A Brief Guide to **Getting the Most** from This Book

1 Read the Book

Feature	Description	Benefit
Applications Using Real-World Data	From the chapter and section openers through the examples and exercises, interesting applications from nearly every discipline, supported by up-to-date real-world data, are included in every section.	Ever wondered how you'll use algebra? This feature will show you how algebra can solve real problems.
Detailed Worked-Out Examples	Examples are clearly written and provide step-by-step solutions. No steps are omitted, and key steps are thoroughly explained to the right of the mathematics.	The blue annotations will help you to understand the solutions by providing the reason why the algebraic steps are true.
Explanatory Voice Balloons	Voice balloons help to demystify algebra. They translate algebraic language into plain English, clarify problem-solving procedures, and present alternative ways of understanding.	Does math ever look foreign to you? This feature translates math into everyday English.
Great Question!	Answers to students' questions offer suggestions for problem solving, point out common errors to avoid, and provide informal hints and suggestions.	This feature should help you not to feel anxious or threatened when asking questions in class.
Achieving Success	The book's Achieving Success boxes offer strategies for success in learning algebra.	Follow these suggestions to help achieve your full academic potential in mathematics.

2 Work the Problems

Feature	Description	Benefit
Check Point Examples	Each example is followed by a similar problem, called a Check Point, that offers you the opportunity to work a similar exercise. Answers to all Check Points are provided in the answer section.	You learn best by doing. You'll solidify your understanding of worked examples if you try a similar problem right away to be sure you understand what you've just read.
Concept and Vocabulary Checks	These short-answer questions, mainly fill-in-the blank and true/false items, assess your understanding of the definitions and concepts presented in each section.	It is difficult to learn algebra without knowing its special language. These exercises test your understanding of the vocabulary and concepts.
Extensive and Varied Exercise Sets	An abundant collection of exercises is included in an Exercise Set at the end of each section. Exercises are organized within several categories. Practice Exercises follow the same order as the section's worked examples. Practice PLUS Exercises contain more challenging problems that often require you to combine several skills or concepts.	The parallel order of the Practice Exercises lets you refer to the worked examples and use them as models for solving these problems. Practice PLUS provides you with ample opportunity to dig in and develop your problem-solving skills.

3 Review for Quizzes and Tests

Feature	Description	Benefit
Mid-Chapter Check Points	Near the midway point in the chapter, an integrated set of review exercises allows you to review the skills and concepts you learned separately over several sections.	Combining exercises from the first half of the chapter gives you a comprehensive review before you continue on.
Chapter Review Charts	Each chapter contains a review chart that summarizes the definitions and concepts in every section of the chapter, complete with examples.	Review this chart and you'll know the most important material in the chapter.
Chapter Tests	Each chapter contains a practice test with problems that cover the important concepts in the chapter. Take the test, check your answers, and then watch the Chapter Test Prep Videos.	You can use the Chapter Test to determine whether you have mastered the material covered in the chapter.
Chapter Test Prep Videos	These videos contain worked-out solutions to every exercise in each chapter test.	These videos let you review any exercises you miss on the chapter test.
Lecture Series	These interactive lecture videos highlight key examples from every section of the textbook.	These videos let you review each objective from the textbook that you need extra help on.

Get the most out of MyMathLab®

MyMathLab, Pearson's online learning management system, creates personalized experiences for students and provides powerful tools for instructors. With a wealth of tested and proven resources, each course can be tailored to fit your specific needs. Talk to your Pearson Representative about ways to integrate MyMathLab into your course for the best results.

Data-Driven Reporting for Instructors

- MyMathLab's comprehensive online gradebook automatically tracks students' results to tests, quizzes, homework, and work in the study plan.

- The Reporting Dashboard, found under More Gradebook Tools, makes it easier than ever to identify topics where students are struggling, or specific students who may need extra help.

Learning in Any Environment

- Because classroom formats and student needs continually change and evolve, MyMathLab has built-in flexibility to accommodate various course designs and formats.

- With a new, streamlined, mobile-friendly design, students and instructors can access courses from most mobile devices to work on exercises and review completed assignments.

SEVENTH EDITION

Introductory Algebra
for College Students

Robert Blitzer
Miami Dade College

PEARSON

Boston • Columbus • Indianapolis • New York • San Francisco
Amsterdam • Cape Town • Dubai • London • Madrid • Milan • Munich • Paris • Montréal • Toronto
Delhi • Mexico City • São Paulo • Sydney • Hong Kong • Seoul • Singapore • Taipei • Tokyo

Editorial Director: *Chris Hoag*

Editor-in-Chief: *Michael Hirsch*

Senior Acquisitions Editor: *Dawn Giovanniello*

Editorial Assistant: *Megan Tripp*

Program Manager: *Beth Kaufman*

Project Manager: *Kathleen A. Manley*

Program Management Team Lead: *Karen Wernholm*

Project Management Team Lead: *Christina Lepre*

Media Producer: *Shana Siegmund*

TestGen Content Manager: *Marty Wright*

MathXL Content Developer: *Rebecca Williams and Eric Gregg*

Product Marketing Manager: *Alicia Frankel*

Field Marketing Managers: *Jenny Crum and Lauren Schur*

Marketing Assistant: *Alexandra Habashi*

Senior Author Support/Technology Specialist: *Joe Vetere*

Rights and Permissions Project Manager: *Gina Cheselka*

Procurement Specialist: *Carol Melville*

Associate Director of Design: *Andrea Nix*

Program Design Lead: *Beth Paquin*

Composition: *codeMantra*

Illustrations: *Scientific Illustrators*

Cover Design: *Studio Montage*

Cover Image: blue mailbox: *Holly Kuchera/Shutterstock;* brick wall painted white: *Raimund Linke/Getty;* cherries: *Annabelle Breakey/Getty;* urban street wall: *Sociologas/ Shutterstock*

Library of Congress Cataloging-in-Publication Data

Blitzer, Robert.
Introductory algebra for college students/Robert F. Blitzer, Miami-Dade College. —7th edition.
    pages cm
ISBN 978-0-13-417805-9
1. Algebra—Textbooks.   I. Title.

QA152.3.B648 2017
512.9—dc23

2015031756

3   18

PEARSON

www.pearsonhighered.com

ISBN 13: 978-0-13-417805-9
ISBN 10: 0-13-417805-X

Table of Contents

Preface

Introductory Algebra for College Students, Seventh Edition, provides comprehensive, in-depth coverage of the topics required in a one-term course in beginning or introductory algebra. The book is written for college students who have no previous experience in algebra and for those who need a review of basic algebra concepts. I wrote the book to help diverse students with different backgrounds and career plans succeed in beginning algebra. *Introductory Algebra for College Students,* Seventh Edition, has two primary goals:

1. To help students acquire a solid foundation in the basic skills of algebra.
2. To show students how algebra can model and solve authentic real-world problems.

One major obstacle in the way of achieving these goals is the fact that very few students actually read their textbook. This has been a regular source of frustration for me and for my colleagues in the classroom. Anecdotal evidence gathered over years highlights two basic reasons students give when asked why they do not take advantage of their textbook:

- "I'll never use this information."
- "I can't follow the explanations."

I've written every page of the Seventh Edition with the intent of eliminating these two objections. The ideas and tools I've used to do so are described in the features that follow. These features and their benefits are highlighted for the student in "A Brief Guide to Getting the Most from This Book," which appears inside the front cover.

What's New in the Seventh Edition?

- **New Applications and Real-World Data.** The Seventh Edition contains 112 new or revised worked-out examples and exercises based on updated and new data sets. Many of these applications involve topics relevant to college students and newsworthy items. Among topics of interest to college students, you'll find new data sets describing student loan debt (Chapter 1 Review, Exercise 103; Exercise Set 3.5, Exercise 39), grade inflation of As and deflation of Cs (Exercise Set 2.1, Exercises 65–66; Exercise Set 2.7, Exercises 105–106), median weekly earnings, by education and gender (Exercise Set 1.8, Exercises 99–100; Section 2.2, Example 8; Exercise Set 2.2, Exercises 71–72), political orientation of college freshmen (Chapter 1 Test, Exercise 20), and the importance of being well-off financially for college freshmen (Section 9.6, Example 6). Among newsworthy topics, new applications include income inequality (Exercise Set 2.4, Exercises 57–58; Exercise Set 4.3, Exercise 70), perceptions of police officers (Exercise Set 2.4, Exercises 59–60), marijuana use among college-age students (Chapter 3 Review, Exercise 44), marriage equality (Chapter 4 Cumulative Review, Exercises 17–20), transformation of the music industry by the Internet (Exercise Set 7.5, Exercises 51–52), and racism, measured by age and political orientation (Section 5.1 opener; Exercise Set 5.1, Exercises 103–104).

- **New Blitzer Bonus Videos with Assessment.** The Blitzer Bonus features throughout the textbook have been turned into animated videos that are built into the MyMathLab course. These videos help students make visual connections to algebra and the world around them. Assignable exercises have been created within the MyMathLab course to assess conceptual understanding and mastery. These videos and exercises can be turned into a media assignment within the Blitzer MyMathLab course.

- **Updated Student Learning Guide.** Organized by the textbook's learning objectives, this updated Learning Guide helps students learn how to make the most of their textbook for test preparation. Projects are now included to give students an opportunity to discover and reinforce the concepts in an active learning environment and are ideal for group work in class.

- **Updated Graphing Calculator Screens.** All screens have been updated using the TI-84 Plus C.

What's New in the Blitzer Developmental Mathematics Series?

Two new textbooks have been added to the series:

- *Developmental Mathematics,* First Edition, is intended for a course sequence covering prealgebra, introductory algebra, and intermediate algebra. The text provides a solid foundation in arithmetic and algebra.

- *Pathways to College Mathematics,* First Edition, provides a general survey of topics to prepare STEM and non-STEM students for success in a variety of college math courses, including college algebra, statistics, liberal arts mathematics, quantitative reasoning, finite mathematics, and mathematics for education majors. The prerequisite is basic math or prealgebra.

- *MyMathLab with Integrated Review* courses are also available for select Blitzer titles. These MyMathLab courses provide the full suite of resources for the core textbook, but also add in study aids and skills check assignments keyed to the prerequisite topics that students need to know, helping them quickly get up to speed.

What Familiar Features Have Been Retained in the Seventh Edition of *Introductory Algebra for College Students*?

- **Learning Objectives.** Learning objectives, framed in the context of a student question (What am I supposed to learn?), are clearly stated at the beginning of each section. These objectives help students recognize and focus on the section's most important ideas. The objectives are restated in the margin at their point of use.

- **Chapter-Opening and Section-Opening Scenarios.** Every chapter and every section open with a scenario presenting a unique application of mathematics in students' lives outside the classroom. These scenarios are revisited in the course of the chapter or section in an example, discussion, or exercise.

- **Innovative Applications.** A wide variety of interesting applications, supported by up-to-date, real-world data, are included in every section.

- **Detailed Worked-Out Examples.** Each example is titled, making the purpose of the example clear. Examples are clearly written and provide students with detailed step-by-step solutions. No steps are omitted and each step is thoroughly explained to the right of the mathematics.

- **Explanatory Voice Balloons.** Voice balloons are used in a variety of ways to demystify mathematics. They translate algebraic ideas into everyday English, help clarify problem-solving procedures, present alternative ways of understanding concepts, and connect problem solving to concepts students have already learned.

- **Check Point Examples.** Each example is followed by a similar matched problem, called a Check Point, offering students the opportunity to test their understanding of the example by working a similar exercise. The answers to the Check Points are provided in the answer section.

- **Concept and Vocabulary Checks.** This feature offers short-answer exercises, mainly fill-in-the-blank and true/false items, that assess students' understanding of the definitions and concepts presented in each section. The Concept and Vocabulary Checks appear as separate features preceding the Exercise Sets.

- **Extensive and Varied Exercise Sets.** An abundant collection of exercises is included in an Exercise Set at the end of each section. Exercises are organized within eight category types: Practice Exercises, Practice Plus Exercises, Application Exercises, Explaining the Concepts, Critical Thinking Exercises, Technology Exercises, Review Exercises, and Preview Exercises. This format makes it easy to create well-rounded homework assignments. The order of the Practice Exercises is exactly the same as the order of the section's worked examples. This parallel order enables students to refer to the titled examples and their detailed explanations to achieve success working the Practice Exercises.

- **Practice Plus Problems.** This category of exercises contains more challenging practice problems that often require students to combine several skills or concepts. With an average of ten Practice Plus problems per Exercise Set, instructors are provided with the option of creating assignments that take Practice Exercises to a more challenging level.

- **Mid-Chapter Check Points.** At approximately the midway point in each chapter, an integrated set of Review Exercises allows students to review and assimilate the skills and concepts they learned separately over several sections.

- **Early Graphing.** Chapter 1 connects formulas and mathematical models to data displayed by bar and line graphs. The rectangular coordinate system is introduced in Chapter 3. Graphs appear in nearly every section and Exercise Set. Examples and exercises use graphs to explore relationships between data and to provide ways of visualizing a problem's solution.

- **Geometric Problem Solving.** Chapter 2 (Linear Equations and Inequalities in One Variable) contains a section that teaches geometric concepts that are important to a student's understanding of algebra. There is frequent emphasis on problem solving in geometric situations, as well as on geometric models that allow students to visualize algebraic formulas.

- **Thorough, Yet Optional, Technology.** Although the use of graphing utilities is optional, they are utilized in Using Technology boxes to enable students to visualize and gain numerical insight into algebraic concepts. The use of graphing utilities is also reinforced in the Technology Exercises appearing in the Exercise Sets for those who want this option. With the book's early introduction to graphing, students can look at the calculator screens in the Using Technology boxes and gain an increased understanding of an example's solution even if they are not using a graphing utility in the course.

- **Great Question!** This feature presents a variety of study tips in the context of students' questions. Answers to questions offer suggestions for problem solving, point out common errors to avoid, and provide informal hints and suggestions. As a secondary benefit, this feature should help students not to feel anxious or threatened when asking questions in class.

- **Achieving Success.** The Achieving Success boxes at the end of most sections offer strategies for persistence and success in college mathematics courses.

- **Chapter Review Grids.** Each chapter contains a review chart that summarizes the definitions and concepts in every section of the chapter. Examples that illustrate these key concepts are also included in the chart.

- **End-of-Chapter Materials.** A comprehensive collection of Review Exercises for each of the chapter's sections follows the review grid. This is followed by a Chapter Test that enables students to test their understanding of the material covered in the chapter. Beginning with Chapter 2, each chapter concludes with a comprehensive collection of mixed Cumulative Review Exercises.

- **Blitzer Bonuses.** These enrichment essays provide historical, interdisciplinary, and otherwise interesting connections to the algebra under study, showing students that math is an interesting and dynamic discipline.

- **Discovery.** Discover for Yourself boxes, found throughout the text, encourage students to further explore algebraic concepts. These explorations are optional and their omission does not interfere with the continuity of the topic under consideration.

I hope that my passion for teaching, as well as my respect for the diversity of students I have taught and learned from over the years, is apparent throughout this new edition. By connecting algebra to the whole spectrum of learning, it is my intent to show students that their world is profoundly mathematical, and indeed, π is in the sky.

Robert Blitzer

Resources for Success
MyMathLab for the Blitzer Developmental Algebra Series

MyMathLab is available to accompany Pearson's market leading text offerings. This text's flavor and approach are tightly integrated throughout the accompanying MyMathLab course, giving students a consistent tone, voice, and teaching method that make learning the material as seamless as possible.

Section Lecture and Chapter Test Prep Videos

An **updated** video program provides a multitude of resources for students. Section Lecture videos walk students through the concepts from every section of the text in a fresh, modern presentation format. Chapter Test Prep videos walk students through the solution of every problem in the text's Chapter Tests, giving students video resources when they might need it most.

Find solutions to the equation $y = 2x + 1$.

x	$y = 2x + 1$	(x, y)
0	$y = 2(0) + 1 = 1$	$(0,1)$
1	$y = 2(1) + 1 = 3$	$(1,3)$
2	$y = 2(2) + 1 = 5$	
−1		

Menu   4/8

09:02/23:53   CC   ESP

Blitzer Bonus Videos

NEW! Animated videos have been created to mirror the Blitzer Bonus features throughout the textbook. Blitzer Bonus features in the text provide interesting real-world connections to the mathematical topics at hand, conveying Bob Blitzer's signature style to engage students. These new Blitzer Bonus videos will help students to connect the topics to the world around them in a visual way. Corresponding assignable exercises in MyMathLab are also available, allowing these new videos to be turned into a media assignment to truly ensure that students have understood what they've watched.

Refresh   Session 78490153   Log out

composite sketch question

Graph the linear equation.

$y = -2$

✖ Clear sketch   Submit response

T Switch to text response

✉ Send a message to the instructor

‹ Join another session

Learning Catalytics

Integrated into MyMathLab, the Learning Catalytics feature uses students' devices in the classroom for an engagement, assessment, and classroom intelligence system that gives instructors real-time feedback on student learning. Learning Catalytics contains Pearson-created content for developmental math topics that allows you to take advantage of this exciting technology immediately.

Student Success Modules

These modules are integrated within the MyMathLab course to help students succeed in college courses and prepare for future professions.

www.mymathlab.com

Resources for Success

Instructor Resources

Annotated Instructor's Edition

This version of the text contains answers to exercises printed on the same page, with graphing answers in a special Graphing Answer Section at the back of the text.

The following resources can be downloaded from www.pearsonhighered.com or in MyMathLab.

PowerPoint® Lecture Slides

Fully editable slides correlated with the textbook include definitions, key concepts, and examples for use in a lecture setting.

Instructor's Solutions Manual

This manual includes fully worked-out solutions to all text exercises.

Instructor's Resource Manual

This manual includes a Mini-Lecture, Skill Builder, and Additional Exercises for every section of the text. It also includes Chapter Test forms, as well as Cumulative and Final Exams, with answers.

TestGen®

TestGen® (www.pearsoned.com/testgen) enables instructors to build, edit, print, and administer tests using a computerized bank of questions developed to cover all the objectives of the text.

Student Resources

The following additional resources are available to support student success:

Learning Guide

UPDATED! Organized by learning objectives, the Learning Guide helps students make the most of their textbook and prepare for tests. Now updated to include projects, students will have the opportunity to discover and reinforce the concepts in an active learning environment. These projects are ideal for group work in class. The Learning Guide is available in MyMathLab, and available as a printed supplement.

Video Lecture Series

Available in MyMathLab, the video program covers every section in the text, providing students with a video tutor at home, in lab, or on the go. The program includes Section Lecture Videos and Chapter Test Prep videos.

Student Solutions Manual

This manual provides detailed, worked-out solutions to odd-numbered section exercises, plus all Check Points, Review/Preview Exercises, Mid-Chapter Check Points, Chapter Reviews, Chapter Tests, and Cumulative Reviews.

www.mymathlab.com

Acknowledgments

An enormous benefit of authoring a successful series is the broad-based feedback I receive from the students, dedicated users, and reviewers. Every change to this edition is the result of their thoughtful comments and suggestions. I would like to express my appreciation to all the reviewers, whose collective insights form the backbone of this revision. In particular, I would like to thank the following people for reviewing *Introductory Algebra for College Students*.

Cindy Adams, *San Jacinto College*

Gwen P. Aldridge, *Northwest Mississippi Community College*

Ronnie Allen, *Central New Mexico Community College*

Dr. Simon Aman, *Harry S. Truman College*

Howard Anderson, *Skagit Valley College*

John Anderson, *Illinois Valley Community College*

Michael H. Andreoli, *Miami-Dade College – North Campus*

Michele Bach, *Kansas City Kansas Community College*

Jana Barnard, *Angelo State University*

Rosanne Benn, *Prince George's Community College*

Christine Brady, *Suffolk County Community College*

Gale Brewer, *Amarillo College*

Carmen Buhler, *Minneapolis Community & Technical College*

Warren J. Burch, *Brevard College*

Alice Burstein, *Middlesex Community College*

Edie Carter, *Amarillo College*

Jerry Chen, *Suffolk County Community College*

Sandra Pryor Clarkson, *Hunter College*

Sally Copeland, *Johnson County Community College*

Valerie Cox, *Calhoun Community College*

Carol Curtis, *Fresno City College*

Robert A. Davies, *Cuyahoga Community College*

Deborah Detrick, *Kansas City Kansas Community College*

Jill DeWitt, *Baker College of Muskegon*

Ben Divers, Jr., *Ferrum College*

Irene Doo, *Austin Community College*

Charles C. Edgar, *Onondaga Community College*

Karen Edwards, *Diablo Valley College*

Scott Fallstrom, *MiraCosta College*

Elise Fischer, *Johnson County Community College*

Susan Forman, *Bronx Community College*

Wendy Fresh, *Portland Community College*

Jennifer Garnes, *Cuyahoga Community College*

Gary Glaze, *Eastern Washington University*

Jay Graening, *University of Arkansas*

Robert B. Hafer, *Brevard College*

Andrea Hendricks, *Georgia Perimeter College*

Donald Herrick, *Northern Illinois University*

Beth Hooper, *Golden West College*

Sandee House, *Georgia Perimeter College*

Tracy Hoy, *College of Lake County*

Laura Hoye, *Trident Community College*

Margaret Huddleston, *Schreiner University*

Marcella Jones, *Minneapolis Community & Technical College*

Shelbra B. Jones, *Wake Technical Community College*

Sharon Keenee, *Georgia Perimeter College*

Regina Keller, *Suffolk County Community College*

Gary Kersting, *North Central Michigan College*

Dennis Kimzey, *Rogue Community College*

Kandace Kling, *Portland Community College*

Gray Knippenberg, *Lansing Community College*

Mary Kochler, *Cuyahoga Community College*

Scot Leavitt, *Portland Community College*

Robert Leibman, *Austin Community College*

Jennifer Lempke, *North Central Michigan College*

Ann M. Loving, *J. Sargent Reynolds Community College*

Kent MacDougall, *Temple College*

Jean-Marie Magnier, *Springfield Technical Community College*

Hank Martel, *Broward College*

Kim Martin, *Southeastern Illinois College*

John Robert Martin, *Tarrant County College*

Lisa McMillen, *Baker College of Auburn Hills*

Irwin Metviner, *State University of New York at Old Westbury*

Jean P. Millen, *Georgia Perimeter College*

Lawrence Morales, *Seattle Central Community College*

Morteza Shafii-Mousavi, *Indiana University South Bend*

Lois Jean Nieme, *Minneapolis Community & Technical College*

Allen R. Newhart, *Parkersburg Community College*

Karen Pain, *Palm Beach State College*

Peg Pankowski, *Community College of Allegheny County – South Campus*

Robert Patenaude, *College of the Canyons*

Matthew Peace, *Florida Gateway College*

Dr. Bernard J. Piña, *New Mexico State University – Doña Ana Community College*

Jill Rafael, *Sierra College*

James Razavi, *Sierra College*

Christopher Reisch, *The State University of New York at Buffalo*

Nancy Ressler, *Oakton Community College*

Katalin Rozsa, *Mesa Community College*

Haazim Sabree, *Georgia Perimeter College*

Chris Schultz, *Iowa State University*

Shannon Schumann, *University of Phoenix*

Barbara Sehr, *Indiana University Kokomo*

Brian Smith, *Northwest Shoals Community College*

Gayle Smith, *Lane Community College*

Dick Spangler, *Tacoma Community College*

Janette Summers, *University of Arkansas*

Robert Thornton, *Loyola University*

Lucy C. Thrower, *Francis Marion College*

Mary Thurow, *Minneapolis Community & Tech College*

Richard Townsend, *North Carolina Central University*

Cindie Wade, *St. Clair County Community College*

Andrew Walker, *North Seattle Community College*

Kathryn Wetzel, *Amarillo College*

Additional acknowledgments are extended to Dan Miller and Kelly Barber for preparing the solutions manuals and the new Learning Guide; Brad Davis, for preparing the answer section and serving as accuracy checker; the codeMantra formatting team for the book's brilliant paging; Brian Morris and Kevin Morris at Scientific Illustrators, for superbly illustrating the book; and Francesca Monaco, project manager, and Kathleen Manley, production editor, whose collective talents kept every aspect of this complex project moving through its many stages.

I would like to thank my editors at Pearson, Dawn Giovanniello and Megan Tripp, who guided and coordinated the book from manuscript through production. Thanks to Beth Paquin and Studio Montage for the quirky cover and interior design. Finally, thanks to marketing manager Alicia Frankel for your innovative marketing efforts, and to the entire Pearson sales force, for your confidence and enthusiasm about the book.

To the Student

The bar graph shows some of the qualities that students say make a great teacher.

It was my goal to incorporate each of the qualities that make a great teacher throughout the pages of this book.

Explains Things Clearly

I understand that your primary purpose in reading *Introductory Algebra for College Students* is to acquire a solid understanding of the required topics in your algebra course. In order to achieve this goal, I've carefully explained each topic. Important definitions and procedures are set off in boxes, and worked-out examples that present solutions in a step-by-step manner appear in every section. Each example is followed by a similar matched problem, called a Check Point, for you to try so that you can actively participate in the learning process as you read the book. (Answers to all Check Points appear in the back of the book.)

Funny/Entertaining

Who says that an algebra textbook can't be entertaining? From our quirky cover to the photos in the chapter and section openers, prepare to expect the unexpected. I hope some of the book's enrichment essays, called Blitzer Bonuses, will put a smile on your face from time to time.

Helpful

I designed the book's features to help you acquire knowledge of introductory algebra, as well as to show you how algebra can solve authentic problems that apply to your life. These helpful features include:

- *Explanatory Voice Balloons:* Voice balloons are used in a variety of ways to make math less intimidating. They translate algebraic language into everyday English, help clarify problem-solving procedures, present alternative ways of understanding concepts, and connect new concepts to concepts you have already learned.
- *Great Question!:* The book's Great Question! boxes are based on questions students ask in class. The answers to these questions give suggestions for problem solving, point out common errors to avoid, and provide informal hints and suggestions.
- *Achieving Success:* The book's Achieving Success boxes give you helpful strategies for success in learning algebra, as well as suggestions that can be applied for achieving your full academic potential in future college coursework.
- *Detailed Chapter Review Charts:* Each chapter contains a review chart that summarizes the definitions and concepts in every section of the chapter. Examples that illustrate these key concepts are also included in the chart. Review these summaries and you'll know the most important material in the chapter!

Passionate about Their Subject

I passionately believe that no other discipline comes close to math in offering a more extensive set of tools for application and development of your mind. I wrote the book in Point Reyes National Seashore, 40 miles north of San Francisco. The park consists of 75,000 acres with miles of pristine surf-washed beaches, forested ridges, and bays bordered by white cliffs. It was my hope to convey the beauty and excitement of mathematics using nature's unspoiled beauty as a source of inspiration and creativity. Enjoy the pages that follow as you empower yourself with the algebra needed to succeed in college, your career, and your life.

Regards,
Bob
Robert Blitzer

About the Author

Bob Blitzer is a native of Manhattan and received a Bachelor of Arts degree with dual majors in mathematics and psychology (minor: English literature) from the City College of New York. His unusual combination of academic interests led him toward a Master of Arts in mathematics from the University of Miami and a doctorate in behavioral sciences from Nova University. Bob's love for teaching mathematics was nourished for nearly 30 years at Miami Dade College, where he received numerous teaching awards, including Innovator of the Year from the League for Innovations in the Community College and an endowed chair based on excellence in the classroom. In addition to *Introductory Algebra for College Students*, Bob has written textbooks covering developmental mathematics, intermediate algebra, college algebra, algebra and trigonometry, precalculus,  and liberal arts mathematics, all published by Pearson. When not secluded in his Northern California writer's cabin, Bob can be found hiking the beaches and trails of Point Reyes National Seashore, and tending to the chores required by his beloved entourage of horses, chickens, and irritable roosters.

Variables, Real Numbers, and Mathematical Models

What can algebra possibly tell me about

- the rising cost of movie ticket prices over the years?
- how I can stretch or shrink my lifespan?
- the widening imbalance between numbers of women and men on college campuses?
- the number of calories I need to maintain energy balance?
- the widening imbalance between salaries of male and female college graduates?

In this chapter, you will learn how the special language of algebra describes your world.

Here's where you'll find these applications:

- Movie ticket prices: Exercise Set 1.1, Exercises 83–84
- Stretching or shrinking my lifespan: Section 1.6, Example 6
- College gender imbalance: Section 1.7, Example 9
- Caloric needs: Section 1.8, Example 12; Exercise Set 1.8, Exercises 97–98
- Gender divide in salaries for college graduates: Exercise Set 1.8, Exercises 99–100.

1958	1967	1980	1990	2000	2010	2013
Cat on a Hot Tin Roof TICKET PRICE $0.68	Cool Hand Luke TICKET PRICE $1.22	Ordinary People TICKET PRICE $2.69	Dances with Wolves TICKET PRICE $4.23	Miss Congeniality TICKET PRICE $5.39	Alice in Wonderland TICKET PRICE $7.89	12 Years a Slave TICKET PRICE $8.38

Sources: Motion Picture Association of America, National Association of Theater Owners (NATO), and Bureau of Labor Statistics (BLS)

SECTION

1.1

Introduction to Algebra: Variables and Mathematical Models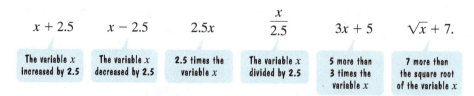

What am I supposed to learn?

After studying this section, you should be able to:

1 Evaluate algebraic expressions.

2 Translate English phrases into algebraic expressions.

3 Determine whether a number is a solution of an equation.

4 Translate English sentences into algebraic equations.

5 Evaluate formulas.

You are thinking about buying a high-definition television. How much distance should you allow between you and the TV for pixels to be undetectable and the image to appear smooth?

Algebraic Expressions

Let's see what the distance between you and your TV has to do with algebra. The biggest difference between arithmetic and algebra is the use of *variables* in algebra. A **variable** is a letter that represents a variety of different numbers. For example, we can let x represent the diagonal length, in inches, of a high-definition television. The industry rule for most of the current HDTVs on the market is to multiply this diagonal length by 2.5 to get the distance, in inches, at which a person with perfect vision can see a smooth image. This can be written $2.5 \cdot x$, but it is usually expressed as $2.5x$. Placing a number and a letter next to one another indicates multiplication.

Notice that $2.5x$ combines the number 2.5 and the variable x using the operation of multiplication. A combination of variables and numbers using the operations of addition, subtraction, multiplication, or division, as well as powers or roots, is called an **algebraic expression**. Here are some examples of algebraic expressions:

$$x + 2.5 \qquad x - 2.5 \qquad 2.5x \qquad \frac{x}{2.5} \qquad 3x + 5 \qquad \sqrt{x} + 7.$$

| The variable x increased by 2.5 | The variable x decreased by 2.5 | 2.5 times the variable x | The variable x divided by 2.5 | 5 more than 3 times the variable x | 7 more than the square root of the variable x |

Great Question!

Are variables always represented by *x*?

No. As you progress through algebra, you will often see x, y, and z used, but any letter can be used to represent a variable. For example, if we use l to represent a TV's diagonal length, your ideal distance from the screen is described by the algebraic expression $2.5l$.

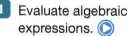

1 Evaluate algebraic expressions.

Evaluating Algebraic Expressions

We can replace a variable that appears in an algebraic expression by a number. We are **substituting** the number for the variable. The process is called **evaluating the expression**. For example, we can evaluate $2.5x$ (the ideal distance between you and your x-inch TV) for $x = 50$. We substitute 50 for x. We obtain $2.5 \cdot 50$, or 125. This means that if the diagonal length of your TV is 50 inches, your distance from the screen should be 125 inches. Because 12 inches = 1 foot, this distance is $\frac{125}{12}$ feet, or approximately 10.4 feet.

Many algebraic expressions involve more than one operation. The order in which we add, subtract, multiply, and divide is important. In Section 1.8, we will discuss the rules for the order in which operations should be done. For now, follow this order:

A First Look at Order of Operations

1. Perform all operations within grouping symbols, such as parentheses.
2. Do all multiplications in the order in which they occur from left to right.
3. Do all additions and subtractions in the order in which they occur from left to right.

EXAMPLE 1   Evaluating Expressions

Evaluate each algebraic expression for $x = 5$:

a. $3 + 4x$ **b.** $4(x + 3)$.

Solution

a. We begin by substituting 5 for x in $3 + 4x$. Then we follow the order of operations: Multiply first, and then add.

$$3 + 4x$$

Replace x with 5.

$$= 3 + 4 \cdot 5$$
$$= 3 + 20 \qquad \text{Perform the multiplication: } 4 \cdot 5 = 20.$$
$$= 23 \qquad \text{Perform the addition.}$$

b. We begin by substituting 5 for x in $4(x + 3)$. Then we follow the order of operations: Perform the addition in parentheses first, and then multiply.

$$4(x + 3)$$

Replace x with 5.

$$= 4(5 + 3)$$
$$= 4(8) \qquad \text{Perform the addition inside the}$$
$$ \qquad \text{parentheses: } 5 + 3 = 8.$$
$$ \qquad 4(8) \text{ can also be written as } 4 \cdot 8.$$
$$= 32 \qquad \text{Multiply.} \quad \blacksquare$$

☑ **CHECK POINT 1**   Evaluate each expression for $x = 10$:

a. $6 + 2x$ **b.** $2(x + 6)$.

Example 2 illustrates that algebraic expressions can contain more than one variable.

EXAMPLE 2   Evaluating Expressions

Evaluate each algebraic expression for $x = 6$ and $y = 4$:

a. $5x - 3y$ **b.** $\dfrac{3x + 5y + 2}{2x - y}$.

Solution

a. $5x - 3y$ This is the given algebraic expression.

Replace x with 6.   Replace y with 4.

$= 5 \cdot 6 - 3 \cdot 4$ We are evaluating the expression for $x = 6$ and $y = 4$.

$= 30 - 12$ Multiply: $5 \cdot 6 = 30$ and $3 \cdot 4 = 12$.

$= 18$ Subtract.

b. $\dfrac{3x + 5y + 2}{2x - y}$ This is the given algebraic expression.

Replace x with 6.   Replace y with 4.

$= \dfrac{3 \cdot 6 + 5 \cdot 4 + 2}{2 \cdot 6 - 4}$ We are evaluating the expression for $x = 6$ and $y = 4$.

$= \dfrac{18 + 20 + 2}{12 - 4}$ Multiply: $3 \cdot 6 = 18$, $5 \cdot 4 = 20$, and $2 \cdot 6 = 12$.

$= \dfrac{40}{8}$ Add in the numerator.
Subtract in the denominator.

$= 5$ Simplify by dividing 40 by 8. ∎

✓ **CHECK POINT 2**   Evaluate each algebraic expression for $x = 3$ and $y = 8$:

a. $7x + 2y$ **b.** $\dfrac{6x - y}{2y - x - 8}$.

---

2 Translate English phrases into algebraic expressions.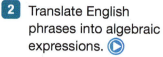

Translating to Algebraic Expressions

Problem solving in algebra often requires the ability to translate word phrases into algebraic expressions. **Table 1.1** lists some key words associated with the operations of addition, subtraction, multiplication, and division.

Table 1.1	Key Words for Addition, Subtraction, Multiplication, and Division			
Operation	**Addition (+)**	**Subtraction (−)**	**Multiplication (·)**	**Division (÷)**
	plus	minus	times	divide
	sum	difference	product	quotient
Key words	more than	less than	twice	ratio
	increased by	decreased by	multiplied by	divided by

EXAMPLE 3   Translating English Phrases into Algebraic Expressions

Write each English phrase as an algebraic expression. Let the variable x represent the number.

a. the sum of a number and 7

b. ten less than a number

c. twice a number, decreased by 6

d. the product of 8 and a number

e. three more than the quotient of a number and 11

Solution

a.

The sum of	a number and 7

$$x + 7$$

The algebraic expression for "the sum of a number and 7" is $x + 7$.

b.

Ten less than	a number

$$x - 10$$

The algebraic expression for "ten less than a number" is $x - 10$.

c.

Twice a number,	decreased by 6

$$2x - 6$$

The algebraic expression for "twice a number, decreased by 6" is $2x - 6$.

d.

The product of	8 and a number

$$8 \cdot x$$

The algebraic expression for "the product of 8 and a number" is $8 \cdot x$, or $8x$.

e.

Three more than	the quotient of a number and 11

$$\frac{x}{11} + 3$$

The algebraic expression for "three more than the quotient of a number and 11" is $\frac{x}{11} + 3$. ■

Great Question!

Are there any helpful hints I can use when writing algebraic expressions for English phrases involving subtraction?

Yes. Pay close attention to order when translating phrases involving subtraction.

Phrase	Translation
A number decreased by 7	$x - 7$
A number subtracted from 7	$7 - x$
Seven less than a number	$x - 7$
Seven less a number	$7 - x$

Think carefully about what is expressed in English before you translate into the language of algebra.

☑ **CHECK POINT 3**   Write each English phrase as an algebraic expression. Let the variable x represent the number.

   a. the product of 6 and a number

   b. a number added to 4

   c. three times a number, increased by 5

   d. twice a number subtracted from 12

   e. the quotient of 15 and a number

3 Determine whether a number is a solution of an equation. ▶

Equations

An **equation** is a statement that two algebraic expressions are equal. **An equation always contains the equality symbol $=$.** Here are some examples of equations:

$$6x + 16 = 46 \qquad 4y + 2 = 2y + 6 \qquad 2(z + 1) = 5(z - 2).$$

Solutions of an equation are values of the variable that make the equation a true statement. To determine whether a number is a solution, substitute that number for the variable and evaluate each side of the equation. If the values on both sides of the equation are the same, the number is a solution.

EXAMPLE 4 Determining Whether Numbers Are Solutions of Equations

Determine whether the given number is a solution of the equation.

a. $6x + 16 = 46; 5$ **b.** $2(z + 1) = 5(z - 2); 7$

Solution

a. $6x + 16 = 46$ This is the given equation.

Is 5 a solution?

$6 \cdot 5 + 16 \overset{?}{=} 46$ To determine whether 5 is a solution, substitute 5 for x. The question mark over the equal sign indicates that we do not yet know if the statement is true.

$30 + 16 \overset{?}{=} 46$ Multiply: $6 \cdot 5 = 30$.

This statement is true. → $46 = 46$ Add: $30 + 16 = 46$.

Because the values on both sides of the equation are the same, the number 5 is a solution of the equation.

b. $2(z + 1) = 5(z - 2)$ This is the given equation.

Is 7 a solution?

$2(7 + 1) \overset{?}{=} 5(7 - 2)$ To determine whether 7 is a solution, substitute 7 for z.

$2(8) \overset{?}{=} 5(5)$ Perform operations in parentheses: $7 + 1 = 8$ and $7 - 2 = 5$.

This statement is false. → $16 = 25$ Multiply: $2 \cdot 8 = 16$ and $5 \cdot 5 = 25$.

Because the values on both sides of the equation are not the same, the number 7 is not a solution of the equation. ∎

☑ **CHECK POINT 4** Determine whether the given number is a solution of the equation.

a. $9x - 3 = 42; 6$ **b.** $2(y + 3) = 5y - 3; 3$

4 Translate English sentences into algebraic equations.

Translating to Equations

Earlier in the section, we translated English phrases into algebraic expressions. Now we will translate English sentences into equations. You'll find that there are a number of different words and phrases for an equation's equality symbol.

equals
gives
yields

$=$

is the same as
is/was/will be

EXAMPLE 5   Translating English Sentences into Algebraic Equations

Write each sentence as an equation. Let the variable x represent the number.

a. The product of 6 and a number is 30.

b. Seven less than 3 times a number gives 17.

Solution

a.

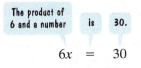

The product of 6 and a number   is   30.

$$6x \; = \; 30$$

The equation for "the product of 6 and a number is 30" is $6x = 30$.

b.

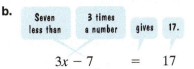

Seven less than   3 times a number   gives   17.

$$3x - 7 \; = \; 17$$

The equation for "seven less than 3 times a number gives 17" is $3x - 7 = 17$. ■

✓ **CHECK POINT 5**   Write each sentence as an equation. Let the variable x represent the number.

a. The quotient of a number and 6 is 5.

b. Seven decreased by twice a number yields 1.

Great Question!

I don't know how to find the solutions of the equations in Example 5. Should I be worried?

Do not be concerned. We'll discuss how to solve these equations in Chapter 2.

Great Question!

When translating English sentences containing commas into algebraic equations, should I pay attention to the commas?

Commas make a difference. In English, sentences and phrases can take on different meanings depending on the way words are grouped with commas. Some examples:

These are meant to be amusing.

- Woman, without her man, is nothing.
  Woman, without her, man is nothing.
- Do not break your bread or roll in your soup.
  Do not break your bread, or roll in your soup.

Algebraically, this is the important item.

- The product of 6 and a number increased by 5 is 30: $6(x + 5) = 30$.
  The product of 6 and a number, increased by 5, is 30: $6x + 5 = 30$.

5 Evaluate formulas.

Formulas and Mathematical Models

One aim of algebra is to provide a compact, symbolic description of the world. These descriptions involve the use of *formulas*. A **formula** is an equation that expresses a relationship between two or more variables. For example, one variety of crickets chirps faster as the temperature rises. You can calculate the temperature by counting the number of times a cricket chirps per minute and applying the following formula:

$$T = 0.3n + 40.$$

In the formula, T is the temperature, in degrees Fahrenheit, and n is the number of cricket chirps per minute. We can use this formula to determine the temperature if you are sitting on your porch and count 80 chirps per minute. Here is how to do so:

$T = 0.3n + 40$	This is the given formula.
$T = 0.3(80) + 40$	Substitute 80 for n.
$T = 24 + 40$	Multiply: 0.3(80) = 24.
$T = 64.$	Add.

When there are 80 cricket chirps per minute, the temperature is 64 degrees.

The process of finding formulas to describe real-world phenomena is called **mathematical modeling**. Such formulas, together with the meaning assigned to the variables, are called **mathematical models**. We often say that these formulas model, or describe, the relationship among the variables.

In creating mathematical models, we strive for both accuracy and simplicity. For example, the formula $T = 0.3n + 40$ is relatively simple to use. However, you should not get upset if you count 80 cricket chirps and the actual temperature is 62 degrees, rather than 64 degrees, as predicted by the formula. Many mathematical models give an approximate, rather than an exact, description of the relationship between variables.

Sometimes a mathematical model gives an estimate that is not a good approximation or is extended to include values of the variable that do not make sense. In these cases, we say that **model breakdown** has occurred. Here is an example:

Use the mathematical model $T = 0.3n + 40$ with $n = 1200$ (1200 cricket chirps per minute).

$$T = 0.3(1200) + 40 = 360 + 40 = 400$$

At 400° F, forget about 1200 chirps per minute! At this temperature, the cricket would "cook" and, alas, all chirping would cease.

EXAMPLE 6   Age at Marriage and the Probability of Divorce

Divorce rates are considerably higher for couples who marry in their teens. The line graphs in **Figure 1.1** show the percentages of marriages ending in divorce based on the wife's age at marriage.

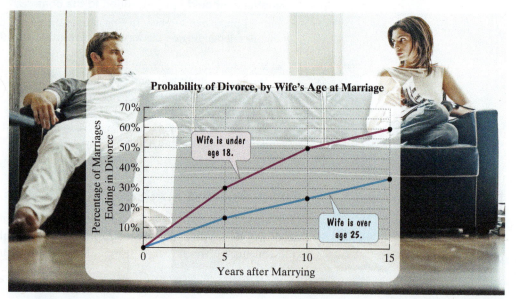

Figure 1.1

Source: B. E. Pruitt et al., *Human Sexuality,* Prentice Hall, 2007

Here are two mathematical models that approximate the data displayed by the line graphs:

Wife is under 18
at time of marriage.

Wife is over 25
at time of marriage.

$$d = 4n + 5 \qquad\qquad d = 2.3n + 1.5.$$

In each model, the variable n is the number of years after marriage and the variable d is the percentage of marriages ending in divorce.

a. Use the appropriate formula to determine the percentage of marriages ending in divorce after 10 years when the wife is over 25 at the time of marriage.

b. Use the appropriate line graph in **Figure 1.1** to determine the percentage of marriages ending in divorce after 10 years when the wife is over 25 at the time of marriage.

c. Does the value given by the mathematical model underestimate or overestimate the actual percentage of marriages ending in divorce after 10 years as shown by the graph? By how much?

Solution

a. Because the wife is over 25 at the time of marriage, we use the formula on the right, $d = 2.3n + 1.5$. To find the percentage of marriages ending in divorce after 10 years, we substitute 10 for n and evaluate the formula.

$$d = 2.3n + 1.5 \qquad \text{This is one of the two given mathematical models.}$$

$$d = 2.3(10) + 1.5 \qquad \text{Replace } n \text{ with 10.}$$

$$d = 23 + 1.5 \qquad \text{Multiply: } 2.3(10) = 23.$$

$$d = 24.5 \qquad \text{Add.}$$

The model indicates that 24.5% of marriages end in divorce after 10 years when the wife is over 25 at the time of marriage.

b. Now let's use the line graph that shows the percentage of marriages ending in divorce when the wife is over 25 at the time of marriage. The graph is shown again in **Figure 1.2**. To find the percentage of marriages ending in divorce after 10 years:

- Locate 10 on the horizontal axis and locate the point above 10.

- Read across to the corresponding percent on the vertical axis.

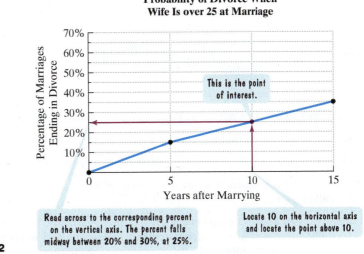

**Probability of Divorce When
Wife Is over 25 at Marriage**

This is the point
of interest.

Read across to the corresponding percent on the vertical axis. The percent falls midway between 20% and 30%, at 25%.

Locate 10 on the horizontal axis and locate the point above 10.

Figure 1.2

The actual data displayed by the graph indicate that 25% of these marriages end in divorce after 10 years.

c. Here's a summary of what we found in parts (a) and (b).

Percentage of Marriages Ending in Divorce after 10 Years (Wife over 25 at Marriage)
Mathematical model: 24.5%
Actual data displayed by graph: 25.0%

The value obtained by evaluating the mathematical model, 24.5%, is close to, but slightly less than, the actual percentage of divorces, 25.0%. The difference between these percents is 25.0% − 24.5%, or 0.5%. The value given by the mathematical model underestimates the actual percent by only 0.5%, providing a fairly accurate description of the data. ■

✓ **CHECK POINT 6**

a. Use the appropriate formula from Example 6 on page 9 to determine the percentage of marriages ending in divorce after 15 years when the wife is under 18 at the time of marriage.

b. Use the appropriate line graph in **Figure 1.1** on page 8 to determine the percentage of marriages ending in divorce after 15 years when the wife is under 18 at the time of marriage.

c. Does the value given by the mathematical model underestimate or overestimate the actual percentage of marriages ending in divorce after 15 years as shown by the graph? By how much?

Achieving Success

Practice! Practice! Practice!

The way to learn algebra is by seeing solutions to examples and **doing exercises**. This means working the Check Points and the assigned exercises in the Exercise Sets. There are no alternatives. It's easy to read a solution, or watch your professor solve an example, and believe you know what to do. However, learning algebra requires that you actually **perform solutions by yourself**. Get in the habit of working exercises every day. The more time you spend solving exercises, the easier the process becomes. It's okay to take a short break after class, but start reviewing and working the assigned homework as soon as possible.

CONCEPT AND VOCABULARY CHECK

Fill in each blank so that the resulting statement is true.

1. A letter that represents a variety of different numbers is called a/an _____.

2. A combination of numbers, letters that represent numbers, and operation symbols is called an algebraic _____.

3. By replacing x with 60 in $2.5x$, we are _____ 60 for x. The process of finding $2.5 \cdot 60$ is called _____ $2.5x$ for $x = 60$.

4. A statement that two algebraic expressions are equal is called a/an _____. A value of the variable that makes such a statement true is called a/an _____.

5. A statement that expresses a relationship between two or more variables, such as $T = 0.3x + 40$, is called a/an _____. The process of finding such statements to describe real-world phenomena is called mathematical _____. Such statements, together with the meaning assigned to the variables, are called mathematical _____.

1.1 EXERCISE SET ▶ MyMathLab®

Practice Exercises

In Exercises 1–14, evaluate each expression for $x = 4$.

1. $x + 8$
2. $x + 10$
3. $12 - x$
4. $16 - x$
5. $5x$
6. $6x$
7. $\dfrac{28}{x}$
8. $\dfrac{36}{x}$
9. $5 + 3x$
10. $3 + 5x$
11. $2(x + 5)$
12. $5(x + 3)$
13. $\dfrac{12x - 8}{2x}$
14. $\dfrac{5x + 52}{3x}$

In Exercises 15–24, evaluate each expression for $x = 7$ and $y = 5$.

15. $2x + y$
16. $3x + y$
17. $2(x + y)$
18. $3(x + y)$
19. $4x - 3y$
20. $5x - 4y$
21. $\dfrac{21}{x} + \dfrac{35}{y}$
22. $\dfrac{50}{y} - \dfrac{14}{x}$
23. $\dfrac{2x - y + 6}{2y - x}$
24. $\dfrac{2y - x + 24}{2x - y}$

In Exercises 25–42, write each English phrase as an algebraic expression. Let the variable x represent the number.

25. four more than a number
26. six more than a number
27. four less than a number
28. six less than a number
29. the sum of a number and 4
30. the sum of a number and 6
31. nine subtracted from a number
32. three subtracted from a number
33. nine decreased by a number
34. three decreased by a number
35. three times a number, decreased by 5
36. five times a number, decreased by 3
37. one less than the product of 12 and a number
38. three less than the product of 13 and a number
39. the sum of 10 divided by a number and that number divided by 10
40. the sum of 20 divided by a number and that number divided by 20
41. six more than the quotient of a number and 30
42. four more than the quotient of 30 and a number

In Exercises 43–58, determine whether the given number is a solution of the equation.

43. $x + 14 = 20; 6$
44. $x + 17 = 22; 5$
45. $30 - y = 10; 20$
46. $50 - y = 20; 30$
47. $4z = 20; 10$
48. $5z = 30; 8$
49. $\dfrac{r}{6} = 8; 48$
50. $\dfrac{r}{9} = 7; 63$
51. $4m + 3 = 23; 6$
52. $3m + 4 = 19; 6$
53. $5a - 4 = 2a + 5; 3$
54. $5a - 3 = 2a + 6; 3$
55. $6(p - 4) = 3p; 8$
56. $4(p + 3) = 6p; 6$
57. $2(w + 1) = 3(w - 1); 7$
58. $3(w + 2) = 4(w - 3); 10$

In Exercises 59–74, write each sentence as an equation. Let the variable x represent the number.

59. Four times a number is 28.
60. Five times a number is 35.
61. The quotient of 14 and a number is $\frac{1}{2}$.
62. The quotient of a number and 8 is $\frac{1}{4}$.
63. The difference between 20 and a number is 5.
64. The difference between 40 and a number is 10.
65. The sum of twice a number and 6 is 16.
66. The sum of twice a number and 9 is 29.
67. Five less than 3 times a number gives 7.
68. Three less than 4 times a number gives 29.
69. The product of 4 and a number, increased by 5, is 33.
70. The product of 6 and a number, increased by 3, is 33.
71. The product of 4 and a number increased by 5 is 33.
72. The product of 6 and a number increased by 3 is 33.
73. Five times a number is equal to 24 decreased by the number.
74. Four times a number is equal to 25 decreased by the number.

Practice PLUS

75. Evaluate $\dfrac{x - y}{4}$ when x is 2 more than 7 times y and $y = 5$.
76. Evaluate $\dfrac{x - y}{3}$ when x is 2 more than 5 times y and $y = 4$.

77. Evaluate $4x + 3(y + 5)$ when x is 1 less than the quotient of y and 4 and $y = 12$.

78. Evaluate $3x + 4(y + 6)$ when x is 1 less than the quotient of y and 3, and $y = 15$.

79. **a.** Evaluate $2(x + 3y)$ for $x = 4$ and $y = 1$.

 b. Is the number you obtained in part (a) a solution of $5z - 30 = 40$?

80. **a.** Evaluate $3(2x + y)$ for $x = 1$ and $y = 5$.

 b. Is the number you obtained in part (a) a solution of $4z - 30 = 54$?

81. **a.** Evaluate $6x - 2y$ for $x = 3$ and $y = 6$.

 b. Is the number you obtained in part (a) a solution of $7w = 45 - 2w$?

82. **a.** Evaluate $5x - 14y$ for $x = 3$ and $y = \frac{1}{2}$.

 b. Is the number you obtained in part (a) a solution of $4w = 54 - 5w$?

Application Exercises

We opened the chapter with what it would have cost to see seven classic movies on the big screen. The bar graph shows the average price of a movie ticket for selected years from 1980 through 2013.

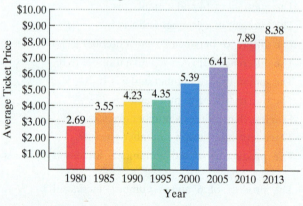

Average Price of a U.S. Movie Ticket

Sources: Motion Picture Association of America, National Association of Theater Owners (NATO), and Bureau of Labor Statistics (BLS)

Here is a mathematical model that approximates the data displayed by the bar graph:

$$T = 0.16n + 2.83.$$

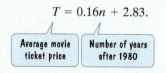

Average movie ticket price | Number of years after 1980

Use this formula to solve Exercises 83–84.

83. **a.** Use the formula to find the average ticket price 10 years after 1980, or in 1990. Does the mathematical model underestimate or overestimate the average ticket price shown by the bar graph for 1990? By how much?

 b. Does the mathematical model underestimate or overestimate the average ticket price shown by the bar graph for 2010? By how much?

84. **a.** Use the formula to find the average ticket price 5 years after 1980, or in 1985. Does the mathematical model underestimate or overestimate the average ticket price shown by the bar graph in the previous column for 1985? By how much?

 b. Does the mathematical model underestimate or overestimate the average ticket price shown by the bar graph for 2005? By how much?

We're just not that into marriage. Among the 2691 American adults surveyed by the Pew Research Center in 2010, 39% said marriage is optional and becoming obsolete, up from 28% who responded to the same survey in 1978. The bar graph shows the percentage of Americans for four selected ages who say that marriage is obsolete.

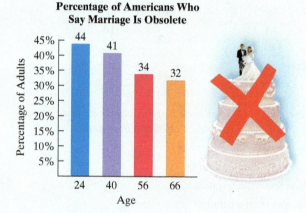

Percentage of Americans Who Say Marriage Is Obsolete

Source: Pew Research Center, "For Millennials, Parenthood Trumps Marriage," March 9, 2011 (www.pewresearch.org)

Here is a mathematical model that approximates the data displayed by the bar graph:

$$p = 52 - 0.3a.$$

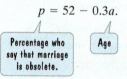

Percentage who say that marriage is obsolete. | Age

Use this formula to solve Exercises 85–86.

85. Does the mathematical model underestimate or overestimate the percentage of 24-year-olds who say that marriage is obsolete? By how much?

86. Does the mathematical model underestimate or overestimate the percentage of 66-year-olds who say that marriage is obsolete? By how much?

A bowler's handicap, H, is often found using the following formula:

$$H = 0.8(200 - A).$$

Bowler's handicap | Bowler's average score

A bowler's final score for a game is the score for that game increased by the handicap.

Use this information and the formula at the bottom of the previous page to solve Exercises 87–88.

87. a. If your average bowling score is 145, what is your handicap?

b. What would your final score be if you bowled 120 in a game?

88. a. If your average bowling score is 165, what is your handicap?

b. What would your final score be if you bowled 140 in a game?

Explaining the Concepts

Achieving Success

An effective way to understand something is to explain it to someone else. You can do this by using the Explaining the Concepts exercises that ask you to respond with verbal or written explanations. Speaking about a new concept uses a different part of your brain than thinking about the concept. Explaining new ideas verbally will quickly reveal any gaps in your understanding. It will also help you to remember new concepts for longer periods of time.

89. What is a variable?

90. What is an algebraic expression?

91. Explain how to evaluate $2 + 5x$ for $x = 3$.

92. If x represents the number, explain the difference between translating the following phrases:
a number decreased by 5
a number subtracted from 5.

93. What is an equation?

94. How do you tell the difference between an algebraic expression and an equation?

95. How do you determine whether a given number is a solution of an equation?

96. What is a mathematical model?

97. The bar graph for Exercises 85–86 shows a decline with increasing age in the percentage of Americans who say that marriage is obsolete. What explanations can you offer for this trend?

98. In Exercises 87–88, we used the formula $H = 0.8(200 - A)$ to find a bowler's handicap, H, where the variable A represents the bowler's average score. Describe what happens to the handicap when the average score is 200.

Critical Thinking Exercises

Make Sense? *In Exercises 99–102, determine whether each statement makes sense or does not make sense, and explain your reasoning.*

99. As I read this book, I write questions in the margins that I might ask in class.

100. I'm solving a problem that requires me to determine if 5 is a solution of $4x + 7$.

101. The model $T = 0.16n + 2.83$ describes the average movie ticket price, T, n years after 1980, so I can use it to estimate the average movie ticket price in 1980.

102. Because there are four quarters in a dollar, I can use the formula $q = 4d$ to determine the number of quarters, q, in d dollars.

In Exercises 103–106, determine whether each statement is true or false. If the statement is false, make the necessary change(s) to produce a true statement.

103. The algebraic expression for "3 less than a number" is the same as the algebraic expression for "a number decreased by 3."

104. Some algebraic expressions contain the equality symbol, =.

105. The algebraic expressions $3 + 2x$ and $(3 + 2)x$ do not mean the same thing.

106. The algebraic expression for "the quotient of a number and 6" is the same as the algebraic expression for "the quotient of 6 and a number."

In Exercises 107–108, define variables and write a formula that describes the relationship in each table.

107.

Number of Hours Worked	Salary
3	$60
4	$80
5	$100
6	$120

108.

Number of Workers	Number of Televisions Assembled
3	30
4	40
5	50
6	60

Preview Exercises

Exercises 109–111 will help you prepare for the material covered in the next section. In each exercise, use the given formula to perform the indicated operation with the two fractions.

109. $\dfrac{a}{b} \cdot \dfrac{c}{d} = \dfrac{a \cdot c}{b \cdot d}$; $\dfrac{3}{7} \cdot \dfrac{2}{5}$

110. $\dfrac{a}{b} \div \dfrac{c}{d} = \dfrac{a}{b} \cdot \dfrac{d}{c} = \dfrac{a \cdot d}{b \cdot c}$; $\dfrac{2}{3} \div \dfrac{7}{5}$

111. $\dfrac{a}{b} - \dfrac{c}{b} = \dfrac{a - c}{b}$; $\dfrac{9}{17} - \dfrac{5}{17}$

SECTION

1.2

Fractions in Algebra

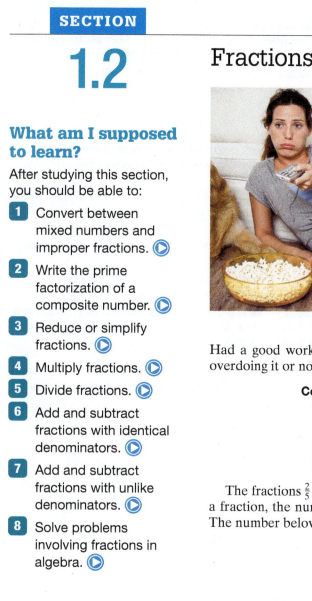

What am I supposed to learn?

After studying this section, you should be able to:

1 Convert between mixed numbers and improper fractions. ▶

2 Write the prime factorization of a composite number. ▶

3 Reduce or simplify fractions. ▶

4 Multiply fractions. ▶

5 Divide fractions. ▶

6 Add and subtract fractions with identical denominators. ▶

7 Add and subtract fractions with unlike denominators. ▶

8 Solve problems involving fractions in algebra. ▶

Had a good workout lately? If so, could you tell from your heart rate if you were overdoing it or not pushing yourself hard enough?

Couch-Potato Exercise

$$H = \frac{2}{5}(220 - a)$$

Heart rate, in beats per minute   Age

Working It

$$H = \frac{9}{10}(220 - a)$$

Heart rate, in beats per minute   Age

The fractions $\frac{2}{5}$ and $\frac{9}{10}$ provide the difference between these formulas. Recall that in a fraction, the number that is written above the fraction bar is called the **numerator**. The number below the fraction bar is called the **denominator**.

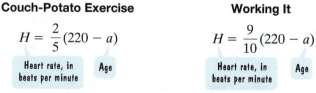

Numerator $\frac{2}{5}$ Fraction bar $\frac{9}{10}$ Numerator

Denominator   Denominator

The numerators and denominators of these fractions, 2, 5, 9, and 10, are examples of *natural numbers*. The **natural numbers** are the numbers that we use for counting.

Natural numbers   1, 2, 3, 4, 5, ...

The three dots after the 5 indicate that the list continues in the same manner without ending.

Fractions appear throughout algebra. The first part of this section provides a review of the arithmetic of fractions. Later in the section, we focus on fractions in algebra.

1 Convert between mixed numbers and improper fractions. ▶

Mixed Numbers and Improper Fractions

A **mixed number** consists of the sum of a natural number and a fraction, expressed without the use of an addition sign. Here is an example of a mixed number:

$$3\frac{4}{5}.$$

The natural number is 3 and the fraction is $\frac{4}{5}$. $3\frac{4}{5}$ means $3 + \frac{4}{5}$.

An **improper fraction** is a fraction whose numerator is greater than its denominator. An example of an improper fraction is $\frac{19}{5}$.

The mixed number $3\frac{4}{5}$ can be converted to the improper fraction $\frac{19}{5}$ using the following procedure:

Converting a Mixed Number to an Improper Fraction

1. Multiply the denominator of the fraction by the natural number and add the numerator to this product.
2. Place the result from step 1 over the denominator of the original mixed number.

EXAMPLE 1   Converting from a Mixed Number to an Improper Fraction

Convert $3\frac{4}{5}$ to an improper fraction.

Solution

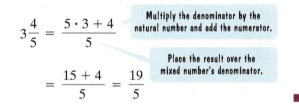

$$3\frac{4}{5} = \frac{5 \cdot 3 + 4}{5}$$

Multiply the denominator by the natural number and add the numerator.

Place the result over the mixed number's denominator.

$$= \frac{15 + 4}{5} = \frac{19}{5}$$

Great Question!

I'm a visual learner. How can I "see" that $3\frac{4}{5}$ and $\frac{19}{5}$ represent the same number?

Figure 1.3 illustrates that shading $3\frac{4}{5}$ circles is the same as shading $\frac{19}{5}$ of the circles.

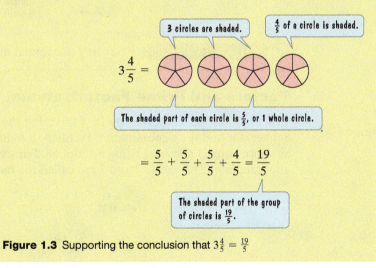

3 circles are shaded.

$\frac{4}{5}$ of a circle is shaded.

$$3\frac{4}{5} =$$

The shaded part of each circle is $\frac{5}{5}$, or 1 whole circle.

$$= \frac{5}{5} + \frac{5}{5} + \frac{5}{5} + \frac{4}{5} = \frac{19}{5}$$

The shaded part of the group of circles is $\frac{19}{5}$.

Figure 1.3 Supporting the conclusion that $3\frac{4}{5} = \frac{19}{5}$

✓ **CHECK POINT 1**   Convert $2\frac{5}{8}$ to an improper fraction.

An **improper fraction** can be converted to a mixed number using the following procedure:

> ### Converting an Improper Fraction to a Mixed Number
>
> 1. Divide the denominator into the numerator. Record the quotient (the result of the division) and the remainder.
> 2. Write the mixed number using the following form:
>
>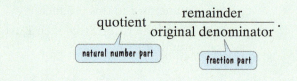
>
> $$\text{quotient } \frac{\text{remainder}}{\text{original denominator}}.$$

EXAMPLE 2   Converting from an Improper Fraction to a Mixed Number

Convert $\frac{42}{5}$ to a mixed number.

Solution   We use two steps to convert $\frac{42}{5}$ to a mixed number.

Step 1. Divide the denominator into the numerator.

$$\begin{array}{r} 8 \\ 5\overline{)42} \\ 40 \\ \hline 2 \end{array}$$

Step 2. Write the mixed number using quotient $\dfrac{\textbf{remainder}}{\textbf{original denominator}}$**. Thus,**

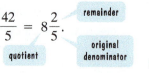

$$\frac{42}{5} = 8\frac{2}{5}.$$

☑ **CHECK POINT 2**   Convert $\frac{5}{3}$ to a mixed number.

---

② Write the prime factorization of a composite number. ▶

Factors and Prime Factorizations

Fractions can be simplified by first factoring the natural numbers that make up the numerator and the denominator. To **factor** a natural number means to write it as two or more natural numbers being multiplied. For example, 21 can be factored as $7 \cdot 3$. In the statement $7 \cdot 3 = 21$, 7 and 3 are called the **factors** and 21 is the **product**.

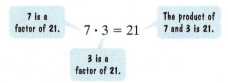

$$7 \cdot 3 = 21$$

7 is a factor of 21.   3 is a factor of 21.   The product of 7 and 3 is 21.

Are 7 and 3 the only factors of 21? The answer is no because 21 can also be factored as $1 \cdot 21$. Thus, 1 and 21 are also factors of 21. The factors of 21 are 1, 3, 7, and 21.

Unlike the number 21, some natural numbers have only two factors: the number itself and 1. For example, the number 7 has only two factors: 7 (the number itself) and 1. The only way to factor 7 is $1 \cdot 7$ or, equivalently, $7 \cdot 1$. For this reason, 7 is called a *prime number*.

Prime Numbers

A **prime number** is a natural number greater than 1 that has only itself and 1 as factors.

The first ten prime numbers are

$$2, 3, 5, 7, 11, 13, 17, 19, 23, \text{ and } 29.$$

Can you see why the natural number 15 is not in this list? In addition to having 15 and 1 as factors ($15 = 1 \cdot 15$), it also has factors of 3 and 5 ($15 = 3 \cdot 5$). The number 15 is an example of a *composite number*.

Composite Numbers

A **composite number** is a natural number greater than 1 that is not a prime number.

Every composite number can be expressed as the product of prime numbers. For example, the composite number 45 can be expressed as

$$45 = 3 \cdot 3 \cdot 5.$$

This product contains only prime numbers: 3 and 5.

Expressing a composite number as the product of prime numbers is called the **prime factorization** of that composite number. The prime factorization of 45 is $3 \cdot 3 \cdot 5$. The order in which we write these factors does not matter. This means that

$$45 = 3 \cdot 3 \cdot 5 \quad \text{or} \quad 45 = 5 \cdot 3 \cdot 3 \quad \text{or} \quad 45 = 3 \cdot 5 \cdot 3.$$

To find the prime factorization of a composite number, begin by selecting any two numbers, excluding 1 and the number itself, whose product is the number to be factored. If one or both of the factors are not prime numbers, continue by factoring each composite number. Stop when all numbers in the factorization are prime.

EXAMPLE 3   Prime Factorization of a Composite Number

Find the prime factorization of 100.

Solution   Begin by selecting any two numbers, excluding 1 and 100, whose product is 100. Here is one possibility:

$$100 = 4 \cdot 25.$$

Because the factors 4 and 25 are not prime, we factor each of these composite numbers.

$$100 = 4 \cdot 25 \qquad \text{This is our first factorization.}$$
$$ = 2 \cdot 2 \cdot 5 \cdot 5 \qquad \text{Factor 4 and 25.}$$

Notice that 2 and 5 are both prime. The prime factorization of 100 is $2 \cdot 2 \cdot 5 \cdot 5$. ∎

✓ **CHECK POINT 3**   Find the prime factorization of 36.

3 Reduce or simplify fractions.

Figure 1.4

Reducing Fractions

Two fractions are **equivalent** if they represent the same value. Writing a fraction as an equivalent fraction with a smaller denominator is called **reducing a fraction**. A fraction is **reduced to its lowest terms**, or **simplified**, when the numerator and denominator have no common factors other than 1.

Look at the rectangle in **Figure 1.4**. Can you see that it is divided into 6 equal parts? Of these 6 parts, 4 of the parts are red. Thus, $\frac{4}{6}$ of the rectangle is red.

The rectangle in **Figure 1.4** is also divided into 3 equal stacks and 2 of the stacks are red. Thus, $\frac{2}{3}$ of the rectangle is red. Because both $\frac{4}{6}$ and $\frac{2}{3}$ of the rectangle are red, we can conclude that $\frac{4}{6}$ and $\frac{2}{3}$ are equivalent fractions.

How can we show that $\frac{4}{6} = \frac{2}{3}$ without using **Figure 1.4**? Prime factorizations of 4 and 6 play an important role in the process. So does the **Fundamental Principle of Fractions**.

Fundamental Principle of Fractions

In words: The value of a fraction does not change if both the numerator and the denominator are divided (or multiplied) by the same nonzero number.

In algebraic language: If $\frac{a}{b}$ is a fraction and c is a nonzero number, then

$$\frac{a \cdot c}{b \cdot c} = \frac{a}{b}.$$

We use prime factorizations and the Fundamental Principle to reduce $\frac{4}{6}$ to its lowest terms as follows:

$$\frac{4}{6} = \frac{2 \cdot 2}{3 \cdot 2} = \frac{2}{3}.$$

Write prime factorizations of 4 and 6.

Divide the numerator and the denominator by the common prime factor, **2.**

Here is a procedure for writing a fraction in lowest terms:

Reducing a Fraction to Its Lowest Terms

1. Write the prime factorizations of the numerator and the denominator.
2. Divide the numerator and the denominator by the greatest common factor, the product of all factors common to both.

Division lines can be used to show dividing out common factors from a fraction's numerator and denominator:

$$\frac{4}{6} = \frac{2 \cdot \cancel{2}}{3 \cdot \cancel{2}} = \frac{2}{3}.$$

Great Question!

When can I divide out numbers that are common to a fraction's numerator and denominator?

When reducing a fraction to its lowest terms, only *factors* that are common to the numerator and the denominator can be divided out. **If you have not factored** and expressed the numerator and denominator in terms of multiplication, **do not divide out.**

Correct:

$$\frac{2 \cdot \cancel{2}}{3 \cdot \cancel{2}} = \frac{2}{3}$$

Incorrect:

$$\frac{2 + \cancel{2}}{3 + \cancel{2}} = \frac{2}{3}$$

Note that $\frac{2+2}{3+2} = \frac{4}{5}$, not $\frac{2}{3}$.

| EXAMPLE 4 | Reducing Fractions |

Reduce each fraction to its lowest terms:

a. $\dfrac{6}{14}$ **b.** $\dfrac{15}{75}$ **c.** $\dfrac{25}{11}$ **d.** $\dfrac{11}{33}$.

Solution   For each fraction, begin with the prime factorization of the numerator and the denominator.

a. $\dfrac{6}{14} = \dfrac{3 \cdot 2}{7 \cdot 2} = \dfrac{3}{7}$ 2 is the greatest common factor of 6 and 14. Divide the numerator and the denominator by 2.

> Including 1 as a factor is helpful when all other factors can be divided out.

b. $\dfrac{15}{75} = \dfrac{3 \cdot 5}{3 \cdot 25} = \dfrac{1 \cdot 3 \cdot 5}{3 \cdot 5 \cdot 5} = \dfrac{1}{5}$ $3 \cdot 5$, or 15, is the greatest common factor of 15 and 75. Divide the numerator and the denominator by $3 \cdot 5$.

c. $\dfrac{25}{11} = \dfrac{5 \cdot 5}{1 \cdot 11}$

Because 25 and 11 share no common factor (other than 1), $\dfrac{25}{11}$ is already reduced to its lowest terms.

d. $\dfrac{11}{33} = \dfrac{1 \cdot 11}{3 \cdot 11} = \dfrac{1}{3}$ 11 is the greatest common factor of 11 and 33. Divide the numerator and denominator by 11.   ■

When reducing fractions, it may not always be necessary to write prime factorizations. In some cases, you can use inspection to find the greatest common factor of the numerator and the denominator. For example, when reducing $\frac{15}{75}$, you can use 15 rather than $3 \cdot 5$:

$$\dfrac{15}{75} = \dfrac{1 \cdot 15}{5 \cdot 15} = \dfrac{1}{5}.$$

| ✓ CHECK POINT 4 | Reduce each fraction to its lowest terms:

a. $\dfrac{10}{15}$ **b.** $\dfrac{42}{24}$ **c.** $\dfrac{13}{15}$ **d.** $\dfrac{9}{45}$.

---

④ Multiply fractions. ▶

Multiplying Fractions

The result of multiplying two fractions is called their **product**.

> ### Multiplying Fractions
>
> In words: The product of two or more fractions is the product of their numerators divided by the product of their denominators.
>
> In algebraic language: If $\frac{a}{b}$ and $\frac{c}{d}$ are fractions, then
>
> $$\dfrac{a}{b} \cdot \dfrac{c}{d} = \dfrac{a \cdot c}{b \cdot d}.$$

Here is an example that illustrates the rule in the previous box:

$$\dfrac{3}{8} \cdot \dfrac{5}{11} = \dfrac{3 \cdot 5}{8 \cdot 11} = \dfrac{15}{88}.$$

> The product of $\frac{3}{8}$ and $\frac{5}{11}$ is $\frac{15}{88}$.

> Multiply numerators and multiply denominators.

EXAMPLE 5 Multiplying Fractions

Multiply. If possible, reduce the product to its lowest terms:

a. $\dfrac{3}{7} \cdot \dfrac{2}{5}$ **b.** $5 \cdot \dfrac{7}{12}$ **c.** $\dfrac{2}{3} \cdot \dfrac{9}{4}$ **d.** $\left(3\dfrac{2}{3}\right)\left(1\dfrac{1}{4}\right)$.

Solution

a. $\dfrac{3}{7} \cdot \dfrac{2}{5} = \dfrac{3 \cdot 2}{7 \cdot 5} = \dfrac{6}{35}$ Multiply numerators and multiply denominators.

b. $5 \cdot \dfrac{7}{12} = \dfrac{5}{1} \cdot \dfrac{7}{12} = \dfrac{5 \cdot 7}{1 \cdot 12} = \dfrac{35}{12}$ or $2\dfrac{11}{12}$ Write 5 as $\frac{5}{1}$. Then multiply numerators and multiply denominators.

c. $\dfrac{2}{3} \cdot \dfrac{9}{4} = \dfrac{2 \cdot 9}{3 \cdot 4} = \dfrac{18}{12} = \dfrac{3 \cdot \cancel{6}}{2 \cdot \cancel{6}} = \dfrac{3}{2}$ or $1\dfrac{1}{2}$

Simplify $\frac{18}{12}$; 6 is the greatest common factor of 18 and 12.

d. $\left(3\dfrac{2}{3}\right)\left(1\dfrac{1}{4}\right) = \dfrac{11}{3} \cdot \dfrac{5}{4} = \dfrac{11 \cdot 5}{3 \cdot 4} = \dfrac{55}{12}$ or $4\dfrac{7}{12}$ ∎

✓ **CHECK POINT 5** Multiply. If possible, reduce the product to its lowest terms:

a. $\dfrac{4}{11} \cdot \dfrac{2}{3}$ **b.** $6 \cdot \dfrac{3}{5}$ **c.** $\dfrac{3}{7} \cdot \dfrac{2}{3}$ **d.** $\left(3\dfrac{2}{5}\right)\left(1\dfrac{1}{2}\right)$.

Great Question!

Can I divide numerators and denominators by common factors before I multiply fractions?

Yes, you can divide numerators and denominators by common factors before performing multiplication. Then multiply the remaining factors in the numerators and multiply the remaining factors in the denominators. For example,

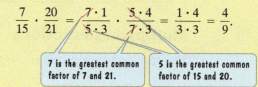

$$\dfrac{7}{15} \cdot \dfrac{20}{21} = \dfrac{7 \cdot 1}{5 \cdot 3} \cdot \dfrac{5 \cdot 4}{7 \cdot 3} = \dfrac{1 \cdot 4}{3 \cdot 3} = \dfrac{4}{9}.$$

7 is the greatest common factor of 7 and 21.   5 is the greatest common factor of 15 and 20.

The divisions involving the common factors, 7 and 5, are often shown as follows:

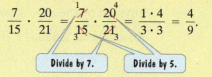

$$\dfrac{7}{15} \cdot \dfrac{20}{21} = \dfrac{\overset{1}{\cancel{7}}}{\underset{3}{\cancel{15}}} \cdot \dfrac{\overset{4}{\cancel{20}}}{\underset{3}{\cancel{21}}} = \dfrac{1 \cdot 4}{3 \cdot 3} = \dfrac{4}{9}.$$

Divide by 7.   Divide by 5.

---

5 Divide fractions. ⊙

Dividing Fractions

The result of dividing two fractions is called their **quotient**. A geometric figure is useful for developing a process for determining the quotient of two fractions.

Consider the division

$$\dfrac{4}{5} \div \dfrac{1}{10}.$$

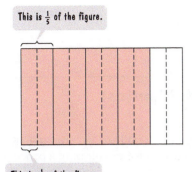

This is $\frac{1}{5}$ of the figure.

This is $\frac{1}{10}$ of the figure.

Figure 1.5

We want to know how many $\frac{1}{10}$'s are in $\frac{4}{5}$. We can use **Figure 1.5** to find this quotient. The rectangle is divided into fifths. The dashed lines further divide the rectangle into tenths.

Figure 1.5 shows that $\frac{4}{5}$ of the rectangle is red. How many $\frac{1}{10}$'s of the rectangle does this include? Can you see that this includes eight of the $\frac{1}{10}$ pieces? Thus, there are eight $\frac{1}{10}$'s in $\frac{4}{5}$:

$$\frac{4}{5} \div \frac{1}{10} = 8.$$

We can obtain the quotient 8 in the following way:

$$\frac{4}{5} \div \frac{1}{10} = \frac{4}{5} \cdot \frac{10}{1} = \frac{4 \cdot 10}{5 \cdot 1} = \frac{40}{5} = 8.$$

Change the division to multiplication.

Invert the divisor, $\frac{1}{10}$.

By inverting the divisor, $\frac{1}{10}$, and obtaining $\frac{10}{1}$, we are writing the divisor's *reciprocal*. Two fractions are **reciprocals** of each other if their product is 1. Thus, $\frac{1}{10}$ and $\frac{10}{1}$ are reciprocals because $\frac{1}{10} \cdot \frac{10}{1} = 1$.

Generalizing from the result above and using the word *reciprocal*, we obtain the following rule for dividing fractions:

Dividing Fractions

In words: The quotient of two fractions is the first fraction multiplied by the reciprocal of the second fraction.

In algebraic language: If $\frac{a}{b}$ and $\frac{c}{d}$ are fractions and $\frac{c}{d}$ is not 0, then

$$\frac{a}{b} \div \frac{c}{d} = \frac{a}{b} \cdot \frac{d}{c}.$$

Change division to multiplication.

Invert $\frac{c}{d}$ and write its reciprocal.

EXAMPLE 6   Dividing Fractions

a. $\dfrac{2}{3} \div \dfrac{7}{15}$ **b.** $\dfrac{3}{4} \div 5$ **c.** $4\dfrac{3}{4} \div 1\dfrac{1}{2}$.

Solution

a. $\dfrac{2}{3} \div \dfrac{7}{15} = \dfrac{2}{3} \cdot \dfrac{15}{7} = \dfrac{2 \cdot 15}{3 \cdot 7} = \dfrac{30}{21} = \dfrac{10 \cdot 3}{7 \cdot 3} = \dfrac{10}{7}$ or $1\dfrac{3}{7}$

Change division to multiplication.

Invert $\frac{7}{15}$ and write its reciprocal.

Simplify: 3 is the greatest common factor of 30 and 21.

b. $\dfrac{3}{4} \div 5 = \dfrac{3}{4} \div \dfrac{5}{1} = \dfrac{3}{4} \cdot \dfrac{1}{5} = \dfrac{3 \cdot 1}{4 \cdot 5} = \dfrac{3}{20}$

Change division to multiplication.

Invert $\frac{5}{1}$ and write its reciprocal.

c. $4\dfrac{3}{4} \div 1\dfrac{1}{2} = \dfrac{19}{4} \div \dfrac{3}{2} = \dfrac{19}{4} \cdot \dfrac{2}{3} = \dfrac{19 \cdot 2}{4 \cdot 3} = \dfrac{38}{12} = \dfrac{19 \cdot 2}{6 \cdot 2} = \dfrac{19}{6}$ or $3\dfrac{1}{6}$ ∎

☑ **CHECK POINT 6**   Divide:

a. $\dfrac{5}{4} \div \dfrac{3}{8}$ **b.** $\dfrac{2}{3} \div 3$ **c.** $3\dfrac{3}{8} \div 2\dfrac{1}{4}$

6 Add and subtract fractions with identical denominators.

Adding and Subtracting Fractions with Identical Denominators

The result of adding two fractions is called their **sum**. The result of subtracting two fractions is called their **difference**. A geometric figure is useful for developing a process for determining the sum or difference of two fractions with identical denominators.

Consider the addition

$$\dfrac{3}{7} + \dfrac{2}{7}.$$

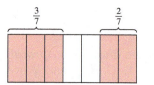

We can use **Figure 1.6** to find this sum. The rectangle is divided into sevenths. On the left, $\frac{3}{7}$ of the rectangle is red. On the right, $\frac{2}{7}$ of the rectangle is red. Including both the left and the right, a total of $\frac{5}{7}$ of the rectangle is red. Thus,

$$\dfrac{3}{7} + \dfrac{2}{7} = \dfrac{5}{7}.$$

Figure 1.6

We can obtain the sum $\frac{5}{7}$ in the following way:

$$\dfrac{3}{7} + \dfrac{2}{7} = \dfrac{3+2}{7} = \dfrac{5}{7}.$$

> Add numerators and put this result over the common denominator.

Generalizing from this result gives us the following rule:

Adding and Subtracting Fractions with Identical Denominators

In words: The sum or difference of two fractions with identical denominators is the sum or difference of their numerators over the common denominator.

In algebraic language: If $\frac{a}{b}$ and $\frac{c}{b}$ are fractions, then

$$\dfrac{a}{b} + \dfrac{c}{b} = \dfrac{a+c}{b} \quad \text{and} \quad \dfrac{a}{b} - \dfrac{c}{b} = \dfrac{a-c}{b}.$$

EXAMPLE 7   Adding and Subtracting Fractions with Identical Denominators

Perform the indicated operations:

a. $\dfrac{3}{11} + \dfrac{4}{11}$ **b.** $\dfrac{11}{12} - \dfrac{5}{12}$ **c.** $5\dfrac{1}{4} - 2\dfrac{3}{4}.$

Solution

a. $\dfrac{3}{11} + \dfrac{4}{11} = \dfrac{3+4}{11} = \dfrac{7}{11}$

b. $\dfrac{11}{12} - \dfrac{5}{12} = \dfrac{11-5}{12} = \dfrac{6}{12} = \dfrac{1 \cdot 6}{2 \cdot 6} = \dfrac{1}{2}$

c. $5\dfrac{1}{4} - 2\dfrac{3}{4} = \dfrac{21}{4} - \dfrac{11}{4} = \dfrac{21-11}{4} = \dfrac{10}{4} = \dfrac{2 \cdot 5}{2 \cdot 2} = \dfrac{5}{2}$ or $2\dfrac{1}{2}$ ∎

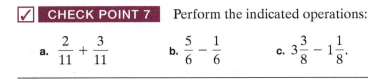

☑ **CHECK POINT 7** Perform the indicated operations:

a. $\dfrac{2}{11} + \dfrac{3}{11}$ **b.** $\dfrac{5}{6} - \dfrac{1}{6}$ **c.** $3\dfrac{3}{8} - 1\dfrac{1}{8}$.

7 Add and subtract fractions with unlike denominators. ▶

Adding and Subtracting Fractions with Unlike Denominators

How do we add or subtract fractions with different denominators? We must first rewrite them as equivalent fractions with the same denominator. We do this by using the Fundamental Principle of Fractions: The value of a fraction does not change if the numerator and the denominator are multiplied by the same nonzero number. Thus, if $\frac{a}{b}$ is a fraction and c is a nonzero number, then

$$\frac{a}{b} = \frac{a \cdot c}{b \cdot c}.$$

EXAMPLE 8   Writing an Equivalent Fraction

Write $\frac{3}{4}$ as an equivalent fraction with a denominator of 16.

Solution   To obtain a denominator of 16, we must multiply the denominator of the given fraction, $\frac{3}{4}$, by 4. So that we do not change the value of the fraction, we also multiply the numerator by 4.

$$\frac{3}{4} = \frac{3 \cdot 4}{4 \cdot 4} = \frac{12}{16} \quad \blacksquare$$

☑ **CHECK POINT 8**   Write $\frac{2}{3}$ as an equivalent fraction with a denominator of 21.

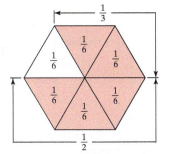

Figure 1.7 $\frac{1}{2} + \frac{1}{3} = \frac{5}{6}$

Equivalent fractions can be used to add fractions with different denominators, such as $\frac{1}{2}$ and $\frac{1}{3}$. **Figure 1.7** indicates that the sum of half the whole figure and one-third of the whole figure results in 5 parts out of 6, or $\frac{5}{6}$, of the figure. Thus,

$$\frac{1}{2} + \frac{1}{3} = \frac{5}{6}.$$

We can obtain the sum $\frac{5}{6}$ if we rewrite each fraction as an equivalent fraction with a denominator of 6.

$$\frac{1}{2} + \frac{1}{3} = \frac{1 \cdot 3}{2 \cdot 3} + \frac{1 \cdot 2}{3 \cdot 2}$$ Rewrite each fraction as an equivalent fraction with a denominator of 6.

$$= \frac{3}{6} + \frac{2}{6}$$ Perform the multiplications. We now have a common denominator.

$$= \frac{3 + 2}{6}$$ Add the numerators and place this sum over the common denominator.

$$= \frac{5}{6}$$ Perform the addition.

When adding $\frac{1}{2}$ and $\frac{1}{3}$, there are many common denominators that we can use, such as 6, 12, 18, and so on. The given denominators, 2 and 3, divide into all of these numbers. However, the denominator 6 is the smallest number that 2 and 3 divide into. For this reason, 6 is called the *least common denominator*, abbreviated LCD.

Adding and Subtracting Fractions with Unlike Denominators

1. Rewrite the fractions as equivalent fractions with the least common denominator.
2. Add or subtract the numerators, putting this result over the common denominator.

EXAMPLE 9 Adding and Subtracting Fractions with Unlike Denominators

Perform the indicated operation:

a. $\dfrac{1}{5} + \dfrac{3}{4}$

b. $\dfrac{3}{4} - \dfrac{1}{6}$

c. $2\dfrac{7}{15} - 1\dfrac{4}{5}.$

Solution

a. Just by looking at $\frac{1}{5} + \frac{3}{4}$, can you tell that the smallest number divisible by both 5 and 4 is 20? Thus, the least common denominator for the denominators 5 and 4 is 20. We rewrite both fractions as equivalent fractions with the least common denominator, 20.

Discover for Yourself

Try Example 9(a), $\frac{1}{5} + \frac{3}{4}$, using a common denominator of 40. Because both 5 and 4 divide into 40, 40 is a common denominator, although not the least common denominator. Describe what happens. What is the advantage of using the least common denominator?

$$\dfrac{1}{5} + \dfrac{3}{4} = \dfrac{1 \cdot 4}{5 \cdot 4} + \dfrac{3 \cdot 5}{4 \cdot 5}$$

To obtain denominators of 20, multiply the numerator and denominator of the first fraction by 4 and the second fraction by 5.

$$= \dfrac{4}{20} + \dfrac{15}{20}$$

Perform the multiplications.

$$= \dfrac{4 + 15}{20}$$

Add the numerators and put this sum over the least common denominator.

$$= \dfrac{19}{20}$$

Perform the addition.

b. By looking at $\frac{3}{4} - \frac{1}{6}$, can you tell that the smallest number divisible by both 4 and 6 is 12? Thus, the least common denominator for the denominators 4 and 6 is 12. We rewrite both fractions as equivalent fractions with the least common denominator, 12.

$$\dfrac{3}{4} - \dfrac{1}{6} = \dfrac{3 \cdot 3}{4 \cdot 3} - \dfrac{1 \cdot 2}{6 \cdot 2}$$

To obtain denominators of 12, multiply the numerator and denominator of the first fraction by 3 and the second fraction by 2.

$$= \dfrac{9}{12} - \dfrac{2}{12}$$

Perform the multiplications.

$$= \dfrac{9 - 2}{12}$$

Subtract the numerators and put this difference over the least common denominator.

$$= \dfrac{7}{12}$$

Perform the subtraction.

c. $2\dfrac{7}{15} - 1\dfrac{4}{5} = \dfrac{37}{15} - \dfrac{9}{5}$ Convert each mixed number to an improper fraction.

> The smallest number divisible by both 15 and 5 is 15. Thus, the least common denominator for the denominators 15 and 5 is 15. Because the first fraction already has a denominator of 15, we only have to rewrite the second fraction.

$$= \dfrac{37}{15} - \dfrac{9 \cdot 3}{5 \cdot 3}$$ To obtain a denominator of 15, multiply the numerator and denominator of the second fraction by 3.

$$= \dfrac{37}{15} - \dfrac{27}{15}$$ Perform the multiplications.

$$= \dfrac{37 - 27}{15}$$ Subtract the numerators and put this difference over the common denominator.

$$= \dfrac{10}{15}$$ Perform the subtraction.

$$= \dfrac{2 \cdot 5}{3 \cdot 5} = \dfrac{2}{3}$$ Reduce to lowest terms. ■

☑ **CHECK POINT 9**   Perform the indicated operation:

a. $\dfrac{1}{2} + \dfrac{3}{5}$ **b.** $\dfrac{4}{3} - \dfrac{3}{4}$ **c.** $3\dfrac{1}{6} - 1\dfrac{11}{12}$.

EXAMPLE 10   Using Prime Factorizations to Find the LCD

Perform the indicated operation: $\dfrac{1}{15} + \dfrac{7}{24}$.

Solution   We need to find the least common denominator first. Using inspection, it is difficult to determine the smallest number divisible by both 15 and 24. We will use their prime factorizations to find the least common denominator:

$$15 = 5 \cdot 3 \quad \text{and} \quad 24 = 8 \cdot 3 = 2 \cdot 2 \cdot 2 \cdot 3.$$

The different prime factors are 5, 3, and 2. The least common denominator is obtained by using the greatest number of times each factor appears in any prime factorization. Because 5 and 3 appear as prime factors and 2 is a factor of 24 three times, the least common denominator is

$$5 \cdot 3 \cdot 2 \cdot 2 \cdot 2 = 5 \cdot 3 \cdot 8 = 120.$$

Now we can rewrite both fractions as equivalent fractions with the least common denominator, 120.

$$\dfrac{1}{15} + \dfrac{7}{24} = \dfrac{1 \cdot 8}{15 \cdot 8} + \dfrac{7 \cdot 5}{24 \cdot 5}$$ To obtain denominators of 120, multiply the numerator and denominator of the first fraction by 8 and the second fraction by 5.

$$= \dfrac{8}{120} + \dfrac{35}{120}$$ Perform the multiplications.

$$= \dfrac{8 + 35}{120}$$ Add the numerators and put this sum over the least common denominator.

$$= \dfrac{43}{120}$$ Perform the addition. ■

Great Question!

I see that least common denominators are used to add and subtract fractions with unlike denominators. Is it ever necessary to use least common denominators when multiplying or dividing fractions?

No. You do not need least common denominators to multiply or divide fractions.

☑ **CHECK POINT 10**   Perform the indicated operation: $\dfrac{3}{10} + \dfrac{7}{12}$.

8 Solve problems involving fractions in algebra. ▶

Fractions in Algebra

Fractions appear throughout algebra. Operations with fractions can be used to determine whether a particular fraction is a solution of an equation.

> **EXAMPLE 11** Determining Whether Fractions Are Solutions of Equations

Determine whether the given number is a solution of the equation.

a. $x + \dfrac{1}{4}x = 7; \, 6\dfrac{2}{5}$ **b.** $\dfrac{1}{7} - w = \dfrac{1}{2}w; \, \dfrac{2}{21}$

Solution

a. To determine whether $6\frac{2}{5}$ is a solution of $x + \frac{1}{4}x = 7$, we begin by converting $6\frac{2}{5}$ from a mixed number to an improper fraction.

$$6\frac{2}{5} = \frac{5 \cdot 6 + 2}{5} = \frac{30 + 2}{5} = \frac{32}{5}$$

Now we substitute $\frac{32}{5}$ for x.

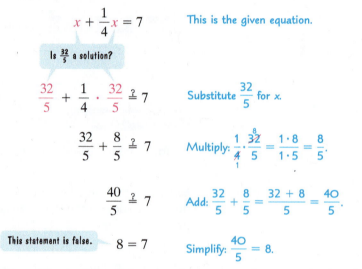

$$x + \frac{1}{4}x = 7 \qquad \text{This is the given equation.}$$

Is $\frac{32}{5}$ a solution?

$$\frac{32}{5} + \frac{1}{4} \cdot \frac{32}{5} \overset{?}{=} 7 \qquad \text{Substitute } \frac{32}{5} \text{ for } x.$$

$$\frac{32}{5} + \frac{8}{5} \overset{?}{=} 7 \qquad \text{Multiply: } \frac{1}{\underset{1}{\cancel{4}}} \cdot \frac{\overset{8}{\cancel{32}}}{5} = \frac{1 \cdot 8}{1 \cdot 5} = \frac{8}{5}.$$

$$\frac{40}{5} \overset{?}{=} 7 \qquad \text{Add: } \frac{32}{5} + \frac{8}{5} = \frac{32 + 8}{5} = \frac{40}{5}.$$

This statement is false. $8 = 7 \qquad \text{Simplify: } \frac{40}{5} = 8.$

Because the values on both sides of the equation are not the same, the fraction $\frac{32}{5}$, or equivalently $6\frac{2}{5}$, is not a solution of the equation.

b.

$$\frac{1}{7} - w = \frac{1}{2}w \qquad \text{This is the given equation.}$$

Is $\frac{2}{21}$ a solution?

$$\frac{1}{7} - \frac{2}{21} \overset{?}{=} \frac{1}{2} \cdot \frac{2}{21} \qquad \text{Substitute } \frac{2}{21} \text{ for } w.$$

$$\frac{1}{7} - \frac{2}{21} \overset{?}{=} \frac{1}{21} \qquad \text{Multiply: } \frac{1}{\underset{1}{\cancel{2}}} \cdot \frac{\overset{1}{\cancel{2}}}{21} = \frac{1 \cdot 1}{1 \cdot 21} = \frac{1}{21}.$$

$$\frac{1 \cdot 3}{7 \cdot 3} - \frac{2}{21} \overset{?}{=} \frac{1}{21} \qquad \text{Turn to the subtraction. The LCD is 21, so multiply the numerator and denominator of the first fraction by 3.}$$

$$\frac{3}{21} - \frac{2}{21} \overset{?}{=} \frac{1}{21} \qquad \text{Perform the multiplications.}$$

This statement is true. $\dfrac{1}{21} = \dfrac{1}{21} \qquad \text{Subtract: } \frac{3}{21} - \frac{2}{21} = \frac{3 - 2}{21} = \frac{1}{21}.$

Because the values on both sides of the equation are the same, the fraction $\frac{2}{21}$ is a solution of the equation. ∎

☑ **CHECK POINT 11** Determine whether the given number is a solution of the equation.

a. $x - \dfrac{2}{9}x = 1; 1\dfrac{2}{7}$

b. $\dfrac{1}{5} - w = \dfrac{1}{3}w; \dfrac{3}{20}$

In Section 1.1, we translated phrases into algebraic expressions and sentences into equations. When these phrases and sentences involve fractions, the word *of* frequently appears. **When used with fractions, the word *of* represents multiplication.** For example, the phrase "$\dfrac{2}{5}$ of a number" can be represented by the algebraic expression $\dfrac{2}{5} \cdot x$, or $\dfrac{2}{5}x$.

EXAMPLE 12 Algebraic Representations of Phrases and Sentences with Fractions

Translate from English to an algebraic expression or equation, whichever is appropriate. Let the variable x represent the number.

a. $\dfrac{1}{3}$ of a number increased by 5 is half of that number.

b. $\dfrac{1}{4}$ of a number, decreased by 7

Solution Part (a) is a sentence, so we will translate into an equation. Part (b) is a phrase, so we will translate into an algebraic expression.

a.

| $\frac{1}{3}$ | of | a number increased by 5 | is | half | of | that number. |

$$\frac{1}{3} \cdot (x+5) = \frac{1}{2} \cdot x$$

The equation for "$\dfrac{1}{3}$ of a number increased by 5 is half of that number" is $\dfrac{1}{3}(x + 5) = \dfrac{1}{2}x$.

b.

| $\frac{1}{4}$ | of | a number | , | decreased by 7 |

$$\frac{1}{4} \cdot x \quad - \quad 7$$

The algebraic expression for "$\dfrac{1}{4}$ of a number, decreased by 7" is $\dfrac{1}{4}x - 7$. ■

☑ **CHECK POINT 12** Translate from English to an algebraic expression or equation, whichever is appropriate. Let the variable x represent the number.

a. $\dfrac{2}{3}$ of a number decreased by 6

b. $\dfrac{3}{4}$ of a number, decreased by 2, is $\dfrac{1}{5}$ of that number.

Many formulas and mathematical models contain fractions. For example, consider temperatures on the Celsius scale and on the Fahrenheit scale, as shown in **Figure 1.8**. The formula $C = \dfrac{5}{9}(F - 32)$ expresses the relationship between Fahrenheit temperature, F, and Celsius temperature, C.

Figure 1.8 The Celsius scale is on the left and the Fahrenheit scale is on the right.

The Formula	What the Formula Tells Us
$C = \dfrac{5}{9}(F - 32)$	If 32 is subtracted from the Fahrenheit temperature, $F - 32$, and this difference is multiplied by $\frac{5}{9}$, the resulting product, $\frac{5}{9}(F - 32)$, gives the Celsius temperature.

EXAMPLE 13   Evaluating a Formula Containing a Fraction

The temperature on a warm summer day is 86°F. Use the formula $C = \frac{5}{9}(F - 32)$ to find the equivalent temperature on the Celsius scale.

Solution   Because the temperature is 86°F, we substitute 86 for F in the given formula. Then we determine the value of C.

$$C = \frac{5}{9}(F - 32) \qquad \text{This is the given formula.}$$

$$C = \frac{5}{9}(86 - 32) \qquad \text{Replace } F \text{ with 86.}$$

$$C = \frac{5}{9}(54) \qquad \text{Work inside parentheses first: } 86 - 32 = 54.$$

$$C = 30 \qquad \text{Multiply: } \frac{5}{9}(54) = \frac{5}{9} \cdot \frac{\overset{6}{\cancel{54}}}{1} = \frac{5 \cdot 6}{1 \cdot 1} = \frac{30}{1} = 30.$$

Thus, 86°F is equivalent to 30°C. ∎

✓ **CHECK POINT 13**   The temperature on a warm spring day is 77°F. Use the formula $C = \frac{5}{9}(F - 32)$ to find the equivalent temperature on the Celsius scale.

CONCEPT AND VOCABULARY CHECK

Fill in each blank so that the resulting statement is true.

1. In the fraction $\frac{2}{5}$, the number 2 is called the _____ and the number 5 is called the _____ .

2. The number $3\frac{2}{5}$ is called a/an _____ number and the number $\frac{17}{5}$ is called a/an _____ fraction.

3. The number $3\frac{2}{5}$ can be converted to $\frac{17}{5}$ by multiplying _____ and _____ , adding _____ , and placing the result over _____ .

4. The numbers that we use for counting $(1, 2, 3, 4, 5, \ldots)$ are called _____ numbers.

5. Among the numbers $1, 2, 3, 4, 5, \ldots$, a number greater than 1 that has only itself and 1 as factors is called a/an _____ number.

6. In $7 \cdot 5 = 35$, the numbers 7 and 5 are called _____ of 35 and the number 35 is called the _____ of 7 and 5.

7. If $\dfrac{a}{b}$ is a fraction and c is a nonzero number, then $\dfrac{a \cdot c}{b \cdot c} = $ _____ .

8. If $\dfrac{a}{b}$ and $\dfrac{c}{d}$ are fractions, then $\dfrac{a}{b} \cdot \dfrac{c}{d} = $ _____ .

9. Two fractions whose product is 1, such as $\frac{2}{3}$ and $\frac{3}{2}$, are called _____ of each other.

10. If $\dfrac{a}{b}$ and $\dfrac{c}{d}$ are fractions and $\dfrac{c}{d}$ is not 0, then $\dfrac{a}{b} \div \dfrac{c}{d} = \dfrac{a}{b} \cdot$ _____ .

11. If $\dfrac{a}{b}$ and $\dfrac{c}{b}$ are fractions, then $\dfrac{a}{b} + \dfrac{c}{b} = $ _____ .

12. In order to add $\frac{1}{5}$ and $\frac{3}{4}$, we use 20 as the _____ .

1.2 EXERCISE SET   ▶ MyMathLab®

Practice Exercises

In Exercises 1–6, convert each mixed number to an improper fraction.

1. $2\dfrac{3}{8}$ **2.** $2\dfrac{7}{9}$ **3.** $7\dfrac{3}{5}$

4. $6\dfrac{2}{5}$ **5.** $8\dfrac{7}{16}$ **6.** $9\dfrac{5}{16}$

In Exercises 7–12, convert each improper fraction to a mixed number.

7. $\dfrac{23}{5}$ **8.** $\dfrac{47}{8}$ **9.** $\dfrac{76}{9}$

10. $\dfrac{59}{9}$ **11.** $\dfrac{711}{20}$ **12.** $\dfrac{788}{25}$

In Exercises 13–28, identify each natural number as prime or composite. If the number is composite, find its prime factorization.

13. 22   **14.** 15   **15.** 20

16. 75   **17.** 37   **18.** 23

19. 36   **20.** 100   **21.** 140

22. 110   **23.** 79   **24.** 83

25. 81   **26.** 64

27. 240   **28.** 360

In Exercises 29–40, simplify each fraction by reducing it to its lowest terms.

29. $\dfrac{10}{16}$ **30.** $\dfrac{8}{14}$ **31.** $\dfrac{15}{18}$ **32.** $\dfrac{18}{45}$

33. $\dfrac{35}{50}$ **34.** $\dfrac{45}{50}$ **35.** $\dfrac{32}{80}$ **36.** $\dfrac{75}{80}$

37. $\dfrac{44}{50}$ **38.** $\dfrac{38}{50}$ **39.** $\dfrac{120}{86}$ **40.** $\dfrac{116}{86}$

In Exercises 41–90, perform the indicated operation. Where possible, reduce the answer to its lowest terms.

41. $\dfrac{2}{5}\cdot\dfrac{1}{3}$ **42.** $\dfrac{3}{7}\cdot\dfrac{1}{4}$ **43.** $\dfrac{3}{8}\cdot\dfrac{7}{11}$

44. $\dfrac{5}{8}\cdot\dfrac{3}{11}$ **45.** $9\cdot\dfrac{4}{7}$ **46.** $8\cdot\dfrac{3}{7}$

47. $\dfrac{1}{10}\cdot\dfrac{5}{6}$ **48.** $\dfrac{1}{8}\cdot\dfrac{2}{3}$ **49.** $\dfrac{5}{4}\cdot\dfrac{6}{7}$

50. $\dfrac{7}{4}\cdot\dfrac{6}{11}$ **51.** $\left(3\dfrac{3}{4}\right)\left(1\dfrac{3}{5}\right)$

52. $\left(2\dfrac{4}{5}\right)\left(1\dfrac{1}{4}\right)$ **53.** $\dfrac{5}{4}\div\dfrac{4}{3}$

54. $\dfrac{7}{8}\div\dfrac{2}{3}$ **55.** $\dfrac{18}{5}\div 2$ **56.** $\dfrac{12}{7}\div 3$

57. $2\div\dfrac{18}{5}$ **58.** $3\div\dfrac{12}{7}$ **59.** $\dfrac{3}{4}\div\dfrac{1}{4}$

60. $\dfrac{3}{7}\div\dfrac{1}{7}$ **61.** $\dfrac{7}{6}\div\dfrac{5}{3}$ **62.** $\dfrac{7}{4}\div\dfrac{3}{8}$

63. $\dfrac{1}{14}\div\dfrac{1}{7}$ **64.** $\dfrac{1}{8}\div\dfrac{1}{4}$ **65.** $6\dfrac{3}{5}\div 1\dfrac{1}{10}$

66. $1\dfrac{3}{4}\div 2\dfrac{5}{8}$ **67.** $\dfrac{2}{11}+\dfrac{4}{11}$ **68.** $\dfrac{5}{13}+\dfrac{2}{13}$

69. $\dfrac{7}{12}+\dfrac{1}{12}$ **70.** $\dfrac{5}{16}+\dfrac{1}{16}$ **71.** $\dfrac{5}{8}+\dfrac{5}{8}$

72. $\dfrac{3}{8}+\dfrac{3}{8}$ **73.** $\dfrac{7}{12}-\dfrac{5}{12}$ **74.** $\dfrac{13}{18}-\dfrac{5}{18}$

75. $\dfrac{16}{7}-\dfrac{2}{7}$ **76.** $\dfrac{17}{5}-\dfrac{2}{5}$ **77.** $\dfrac{1}{2}+\dfrac{1}{5}$

78. $\dfrac{1}{3}+\dfrac{1}{5}$ **79.** $\dfrac{3}{4}+\dfrac{3}{20}$ **80.** $\dfrac{2}{5}+\dfrac{2}{15}$

81. $\dfrac{3}{8}+\dfrac{5}{12}$ **82.** $\dfrac{3}{10}+\dfrac{2}{15}$ **83.** $\dfrac{11}{18}-\dfrac{2}{9}$

84. $\dfrac{17}{18}-\dfrac{4}{9}$ **85.** $\dfrac{4}{3}-\dfrac{3}{4}$ **86.** $\dfrac{3}{2}-\dfrac{2}{3}$

87. $\dfrac{7}{10}-\dfrac{3}{16}$ **88.** $\dfrac{7}{30}-\dfrac{5}{24}$

89. $3\dfrac{3}{4}-2\dfrac{1}{3}$ **90.** $3\dfrac{2}{3}-2\dfrac{1}{2}$

In Exercises 91–102, determine whether the given number is a solution of the equation.

91. $\dfrac{7}{2}x=28;\ 8$ **92.** $\dfrac{5}{3}x=30;\ 18$

93. $w-\dfrac{2}{3}=\dfrac{3}{4};\ 1\dfrac{5}{12}$

94. $w-\dfrac{3}{4}=\dfrac{7}{4};\ 2\dfrac{1}{2}$

95. $20-\dfrac{1}{3}z=\dfrac{1}{2}z;\ 12$

96. $12-\dfrac{1}{4}z=\dfrac{1}{2}z;\ 20$

97. $\dfrac{2}{9}y+\dfrac{1}{3}y=\dfrac{3}{7};\ \dfrac{27}{35}$

98. $\dfrac{2}{3}y+\dfrac{5}{6}y=2;\ 1\dfrac{1}{3}$

99. $\dfrac{1}{3}(x-2)=\dfrac{1}{5}(x+4)+3;\ 26$

100. $\dfrac{1}{2}(x-2)+3=\dfrac{3}{8}(3x-4);\ 4$

101. $(y \div 6) + \frac{2}{3} = (y \div 2) - \frac{7}{9}; 4\frac{1}{3}$

102. $(y \div 6) + \frac{1}{3} = (y \div 2) - \frac{5}{9}; 2\frac{2}{3}$

In Exercises 103–114, translate from English to an algebraic expression or equation, whichever is appropriate. Let the variable x represent the number.

103. $\frac{1}{5}$ of a number   **104.** $\frac{1}{6}$ of a number

105. A number decreased by $\frac{1}{4}$ of itself

106. A number decreased by $\frac{1}{3}$ of itself

107. A number decreased by $\frac{1}{4}$ is half of that number.

108. A number decreased by $\frac{1}{3}$ is half of that number.

109. The sum of $\frac{1}{7}$ of a number and $\frac{1}{8}$ of that number gives 12.

110. The sum of $\frac{1}{9}$ of a number and $\frac{1}{10}$ of that number gives 15.

111. The product of $\frac{2}{3}$ and a number increased by 6

112. The product of $\frac{3}{4}$ and a number increased by 9

113. The product of $\frac{2}{3}$ and a number, increased by 6, is 3 less than the number.

114. The product of $\frac{3}{4}$ and a number, increased by 9, is 2 less than the number.

Practice PLUS

In Exercises 115–118, perform the indicated operation. Write the answer as an algebraic expression.

115. $\frac{3}{4} \cdot \frac{a}{5}$ **116.** $\frac{2}{3} \div \frac{a}{7}$

117. $\frac{11}{x} + \frac{9}{x}$ **118.** $\frac{10}{y} - \frac{6}{y}$

In Exercises 119–120, perform the indicated operations. Begin by performing operations in parentheses.

119. $\left(\frac{1}{2} - \frac{1}{3}\right) \div \frac{5}{8}$

120. $\left(\frac{1}{2} + \frac{1}{4}\right) \div \left(\frac{1}{2} + \frac{1}{3}\right)$

In Exercises 121–122, determine whether the given number is a solution of the equation.

121. $\frac{1}{5}(x + 2) = \frac{1}{2}\left(x - \frac{1}{5}\right); \frac{5}{8}$

122. $12 - 3(x - 2) = 4x - (x + 3); 3\frac{1}{2}$

Application Exercises

The formula

$$C = \frac{5}{9}(F - 32)$$

expresses the relationship between Fahrenheit temperature, F, and Celsius temperature, C. In Exercises 123–124, use the formula to convert the given Fahrenheit temperature to its equivalent temperature on the Celsius scale.

123. 68°F   **124.** 41°F

The maximum heart rate, in beats per minute, that you should achieve during exercise is 220 minus your age:

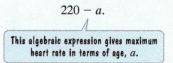

$$220 - a.$$

This algebraic expression gives maximum heart rate in terms of age, *a*.

The bar graph shows the target heart rate ranges for four types of exercise goals. The lower and upper limits of these ranges are fractions of the maximum heart rate, 220 − a. Exercises 125–128 are based on the information in the graph.

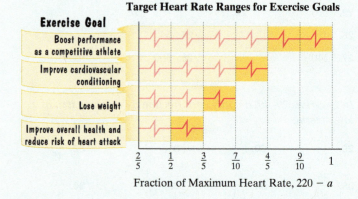

Target Heart Rate Ranges for Exercise Goals

Fraction of Maximum Heart Rate, 220 − *a*

125. If your exercise goal is to improve cardiovascular conditioning, the graph shows the following range for target heart rate, *H*, in beats per minute:

Lower limit of range — $H = \frac{7}{10}(220 - a)$

Upper limit of range — $H = \frac{4}{5}(220 - a).$

a. What is the lower limit of the heart range, in beats per minute, for a 20-year-old with this exercise goal?

b. What is the upper limit of the heart range, in beats per minute, for a 20-year-old with this exercise goal?

126. If your exercise goal is to improve overall health, the graph on the previous page shows the following range for target heart rate, H, in beats per minute:

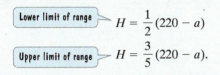

Lower limit of range $\quad H = \dfrac{1}{2}(220 - a)$

Upper limit of range $\quad H = \dfrac{3}{5}(220 - a)$.

a. What is the lower limit of the heart range, in beats per minute, for a 30-year-old with this exercise goal?

b. What is the upper limit of the heart range, in beats per minute, for a 30-year-old with this exercise goal?

127. a. Write a formula that models the heart rate, H, in beats per minute, for a person who is a years old and would like to achieve $\frac{9}{10}$ of maximum heart rate during exercise.

b. Use your formula from part (a) to find the heart rate during exercise for a 40-year-old with this goal.

128. a. Write a formula that models the heart rate, H, in beats per minute, for a person who is a years old and would like to achieve $\frac{7}{8}$ of maximum heart rate during exercise.

b. Use your formula from part (a) to find the heart rate during exercise for a 20-year-old with this goal.

The bar graph shows that the fraction of U.S. adults who use the Internet has continued to increase.

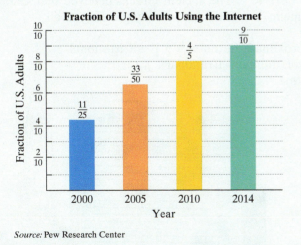

Fraction of U.S. Adults Using the Internet

Source: Pew Research Center

The data can be modeled by

$$I = \frac{3}{100}x + \frac{1}{2},$$

where I is the fraction of adults who used the Internet x years after 2000. Use this information to solve Exercises 129–130.

129. a. Use the formula to find the fraction of adults who used the Internet in 2010.

b. Use the least common denominator to compare the fraction of adults shown by the graph who used the Internet in 2010 with the fraction obtained by the mathematical model. What do you observe?

c. Use the formula to project the fraction of adults who will use the Internet in 2016.

130. a. Use the formula to find the fraction of adults who used the Internet in 2014.

b. Use the least common denominator to compare the fraction of adults shown by the graph who used the Internet in 2014 with the fraction obtained by the mathematical model in part (a). Does the formula underestimate or overestimate the fraction shown by the graph? By how much?

c. Use the formula to project the fraction of adults who will use the Internet in 2015.

Explaining the Concepts

131. Explain how to convert a mixed number to an improper fraction and give an example.

132. Explain how to convert an improper fraction to a mixed number and give an example.

133. Describe the difference between a prime number and a composite number.

134. What is meant by the prime factorization of a composite number?

135. What is the Fundamental Principle of Fractions?

136. Explain how to reduce a fraction to its lowest terms. Give an example with your explanation.

137. Explain how to multiply fractions and give an example.

138. Explain how to divide fractions and give an example.

139. Describe how to add or subtract fractions with identical denominators. Provide an example with your description.

140. Explain how to add fractions with different denominators. Use $\frac{5}{6} + \frac{1}{2}$ as an example.

Critical Thinking Exercises

Make Sense? In Exercises 141–144, determine whether each statement makes sense or does not make sense, and explain your reasoning.

141. I find it easier to multiply $\frac{1}{5}$ and $\frac{3}{4}$ than to add them.

142. Fractions frustrated me in arithmetic, so I'm glad I won't have to use them in algebra.

143. I need to be able to perform operations with fractions to determine whether $\frac{3}{2}$ is a solution of $8x = 12\left(x - \frac{1}{2}\right)$.

144. I saved money by buying a computer for $\frac{3}{2}$ of its original price.

In Exercises 145–148, determine whether each statement is true or false. If the statement is false, make the necessary change(s) to produce a true statement.

145. $\frac{1}{2} + \frac{1}{5} = \frac{2}{7}$

146. $\frac{1}{2} \div 4 = 2$

147. Every fraction has infinitely many equivalent fractions.

148. $\frac{3 + 7}{30} = \frac{\overset{1}{3} + 7}{\underset{10}{30}} = \frac{8}{10} = \frac{4}{5}$

149. Shown at the top of the next column is a short excerpt from "The Star-Spangled Banner." The time is $\frac{3}{4}$, which means that each measure must contain notes that add up to $\frac{3}{4}$. The values of the different notes tell musicians how long to hold each note.

$$\circ = 1 \quad \mathrel{\rlap{\textstyle\text{\j}}} = \frac{1}{2} \quad \mathrel{\rlap{\textstyle\text{\j}}} = \frac{1}{4} \quad \mathrel{\rlap{\textstyle\text{\j}}} = \frac{1}{8}$$

Use vertical lines to divide this line of "The Star-Spangled Banner" into measures.

say does that Star-span-gled Ban-ner yet wave O'er the

Preview Exercises

Exercises 150–152 will help you prepare for the material covered in the next section. Consider the following "infinite ruler" that shows numbers that lie to the left and to the right of zero.

150. What number is represented by point (a)?

151. What number is represented by point (b)? Express the number as an improper fraction.

152. What number is represented by point (c)?

SECTION

1.3

The Real Numbers

The United Nations Building in New York was designed to represent its mission of promoting world harmony. Viewed from the front, the building looks like three rectangles stacked upon each other. In each rectangle, the ratio of the width to height is $\sqrt{5} + 1$ to 2, approximately 1.618 to 1. The ancient Greeks believed that such a rectangle, called a **golden rectangle**, was the most visually pleasing of all rectangles.

The ratio 1.618 to 1 is approximate because $\sqrt{5}$ is an irrational number, a special kind of real number. Irrational? Real? Let's make sense of all this by describing the kinds of numbers you will encounter in this course.

What am I supposed to learn?

After studying this section, you should be able to:

1 Define the sets that make up the real numbers.

2 Graph numbers on a number line.

3 Express rational numbers as decimals.

4 Classify numbers as belonging to one or more sets of the real numbers.

5 Understand and use inequality symbols.

6 Find the absolute value of a real number.

The U.N. building is designed with three golden rectangles.

1 Define the sets that make up the real numbers.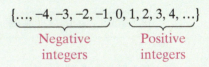

Natural Numbers and Whole Numbers

Before we describe the set of real numbers, let's be sure you are familiar with some basic ideas about sets. A **set** is a collection of objects whose contents can be clearly determined. The objects in a set are called the **elements** of the set. For example, the set of numbers used for counting can be represented by

$$\{1, 2, 3, 4, 5, \dots\}.$$

The braces, $\{\ \}$, indicate that we are representing a set. This form of representing a set uses commas to separate the elements of the set. Remember that the three dots after the 5 indicate that there is no final element and that the listing goes on forever.

We have seen that the set of numbers used for counting is called the set of **natural numbers**. When we combine the number 0 with the natural numbers, we obtain the set of **whole numbers**.

> ### Natural Numbers and Whole Numbers
> The set of **natural numbers** is $\{1, 2, 3, 4, 5, \dots\}$.
> The set of **whole numbers** is $\{0, 1, 2, 3, 4, 5, \dots\}$.

Integers and the Number Line

The whole numbers do not allow us to describe certain everyday situations. For example, if the balance in your checking account is $30 and you write a check for $35, your checking account is overdrawn by $5. We can write this as -5, read *negative* 5. The set consisting of the natural numbers, 0, and the negatives of the natural numbers is called the set of **integers**.

> ### Integers
> The set of **integers** is
>
> $$\underbrace{\{\dots, -4, -3, -2, -1,}_{\substack{\text{Negative} \\ \text{integers}}} 0, \underbrace{1, 2, 3, 4, \dots\}}_{\substack{\text{Positive} \\ \text{integers}}}$$

Notice that the term **positive integers** is another name for the natural numbers. The positive integers can be written in two ways:

1. Use a "+" sign. For example, $+4$ is "positive four."
2. Do not write any sign. For example, 4 is assumed to be "positive four."

EXAMPLE 1   Practical Examples of Negative Integers

Write a negative integer that describes each of the following situations:

a. A debt of $10
b. The shore surrounding the Dead Sea is 1312 feet below sea level.

Solution

a. A debt of $10 can be expressed by the negative integer -10 (negative ten).
b. The shore surrounding the Dead Sea is 1312 feet below sea level, expressed as -1312. ∎

✓ **CHECK POINT 1**   Write a negative integer that describes each of the following situations:

a. A debt of $500

b. Death Valley, the lowest point in North America, is 282 feet below sea level.

---

2 Graph numbers on a number line.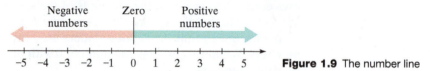

The **number line** is a graph we use to visualize the set of integers, as well as other sets of numbers. The number line is shown in **Figure 1.9**.

Negative numbers   Zero   Positive numbers

$$-5 \quad -4 \quad -3 \quad -2 \quad -1 \quad 0 \quad 1 \quad 2 \quad 3 \quad 4 \quad 5$$

Figure 1.9 The number line

The number line extends indefinitely in both directions. Zero separates the positive numbers from the negative numbers on the number line. The positive integers are located to the right of 0, and the negative integers are located to the left of 0. Zero is neither positive nor negative. For every positive integer on a number line, there is a corresponding negative integer on the opposite side of 0.

Integers are graphed on a number line by placing a dot at the correct location for each number.

EXAMPLE 2   Graphing Integers on a Number Line

Graph:

a. −3   **b.** 4   **c.** 0.

Solution   Place a dot at the correct location for each integer.

                (a)   (c)   (b)

$$-5 \quad -4 \quad -3 \quad -2 \quad -1 \quad 0 \quad 1 \quad 2 \quad 3 \quad 4 \quad 5$$

■

✓ **CHECK POINT 2**   Graph: **a.** −4   **b.** 0   **c.** 3.

Rational Numbers

If two integers are added, subtracted, or multiplied, the result is always another integer. This, however, is not always the case with division. For example, 10 divided by 5 is the integer 2. By contrast, 5 divided by 10 is $\frac{1}{2}$, and $\frac{1}{2}$ is not an integer. To permit divisions such as $\frac{5}{10}$, we enlarge the set of integers, calling the new collection the *rational numbers*. The set of **rational numbers** consists of all the numbers that can be expressed as a quotient of two integers, with the denominator not 0.

Great Question!

Is there another way to express the rational number $\frac{-3}{4}$?

In Section 1.7, you will learn that a negative number divided by a positive number gives a negative result. Thus, $\frac{-3}{4}$ can also be written as $-\frac{3}{4}$.

The Rational Numbers

The set of **rational numbers** is the set of all numbers that can be expressed in the form $\frac{a}{b}$, where a and b are integers and b is not equal to 0, written $b \neq 0$. The integer a is called the **numerator** and the integer b is called the **denominator**.

Here are two examples of rational numbers:

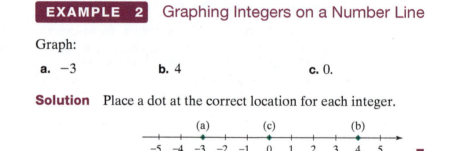

- $\frac{1}{2}$ $a = 1$ $b = 2$

- $\frac{-3}{4}$ $a = -3$ $b = 4$

Is the integer 5 another example of a rational number? Yes. The integer 5 can be written with a denominator of 1.

$$5 = \frac{5}{1}$$

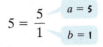

$a = 5$
$b = 1$

All integers are also rational numbers because they can be written with a denominator of 1.

How can we express a negative mixed number, such as $-2\frac{3}{4}$, in the form $\frac{a}{b}$? Copy the negative sign and then follow the procedure discussed in the previous section:

$$-2\frac{3}{4} = -\frac{4 \cdot 2 + 3}{4} = -\frac{8 + 3}{4} = -\frac{11}{4} = \frac{-11}{4}.$$

$a = -11$
$b = 4$

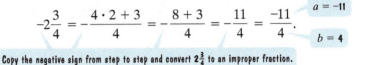
Copy the negative sign from step to step and convert $2\frac{3}{4}$ to an improper fraction.

Rational numbers are graphed on a number line by placing a dot at the correct location for each number.

EXAMPLE 3 Graphing Rational Numbers on a Number Line

Graph: **a.** $\frac{7}{2}$ **b.** -4.6.

Solution Place a dot at the correct location for each rational number.

a. Because $\frac{7}{2} = 3\frac{1}{2}$, its graph is midway between 3 and 4.

b. Because $-4.6 = -4\frac{6}{10}$, its graph is $\frac{6}{10}$, or $\frac{3}{5}$, of a unit to the left of -4.

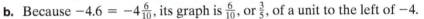

☑ **CHECK POINT 3** Graph: **a.** $\frac{9}{2}$ **b.** -1.2.

---

③ Express rational numbers as decimals. ▶

Every rational number can be expressed as a fraction and as a decimal. To express the fraction $\frac{a}{b}$ as a decimal, divide the denominator, b, into the numerator, a.

EXAMPLE 4 Expressing Rational Numbers as Decimals

Express each rational number as a decimal:

a. $\frac{5}{8}$ **b.** $\frac{7}{11}$.

Solution In each case, divide the denominator into the numerator.

a.
$$\begin{array}{r} 0.625 \\ 8\overline{)5.000} \\ \underline{4\,8} \\ 20 \\ \underline{16} \\ 40 \\ \underline{40} \\ 0 \end{array}$$

b.
$$\begin{array}{r} 0.6363\ldots \\ 11\overline{)7.0000\ldots} \\ \underline{6\,6} \\ 40 \\ \underline{33} \\ 70 \\ \underline{66} \\ 40 \\ \underline{33} \\ 70 \\ \vdots \end{array}$$

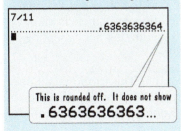

In Example 4, the decimal for $\frac{5}{8}$, namely 0.625, stops and is called a **terminating decimal**. Other examples of terminating decimals are

$$\frac{1}{4} = 0.25, \qquad \frac{2}{5} = 0.4, \qquad \text{and} \qquad \frac{7}{8} = 0.875.$$

By contrast, the division process for $\frac{7}{11}$ results in 0.6363. . . , with the digits 63 repeating over and over indefinitely. To indicate this, write a bar over the digits that repeat. Thus,

$$\frac{7}{11} = 0.\overline{63}.$$

The decimal for $\frac{7}{11}$, $0.\overline{63}$, is called a **repeating decimal**. Other examples of repeating decimals are

$$\frac{1}{3} = 0.333 \ldots = 0.\overline{3} \qquad \text{and} \qquad \frac{2}{3} = 0.666 \ldots = 0.\overline{6}.$$

Rational Numbers and Decimals

Any rational number can be expressed as a decimal. The resulting decimal will either terminate (stop), or it will have a digit that repeats or a block of digits that repeat.

✓ **CHECK POINT 4**   Express each rational number as a decimal:

a. $\dfrac{3}{8}$ **b.** $\dfrac{5}{11}$.

Irrational Numbers

Can you think of a number that, when written in decimal form, neither terminates nor repeats? An example of such a number is $\sqrt{2}$ (read: "the square root of 2"). The number $\sqrt{2}$ is a number that can be multiplied by itself to obtain 2. No terminating or repeating decimal can be multiplied by itself to get 2. However, some approximations come close to 2.

- 1.4 is an approximation of $\sqrt{2}$:

$$1.4 \times 1.4 = 1.96.$$

- 1.41 is an approximation of $\sqrt{2}$:

$$1.41 \times 1.41 = 1.9881.$$

- 1.4142 is an approximation of $\sqrt{2}$:

$$1.4142 \times 1.4142 = 1.99996164.$$

Can you see how each approximation in the list is getting better? This is because the products are getting closer and closer to 2.

The number $\sqrt{2}$, whose decimal representation does not come to an end and does not have a block of repeating digits, is an example of an **irrational number**.

The Irrational Numbers

Any number that can be represented on the number line that is not a rational number is called an **irrational number**. Thus, the set of irrational numbers is the set of numbers whose decimal representations are neither terminating nor repeating.

Perhaps the best known of all the irrational numbers is π (pi). This irrational number represents the distance around a circle (its circumference) divided by the diameter of the circle. In the *Star Trek* episode "Wolf in the Fold," Spock foils an evil computer

by telling it to "compute the last digit in the value of π." Because π is an irrational number, there is no last digit in its decimal representation:

$$\pi = 3.14159265358979323846264433832795\ldots .$$

Because irrational numbers cannot be represented by decimals that come to an end, mathematicians use symbols such as $\sqrt{2}$, $\sqrt{3}$, and π to represent these numbers. However, **not all square roots are irrational**. For example, $\sqrt{25} = 5$ because 5 multiplied by itself is 25. Thus, $\sqrt{25}$ is a natural number, a whole number, an integer, and a rational number $\left(\sqrt{25} = \frac{5}{1}\right)$.

Using Technology

You can obtain decimal approximations for irrational numbers using a calculator. For example, to approximate $\sqrt{2}$, use the following keystrokes:

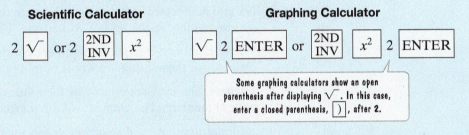

The display may read 1.41421356237, although your calculator may show more or fewer digits. Between which two integers would you graph $\sqrt{2}$ on a number line?

4 Classify numbers as belonging to one or more sets of the real numbers.

The Set of Real Numbers

All numbers that can be represented by points on the number line are called **real numbers**. Thus, the set of real numbers is formed by combining the rational numbers and the irrational numbers. Every real number is either rational or irrational.

The sets that make up the real numbers are summarized in **Table 1.2**. Notice the use of the symbol $\approx$ in the examples of irrational numbers. The symbol $\approx$ means "is approximately equal to."

Real numbers

Rational numbers		Irrational numbers
Integers		
Whole numbers		
Natural numbers		

This diagram shows that every real number is rational or irrational.

Table 1.2	The Sets That Make Up the Real Numbers	
Name	**Description**	**Examples**
Natural numbers	$\{1, 2, 3, 4, 5, \ldots\}$ These numbers are used for counting.	$2, 3, 5, 17$
Whole numbers	$\{0, 1, 2, 3, 4, 5, \ldots\}$ The set of whole numbers is formed by adding 0 to the set of natural numbers.	$0, 2, 3, 5, 17$
Integers	$\{\ldots, -5, -4, -3, -2, -1, 0, 1, 2, 3, 4, 5, \ldots\}$ The set of integers is formed by adding the negatives of the natural numbers to the set of whole numbers.	$-17, -5, -3, -2, 0,$ $2, 3, 5, 17$
Rational numbers	The set of rational numbers is the set of all numbers that can be expressed in the form $\frac{a}{b}$, where a and b are integers and b is not equal to 0, written $b \neq 0$. Rational numbers can be expressed as terminating or repeating decimals.	$-17 = \frac{-17}{1}, -5 = \frac{-5}{1}, -3, -2,$ $0, 2, 3, 5, 17,$ $\frac{2}{5} = 0.4,$ $\frac{-2}{3} = -0.6666\cdots = -0.\overline{6}$
Irrational numbers	The set of irrational numbers is the set of all numbers whose decimal representations are neither terminating nor repeating. Irrational numbers cannot be expressed as a quotient of integers.	$\sqrt{2} \approx 1.414214$ $-\sqrt{3} \approx -1.73205$ $\pi \approx 3.142$ $-\frac{\pi}{2} \approx -1.571$

EXAMPLE 5   Classifying Real Numbers

Consider the following set of numbers:

$$\left\{ -7, -\frac{3}{4}, 0, 0.\overline{6}, \sqrt{5}, \pi, 7.3, \sqrt{81} \right\}.$$

List the numbers in the set that are

a. natural numbers.   **b.** whole numbers.   **c.** integers.

d. rational numbers.   **e.** irrational numbers.   **f.** real numbers.

Solution

a. Natural numbers: The natural numbers are the numbers used for counting. The only natural number in the set is $\sqrt{81}$ because $\sqrt{81} = 9$. (9 multiplied by itself is 81.)

b. Whole numbers: The whole numbers consist of the natural numbers and 0. The elements of the set that are whole numbers are 0 and $\sqrt{81}$.

c. Integers: The integers consist of the natural numbers, 0, and the negatives of the natural numbers. The elements of the set that are integers are $\sqrt{81}, 0$, and -7.

d. Rational numbers: All numbers in the set that can be expressed as the quotient of integers are rational numbers. These include $-7 \left(-7 = \frac{-7}{1} \right), -\frac{3}{4}, 0 \left(0 = \frac{0}{1} \right),$ and $\sqrt{81} \left(\sqrt{81} = \frac{9}{1} \right)$. Furthermore, all numbers in the set that are terminating or repeating decimals are also rational numbers. These include $0.\overline{6}$ and 7.3.

e. Irrational numbers: The irrational numbers in the set are $\sqrt{5} \left(\sqrt{5} \approx 2.236 \right)$ and $\pi \left(\pi \approx 3.14 \right)$. Both $\sqrt{5}$ and π are only approximately equal to 2.236 and 3.14, respectively. In decimal form, $\sqrt{5}$ and π neither terminate nor have blocks of repeating digits.

f. Real numbers: All the numbers in the given set are real numbers. ∎

✓ **CHECK POINT 5**   Consider the following set of numbers:

$$\left\{ -9, -1.3, 0, 0.\overline{3}, \frac{\pi}{2}, \sqrt{9}, \sqrt{10} \right\}.$$

List the numbers in the set that are

a. natural numbers.   **b.** whole numbers.

c. integers.   **d.** rational numbers.

e. irrational numbers.   **f.** real numbers.

5 Understand and use inequality symbols. ▶

Ordering the Real Numbers

On the real number line, the real numbers increase from left to right. The lesser of two real numbers is the one farther to the left on a number line. The greater of two real numbers is the one farther to the right on a number line.

Look at the number line in **Figure 1.10**. The integers 2 and 5 are graphed. Observe that 2 is to the left of 5 on the number line. This means that 2 is less than 5.

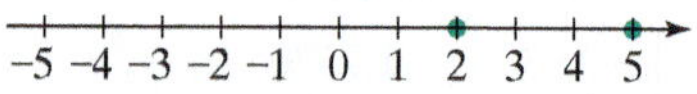

Figure 1.10

$2 < 5$: 2 is less than 5 because 2 is to the *left* of 5 on the number line.

In **Figure 1.10**, we can also observe that 5 is to the *right* of 2 on the number line. This means that 5 is greater than 2.

> 5 > 2: 5 is greater than 2 because 5 is to the *right* of 2 on the number line.

The symbols < and > are called **inequality symbols**. These symbols always point to the lesser of the two real numbers when the inequality is true.

2 is less than 5.   2 < 5   The symbol points to 2, the lesser number.
5 is greater than 2.   5 > 2   The symbol points to 2, the lesser number.

EXAMPLE 6   Using Inequality Symbols

Insert either < or > in the shaded area between each pair of numbers to make a true statement:

a. 3 ▮ 17   **b.** −4.5 ▮ 1.2   **c.** −5 ▮ −83   **d.** $\frac{4}{5}$ ▮ $\frac{2}{3}$.

Solution In each case, mentally compare the graph of the first number to the graph of the second number. If the first number is to the left of the second number, insert the symbol < for "is less than." If the first number is to the right of the second number, insert the symbol > for "is greater than."

a. Compare the graphs of 3 and 17 on the number line.

Because 3 is to the left of 17, this means that 3 is less than 17: 3 < 17.

b. Compare the graphs of −4.5 and 1.2.

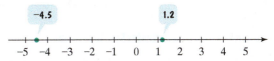

Because −4.5 is to the left of 1.2, this means that −4.5 is less than 1.2: −4.5 < 1.2.

c. Compare the graphs of −5 and −83.

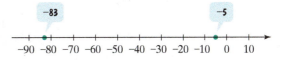

Because −5 is to the right of −83, this means that −5 is greater than −83: −5 > −83.

d. Compare the graphs of $\frac{4}{5}$ and $\frac{2}{3}$. To do so, convert to decimal notation or use a common denominator. Using decimal notation, $\frac{4}{5} = 0.8$ and $\frac{2}{3} = 0.\overline{6}$.

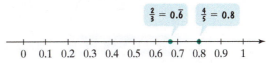

Because 0.8 is to the right of $0.\overline{6}$, this means that $\frac{4}{5}$ is greater than $\frac{2}{3}$: $\frac{4}{5} > \frac{2}{3}$. ■

✓ **CHECK POINT 6** Insert either $<$ or $>$ in the shaded area between each pair of numbers to make a true statement:

a. 14 ▮ 5   **b.** -5.4 ▮ 2.3   **c.** -19 ▮ -6 **d.** $\dfrac{1}{4}$ ▮ $\dfrac{1}{2}$.

The symbols $<$ and $>$ may be combined with an equal sign, as shown in the table.

	Symbols	Meaning	Examples	Explanation
This inequality is true if either the $<$ part or the $=$ part is true.	$a \le b$	a is less than or equal to b.	$3 \le 7$ $7 \le 7$	Because $3 < 7$ Because $7 = 7$
This inequality is true if either the $>$ part or the $=$ part is true.	$b \ge a$	b is greater than or equal to a.	$7 \ge 3$ $-5 \ge -5$	Because $7 > 3$ Because $-5 = -5$

When using the symbol $\le$ (is less than or equal to), the inequality is a true statement if either the $<$ part or the $=$ part is true. When using the symbol $\ge$ (is greater than or equal to), the inequality is a true statement if either the $>$ part or the $=$ part is true.

EXAMPLE 7 Using Inequality Symbols

Determine whether each inequality is true or false:

a. $-7 \le 4$ **b.** $-7 \le -7$ **c.** $-9 \ge 6$.

Solution

a. $-7 \le 4$ is true because $-7 < 4$ is true.

b. $-7 \le -7$ is true because $-7 = -7$ is true.

c. $-9 \ge 6$ is false because neither $-9 > 6$ nor $-9 = 6$ is true. ∎

✓ **CHECK POINT 7** Determine whether each inequality is true or false:

a. $-2 \le 3$ **b.** $-2 \ge -2$ **c.** $-4 \ge 1$.

6 Find the absolute value of a real number.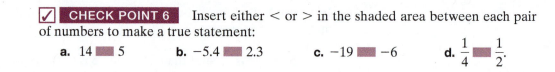

Absolute Value

Absolute value describes distance from 0 on a number line. If a represents a real number, the symbol $|a|$ represents its absolute value, read "the absolute value of a." For example,

$$|-5| = 5.$$

The absolute value of -5 is 5 because -5 is 5 units from 0 on a number line.

Absolute Value

The **absolute value** of a real number a, denoted by $|a|$, is the distance from 0 to a on a number line. Because absolute value describes a distance, it is never negative.

EXAMPLE 8 Finding Absolute Value

Find the absolute value:

a. $|-3|$ **b.** $|5|$ **c.** $|0|$.

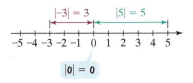

Figure 1.11 Absolute value describes distance from 0 on a number line.

Solution   The solution is illustrated in **Figure 1.11**.

a. $|-3| = 3$ The absolute value of −3 is 3 because −3 is 3 units from O.

b. $|5| = 5$ 5 is 5 units from O.

c. $|0| = 0$ O is O units from itself. ∎

Example 8 illustrates that the absolute value of a positive real number or 0 is the number itself. The absolute value of a negative real number, such as −3, is the number without the negative sign. Zero is the only real number whose absolute value is 0: $|0| = 0$. **The absolute value of any real number other than 0 is always positive.**

✓ **CHECK POINT 8**   Find the absolute value:

a. $|-4|$ b. $|6|$ c. $|-\sqrt{2}|$.

Achieving Success

Check! Check! Check!

After completing each Check Point or odd-numbered exercise, compare your answer with the one given in the answer section at the back of the book. To make this process more convenient, place a Post-it® or some other marker at the appropriate page of the answer section. If your answer is different from the one given in the answer section, try to figure out your mistake. Then correct the error. If you cannot determine what went wrong, show your work to your professor. **By recording each step neatly and using as much paper as you need**, your professor will find it easier to determine where you had trouble.

CONCEPT AND VOCABULARY CHECK

Fill in each blank so that the resulting statement is true.

1. The set $\{1, 2, 3, 4, 5, \dots\}$ is called the set of _____ numbers.

2. The set $\{0, 1, 2, 3, 4, 5, \dots\}$ is called the set of _____ numbers.

3. The set $\{\dots, -4, -3, -2, -1, 0, 1, 2, 3, 4, \dots\}$ is called the set of _____ .

4. The set of numbers in the form $\frac{a}{b}$, where a and b belong to the set in statement 3 above and $b \neq 0$, is called the set of _____ numbers.

5. The set of numbers whose decimal representations are neither terminating nor repeating is called the set of _____ numbers.

6. Every real number is either a /an _____ number or a/an _____ number.

7. The notation $2 < 5$ means that 2 is to the _____ of 5 on a number line.

8. The distance from 0 to a on a number line is called the _____ of a, denoted by _____ .

1.3 EXERCISE SET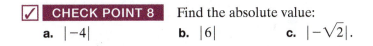

Practice Exercises

In Exercises 1–8, write a positive or negative integer that describes each situation.

1. Meteorology: 20° below zero

2. Navigation: 65 feet above sea level

3. Health: A gain of 8 pounds

4. Economics: A loss of $12,500.00

5. Banking: A withdrawal of $3000.00

6. Physics: An automobile slowing down at a rate of 3 meters per second each second

7. Economics: A budget deficit of 4 billion dollars

8. Football: A 14-yard loss

In Exercises 9–20, start by drawing a number line that shows integers from −5 to 5. Then graph each real number on your number line.

9. 2 **10.** 5 **11.** −5 **12.** −2 **13.** $3\frac{1}{2}$ **14.** $2\frac{1}{4}$

15. $\frac{11}{3}$ **16.** $\frac{7}{3}$ **17.** −1.8 **18.** −3.4 **19.** $-\frac{16}{5}$ **20.** $-\frac{11}{5}$

In Exercises 21–32, express each rational number as a decimal.

21. $\frac{3}{4}$ **22.** $\frac{3}{5}$ **23.** $\frac{7}{20}$

24. $\frac{3}{20}$ **25.** $\frac{7}{8}$ **26.** $\frac{5}{16}$

27. $\frac{9}{11}$ **28.** $\frac{3}{11}$ **29.** $-\frac{1}{2}$

30. $-\frac{1}{4}$ **31.** $-\frac{5}{6}$ **32.** $-\frac{7}{6}$

In Exercises 33–36, list all numbers from the given set that are: **a.** *natural numbers,* **b.** *whole numbers,* **c.** *integers,* **d.** *rational numbers,* **e.** *irrational numbers,* **f.** *real numbers.*

33. $\left\{-9, -\frac{4}{5}, 0, 0.25, \sqrt{3}, 9.2, \sqrt{100}\right\}$

34. $\left\{-7, -0.\overline{6}, 0, \sqrt{49}, \sqrt{50}\right\}$

35. $\left\{-11, -\frac{5}{6}, 0, 0.75, \sqrt{5}, \pi, \sqrt{64}\right\}$

36. $\left\{-5, -0.\overline{3}, 0, \sqrt{2}, \sqrt{4}\right\}$

37. Give an example of a whole number that is not a natural number.

38. Give an example of an integer that is not a whole number.

39. Give an example of a rational number that is not an integer.

40. Give an example of a rational number that is not a natural number.

41. Give an example of a number that is an integer, a whole number, and a natural number.

42. Give an example of a number that is a rational number, an integer, and a real number.

43. Give an example of a number that is an irrational number and a real number.

44. Give an example of a number that is a real number, but not an irrational number.

In Exercises 45–62, insert either < or > in the shaded area between each pair of numbers to make a true statement.

45. $\frac{1}{2}$  2   **46.** 4 ▨ −3

47. 3 ▨ $-\frac{5}{2}$ **48.** 3 ▨ $\frac{3}{2}$

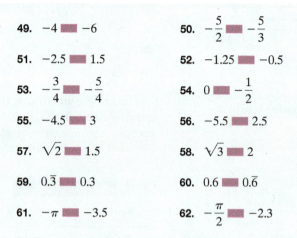

49. −4 ▨ −6   **50.** $-\frac{5}{2}$ ▨ $-\frac{5}{3}$

51. −2.5 ▨ 1.5   **52.** −1.25 ▨ −0.5

53. $-\frac{3}{4}$ ▨ $-\frac{5}{4}$ **54.** 0 ▨ $-\frac{1}{2}$

55. −4.5 ▨ 3   **56.** −5.5 ▨ 2.5

57. $\sqrt{2}$ ▨ 1.5   **58.** $\sqrt{3}$ ▨ 2

59. $0.\overline{3}$ ▨ 0.3   **60.** 0.6 ▨ $0.\overline{6}$

61. $-\pi$ ▨ −3.5   **62.** $-\frac{\pi}{2}$ ▨ −2.3

In Exercises 63–70, determine whether each inequality is true or false.

63. $-5 \geq -13$ **64.** $-5 \leq -8$

65. $-9 \geq -9$ **66.** $-14 \leq -14$

67. $0 \geq -6$ **68.** $0 \geq -13$

69. $-17 \geq 6$ **70.** $-14 \geq 8$

In Exercises 71–78, find each absolute value.

71. $|6|$ **72.** $|3|$ **73.** $|-7|$

74. $|-9|$ **75.** $\left|\frac{5}{6}\right|$ **76.** $\left|\frac{4}{5}\right|$

77. $|-\sqrt{11}|$ **78.** $|-\sqrt{29}|$

Practice PLUS

In Exercises 79–86, insert either <, >, or = in the shaded area to make a true statement.

79. $|-6|$ ▨ $|-3|$ **80.** $|-20|$ ▨ $|-50|$

81. $\left|\frac{3}{5}\right|$ ▨ $|-0.6|$ **82.** $\left|\frac{5}{2}\right|$ ▨ $|-2.5|$

83. $\frac{30}{40} - \frac{3}{4}$ ▨ $\frac{14}{15} \cdot \frac{15}{14}$ **84.** $\frac{17}{18} \cdot \frac{18}{17}$ ▨ $\frac{50}{60} - \frac{5}{6}$

85. $\frac{8}{13} \div \frac{8}{13}$ ▨ $|-1|$ **86.** $|-2|$ ▨ $\frac{4}{17} \div \frac{4}{17}$

Application Exercises

In Exercises 87–94, determine whether natural numbers, whole numbers, integers, rational numbers, or all real numbers are appropriate for each situation.

87. Shoe sizes of students on campus

88. Recorded heights of students on campus

89. Temperatures in weather reports

90. Class sizes of algebra courses

91. Values of d given by the formula $d = \sqrt{1.5h}$, where d is the distance, in miles, that you can see to the horizon from a height of h feet

92. Values of C given by the formula $C = 2\pi r$, where C is the circumference of a circle with radius r

93. The number of pets a person has

94. The number of siblings a person has

95. The table shows the record low temperatures for five U.S. states.

State	Record Low (°F)	Date
Florida	−2	Feb. 13, 1899
Georgia	−17	Jan. 27, 1940
Hawaii	12	May 17, 1979
Louisiana	−16	Feb. 13, 1899
Rhode Island	−25	Feb. 5, 1996

Source: National Climatic Data Center

a. Graph the five record low temperatures on a number line.

b. Write the names of the states in order from the coldest record low to the warmest record low.

96. The table shows the record low temperatures for five U.S. states.

State	Record Low (°F)	Date
Virginia	−30	Jan. 22, 1985
Washington	−48	Dec. 30, 1968
West Virginia	−37	Dec. 30, 1917
Wisconsin	−55	Feb. 4, 1996
Wyoming	−66	Feb. 9, 1933

Source: National Climatic Data Center

a. Graph the five record low temperatures on a number line.

b. Write the names of the states in order from the coldest record low to the warmest record low.

Explaining the Concepts

97. What is a set?

98. What are the natural numbers?

99. What are the whole numbers?

100. What are the integers?

101. How does the set of integers differ from the set of whole numbers?

102. Describe how to graph a number on the number line.

103. What is a rational number?

104. Explain how to express $\frac{3}{8}$ as a decimal.

105. Describe the difference between a rational number and an irrational number.

106. If you are given two different real numbers, explain how to determine which one is the lesser.

107. Describe what is meant by the absolute value of a number. Give an example with your explanation.

Critical Thinking Exercises

Make Sense?   *In Exercises 108–111, determine whether each statement makes sense or does not make sense, and explain your reasoning.*

108. The humor in this joke is based on the fact that the football will never be hiked.

Foxtrot copyright © 2003, 2009 by Bill Amend/Distributed by Universal Uclick

109. *Titanic* came to rest 12,500 feet below sea level and *Bismarck* came to rest 15,617 feet below sea level, so *Bismarck*'s resting place is higher than *Titanic*'s.

110. I expressed a rational number as a decimal and the decimal neither terminated nor repeated.

111. I evaluated the formula $d = \sqrt{1.5h}$ for a value of h that resulted in a rational number for d.

In Exercises 112–117, determine whether each statement is true or false. If the statement is false, make the necessary change(s) to produce a true statement.

112. Every rational number is an integer.

113. Some whole numbers are not integers.

114. Some rational numbers are not positive.

115. Irrational numbers cannot be negative.

116. Some real numbers are not rational numbers.

117. Some integers are not rational numbers.

In Exercises 118–119, write each phrase as an algebraic expression.

118. a loss of $\frac{1}{3}$ of an investment of d dollars

119. a loss of half of an investment of d dollars

Technology Exercises

In Exercises 120–123, use a calculator to find a decimal approximation for each irrational number, correct to three decimal places. Between which two integers should you graph each of these numbers on the number line?

120. $\sqrt{3}$

121. $-\sqrt{12}$

122. $1 - \sqrt{2}$

123. $2 - \sqrt{5}$

Preview Exercises

Exercises 124–126 will help you prepare for the material covered in the next section. In each exercise, evaluate both expressions for $x = 4$. What do you observe?

124. $3(x + 5); 3x + 15$

125. $3x + 5x; 8x$

126. $9x - 2x; 7x$

1.4

Basic Rules of Algebra

Starting as a link among U.S. research scientists, more than 2.5 billion people worldwide now use the Internet. Some random Internet factoids:

What am I supposed to learn?

After studying this section, you should be able to:

1. Understand and use the vocabulary of algebraic expressions.

2. Use commutative properties.

3. Use associative properties.

4. Use the distributive property.

5. Combine like terms.

6. Simplify algebraic expressions.

- Fraction of all Internet searches that are for pornography: $\frac{1}{3}$
- Peak time for sex-related searches: 11 P.M.
- Fraction of people who use the word "password" as their password: $\frac{1}{8}$
- Vanity searchers: Fraction of people who typed their own name into a search engine: $\frac{1}{4}$
- What happens in an Internet minute: 204 million emails sent, 2 million Google searches, 1.8 million Facebook "likes," 278,000 tweets

(*Sources*: Paul Grobman, *Vital Statistics*, Plume, 2005, Intel and qmee.com, 2013)

In this section, we move from these quirky tidbits to mathematical models that describe the remarkable growth of the Internet in the United States and worldwide. To use these models efficiently (you'll work with them in the Exercise Set), they should be simplified using basic rules of algebra. Before turning to these rules, we open the section with a closer look at algebraic expressions.

1. Understand and use the vocabulary of algebraic expressions.

The Vocabulary of Algebraic Expressions

We have seen that an algebraic expression combines numbers and variables. Here is an example of an algebraic expression:

$$7x + 3.$$

The **terms** of an algebraic expression are those parts that are separated by addition. For example, the algebraic expression $7x + 3$ contains two terms, namely $7x$ and 3. Notice that a term is a number, a variable, or a number multiplied by one or more variables.

The numerical part of a term is called its **coefficient**. In the term $7x$, the 7 is the coefficient. If a term containing one or more variables is written without a coefficient, the coefficient is understood to be 1. Thus, x means $1x$ and ab means $1ab$.

A term that consists of just a number is called a **constant term**. The constant term of $7x + 3$ is 3.

The parts of each term that are multiplied are called the **factors of the term**. The factors of the term $7x$ are 7 and x.

Like terms are terms that have exactly the same variable factors. Here are two examples of like terms:

$7x$ and $3x$ These terms have the same variable factor, *x*.

$4y$ and $9y$.   These terms have the same variable factor, *y*.

By contrast, here are some examples of terms that are not like terms. These terms do not have the same variable factor.

$7x$ and 3 The variable factor of the first term is *x*.
            The second term has no variable factor.

$7x$ and $3y$ The variable factor of the first term is *x*.
            The variable factor of the second term is *y*.

Constant terms are like terms. Thus, the constant terms 7 and -12 are like terms.

| EXAMPLE 1 | Using the Vocabulary of Algebraic Expressions |

Use the algebraic expression

$$4x + 7 + 5x$$

to answer the following questions:

a. How many terms are there in the algebraic expression?

b. What is the coefficient of the first term?

c. What is the constant term?

d. What are the like terms in the algebraic expression?

Solution

a. Because terms are separated by addition, the algebraic expression $4x + 7 + 5x$ contains three terms.

$$4x + 7 + 5x$$

First   Second   Third
term   term   term

b. The coefficient of the first term, $4x$, is 4.

c. The constant term in $4x + 7 + 5x$ is 7.

d. The like terms in $4x + 7 + 5x$ are $4x$ and $5x$. These terms have the same variable factor, x. ∎

| ✓ CHECK POINT 1 | Use the algebraic expression $6x + 2x + 11$ to answer each of the four questions in Example 1.

Equivalent Algebraic Expressions

In Example 1, we considered the algebraic expression

$$4x + 7 + 5x.$$

Let's compare this expression with a second algebraic expression

$$9x + 7.$$

Evaluate each expression for some choice of x. We will select $x = 2$.

$$4x + 7 + 5x \qquad\qquad 9x + 7$$

Replace x with 2.   Replace x with 2.

$$= 4 \cdot 2 + 7 + 5 \cdot 2 \qquad = 9 \cdot 2 + 7$$
$$= 8 + 7 + 10 \qquad\qquad = 18 + 7$$
$$= 25 \qquad\qquad\qquad\quad = 25$$

Both algebraic expressions have the same value when $x = 2$. Regardless of what number you select for x, the algebraic expressions $4x + 7 + 5x$ and $9x + 7$ will have the same value. These expressions are called *equivalent algebraic expressions*. Two algebraic expressions that have the same value for all replacements are called **equivalent algebraic expressions**. Because $4x + 7 + 5x$ and $9x + 7$ are equivalent algebraic expressions, we write

$$4x + 7 + 5x = 9x + 7.$$

Properties of Real Numbers and Algebraic Expressions

We now turn to basic properties, or rules, that you know from past experiences in working with whole numbers and fractions. These properties will be extended to include all real numbers and algebraic expressions. We will give each property a name so that we can refer to it throughout the study of algebra.

2 Use commutative properties. ▶

The Commutative Properties

The addition or multiplication of two real numbers can be done in any order. For example, $3 + 5 = 5 + 3$ and $3 \cdot 5 = 5 \cdot 3$. Changing the order does not change the answer of a sum or a product. These facts are called **commutative properties**.

Great Question!

Are there commutative properties for subtraction and division?

No. The commutative property does not hold for subtraction or division.

$$6 - 1 \neq 1 - 6$$
$$8 \div 4 \neq 4 \div 8$$

The Commutative Properties

Let a and b represent real numbers, variables, or algebraic expressions.

Commutative Property of Addition

$$a + b = b + a$$

Changing order when adding does not affect the sum.

Commutative Property of Multiplication

$$ab = ba$$

Changing order when multiplying does not affect the product.

EXAMPLE 2   Using the Commutative Properties

Use the commutative properties to write an algebraic expression equivalent to each of the following:

a. $y + 6$ **b.** $5x$.

Solution

a. By the commutative property of addition, an algebraic expression equivalent to $y + 6$ is $6 + y$. Thus,

$$y + 6 = 6 + y.$$

b. By the commutative property of multiplication, an algebraic expression equivalent to $5x$ is $x5$. Thus,

$$5x = x5. \quad \blacksquare$$

✓ **CHECK POINT 2**   Use the commutative properties to write an algebraic expression equivalent to each of the following:

a. $x + 14$ **b.** $7y$.

EXAMPLE 3   Using the Commutative Properties

Write an algebraic expression equivalent to $13x + 8$ using

a. the commutative property of addition.

b. the commutative property of multiplication.

Solution

a. By the commutative property of addition, we change the order of the terms being added. This means that an algebraic expression equivalent to $13x + 8$ is $8 + 13x$:

$$13x + 8 = 8 + 13x.$$

b. By the commutative property of multiplication, we change the order of the factors being multiplied. This means that an algebraic expression equivalent to $13x + 8$ is $x13 + 8$:

$$13x + 8 = x13 + 8. \quad \blacksquare$$

✓ **CHECK POINT 3**   Write an algebraic expression equivalent to $5x + 17$ using

a. the commutative property of addition.

b. the commutative property of multiplication.

Blitzer Bonus ◉

Commutative Words and Sentences

The commutative property states that a change in order produces no change in the answer. The words and sentences listed here suggest a characteristic of the commutative property; they read the same from left to right and from right to left!

- dad
- repaper
- never odd or even
- Six is a six is a six is a six is a six is . . . .

- Go deliver a dare, vile dog!
- May a moody baby doom a yam?
- Madam, in Eden I'm Adam.

- Ma is a nun, as I am.
- A man, a plan, a canal: Panama
- Was it a rat I saw?

- Deb sat in Anita's bed.
  Ned sat in Anita's den.
  But Anita sat in a tub.

- Ed, is Nik inside?
  Ed, is Deb bedside?
  Ed is busy—subside!

3 Use associative properties. ◉

The Associative Properties

Parentheses indicate groupings. As we have seen, we perform operations within the parentheses first. For example,

$$(2 + 5) + 10 = 7 + 10 = 17$$

and

$$2 + (5 + 10) = 2 + 15 = 17.$$

In general, the way in which three numbers are grouped does not change their sum. It also does not change their product. These facts are called the **associative properties**.

Great Question!

Are there associative properties for subtraction and division?

No. The associative property does not hold for subtraction or division.

$$(6 - 3) - 1 \neq 6 - (3 - 1)$$
$$(8 \div 4) \div 2 \neq 8 \div (4 \div 2)$$

The Associative Properties

Let a, b, and c represent real numbers, variables, or algebraic expressions.

Associative Property of Addition

$$(a + b) + c = a + (b + c)$$

Changing grouping when adding does not affect the sum.

Associative Property of Multiplication

$$(ab)c = a(bc)$$

Changing grouping when multiplying does not affect the product.

The associative properties can be used to simplify some algebraic expressions by removing the parentheses.

EXAMPLE 4 Simplifying Using the Associative Properties

Simplify:

a. $3 + (8 + x)$ **b.** $8(4x)$.

Solution

a. $3 + (8 + x)$ This is the given algebraic expression.

$= (3 + 8) + x$ Use the associative property of addition to group the first two numbers.

$= 11 + x$ Add within parentheses.

Using the commutative property of addition, this simplified algebraic expression can also be written as $x + 11$.

b. $8(4x)$ This is the given algebraic expression.

$= (8 \cdot 4)x$ Use the associative property of multiplication to group the first two numbers.

$= 32x$ Multiply within parentheses.

We can use the commutative property of multiplication to write this simplified algebraic expression as $x32$ or $x \cdot 32$. However, it is customary to express a term with its coefficient on the left. Thus, we use $32x$ as the simplified form of the algebraic expression. ∎

✓ **CHECK POINT 4** Simplify:

a. $8 + (12 + x)$ **b.** $6(5x)$.

The next example involves the use of both basic properties to simplify an algebraic expression.

Great Question!

Is there an easy way to distinguish between the commutative and associative properties?

Commutative: changes *order*

Associative: changes *grouping*

EXAMPLE 5 Using the Commutative and Associative Properties

Simplify: $7 + (x + 2)$.

Solution

$7 + (x + 2)$ This is the given algebraic expression.

$= 7 + (2 + x)$ Use the commutative property to change the order of the addition.

$= (7 + 2) + x$ Use the associative property to group the first two numbers.

$= 9 + x$ Add within parentheses.

Using the commutative property of addition, an equivalent algebraic expression is $x + 9$. ∎

✓ **CHECK POINT 5** Simplify: $8 + (x + 4)$.

4 Use the distributive property.

The Distributive Property

The **distributive property** involves both multiplication and addition. The property shows how to multiply the sum of two numbers by a third number. Consider, for example, $4(7 + 3)$, which can be calculated in two ways. One way is to perform the addition within the grouping symbols and then multiply:

$$4(7 + 3) = 4(10) = 40.$$

The other way to find $4(7 + 3)$ is to *distribute* the multiplication by 4 over the addition: First multiply each number within the parentheses by 4 and then add:

$$4 \cdot 7 + 4 \cdot 3 = 28 + 12 = 40.$$

The result in both cases is 40. Thus,

$$\overparen{4(7 + 3)} = 4 \cdot 7 + 4 \cdot 3. \qquad \text{Multiplication distributes over addition.}$$

The distributive property allows us to rewrite the product of a number and a sum as the sum of two products.

The Distributive Property

Let a, b, and c represent real numbers, variables, or algebraic expressions.

$$\overparen{a(b + c)} = ab + ac$$

Multiplication distributes over addition.

Great Question!

Can you give examples so that I don't confuse the distributive property with the associative property of multiplication?

Distributive:

$$4(5 + x) = 4 \cdot 5 + 4x$$
$$= 20 + 4x$$

Associative:

$$4(5 \cdot x) = (4 \cdot 5)x$$
$$= 20x$$

EXAMPLE 6   Using the Distributive Property

Multiply: $6(x + 4)$.

Solution Multiply *each term* inside the parentheses, x and 4, by the multiplier outside, 6.

$$6(x + 4) = 6x + 6 \cdot 4 \qquad \text{Use the distributive property to remove parentheses.}$$
$$= 6x + 24 \qquad \text{Multiply: } 6 \cdot 4 = 24. \blacksquare$$

✓ **CHECK POINT 6**   Multiply: $5(x + 3)$.

EXAMPLE 7   Using the Distributive Property

Multiply: $5(3y + 7)$.

Solution Multiply *each term* inside the parentheses, $3y$ and 7, by the multiplier outside, 5.

$$5(3y + 7) = 5 \cdot 3y + 5 \cdot 7 \qquad \text{Use the distributive property to remove parentheses.}$$
$$= 15y + 35 \qquad \text{Multiply. Use the associative property of multiplication}$$
$$\text{to find } 5 \cdot 3y: 5 \cdot 3y = (5 \cdot 3)y = 15y. \blacksquare$$

✓ **CHECK POINT 7**   Multiply: $6(4y + 7)$.

Great Question!

What's the bottom line when using the distributive property?

When using the distributive property to remove parentheses, be sure to multiply *each term* inside the parentheses by the multiplier outside.

Incorrect!

$$5(3y + 7) = 5 \cdot 3y + 7$$
$$= 15y + 7$$

> 7 must also be multiplied by 5.

Table 1.3 shows a number of other forms of the distributive property.

Table 1.3 Other Forms of the Distributive Property

Property	Meaning	Example
$a(b - c) = ab - ac$	Multiplication distributes over subtraction.	$5(4x - 3) = 5 \cdot 4x - 5 \cdot 3$ $= 20x - 15$
$a(b + c + d) = ab + ac + ad$	Multiplication distributes over three or more terms in parentheses.	$4(x + 10 + 3y)$ $= 4x + 4 \cdot 10 + 4 \cdot 3y$ $= 4x + 40 + 12y$
$(b + c)a = ba + ca$	Multiplication on the right distributes over addition (or subtraction).	$(x + 7)9 = x \cdot 9 + 7 \cdot 9$ $= 9x + 63$

5 Combine like terms. ▶

Combining Like Terms

The distributive property

$$a(b + c) = ab + ac$$

lets us add and subtract like terms. To do this, we will usually apply the property in the form

$$ax + bx = (a + b)x$$

and then combine a and b. For example,

$$3x + 7x = (3 + 7)x = 10x.$$

This process is called **combining like terms**.

EXAMPLE 8 Combining Like Terms

Combine like terms:

a. $4x + 15x$ **b.** $7a - 2a$.

Solution

a. $4x + 15x$ These are like terms because $4x$ and $15x$ have identical variable factors.
   $= (4 + 15)x$ Apply the distributive property.
   $= 19x$ Add within parentheses.

b. $7a - 2a$ These are like terms because $7a$ and $2a$ have identical variable factors.
   $= (7 - 2)a$ Apply the distributive property.
   $= 5a$ Subtract within parentheses. ■

☑ **CHECK POINT 8** Combine like terms:
   a. $7x + 3x$ **b.** $9a - 4a$.

When combining like terms, you may find yourself leaving out the details of the distributive property. For example, you may simply write

$$7x + 3x = 10x.$$

It might be useful to think along these lines: Seven things plus three of the (same) things give ten of those things. To add like terms, add the coefficients and copy the common variable.

Combining Like Terms Mentally

1. Add or subtract the coefficients of the terms.
2. Use the result of step 1 as the coefficient of the terms' variable factor.

When an expression contains three or more terms, use the commutative and associative properties to group like terms. Then combine the like terms.

EXAMPLE 9   Grouping and Combining Like Terms

Simplify:

a. $7x + 5 + 3x + 8$ **b.** $4x + 7y + 2x + 3y.$

Solution

a. $7x + 5 + 3x + 8$
$= (7x + 3x) + (5 + 8)$ Rearrange terms and group the like terms using the commutative and associative properties. This step is often done mentally.

$= 10x + 13$ Combine like terms: $7x + 3x = 10x$. Combine constant terms: $5 + 8 = 13$.

b. $4x + 7y + 2x + 3y$
$= (4x + 2x) + (7y + 3y)$ Group like terms.
$= 6x + 10y$ Combine like terms by adding coefficients and keeping the variable factor. ■

✓ **CHECK POINT 9**   Simplify:
a. $8x + 7 + 10x + 3$ **b.** $9x + 6y + 5x + 2y.$

Great Question!

What do like objects, such as apples and apples, have to do with like terms?

Combining like terms should remind you of adding and subtracting numbers of like objects.

7 apples + 3 apples = 10 apples — $7a + 3a = 10a$

9 feet − 5 feet = 4 feet — $9f - 5f = 4f$

6 apples + 10 feet = ?
$6a$ and $10f$ are not like terms and cannot be added.

---

6 Simplify algebraic expressions.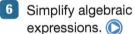

Simplifying Algebraic Expressions

An algebraic expression is **simplified** when parentheses have been removed and like terms have been combined.

Simplifying Algebraic Expressions

1. Use the distributive property to remove parentheses.
2. Rearrange terms and group like terms using the commutative and associative properties. This step may be done mentally.
3. Combine like terms by combining the coefficients of the terms and keeping the same variable factor.

EXAMPLE 10 Simplifying an Algebraic Expression

Simplify: $5(3x + 7) + 6x$.

Solution

$$5(3x + 7) + 6x$$
$$= 5 \cdot 3x + 5 \cdot 7 + 6x \qquad \text{Use the distributive property to remove the parentheses.}$$
$$= 15x + 35 + 6x \qquad \text{Multiply.}$$
$$= (15x + 6x) + 35 \qquad \text{Group like terms.}$$
$$= 21x + 35 \qquad \text{Combine like terms.} \blacksquare$$

✓ **CHECK POINT 10** Simplify: $7(2x + 3) + 11x$.

EXAMPLE 11 Simplifying an Algebraic Expression

Simplify: $6(2x + 4y) + 10(4x + 3y)$.

Solution

$$6(2x + 4y) + 10(4x + 3y)$$
$$= 6 \cdot 2x + 6 \cdot 4y + 10 \cdot 4x + 10 \cdot 3y \qquad \text{Use the distributive property to remove the parentheses.}$$
$$= 12x + 24y + 40x + 30y \qquad \text{Multiply.}$$
$$= (12x + 40x) + (24y + 30y) \qquad \text{Group like terms.}$$
$$= 52x + 54y \qquad \text{Combine like terms.} \blacksquare$$

✓ **CHECK POINT 11** Simplify: $7(4x + 3y) + 2(5x + y)$.

Great Question!

Do I need to use the distributive property to simplify an algebraic expression such as $4(3x + 5x)$ that contains like terms within the parentheses?

Use the distributive property to remove parentheses when the terms inside parentheses are not like terms:

$$4(3x + 5y) = 4 \cdot 3x + 4 \cdot 5y = 12x + 20y.$$

> 3x and 5y are not like terms.

It is not necessary to use the distributive property to remove parentheses when the terms inside parentheses are like terms:

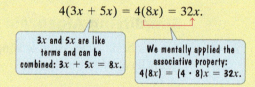

> 3x and 5x are like terms and can be combined: $3x + 5x = 8x$.

> We mentally applied the associative property: $4(8x) = (4 \cdot 8)x = 32x$.

Achieving Success

Read ahead. You might find it helpful to use some of your homework time to read (or skim) the section in the textbook that will be covered in your professor's next lecture. Having a clear idea of the new material that will be discussed will help you to understand the class a whole lot better.

CONCEPT AND VOCABULARY CHECK

Fill in each blank so that the resulting statement is true.

1. Terms that have the same variable factors, such as $7x$ and $5x$, are called _____ terms.

2. If a and b are real numbers, the commutative property of addition states that $a + b =$ _____.

3. If a and b are real numbers, the commutative property of multiplication states that _____ $= ba$.

4. If a, b, and c are real numbers, the associative property of addition states that $(a + b) + c =$ _____.

5. If a, b, and c are real numbers, the associative property of multiplication states that _____ $= a(bc)$.

6. If a, b, and c are real numbers, the distributive property states that $a(b + c) =$ _____.

7. An algebraic expression is _____ when parentheses have been removed and like terms have been combined.

1.4 EXERCISE SET   ▶ MyMathLab®

Practice Exercises

In Exercises 1–6, an algebraic expression is given. Use each expression to answer the following questions.
  a. *How many terms are there in the algebraic expression?*
  b. *What is the numerical coefficient of the first term?*
  c. *What is the constant term?*
  d. *Does the algebraic expression contain like terms? If so, what are the like terms?*

1. $3x + 5$
2. $9x + 4$
3. $x + 2 + 5x$
4. $x + 6 + 7x$
5. $4y + 1 + 3x$
6. $8y + 1 + 10x$

In Exercises 7–14, use the commutative property of addition to write an equivalent algebraic expression.

7. $y + 4$
8. $x + 7$
9. $5 + 3x$
10. $4 + 9x$
11. $4x + 5y$
12. $10x + 9y$
13. $5(x + 3)$
14. $6(x + 4)$

In Exercises 15–22, use the commutative property of multiplication to write an equivalent algebraic expression.

15. $9x$
16. $8x$
17. $x + y6$
18. $x + y7$
19. $7x + 23$
20. $13x + 11$
21. $5(x + 3)$
22. $6(x + 4)$

In Exercises 23–26, use an associative property to rewrite each algebraic expression. Once the grouping has been changed, simplify the resulting algebraic expression.

23. $7 + (5 + x)$
24. $9 + (3 + x)$
25. $7(4x)$
26. $8(5x)$

In Exercises 27–46, use a form of the distributive property to rewrite each algebraic expression without parentheses.

27. $3(x + 5)$
28. $4(x + 6)$
29. $8(2x + 3)$
30. $9(2x + 5)$
31. $\frac{1}{3}(12 + 6r)$
32. $\frac{1}{4}(12 + 8r)$
33. $5(x + y)$
34. $7(x + y)$
35. $3(x - 2)$
36. $4(x - 5)$
37. $2(4x - 5)$
38. $6(3x - 2)$
39. $\frac{1}{2}(5x - 12)$
40. $\frac{1}{3}(7x - 21)$
41. $(2x + 7)4$
42. $(5x + 3)6$
43. $6(x + 3 + 2y)$
44. $7(2x + 4 + y)$
45. $5(3x - 2 + 4y)$
46. $4(5x - 3 + 7y)$

In Exercises 47–64, simplify each algebraic expression.

47. $7x + 10x$
48. $5x + 13x$
49. $11a - 3a$
50. $14b - 5b$
51. $3 + (x + 11)$
52. $7 + (x + 10)$
53. $5y + 3 + 6y$
54. $8y + 7 + 10y$
55. $2x + 5 + 7x - 4$
56. $7x + 8 + 2x - 3$
57. $11a + 12 + 3a + 2$
58. $13a + 15 + 2a + 11$
59. $5(3x + 2) - 4$
60. $2(5x + 4) - 3$
61. $12 + 5(3x - 2)$
62. $14 + 2(5x - 1)$
63. $7(3a + 2b) + 5(4a + 2b)$
64. $11(6a + 3b) + 4(12a + 5b)$

Practice PLUS

In Exercises 65–66, name the property used to go from step to step each time that "(why?)" occurs.

65. $7 + 2(x + 9)$
$= 7 + (2x + 18)$ (why?)
$= 7 + (18 + 2x)$ (why?)
$= (7 + 18) + 2x$ (why?)
$= 25 + 2x$
$= 2x + 25$ (why?)

66. $5(x + 4) + 3x$
$= (5x + 20) + 3x$ (why?)
$= (20 + 5x) + 3x$ (why?)
$= 20 + (5x + 3x)$ (why?)
$= 20 + (5 + 3)x$ (why?)
$= 20 + 8x$
$= 8x + 20$ (why?)

In Exercises 67–76, write each English phrase as an algebraic expression. Then simplify the expression. Let x represent the number.

67. the sum of 7 times a number and twice the number

68. the sum of 8 times a number and twice the number

69. the product of 3 and a number, which is then subtracted from the product of 12 and a number

70. the product of 5 and a number, which is then subtracted from the product of 11 and a number

71. six times the product of 4 and a number

72. nine times the product of 3 and a number

73. six times the sum of 4 and a number

74. nine times the sum of 3 and a number

75. eight increased by the product of 5 and one less than a number

76. nine increased by the product of 3 and 2 less than a number

Application Exercises

The graph shows the number of Internet users in the United States and worldwide for five selected years. Exercises 77–78 involve mathematical models for the data.

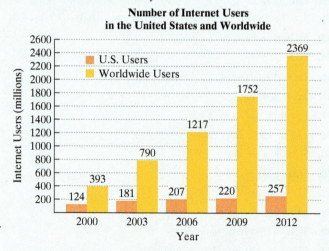

Number of Internet Users in the United States and Worldwide

Source: International Telecommunication Union

77. The number of Internet users in the United States, U, in millions, n years after 2000 can be modeled by the formula

$$U = 4(2n + 32) + 2\left(n + \frac{9}{2}\right).$$

a. Simplify the formula.

b. Use the simplified form of the mathematical model to find the number of Internet users in the United States in 2012. Does the model underestimate, overestimate, or give the exact value for the actual number shown by the bar graph for 2012?

78. The number of worldwide Internet users, W, in millions, n years after 2000 can be modeled by the formula

$$W = 3(50n + 95) + 2\left(\frac{23}{2}n + 4\right).$$

a. Simplify the formula.

b. Use the simplified form of the mathematical model to find the number of worldwide Internet users in 2009. Does the model underestimate or overestimate the actual number shown by the bar graph for 2009? By how much?

Explaining the Concepts

79. What is a term? Provide an example with your description.

80. What are like terms? Provide an example with your description.

81. What are equivalent algebraic expressions?

82. State a commutative property and give an example.

83. State an associative property and give an example.

84. State a form of the distributive property and give an example.

85. Explain how to add like terms. Give an example.

86. What does it mean to simplify an algebraic expression?

87. An algebra student incorrectly used the distributive property and wrote $3(5x + 7) = 15x + 7$. If you were that student's teacher, what would you say to help the student avoid this kind of error?

88. You can transpose the letters in the word "conversation" to form the phrase "voices rant on." From "total abstainers" we can form "sit not at ale bars." What two algebraic properties do each of these transpositions (called *anagrams*) remind you of? Explain your answer.

Critical Thinking Exercises

Make Sense? *In Exercises 89–92, determine whether each statement makes sense or does not make sense, and explain your reasoning.*

89. I applied the commutative property and rewrote $x - 4$ as $4 - x$.

90. Just as the commutative properties change groupings, the associative properties change order.

91. I did not use the distributive property to simplify $3(2x + 5x)$.

92. The commutative, associative, and distributive properties remind me of the rules of a game.

In Exercises 93–96, determine whether each statement is true or false. If the statement is false, make the necessary change(s) to produce a true statement.

93. $(24 \div 6) \div 2 = 24 \div (6 \div 2)$

94. $2x + 5 = 5x + 2$

95. $a + (bc) = (a + b)(a + c)$; in words, addition can be distributed over multiplication.

96. Like terms contain the same coefficients.

Exercises 97–99 will help you prepare for the material covered in the next section. In each exercise, write an integer that is the result of the given situation.

97. You earn $150, but then you misplace $90.

98. You lose $50 and then you misplace $10.

99. The temperature is 30 degrees, and then it drops by 35 degrees.

MID-CHAPTER CHECK POINT   Section 1.1–Section 1.4

✓ **What You Know:** Algebra uses variables, or letters, that represent a variety of different numbers. These variables appear in algebraic expressions, equations, and formulas. Mathematical models use variables to describe real-world phenomena. We reviewed operations with fractions and saw how fractions are used in algebra. We defined the real numbers and represented them as points on a number line. Finally, we introduced some basic rules of algebra and used the commutative, associative, and distributive properties to simplify algebraic expressions.

1. Evaluate for $x = 6$: $2 + 10x$

2. Evaluate for $x = \frac{3}{5}$: $10x - 4$.

3. Evaluate for $x = 3$ and $y = 10$:
$$\frac{xy}{2} + 4(y - x).$$

In Exercises 4–5, translate from English to an algebraic expression or equation, whichever is appropriate. Let the variable x represent the number.

4. Two less than $\frac{1}{4}$ of a number

5. Five more than the quotient of a number and 6 gives 19.

In Exercises 6–7, determine whether the given number is a solution of the equation.

6. $3(x + 2) = 4x - 1$; 6

7. $8y = 12\left(y - \frac{1}{2}\right)$; $\frac{3}{4}$

8. The bar graph shows the average price of expanded basic cable television service in the United States for four selected years. (Expanded basic cable is a step up from the entry-level package offered by most providers.)

Average Price of Expanded Basic Cable Service

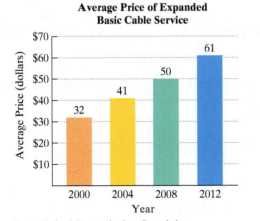

Here is a mathematical model that approximates the data displayed by the bar graph:

$$C = 2n + 32.$$

Average price of expanded basic cable service ⟵ C

Number of years after 2000 ⟵ n

a. Use the formula to find the average price of expanded basic cable service in 2008. Does the mathematical model underestimate or overestimate the actual average price shown by the bar graph for 2008? By how much?

b. If trends from 2000 through 2012 continue, use the formula to project the average price of expanded basic cable service in 2020.

In Exercises 9–15, perform the indicated operation. Where possible, reduce answers to lowest terms.

9. $\dfrac{7}{10} - \dfrac{8}{15}$

10. $\dfrac{2}{3} \cdot \dfrac{3}{4}$

11. $\dfrac{5}{22} + \dfrac{5}{33}$

12. $\dfrac{3}{5} \div \dfrac{9}{10}$

13. $\dfrac{23}{105} - \dfrac{2}{105}$

14. $2\dfrac{7}{9} \div 3$

15. $5\dfrac{2}{9} - 3\dfrac{1}{6}$

16. The formula $C = \frac{5}{9}(F - 32)$ expresses the relationship between Fahrenheit temperature, F, and Celsius temperature, C. If the temperature is 50°F, use the formula to find the equivalent temperature on the Celsius scale.

17. Insert either $<$ or $>$ in the shaded area to make a true statement:
$$-8000 \ \blacksquare \ -8\dfrac{1}{4}.$$

18. Express $\frac{1}{11}$ as a decimal.

19. Find the absolute value: $|-19.3|$.

20. List all the rational numbers in this set:
$$\left\{-11, -\dfrac{3}{7}, 0, 0.45, \sqrt{23}, \sqrt{25}\right\}.$$

In Exercises 21–23, rewrite 5(x + 3) as an equivalent expression using the given property.

21. the commutative property of multiplication

22. the commutative property of addition

23. the distributive property

In Exercises 24–25, simplify each algebraic expression.

24. $7(9x + 3) + \dfrac{1}{3}(6x)$

25. $2(3x + 5y) + 4(x + 6y)$

SECTION

1.5

Addition of Real Numbers

What am I supposed to learn?

After studying this section, you should be able to:

1 Add numbers with a number line.

2 Find sums using identity and inverse properties.

3 Add numbers without a number line.

4 Use addition rules to simplify algebraic expressions.

5 Solve applied problems using a series of additions.

1 Add numbers with a number line.

It has not been a good day! First, you lost your wallet with $50 in it. Then, to get through the day, you borrowed $10, which you somehow misplaced. Your loss of $50 followed by a loss of $10 is an overall loss of $60. This can be written

$$-50 + (-10) = -60.$$

The result of adding two or more numbers is called the **sum** of the numbers. The sum of -50 and -10 is -60. You can think of gains and losses of money to find sums. For example, to find $17 + (-13)$, think of a gain of $17 followed by a loss of $13. There is an overall gain of $4. Thus, $17 + (-13) = 4$. In the same way, to find $-17 + 13$, think of a loss of $17 followed by a gain of $13. There is an overall loss of $4, so $-17 + 13 = -4$.

Adding with a Number Line

We use the number line to help picture the addition of real numbers. Here is the procedure for finding $a + b$, the sum of a and b, using the number line:

> **Using the Number Line to Find a Sum**
>
> Let a and b represent real numbers. To find $a + b$ using a number line,
>
> 1. Start at a.
> 2. a. If b is **positive**, move b units to the **right**.
>   b. If b is **negative**, move $|b|$ units to the **left**.
>   c. If b is **0, stay** at a.
> 3. The number where we finish on the number line represents the sum of a and b.

This procedure is illustrated in Examples 1 and 2. Think of moving to the right as a gain and moving to the left as a loss.

EXAMPLE 1 Adding Real Numbers Using a Number Line

Find the sum using a number line:

$$3 + (-5).$$

Solution We find $3 + (-5)$ using the number line in **Figure 1.12**.

Step 1. We consider 3 to be the first number, represented by a in the box on the previous page. We start at a, or 3.

Step 2. We consider -5 to be the second number, represented by b. Because this number is negative, we move 5 units to the left.

Step 3. We finish at -2 on the number line. The number where we finish represents the sum of 3 and -5. Thus,

$$3 + (-5) = -2.$$

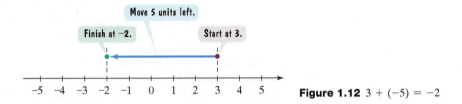

Figure 1.12 $3 + (-5) = -2$

Observe that if there is a gain of \$3 followed by a loss of \$5, there is an overall loss of \$2. ∎

☑ **CHECK POINT 1** Find the sum using a number line:

$$4 + (-7).$$

EXAMPLE 2 Adding Real Numbers Using a Number Line

Find each sum using a number line:

a. $-3 + (-4)$ **b.** $-6 + 2$.

Solution

a. To find $-3 + (-4)$, start at -3. Move 4 units to the left. We finish at -7. Thus,

$$-3 + (-4) = -7.$$

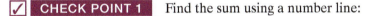

Observe that if there is a loss of \$3 followed by a loss of \$4, there is an overall loss of \$7.

b. To find $-6 + 2$, start at -6. Move 2 units to the right because 2 is positive. We finish at -4. Thus,

$$-6 + 2 = -4.$$

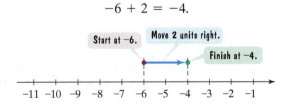

Observe that if there is a loss of \$6 followed by a gain of \$2, there is an overall loss of \$4. ∎

☑ **CHECK POINT 2** Find each sum using a number line:

a. $-1 + (-3)$ **b.** $-5 + 3$.

2 Find sums using identity and inverse properties.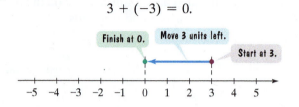

The Number Line and Properties of Addition

The number line can be used to picture some useful properties of addition. For example, let's see what happens if we add two numbers with different signs but the same absolute value. Two such numbers are 3 and −3. To find $3 + (−3)$ on a number line, we start at 3 and move 3 units to the left. We finish at 0. Thus,

$$3 + (−3) = 0.$$

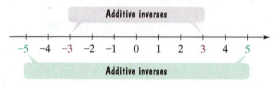

Numbers that differ only in sign, such as 3 and −3, are called *additive inverses*. **Additive inverses**, also called **opposites**, are pairs of real numbers that are the same number of units from zero on the number line, but are on opposite sides of zero. Thus, −3 is the additive inverse, or opposite, of 3, and 5 is the additive inverse, or opposite, of −5. The additive inverse of 0 is 0. Other additive inverses come in pairs.

In general, the sum of any real number, denoted by a, and its additive inverse, denoted by $−a$, is zero:

$$a + (−a) = 0.$$

This property is called the **inverse property of addition**.

Table 1.4 shows the identity and inverse properties of addition.

Table 1.4   Identity and Inverse Properties of Addition

Let a be a real number, a variable, or an algebraic expression.

Property	Meaning	Examples
Identity Property of Addition	Zero can be deleted from a sum. $a + 0 = a$ $0 + a = a$	• $4 + 0 = 4$ • $−3x + 0 = −3x$ • $0 + (5a + b) = 5a + b$
Inverse Property of Addition	The sum of a real number and its additive inverse, or opposite, gives 0, the additive identity. $a + (−a) = 0$ $(−a) + a = 0$	• $6 + (−6) = 0$ • $3x + (−3x) = 0$ • $[−(2y + 1)] + (2y + 1) = 0$

3 Add numbers without a number line.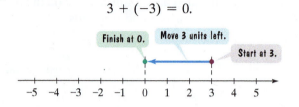

Adding without a Number Line

Now that we can picture the addition of real numbers, we look at two rules for using absolute value to add signed numbers.

Adding Two Numbers with the Same Sign

1. Add the absolute values.
2. Use the common sign as the sign of the sum.

EXAMPLE 3 Adding Real Numbers

Add without using a number line:

a. $-11 + (-15)$ **b.** $-0.2 + (-0.8)$ **c.** $-\dfrac{3}{4} + \left(-\dfrac{1}{2}\right)$.

Solution In each part of this example, we are adding numbers with the same sign.

a. $-11 + (-15) = -26$
> Add absolute values: 11 + 15 = 26.

> Use the common sign.

b. $-0.2 + (-0.8) = -1$
> Add absolute values: 0.2 + 0.8 = 1.0 or 1.

> Use the common sign.

c. $-\dfrac{3}{4} + \left(-\dfrac{1}{2}\right) = -\dfrac{5}{4}$
> Add absolute values: $\frac{3}{4} + \frac{1}{2} = \frac{3}{4} + \frac{2}{4} = \frac{5}{4}$.

> Use the common sign.

☑ **CHECK POINT 3** Add without using a number line:

a. $-10 + (-25)$ **b.** $-0.3 + (-1.2)$ **c.** $-\dfrac{2}{3} + \left(-\dfrac{1}{6}\right)$.

We also use absolute value to add two real numbers with different signs.

Adding Two Numbers with Different Signs

1. Subtract the smaller absolute value from the greater absolute value.
2. Use the sign of the number with the greater absolute value as the sign of the sum.

EXAMPLE 4 Adding Real Numbers

Add without using a number line:

a. $-13 + 4$ **b.** $-0.2 + 0.8$ **c.** $-\dfrac{3}{4} + \dfrac{1}{2}$.

Solution In each part of this example, we are adding numbers with different signs.

a. $-13 + 4 = -9$
> Subtract absolute values: 13 − 4 = 9.

> Use the sign of the number with the greater absolute value.

b. $-0.2 + 0.8 = 0.6$
> Subtract absolute values: 0.8 − 0.2 = 0.6.

> Use the sign of the number with the greater absolute value. The sign is assumed to be positive.

c. $-\dfrac{3}{4} + \dfrac{1}{2} = -\dfrac{1}{4}$
> Subtract absolute values: $\frac{3}{4} - \frac{1}{2} = \frac{3}{4} - \frac{2}{4} = \frac{1}{4}$.

> Use the sign of the number with the greater absolute value.

✓ **CHECK POINT 4** Add without using a number line:

a. $-15 + 2$ **b.** $-0.4 + 1.6$ **c.** $-\dfrac{2}{3} + \dfrac{1}{6}$.

Great Question!

What's the bottom line for determining the sign of a sum?

- The sum of two positive numbers is always positive.
  Example: $0.8 + 0.3$ is a positive number.

- The sum of two negative numbers is always negative.
  Example: $-0.8 + (-0.3)$ is a negative number.

- The sum of two numbers with different signs may be positive or negative. The sign of the sum is the sign of the number with the greater absolute value.
  Examples: $-0.8 + 0.3$ is a negative number.
  $\qquad\qquad\ \ 0.8 + (-0.3)$ is a positive number.

- The sum of a number and its additive inverse, or opposite, is always zero.
  Example: $0.8 + (-0.8) = 0$.

4 Use addition rules to simplify algebraic expressions. ▶

Algebraic Expressions

The rules for adding real numbers can be used to simplify certain algebraic expressions.

EXAMPLE 5 Simplifying Algebraic Expressions

Simplify:

a. $-11x + 7x$ **b.** $7y + (-12z) + (-9y) + 15z$ **c.** $3(7x + 8) + (-25x)$.

Solution

a. $-11x + 7x$ The given algebraic expression has two like terms: $-11x$ and $7x$ have identical variable factors.

$\quad = (-11 + 7)x$ Apply the distributive property.
$\quad = -4x$ Add within parentheses: $-11 + 7 = -4$.

b. $7y + (-12z) + (-9y) + 15z$ The colors indicate that there are two pairs of like terms.

$\quad = 7y + (-9y) + (-12z) + 15z$ Arrange like terms so that they are next to one another.

$\quad = [7 + (-9)]y + [(-12) + 15]z$ Apply the distributive property.

$\quad = -2y + 3z$ Add within the grouping symbols: $7 + (-9) = -2$ and $-12 + 15 = 3$.

c. $3(7x + 8) + (-25x)$

$\quad = 3 \cdot 7x + 3 \cdot 8 + (-25x)$ Use the distributive property to remove the parentheses.

$\quad = 21x + 24 + (-25x)$ Multiply.

$\quad = 21x + (-25x) + 24$ Arrange like terms so that they are next to one another.

$\quad = [21 + (-25)]x + 24$ Apply the distributive property.

$\quad = -4x + 24$ Add within the grouping symbols: $21 + (-25) = -4$. ∎

✓ **CHECK POINT 5** Simplify:

a. $-20x + 3x$

b. $3y + (-10z) + (-10y) + 16z$

c. $5(2x + 3) + (-30x)$.

5 Solve applied problems using a series of additions.

Applications

Positive and negative numbers are used in everyday life to represent such things as gains and losses in the stock market, rising and falling temperatures, deposits and withdrawals on bank statements, and ascending and descending motion. Positive and negative numbers are used to solve applied problems involving a series of additions.

One way to add a series of positive and negative numbers is to use the commutative and associative properties.

- Add all the positive numbers.
- Add all the negative numbers.
- Add the sums obtained in the first two steps.

The next example illustrates this idea.

EXAMPLE 6 An Application of Adding Signed Numbers

A glider was towed 1000 meters into the air and then let go. It descended 70 meters into a thermal (rising bubble of warm air), which took it up 2100 meters. At this point it dropped 230 meters into a second thermal. Then it rose 1200 meters. What was its altitude at that point?

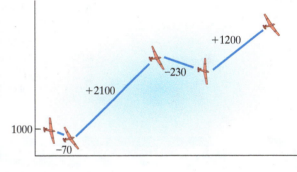

Solution We use the problem's conditions to write a sum. The altitude of the glider is expressed by the following sum:

Towed to 1000 meters	then	Descended 70 meters	then	Taken up 2100 meters	then	Dropped 230 meters	then	Rose 1200 meters
1000	+	(−70)	+	2100	+	(−230)	+	1200.

Discover for Yourself

Try working Example 6 by adding from left to right. You should still obtain 4000 for the sum. Which method do you find easier?

$1000 + (-70) + 2100 + (-230) + 1200$ This is the sum arising from the problem's conditions.

$= (1000 + 2100 + 1200) + [(-70) + (-230)]$ Use the commutative and associative properties to group the positive and negative numbers.

$= 4300 + (-300)$ Add the positive numbers.
Add the negative numbers.

$= 4000$ Add the results.

The altitude of the glider is 4000 meters. ∎

☑ **CHECK POINT 6** The water level of a reservoir is measured over a five-month period. During this time, the level rose 2 feet, then fell 4 feet, then rose 1 foot, then fell 5 feet, and then rose 3 feet. What was the change in the water level at the end of the five months?

Achieving Success

Some of the difficulties facing students who take a college math class for the first time are a result of the differences between a high school math class and a college math course. It is helpful to understand these differences throughout your college math courses.

High School Math Class	College Math Course
Attendance is required.	Attendance may be optional.
Teachers monitor progress and performance closely.	Students receive grades, but may not be informed by the professor if they are in trouble.
There are frequent tests, as well as make-up tests if grades are poor.	There are usually no more than three or four tests per semester. Make-up tests are rarely allowed.
Grades are often based on participation and effort.	Grades are usually based exclusively on test grades.
Students have contact with their instructor every day.	Students usually meet with their instructor two or three times per week.
Teachers cover all material for tests in class through lectures and activities.	Students are responsible for information whether or not it is covered in class.
A course is covered over the school year, usually ten months.	A course is covered in a semester, usually four months.
Extra credit is often available for struggling students.	Extra credit is almost never offered.

Source: BASS, ALAN, *MATH STUDY SKILLS*, 1st, © 2008. Printed and Electronically reproduced by permission of Pearson Education, Inc., Upper Saddle River, New Jersey.

CONCEPT AND VOCABULARY CHECK

Fill in each blank so that the resulting statement is true.

1. Pairs of real numbers that are the same number of units from zero on the number line, but are on opposite sides of zero, are called opposites or _____ .

2. The sum of any real number and its opposite is always _____ .

In the remaining items, state whether each sum is a positive number, a negative number, or 0. Do not actually find the sum.

3. $-20 + (-18)$ _____

4. $-20 + 28$ _____

5. $-0.8 + 0.8$ _____

6. $-\dfrac{4}{5} + \dfrac{1}{2}$ _____

7. $-1.3 + 2.7$ _____

8. $-\dfrac{2}{5} + \dfrac{2}{5}$ _____

1.5 EXERCISE SET ▶ MyMathLab®

Practice Exercises

In Exercises 1–8, find each sum using a number line.

1. $7 + (-3)$
2. $7 + (-2)$
3. $-2 + (-5)$
4. $-1 + (-5)$
5. $-6 + 2$
6. $-8 + 3$
7. $3 + (-3)$
8. $5 + (-5)$

In Exercises 9–46, find each sum without the use of a number line.

9. $-7 + 0$
10. $-5 + 0$

11. $30 + (-30)$

12. $15 + (-15)$

13. $-30 + (-30)$

14. $-15 + (-15)$

15. $-8 + (-10)$

16. $-4 + (-6)$

17. $-0.4 + (-0.9)$

18. $-1.5 + (-5.3)$

19. $-\dfrac{7}{10} + \left(-\dfrac{3}{10}\right)$

20. $-\dfrac{7}{8} + \left(-\dfrac{1}{8}\right)$

21. $-9 + 4$

22. $-7 + 3$

23. $12 + (-8)$

24. $13 + (-5)$

25. $6 + (-9)$

26. $3 + (-11)$

27. $-3.6 + 2.1$

28. $-6.3 + 5.2$

29. $-3.6 + (-2.1)$

30. $-6.3 + (-5.2)$

31. $\dfrac{9}{10} + \left(-\dfrac{3}{5}\right)$

32. $\dfrac{7}{10} + \left(-\dfrac{2}{5}\right)$

33. $-\dfrac{5}{8} + \dfrac{3}{4}$

34. $-\dfrac{5}{6} + \dfrac{1}{3}$

35. $-\dfrac{3}{7} + \left(-\dfrac{4}{5}\right)$

36. $-\dfrac{3}{8} + \left(-\dfrac{2}{3}\right)$

37. $4 + (-7) + (-5)$

38. $10 + (-3) + (-8)$

39. $85 + (-15) + (-20) + 12$

40. $60 + (-50) + (-30) + 25$

41. $17 + (-4) + 2 + 3 + (-10)$

42. $19 + (-5) + 1 + 8 + (-13)$

43. $-45 + \left(-\dfrac{3}{7}\right) + 25 + \left(-\dfrac{4}{7}\right)$

44. $-50 + \left(-\dfrac{7}{9}\right) + 35 + \left(-\dfrac{11}{9}\right)$

45. $3.5 + (-45) + (-8.4) + 72$

46. $6.4 + (-35) + (-2.6) + 14$

In Exercises 47–60, simplify each algebraic expression.

47. $-10x + 2x$

48. $-19x + 10x$

49. $25y + (-12y)$

50. $26y + (-14y)$

51. $-8a + (-15a)$

52. $-9a + (-13a)$

53. $4y + (-13z) + (-10y) + 17z$

54. $5y + (-11z) + (-15y) + 20z$

55. $-7b + 10 + (-b) + (-6)$

56. $-10b + 13 + (-b) + (-4)$

57. $7x + (-5y) + (-9x) + 19y$

58. $13x + (-9y) + (-17x) + 20y$

59. $8(4y + 3) + (-35y)$

60. $7(3y + 5) + (-25y)$

Practice PLUS

In Exercises 61–64, find each sum.

61. $|-3 + (-5)| + |2 + (-6)|$

62. $|4 + (-11)| + |-3 + (-4)|$

63. $-20 + \left[-|15 + (-25)|\right]$

64. $-25 + \left[-|18 + (-26)|\right]$

In Exercises 65–66, insert either $<$, $>$, or $=$ in the shaded area to make a true statement.

65. $6 + [2 + (-13)]$ ▉ $-3 + [4 + (-8)]$

66. $[(-8) + (-6)] + 10$ ▉ $-8 + [9 + (-2)]$

In Exercises 67–70, write each English phrase as an algebraic expression. Then simplify the expression. Let x represent the number.

67. The product of -6 and a number, which is then increased by the product of -13 and the number

68. The product of -9 and a number, which is then increased by the product of -11 and the number

69. The quotient of -20 and a number, increased by the quotient of 3 and the number

70. The quotient of -15 and a number, increased by the quotient of 4 and the number

Application Exercises

Solve Exercises 71–78 by writing a sum of signed numbers and adding.

71. The greatest temperature variation recorded in a day is 100 degrees in Browning, Montana, on January 23, 1916. The low temperature was $-56°F$. What was the high temperature?

72. In Spearfish, South Dakota, on January 22, 1943, the temperature rose 49 degrees in two minutes. If the initial temperature was $-4°F$, what was the temperature two minutes later?

73. The Dead Sea is the lowest elevation on Earth, 1312 feet below sea level. What is the elevation of a person standing 712 feet above the Dead Sea?

74. Lake Assal in Africa is 512 feet below sea level. What is the elevation of a person standing 642 feet above Lake Assal?

75. The temperature at 8:00 A.M. was $-7°F$. By noon it had risen 15°F, but by 4:00 P.M. it had fallen 5°F. What was the temperature at 4:00 P.M.?

76. On three successive plays, a football team lost 15 yards, gained 13 yards, and then lost 4 yards. What was the team's total gain or loss for the three plays?

77. A football team started with the football at the 27-yard line, advancing toward the center of the field (the 50-yard line). Four successive plays resulted in a 4-yard gain, a 2-yard loss, an 8-yard gain, and a 12-yard loss. What was the location of the football at the end of the fourth play?

78. The water level of a reservoir is measured over a five-month period. At the beginning, the level is 20 feet. During this time, the level rose 3 feet, then fell 2 feet, then fell 1 foot, then fell 4 feet, and then rose 2 feet. What is the reservoir's water level at the end of the five months?

The bar graph shows that in 2001, the U.S. government collected more in taxes than it spent, so there was a budget surplus for that year. By contrast, in 2003 through 2013, the government spent more than it collected, resulting in budget deficits. Exercises 79–80 involve these deficits.

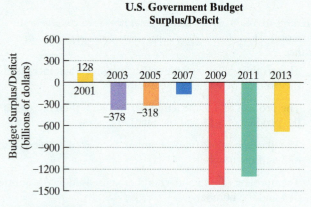

U.S. Government Budget Surplus/Deficit

Source: Budget of the U.S. Government

79. a. In 2011, the government collected $2303 billion and spent $3603 billion. Find $2303 + (-3603)$ and determine the deficit, in billions of dollars, for 2011.

b. In 2013, the government collected $2775 billion and spent $3455 billion. Find the deficit, in billions of dollars, for 2013.

c. Use your answers from parts (a) and (b) to determine the combined deficit, in billions of dollars, for 2011 and 2013.

80. a. In 2007, the government collected $2568 billion and spent $2729 billion. Find $2568 + (-2729)$ and determine the deficit, in billions of dollars, for 2007.

b. In 2009, the government collected $2105 billion and spent $3518 billion. Find the deficit, in billions of dollars, for 2009.

c. Use your answers from parts (a) and (b) to determine the combined deficit, in billions of dollars, for 2007 and 2009.

Explaining the Concepts

81. Explain how to add two numbers with a number line. Provide an example with your explanation.

82. What are additive inverses?

83. Describe how the inverse property of addition

$$a + (-a) = 0$$

can be shown on a number line.

84. Without using a number line, describe how to add two numbers with the same sign. Give an example.

85. Without using a number line, describe how to add two numbers with different signs. Give an example.

86. Write a problem that can be solved by finding the sum of at least three numbers, some positive and some negative. Then explain how to solve the problem.

87. Without a calculator, you can add numbers using a number line, using absolute value, or using gains and losses. Which method do you find most helpful? Why is this so?

Critical Thinking Exercises

Make Sense?   *In Exercises 88–91, determine whether each statement makes sense or does not make sense, and explain your reasoning.*

88. It takes me too much time to add real numbers with a number line.

89. I found the sum of -13 and 4 by thinking of temperatures above and below zero: If it's 13 below zero and the temperature rises 4 degrees, the new temperature will be 9 below zero, so $-13 + 4 = -9$.

90. I added two negative numbers and obtained a positive sum.

91. Without adding numbers, I can see that the sum of -227 and 319 is greater than the sum of 227 and -319.

In Exercises 92–95, determine whether each statement is true or false. If the statement is false, make the necessary change(s) to produce a true statement.

92. $\dfrac{3}{4} + \left(-\dfrac{3}{5}\right) = -\dfrac{3}{20}$

93. The sum of zero and a negative number is always a negative number.

94. If one number is positive and the other negative, then the absolute value of their sum equals the sum of their absolute values.

95. The sum of a positive number and a negative number is always a positive number.

In Exercises 96–99, use the number line to determine whether each expression is positive or negative.

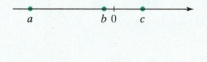

96. $a + b$

97. $a + c$

98. $b + c$

99. $|a + c|$

Technology Exercises

100. Use a calculator to verify any five of the sums that you found in Exercises 9–46.

101. Use a calculator to verify any three of the answers that you obtained in Application Exercises 71–78.

Review Exercises

From here on, each Exercise Set will contain three review exercises. It is essential to review previously covered topics to improve your understanding of the topics and to help maintain your mastery of the material. If you are not certain how to solve a review exercise, turn to the section and the worked example given in parentheses at the end of each exercise.

102. Consider the set

$$\{-6, -\pi, 0, 0.\overline{7}, \sqrt{3}, \sqrt{4}\}.$$

List all numbers from the set that are **a.** natural numbers, **b.** whole numbers, **c.** integers, **d.** rational numbers, **e.** irrational numbers, **f.** real numbers. (Section 1.3, Example 5)

103. Determine whether this inequality is true or false: $19 \geq -18$. (Section 1.3, Example 7)

104. Determine whether 18 is a solution of $16 = 2(x - 1) - x$. (Section 1.1, Example 4)

Preview Exercises

Exercises 105–107 will help you prepare for the material covered in the next section. In each exercise, a subtraction is expressed as addition of an opposite. Find this sum, indicated by a question mark.

105. $7 - 10 = 7 + (-10) = ?$

106. $-8 - 13 = -8 + (-13) = ?$

107. $-8 - (-13) = -8 + 13 = ?$

SECTION

1.6

Subtraction of Real Numbers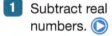

What am I supposed to learn?

After studying this section, you should be able to:

1. Subtract real numbers.
2. Simplify a series of additions and subtractions.
3. Use the definition of subtraction to identify terms.
4. Use the subtraction definition to simplify algebraic expressions.
5. Solve problems involving subtraction.

Everybody complains about the weather, but where is it really the worst? **Table 1.5** shows that in the United States, Alaska and Montana are the record-holders for the coldest temperatures.

Table 1.5 Lowest Recorded U.S. Temperatures

State	Year	Temperature (°F)
Alaska	1971	−80
Montana	1954	−70

Source: National Climatic Data Center

The table shows that Montana's record low temperature, −70, decreased by 10 degrees gives Alaska's record low temperature, −80. This can be expressed in two ways:

$$-70 - 10 = -80 \quad \text{or} \quad -70 + (-10) = -80.$$

This means that

$$-70 - 10 = -70 + (-10).$$

To subtract 10 from −70, we add −70 and the opposite, or additive inverse, of 10. Generalizing from this situation, we define subtraction as addition of an opposite.

Definition of Subtraction

For all real numbers a and b,

$$a - b = a + (-b).$$

In words: To subtract b from a, add the opposite, or additive inverse, of b to a. The result of subtraction is called the **difference**.

1 Subtract real numbers.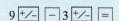

A Procedure for Subtracting Real Numbers

The definition of subtraction gives us a procedure for subtracting real numbers.

Subtracting Real Numbers

1. Change the subtraction operation to addition.
2. Change the sign of the number being subtracted.
3. Add, using one of the rules for adding numbers with the same sign or different signs.

Using Technology

You can use a calculator to subtract signed numbers. Here are the keystrokes for finding $5 - (-6)$:

Scientific Calculator

$5 \boxed{-} 6 \boxed{+/-} \boxed{=}$

Graphing Calculator

$5 \boxed{-} \boxed{(-)} 6 \boxed{\text{ENTER}}$

Here are the keystrokes for finding $-9 - (-3)$:

Scientific Calculator

$9 \boxed{+/-} \boxed{-} 3 \boxed{+/-} \boxed{=}$

Graphing Calculator

$\boxed{(-)} 9 \boxed{-} \boxed{(-)} 3 \boxed{\text{ENTER}}$

Don't confuse the subtraction key on a graphing calculator, $\boxed{-}$, with the sign change or additive inverse key, $\boxed{(-)}$. What happens if you do?

EXAMPLE 1 Using the Definition of Subtraction

Subtract:

a. $7 - 10$ **b.** $5 - (-6)$ **c.** $-9 - (-3)$.

Solution

a. $7 - 10 = 7 + (-10) = -3$

> Change the subtraction to addition. Replace 10 with its opposite.

b. $5 - (-6) = 5 + 6 = 11$

> Change the subtraction to addition. Replace −6 with its opposite.

c. $-9 - (-3) = -9 + 3 = -6$

> Change the subtraction to addition. Replace −3 with its opposite.

■

✓ **CHECK POINT 1** Subtract:

a. $3 - 11$ **b.** $4 - (-5)$ **c.** $-7 - (-2)$.

The definition of subtraction can be applied to real numbers that are not integers.

EXAMPLE 2 Using the Definition of Subtraction

Subtract:

a. $-5.2 - (-11.4)$ **b.** $-\dfrac{3}{4} - \dfrac{2}{3}$ **c.** $4\pi - (-9\pi)$.

Solution

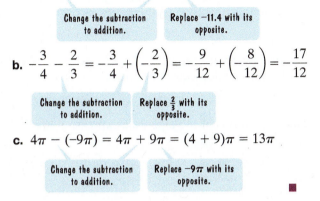

a. $-5.2 - (-11.4) = -5.2 + 11.4 = 6.2$

> Change the subtraction to addition.
>
> Replace -11.4 with its opposite.

b. $-\dfrac{3}{4} - \dfrac{2}{3} = -\dfrac{3}{4} + \left(-\dfrac{2}{3}\right) = -\dfrac{9}{12} + \left(-\dfrac{8}{12}\right) = -\dfrac{17}{12}$

> Change the subtraction to addition.
>
> Replace $\frac{2}{3}$ with its opposite.

c. $4\pi - (-9\pi) = 4\pi + 9\pi = (4+9)\pi = 13\pi$

> Change the subtraction to addition.
>
> Replace -9π with its opposite.

■

Reading the symbol "−" can be a bit tricky. The way you read it depends on where it appears. For example,

$$-5.2 - (-11.4)$$

is read "negative five point two minus negative eleven point four." Read parts (b) and (c) of Example 2 aloud. When is "−" read "negative" and when is it read "minus"?

✓ **CHECK POINT 2** Subtract:

a. $-3.4 - (-12.6)$ **b.** $-\dfrac{3}{5} - \dfrac{1}{3}$ **c.** $5\pi - (-2\pi)$.

---

2 Simplify a series of additions and subtractions. ▶

Problems Containing a Series of Additions and Subtractions

In some problems, several additions and subtractions occur together. We begin by converting all subtractions to additions of opposites, or additive inverses.

> ### Simplifying a Series of Additions and Subtractions
>
> 1. Change all subtractions to additions of opposites.
> 2. Group and then add all the positive numbers.
> 3. Group and then add all the negative numbers.
> 4. Add the results of steps 2 and 3.

EXAMPLE 3 Simplifying a Series of Additions and Subtractions

Simplify: $7 - (-5) - 11 - (-6) - 19$.

Solution

$$7 - (-5) - 11 - (-6) - 19$$

$= 7 + 5 + (-11) + 6 + (-19)$ Write subtractions as additions of opposites.

$= (7 + 5 + 6) + [(-11) + (-19)]$ Group the positive numbers. Group the negative numbers.

$= 18 + (-30)$ Add the positive numbers. Add the negative numbers.

$= -12$ Add the results. ■

✓ **CHECK POINT 3** Simplify: $10 - (-12) - 4 - (-3) - 6$.

3 Use the definition of subtraction to identify terms. ▶

Subtraction and Algebraic Expressions

We know that the terms of an algebraic expression are separated by addition signs. Let's use this idea to identify the terms of the following algebraic expression:

$$9x - 4y - 5.$$

Because terms are separated by addition, we rewrite the subtractions in the algebraic expression as additions of opposites. Thus,

$$9x - 4y - 5 = 9x + (-4y) + (-5).$$

The three terms of the algebraic expression are $9x$, $-4y$, and -5.

EXAMPLE 4 Using the Definition of Subtraction to Identify Terms

Identify the terms of the algebraic expression:

$$2xy - 13y - 6.$$

Solution Rewrite the subtractions in the algebraic expression as additions of opposites.

$$2xy - 13y - 6 = 2xy + (-13y) + (-6)$$

| First term | Second term | Third term |

Because terms are separated by addition, the terms are $2xy$, $-13y$, and -6. ■

✓ **CHECK POINT 4** Identify the terms of the algebraic expression:

$$-6 + 4a - 7ab.$$

---

4 Use the subtraction definition to simplify algebraic expressions. ▶

The procedure for subtracting real numbers can be used to simplify certain algebraic expressions that involve subtraction.

EXAMPLE 5 Simplifying Algebraic Expressions

Simplify:

a. $2 + 3x - 8x$ **b.** $-4x - 9y - 2x + 12y.$

Solution

a. $2 + 3x - 8x$ This is the given algebraic expression.

$= 2 + 3x + (-8x)$ Write the subtraction as the addition of an opposite.

$= 2 + [3 + (-8)]x$ Apply the distributive property.

$= 2 + (-5x)$ Add within the grouping symbols.

$= 2 - 5x$ Be concise and express as subtraction.

b. $-4x - 9y - 2x + 12y$ This is the given algebraic expression.

$= -4x + (-9y) + (-2x) + 12y$ Write the subtractions as the additions of opposites.

$= -4x + (-2x) + (-9y) + 12y$ Arrange like terms so that they are next to one another.

$= [-4 + (-2)]x + (-9 + 12)y$ Apply the distributive property.

$= -6x + 3y$ Add within the grouping symbols. ■

Great Question!

How can I speed up the process of subtracting like terms?

You can think of gains and losses of money to work some of the steps in Example 5 mentally:

- $3x - 8x = -5x$ A gain of 3 dollars followed by a loss of 8 dollars is a net loss of 5 dollars.
- $-9y + 12y = 3y$ A loss of 9 dollars followed by a gain of 12 dollars is a net gain of 3 dollars.

✓ **CHECK POINT 5** Simplify:

a. $4 + 2x - 9x$ **b.** $-3x - 10y - 6x + 14y.$

---

5 Solve problems involving subtraction.

Applications

Subtraction is used to solve problems in which the word *difference* appears. The difference between real numbers a and b is expressed as $a - b$.

EXAMPLE 6	An Application of Subtraction Using the Word *Difference*

Life expectancy for the average American man is 75.2 years; for a woman, it's 80.4 years. The number line in **Figure 1.13**, with points representing eight integers, indicates factors, many within our control, that can stretch or shrink one's probable lifespan.

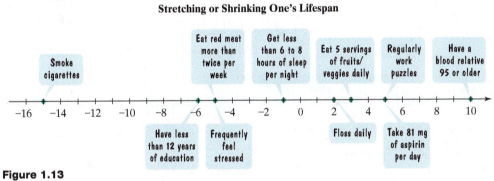

Figure 1.13

Source: Newsweek

Years of Life Gained or Lost

a. What is the difference in the lifespan between a person who regularly works puzzles and a person who eats red meat more than twice per week?

b. What is the difference in the lifespan between a person with less than 12 years of education and a person who smokes cigarettes?

Solution

a. We begin with the difference in the lifespan between a person who regularly works puzzles and a person who eats red meat more than twice per week. Refer to **Figure 1.13** to determine years of life gained or lost.

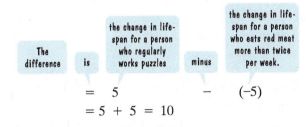

$$= 5 - (-5)$$
$$= 5 + 5 = 10$$

The difference in the lifespan is 10 years.

b. Now we consider the difference in the lifespan between a person with less than 12 years of education and a person who smokes cigarettes.

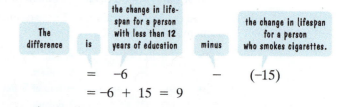

$$= -6 \quad - \quad (-15)$$
$$= -6 + 15 = 9$$

The difference in the lifespan is 9 years. ∎

✓ **CHECK POINT 6**   Use the number line in **Figure 1.13** to answer the following questions:

a. What is the difference in the lifespan between a person who eats five servings of fruits/veggies daily and a person who frequently feels stressed?

b. What is the difference in the lifespan between a person who gets less than 6 to 8 hours of sleep per night and a person who smokes cigarettes?

Stretching or Shrinking One's Lifespan

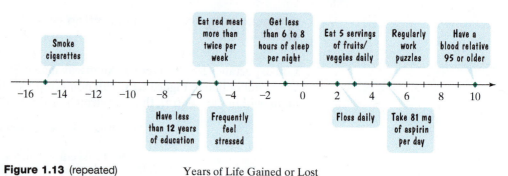

Figure 1.13 (repeated)   Years of Life Gained or Lost

CONCEPT AND VOCABULARY CHECK

Fill in each blank so that the resulting statement is true.

1. $8 - 14 = 8 +$ _____

2. $8 - (-14) = 8 +$ _____

3. $-8 - (-14) = -8 +$ _____

4. $-8 - 14 =$ _____ $+$ _____

5. $7 - (-3) - 12 - 23 = 7 +$ _____ $+$ _____ $+$ _____

6. $9x - 4y - (-6) = 9x +$ _____ $+$ _____

7. The algebraic expression in Concept Check 6 contains _____ terms because terms are separated by _____ .

1.6 EXERCISE SET   ▶ MyMathLab®

Practice Exercises

1. Consider the subtraction $5 - 12$.

   a. Find the opposite, or additive inverse, of 12.

   b. Rewrite the subtraction as the addition of the opposite of 12.

2. Consider the subtraction $4 - 10$.

   a. Find the opposite, or additive inverse, of 10.

   b. Rewrite the subtraction as the addition of the opposite of 10.

3. Consider the subtraction $5 - (-7)$.

   a. Find the opposite, or additive inverse, of -7.

   b. Rewrite the subtraction as the addition of the opposite of -7.

4. Consider the subtraction $2 - (-8)$.

   a. Find the opposite, or additive inverse, of -8.

   b. Rewrite the subtraction as the addition of the opposite of -8.

In Exercises 5–50, perform the indicated subtraction.

5. $14 - 8$

6. $15 - 2$

7. $8 - 14$

8. $2 - 15$

9. $3 - (-20)$

10. $5 - (-17)$

11. $-7 - (-18)$

12. $-5 - (-19)$

13. $-13 - (-2)$

14. $-21 - (-3)$

15. $-21 - 17$

16. $-29 - 21$

17. $-45 - (-45)$

18. $-65 - (-65)$

19. $23 - 23$

20. $26 - 26$

21. $13 - (-13)$

22. $15 - (-15)$

23. $0 - 13$

24. $0 - 15$

25. $0 - (-13)$

26. $0 - (-15)$

27. $\frac{3}{7} - \frac{5}{7}$

28. $\frac{4}{9} - \frac{7}{9}$

29. $\frac{1}{5} - \left(-\frac{3}{5}\right)$

30. $\frac{1}{7} - \left(-\frac{3}{7}\right)$

31. $-\frac{4}{5} - \frac{1}{5}$

32. $-\frac{4}{9} - \frac{1}{9}$

33. $-\frac{4}{5} - \left(-\frac{1}{5}\right)$

34. $-\frac{4}{9} - \left(-\frac{1}{9}\right)$

35. $\frac{1}{2} - \left(-\frac{1}{4}\right)$

36. $\frac{2}{5} - \left(-\frac{1}{10}\right)$

37. $\frac{1}{2} - \frac{1}{4}$

38. $\frac{2}{5} - \frac{1}{10}$

39. $9.8 - 2.2$

40. $5.7 - 3.3$

41. $-3.1 - (-1.1)$

42. $-4.6 - (-1.1)$

43. $1.3 - (-1.3)$

44. $1.4 - (-1.4)$

45. $-2.06 - (-2.06)$

46. $-3.47 - (-3.47)$

47. $5\pi - 2\pi$

48. $9\pi - 7\pi$

49. $3\pi - (-10\pi)$

50. $4\pi - (-12\pi)$

In Exercises 51–68, simplify each series of additions and subtractions.

51. $13 - 2 - (-8)$

52. $14 - 3 - (-7)$

53. $9 - 8 + 3 - 7$

54. $8 - 2 + 5 - 13$

55. $-6 - 2 + 3 - 10$

56. $-9 - 5 + 4 - 17$

57. $-10 - (-5) + 7 - 2$

58. $-6 - (-3) + 8 - 11$

59. $-23 - 11 - (-7) + (-25)$

60. $-19 - 8 - (-6) + (-21)$

61. $-823 - 146 - 50 - (-832)$

62. $-726 - 422 - 921 - (-816)$

63. $1 - \frac{2}{3} - \left(-\frac{5}{6}\right)$

64. $2 - \frac{3}{4} - \left(-\frac{7}{8}\right)$

65. $-0.16 - 5.2 - (-0.87)$

66. $-1.9 - 3 - (-0.26)$

67. $-\frac{3}{4} - \frac{1}{4} - \left(-\frac{5}{8}\right)$

68. $-\frac{1}{2} - \frac{2}{3} - \left(-\frac{1}{3}\right)$

In Exercises 69–72, identify the terms in each algebraic expression.

69. $-3x - 8y$

70. $-9a - 4b$

71. $12x - 5xy - 4$

72. $8a - 7ab - 13$

In Exercises 73–84, simplify each algebraic expression.

73. $3x - 9x$

74. $2x - 10x$

75. $4 + 7y - 17y$

76. $5 + 9y - 29y$

77. $2a + 5 - 9a$

78. $3a + 7 - 11a$

79. $4 - 6b - 8 - 3b$

80. $5 - 7b - 13 - 4b$

81. $13 - (-7x) + 4x - (-11)$

82. $15 - (-3x) + 8x - (-10)$

83. $-5x - 10y - 3x + 13y$

84. $-6x - 9y - 4x + 15y$

Practice PLUS

In Exercises 85–90, find the value of each expression.

85. $-|-9 - (-6)| - (-12)$

86. $-|-8 - (-2)| - (-6)$

87. $\frac{5}{8} - \left(\frac{1}{2} - \frac{3}{4}\right)$

88. $\frac{9}{10} - \left(\frac{1}{4} - \frac{7}{10}\right)$

89. $|-9 - (-3 + 7)| - |-17 - (-2)|$

90. $|24 - (-16)| - |-51 - (-31 + 2)|$

In Exercises 91–94, write each English phrase as an algebraic expression. Then simplify the expression. Let x represent the number.

91. The difference between 6 times a number and -5 times the number

92. The difference between 9 times a number and -4 times the number

93. The quotient of -2 and a number, subtracted from the quotient of -5 and the number

94. The quotient of -7 and a number, subtracted from the quotient of -12 and the number

Application Exercises

95. The peak of Mount Kilimanjaro, the highest point in Africa, is 19,321 feet above sea level. Qattara Depression, Egypt, one of the lowest points in Africa, is 436 feet below sea level. What is the difference in elevation between the peak of Mount Kilimanjaro and the Qattara Depression?

96. The peak of Mount Whitney is 14,494 feet above sea level. Mount Whitney can be seen directly above Death Valley, which is 282 feet below sea level. What is the difference in elevation between these geographic locations?

In Exercises 97–106, we return to the number line that shows factors that can stretch or shrink one's probable lifespan.

Stretching or Shrinking One's Lifespan

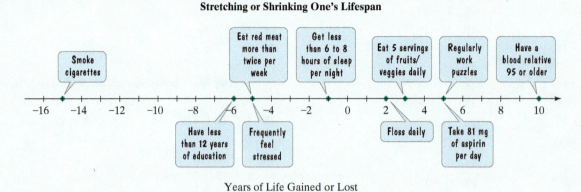

Years of Life Gained or Lost

Source: Newsweek

97. If you have a blood relative 95 or older and you smoke cigarettes, do you stretch or shrink your lifespan? By how many years?

98. If you floss daily and eat red meat more than twice per week, do you stretch or shrink your lifespan? By how many years?

99. If you frequently feel stressed and have less than 12 years of education, do you stretch or shrink your lifespan? By how many years?

100. If you get less than 6 to 8 hours of sleep per night and smoke cigarettes, do you stretch or shrink your lifespan? By how many years?

101. What happens to the lifespan for a person who takes 81 mg of aspirin per day and eats red meat more than twice per week?

102. What happens to the lifespan for a person who regularly works puzzles and a person who frequently feels stressed?

103. What is the difference in the lifespan between a person who has a blood relative 95 or older and a person who smokes cigarettes?

104. What is the difference in the lifespan between a person who has a blood relative 95 or older and a person who has less than 12 years of education?

105. What is the difference in the lifespan between a person who frequently feels stressed and a person who has less than 12 years of education?

106. What is the difference in the lifespan between a person who gets less than 6 to 8 hours of sleep per night and a person who frequently feels stressed?

The number line shows the depths below the ocean surface, in meters, at which scientists have recorded different animals. Use this information to solve Exercises 107–110.

Deepest Dives

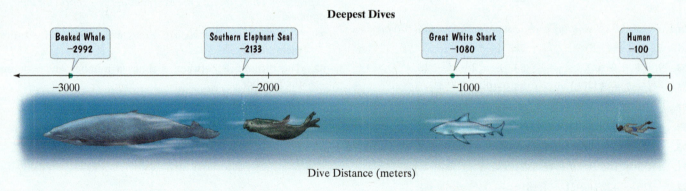

Dive Distance (meters)

Source: Scholastic Math

107. How much deeper can a beaked whale dive than a human?

108. How much deeper can a southern elephant seal dive than a great white shark?

109. A southern elephant seal dove down to −2000 meters. Then it headed up 400 meters toward the surface. Then it dove another 500 meters. What was the seal's final distance from the surface?

110. A great white shark dove down to −900 meters. Then it headed up 300 meters toward the surface. Then it dove another 200 meters. What was the shark's final distance from the surface?

Explaining the Concepts

111. Explain how to subtract real numbers.

112. How is $4 - (-2)$ read?

113. Explain how to simplify a series of additions and subtractions. Provide an example with your explanation.

114. Explain how to find the terms of the algebraic expression $5x - 2y - 7$.

115. Write a problem that can be solved by finding the difference between two numbers. At least one of the numbers should be negative. Then explain how to solve the problem.

Critical Thinking Exercises

Make Sense? *In Exercises 116–119, determine whether each statement makes sense or does not make sense, and explain your reasoning.*

116. I already knew how to add positive and negative numbers, so there was not that much new to learn when it came to subtracting them.

117. I can find the closing price of stock PQR on Wednesday by subtracting the change in price, -1.23, from the closing price on Thursday, 47.19.

118. I found the variation in elevation between two heights by taking the difference between the high point and the low point.

119. I found the variation in U.S. temperature by subtracting the record low temperature, a negative number, from the record high temperature, a positive number.

In Exercises 120–123, determine whether each statement is true or false. If the statement is false, make the necessary change(s) to produce a true statement.

120. If a and b are negative numbers, then $a - b$ is sometimes a negative number.

121. $7 - (-2) = 5$

122. The difference between 0 and a negative number is always a positive number.

123. $|a - b| = |b - a|$

In Exercises 124–127, use the number line to determine whether each difference is positive or negative.

124. $c - a$

125. $a - b$

126. $b - c$

127. $0 - b$

128. Order the expressions $|x - y|$, $|x| - |y|$, and $|x + y|$ from least to greatest for $x = -6$ and $y = -8$.

Technology Exercises

129. Use a calculator to verify any five of the differences that you found in Exercises 5–46.

130. Use a calculator to verify any three of the answers that you found in Exercises 51–68.

Review Exercises

131. Determine whether 2 is a solution of $13x + 3 = 3(5x - 1)$. (Section 1.1, Example 4)

132. Simplify: $5(3x + 2y) + 6(5y)$. (Section 1.4, Example 11)

133. Give an example of an integer that is not a natural number. (Section 1.3, Example 5)

Preview Exercises

Exercises 134–136 will help you prepare for the material covered in the next section.

In Exercises 134–135, a multiplication is expressed as a repeated addition. Find this sum, indicated by a question mark.

134. $4(-3) = (-3) + (-3) + (-3) + (-3) = ?$

135. $3(-3) = (-3) + (-3) + (-3) = ?$

136. The list shows a pattern for various products.

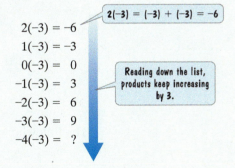

Use this pattern to find $-4(-3)$.

SECTION

1.7

Multiplication and Division of Real Numbers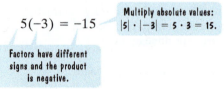

What am I Supposed to Learn?

After studying this section, you should be able to:

1. Multiply real numbers.
2. Multiply more than two real numbers.
3. Find multiplicative inverses.
4. Use the definition of division.
5. Divide real numbers.
6. Simplify algebraic expressions involving multiplication.
7. Determine whether a number is a solution of an equation.
8. Use mathematical models involving multiplication and division.

Where the Boys Are, released in 1960, was the first college spring-break movie. Today, visitors to college campuses can't help but ask: Where are the boys?

In 2007, 135 women received bachelor's degrees for every 100 men. In this section, we use data from a report by the U.S. Department of Education to develop mathematical models showing that this gender imbalance will widen in the coming years. Working with these models requires that we know how to multiply and divide real numbers.

Multiplying Real Numbers

Multiplication is repeated addition. For example, $5(-3)$ means that -3 is added five times:

$$5(-3) = (-3) + (-3) + (-3) + (-3) + (-3) = -15.$$

The result of the multiplication, -15, is called the **product** of 5 and -3. The numbers being multiplied, 5 and -3, are called the **factors** of the product.

Rules for multiplying real numbers are described in terms of absolute value. For example, $5(-3) = -15$ illustrates that the product of numbers with different signs is found by multiplying their absolute values. The product is negative.

$$5(-3) = -15$$

Multiply absolute values:
$|5| \cdot |-3| = 5 \cdot 3 = 15.$

Factors have different signs and the product is negative.

The following rules are used to determine the sign of the product of two numbers:

1. Multiply real numbers.

The Product of Two Real Numbers

- The product of two real numbers with **different signs** is found by multiplying their absolute values. The product is **negative**.
- The product of two real numbers with the **same sign** is found by multiplying their absolute values. The product is **positive**.
- The product of 0 and any real number is 0. Thus, for any real number a,

$$a \cdot 0 = 0 \quad \text{and} \quad 0 \cdot a = 0.$$

EXAMPLE 1   Multiplying Real Numbers

Multiply:

a. $6(-3)$ **b.** $-\dfrac{1}{5} \cdot \dfrac{2}{3}$ **c.** $(-9)(-10)$ **d.** $(-1.4)(-2)$ **e.** $(-372)(0)$.

Solution

a. $6(-3) = -18$ ⟵ Multiply absolute values: $6 \cdot 3 = 18$.

Different signs: negative product

b. $-\dfrac{1}{5} \cdot \dfrac{2}{3} = -\dfrac{2}{15}$ ⟵ Multiply absolute values: $\frac{1}{5} \cdot \frac{2}{3} = \frac{1 \cdot 2}{5 \cdot 3} = \frac{2}{15}$.

Different signs: negative product

c. $(-9)(-10) = 90$ ⟵ Multiply absolute values: $9 \cdot 10 = 90$.

Same sign: positive product

d. $(-1.4)(-2) = 2.8$ ⟵ Multiply absolute values: $(1.4)(2) = 2.8$.

Same sign: positive product

e. $(-372)(0) = 0$ ⟵ The product of 0 and any real number is 0: $a \cdot 0 = 0$.

■

✓ **CHECK POINT 1**   Multiply:

a. $8(-5)$ **b.** $-\dfrac{1}{3} \cdot \dfrac{4}{7}$ **c.** $(-12)(-3)$

d. $(-1.1)(-5)$ **e.** $(-543)(0)$.

---

2 Multiply more than two real numbers. ⓟ

Multiplying More Than Two Numbers

How do we perform more than one multiplication, such as

$$-4(-3)(-2)?$$

Because of the associative and commutative properties, we can order and group the numbers in any manner. Each pair of negative numbers will produce a positive product. Thus, the product of an even number of negative numbers is always positive. By contrast, the product of an odd number of negative numbers is always negative.

$$-4(-3)(-2) = -24$$ ⟵ Multiply absolute values: $4 \cdot 3 \cdot 2 = 24$.

Odd number of negative numbers (three): negative product

> **Multiplying More Than Two Numbers**
>
> 1. Assuming that no factor is zero,
>   - The product of an **even** number of **negative numbers** is **positive**.
>   - The product of an **odd** number of **negative numbers** is **negative**.
>
>   The multiplication is performed by multiplying the absolute values of the given numbers.
>
> 2. If any factor is 0, the product is 0.

EXAMPLE 2   Multiplying More Than Two Numbers

Multiply:

a. $(-3)(-1)(2)(-2)$ **b.** $(-1)(-2)(-2)(3)(-4)$.

Solution

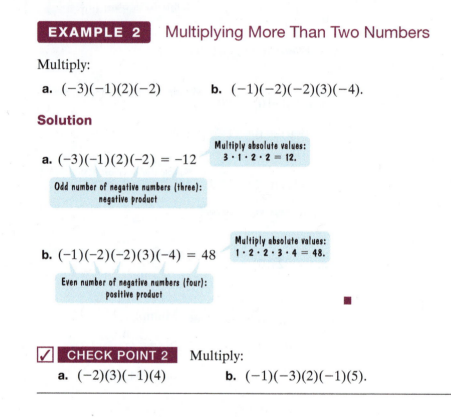

a. $(-3)(-1)(2)(-2) = -12$ Multiply absolute values: $3 \cdot 1 \cdot 2 \cdot 2 = 12.$

Odd number of negative numbers (three): negative product

b. $(-1)(-2)(-2)(3)(-4) = 48$ Multiply absolute values: $1 \cdot 2 \cdot 2 \cdot 3 \cdot 4 = 48.$

Even number of negative numbers (four): positive product

☑ **CHECK POINT 2**   Multiply:

a. $(-2)(3)(-1)(4)$ **b.** $(-1)(-3)(2)(-1)(5)$.

Is it always necessary to count the number of negative factors when multiplying more than two numbers? No. If any factor is 0, you can immediately write 0 for the product. For example,

$$(-37)(423)(0)(-55)(-3.7) = 0.$$

If any factor is 0, the product is 0.

③ Find multiplicative inverses. ▶

The Meaning of Division

The result of dividing the real number a by the nonzero real number b is called the **quotient** of a and b. We can write this quotient as $a \div b$ or $\frac{a}{b}$.

We know that subtraction is defined in terms of addition of an additive inverse, or opposite:

$$a - b = a + (-b).$$

In a similar way, we can define division in terms of multiplication. For example, the quotient of 8 and 2 can be written as multiplication:

$$8 \div 2 = 8 \cdot \frac{1}{2}.$$

We call $\frac{1}{2}$ the *multiplicative inverse*, or *reciprocal*, of 2. Two numbers whose product is 1 are called **multiplicative inverses** or **reciprocals** of each other. Thus, the multiplicative inverse of 2 is $\frac{1}{2}$ and the multiplicative inverse of $\frac{1}{2}$ is 2 because $2 \cdot \frac{1}{2} = 1$.

EXAMPLE 3   Finding Multiplicative Inverses

Find the multiplicative inverse of each number:

a. 5   **b.** $\frac{1}{3}$ **c.** -4 **d.** $-\frac{4}{5}$.

Solution

a. The multiplicative inverse of 5 is $\frac{1}{5}$ because $5 \cdot \frac{1}{5} = 1$.

b. The multiplicative inverse of $\frac{1}{3}$ is 3 because $\frac{1}{3} \cdot 3 = 1$.

c. The multiplicative inverse of -4 is $-\frac{1}{4}$ because $(-4)\left(-\frac{1}{4}\right) = 1$.

d. The multiplicative inverse of $-\frac{4}{5}$ is $-\frac{5}{4}$ because $\left(-\frac{4}{5}\right)\left(-\frac{5}{4}\right) = 1$. ■

✓ **CHECK POINT 3**   Find the multiplicative inverse of each number:

a. 7   **b.** $\frac{1}{8}$ **c.** -6 **d.** $-\frac{7}{13}$.

Can you think of a real number that has no multiplicative inverse? The number **0 has no multiplicative inverse** because 0 multiplied by any number is never 1, but always 0. We now define division in terms of multiplication by a multiplicative inverse.

4 Use the definition of division. ▶

Definition of Division

If a and b are real numbers and b is not 0, then the quotient of a and b is defined as

$$a \div b = a \cdot \frac{1}{b}.$$

In words: The quotient of two real numbers is the product of the first number and the multiplicative inverse of the second number.

EXAMPLE 4   Using the Definition of Division

Use the definition of division to find each quotient:

a. $-15 \div 3$ **b.** $\dfrac{-20}{-4}$.

5 Divide real numbers.

Solution

a. $-15 \div 3 = -15 \cdot \dfrac{1}{3} = -5$

> Change the division to multiplication.

> Replace 3 with its multiplicative inverse.

b. $\dfrac{-20}{-4} = -20 \cdot \left(-\dfrac{1}{4}\right) = 5$

> Change the division to multiplication.

> Replace −4 with its multiplicative inverse.

✓ **CHECK POINT 4** Use the definition of division to find each quotient:

a. $-28 \div 7$ **b.** $\dfrac{-16}{-2}.$

A Procedure for Dividing Real Numbers

Because the quotient $a \div b$ is defined as the product $a \cdot \frac{1}{b}$, the sign rules for dividing numbers are the same as the sign rules for multiplying them.

The Quotient of Two Real Numbers

- The quotient of two real numbers with **different signs** is found by dividing their absolute values. The quotient is **negative**.
- The quotient of two real numbers with the **same sign** is found by dividing their absolute values. The quotient is **positive**.
- Division of any number by zero is undefined.
- Any nonzero number divided into 0 is 0.

EXAMPLE 5 Dividing Real Numbers

Divide:

a. $\dfrac{8}{-2}$ **b.** $-\dfrac{3}{4} \div \left(-\dfrac{5}{9}\right)$ **c.** $\dfrac{-20.8}{4}$ **d.** $\dfrac{0}{-7}.$

Solution

a. $\dfrac{8}{-2} = -4$

> Divide absolute values: $\frac{8}{2} = 4$.

> Different signs: negative quotient

b. $-\dfrac{3}{4} \div \left(-\dfrac{5}{9}\right) = \dfrac{27}{20}$

> Divide absolute values: $\frac{3}{4} \div \frac{5}{9} = \frac{3}{4} \cdot \frac{9}{5} = \frac{27}{20}$.

> Same sign: positive quotient

c. $\dfrac{-20.8}{4} = -5.2$

> Divide absolute values: $4\overline{)20.8}$ $= 5.2$.

> Different signs: negative quotient

d. $\dfrac{0}{-7} = 0$

> Any nonzero number divided into 0 is 0.

Great Question!

Why is zero under the line undefined?

Here's another reason why division by zero is undefined. We know that

$$\frac{12}{4} = 3 \text{ because } 3 \cdot 4 = 12.$$

Now think about $\frac{-7}{0}$. If

$$\frac{-7}{0} = ? \text{ then } ? \cdot 0 = -7.$$

However, any real number multiplied by 0 is 0 and not -7. No matter what number we use to replace the question mark in $? \cdot 0 = -7$, we can never obtain -7.

Can you see why $\frac{0}{-7}$ must be 0? The definition of division tells us that

$$\frac{0}{-7} = 0 \cdot \left(-\frac{1}{7}\right)$$

and the product of 0 and any real number is 0. By contrast, the definition of division does not allow for division by 0 because 0 does not have a multiplicative inverse. It is incorrect to write

$$\frac{-7}{0} = -7 \cdot \frac{1}{0}.$$

> *O does not have a multiplicative inverse.*

Division by zero is not allowed, or not defined. Thus, $\frac{-7}{0}$ does not represent a real number. A real number can never have a denominator of 0.

✓ **CHECK POINT 5**   Divide:

a. $\dfrac{-32}{-4}$ **b.** $-\dfrac{2}{3} \div \dfrac{5}{4}$ **c.** $\dfrac{21.9}{-3}$ **d.** $\dfrac{0}{-5}.$

6 Simplify algebraic expressions involving multiplication. ▶

Multiplication and Algebraic Expressions

In Section 1.4, we discussed the commutative and associative properties of multiplication. We also know that multiplication distributes over addition and subtraction. We now add some additional properties to our previous list (**Table 1.6**). These properties are frequently helpful in simplifying algebraic expressions.

Table 1.6   Additional Properties of Multiplication

Let a be a real number, a variable, or an algebraic expression.		
Property	**Meaning**	**Examples**
Identity Property of Multiplication	1 can be deleted from a product. $a \cdot 1 = a$ $1 \cdot a = a$	• $\sqrt{3} \cdot 1 = \sqrt{3}$ • $1x = x$ • $1(2x + 3) = 2x + 3$
Inverse Property of Multiplication	If a is not 0: $a \cdot \dfrac{1}{a} = 1$ $\dfrac{1}{a} \cdot a = 1$ The product of a nonzero number and its multiplicative inverse, or reciprocal, gives 1, the multiplicative identity.	• $6 \cdot \dfrac{1}{6} = 1$ • $3x \cdot \dfrac{1}{3x} = 1$ (x is not 0.) • $\dfrac{1}{(y-2)} \cdot (y-2) = 1$ (y is not 2.)
Multiplication Property of -1	Negative 1 times a is the opposite, or additive inverse, of a. $-1 \cdot a = -a$ $a(-1) = -a$	• $-1 \cdot \sqrt{3} = -\sqrt{3}$ • $-1\left(-\dfrac{3}{4}\right) = \dfrac{3}{4}$ • $-1x = -x$ • $-(x + 4) = -1(x + 4)$ $\qquad\qquad\quad = -x - 4$
Double Negative Property	The opposite of $-a$ is a. $-(-a) = a$	• $-(-4) = 4$ • $-(-6y) = 6y$

In the table on the previous page, we used two steps to remove the parentheses from $-(x + 4)$. First, we used the multiplication property of -1.

$$-(x + 4) = -1(x + 4)$$

Then we used the distributive property, distributing -1 to each term in parentheses.

$$-1(x + 4) = (-1)x + (-1)4 = -x + (-4) = -x - 4$$

There is a fast way to obtain $-(x + 4) = -x - 4$ in just one step.

> ### Negative Signs and Parentheses
>
> If a negative sign precedes parentheses, remove the parentheses and change the sign of every term within the parentheses.

Here are some examples that illustrate this method.

$$-(11x + 5) = -11x - 5$$
$$-(11x - 5) = -11x + 5$$
$$-(-11x + 5) = 11x - 5$$
$$-(-11x - 5) = 11x + 5$$

EXAMPLE 6   Simplifying Algebraic Expressions

Simplify:

a. $-2(3x)$ **b.** $6x + x$ **c.** $8a - 9a$ **d.** $-3(2x - 5)$ **e.** $-(3y - 8)$.

Solution   We will show all steps in the solution process. However, you probably are working many of these steps mentally.

a. $-2(3x)$ This is the given algebraic expression.

$\quad = (-2 \cdot 3)x$ Use the associative property and group the first two numbers.

$\quad = -6x$ Numbers with opposite signs have a negative product.

b. $6x + x$ This is the given algebraic expression.

$\quad = 6x + 1x$ Use the multiplication property of 1.

$\quad = (6 + 1)x$ Apply the distributive property.

$\quad = 7x$ Add within parentheses.

c. $8a - 9a$ This is the given algebraic expression.

$\quad = (8 - 9)a$ Apply the distributive property.

$\quad = -1a$ Subtract within parentheses: $8 - 9 = 8 + (-9) = -1$.

$\quad = -a$ Apply the multiplication property of -1.

d. $-3(2x - 5)$ This is the given algebraic expression.

$\quad = -3(2x) - (-3) \cdot (5)$ Apply the distributive property.

$\quad = -6x - (-15)$ Multiply.

$\quad = -6x + 15$ Subtraction is the addition of an opposite.

e. $-(3y - 8)$ This is the given algebraic expression.

$\quad = -3y + 8$ Remove parentheses by changing the sign of every term inside the parentheses. ∎

✓ **CHECK POINT 6**   Simplify:

a. $-4(5x)$ **b.** $9x + x$ **c.** $13b - 14b$

d. $-7(3x - 4)$ **e.** $-(7y - 6)$.

EXAMPLE 7   Simplifying an Algebraic Expression

Simplify: $5(2y - 9) - (9y - 8)$.

Solution

$5(2y - 9) - (9y - 8)$	This is the given algebraic expression.
$= 5 \cdot 2y - 5 \cdot 9 - (9y - 8)$	Apply the distributive property over the first parentheses.
$= 10y - 45 - (9y - 8)$	Multiply.
$= 10y - 45 - 9y + 8$	Remove the remaining parentheses by changing the sign of each term within parentheses.
$= (10y - 9y) + (-45 + 8)$	Group like terms.
$= 1y + (-37)$	Combine like terms. For the variable terms, $10y - 9y = 10y + (-9y) = [10 + (-9)]y = 1y$.
$= y + (-37)$	Use the multiplication property of 1: $1y = y$.
$= y - 37$	Express addition of an opposite as subtraction. ∎

✓ **CHECK POINT 7**   Simplify: $4(3y - 7) - (13y - 2)$.

A Summary of Operations with Real Numbers

Operations with real numbers are summarized in **Table 1.7**.

Table 1.7   Summary of Operations with Real Numbers

Signs of Numbers	Addition	Subtraction	Multiplication	Division
Both numbers are positive.		Difference may be either positive or negative.	Product is always positive.	Quotient is always positive.
Sum is always positive.				
Examples				
8 and 2	$8 + 2 = 10$	$8 - 2 = 6$	$8 \cdot 2 = 16$	$8 \div 2 = 4$
2 and 8	$2 + 8 = 10$	$2 - 8 = -6$	$2 \cdot 8 = 16$	$2 \div 8 = \frac{1}{4}$
One number is positive and the other number is negative.	Sum may be either positive or negative.	Difference may be either positive or negative.	Product is always negative.	Quotient is always negative.
Examples				
8 and -2	$8 + (-2) = 6$	$8 - (-2) = 10$	$8(-2) = -16$	$8 \div (-2) = -4$
-8 and 2	$-8 + 2 = -6$	$-8 - 2 = -10$	$-8(2) = -16$	$-8 \div 2 = -4$
Both numbers are negative.		Difference may be either positive or negative.	Product is always positive.	Quotient is always positive.
Sum is always negative.				
Examples				
-8 and -2	$-8 + (-2) = -10$	$-8 - (-2) = -6$	$-8(-2) = 16$	$-8 \div (-2) = 4$
-2 and -8	$-2 + (-8) = -10$	$-2 - (-8) = 6$	$-2(-8) = 16$	$-2 \div (-8) = \frac{1}{4}$

7 Determine whether a number is a solution of an equation. ⏵

To determine if a number is a solution of an equation, it is often necessary to perform more than one operation with real numbers.

EXAMPLE 8   Determining Whether a Number Is a Solution of an Equation

Determine whether -3 is a solution of

$$6x + 14 = -7 - x.$$

Solution

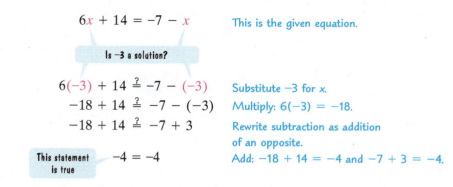

$$6x + 14 = -7 - x \qquad \text{This is the given equation.}$$

Is −3 a solution?

$$6(-3) + 14 \overset{?}{=} -7 - (-3) \qquad \text{Substitute −3 for } x.$$

$$-18 + 14 \overset{?}{=} -7 - (-3) \qquad \text{Multiply: } 6(-3) = -18.$$

$$-18 + 14 \overset{?}{=} -7 + 3 \qquad \text{Rewrite subtraction as addition of an opposite.}$$

This statement is true
$$-4 = -4 \qquad \text{Add: } -18 + 14 = -4 \text{ and } -7 + 3 = -4.$$

Because the values on both sides of the equation are the same, −3 is a solution of the equation. ∎

✓ **CHECK POINT 8**   Determine whether −3 is a solution of $2x - 5 = 8x + 7$.

---

8 Use mathematical models involving multiplication and division.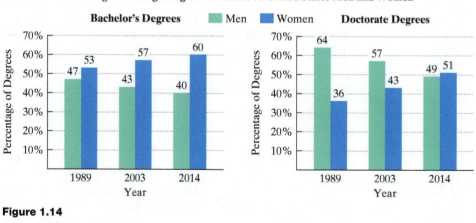

Applications

Mathematical models frequently involve multiplication and division.

EXAMPLE 9   Big (Lack of) Men on Campus

In 2007, 135 women received bachelor's degrees for every 100 men. According to the U.S. Department of Education, that gender imbalance will widen in the coming years, as shown by the bar graphs in **Figure 1.14**.

Percentage of College Degrees Awarded to United States Men and Women

Figure 1.14

Source: U.S. Department of Education

The data for bachelor's degrees can be described by the following mathematical models:

Percentage of bachelor's degrees awarded to men
$$M = -0.28n + 47$$

Number of years after 1989

Percentage of bachelor's degrees awarded to women
$$W = 0.28n + 53.$$

a. According to the first formula, $M = -0.28\,n + 47$, what percentage of bachelor's degrees were awarded to men in 2003? Does this underestimate or overestimate the actual percent shown in **Figure 1.14** for 2003? By how much?

b. A **ratio** is a comparison by division. Write a formula that describes the ratio of the percentage of bachelor's degrees received by men, $M = -0.28n + 47$, to the percentage of bachelor's degrees received by women, $W = 0.28n + 53$. Name this new mathematical model R, for ratio.

c. Use the formula for R to find the ratio of bachelor's degrees received by men to degrees received by women in 2014.

Solution

a. We are interested in the percentage of bachelor's degrees awarded to men in 2003. The year 2003 is 14 years after 1989: $2003 - 1989 = 14$. We substitute 14 for n in the formula for men and evaluate the formula.

$M = -0.28n + 47$ This is the model for the percentage of degrees awarded to men.

$M = -0.28(14) + 47$ Replace n with 14.

$M = -3.92 + 47$ Multiply: $-0.28(14) = -3.92$.

$M = 43.08$ Add. Use the sign of 47 and subtract absolute values:

$$\begin{array}{r} {\scriptstyle 6\ 9\,10} \\ 47.\cancel{0}\cancel{0} \\ -\ \ 3.92. \\ \hline 43.08 \end{array}$$

The model indicates that 43.08% of bachelor's degrees were awarded to men in 2003. Because the number displayed in **Figure 1.14** is 43%, the formula overestimates the actual percent by 0.08%.

b. Now we use division to write a model comparing the ratio, R, of bachelor's degrees received by men to degrees received by women.

$$R = \frac{M}{W} = \frac{-0.28n + 47}{0.28n + 53}$$

Formula for percentage of degrees awarded to men

Formula for percentage of degrees awarded to women

c. Let's see what happens to the ratio, R, in 2014. The year 2014 is 25 years after 1989 ($2014 - 1989 = 25$), so substitute 25 for n.

$R = \dfrac{-0.28n + 47}{0.28n + 53}$ This is the model for the ratio of the percentage of degrees awarded to men and to women.

$R = \dfrac{-0.28(25) + 47}{0.28(25) + 53}$ Replace each occurrence of n with 25.

$R = \dfrac{-7 + 47}{7 + 53}$ Multiply: $-0.28(25) = -7$ and $0.28(25) = 7$.

The models for M and W give the projected percents for 2014 in **Figure 1.14**.

$R = \dfrac{40}{60}$ Add: $-7 + 47 = 40$ and $7 + 53 = 60$.

$R = \dfrac{2}{3}$ Reduce: $\dfrac{40}{60} = \dfrac{2 \cdot 20}{3 \cdot 20} = \dfrac{2}{3}$.

Our model for R, consistent with the actual data, shows that in 2014, the ratio of bachelor's degrees received by men to degrees received by women is 2 to 3. Three women received bachelor's degrees for every two men. ∎

Achieving Success

Ask! Ask! Ask! Do not be afraid to ask questions in class. Your professor may not realize that a concept is unclear until you raise your hand and ask a question. Other students who have problems asking questions in class will be appreciative that you have spoken up. Be polite and professional, but ask as many questions as required.

✓ **CHECK POINT 9** The data for doctorate degrees shown in **Figure 1.14** can be described by $M = -0.6n + 64.4$, where M is the percentage of doctorate degrees awarded to men n years after 1989. According to this mathematical model, what percentage of doctorate degrees were received by men in 2014? Does this underestimate or overestimate the percentage shown in **Figure 1.14**? By how much?

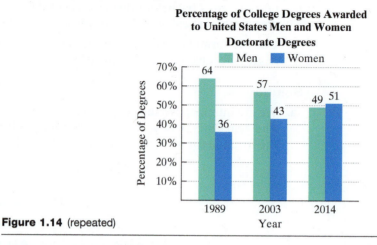

Figure 1.14 (repeated)

CONCEPT AND VOCABULARY CHECK

Use the choices below to fill in each blank so that the resulting statement is true:

positive   negative   0   undefined.

1. The product of two negative numbers is a/an _____ number.

2. The product of a negative number and a positive number is a/an _____ number.

3. The product of three negative numbers is a/an _____ number.

4. The product of an even number of negative numbers is a/an _____ number.

5. The product of two negative numbers and 0 is _____ .

6. The multiplicative inverse, or reciprocal, of a negative number is a/an _____ number.

7. The quotient of a positive number and a negative number is a/an _____ number.

8. The quotient of two negative numbers is a/an _____ number.

9. The quotient of 0 and a negative number is _____ .

10. The quotient of a negative number and 0 is _____ .

11. The opposite of a negative number is a/an _____ number.

1.7 EXERCISE SET   ▶ MyMathLab®

Practice Exercises

In Exercises 1–34, perform the indicated multiplication.

1. $5(-9)$
2. $10(-7)$
3. $(-8)(-3)$
4. $(-9)(-5)$
5. $(-3)(7)$
6. $(-4)(8)$
7. $(-19)(-1)$
8. $(-11)(-1)$
9. $0(-19)$
10. $0(-11)$
11. $\frac{1}{2}(-24)$
12. $\frac{1}{3}(-21)$
13. $\left(-\frac{3}{4}\right)(-12)$
14. $\left(-\frac{4}{5}\right)(-30)$
15. $-\frac{3}{5} \cdot \left(-\frac{4}{7}\right)$
16. $-\frac{5}{7} \cdot \left(-\frac{3}{8}\right)$
17. $-\frac{7}{9} \cdot \frac{2}{3}$
18. $-\frac{5}{11} \cdot \frac{2}{7}$
19. $3(-1.2)$
20. $4(-1.2)$
21. $-0.2(-0.6)$
22. $-0.3(-0.7)$
23. $(-5)(-2)(3)$
24. $(-6)(-3)(10)$
25. $(-4)(-3)(-1)(6)$
26. $(-2)(-7)(-1)(3)$
27. $-2(-3)(-4)(-1)$
28. $-3(-2)(-5)(-1)$
29. $(-3)(-3)(-3)$
30. $(-4)(-4)(-4)$

31. $5(-3)(-1)(2)(3)$

32. $2(-5)(-2)(3)(1)$

33. $(-8)(-4)(0)(-17)(-6)$

34. $(-9)(-12)(-18)(0)(-3)$

In Exercises 35–42, find the multiplicative inverse of each number.

35. 4

36. 3

37. $\dfrac{1}{5}$

38. $\dfrac{1}{7}$

39. -10

40. -12

41. $-\dfrac{2}{5}$

42. $-\dfrac{4}{9}$

In Exercises 43–46,
  a. *Rewrite the division as multiplication involving a multiplicative inverse.*
  b. *Use the multiplication from part (a) to find the given quotient.*

43. $-32 \div 4$

44. $-18 \div 6$

45. $\dfrac{-60}{-5}$

46. $\dfrac{-30}{-5}$

In Exercises 47–76, perform the indicated division or state that the expression is undefined.

47. $\dfrac{12}{-4}$

48. $\dfrac{40}{-5}$

49. $\dfrac{-21}{3}$

50. $\dfrac{-60}{6}$

51. $\dfrac{-90}{-3}$

52. $\dfrac{-66}{-6}$

53. $\dfrac{0}{-7}$

54. $\dfrac{0}{-8}$

55. $\dfrac{7}{0}$

56. $\dfrac{-8}{0}$

57. $-15 \div 3$

58. $-80 \div 8$

59. $120 \div (-10)$

60. $130 \div (-10)$

61. $(-180) \div (-30)$

62. $(-150) \div (-25)$

63. $0 \div (-4)$

64. $0 \div (-10)$

65. $-4 \div 0$

66. $-10 \div 0$

67. $\dfrac{-12.9}{3}$

68. $\dfrac{-21.6}{3}$

69. $-\dfrac{1}{2} \div \left(-\dfrac{3}{5}\right)$

70. $-\dfrac{1}{2} \div \left(-\dfrac{7}{9}\right)$

71. $-\dfrac{14}{9} \div \dfrac{7}{8}$

72. $-\dfrac{5}{16} \div \dfrac{25}{8}$

73. $\dfrac{1}{3} \div \left(-\dfrac{1}{3}\right)$

74. $\dfrac{1}{5} \div \left(-\dfrac{1}{5}\right)$

75. $6 \div \left(-\dfrac{2}{5}\right)$

76. $8 \div \left(-\dfrac{2}{9}\right)$

In Exercises 77–96, simplify each algebraic expression.

77. $-5(2x)$

78. $-9(3x)$

79. $-4\left(-\dfrac{3}{4}y\right)$

80. $-5\left(-\dfrac{3}{5}y\right)$

81. $8x + x$

82. $12x + x$

83. $-5x + x$

84. $-6x + x$

85. $6b - 7b$

86. $12b - 13b$

87. $-y + 4y$

88. $-y + 9y$

89. $-4(2x - 3)$

90. $-3(4x - 5)$

91. $-3(-2x + 4)$

92. $-4(-3x + 2)$

93. $-(2y - 5)$

94. $-(3y - 1)$

95. $4(2y - 3) - (7y + 2)$

96. $5(3y - 1) - (14y - 2)$

In Exercises 97–108, determine whether the given number is a solution of the equation.

97. $4x = 2x - 10; -5$

98. $5x = 3x - 6; -3$

99. $-7y + 18 = -10y + 6; -4$

100. $-4y + 21 = -7y + 15; -2$

101. $5(w + 3) = 2w - 21; -10$

102. $6(w + 2) = 4w - 10; -9$

103. $4(6 - z) + 7z = 0; -8$

104. $5(7 - z) + 12z = 0; -5$

105. $14 - 2x = -4x + 7; -2\dfrac{1}{2}$

106. $16 - 4x = -2x + 21; -3\dfrac{1}{2}$

107. $\dfrac{5m - 1}{6} = \dfrac{3m - 2}{4}; -4$

108. $\dfrac{6m - 5}{11} = \dfrac{3m - 2}{5}; -1$

Practice PLUS

In Exercises 109–116, write a numerical expression for each phrase. Then simplify the numerical expression by performing the given operations.

109. 8 added to the product of 4 and -10

110. 14 added to the product of 3 and -15

111. The product of -9 and -3, decreased by -2

112. The product of -6 and -4, decreased by -5

113. The quotient of -18 and the sum of -15 and 12

114. The quotient of -25 and the sum of -21 and 16

115. The difference between -6 and the quotient of 12 and -4

116. The difference between -11 and the quotient of 20 and -5

Application Exercises

In Exercises 117–118, use the formula $C = \dfrac{5}{9}(F - 32)$ to express each Fahrenheit temperature, F, as its equivalent Celsius temperature, C.

117. $-22°\text{F}$

118. $-31°\text{F}$

In the years after warning labels were put on cigarette packs, the number of smokers dropped from approximately two in five adults to less than one in five. The bar graph shows the percentage of American adults who smoked cigarettes for selected years from 1965 through 2012.

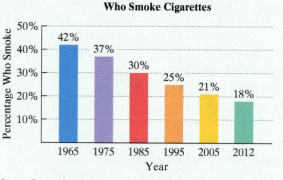

Percentage of American Adults Who Smoke Cigarettes

Source: Centers for Disease Control and Prevention

Here is a mathematical model that approximates the data displayed by the bar graph:

$$C = -0.5x + 42.$$

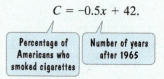

Use this formula to solve Exercises 119–120.

119. a. Does the mathematical model underestimate or overestimate the percentage of American adults who smoked cigarettes in 2012? By how much?

b. Use the mathematical model to project the percentage of American adults who will smoke cigarettes in 2019.

120. a. Does the mathematical model underestimate or overestimate the percentage of American adults who smoked cigarettes in 2005? By how much?

b. Use the mathematical model to project the percentage of American adults who will smoke cigarettes in 2021.

The line graphs show the percentage of Americans who preferred smaller or larger families for four selected years from 1997 through 2011.

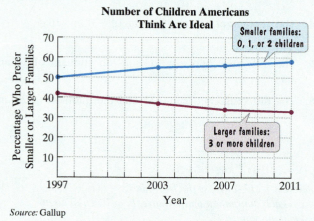

Number of Children Americans Think Are Ideal

Source: Gallup

The data can be described by the following mathematical models:

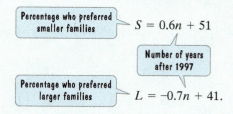

$$S = 0.6n + 51$$
$$L = -0.7n + 41.$$

Use this information to solve Exercises 121–122.

121. a. Use the appropriate graph to estimate the percentage of Americans in 2007 who preferred larger families.

b. Use the appropriate formula to determine the percentage of Americans in 2007 who preferred larger families. How does this compare with your estimate in part (a)?

c. Write a formula that describes the ratio of the percentage of Americans who preferred larger families to the percentage who preferred smaller families n years after 1997. Name this new model R for ratio.

d. Use the mathematical model from part (c) to write a fraction for 2007 comparing the ratio of the percentage who preferred larger families to the percentage who preferred smaller families. If necessary, reduce the fraction to its lowest terms.

122. a. Use the appropriate graph to estimate the percentage of Americans in 2007 who preferred smaller families.

b. Use the appropriate formula to determine the percentage of Americans in 2007 who preferred smaller families. How does this compare with your estimate in part (a)?

c. Write a formula that describes the difference between the percentage of Americans who preferred smaller families and the percentage who preferred larger families n years after 1997. Name this new mathematical model D for difference. Then simplify the algebraic expression in the model.

d. Use the simplified form of the mathematical model from part (c) to determine the difference in 2007 between the percentage of Americans who preferred smaller families and the percentage who preferred larger families. How does this compare with the difference displayed by the graphs for this year?

Explaining the Concepts

123. Explain how to multiply two real numbers. Provide examples with your explanation.

124. Explain how to determine the sign of a product that involves more than two numbers.

125. Explain how to find the multiplicative inverse of a number.

126. Why is it that 0 has no multiplicative inverse?

127. Explain how to divide real numbers.

128. Why is division by zero undefined?

129. Explain how to simplify an algebraic expression in which a negative sign precedes parentheses.

130. Do you believe that the trend in the graphs showing a decline in male college attendance (**Figure 1.14** on page 82) should be reversed by providing admissions preferences for men? Explain your position on this issue.

Critical Thinking Exercises

Make Sense?   In Exercises 131–134, determine whether each statement makes sense or does not make sense, and explain your reasoning.

131. I've noticed that the sign rules for dividing real numbers are slightly different than the sign rules for multiplying real numbers.

132. Just as two negative factors give a positive product, I've seen the same thing occur with double negatives in English:

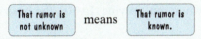

133. This pattern suggests that multiplying two negative numbers results in a positive answer:

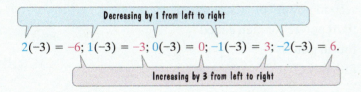

$2(-3) = -6$; $1(-3) = -3$; $0(-3) = 0$; $-1(-3) = 3$; $-2(-3) = 6$.

134. When I used

$$R = \frac{M}{W} = \frac{-0.28n + 47}{0.28n + 53}$$

Number of years after 1989

to project the ratio of bachelor's degrees received by men to degrees received by women in 2020, I had to find the quotient of two negative numbers.

In Exercises 135–138, determine whether each statement is true or false. If the statement is false, make the necessary change(s) to produce a true statement.

135. Both the addition and the multiplication of two negative numbers result in a positive number.

136. Multiplying a negative number by a nonnegative number will always give a negative number.

137. $0 \div (-\sqrt{2})$ is undefined.

138. If a is negative, b is positive, and c is negative, then $\dfrac{a}{bc}$ is positive.

In Exercises 139–142, write an algebraic expression for the given English phrase.

139. The value, in cents, of x nickels

140. The distance covered by a car traveling at 50 miles per hour for x hours

141. The monthly salary, in dollars, for a person earning x dollars per year

142. The fraction of people in a room who are women if there are 40 women and x men in the room

Technology Exercises

143. Use a calculator to verify any five of the products that you found in Exercises 1–34.

144. Use a calculator to verify any five of the quotients that you found in Exercises 47–76.

145. Simplify using a calculator:

$$0.3(4.7x - 5.9) - 0.07(3.8x - 61).$$

146. Use your calculator to attempt to find the quotient of -3 and 0. Describe what happens. Does the same thing occur when finding the quotient of 0 and -3? Explain the difference. Finally, what happens when you enter the quotient of 0 and itself?

Review Exercises

In Exercises 147–149, perform the indicated operation.

147. $-6 + (-3)$ (Section 1.5, Example 3)

148. $-6 - (-3)$ (Section 1.6, Example 1)

149. $-6 \div (-3)$ (Section 1.7, Example 4)

Preview Exercises

Exercises 150–152 will help you prepare for the material covered in the next section. In each exercise, an expression with an exponent is written as a repeated multiplication. Find this product, indicated by a question mark.

150. $(-6)^2 = (-6)(-6) = ?$

151. $(-5)^3 = (-5)(-5)(-5) = ?$

152. $(-2)^4 = (-2)(-2)(-2)(-2) = ?$

1.8

Exponents and Order of Operations ▶

What am I supposed to learn?

After studying this section, you should be able to:

1. Evaluate exponential expressions. ▶

2. Simplify algebraic expressions with exponents. ▶

3. Use the order of operations agreement. ▶

4. Evaluate mathematical models. ▶

1. Evaluate exponential expressions. ▶

Eat, drink, and be merry—for tomorrow we diet. But is there a way to avoid the extremes of merriment and dieting? In this section, we continue to see how mathematical models describe your world, including formulas that provide daily caloric needs for maintaining energy balance based on age and lifestyle.

Natural Number Exponents

Although people do a great deal of talking, the total output since the beginning of gabble to the present day, including all baby talk, love songs, and congressional debates, only amounts to about 10 million billion words. This can be expressed as 16 factors of 10, or 10^{16} words.

Exponents such as 2, 3, 4, and so on are used to indicate repeated multiplication. For example,

$$10^2 = 10 \cdot 10 = 100,$$

$$10^3 = 10 \cdot 10 \cdot 10 = 1000, \quad 10^4 = 10 \cdot 10 \cdot 10 \cdot 10 = 10,000.$$

The 10 that is repeated when multiplying is called the **base**. The small numbers above and to the right of the base are called **exponents** or **powers**. The exponent tells the number of times the base is to be used when multiplying. In 10^3, the base is 10 and the exponent is 3.

Any number with an exponent of 1 is the number itself. Thus, $10^1 = 10$.

Multiplications that are expressed in exponential notation are read as follows:

10^1: "ten to the first power"

10^2: "ten to the second power" or "ten squared"

10^3: "ten to the third power" or "ten cubed"

10^4: "ten to the fourth power"

10^5: "ten to the fifth power"

etc.

Any real number can be used as the base. Thus,

$$7^2 = 7 \cdot 7 = 49 \quad \text{and} \quad (-3)^4 = (-3)(-3)(-3)(-3) = 81.$$

The bases are 7 and -3, respectively. Do not confuse $(-3)^4$ and -3^4.

$$-3^4 = -(3 \cdot 3 \cdot 3 \cdot 3) = -81$$

The negative is not taken to the power because it is not inside parentheses.

An exponent applies only to a base. A negative sign is not part of a base unless it appears in parentheses.

Using Technology

You can use a calculator to evaluate exponential expressions. For example, to evaluate 5^3, press the following keys:

Many Scientific Calculators

5 $\boxed{y^x}$ 3 $\boxed{=}$

Many Graphing Calculators

5 $\boxed{\wedge}$ 3 $\boxed{\text{ENTER}}$

Although calculators have special keys to evaluate powers of ten and squaring bases, you can always use one of the sequences shown here.

EXAMPLE 1 Evaluating Exponential Expressions

Evaluate:

a. 4^2 **b.** $(-5)^3$ **c.** $(-2)^4$ **d.** -2^4.

Solution

Exponent is 2.

a. $4^2 = 4 \cdot 4 = 16$ The exponent indicates that the base is used as a factor two times.

Base is 4.

We read $4^2 = 16$ as "4 to the second power is 16" or "4 squared is 16."

Exponent is 3.

b. $(-5)^3 = (-5)(-5)(-5)$ The exponent indicates that the base is used as a factor three times.

Base is –5.

$\qquad = -125$ An odd number of negative factors yields a negative product.

We read $(-5)^3 = -125$ as "the number negative 5 to the third power is negative 125" or "negative 5 cubed is negative 125."

Exponent is 4.

c. $(-2)^4 = (-2)(-2)(-2)(-2)$ The exponent indicates that the base is used as a factor four times.

Base is –2.

$\qquad = 16$ An even number of negative factors yields a positive product.

We read $(-2)^4 = 16$ as "the number negative 2 to the fourth power is 16."

Exponent is 4.

d. $-2^4 = -(2 \cdot 2 \cdot 2 \cdot 2)$ The negative is not inside parentheses and is not taken to the fourth power.

Base is 2.

$\qquad = -16$ Multiply the twos and copy the negative.

We read $-2^4 = -16$ as "the negative of 2 raised to the fourth power is negative 16" or "the opposite, or additive inverse, of 2 raised to the fourth power is negative 16." ■

✓ CHECK POINT 1 Evaluate:

   a. 6^2 **b.** $(-4)^3$ **c.** $(-1)^4$ **d.** -1^4.

The formal algebraic definition of a natural number exponent summarizes our discussion:

Definition of a Natural Number Exponent

If b is a real number and n is a natural number,

$$\underset{\text{Base}}{b}{}^{\overset{\text{Exponent}}{n}} = \underbrace{b \cdot b \cdot b \cdot \ldots \cdot b}_{\substack{b \text{ appears as a} \\ \text{factor } n \text{ times.}}}.$$

b^n is read "the nth power of b" or "b to the nth power." Thus, the nth power of b is defined as the product of n factors of b. The expression b^n is called an **exponential expression**.

Furthermore, $b^1 = b$.

Blitzer Bonus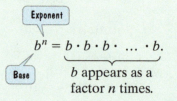

Integers, Karma, and Exponents

On Friday the 13th, are you a bit more careful crossing the street even if you don't consider yourself superstitious? Numerology, the belief that certain integers have greater significance and can be lucky or unlucky, is widespread in many cultures.

Integer	Connotation	Culture	Origin	Example
4	Negative	Chinese	The word for the number 4 sounds like the word for death.	Many buildings in China have floor-numbering systems that skip 40–49.
7	Positive	United States	In dice games, this prime number is the most frequently rolled number with two dice.	There was a spike in the number of couples getting married on 7/7/07.
8	Positive	Chinese	It's considered a sign of prosperity.	The Beijing Olympics began at 8 p.m. on 8/8/08.
13	Negative	Various	Various reasons, including the number of people at the Last Supper	Many buildings around the world do not label any floor "13."
18	Positive	Jewish	The Hebrew letters spelling *chai*, or living, are the 8th and 10th in the alphabet, adding up to 18.	Monetary gifts for celebrations are often given in multiples of 18.
666	Negative	Christian	The New Testament's Book of Revelation identifies 666 as the "number of the beast," which some say refers to Satan.	In 2008, Reeves, Louisiana, eliminated 666 as the prefix of its phone numbers.

Although your author is not a numerologist, he is intrigued by curious exponential representations for 666:

$$666 = 6 + 6 + 6 + 6^3 + 6^3 + 6^3$$
$$666 = 1^3 + 2^3 + 3^3 + 4^3 + 5^3 + 6^3 + 5^3 + 4^3 + 3^3 + 2^3 + 1^3$$
$$666 = 2^2 + 3^2 + 5^2 + 7^2 + 11^2 + 13^2 + 17^2$$
$$666 = 1^6 - 2^6 + 3^6.$$

Sum of the squares of the first seven prime numbers

2 Simplify algebraic expressions with exponents. ▶

Exponents and Algebraic Expressions

The distributive property can be used to simplify certain algebraic expressions that contain exponents. For example, we can use the distributive property to combine like terms in the algebraic expression $4x^2 + 6x^2$:

$$4x^2 + 6x^2 \quad = \quad (4 + 6)x^2 = 10x^2.$$

| First term with variable factor x^2 | Second term with variable factor x^2 | The common variable factor is x^2. |

Great Question!

When I add like terms, do I add exponents?

When adding algebraic expressions, if you have like terms you add only the numerical coefficients—not the exponents. **Exponents are never added when the operation is addition.** Avoid these common errors.

Incorrect

- $7x^3 + 2x^3 + 9x^6$
- $5x^2 + x^2 + 6x^4$
- $3x^2 + 4x^3 + 7x^5$

EXAMPLE 2   Simplifying Algebraic Expressions

Simplify, if possible:

a. $7x^3 + 2x^3$ **b.** $5x^2 + x^2$ **c.** $3x^2 + 4x^3$.

Solution

a. $7x^3 + 2x^3$ There are two like terms with the same variable factor, namely x^3.
$= (7 + 2)x^3$ Apply the distributive property.
$= 9x^3$ Add within parentheses.

b. $5x^2 + x^2$ There are two like terms with the same variable factor, namely x^2.
$= 5x^2 + 1x^2$ Use the multiplication property of 1.
$= (5 + 1)x^2$ Apply the distributive property.
$= 6x^2$ Add within parentheses.

c. $3x^2 + 4x^3$ cannot be simplified. The terms $3x^2$ and $4x^3$ are not like terms because they have different variable factors, namely x^2 and x^3. ■

✓ **CHECK POINT 2**   Simplify, if possible:
    a. $16x^2 + 5x^2$ **b.** $7x^3 + x^3$ **c.** $10x^2 + 8x^3$.

3 Use the order of operations agreement. ▶

Order of Operations

Suppose that you want to find the value of $3 + 7 \cdot 5$. Which procedure shown is correct?

$$3 + 7 \cdot 5 = 3 + 35 = 38 \quad \text{or} \quad 3 + 7 \cdot 5 = 10 \cdot 5 = 50$$

You know the answer because we introduced certain rules, called the **order of operations**, at the beginning of the chapter. One of these rules stated that if a problem contains no parentheses or other grouping symbols, perform multiplication before addition. Thus, the procedure on the left is correct because the multiplication of 7 and 5 is done first. Then the addition is performed. The correct answer is 38.

Some problems contain grouping symbols, such as parentheses, $(\ )$; brackets, $[\ ]$; braces, $\{\ \}$; absolute value symbols, $|\ |$; or fraction bars. These grouping symbols tell us what to do first. Here are two examples:

- $(3 + 7) \cdot 5 = 10 \cdot 5 = 50$

First, perform operations in grouping symbols.

- $8|6 - 16| = 8|-10| = 8 \cdot 10 = 80$.

Here are the rules for determining the order in which operations should be performed:

Order of Operations

1. Perform all operations within grouping symbols.
2. Evaluate all exponential expressions.
3. Do all multiplications and divisions in the order in which they occur, working from left to right.
4. Finally, do all additions and subtractions using one of the following procedures:
   a. Work from left to right and do additions and subtractions in the order in which they occur.

   or

   b. Rewrite subtractions as additions of opposites. Combine positive and negative numbers separately, and then add these results.

The last step in the order of operations indicates that you have a choice when working with additions and subtractions, although in this section we will perform these operations from left to right. However, when working with multiplications and divisions, you must perform these operations *as they occur* from left to right. For example,

$$8 \div 4 \cdot 2 = 2 \cdot 2 = 4 \qquad \text{Do the division first because it occurs first.}$$

$$8 \cdot 4 \div 2 = 32 \div 2 = 16. \qquad \text{Do the multiplication first because it occurs first.}$$

EXAMPLE 3 Using the Order of Operations

Simplify: $18 + 2 \cdot 3 - 10$.

Solution There are no grouping symbols or exponential expressions. In cases like this, we multiply and divide before adding and subtracting.

$$
\begin{aligned}
18 + 2 \cdot 3 - 10 &= 18 + 6 - 10 \qquad &&\text{Multiply: } 2 \cdot 3 = 6. \\
&= 24 - 10 \qquad &&\text{Add and subtract from left to right:} \\
& &&18 + 6 = 24. \\
&= 14 \qquad &&\text{Subtract: } 24 - 10 = 14. \quad\blacksquare
\end{aligned}
$$

✓ **CHECK POINT 3** Simplify: $20 + 4 \cdot 3 - 17$.

EXAMPLE 4 Using the Order of Operations

Simplify: $6^2 - 24 \div 2^2 \cdot 3 - 1$.

Solution There are no grouping symbols. Thus, we begin by evaluating exponential expressions. Then we multiply or divide. Finally, we add or subtract.

$$
\begin{aligned}
6^2 - 24 &\div 2^2 \cdot 3 - 1 \\
&= 36 - 24 \div 4 \cdot 3 - 1 \qquad &&\text{Evaluate exponential expressions:} \\
& &&6^2 = 6 \cdot 6 = 36 \text{ and } 2^2 = 2 \cdot 2 = 4. \\
&= 36 - 6 \cdot 3 - 1 \qquad &&\text{Perform the multiplications and divisions from left to right. Start with} \\
& &&24 \div 4 = 6. \\
&= 36 - 18 - 1 \qquad &&\text{Now do the multiplication: } 6 \cdot 3 = 18. \\
&= 18 - 1 \qquad &&\text{Finally, perform the subtraction from left to right:} \\
& &&36 - 18 = 18. \\
&= 17 \qquad &&\text{Complete the subtraction: } 18 - 1 = 17. \quad\blacksquare
\end{aligned}
$$

✓ **CHECK POINT 4** Simplify: $7^2 - 48 \div 4^2 \cdot 5 - 2$.

EXAMPLE 5   Using the Order of Operations

Simplify:

a. $(2 \cdot 5)^2$ **b.** $2 \cdot 5^2$.

Solution

a. Because $(2 \cdot 5)^2$ contains grouping symbols, namely parentheses, we perform the operation within parentheses first.

$$(2 \cdot 5)^2 = 10^2 \qquad \text{Multiply within parentheses: } 2 \cdot 5 = 10.$$
$$= 100 \qquad \text{Evaluate the exponential expression: } 10^2 = 10 \cdot 10 = 100.$$

b. Because $2 \cdot 5^2$ does not contain grouping symbols, we begin by evaluating the exponential expression.

$$2 \cdot 5^2 = 2 \cdot 25 \qquad \text{Evaluate the exponential expression: } 5^2 = 5 \cdot 5 = 25.$$
$$= 50 \qquad \text{Now do the multiplication: } 2 \cdot 25 = 50. \quad \blacksquare$$

✓ **CHECK POINT 5**   Simplify:
a. $(3 \cdot 2)^2$ **b.** $3 \cdot 2^2$.

EXAMPLE 6   Using the Order of Operations

Simplify: $\left(\dfrac{1}{2}\right)^3 - \left(\dfrac{1}{2} - \dfrac{3}{4}\right)^2 (-4)$.

Solution   Because grouping symbols appear, we perform the operation within parentheses first.

$$\left(\frac{1}{2}\right)^3 - \left(\frac{1}{2} - \frac{3}{4}\right)^2 (-4)$$

$$= \left(\frac{1}{2}\right)^3 - \left(-\frac{1}{4}\right)^2 (-4) \qquad \begin{array}{l} \text{Work inside parentheses first:} \\ \dfrac{1}{2} - \dfrac{3}{4} = \dfrac{2}{4} - \dfrac{3}{4} = \dfrac{2}{4} + \left(-\dfrac{3}{4}\right) = -\dfrac{1}{4}. \end{array}$$

$$= \frac{1}{8} - \frac{1}{16}(-4) \qquad \begin{array}{l} \text{Evaluate exponential expressions:} \\ \left(\dfrac{1}{2}\right)^3 = \dfrac{1}{2} \cdot \dfrac{1}{2} \cdot \dfrac{1}{2} = \dfrac{1}{8} \text{ and } \left(-\dfrac{1}{4}\right)^2 = \left(-\dfrac{1}{4}\right)\left(-\dfrac{1}{4}\right) = \dfrac{1}{16}. \end{array}$$

$$= \frac{1}{8} - \left(-\frac{1}{4}\right) \qquad \text{Multiply: } \dfrac{1}{16} \cdot \left(\dfrac{-4}{1}\right) = -\dfrac{4}{16} = -\dfrac{1}{4}.$$

$$= \frac{3}{8} \qquad \text{Subtract: } \dfrac{1}{8} - \left(-\dfrac{1}{4}\right) = \dfrac{1}{8} + \dfrac{1}{4} = \dfrac{1}{8} + \dfrac{2}{8} = \dfrac{3}{8}. \quad \blacksquare$$

✓ **CHECK POINT 6**   Simplify: $\left(-\dfrac{1}{2}\right)^2 - \left(\dfrac{7}{10} - \dfrac{8}{15}\right)^2 (-18)$.

Some expressions contain many grouping symbols. An example of such an expression is $2[5(4 - 7) + 9]$. The grouping symbols are the parentheses and the brackets.

The parentheses, the innermost grouping symbols, group $4 - 7$.

$$2[5(4 - 7) + 9]$$

The brackets, the outermost grouping symbols, group $5(4 - 7) + 9$.

When combinations of grouping symbols appear, **perform operations within the innermost grouping symbols first**. Then work to the outside, performing operations within the outermost grouping symbols.

EXAMPLE 7 Using the Order of Operations

Simplify: $2[5(4 - 7) + 9]$.

Solution

$2[5(4 - 7) + 9]$

$= 2[5(-3) + 9]$ Work inside parentheses first: $4 - 7 = 4 + (-7) = -3$.

$= 2[-15 + 9]$ Work inside brackets and multiply: $5(-3) = -15$.

$= 2[-6]$ Add inside brackets: $-15 + 9 = -6$. The resulting problem can also be expressed as $2(-6)$.

$= -12$ Multiply: $2[-6] = -12$. ∎

Parentheses can be used for both innermost and outermost grouping symbols. For example, the expression $2[5(4 - 7) + 9]$ can also be written $2(5(4 - 7) + 9)$. However, too many parentheses can be confusing. The use of both parentheses and brackets makes it easier to identify inner and outer groupings.

☑ **CHECK POINT 7** Simplify: $4[3(6 - 11) + 5]$.

EXAMPLE 8 Using the Order of Operations

Simplify: $18 \div 6 + 4[5 + 2(8 - 10)^3]$.

Solution

$18 \div 6 + 4[5 + 2(8 - 10)^3]$

$= 18 \div 6 + 4[5 + 2(-2)^3]$ Work inside parentheses first: $8 - 10 = 8 + (-10) = -2$.

$= 18 \div 6 + 4[5 + 2(-8)]$ Work inside brackets and evaluate the exponential expression: $(-2)^3 = (-2)(-2)(-2) = -8$.

$= 18 \div 6 + 4[5 + (-16)]$ Work inside brackets and multiply: $2(-8) = -16$.

$= 18 \div 6 + 4[-11]$ Work inside brackets and add: $5 + (-16) = -11$.

$= 3 + 4[-11]$ Perform the multiplications and divisions from left to right. Start with $18 \div 6 = 3$.

$= 3 + (-44)$ Now do the multiplication: $4(-11) = -44$.

$= -41$ Finally, perform the addition: $3 + (-44) = -41$. ∎

☑ **CHECK POINT 8** Simplify: $25 \div 5 + 3[4 + 2(7 - 9)^3]$.

Fraction bars are grouping symbols that separate expressions into two parts, the numerator and the denominator. Consider, for example,

The numerator is one part of the expression.

The fraction bar is the grouping symbol.

$$\frac{2(3 - 12) + 6 \cdot 4}{2^4 + 1}.$$

The denominator is the other part of the expression.

We can use brackets instead of the fraction bar. An equivalent expression for $\dfrac{2(3 - 12) + 6 \cdot 4}{2^4 + 1}$ is

$$[2(3 - 12) + 6 \cdot 4] \div [2^4 + 1].$$

The grouping suggests a method for simplifying expressions with fraction bars as grouping symbols:

- Simplify the numerator.
- Simplify the denominator.
- If possible, simplify the fraction.

EXAMPLE 9   Using the Order of Operations

Simplify: $\dfrac{2(3 - 12) + 6 \cdot 4}{2^4 + 1}$.

Solution

$$\dfrac{2(3 - 12) + 6 \cdot 4}{2^4 + 1}$$

$$= \dfrac{2(-9) + 6 \cdot 4}{16 + 1}$$
Work inside parentheses in the numerator: $3 - 12 = 3 + (-12) = -9$. Evaluate the exponential expression in the denominator: $2^4 = 2 \cdot 2 \cdot 2 \cdot 2 = 16$.

$$= \dfrac{-18 + 24}{16 + 1}$$
Multiply in the numerator: $2(-9) = -18$ and $6 \cdot 4 = 24$.

$$= \dfrac{6}{17}$$
Perform the addition in the numerator and in the denominator. ∎

✓ **CHECK POINT 9**   Simplify: $\dfrac{5(4 - 9) + 10 \cdot 3}{2^3 - 1}$.

EXAMPLE 10   Using the Order of Operations

Evaluate: $-x^2 - 7x$ for $x = -2$.

Solution   We begin by substituting -2 for each occurrence of x in the algebraic expression. Then we use the order of operations to evaluate the expression.

$$-x^2 - 7x$$

Replace x with -2. Place parentheses around -2.

$$= -(-2)^2 - 7(-2)$$

$$= -4 - 7(-2)$$
Evaluate the exponential expression: $(-2)^2 = (-2)(-2) = 4$.

$$= -4 - (-14)$$
Multiply: $7(-2) = -14$.

$$= 10$$
Subtract: $-4 - (-14) = -4 + 14 = 10$. ∎

✓ **CHECK POINT 10**   Evaluate: $-x^2 - 4x$ for $x = -5$.

Some algebraic expressions contain two sets of grouping symbols. Using the order of operations, grouping symbols are removed from innermost (parentheses) to outermost (brackets).

EXAMPLE 11 Simplifying an Algebraic Expression

Simplify: $18x^2 + 4 - [6(x^2 - 2) + 5]$.

Solution

$18x^2 + 4 - [6(x^2 - 2) + 5]$

$= 18x^2 + 4 - [6x^2 - 12 + 5]$ Use the distributive property to remove parentheses: $6(x^2 - 2) = 6x^2 - 6 \cdot 2 = 6x^2 - 12$.

$= 18x^2 + 4 - [6x^2 - 7]$ Add inside brackets: $-12 + 5 = -7$.

$= 18x^2 + 4 - 6x^2 + 7$ Remove brackets by changing the sign of each term within brackets.

$= (18x^2 - 6x^2) + 4 + 7$ Group like terms.

$= 12x^2 + 11$ Combine like terms. ∎

✓ **CHECK POINT 11** Simplify: $14x^2 + 5 - [7(x^2 - 2) + 4]$.

4 Evaluate mathematical models. ▶

Applications

In Examples 12 and 13, we use the order of operations to evaluate mathematical models.

EXAMPLE 12 Modeling Caloric Needs

The bar graph in **Figure 1.15** shows the estimated number of calories per day needed to maintain energy balance for various gender and age groups for moderately active lifestyles. (Moderately active means a lifestyle that includes physical activity equivalent to walking 1.5 to 3 miles per day at 3 to 4 miles per hour, in addition to the light physical activity associated with typical day-to-day life.)

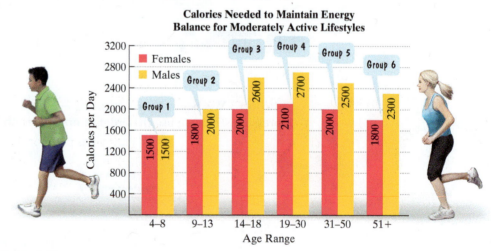

Figure 1.15

Source: U.S.D.A.

The mathematical model

$$F = -66x^2 + 526x + 1030$$

describes the number of calories needed per day, F, by females in age group x with moderately active lifestyles. According to the model, how many calories per day are needed by females between the ages of 19 and 30, inclusive, with this lifestyle? Does this underestimate or overestimate the number shown by the graph in **Figure 1.15**? By how much?

Solution   Because the 19–30 age range is designated as group 4, we substitute 4 for x in the given model. Then we use the order of operations to find F, the number of calories needed per day by females between the ages of 19 and 30.

$F = -66x^2 + 526x + 1030$	This is the given mathematical model.
$F = -66 \cdot 4^2 + 526 \cdot 4 + 1030$	Replace each occurrence of x with 4.
$F = -66 \cdot 16 + 526 \cdot 4 + 1030$	Evaluate the exponential expression: $4^2 = 4 \cdot 4 = 16$.
$F = -1056 + 2104 + 1030$	Multiply from left to right: $-66 \cdot 16 = -1056$ and $526 \cdot 4 = 2104$.
$F = 2078$	Add: $-1056 + 2104 = 1048$ and $1048 + 1030 = 2078$.

The formula indicates that females in the 19–30 age range with moderately active lifestyles need 2078 calories per day. **Figure 1.15** on the previous page indicates that 2100 calories are needed. Thus, the mathematical model underestimates caloric needs by 2100 – 2078 calories, or by 22 calories per day. ∎

Great Question!

In the solution to Example 12, −1056 + 2104 + 1030 was simplified by performing the addition from left to right. Do I have to do it that way?

No. Here's another way to simplify the expression. First add the positive numbers

$$2104 + 1030 = 3134$$

and then add the negative number to that result:

$$3134 + (-1056) = 2078.$$

✓ **CHECK POINT 12**   The mathematical model

$$M = -120x^2 + 998x + 590$$

describes the number of calories needed per day, M, by males in age group x with moderately active lifestyles. According to the model, how many calories per day are needed by males between the ages of 19 and 30, inclusive, with this lifestyle? Does this underestimate or overestimate the number shown by the graph in **Figure 1.15** on the previous page? By how much?

EXAMPLE 13   Shrinky-Dink Size without Rinky-Dink Style

A company has decided to jump into the pedal game with foldable, ultraportable bikes. Its fixed monthly costs are $500,000 and it will cost $400 to manufacture each bike. The average cost per bike, $\overline{C}$, for the company to manufacture x foldable bikes per month is modeled by

$$\overline{C} = \frac{400x + 500,000}{x}.$$

Find the average cost per bike for the company to manufacture

    a. 10,000 bikes per month.

    b. 50,000 bikes per month.

    c. 100,000 bikes per month.

What happens to the average cost per bike as the production level increases?

Solution

a. We are interested in the average cost per bike for the company if 10,000 bikes are manufactured per month. Because x represents the number of bikes manufactured per month, we substitute 10,000 for x in the given mathematical model.

$$\overline{C} = \frac{400x + 500,000}{x} = \frac{400(10,000) + 500,000}{10,000} = \frac{4,000,000 + 500,000}{10,000}$$

$$= \frac{4,500,000}{10,000} = 450$$

The average cost per bike of producing 10,000 bikes per month is $450.

b. Now, 50,000 bikes are manufactured per month. We find the average cost per bike by substituting 50,000 for x in the given mathematical model.

$$\overline{C} = \frac{400x + 500,000}{x} = \frac{400(50,000) + 500,000}{50,000} = \frac{20,000,000 + 500,000}{50,000}$$

$$= \frac{20,500,000}{50,000} = 410$$

The average cost per bike of producing 50,000 bikes per month is $410.

c. Finally, the production level has increased to 100,000 bikes per month. We find the average cost per bike for the company by substituting 100,000 for x in the given mathematical model.

$$\overline{C} = \frac{400x + 500,000}{x} = \frac{400(100,000) + 500,000}{100,000} = \frac{40,000,000 + 500,000}{100,000}$$

$$= \frac{40,500,000}{100,000} = 405$$

The average cost per bike of producing 100,000 bikes per month is $405.

As the production level increases, the average cost of producing each foldable bike decreases. This illustrates the difficulty with small businesses. It is nearly impossible to have competitively low prices when production levels are low. ■

The graph in **Figure 1.16** shows the relationship between production level and cost. The three points with the voice balloons illustrate our evaluations in Example 13. The other unlabeled points along the smooth blue curve represent the company's average costs for various production levels. The symbol ╪ on the vertical axis indicates that there is a break in the values between 0 and 400. Thus, values for the average cost per bike begin at $400.

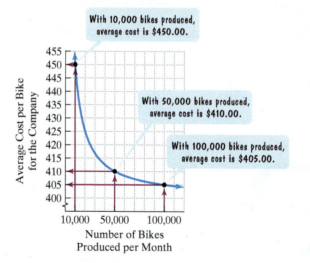

The points that lie along the blue curve in **Figure 1.16** are falling from left to right. Can you see how this shows that the company's cost per foldable bike is decreasing as the production level increases?

Figure 1.16

✓ **CHECK POINT 13**   A company that manufactures running shoes has weekly fixed costs of $300,000 and it costs $30 to manufacture each pair of running shoes. The average cost per pair of running shoes, $\overline{C}$, for the company to manufacture x pairs per week is modeled by

$$\overline{C} = \frac{30x + 300,000}{x}.$$

Find the average cost per pair of running shoes for the company to manufacture

a. 1000 pairs per week.   **b.** 10,000 pairs per week.

c. 100,000 pairs per week.

Achieving Success

Do not wait until the last minute to study for an exam. Cramming is a high-stress activity that forces your brain to make a lot of weak connections. No wonder crammers tend to forget everything they learned minutes after taking a test.

Preparing for Tests Using the Book

- Study the appropriate sections from the review chart in the Chapter Summary. The chart contains definitions, concepts, procedures, and examples. Review this chart and you'll know the most important material in each section!

- Work the assigned exercises from the Review Exercises. The Review Exercises contain the most significant problems for each of the chapter's sections.

- Find a quiet place to take the Chapter Test. Do not use notes, index cards, or any other resources. Check your answers and ask your professor to review any exercises you missed.

CONCEPT AND VOCABULARY CHECK

Fill in each blank so that the resulting statement is true.

1. In the expression b^n, b is called the _____ and n is called the _____.

2. The expression b^n is read _____.

In the remaining items, use the choices below to fill in each blank:

add   subtract   multiply   divide.

3. To simplify the expression $10 + 4 \cdot 5 - 30$, first _____.

4. To simplify the expression $(10 + 4) \cdot 5 - 30$, first _____.

5. To simplify the expression $36 - 24 \div 4 \cdot 3$, first _____.

6. To simplify the expression $4[8 + 3(2 - 6)^2]$, first _____.

7. To simplify the expression $8|5 - 4 \cdot 6|$, first _____.

1.8 EXERCISE SET   ▶ MyMathLab®

Practice Exercises

In Exercises 1–14, evaluate each exponential expression.

1. 9^2
2. 3^2
3. 4^3
4. 6^3
5. $(-4)^2$
6. $(-10)^2$
7. $(-4)^3$
8. $(-10)^3$
9. $(-5)^4$
10. $(-1)^6$
11. -5^4
12. -1^6
13. -10^2
14. -8^2

In Exercises 15–28, simplify each algebraic expression, or explain why the expression cannot be simplified.

15. $7x^2 + 12x^2$
16. $6x^2 + 18x^2$
17. $10x^3 + 5x^3$
18. $14x^3 + 8x^3$
19. $8x^4 + x^4$
20. $14x^4 + x^4$

21. $26x^2 - 27x^2$
22. $29x^2 - 30x^2$
23. $27x^3 - 26x^3$
24. $30x^3 - 29x^3$
25. $5x^2 + 5x^3$
26. $8x^2 + 8x^3$
27. $16x^2 - 16x^2$
28. $34x^2 - x^2$

In Exercises 29–72, use the order of operations to simplify each expression.

29. $7 + 6 \cdot 3$
30. $3 + 4 \cdot 5$
31. $45 \div 5 \cdot 3$
32. $40 \div 4 \cdot 2$
33. $6 \cdot 8 \div 4$
34. $8 \cdot 6 \div 2$
35. $14 - 2 \cdot 6 + 3$
36. $36 - 12 \div 4 + 2$
37. $8^2 - 16 \div 2^2 \cdot 4 - 3$
38. $10^2 - 100 \div 5^2 \cdot 2 - 1$

39. $3(-2)^2 - 4(-3)^2$

40. $5(-3)^2 - 2(-4)^2$

41. $(4 \cdot 5)^2 - 4 \cdot 5^2$

42. $(3 \cdot 5)^2 - 3 \cdot 5^2$

43. $(2 - 6)^2 - (3 - 7)^2$

44. $(4 - 6)^2 - (5 - 9)^2$

45. $6(3 - 5)^3 - 2(1 - 3)^3$

46. $-3(-6 + 8)^3 - 5(-3 + 5)^3$

47. $[2(6 - 2)]^2$

48. $[3(4 - 6)]^3$

49. $2[5 + 2(9 - 4)]$

50. $3[4 + 3(10 - 8)]$

51. $[7 + 3(2^3 - 1)] \div 21$

52. $[11 - 4(2 - 3^3)] \div 37$

53. $\dfrac{10 + 8}{5^2 - 4^2}$

54. $\dfrac{6^2 - 4^2}{2 - (-8)}$

55. $\dfrac{37 + 15 \div (-3)}{2^4}$

56. $\dfrac{22 + 20 \div (-5)}{3^2}$

57. $\dfrac{(-11)(-4) + 2(-7)}{7 - (-3)}$

58. $\dfrac{-5(7 - 2) - 3(4 - 7)}{-13 - (-5)}$

59. $4|10 - (8 - 20)|$

60. $6|7 - 4 \cdot 3|$

61. $8(-10) + |4(-5)|$

62. $4(-15) + |3(-10)|$

63. $-2^2 + 4[16 \div (3 - 5)]$

64. $-3^2 + 2[20 \div (7 - 11)]$

65. $24 \div \dfrac{3^2}{8 - 5} - (-6)$

66. $30 \div \dfrac{5^2}{7 - 12} - (-9)$

67. $\dfrac{\frac{1}{4} - \frac{1}{2}}{\frac{1}{3}}$

68. $\dfrac{\frac{3}{5} - \frac{7}{10}}{\frac{1}{2}}$

69. $-\dfrac{9}{4}\left(\dfrac{1}{2}\right) + \dfrac{3}{4} \div \dfrac{5}{6}$

70. $\left[-\dfrac{4}{7} - \left(-\dfrac{2}{5}\right)\right]\left[-\dfrac{3}{8} + \left(-\dfrac{1}{9}\right)\right]$

71. $\dfrac{\frac{7}{9} - 3}{\frac{5}{6}} \div \dfrac{3}{2} + \dfrac{3}{4}$

72. $\dfrac{\frac{17}{25}}{\frac{3}{5} - 4} \div \dfrac{1}{5} + \dfrac{1}{2}$

In Exercises 73–80, evaluate each algebraic expression for the given value of the variable.

73. $x^2 + 5x; x = 3$

74. $x^2 - 2x; x = 6$

75. $3x^2 - 8x; x = -2$

76. $4x^2 - 2x; x = -3$

77. $-x^2 - 10x; x = -1$

78. $-x^2 - 14x; x = -1$

79. $\dfrac{6y - 4y^2}{y^2 - 15}; y = 5$

80. $\dfrac{3y - 2y^2}{y(y - 2)}; y = 5$

In Exercises 81–88, simplify each algebraic expression by removing parentheses and brackets.

81. $3[5(x - 2) + 1]$

82. $4[6(x - 3) + 1]$

83. $3[6 - (y + 1)]$

84. $5[2 - (y + 3)]$

85. $7 - 4[3 - (4y - 5)]$

86. $6 - 5[8 - (2y - 4)]$

87. $2(3x^2 - 5) - [4(2x^2 - 1) + 3]$

88. $4(6x^2 - 3) - [2(5x^2 - 1) + 1]$

Practice PLUS

In Exercises 89–92, express each sentence as a single numerical expression. Then use the order of operations to simplify the expression.

89. Cube -2. Subtract this exponential expression from -10.

90. Cube -5. Subtract this exponential expression from -100.

91. Subtract 10 from 7. Multiply this difference by 2. Square this product.

92. Subtract 11 from 9. Multiply this difference by 2. Raise this product to the fourth power.

In Exercises 93–96, let x represent the number. Express each sentence as a single algebraic expression. Then simplify the expression.

93. Multiply a number by 5. Add 8 to this product. Subtract this sum from the number.

94. Multiply a number by 3. Add 9 to this product. Subtract this sum from the number.

95. Cube a number. Subtract 4 from this exponential expression. Multiply this difference by 5.

96. Cube a number. Subtract 6 from this exponential expression. Multiply this difference by 4.

Application Exercises

The bar graph shows the estimated number of calories per day needed to maintain energy balance for various gender and age groups for sedentary lifestyles. (Sedentary means a lifestyle that includes only the light physical activity associated with typical day-to-day life.)

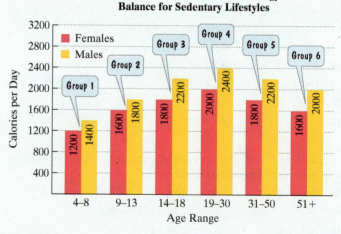

Calories Needed to Maintain Energy Balance for Sedentary Lifestyles

Source: U.S.D.A.

Use the appropriate information displayed by the graph at the bottom of the previous page to solve Exercises 97–98.

97. The mathematical model

$$F = -82x^2 + 654x + 620$$

describes the number of calories needed per day, F, by females in age group x with sedentary lifestyles. According to the model, how many calories per day are needed by females between the ages of 19 and 30, inclusive, with this lifestyle? Does this underestimate or overestimate the number shown by the graph? By how much?

98. The mathematical model

$$M = -96x^2 + 802x + 660$$

describes the number of calories needed per day, M, by males in age group x with sedentary lifestyles. According to the model, how many calories per day are needed by males between the ages of 19 and 30, inclusive, with this lifestyle? Does this underestimate or overestimate the number shown by the graph? By how much?

Gender Divide in Salaries for College Graduates. The graph shows the median, or middlemost, weekly earnings of male and female college graduates for six selected years from 2000 through 2011.

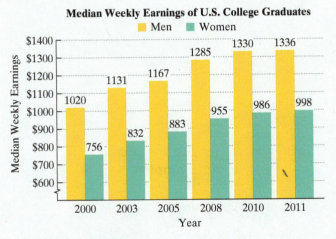

Median Weekly Earnings of U.S. College Graduates

Source: Bureau of Labor Statistics

The data can be described by the following mathematical models:

Median weekly earnings of male college graduates
$$M = \frac{-2n^2 + 170n + 5115}{5}$$

Median weekly earnings of female college graduates
$$F = \frac{-3n^2 + 145n + 3775}{5}.$$

In each mathematical model, n represents the number of years after 2000. Use this information to solve Exercises 99–100.

99. a. Use the appropriate formula to find the median weekly earnings of male college graduates in 2010. Does this value underestimate or overestimate the earnings shown by the bar graph? By how much?

b. Use the appropriate formula to find the median weekly earnings of female college graduates in 2010. Does this value underestimate or overestimate the earnings shown by the bar graph? By how much?

c. Use the values given by the mathematical models in parts (a) and (b) to find the difference between men's weekly earnings and women's weekly earnings in 2010. How does this compare with the difference in earnings shown by the graph?

100. a. Use the appropriate formula to find the median weekly earnings of male college graduates in 2005. Does this value underestimate or overestimate the earnings shown by the bar graph? By how much?

b. Use the appropriate formula to find the median weekly earnings of female college graduates in 2005. Does this value underestimate or overestimate the earnings shown by the bar graph? By how much?

c. Use the values given by the mathematical models in parts (a) and (b) to find the difference between men's weekly earnings and women's weekly earnings in 2005. How does this compare with the difference in earnings shown by the graph?

In Palo Alto, California, a government agency ordered computer-related companies to contribute to a pool of money to clean up underground water supplies. (The companies had stored toxic chemicals in leaking underground containers.) The mathematical model

$$C = \frac{200x}{100 - x}$$

describes the cost, C, in tens of thousands of dollars, for removing x percent of the contaminants. Use this formula to solve Exercises 101–102.

101. a. Find the cost, in tens of thousands of dollars, for removing 50% of the contaminants.

b. Find the cost, in tens of thousands of dollars, for removing 80% of the contaminants.

c. Describe what is happening to the cost of the cleanup as the percentage of contaminant removed increases.

102. a. Find the cost, in tens of thousands of dollars, for removing 60% of the contaminants.

b. Find the cost, in tens of thousands of dollars, for removing 90% of the contaminants.

c. Describe what is happening to the cost of the cleanup as the percentage of contaminants removed increases.

Explaining the Concepts

103. Describe what it means to raise a number to a power. In your description, include a discussion of the difference between -5^2 and $(-5)^2$.

104. Explain how to simplify $4x^2 + 6x^2$. Why is the sum not equal to $10x^4$?

105. Why is the order of operations agreement needed?

Critical Thinking Exercises

Make Sense? *In Exercises 106–109, determine whether each statement makes sense or does not make sense, and explain your reasoning.*

106. Without parentheses, an exponent has only the number next to it as its base.

107. I read that a certain star is 10^4 light-years from Earth, which means 100,000 light-years.

108. When I evaluated $(-1)^n$, I obtained positive numbers when n was even and negative numbers when n was odd.

109. The rules for the order of operations avoid the confusion of obtaining different results when I simplify the same expression.

In Exercises 110–113, determine whether each statement is true or false. If the statement is false, make the necessary change(s) to produce a true statement.

110. If x is -3, then the value of $-3x - 9$ is -18.

111. The algebraic expression $\dfrac{6x + 6}{x + 1}$ cannot have the same value when two different replacements are made for x such as $x = -3$ and $x = 2$.

112. The value of $\dfrac{|3 - 7| - 2^3}{(-2)(-3)}$ is the fraction that results when $\frac{1}{3}$ is subtracted from $-\frac{1}{3}$.

113. $-2(6 - 4^2)^3 = -2(6 - 16)^3$
$= -2(-10)^3 = (-20)^3 = -8000$

114. Simplify: $\dfrac{1}{4} - 6(2 + 8) \div \left(-\dfrac{1}{3}\right)\left(-\dfrac{1}{9}\right).$

In Exercises 115–116, insert parentheses in each expression so that the resulting value is 45.

115. $2 \cdot 3 + 3 \cdot 5$

116. $2 \cdot 5 - \dfrac{1}{2} \cdot 10 \cdot 9$

Review Exercises

117. Simplify: $-8 - 2 - (-5) + 11$. (Section 1.6, Example 3)

118. Multiply: $-4(-1)(-3)(2)$. (Section 1.7, Example 2)

119. Give an example of a real number that is not an irrational number. (Section 1.3, Example 5).

Preview Exercises

Exercises 120–122 will help you prepare for the material covered in the first section of the next chapter. In each exercise, determine whether the given number is a solution of the equation.

120. $-\dfrac{1}{2} = x - \dfrac{2}{3}; \dfrac{1}{6}$

121. $5y + 3 - 4y - 8 = 15; 20$

122. $4x + 2 = 3(x - 6) + 8; -11$

Chapter 1 Summary

Definitions and Concepts	**Examples**

Section 1.1 Introduction to Algebra: Variables and Mathematical Models

A letter that represents a variety of different numbers is called a variable. An algebraic expression is a combination of variables, numbers, and operation symbols. To evaluate an algebraic expression, substitute a given number for the variable and simplify:

1. Perform calculations within parentheses first.
2. Perform multiplication before addition or subtraction.

(A more detailed order of operations is given in Section 1.8.)

Evaluate $6(x - 3) + 4x$ for $x = 5$.

Replace x with 5.

$6(5 - 3) + 4 \cdot 5$
$= 6(2) + 4 \cdot 5$
$= 12 + 20$
$= 32$

Here are some key words for translating into algebraic expressions:

- Addition: plus, sum, more than, increased by
- Subtraction: minus, difference, less than, decreased by
- Multiplication: times, product, twice, multiplied by
- Division: divide, quotient, ratio, divided by

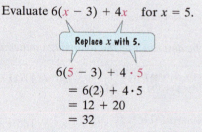

Two less than · 3 times a number
$3x - 2$

The product of 4 and · a number increased by 6
$4(x + 6)$

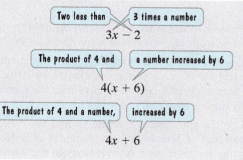

The product of 4 and a number, · increased by 6
$4x + 6$

Definitions and Concepts	**Examples**
An equation is a statement that two algebraic expressions are equal. Solutions of an equation are values of the variable that make the equation a true statement. To determine whether a number is a solution, substitute that number for the variable and evaluate each side of the equation. If the values on both sides of the equation are the same, the number is a solution.	Is 5 a solution of $$3(3x + 5) = 10x + 10?$$ *Substitute 5 for x.* $$3(3 \cdot 5 + 5) \stackrel{?}{=} 10 \cdot 5 + 10$$ $$3(15 + 5) \stackrel{?}{=} 50 + 10$$ $$3(20) \stackrel{?}{=} 60$$ $$60 = 60,$$ The true statement indicates that 5 is a solution.
A formula is an equation that expresses a relationship between two or more variables. The process of finding formulas to describe real-world phenomena is called mathematical modeling. Such formulas, together with the meaning assigned to the variables, are called mathematical models. These formulas model, or describe, the relationship among the variables.	The formulas $$M = 0.2n + 12$$ and $$W = -0.4n + 39$$ model the time each week that men, M, and women, W, devoted to housework n years after 1965.

Section 1.2 Fractions in Algebra

Mixed Numbers and Improper Fractions A mixed number consists of the addition of a natural number $(1, 2, 3, \dots)$ and a fraction, expressed without the use of an addition sign. An improper fraction has a numerator that is greater than its denominator. To convert a mixed number to an improper fraction, multiply the denominator by the natural number and add the numerator. Then place this result over the original denominator. To convert an improper fraction to a mixed number, divide the denominator into the numerator and write the mixed number using $$\text{quotient } \frac{\text{remainder}}{\text{original denominator}}.$$	Convert $5\frac{3}{7}$ to an improper fraction. $$5\frac{3}{7} = \frac{7 \cdot 5 + 3}{7} = \frac{35 + 3}{7} = \frac{38}{7}$$ Convert $\frac{14}{3}$ to a mixed number. $$\frac{14}{3} = 4\frac{2}{3} \qquad \begin{array}{r} 4 \\ 3\overline{)14} \\ \underline{12} \\ 2 \end{array}$$
A prime number is a natural number greater than 1 that has only itself and 1 as factors. A composite number is a natural number greater than 1 that is not a prime number. The prime factorization of a composite number is the expression of the composite number as the product of prime numbers.	Find the prime factorization: $$60 = 6 \cdot 10$$ $$= 2 \cdot 3 \cdot 2 \cdot 5$$
A fraction is reduced to its lowest terms when the numerator and denominator have no common factors other than 1. To reduce a fraction to its lowest terms, divide both the numerator and the denominator by their greatest common factor. The greatest common factor can be found by inspection or prime factorizations of the numerator and the denominator.	Reduce to lowest terms: $$\frac{8}{14} = \frac{2 \cdot 4}{2 \cdot 7} = \frac{4}{7}$$
Multiplying Fractions The product of two or more fractions is the product of their numerators divided by the product of their denominators.	Multiply: $$\frac{2}{7} \cdot \frac{5}{9} = \frac{2 \cdot 5}{7 \cdot 9} = \frac{10}{63}$$
Dividing Fractions The quotient of two fractions is the first multiplied by the reciprocal (or multiplicative inverse) of the second.	Divide: $$\frac{4}{9} \div \frac{3}{7} = \frac{4}{9} \cdot \frac{7}{3} = \frac{4 \cdot 7}{9 \cdot 3} = \frac{28}{27}$$

Definitions and Concepts	Examples

Section 1.2 Fractions in Algebra (continued)

Adding and Subtracting Fractions with Identical Denominators

Add or subtract numerators. Put this result over the common denominator.

Subtract:

$$\frac{5}{8} - \frac{3}{8} = \frac{5-3}{8} = \frac{2}{8} = \frac{2 \cdot 1}{2 \cdot 4} = \frac{1}{4}$$

Adding and Subtracting Fractions with Unlike Denominators

Rewrite the fractions as equivalent fractions with the least common denominator. Then add or subtract numerators, putting this result over the common denominator.

Add:

$$\frac{3}{8} + \frac{5}{12} = \frac{3 \cdot 3}{8 \cdot 3} + \frac{5 \cdot 2}{12 \cdot 2}$$

The LCD is 24. $= \dfrac{9}{24} + \dfrac{10}{24} = \dfrac{19}{24}$

Fractions appear throughout algebra. Many formulas and mathematical models contain fractions. Operations with fractions can be used to determine whether a particular fraction is a solution of an equation. When used with fractions, the word *of* represents multiplication.

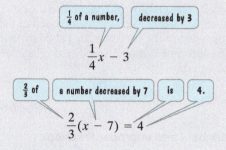

$\frac{1}{4}$ of a number, | decreased by 3

$$\frac{1}{4}x - 3$$

$\frac{2}{3}$ of | a number decreased by 7 | is | 4.

$$\frac{2}{3}(x - 7) = 4$$

Section 1.3 The Real Numbers

A set is a collection of objects, called elements, whose contents can be clearly determined.

$$\{a, b, c\}$$

A line used to visualize numbers is called a number line.

$$\xleftarrow{\quad}\overset{\begin{array}{cccccccccc} & & & & & & & & & \\ -4 & -3 & -2 & -1 & 0 & 1 & 2 & 3 & 4 \end{array}}{\xrightarrow{\qquad\qquad\qquad\qquad}}$$

Real Numbers: the set of all numbers that can be represented by points on the number line

The Sets That Make Up the Real Numbers

- Natural Numbers: $\{1, 2, 3, 4, \dots\}$
- Whole Numbers: $\{0, 1, 2, 3, 4, \dots\}$
- Integers: $\{\dots, -3, -2, -1, 0, 1, 2, 3, \dots\}$
- Rational Numbers: the set of numbers that can be expressed as the quotient of an integer and a nonzero integer; can be expressed as terminating or repeating decimals
- Irrational Numbers: the set of numbers that cannot be expressed as the quotient of integers; decimal representations neither terminate nor repeat.

Given the set

$$\left\{-1.4, 0, 0.\overline{7}, \frac{9}{10}, \sqrt{2}, \sqrt{4}\right\}$$

list the

- natural numbers: $\sqrt{4}$, or 2
- whole numbers: $0, \sqrt{4}$
- integers: $0, \sqrt{4}$
- rational numbers: $-1.4, 0, 0.\overline{7}, \dfrac{9}{10}, \sqrt{4}$
- irrational numbers: $\sqrt{2}$
- real numbers:

$$-1.4, 0, 0.\overline{7}, \frac{9}{10}, \sqrt{2}, \sqrt{4}$$

For any two real numbers, a and b, a is less than b if a is to the left of b on the number line.

Inequality Symbols

$<$: is less than $>$: is greater than
$\leq$: is less than or equal to $\geq$: is greater than or equal to

$$\xleftarrow{\quad}\overset{\begin{array}{cccccccccc} & & & & & & & & & \\ -4 & -3 & -2 & -1 & 0 & 1 & 2 & 3 & 4 \end{array}}{\xrightarrow{\qquad\qquad\qquad\qquad}}$$

$-2 < 0 \qquad 0 > -2$

$0 < 2.5 \qquad 2.5 > 0$

The absolute value of a, written $|a|$, is the distance from 0 to a on the number line.

$|4| = 4 \quad |0| = 0 \quad |-6| = 6$

Definitions and Concepts	Examples

Section 1.4 Basic Rules of Algebra

Terms of an algebraic expression are separated by addition. The parts of each term that are multiplied are its factors. The numerical part of a term is its coefficient. Like terms have the same variable factors raised to the same powers.

- An expression with three terms:

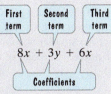

The like terms are $8x$ and $6x$.

Properties of Real Numbers and Algebraic Expressions

- Commutative Properties:
$$a + b = b + a$$
$$ab = ba$$
- Associative Properties:
$$(a + b) + c = a + (b + c)$$
$$(ab)c = a(bc)$$
- Distributive Properties:
$$a(b + c) = ab + ac$$
$$(b + c)a = ba + ca$$
$$a(b - c) = ab - ac$$
$$(b - c)a = ba - ca$$
$$a(b + c + d) = ab + ac + ad$$

Commutative of Addition:
$$5x + 4 = 4 + 5x$$
Commutative of Multiplication:
$$5x + 4 = x5 + 4$$
Associative of Addition:
$$6 + (4 + x) = (6 + 4) + x = 10 + x$$
Associative of Multiplication:
$$7(10x) = (7 \cdot 10)x = 70x$$
Distributive:
$$8(x + 5 + 4y) = 8x + 40 + 32y$$
Distributive to Combine Like Terms:
$$8x + 12x = (8 + 12)x = 20x$$

Simplifying Algebraic Expressions

Use the distributive property to remove grouping symbols. Then combine like terms.

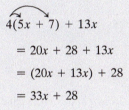

$$= 20x + 28 + 13x$$
$$= (20x + 13x) + 28$$
$$= 33x + 28$$

Section 1.5 Addition of Real Numbers

Sums on a Number Line

To find $a + b$, the sum of a and b, on a number line, start at a. If b is positive, move b units to the right. If b is negative, move $|b|$ units to the left. If b is 0, stay at a. The number where we finish on the number line represents $a + b$.

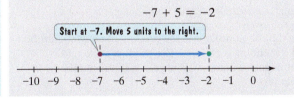

Additive inverses, or opposites, are pairs of real numbers that are the same number of units from zero on the number line, but on opposite sides of zero.

- Identity Property of Addition:
$$a + 0 = 0 \qquad 0 + a = a$$
- Inverse Property of Addition:
$$a + (-a) = 0 \qquad (-a) + a = 0$$

The additive inverse (or opposite) of 4 is -4.
The additive inverse of -1.7 is 1.7.

Identity Property of Addition:
$$4x + 0 = 4x$$
Inverse Property of Addition:
$$4x + (-4x) = 0$$

Definitions and Concepts	**Examples**

Section 1.5 Addition of Real Numbers (continued)

Addition without a Number Line

To add two numbers with the same sign, add their absolute values and use their common sign. To add two numbers with different signs, subtract the smaller absolute value from the greater absolute value and use the sign of the number with the greater absolute value.

To add a series of positive and negative numbers, add all the positive numbers and add all the negative numbers. Then add the resulting positive and negative sums.

Add:

$$10 + 4 = 14$$
$$-4 + (-6) = -10$$
$$-30 + 5 = -25$$
$$12 + (-8) = 4$$

$$5 + (-3) + (-7) + 2$$
$$= (5 + 2) + [(-3) + (-7)]$$
$$= 7 + (-10)$$
$$= -3$$

Section 1.6 Subtraction of Real Numbers

To subtract b from a, add the opposite, or additive inverse, of b to a:
$$a - b = a + (-b).$$
The result is called the difference between a and b.

Subtract:

$$-7 - (-5) = -7 + 5 = -2$$
$$-\frac{3}{4} - \frac{1}{2} = -\frac{3}{4} + \left(-\frac{1}{2}\right)$$
$$= -\frac{3}{4} + \left(-\frac{2}{4}\right) = -\frac{5}{4}$$

To simplify a series of additions and subtractions, change all subtractions to additions of opposites. Then use the procedure for adding a series of positive and negative numbers.

Simplify:

$$-6 - 2 - (-3) + 10$$
$$= -6 + (-2) + 3 + 10$$
$$= -8 + 13$$
$$= 5$$

Section 1.7 Multiplication and Division of Real Numbers

The result of multiplying a and b, ab, is called the product of a and b. If the two numbers have different signs, the product is negative. If the two numbers have the same sign, the product is positive. If either number is 0, the product is 0.

Multiply:

$$-5(-10) = 50$$
$$\frac{3}{4}\left(-\frac{5}{7}\right) = -\frac{3}{4} \cdot \frac{5}{7} = -\frac{15}{28}$$

Assuming that no number is 0, the product of an even number of negative numbers is positive. The product of an odd number of negative numbers is negative. If any number is 0, the product is 0.

Multiply:

$$(-3)(-2)(-1)(-4) = 24$$
$$(-3)(2)(-1)(-4) = -24$$

The result of dividing the real number a by the nonzero real number b is called the quotient of a and b. If two numbers have different signs, their quotient is negative. If two numbers have the same sign, their quotient is positive. Division by zero is undefined.

Divide:

$$\frac{21}{-3} = -7$$
$$-\frac{1}{3} \div (-3) = \frac{1}{3} \cdot \frac{1}{3} = \frac{1}{9}$$

Definitions and Concepts	**Examples**

Section 1.7 Multiplication and Division of Real Numbers (continued)

Two numbers whose product is 1 are called multiplicative inverses, or reciprocals, of each other. The number 0 has no multiplicative inverse.	The multiplicative inverse of 4 is $\frac{1}{4}$. The multiplicative inverse of $-\frac{1}{3}$ is -3.

- Identity Property of Multiplication
$$a \cdot 1 = a \qquad 1 \cdot a = a$$
- Inverse Property of Multiplication
  If a is not 0:
$$a \cdot \frac{1}{a} = 1 \qquad \frac{1}{a} \cdot a = 1$$
- Multiplication Property of -1
$$-1a = -a \qquad a(-1) = -a$$
- Double Negative Property
$$-(-a) = a$$

Simplify:
$$1x = x$$
$$7x \cdot \frac{1}{7x} = 1, x \neq 0$$
$$4x - 5x = -1x = -x$$
$$-(-7y) = 7y$$

If a negative sign precedes parentheses, remove parentheses and change the sign of every term within parentheses.	Simplify: $$-(7x - 3y + 2) = -7x + 3y - 2$$

Section 1.8 Exponents and Order of Operations

If b is a real number and n is a natural number, b^n, the nth power of b, is the product of n factors of b. Furthermore, $b^1 = b$.	Evaluate: $$8^2 = 8 \cdot 8 = 64$$ $$(-5)^3 = (-5)(-5)(-5) = -125$$

Order of Operations

1. Perform operations within grouping symbols, starting with the innermost grouping symbols. Grouping symbols include parentheses, brackets, fraction bars, and absolute value symbols.
2. Evaluate exponential expressions.
3. Multiply and divide, from left to right.
4. Add and subtract. In this step, you have a choice. You can add and subtract in order from left to right. You can also rewrite subtractions as additions of opposites, combine positive and negative numbers separately, and then add these results.

Simplify:
$$5(4 - 6)^2 - 2(1 - 3)^3$$
$$= 5(-2)^2 - 2(-2)^3$$
$$= 5(4) - 2(-8)$$
$$= 20 - (-16)$$
$$= 20 + 16 = 36$$

Some algebraic expressions contain two sets of grouping symbols: parentheses, the inner grouping symbols, and brackets, the outer grouping symbols. To simplify such expressions, use the order of operations and remove grouping symbols from innermost (parentheses) to outermost (brackets).	Simplify: $$5 - 3[2(x + 1) - 7]$$ $$= 5 - 3[2x + 2 - 7]$$ $$= 5 - 3[2x - 5]$$ $$= 5 - 6x + 15$$ $$= -6x + 20$$

CHAPTER 1 REVIEW EXERCISES

1.1 *In Exercises 1–2, evaluate each expression for x = 6.*

1. $10 + 5x$

2. $8(x - 2) + 3x$

In Exercises 3–4, evaluate each expression for x = 8 and y = 10.

3. $\dfrac{40}{x} - \dfrac{y}{5}$

4. $3(2y + x)$

In Exercises 5–8, translate from English to an algebraic expression or equation, whichever is appropriate. Let the variable x represent the number.

5. Six subtracted from the product of 7 and a number

6. The quotient of a number and 5, decreased by 2, is 18.

7. Nine less twice a number is 14.

8. The product of 3 and 7 more than a number

In Exercises 9–11, determine whether the given number is a solution of the equation.

9. $4x + 5 = 13; 3$

10. $2y + 7 = 4y - 5; 6$

11. $3(w + 1) + 11 = 2(w + 8); 2$

Exercises 12–13 involve an experiment on memory. Students in a language class are asked to memorize 40 vocabulary words in Latin, a language with which they are not familiar. After studying the words for one day, students are tested each day after to see how many words they remember. The class average is taken and the results are shown as ten points on a line graph.

**Average Number of Words
Remembered Over Time**

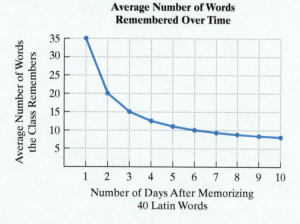

Number of Days After Memorizing
40 Latin Words

12. Use the line graph to estimate the average number of Latin words the class remembered after 5 days.

13. The mathematical model

$$L = \dfrac{5n + 30}{n}$$

describes the average number of Latin words remembered by the students, L, after n days. Use the formula to find the average number of words remembered after 5 days. How well does this compare with your estimate from Exercise 12?

14. **Cost per Day to Punish: $80.25.** In 2013, most of the 215,866 federal inmates in the United States were in prison on drug charges. The bar graph shows the average annual cost per inmate for three selected years from 2000 through 2013.

**Average Annual Cost per Federal Prisoner
in the United States**

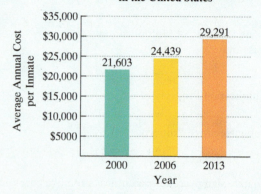

Source: Bureau of Justice Statistics

Here is a mathematical model that approximates the data displayed by the bar graph:

$$C = 594n + 21{,}348.$$

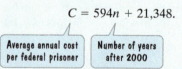

Average annual cost per federal prisoner   Number of years after 2000

a. Use the formula to find the average annual cost per federal prisoner in 2013. Does the mathematical model underestimate or overestimate the actual cost shown by the bar graph? By how much?

b. If trends from 2000 through 2013 continue, use the formula to project the average annual cost per federal prisoner in 2020.

1.2 *In Exercises 15–16, convert each mixed number to an improper fraction.*

15. $3\dfrac{2}{7}$

16. $5\dfrac{9}{11}$

In Exercises 17–18, convert each improper fraction to a mixed number.

17. $\dfrac{17}{9}$

18. $\dfrac{27}{5}$

In Exercises 19–21, identify each natural number as prime or composite. If the number is composite, find its prime factorization.

19. 60

20. 63

21. 67

In Exercises 22–23, simplify each fraction by reducing it to its lowest terms.

22. $\dfrac{15}{33}$

23. $\dfrac{40}{75}$

In Exercises 24–29, perform the indicated operation. Where possible, reduce the answer to its lowest terms.

24. $\dfrac{3}{5} \cdot \dfrac{7}{10}$ **25.** $\dfrac{4}{5} \div \dfrac{3}{10}$ **26.** $1\dfrac{2}{3} \div 6\dfrac{2}{3}$

27. $\dfrac{2}{9} + \dfrac{4}{9}$ **28.** $\dfrac{5}{6} + \dfrac{7}{9}$ **29.** $\dfrac{3}{4} - \dfrac{2}{15}$

In Exercises 30–31, determine whether the given number is a solution of the equation.

30. $x - \dfrac{3}{4} = \dfrac{7}{4}; 2\dfrac{1}{2}$

31. $\dfrac{2}{3}w = \dfrac{1}{15}w + \dfrac{3}{5}; 2$

In Exercises 32–33, translate from English to an algebraic expression or equation, whichever is appropriate. Let the variable x represent the number.

32. Two decreased by half of a number is $\frac{1}{4}$ of the number.

33. $\frac{3}{5}$ of a number increased by 6

34. Suppose that the target heart rate, H, in beats per minute, for your exercise goal is given by

$$H = \frac{4}{5}(220 - a),$$

where a is your age. If you are 30 years old, what is your target heart rate?

1.3   *In Exercises 35–36, graph each real number on a number line.*

35. -2.5 **36.** $4\dfrac{3}{4}$

In Exercises 37–38, express each rational number as a decimal.

37. $\dfrac{5}{8}$ **38.** $\dfrac{3}{11}$

39. Consider the set

$$\left\{ -17, -\frac{9}{13}, 0, 0.75, \sqrt{2}, \pi, \sqrt{81} \right\}.$$

List all numbers from the set that are: **a.** natural numbers, **b.** whole numbers, **c.** integers, **d.** rational numbers, **e.** irrational numbers, **f.** real numbers.

40. Give an example of an integer that is not a natural number.

41. Give an example of a rational number that is not an integer.

42. Give an example of a real number that is not a rational number.

In Exercises 43–46, insert either $<$ or $>$ in the shaded area between each pair of numbers to make a true statement.

43. $-93 \ \blacksquare\ 17$ **44.** $-2 \ \blacksquare\ -200$

45. $0 \ \blacksquare\ -\dfrac{1}{3}$ **46.** $-\dfrac{1}{4} \ \blacksquare\ -\dfrac{1}{5}$

In Exercises 47–48, determine whether each inequality is true or false.

47. $-13 \geq -11$ **48.** $-126 \leq -126$

In Exercises 49–50, find each absolute value.

49. $|-58|$ **50.** $|2.75|$

1.4

51. Use the commutative property of addition to write an equivalent algebraic expression: $7 + 13y$.

52. Use the commutative property of multiplication to write an equivalent algebraic expression: $9(x + 7)$.

In Exercises 53–54, use an associative property to rewrite each algebraic expression. Then simplify the resulting algebraic expression.

53. $6 + (4 + y)$

54. $7(10x)$

55. Use the distributive property to rewrite without parentheses: $6(4x - 2 + 5y)$.

In Exercises 56–57, simplify each algebraic expression.

56. $4a + 9 + 3a - 7$

57. $6(3x + 4) + 5(2x - 1)$

1.5

58. Use a number line to find the sum: $-6 + 8$.

In Exercises 59–61, find each sum without the use of a number line.

59. $8 + (-11)$

60. $-\dfrac{3}{4} + \dfrac{1}{5}$

61. $7 + (-5) + (-13) + 4$

In Exercises 62–63, simplify each algebraic expression.

62. $8x + (-6y) + (-12x) + 11y$

63. $10(3y + 4) + (-40y)$

64. The Dead Sea is the lowest elevation on Earth, 1312 feet below sea level. If a person is standing 512 feet above the Dead Sea, what is that person's elevation?

65. The water level of a reservoir is measured over a five-month period. At the beginning, the level is 25 feet. During this time, the level fell 3 feet, then rose 2 feet, then rose 1 foot, then fell 4 feet, and then rose 2 feet. What is the reservoir's water level at the end of the five months?

1.6

66. Rewrite $9 - 13$ as the addition of an opposite.

In Exercises 67–69, perform the indicated subtraction.

67. $-9 - (-13)$ **68.** $-\dfrac{7}{10} - \dfrac{1}{2}$

69. $-3.6 - (-2.1)$

In Exercises 70–71, simplify each series of additions and subtractions.

70. $-7 - (-5) + 11 - 16$

71. $-25 - 4 - (-10) + 16$

72. Simplify: $3 - 6a - 8 - 2a$.

73. What is the difference in elevation between a plane flying 26,500 feet above sea level and a submarine traveling 650 feet below sea level?

1.7 *In Exercises 74–76, perform the indicated multiplication.*

74. $-7(-12)$

75. $\dfrac{3}{5}\left(-\dfrac{5}{11}\right)$

76. $5(-3)(-2)(-4)$

In Exercises 77–79, perform the indicated division or state that the expression is undefined.

77. $\dfrac{45}{-5}$

78. $-17 \div 0$

79. $-\dfrac{4}{5} \div \left(-\dfrac{2}{5}\right)$

In Exercises 80–81, simplify each algebraic expression.

80. $-4\left(-\dfrac{3}{4}x\right)$

81. $-3(2x - 1) - (4 - 5x)$

In Exercises 82–83, determine whether the given number is a solution of the equation.

82. $5x + 16 = -8 - x; -6$

83. $2(x + 3) - 18 = 5x; -4$

84. The bar graph shows the number of Americans, in millions, ages 7 and older who rode a bicycle six or more days during three selected years.

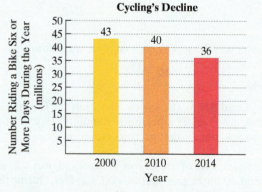

Cycling's Decline

Source: National Sporting Goods Association

Here is a mathematical model for the data displayed by the bar graph:

$$B = -\frac{1}{2}n + 43.$$

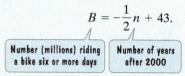

For which year or years shown by the graph does the mathematical model give the exact number of Americans riding a bike six or more days?

1.8 *In Exercises 85–87, evaluate each exponential expression.*

85. $(-6)^2$

86. -6^2

87. $(-2)^5$

In Exercises 88–89, simplify each algebraic expression, or explain why the expression cannot be simplified.

88. $4x^3 + 2x^3$

89. $4x^3 + 4x^2$

In Exercises 90–98, use the order of operations to simplify each expression.

90. $-40 \div 5 \cdot 2$

91. $-6 + (-2) \cdot 5$

92. $6 - 4(-3 + 2)$

93. $28 \div (2 - 4^2)$

94. $36 - 24 \div 4 \cdot 3 - 1$

95. $-8[-4 - 5(-3)]$

96. $\dfrac{6(-10 + 3)}{2(-15) - 9(-3)}$

97. $\left(\dfrac{1}{2} + \dfrac{1}{3}\right) \div \left(\dfrac{1}{4} - \dfrac{3}{8}\right)$

98. $\dfrac{1}{2} - \dfrac{2}{3} \div \dfrac{5}{9} + \dfrac{3}{10}$

In Exercises 99–100, evaluate each algebraic expression for the given value of the variable.

99. $x^2 - 2x + 3; x = -1$

100. $-x^2 - 7x; x = -2$

In Exercises 101–102, simplify each algebraic expression.

101. $4[7(a - 1) + 2]$

102. $-6[4 - (y + 2)]$

103. The bar graph shows the average student loan debt in the United States for four selected years from 2001 through 2010.

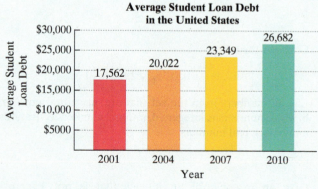

Average Student Loan Debt in the United States

Source: Pew Research Center

Average student loan debt, D, can be described by the mathematical model

$$D = 24n^2 + 756n + 16{,}739,$$

where n is the number of years after 2000.

a. Does the mathematical model underestimate or overestimate average student loan debt in 2010? By how much?

b. If the trend shown by the graph continues, use the formula to project average student loan debt in 2020.

CHAPTER 1 TEST

Step-by-step test solutions are found on the Chapter Test Prep Videos available in MyMathLab® or on YouTube (search "BlitzerIntroAlg7e" and click on "Channels").

In Exercises 1–10, perform the indicated operation or operations.

1. $1.4 - (-2.6)$

2. $-9 + 3 + (-11) + 6$

3. $3(-17)$

4. $\left(-\dfrac{3}{7}\right) \div \left(-\dfrac{15}{7}\right)$

5. $\left(3\dfrac{1}{3}\right)\left(-1\dfrac{3}{4}\right)$

6. $-50 \div 10$

7. $-6 - (5 - 12)$

8. $(-3)(-4) \div (7 - 10)$

9. $(6 - 8)^2(5 - 7)^3$

10. $\dfrac{3(-2) - 2(2)}{-2(8 - 3)}$

In Exercises 11–13, simplify each algebraic expression.

11. $11x - (7x - 4)$

12. $5(3x - 4y) - (2x - y)$

13. $6 - 2[3(x + 1) - 5]$

14. List all the rational numbers in this set.

$$\left\{-7, -\dfrac{4}{5}, 0, 0.25, \sqrt{3}, \sqrt{4}, \dfrac{22}{7}, \pi\right\}$$

15. Insert either $<$ or $>$ in the shaded area to make a true statement: $-1 \;\blacksquare\; -100$.

16. Find the absolute value: $|-12.8|$.

In Exercises 17–18, evaluate each algebraic expression for the given value of the variable.

17. $5(x - 7); x = 4$

18. $x^2 - 5x; x = -10$

19. Use the commutative property of addition to write an equivalent algebraic expression: $2(x + 3)$.

20. Use the associative property of multiplication to rewrite $-6(4x)$. Then simplify the expression.

21. Use the distributive property to rewrite without parentheses: $7(5x - 1 + 2y)$.

22. What is the difference in elevation between a plane flying 16,200 feet above sea level and a submarine traveling 830 feet below sea level?

In Exercises 23–24, determine whether the given number is a solution of the equation.

23. $\dfrac{1}{5}(x + 2) = \dfrac{1}{10}x + \dfrac{3}{5}; 3$

24. $3(x + 2) - 15 = 4x; -9$

In Exercises 25–26, translate from English to an algebraic expression or equation, whichever is appropriate. Let the variable x represent the number.

25. $\frac{1}{4}$ of a number, decreased by 5, is 32.

26. Seven subtracted from the product of 5 and 4 more than a number

27. On average, Americans look at their smartphones more than 150 times per day, spending approximately 2.5 hours with their phones. (Source: *Scholastic Math*) The bar graph shows the percentage of Americans owning smartphones, by age group.

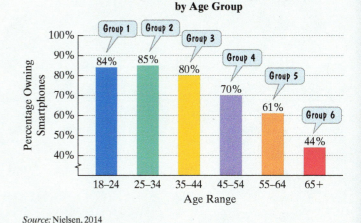

Percentage of Americans Owning Smartphones, by Age Group

Source: Nielsen, 2014

The mathematical model

$$S = -2x^2 + 5x + 81$$

describes the percentage of Americans owning smartphones, S, in age group x. Does the model underestimate or overestimate the actual percentage of Americans owning smartphones in the 35–44 age range? By how much?

28. Electrocardiograms are used in exercise stress tests to determine a person's fitness for strenuous exercise. The target heart rate, in beats per minute, for such tests depends on a person's age. The line graph shows target heart rates for stress tests for people of various ages.

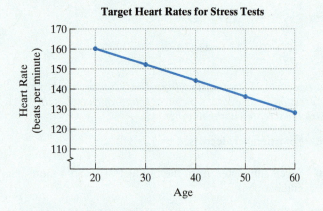

Target Heart Rates for Stress Tests

Use the graph to estimate the target heart rate for a 40-year-old taking a stress test.

29. The formula $H = \frac{4}{5}(220 - a)$ gives the target heart rate, H, in beats per minute, on a stress test for a person of age a. Use the formula to find the target heart rate for a 40-year-old. How does this compare with your estimate from Exercise 28?

Linear Equations and Inequalities in One Variable

The belief that humor and laughter can have positive effects on our lives is not new. The Bible tells us, "A merry heart doeth good like a medicine, but a broken spirit drieth the bones." (**Proverbs 17:22**)

Some random humor factoids: ■ The average adult laughs 15 times each day. (*Newhouse News Service*) ■ Forty-six percent of people who are telling a joke laugh more than the people they are telling it to. (*U.S. News and World Report*) ■ Eighty percent of adult laughter does not occur in response to jokes or funny situations. (*Independent*) ■ Algebra can be used to model the influence that humor plays in our responses to negative life events. (**Bob Blitzer,** *Introductory Algebra*)

That last tidbit that your author threw into the list is true. Based on our sense of humor, there is actually a formula that predicts how we will respond to difficult life events.

Formulas can be used to explain what is happening in the present and to make predictions about what might occur in the future. In this chapter, you will learn to use formulas in new ways that will help you to recognize patterns, logic, and order in a world that can appear chaotic to the untrained eye.

Here's where you'll find this application:

- A mathematical model that includes sense of humor as a variable is developed in Example 8 of Section 2.3.

2.1

The Addition Property of Equality

What am I supposed to learn?

After studying this section, you should be able to:

1. Identify linear equations in one variable. ▶

2. Use the addition property of equality to solve equations. ▶

3. Solve applied problems using formulas. ▶

Language is one of the few areas in which children are more efficient learners than adults. Although most infants don't clearly start comprehending words until nine or ten months, they recognize their own names by as early as six months. They begin to recognize other commonly used and important (to them) words like "bottle," "mama," and "doggie" by ten to 12 months. By the time they turn two, most childern can produce several hundred words. By kindergarten, their vocabularies have ballooned to several thousand words.

In this section, we will work with a mathematical model (Example 7) that describes the number of words in a child's vocabulary for ages between 15 and 50 months. Your work with this model will involve the ability to *solve equations*.

Linear Equations in One Variable

In Chapter 1, we learned that an equation is a statement that two algebraic expressions are equal. We determined whether a given number is an equation's solution by substituting that number for each occurrence of the variable. When the substitution resulted in a true statement, that number was a solution. When the substitution resulted in a false statement, that number was not a solution.

In the next three sections, we will study how to solve equations in one variable. **Solving an equation** is the process of finding the number (or numbers) that make the equation a true statement. These numbers are called the **solutions**, or **roots**, of the equation, and we say that they **satisfy** the equation.

In this chapter, you will learn to solve the simplest type of equation, called a *linear equation in one variable*.

1. Identify linear equations in one variable. ▶

> ### Definition of a Linear Equation in One Variable
>
> A **linear equation in one variable** x is an equation that can be written in the form
>
> $$ax + b = c,$$
>
> where a, b, and c are real numbers, and $a \neq 0$ (a is not equal to 0).

Linear Equations in One Variable (x)

$$3x + 7 = 9 \qquad\qquad -15x = 45 \qquad\qquad x = 6.8$$

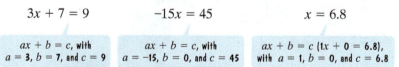

$ax + b = c$, with $a = 3$, $b = 7$, and $c = 9$ | $ax + b = c$, with $a = -15$, $b = 0$, and $c = 45$ | $ax + b = c$ ($1x + 0 = 6.8$), with $a = 1$, $b = 0$, and $c = 6.8$

Nonlinear Equations in One Variable (x)

$$3x^2 + 7 = 9 \qquad\qquad -\frac{15}{x} = 45 \qquad\qquad |x| = 6.8$$

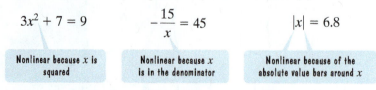

Nonlinear because x is squared | Nonlinear because x is in the denominator | Nonlinear because of the absolute value bars around x

The adjective *linear* contains the word *line*. As we shall see, linear equations are related to graphs whose points lie along a straight line.

2 Use the addition property of equality to solve equations. ▶

Using the Addition Property of Equality to Solve Equations

Consider the equation

$$x = 11.$$

By inspection, we can see that the solution to this equation is 11. If we substitute 11 for x, we obtain the true statement $11 = 11$.

Now consider the equation

$$x - 3 = 8.$$

If we substitute 11 for x, we obtain $11 - 3 \overset{?}{=} 8$. Subtracting on the left side, we get the true statement $8 = 8$.

The equations $x - 3 = 8$ and $x = 11$ both have the same solution, namely 11, and are called *equivalent equations*. **Equivalent equations** are equations that have the same solution.

The idea in solving a linear equation is to get an equivalent equation with the variable (the letter) by itself on one side of the equal sign and a number by itself on the other side. For example, consider the equation $x - 3 = 8$. To get x by itself on the left side, add 3 to the left side, because $x - 3 + 3$ gives $x + 0$, or just x. You must then add 3 to the right side also. By doing this, we are using the **addition property of equality**.

The Addition Property of Equality

The same real number (or algebraic expression) may be added to both sides of an equation without changing the equation's solution. This can be expressed symbolically as follows:

$$\text{If } a = b, \text{ then } a + c = b + c.$$

EXAMPLE 1   Solving an Equation Using the Addition Property

Solve the equation: $x - 3 = 8$.

Solution   We can isolate the variable, x, by adding 3 to both sides of the equation.

$x - 3 = 8$	This is the given equation.
$x - 3 + 3 = 8 + 3$	Add 3 to both sides.
$x + 0 = 11$	This step is often done mentally and not listed.
$x = 11$	

By inspection, we can see that the solution to $x = 11$ is 11. To check this proposed solution, replace x with 11 in the original equation.

Check

$x - 3 = 8$	This is the original equation.
$11 - 3 \overset{?}{=} 8$	Substitute 11 for x.
$8 = 8$	Subtract: $11 - 3 = 8$.

This statement is true.

Because the check results in a true statement, we conclude that the solution to the given equation is 11. ∎

Great Question!

Is there another way to show that I'm adding the same number to both sides of an equation?

Some people prefer to show the number below the equation:

$$\begin{aligned} x - 3 &= 8 \\ +3 \quad &\;\; +3 \\ \hline x \quad &= 11. \end{aligned}$$

Great Question!

In a high school algebra course, I remember my teacher talking about balancing an equation. What does the addition property of equality have to do with a balanced equation?

You can think of an equation as a balanced scale—balanced because its two sides are equal. To maintain this balance, whatever you do to one side must also be done to the other side.

The set of an equation's solutions is called its **solution set**. Thus, the solution set of the equation in Example 1, $x - 3 = 8$, is {11}. The solution can be expressed as 11 or, using set notation, {11}.

✓ **CHECK POINT 1**   Solve the equation and check your proposed solution:

$$x - 5 = 12.$$

When we use the addition property of equality, we add the same number to both sides of an equation. We know that subtraction is the addition of an opposite, or additive inverse. Thus, the addition property also lets us subtract the same number from both sides of an equation without changing the equation's solution.

EXAMPLE 2   Subtracting the Same Number from Both Sides

Solve and check: $z + 1.4 = 2.06$.

Solution

$z + 1.4 = 2.06$	This is the given equation.
$z + 1.4 - 1.4 = 2.06 - 1.4$	Subtract 1.4 from both sides. This is equivalent to adding −1.4 to both sides.
$z = 0.66$	Subtracting 1.4 from both sides eliminates 1.4 on the left.

Can you see that the solution to $z = 0.66$ is 0.66? To check this proposed solution, replace z with 0.66 in the original equation.

Check	$z + 1.4 = 2.06$	This is the original equation.
	$0.66 + 1.4 \stackrel{?}{=} 2.06$	Substitute 0.66 for z.
	$2.06 = 2.06$	This statement is true.

This true statement indicates that the solution is 0.66, or the solution set is {0.66}. ■

✓ **CHECK POINT 2**   Solve and check: $z + 2.8 = 5.09$.

When isolating the variable, we can isolate it on either the left side or the right side of an equation.

EXAMPLE 3   Isolating the Variable on the Right

Solve and check: $-\dfrac{1}{2} = x - \dfrac{2}{3}.$

Solution   We can isolate the variable, x, on the right side by adding $\frac{2}{3}$ to both sides of the equation.

$$-\frac{1}{2} = x - \frac{2}{3}$$ This is the given equation.

$$-\frac{1}{2} + \frac{2}{3} = x - \frac{2}{3} + \frac{2}{3}$$ Add $\frac{2}{3}$ to both sides, isolating x on the right.

$$-\frac{3}{6} + \frac{4}{6} = x$$ Rewrite each fraction as an equivalent fraction with a denominator of 6: $-\frac{1}{2} + \frac{2}{3} = -\frac{1}{2} \cdot \frac{3}{3} + \frac{2}{3} \cdot \frac{2}{2} = -\frac{3}{6} + \frac{4}{6}.$

$$\frac{1}{6} = x$$ Add on the left side: $-\frac{3}{6} + \frac{4}{6} = \frac{-3+4}{6} = \frac{1}{6}.$

Take a moment to check the proposed solution, $\frac{1}{6}$. Substitute $\frac{1}{6}$ for x in the original equation. You should obtain $-\frac{1}{2} = -\frac{1}{2}$. This true statement indicates that the solution is $\frac{1}{6}$, or the solution set is $\left\{\frac{1}{6}\right\}$. ■

✓ **CHECK POINT 3**   Solve and check: $-\dfrac{1}{2} = x - \dfrac{3}{4}.$

> ### Great Question!
>
> **Am I allowed to "flip" the two sides of an equation?**
>
> The equations $a = b$ and $b = a$ have the same meaning. If you prefer, you can solve
>
> $$-\frac{1}{2} = x - \frac{2}{3}$$
>
> by reversing the two sides and solving
>
> $$x - \frac{2}{3} = -\frac{1}{2}.$$

In Example 4, we combine like terms before using the addition property.

EXAMPLE 4   Combining Like Terms before Using the Addition Property

Solve and check: $5y + 3 - 4y - 8 = 6 + 9.$

Solution

$$5y + 3 - 4y - 8 = 6 + 9$$ This is the given equation.

$$y - 5 = 15$$ Combine like terms:
$5y - 4y = y, 3 - 8 = -5,$ and $6 + 9 = 15.$

$$y - 5 + 5 = 15 + 5$$ Add 5 to both sides.

$$y = 20$$ Simplify.

To check the proposed solution, 20, replace y with 20 in the original equation.

Check $5y + 3 - 4y - 8 = 6 + 9$ Be sure to use the original equation and not the simplified form from the second step above. (Why?)

$$5(20) + 3 - 4(20) - 8 \overset{?}{=} 6 + 9$$ Substitute 20 for y.

$$100 + 3 - 80 - 8 \overset{?}{=} 6 + 9$$ Multiply on the left.

$$103 - 88 \overset{?}{=} 6 + 9$$ Combine positive numbers and combine negative numbers on the left.

$$15 = 15$$ This statement is true.

This true statement verifies that the solution is 20, or the solution set is {20}. ■

✓ **CHECK POINT 4**   Solve and check: $8y + 7 - 7y - 10 = 6 + 4.$

Adding and Subtracting Variable Terms on Both Sides of an Equation

In some equations, variable terms appear on both sides. Here is an example:

$$4x = 7 + 3x.$$

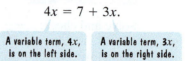

Our goal is to isolate all the variable terms on one side of the equation. We can use the addition property of equality to do this. The property allows us to add or subtract the same variable term on both sides of an equation without changing the solution. Let's see how we can use this idea to solve $4x = 7 + 3x$.

EXAMPLE 5   Using the Addition Property to Isolate Variable Terms

Solve and check: $4x = 7 + 3x$.

Solution   In the given equation, variable terms appear on both sides. We can isolate them on one side by subtracting $3x$ from both sides of the equation.

$4x = 7 + 3x$	This is the given equation.
$4x - 3x = 7 + 3x - 3x$	Subtract $3x$ from both sides and isolate variable terms on the left.
$x = 7$	Subtracting $3x$ from both sides eliminates $3x$ on the right. On the left, $4x - 3x = 1x = x$.

To check the proposed solution, 7, replace x with 7 in the original equation.

Check	$4x = 7 + 3x$	Use the original equation.
	$4(7) \overset{?}{=} 7 + 3(7)$	Substitute 7 for x.
	$28 \overset{?}{=} 7 + 21$	Multiply: $4(7) = 28$ and $3(7) = 21$.
	$28 = 28$	This statement is true.

This true statement verifies that the solution is 7, or the solution set is {7}. ∎

☑ **CHECK POINT 5**   Solve and check: $7x = 12 + 6x$.

EXAMPLE 6   Solving an Equation by Isolating the Variable

Solve and check: $3y - 9 = 2y + 6$.

Solution   Our goal is to isolate variable terms on one side and constant terms on the other side. Let's begin by isolating the variable on the left.

$3y - 9 = 2y + 6$	This is the given equation.
$3y - 2y - 9 = 2y - 2y + 6$	Isolate the variable terms on the left by subtracting $2y$ from both sides.
$y - 9 = 6$	Subtracting $2y$ from both sides eliminates $2y$ on the right. On the left, $3y - 2y = 1y = y$.

Now we isolate the constant terms on the right by adding 9 to both sides.

$y - 9 + 9 = 6 + 9$	Add 9 to both sides.
$y = 15$	Simplify.

Check $3y - 9 = 2y + 6$ Use the original equation.

$3(15) - 9 \stackrel{?}{=} 2(15) + 6$ Substitute 15 for y.

$45 - 9 \stackrel{?}{=} 30 + 6$ Multiply: $3(15) = 45$ and $2(15) = 30$.

$36 = 36$ This statement is true.

The solution is 15, or the solution set is {15}. ■

> ✓ **CHECK POINT 6**   Solve and check: $3x - 6 = 2x + 5$.

3 Solve applied problems using formulas. ▶

Applications

Our next example shows how the addition property of equality can be used to find the value of a variable in a mathematical model.

> **EXAMPLE 7**   An Application: Vocabulary and Age

There is a relationship between the number of words in a child's vocabulary, V, and the child's age, A, in months, for ages between 15 and 50 months, inclusive. This relationship can be modeled by the formula

$$V + 900 = 60A.$$

Use the formula to find the number of words in a child's vocabulary at the age of 30 months.

Solution   In the formula, A represents the child's age, in months. We are interested in a 30-month-old child. Thus, we substitute 30 for A. Then we use the addition property of equality to find V, the number of words in the child's vocabulary.

$V + 900 = 60A$ This is the given formula.

$V + 900 = 60(30)$ Substitute 30 for A.

$V + 900 = 1800$ Multiply: $60(30) = 1800$.

$V + 900 - 900 = 1800 - 900$ Subtract 900 from both sides and solve for V.

$V = 900$

At the age of 30 months, a child has a vocabulary of 900 words. ■

> ✓ **CHECK POINT 7**   Use the formula $V + 900 = 60A$ to find the number of words in a child's vocabulary at the age of 50 months.

The line graph in **Figure 2.1** allows us to "see" the formula $V + 900 = 60A$. The two points labeled with voice balloons illustrate what we learned about vocabulary and age by solving equations in Example 7 and Check Point 7.

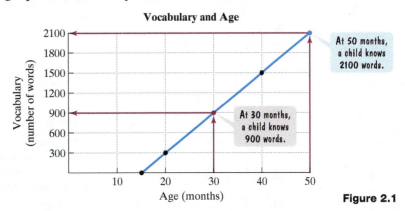

Vocabulary and Age

At 50 months, a child knows 2100 words.

At 30 months, a child knows 900 words.

Vocabulary (number of words)

Age (months)

Figure 2.1

Later in the book, you will learn to create graphs for mathematical models and formulas. Linear equations are related to models whose graphs are straight lines, like the one in **Figure 2.1**. For this reason, these equations are called *linear* equations.

Achieving Success

Algebra is cumulative. This means that the topics build on one another. Understanding each topic depends on understanding the previous material. Do not let yourself fall behind.

CONCEPT AND VOCABULARY CHECK

Fill in each blank so that the resulting statement is true.

1. The process of finding the number or numbers that make an equation a true statement is called _____ the equation.

2. An equation in the form $ax + b = c$, such as $7x + 9 = 13$, is called a/an _____ equation in one variable.

3. Equations that have the same solution are called _____ equations.

4. The addition property of equality states that if $a = b$, then $a + c =$ _____.

5. The addition property of equality lets us add or _____ the same number on both sides of an equation without changing the equation's _____.

6. The equation $x - 7 = 13$ can be solved by _____ to both sides.

7. The equation $7x = 5 + 6x$ can be solved by _____ from both sides.

2.1 EXERCISE SET ▶ MyMathLab®

Practice Exercises

In Exercises 1–10, identify the linear equations in one variable.

1. $x - 9 = 13$
2. $x - 15 = 20$
3. $x^2 - 9 = 13$
4. $x^2 - 15 = 20$
5. $\dfrac{9}{x} = 13$
6. $\dfrac{15}{x} = 20$
7. $\sqrt{2}x + \pi = 0.\overline{3}$
8. $\sqrt{3}x + \pi = 0.\overline{6}$
9. $|x + 2| = 5$
10. $|x + 5| = 8$

Solve each equation in Exercises 11–54 using the addition property of equality. Be sure to check your proposed solutions.

11. $x - 4 = 19$
12. $y - 5 = -18$
13. $z + 8 = -12$
14. $z + 13 = -15$
15. $-2 = x + 14$
16. $-13 = x + 11$
17. $-17 = y - 5$
18. $-21 = y - 4$
19. $7 + z = 11$
20. $18 + z = 14$
21. $-6 + y = -17$
22. $-8 + y = -29$
23. $x + \dfrac{1}{3} = \dfrac{7}{3}$
24. $x + \dfrac{7}{8} = \dfrac{9}{8}$
25. $t + \dfrac{5}{6} = -\dfrac{7}{12}$
26. $t + \dfrac{2}{3} = -\dfrac{7}{6}$
27. $x - \dfrac{3}{4} = \dfrac{9}{2}$
28. $x - \dfrac{3}{5} = \dfrac{7}{10}$
29. $-\dfrac{1}{5} + y = -\dfrac{3}{4}$
30. $-\dfrac{1}{8} + y = -\dfrac{1}{4}$
31. $3.2 + x = 7.5$
32. $-2.7 + w = -5.3$
33. $x + \dfrac{3}{4} = -\dfrac{9}{2}$
34. $r + \dfrac{3}{5} = -\dfrac{7}{10}$

35. $5 = -13 + y$
36. $-11 = 8 + x$
37. $-\dfrac{3}{5} = -\dfrac{3}{2} + s$
38. $\dfrac{7}{3} = -\dfrac{5}{2} + z$
39. $830 + y = 520$
40. $-90 + t = -35$
41. $r + 3.7 = 8$
42. $x + 10.6 = -9$
43. $-3.7 + m = -3.7$
44. $y + \dfrac{7}{11} = \dfrac{7}{11}$
45. $6y + 3 - 5y = 14$
46. $-3x - 5 + 4x = 9$
47. $7 - 5x + 8 + 2x + 4x - 3 = 2 + 3 \cdot 5$
48. $13 - 3r + 2 + 6r - 2r - 1 = 3 + 2 \cdot 9$
49. $7y + 4 = 6y - 9$
50. $4r - 3 = 5 + 3r$
51. $12 - 6x = 18 - 7x$
52. $20 - 7s = 26 - 8s$
53. $4x + 2 = 3(x - 6) + 8$
54. $7x + 3 = 6(x - 1) + 9$

Practice PLUS

The equations in Exercises 55–58 contain small geometric figures that represent real numbers. Use the addition property of equality to isolate x on one side of the equation and the geometric figures on the other side.

55. $x - \square = \triangle$
56. $x + \square = \triangle$
57. $2x + \triangle = 3x + \square$
58. $6x - \triangle = 7x - \square$

In Exercises 59–62, use the given information to write an equation. Let x represent the number described in each exercise. Then solve the equation and find the number.

59. If 12 is subtracted from a number, the result is −2. Find the number.

60. If 23 is subtracted from a number, the result is −8. Find the number.

61. The difference between $\frac{2}{5}$ of a number and 8 is $\frac{7}{5}$ of that number. Find the number.

62. The difference between 3 and $\frac{2}{7}$ of a number is $\frac{5}{7}$ of that number. Find the number.

Application Exercises

Formulas frequently appear in the business world. For example, the cost, C, of an item (the price paid by a retailer) plus the markup, M, on that item (the retailer's profit) equals the selling price, S, of the item. The formula is

$$C + M = S.$$

Use the formula to solve Exercises 63–64.

63. The selling price of a computer is $1850. If the markup on the computer is $150, find the cost to the retailer for the computer.

64. The selling price of a television is $650. If the cost to the retailer for the television is $520, find the markup.

Grade Inflation. *The bar graph shows the percentage of U.S. college freshmen with an average grade of A in high school.*

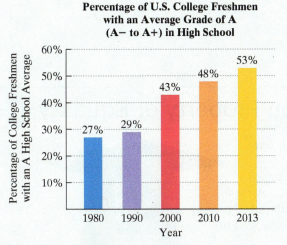

Percentage of U.S. College Freshmen with an Average Grade of A (A− to A+) in High School

Source: Higher Education Research Institute

The data displayed by the bar graph can be described by the mathematical model

$$p - 0.8x = 25,$$

where x is the number of years after 1980 and p is the percentage of U.S. college freshmen who had an average grade of A in high school. Use this information to solve Exercises 65–66.

65. a. According to the formula, in 2010, what percentage of U.S. college freshmen had an average grade of A in high school? Does this underestimate or overestimate the percent displayed by the bar graph? By how much?

b. Use the formula to project the percentage of U.S. college freshmen in 2020 who will have had an average grade of A in high school.

66. a. According to the formula, in 2000, what percentage of U.S. college freshmen had an average grade of A in high school? Does this underestimate or overestimate the percent displayed by the bar graph? By how much?

b. Use the formula to project the percentage of U.S. college freshmen in 2030 who will have had an average grade of A in high school.

Diversity Index. *The diversity index, from 0 (no diversity) to 100, measures the chance that two randomly selected people are a different race or ethnicity. The diversity index in the United States varies widely from region to region, from as high as 81 in Hawaii to as low as 11 in Vermont. The line graph shows the national diversity index for the United States for four years in the period from 1980 through 2010.*

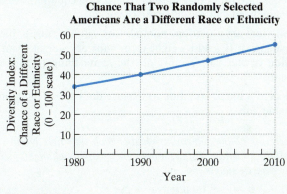

Chance That Two Randomly Selected Americans Are a Different Race or Ethnicity

Source: USA Today

The data displayed by the line graph can be described by the mathematical model

$$I - 0.7x = 34,$$

where I is the national diversity index in the United States x years after 1980. Use this information to solve Exercises 67–68.

67. a. Use the line graph to estimate the U.S. diversity index in 2010.

b. Use the formula to determine the U.S. diversity index in 2010. How does this compare with your graphical estimate from part (a)?

68. a. Use the line graph to estimate the U.S. diversity index in 2000.

b. Use the formula to determine the U.S. diversity index in 2000. How does this compare with your graphical estimate from part (a)?

Explaining the Concepts

69. State the addition property of equality and give an example.

70. Explain why $x + 2 = 9$ and $x + 2 = -6$ are not equivalent equations.

71. What is the difference between solving an equation such as

$$5y + 3 - 4y - 8 = 6 + 9$$

and simplifying an algebraic expression such as

$$5y + 3 - 4y - 8?$$

If there is a difference, which topic should be taught first? Why?

72. Look, again, at the graph for Exercises 67–68 that shows the diversity index in the United States over time. You used a *linear* equation to solve Exercise 67 or 68. What does the adjective *linear* have to do with the relationship among the four data points in the graph?

Critical Thinking Exercises

Make Sense?   *In Exercises 73–76, determine whether each statement makes sense or does not make sense, and explain your reasoning.*

73. The book is teaching me totally different things than my instructor: The book adds the number to both sides beside the equation, but my instructor adds the number underneath.

74. There are times that I prefer to check an equation's solution in my head and not show the check.

75. Solving an equation reminds me of keeping a barbell balanced: If I add weight to or subtract weight from one side of the bar, I must do the same thing to the other side.

76. I used a linear equation to explore data points lying on the same line.

In Exercises 77–80, determine whether each statement is true or false. If the statement is false, make the necessary change(s) to produce a true statement.

77. If $y - a = -b$, then $y = a + b$.

78. If $y + 7 = 0$, then $y = 7$.

79. If $2x - 5 = 3x$, then $x = -5$.

80. If $3x = 18$, then $x = 18 - 3$.

81. Write an equation with a negative solution that can be solved by adding 100 to both sides.

Technology Exercises

Use a calculator to solve each equation in Exercises 82–83.

82. $x - 7.0463 = -9.2714$

83. $6.9825 = 4.2296 + y$

Review Exercises

84. Write an algebraic expression in which x represents the number: the quotient of 9 and a number, decreased by 4 times the number. (Section 1.1, Example 3)

85. Simplify: $-16 - 8 \div 4 \cdot (-2)$. (Section 1.8, Example 4)

86. Simplify: $3[7x - 2(5x - 1)]$. (Section 1.8, Example 11)

Preview Exercises

Exercises 87–89 will help you prepare for the material covered in the next section.

87. Multiply and simplify: $5 \cdot \dfrac{x}{5}$.

88. Divide and simplify: $\dfrac{-7y}{-7}$.

89. Is 4 a solution of $3x - 14 = -2x + 6$?

SECTION

2.2

The Multiplication Property of Equality ▶

What am I supposed to learn?

After studying this section, you should be able to:

1. Use the multiplication property of equality to solve equations. ▶

2. Solve equations in the form $-x = c$. ▶

3. Use the addition and multiplication properties to solve equations. ▶

4. Solve applied problems using formulas. ▶

Is a college education worth the time and money?
Fifty members of the Forbes World's Billionaires List were asked this question. The results of the anonymous poll: 92% said yes; 8% said no. (Source: *Forbes*)

In this section (Example 8), we will work with mathematical models that show people with college degrees earn considerably more than people with only high school diplomas. Our work with these models is based on another property for solving equations, called the *multiplication property of equality*.

1 Use the multiplication property of equality to solve equations. ▶

Using the Multiplication Property of Equality to Solve Equations

Can the addition property of equality be used to solve every linear equation in one variable? No. For example, consider the equation

$$\frac{x}{5} = 9.$$

We cannot isolate the variable x by adding or subtracting 5 on both sides. To get x by itself on the left side, multiply the left side by 5:

$$5 \cdot \frac{x}{5} = \left(5 \cdot \frac{1}{5}\right)x = 1x = x.$$

> **5 is the multiplicative inverse of $\frac{1}{5}$.**

You must then multiply the right side by 5 also. By doing this, we are using the **multiplication property of equality**.

> ### The Multiplication Property of Equality
>
> The same nonzero real number (or algebraic expression) may multiply both sides of an equation without changing the solution. This can be expressed symbolically as follows:
>
> If $a = b$ and $c \neq 0$, then $ac = bc$.

EXAMPLE 1 Solving an Equation Using the Multiplication Property

Solve the equation: $\dfrac{x}{5} = 9$.

Solution We can isolate the variable, x, by multiplying both sides of the equation by 5.

$\dfrac{x}{5} = 9$	This is the given equation.
$5 \cdot \dfrac{x}{5} = 5 \cdot 9$	Multiply both sides by 5.
$1x = 45$	Simplify.
$x = 45$	$1x = x$

By substituting 45 for x in the original equation, we obtain the true statement $9 = 9$. This verifies that the solution is 45, or the solution set is {45}. ∎

☑ **CHECK POINT 1** Solve the equation: $\dfrac{x}{3} = 12$.

When we use the multiplication property of equality, we multiply both sides of an equation by the same nonzero number. We know that division is multiplication by a multiplicative inverse. Thus, the multiplication property also lets us divide both sides of an equation by a nonzero number without changing the solution.

EXAMPLE 2 Dividing Both Sides by the Same Nonzero Number

Solve: **a.** $6x = 30$ **b.** $-7y = 56$ **c.** $-18.9 = 3z$.

Solution In each equation, the variable is multiplied by a number. We can isolate the variable by dividing both sides of the equation by that number.

a. $6x = 30$ This is the given equation.

$$\frac{6x}{6} = \frac{30}{6}$$ Divide both sides by 6.

$1x = 5$ Simplify.

$x = 5$ $1x = x$

By substituting 5 for x in the original equation, we obtain the true statement $30 = 30$. The solution is 5, or the solution set is $\{5\}$.

b. $-7y = 56$ This is the given equation.

$$\frac{-7y}{-7} = \frac{56}{-7}$$ Divide both sides by -7.

$1y = -8$ Simplify.

$y = -8$ $1y = y$

By substituting -8 for y in the original equation, we obtain the true statement $56 = 56$. The solution is -8, or the solution set is $\{-8\}$.

c. $-18.9 = 3z$ This is the given equation.

$$\frac{-18.9}{3} = \frac{3z}{3}$$ Divide both sides by 3.

$-6.3 = 1z$ Simplify.

$-6.3 = z$ $1z = z$

By substituting -6.3 for z in the original equation, we obtain the true statement $-18.9 = -18.9$. The solution is -6.3, or the solution set is $\{-6.3\}$. ∎

✓ **CHECK POINT 2** Solve:

a. $4x = 84$

b. $-11y = 44$

c. $-15.5 = 5z$.

Some equations have a variable term with a fractional coefficient. Here is an example:

The coefficient of the term $\frac{3}{4}y$ is $\frac{3}{4}$. $\dfrac{3}{4}y = 12$.

To isolate the variable, multiply both sides of the equation by the multiplicative inverse of the fraction. For the equation $\frac{3}{4}y = 12$, the multiplicative inverse of $\frac{3}{4}$ is $\frac{4}{3}$. Thus, we solve $\frac{3}{4}y = 12$ by multiplying both sides by $\frac{4}{3}$.

EXAMPLE 3 Using the Multiplication Property to Eliminate a Fractional Coefficient

Solve: **a.** $\dfrac{3}{4}y = 12$ **b.** $9 = -\dfrac{3}{5}x$.

Solution

a. $\dfrac{3}{4}y = 12$ This is the given equation.

$\dfrac{4}{3}\left(\dfrac{3}{4}y\right) = \dfrac{4}{3} \cdot 12$ Multiply both sides by $\frac{4}{3}$, the multiplicative inverse of $\frac{3}{4}$.

$1y = 16$ On the left, $\frac{4}{3}\left(\frac{3}{4}y\right) = \left(\frac{4}{3} \cdot \frac{3}{4}\right)y = 1y$.

   On the right, $\frac{4}{3} \cdot \frac{12}{1} = \frac{48}{3} = 16$.

$y = 16$ $1y = y$

By substituting 16 for y in the original equation, we obtain the true statement $12 = 12$. The solution is 16, or the solution set is $\{16\}$.

b. $9 = -\dfrac{3}{5}x$ This is the given equation.

$-\dfrac{5}{3} \cdot 9 = -\dfrac{5}{3}\left(-\dfrac{3}{5}x\right)$ Multiply both sides by $-\frac{5}{3}$, the multiplicative inverse of $-\frac{3}{5}$.

$-15 = 1x$ Simplify.

$-15 = x$ $1x = x$

By substituting -15 for x in the original equation, we obtain the true statement $9 = 9$. The solution is -15, or the solution set is $\{-15\}$. ∎

☑ **CHECK POINT 3**   Solve:

a. $\dfrac{2}{3}y = 16$ **b.** $28 = -\dfrac{7}{4}x.$

---

2 Solve equations in the form $-x = c$.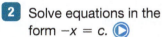

Equations and Coefficients of -1

How do we solve an equation in the form $-x = c$, such as $-x = 4$? Because the equation means $-1x = 4$, we have not yet obtained a solution. The solution of an equation is obtained from the form $x =$ some number. The equation $-x = 4$ is not yet in this form. We still need to isolate x. We can do this by multiplying or dividing both sides of the equation by -1. We will multiply by -1.

EXAMPLE 4   Solving Equations in the Form $-x = c$

Solve: **a.** $-x = 4$ **b.** $-x = -7.$

Solution   We multiply both sides of each equation by -1. This will isolate x on the left side.

a. $-x = 4$ This is the given equation.

$-1x = 4$ Rewrite $-x$ as $-1x$.

$(-1)(-1x) = (-1)(4)$ Multiply both sides by -1.

$1x = -4$ On the left, $(-1)(-1) = 1$. On the right, $(-1)(4) = -4$.

$x = -4$ $1x = x$

Check $-x = 4$ This is the original equation.

$-(-4) \overset{?}{=} 4$ Substitute -4 for x.

$4 = 4$ $-(-a) = a$, so $-(-4) = 4$.

This true statement indicates that the solution is -4, or the solution set is $\{-4\}$.

Great Question!

Is there a fast way to solve equations in the form −x = c?

If −x = c, then the equation's solution is the opposite, or additive inverse, of c. For example, the solution of −x = −7 is the opposite of −7, which is 7.

b.

$-x = -7$	This is the given equation.
$-1x = -7$	Rewrite $-x$ as $-1x$.
$(-1)(-1x) = (-1)(-7)$	Multiply both sides by -1.
$1x = 7$	$(-1)(-1) = 1$ and $(-1)(-7) = 7$.
$x = 7$	$1x = x$

By substituting 7 for x in the original equation, we obtain the true statement $-7 = -7$. The solution is 7, or the solution set is {7}. ■

✓ CHECK POINT 4 Solve: **a.** $-x = 5$ **b.** $-x = -3$.

3 Use the addition and multiplication properties to solve equations. ⊙

Equations Requiring Both the Addition and Multiplication Properties

When an equation does not contain fractions, we will often use the addition property of equality before the multiplication property of equality. Our overall goal is to isolate the variable with a coefficient of 1 on either the left or right side of the equation.

Here is the procedure that we will be using to solve the equations in the next three examples:

- Use the addition property of equality to isolate the variable term.
- Use the multiplication property of equality to isolate the variable.

EXAMPLE 5 Using Both the Addition and Multiplication Properties

Solve: $3x + 1 = 7$.

Solution We first isolate the variable term, $3x$, by subtracting 1 from both sides. Then we isolate the variable, x, by dividing both sides by 3.

- **Use the addition property of equality to isolate the variable term.**

$3x + 1 = 7$	This is the given equation.
$3x + 1 - 1 = 7 - 1$	Use the addition property, subtracting 1 from both sides.
$3x = 6$	Simplify.

- **Use the multiplication property of equality to isolate the variable.**

$\dfrac{3x}{3} = \dfrac{6}{3}$	Divide both sides of $3x = 6$ by 3.
$x = 2$	Simplify.

By substituting 2 for x in the original equation, $3x + 1 = 7$, we obtain the true statement $7 = 7$. The solution is 2, or the solution set is {2}. ■

✓ CHECK POINT 5 Solve: $4x + 3 = 27$.

EXAMPLE 6 Using Both the Addition and Multiplication Properties

Solve: $-2y - 28 = 4$.

Solution We first isolate the variable term, $-2y$, by adding 28 to both sides. Then we isolate the variable, y, by dividing both sides by -2.

- **Use the addition property of equality to isolate the variable term.**

$$-2y - 28 = 4 \qquad \text{This is the given equation.}$$

$$-2y - 28 + 28 = 4 + 28 \qquad \text{Use the addition property, adding 28 to both sides.}$$

$$-2y = 32 \qquad \text{Simplify.}$$

- **Use the multiplication property of equality to isolate the variable.**

$$\frac{-2y}{-2} = \frac{32}{-2} \qquad \text{Divide both sides by } -2.$$

$$y = -16 \qquad \text{Simplify.}$$

Take a moment to substitute -16 for y in the given equation. Do you obtain the true statement $4 = 4$? The solution is -16, or the solution set is $\{-16\}$. ■

✓ **CHECK POINT 6**   Solve: $-4y - 15 = 25$.

EXAMPLE 7   Using Both the Addition and Multiplication Properties

Solve: $3x - 14 = -2x + 6$.

Solution   We will use the addition property to collect all terms involving x on the left and all numerical terms on the right. Then we will isolate the variable, x, by dividing both sides by its coefficient.

- **Use the addition property of equality to isolate the variable term.**

$$3x - 14 = -2x + 6 \qquad \text{This is the given equation.}$$

$$3x + 2x - 14 = -2x + 2x + 6 \qquad \text{Add } 2x \text{ to both sides.}$$

$$5x - 14 = 6 \qquad \text{Simplify.}$$

$$5x - 14 + 14 = 6 + 14 \qquad \text{Add 14 to both sides.}$$

$$5x = 20 \qquad \text{Simplify. The variable term, } 5x \text{, is isolated on the left. The numerical term, 20, is isolated on the right.}$$

- **Use the multiplication property of equality to isolate the variable.**

$$\frac{5x}{5} = \frac{20}{5} \qquad \text{Divide both sides by 5.}$$

$$x = 4 \qquad \text{Simplify.}$$

Check
$$3x - 14 = -2x + 6 \qquad \text{Use the original equation.}$$
$$3(4) - 14 \stackrel{?}{=} -2(4) + 6 \qquad \text{Substitute the proposed solution, 4, for } x.$$
$$12 - 14 \stackrel{?}{=} -8 + 6 \qquad \text{Multiply.}$$
$$-2 = -2 \qquad \text{Simplify.}$$

The true statement $-2 = -2$ verifies that the solution is 4, or the solution set is $\{4\}$. ■

✓ **CHECK POINT 7**   Solve: $2x - 15 = -4x + 21$.

4 Solve applied problems using formulas.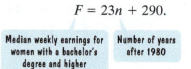

Applications

People with college degrees earn a lot more than people with only high school diplomas, and the gap is widening. The graph in **Figure 2.2** shows the median, or middlemost, weekly earnings in the United States, by education and gender, for 1980 and 2013. Data are expressed in current dollars for full-time workers 25 years old and over.

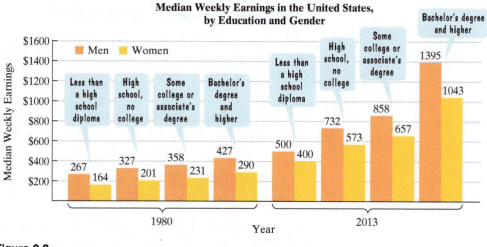

Figure 2.2

Source: Bureau of Labor Statistics

Our next example, as well as forthcoming exercises, shows how solving linear equations can be used to explore mathematical models based on some of the data displayed by the graph.

EXAMPLE 8   The Value of a College Degree

The data in **Figure 2.2** showing the median weekly earnings for women with a bachelor's degree and higher can be described by the following mathematical model:

$$F = 23n + 290.$$

| Median weekly earnings for women with a bachelor's degree and higher | Number of years after 1980 |

a. Does the formula underestimate or overestimate median weekly earnings in 2013? By how much?

b. If trends shown by the formula continue, when will median weekly earnings for women with a bachelor's degree and higher be $1210?

Solution

a. **Figure 2.2** indicates that median weekly earnings for women with a bachelor's degree and higher were $1043 in 2013. Let's see if the formula underestimates or overestimates this value. The year 2013 is 33 years after 1980 (2013 − 1980 = 33), so we substitute 33 for *n*.

$F = 23n + 290$	This is the given formula.
$F = 23(33) + 290$	Replace *n* with 33.
$F = 759 + 290$	Multiply: 23(33) = 759.
$F = 1049$	Add: 759 + 290 = 1049.

The model indicates that median weekly earnings for women with a bachelor's degree and higher were $1049 in 2013. This overestimates the number displayed by the graph, $1043, by $6.

b. We are interested in when the median weekly earnings for women with a bachelor's degree and higher will be $1210. We substitute 1210 for the median weekly earnings, F, in the given formula. Then we solve for n, the number of years after 1980.

$$F = 23n + 290$$ This is the given formula.

$$1210 = 23n + 290$$ Replace F with 1210.

Our goal is to isolate n.

$$1210 - 290 = 23n + 290 - 290$$ Isolate the term containing n by subtracting 290 from both sides.

$$920 = 23n$$ Simplify.

$$\frac{920}{23} = \frac{23n}{23}$$ Divide both sides by 23.

$$40 = n$$ Simplify: $\frac{920}{23} = 40$.

The formula indicates that 40 years after 1980, or in 2020, the median weekly earnings for women with a bachelor's degree and higher will be $1210. ■

Great Question!

In Example 8, did you use the formula $F = 23n + 290$ in two different ways?

Good observation! In Example 8(a), we substituted a value for n and used the order of operations to determine F. In Example 8(b), we substituted a value for F and solved a linear equation to determine n.

✓ **CHECK POINT 8** The data in **Figure 2.2** showing the median weekly earnings for men with a bachelor's degree and higher can be described by the mathematical model

$$M = 29n + 427,$$

where M represents median weekly earnings n years after 1980.

a. Does the formula underestimate or overestimate median weekly earnings for men with a bachelor's degree and higher in 2013? By how much?

b. If trends shown by the formula continue, when will median weekly earnings for men with a bachelor's degree and higher be $1442?

CONCEPT AND VOCABULARY CHECK

Fill in each blank so that the resulting statement is true.

1. The multiplication property of equality states that if $a = b$ and $c \neq 0$, then $ac =$ _____.

2. The multiplication property of equality lets us multiply or _____ both sides of an equation by the same nonzero number.

3. The equation $\frac{x}{7} = 13$ can be solved by _____ both sides by _____.

4. The equation $-8y = 32$ can be solved by _____ both sides by _____.

5. The equation $\frac{3}{5}x = 20$ can be solved by _____ both sides by _____.

6. The equation $-x = 12$ can be solved by _____ both sides by _____.

7. The equation $5x + 2 = 17$ can be solved by first _____ from both sides and then _____ both sides by _____.

2.2 EXERCISE SET ▶ MyMathLab®

Practice Exercises

Solve each equation in Exercises 1–28 using the multiplication property of equality. Be sure to check your proposed solutions.

1. $\dfrac{x}{6} = 5$

2. $\dfrac{x}{7} = 4$

3. $\dfrac{x}{-3} = 11$

4. $\dfrac{x}{-5} = 8$

5. $5y = 35$

6. $6y = 42$

7. $-7y = 63$

8. $-4y = 32$

9. $-28 = 8z$

10. $-36 = 8z$

11. $-18 = -3z$

12. $-54 = -9z$

13. $-8x = 6$

14. $-8x = 4$

15. $17y = 0$

16. $-16y = 0$

17. $\dfrac{2}{3}y = 12$

18. $\dfrac{3}{4}y = 15$

19. $28 = -\dfrac{7}{2}x$

20. $20 = -\dfrac{5}{8}x$

21. $-x = 17$

22. $-x = 23$

23. $-47 = -y$

24. $-51 = -y$

25. $-\dfrac{x}{5} = -9$

26. $-\dfrac{x}{5} = -1$

27. $2x - 12x = 50$

28. $8x - 3x = -45$

Solve each equation in Exercises 29–54 using both the addition and multiplication properties of equality. Check proposed solutions.

29. $2x + 1 = 11$

30. $2x + 5 = 13$

31. $2x - 3 = 9$

32. $3x - 2 = 9$

33. $-2y + 5 = 7$

34. $-3y + 4 = 13$

35. $-3y - 7 = -1$

36. $-2y - 5 = 7$

37. $12 = 4z + 3$

38. $14 = 5z - 21$

39. $-x - 3 = 3$

40. $-x - 5 = 5$

41. $6y = 2y - 12$

42. $8y = 3y - 10$

43. $3z = -2z - 15$

44. $2z = -4z + 18$

45. $-5x = -2x - 12$

46. $-7x = -3x - 8$

47. $8y + 4 = 2y - 5$

48. $5y + 6 = 3y - 6$

49. $6z - 5 = z + 5$

50. $6z - 3 = z + 2$

51. $6x + 14 = 2x - 2$

52. $9x + 2 = 6x - 4$

53. $-3y - 1 = 5 - 2y$

54. $-3y - 2 = -5 - 4y$

Practice PLUS

The equations in Exercises 55–58 contain small geometric figures that represent nonzero real numbers. Use the multiplication property of equality to isolate x on one side of the equation and the geometric figures on the other side.

55. $\dfrac{x}{\square} = \triangle$

56. $\triangle = \square x$

57. $\triangle = -x$

58. $\dfrac{-x}{\square} = \triangle$

In Exercises 59–66, use the given information to write an equation. Let x represent the number described in each exercise. Then solve the equation and find the number.

59. If a number is multiplied by 6, the result is 10. Find the number.

60. If a number is multiplied by -6, the result is 20. Find the number.

61. If a number is divided by -9, the result is 5. Find the number.

62. If a number is divided by -7, the result is 8. Find the number.

63. Eight subtracted from the product of 4 and a number is 56.

64. Ten subtracted from the product of 3 and a number is 23.

65. Negative three times a number, increased by 15, is -6.

66. Negative five times a number, increased by 11, is -29.

Application Exercises

The formula

$$M = \frac{n}{5}$$

models your distance, M, in miles, from a lightning strike in a thunderstorm if it takes n seconds to hear thunder after seeing the lightning. Use this formula to solve Exercises 67–68.

67. If you are 2 miles away from the lightning flash, how long will it take the sound of thunder to reach you?

68. If you are 3 miles away from the lightning flash, how long will it take the sound of thunder to reach you?

The Mach number is a measurement of speed, named after the man who suggested it, Ernst Mach (1838–1916). The formula

$$M = \frac{A}{740}$$

indicates that the speed of an aircraft, A, in miles per hour, divided by the speed of sound, approximately 740 miles per hour, results in the Mach number, M. Use the formula to determine the speed, in miles per hour, of the aircrafts in Exercises 69–70. (Note: When an aircraft's speed increases beyond Mach 1, it is said to have broken the sound barrier.)

69.   **70.**

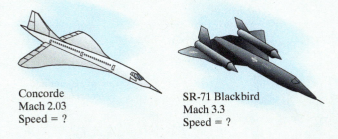

Concorde
Mach 2.03
Speed = ?

SR-71 Blackbird
Mach 3.3
Speed = ?

In Exercises 71–72, we continue using the data for median weekly earnings, by education and gender.

71.   **Median Weekly Earnings for Men with Some College or an Associate's Degree**

| 1980 | $358 |
| 2013 | $858 |

Using today's dollars, the data in the bar graph can be described by the mathematical model

$$M = 15n + 358,$$

where M represents median weekly earnings n years after 1980.

a. Does the formula underestimate or overestimate median weekly earnings in 2013? By how much?

b. If trends shown by the formula continue, when will median weekly earnings for men with some college or an associate's degree be $1033?

72.   **Median Weekly Earnings for Women with Some College or an Associate's Degree**

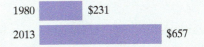

| 1980 | $231 |
| 2013 | $657 |

Using today's dollars, the data in the bar graph can be described by the mathematical model

$$W = 13n + 231,$$

where W represents median weekly earnings n years after 1980.

a. Does the formula underestimate or overestimate the median weekly earnings in 2013? By how much?

b. If trends shown by the formula continue, when will median weekly earnings for women with some college or an associate's degree be $777?

Explaining the Concepts

73. State the multiplication property of equality and give an example.

74. Explain how to solve the equation $-x = -50$.

75. Explain how to solve the equation $2x + 8 = 5x - 3$.

Critical Thinking Exercises

Make Sense? *In Exercises 76–79, determine whether each statement makes sense or does not make sense, and explain your reasoning.*

76. I used the addition and multiplication properties of equality to solve $3x = 20 + 4$.

77. I used the addition and multiplication properties of equality to solve $12 - 3x = 15$ as follows:

$$12 - 3x = 15$$
$$3x = 3$$
$$x = 1.$$

78. When I use the addition and multiplication properties to solve $2x + 5 = 17$, I undo the operations in the opposite order in which they are performed.

79. The model $F = 23n + 290$ describes the median weekly earnings, F, for women with a bachelor's degree and higher n years after 1980, so I have to solve a linear equation to determine the median weekly earnings in 2013.

In Exercises 80–83, determine whether each statement is true or false. If the statement is false, make the necessary change(s) to produce a true statement.

80. If $7x = 21$, then $x = 21 - 7$.

81. If $3x - 4 = 16$, then $3x = 12$.

82. If $3x + 7 = 0$, then $x = \frac{7}{3}$.

83. The solution of $6x = 0$ is not a natural number.

In Exercises 84–85, write an equation with the given characteristics.

84. The solution is a positive integer and the equation can be solved by dividing both sides by -60.

85. The solution is a negative integer and the equation can be solved by multiplying both sides by $\frac{4}{5}$.

Technology Exercises

Solve each equation in Exercises 86–87. Use a calculator to help with the arithmetic. Check your solution using the calculator.

86. $3.7x - 19.46 = -9.988$

87. $-72.8y - 14.6 = -455.43 - 4.98y$

Review Exercises

88. Evaluate: $(-10)^2$. (Section 1.8, Example 1)

89. Evaluate: -10^2. (Section 1.8, Example 1)

90. Evaluate $x^3 - 4x$ for $x = -1$. (Section 1.8, Example 10)

Preview Exercises

Exercises 91–93 will help you prepare for the material covered in the next section.

91. Simplify: $13 - 3(x + 2)$.

92. Is 6 a solution of $2(x - 3) - 17 = 13 - 3(x + 2)$?

93. Multiply and simplify: $10\left(\dfrac{x}{5} - \dfrac{39}{5}\right)$.

SECTION

2.3

Solving Linear Equations

What am I supposed to learn?

After studying this section, you should be able to:

1 Solve linear equations.

2 Solve linear equations containing fractions.

3 Solve linear equations containing decimals.

4 Identify equations with no solution or infinitely many solutions.

5 Solve applied problems using formulas.

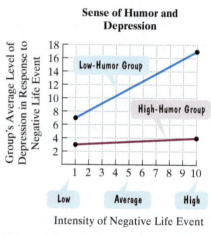

Sense of Humor and Depression

Figure 2.3

Source: Steven Davis and Joseph Palladino. *Psychology,* Fifth Edition, Prentice Hall, 2007.

The belief that humor and laughter can have positive benefits on our lives is not new. The graphs in **Figure 2.3** indicate that persons with a low sense of humor have higher levels of depression in response to negative life events than those with a high sense of humor. In this section, we will see how algebra models these relationships. To use these mathematical models, it would be helpful to have a systematic procedure for solving linear equations. We open this section with such a procedure.

1 Solve linear equations.

A Step-by-Step Procedure for Solving Linear Equations

Here is a step-by-step procedure for solving a linear equation in one variable. Not all of these steps are necessary to solve every equation.

Solving a Linear Equation

1. Simplify the algebraic expression on each side.

2. Collect all the variable terms on one side and all the constant terms on the other side.

3. Isolate the variable and solve.

4. Check the proposed solution in the original equation.

EXAMPLE 1   Solving a Linear Equation

Solve and check: $2x - 8x + 40 = 13 - 3x - 3$.

Solution
Step 1. Simplify the algebraic expression on each side.

$2x - 8x + 40 = 13 - 3x - 3$	This is the given equation.
$-6x + 40 = 10 - 3x$	Combine like terms: $2x - 8x = -6x$ and $13 - 3 = 10$.

Discover for Yourself

Solve the equation in Example 1 by collecting terms with the variable on the right and numbers on the left. What do you observe?

Step 2. Collect variable terms on one side and constant terms on the other side. The simplified equation is $-6x + 40 = 10 - 3x$. We will collect variable terms on the left by adding $3x$ to both sides. We will collect the numbers on the right by subtracting 40 from both sides.

$-6x + 40 + 3x = 10 - 3x + 3x$	Add $3x$ to both sides.
$-3x + 40 = 10$	Simplify: $-6x + 3x = -3x$.
$-3x + 40 - 40 = 10 - 40$	Subtract 40 from both sides.
$-3x = -30$	Simplify.

Step 3. Isolate the variable and solve. We isolate the variable, x, by dividing both sides by -3.

$\dfrac{-3x}{-3} = \dfrac{-30}{-3}$	Divide both sides by -3.
$x = 10$	Simplify.

Step 4. Check the proposed solution in the original equation. Substitute 10 for x in the original equation.

$2x - 8x + 40 = 13 - 3x - 3$	This is the original equation.
$2 \cdot 10 - 8 \cdot 10 + 40 \overset{?}{=} 13 - 3 \cdot 10 - 3$	Substitute 10 for x.
$20 - 80 + 40 \overset{?}{=} 13 - 30 - 3$	Perform the indicated multiplications.
$-60 + 40 \overset{?}{=} -17 - 3$	Subtract: $20 - 80 = -60$ and $13 - 30 = -17$.
$-20 = -20$	Simplify.

By substituting 10 for x in the original equation, we obtain the true statement $-20 = -20$. This verifies that the solution is 10, or the solution set is $\{10\}$. ∎

✓ **CHECK POINT 1**   Solve and check: $-7x + 25 + 3x = 16 - 2x - 3$.

EXAMPLE 2   Solving a Linear Equation

Solve and check: $5x = 8(x + 3)$.

Solution
Step 1. Simplify the algebraic expression on each side. Use the distributive property to remove parentheses on the right.

$5x = 8(x + 3)$	This is the given equation.
$5x = 8x + 24$	Use the distributive property.

Step 2. Collect variable terms on one side and constant terms on the other side. We will work with $5x = 8x + 24$ and collect variable terms on the left by subtracting $8x$ from both sides. The only constant term, 24, is already on the right.

$5x - 8x = 8x + 24 - 8x$	Subtract $8x$ from both sides.
$-3x = 24$	Simplify: $5x - 8x = -3x$.

Step 3. Isolate the variable and solve. We isolate the variable, x, by dividing both sides of $-3x = 24$ by -3.

$$\frac{-3x}{-3} = \frac{24}{-3} \qquad \text{Divide both sides by } -3.$$

$$x = -8 \qquad \text{Simplify.}$$

Step 4. Check the proposed solution in the original equation. Substitute -8 for x in the original equation.

$$5x = 8(x + 3) \qquad \text{This is the original equation.}$$
$$5(-8) \overset{?}{=} 8(-8 + 3) \qquad \text{Substitute } -8 \text{ for } x.$$
$$5(-8) \overset{?}{=} 8(-5) \qquad \text{Perform the addition in parentheses: } -8 + 3 = -5.$$
$$-40 = -40 \qquad \text{Multiply.}$$

The true statement $-40 = -40$ verifies that -8 is the solution, or the solution set is $\{-8\}$. ∎

✓ **CHECK POINT 2** Solve and check: $8x = 2(x + 6)$.

EXAMPLE 3 Solving a Linear Equation

Solve and check: $2(x - 3) - 17 = 13 - 3(x + 2)$.

Solution
Step 1. Simplify the algebraic expression on each side.

> Do not begin with $13 - 3$. Multiplication (the distributive property) is applied before subtraction.

$$2(x - 3) - 17 = 13 - 3(x + 2) \qquad \text{This is the given equation.}$$
$$2x - 6 - 17 = 13 - 3x - 6 \qquad \text{Use the distributive property.}$$
$$2x - 23 = -3x + 7 \qquad \text{Combine like terms.}$$

Step 2. Collect variable terms on one side and constant terms on the other side. We will collect variable terms on the left by adding $3x$ to both sides. We will collect the numbers on the right by adding 23 to both sides.

$$2x - 23 + 3x = -3x + 7 + 3x \qquad \text{Add } 3x \text{ to both sides.}$$
$$5x - 23 = 7 \qquad \text{Simplify: } 2x + 3x = 5x.$$
$$5x - 23 + 23 = 7 + 23 \qquad \text{Add 23 to both sides.}$$
$$5x = 30 \qquad \text{Simplify.}$$

Step 3. Isolate the variable and solve. We isolate the variable, x, by dividing both sides by 5.

$$\frac{5x}{5} = \frac{30}{5} \qquad \text{Divide both sides by 5.}$$

$$x = 6 \qquad \text{Simplify.}$$

Step 4. Check the proposed solution in the original equation. Substitute 6 for x in the original equation.

$$2(x - 3) - 17 = 13 - 3(x + 2) \qquad \text{This is the original equation.}$$
$$2(6 - 3) - 17 \overset{?}{=} 13 - 3(6 + 2) \qquad \text{Substitute 6 for } x.$$
$$2(3) - 17 \overset{?}{=} 13 - 3(8) \qquad \text{Simplify inside parentheses.}$$
$$6 - 17 \overset{?}{=} 13 - 24 \qquad \text{Multiply.}$$
$$-11 = -11 \qquad \text{Subtract.}$$

The true statement $-11 = -11$ verifies that 6 is the solution, or the solution set is $\{6\}$. ∎

> ✓ **CHECK POINT 3** Solve and check: $4(2x + 1) - 29 = 3(2x - 5)$.

---

2 Solve linear equations containing fractions. ▶

Linear Equations with Fractions

Equations are easier to solve when they do not contain fractions. How do we remove fractions from an equation? We begin by multiplying both sides of the equation by the least common denominator of any fractions in the equation. The least common denominator is the smallest number that all denominators will divide into. Multiplying every term on both sides of the equation by the least common denominator will eliminate the fractions in the equation. Example 4 shows how we "clear an equation of fractions."

EXAMPLE 4 Solving a Linear Equation Involving Fractions

Solve and check: $\dfrac{3x}{2} = \dfrac{x}{5} - \dfrac{39}{5}$.

Solution The denominators are 2, 5, and 5. The smallest number that is divisible by 2, 5, and 5 is 10. We begin by multiplying both sides of the equation by 10, the least common denominator.

$$\frac{3x}{2} = \frac{x}{5} - \frac{39}{5}$$
This is the given equation.

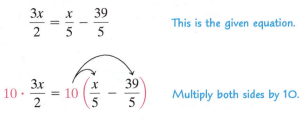

$$10 \cdot \frac{3x}{2} = 10\left(\frac{x}{5} - \frac{39}{5}\right)$$
Multiply both sides by 10.

$$10 \cdot \frac{3x}{2} = 10 \cdot \frac{x}{5} - 10 \cdot \frac{39}{5}$$
Use the distributive property. Be sure to multiply all terms by 10.

$$\overset{5}{\cancel{10}} \cdot \frac{3x}{\underset{1}{2}} = \overset{2}{\cancel{10}} \cdot \frac{x}{\underset{1}{5}} - \overset{2}{\cancel{10}} \cdot \frac{39}{\underset{1}{5}}$$
Divide out common factors in the multiplications.

$$15x = 2x - 78$$
Complete the multiplications. The fractions are now cleared.

At this point, we have an equation similar to those we previously have solved. Collect the variable terms on one side and the constant terms on the other side.

$$15x - 2x = 2x - 2x - 78$$
Subtract 2x to get the variable terms on the left.

$$13x = -78$$
Simplify.

Isolate x by dividing both sides by 13.

$$\frac{13x}{13} = \frac{-78}{13}$$
Divide both sides by 13.

$$x = -6$$
Simplify.

Check $\dfrac{3x}{2} = \dfrac{x}{5} - \dfrac{39}{5}$ This is the original equation.

$\dfrac{3(-6)}{2} \overset{?}{=} \dfrac{-6}{5} - \dfrac{39}{5}$ Substitute -6 for x.

$-9 \overset{?}{=} \dfrac{-6}{5} - \dfrac{39}{5}$ Simplify the left side: $\dfrac{3(-6)}{2} = \dfrac{-18}{2} = -9$.

$-9 \overset{?}{=} \dfrac{-45}{5}$ Subtract on the right side: $\dfrac{-6}{5} - \dfrac{39}{5} = \dfrac{-6}{5} + \left(\dfrac{-39}{5}\right) = \dfrac{-45}{5}$.

$-9 = -9$ Simplify: $\dfrac{-45}{5} = -9$.

The true statement $-9 = -9$ verifies that -6 is the solution, or the solution set is $\{-6\}$. ∎

✓ **CHECK POINT 4**   Solve and check: $\dfrac{x}{4} = \dfrac{2x}{3} + \dfrac{5}{6}$.

---

3 Solve linear equations containing decimals. ▶

Linear Equations with Decimals

It is not a requirement to clear decimals in an equation. However, if you are not using a calculator, eliminating decimal numbers can make calculations easier.

Multiplying a decimal number by 10^n, where n is a positive integer, has the effect of moving the decimal point n places to the right. For example,

$$0.3 \times 10 = 3 \qquad 0.37 \times 10^2 = 37 \qquad 0.408 \times 10^3 = 408.$$

One decimal place   Two decimal places   Three decimal places

These numerical examples suggest a procedure for clearing decimals in an equation.

Clearing an Equation of Decimals

Multiply every term on both sides of the equation by a power of 10. The exponent on 10 will equal the greatest number of decimal places in the equation.

EXAMPLE 5   Solving a Linear Equation Involving Decimals

Solve and check: $0.3(x - 6) = 0.37x - 1.1$.

Solution   We will first apply the distributive property to remove parentheses on the left. Then we will clear the equation of decimals.

$0.3(x - 6) = 0.37x - 1.1$ This is the given equation.

$0.3x - 0.3(6) = 0.37x - 1.1$ Use the distributive property.

$0.3x - 1.8 = 0.37x - 1.1$ Multiply: $0.3(6) = 1.8$.

The number 0.37 has two decimal places, the greatest number of decimal places in the equation. Multiply both sides by 10^2, or 100, to clear the decimals.

$100(0.3x - 1.8) = 100(0.37x - 1.1)$ Multiply both sides by 100.

$100(0.3x) - 100(1.8) = 100(0.37x) - 100(1.1)$ Use the distributive property.

$30x - 180 = 37x - 110$ Simplify by moving decimal points two places to the right.

At this point, we have an equation similar to those we previously solved. We will solve $30x - 180 = 37x - 110$ by collecting the variable terms on one side and the constant terms on the other side.

$30x - 37x - 180 = 37x - 37x - 110$	Subtract $37x$ to get the variable terms on the left.
$-7x - 180 = -110$	Simplify.
$-7x - 180 + 180 = -110 + 180$	Add 180 to get the constant terms on the right.
$-7x = 70$	Simplify.
$\dfrac{-7x}{-7} = \dfrac{70}{-7}$	Isolate x by dividing both sides by -7.
$x = -10$	Simplify.

Check	$0.3(x - 6) = 0.37x - 1.1$	This is the original equation.
	$0.3(-10 - 6) = 0.37(-10) - 1.1$	Substitute -10 for x.
	$0.3(-16) = 0.37(-10) - 1.1$	Simplify the left side: $-10 - 6 = -16$.
	$-4.8 = -3.7 - 1.1$	Multiply: $0.3(-16) = -4.8$ and $0.37(-10) = -3.7$.
	$-4.8 = -4.8$	Subtract on the right side: $-3.7 - 1.1 = -3.7 + (-1.1) = -4.8$.

The true statement $-4.8 = -4.8$ verifies that -10 is the solution, or the solution set is $\{-10\}$. ■

Great Question!

In the solution to Example 5, you simplified using the distributive property. Then you multiplied by 100 and cleared the equation of decimals. Can I clear the equation of decimals before using the distributive property?

Yes. Decimals can be cleared at any step in the process of solving an equation. Here's how it works for the equation in Example 5:

$0.3(x - 6) = 0.37x - 1.1$	This is the given equation.
$100[0.3(x - 6)] = 100(0.37x - 1.1)$	Multiply both sides by 100.

> $0.3(x - 6)$ is one term with two factors, 0.3 and $x - 6$. To multiply the term by 100, we need only multiply one factor, in this case 0.3, by 100.

$[100(0.3)](x - 6) = 100(0.37x) - 100(1.1)$	Use the associative property on the left and the distributive property on the right.
$30(x - 6) = 37x - 110$	Simplify by moving decimal points two places to the right.
$30x - 180 = 37x - 110.$	Use the distributive property.

This last equation, $30x - 180 = 37x - 110$, is the same equation cleared of decimals that we obtained in the solution to Example 5.

☑ **CHECK POINT 5**   Solve and check: $0.48x + 3 = 0.2(x - 6)$.

---

4 Identify equations with no solution or infinitely many solutions. ▶

Equations with No Solution or Infinitely Many Solutions

Thus far, each equation that we have solved has had a single solution. However, some equations are not true for even one real number. Such an equation is called an **inconsistent equation**. Here is an example of such an equation:

$$x = x + 4.$$

There is no number that is equal to itself plus 4. This equation has no solution.

You can express the fact that an equation has no solution using words or set notation.

- Use the phrase "no solution."
- Use set notation: $\varnothing$.

> This symbol stands for the empty set, a set with no elements.

An equation that is true for all real numbers is called an **identity**. An example of an identity is

$$x + 3 = x + 2 + 1.$$

Every number plus 3 is equal to that number plus 2 plus 1. Every real number is a solution to this equation.

You can express the fact that every real number is a solution of an equation using words or set notation.

- Use the phrase "all real numbers."
- Use set notation: $\{x \mid x \text{ is a real number}\}$.

> The set of / all x / such that / x is a real number

Recognizing Inconsistent Equations and Identities

If you attempt to solve an equation with no solution or one that is true for every real number, you will eliminate the variable.

- An inconsistent equation with no solution results in a false statement, such as $2 = 5$.

- An identity that is true for every real number results in a true statement, such as $4 = 4$.

EXAMPLE 6 Solving an Equation

Solve: $2x + 6 = 2(x + 4)$.

Solution

$$
\begin{array}{ll}
2x + 6 = 2(x + 4) & \text{This is the given equation.} \\
2x + 6 = 2x + 8 & \text{Use the distributive property.} \\
2x + 6 - 2x = 2x + 8 - 2x & \text{Subtract } 2x \text{ from both sides.} \\
6 = 8 & \text{Simplify.}
\end{array}
$$

> Keep reading. 6 = 8 is not the solution.

The original equation is equivalent to the statement $6 = 8$, which is false for every value of x. The equation is inconsistent and has no solution. You can express this by writing "no solution" or using the symbol for the empty set, $\varnothing$. ∎

✓ **CHECK POINT 6** Solve: $3x + 7 = 3(x + 1)$.

EXAMPLE 7 Solving an Equation

Solve: $-3x + 5 + 5x = 4x - 2x + 5$.

Solution

$$-3x + 5 + 5x = 4x - 2x + 5$$ This is the given equation.

$$2x + 5 = 2x + 5$$ Combine like terms: $-3x + 5x = 2x$ and $4x - 2x = 2x$.

$$2x + 5 - 2x = 2x + 5 - 2x$$ Subtract $2x$ from both sides.

Keep reading. 5 = 5 is not the solution. $$5 = 5$$ Simplify.

The original equation is equivalent to the statement $5 = 5$, which is true for every value of x. The equation is an identity and all real numbers are solutions. You can express this by writing "all real numbers" or using set notation: $\{x \mid x \text{ is a real number}\}$. ∎

✓ **CHECK POINT 7** Solve: $3(x - 1) + 9 = 8x + 6 - 5x$.

---

5 Solve applied problems using formulas.

Applications

The next example shows how our procedure for solving equations with fractions can be used to find the value of a variable in a mathematical model.

EXAMPLE 8 An Application: Responding to Negative Life Events

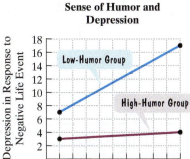

Sense of Humor and Depression

Low-Humor Group

High-Humor Group

Group's Average Level of Depression in Response to Negative Life Event

Low   Average   High

Intensity of Negative Life Event

Figure 2.3 (repeated)

In the section opener, we introduced line graphs, repeated in **Figure 2.3**, indicating that persons with a low sense of humor have higher levels of depression in response to negative life events than those with a high sense of humor. These graphs can be modeled by the following formulas:

Low-Humor Group   High-Humor Group

$$D = \frac{10}{9}x + \frac{53}{9} \qquad\qquad D = \frac{1}{9}x + \frac{26}{9}.$$

In each formula, x represents the intensity of a negative life event (from 1, low, to 10, high) and D is the level of depression in response to that event. If the high-humor group averages a level of depression of 3.5, or $\frac{7}{2}$, in response to a negative life event, what is the intensity of that event? How is the solution shown on the red line graph in **Figure 2.3**?

Solution We are interested in the intensity of a negative life event with an average level of depression of $\frac{7}{2}$ for the high-humor group. We substitute $\frac{7}{2}$ for D in the high-humor model and solve for x, the intensity of the negative life event.

$$D = \frac{1}{9}x + \frac{26}{9}$$ This is the given formula for the high-humor group.

$$\frac{7}{2} = \frac{1}{9}x + \frac{26}{9}$$ Replace D with $\frac{7}{2}$.

$$18 \cdot \frac{7}{2} = 18\left(\frac{1}{9}x + \frac{26}{9}\right)$$ Multiply both sides by 18, the least common denominator.

$$18 \cdot \frac{7}{2} = 18 \cdot \frac{1}{9}x + 18 \cdot \frac{26}{9}$$ Use the distributive property.

$$\overset{9}{18} \cdot \frac{7}{\underset{1}{2}} = \overset{2}{18} \cdot \frac{1}{\underset{1}{9}}x + \overset{2}{18} \cdot \frac{26}{\underset{1}{9}}$$ Divide out common factors in the multiplications.

$$63 = 2x + 52$$

Complete the multiplications. The fractions are now cleared.

$$63 - 52 = 2x + 52 - 52$$

Subtract 52 from both sides to get constants on the left.

$$11 = 2x$$

Simplify.

$$\frac{11}{2} = \frac{2x}{2}$$

Divide both sides by 2.

$$\frac{11}{2} = x$$

Simplify.

The formula indicates that if the high-humor group averages a level of depression of 3.5 in response to a negative life event, the intensity of that event is $\frac{11}{2}$, or 5.5. This is illustrated on the line graph for the high-humor group in **Figure 2.4**.

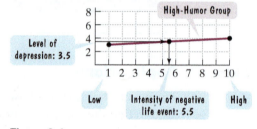

Figure 2.4

☑ **CHECK POINT 8**   Use the model for the low-humor group given in Example 8 to solve this problem. If the low-humor group averages a level of depression of 10 in response to a negative life event, what is the intensity of that event? How is the solution shown on the blue line graph in **Figure 2.3**?

Achieving Success

FoxTrot

Foxtrot copyright © 2003, 2009 by Bill Amend/Distributed by Universal Uclick

Because concepts in mathematics build on each other, **it is extremely important that you complete all homework assignments**. This requires more than attempting a few of the assigned exercises. When it comes to assigned homework, you need to do four things and to do these things consistently throughout any math course:

1. Attempt to work *every* assigned problem.
2. Check your answers.
3. Correct your errors.
4. Ask for help with the problems you have attempted but do not understand.

Having said this, **don't panic at the length of the Exercise Sets**. You are not expected to work all, or even most, of the problems. Your professor will provide guidance on which exercises to work by assigning those problems that are consistent with the goals and objectives of your course.

CONCEPT AND VOCABULARY CHECK

Fill in each blank so that the resulting statement is true.

1. The first step in solving $3x - 9x + 30 = 15 - 2x - 4$ is to _____.

2. The equation $\frac{x}{5} - \frac{1}{2} = \frac{x}{6}$ can be cleared of fractions by multiplying both sides by _____.

3. The equation $0.9x - 4.3 = 0.47$ can be cleared of decimals by multiplying both sides by _____.

4. A linear equation that is not true for even one real number, and therefore has no solution, is called a/an _____ equation.

5. A linear equation that is true for all real numbers is called a/an _____.

6. In solving an equation, if you eliminate the variable and obtain a false statement such as $2 = 5$, the equation is a/an _____ equation.

7. In solving an equation, if you eliminate the variable and obtain a true statement such as $5 = 5$, the equation is a/an _____.

2.3 EXERCISE SET ▶ MyMathLab®

Practice Exercises

In Exercises 1–30, solve each equation. Be sure to check your proposed solution by substituting it for the variable in the original equation.

1. $5x + 3x - 4x = 10 + 2$

2. $4x + 8x - 2x = 20 - 15$

3. $4x - 9x + 22 = 3x + 30$

4. $3x + 2x + 64 = 40 - 7x$

5. $3x + 6 - x = 8 + 3x - 6$

6. $3x + 2 - x = 6 + 3x - 8$

7. $4(x + 1) = 20$

8. $3(x - 2) = -6$

9. $7(2x - 1) = 42$

10. $4(2x - 3) = 32$

11. $38 = 30 - 2(x - 1)$

12. $20 = 44 - 8(2 - x)$

13. $2(4z + 3) - 8 = 46$

14. $3(3z + 5) - 7 = 89$

15. $6x - (3x + 10) = 14$

16. $5x - (2x + 14) = 10$

17. $5(2x + 1) = 12x - 3$

18. $3(x + 2) = x + 30$

19. $3(5 - x) = 4(2x + 1)$

20. $3(3x - 1) = 4(3 + 3x)$

21. $8(y + 2) = 2(3y + 4)$

22. $8(y + 3) = 3(2y + 12)$

23. $3(x + 1) = 7(x - 2) - 3$

24. $5x - 4(x + 9) = 2x - 3$

25. $5(2x - 8) - 2 = 5(x - 3) + 3$

26. $7(3x - 2) + 5 = 6(2x - 1) + 24$

27. $6 = -4(1 - x) + 3(x + 1)$

28. $100 = -(x - 1) + 4(x - 6)$

29. $10(z + 4) - 4(z - 2) = 3(z - 1) + 2(z - 3)$

30. $-2(z - 4) - (3z - 2) = -2 - (6z - 2)$

Solve each equation and check your proposed solution in Exercises 31–46. Begin your work by rewriting each equation without fractions.

31. $\dfrac{x}{5} - 4 = -6$

32. $\dfrac{x}{2} + 13 = -22$

33. $\dfrac{2x}{3} - 5 = 7$

34. $\dfrac{3x}{4} - 9 = -6$

35. $\dfrac{2y}{3} - \dfrac{3}{4} = \dfrac{5}{12}$

36. $\dfrac{3y}{4} - \dfrac{2}{3} = \dfrac{7}{12}$

37. $\dfrac{x}{3} + \dfrac{x}{2} = \dfrac{5}{6}$

38. $\dfrac{x}{4} - \dfrac{x}{5} = 1$

39. $20 - \dfrac{z}{3} = \dfrac{z}{2}$

40. $\dfrac{z}{5} - \dfrac{1}{2} = \dfrac{z}{6}$

41. $\dfrac{y}{3} + \dfrac{2}{5} = \dfrac{y}{5} - \dfrac{2}{5}$

42. $\dfrac{y}{12} + \dfrac{1}{6} = \dfrac{y}{2} - \dfrac{1}{4}$

43. $\dfrac{3x}{4} - 3 = \dfrac{x}{2} + 2$

44. $\dfrac{3x}{5} - \dfrac{2}{5} = \dfrac{x}{3} + \dfrac{2}{5}$

45. $\dfrac{x - 3}{5} - 1 = \dfrac{x - 5}{4}$

46. $\dfrac{x - 2}{3} - 4 = \dfrac{x + 1}{4}$

Solve each equation and check your proposed solution in Exercises 47–58.

47. $3.6x = 2.9x + 6.3$

48. $1.2x - 3.6 = 2.4 - 0.3x$

49. $0.92y + 2 = y - 0.4$

50. $0.15y - 0.1 = 2.5y - 1.04$

51. $0.3x - 4 = 0.1(x + 10)$

52. $0.1(x + 80) = 14 - 0.2x$

53. $0.4(2z + 6) + 0.1 = 0.5(2z - 3)$

54. $1.4(z - 5) - 0.2 = 0.5(6z - 8)$

55. $0.01(x + 4) - 0.04 = 0.01(5x + 4)$

56. $0.02(x - 2) = 0.06 - 0.01(x + 1)$

57. $0.6(x + 300) = 0.65x - 205$

58. $0.05(7x + 36) = 0.4x + 1.2$

In Exercises 59–78, solve each equation. Use words or set notation to identify equations that have no solution, or equations that are true for all real numbers.

59. $3x - 7 = 3(x + 1)$

60. $2(x - 5) = 2x + 10$

61. $2(x + 4) = 4x + 5 - 2x + 3$

62. $3(x - 1) = 8x + 6 - 5x - 9$

63. $7 + 2(3x - 5) = 8 - 3(2x + 1)$

64. $2 + 3(2x - 7) = 9 - 4(3x + 1)$

65. $4x + 1 - 5x = 5 - (x + 4)$

66. $5x - 5 = 3x - 7 + 2(x + 1)$

67. $4(x + 2) + 1 = 7x - 3(x - 2)$

68. $5x - 3(x + 1) = 2(x + 3) - 5$

69. $3 - x = 2x + 3$

70. $5 - x = 4x + 5$

71. $\dfrac{x}{3} + 2 = \dfrac{x}{3}$

72. $\dfrac{x}{4} + 3 = \dfrac{x}{4}$

73. $\dfrac{x}{2} - \dfrac{x}{4} + 4 = x + 4$

74. $\dfrac{x}{2} + \dfrac{2x}{3} + 3 = x + 3$

75. $\dfrac{2}{3}x = 2 - \dfrac{5}{6}x$

76. $\dfrac{2}{3}x = \dfrac{1}{4}x - 8$

77. $0.06(x + 5) = 0.03(2x + 7) + 0.09$

78. $0.04(x - 2) = 0.02(6x - 3) - 0.02$

Practice PLUS

The equations in Exercises 79–80 contain small figures ($\square$, $\triangle$, and $\$$) that represent nonzero real numbers. Use properties of equality to isolate x on one side of the equation and the small figures on the other side.

79. $\dfrac{x}{\square} + \triangle = \$$

80. $\dfrac{x}{\square} - \triangle = -\$$

81. If $\dfrac{x}{5} - 2 = \dfrac{x}{3}$, evaluate $x^2 - x$.

82. If $\dfrac{3x}{2} + \dfrac{3x}{4} = \dfrac{x}{4} - 4$, evaluate $x^2 - x$.

In Exercises 83–86, use the given information to write an equation. Let x represent the number described in each exercise. Then solve the equation and find the number.

83. When one-third of a number is added to one-fifth of the number, the sum is 16. What is the number?

84. When two-fifths of a number is added to one-fourth of the number, the sum is 13. What is the number?

85. When 3 is subtracted from three-fourths of a number, the result is equal to one-half of the number. What is the number?

86. When 30 is subtracted from seven-eighths of a number, the result is equal to one-half of the number. What is the number?

Application Exercises

In Massachusetts, speeding fines are determined by the formula

$$F = 10(x - 65) + 50,$$

where F is the cost, in dollars, of the fine if a person is caught driving x miles per hour. Use this formula to solve Exercises 87–88.

87. If a fine comes to $250, how fast was that person driving?

88. If a fine comes to $400, how fast was that person driving?

The latest guidelines, which apply to both men and women, give healthy weight ranges, rather than specific weights, for your height. The further you are above the upper limit of your range, the greater are the risks of developing weight-related health problems. The bar graph shows these ranges for various heights for people between the ages of 19 and 34, inclusive.

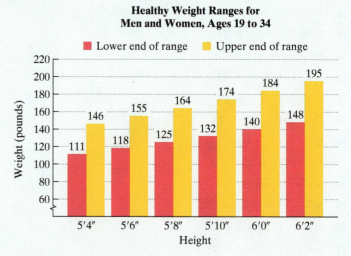

Healthy Weight Ranges for Men and Women, Ages 19 to 34

Source: U.S. Department of Health and Human Services

The mathematical model

$$\frac{W}{2} - 3H = 53$$

describes a weight, W, in pounds, that lies within the healthy weight range for a person whose height is H inches over 5 feet. Use this information to solve Exercises 89–90.

89. Use the formula to find a healthy weight for a person whose height is 5'6". (*Hint: H* = 6 because this person's height is 6 inches over 5 feet.) How many pounds is this healthy weight below the upper end of the range shown by the bar graph?

90. Use the formula to find a healthy weight for a person whose height is 6'0". (*Hint: H* = 12 because this person's height is 12 inches over 5 feet.) How many pounds is this healthy weight below the upper end of the range shown by the bar graph?

The formula

$$p = 15 + \frac{5d}{11}$$

describes the pressure of sea water, p, in pounds per square foot, at a depth of d feet below the surface. Use the formula to solve Exercises 91–92.

91. The record depth for breath-held diving, by Francisco Ferreras (Cuba) off Grand Bahama Island, on November 14, 1993, involved pressure of 201 pounds per square foot. To what depth did Ferreras descend on this ill-advised venture? (He was underwater for 2 minutes and 9 seconds!)

92. At what depth is the pressure 20 pounds per square foot?

Explaining the Concepts

93. In your own words, describe how to solve a linear equation.

94. Explain how to solve a linear equation containing fractions.

95. Suppose that you solve $\frac{x}{5} - \frac{x}{2} = 1$ by multiplying both sides by 20, rather than the least common denominator, 10. Describe what happens. If you get the correct solution, why do you think we clear the equation of fractions by multiplying by the *least* common denominator?

96. Explain how to clear decimals in a linear equation.

97. Suppose you are an algebra teacher grading the following solution on an examination:

Solve: $-3(x - 6) = 2 - x$

Solution: $-3x - 18 = 2 - x$

$-2x - 18 = 2$

$-2x = -16$

$x = 8.$

You should note that 8 checks, and the solution is 8. The student who worked the problem therefore wants full credit. Can you find any errors in the solution? If full credit is 10 points, how many points should you give the student? Justify your position.

Critical Thinking Exercises

Make Sense? *In Exercises 98–101, determine whether each statement makes sense or does not make sense, and explain your reasoning.*

98. Although I can solve $3x + \frac{1}{5} = \frac{1}{4}$ by first subtracting $\frac{1}{5}$ from both sides, I find it easier to begin by multiplying both sides by 20, the least common denominator.

99. I can use any common denominator to clear an equation of fractions, but using the least common denominator makes the arithmetic easier.

100. When I substituted 5 for *x* in the equation

$$4x + 6 = 6(x + 1) - 2x$$

I obtained a true statement, so the equation's solution is 5.

101. I cleared the equation $0.5x + 8.3 = 12.4$ of decimals by multiplying both sides by 100.

In Exercises 102–105, determine whether each statement is true or false. If the statement is false, make the necessary change(s) to produce a true statement.

102. The equation $3(x + 4) = 3(4 + x)$ has precisely one solution.

103. The equation $2y + 5 = 0$ is equivalent to $2y = 5$.

104. If $2 - 3y = 11$ and the solution to the equation is substituted into $y^2 + 2y - 3$, a number results that is neither positive nor negative.

105. The equation $x + \frac{1}{3} = \frac{1}{2}$ is equivalent to $x + 2 = 3$.

106. A woman's height, h, is related to the length of her femur, f (the bone from the knee to the hip socket), by the formula $f = 0.432h - 10.44$. Both h and f are measured in inches. A partial skeleton is found of a woman in which the femur is 16 inches long. Police find the skeleton in an area where a woman slightly over 5 feet tall has been missing for over a year. Can the partial skeleton be that of the missing woman? Explain.

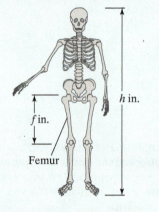

h in.

f in.

Femur

Solve each equation in Exercises 107–108.

107. $\dfrac{2x - 3}{9} + \dfrac{x - 3}{2} = \dfrac{x + 5}{6} - 1$

108. $2(3x + 4) = 3x + 2[3(x - 1) + 2]$

Review Exercises

In Exercises 109–110, insert either $<$ or $>$ in the shaded area between each pair of numbers to make a true statement.

109. -24 ▮ -20 (Section 1.3, Example 6)

110. $-\dfrac{1}{3}$ ▮ $-\dfrac{1}{5}$ (Section 1.3, Example 6)

111. Simplify: $-9 - 11 + 7 - (-3)$. (Section 1.6, Example 3)

Preview Exercises

Exercises 112–114 will help you prepare for the material covered in the next section.

112. Consider the formula

$$T = D + pm.$$

a. Subtract D from both sides and write the resulting formula.

b. Divide both sides of your formula from part (a) by p and write the resulting formula.

113. Solve: $4 = 0.25B$.

114. Solve: $1.3 = P \cdot 26$.

SECTION

2.4

Formulas and Percents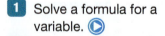

What am I supposed to learn?

After studying this section, you should be able to:

1 Solve a formula for a variable. ▶

2 Use the percent formula. ▶

3 Solve applied problems involving percent change. ▶

"And, if elected, it is my solemn pledge to cut your taxes by 10% for each of my first three years in office, for a total cut of 30%."

Did you know that one of the most common ways that you are given numerical information is with percents? In this section, you will learn to use a formula that will help you to understand percents from an algebraic perspective, enabling you to make sense of the politician's promise.

1 Solve a formula for a variable.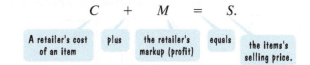

Solving a Formula for One of Its Variables

We know that solving an equation is the process of finding the number (or numbers) that make the equation a true statement. All of the equations we have solved contained only one letter, such as x or y.

By contrast, the formulas we have seen contain two or more letters, representing two or more variables. Here is an example:

$$C \quad + \quad M \quad = \quad S.$$

| A retailer's cost of an item | plus | the retailer's markup (profit) | equals | the items's selling price. |

We say that this formula is solved for the variable S because S is alone on one side of the equation and the other side does not contain an S.

Solving a formula for a variable means rewriting the formula so that the variable is isolated on one side of the equation. It does not mean obtaining a numerical value for that variable.

The addition and multiplication properties of equality are used to solve a formula for one of its variables. Consider the retailer's formula, $C + M = S$. How do we solve this formula for C? Use the addition property to isolate C by subtracting M from both sides:

We need to isolate C.

$C + M = S$	This is the given formula.
$C + M - M = S - M$	Subtract M from both sides.
$C = S - M$	Simplify.

Solved for C, the formula $C = S - M$ tells us that the cost of an item for a retailer is the item's selling price minus its markup.

To solve a formula for one of its variables, treat that variable as if it were the only variable in the equation. Think of the other variables as if they were numbers. Use the addition property of equality to isolate all terms with the specified variable on one side of the equation and all terms without the specified variable on the other side. Then use the multiplication property of equality to get the specified variable alone.

Our first example involves the formula for the area of a rectangle. The **area of a two-dimensional figure** is the number of square units it takes to fill the interior of the figure. A **square unit** is a square, each of whose sides is one unit in length, as illustrated in **Figure 2.5**. The figure shows that there are 12 square units contained within the rectangle. The area of the rectangle is 12 square units. Notice that the area can be determined in the following manner:

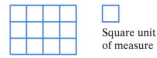

Square unit of measure

Figure 2.5

| Across | Down |

$$4 \text{ units} \cdot 3 \text{ units} = 4 \cdot 3 \text{ units} \cdot \text{units} = 12 \text{ square units.}$$

The area of a rectangle is the product of the distance across, its length, and the distance down, its width.

Area of a Rectangle

The area, A, of a rectangle with length l and width w is given by the formula

$$A = lw.$$

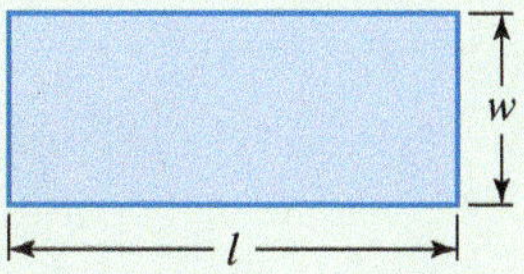

EXAMPLE 1 Solving a Formula for a Variable

Solve the formula $A = lw$ for w.

Solution Our goal is to get w by itself on one side of the formula. There is only one term with w, lw, and it is already isolated on the right side. We isolate w on the right by using the multiplication property of equality and dividing both sides by l.

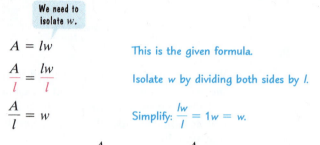

We need to isolate w.

$A = lw$	This is the given formula.
$\dfrac{A}{l} = \dfrac{lw}{l}$	Isolate w by dividing both sides by l.
$\dfrac{A}{l} = w$	Simplify: $\dfrac{lw}{l} = 1w = w.$

The formula solved for w is $\dfrac{A}{l} = w$ or $w = \dfrac{A}{l}$. Thus, the area of a rectangle divided by its length is equal to its width. ■

✓ **CHECK POINT 1** Solve the formula $A = lw$ for l.

The **perimeter of a two-dimensional figure** is the sum of the lengths of its sides. Perimeter is measured in linear units, such as inches, feet, yards, meters, or kilometers.

Example 2 involves the perimeter, P, of a rectangle. Because perimeter is the sum of the lengths of the sides, the perimeter of the rectangle shown in **Figure 2.6** is $l + w + l + w$. This can be expressed as

$$P = 2l + 2w.$$

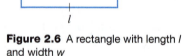

Figure 2.6 A rectangle with length l and width w

Perimeter of a Rectangle

The perimeter, P, of a rectangle with length l and width w is given by the formula

$$P = 2l + 2w.$$

The perimeter of a rectangle is the sum of twice the length and twice the width.

EXAMPLE 2 Solving a Formula for a Variable

Solve the formula $2l + 2w = P$ for w.

Solution First, isolate $2w$ on the left by subtracting $2l$ from both sides. Then solve for w by dividing both sides by 2.

We need to isolate w.

$2l + 2w = P$	This is the given formula.
$2l - 2l + 2w = P - 2l$	Isolate $2w$ by subtracting $2l$ from both sides.
$2w = P - 2l$	Simplify.
$\dfrac{2w}{2} = \dfrac{P - 2l}{2}$	Isolate w by dividing both sides by 2.
$w = \dfrac{P - 2l}{2}$	Simplify. ■

✓ **CHECK POINT 2** Solve the formula $2l + 2w = P$ for l.

EXAMPLE 3 Solving a Formula for a Variable

The total price of an article purchased on a monthly deferred payment plan is described by the following formula:

$$T = D + pm.$$

In this formula, T is the total price, D is the down payment, p is the amount of the monthly payment, and m is the number of payments. Solve the formula for p.

Solution First, isolate pm on the right by subtracting D from both sides. Then isolate p from pm by dividing both sides of the formula by m.

> We need to isolate p.

$T = D + pm$	This is the given formula. We want p alone.
$T - D = D - D + pm$	Isolate pm by subtracting D from both sides.
$T - D = pm$	Simplify.
$\dfrac{T - D}{m} = \dfrac{pm}{m}$	Now isolate p by dividing both sides by m.
$\dfrac{T - D}{m} = p$	Simplify: $\dfrac{pm}{m} = \dfrac{p\overset{1}{\cancel{m}}}{\underset{1}{\cancel{m}}} = p \cdot 1 = p.$ ■

✓ **CHECK POINT 3** Solve the formula $T = D + pm$ for m.

The next example has a formula that contains a fraction. To solve for a variable in a formula involving fractions, we begin by multiplying both sides by the least common denominator of all fractions in the formula. This will eliminate the fractions. Then we solve for the specified variable.

EXAMPLE 4 Solving a Formula Containing a Fraction for a Variable

Solve the formula $\dfrac{W}{2} - 3H = 53$ for W.

Solution Do you remember seeing this formula in the last section's Exercise Set? It models a person's healthy weight, W, where H represents that person's height, in inches, in excess of 5 feet. We begin by multiplying both sides of the formula by 2 to eliminate the fraction. Then we isolate the variable W.

$\dfrac{W}{2} - 3H = 53$	This is the given formula.
$2\left(\dfrac{W}{2} - 3H\right) = 2 \cdot 53$	Multiply both sides by 2.
$2 \cdot \dfrac{W}{2} - 2 \cdot 3H = 2 \cdot 53$	Use the distributive property.

> We need to isolate W.

$W - 6H = 106$	Simplify.
$W - 6H + 6H = 106 + 6H$	Isolate W by adding $6H$ to both sides.
$W = 106 + 6H$	Simplify.

This form of the formula makes it easy to find a person's healthy weight, W, if we know that person's height, H, in inches, in excess of 5 feet. ■

> ✓ **CHECK POINT 4** Solve for x: $\dfrac{x}{3} - 4y = 5$.

2 Use the percent formula. ▶

A Formula Involving Percent

Great Question!

Before presenting this formula involving percent, can you briefly review what I should already know about the basics of percent?

- **Percents** are the result of expressing numbers as part of 100. The word *percent* means *per hundred*. For example, the bar graph in **Figure 2.7** shows that 47% of Americans find "whatever" most annoying in conversation. Thus, 47 out of every 100 Americans are most annoyed by "whatever" in conversation: $47\% = \frac{47}{100}$.

Which Is Most Annoying in Conversation?

"Whatever" — 47%
"You know" — 25%
"It is what it is" — 11%
"Anyway" — 7%
"At the end of the day" — 2%

Figure 2.7

Source: Marist Poll Survey of 938 Americans ages 16 and older, August 3–6, 2010.

- To convert a number from percent form to decimal form, move the decimal point two places to the left and drop the percent sign. Example:

$$47\% = 47.\% = 0.47\,\cancel{\%}$$

Thus, $47\% = 0.47$.

- To convert a number from decimal form to a percent, move the decimal point two places to the right and attach a percent sign.

$$0.86 = 086.\%$$

Thus, $0.86 = 86\%$.

- Dictionaries indicate that the word *percentage* has the same meaning as the word *percent*. Use the word that sounds better in a given circumstance.

Percents are useful in comparing two numbers. To compare the number A to the number B using a percent P, the following formula is used:

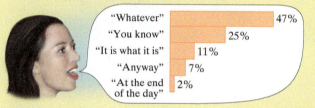

$$A \text{ is } P \text{ percent of } B.$$
$$A = P \cdot B.$$

In the formula

$$A = PB,$$

$B =$ the base number, $P =$ the percent (in decimal form), and $A =$ the number being compared to B.

There are three basic types of percent problems that can be solved using the percent formula

$$A = PB. \quad \text{A is } P \text{ percent of } B.$$

Question	Given	Percent Formula
What number is P percent of B?	P and B	Solve for A.
A is P percent of what number?	A and P	Solve for B.
A is what percent of B?	A and B	Solve for P.

Let's look at an example of each type of problem.

EXAMPLE 5   Using the Percent Formula: What Number Is P Percent of B?

What number is 8% of 20?

Solution   We use the formula $A = PB$: A is P percent of B. We are interested in finding the quantity A in this formula.

What number   is   8%   of   20?

$$A = 0.08 \cdot 20 \qquad \text{Express 8\% as 0.08.}$$
$$A = 1.6 \qquad \text{Multiply:} \quad \begin{array}{r} 0.08 \\ \times\ 20 \\ \hline 1.60 \end{array}$$

Thus, 1.6 is 8% of 20. The answer is 1.6. ■

✓ **CHECK POINT 5**   What number is 9% of 50?

EXAMPLE 6   Using the Percent Formula: A Is P Percent of What Number?

4 is 25% of what number?

Solution   We use the formula $A = PB$: A is P percent of B. We are interested in finding the quantity B in this formula.

4   is   25%   of   what number?

$$4 = 0.25 \cdot B \qquad \text{Express 25\% as 0.25.}$$
$$\frac{4}{0.25} = \frac{0.25B}{0.25} \qquad \text{Divide both sides by 0.25.}$$
$$16 = B \qquad \text{Simplify:} \quad 0.25\overline{)4.00} \;\; 16.$$

Thus, 4 is 25% of 16. The answer is 16. ■

✓ **CHECK POINT 6**   9 is 60% of what number?

EXAMPLE 7 Using the Percent Formula:
A Is What Percent of *B*?

1.3 is what percent of 26?

Solution We use the formula $A = PB$: *A* is *P* percent of *B*. We are interested in finding the quantity *P* in this formula.

| 1.3 | is | what percent | of | 26? |

$$1.3 = P \cdot 26$$

$$\frac{1.3}{26} = \frac{P \cdot 26}{26} \qquad \text{Divide both sides by 26.}$$

$$0.05 = P \qquad \text{Simplify: } 26\overline{)1.30} = 0.05$$

We change 0.05 to a percent by moving the decimal point two places to the right and adding a percent sign: $0.05 = 5\%$. Thus, 1.3 is 5% of 26. The answer is 5%. ∎

✓ **CHECK POINT 7** 18 is what percent of 50?

---

3 Solve applied problems involving percent change. ▶

Applications

Percents are used for comparing changes, such as increases or decreases in sales, population, prices, and production. If a quantity changes, its percent increase or percent decrease can be determined by asking the following question:

The change is what percent of the original amount?

The question is answered using the percent formula as follows:

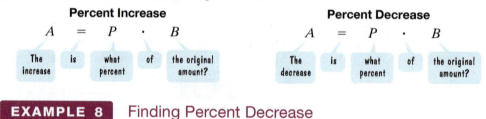

Percent Increase

$$A = P \cdot B$$

| The increase | is | what percent | of | the original amount? |

Percent Decrease

$$A = P \cdot B$$

| The decrease | is | what percent | of | the original amount? |

EXAMPLE 8 Finding Percent Decrease

A jacket regularly sells for $135.00. The sale price is $60.75. Find the percent decrease in the jacket's price.

Solution The percent decrease in price can be determined by asking the following question:

The price decrease is what percent of the original price ($135.00)?

The price decrease is the difference between the original price and the sale price ($60.75):

$$\$135.00 - \$60.75 = \$74.25.$$

Now we use the percent formula to find the percent decrease.

| The price decrease | is | what percent | of | the original price? |

$$74.25 = P \cdot 135$$

$$\frac{74.25}{135} = \frac{P \cdot 135}{135} \qquad \text{Divide both sides by 135.}$$

$$0.55 = P \qquad \text{Simplify: } 135\overline{)74.25} = 0.55$$

We change 0.55 to a percent by moving the decimal point two places to the right and adding a percent sign: $0.55 = 55\%$. Thus, the percent decrease in the jacket's price is 55%. ∎

✓ **CHECK POINT 8**   A television regularly sells for $940. The sale price is $611. Find the percent decrease in the television's price.

In our next example, we look at one of the many ways that percent can be used incorrectly.

EXAMPLE 9   Promises of a Politician

A politician states, "If you elect me to office, I promise to cut your taxes for each of my first three years in office by 10% each year, for a total reduction of 30%." Evaluate the accuracy of the politician's statement.

Solution   To make things simple, let's assume that a taxpayer paid $100 in taxes in the year previous to the politician's election. A 10% reduction during year 1 is 10% of $100.

$$10\% \text{ of previous year's tax} = 10\% \text{ of } \$100 = 0.10 \cdot \$100 = \$10$$

With a 10% reduction the first year, the taxpayer will pay only $100 − $10, or $90, in taxes during the politician's first year in office.

The table below shows how we calculate the new, reduced tax for each of the first three years in office.

Year	Tax Paid the Year Before	10% Reduction	Taxes Paid This Year
1	$100	$0.10 \cdot \$100 = \$10$	$\$100 - \$10 = \$90$
2	$ 90	$0.10 \cdot \$90 = \$9$	$\$90 - \$9 = \$81$
3	$ 81	$0.10 \cdot \$81 = \$8.10$	$\$81 - \$8.10 = \$72.90$

The tax reduction is the amount originally paid, $100.00, minus the amount paid during the politician's third year in office, $72.90:

$$\$100.00 - \$72.90 = \$27.10.$$

Now we use the percent formula to determine the percent decrease in taxes over the three years.

The tax decrease	is	what percent	of	the original tax?

$$27.1 = P \cdot 100$$

$$\frac{27.1}{100} = \frac{P \cdot 100}{100} \qquad \text{Divide both sides by 100.}$$

$$0.271 = P \qquad \text{Simplify.}$$

Change 0.271 to a percent: 0.271 = 27.1%. The percent decrease is 27.1%. The taxes decline by 27.1%, not by 30%. The politician is ill-informed in saying that three consecutive 10% cuts add up to a total tax cut of 30%. In our calculation, which serves as a counterexample to the promise, the total tax cut is only 27.1%. ∎

✓ **CHECK POINT 9**   Suppose you paid $1200 in taxes. During year 1, taxes decrease by 20%. During year 2, taxes increase by 20%.

   a. What do you pay in taxes for year 2?

   b. How do your taxes for year 2 compare with what you originally paid, namely $1200? If the taxes are not the same, find the percent increase or decrease.

Achieving Success

Write down all the steps. In this textbook, examples are written that provide step-by-step solutions. No steps are omitted and each step is explained to the right of the mathematics. Some professors are careful to write down every step of a problem as they are doing it in class; others aren't so fastidious. In either case, write down what the professor puts up and when you get home, fill in whatever steps have been omitted (if any). In your math work, including homework and tests, show clear step-by-step solutions. Detailed solutions help organize your thoughts and enhance understanding. Doing too many steps mentally often results in preventable mistakes.

CONCEPT AND VOCABULARY CHECK

Fill in each blank so that the resulting statement is true.

1. Solving a formula for a variable means rewriting the formula so that the variable is _____.

2. The area, A, of a rectangle with length l and width w is given by the formula _____.

3. The perimeter, P, of a rectangle with length l and width w is given by the formula _____.

4. The sentence "A is P percent of B" is expressed by the formula _____.

5. In order to solve $y = mx + b$ for x, we first _____ and then _____.

2.4 EXERCISE SET ▶ MyMathLab®

Practice Exercises

In Exercises 1–26, solve each formula for the specified variable. Do you recognize the formula? If so, what does it describe?

1. $d = rt$ for r
2. $d = rt$ for t
3. $I = Prt$ for P
4. $I = Prt$ for r
5. $C = 2\pi r$ for r
6. $C = \pi d$ for d
7. $E = mc^2$ for m
8. $V = \pi r^2 h$ for h
9. $y = mx + b$ for m
10. $y = mx + b$ for x
11. $T = D + pm$ for D
12. $P = C + MC$ for M
13. $A = \dfrac{1}{2}bh$ for b
14. $A = \dfrac{1}{2}bh$ for h
15. $M = \dfrac{n}{5}$ for n
16. $M = \dfrac{A}{740}$ for A
17. $\dfrac{c}{2} + 80 = 2F$ for c
18. $p = 15 + \dfrac{5d}{11}$ for d
19. $A = \dfrac{1}{2}(a + b)$ for a
20. $A = \dfrac{1}{2}(a + b)$ for b
21. $S = P + Prt$ for r
22. $S = P + Prt$ for t
23. $A = \dfrac{1}{2}h(a + b)$ for b
24. $A = \dfrac{1}{2}h(a + b)$ for a
25. $Ax + By = C$ for x
26. $Ax + By = C$ for y

Use the percent formula, $A = PB$: A is P percent of B, to solve Exercises 27–42.

27. What number is 3% of 200?
28. What number is 8% of 300?
29. What number is 18% of 40?
30. What number is 16% of 90?
31. 3 is 60% of what number?
32. 8 is 40% of what number?
33. 24% of what number is 40.8?
34. 32% of what number is 51.2?
35. 3 is what percent of 15?
36. 18 is what percent of 90?
37. What percent of 2.5 is 0.3?
38. What percent of 7.5 is 0.6?
39. If 5 is increased to 8, the increase is what percent of the original number?
40. If 5 is increased to 9, the increase is what percent of the original number?
41. If 4 is decreased to 1, the decrease is what percent of the original number?
42. If 8 is decreased to 6, the decrease is what percent of the original number?

Practice PLUS

In Exercises 43–50, solve each equation for x.

43. $y = (a + b)x$

44. $y = (a - b)x$

45. $y = (a - b)x + 5$

46. $y = (a + b)x - 8$

47. $y = cx + dx$

48. $y = cx - dx$

49. $y = Ax - Bx - C$

50. $y = Ax + Bx + C$

Application Exercises

51. The average, or mean, A, of three exam grades, x, y, and z, is given by the formula

$$A = \frac{x + y + z}{3}.$$

a. Solve the formula for z.

b. Use the formula in part (a) to solve this problem. On your first two exams, your grades are 86% and 88%: $x = 86$ and $y = 88$. What must you get on the third exam to have an average of 90%?

52. The average, or mean, A, of four exam grades, x, y, z, and w, is given by the formula

$$A = \frac{x + y + z + w}{4}.$$

a. Solve the formula for w.

b. Use the formula in part (a) to solve this problem. On your first three exams, your grades are 76%, 78%, and 79%: $x = 76$, $y = 78$, and $z = 79$. What must you get on the fourth exam to have an average of 80%?

53. If you are traveling in your car at an average rate of r miles per hour for t hours, then the distance, d, in miles, that you travel is described by the formula $d = rt$: distance equals rate times time.

a. Solve the formula for t.

b. Use the formula in part (a) to find the time that you travel if you cover a distance of 100 miles at an average rate of 40 miles per hour.

54. The formula $F = \frac{9}{5}C + 32$ expresses the relationship between Celsius temperature, C, and Fahrenheit temperature, F.

a. Solve the formula for C.

b. Use the formula from part (a) to find the equivalent Celsius temperature for a Fahrenheit temperature of 59°.

An Adecco survey of 1800 workers asked participants about taboo topics to discuss at work. The circle graph shows the results. Use this information to solve Exercises 55–56.

What Is the Most Taboo Topic to Discuss at Work?

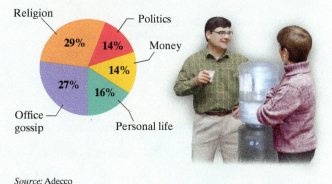

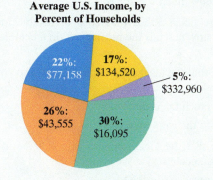

Religion 29%
Politics 14%
Money 14%
Personal life 16%
Office gossip 27%

Source: Adecco

55. Among the 1800 workers who participated in the poll, how many stated that religion is the most taboo topic to discuss at work?

56. Among the 1800 workers who participated in the poll, how many stated that politics is the most taboo topic to discuss at work?

In 2014, America's income gap was wider than ever. The circle graph shows the average household income from the richest 5% of households to the poorest 30% of households. Use this information to solve Exercises 57–58.

Average U.S. Income, by Percent of Households

22%: $77,158
17%: $134,520
5%: $332,960
26%: $43,555
30%: $16,095

Source: Time, September 8–15, 2014

57. a. If the richest 5% includes 5.85 million households, how many households, in millions, are there in the United States?

b. The average income in 2014 for the richest 5% was 180% of what this group's average income had been in 1975. What was the average income in 1975, to the nearest dollar, for the richest 5%?

58. a. If the poorest 30% includes 35.1 million households, how many households, in millions, are there in the United States?

b. The average income in 2014 for the poorest 30% was 107% of what this group's average income had been in 1975. What was the average income in 1975, to the nearest dollar, for the poorest 30%?

Views of Police Officers. According to a poll of 1500 adults conducted in 2014 by USA Today/Pew Research Center, a majority of Americans said police failed to do a good job at holding officers accountable. The circle graph shows a breakdown of the 1500 respondents who were surveyed. Use this information to solve Exercises 59–60.

How Well Are Police Departments Holding Officers Accountable?

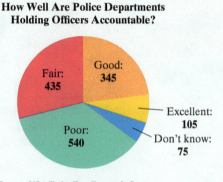

Source: USA Today/Pew Research Center

59. What percentage of those surveyed said police departments did a poor job at holding officers accountable?

60. What percentage of those surveyed said police departments did an excellent job at holding officers accountable?

61. A charity has raised $7500, with a goal of raising $60,000. What percent of the goal has been raised?

62. A charity has raised $225,000, with a goal of raising $500,000. What percent of the goal has been raised?

63. A restaurant bill came to $60. If 15% of this amount was left as a tip, how much was the tip?

64. If income tax is $3502 plus 28% of taxable income over $23,000, how much is the income tax on a taxable income of $35,000?

65. Suppose that the local sales tax rate is 6% and you buy a car for $16,800.

  a. How much tax is due?

  b. What is the car's total cost?

66. Suppose that the local sales tax rate is 7% and you buy a graphing calculator for $96.

  a. How much tax is due?

  b. What is the calculator's total cost?

67. An exercise machine with an original price of $860 is on sale at 12% off.

  a. What is the discount amount?

  b. What is the exercise machine's sale price?

68. A book that normally sells for $16.50 is on sale at 40% off.

  a. What is the discount amount?

  b. What is the book's sale price?

69. A sofa regularly sells for $840. The sale price is $714. Find the percent decrease in the sofa's price.

70. A fax machine regularly sells for $380. The sale price is $266. Find the percent decrease in the machine's price.

71. Suppose that you put $10,000 in a rather risky investment recommended by your financial advisor. During the first year, your investment decreases by 30% of its original value. During the second year, your investment increases by 40% of its first-year value. Your advisor tells you that there must have been a 10% overall increase of your original $10,000 investment. Is your financial advisor using percentages properly? If not, what is the actual percent gain or loss on your original $10,000 investment?

72. The price of a color printer is reduced by 30% of its original price. When it still does not sell, its price is reduced by 20% of the reduced price. The salesperson informs you that there has been a total reduction of 50%. Is the salesperson using percentages properly? If not, what is the actual percent reduction from the original price?

Explaining the Concepts

73. Explain what it means to solve a formula for a variable.

74. What does the percent formula, $A = PB$, describe? Give an example of how the formula is used.

Critical Thinking Exercises

Make Sense? In Exercises 75–78, determine whether each statement makes sense or does not make sense, and explain your reasoning.

75. To help me get started, I circle the variable that I need to solve for in a formula.

76. By solving a formula for one of its variables, I find a numerical value for that variable.

77. I have $100 and my restaurant bill comes to $80, which is not enough to leave a 20% tip.

78. I found the percent decrease in a jacket's price to be 120%.

In Exercises 79–82, determine whether each statement is true or false. If the statement is false, make the necessary change(s) to produce a true statement.

79. If $ax + b = 0$, then $x = \dfrac{b}{a}$.

80. If $A = lw$, then $w = \dfrac{l}{A}$.

81. If $A = \dfrac{1}{2}bh$, then $b = \dfrac{A}{2h}$.

82. Solving $x - y = -7$ for y gives $y = x + 7$.

83. In psychology, an intelligence quotient, Q, also called IQ, is measured by the formula

$$Q = \frac{100M}{C},$$

where M = mental age and C = chronological age. Solve the formula for C.

Review Exercises

84. Solve and check: $5x + 20 = 8x - 16$.
(Section 2.2, Example 7)

85. Solve and check: $5(2y - 3) - 1 = 4(6 + 2y)$.
(Section 2.3, Example 3)

86. Simplify: $x - 0.3x$.
(Section 1.4, Example 8)

Preview Exercises

Exercises 87–89 will help you prepare for the material covered in the next section. In each exercise, let x represent the number and write the phrase as an algebraic expression.

87. The quotient of 13 and a number, decreased by 7 times the number

88. Eight times the sum of a number and 14

89. Nine times the difference of a number and 5

MID-CHAPTER CHECK POINT   Section 2.1–Section 2.4

✓ **What You Know:** We learned a step-by-step procedure for solving linear equations, including equations with fractions and decimals. We saw that some equations have no solution, whereas others have all real numbers as solutions. We used the addition and multiplication properties of equality to solve formulas for variables. Finally, we worked with the percent formula $A = PB$: A is P percent of B.

1. Solve: $\dfrac{x}{2} = 12 - \dfrac{x}{4}$.

2. Solve: $5x - 42 = -57$.

3. Solve for C: $H = \dfrac{EC}{825}$.

4. What number is 6% of 140?

5. Solve: $\dfrac{-x}{10} = -3$.

6. Solve: $1 - 3(y - 5) = 4(2 - 3y)$.

7. Solve for r: $S = 2\pi rh$.

8. 12 is 30% of what number?

9. Solve: $\dfrac{3y}{5} + \dfrac{y}{2} = \dfrac{5y}{4} - 3$.

10. Solve: $2.4x + 6 = 1.4x + 0.5(6x - 9)$

11. Solve: $5z + 7 = 6(z - 2) - 4(2z - 3)$.

12. Solve for x: $Ax - By = C$.

13. Solve: $6y + 7 + 3y = 3(3y - 1)$.

14. Solve: $10\left(\dfrac{1}{2}x + 3\right) = 10\left(\dfrac{3}{5}x - 1\right)$.

15. 50 is what percent of 400?

16. Solve: $\dfrac{3(m + 2)}{4} = 2m + 3$.

17. If 40 is increased to 50, the increase is what percent of the original number?

18. Solve: $12x - 4 + 8x - 4 = 4(5x - 2)$.

19. The bar graph indicates that reading books for fun loses value as kids age.

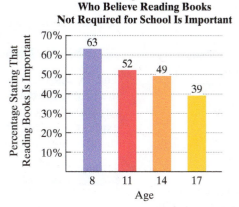

Percentage of U.S. Kids Who Believe Reading Books Not Required for School Is Important

Source: Scholastic Kids and Family Reading Report 2010, *Scholastic Inc.*

The data can be described by the mathematical model

$$B = -\frac{5}{2}a + 82,$$

where B is the percentage at age a who believe that reading books is important.

a. Does the formula underestimate or overestimate the percentage of 14-year-olds who believe that reading books is important? By how much?

b. If trends shown by the formula continue, at which age will 22% believe that reading books is important?

2.5

An Introduction to Problem Solving ▶

What am I supposed to learn?

After studying this section, you should be able to:

1 Translate English phrases into algebraic expressions. ▶

2 Solve algebraic word problems using linear equations. ▶

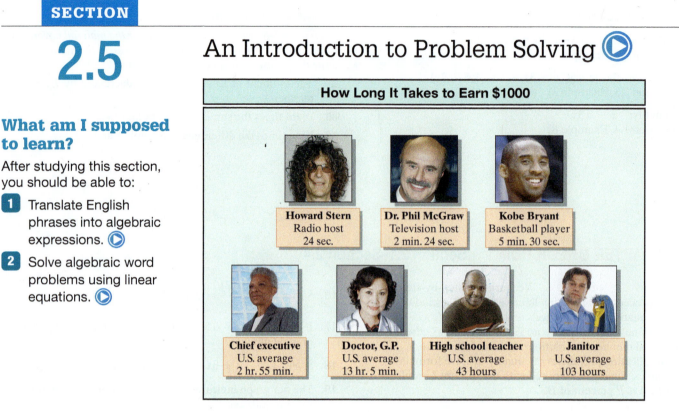

How Long It Takes to Earn $1000

Howard Stern
Radio host
24 sec.

Dr. Phil McGraw
Television host
2 min. 24 sec.

Kobe Bryant
Basketball player
5 min. 30 sec.

Chief executive
U.S. average
2 hr. 55 min.

Doctor, G.P.
U.S. average
13 hr. 5 min.

High school teacher
U.S. average
43 hours

Janitor
U.S. average
103 hours

Source: *Time*

In this section, you'll see examples and exercises focused on how much money Americans earn. These situations illustrate a step-by-step strategy for solving problems. As you become familiar with this strategy, you will learn to solve a wide variety of problems.

1 Translate English phrases into algebraic expressions. ▶

Algebraic Expressions for English Phrases

The hardest thing about word problems is writing an equation that translates, or models, the problem's conditions. Throughout Chapter 1, you wrote algebraic expressions and equations for conditions about numbers. **Table 2.1** summarizes many of the algebraic expressions that you wrote for these conditions. We are using x to represent the variable, but we can use any letter.

Great Question!

Table 2.1 looks long and intimidating. What's the best way to get through the table?

Cover the right column with a sheet of paper and attempt to formulate the algebraic expression for the English phrase in the left column on your own. Then slide the paper down and check your answer. Work through the entire table in this manner.

Table 2.1 Algebraic Translations of English Phrases

English Phrase	Algebraic Expression
Addition	
The sum of a number and 7	$x + 7$
Five more than a number; a number plus 5	$x + 5$
A number increased by 6; 6 added to a number	$x + 6$
Subtraction	
A number minus 4	$x - 4$
A number decreased by 5	$x - 5$
A number subtracted from 8	$8 - x$
The difference between a number and 6	$x - 6$
The difference between 6 and a number	$6 - x$
Seven less than a number	$x - 7$
Seven minus a number	$7 - x$
Nine fewer than a number	$x - 9$

Table 2.1 continued

English Phrase	Algebraic Expression
Multiplication	
Five times a number	$5x$
The product of 3 and a number	$3x$
Two-thirds of a number (used with fractions)	$\frac{2}{3}x$
Seventy-five percent of a number (used with decimals)	$0.75x$
Thirteen multiplied by a number	$13x$
A number multiplied by 13	$13x$
Twice a number	$2x$
Division	
A number divided by 3	$\dfrac{x}{3}$
The quotient of 7 and a number	$\dfrac{7}{x}$
The quotient of a number and 7	$\dfrac{x}{7}$
The reciprocal of a number	$\dfrac{1}{x}$
More than one operation	
The sum of twice a number and 7	$2x + 7$
Twice the sum of a number and 7	$2(x + 7)$
Three times the sum of 1 and twice a number	$3(1 + 2x)$
Nine subtracted from 8 times a number	$8x - 9$
Twenty-five percent of the sum of 3 times a number and 14	$0.25(3x + 14)$
Seven times a number, increased by 24	$7x + 24$
Seven times the sum of a number and 24	$7(x + 24)$

A Strategy for Solving Word Problems Using Equations

Here are some general steps we will follow in solving word problems:

Strategy for Solving Word Problems

Step 1. Read the problem carefully several times until you can state in your own words what is given and what the problem is looking for. Let x (or any variable) represent one of the unknown quantities in the problem.

Step 2. If necessary, write expressions for any other unknown quantities in the problem in terms of x.

Step 3. Write an equation in x that translates, or models, the conditions of the problem.

Step 4. Solve the equation and answer the problem's question.

Step 5. Check the solution *in the original wording* of the problem, not in the equation obtained from the words.

Great Question!

Why are word problems important?

There is great value in reasoning through the steps for solving a word problem. This value comes from the problem-solving skills that you will attain and is often more important than the specific problem or its solution.

2 Solve algebraic word problems using linear equations. ▶

Applying the Strategy for Solving Word Problems

Now that you've read why word problems are important, let's apply our five-step strategy for solving these problems.

EXAMPLE 1 Solving a Word Problem

Nine subtracted from eight times a number is 39. Find the number.

Solution

Step 1. Let *x* represent one of the quantities. Because we are asked to find a number, let

$$x = \text{the number.}$$

Step 2. Represent other unknown quantities in terms of *x*. There are no other unknown quantities to find, so we can skip this step.

Step 3. Write an equation in *x* that models the conditions.

| Nine subtracted from | eight times a number | is | 39. |

$$8x \quad - \quad 9 \quad = \quad 39$$

Step 4. Solve the equation and answer the question.

$$8x - 9 = 39 \qquad \text{This is the equation that models the problem's conditions.}$$
$$8x - 9 + 9 = 39 + 9 \qquad \text{Add 9 to both sides.}$$
$$8x = 48 \qquad \text{Simplify.}$$
$$\frac{8x}{8} = \frac{48}{8} \qquad \text{Divide both sides by 8.}$$
$$x = 6 \qquad \text{Simplify.}$$

The number is 6.

Step 5. Check the proposed solution in the original wording of the problem. "Nine subtracted from eight times a number is 39." The proposed number is 6. Eight times 6 is $8 \cdot 6$, or 48. Nine subtracted from 48 is $48 - 9$, or 39. The proposed solution checks in the problem's wording, verifying that the number is 6. ∎

☑ **CHECK POINT 1** Four subtracted from six times a number is 68. Find the number.

EXAMPLE 2 Starting Salaries for College Graduates with Undergraduate Degrees

The bar graph in **Figure 2.8** shows the ten most popular college majors with median, or middlemost, starting salaries for recent college graduates.

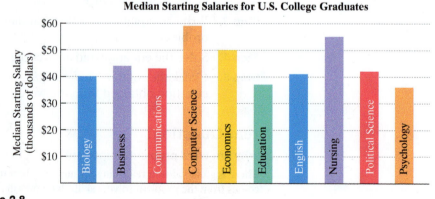

Median Starting Salaries for U.S. College Graduates

Figure 2.8
Source: PayScale (2013 data)

The median starting salary of a nursing major exceeds that of a business major by $11 thousand. Combined, their median starting salaries are $99 thousand. Determine the median starting salaries of business and nursing majors with bachelor's degrees.

Solution

Step 1. Let x represent one of the quantities. We know something about the median starting salary of a nursing major: It exceeds that of a business major by $11 thousand. This means that nursing majors with bachelor's degrees earn $11 thousand more than business majors. We will let

x = the median starting salary, in thousands of dollars, of business majors.

Step 2. Represent other unknown quantities in terms of x. The other unknown quantity is the median starting salary of nursing majors. Because it is $11 thousand more than that of business majors, let

$x + 11$ = the median starting salary, in thousands of dollars, of nursing majors.

Step 3. Write an equation in x that models the conditions. Combined, the median starting salaries for business and nursing majors are $99 thousand.

The median starting salary for business majors	plus	the median starting salary for nursing majors	is	$99 thousand.
x	$+$	$(x + 11)$	$=$	99

Step 4. Solve the equation and answer the question.

$$x + (x + 11) = 99 \quad \text{This is the equation that models the problem's conditions.}$$
$$2x + 11 = 99 \quad \text{Regroup and combine like terms on the left side.}$$
$$2x + 11 - 11 = 99 - 11 \quad \text{Subtract 11 from both sides.}$$
$$2x = 88 \quad \text{Simplify.}$$
$$\frac{2x}{2} = \frac{88}{2} \quad \text{Divide both sides by 2.}$$
$$x = 44 \quad \text{Simplify.}$$

Because x represents the median starting salary, in thousands of dollars, of business majors, we see that business majors with bachelor's degrees have a median starting salary of $44 thousand, or $44,000. Because $x + 11$ represents the median starting salary, in thousands of dollars, of nursing majors, their median starting salary is $44 + 11$, or $55 thousand ($55,000).

Step 5. Check the proposed solution in the original wording of the problem. The problem states that combined, the median starting salaries are $99 thousand. By adding $44 thousand, the median starting salary of business majors, and $55 thousand, the median starting salary of nursing majors, we do, indeed, obtain a sum of $99 thousand. ∎

Great Question!

Example 2 involves using the word *exceeds* to represent one of the unknown quantities. Can you help me to write algebraic expressions for quantities described using *exceeds*?

Modeling with the word *exceeds* can be a bit tricky. It's helpful to identify the smaller quantity. Then add to this quantity to represent the larger quantity. For example, suppose that Tim's height exceeds Tom's height by a inches. Tom is the shorter person. If Tom's height is represented by x, then Tim's height is represented by $x + a$.

✓ **CHECK POINT 2**   Two of the bars in **Figure 2.8** on page 158 represent starting salaries of computer science and English majors. The median starting salary of a computer science major exceeds that of an English major by $18 thousand. Combined, their median starting salaries are $100 thousand. Determine the median starting salaries of English and computer science majors with bachelor's degrees.

EXAMPLE 3   Consecutive Integers and the Super Bowl

Only once, in 1991, were the winning and losing scores in the Super Bowl consecutive integers. The New York Giants beat the Buffalo Bills in a nearly error-free game. If the sum of the scores was 39, determine the points scored by the losing team, the Bills, and the winning team, the Giants.

Solution

Step 1. Let x represent one of the quantities. We will let

$$x = \text{points scored by the losing team, the Bills.}$$

Step 2. Represent other unknown quantities in terms of x. The other unknown quantity involves points scored by the winning team, the Giants. Because the scores were consecutive integers, the winning team scored one point more than the losing team. Thus,

$$x + 1 = \text{points scored by the winning team, the Giants.}$$

Step 3. Write an equation in x that models the conditions. We are told that the sum of the scores was 39.

Points scored by the losing team	plus	points scored by the winning team	result in	39 points.
x	$+$	$(x + 1)$	$=$	39

Step 4. Solve the equation and answer the question.

$$x + (x + 1) = 39 \qquad \text{This is the equation that models the problem's conditions.}$$
$$2x + 1 = 39 \qquad \text{Regroup and combine like terms.}$$
$$2x + 1 - 1 = 39 - 1 \qquad \text{Subtract 1 from both sides.}$$
$$2x = 38 \qquad \text{Simplify.}$$
$$\frac{2x}{2} = \frac{38}{2} \qquad \text{Divide both sides by 2.}$$
$$x = 19 \qquad \text{Simplify.}$$

Thus,

$$\text{points scored by the losing team, the Bills} = x = 19$$

and

$$\text{points scored by the winning team, the Giants} = x + 1 = 20.$$

With the closest final score in Super Bowl history, the Giants scored 20 points and the Bills scored 19 points.

Step 5. Check the proposed solution in the original wording of the problem. The problem states that the sum of the scores was 39. With a final score of 20 to 19, we see that the sum of these numbers is, indeed, 39. ∎

✓ **CHECK POINT 3**   Page numbers on facing pages of a book are consecutive integers. Two pages that face each other have 145 as the sum of their page numbers. What are the page numbers?

Example 3 and Check Point 3 involved consecutive integers. By contrast, some word problems involve consecutive odd integers, such as 5, 7, and 9. Other word problems involve consecutive even integers, such as 6, 8, and 10. When working with consecutive even or consecutive odd integers, we must continually add 2 to move from one integer to the next successive integer in the list.

Table 2.2 should be helpful in solving consecutive integer problems.

Table 2.2	Consecutive Integers	
English Phrase	**Algebraic Expressions**	**Example**
Two consecutive integers	$x, x + 1$	13, 14
Three consecutive integers	$x, x + 1, x + 2$	$-8, -7, -6$
Two consecutive even integers	$x, x + 2$	40, 42
Two consecutive odd integers	$x, x + 2$	$-37, -35$
Three consecutive even integers	$x, x + 2, x + 4$	30, 32, 34
Three consecutive odd integers	$x, x + 2, x + 4$	9, 11, 13

EXAMPLE 4 Renting a Car

Rent-a-Heap Agency charges $125 per week plus $0.20 per mile to rent a small car. How many miles can you travel for $335?

Solution

Step 1. Let x represent one of the quantities. Because we are asked to find the number of miles we can travel for $335, let

$$x = \text{the number of miles.}$$

Step 2. Represent other unknown quantities in terms of x. There are no other unknown quantities to find, so we can skip this step.

Step 3. Write an equation in x that models the conditions. Before writing the equation, let us consider a few specific values for the number of miles traveled. The rental charge is $125 plus $0.20 for each mile.

3 miles: The rental charge is $125 + $0.20(3).

30 miles: The rental charge is $125 + $0.20(30).

100 miles: The rental charge is $125 + $0.20(100).

x miles: The rental charge is $125 + 0.20x$.

The weekly charge of $125	plus	the charge of $0.20 per mile for x miles	equals	the total $335 rental charge.
125	+	0.20x	=	335

Step 4. Solve the equation and answer the question.

$$125 + 0.20x = 335 \qquad \text{This is the equation that models the conditions of the problem.}$$

$$125 + 0.20x - 125 = 335 - 125 \qquad \text{Subtract 125 from both sides.}$$

$$0.20x = 210 \qquad \text{Simplify.}$$

$$\frac{0.20x}{0.20} = \frac{210}{0.20} \qquad \text{Divide both sides by 0.20.}$$

$$x = 1050 \qquad \text{Simplify.}$$

You can travel 1050 miles for $335.

Step 5. Check the proposed solution in the original wording of the problem. Traveling 1050 miles should result in a total rental charge of $335. The mileage charge of $0.20 per mile is

$$\$0.20(1050) = \$210.$$

Adding this to the $125 weekly charge gives a total rental charge of

$$\$125 + \$210 = \$335.$$

Because this results in the given rental charge of $335, this verifies that you can travel 1050 miles. ■

☑ **CHECK POINT 4**   A taxi charges $2.00 to turn on the meter plus $0.25 for each eighth of a mile. If you have $10.00, how many eighths of a mile can you go? How many miles is that?

We will be using the formula for the perimeter of a rectangle, $P = 2l + 2w$, in our next example. Twice the rectangle's length plus twice the rectangle's width is its perimeter.

EXAMPLE 5   Finding the Dimensions of a Soccer Field

A rectangular soccer field is twice as long as it is wide. If the perimeter of a soccer field is 300 yards, what are the field's dimensions?

Solution

Step 1. Let x represent one of the quantities. We know something about the length; the field is twice as long as it is wide. We will let

$$x = \text{the width.}$$

Step 2. Represent other unknown quantities in terms of x. Because the field is twice as long as it is wide, let

$$2x = \text{the length.}$$

Great Question!

Should I draw pictures like Figure 2.9 when solving geometry problems?

When solving word problems, particularly problems involving geometric figures, drawing a picture of the situation is often helpful. Label x on your drawing and, where appropriate, label other parts of the drawing in terms of x.

Figure 2.9 illustrates the soccer field and its dimensions.

Width x
$2x$
Length
Figure 2.9

Step 3. Write an equation in x that models the conditions. Because the perimeter of a soccer field is 300 yards,

Twice the length	plus	twice the width	is	the perimeter.
$2 \cdot 2x$	$+$	$2 \cdot x$	$=$	$300.$

Step 4. Solve the equation and answer the question.

$$2 \cdot 2x + 2 \cdot x = 300 \quad \text{This is the equation that models the problem's conditions.}$$

$$4x + 2x = 300 \quad \text{Multiply.}$$

$$6x = 300 \quad \text{Combine like terms.}$$

$$\frac{6x}{6} = \frac{300}{6} \quad \text{Divide both sides by 6.}$$

$$x = 50 \quad \text{Simplify.}$$

Thus,   width $= x = 50$

length $= 2x = 2(50) = 100.$

The dimensions of a soccer field are 50 yards by 100 yards.

Step 5. Check the proposed solution in the original wording of the problem. The perimeter of the soccer field, using the dimensions that we found, is 2(50 yards) + 2(100 yards) = 100 yards + 200 yards, or 300 yards. Because the problem's wording tells us that the perimeter is 300 yards, our dimensions are correct. ■

☑ **CHECK POINT 5**   A rectangular swimming pool is three times as long as it is wide. If the perimeter of the pool is 320 feet, what are the pool's dimensions?

EXAMPLE 6   A Price Reduction

Your local computer store is having a sale. After a 30% price reduction, you purchase a new computer for $980. What was the computer's price before the reduction?

Solution

Step 1. Let x represent one of the quantities. We will let

$x =$ the original price of the computer.

Step 2. Represent other unknown quantities in terms of x. There are no other unknown quantities to find, so we can skip this step.

Step 3. Write an equation in x that models the conditions. The computer's original price minus the 30% reduction is the reduced price, $980.

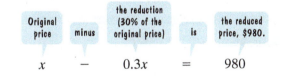

Step 4. Solve the equation and answer the question.

$$x - 0.3x = 980 \quad \text{This is the equation that models the problem's conditions.}$$

$$0.7x = 980 \quad \text{Combine like terms: } x - 0.3x = 1x - 0.3x = 0.7x.$$

$$\frac{0.7x}{0.7} = \frac{980}{0.7} \quad \text{Divide both sides by 0.7.}$$

$$x = 1400 \quad \text{Simplify: } 0.7\overline{)980.0}^{\,1400}$$

The computer's price before the reduction was $1400.

Step 5. Check the proposed solution in the original wording of the problem. The price before the reduction, $1400, minus the reduction in price should equal the reduced price given in the original wording, $980. The reduction in price is equal to 30% of

Great Question!

Is there a common error that I can avoid when solving problems about price reductions?

Yes. In Example 6, notice that the original price, x, reduced by 30% is $x - 0.3x$ and *not* $x - 0.3$.

the price before the reduction, $1400. To find the reduction, we multiply the decimal equivalent of 30%, 0.30 or 0.3, by the original price, $1400:

$$30\% \text{ of } \$1400 = (0.3)(\$1400) = \$420.$$

Now we can determine whether the calculation for the price before the reduction, $1400, minus the reduction, $420, is equal to the reduced price given in the problem, $980. We subtract:

$$\$1400 - \$420 = \$980.$$

This verifies that the price of the computer before the reduction was $1400. ■

Great Question!

Can I solve the equation in Example 6, $x - 0.3x = 980$, by first clearing the decimal?

Yes. Because 0.3 has one decimal place, the greatest number of decimal places in the equation, multiply both sides by 10^1, or 10, to clear the decimal.

$x - 0.3x = 980$	This is the equation that models the conditions in Example 6.
$10(x - 0.3x) = 10(980)$	Multiply both sides by 10.
$10x - 10(0.3x) = 10(980)$	Use the distributive property.
$10x - 3x = 9800$	Simplify by moving decimal points one place to the right.
$7x = 9800$	Combine like terms.
$\dfrac{7x}{7} = \dfrac{9800}{7}$	Divide both sides by 7.
$x = 1400$	Simplify.

This is the same value for x that we obtained in Example 6 when we did not clear the decimal. Which method do you prefer?

✓ **CHECK POINT 6**   After a 40% price reduction, an exercise machine sold for $564. What was the exercise machine's price before this reduction?

Achieving Success

Do not expect to solve every word problem immediately. As you read each problem, underline the important parts. It's a good idea to read the problem at least twice. Be persistent, but use the **"Ten Minutes of Frustration" Rule**. If you have exhausted every possible means for solving a problem and you are still bogged down, stop after ten minutes. Put a question mark by the exercise and move on. When you return to class, ask your professor for assistance.

CONCEPT AND VOCABULARY CHECK

Fill in each blank so that the resulting statement is true.

1. If x represents a number, six subtracted from four times the number can be represented by _____.

2. According to *Forbes* magazine, the top-earning dead celebrities in 2010 were Michael Jackson and Elvis Presley. Jackson's earnings exceeded Presley's earnings by $215 million. If x represents Presley's earnings, in millions of dollars, Jackson's earnings can be represented by _____.

3. If the number of any left-hand page in this book is represented by x, the number on the facing page can be represented by _____.

4. I can rent a car for $125 per week plus $0.15 for each mile driven. If I drive x miles in a week, the cost for the rental can be represented by _____.

5. If x represents a rectangle's width and $4x$ represents its length, the perimeter of the rectangle can be represented by _____.

6. I purchased a computer after a 35% price reduction. If x represents the computer's original price, the reduced price can be represented by _____.

2.5 EXERCISE SET   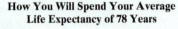 MyMathLab®

Practice Exercises

In Exercises 1–20, let x represent the number. Use the given conditions to write an equation. Solve the equation and find the number.

1. A number increased by 60 is equal to 410. Find the number.
2. The sum of a number and 43 is 107. Find the number.
3. A number decreased by 23 is equal to 214. Find the number.
4. The difference between a number and 17 is 96. Find the number.
5. The product of 7 and a number is 126. Find the number.
6. The product of 8 and a number is 272. Find the number.
7. The quotient of a number and 19 is 5. Find the number.
8. The quotient of a number and 14 is 8. Find the number.
9. The sum of four and twice a number is 56. Find the number.
10. The sum of five and three times a number is 59. Find the number.
11. Seven subtracted from five times a number is 178. Find the number.
12. Eight subtracted from six times a number is 298. Find the number.
13. A number increased by 5 is two times the number. Find the number.
14. A number increased by 12 is four times the number. Find the number.
15. Twice the sum of four and a number is 36. Find the number.
16. Three times the sum of five and a number is 48. Find the number.
17. Nine times a number is 30 more than three times that number. Find the number.
18. Five more than four times a number is that number increased by 35. Find the number.
19. If the quotient of three times a number and five is increased by four, the result is 34. Find the number.
20. If the quotient of three times a number and four is decreased by three, the result is nine. Find the number.

Application Exercises

In Exercises 21–46, use the five-step strategy to solve each problem.

How will you spend your average life expectancy of 78 years? The bar graph shows the average number of years you will devote to each of your most time-consuming activities. Exercises 21–22 are based on the data displayed by the graph.

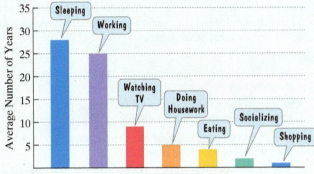

Source: U.S. Bureau of Labor Statistics

21. According to the U.S. Bureau of Labor Statistics, you will devote 37 years to sleeping and watching TV. The number of years sleeping will exceed the number of years watching TV by 19. Over your lifetime, how many years will you spend on each of these activities?

22. According to the U.S. Bureau of Labor Statistics, you will devote 32 years to sleeping and eating. The number of years sleeping will exceed the number of years eating by 24. Over your lifetime, how many years will you spend on each of these activities?

The bar graph shows median yearly earnings of full-time workers in the United States for people 25 years and over with a college education, by final degree earned. Exercises 23–24 are based on the data displayed by the graph.

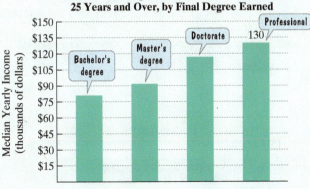

Source: U.S. Census Bureau

(Exercises 23–24 are based on the bar graph at the bottom of the previous page.)

23. The median yearly salary of an American whose final degree is a master's is $70 thousand less than twice that of an American whose final degree is a bachelor's. Combined, two people with each of these educational attainments earn $173 thousand. Find the median yearly salary of Americans with each of these final degrees.

24. The median yearly salary of an American whose final degree is a doctorate is $45 thousand less than twice that of an American whose final degree is a bachelor's. Combined, two people with each of these educational attainments earn $198 thousand. Find the median yearly salary of Americans with each of these final degrees.

In Exercises 25–26, use the fact that page numbers on facing pages of a book are consecutive integers.

25. The sum of the page numbers on the facing pages of a book is 629. What are the page numbers?

26. The sum of the page numbers on the facing pages of a book is 525. What are the page numbers?

27. Roger Maris and Babe Ruth are among the ten baseball players with the most home runs in a major league baseball season.

The number of home runs by these players for the seasons shown are consecutive odd integers whose sum is 120. Determine the number of homers hit by Ruth and by Maris.

28. Babe Ruth and Hank Greenberg are among the ten baseball players with the most home runs in a major league baseball season.

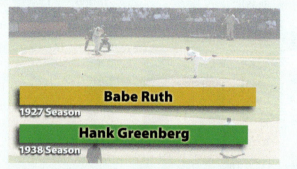

The number of home runs by these players for the seasons shown are consecutive even integers whose sum is 118. Determine the number of homers hit by Greenberg and by Ruth.

29. A car rental agency charges $200 per week plus $0.15 per mile to rent a car. How many miles can you travel in one week for $320?

30. A car rental agency charges $180 per week plus $0.25 per mile to rent a car. How many miles can you travel in one week for $395?

Despite booming new car sales with their cha-ching sounds, the average age of vehicles on U.S. roads is not going down. The bar graph shows the average price of new cars in the United States and the average age of cars on U.S. roads for two selected years. Exercises 31–32 are based on the information displayed by the graph.

**Average Price of New Cars and
Average Age of Cars on U.S. Roads**

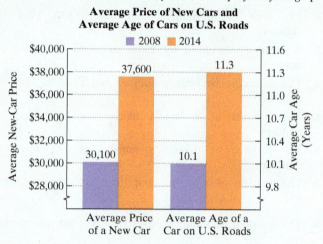

Source: Kelley Blue Book, IHS Automotive/Polk

31. In 2014, the average price of a new car was $37,600. For the period shown, new-car prices increased by approximately $1250 per year. If this trend continues, how many years after 2014 will the price of a new car average $46,350? In which year will this occur?

32. In 2014, the average age of cars on U.S. roads was 11.3 years. For the period shown, this average age increased by approximately 0.2 year per year. If this trend continues, how many years after 2014 will the average age of vehicles on U.S. roads be 12.3 years? In which year will this occur?

33. A rectangular field is four times as long as it is wide. If the perimeter of the field is 500 yards, what are the field's dimensions?

34. A rectangular field is five times as long as it is wide. If the perimeter of the field is 288 yards, what are the field's dimensions?

35. An American football field is a rectangle with a perimeter of 1040 feet. The length is 200 feet more than the width. Find the width and length of the rectangular field.

36. A basketball court is a rectangle with a perimeter of 86 meters. The length is 13 meters more than the width. Find the width and length of the basketball court.

37. A bookcase is to be constructed as shown in the figure. The length is to be 3 times the height. If 60 feet of lumber is available for the entire unit, find the length and height of the bookcase.

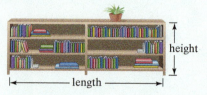

height

length

38. The height of the bookcase in the figure is 3 feet longer than the length of a shelf. If 18 feet of lumber is available for the entire unit, find the length and height of the unit.

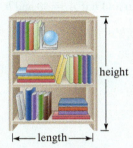

39. After a 20% reduction, you purchase a television for $320. What was the television's price before the reduction?

40. After a 30% reduction, you purchase a DVD player for $98. What was the price before the reduction?

41. This year's salary, $50,220, is an 8% increase over last year's salary. What was last year's salary?

42. This year's salary, $42,074, is a 9% increase over last year's salary. What was last year's salary?

43. Including 6% sales tax, a car sold for $23,850. Find the price of the car before the tax was added.

44. Including 8% sales tax, a bed-and-breakfast inn charges $172.80 per night. Find the inn's nightly cost before the tax is added.

45. An automobile repair shop charged a customer $448, listing $63 for parts and the remainder for labor. If the cost of labor is $35 per hour, how many hours of labor did it take to repair the car?

46. A repair bill on a sailboat came to $1603, including $532 for parts and the remainder for labor. If the cost of labor is $63 per hour, how many hours of labor did it take to repair the sailboat?

Explaining the Concepts

47. In your own words, describe a step-by-step approach for solving algebraic word problems.

48. Many students find solving linear equations much easier than solving algebraic word problems. Discuss some of the reasons why this is the case.

49. Did you have some difficulties solving some of the problems that were assigned in this Exercise Set? Discuss what you did if this happened to you. Did your course of action enhance your ability to solve algebraic word problems?

50. Write an original word problem that can be solved using a linear equation. Then solve the problem.

Critical Thinking Exercises

Make Sense?   In Exercises 51–54, determine whether each statement makes sense or does not make sense, and explain your reasoning.

51. Rather than struggling with the assigned word problems, I'll ask my instructor to solve them all in class and then study the solutions.

52. By reasoning through word problems, I can increase my problem-solving skills in general.

53. I find the hardest part in solving a word problem is writing the equation that models the verbal conditions.

54. I made a mistake when I used x and $x + 2$ to represent two consecutive odd integers, because 2 is even.

In Exercises 55–58, determine whether each statement is true or false. If the statement is false, make the necessary change(s) to produce a true statement.

55. Ten pounds less than Bill's weight, x, equals 160 pounds is modeled by $10 - x = 160$.

56. After a 35% reduction, a computer's price is $780, so its original price, x, can be found by solving $x - 0.35 = 780$.

57. If the length of a rectangle is 6 inches more than its width, and its perimeter is 24 inches, the distributive property must be used to solve the equation that determines the length.

58. On a number line, consecutive integers do not have any other integers between them.

59. An HMO pamphlet contains the following recommended weight for women: "Give yourself 100 pounds for the first 5 feet plus 5 pounds for every inch over 5 feet tall." Using this description, which height corresponds to an ideal weight of 135 pounds?

60. The rate for a particular international telephone call is $0.55 for the first minute and $0.40 for each additional minute. Determine the length of a call that costs $6.95.

61. In a film, the actor Charles Coburn played an elderly "uncle" character criticized for marrying a woman when he is 3 times her age. He wittily replies, "Ah, but in 20 years time I shall only be twice her age." How old is the "uncle" and the woman?

62. Answer the question in the following *Peanuts* cartoon strip. (*Note*: You may not use the answer given in the cartoon!)

Peanuts copyright © 1979, 2011 by United Features Syndicate, Inc.

Review Exercises

63. Solve and check: $\frac{4}{5}x = -16$.

    (Section 2.2, Example 3)

64. Solve and check: $6(y - 1) + 7 = 9y - y + 1$.
    (Section 2.3, Example 3)

65. Solve for w: $V = \frac{1}{3}lwh$. (Section 2.4, Example 4)

Preview Exercises

Exercises 66–68 will help you prepare for the material covered in the next section.

66. Use $A = \frac{1}{2}bh$ to find h if $A = 30$ and $b = 12$.

67. Evaluate $A = \frac{1}{2}h(a + b)$ for $a = 10$, $b = 16$, and $h = 7$.

68. Solve: $x = 4(90 - x) - 40$.

SECTION

2.6

What am I supposed to learn?

After studying this section, you should be able to:

1 Solve problems using formulas for perimeter and area.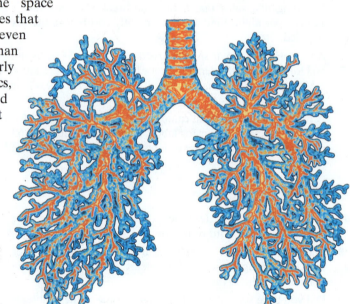

2 Solve problems using formulas for a circle's area and circumference.

3 Solve problems using formulas for volume.

4 Solve problems involving the angles of a triangle.

5 Solve problems involving complementary and supplementary angles.

1 Solve problems using formulas for perimeter and area.

Problem Solving in Geometry

Geometry is about the space you live in and the shapes that surround you. You're even made of it. The human lung consists of nearly 300 spherical air sacs, geometrically designed to provide the greatest surface area within the limited volume of our bodies. Viewed in this way, geometry becomes an intimate experience.

For thousands of years, people have studied geometry in some form to obtain a better understanding of the world in which they live. A study of the shape of your world will provide you with many practical applications that will help to increase your problem-solving skills.

Geometric Formulas for Perimeter and Area

Solving geometry problems often requires using basic geometric formulas. Formulas for perimeter and area are summarized in **Table 2.3**. Remember that perimeter is measured in linear units, such as feet or meters, and area is measured in square units, such as square feet, ft², or square meters, m².

Table 2.3	Common Formulas for Perimeter and Area		
Square $A = s^2$ $P = 4s$	**Rectangle** $A = lw$ $P = 2l + 2w$	**Triangle** $A = \frac{1}{2}bh$	**Trapezoid** $A = \frac{1}{2}h(a + b)$

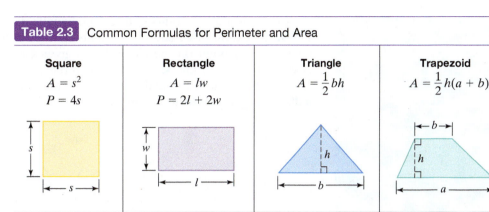

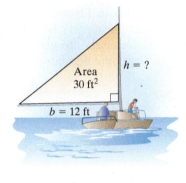

Figure 2.10 Finding the height of a triangular sail

EXAMPLE 1   Using the Formula for the Area of a Triangle

A sailboat has a triangular sail with an area of 30 square feet and a base that is 12 feet long. (See **Figure 2.10**.) Find the height of the sail.

Solution   We begin with the formula for the area of a triangle given in **Table 2.3**.

$$A = \frac{1}{2}bh \qquad \text{The area of a triangle is } \tfrac{1}{2} \text{ the product of its base and height.}$$

$$30 = \frac{1}{2}(12)h \qquad \text{Substitute 30 for A and 12 for } b.$$

$$30 = 6h \qquad \text{Simplify.}$$

$$\frac{30}{6} = \frac{6h}{6} \qquad \text{Divide both sides by 6.}$$

$$5 = h \qquad \text{Simplify.}$$

The height of the sail is 5 feet.

Check
The area is $A = \frac{1}{2}bh = \frac{1}{2}(12 \text{ feet})(5 \text{ feet}) = 30$ square feet. ∎

☑ **CHECK POINT 1**   A sailboat has a triangular sail with an area of 24 square feet and a base that is 4 feet long. Find the height of the sail.

---

2 Solve problems using formulas for a circle's area and circumference. ▶

Geometric Formulas for Circumference and Area of a Circle

It's a good idea to know your way around a circle. Clocks, angles, maps, and compasses are based on circles. Circles occur everywhere in nature: in ripples on water, patterns on a butterfly's wings, and cross sections of trees. Some consider the circle to be the most pleasing of all shapes.

A **circle** is a set of points in the plane equally distant from a given point, its center. **Figure 2.11** shows two circles. A **radius** (plural: radii), *r*, is a line segment from the center to any point on the circle. For a given circle, all radii have the same length. A **diameter**, *d*, is a line segment through the center whose endpoints both lie on the circle. For a given circle, all diameters have the same length. In any circle, **the length of a diameter is twice the length of a radius**.

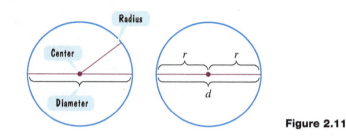

Figure 2.11

The point at which a pebble hits a flat surface of water becomes the center of a number of circular ripples.

The words *radius* and *diameter* refer to both the line segments in **Figure 2.11** as well as their linear measures. The distance around a circle (its perimeter) is called its **circumference**. Formulas for the area and circumference of a circle are given in terms of π and appear in **Table 2.4**. We have seen that π is an irrational number and is approximately equal to 3.14.

Table 2.4   Formulas for Circles

Circle	Area	Circumference
	$A = \pi r^2$	$C = 2\pi r$

When computing a circle's area or circumference by hand, round π to 3.14. When using a calculator, use the $\boxed{\pi}$ key, which gives the value of π rounded to approximately 11 decimal places. In either case, calculations involving π give approximate answers. These answers can vary slightly depending on how π is rounded. The symbol $\approx$ (is approximately equal to) will be written in these calculations.

EXAMPLE 2   Finding the Area and Circumference of a Circle

Find the area and circumference of a circle with a diameter measuring 20 inches.

Solution   The radius is half the diameter, so $r = \frac{20}{2} = 10$ inches.

$A = \pi r^2$ $C = 2\pi r$ Use the formulas for area and circumference of a circle.

$A = \pi(10)^2$ $C = 2\pi(10)$ Substitute 10 for *r*.

$A = 100\pi$ $C = 20\pi$

The area of the circle is 100π square inches and the circumference is 20π inches. Using the fact that $\pi \approx 3.14$, the area is approximately $100(3.14)$, or 314 square inches. The circumference is approximately $20(3.14)$, or 62.8 inches. ∎

✓ **CHECK POINT 2**   The diameter of a circular landing pad for helicopters is 40 feet. Find the area and circumference of the landing pad. Express answers in terms of π. Then round answers to the nearest square foot and foot, respectively.

EXAMPLE 3	Problem Solving Using the Formula for a Circle's Area

Which one of the following is the better buy: a large pizza with a 16-inch diameter for $15.00 or a medium pizza with an 8-inch diameter for $7.50?

Solution   The better buy is the pizza with the lower price per square inch. The radius of the large pizza is $\frac{1}{2} \cdot 16$ inches, or 8 inches, and the radius of the medium pizza is $\frac{1}{2} \cdot 8$ inches, or 4 inches. The area of the surface of each circular pizza is determined using the formula for the area of a circle.

$$\text{Large pizza:}\quad A = \pi r^2 = \pi(8\text{ in.})^2 = 64\pi\text{ in.}^2 \approx 201\text{ in.}^2$$

$$\text{Medium pizza:}\ A = \pi r^2 = \pi(4\text{ in.})^2 = 16\pi\text{ in.}^2 \approx 50\text{ in.}^2$$

For each pizza, the price per square inch is found by dividing the price by the area:

$$\text{Price per square inch for large pizza} = \frac{\$15.00}{64\pi\text{ in.}^2} \approx \frac{\$15.00}{201\text{ in.}^2} \approx \frac{\$0.07}{\text{in.}^2}$$

$$\text{Price per square inch for medium pizza} = \frac{\$7.50}{16\pi\text{ in.}^2} \approx \frac{\$7.50}{50\text{ in.}^2} = \frac{\$0.15}{\text{in.}^2}.$$

The large pizza costs approximately $0.07 per square inch and the medium pizza costs approximately $0.15 per square inch. Thus, the large pizza is the better buy.   ■

In Example 3, did you at first think that the price per square inch would be the same for the large and the medium pizzas? After all, the radius of the large pizza is twice that of the medium pizza, and the cost of the large is twice that of the medium. However, the large pizza's area, 64π square inches, is *four times the area* of the medium pizza, 16π square inches. Doubling the radius of a circle increases its area by four times the original amount.

☑ **CHECK POINT 3**   Which one of the following is the better buy: a large pizza with an 18-inch diameter for $20.00 or a medium pizza with a 14-inch diameter for $14.00?

Using Technology

You can use your calculator to obtain the price per square inch for each pizza in Example 3. The price per square inch for the large pizza, $\frac{15}{64\pi}$, is approximated by one of the following keystrokes:

Many Scientific Calculators

15 ÷ ((64 × π)) =

Many Graphing Calculators

15 ÷ ((64 π)) ENTER

3 Solve problems using formulas for volume.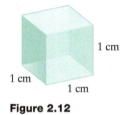

Geometric Formulas for Volume

A shoe box and a basketball are examples of three-dimensional figures. **Volume** refers to the amount of space occupied by such a figure. To measure this space, we begin by selecting a cubic unit. One such cubic unit, 1 cubic centimeter (cm³), is shown in **Figure 2.12**.

The edges of a cube all have the same length. Other cubic units used to measure volume include 1 cubic inch (in.³) and 1 cubic foot (ft³). The volume of a solid is the number of cubic units that can be contained in the solid.

Formulas for volumes of three-dimensional figures are given in **Table 2.5**.

Figure 2.12

Table 2.5	Common Formulas for Volume				

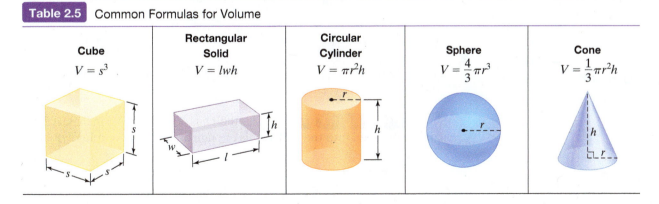

| Cube $V = s^3$ | Rectangular Solid $V = lwh$ | Circular Cylinder $V = \pi r^2 h$ | Sphere $V = \frac{4}{3}\pi r^3$ | Cone $V = \frac{1}{3}\pi r^2 h$ |

Radius: 2 inches
Height: 6 inches

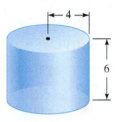

Radius: 4 inches
Height: 6 inches

Figure 2.13 Doubling a cylinder's radius

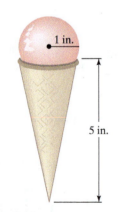

Figure 2.14

EXAMPLE 4 Using the Formula for the Volume of a Cylinder

A cylinder with a radius of 2 inches and a height of 6 inches has its radius doubled. (See **Figure 2.13.**) How many times greater is the volume of the larger cylinder than the volume of the smaller cylinder?

Solution We begin with the formula for the volume of a cylinder, $V = \pi r^2 h$, given on the previous page in **Table 2.5**. Find the volume of the smaller cylinder and the volume of the larger cylinder. To compare the volumes, divide the volume of the larger cylinder by the volume of the smaller cylinder.

$$V = \pi r^2 h \qquad \text{Use the formula for the volume of a cylinder.}$$

Radius is doubled.

$$V_{\text{Smaller}} = \pi(2)^2(6) \quad V_{\text{Larger}} = \pi(4)^2(6) \qquad \text{Substitute the given values.}$$
$$V_{\text{Smaller}} = \pi(4)(6) \quad V_{\text{Larger}} = \pi(16)(6)$$
$$V_{\text{Smaller}} = 24\pi \qquad V_{\text{Larger}} = 96\pi$$

The volume of the smaller cylinder is 24π cubic inches. The volume of the larger cylinder is 96π cubic inches. We use division to compare the volumes:

$$\frac{V_{\text{Larger}}}{V_{\text{Smaller}}} = \frac{96\pi}{24\pi} = \frac{4}{1}.$$

Thus, the volume of the larger cylinder is 4 times the volume of the smaller cylinder. ∎

✓ CHECK POINT 4 A cylinder with a radius of 3 inches and a height of 5 inches has its height doubled. How many times greater is the volume of the larger cylinder than the volume of the smaller cylinder?

EXAMPLE 5 Applying Volume Formulas

An ice cream cone is 5 inches deep and has a radius of 1 inch. A spherical scoop of ice cream also has a radius of 1 inch. (See **Figure 2.14.**) If the ice cream melts into the cone, will it overflow?

Solution The ice cream will overflow if the volume of the ice cream, a sphere, is greater than the volume of the cone. Find the volume of each.

$$V_{\text{cone}} = \frac{1}{3}\pi r^2 h = \frac{1}{3}\pi(1 \text{ in.})^2 \cdot 5 \text{ in.} = \frac{5\pi}{3}\text{in.}^3 \approx 5 \text{ in.}^3$$

$$V_{\text{sphere}} = \frac{4}{3}\pi r^3 = \frac{4}{3}\pi(1 \text{ in.})^3 = \frac{4\pi}{3}\text{in.}^3 \approx 4 \text{ in.}^3$$

The volume of the spherical scoop of ice cream is less than the volume of the cone, so there will be no overflow. ∎

✓ CHECK POINT 5 A basketball has a radius of 4.5 inches. If 350 cubic inches of air are pumped into the ball, is this enough air to fill it completely?

Blitzer Bonus ⓒ

Deceptions in Visual Displays of Data

Graphs can be used to distort data, making it difficult for the viewer to learn the truth. One potential source of misunderstanding involves geometric figures whose lengths are in the correct ratios for the displayed data, but whose areas or volumes are then varied to create a misimpression about how the data are changing over time. Here are two examples of misleading visual displays.

Graphic Display	Presentation Problems
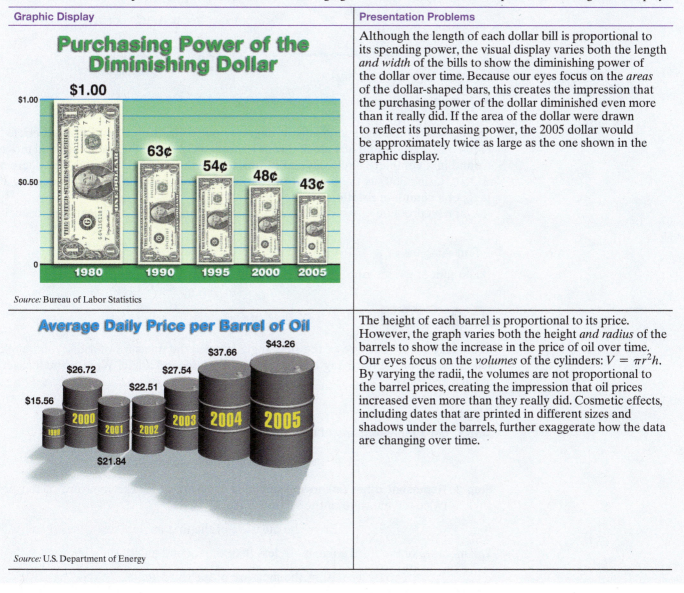 *Source:* Bureau of Labor Statistics	Although the length of each dollar bill is proportional to its spending power, the visual display varies both the length *and width* of the bills to show the diminishing power of the dollar over time. Because our eyes focus on the *areas* of the dollar-shaped bars, this creates the impression that the purchasing power of the dollar diminished even more than it really did. If the area of the dollar were drawn to reflect its purchasing power, the 2005 dollar would be approximately twice as large as the one shown in the graphic display.
Source: U.S. Department of Energy	The height of each barrel is proportional to its price. However, the graph varies both the height *and radius* of the barrels to show the increase in the price of oil over time. Our eyes focus on the *volumes* of the cylinders: $V = \pi r^2 h$. By varying the radii, the volumes are not proportional to the barrel prices, creating the impression that oil prices increased even more than they really did. Cosmetic effects, including dates that are printed in different sizes and shadows under the barrels, further exaggerate how the data are changing over time.

④ Solve problems involving the angles of a triangle. ⓒ

The Angles of a Triangle

The hour hand of a clock moves from 12 to 2. The hour hand suggests a **ray**, a part of a line that has only one endpoint and extends forever in the opposite direction. An *angle* is formed as the ray in **Figure 2.15** rotates from 12 to 2.

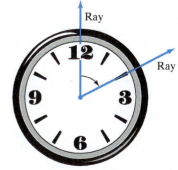

Figure 2.15 Clock with a ray rotating to form an angle

An **angle**, symbolized ∢, is made up of two rays that have a common endpoint. **Figure 2.16** shows an angle. The common endpoint, B in the figure, is called the **vertex** of the angle. The two rays that form the angle are called its **sides**. The four ways of naming the angle are shown to the right of **Figure 2.16**.

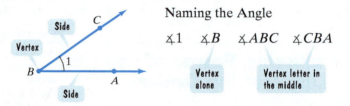

Naming the Angle

∢1 ∢B ∢ABC ∢CBA

Figure 2.16 An angle: two rays with a common endpoint

One way to measure angles is in **degrees**, symbolized by a small, raised circle °. Think of the hour hand of a clock. From 12 noon to 12 midnight, the hour hand moves around in a complete circle. By definition, the ray has rotated through 360 degrees, or 360°. Using 360° as the amount of rotation of a ray back onto itself, **a degree, 1°, is $\frac{1}{360}$ of a complete rotation.**

Our next problem is based on the relationship among the three angles of any triangle.

The Angles of a Triangle

The sum of the measures of the three angles of any triangle is 180°.

EXAMPLE 6 Angles of a Triangle

In a triangle, the measure of the first angle is twice the measure of the second angle. The measure of the third angle is 20° less than the second angle. What is the measure of each angle?

Solution

Step 1. Let x represent one of the quantities. Let

$$x = \text{the measure of the second angle.}$$

Step 2. Represent other unknown quantities in terms of x. The measure of the first angle is twice the measure of the second angle. Thus, let

$$2x = \text{the measure of the first angle.}$$

The measure of the third angle is 20° less than the second angle. Thus, let

$$x - 20 = \text{the measure of the third angle.}$$

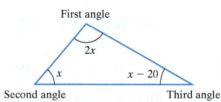

First angle

$2x$

x $x - 20$

Second angle   Third angle

Step 3. Write an equation in x that models the conditions. Because we are working with a triangle, the sum of the measures of its three angles is 180°.

$$2x \quad + \quad x \quad + \quad (x - 20) \quad = \quad 180$$

Step 4. Solve the equation and answer the question.

$$2x + x + (x - 20) = 180 \qquad \text{This is the equation that models the sum of the measures of the angles.}$$

$$4x - 20 = 180 \qquad \text{Regroup and combine like terms.}$$

$$4x - 20 + 20 = 180 + 20 \qquad \text{Add 20 to both sides.}$$

$$4x = 200 \qquad \text{Simplify.}$$

$$\frac{4x}{4} = \frac{200}{4} \qquad \text{Divide both sides by 4.}$$

$$x = 50 \qquad \text{Simplify.}$$

Measure of first angle $= 2x = 2 \cdot 50 = 100$

Measure of second angle $= x = 50$

Measure of third angle $= x - 20 = 50 - 20 = 30$

The angles measure 100°, 50°, and 30°.

Step 5. Check the proposed solution in the original wording of the problem. The problem tells us that we are working with a triangle's angles. Thus, the sum of the measures should be 180°. Adding the three measures, we obtain $100° + 50° + 30°$, giving the required sum of 180°. ■

☑ **CHECK POINT 6** In a triangle, the measure of the first angle is three times the measure of the second angle. The measure of the third angle is 20° less than the second angle. What is the measure of each angle?

---

5 Solve problems involving complementary and supplementary angles. ▶

Complementary and Supplementary Angles

Two angles with measures having a sum of 90° are called **complementary angles**. For example, angles measuring 70° and 20° are complementary angles because $70° + 20° = 90°$. For angles such as those measuring 70° and 20°, each angle is a **complement** of the other: The 70° angle is the complement of the 20° angle and the 20° angle is the complement of the 70° angle. The measure of the complement can be found by subtracting the angle's measure from 90°. For example, we can find the complement of a 25° angle by subtracting 25° from 90°: $90° - 25° = 65°$. Thus, an angle measuring 65° is the complement of one measuring 25°.

Two angles with measures having a sum of 180° are called **supplementary angles**. For example, angles measuring 110° and 70° are supplementary angles because $110° + 70° = 180°$. For angles such as those measuring 110° and 70°, each angle is a **supplement** of the other: The 110° angle is the supplement of the 70° angle and the 70° angle is the supplement of the 110° angle. The measure of the supplement can be found by subtracting the angle's measure from 180°. For example, we can find the supplement of a 25° angle by subtracting 25° from 180°: $180° - 25° = 155°$. Thus, an angle measuring 155° is the supplement of one measuring 25°.

Algebraic Expressions for Complements and Supplements

Measure of an angle: x

Measure of the angle's complement: $90 - x$

Measure of the angle's supplement: $180 - x$

EXAMPLE 7 Angle Measures and Complements

The measure of an angle is 40° less than four times the measure of its complement. What is the angle's measure?

Solution

Step 1. Let x represent one of the quantities. Let

$$x = \text{the measure of the angle.}$$

Step 2. Represent other unknown quantities in terms of x. Because this problem involves an angle and its complement, let

$$90 - x = \text{the measure of the complement.}$$

Step 3. Write an equation in x that models the conditions.

The angle's measure	is	40° less than	four times the measure of the complement.
x	$=$	$4(90 - x)$	$- \ 40$

Step 4. Solve the equation and answer the question.

$$x = 4(90 - x) - 40 \qquad \text{This is the equation that models the problem's conditions.}$$

$$x = 360 - 4x - 40 \qquad \text{Use the distributive property.}$$

$$x = 320 - 4x \qquad \text{Simplify: } 360 - 40 = 320.$$

$$x + 4x = 320 - 4x + 4x \qquad \text{Add } 4x \text{ to both sides.}$$

$$5x = 320 \qquad \text{Simplify.}$$

$$\frac{5x}{5} = \frac{320}{5} \qquad \text{Divide both sides by 5.}$$

$$x = 64 \qquad \text{Simplify.}$$

The angle measures 64°.

Step 5. Check the proposed solution in the original wording of the problem. The measure of the complement is $90° - 64° = 26°$. Four times the measure of the complement is $4 \cdot 26°$, or $104°$. The angle's measure, 64°, is 40° less than 104°: $104° - 40° = 64°$. As specified by the problem's wording, the angle's measure is 40° less than four times the measure of its complement. ∎

☑ **CHECK POINT 7** The measure of an angle is twice the measure of its complement. What is the angle's measure?

CONCEPT AND VOCABULARY CHECK

Fill in each blank so that the resulting statement is true.

1. The area, A, of a triangle with base b and height h is given by the formula _____.

2. The area, A, of a circle with radius r is given by the formula _____.

3. The circumference, C, of a circle with radius r is given by the formula _____.

4. In any circle, twice the length of the _____ is the length of the _____.

5. The volume, V, of a rectangular solid with length l, width w, and height h is given by the formula _____.

6. The volume, V, of a circular cylinder with radius r and height h is given by the formula _____.

7. The sum of the measures of the three angles of any triangle is _____.

8. Two angles with measures having a sum of 90° are called _____ angles.

9. Two angles with measures having a sum of 180° are called _____ angles.

10. If the measure of an angle is represented by x, the measure of its complement is represented by _____ and the measure of its supplement is represented by _____.

2.6 EXERCISE SET   ▶ MyMathLab®

Practice Exercises

Use the formulas for perimeter and area in **Table 2.3** on page 169 to solve Exercises 1–12.

In Exercises 1–2, find the perimeter and area of each rectangle.

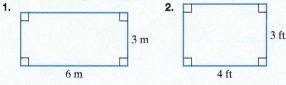

1.   3 m   6 m

2.   3 ft   4 ft

In Exercises 3–4, find the area of each triangle.

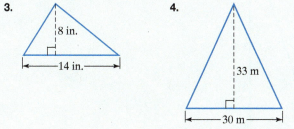

3.   8 in.   14 in.

4.   33 m   30 m

In Exercises 5–6, find the area of each trapezoid.

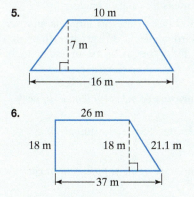

5.   10 m   7 m   16 m

6.   26 m   18 m   18 m   21.1 m   37 m

7. A rectangular swimming pool has a width of 25 feet and an area of 1250 square feet. What is the pool's length?

8. A rectangular swimming pool has a width of 35 feet and an area of 2450 square feet. What is the pool's length?

9. A triangle has a base of 5 feet and an area of 20 square feet. Find the triangle's height.

10. A triangle has a base of 6 feet and an area of 30 square feet. Find the triangle's height.

11. A rectangle has a width of 44 centimeters and a perimeter of 188 centimeters. What is the rectangle's length?

12. A rectangle has a width of 46 centimeters and a perimeter of 208 centimeters. What is the rectangle's length?

Use the formulas for the area and circumference of a circle in **Table 2.4** on page 170 to solve Exercises 13–18.

In Exercises 13–16, find the area and circumference of each circle. Express answers in terms of π. Then round to the nearest whole number.

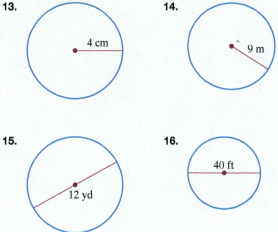

13.   4 cm

14.   9 m

15.   12 yd

16.   40 ft

17. The circumference of a circle is 14π inches. Find the circle's radius and diameter.

18. The circumference of a circle is 16π inches. Find the circle's radius and diameter.

Use the formulas for volume in **Table 2.5** on page 171 to solve Exercises 19–30.

In Exercises 19–26, find the volume of each figure. Where applicable, express answers in terms of π. Then round to the nearest whole number.

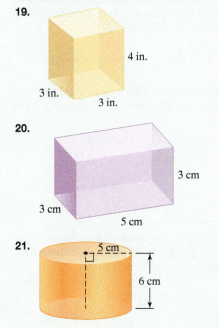

19.   4 in.   3 in.   3 in.

20.   3 cm   3 cm   5 cm

21.   5 cm   6 cm

22.

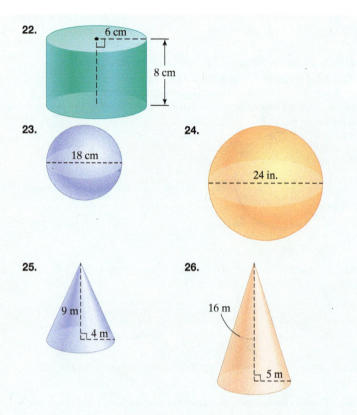

6 cm

8 cm

23.

18 cm

24.

24 in.

25.

9 m

4 m

26.

16 m

5 m

27. Solve the formula for the volume of a circular cylinder for h.

28. Solve the formula for the volume of a cone for h.

29. A cylinder with radius 3 inches and height 4 inches has its radius tripled. How many times greater is the volume of the larger cylinder than the smaller cylinder?

30. A cylinder with radius 2 inches and height 3 inches has its radius quadrupled. How many times greater is the volume of the larger cylinder than the smaller cylinder?

Use the relationship among the three angles of any triangle to solve Exercises 31–36.

31. Two angles of a triangle have the same measure and the third angle is 30° greater than the measure of the other two. Find the measure of each angle.

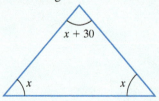

$x + 30$

x x

32. One angle of a triangle is three times as large as another. The measure of the third angle is 40° more than that of the smallest angle. Find the measure of each angle.

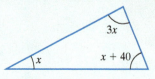

$3x$

x $x + 40$

Find the measure of each angle whose degree measure is represented in terms of x in the triangles in Exercises 33–34.

33.

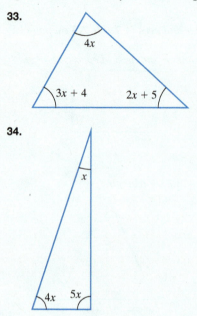

$4x$

$3x + 4$ $2x + 5$

34.

x

$4x$ $5x$

35. One angle of a triangle is twice as large as another. The measure of the third angle is 20° more than that of the smallest angle. Find the measure of each angle.

36. One angle of a triangle is three times as large as another. The measure of the third angle is 30° greater than that of the smallest angle. Find the measure of each angle.

In Exercises 37–40, find the measure of the complement of each angle.

37. 58°   **38.** 41°   **39.** 88°   **40.** 2°

In Exercises 41–44, find the measure of the supplement of each angle.

41. 132°   **42.** 93°

43. 90°   **44.** 179.5°

In Exercises 45–50, use the five-step problem-solving strategy to find the measure of the angle described.

45. The angle's measure is 60° more than that of its complement.

46. The angle's measure is 78° less than that of its complement.

47. The angle's measure is three times that of its supplement.

48. The angle's measure is 16° more than triple that of its supplement.

49. The measure of the angle's supplement is 10° more than three times that of its complement.

50. The measure of the angle's supplement is 52° more than twice that of its complement.

Practice PLUS

In Exercises 51–53, find the area of each figure.

51.

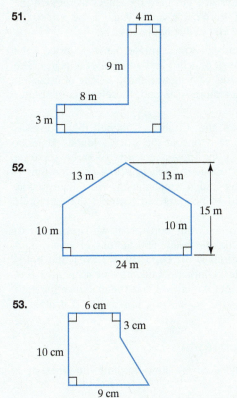

52.

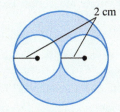

53.

54. Find the area of the shaded region in terms of π.

55. Find the volume of the cement block in the figure shown.

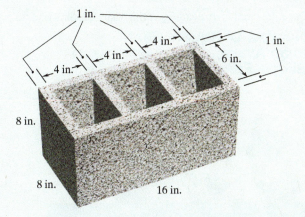

56. Find the volume of the darkly shaded region. Express the answer in terms of π.

Application Exercises

*Use the formulas for perimeter and area in **Table 2.3** on page 169 to solve Exercises 57–58.*

57. Taxpayers with an office in their home may deduct a percentage of their home-related expenses. This percentage is based on the ratio of the office's area to the area of the home. A taxpayer with a 2200-square-foot home maintains a 20-foot by 16-foot office. If the yearly electricity bills for the home come to $4800, how much of this is deductible?

58. The lot in the figure shown, except for the house, shed, and driveway, is lawn. One bag of lawn fertilizer costs $25.00 and covers 4000 square feet.

   a. Determine the minimum number of bags of fertilizer needed for the lawn.

   b. Find the total cost of the fertilizer.

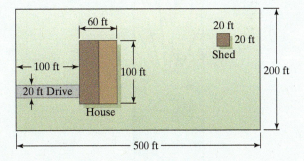

*Use the formulas for the area and the circumference of a circle in **Table 2.4** on page 170 to solve Exercises 59–64. Unless otherwise indicated, round all circumference and area calculations to the nearest whole number.*

59. Which one of the following is a better buy: a large pizza with a 14-inch diameter for $12.00 or a medium pizza with a 7-inch diameter for $5.00?

60. Which one of the following is a better buy: a large pizza with a 16-inch diameter for $12.00 or two small pizzas, each with a 10-inch diameter, for $12.00?

61. If asphalt pavement costs $0.80 per square foot, find the cost to pave the circular road in the figure shown. Round to the nearest dollar.

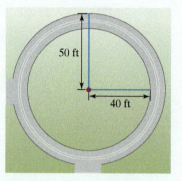

62. Hardwood flooring costs $10.00 per square foot. How much will it cost (to the nearest dollar) to cover the dance floor shown in the figure with hardwood flooring?

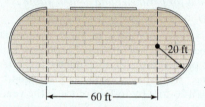

63. A glass window is to be placed in a house. The window consists of a rectangle, 6 feet high by 3 feet wide, with a semicircle at the top. Approximately how many feet of stripping, to the nearest tenth of a foot, will be needed to frame the window?

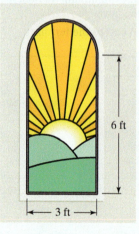

64. How many plants spaced every 6 inches are needed to surround a circular garden with a 30-foot radius?

*Use the formulas for volume in **Table 2.5** on page 171 to solve Exercises 65–69. When necessary, round all volume calculations to the nearest whole number.*

65. A water reservoir is shaped like a rectangular solid with a base that is 50 yards by 30 yards, and a vertical height of 20 yards. At the start of a three-month period of no rain, the reservoir was completely full. At the end of this period, the height of the water was down to 6 yards. How much water was used in the three-month period?

66. A building contractor is to dig a foundation 4 yards long, 3 yards wide, and 2 yards deep for a toll booth's foundation. The contractor pays $10 per load for trucks to remove the dirt. Each truck holds 6 cubic yards. What is the cost to the contractor to have all the dirt hauled away?

67. Two cylindrical cans of soup sell for the same price. One can has a diameter of 6 inches and a height of 5 inches. The other has a diameter of 5 inches and a height of 6 inches. Which can contains more soup and, therefore, is the better buy?

68. The tunnel under the English Channel that connects England and France is one of the world's longest tunnels. The Chunnel, as it is known, consists of three separate tunnels built side by side. Each is a half-cylinder that is 50,000 meters long and 4 meters high. How many cubic meters of dirt had to be removed to build the Chunnel?

69. You are about to sue your contractor who promised to install a water tank that holds 500 gallons of water. You know that 500 gallons is the capacity of a tank that holds 67 cubic feet. The cylindrical tank has a radius of 3 feet and a height of 2 feet 4 inches. Does the evidence indicate you can win the case against the contractor if it goes to court?

Explaining the Concepts

70. Using words only, describe how to find the area of a triangle.

71. Describe the difference between the following problems: How much fencing is needed to enclose a garden? How much fertilizer is needed for the garden?

72. Describe how volume is measured. Explain why linear or square units cannot be used.

73. What is an angle?

74. If the measures of two angles of a triangle are known, explain how to find the measure of the third angle.

75. Can a triangle contain two 90° angles? Explain your answer.

76. What are complementary angles? Describe how to find the measure of an angle's complement.

77. What are supplementary angles? Describe how to find the measure of an angle's supplement.

78. Describe what is misleading in this visual display of data.

Number of Square Feet in an Average U.S. Single-Family Home

Source: National Association of Home Builders

Critical Thinking Exercises

Make Sense? In Exercises 79–82, determine whether each statement makes sense or does not make sense, and explain your reasoning.

79. There is nothing that is misleading in this visual display of data.

Book Title Output in the United States

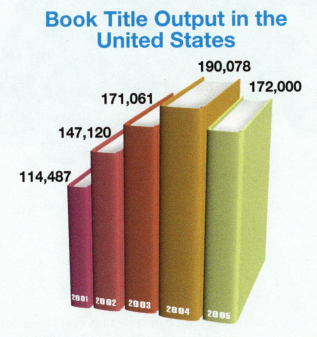

190,078
172,000
171,061
147,120
114,487

2001 2002 2003 2004 2005

Source: R. R. Bowker

80. I solved a word problem and determined that a triangle had angles measuring 37°, 58°, and 86°.

81. I paid $10 for a pizza, so I would expect to pay approximately $20 for the same kind of pizza with twice the radius.

82. I find that my answers involving π can vary slightly depending on whether I round π mid-calculation or use the π key on my calculator and then round at the very end.

In Exercises 83–86, determine whether each statement is true or false. If the statement is false, make the necessary change(s) to produce a true statement.

83. It is possible to have a circle whose circumference is numerically equal to its area.

84. When the measure of a given angle is added to three times the measure of its complement, the sum equals the sum of the measures of the complement and supplement of the angle.

85. The complement of an angle that measures less than 90° is an angle that measures more than 90°.

86. Two complementary angles can be equal in measure.

87. Suppose you know the cost for building a rectangular deck measuring 8 feet by 10 feet. If you decide to increase the dimensions to 12 feet by 15 feet, by how many times will the cost increase?

88. A rectangular swimming pool measures 14 feet by 30 feet. The pool is surrounded on all four sides by a path that is 3 feet wide. If the cost to resurface the path is $2 per square foot, what is the total cost of resurfacing the path?

89. What happens to the volume of a sphere if its radius is doubled?

90. A scale model of a car is constructed so that its length, width, and height are each $\frac{1}{10}$ the length, width, and height of the actual car. By how many times does the volume of the car exceed its scale model?

91. Find the measure of the angle of inclination, denoted by x in the figure, for the road leading to the bridge.

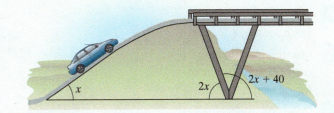

x $2x$ $2x + 40$

Review Exercises

92. Solve for s: $P = 2s + b$.
(Section 2.4, Example 3)

93. Solve for x: $\dfrac{x}{2} + 7 = 13 - \dfrac{x}{4}$.
(Section 2.3, Example 4)

94. Simplify: $\left[3\left(12 \div 2^2 - 3\right)^2\right]^2$.
(Section 1.8, Example 8)

Preview Exercises

Exercises 95–97 will help you prepare for the material covered in the next section.

95. Is 2 a solution of $x + 3 < 8$?

96. Is 6 a solution of $4y - 7 \geq 5$?

97. Solve: $2(x - 3) + 5x = 8(x - 1)$.

2.7

Solving Linear Inequalities

What am I supposed to learn?

After studying this section, you should be able to:

1. Graph the solutions of an inequality on a number line. ▶

2. Use interval notation. ▶

3. Understand properties used to solve linear inequalities. ▶

4. Solve linear inequalities. ▶

5. Identify inequalities with no solution or true for all real numbers. ▶

6. Solve problems using linear inequalities. ▶

Do you remember Rent-a-Heap, the car rental company that charged $125 per week plus $0.20 per mile to rent a small car? In Example 4 on page 161, we asked the question: How many miles can you travel for $335? We let x represent the number of miles and set up a linear equation as follows:

The weekly charge of $125	plus	the charge of $0.20 per mile for x miles	equals	the total $335 rental charge.
125	+	0.20x	=	335.

Because we are limited by how much money we can spend on everything from buying clothing to renting a car, it is also possible to ask: How many miles can you travel if you can spend *at most* $335? We again let x represent the number of miles. Spending *at most* $335 means that the amount spent on the weekly rental must be *less than or equal to* $335:

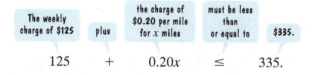

The weekly charge of **$125**	plus	the charge of **$0.20** per mile for x miles	must be less than or equal to	**$335.**
125	+	0.20x	≤	335.

Using the commutative property of addition, we can express this inequality as

$$0.20x + 125 \le 335.$$

The form of this inequality is $ax + b \le c$, with $a = 0.20$, $b = 125$, and $c = 335$. Any inequality in this form is called a **linear inequality in one variable**. The symbol between $ax + b$ and c can be ≤ (is less than or equal to), < (is less than), ≥ (is greater than or equal to), or > (is greater than). The greatest exponent on the variable in such an inequality is 1.

In this section, we will study how to solve linear inequalities such as $0.20x + 125 \le 335$. **Solving an inequality** is the process of finding the set of numbers that will make the inequality a true statement. These numbers are called the **solutions** of the inequality, and we say that they **satisfy** the inequality. The set of all solutions is called the **solution set** of the inequality. We begin by discussing how to graph and how to represent these solution sets.

① Graph the solutions of an inequality on a number line.

Graphs of Inequalities

There are infinitely many solutions to the inequality $x < 3$, namely, all real numbers that are less than 3. Although we cannot list all the solutions, we can make a drawing on a number line that represents these solutions. Such a drawing is called the **graph of the inequality**.

Graphs of solutions to linear inequalities are shown on a number line by shading all points representing numbers that are solutions. **Square brackets, [ ], indicate endpoints that are solutions. Parentheses, ( ), indicate endpoints that are not solutions.**

EXAMPLE 1   Graphing Inequalities

Great Question!

Is there another way that I can write $x < 3$?

Because an inequality symbol points to the smaller number, $x < 3$ (x is less than 3) may be expressed as $3 > x$ (3 is greater than x).

Graph the solutions of each inequality:

a. $x < 3$ **b.** $x \geq -1$ **c.** $-1 < x \leq 3$.

Solution

a. The solutions of $x < 3$ are all real numbers that are less than 3. They are graphed on a number line by shading all points to the left of 3. The parenthesis at 3 indicates that 3 is not a solution, but numbers such as 2.9999 and 2.6 are. The arrow shows that the graph extends indefinitely to the left.

$$\xleftarrow{\qquad} \overset{-4 \quad -3 \quad -2 \quad -1 \quad 0 \quad 1 \quad 2 \quad 3 \quad 4}{\longrightarrow}$$

b. The solutions of $x \geq -1$ are all real numbers that are greater than or equal to -1. We shade all points to the right of -1 and the point for -1 itself. The square bracket at -1 shows that -1 is a solution of the given inequality. The arrow shows that the graph extends indefinitely to the right.

$$\overset{-4 \quad -3 \quad -2 \quad -1 \quad 0 \quad 1 \quad 2 \quad 3 \quad 4}{\xrightarrow{\qquad}}$$

c. The inequality $-1 < x \leq 3$ is read "-1 is less than x *and* x is less than or equal to 3," or "x is greater than -1 *and* less than or equal to 3." The solutions of $-1 < x \leq 3$ are all real numbers between -1 and 3, not including -1 but including 3. The parenthesis at -1 indicates that -1 is not a solution. The square bracket at 3 shows that 3 is a solution. Shading indicates the other solutions.

$$\overset{-4 \quad -3 \quad -2 \quad -1 \quad 0 \quad 1 \quad 2 \quad 3 \quad 4}{\longrightarrow}$$

✓ **CHECK POINT 1**   Graph the solutions of each inequality:

a. $x < 4$ **b.** $x \geq -2$ **c.** $-4 \leq x < 1$.

---

② Use interval notation.

Interval Notation

The solutions of $x < 3$ are all real numbers that are less than 3:

$$\xleftarrow{\qquad} \overset{-4 \quad -3 \quad -2 \quad -1 \quad 0 \quad 1 \quad 2 \quad 3 \quad 4}{\longrightarrow}$$

These numbers form an *interval* on the number line. The solution set of $x < 3$ can be expressed in **interval notation** as

$$(-\infty, 3).$$

The negative infinity symbol indicates that the interval extends indefinitely to the left.

The parenthesis indicates that 3 is excluded from the interval.

The solution set of $x < 3$ can also be expressed in set-builder notation as

$$\{ x \mid x < 3 \}.$$

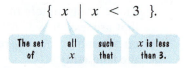

| The set of | all x | such that | x is less than 3. |

Table 2.6 shows four inequalities, their solution sets using interval and set-builder notations, and graphs of the solution sets.

Table 2.6 Solution Sets of Inequalities

Let a and b be real numbers.

Inequality	Interval Notation	Set-Builder Notation	Graph
$x > a$	(a, ∞)	$\{x \mid x > a\}$	⟵(———▶ at a
$x \geq a$	$[a, \infty)$	$\{x \mid x \geq a\}$	⟵[———▶ at a
$x < b$	$(-\infty, b)$	$\{x \mid x < b\}$	◀———) at b
$x \leq b$	$(-\infty, b]$	$\{x \mid x \leq b\}$	◀———] at b

Parentheses and Brackets in Interval Notation

Parentheses indicate endpoints that are not included in an interval. Square brackets indicate endpoints that are included in an interval. Parentheses are always used with ∞ or $-\infty$.

EXAMPLE 2 Using Interval Notation

Express the solution set of each inequality in interval notation and graph the interval:

a. $x \leq -1$ **b.** $x > 2$.

Solution

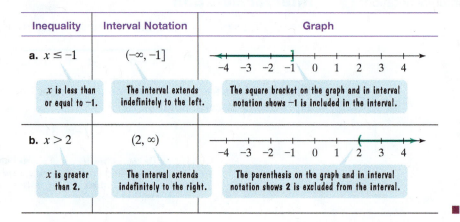

Inequality	Interval Notation	Graph
a. $x \leq -1$	$(-\infty, -1]$	number line from −4 to 4 with square bracket at −1
	x is less than or equal to −1.	The interval extends indefinitely to the left.	The square bracket on the graph and in interval notation shows −1 is included in the interval.
b. $x > 2$	$(2, \infty)$	number line from −4 to 4 with parenthesis at 2
	x is greater than 2.	The interval extends indefinitely to the right.	The parenthesis on the graph and in interval notation shows 2 is excluded from the interval.

✓ **CHECK POINT 2** Express the solution set of each inequality in interval notation and graph the interval:

a. $x \geq 0$ **b.** $x < 5$.

---

③ Understand properties used to solve linear inequalities. ▶

Properties Used to Solve Linear Inequalities

Back to our question that opened this section: How many miles can you drive your Rent-a-Heap car if you can spend at most $335 per week? We answer the question by solving

$$0.20x + 125 \leq 335$$

for x. The solution procedure is nearly identical to that for solving

$$0.20x + 125 = 335.$$

Our goal is to get x by itself on the left side. We do this by first subtracting 125 from both sides to isolate $0.20x$:

$0.20x + 125 \leq 335$	This is the given inequality.
$0.20x + 125 - 125 \leq 335 - 125$	Subtract 125 from both sides.
$0.20x \leq 210.$	Simplify.

Finally, we isolate x from $0.20x$ by dividing both sides of the inequality by 0.20:

$\dfrac{0.20x}{0.20} \leq \dfrac{210}{0.20}$	Divide both sides by 0.20.
$x \leq 1050.$	Simplify.

With at most $335 per week to spend, you can travel at most 1050 miles.

We started with the inequality $0.20x + 125 \leq 335$ and obtained the inequality $x \leq 1050$ in the final step. Both of these inequalities have the same solution set, namely $\{x \,|\, x \leq 1050\}$. Inequalities such as these, with the same solution set, are said to be **equivalent**.

We isolated x from $0.20x$ by dividing both sides of $0.20x \leq 210$ by 0.20, a positive number. Let's see what happens if we divide both sides of an inequality by a negative number. Consider the inequality $10 < 14$. Divide both 10 and 14 by -2:

$$\frac{10}{-2} = -5 \quad \text{and} \quad \frac{14}{-2} = -7.$$

Because -5 lies to the right of -7 on the number line, -5 is greater than -7:

$$-5 > -7.$$

Notice that the direction of the inequality symbol is reversed:

$$10 < 14$$
$$\updownarrow$$
$$-5 > -7.$$

Dividing by -2 changes the direction of the inequality symbol.

In general, **when we multiply or divide both sides of an inequality by a negative number, the direction of the inequality symbol is reversed**. When we reverse the direction of the inequality symbol, we say that we change the *sense* of the inequality.

Great Question!

What are some common English phrases and sentences that I can model with linear inequalities?

English phrases such as "at least" and "at most" can be represented by inequalities.

English Sentence	Inequality
x is at least 5.	$x \geq 5$
x is at most 5.	$x \leq 5$
x is no more than 5.	$x \leq 5$
x is no less than 5.	$x \geq 5$

We can isolate a variable in a linear inequality the same way we can isolate a variable in a linear equation. The following properties are used to create equivalent inequalities:

Properties of Inequalities

Property	The Property in Words	Example
The Addition Property of Inequality If $a < b$, then $a + c < b + c$. If $a < b$, then $a - c < b - c$.	If the same quantity is added to or subtracted from both sides of an inequality, the resulting inequality is equivalent to the original one.	$2x + 3 < 7$ Subtract 3: $2x + 3 - 3 < 7 - 3$. Simplify: $2x < 4$.
The Positive Multiplication Property of Inequality If $a < b$ and c is positive, then $ac < bc$. If $a < b$ and c is positive, then $\dfrac{a}{c} < \dfrac{b}{c}$.	If we multiply or divide both sides of an inequality by the same positive quantity, the resulting inequality is equivalent to the original one.	$2x < 4$ Divide by 2: $\dfrac{2x}{2} < \dfrac{4}{2}$. Simplify: $x < 2$.
The Negative Multiplication Property of Inequality If $a < b$ and c is negative, then $ac > bc$. If $a < b$ and c is negative, then $\dfrac{a}{c} > \dfrac{b}{c}$.	If we multiply or divide both sides of an inequality by the same negative quantity and reverse the direction of the inequality symbol, the resulting inequality is equivalent to the original one.	$-4x < 20$ Divide by -4 and reverse the direction of the inequality symbol: $\dfrac{-4x}{-4} > \dfrac{20}{-4}$. Simplify: $x > -5$.

4 Solve linear inequalities. ▶

Solving Linear Inequalities Involving Only One Property of Inequality

If you can solve a linear equation, it is likely that you can solve a linear inequality. Why? The procedure for solving linear inequalities is nearly the same as the procedure for solving linear equations, with one important exception: **When multiplying or dividing by a negative number, reverse the direction of the inequality symbol, changing the sense of the inequality.**

EXAMPLE 3 Solving a Linear Inequality

Solve and graph the solution set on a number line:

$$x + 3 < 8.$$

Solution Our goal is to isolate x. We can do this by using the addition property, subtracting 3 from both sides.

$x + 3 < 8$	This is the given inequality.
$x + 3 - 3 < 8 - 3$	Subtract 3 from both sides.
$x < 5$	Simplify.

The solution set consists of all real numbers that are less than 5. We express this in interval notation as $(-\infty, 5)$, or in set-builder notation as $\{x \mid x < 5\}$. The graph of the solution set is shown as follows:

Discover for Yourself

Can you check all solutions to Example 3 in the given inequality? Is a partial check possible? Select a real number that is less than 5 and show that it satisfies $x + 3 < 8$.

☑ **CHECK POINT 3** Solve and graph the solution set on a number line:

$$x + 6 < 9.$$

EXAMPLE 4 Solving a Linear Inequality

Solve and graph the solution set on a number line:

$$4x - 1 \geq 3x - 6.$$

Solution Our goal is to isolate all terms involving x on one side and all numerical terms on the other side, exactly as we did when solving equations. Let's begin by using the addition property to isolate variable terms on the left.

$4x - 1 \geq 3x - 6$	This is the given inequality.
$4x - 3x - 1 \geq 3x - 3x - 6$	Subtract 3x from both sides.
$x - 1 \geq -6$	Simplify.

Now we isolate the numerical terms on the right. Use the addition property and add 1 to both sides.

$x - 1 + 1 \geq -6 + 1$	Add 1 to both sides.
$x \geq -5$	Simplify.

The solution set consists of all real numbers that are greater than or equal to -5. We express this in interval notation as $[-5, \infty)$, or in set-builder notation as $\{x \mid x \geq -5\}$.

The graph of the solution set is shown as follows:

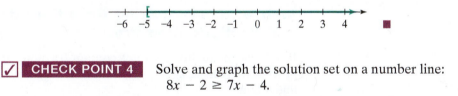

☑ **CHECK POINT 4** Solve and graph the solution set on a number line:

$$8x - 2 \geq 7x - 4.$$

We solved the inequalities in Examples 3 and 4 using the addition property of inequality. Now let's practice using the multiplication property of inequality. Do not forget to reverse the direction of the inequality symbol when multiplying or dividing both sides by a negative number.

EXAMPLE 5 Solving Linear Inequalities

Solve and graph the solution set on a number line:

a. $\frac{1}{3}x < 5$ **b.** $-3x < 21.$

Solution In each case, our goal is to isolate x. In the first inequality, this is accomplished by multiplying both sides by 3. In the second inequality, we can do this by dividing both sides by -3.

a. $\frac{1}{3}x < 5$ This is the given inequality.

$3 \cdot \frac{1}{3}x < 3 \cdot 5$ Isolate x by multiplying by 3 on both sides.
└──── The symbol $<$ stays the same because we are multiplying both sides by a positive number.

$x < 15$ Simplify.

The solution set consists of all real numbers that are less than 15. We express this in interval notation as $(-\infty, 15)$, or in set-builder notation as $\{x \mid x < 15\}$. The graph of the solution set is shown as follows:

b. $-3x < 21$ This is the given inequality.

$$\dfrac{-3x}{-3} > \dfrac{21}{-3}$$ Isolate x by dividing by -3 on both sides. The symbol $<$ must be reversed because we are dividing both sides by a negative number.

$x > -7$ Simplify.

The solution set consists of all real numbers that are greater than -7. We express this in interval notation as $(-7, \infty)$, or in set-builder notation as $\{x \mid x > -7\}$. The graph of the solution set is shown as follows:

✓ **CHECK POINT 5**   Solve and graph the solution set on a number line:

a. $\dfrac{1}{4}x < 2$ **b.** $-6x < 18$.

Inequalities Requiring Both the Addition and Multiplication Properties

If a linear inequality does not contain fractions, it can be solved using the following procedure. Notice, again, how similar this procedure is to the procedure for solving a linear equation.

Solving a Linear Inequality

1. Simplify the algebraic expression on each side.
2. Use the addition property of inequality to collect all the variable terms on one side and all the constant terms on the other side.
3. Use the multiplication property of inequality to isolate the variable and solve. Change the sense of the inequality when multiplying or dividing both sides by a negative number.
4. Express the solution set in interval or set-builder notation, and graph the solution set on a number line.

EXAMPLE 6   Solving a Linear Inequality

Solve and graph the solution set on a number line:

$$4y - 7 \geq 5.$$

Solution

Step 1. Simplify each side. Because each side is already simplified, we can skip this step.

Step 2. Collect variable terms on one side and constant terms on the other side. The only variable term, $4y$, is already on the left side of $4y - 7 \geq 5$. We will collect constant terms on the right by adding 7 to both sides.

$$4y - 7 \geq 5 \qquad \text{This is the given inequality.}$$
$$4y - 7 + 7 \geq 5 + 7 \qquad \text{Add 7 to both sides.}$$
$$4y \geq 12 \qquad \text{Simplify.}$$

Step 3. Isolate the variable and solve. We isolate the variable, y, by dividing both sides by 4. Because we are dividing by a positive number, we do not reverse the inequality symbol.

$$\frac{4y}{4} \geq \frac{12}{4} \qquad \text{Divide both sides by 4.}$$
$$y \geq 3 \qquad \text{Simplify.}$$

Step 4. Express the solution set in interval or set-builder notation, and graph the set on a number line. The solution set consists of all real numbers that are greater than or equal to 3, expressed in interval notation as $[3, \infty)$, or in set-builder notation as $\{y \mid y \geq 3\}$. The graph of the solution set is shown as follows:

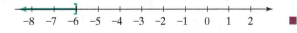

✓ **CHECK POINT 6** Solve and graph the solution set on a number line:
$$5y - 3 \geq 17.$$

EXAMPLE 7 Solving a Linear Inequality

Solve and graph the solution set on a number line:
$$7x + 15 \geq 13x + 51.$$

Solution

Step 1. Simplify each side. Because each side is already simplified, we can skip this step.

Step 2. Collect variable terms on one side and constant terms on the other side. We will collect variable terms on the left and constant terms on the right.

$$7x + 15 \geq 13x + 51 \qquad \text{This is the given inequality.}$$
$$7x + 15 - 13x \geq 13x + 51 - 13x \qquad \text{Subtract } 13x \text{ from both sides.}$$
$$-6x + 15 \geq 51 \qquad \text{Simplify.}$$
$$-6x + 15 - 15 \geq 51 - 15 \qquad \text{Subtract 15 from both sides.}$$
$$-6x \geq 36 \qquad \text{Simplify.}$$

Step 3. Isolate the variable and solve. We isolate the variable, x, by dividing both sides by -6. Because we are dividing by a negative number, we must reverse the inequality symbol.

$$\frac{-6x}{-6} \leq \frac{36}{-6} \qquad \text{Divide both sides by } -6 \text{ and change the sense of the inequality.}$$

$$x \leq -6 \qquad \text{Simplify.}$$

Step 4. Express the solution set in interval or set-builder notation, and graph the set on a number line. The solution set consists of all real numbers that are less than or equal to -6, expressed in interval notation as $(-\infty, -6]$, or in set-builder notation as $\{x \mid x \leq -6\}$. The graph of the solution set is shown as follows:

☑ **CHECK POINT 7** Solve and graph the solution set: $6 - 3x \le 5x - 2$.

EXAMPLE 8 Solving a Linear Inequality

Solve and graph the solution set on a number line:

$$2(x - 3) + 5x \le 8(x - 1).$$

Solution

Step 1. Simplify each side. We use the distributive property to remove parentheses. Then we combine like terms.

$2(x - 3) + 5x \le 8(x - 1)$	This is the given inequality.
$2x - 6 + 5x \le 8x - 8$	Use the distributive property.
$7x - 6 \le 8x - 8$	Add like terms on the left.

Step 2. Collect variable terms on one side and constant terms on the other side. We will collect variable terms on the left and constant terms on the right.

$7x - 8x - 6 \le 8x - 8x - 8$	Subtract $8x$ from both sides.
$-x - 6 \le -8$	Simplify.
$-x - 6 + 6 \le -8 + 6$	Add 6 to both sides.
$-x \le -2$	Simplify.

Step 3. Isolate the variable and solve. To isolate x in $-x \le -2$, we must eliminate the negative sign in front of the x. Because $-x$ means $-1x$, we can do this by multiplying (or dividing) both sides of the inequality by -1. We are multiplying by a negative number. Thus, we must reverse the inequality symbol.

$(-1)(-x) \ge (-1)(-2)$	Multiply both sides of $-x \le -2$ by -1 and change the sense of the inequality.
$x \ge 2$	Simplify.

Step 4. Express the solution set in interval or set-builder notation, and graph the set on a number line. The solution set consists of all real numbers that are greater than or equal to 2, expressed in interval notation as $[2, \infty)$, or in set-builder notation as $\{x \mid x \ge 2\}$. The graph of the solution set is shown as follows:

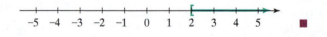

☑ **CHECK POINT 8** Solve and graph the solution set on a number line:

$$2(x - 3) - 1 \le 3(x + 2) - 14.$$

---

5 Identify inequalities with no solution or true for all real numbers. ⊙

Inequalities with Unusual Solution Sets

We have seen that some equations have no solution. This is also true for some inequalities. An example of such an inequality is

$$x > x + 1.$$

There is no number that is greater than itself plus 1. This inequality has no solution. Its solution set is $\varnothing$, the empty set.

By contrast, some inequalities are true for all real numbers. An example of such an inequality is

$$x < x + 1.$$

Every real number is less than itself plus 1. The solution set is expressed in interval notation as $(-\infty, \infty)$, or in set-builder notation as $\{x \mid x \text{ is a real number}\}$.

Recognizing Inequalities with No Solution or True for All Real Numbers

If you attempt to solve an inequality with no solution or one that is true for every real number, you will eliminate the variable.

- An inequality with no solution results in a false statement, such as $0 > 1$. The solution set is $\varnothing$, the empty set.

- An inequality that is true for every real number results in a true statement, such as $0 < 1$. The solution set is $(-\infty, \infty)$ or $\{x \mid x \text{ is a real number}\}$.

EXAMPLE 9 Solving a Linear Inequality

Solve: $3(x + 1) > 3x + 5$.

Solution

$3(x + 1) > 3x + 5$	This is the given inequality.
$3x + 3 > 3x + 5$	Apply the distributive property.
$3x + 3 - 3x > 3x + 5 - 3x$	Subtract $3x$ from both sides.
$3 > 5$	Simplify.

Keep reading. **3 > 5** is not the solution.

The original inequality is equivalent to the statement $3 > 5$, which is false for every value of x. The inequality has no solution. The solution set is $\varnothing$, the empty set. ∎

☑ **CHECK POINT 9**   Solve: $4(x + 2) > 4x + 15$.

EXAMPLE 10 Solving a Linear Inequality

Solve: $2(x + 5) \le 5x - 3x + 14$.

Solution

$2(x + 5) \le 5x - 3x + 14$	This is the given inequality.
$2x + 10 \le 5x - 3x + 14$	Apply the distributive property.
$2x + 10 \le 2x + 14$	Combine like terms.
$2x + 10 - 2x \le 2x + 14 - 2x$	Subtract $2x$ from both sides.
$10 \le 14$	Simplify.

Keep reading. **10 ≤ 14** is not the solution.

The original inequality is equivalent to the statement $10 \le 14$, which is true for every value of x. The solution is the set of all real numbers, expressed in interval notation as $(-\infty, \infty)$, or in set-builder notation as $\{x \mid x \text{ is a real number}\}$. ∎

☑ **CHECK POINT 10**   Solve: $3(x + 1) \ge 2x + 1 + x$.

6 Solve problems using linear inequalities. ▶

Applications

As you know, different professors may use different grading systems to determine your final course grade. Some professors require a final examination; others do not. In our next example, a final exam is required *and* it also counts as two grades.

EXAMPLE 11 An Application: Final Course Grade

To earn an A in a course, you must have a final average of at least 90%. On the first four examinations, you have grades of 86%, 88%, 92%, and 84%. If the final examination counts as two grades, what must you get on the final to earn an A in the course?

Solution We will use our five-step strategy for solving algebraic word problems.

Steps 1 and 2. Represent unknown quantities in terms of *x*. Let

$$x = \text{your grade on the final examination.}$$

Step 3. Write an inequality in *x* that models the conditions. The average of the six grades is found by adding the grades and dividing the sum by 6.

$$\text{Average} = \frac{86 + 88 + 92 + 84 + x + x}{6}$$

Because the final counts as two grades, the *x* (your grade on the final examination) is added twice. This is also why the sum is divided by 6.

To get an A, your average must be at least 90. This means that your average must be greater than or equal to 90.

$$\frac{86 + 88 + 92 + 84 + x + x}{6} \geq 90$$

Step 4. Solve the inequality and answer the problem's question.

$$\frac{86 + 88 + 92 + 84 + x + x}{6} \geq 90 \quad \text{This is the inequality that models the given conditions.}$$

$$\frac{350 + 2x}{6} \geq 90 \quad \text{Combine like terms in the numerator.}$$

$$6\left(\frac{350 + 2x}{6}\right) \geq 6(90) \quad \text{Multiply both sides by 6, clearing the fraction.}$$

$$350 + 2x \geq 540 \quad \text{Multiply.}$$

$$350 + 2x - 350 \geq 540 - 350 \quad \text{Subtract 350 from both sides.}$$

$$2x \geq 190 \quad \text{Simplify.}$$

$$\frac{2x}{2} \geq \frac{190}{2} \quad \text{Divide both sides by 2.}$$

$$x \geq 95 \quad \text{Simplify.}$$

You must get at least 95% on the final examination to earn an A in the course.

Step 5. Check. We can perform a partial check by computing the average with any grade that is at least 95. We will use 96. If you get 96% on the final examination, your average is

$$\frac{86 + 88 + 92 + 84 + 96 + 96}{6} = \frac{542}{6} = 90\frac{1}{3}.$$

Because $90\frac{1}{3} > 90$, you earn an A in the course. ∎

☑ **CHECK POINT 11** To earn a B in a course, you must have a final average of at least 80%. On the first three examinations, you have grades of 82%, 74%, and 78%. If the final examination counts as two grades, what must you get on the final to earn a B in the course?

EXAMPLE 12 An Application: Staying within a Budget

You can spend at most $1000 to have a picnic catered. The caterer charges a $150 setup fee and $25 per person. How many people can you invite while staying within your budget?

Solution

Steps 1 and 2. Represent unknown quantities in terms of x. Let

$$x = \text{the number of people you invite to the picnic.}$$

Step 3. Write an inequality in x that models the conditions. You can spend at most $1000. This means that the caterer's setup fee plus the cost of the meals must be less than or equal to $1000.

The setup fee: $150	plus	the cost of the meals: $25 per person for x people	must be less than or equal to	$1000.
150	+	25x	≤	1000

Step 4. Solve the inequality and answer the problem's question.

$$150 + 25x \le 1000 \qquad \text{This is the inequality that models the given conditions.}$$

$$150 + 25x - 150 \le 1000 - 150 \qquad \text{Subtract 150 from both sides.}$$

$$25x \le 850 \qquad \text{Simplify.}$$

$$\frac{25x}{25} \le \frac{850}{25} \qquad \text{Divide both sides by 25.}$$

$$x \le 34 \qquad \text{Simplify.}$$

You can invite at most 34 people to the picnic and still stay within your budget.

Step 5. Check. We can perform a partial check. Because $x \le 34$, let's see if you stay within your $1000 budget by inviting 33 guests.

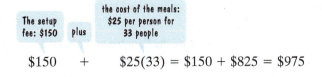

The setup fee: $150	plus	the cost of the meals: $25 per person for 33 people
$150	+	$25(33) = $150 + $825 = $975

Inviting 33 people results in catering costs of $975, which is within your $1000 budget. ■

☑ **CHECK POINT 12** You can spend at most $1600 to have a picnic catered. The caterer charges a $95 setup fee and $35 per person. How many people can you invite while staying within your budget?

Achieving Success

Assuming that you have done very well preparing for an exam, **there are certain things you can do that will make you a better test taker.**

- Get a good sleep the night before the exam.
- Have a good breakfast that balances protein, carbohydrates, and fruit.
- Just before the exam, briefly review the relevant material in the chapter summary.
- Bring everything you need to the exam, including two pencils, an eraser, scratch paper (if permitted), a calculator (if you're allowed to use one), water, and a watch.
- Survey the entire exam quickly to get an idea of its length.
- Read the directions to each problem carefully. Make sure that you have answered the specific question asked.
- Work the easy problems first. Then return to the hard problems you are not sure of. Doing the easy problems first will build your confidence. If you get bogged down on any one problem, you may not be able to complete the exam and receive credit for the questions you can easily answer.
- Attempt every problem. There may be partial credit even if you do not obtain the correct answer.
- Work carefully. Show your step-by-step solutions neatly. Check your work and answers.
- Watch the time. Pace yourself and be aware of when half the time is up. Determine how much of the exam you have completed. This will indicate if you're moving at a good pace or need to speed up. Prepare to spend more time on problems worth more points.
- Never turn in a test early. Use every available minute you are given for the test. If you have extra time, double check your arithmetic and look over your solutions.

CONCEPT AND VOCABULARY CHECK

Fill in each blank so that the resulting statement is true.

1. The solution set of $x < 5$ can be expressed in interval notation as _____.

2. The solution set of $x \geq 2$ can be expressed in interval notation as _____.

3. The addition property of inequality states that if $a < b$, then $a + c$ _____.

4. The positive multiplication property of inequality states that if $a < b$ and c is positive, then ac _____.

5. The negative multiplication property of inequality states that if $a < b$ and c is negative, then ac _____.

6. The linear inequality $-3x + 4 > 13$ can be solved by first _____ from both sides and then _____ both sides by _____, which changes the _____ of the inequality symbol from _____ to _____.

7. In solving an inequality, if you eliminate the variable and obtain a false statement such as $0 > 1$, the solution set is _____.

8. In solving an inequality, if you eliminate the variable and obtain a true statement such as $0 < 1$, the solution set in interval notation is _____.

2.7 EXERCISE SET ▶ MyMathLab®

Practice Exercises

In Exercises 1–12, graph the solutions of each inequality on a number line.

1. $x > 5$
2. $x > -3$
3. $x < -2$
4. $x < 0$
5. $x \geq -4$
6. $x \geq -6$
7. $x \leq 4.5$
8. $x \leq 7.5$
9. $-2 < x \leq 6$
10. $-3 \leq x < 6$
11. $-1 < x < 3$
12. $-2 \leq x \leq 0$

In Exercises 13–20, express the solution set of each inequality in interval notation and graph the interval.

13. $x \leq 3$
14. $x \leq 5$
15. $x > \dfrac{5}{2}$
16. $x > \dfrac{7}{2}$
17. $x \leq 0$
18. $x \leq 1$
19. $x < 4$
20. $x < 5$

Use the addition property of inequality to solve each inequality in Exercises 21–38 and graph the solution set on a number line.

21. $x - 3 > 4$
22. $x + 1 < 6$
23. $x + 4 \leq 10$
24. $x - 5 \geq 2$
25. $y - 2 < 0$
26. $y + 3 \geq 0$
27. $3x + 4 \leq 2x + 7$
28. $2x + 9 \leq x + 2$
29. $5x - 9 < 4x + 7$
30. $3x - 8 < 2x + 11$
31. $7x - 7 > 6x - 3$
32. $8x - 9 > 7x - 3$
33. $x - \dfrac{2}{3} > \dfrac{1}{2}$
34. $x - \dfrac{1}{3} \geq \dfrac{5}{6}$
35. $y + \dfrac{7}{8} \leq \dfrac{1}{2}$
36. $y + \dfrac{1}{3} \leq \dfrac{3}{4}$
37. $-15y + 13 > 13 - 16y$
38. $-12y + 17 > 20 - 13y$

Use the multiplication property of inequality to solve each inequality in Exercises 39–56 and graph the solution set on a number line.

39. $\dfrac{1}{2}x < 4$
40. $\dfrac{1}{2}x > 3$
41. $\dfrac{x}{3} > -2$
42. $\dfrac{x}{4} < -1$
43. $4x < 20$
44. $6x < 18$
45. $3x \geq -21$
46. $7x \geq -56$
47. $-3x < 15$
48. $-7x > 21$
49. $-3x \geq 15$
50. $-7x \leq 21$
51. $-16x > -48$
52. $-20x > -140$
53. $-4y \leq \dfrac{1}{2}$
54. $-2y \leq \dfrac{1}{2}$
55. $-x < 4$
56. $-x > -3$

Use both the addition and multiplication properties of inequality to solve each inequality in Exercises 57–80 and graph the solution set on a number line.

57. $2x - 3 > 7$
58. $3x + 2 \leq 14$
59. $3x + 3 < 18$
60. $8x - 4 > 12$
61. $3 - 7x \leq 17$
62. $5 - 3x \geq 20$
63. $-2x - 3 < 3$
64. $-3x + 14 < 5$
65. $5 - x \leq 1$
66. $3 - x \geq -3$
67. $2x - 5 > -x + 6$
68. $6x - 2 \geq 4x + 6$
69. $2y - 5 < 5y - 11$
70. $4y - 7 > 9y - 2$
71. $3(2y - 1) < 9$
72. $4(2y - 1) > 12$
73. $3(x + 1) - 5 < 2x + 1$
74. $4(x + 1) + 2 \geq 3x + 6$
75. $8x + 3 > 3(2x + 1) - x + 5$
76. $7 - 2(x - 4) < 5(1 - 2x)$
77. $\dfrac{x}{3} - 2 \geq 1$
78. $\dfrac{x}{4} - 3 \geq 1$
79. $1 - \dfrac{x}{2} > 4$
80. $1 - \dfrac{x}{2} < 5$

In Exercises 81–90, solve each inequality.

81. $4x - 4 < 4(x - 5)$
82. $3x - 5 < 3(x - 2)$

83. $x + 3 < x + 7$

84. $x + 4 < x + 10$

85. $7x \le 7(x - 2)$

86. $3x + 1 \le 3(x - 2)$

87. $2(x + 3) > 2x + 1$

88. $5(x + 4) > 5x + 10$

89. $5x - 4 \le 4(x - 1)$

90. $6x - 3 \le 3(x - 1)$

Practice PLUS

In Exercises 91–94, use properties of inequality to rewrite each inequality so that x is isolated on one side.

91. $3x + a > b$

92. $-2x - a \le b$

93. $y \le mx + b$ and $m < 0$

94. $y > mx + b$ and $m > 0$

We know that $|x|$ represents the distance from 0 to x on a number line. In Exercises 95–98, use each sentence to describe all possible locations of x on a number line. Then rewrite the given sentence as an inequality involving $|x|$.

95. The distance from 0 to x on a number line is less than 2.

96. The distance from 0 to x on a number line is less than 3.

97. The distance from 0 to x on a number line is greater than 2.

98. The distance from 0 to x on a number line is greater than 3.

Application Exercises

An online test of English spelling looked at how well people spelled difficult words. The bar graph shows the percentage of people who spelled each word correctly. Let x represent the percentage who spelled a word correctly. In Exercises 99–104, write the word or words described by the given inequality. (Yes, spelling counts!)

Percentage of People Spelling Various Words Correctly

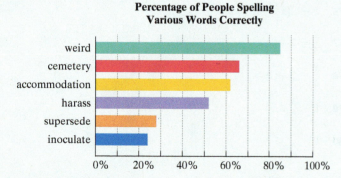

Source: Vivian Cook. Accomodating Brocolli in the Cemetary or Why Can't Anybody Spell?, Simon and Schuster, 2004.

99. $x > 55\%$

100. $x \ge 70\%$

101. $x \le 30\%$

102. $x \le 50\%$

103. $40\% \le x < 60\%$

104. $50\% < x \le 70\%$

Mediocre Grade Deflation. The bar graph shows the percentage of U.S. college freshmen with an average grade of C in high school.

Percentage of U.S. College Freshmen with an Average Grade of C (C− to C+) in High School

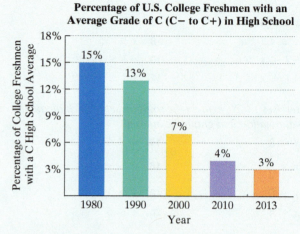

Source: Higher Education Research Institute

The data displayed by the bar graph can be described by the mathematical model

$$p = -0.4x + 16,$$

where x is the number of years after 1980 and p is the percentage of U.S. college freshmen who had an average grade of C in high school. Use this information to solve Exercises 105–106.

105. **a.** According to the formula, in 2010, what percentage of U.S. college freshmen had an average grade of C in high school? How does this compare with the percent displayed by the bar graph?

   b. According to the formula, how many years after 1980 will no more than 1.2% of college freshmen have an average grade of C in high school? Which years does this include?

106. **a.** According to the formula, in 2000, what percentage of U.S. college freshmen had an average grade of C in high school? Does this underestimate or overestimate the percent displayed by the bar graph? By how much?

   b. According to the formula, how many years after 1980 will no more than 0.8% of college freshmen have an average grade of C in high school? Which years does this include?

107. On two examinations, you have grades of 86 and 88. There is an optional final examination, which counts as one grade. You decide to take the final in order to get a course grade of A, meaning a final average of at least 90.

   a. What must you get on the final to earn an A in the course?

   b. By taking the final, if you do poorly, you might risk the B that you have in the course based on the first two exam grades. If your final average is less than 80, you will lose your B in the course. Describe the grades on the final that will cause this to happen.

108. On three examinations, you have grades of 88, 78, and 86. There is still a final examination, which counts as one grade.

    a. In order to get an A, your average must be at least 90. If you get 100 on the final, compute your average and determine if an A in the course is possible.

    b. To earn a B in the course, you must have a final average of at least 80. What must you get on the final to earn a B in the course?

109. A car can be rented from Continental Rental for $80 per week plus 25 cents for each mile driven. How many miles can you travel if you can spend at most $400 for the week?

110. A car can be rented from Basic Rental for $60 per week plus 50 cents for each mile driven. How many miles can you travel if you can spend at most $600 for the week?

111. An elevator at a construction site has a maximum capacity of 3000 pounds. If the elevator operator weighs 245 pounds and each cement bag weighs 95 pounds, how many bags of cement can be safely lifted on the elevator in one trip?

112. An elevator at a construction site has a maximum capacity of 2800 pounds. If the elevator operator weighs 265 pounds and each cement bag weighs 65 pounds, how many bags of cement can be safely lifted on the elevator in one trip?

Explaining the Concepts

113. When graphing the solutions of an inequality, what is the difference between a parenthesis and a bracket?

114. When solving an inequality, when is it necessary to change the direction of the inequality symbol? Give an example.

115. Describe ways in which solving a linear inequality is similar to solving a linear equation.

116. Describe ways in which solving a linear inequality is different from solving a linear equation.

Critical Thinking Exercises

Make Sense? *In Exercises 117–120, determine whether each statement makes sense or does not make sense, and explain your reasoning.*

117. I prefer interval notation over set-builder notation because it takes less space to write solution sets.

118. I can check inequalities by substituting 0 for the variable: When 0 belongs to the solution set, I should obtain a true statement, and when 0 does not belong to the solution set, I should obtain a false statement.

119. In an inequality such as $5x + 4 < 8x - 5$, I can avoid division by a negative number depending on which side I collect the variable terms and on which side I collect the constant terms.

120. I solved $-2x + 5 \geq 13$ and concluded that -4 is the greatest integer in the solution set.

In Exercises 121–124, determine whether each statement is true or false. If the statement is false, make the necessary change(s) to produce a true statement.

121. The inequality $x - 3 > 0$ is equivalent to $x < 3$.

122. The statement "x is at most 5" is written $x < 5$.

123. The inequality $-4x < -20$ is equivalent to $x > -5$.

124. The statement "the sum of x and 6% of x is at least 80" is modeled by $x + 0.06x \geq 80$.

125. A car can be rented from Basic Rental for $260 per week with no extra charge for mileage. Continental charges $80 per week plus 25 cents for each mile driven to rent the same car. How many miles should be driven in a week to make the rental cost for Basic Rental a better deal than Continental's?

126. Membership in a fitness club costs $500 yearly plus $1 per hour spent working out. A competing club charges $440 yearly plus $1.75 per hour for use of their equipment. How many hours must a person work out yearly to make membership in the first club cheaper than membership in the second club?

Technology Exercises

Solve each inequality in Exercises 127–128. Use a calculator to help with the arithmetic.

127. $1.45 - 7.23x > -1.442$

128. $126.8 - 9.4y \leq 4.8y + 34.5$

Review Exercises

129. 8 is 40% of what number? (Section 2.4, Example 6)

130. The length of a rectangle exceeds the width by 5 inches. The perimeter is 34 inches. What are the rectangle's dimensions? (Section 2.5, Example 5)

131. Solve and check: $5x + 16 = 3(x + 8)$. (Section 2.3, Example 2)

Preview Exercises

Exercises 132–134 will help you prepare for the material covered in the first section of the next chapter.

132. Is $x - 4y = 14$ a true statement for $x = 2$ and $y = -3$?

133. Is $x - 4y = 14$ a true statement for $x = 12$ and $y = 1$?

134. If $y = \frac{2}{3}x + 1$, find the value of y for $x = -6$.

Chapter 2 Summary

Definitions and Concepts	**Examples**

Section 2.1 The Addition Property of Equality

A linear equation in one variable can be written in the form $ax + b = c$, where a is not zero.

$3x + 7 = 9$ is a linear equation.

Equivalent equations have the same solution.

$2x - 4 = 6$, $2x = 10$, and $x = 5$ are equivalent equations.

The Addition Property of Equality
Adding the same number (or algebraic expression) to both sides of an equation or subtracting the same number (or algebraic expression) from both sides of an equation does not change its solution.

$$x - 3 = 8 \qquad\qquad x + 4 = 10$$
$$x - 3 + 3 = 8 + 3 \qquad x + 4 - 4 = 10 - 4$$
$$x = 11 \qquad\qquad x = 6$$

Section 2.2 The Multiplication Property of Equality

The Multiplication Property of Equality
Multiplying both sides of an equation or dividing both sides of an equation by the same nonzero real number (or algebraic expression) does not change the solution.

$$\frac{x}{-5} = 6 \qquad\qquad -50 = -5y$$
$$-5\left(\frac{x}{-5}\right) = -5(6) \qquad \frac{-50}{-5} = \frac{-5y}{-5}$$
$$x = -30 \qquad\qquad 10 = y$$

Equations and Coefficients of -1
If $-x = c$, multiply both sides by -1 to solve for x. The solution is the opposite, or additive inverse, of c.

$$-x = -12$$
$$(-1)(-x) = (-1)(-12)$$
$$x = 12$$

Using the Addition and Multiplication Properties
If an equation does not contain fractions,

• Use the addition property to isolate the variable term.
• Use the multiplication property to isolate the variable.

$$-2x - 5 = 11$$
$$-2x - 5 + 5 = 11 + 5$$
$$-2x = 16$$
$$\frac{-2x}{-2} = \frac{16}{-2}$$
$$x = -8$$

Section 2.3 Solving Linear Equations

Solving a Linear Equation

1. Simplify each side.

2. Collect all the variable terms on one side and all the constant terms on the other side.

3. Isolate the variable and solve. (If the variable is eliminated and a false statement results, the inconsistent equation has no solution. If a true statement results, all real numbers are solutions of the identity.)

4. Check the proposed solution in the original equation.

Solve: $\qquad 7 - 4(x - 1) = x + 1.$
$$7 - 4x + 4 = x + 1$$
$$-4x + 11 = x + 1$$
$$-4x - x + 11 = x - x + 1$$
$$-5x + 11 = 1$$
$$-5x + 11 - 11 = 1 - 11$$
$$-5x = -10$$
$$\frac{-5x}{-5} = \frac{-10}{-5}$$
$$x = 2$$

$$7 - 4(x - 1) = x + 1$$
$$7 - 4(2 - 1) \stackrel{?}{=} 2 + 1$$
$$7 - 4(1) \stackrel{?}{=} 2 + 1$$
$$7 - 4 \stackrel{?}{=} 2 + 1$$
$$3 = 3, \text{ true}$$

The solution is 2, or the solution set is $\{2\}$.

Definitions and Concepts	**Examples**

Section 2.3 Solving Linear Equations (continued)

Equations Containing Fractions
Multiply both sides (all terms) by the least common denominator. This clears the equation of fractions.

Solve: $\dfrac{x}{5} + \dfrac{1}{2} = \dfrac{x}{2} - 1.$

$$10\left(\dfrac{x}{5} + \dfrac{1}{2}\right) = 10\left(\dfrac{x}{2} - 1\right)$$

$$10 \cdot \dfrac{x}{5} + 10 \cdot \dfrac{1}{2} = 10 \cdot \dfrac{x}{2} - 10 \cdot 1$$

$$2x + 5 = 5x - 10$$

$$-3x = -15$$

$$x = 5$$

The solution is 5, or the solution set is $\{5\}$.

Equations Containing Decimals
An equation may be cleared of decimals by multiplying every term on both sides by a power of 10. The exponent on 10 will equal the greatest number of decimal places in the equation.

Solve: $1.4(x - 5) = 3x - 3.8$

$$1.4x - 7 = 3x - 3.8$$

$$10(1.4x) - 10(7) = 10(3x) - 10(3.8)$$

$$14x - 70 = 30x - 38$$

$$-16x = 32$$

$$x = -2$$

The solution is -2, or the solution set is $\{-2\}$.

Inconsistent equations with no solution (solution set: $\varnothing$) result in false statements in the solution process. Identities, true for every real number (solution set: $\{x \mid x \text{ is a real number}\}$), result in true statements in the solution process.

Solve: $3x + 2 = 3(x + 5).$

$$3x + 2 = 3x + 15$$

$$3x + 2 - 3x = 3x + 15 - 3x$$

$$2 = 15 \quad \text{(false)}$$

No solution: $\varnothing$

Solve: $2(x + 4) = x + x + 8.$

$$2x + 8 = 2x + 8$$

$$2x + 8 - 2x = 2x + 8 - 2x$$

$$8 = 8 \quad \text{(true)}$$

$\{x \mid x \text{ is a real number}\}$

Section 2.4 Formulas and Percents

To solve a formula for one of its variables, use the steps for solving a linear equation and isolate the specified variable on one side of the equation.

Solve for l:

$$w = \dfrac{P - 2l}{2}.$$

$$2w = 2\left(\dfrac{P - 2l}{2}\right)$$

$$2w = P - 2l$$

$$2w - P = P - P - 2l$$

$$2w - P = -2l$$

$$\dfrac{2w - P}{-2} = \dfrac{-2l}{-2}$$

$$\dfrac{2w - P}{-2} = l$$

Definitions and Concepts	Examples

Section 2.4 Formulas and Percents (continued)

A Formula Involving Percent

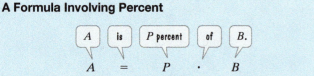

In the formula $A = PB$, P is expressed as a decimal.

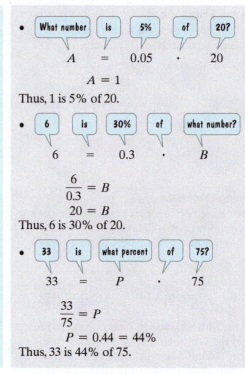

$A = 0.05 \cdot 20$

$A = 1$

Thus, 1 is 5% of 20.

$6 = 0.3 \cdot B$

$\dfrac{6}{0.3} = B$

$20 = B$

Thus, 6 is 30% of 20.

$33 = P \cdot 75$

$\dfrac{33}{75} = P$

$P = 0.44 = 44\%$

Thus, 33 is 44% of 75.

Section 2.5 An Introduction to Problem Solving

Strategy for Solving Word Problems
Step 1. Let x represent one of the quantities.

Step 2. Represent other unknown quantities in terms of x.

Step 3. Write an equation that models the conditions.

Step 4. Solve the equation and answer the question.

Step 5. Check the proposed solution in the original wording of the problem.

The length of a rectangle exceeds the width by 3 inches. The perimeter is 26 inches. What are the rectangle's dimensions?

Let x = the width.

$x + 3$ = the length

Twice length plus twice width is perimeter.

$2(x + 3) + 2x = 26$

$2(x + 3) + 2x = 26$

$2x + 6 + 2x = 26$

$4x + 6 = 26$

$4x = 20$

$x = 5$

The width (x) is 5 inches and the length ($x + 3$) is $5 + 3$, or 8 inches.

$\text{Perimeter} = 2(8 \text{ in.}) + 2(5 \text{ in.})$

$= 16 \text{ in.} + 10 \text{ in.} = 26 \text{ in.}$

This checks with the given perimeter.

Definitions and Concepts	**Examples**

Section 2.6 Problem Solving in Geometry

Solving geometry problems often requires using basic geometric formulas. Formulas for perimeter, area, circumference, and volume are given in **Table 2.3** (page 169), **Table 2.4** (page 170), and **Table 2.5** (page 171) in Section 2.6.

A sailboat's triangular sail has an area of 24 ft^2 and a base of 8 ft. Find its height.

$$A = \frac{1}{2}bh$$

$$24 = \frac{1}{2}(8)h$$

$$24 = 4h$$

$$6 = h$$

The sail's height is 6 ft.

The sum of the measures of the three angles of any triangle is 180°.

In a triangle, the first angle measures 3 times the second and the third measures 40° less than the second. Find each angle's measure.

$$\text{Second angle} = x$$

$$\text{First angle} = 3x$$

$$\text{Third angle} = x - 40$$

Sum of measures is 180°.

$$x + 3x + (x - 40) = 180$$

$$5x - 40 = 180$$

$$5x = 220$$

$$x = 44$$

The angles measure $x = 44$, $3x = 3 \cdot 44 = 132$, and $x - 40 = 44 - 40 = 4$.
The angles measure 44°, 132°, and 4°.

Two complementary angles have measures whose sum is 90°. Two supplementary angles have measures whose sum is 180°. If an angle measures x, its complement measures $90 - x$, and its supplement measures $180 - x$.

An angle measures five times its complement. Find the angle's measure.

$$x = \text{angle's measure}$$

$$90 - x = \text{measure of complement}$$

$$x = 5(90 - x)$$

$$x = 450 - 5x$$

$$6x = 450$$

$$x = 75$$

The angle measures 75°.

Section 2.7 Solving Linear Inequalities

A linear inequality in one variable can be written in one of these forms:

$$ax + b < c \qquad ax + b \le c.$$

$$ax + b > c \qquad ax + b \ge c.$$

$3x + 6 > 12$ is a linear inequality.

Definitions and Concepts	**Examples**

Section 2.7 Solving Linear Inequalities (continued)

Graphs of solutions to linear inequalities are shown on a number line by shading all points representing numbers that are solutions. Square brackets, [ ], indicate endpoints that are solutions. Parentheses, ( ), indicate endpoints that are not solutions.

- Graph the solutions of $x < 4$.

- Graph the solutions of $-2 < x \le 1$.

Solutions of inequalities can be expressed in interval notation or set-builder notation.

Inequality	Interval Notation	Set-Builder Notation	Graph
$x > b$	(b, ∞)	$\{x \mid x > b\}$	
$x \le a$	$(-\infty, a]$	$\{x \mid x \le a\}$	

Express the solution set in interval notation and graph the interval:

- $x \le 1$
  $(-\infty, 1]$

- $x > -2$
  $(-2, \infty)$

The Addition Property of Inequality
Adding the same number to both sides of an inequality or subtracting the same number from both sides of an inequality does not change the solutions.

$$x + 3 < 8$$
$$x + 3 - 3 < 8 - 3$$
$$x < 5$$

The Positive Multiplication Property of Inequality
Multiplying or dividing both sides of an inequality by the same positive number does not change the solutions.

$$\frac{x}{6} \ge 5$$
$$6 \cdot \frac{x}{6} \ge 6 \cdot 5$$
$$x \ge 30$$

The Negative Multiplication Property of Inequality
Multiplying or dividing both sides of an inequality by the same negative number and reversing the direction of the inequality sign does not change the solutions.

$$-3x \le 12$$
$$\frac{-3x}{-3} \ge \frac{12}{-3}$$
$$x \ge -4$$

Solving Linear Inequalities
Use the procedure for solving linear equations. When multiplying or dividing by a negative number, reverse the direction of the inequality symbol. Express the solution set in interval or set-builder notation, and graph the set on a number line. If the variable is eliminated and a false statement results, the inequality has no solution. The solution set is $\varnothing$, the empty set. If a true statement results, the solution is the set of all real numbers: $(-\infty, \infty)$ or $\{x \mid x$ is a real number$\}$.

Solve:
$$x + 4 \ge 6x - 16$$
$$x + 4 - 6x \ge 6x - 16 - 6x$$
$$-5x + 4 \ge -16$$
$$-5x + 4 - 4 \ge -16 - 4$$
$$-5x \ge -20$$
$$\frac{-5x}{-5} \le \frac{-20}{-5}$$
$$x \le 4$$

Solution set: $(-\infty, 4]$ or $\{x \mid x \le 4\}$

CHAPTER 2 REVIEW EXERCISES

2.1 *Solve each equation in Exercises 1–5 using the addition property of equality. Be sure to check proposed solutions.*

1. $x - 10 = 22$

2. $-14 = y + 8$

3. $7z - 3 = 6z + 9$

4. $4(x + 3) = 3x - 10$

5. $6x - 3x - 9 + 1 = -5x + 7x - 3$

2.2 *Solve each equation in Exercises 6–13 using the multiplication property of equality. Be sure to check proposed solutions.*

6. $\dfrac{x}{8} = 10$

7. $\dfrac{y}{-8} = 7$

8. $7z = 77$

9. $-36 = -9y$

10. $\dfrac{3}{5}x = -9$

11. $30 = -\dfrac{5}{2}y$

12. $-x = 25$

13. $\dfrac{-x}{10} = -1$

Solve each equation in Exercises 14–18 using both the addition and multiplication properties of equality. Check proposed solutions.

14. $4x + 9 = 33$

15. $-3y - 2 = 13$

16. $5z + 20 = 3z$

17. $5x - 3 = x + 5$

18. $3 - 2x = 9 - 8x$

19. The number of Americans who consider themselves nonreligious is growing. The bar graph shows the percentage of Americans who were religiously unaffiliated from 2007 through 2012.

Percentage of Americans Who Are Religiously Unaffiliated*

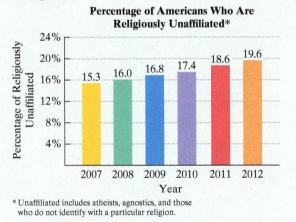

*Unaffiliated includes atheists, agnostics, and those who do not identify with a particular religion.

Source: Pew Research Center

The mathematical model

$$p = 0.9n + 15$$

describes the percentage of Americans who were religiously unaffiliated, p, n years after 2007.

a. Does the formula underestimate or overestimate the percentage who were religiously unaffiliated in 2012? By how much?

b. If trends shown by the formula continue, when will 24% of Americans be religiously unaffiliated?

2.3 *Solve and check each equation in Exercises 20–30.*

20. $5x + 9 - 7x + 6 = x + 18$

21. $3(x + 4) = 5x - 12$

22. $1 - 2(6 - y) = 3y + 2$

23. $2(x - 4) + 3(x + 5) = 2x - 2$

24. $-2(y - 4) - (3y - 2) = -2 - (6y - 2)$

25. $\dfrac{2x}{3} = \dfrac{x}{6} + 1$

26. $\dfrac{x}{2} - \dfrac{1}{10} = \dfrac{x}{5} + \dfrac{1}{2}$

27. $0.5x + 8.75 = 13.25$

28. $0.1(x - 3) = 1.1 - 0.25x$

29. $3(8x - 1) = 6(5 + 4x)$

30. $4(2x - 3) + 4 = 8x - 8$

31. The formula $H = 0.7(220 - a)$ can be used to determine target heart rate during exercise, H, in beats per minute, by a person of age a. If the target heart rate is 133 beats per minute, how old is that person?

2.4 *In Exercises 32–36, solve each formula for the specified variable.*

32. $I = Pr$ for r

33. $V = \dfrac{1}{3}Bh$ for h

34. $P = 2l + 2w$ for w

35. $A = \dfrac{B + C}{2}$ for B

36. $T = D + pm$ for m

37. What number is 8% of 120?

38. 90 is 45% of what number?

39. 36 is what percent of 75?

40. If 6 is increased to 12, the increase is what percent of the original number?

41. If 5 is decreased to 3, the decrease is what percent of the original number?

42. A college that had 40 students for each lecture course increased the number to 45 students. What is the percent increase in the number of students in a lecture course?

43. Consider the following statement:

> My portfolio fell 10% last year, but then it rose 10% this year, so at least I recouped my losses.

Is this statement true? In particular, suppose you invested $10,000 in the stock market last year. How much money would be left in your portfolio with a 10% fall and then a 10% rise? If there is a loss, what is the percent decrease in your portfolio?

44. The radius is one of two bones that connect the elbow and the wrist. The formula $r = \dfrac{h}{7}$ models the length of a woman's radius, r, in inches, and her height, h, in inches.

a. Solve the formula for h.

b. Use the formula in part (a) to find a woman's height if her radius is 9 inches long.

45. Every day, the average U.S. household uses 91 gallons of water flushing toilets. The circle graph shows that this represents 26% of the total number of gallons of water used per day. How many gallons of water does the average U.S. household use per day?

Where United States Households Use Water

Baths 2%

Other domestic 2%

Dishwashers 1%

Leaks 14%

Toilets 26%

Faucets 16%

Showers 17%

Clothes washers 22%

Source: American Water Works Association

2.5 *In Exercises 46–53, use the five-step strategy to solve each problem.*

46. Six times a number, decreased by 20, is four times the number. Find the number.

47. In 2014, the wealthiest Americans were Bill Gates and Warren Buffett. At that time, Gates's net worth exceeded that of Buffett's by $14 billion. Combined, the two men were worth $148 billion. Determine each man's net worth in 2010. *(Source: Forbes)*

48. Two pages that face each other in a book have 93 as the sum of their page numbers. What are the page numbers?

49. The graph shows the gender breakdown of the U.S. population at each end of the age spectrum.

Gender Breakdown of the American Population

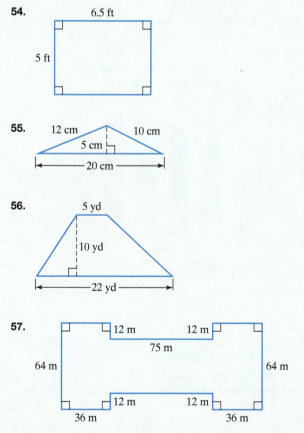

☐ Male ☐ Female

60%

40%

Younger than 20   70 and Older

Source: U.S. Census Bureau

For Americans under 20, the percentage of males is greater than the percentage of females and these percents are consecutive odd integers. What percentage of Americans younger than 20 are females and what percentage are males?

50. In 2001, the average spent per student on K–12 public education in the United States was $7284, increasing by approximately $328 per year for the period shown by the bar graph. If this trend continues, in how many years after 2001 will per-pupil school spending be $12,204? In which year will that be?

Per-Pupil School Spending

2001 $7284

2011 $10,560

Source: U.S. Census Bureau

51. A bank's total monthly charge for a checking account is $6 plus $0.05 per check. If your total monthly charge is $6.90, how many checks did you write during that month?

52. A rectangular field is three times as long as it is wide. If the perimeter of the field is 400 yards, what are the field's dimensions?

53. After a 25% reduction, you purchase a table for $180. What was the table's price before the reduction?

2.6 *Use a formula for area to find the area of each figure in Exercises 54–57.*

54.

6.5 ft

5 ft

55.

12 cm   10 cm

5 cm

20 cm

56.

5 yd

10 yd

22 yd

57.

12 m   12 m

75 m

64 m   64 m

12 m   12 m

36 m   36 m

58. Find the circumference and the area of a circle with a diameter of 20 meters. Round answers to the nearest whole number.

59. A sailboat has a triangular sail with an area of 42 square feet and a base that measures 14 feet. Find the height of the sail.

60. A rectangular kitchen floor measures 12 feet by 15 feet. A stove on the floor has a rectangular base measuring 3 feet by 4 feet, and a refrigerator covers a rectangular area of the floor measuring 3 feet by 4 feet. How many square feet of tile will be needed to cover the kitchen floor not counting the area used by the stove and the refrigerator?

61. A yard that is to be covered with mats of grass is shaped like a trapezoid. The bases are 80 feet and 100 feet, and the height is 60 feet. What is the cost of putting the grass mats on the yard if the landscaper charges $0.35 per square foot?

62. Which one of the following is a better buy: a medium pizza with a 14-inch diameter for $6.00 or two small pizzas, each with an 8-inch diameter, for $6.00?

Use a formula for volume to find the volume of each figure in Exercises 63–65. Where applicable, express answers in terms of π. Then round to the nearest whole number.

63.

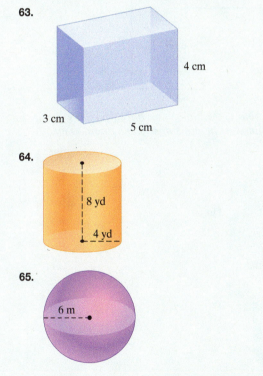

4 cm

3 cm

5 cm

64.

8 yd

4 yd

65.

6 m

66. A train is being loaded with freight containers. Each box is 8 meters long, 4 meters wide, and 3 meters high. If there are 50 freight containers, how much space is needed?

67. A cylindrical fish tank has a diameter of 6 feet and a height of 3 feet. How many tropical fish can be put in the tank if each fish needs 5 cubic feet of water?

68. Find the measure of each angle of the triangle shown in the figure.

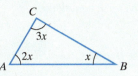

69. In a triangle, the measure of the first angle is 15° more than twice the measure of the second angle. The measure of the third angle exceeds that of the second angle by 25°. What is the measure of each angle?

70. Find the measure of the complement of a 57° angle.

71. Find the measure of the supplement of a 75° angle.

72. How many degrees are there in an angle that measures 25° more than the measure of its complement?

73. The measure of the supplement of an angle is 45° less than four times the measure of the angle. Find the measure of the angle and its supplement.

2.7 *In Exercises 74–75, graph the solution of each inequality on a number line.*

74. $x < -1$

75. $-2 < x \le 4$

In Exercises 76–77, express the solution set of each inequality in interval notation and graph the interval.

76. $x \ge \dfrac{3}{2}$

77. $x < 0$

In Exercises 78–85, solve each inequality and graph the solution set on a number line. It is not necessary to provide graphs if the inequality has no solution or is true for all real numbers.

78. $2x - 5 < 3$

79. $\dfrac{x}{2} > -4$

80. $3 - 5x \le 18$

81. $4x + 6 < 5x$

82. $6x - 10 \ge 2(x + 3)$

83. $4x + 3(2x - 7) \le x - 3$

84. $2(2x + 4) > 4(x + 2) - 6$

85. $-2(x - 4) \le 3x + 1 - 5x$

86. To pass a course, a student must have an average on three examinations of at least 60. If a student scores 42 and 74 on the first two tests, what must be earned on the third test to pass the course?

87. You can spend at most $2000 for a catered party. The caterer charges a setup fee of $350 and $55 per person. How many people can you invite while staying within your budget?

In Exercises 1–7, solve each equation.

1. $4x - 5 = 13$

2. $12x + 4 = 7x - 21$

3. $8 - 5(x - 2) = x + 26$

4. $3(2y - 4) = 9 - 3(y + 1)$

5. $\frac{3}{4}x = -15$

6. $\frac{x}{10} + \frac{1}{3} = \frac{x}{5} + \frac{1}{2}$

7. $9.2x - 80.1 = 21.3x - 19.6$

8. The formula $P = 2.4x + 180$ models U.S. population, P, in millions x years after 1960. How many years after 1960 is the U.S. population expected to reach 324 million? In which year is this expected to occur?

In Exercises 9–10, solve each formula for the specified variable.

9. $V = \pi r^2 h$ for h

10. $l = \dfrac{P - 2w}{2}$ for w

11. What number is 6% of 140?

12. 120 is 80% of what number?

13. 12 is what percent of 240?

In Exercises 14–18, solve each problem.

14. The product of 5 and a number, decreased by 9, is 306. What is the number?

15. Compared with other major countries, American employees have less vacation time. The bar graph shows the average number of vacation days per person for selected countries.

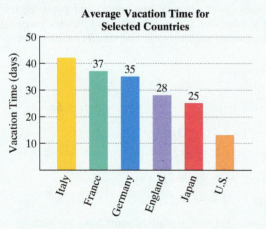

Average Vacation Time for Selected Countries

Source: World Development Indicators

The average time Italians spend on vacation exceeds the average American vacation time by 29 days. The combined average vacation time for Americans and Italians is 55 days. On average, how many days do Americans spend on vacation and how many days do Italians spend on vacation?

16. A texting plan has a monthly fee of $15.00 and a charge of $0.05 per text. How many text messages can you send in a month for a total cost, including the $15.00, of $45.00?

17. A rectangular field is twice as long as it is wide. If the perimeter of the field is 450 yards, what are the field's dimensions?

18. After a 20% reduction, you purchase a new Stephen King novel for $28. What was the book's price before the reduction?

In Exercises 19–21, find the area of each figure.

19.

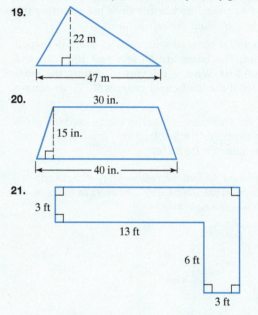

22 m

47 m

20.

30 in.

15 in.

40 in.

21.

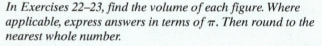

3 ft

13 ft

6 ft

3 ft

In Exercises 22–23, find the volume of each figure. Where applicable, express answers in terms of π. Then round to the nearest whole number.

22.

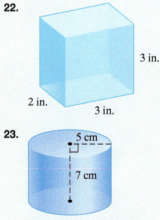

3 in.

2 in.

3 in.

23.

5 cm

7 cm

24. What will it cost to cover a rectangular floor measuring 40 feet by 50 feet with square tiles that measure 2 feet on each side if a package of 10 tiles costs $13 per package?

25. A sailboat has a triangular sail with an area of 56 square feet and a base that measures 8 feet. Find the height of the sail.

26. In a triangle, the measure of the first angle is three times that of the second angle. The measure of the third angle is 30° less than the measure of the second angle. What is the measure of each angle?

27. How many degrees are there in an angle that measures 16° more than the measure of its complement?

In Exercises 28–29, express the solution set of each inequality in interval notation and graph the interval.

28. $x > -2$ 29. $x \leq 3$

Solve each inequality in Exercises 30–32 and graph the solution set on a number line.

30. $\dfrac{x}{2} < -3$

31. $6 - 9x \geq 33$

32. $4x - 2 > 2(x + 6)$

33. A student has grades on three examinations of 76, 80, and 72. What must the student earn on a fourth examination to have an average of at least 80?

34. The length of a rectangle is 20 inches. For what widths is the perimeter greater than 56 inches?

CUMULATIVE REVIEW EXERCISES   (CHAPTERS 1–2)

In Exercises 1–3, perform the indicated operation or operations.

1. $-8 - (12 - 16)$ 2. $(-3)(-2) + (-2)(4)$

3. $(8 - 10)^3(7 - 11)^2$

4. Simplify: $2 - 5[x + 3(x + 7)]$.

5. List all the rational numbers in this set:
$$\left\{ -4, -\frac{1}{3}, 0, \sqrt{2}, \sqrt{4}, \frac{\pi}{2}, 1063 \right\}.$$

6. Write as an algebraic expression, using x to represent the number:
The quotient of 5 and a number, decreased by 2 more than the number.

7. Insert either $<$ or $>$ in the shaded area to make a true statement:
$$-10,000 \quad \blacksquare \quad -2.$$

8. Use the distributive property to rewrite without parentheses:
$$6(4x - 1 - 5y).$$

The bar graph shows the percentage of U.S. high school seniors who had ever used alcohol for four selected years from 2000 through 2013.

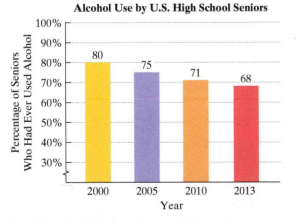

Alcohol Use by U.S. High School Seniors

Source: University of Michigan Institute for Social Research

The mathematical model
$$A = -0.9n + 80$$
describes the percentage of seniors who had ever used alcohol, A, n years after 2000. Use this information to solve Exercises 9–10.

9. Use the formula to determine the percentage of seniors who had ever used alcohol in 2000. How does this compare with the percent displayed by the graph?

10. If trends shown by the formula continue, when will 62% of high school seniors ever have used alcohol?

In Exercises 11–12, solve each equation.

11. $5 - 6(x + 2) = x - 14$

12. $\dfrac{x}{5} - 2 = \dfrac{x}{3}$

13. Solve for A: $V = \dfrac{1}{3}Ah$.

14. 48 is 30% of what number?

15. The length of a rectangular parking lot is 10 yards less than twice its width. If the perimeter of the lot is 400 yards, what are its dimensions?

16. A gas station owner makes a profit of 40 cents per gallon of gasoline sold. How many gallons of gasoline must be sold in a year to make a profit of $30,000 from gasoline sales?

17. Express the solution set of $x \leq \frac{1}{2}$ in interval notation and graph the interval.

Solve each inequality in Exercises 18–19 and graph the solution set on a number line.

18. $3 - 3x > 12$

19. $5 - 2(3 - x) \leq 2(2x + 5) + 1$

20. You take a summer job selling medical supplies. You are paid $600 per month plus 4% of the sales price of all the supplies you sell. If you want to earn more than $2500 per month, what value of medical supplies must you sell?

Linear Equations and Inequalities in Two Variables

These photos of presidential puffing indicate that the White House was not always a no-smoking zone. According to *Cigar Aficionado*, nearly half of U.S. presidents have had a nicotine habit, from cigarettes to pipes to cigars. Franklin Roosevelt's stylish way with a cigarette holder was part of his mystique. Although Dwight Eisenhower quit his wartime four-pack-a-day habit before taking office, smoking in the residence was still common, with ashtrays on the tables at state dinners and free cigarettes for guests. In 1993, Hillary Clinton banned smoking in the White House, although Bill Clinton's cigars later made a sordid cameo in the Lewinsky scandal. Barack Obama quit smoking before entering the White House, but had "fallen off the wagon occasionally" as he admitted in a *Meet the Press* interview.

Changing attitudes toward smoking, both inside and outside the White House, date back to 1964 and an equation in two variables. In this chapter, you will learn graphing and modeling methods for situations involving two variables that will enable you to understand the mathematics behind this turning point in public health.

Here's where you'll find this application:

- Got a light, Mr. President? It's unlikely you'll ask after reading the Blitzer Bonus on page 259.

SECTION

3.1

Graphing Linear Equations in Two Variables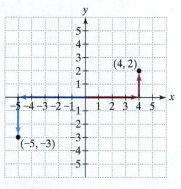

What am I supposed to learn?

After studying this section, you should be able to:

1 Plot ordered pairs in the rectangular coordinate system.

2 Find coordinates of points in the rectangular coordinate system.

3 Determine whether an ordered pair is a solution of an equation.

4 Find solutions of an equation in two variables.

5 Use point plotting to graph linear equations.

6 Use graphs of linear equations to solve problems.

1 Plot ordered pairs in the rectangular coordinate system.

Credit cards. Are they convenient tools or snakes in your wallet? Despite high interest rates, fees, and penalties, American consumers are paying more with credit cards, and less with checks and cash. In this section (Example 8), we will use geometric figures to visualize mathematical models describing our primary spending methods. The idea of visualizing equations as geometric figures was developed by the French philosopher and mathematician René Descartes (1596–1650). We begin by looking at Descartes's idea that brought together algebra and geometry, called the **rectangular coordinate system** or (in his honor) the **Cartesian coordinate system**.

Points and Ordered Pairs

The rectangular coordinate system consists of two number lines that intersect at right angles at their zero points, as shown in **Figure 3.1**. The horizontal number line is the **x-axis.** The vertical number line is the **y-axis.** The point of intersection of these axes is their zero points, called the **origin.** Positive numbers are shown to the right and above the origin. Negative numbers are shown to the left and below the origin. The axes divide the plane into four regions, called **quadrants**. The points located on the axes are not in any quadrant.

Figure 3.1 The rectangular coordinate system

Each point in the rectangular coordinate system corresponds to an **ordered pair** of real numbers, (x, y). Examples of such pairs are $(4, 2)$ and $(-5, -3)$. The first number in each pair, called the **x-coordinate**, denotes the distance and direction from the origin along the x-axis. The second number, called the **y-coordinate**, denotes vertical distance and direction along a line parallel to the y-axis or along the y-axis itself.

Figure 3.2 shows how we **plot**, or locate, the points corresponding to the ordered pairs $(4, 2)$ and $(-5, -3)$. We plot $(4, 2)$ by going 4 units from 0 to the right along the x-axis. Then we go 2 units up parallel to the y-axis. We plot $(-5, -3)$ by going 5 units from 0 to the left along the x-axis and 3 units down parallel to the y-axis. The phrase "the point corresponding to the ordered pair $(-5, -3)$" is often abbreviated as "the point $(-5, -3)$."

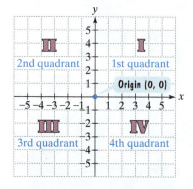

Figure 3.2 Plotting $(4, 2)$ and $(-5, -3)$

EXAMPLE 1   Plotting Points in the Rectangular Coordinate System

Plot the points: $A(-3, 5)$, $B(2, -4)$, $C(5, 0)$, $D(-5, -2)$, $E(0, 4)$, and $F(0, 0)$.

Solution   See **Figure 3.3**. We move from the origin and plot the points in the following way:

$A(-3, 5)$:   3 units left, 5 units up

$B(2, -4)$:   2 units right, 4 units down

$C(5, 0)$:   5 units right, 0 units up or down

$D(-5, -2)$:   5 units left, 2 units down

$E(0, 4)$:   0 units right or left, 4 units up

$F(0, 0)$:   0 units right or left, 0 units up or down

Notice that the origin is represented by (0, 0).

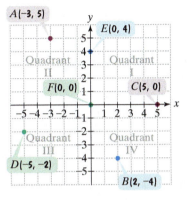

Figure 3.3   Plotting points   ■

Great Question!

If a point is located on an axis, is one of its coordinates always zero?

Any point on the x-axis, such as $(5, 0)$, has a y-coordinate of 0.
Any point on the y-axis, such as $(0, 4)$, has an x-coordinate of 0.

The phrase *ordered pair* is used because **order is important**. For example, the points $(2, 5)$ and $(5, 2)$ are not the same. To plot $(2, 5)$, move 2 units right and 5 units up. To plot $(5, 2)$, move 5 units right and 2 units up. The points $(2, 5)$ and $(5, 2)$ are in different locations. **The order in which coordinates appear makes a difference in a point's location.**

✓ **CHECK POINT 1**   Plot the points: $A(-2, 4)$, $B(4, -2)$, $C(-3, 0)$, and $D(0, -3)$.

---

2 Find coordinates of points in the rectangular coordinate system. ▶

In the rectangular coordinate system, each ordered pair corresponds to exactly one point. Example 2 illustrates that each point in the rectangular coordinate system corresponds to exactly one ordered pair.

EXAMPLE 2   Finding Coordinates of Points

Determine the coordinates of points A, B, C, and D shown in **Figure 3.4**.

Solution

Point	Position from the Origin	Coordinates
A	6 units left, 0 units up or down	$(-6, 0)$
B	0 units right or left, 2 units up	$(0, 2)$
C	2 units right, 0 units up or down	$(2, 0)$
D	4 units right, 2 units down	$(4, -2)$

■

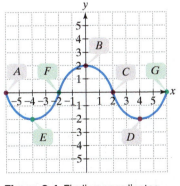

Figure 3.4 Finding coordinates of points

✓ **CHECK POINT 2**   Determine the coordinates of points E, F, and G shown in **Figure 3.4**.

3 Determine whether an ordered pair is a solution of an equation.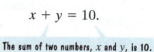

Solutions of Equations in Two Variables

The rectangular coordinate system allows us to visualize relationships between two variables by connecting any equation in two variables with a geometric figure. Consider, for example, the following equation in two variables:

$$x + y = 10.$$

The sum of two numbers, x and y, is 10.

Many pairs of numbers fit the description in the voice balloon, such as $x = 1$ and $y = 9$, or $x = 3$ and $y = 7$. The phrase "$x = 1$ and $y = 9$" is abbreviated using the ordered pair $(1, 9)$. Similarly, the phrase "$x = 3$ and $y = 7$" is abbreviated using the ordered pair $(3, 7)$.

A **solution of an equation in two variables**, x and y, is an ordered pair of real numbers with the following property: When the x-coordinate is substituted for x and the y-coordinate is substituted for y in the equation, we obtain a true statement. For example, $(1, 9)$ is a solution of the equation $x + y = 10$. When 1 is substituted for x and 9 is substituted for y, we obtain the true statement $1 + 9 = 10$, or $10 = 10$. Because there are infinitely many pairs of numbers that have a sum of 10, the equation $x + y = 10$ has infinitely many solutions. Each ordered-pair solution is said to **satisfy** the equation. Thus, $(1, 9)$ satisfies the equation $x + y = 10$.

EXAMPLE 3 Deciding Whether an Ordered Pair Satisfies an Equation

Determine whether each ordered pair is a solution of the equation

$$x - 4y = 14:$$

a. $(2, -3)$ **b.** $(12, 1)$.

Solution

a. To determine whether $(2, -3)$ is a solution of the equation, we substitute 2 for x and -3 for y.

$x - 4y = 14$	This is the given equation.
$2 - 4(-3) \stackrel{?}{=} 14$	Substitute 2 for x and -3 for y.
$2 - (-12) \stackrel{?}{=} 14$	Multiply: $4(-3) = -12$.
$14 = 14$	Subtract: $2 - (-12) = 2 + 12 = 14$.

This statement is true.

Because we obtain a true statement, we conclude that $(2, -3)$ is a solution of the equation $x - 4y = 14$. Thus, $(2, -3)$ satisfies the equation.

b. To determine whether $(12, 1)$ is a solution of the equation, we substitute 12 for x and 1 for y.

$x - 4y = 14$	This is the given equation.
$12 - 4(1) \stackrel{?}{=} 14$	Substitute 12 for x and 1 for y.
$12 - 4 \stackrel{?}{=} 14$	Multiply: $4(1) = 4$.
$8 = 14$	Subtract: $12 - 4 = 8$.

This statement is false.

Because we obtain a false statement, we conclude that $(12, 1)$ is not a solution of $x - 4y = 14$. The ordered pair $(12, 1)$ does not satisfy the equation. ■

✓ **CHECK POINT 3**   Determine whether each ordered pair is a solution of the equation $x - 3y = 9$:

a. $(3, -2)$ **b.** $(-2, 3)$.

---

4 Find solutions of an equation in two variables.

In this chapter, we will use x and y to represent the variables of an equation in two variables. However, any two letters can be used. Solutions are still ordered pairs. Although it depends on the role variables play in equations, the first number in an ordered pair often replaces the variable that occurs first alphabetically. The second number in an ordered pair often replaces the variable that occurs last alphabetically.

How do we find ordered pairs that are solutions of an equation in two variables, x and y?

- Select a value for one of the variables.
- Substitute that value into the equation and find the corresponding value of the other variable.
- Use the values of the two variables to form an ordered pair (x, y). This pair is a solution of the equation.

EXAMPLE 4   Finding Solutions of an Equation

Find five solutions of

$$y = 2x - 1.$$

Select integers for x, starting with -2 and ending with 2.

Solution   We organize the process of finding solutions in the following table of values.

	Start with these values of x.	Substitute x into $y = 2x - 1$ and compute y.	Use values for x and y to form an ordered-pair solution.
	x	$y = 2x - 1$	(x, y)
Any numbers can be selected for x. There is nothing special about integers from −2 to 2, inclusive. We chose these values to include two negative numbers, 0, and two positive numbers. We also wanted to keep the resulting computations for y relatively simple.	-2	$y = 2(-2) - 1 = -4 - 1 = -5$	$(-2, -5)$
	-1	$y = 2(-1) - 1 = -2 - 1 = -3$	$(-1, -3)$
	0	$y = 2 \cdot 0 - 1 = 0 - 1 = -1$	$(0, -1)$
	1	$y = 2 \cdot 1 - 1 = 2 - 1 = 1$	$(1, 1)$
	2	$y = 2 \cdot 2 - 1 = 4 - 1 = 3$	$(2, 3)$

Look at the ordered pairs in the last column. Five solutions of $y = 2x - 1$ are $(-2, -5), (-1, -3), (0, -1), (1, 1)$, and $(2, 3)$. ∎

✓ **CHECK POINT 4**   Find five solutions of $y = 3x + 2$. Select integers for x, starting with -2 and ending with 2.

---

5 Use point plotting to graph linear equations.

Graphing Linear Equations in the Form $y = mx + b$

In Example 4, we found five solutions of $y = 2x - 1$. We can generate as many ordered-pair solutions as desired to $y = 2x - 1$ by substituting numbers for x and then finding the corresponding values for y. The **graph of the equation** is the set of all points whose coordinates satisfy the equation.

One method for graphing an equation such as $y = 2x - 1$ is the **point-plotting method**.

The Point-Plotting Method for Graphing an Equation in Two Variables

1. Find several ordered pairs that are solutions of the equation.
2. Plot these ordered pairs as points in the rectangular coordinate system.
3. Connect the points with a smooth curve or line.

 **EXAMPLE 5**   Graphing an Equation Using the Point-Plotting Method

Graph the equation: $y = 3x$.

Solution

Step 1. Find several ordered pairs that are solutions of the equation. Because there are infinitely many solutions, we cannot list them all. To find a few solutions of the equation, we select integers for x, starting with -2 and ending with 2.

Start with these values of x.	Substitute x into $y = 3x$ and compute y.	These are solutions of $y = 3x$.
x	$y = 3x$	(x, y)
-2	$y = 3(-2) = -6$	$(-2, -6)$
-1	$y = 3(-1) = -3$	$(-1, -3)$
0	$y = 3 \cdot 0 = 0$	$(0, 0)$
1	$y = 3 \cdot 1 = 3$	$(1, 3)$
2	$y = 3 \cdot 2 = 6$	$(2, 6)$

Step 2. Plot these ordered pairs as points in the rectangular coordinate system. The five ordered pairs in the table of values are plotted in **Figure 3.5(a)**.

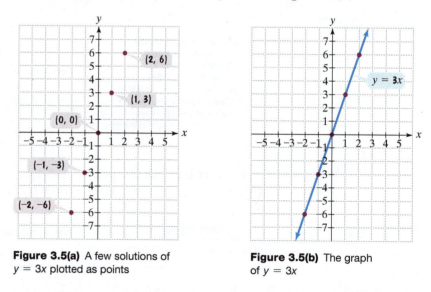

Figure 3.5(a) A few solutions of $y = 3x$ plotted as points

Figure 3.5(b) The graph of $y = 3x$

Step 3. Connect the points with a smooth curve or line. The points lie along a straight line. The graph of $y = 3x$ is shown in **Figure 3.5(b)**. The arrows on both ends of the line indicate that it extends indefinitely in both directions. ∎

✓ **CHECK POINT 5**   Graph the equation: $y = 2x$.

Equations like $y = 3x$ and $y = 2x$ are called **linear equations in two variables** because the graph of each equation is a line. Any equation that can be written in the form $y = mx + b$, where m and b are constants, is a linear equation in two variables. Here are examples of linear equations in two variables:

$$y = 3x \qquad\qquad y = 3x - 2$$

$$\text{or} \quad y = 3x + 0 \qquad \text{or} \quad y = 3x + (-2).$$

> This is in the form of $y = mx + b$, with $m = 3$ and $b = 0$.

> This is in the form of $y = mx + b$, with $m = 3$ and $b = -2$.

Great Question!

What's the big deal about the constants m and b in the linear equation $y = mx + b$?

In Section 3.4, you will see how these numbers affect the equation's graph in special ways. The number m determines the line's steepness. The number b determines the point where the line crosses the y-axis. Any real numbers can be used for m and b.

Can you guess how the graph of the linear equation $y = 3x - 2$ compares with the graph of $y = 3x$? In Example 5, we graphed $y = 3x$. Now, let's graph the equation $y = 3x - 2$.

EXAMPLE 6   Graphing a Linear Equation in Two Variables

Graph the equation: $y = 3x - 2$.

Solution

Step 1. Find several ordered pairs that are solutions of the equation. To find a few solutions, we select integers for x, starting with -2 and ending with 2.

Start with x.	Compute y.	Form the ordered pair (x, y).
x	$y = 3x - 2$	(x, y)
-2	$y = 3(-2) - 2 = -6 - 2 = -8$	$(-2, -8)$
-1	$y = 3(-1) - 2 = -3 - 2 = -5$	$(-1, -5)$
0	$y = 3 \cdot 0 - 2 = 0 - 2 = -2$	$(0, -2)$
1	$y = 3 \cdot 1 - 2 = 3 - 2 = 1$	$(1, 1)$
2	$y = 3 \cdot 2 - 2 = 6 - 2 = 4$	$(2, 4)$

Step 2. Plot these ordered pairs as points in the rectangular coordinate system. The five ordered pairs in the table of values are plotted in **Figure 3.6(a)**.

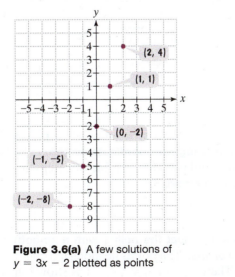

Figure 3.6(a) A few solutions of $y = 3x - 2$ plotted as points

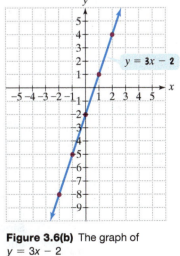

Figure 3.6(b) The graph of $y = 3x - 2$

Step 3. Connect the points with a smooth curve or line. The points lie along a straight line. The graph of $y = 3x - 2$ is shown in **Figure 3.6(b)**. ■

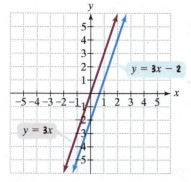

Figure 3.7

Now we are ready to compare the graphs of $y = 3x - 2$ and $y = 3x$. The graphs of both linear equations are shown in the same rectangular coordinate system in **Figure 3.7**. Can you see that the blue graph of $y = 3x - 2$ is parallel to the red graph of $y = 3x$ and 2 units lower than the graph of $y = 3x$? Instead of crossing the y-axis at $(0, 0)$, the graph of $y = 3x - 2$ crosses the y-axis at $(0, -2)$.

> ### Comparing Graphs of Linear Equations
>
> If the value of m does not change,
>
> - The graph of $y = mx + b$ is the graph of $y = mx$ shifted b units up when b is a positive number.
> - The graph of $y = mx + b$ is the graph of $y = mx$ shifted $|b|$ units down when b is a negative number.

✓ **CHECK POINT 6** Graph the equation: $y = 2x - 2$.

EXAMPLE 7 Graphing a Linear Equation in Two Variables

Graph the equation: $y = \dfrac{2}{3}x + 1$.

Solution

Step 1. Find several ordered pairs that are solutions of the equation. Notice that m, the coefficient of x, is $\frac{2}{3}$. When m is a fraction, we will select values of x that are multiples of the denominator. In this way, we can avoid values of y that are fractions. Because the denominator of $\frac{2}{3}$ is 3, we select multiples of 3 for x. Let's use $-6, -3, 0, 3,$ and 6.

Start with multiples of 3 for *x*.	Compute *y*.	Form the ordered pair (*x, y*).
x	$y = \frac{2}{3}x + 1$	**(x, y)**
−6	$y = \frac{2}{3}(-6) + 1 = -4 + 1 = -3$	(−6, −3)
−3	$y = \frac{2}{3}(-3) + 1 = -2 + 1 = -1$	(−3, −1)
0	$y = \frac{2}{3} \cdot 0 + 1 = 0 + 1 = 1$	(0, 1)
3	$y = \frac{2}{3} \cdot 3 + 1 = 2 + 1 = 3$	(3, 3)
6	$y = \frac{2}{3} \cdot 6 + 1 = 4 + 1 = 5$	(6, 5)

Step 2. Plot these ordered pairs as points in the rectangular coordinate system. The five ordered pairs in the table of values are plotted in **Figure 3.8**.

Step 3. Connect the points with a smooth curve or line. The points lie along a straight line. The graph of $y = \frac{2}{3}x + 1$ is shown in **Figure 3.8**.

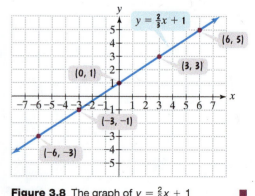

Figure 3.8 The graph of $y = \frac{2}{3}x + 1$

✓ **CHECK POINT 7**   Graph the equation: $y = \frac{1}{2}x + 2$.

6 Use graphs of linear equations to solve problems. ▶

Applications

Part of the beauty of the rectangular coordinate system is that it allows us to "see" mathematical models and visualize the solution to a problem. This idea is demonstrated in Example 8.

EXAMPLE 8   Graphing a Mathematical Model

The graph in **Figure 3.9** shows that over time fewer consumers are paying with checks and cash, and more are paying with plastic.

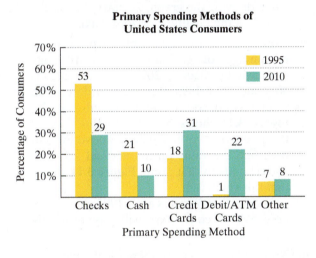

Primary Spending Methods of United States Consumers

Figure 3.9

Source: Mintel/Visa USA Research Services

The mathematical model

$$C = -1.6n + 53$$

describes the percentage of consumers, C, who paid primarily with checks n years after 1995.

a. Let $n = 0, 5, 10, 15,$ and 20. Make a table of values showing five solutions of the equation.

b. Graph the formula in a rectangular coordinate system.

c. Use the graph to estimate the percentage of consumers who will pay primarily by check in 2025.

d. Use the formula to project the percentage of consumers who will pay primarily by check in 2025.

Solution

a. The table of values shows five solutions of $C = -1.6n + 53$.

Start with these values of n.	Substitute n into $C = -1.6n + 53$ and compute C.	Form the ordered pair (n, C).
n	$C = -1.6n + 53$	(n, C)
0	$C = -1.6(0) + 53 = 0 + 53 = 53$	$(0, 53)$
5	$C = -1.6(5) + 53 = -8 + 53 = 45$	$(5, 45)$
10	$C = -1.6(10) + 53 = -16 + 53 = 37$	$(10, 37)$
15	$C = -1.6(15) + 53 = -24 + 53 = 29$	$(15, 29)$
20	$C = -1.6(20) + 53 = -32 + 53 = 21$	$(20, 21)$

b. Now we are ready to graph $C = -1.6n + 53$. Because of the letters used to represent the variables, we label the horizontal axis n and the vertical axis C. We plot the five ordered pairs from the table of values and connect them with a line. The graph of $C = -1.6n + 53$ is shown in **Figure 3.10**.

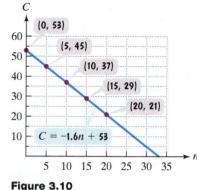

Figure 3.10

As you examine **Figure 3.10**, are you wondering why we did not extend the graph into the other quadrants of the rectangular coordinate system? Keep in mind that n represents the number of years after 1995 and C represents the percentage of consumers paying by check. Only nonnegative values of each variable are meaningful. Thus, when drawing the graph, we started on the positive portion of the vertical axis and ended on the positive horizontal axis.

c. Now we can use the graph to project the percentage of consumers who will pay primarily by check in 2025. The year 2025 is 30 years after 1995. We need to determine what second coordinate is paired with 30. **Figure 3.11** shows how to do this. Locate the point on the line that is above 30 and then find the value on the vertical axis that corresponds to that point. **Figure 3.11** shows that this value is 5.

We see that 30 years after 1995, or in 2025, only 5% of consumers will pay primarily by check.

d. Now let's use the mathematical model to verify our graphical observation in part (c).

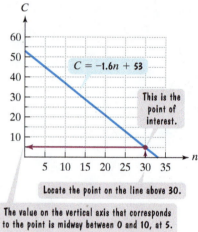

Locate the point on the line above 30.

The value on the vertical axis that corresponds to the point is midway between 0 and 10, at 5.

Figure 3.11

$$C = -1.6n + 53 \qquad \text{This is the given mathematical model.}$$
$$C = -1.6(30) + 53 \qquad \text{Substitute 30 for } n.$$
$$C = -48 + 53 \qquad \text{Multiply: } -1.6(30) = -48.$$
$$C = 5 \qquad \text{Add.}$$

This verifies that 30 years after 1995, in 2025, 5% of consumers will pay primarily by check. ∎

✓ **CHECK POINT 8**   The mathematical model

$$D = 1.4n + 1$$

describes the percentage of consumers, D, who paid primarily with debit cards n years after 1995.

a. Let $n = 0, 5, 10,$ and 15. Make a table of values showing four solutions of the equation.

b. Graph the formula in a rectangular coordinate system.

c. Use the graph to estimate the percentage of consumers who paid primarily with debit cards in 2015.

d. Use the formula to project the percentage of consumers who paid primarily with debit cards in 2015.

Using Technology

Graphing calculators or graphing software packages for computers are referred to as **graphing utilities** or graphers. A graphing utility is a powerful tool that quickly generates the graph of an equation in two variables. **Figure 3.12** shows two such graphs for the equations in Examples 5 and 6.

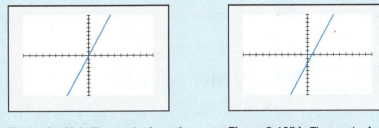

Figure 3.12(a) The graph of $y = 3x$ **Figure 3.12(b)** The graph of $y = 3x - 2$

What differences do you notice between these graphs and the graphs that we drew by hand? They do seem a bit "jittery." Arrows do not appear on both ends of the graphs. Furthermore, numbers are not given along the axes. For both graphs in **Figure 3.12**, the x-axis extends from -10 to 10 and the y-axis also extends from -10 to 10. The distance represented by each consecutive tick mark is one unit. We say that the **viewing rectangle**, or the **viewing window**, is $[-10, 10, 1]$ by $[-10, 10, 1]$.

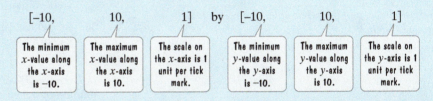

$[-10,$	$10,$	$1]$	by	$[-10,$	$10,$	$1]$
The minimum x-value along the x-axis is -10.	The maximum x-value along the x-axis is 10.	The scale on the x-axis is 1 unit per tick mark.		The minimum y-value along the y-axis is -10.	The maximum y-value along the y-axis is 10.	The scale on the y-axis is 1 unit per tick mark.

To graph an equation in x and y using a graphing utility, enter the equation and specify the size of the viewing rectangle. The size of the viewing rectangle sets minimum and maximum values for both the x- and y-axes. Enter these values, as well as the values for the distances between consecutive tick marks on the respective axes. The $[-10, 10, 1]$ by $[-10, 10, 1]$ viewing rectangle used in **Figure 3.12** is called the **standard viewing rectangle**.

On many graphing utilities, the display screen is five-eighths as high as it is wide. By using a square setting, you can equally space the x and y tick marks. (This does not occur in the standard viewing rectangle.) Graphing utilities can also *zoom in* and *zoom out*. When you zoom in, you see a smaller portion of the graph, but you do so in greater detail. When you zoom out, you see a larger portion of the graph. Thus, zooming out may help you to develop a better understanding of the overall character of the graph. With practice, you will become more comfortable with graphing equations in two variables using your graphing utility. You will also develop a better sense of the size of the viewing rectangle that will reveal needed information about a particular graph.

CONCEPT AND VOCABULARY CHECK

Fill in each blank so that the resulting statement is true.

1. In the rectangular coordinate system, the horizontal number line is called the _____.

2. In the rectangular coordinate system, the vertical number line is called the _____.

3. In the rectangular coordinate system, the point of intersection of the horizontal axis and the vertical axis is called the _____.

4. The axes of the rectangular coordinate system divide the plane into regions, called _____. There are _____ of these regions.

5. The first number in an ordered pair such as (3, 8) is called the _____. The second number in such an ordered pair is called the _____.

6. The ordered pair (1, 3) is a/an _____ of the equation $y = 5x - 2$ because when 1 is substituted for x and 3 is substituted for y, we obtain a true statement. We also say that (1, 3) _____ the equation.

7. Each ordered pair of numbers corresponds to _____ point in the rectangular coordinate system.

8. A linear equation in two variables can be written in the form $y =$ _____, where m and b are constants.

3.1 EXERCISE SET ▶ MyMathLab®

Practice Exercises

In Exercises 1–8, plot the given point in a rectangular coordinate system. Indicate in which quadrant each point lies.

1. (3, 5)
2. (5, 3)
3. (−5, 1)
4. (1, −5)
5. (−3, −1)
6. (−1, −3)
7. (6, −3.5)
8. (−3.5, 6)

In Exercises 9–24, plot the given point in a rectangular coordinate system.

9. (−3, −3)
10. (−5, −5)
11. (−2, 0)
12. (−5, 0)
13. (0, 2)
14. (0, 5)
15. (0, −3)
16. (0, −5)
17. $\left(\frac{5}{2}, \frac{7}{2}\right)$
18. $\left(\frac{7}{2}, \frac{5}{2}\right)$
19. $\left(-5, \frac{3}{2}\right)$
20. $\left(-\frac{9}{2}, -4\right)$
21. (0, 0)
22. $\left(-\frac{5}{2}, 0\right)$
23. $\left(0, -\frac{5}{2}\right)$
24. $\left(0, \frac{7}{2}\right)$

In Exercises 25–32, give the ordered pairs that correspond to the points labeled in the figure.

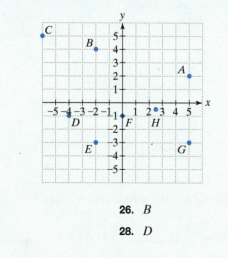

25. *A*
26. *B*
27. *C*
28. *D*
29. *E*
30. *F*
31. *G*
32. *H*

33. In which quadrants are the *y*-coordinates positive?

34. In which quadrants are the *x*-coordinates negative?

35. In which quadrants do the *x*-coordinates and the *y*-coordinates have the same sign?

36. In which quadrants do the *x*-coordinates and the *y*-coordinates have opposite signs?

In Exercises 37–48, determine whether each ordered pair is a solution of the given equation.

37. $y = 3x$ (2, 3), (3, 2), (−4, −12)

38. $y = 4x$ (3, 12), (12, 3), (−5, −20)

39. $y = -4x$ (−5, −20), (0, 0), (9, −36)

40. $y = -3x$ (−5, 15), (0, 0), (7, −21)

41. $y = 2x + 6$ (0, 6), (−3, 0), (2, −2)

42. $y = 8 - 4x$ (8, 0), (16, −2), (3, −4)

43. $3x + 5y = 15$ (−5, 6), (0, 5), (10, −3)

44. $2x - 5y = 0$ (−2, 0), (−10, 6), (5, 0)

45. $x + 3y = 0$ (0, 0), $\left(1, \frac{1}{3}\right)$, $\left(2, -\frac{2}{3}\right)$

46. $x + 5y = 0$ (0, 0), $\left(1, \frac{1}{5}\right)$, $\left(2, -\frac{2}{5}\right)$

47. $x - 4 = 0$ (4, 7), (3, 4), (0, −4)

48. $y + 2 = 0$ (0, 2), (2, 0), (0, −2)

In Exercises 49–56, find five solutions of each equation. Select integers for x, starting with −2 and ending with 2. Organize your work in a table of values.

49. $y = 12x$

50. $y = 14x$

51. $y = -10x$

52. $y = -20x$

53. $y = 8x - 5$

54. $y = 6x - 4$

55. $y = -3x + 7$

56. $y = -5x + 9$

In Exercises 57–80, graph each linear equation in two variables. Find at least five solutions in your table of values for each equation.

57. $y = x$

58. $y = x + 1$

59. $y = x - 1$

60. $y = x - 2$

61. $y = 2x + 1$

62. $y = 2x - 1$

63. $y = -x + 2$

64. $y = -x + 3$

65. $y = -3x - 1$

66. $y = -3x - 2$

67. $y = \dfrac{1}{2}x$

68. $y = -\dfrac{1}{2}x$

69. $y = -\dfrac{1}{4}x$

70. $y = \dfrac{1}{4}x$

71. $y = \dfrac{1}{3}x + 1$

72. $y = \dfrac{1}{3}x - 1$

73. $y = -\dfrac{3}{2}x + 1$

74. $y = -\dfrac{3}{2}x + 2$

75. $y = -\dfrac{5}{2}x - 1$

76. $y = -\dfrac{5}{2}x + 1$

77. $y = x + \dfrac{1}{2}$

78. $y = x - \dfrac{1}{2}$

79. $y = 4$, or $y = 0x + 4$

80. $y = 3$, or $y = 0x + 3$

Practice PLUS

In Exercises 81–84, write each sentence as a linear equation in two variables. Then graph the equation.

81. The y-variable is 3 more than the x-variable.

82. The y-variable exceeds the x-variable by 4.

83. The y-variable exceeds twice the x-variable by 5.

84. The y-variable is 2 less than 3 times the x-variable.

85. At the beginning of a semester, a student purchased eight pens and six pads for a total cost of $14.50.

   a. If x represents the cost of one pen and y represents the cost of one pad, write an equation in two variables that reflects the given conditions.

   b. If pads cost $0.75 each, find the cost of one pen.

86. A nursery offers a package of three small orange trees and four small grapefruit trees for $22.

   a. If x represents the cost of one orange tree and y represents the cost of one grapefruit tree, write an equation in two variables that reflects the given conditions.

   b. If a grapefruit tree costs $2.50, find the cost of an orange tree.

Application Exercises

A football is thrown by a quarterback to a receiver. The points in the figure show the height of the football, in feet, above the ground in terms of its distance, in yards, from the quarterback. Use this information to solve Exercises 87–92.

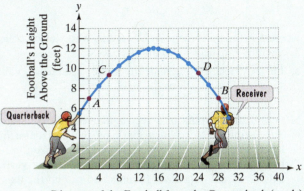

Distance of the Football from the Quarterback (yards)

87. Find the coordinates of point A. Then interpret the coordinates in terms of the information given.

88. Find the coordinates of point B. Then interpret the coordinates in terms of the information given.

89. Estimate the coordinates of point C.

90. Estimate the coordinates of point D.

91. What is the football's maximum height? What is its distance from the quarterback when it reaches its maximum height?

92. What is the football's height when it is caught by the receiver? What is the receiver's distance from the quarterback when he catches the football?

Even as Americans increasingly view a college education as essential for success, many believe that a college education is becoming less available to qualified students. Exercises 93–94 are based on the data displayed by the graph.

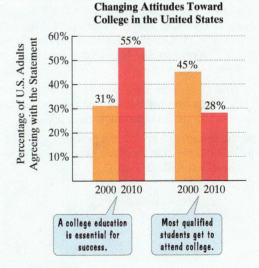

Source: Public Agenda

93. The graph at the bottom of the previous page shows that in 2000, 31% of U.S. adults viewed a college education as essential for success. For the period from 2000 through 2010, the percentage viewing a college education as essential for success increased on average by approximately 2.4 each year. These conditions can be described by the mathematical model

$$S = 2.4n + 31,$$

where S is the percentage of U.S. adults who viewed college as essential for success n years after 2000.

a. Let $n = 0, 5, 10, 15,$ and 20. Make a table of values showing five solutions of the equation.

b. Graph the formula in a rectangular coordinate system. Suggestion: Let each tick mark on the horizontal axis, labeled n, represent 5 units. Extend the horizontal axis to include $n = 25$. Let each tick mark on the vertical axis, labeled S, represent 10 units and extend the axis to include $S = 100$.

c. Use your graph from part (b) to estimate the percentage of U.S. adults who will view college as essential for success in 2018.

d. Use the formula to project the percentage of U.S. adults who will view college as essential for success in 2018.

94. The graph at the bottom of the previous page shows that in 2000, 45% of U.S. adults believed that most qualified students get to attend college. For the period from 2000 through 2010, the percentage who believed that a college education is available to most qualified students decreased by approximately 1.7 each year. These conditions can be described by the mathematical model

$$Q = -1.7n + 45,$$

where Q is the percentage believing that a college education is available to most qualified students n years after 2000.

a. Let $n = 0, 5, 10, 15,$ and 20. Make a table of values showing five solutions of the equation.

b. Graph the formula in a rectangular coordinate system. Suggestion: Let each tick mark on the horizontal axis, labeled n, represent 5 units. Extend the horizontal axis to include $n = 25$. Let each tick mark on the vertical axis, labeled Q, represent 5 units and extend the axis to include $Q = 50$.

c. Use your graph from part (b) to estimate the percentage of U.S. adults who will believe that a college education is available to most qualified students in 2018.

d. Use the formula to project the percentage of U.S. adults who will believe that a college education is available to most qualified students in 2018.

Explaining the Concepts

95. What is the rectangular coordinate system?

96. Explain how to plot a point in the rectangular coordinate system. Give an example with your explanation.

97. Explain why $(5, -2)$ and $(-2, 5)$ do not represent the same point.

98. Explain how to find the coordinates of a point in the rectangular coordinate system.

99. How do you determine whether an ordered pair is a solution of an equation in two variables, x and y?

100. Explain how to find ordered pairs that are solutions of an equation in two variables, x and y.

101. What is the graph of an equation?

102. Explain how to graph an equation in two variables in the rectangular coordinate system.

Critical Thinking Exercises

Make Sense?   *In Exercises 103–106, determine whether each statement makes sense or does not make sense, and explain your reasoning.*

103. When I know that an equation's graph is a straight line, I don't need to plot more than two points, although I sometimes plot three just to check that the points line up.

104. The graph that I'm looking at is U-shaped, so its equation cannot be of the form $y = mx + b$.

105. I'm working with a linear equation in two variables and found that $(-2, 2)$, $(0, 0)$, and $(2, 2)$ are solutions.

106. When a real-world situation is modeled with a linear equation in two variables, I can use its graph to predict specific information about the situation.

In Exercises 107–110, determine whether each statement is true or false. If the statement is false, make the necessary change(s) to produce a true statement.

107. The graph of $y = 3x + 1$ is the graph of $y = 2x$ shifted up 1 unit.

108. The graph of any equation in the form $y = mx + b$ passes through the point $(0, b)$.

109. The ordered pair $(3, 4)$ satisfies the equation

$$2y - 3x = -6.$$

110. If $(2, 5)$ satisfies an equation, then $(5, 2)$ also satisfies the equation.

111. a. Graph each of the following points:

$$\left(1, \frac{1}{2}\right), \ (2, 1), \ \left(3, \frac{3}{2}\right), \ (4, 2).$$

Parts (b)–(d) can be answered by changing the sign of one or both coordinates of the points in part (a).

b. What must be done to the coordinates so that the resulting graph is a mirror-image reflection about the *y*-axis of your graph in part (a)?

c. What must be done to the coordinates so that the resulting graph is a mirror-image reflection about the *x*-axis of your graph in part (a)?

d. What must be done to the coordinates so that the resulting graph is a straight-line extension of your graph in part (a)?

Technology Exercises

Use a graphing utility to graph each equation in Exercises 112–115 in a standard viewing rectangle, $[-10, 10, 1]$ by $[-10, 10, 1]$. Then use the $\boxed{\text{TRACE}}$ *feature to trace along the line and find the coordinates of two points.*

112. $y = 2x - 1$ **113.** $y = -3x + 2$

114. $y = \dfrac{1}{2}x$ **115.** $y = \dfrac{3}{4}x - 2$

116. Use a graphing utility to verify any five of your hand-drawn graphs in Exercises 57–80. Use an appropriate viewing rectangle and the $\boxed{\text{ZOOM SQUARE}}$ feature to make the graph look like the one you drew by hand.

Review Exercises

117. Solve: $3x + 5 = 4(2x - 3) + 7$.
(Section 2.3, Example 3)

118. Simplify: $3(1 - 2 \cdot 5) - (-28)$.
(Section 1.8, Example 7)

119. Solve for *h*: $V = \dfrac{1}{3}Ah$. (Section 2.4, Example 4)

Preview Exercises

Exercises 120–122 will help you prepare for the material covered in the next section. Remember that a solution of an equation in two variables is an ordered pair.

120. Let $y = 0$ and find a solution of $3x - 4y = 24$.

121. Let $x = 0$ and find a solution of $3x - 4y = 24$.

122. Let $x = 0$ and find a solution of $x + 2y = 0$.

SECTION

3.2

Graphing Linear Equations Using Intercepts

What am I supposed to learn?

After studying this section, you should be able to:

1 Use a graph to identify intercepts.

2 Graph a linear equation in two variables using intercepts.

3 Graph horizontal or vertical lines.

It's hard to believe that this gas-guzzler, with its huge fins and overstated design, was available in 1957 for approximately $1800. Sadly, its elegance quickly faded, depreciating linearly by $300 per year, often sold for scrap just six years after its glorious emergence from the dealer's showroom.

From these casual observations, we can obtain a mathematical model and its graph. The model is

$$y = -300x + 1800.$$

The car depreciated by $300 per year for *x* years.

The new car was worth $1800.

In the model $y = -300x + 1800$, y is the car's value after x years. **Figure 3.13** shows the equation's graph.

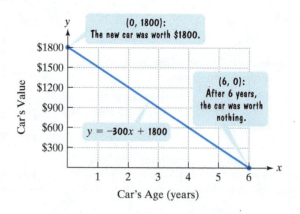

Figure 3.13

Here are some important observations about the model and its graph:

- The points in **Figure 3.13** show where the graph intersects the y-axis and where it intersects the x-axis. These important points are the topic of this section.
- We can rewrite the model, $y = -300x + 1800$, by adding $300x$ to both sides:

$$300x + y = 1800.$$

The form of this equation is $Ax + By = C$.

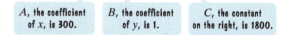

$$300x + y = 1800$$

| A, the coefficient of x, is 300. | B, the coefficient of y, is 1. | C, the constant on the right, is 1800. |

All equations of the form $Ax + By = C$ are straight lines when graphed as long as A and B are not both zero. To graph linear equations of this form, we will use the *intercepts*.

1 Use a graph to identify intercepts.

Intercepts

An **x-intercept** of a graph is the x-coordinate of a point where the graph intersects the x-axis. For example, look at the graph of $2x - 4y = 8$ in **Figure 3.14**. The graph crosses the x-axis at $(4, 0)$. Thus, the x-intercept is 4. **The y-coordinate corresponding to an x-intercept is always zero.**

A **y-intercept** of a graph is the y-coordinate of a point where the graph intersects the y-axis. The graph of $2x - 4y = 8$ in **Figure 3.14** shows that the graph crosses the y-axis at $(0, -2)$. Thus, the y-intercept is -2. **The x-coordinate corresponding to a y-intercept is always zero.**

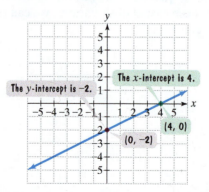

Figure 3.14 The graph of $2x - 4y = 8$

Great Question!

Are single numbers the only way to represent intercepts? Can ordered pairs also be used?

Mathematicians tend to use two ways to describe intercepts. Did you notice that we are using single numbers? If a graph's x-intercept is a, it passes through the point $(a, 0)$. If a graph's y-intercept is b, it passes through the point $(0, b)$.

Some books state that the x-intercept is the *point* $(a, 0)$ and the x-intercept is *at a* on the x-axis. Similarly, the y-intercept is the *point* $(0, b)$ and the y-intercept is *at b* on the y-axis. In these descriptions, the intercepts are the actual points where a graph intersects the axes.

Although we'll describe intercepts as single numbers, we'll immediately state the point on the x- or y-axis that the graph passes through. Here's the important thing to keep in mind:

x-intercept: The corresponding y is 0.

y-intercept: The corresponding x is 0.

EXAMPLE 1 Identifying Intercepts

Identify the x- and y-intercepts.

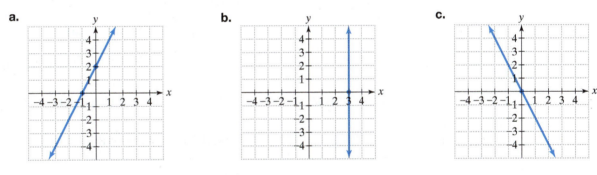

a. **b.** **c.**

Solution

a. The graph crosses the x-axis at $(-1, 0)$. Thus, the x-intercept is -1. The graph crosses the y-axis at $(0, 2)$. Thus, the y-intercept is 2.

b. The graph crosses the x-axis at $(3, 0)$, so the x-intercept is 3. This vertical line does not cross the y-axis. Thus, there is no y-intercept.

c. This graph crosses the x- and y-axes at the same point, the origin. Because the graph crosses both axes at $(0, 0)$, the x-intercept is 0 and the y-intercept is 0. ∎

✓ **CHECK POINT 1** Identify the x- and y-intercepts.

a. **b.** **c.**

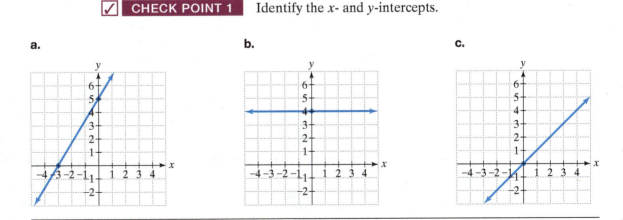

2 Graph a linear equation in two variables using intercepts. ▶

Great Question!

Can *x* or *y* have exponents in the standard form of the equation of a line?

In the form $Ax + By = C$, the exponent that is understood on both x and y is 1. Neither x nor y can have exponents other than 1.

Forms of Equations That Do Not Represent Lines

$$Ax^2 + By^2 = C \quad Ax + By^2 = C$$

Exponents on x and y are not both 1.

Graphing Using Intercepts

An equation of the form $Ax + By = C$, where $A, B,$ and C are integers, is called the **standard form** of the equation of a line. The equation can be graphed by finding the x- and y-intercepts, plotting the intercepts, and drawing a straight line through these points. How do we find the intercepts of a line, given its equation? Because the y-coordinate of the x-intercept is 0, to find the x-intercept,

- Substitute 0 for y in the equation.
- Solve for x.

EXAMPLE 2   Finding the *x*-Intercept

Find the x-intercept of the graph of $3x - 4y = 24$.

Solution   To find the x-intercept, let $y = 0$ and solve for x.

$$3x - 4y = 24 \qquad \text{This is the given equation.}$$
$$3x - 4 \cdot 0 = 24 \qquad \text{Let } y = 0.$$
$$3x = 24 \qquad \text{Simplify: } 4 \cdot 0 = 0 \text{ and } 3x - 0 = 3x.$$
$$x = 8 \qquad \text{Divide both sides by 3.}$$

The x-intercept is 8. The graph of $3x - 4y = 24$ passes through the point $(8, 0)$. ∎

☑ **CHECK POINT 2**   Find the x-intercept of the graph of $4x - 3y = 12$.

Because the x-coordinate of the y-intercept is 0, to find the y-intercept,

- Substitute 0 for x in the equation.
- Solve for y.

EXAMPLE 3   Finding the *y*-Intercept

Find the y-intercept of the graph of $3x - 4y = 24$.

Solution   To find the y-intercept, let $x = 0$ and solve for y.

$$3x - 4y = 24 \qquad \text{This is the given equation.}$$
$$3 \cdot 0 - 4y = 24 \qquad \text{Let } x = 0.$$
$$-4y = 24 \qquad \text{Simplify: } 3 \cdot 0 = 0 \text{ and } 0 - 4y = -4y.$$
$$y = -6 \qquad \text{Divide both sides by } -4.$$

The y-intercept is -6. The graph of $3x - 4y = 24$ passes through the point $(0, -6)$. ∎

☑ **CHECK POINT 3**   Find the y-intercept of the graph of $4x - 3y = 12$.

When graphing using intercepts, it is a good idea to use a third point, a checkpoint, before drawing the line. A checkpoint can be obtained by selecting a value for either variable, other than 0, and finding the corresponding value for the other variable. The checkpoint should lie on the same line as the x- and y-intercepts. If it does not, recheck your work and find the error.

> ### Using Intercepts to Graph $Ax + By = C$
>
> 1. Find the x-intercept. Let $y = 0$ and solve for x.
> 2. Find the y-intercept. Let $x = 0$ and solve for y.
> 3. Find a checkpoint, a third ordered-pair solution.
> 4. Graph the equation by drawing a line through the three points.

EXAMPLE 4 Using Intercepts to Graph a Linear Equation

Graph: $3x + 2y = 6$.

Solution
Step 1. Find the x-intercept. Let $y = 0$ and solve for x.

$$3x + 2 \cdot 0 = 6 \qquad \text{Replace } y \text{ with 0 in } 3x + 2y = 6.$$
$$3x = 6 \qquad \text{Simplify.}$$
$$x = 2 \qquad \text{Divide both sides by 3.}$$

The x-intercept is 2, so the line passes through $(2, 0)$.

Step 2. Find the y-intercept. Let $x = 0$ and solve for y.

$$3 \cdot 0 + 2y = 6 \qquad \text{Replace } x \text{ with 0 in } 3x + 2y = 6.$$
$$2y = 6 \qquad \text{Simplify.}$$
$$y = 3 \qquad \text{Divide both sides by 2.}$$

The y-intercept is 3, so the line passes through $(0, 3)$.

Step 3. Find a checkpoint, a third ordered-pair solution. For our checkpoint, we will let $x = 1$ and find the corresponding value for y.

$$3x + 2y = 6 \qquad \text{This is the given equation.}$$
$$3 \cdot 1 + 2y = 6 \qquad \text{Substitute 1 for } x.$$
$$3 + 2y = 6 \qquad \text{Simplify.}$$
$$2y = 3 \qquad \text{Subtract 3 from both sides.}$$
$$y = \frac{3}{2} \qquad \text{Divide both sides by 2.}$$

The checkpoint is the ordered pair $\left(1, \dfrac{3}{2}\right)$, or $(1, 1.5)$.

Step 4. Graph the equation by drawing a line through the three points. The three points in **Figure 3.15** lie along the same line. Drawing a line through the three points results in the graph of $3x + 2y = 6$. ∎

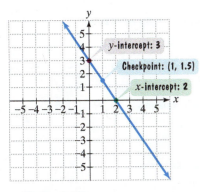

Figure 3.15
The graph of $3x + 2y = 6$

Using Technology

You can use a graphing utility to graph equations of the form $Ax + By = C$. Begin by solving the equation for y. For example, to graph $3x + 2y = 6$, solve the equation for y.

$$3x + 2y = 6 \qquad \text{This is the equation to be graphed.}$$

$$3x - 3x + 2y = -3x + 6 \qquad \text{Subtract } 3x \text{ from both sides.}$$

$$2y = -3x + 6 \qquad \text{Simplify.}$$

$$\frac{2y}{2} = \frac{-3x + 6}{2} \qquad \text{Divide both sides by 2.}$$

$$y = -\frac{3}{2}x + 3 \qquad \text{Divide each term of } -3x + 6 \text{ by 2.}$$

This is the equation to enter into your graphing utility. The graph of $y = -\frac{3}{2}x + 3$ or, equivalently, $3x + 2y = 6$, is shown below in a $[-6, 6, 1]$ by $[-6, 6, 1]$ viewing rectangle.

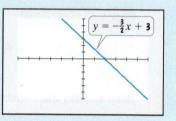

✅ **CHECK POINT 4** Graph: $2x + 3y = 6$.

EXAMPLE 5 Using Intercepts to Graph a Linear Equation

Graph: $2x - y = 4$.

Solution

Step 1. Find the x-intercept. Let $y = 0$ and solve for x.

$$2x - 0 = 4 \qquad \text{Replace } y \text{ with 0 in } 2x - y = 4.$$

$$2x = 4 \qquad \text{Simplify.}$$

$$x = 2 \qquad \text{Divide both sides by 2.}$$

The x-intercept is 2, so the line passes through $(2, 0)$.

Step 2. Find the y-intercept. Let $x = 0$ and solve for y.

$$2 \cdot 0 - y = 4 \qquad \text{Replace } x \text{ with 0 in } 2x - y = 4.$$

$$-y = 4 \qquad \text{Simplify.}$$

$$y = -4 \qquad \text{Divide (or multiply) both sides by } -1.$$

The y-intercept is -4, so the line passes through $(0, -4)$.

Step 3. Find a checkpoint, a third ordered-pair solution. For our checkpoint, we will let $x = 1$ and find the corresponding value for y.

$$2x - y = 4 \qquad \text{This is the given equation.}$$

$$2 \cdot 1 - y = 4 \qquad \text{Substitute 1 for } x.$$

$$2 - y = 4 \qquad \text{Simplify.}$$

$$-y = 2 \qquad \text{Subtract 2 from both sides.}$$

$$y = -2 \qquad \text{Divide (or multiply) both sides by } -1.$$

The checkpoint is $(1, -2)$.

Figure 3.16
The graph of $2x - y = 4$

x-intercept: 2

Checkpoint: (1, −2)

y-intercept: −4

Step 4. Graph the equation by drawing a line through the three points. The three points in **Figure 3.16** lie along the same line. Drawing a line through the three points results in the graph of $2x - y = 4$. ■

☑ **CHECK POINT 5** Graph: $x - 2y = 4$.

We have seen that not all lines have two different intercepts. Some lines pass through the origin. Thus, they have an x-intercept of 0 and a y-intercept of 0. Is it possible to recognize these lines by their equations? Yes. **The graph of the linear equation $Ax + By = 0$ passes through the origin.** Notice that the constant on the right side of this equation is 0.

An equation of the form $Ax + By = 0$ can be graphed by using the origin as one point on the line. Find two other points by finding two other solutions of the equation. Select values for either variable, other than 0, and find the corresponding values for the other variable.

EXAMPLE 6 Graphing a Linear Equation of the Form $Ax + By = 0$

Graph: $x + 2y = 0$.

Solution Because the constant on the right is 0, the graph passes through the origin. The x- and y-intercepts are both 0. Remember that we are using two points and a checkpoint to graph a line. Thus, we still want to find two other points. Let $y = -1$ to find a second ordered-pair solution. Let $y = 1$ to find a third ordered-pair (checkpoint) solution.

$$x + 2y = 0 \qquad\qquad x + 2y = 0$$

Let $y = -1$. Let $y = 1$.

$$x + 2(-1) = 0 \qquad\qquad x + 2 \cdot 1 = 0$$
$$x + (-2) = 0 \qquad\qquad x + 2 = 0$$
$$x = 2 \qquad\qquad\qquad x = -2$$

The solutions are $(2, -1)$ and $(-2, 1)$. Plot these two points, as well as the origin—that is, $(0, 0)$. The three points in **Figure 3.17** lie along the same line. Drawing a line through the three points results in the graph of $x + 2y = 0$. ■

☑ **CHECK POINT 6** Graph: $x + 3y = 0$.

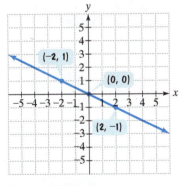

Figure 3.17
The graph of $x + 2y = 0$

3 Graph horizontal or vertical lines.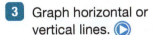

Equations of Horizontal and Vertical Lines

We know that the graph of any equation of the form $Ax + By = C$ is a line as long as A and B are not both zero. What happens if A or B, but not both, is zero? Example 7 shows that if $A = 0$, the equation $Ax + By = C$ has no x-term and the graph is a horizontal line.

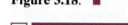

Figure 3.18
The graph of $y = -3$

EXAMPLE 7 Graphing a Horizontal Line

Graph: $y = -3$.

Solution All ordered pairs that are solutions of $y = -3$ have a value of y that is -3. Any value can be used for x. In the table on the right, we have selected three of the possible values for x: $-2, 0$, and 3. The table shows that three ordered pairs that are solutions of $y = -3$ are $(-2, -3)$, $(0, -3)$, and $(3, -3)$. Drawing a line that passes through the three points gives the horizontal line shown in **Figure 3.18**. ∎

x	$y = -3$	(x, y)
-2	-3	$(-2, -3)$
0	-3	$(0, -3)$
3	-3	$(3, -3)$

For all choices of x, y is a constant -3.

☑ **CHECK POINT 7** Graph: $y = 3$.

Example 8 shows that if $B = 0$, the equation $Ax + By = C$ has no y-term and the graph is a vertical line.

EXAMPLE 8 Graphing a Vertical Line

Graph: $x = 5$.

Solution All ordered pairs that are solutions of $x = 5$ have a value of x that is 5. Any value can be used for y. In the table on the right, we have selected three of the possible values for y: $-2, 0$, and 3. The table shows that three ordered pairs that are solutions of $x = 5$ are $(5, -2)$, $(5, 0)$, and $(5, 3)$. Drawing a line that passes through the three points gives the vertical line shown in **Figure 3.19**. ∎

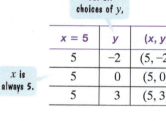

For all choices of y, x is always 5.

$x = 5$	y	(x, y)
5	-2	$(5, -2)$
5	0	$(5, 0)$
5	3	$(5, 3)$

Figure 3.19
The graph of $x = 5$

Great Question!

Why isn't the graph of $x = 5$ just a single point at 5 on a number line?

Do not confuse two-dimensional graphing and one-dimensional graphing of $x = 5$. The graph of $x = 5$ in a two-dimensional rectangular coordinate system is the vertical line in **Figure 3.19**. By contrast, the graph of $x = 5$ on a one-dimensional number line representing values of x is a single point at 5:

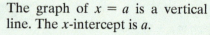

☑ **CHECK POINT 8** Graph: $x = -2$.

Horizontal and Vertical Lines

The graph of $y = b$ is a horizontal line. The y-intercept is b.

The graph of $x = a$ is a vertical line. The x-intercept is a.

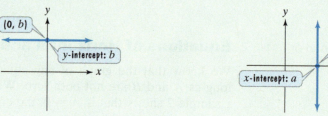

CONCEPT AND VOCABULARY CHECK

Fill in each blank so that the resulting statement is true.

1. The *x*-coordinate of a point where a graph crosses the *x*-axis is called a/an _____.

2. The *y*-coordinate of a point where a graph crosses the *y*-axis is called a/an _____.

3. The point (4, 0) lies along a line, so 4 is the _____ of that line.

4. The point (0, 3) lies along a line, so 3 is the _____ of that line.

5. An equation that can be written in the form $Ax + By = C$, where A and B are not both zero, is called the _____ form of the equation of a line.

6. Given the equation $Ax + By = C$, to find the *x*-intercept (if there is one), let _____ = 0 and solve for _____.

7. Given the equation $Ax + By = C$, to find the *y*-intercept (if there is one), let _____ = 0 and solve for _____.

8. The graph of the equation $y = 3$ is a/an _____ line.

9. The graph of the equation $x = -2$ is a/an _____ line.

3.2 EXERCISE SET   ▶ MyMathLab®

Practice Exercises

In Exercises 1–8, use the graph to identify the

a. *x-intercept, or state that there is no x-intercept;*

b. *y-intercept, or state that there is no y-intercept.*

1.
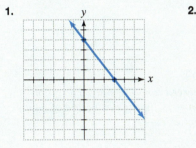

2.
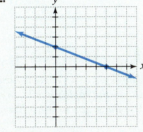

3.
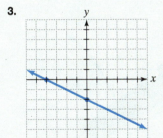

4.
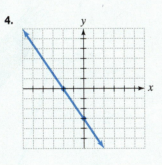

5.
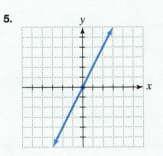

6.
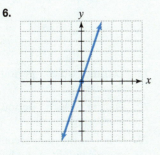

7.   8.
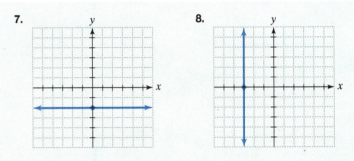

In Exercises 9–18, find the x-intercept and the y-intercept of the graph of each equation. Do not graph the equation.

9. $2x + 5y = 20$

10. $2x + 6y = 30$

11. $2x - 3y = 15$

12. $4x - 5y = 10$

13. $-x + 3y = -8$

14. $-x + 3y = -10$

15. $7x - 9y = 0$

16. $8x - 11y = 0$

17. $2x = 3y - 11$

18. $2x = 4y - 13$

In Exercises 19–40, use intercepts and a checkpoint to graph each equation.

19. $x + y = 5$

20. $x + y = 6$

21. $x + 3y = 6$

22. $2x + y = 4$

23. $6x - 9y = 18$

24. $6x - 2y = 12$

25. $-x + 4y = 6$

26. $-x + 3y = 10$

27. $2x - y = 7$

28. $2x - y = 5$

29. $3x = 5y - 15$ **30.** $2x = 3y + 6$

31. $25y = 100 - 50x$ **32.** $10y = 60 - 40x$

33. $2x - 8y = 12$ **34.** $3x - 6y = 15$

35. $x + 2y = 0$ **36.** $2x + y = 0$

37. $y - 3x = 0$ **38.** $y - 4x = 0$

39. $2x - 3y = -11$ **40.** $3x - 2y = -7$

In Exercises 41–46, write an equation for each graph.

41.

42.

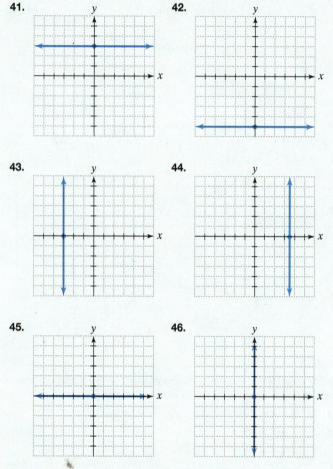

43.

44.

45.

46.

In Exercises 47–62, graph each equation.

47. $y = 4$ **48.** $y = 2$

49. $y = -2$ **50.** $y = -3$

51. $x = 2$ **52.** $x = 4$

53. $x + 1 = 0$ **54.** $x + 5 = 0$

55. $y - 3.5 = 0$ **56.** $y - 2.5 = 0$

57. $x = 0$ **58.** $y = 0$

59. $3y = 9$ **60.** $5y = 20$

61. $12 - 3x = 0$ **62.** $12 - 4x = 0$

Practice PLUS

In Exercises 63–68, match each equation with one of the graphs shown in Exercises 1–8.

63. $3x + 2y = -6$ **64.** $x + 2y = -4$

65. $y = -2$ **66.** $x = -3$

67. $4x + 3y = 12$ **68.** $2x + 5y = 10$

In Exercises 69–70,

  a. *Write a linear equation in standard form satisfying the given condition. Assume that all measurements shown in each figure are in feet.*

  b. *Graph the equation in part (a). Because x and y must be nonnegative (why?), limit your final graph to quadrant I and its boundaries.*

69. The perimeter of the larger rectangle is 58 feet.

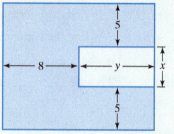

70. The perimeter of the shaded trapezoid is 84 feet.

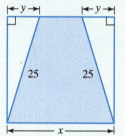

Application Exercises

The flight of an eagle is observed for 30 seconds. The graph shows the eagle's height, in meters, during this period of time. Use the graph to solve Exercises 71–75.

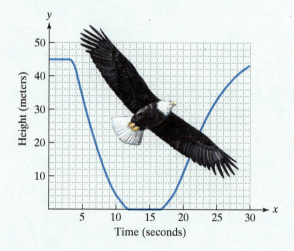

71. During which period of time is the eagle's height decreasing?

72. During which period of time is the eagle's height increasing?

73. What is the y-intercept? What does this mean about the eagle's height at the beginning of the observation?

(In Exercises 74–75, be sure to refer to the graph at the bottom of the previous page.)

74. During the first three seconds of observation, the eagle's flight is graphed as a horizontal line. Write the equation of the line. What does this mean about the eagle's flight pattern during this time?

75. Use integers to write five *x*-intercepts of the graph. What is the eagle doing during these times?

76. A new car worth $24,000 is depreciating in value by $3000 per year. The mathematical model

$$y = -3000x + 24,000$$

describes the car's value, *y*, in dollars, after *x* years.

   a. Find the *x*-intercept. Describe what this means in terms of the car's value.

   b. Find the *y*-intercept. Describe what this means in terms of the car's value.

   c. Use the intercepts to graph the linear equation. Because *x* and *y* must be nonnegative (why?), limit your graph to quadrant I and its boundaries.

   d. Use your graph to estimate the car's value after five years.

77. A new car worth $45,000 is depreciating in value by $5000 per year. The mathematical model

$$y = -5000x + 45,000$$

describes the car's value, *y*, in dollars, after *x* years.

   a. Find the *x*-intercept. Describe what this means in terms of the car's value.

   b. Find the *y*-intercept. Describe what this means in terms of the car's value.

   c. Use the intercepts to graph the linear equation. Because *x* and *y* must be nonnegative (why?), limit your graph to quadrant I and its boundaries.

   d. Use your graph to estimate the car's value after five years.

Explaining the Concepts

78. What is an *x*-intercept of a graph?

79. What is a *y*-intercept of a graph?

80. If you are given an equation of the form $Ax + By = C$, explain how to find the *x*-intercept.

81. If you are given an equation of the form $Ax + By = C$, explain how to find the *y*-intercept.

82. Explain how to graph $Ax + By = C$ if *C* is not equal to zero.

83. Explain how to graph a linear equation of the form $Ax + By = 0$.

84. How many points are needed to graph a line? How many should actually be used? Explain.

85. Describe the graph of $y = 200$.

86. Describe the graph of $x = -100$.

Critical Thinking Exercises

Make Sense? *In Exercises 87–90, determine whether each statement makes sense or does not make sense, and explain your reasoning.*

87. If I could be absolutely certain that I have not made an algebraic error in obtaining intercepts, I would not need to use checkpoints.

88. I like to select a point represented by one of the intercepts as my checkpoint.

89. The graphs of $2x - 3y = -18$ and $-2x + 3y = 18$ must have the same intercepts because I can see that the equations are equivalent.

90. From 1997 through 2007, the federal minimum wage remained constant at $5.15 per hour, so I modeled the situation with $y = 5.15$ and the graph of a vertical line.

In Exercises 91–92, find the coefficients that must be placed in each shaded area so that the equation's graph will be a line with the specified intercepts.

91. ▮ $x +$ ▮ $y = 10$; *x*-intercept = 5; *y*-intercept = 2

92. ▮ $x +$ ▮ $y = 12$; *x*-intercept = −2; *y*-intercept = 4

Technology Exercises

93. Use a graphing utility to verify any five of your hand-drawn graphs in Exercises 19–40. Solve the equation for *y* before entering it.

In Exercises 94–97, use a graphing utility to graph each equation. You will need to solve the equation for y before entering it. Use the graph displayed on the screen to identify the x-intercept and the y-intercept.

94. $2x + y = 4$

95. $3x - y = 9$

96. $2x + 3y = 30$

97. $4x - 2y = -40$

Review Exercises

98. Find the absolute value: $|-13.4|$.
    (Section 1.3, Example 8)

99. Simplify: $7x - (3x - 5)$.
    (Section 1.7, Example 7)

100. Solve: $8(x - 2) - 2(x - 3) \le 8x$.
     (Section 2.7, Example 8)

Preview Exercises

Exercises 101–103 will help you prepare for the material covered in the next section. In each exercise, evaluate

$$\frac{y_2 - y_1}{x_2 - x_1}$$

for the given ordered pairs (x_1, y_1) and (x_2, y_2).

101. $(x_1, y_1) = (1, 3)$; $(x_2, y_2) = (6, 13)$

102. $(x_1, y_1) = (4, -2)$; $(x_2, y_2) = (6, -4)$

103. $(x_1, y_1) = (3, 4)$; $(x_2, y_2) = (5, 4)$

SECTION

3.3

Slope

What am I supposed to learn?

After studying this section, you should be able to:

1 Compute a line's slope.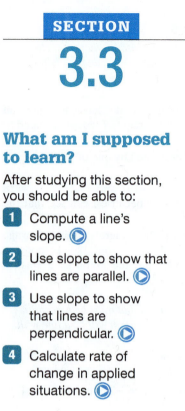

2 Use slope to show that lines are parallel.

3 Use slope to show that lines are perpendicular.

4 Calculate rate of change in applied situations.

A best guess at the future of our nation indicates that the numbers of men and women living alone will increase each year. **Figure 3.20** shows that in 2013, 15.0 million men and 18.6 million women lived alone, an increase over the numbers displayed in the graph for 2000.

By looking at **Figure 3.20**, can you tell that the green graph representing men is steeper than the red graph representing women? This indicates a greater rate of increase in the millions of men living alone than in the millions of women living alone over the period from 2000 through 2013. In this section, we will study the idea of a line's steepness and see what this has to do with how its variables are changing.

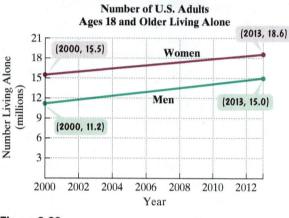

Number of U.S. Adults Ages 18 and Older Living Alone

Figure 3.20

Source: U.S. Census Bureau

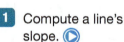
Compute a line's slope.

The Slope of a Line

Mathematicians have developed a useful measure of the steepness of a line, called the **slope** of the line. Slope compares the vertical change (the **rise**) to the horizontal change (the **run**) when moving from one fixed point to another along the line. To calculate the slope of a line, we use a ratio that compares the change in y (the rise) to the change in x (the run).

Definition of Slope

The **slope** of the line through the distinct points (x_1, y_1) and (x_2, y_2) is

$$\frac{\text{Change in } y}{\text{Change in } x} = \frac{\text{Rise}}{\text{Run}} \quad \begin{array}{l}\text{Vertical change}\\\text{Horizontal change}\end{array}$$

$$= \frac{y_2 - y_1}{x_2 - x_1}$$

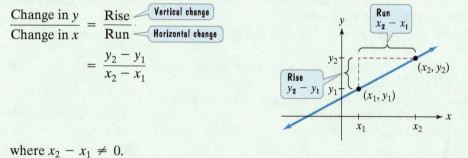

where $x_2 - x_1 \neq 0$.

It is common notation to let the letter m represent the slope of a line. The letter m is used because it is the first letter of the French verb *monter*, meaning to rise, or to ascend.

EXAMPLE 1   Using the Definition of Slope

Find the slope of the line passing through each pair of points:

a. $(-3, -1)$ and $(-2, 4)$

b. $(-3, 4)$ and $(2, -2)$.

Solution

a. To find the slope of the line passing through $(-3, -1)$ and $(-2, 4)$, we let $(x_1, y_1) = (-3, -1)$ and $(x_2, y_2) = (-2, 4)$. We obtain the slope as follows:

$$m = \frac{\text{Change in } y}{\text{Change in } x} = \frac{y_2 - y_1}{x_2 - x_1} = \frac{4 - (-1)}{-2 - (-3)} = \frac{5}{1} = 5.$$

The situation is illustrated in **Figure 3.21**. The slope of the line is 5, indicating that there is a vertical change, a rise, of 5 units for each horizontal change, a run, of 1 unit. The slope is positive and the line rises from left to right.

Great Question!

When using the definition of slope, how do I know which point to call (x_1, y_1) and which point to call (x_2, y_2)?

When computing slope, it makes no difference which point you call (x_1, y_1) and which point you call (x_2, y_2). If we let $(x_1, y_1) = (-2, 4)$ and $(x_2, y_2) = (-3, -1)$, the slope is still 5:

$$m = \frac{\text{Change in } y}{\text{Change in } x} = \frac{y_2 - y_1}{x_2 - x_1} = \frac{-1 - 4}{-3 - (-2)} = \frac{-5}{-1} = 5.$$

However, you should not subtract in one order in the numerator $(y_2 - y_1)$ and then in the opposite order in the denominator $(x_1 - x_2)$.

$$\frac{-1 - 4}{-2 - (-3)} = \frac{-5}{1} = -5 \qquad \text{Incorrect! The slope is not } -5.$$

b. To find the slope of the line passing through $(-3, 4)$ and $(2, -2)$, we can let $(x_1, y_1) = (-3, 4)$ and $(x_2, y_2) = (2, -2)$. The slope is computed as follows:

$$m = \frac{\text{Change in } y}{\text{Change in } x} = \frac{y_2 - y_1}{x_2 - x_1} = \frac{-2 - 4}{2 - (-3)} = \frac{-6}{5} = -\frac{6}{5}.$$

The situation is illustrated in **Figure 3.22**. The slope of the line is $-\frac{6}{5}$. For every vertical change of -6 units (6 units down), there is a corresponding horizontal change of 5 units. The slope is negative and the line falls from left to right.

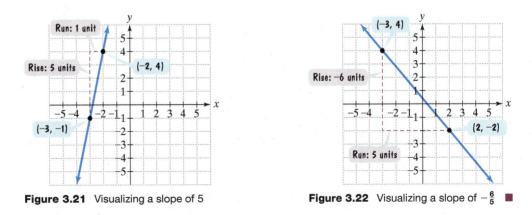

Figure 3.21 Visualizing a slope of 5

Figure 3.22 Visualizing a slope of $-\frac{6}{5}$ ∎

☑ **CHECK POINT 1** Find the slope of the line passing through each pair of points:
a. $(-3, 4)$ and $(-4, -2)$ **b.** $(4, -2)$ and $(-1, 5)$.

EXAMPLE 2   Using the Definition of Slope for Horizontal and Vertical Lines

Find the slope of the line passing through each pair of points:

a. $(5, 4)$ and $(3, 4)$ **b.** $(2, 5)$ and $(2, 1)$.

Solution

a. Let $(x_1, y_1) = (5, 4)$ and $(x_2, y_2) = (3, 4)$. We obtain the slope as follows:

$$m = \frac{\text{Change in } y}{\text{Change in } x} = \frac{y_2 - y_1}{x_2 - x_1} = \frac{4 - 4}{3 - 5} = \frac{0}{-2} = 0.$$

The situation is illustrated in **Figure 3.23**. Can you see that the line is horizontal? Because any two points on a horizontal line have the same y-coordinate, these lines neither rise nor fall from left to right. The change in y, $y_2 - y_1$, is always zero. Thus, **the slope of any horizontal line is zero**.

b. We can let $(x_1, y_1) = (2, 5)$ and $(x_2, y_2) = (2, 1)$. **Figure 3.24** shows that these points are on a vertical line. We attempt to compute the slope as follows:

$$m = \frac{\text{Change in } y}{\text{Change in } x} = \frac{1 - 5}{2 - 2} = \frac{-4}{0}. \quad \text{Division by zero is undefined.}$$

Because division by zero is undefined, the slope of the vertical line in **Figure 3.24** is undefined. In general, **the slope of any vertical line is undefined**.

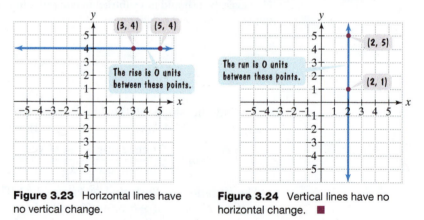

Figure 3.23 Horizontal lines have no vertical change.

Figure 3.24 Vertical lines have no horizontal change. ■

Table 3.1 summarizes four possibilities for the slope of a line.

Table 3.1 Possibilities for a Line's Slope

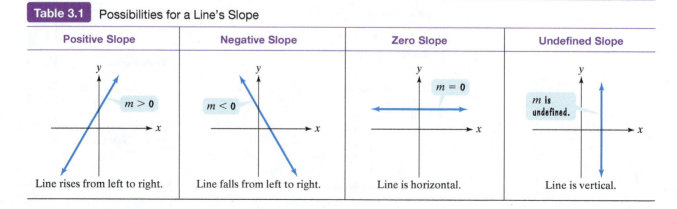

Positive Slope	Negative Slope	Zero Slope	Undefined Slope
$m > 0$	$m < 0$	$m = 0$	m is undefined.
Line rises from left to right.	Line falls from left to right.	Line is horizontal.	Line is vertical.

✓ **CHECK POINT 2** Find the slope of the line passing through each pair of points or state that the slope is undefined:

a. (6, 5) and (2, 5) **b.** (1, 6) and (1, 4).

2 Use slope to show that lines are parallel.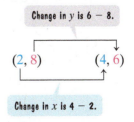

Slope and Parallel Lines

Two nonintersecting lines that lie in the same plane are **parallel**. If two lines do not intersect, the ratio of the vertical change to the horizontal change is the same for each line. Because two parallel lines have the same "steepness," they must have the same slope.

> **Slope and Parallel Lines**
>
> 1. If two nonvertical lines are parallel, then they have the same slope.
> 2. If two distinct nonvertical lines have the same slope, then they are parallel.
> 3. Two distinct vertical lines, each with undefined slope, are parallel.

EXAMPLE 3 Using Slope to Show That Lines Are Parallel

Show that the line passing through (1, 4) and (3, 2) is parallel to the line passing through (2, 8) and (4, 6).

Solution The situation is illustrated in **Figure 3.25**. The lines certainly look like they are parallel. Let's use equal slopes to confirm this fact. For each line, we compute the ratio of the difference in y-coordinates to the difference in x-coordinates. Be sure to subtract the coordinates in the same order.

We begin with the slope of the blue line through (1, 4) and (3, 2).

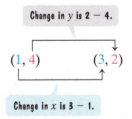

$$m = \frac{\text{Change in } y}{\text{Change in } x} = \frac{2 - 4}{3 - 1} = \frac{-2}{2} = -1$$

Now we find the slope of the red line through (2, 8) and (4, 6).

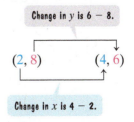

$$m = \frac{\text{Change in } y}{\text{Change in } x} = \frac{6 - 8}{4 - 2} = \frac{-2}{2} = -1$$

Because the slopes are equal, the lines are parallel. ■

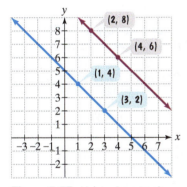

Figure 3.25 Using slope to show that lines are parallel

✓ **CHECK POINT 3** Show that the line passing through $(4, 2)$ and $(6, 6)$ is parallel to the line passing through $(0, -2)$ and $(1, 0)$.

3 Use slope to show that lines are perpendicular.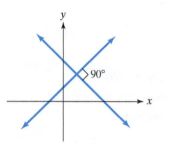

Slope and Perpendicular Lines

Two lines that intersect at a right angle (90°) are said to be **perpendicular**, as shown in **Figure 3.26**. There is a relationship between the slopes of perpendicular lines.

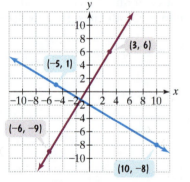

Figure 3.26 Perpendicular lines

Slope and Perpendicular Lines

1. If two nonvertical lines are perpendicular, then the product of their slopes is -1.
2. If the product of the slopes of two lines is -1, then the lines are perpendicular.
3. A horizontal line having zero slope is perpendicular to a vertical line having undefined slope.

EXAMPLE 4 Using Slope to Show That Lines Are Perpendicular

Show that the line passing through $(-6, -9)$ and $(3, 6)$ is perpendicular to the line passing through $(10, -8)$ and $(-5, 1)$.

Solution The situation is illustrated in **Figure 3.27**. The lines certainly look like they intersect at a right angle. Let's show that the product of their slopes is -1 to confirm this fact.

We begin with the slope of the red line through $(-6, -9)$ and $(3, 6)$.

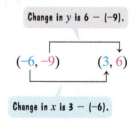

Change in y is $6 - (-9)$.

$$(-6, -9) \qquad (3, 6)$$

Change in x is $3 - (-6)$.

Figure 3.27 Using slope to show that lines are perpendicular

$$m = \frac{\text{Change in } y}{\text{Change in } x} = \frac{6 - (-9)}{3 - (-6)} = \frac{6 + 9}{3 + 6} = \frac{15}{9} = \frac{5}{3}$$

Now we find the slope of the blue line through $(10, -8)$ and $(-5, 1)$.

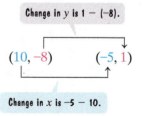

Change in y is $1 - (-8)$.

$$(10, -8) \qquad (-5, 1)$$

Change in x is $-5 - 10$.

$$m = \frac{\text{Change in } y}{\text{Change in } x} = \frac{1 - (-8)}{-5 - 10} = \frac{1 + 8}{-5 - 10} = \frac{9}{-15} = -\frac{3}{5}$$

We have determined that the slope of the red line in **Figure 3.27** is $\frac{5}{3}$ and the slope of the blue line is $-\frac{3}{5}$. Now we can find the product of the slopes:

$$\frac{5}{3}\left(-\frac{3}{5}\right) = -1.$$

Because the product of the slopes is -1, the lines are perpendicular. ■

In this example, the slopes of the perpendicular lines, $\frac{5}{3}$ and $-\frac{3}{5}$, are reciprocals with opposite signs, called **negative reciprocals**. This gives us an equivalent way of stating the relationship between slope and perpendicular lines:

Two nonvertical lines are perpendicular if the slope of one is the negative reciprocal of the slope of the other.

✓ **CHECK POINT 4**   Show that the line passing through $(-1, 4)$ and $(3, 2)$ is perpendicular to the line passing through $(-2, -1)$ and $(2, 7)$.

---

4 Calculate rate of change in applied situations.

Slope as Rate of Change

Slope is defined as the ratio of a change in y to a corresponding change in x. It tells how fast y is changing with respect to x. Thus, the slope of a line represents its rate of change.

Our next example shows how slope can be interpreted as a rate of change in an applied situation. When calculating slope in applied problems, keep track of the units in the numerator and the denominator.

EXAMPLE 5   Slope as a Rate of Change

The line graphs for the numbers of women and men living alone are shown again in **Figure 3.28**. Find the slope of the line segment for the women. Describe what this slope represents.

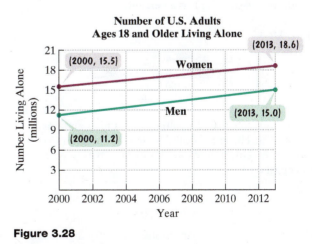

Figure 3.28

Source: U.S. Census Bureau

Solution   We let x represent a year and y the number of women living alone in that year. Use the two points shown on the red line segment for women.

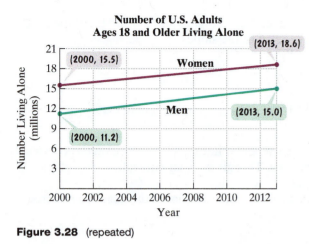

**Number of U.S. Adults
Ages 18 and Older Living Alone**

Figure 3.28 (repeated)

The two points shown on the line segment for women have the following coordinates:

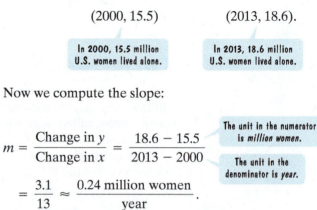

$$(2000, 15.5) \qquad (2013, 18.6).$$

In 2000, 15.5 million U.S. women lived alone.

In 2013, 18.6 million U.S. women lived alone.

Now we compute the slope:

$$m = \frac{\text{Change in } y}{\text{Change in } x} = \frac{18.6 - 15.5}{2013 - 2000}$$

The unit in the numerator is *million women.*

The unit in the denominator is *year.*

$$= \frac{3.1}{13} \approx \frac{0.24 \text{ million women}}{\text{year}}.$$

The slope indicates that the number of American women living alone increased at a rate of approximately 0.24 million each year for the period from 2000 through 2013. The rate of change is 0.24 million women per year. ■

✓ **CHECK POINT 5**   Use the ordered pairs in **Figure 3.28** to find the slope of the green line segment for the men. Express the slope correct to two decimal places and describe what it represents.

In Check Point 5, did you find that the slope of the line segment for the men is different from that of the women? The rate of change for men living alone is greater than the rate of change for women living alone. The line segment representing men in **Figure 3.28** is steeper than the line segment representing women. If you extend the line segments far enough, the resulting lines will intersect. They are not parallel.

Blitzer Bonus ⊙

Railroads and Highways

The steepest part of Mount Washington Cog Railway in New Hampshire has a 37% grade. This is equivalent to a slope of $\frac{37}{100}$. For every horizontal change of 100 feet, the railroad ascends 37 feet vertically. Engineers denote slope by grade, expressing slope as a percent.

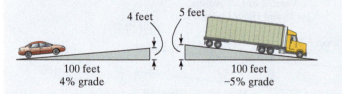

4 feet

5 feet

100 feet
4% grade

100 feet
−5% grade

Railroad grades are usually less than 2%, although in the mountains they may go as high as 4%. The grade of the Mount Washington Cog Railway is phenomenal, making it necessary for locomotives to *push* single cars up its steepest part.

A Mount Washington Cog Railway locomotive pushing a single car up the steepest part of the railroad. The locomotive is about 120 years old.

Achieving Success

According to the Ebbinghaus retention model, you forget 50% of processed information within one hour of leaving the classroom. You lose 60% within 24 hours. After 30 days, 70% is gone. Reviewing and rewriting class notes is an effective way to counteract this phenomenon. At the very least, read your lecture notes at the end of each day. The more you engage with the material, the more you retain.

CONCEPT AND VOCABULARY CHECK

Fill in each blank so that the resulting statement is true.

1. The slope, m, of the line through the distinct points (x_1, y_1) and (x_2, y_2) is given by the formula $m =$ _____.

2. The slope of the line through the distinct points (x_1, y_1) and (x_2, y_2) can be interpreted as the rate of change in _____ with respect to _____.

3. If a line rises from left to right, the line has _____ slope.

4. If a line falls from left to right, the line has _____ slope.

5. The slope of a horizontal line is _____.

6. The slope of a vertical line is _____.

7. If two distinct nonvertical lines have the same slope, then the lines are _____.

8. If the product of the slopes of two lines is -1, then the lines are _____.

3.3 EXERCISE SET   ▶ MyMathLab®

Practice Exercises

In Exercises 1–10, find the slope of the line passing through each pair of points or state that the slope is undefined. Then indicate whether the line through the points rises, falls, is horizontal, or is vertical.

1. $(4, 7)$ and $(8, 10)$

2. $(2, 1)$ and $(3, 4)$

3. $(-2, 1)$ and $(2, 2)$

4. $(-1, 3)$ and $(2, 4)$

5. $(4, -2)$ and $(3, -2)$

6. $(4, -1)$ and $(3, -1)$

7. $(-2, 4)$ and $(-1, -1)$

8. $(6, -4)$ and $(4, -2)$

9. $(5, 3)$ and $(5, -2)$

10. $(3, -4)$ and $(3, 5)$

In Exercises 11–22, find the slope of each line, or state that the slope is undefined.

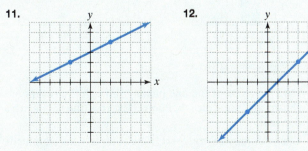

11.

12.

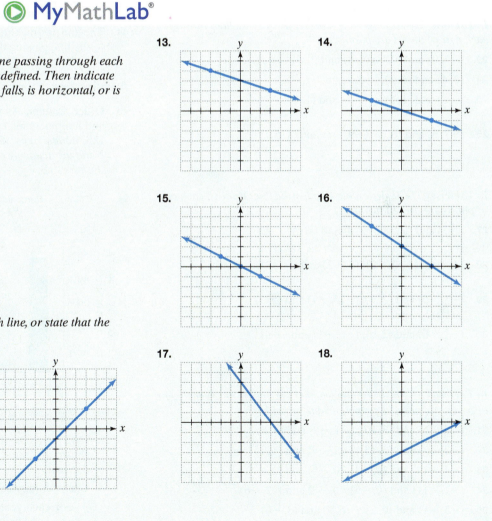

13.

14.

15.

16.

17.

18.

19.   **20.**

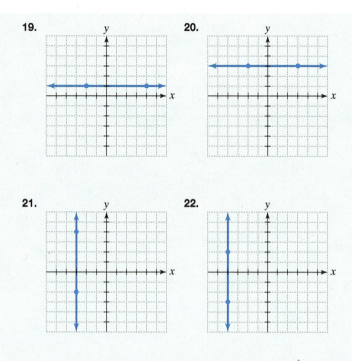

21.   **22.**

In Exercises 23–26, determine whether the distinct lines through each pair of points are parallel.

23. $(-2, 0)$ and $(0, 6)$; $(1, 8)$ and $(0, 5)$

24. $(2, 4)$ and $(6, 1)$; $(-3, 1)$ and $(1, -2)$

25. $(0, 3)$ and $(1, 5)$; $(-1, 7)$ and $(1, 10)$

26. $(-7, 6)$ and $(0, 4)$; $(-9, -3)$ and $(1, 5)$

In Exercises 27–30, determine whether the lines through each pair of points are perpendicular.

27. $(1, 5)$ and $(0, 3)$; $(-2, 8)$ and $(2, 6)$

28. $(3, 2)$ and $(-2, -2)$; $(3, -2)$ and $(-1, 3)$

29. $(-1, -6)$ and $(2, 9)$; $(-15, -1)$ and $(5, 3)$

30. $(-1, -6)$ and $(2, 6)$; $(-8, -1)$ and $(4, 2)$

In Exercises 31–36, determine whether the lines through each pair of points are parallel, perpendicular, or neither.

31. $(-2, -5)$ and $(3, 10)$; $(-1, -9)$ and $(4, 6)$

32. $(-2, -7)$ and $(3, 13)$; $(-1, -9)$ and $(5, 15)$

33. $(-4, -12)$ and $(0, -4)$; $(0, -5)$ and $(2, -4)$

34. $(-1, -11)$ and $(0, -5)$; $(0, -8)$ and $(12, -6)$

35. $(-5, -1)$ and $(0, 2)$; $(-6, 9)$ and $(3, -6)$

36. $(-2, -15)$ and $(0, -3)$; $(-12, 6)$ and $(6, 3)$

Practice PLUS

37. On the same set of axes, draw lines passing through the origin with slopes -1, $-\frac{1}{2}$, 0, $\frac{1}{3}$, and 2.

38. On the same set of axes, draw lines with y-intercept 4 and slopes -1, $-\frac{1}{2}$, 0, $\frac{1}{3}$, and 2.

Use slopes to solve Exercises 39–40.

39. Show that the points whose coordinates are $(-3, -3)$, $(2, -5)$, $(5, -1)$, and $(0, 1)$ are the vertices of a four-sided figure whose opposite sides are parallel. (Such a figure is called a *parallelogram*.)

40. Show that the points whose coordinates are $(-3, 6)$, $(2, -3)$, $(11, 2)$, and $(6, 11)$ are the vertices of a four-sided figure whose opposite sides are parallel.

41. The line passing through $(5, y)$ and $(1, 0)$ is parallel to the line joining $(2, 3)$ and $(-2, 1)$. Find y.

42. The line passing through $(1, y)$ and $(7, 12)$ is parallel to the line joining $(-3, 4)$ and $(-5, -2)$. Find y.

43. The line passing through $(-1, y)$ and $(1, 0)$ is perpendicular to the line joining $(2, 3)$ and $(-2, 1)$. Find y.

44. The line passing through $(-2, y)$ and $(-4, 4)$ is perpendicular to the line passing through $(-1, -2)$ and $(4, -1)$. Find y.

Application Exercises

45. **Older, Calmer.** As we age, daily stress and worry decrease and happiness increases, according to an analysis of 340,847 U.S. adults, ages 18–85, in the journal *Proceedings of the National Academy of Sciences*. The graphs show a portion of the research.

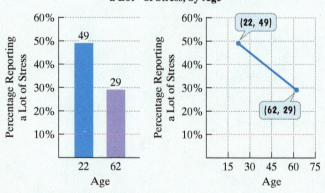

Percentage of Americans Reporting "a Lot" of Stress, by Age

Source: National Academy of Sciences

a. Find the slope of the line passing through the two points shown by the voice balloons. Express the slope as a decimal.

b. Use your answer from part (a) to complete the statement:

For each year of aging, the percentage of Americans reporting "a lot" of stress decreases by _____%. The rate of change is _____% per _____.

46. Exercise is useful not only in preventing depression, but also as a treatment. The graphs show the percentage of patients with depression in remission when exercise (brisk walking) was used as a treatment. (The control group that engaged in no exercise had 11% of the patients in remission.)

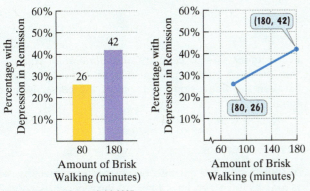

Exercise and Percentage of Patients with Depression in Remission

Source: Newsweek, March 26, 2007

a. Find the slope of the line passing through the two points shown by the voice balloons. Express the slope as a decimal.

b. Use your answer from part (a) to complete this statement: For each minute of brisk walking, the percentage of patients with depression in remission increased by _____ %. The rate of change is _____ % per _____.

Hitting the Slopes. *Many factors determine a ski trail's rating, including its width, the presence of obstacles, and the trail's steepness.*

Most resorts use the same general classifications: green circles for beginners, blue squares for intermediate skiers, and black diamonds for experienced skiers. In general, the following slopes, expressed as percents, are typical for each type of trail. Use this information to solve Exercises 47–48. Round answers to the nearest tenth of a percent.

Ski Trail Rating	Type of Trail	Absolute Value of the Slope (as a percent)
🟢 Green Circle	Beginner	5% to 20%
🟦 Blue Square	Intermediate	21% to 35%
◆ Black Diamond	Expert	Greater than 35%

47. At the Keystone Resort in Colorado, the Schoolmarm Trail starts at an elevation of 11,640 feet and ends at an elevation of 9309 feet. Its run is 15,662 feet. What is the slope of the Schoolmarm Trail and what is the ski trail's rating?

48. At the Keystone Resort in Colorado, the Wild Irishman Trail starts at an elevation of 11,640 feet and ends at an elevation of 10,414 feet. Its run is 4979 feet. What is the slope of the Wild Irishman Trail and what is the ski trail's rating?

The grade of a road or ramp refers to its slope expressed as a percent. Use this information to solve Exercises 49–50.

49. Construction laws are very specific when it comes to access ramps for the disabled. Every vertical rise of 1 foot requires a horizontal run of 12 feet. What is the grade of such a ramp? Round to the nearest tenth of a percent.

50. A college campus goes beyond the standards described in Exercise 49. All wheelchair ramps on campus are designed so that every vertical rise of 1 foot is accompanied by a horizontal run of 14 feet. What is the grade of such a ramp? Round to the nearest tenth of a percent.

Explaining the Concepts

51. What is the slope of a line?

52. Describe how to calculate the slope of a line passing through two points.

53. What does it mean if the slope of a line is zero?

54. What does it mean if the slope of a line is undefined?

55. If two lines are parallel, describe the relationship between their slopes.

56. If two lines are perpendicular, describe the relationship between their slopes.

Critical Thinking Exercises

Make Sense? *In Exercises 57–60, determine whether each statement makes sense or does not make sense, and explain your reasoning.*

57. When finding the slope of the line passing through $(-1, 5)$ and $(2, -3)$, I must let (x_1, y_1) be $(-1, 5)$ and (x_2, y_2) be $(2, -3)$.

58. When applying the slope formula, it is important to subtract corresponding coordinates in the same order.

59. I visualize slope as walking along a line from left to right. If I'm walking uphill, the slope is positive, and if I'm walking downhill, the slope is negative.

60. I computed the slope of one line to be $-\frac{3}{5}$ and the slope of a second line to be $-\frac{5}{3}$, so the lines must be perpendicular.

In Exercises 61–64, determine whether each statement is true or false. If the statement is false, make the necessary change(s) to produce a true statement.

61. Slope is run divided by rise.

62. The line through $(2, 2)$ and the origin has slope 1.

63. A line with slope 3 can be parallel to a line with slope -3.

64. The line through $(3, 1)$ and $(3, -5)$ has zero slope.

In Exercises 65–66, use the figure shown to make the indicated list.

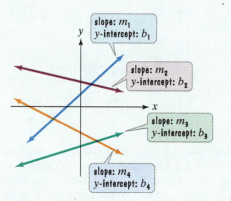

65. List the slopes $m_1, m_2, m_3,$ and m_4 in order of decreasing size.

66. List the y-intercepts $b_1, b_2, b_3,$ and b_4 in order of decreasing size.

Technology Exercises

Use a graphing utility to graph each equation in Exercises 67–70. Then use the | TRACE | *feature to trace along the line and find the coordinates of two points. Use these points to compute the line's slope.*

67. $y = 2x + 4$

68. $y = -3x + 6$

69. $y = -\frac{1}{2}x - 5$

70. $y = \frac{3}{4}x - 2$

71. In Exercises 67–70, compare the slope that you found with the line's equation. What relationship do you observe between the line's slope and one of the constants in the equation?

Review Exercises

72. A 36-inch board is cut into two pieces. One piece is twice as long as the other. How long are the pieces? (Section 2.5, Example 2)

73. Simplify: $-10 + 16 \div 2(-4)$. (Section 1.8, Example 4)

74. Solve and graph the solution set on a number line: $2x - 3 \le 5$. (Section 2.7, Example 6)

Preview Exercises

Exercises 75–77 will help you prepare for the material covered in the next section.

75. From $(0, -3)$, move 4 units up and 1 unit to the right. What point do you obtain?

76. From $(0, 1)$, move 2 units down and 3 units to the right. What point do you obtain?

77. Solve for y: $2x + 5y = 0$.

SECTION

3.4

The Slope-Intercept Form of the Equation of a Line

What am I supposed to learn?

After studying this section, you should be able to:

1. Find a line's slope and *y*-intercept from its equation. ▶

2. Graph lines in slope-intercept form. ▶

3. Use slope and *y*-intercept to graph $Ax + By = C$. ▶

4. Use slope and *y*-intercept to model data. ▶

Got joy? According to polls conducted by *Time* magazine, in 2003, 79% of Americans ages 18 and older considered themselves optimists. By 2013, this number had decreased to 50%. In the period from 2003 to 2013, fewer Americans shopped, helped others, and prayed/meditated to improve their mood. This section (Example 5) contains two mathematical models based on these joy-deflating observations. To develop these models, we turn to a new form for a line's equation.

---

1. Find a line's slope and *y*-intercept from its equation. ▶

The Slope-Intercept Form of the Equation of a Line

Let's begin with an example that shows how easy it is to find a line's slope and *y*-intercept from its equation.

Figure 3.29 shows the graph of $y = 2x + 4$. Verify that the *x*-intercept is −2 by setting *y* equal to 0 and solving for *x*. Similarly, verify that the *y*-intercept is 4 by setting *x* equal to 0 and solving for *y*.

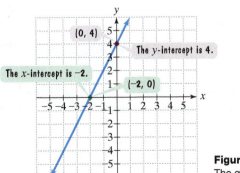

Figure 3.29
The graph of $y = 2x + 4$

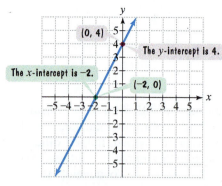

Figure 3.29 The graph of $y = 2x + 4$ (repeated)

Now that we have two points on the line, $(-2, 0)$ and $(0, 4)$, we can calculate the slope of the graph of $y = 2x + 4$.

$$\text{Slope} = \frac{\text{Change in } y}{\text{Change in } x}$$

$$= \frac{4 - 0}{0 - (-2)} = \frac{4}{2} = 2$$

We see that the slope of the line is 2, the same as the coefficient of x in the equation $y = 2x + 4$. The y-intercept is 4, the same as the constant in the equation $y = 2x + 4$.

$$y = 2x + 4$$

The slope is 2.   The y-intercept is 4.

A linear equation like $y = 2x + 4$ that is solved for y is said to be in *slope-intercept form*. This is because the slope and the y-intercept can be immediately determined from the equation. The x-coefficient is the line's slope and the constant term is the y-intercept.

Slope-Intercept Form of the Equation of a Line

The **slope-intercept form of the equation** of a nonvertical line with slope m and y-intercept b is

$$y = mx + b.$$

EXAMPLE 1 Finding a Line's Slope and y-Intercept from Its Equation

Find the slope and the y-intercept of the line with the given equation:

a. $y = 2x - 4$ **b.** $y = \dfrac{1}{2}x + 2$ **c.** $5x + y = 4$.

Great Question!

Which are the constants and which are the variables in $y = mx + b$?

The variables in $y = mx + b$ vary in different ways. The variables for slope, m, and y-intercept, b, vary from one line's equation to another. However, they remain constant in the equation of a single line. By contrast, the variables x and y represent the infinitely many points, (x, y), on a single line. Thus, these variables vary in both the equation of a single line, as well as from one equation to another.

Solution

a. We write $y = 2x - 4$ as $y = 2x + (-4)$. The slope is the x-coefficient and the y-intercept is the constant term.

$$y = 2x + (-4)$$

The slope is 2.   The y-intercept is −4.

b. The equation $y = \frac{1}{2}x + 2$ is in the form $y = mx + b$. We can find the slope, m, by identifying the coefficient of x. We can find the y-intercept, b, by identifying the constant term.

$$y = \frac{1}{2}x + 2$$

The slope is $\frac{1}{2}$.   The y-intercept is 2.

c. The equation $5x + y = 4$ is not in the form $y = mx + b$. We can obtain this form by isolating y on one side. We isolate y on the left side by subtracting $5x$ from both sides.

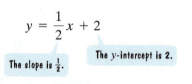

$$5x + y = 4 \qquad \text{This is the given equation.}$$
$$5x - 5x + y = -5x + 4 \qquad \text{Subtract } 5x \text{ from both sides.}$$
$$y = -5x + 4 \qquad \text{Simplify.}$$

Now, the equation is in the form $y = mx + b$. The slope is the coefficient of x and the y-intercept is the constant term.

$$y = -5x + 4$$

The slope is –5.   The y-intercept is 4.   ■

✓ **CHECK POINT 1**   Find the slope and the y-intercept of the line with the given equation:

a. $y = 5x - 3$

b. $y = \dfrac{2}{3}x + 4$

c. $7x + y = 6$.

---

2 Graph lines in slope-intercept form.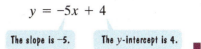

Graphing $y = mx + b$ by Using the Slope and y-Intercept

If a line's equation is written with y isolated on one side, we can use the y-intercept and the slope to obtain its graph.

Graphing $y = mx + b$ by Using the Slope and y-Intercept

1. Plot the point containing the y-intercept on the y-axis. This is the point $(0, b)$.

2. Obtain a second point using the slope, m. Write m as a fraction, and use rise over run, starting at the point on the y-axis, to plot this point.

3. Use a straightedge to draw a line through the two points. Draw arrowheads at the ends of the line to show that the line continues indefinitely in both directions.

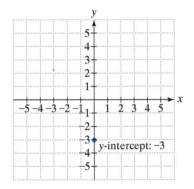

Figure 3.30(a)
The y-intercept is –3, so $(0, -3)$ is a point on the line.

EXAMPLE 2   Graphing by Using the Slope and y-Intercept

Graph the line whose equation is $y = 4x - 3$.

Solution   We write $y = 4x - 3$ in the form $y = mx + b$.

$$y = 4x + (-3)$$

The slope is 4.   The y-intercept is –3.

Now that we have identified the slope and the y-intercept, we use the three steps in the preceding box to graph the equation.

Step 1. Plot the point containing the y-intercept on the y-axis.   The y-intercept is -3. We plot the point $(0, -3)$, shown in **Figure 3.30(a)**.

Step 2. Obtain a second point using the slope, m. Write m as a fraction, and use rise over run, starting at the point on the y-axis, to plot this point.   The slope, 4, written as a fraction is $\frac{4}{1}$.

$$m = \frac{4}{1} = \frac{\text{Rise}}{\text{Run}}$$

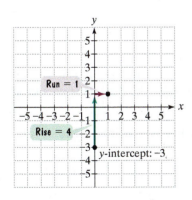

Figure 3.30(b)
The slope is 4.

We plot the second point on the line by starting at $(0, -3)$, the first point. Based on the slope, we move 4 units *up* (the rise) and 1 unit to the *right* (the run). This puts us at a second point on the line, $(1, 1)$, shown in **Figure 3.30(b)**.

Step 3. Use a straightedge to draw a line through the two points. The graph of $y = 4x - 3$ is shown in **Figure 3.30(c)**.

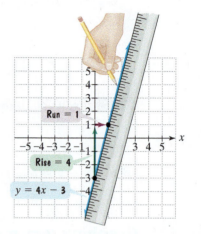

Figure 3.30(c) Graphing $y = 4x - 3$ using the y-intercept and slope ■

✓ **CHECK POINT 2** Graph the line whose equation is $y = 3x - 2$.

EXAMPLE 3 Graphing by Using the Slope and y-Intercept

Graph the line whose equation is $y = \dfrac{2}{3}x + 2$.

Solution The equation of the line is in the form $y = mx + b$. We can find the slope, m, by identifying the coefficient of x. We can find the y-intercept, b, by identifying the constant term.

$$y = \frac{2}{3}x + 2$$

The slope is $\frac{2}{3}$.   The y-intercept is 2.

Now that we have identified the slope and the y-intercept, we use the three-step procedure to graph the equation.

Step 1. Plot the point containing the y-intercept on the y-axis. The y-intercept is 2. We plot $(0, 2)$, shown in **Figure 3.31**.

Step 2. Obtain a second point using the slope, m. Write m as a fraction, and use rise over run, starting at the point on the y-axis, to plot this point. The slope, $\frac{2}{3}$, is already written as a fraction.

$$m = \frac{2}{3} = \frac{\text{Rise}}{\text{Run}}$$

We plot the second point on the line by starting at $(0, 2)$, the first point. Based on the slope, we move 2 units *up* (the rise) and 3 units to the *right* (the run). This puts us at a second point on the line, $(3, 4)$, shown in **Figure 3.31**.

Step 3. Use a straightedge to draw a line through the two points. The graph of $y = \frac{2}{3}x + 2$ is shown in **Figure 3.31**. ■

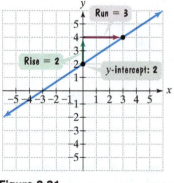

Figure 3.31
The graph of $y = \frac{2}{3}x + 2$

✓ **CHECK POINT 3** Graph the line whose equation is $y = \dfrac{3}{5}x + 1$.

3 Use slope and y-intercept to graph $Ax + By = C$. ▶

Graphing *Ax + By = C* by Using the Slope and *y*-Intercept

Earlier in this chapter, we considered linear equations of the form $Ax + By = C$. We used x- and y-intercepts, as well as checkpoints, to graph these equations. It is also possible to obtain the graphs by using the slope and y-intercept. To do this, begin by solving $Ax + By = C$ for y. This will put the equation in slope-intercept form. Then use the three-step procedure to graph the equation. This is illustrated in Example 4.

EXAMPLE 4   Graphing by Using the Slope and *y*-Intercept

Graph the linear equation $2x + 5y = 0$ by using the slope and y-intercept.

Solution   We put the equation in slope-intercept form by solving for y.

$2x + 5y = 0$	This is the given equation.
$2x - 2x + 5y = -2x + 0$	Subtract 2x from both sides.
$5y = -2x + 0$	Simplify.
$\dfrac{5y}{5} = \dfrac{-2x + 0}{5}$	Divide both sides by 5.
$y = \dfrac{-2x}{5} + \dfrac{0}{5}$	Divide each term in the numerator by 5.
$y = -\dfrac{2}{5}x + 0$	Simplify.

Now that the equation is in slope-intercept form, we can use the slope and y-intercept to obtain its graph. Examine the slope-intercept form:

$$y = -\frac{2}{5}x + 0.$$

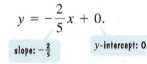

slope: $-\frac{2}{5}$ y-intercept: 0

Note that the slope is $-\frac{2}{5}$ and the y-intercept is 0. Use the y-intercept to plot $(0, 0)$ on the y-axis. Then locate a second point by using the slope.

$$m = -\frac{2}{5} = \frac{-2}{5} = \frac{\text{Rise}}{\text{Run}}$$

Because the rise is -2 and the run is 5, move *down* 2 units and to the *right* 5 units, starting at the point $(0, 0)$. This puts us at a second point on the line, $(5, -2)$. The graph of $2x + 5y = 0$ is the line drawn through these points, shown in **Figure 3.32**. ∎

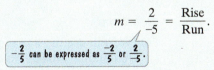

Figure 3.32 The graph of $2x + 5y = 0$, or $y = -\frac{2}{5}x + 0$

Discover for Yourself

You can obtain a second point in Example 4 by writing the slope as follows:

$$m = \frac{2}{-5} = \frac{\text{Rise}}{\text{Run}}.$$

$-\frac{2}{5}$ can be expressed as $\frac{-2}{5}$ or $\frac{2}{-5}$.

Now obtain this second point in **Figure 3.32** by moving *up* 2 units and to the *left* 5 units, starting at $(0, 0)$. What do you observe once you graph the line?

✓ **CHECK POINT 4**   Graph the linear equation $3x + 4y = 0$ by using the slope and y-intercept.

4 Use slope and *y*-intercept to model data. ▶

Modeling with the Slope-Intercept Form of the Equation of a Line

The slope-intercept form for equations of lines is useful for obtaining mathematical models for data that fall on or near a line. For example, the bar graph in **Figure 3.33(a)** shows the percentage of Americans who did various things in 2003 and 2013 to improve their mood. The data for helping others and shopping are displayed as points in a rectangular coordinate system in **Figure 3.33(b)**.

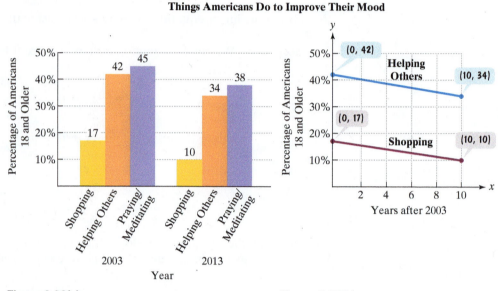

Figure 3.33(a)

Source: Time

Figure 3.33(b)

Example 5 illustrates how we can use the equation $y = mx + b$ to obtain a model for the data and make predictions about what might occur in the future.

EXAMPLE 5   Modeling with the Slope-Intercept Form of the Equation

a. Use the two points for helping others in **Figure 3.33(b)** to find an equation in the form $y = mx + b$ that models the percentage of Americans who improved their mood by helping others, y, x years after 2003.

b. If trends shown by the data continue, use the model to project the percentage of Americans who will improve their mood by helping others in 2018.

Solution

a. We will use the line segment for helping others with the points $(0, 42)$ and $(10, 34)$ to obtain a model. We need values for m, the slope, and b, the y-intercept.

$$y = mx + b$$

$$m = \frac{\text{Change in } y}{\text{Change in } x}$$

The point $(0, 42)$ lies on the line segment, so the y-intercept is 42: $b = 42$.

$$= \frac{34 - 42}{10 - 0}$$

$$= \frac{-8}{10} = -0.8$$

The percentage of Americans who improved their mood by helping others, y, x years after 2003 can be modeled by the linear equation

$$y = -0.8x + 42.$$

The slope, -0.8, indicates a decrease in the percentage of Americans who improved their mood by helping others by 0.8% per year from 2003 through 2013.

b. Now let's use this model to project the percentage of Americans who will improve their mood by helping others in 2018. Because 2018 is 15 years after 2003, substitute 15 for x in $y = -0.8x + 42$ and evaluate the formula.

$$y = -0.8(15) + 42 = -12 + 42 = 30$$

Our model projects that 30% of Americans will improve their mood by helping others in 2018. ■

✓ CHECK POINT 5

a. Use the two points for shopping in **Figure 3.33(b)** to find an equation in the form $y = mx + b$ that models the percentage of Americans who improved their mood by shopping, y, x years after 2003.

b. If trends shown by the data continue, use the model to project the percentage of Americans who will improve their mood by shopping in 2023.

Achieving Success

Use index cards to help learn new terms.

Many of the terms, notations, and formulas used in this book will be new to you. Buy a pack of 3×5 index cards. On each card, list a new vocabulary word, symbol, or title of a formula. On the other side of the card, put the definition or formula. Here are two examples:

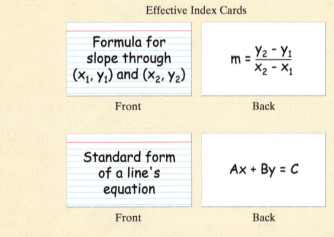

Effective Index Cards

Front   Back

Review these cards frequently. Use the cards to quiz yourself and prepare for exams.

CONCEPT AND VOCABULARY CHECK

Fill in each blank so that the resulting statement is true.

1. The slope-intercept form of the equation of a line is _____ , where m represents the _____ and b represents the _____ .

2. In order to graph the line whose equation is $y = \frac{2}{5}x + 3$, begin by plotting the point _____ . From this point, we move _____ units up (the rise) and _____ units to the right (the run).

3. In order to graph equations of the form $Ax + By = C$ using the slope and y-intercept, we begin by solving the equation for _____ .

3.4 EXERCISE SET ▶ MyMathLab®

Practice Exercises

In Exercises 1–12, find the slope and the y-intercept of the line with the given equation.

1. $y = 3x + 2$ **2.** $y = 9x + 4$

3. $y = 3x - 5$ **4.** $y = 4x - 2$

5. $y = -\frac{1}{2}x + 5$ **6.** $y = -\frac{3}{4}x + 6$

7. $y = 7x$ **8.** $y = 10x$

9. $y = 10$ **10.** $y = 7$

11. $y = 4 - x$ **12.** $y = 5 - x$

In Exercises 13–26, begin by solving the linear equation for y. This will put the equation in slope-intercept form. Then find the slope and the y-intercept of the line with this equation.

13. $-5x + y = 7$

14. $-9x + y = 5$

15. $x + y = 6$

16. $x + y = 8$

17. $6x + y = 0$

18. $8x + y = 0$

19. $3y = 6x$

20. $3y = -9x$

21. $2x + 7y = 0$

22. $2x + 9y = 0$

23. $3x + 2y = 3$

24. $4x + 3y = 4$

25. $3x - 4y = 12$

26. $5x - 2y = 10$

In Exercises 27–38, graph each linear equation using the slope and y-intercept

27. $y = 2x + 4$ **28.** $y = 3x + 1$

29. $y = -3x + 5$ **30.** $y = -2x + 4$

31. $y = \frac{1}{2}x + 1$ **32.** $y = \frac{1}{3}x + 2$

33. $y = \frac{2}{3}x - 5$ **34.** $y = \frac{3}{4}x - 4$

35. $y = -\frac{3}{4}x + 2$ **36.** $y = -\frac{2}{3}x + 4$

37. $y = -\frac{5}{3}x$ **38.** $y = -\frac{4}{3}x$

In Exercises 39–46,

a. *Put the equation in slope-intercept form by solving for y.*
b. *Identify the slope and the y-intercept.*
c. *Use the slope and y-intercept to graph the equation.*

39. $3x + y = 0$ **40.** $2x + y = 0$

41. $3y = 4x$ **42.** $4y = 5x$

43. $2x + y = 3$ **44.** $3x + y = 4$

45. $7x + 2y = 14$

46. $5x + 3y = 15$

In Exercises 47–56, graph both linear equations in the same rectangular coordinate system. If the lines are parallel or perpendicular, explain why.

47. $y = 3x + 1$
$y = 3x - 3$

48. $y = 2x + 4$
$y = 2x - 3$

49. $y = -3x + 2$
$y = 3x + 2$

50. $y = -2x + 1$
$y = 2x + 1$

51. $y = x + 3$
$y = -x + 1$

52. $y = x + 2$
$y = -x - 1$

53. $x - 2y = 2$
$2x - 4y = 3$

54. $x - 3y = 9$
$3x - 9y = 18$

55. $2x - y = -1$
$x + 2y = -6$

56. $3x - y = -2$
$x + 3y = -9$

Practice PLUS

In Exercises 57–64, write an equation in the form $y = mx + b$ of the line that is described.

57. The y-intercept is 5 and the line is parallel to the line whose equation is $3x + y = 6$.

58. The y-intercept is -4 and the line is parallel to the line whose equation is $2x + y = 8$.

59. The y-intercept is 6 and the line is perpendicular to the line whose equation is $y = 5x - 1$.

60. The y-intercept is 7 and the line is perpendicular to the line whose equation is $y = 8x - 3$.

61. The line has the same y-intercept as the line whose equation is $16y = 8x + 32$ and is parallel to the line whose equation is $3x + 3y = 9$.

62. The line has the same y-intercept as the line whose equation is $2y = 6x + 8$ and is parallel to the line whose equation is $4x + 4y = 20$.

63. The line rises from left to right. It passes through the origin and a second point with equal x- and y-coordinates.

64. The line falls from left to right. It passes through the origin and a second point with opposite x- and y-coordinates.

Application Exercises

The bar graph breaks down the U.S. population by race/ethnicity for 2010, with projections by the U.S. Census Bureau for 2050. Exercises 65–66 involve the graphs shown in the rectangular coordinate system for two of the racial/ethnic groups.

United States Population by Race/Ethnicity

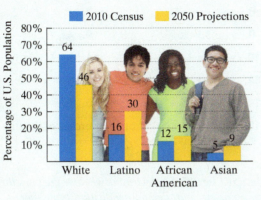

Source: U.S. Census Bureau

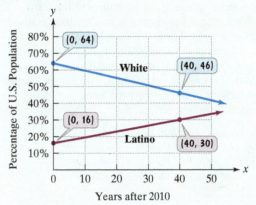

Years after 2010

65. a. Use the two points for whites to find an equation in the form $y = mx + b$ that models the percentage of whites, y, in the United States population x years after 2010.

   b. Use the model from part (a) to project the percentage of whites in the United States in 2110.

66. a. Use the two points for Latinos to find an equation in the form $y = mx + b$ that models the percentage of Latinos, y, in the United States population x years after 2010.

   b. Use the model from part (a) to project the percentage of Latinos in the United States in 2110.

Explaining the Concepts

67. Describe how to find the slope and the y-intercept of a line whose equation is given.

68. Describe how to graph a line using the slope and y-intercept. Provide an original example with your description.

69. A formula in the form $y = mx + b$ models the cost, y, of a four-year college x years after 2010. Would you expect m to be positive, negative, or zero? Explain your answer.

Critical Thinking Exercises

Make Sense? *In Exercises 70–73, determine whether each statement makes sense or does not make sense, and explain your reasoning.*

70. The slope-intercept form of a line's equation makes it possible for me to determine immediately the slope and the y-intercept.

71. Because the variable m does not appear in $Ax + By = C$, equations in this form make it impossible to determine the line's slope.

72. If I drive m miles in a year, the formula $c = 0.25m + 3500$ models the annual cost, c, in dollars, of operating my car, so the equation shows that with no driving at all, the cost is $3500, and the rate of increase in this cost is $0.25 for each mile that I drive.

73. Some lines are impossible to graph because when obtaining a second point using the slope, this point lands outside the edge of my graph paper.

In Exercises 74–77, determine whether each statement is true or false. If the statement is false, make the necessary change(s) to produce a true statement.

74. The equation $y = mx + b$ shows that no line can have a y-intercept that is numerically equal to its slope.

75. Every line in the rectangular coordinate system has an equation that can be expressed in slope-intercept form.

76. The line $3x + 2y = 5$ has slope $-\frac{3}{2}$.

77. The line $2y = 3x + 7$ has a y-intercept of 7.

78. The relationship between Celsius temperature, C, and Fahrenheit temperature, F, can be described by a linear equation in the form $F = mC + b$. The graph of this equation contains the point $(0, 32)$: Water freezes at $0°C$ or at $32°F$. The line also contains the point $(100, 212)$: Water boils at $100°C$ or at $212°F$. Write the linear equation expressing Fahrenheit temperature in terms of Celsius temperature.

Review Exercises

79. Solve: $\frac{x}{2} + 7 = 13 - \frac{x}{4}$. (Section 2.3, Example 4)

80. Simplify: $3(12 \div 2^2 - 3)^2$. (Section 1.8, Example 6)

81. 14 is 25% of what number? (Section 2.4, Example 6)

Preview Exercises

Exercises 82–84 will help you prepare for the material covered in the next section. In each exercise, solve for y and put the equation in slope-intercept form.

82. $y - 3 = 4(x + 1)$

83. $y + 3 = -\frac{3}{2}(x - 4)$

84. $y - 30.0 = 0.265(x - 10)$

MID-CHAPTER CHECK POINT   Section 3.1–Section 3.4

✓ **What You Know:** We learned to graph equations in two variables using point plotting, as well as a variety of other techniques. We used intercepts and a checkpoint to graph linear equations in the form $Ax + By = C$. We saw that $y = b$ graphs as a horizontal line and $x = a$ graphs as a vertical line. We determined a line's steepness, or rate of change, by computing its slope. We saw that lines with the same slope are parallel and lines with slopes that have a product of -1 are perpendicular. Finally, we learned to graph linear equations in slope-intercept form, $y = mx + b$, using the slope, m, and the y-intercept, b.

In Exercises 1–3, use each graph to determine
  a. *the x-intercept, or state that there is no x-intercept.*
  b. *the y-intercept, or state that there is no y-intercept.*
  c. *the line's slope, or state that the slope is undefined.*

1.

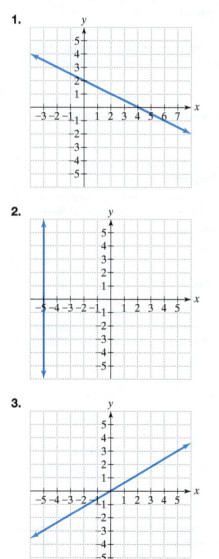

2.

3.

In Exercises 4–15, graph each equation in a rectangular coordinate system.

4. $y = -2x$

5. $y = -2$

6. $x + y = -2$

7. $y = \frac{1}{3}x - 2$

8. $x = 3.5$

9. $4x - 2y = 8$

10. $y = 3x + 2$

11. $3x + y = 0$

12. $y = -x + 4$

13. $y = x - 4$

14. $5y = -3x$

15. $5y = 20$

16. Find the slope and the y-intercept of the line whose equation is $5x - 2y = 10$.

In Exercises 17–19, determine whether the lines through each pair of points are parallel, perpendicular, or neither.

17. $(-5, -3)$ and $(0, -4)$; $(-2, -8)$ and $(1, 7)$

18. $(-4, 1)$ and $(2, 7)$; $(-5, 13)$ and $(4, -5)$

19. $(2, -4)$ and $(7, 0)$; $(-4, 2)$ and $(1, 6)$

20. Surfing While Driving. The graphs indicate that a greater percentage of U.S. drivers ages 18–29 used their cellphones to access the Internet while behind the wheel in 2013 than in 2009.

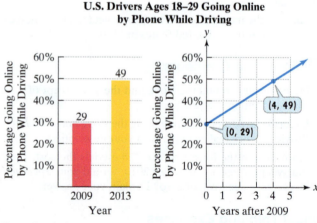

U.S. Drivers Ages 18–29 Going Online by Phone While Driving

Source: State Farm

a. Use the two points labeled by the voice balloons to find an equation in the form $y = mx + b$ that models the percentage of drivers ages 18–29 who went online by phone while driving, y, x years after 2009.

b. If trends shown by the graphs continue, use the model in part (a) to project the percentage of drivers ages 18–29 who will surf while driving in 2018.

3.5

The Point-Slope Form of the Equation of a Line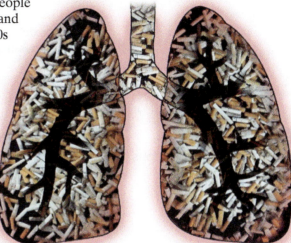

What am I supposed to learn?

After studying this section, you should be able to:

1. Use the point-slope form to write equations of a line.

2. Write linear equations that model data and make predictions.

Surprised by the number of people smoking cigarettes in movies and television shows made in the 1940s and 1950s? At that time, there was little awareness of the relationship between tobacco use and numerous diseases. Cigarette smoking was seen as a healthy way to relax and help digest a hearty meal. Then, in 1964, a linear equation changed everything. To understand the mathematics behind this turning point in public health, we explore another form of a line's equation.

Point-Slope Form

We can use the slope of a line to obtain another useful form of the line's equation. Consider a nonvertical line that has slope m and contains the point (x_1, y_1). Now let (x, y) represent any other point on the line, shown in **Figure 3.34**. Keep in mind that the point (x, y) is arbitrary and is not in one fixed position. By contrast, the point (x_1, y_1) is fixed.

Regardless of where the point (x, y) is located on the line, the steepness of the line in **Figure 3.34** remains the same. Thus, the ratio for slope stays a constant m. This means that for all points (x, y) along the line,

$$m = \frac{\text{Change in } y}{\text{Change in } x} = \frac{y - y_1}{x - x_1}.$$

We can clear the fraction by multiplying both sides by $x - x_1$, the least common denominator, where $x - x_1 \neq 0$.

$$m = \frac{y - y_1}{x - x_1} \qquad \text{This is the slope of the line in Figure 3.34.}$$

$$m(x - x_1) = \frac{y - y_1}{x - x_1} \cdot (x - x_1) \qquad \text{Multiply both sides by } x - x_1.$$

$$m(x - x_1) = y - y_1 \qquad \text{Simplify: } \frac{y - y_1}{x - x_1} \cdot (x - x_1) = y - y_1.$$

Now, if we reverse the two sides, we obtain the **point-slope form** of the equation of a line.

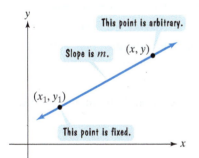

Figure 3.34 A line passing through (x_1, y_1) with slope m

Within the figure: This point is arbitrary. Slope is m. (x, y). (x_1, y_1). This point is fixed.

Point-Slope Form of the Equation of a Line

The **point-slope form of the equation** of a nonvertical line with slope m that passes through the point (x_1, y_1) is

$$y - y_1 = m(x - x_1).$$

Great Question!

When using $y - y_1 = m(x - x_1)$, for which variables do I substitute numbers?

When writing the point-slope form of a line's equation, you will never substitute numbers for x and y. You will substitute values for x_1, y_1, and m.

For example, using $y - y_1 = m(x - x_1)$, the point-slope form of the equation of the line passing through $(1, 4)$ with slope 2 $(m = 2)$ is

$$y - 4 = 2(x - 1).$$

1 Use the point-slope form to write equations of a line. ▶

Using the Point-Slope Form to Write a Line's Equation

If we know the slope of a line and a point not containing the y-intercept through which the line passes, the point-slope form is the equation that we should use. Once we have obtained this equation, it is customary to solve for y and write the equation in slope-intercept form. Examples 1 and 2 illustrate these ideas.

EXAMPLE 1 Writing the Point-Slope Form and the Slope-Intercept Form

Write the point-slope form and the slope-intercept form of the equation of the line with slope 4 that passes through the point $(-1, 3)$.

Solution We begin with the point-slope form of the equation of a line with $m = 4$, $x_1 = -1$, and $y_1 = 3$.

$y - y_1 = m(x - x_1)$	This is the point-slope form of the equation.
$y - 3 = 4[x - (-1)]$	Substitute: $(x_1, y_1) = (-1, 3)$ and $m = 4$.
$y - 3 = 4(x + 1)$	We now have the point-slope form of the equation of the given line.

Now we solve the equation $y - 3 = 4(x + 1)$ for y and write an equivalent equation in slope-intercept form $(y = mx + b)$.

We need to isolate y.

$y - 3 = 4(x + 1)$	This is the point-slope form of the equation.
$y - 3 = 4x + 4$	Use the distributive property.
$y = 4x + 7$	Add 3 to both sides.

The slope-intercept form of the line's equation is $y = 4x + 7$. ∎

✓ **CHECK POINT 1** Write the point-slope form and the slope-intercept form of the equation of the line with slope 6 that passes through the point $(2, -5)$.

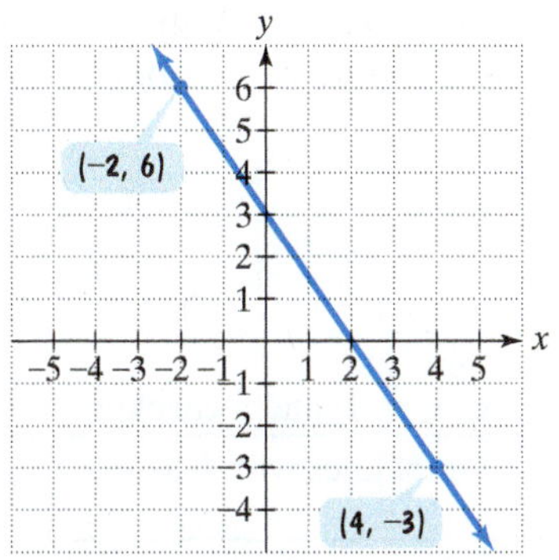

Figure 3.35

EXAMPLE 2 Writing the Point-Slope Form and the Slope-Intercept Form

A line passes through the points $(4, -3)$ and $(-2, 6)$. (See **Figure 3.35**.) Find the equation of the line

a. in point-slope form.   **b.** in slope-intercept form.

Solution

a. To use the point-slope form, we need to find the slope. The slope is the change in the y-coordinates divided by the corresponding change in the x-coordinates.

$$m = \frac{6 - (-3)}{-2 - 4} = \frac{9}{-6} = -\frac{3}{2} \qquad \text{This is the definition of slope using } (4, -3) \text{ and } (-2, 6).$$

We can take either point on the line to be (x_1, y_1). Let's use $(x_1, y_1) = (4, -3)$. Now, we are ready to write the point-slope form of the equation.

$$y - y_1 = m(x - x_1) \qquad \text{This is the point-slope form of the equation.}$$

$$y - (-3) = -\frac{3}{2}(x - 4) \qquad \text{Substitute: } (x_1, y_1) = (4, -3) \text{ and } m = -\frac{3}{2}.$$

$$y + 3 = -\frac{3}{2}(x - 4) \qquad \text{Simplify.}$$

This equation is one point-slope form of the equation of the line shown in **Figure 3.35**.

b. Now, we solve this equation for y and write an equivalent equation in slope-intercept form ($y = mx + b$).

We need to isolate y.

$$y + 3 = -\frac{3}{2}(x - 4) \qquad \text{This is the point-slope form of the equation.}$$

$$y + 3 = -\frac{3}{2}x + 6 \qquad \text{Use the distributive property.}$$

$$y = -\frac{3}{2}x + 3 \qquad \text{Subtract 3 from both sides.}$$

This equation is the slope-intercept form of the equation of the line shown in **Figure 3.35**. ∎

Discover for Yourself

If you are given two points on a line, you can use either point for (x_1, y_1) when you write its point-slope equation. Rework Example 2 using $(-2, 6)$ for (x_1, y_1). Once you solve for y, you should obtain the same slope-intercept equation as the one shown in the last line of the solution to Example 2.

✓ **CHECK POINT 2**   A line passes through the points $(-2, -1)$ and $(-1, -6)$. Find the equation of the line

a. in point-slope form.

b. in slope-intercept form.

The many forms of a line's equation can be a bit overwhelming. **Table 3.2** summarizes the various forms and contains the most important things you should remember about each form.

Table 3.2 Equations of Lines	
Form	**What You Should Know**
Standard Form $Ax + By = C$	• Graph equations in this form using intercepts (x-intercept: set $y = 0$; y-intercept: set $x = 0$) and a checkpoint.
$y = b$	• Graph equations in this form as horizontal lines with b as the y-intercept.
$x = a$	• Graph equations in this form as vertical lines with a as the x-intercept.
Slope-Intercept Form $y = mx + b$	• Graph equations in this form using the y-intercept, b, and the slope, m. • Start with this form when writing a linear equation if you know a line's slope and y-intercept.
Point-Slope Form $y - y_1 = m(x - x_1)$	• Start with this form when writing a linear equation if you know the slope of the line and a point on the line not containing the y-intercept or two points on the line, neither of which contains the y-intercept. Calculate the slope using $$m = \frac{\text{Change in } y}{\text{Change in } x} = \frac{y_2 - y_1}{x_2 - x_1}.$$ Although you begin with point-slope form, you usually solve for y and convert to slope-intercept form.

2 Write linear equations that model data and make predictions.

Applications

Linear equations are useful for modeling data that fall on or near a line. For example, the bar graph in **Figure 3.36(a)** gives the median age of the U.S. population in the indicated year. (The median age is the age in the middle when all the ages of the U.S. population are arranged from youngest to oldest.) The data are displayed as a set of five points in a rectangular coordinate system in **Figure 3.36(b).**

The Graying of America: Median Age of the United States Population

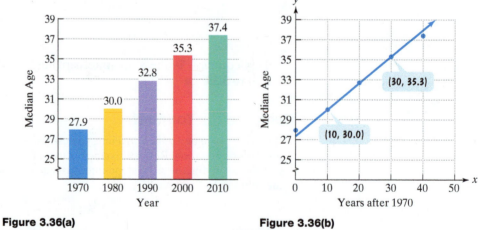

Figure 3.36(a)

Source: U.S. Census Bureau

Figure 3.36(b)

A set of points representing data is called a **scatter plot**. Also shown on the scatter plot in **Figure 3.36(b)** is a line that passes through or near the five points. By writing the equation of this line, we can obtain a model of the data and make predictions about the median age of the U.S. population in the future.

EXAMPLE 3   Modeling the Graying of America

Write the slope-intercept form of the equation of the line shown in **Figure 3.36(b)**. Use the equation to predict the median age of the U.S. population in 2020.

Solution   The line in **Figure 3.36(b)** passes through $(10, 30.0)$ and $(30, 35.3)$. We start by finding its slope.

$$m = \frac{\text{Change in } y}{\text{Change in } x} = \frac{35.3 - 30.0}{30 - 10} = \frac{5.3}{20} = 0.265$$

The slope indicates that each year the median age of the U.S. population is increasing by 0.265 year.

Now, we write the slope-intercept form of the equation for the line.

$y - y_1 = m(x - x_1)$	Begin with the point-slope form.
$y - 30.0 = 0.265(x - 10)$	Either ordered pair can be (x_1, y_1). Let $(x_1, y_1) = (10, 30.0)$. From above, $m = 0.265$.
$y - 30.0 = 0.265x - 2.65$	Apply the distributive property.
$y = 0.265x + 27.35$	Add 30 to both sides and solve for y.

A linear equation that models the median age of the U.S. population, y, x years after 1970 is

$$y = 0.265x + 27.35.$$

Now, let's use this equation to predict the median age in 2020. Because 2020 is 50 years after 1970, substitute 50 for x and compute y.

$$y = 0.265(50) + 27.35 = 40.6$$

Our model predicts that the median age of the U.S. population in 2020 will be 40.6.   ∎

☑ **CHECK POINT 3**   Use the data points $(10, 30.0)$ and $(20, 32.8)$ from **Figure 3.36(b)** to write the slope-intercept form of an equation that models the median age of the U.S. population x years after 1970. Use this model to predict the median age in 2020.

Using Technology

You can use a graphing utility to obtain a model for a scatter plot in which the data points fall on or near a straight line. The line that best fits the data is called the **regression line**. After entering the data in **Figure 3.36(b)**, a graphing utility displays a scatter plot of the data and the regression line.

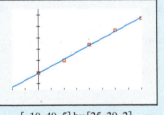

[−10, 40, 5] by [25, 39, 2]

Also displayed is the regression line's equation.

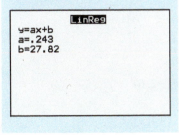

LinReg
y=ax+b
a=.243
b=27.82

Blitzer Bonus ⏵

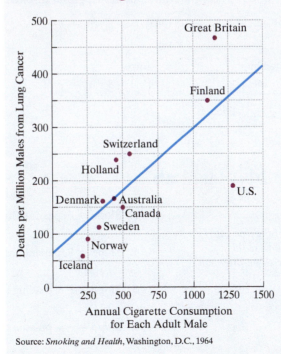

Deaths per Million Males from Lung Cancer

Annual Cigarette Consumption for Each Adult Male

Source: *Smoking and Health*, Washington, D.C., 1964

Cigarettes and Lung Cancer
This scatter plot shows a relationship between cigarette consumption among males and deaths due to lung cancer per million males. The data are from 11 countries and date back to a 1964 report by the U.S. Surgeon General. The scatter plot can be modeled by a line whose slope indicates an increasing death rate from lung cancer with increased cigarette consumption. At that time, the tobacco industry argued that in spite of this regression line, tobacco use is not the cause of cancer. Recent data do, indeed, show a causal effect between tobacco use and numerous diseases.

Achieving Success

The Secret of Math Success

What's the secret of math success? The bar graph in **Figure 3.37** shows that Japanese teachers and students are more likely than their American counterparts to believe that the key to doing well in math is working hard. Americans tend to think that either you have mathematical intelligence or you don't. Alan Bass, author of *Math Study Skills* (Pearson Education, 2008), strongly disagrees with this American perspective:

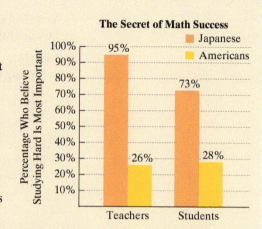

Figure 3.37

Source: Wade and Tavris, *Psychology*, Ninth Edition, Pearson, 2008.

"Human beings are easily intelligent enough to understand the basic principles of math. I cannot repeat this enough, but I'll try … **Poor performance in math is not due to a lack of intelligence!** The fact is that the key to success in math is in taking an intelligent approach. Students come up to me and say, 'I'm just not good at math.' Then I ask to see their class notebooks and they show me a chaotic mess of papers jammed into a folder. In math, that's a lot like taking apart your car's engine, dumping the heap of disconnected parts back under the hood, and then going to a mechanic to ask why it won't run. Students come to me and say, 'I'm just not good at math.' Then I ask them about their study habits and they say, 'I have to do all my studying on the weekend.' In math, that's a lot like trying to do all your eating or sleeping on the weekends and wondering why you're so tired and hungry all the time. **How you approach math is much more important than how smart you are.**"

— Alan Bass

CONCEPT AND VOCABULARY CHECK

Fill in each blank so that the resulting statement is true.

1. The point-slope form of the equation of a nonvertical line with slope m that passes through the point (x_1, y_1) is _____ .

2. The equation $5x + 3y = 10$ is written in _____ form.

3. The equation $y = 3x + 7$ is written in _____ form.

4. The equation $y - 6 = 4(x + 1)$ is written in _____ form.

5. The graph of $y = 3$ is a/an _____ line.

6. The graph of $x = -1$ is a/an _____ line.

3.5 EXERCISE SET ▶ MyMathLab®

Practice Exercises

Write the point-slope form of the equation of the line satisfying each of the conditions in Exercises 1–28. Then use the point-slope form of the equation to write the slope-intercept form of the equation.

1. Slope = 3, passing through $(2, 5)$

2. Slope = 6, passing through $(3, 1)$

3. Slope = 5, passing through $(-2, 6)$

4. Slope = 7, passing through $(-4, 9)$

5. Slope = -8, passing through $(-3, -2)$

6. Slope = -4, passing through $(-5, -2)$

7. Slope $= -12$, passing through $(-8, 0)$

8. Slope $= -11$, passing through $(0, -3)$

9. Slope $= -1$, passing through $\left(-\frac{1}{2}, -2\right)$

10. Slope $= -1$, passing through $\left(-4, -\frac{1}{4}\right)$

11. Slope $= \frac{1}{2}$, passing through the origin

12. Slope $= \frac{1}{3}$, passing through the origin

13. Slope $= -\frac{2}{3}$, passing through $(6, -2)$

14. Slope $= -\frac{3}{5}$, passing through $(10, -4)$

15. Passing through $(1, 2)$ and $(5, 10)$

16. Passing through $(3, 5)$ and $(8, 15)$

17. Passing through $(-3, 0)$ and $(0, 3)$

18. Passing through $(-2, 0)$ and $(0, 2)$

19. Passing through $(-3, -1)$ and $(2, 4)$

20. Passing through $(-2, -4)$ and $(1, -1)$

21. Passing through $(-4, -1)$ and $(3, 4)$

22. Passing through $(-6, 1)$ and $(2, -5)$

23. Passing through $(-3, -1)$ and $(4, -1)$

24. Passing through $(-2, -5)$ and $(6, -5)$

25. Passing through $(2, 4)$ with x-intercept $= -2$

26. Passing through $(1, -3)$ with x-intercept $= -1$

27. x-intercept $= -\frac{1}{2}$ and y-intercept $= 4$

28. x-intercept $= 4$ and y-intercept $= -2$

Practice PLUS

In Exercises 29–38, write an equation in slope-intercept form of the line satisfying the given conditions.

29. The line passes through $(-3, 2)$ and is parallel to the line whose equation is $y = 4x + 1$.

30. The line passes through $(5, -3)$ and is parallel to the line whose equation is $y = 2x + 1$.

31. The line passes through $(-1, -5)$ and is parallel to the line whose equation is $3x + y = 6$.

32. The line passes through $(-4, -7)$ and is parallel to the line whose equation is $6x + y = 8$.

33. The line passes through $(4, -7)$ and is perpendicular to the line whose equation is $x - 2y = 3$.

34. The line passes through $(5, -9)$ and is perpendicular to the line whose equation is $x + 7y = 12$.

35. The line passes through $(2, 4)$ and has the same y-intercept as the line whose equation is $x - 4y = 8$.

36. The line passes through $(2, 6)$ and has the same y-intercept as the line whose equation is $x - 3y = 18$.

37. The line has an x-intercept at -4 and is parallel to the line containing $(3, 1)$ and $(2, 6)$.

38. The line has an x-intercept at -6 and is parallel to the line containing $(4, -3)$ and $(2, 2)$.

Application Exercises

39. The bar graph shows the mean, or average, student loan debt in the United States, in 2011 dollars, for four selected years from 2001 through 2010.

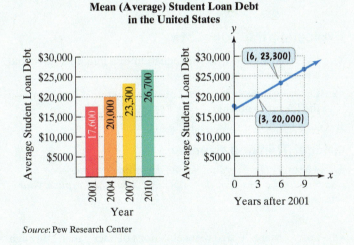

Mean (Average) Student Loan Debt in the United States

Source: Pew Research Center

a. Shown to the right of the bar graph is a scatter plot with a line passing through two of the data points. Use the two points whose coordinates are show in by the voice balloons to write the slope-intercept form of an equation that models the average student loan debt, y, in the United States x years after 2001.

b. If trends shown by the data continue, use the model from part (a) to project the average student loan debt in 2021.

40. In Section 3.3 (Example 5), we worked with data involving the increasing number of U.S. adults ages 18 and older living alone. The bar graph reinforces the fact that one-person households are growing more common. It shows one-person households as a percentage of the U.S. total for five selected years from 1980 through 2012.

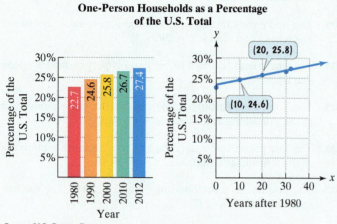

One-Person Households as a Percentage of the U.S. Total

Source: U.S. Census Bureau

a. Shown to the right of the bar graph is a scatter plot with a line passing through two of the data points. Use the two points whose coordinates are shown by the voice balloons to write the slope-intercept form of an equation that models one-person households as a percentage of the U.S. total, y, x years after 1980.

b. If trends shown by the data continue, use the model from part (a) to project one-person households as a percentage of the U.S. total in 2020.

Explaining the Concepts

41. Describe how to write the equation of a line if its slope and a point on the line are known.

42. Describe how to write the equation of a line if two points on the line are known.

Critical Thinking Exercises

Make Sense?   *In Exercises 43–46, determine whether each statement makes sense or does not make sense, and explain your reasoning.*

43. I use $y = mx + b$ to write equations of lines passing through two points when neither contains the y-intercept.

44. In many examples, I use the slope-intercept form of a line's equation to obtain an equivalent equation in point-slope form.

45. I have linear models that describe changes for men and women over the same time period. The models have the same slope, so the graphs are parallel lines, indicating that the rate of change for men is the same as the rate of change for women.

46. The shape of this scatter plot showing the average box office revenue of Oscar best picture nominees suggests that I should not use a line or any of its equations to obtain a model.

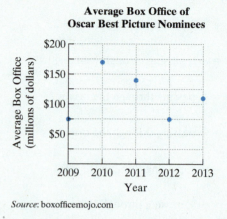

Average Box Office of Oscar Best Picture Nominees

Source: boxofficemojo.com

In Exercises 47–50, determine whether each statement is true or false. If the statement is false, make the necessary change(s) to produce a true statement.

47. If a line has undefined slope, then it has no equation.

48. The line whose equation is $y - 3 = 7(x + 2)$ passes through $(-3, 2)$.

49. The point-slope form can be applied to obtain the equation of the line through the points $(2, -5)$ and $(2, 6)$.

50. The slope of the line whose equation is $3x + y = 7$ is 3.

51. Excited about the success of celebrity stamps, post office officials were rumored to have put forth a plan to institute two new types of thermometers. On these new scales, $°E$ represents degrees Elvis and $°M$ represents degrees Madonna. If it is known that $40°E = 25°M$, $280°E = 125°M$, and degrees Elvis is linearly related to degrees Madonna, write an equation expressing E in terms of M.

Technology Exercises

52. Use a graphing utility to graph $y = 1.75x - 2$. Select the best viewing rectangle possible by experimenting with the range settings to show that the line's slope is $\frac{7}{4}$.

53. a. Use the statistical menu of a graphing utility to enter the five data points shown in the scatter plot in Exercise 40. Refer to the bar graph to obtain coordinates of points that are not given in the scatter plot.

b. Use the scatter plot capability to draw a scatter plot of the data points like the one shown in Exercise 40.

c. Select the linear regression option. Use your utility to obtain values for a and b for the equation of the regression line, $y = ax + b$. You may also be given a *correlation coefficient*, r. Values of r close to 1 indicate that the points can be described by a linear relationship and the regression line has a positive slope. Values of r close to -1 indicate that the points can be described by a linear relationship and the regression line has a negative slope. Values of r close to 0 indicate no linear relationship between the variables.

d. Use the appropriate sequence (consult your manual) to graph the regression equation on top of the points in the scatter plot.

Review Exercises

54. How many sheets of paper, weighing 2 grams each, can be put in an envelope weighing 4 grams if the total weight must not exceed 29 grams? (Section 2.7, Example 11)

55. List all the natural numbers in this set:

$$\left\{ -2, 0, \frac{1}{2}, 1, \sqrt{3}, \sqrt{4} \right\}.$$

(Section 1.3, Example 5)

56. Use intercepts to graph $3x - 5y = 15$. (Section 3.2, Example 4)

Preview Exercises

Exercises 57–59 will help you prepare for the material covered in the next section.

57. Is $2x - 3y \geq 6$ a true statement for $x = 3$ and $y = -1$?

58. Is $2x - 3y \geq 6$ a true statement for $x = 0$ and $y = 0$?

59. Is $y \leq \frac{2}{3}x$ a true statement for $x = 1$ and $y = 1$?

SECTION

3.6   Linear Inequalities in Two Variables

What am I supposed to learn?

After studying this section, you should be able to:

1 Determine whether an ordered pair is a solution of an inequality.

2 Graph a linear inequality in two variables.

3 Solve applied problems involving linear inequalities in two variables.

Temperature and precipitation are two variables that affect whether regions are forests, grasslands, or deserts. In this section, you will see how linear inequalities in two variables describe some of the most magnificent places in our nation's landscape.

1 Determine whether an ordered pair is a solution of an inequality.

Linear Inequalities in Two Variables and Their Solutions

We have seen that equations in the form $Ax + By = C$ are straight lines when graphed. If we change the = sign to $>, <, \geq,$ or $\leq$, we obtain a **linear inequality in two variables**. Some examples of linear inequalities in two variables are $x + y > 2$, $3x - 5y \leq 15$, and $2x - y < 4$.

A **solution of an inequality in two variables**, x and y, is an ordered pair of real numbers with the following property: When the x-coordinate is substituted for x and the y-coordinate is substituted for y in the inequality, we obtain a true statement. For example, $(3, 2)$ is a solution of the inequality $x + y > 1$. When 3 is substituted for x and 2 is substituted for y, we obtain the true statement $3 + 2 > 1$, or $5 > 1$. Because there are infinitely many pairs of numbers that have a sum greater than 1, the inequality $x + y > 1$ has infinitely many solutions. Each ordered-pair solution is said to **satisfy** the inequality. Thus, $(3, 2)$ satisfies the inequality $x + y > 1$.

EXAMPLE 1   Deciding Whether an Ordered Pair Satisfies an Inequality

Determine whether each ordered pair is a solution of the inequality

$$2x - 3y \geq 6:$$

a. $(0, 0)$ **b.** $(3, -1)$.

Solution

a. To determine whether $(0, 0)$ is a solution of the inequality, we substitute 0 for x and 0 for y.

$$2x - 3y \geq 6 \qquad \text{This is the given inequality.}$$

$$2 \cdot 0 - 3 \cdot 0 \overset{?}{\geq} 6 \qquad \text{Substitute 0 for } x \text{ and 0 for } y.$$

$$0 - 0 \overset{?}{\geq} 6 \qquad \text{Multiply: } 2 \cdot 0 = 0 \text{ and } 3 \cdot 0 = 0.$$

$$\boxed{\text{This statement is false.}} \quad 0 \geq 6 \qquad \text{Subtract: } 0 - 0 = 0.$$

Because we obtain a false statement, we conclude that $(0, 0)$ is not a solution of $2x - 3y \geq 6$. The ordered pair $(0, 0)$ does not satisfy the inequality.

b. To determine whether $(3, -1)$ is a solution of the inequality, we substitute 3 for x and -1 for y.

$$2x - 3y \geq 6 \qquad \text{This is the given inequality.}$$

$$2 \cdot 3 - 3(-1) \overset{?}{\geq} 6 \qquad \text{Substitute 3 for } x \text{ and } -1 \text{ for } y.$$

$$6 - (-3) \overset{?}{\geq} 6 \qquad \text{Multiply: } 2 \cdot 3 = 6 \text{ and } 3(-1) = -3.$$

$$\boxed{\text{This statement is true.}} \quad 9 \geq 6 \qquad \text{Subtract: } 6 - (-3) = 6 + 3 = 9.$$

Because we obtain a true statement, we conclude that $(3, -1)$ is a solution of $2x - 3y \geq 6$. The ordered pair $(3, -1)$ satisfies the inequality. ■

✓ **CHECK POINT 1** Determine whether each ordered pair is a solution of the inequality $5x + 4y \leq 20$:

a. $(0, 0)$

b. $(6, 2)$.

---

2 Graph a linear inequality in two variables. ▶

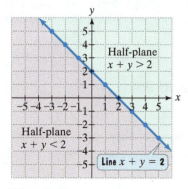

Figure 3.38

The Graph of a Linear Inequality in Two Variables

We know that the graph of an equation in two variables is the set of all points whose coordinates satisfy the equation. Similarly, the **graph of an inequality in two variables** is the set of all points whose coordinates satisfy the inequality.

Let's use **Figure 3.38** to get an idea of what the graph of a linear inequality in two variables looks like. Part of the figure shows the graph of the linear equation $x + y = 2$. The line divides the points in the rectangular coordinate system into three sets. First, there is the set of points along the line, satisfying $x + y = 2$. Next, there is the set of points in the green region above the line. Points in the green region satisfy the linear inequality $x + y > 2$. Finally, there is the set of points in the purple region below the line. Points in the purple region satisfy the linear inequality $x + y < 2$.

A **half-plane** is the set of all the points on one side of a line. In **Figure 3.38**, the green region is a half-plane. The purple region is also a half-plane. A half-plane is the graph of a linear inequality that involves $>$ or $<$. The graph of an inequality that involves $\geq$ or $\leq$ is a half-plane and a line. A solid line is used to show that the line is part of the graph. A dashed line is used to show that a line is not part of a graph.

Graphing a Linear Inequality in Two Variables

1. Replace the inequality symbol with an equal sign and graph the corresponding linear equation. Draw a solid line if the original inequality contains a $\leq$ or $\geq$ symbol. Draw a dashed line if the original inequality contains a $<$ or $>$ symbol.

2. Choose a test point from one of the half-planes. (Do not choose a point that is on the line.) Substitute the coordinates of the test point into the inequality.

3. If a true statement results, shade the half-plane containing this test point. If a false statement results, shade the half-plane not containing this test point.

EXAMPLE 2   Graphing a Linear Inequality in Two Variables

Graph: $3x - 5y < 15$.

Solution

Step 1. Replace the inequality symbol with = and graph the linear equation. We need to graph $3x - 5y = 15$. We can use intercepts to graph this line.

We set $y = 0$ to find the x-intercept:	We set $x = 0$ to find the y-intercept:
$3x - 5y = 15$	$3x - 5y = 15$
$3x - 5 \cdot 0 = 15$	$3 \cdot 0 - 5y = 15$
$3x = 15$	$-5y = 15$
$x = 5.$	$y = -3.$

The x-intercept is 5, so the line passes through $(5, 0)$. The y-intercept is -3, so the line passes through $(0, -3)$. The graph of the equation is indicated by a dashed line because the inequality $3x - 5y < 15$ contains a $<$ symbol, rather than $\leq$. The graph of the line is shown in **Figure 3.39**.

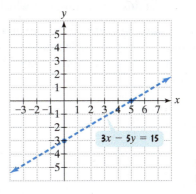

$3x - 5y = 15$

Figure 3.39 Preparing to graph $3x - 5y < 15$

Step 2. Choose a test point from one of the half-planes and not from the line. Substitute its coordinates into the inequality. The line $3x - 5y = 15$ divides the plane into three parts—the line itself and two half-planes. The points in one half-plane satisfy $3x - 5y > 15$. The points in the other half-plane satisfy $3x - 5y < 15$. We need to find which half-plane is the solution. To do so, we test a point from either half-plane. The origin, $(0, 0)$, is the easiest point to test.

$$3x - 5y < 15 \qquad \text{This is the given inequality.}$$
$$3 \cdot 0 - 5 \cdot 0 \stackrel{?}{<} 15 \qquad \text{Test (O, O) by substituting O for } x \text{ and O for } y.$$
$$0 - 0 \stackrel{?}{<} 15 \qquad \text{Multiply.}$$
$$0 < 15 \qquad \text{This statement is true.}$$

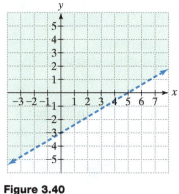

Figure 3.40
The graph of $3x - 5y < 15$

Step 3. If a true statement results, shade the half-plane containing the test point. Because 0 is less than 15, the test point $(0, 0)$ is part of the solution set. All the points on the same side of the line $3x - 5y = 15$ as the point $(0, 0)$ are members of the solution set. The solution set is the half-plane that contains the point $(0, 0)$, indicated by shading this half-plane. The graph is shown using green shading and a dashed blue line in **Figure 3.40**. ∎

✓ **CHECK POINT 2** Graph: $2x - 4y < 8$.

When graphing a linear inequality, test a point that lies in one of the half-planes and *not on the line dividing the half-planes*. The test point $(0, 0)$ is convenient because it is easy to calculate when 0 is substituted for each variable. However, if $(0, 0)$ lies on the dividing line and not in a half-plane, a different test point must be selected.

EXAMPLE 3 Graphing a Linear Inequality

Graph: $y \le \dfrac{2}{3}x$.

Solution

Step 1. Replace the inequality symbol with = and graph the linear equation. Because we are interested in graphing $y \le \frac{2}{3}x$, we begin by graphing $y = \frac{2}{3}x$. We can use the slope and the y-intercept to graph this line.

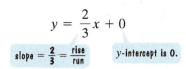

$$y = \frac{2}{3}x + 0$$

slope $= \dfrac{2}{3} = \dfrac{\text{rise}}{\text{run}}$ y-intercept is 0.

The y-intercept of $y = \frac{2}{3}x$ is 0, so the line passes through $(0, 0)$. Using the y-intercept and the slope, $\frac{2}{3}$, the line is shown in **Figure 3.41** as a solid line. This is because the inequality $y \le \frac{2}{3}x$ contains a $\le$ symbol, in which equality is included.

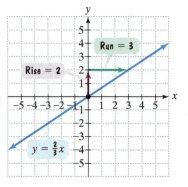

Figure 3.41
Preparing to graph $y \le \frac{2}{3}x$

Step 2. Choose a test point from one of the half-planes and not from the line. Substitute its coordinates into the inequality. We cannot use $(0, 0)$ as a test point because it lies on the line and not in a half-plane. Let's use $(1, 1)$, which lies in the half-plane above the line.

$$y \le \frac{2}{3}x \qquad \text{This is the given inequality.}$$

$$1 \overset{?}{\le} \frac{2}{3} \cdot 1 \qquad \text{Test } (1, 1) \text{ by substituting 1 for } x \text{ and 1 for } y.$$

$$1 \le \frac{2}{3} \qquad \text{This statement is false.}$$

Step 3. If a false statement results, shade the half-plane not containing the test point. Because 1 is not less than or equal to $\frac{2}{3}$, the test point $(1, 1)$ is not part of the solution set. Thus, the half-plane below the solid line $y = \frac{2}{3}x$ is part of the solution set. The solution set is the line and the half-plane that does not contain the point $(1, 1)$, indicated by shading this half-plane. The graph is shown using green shading and a blue line in **Figure 3.42**. ∎

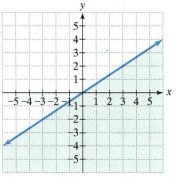

Figure 3.42
The graph of $y \le \frac{2}{3}x$

✓ **CHECK POINT 3** Graph: $y \ge \dfrac{1}{2}x$.

Graphing Linear Inequalities without Using Test Points

You can graph inequalities in the form $y > mx + b$ or $y < mx + b$ without using test points. The inequality symbol indicates which half-plane to shade.

- If $y > mx + b$, shade the half-plane above the line $y = mx + b$.
- If $y < mx + b$, shade the half-plane below the line $y = mx + b$.

Observe how this is illustrated in **Figure 3.42**. The graph of $y \le \frac{2}{3}x$ contains the half-plane below the line $y = \frac{2}{3}x$.

It is also not necessary to use test points when graphing inequalities involving half-planes on one side of a vertical or a horizontal line.

For the Vertical Line $x = a$:

- If $x > a$, shade the half-plane to the right of $x = a$.
- If $x < a$, shade the half-plane to the left of $x = a$.

For the Horizontal Line $y = b$:

- If $y > b$, shade the half-plane above $y = b$.
- If $y < b$, shade the half-plane below $y = b$.

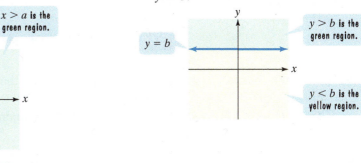

Great Question!

When is it important to use test points to graph linear inequalities?

Continue using test points to graph inequalities in the form $Ax + By > C$ or $Ax + By < C$. The graph of $Ax + By > C$ can lie above or below the line of $Ax + By = C$, depending on the value of B. The same comment applies to the graph of $Ax + By < C$.

EXAMPLE 4   Graphing Inequalities without Using Test Points

Graph each inequality in a rectangular coordinate system:

a. $y \le -3$ **b.** $x > 2$.

Solution

a. $y \le -3$

Graph $y = -3$, a horizontal line with y-intercept -3. The line is solid because equality is included in $y \le -3$. Because of the less than part of $\le$, shade the half-plane below the horizontal line.

b. $x > 2$

Graph $x = 2$, a vertical line with x-intercept 2. The line is dashed because equality is not included in $x > 2$. Because of $>$, the greater than symbol, shade the half-plane to the right of the vertical line.

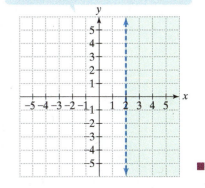

✓ **CHECK POINT 4**   Graph each inequality in a rectangular coordinate system:

a. $y > 1$ **b.** $x \le -2$.

3 Solve applied problems involving linear inequalities in two variables.

Applications

Temperature and precipitation affect whether or not trees and forests can grow. At certain levels of precipitation and temperature, only grasslands and deserts will exist. **Figure 3.43** shows three kinds of regions—deserts, grasslands, and forests—that result from various ranges of temperature and precipitation. Notice that the horizontal axis is labeled T, for temperature, rather than x. The vertical axis is labeled P, for precipitation, rather than y. We can use inequalities in two variables, T and P, to describe the regions in the figure.

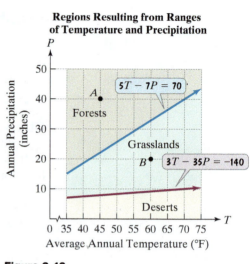

Regions Resulting from Ranges of Temperature and Precipitation

Figure 3.43

Source: A. Miller and J. Thompson, *Elements of Meteorology*

EXAMPLE 5   Forests and Inequalities

a. Use **Figure 3.43** to find the coordinates of point A. What does this mean in terms of the kind of region that occurs?

b. For average annual temperatures that exceed 35°F, the inequality

$$5T - 7P < 70$$

models where forests occur. Show that the coordinates of point A satisfy this inequality.

Solution

a. Point A has coordinates $(45, 40)$. This means that if a region has an average annual temperature of 45°F and an average annual precipitation of 40 inches, then a forest occurs.

b. We can show that $(45, 40)$ satisfies the inequality for forests by substituting 45 for T and 40 for P.

$5T - 7P < 70$	This is the given inequality.
$5 \cdot 45 - 7 \cdot 40 \overset{?}{<} 70$	Substitute 45 for T and 40 for P.
$225 - 280 \overset{?}{<} 70$	Multiply: $5 \cdot 45 = 225$ and $7 \cdot 40 = 280$.
$-55 < 70$	Subtract: $225 - 280 = -55$.

This statement is true.

The coordinates $(45, 40)$ make the inequality true. Thus, $(45, 40)$ satisfies the inequality. ∎

✓ CHECK POINT 5

a. Use **Figure 3.43** to find the coordinates of point B. What does this mean in terms of the kind of region that occurs?

b. For average annual temperatures that exceed 35°F, the inequalities $5T - 7P \geq 70$ and $3T - 35P \leq -140$ model where grasslands occur. Show that the coordinates of point B satisfy both of these inequalities.

Achieving Success

Organizing and creating your own compact chapter summaries can reinforce what you know and help with the retention of this information. Imagine that your professor will permit two index cards of notes (3 by 5; front and back) on all exams. Organize and create such a two-card summary for the test on this chapter. Begin by determining what information you would find most helpful to include on the cards. Take as long as you need to create the summary. Based on how effective you find this strategy, you may decide to use the technique to help prepare for future exams.

CONCEPT AND VOCABULARY CHECK

Fill in each blank so that the resulting statement is true.

1. The ordered pair (5, 4) is a/an _____ of the inequality $x + y > 2$ because when 5 is substituted for _____ and 4 is substituted for _____, the true statement _____ is obtained.

2. The set of all points that satisfy an inequality is called the _____ of the inequality.

3. The set of all points on one side of a line is called a/an _____.

4. True or false: The graph of $5x - 3y > 15$ includes the line $5x - 3y = 15$. _____

5. True or false: The graph of the linear equation $5x - 3y = 15$ is used to graph the linear inequality $5x - 3y > 15$. _____

6. True or false: When graphing $5x - 3y > 15$, to determine which side of the line to shade, choose a test point on $5x - 3y = 15$. _____

3.6 EXERCISE SET ▶ MyMathLab®

Practice Exercises

In Exercises 1–8, determine whether each ordered pair is a solution of the given inequality.

1. $x + y > 4$: (2, 2), (3, 2), (−3, 8)
2. $2x - y < 3$: (0, 0), (3, 0), (−4, −15)
3. $2x + y \geq 5$: (4, 0), (1, 3), (0, 0)
4. $3x - 5y \geq -12$: (2, −3), (2, 8), (0, 0)
5. $y \geq -2x + 4$: (4, 0), (1, 3), (−2, −4)
6. $y \leq -x + 5$: (5, 0), (0, 5), (8, −4)
7. $y > -2x + 1$: (2, 3), (0, 0), (0, 5)
8. $x < -y - 2$: (−1, −1), (0, 0), (4, −5)

In Exercises 9–36, graph each inequality.

9. $x + y \geq 3$
10. $x + y \geq 4$
11. $x - y < 5$
12. $x - y < 2$
13. $x + 2y > 4$
14. $2x + y > 6$
15. $3x - y \leq 6$
16. $x - 3y \leq -6$
17. $3x - 2y \leq 8$
18. $2x - 3y \geq 8$
19. $4x + 3y > 15$
20. $5x + 10y > 15$
21. $5x - y < -7$
22. $x - 5y < -7$
23. $y \leq \frac{1}{3}x$
24. $y \leq \frac{1}{4}x$
25. $y > 2x$
26. $y > 4x$
27. $y > 3x + 2$
28. $y > 2x - 1$
29. $y < \frac{3}{4}x - 3$
30. $y < \frac{2}{3}x - 1$
31. $x \leq 1$
32. $x \leq -3$
33. $y > 1$
34. $y > -3$
35. $x \geq 0$
36. $y \leq 0$

Practice PLUS

In Exercises 37–44, write each sentence as a linear inequality in two variables. Then graph the inequality.

37. The sum of the x-variable and the y-variable is at least 2.

38. The difference between the x-variable and the y-variable is at least 3.

39. The difference between 5 times the x-variable and 2 times the y-variable is at most 10.

40. The sum of 4 times the x-variable and 2 times the y-variable is at most 8.

41. The y-variable is no less than $\frac{1}{2}$ of the x-variable.

42. The y-variable is no less than $\frac{1}{4}$ of the x-variable.

43. The y-variable is no more than -1.

44. The y-variable is no more than -2.

Application Exercises

45. Bottled water and medical supplies are to be shipped to survivors of a hurricane by plane. Each plane can carry no more than 80,000 pounds. The bottled water weighs 20 pounds per container and each medical kit weighs 10 pounds. Let x represent the number of bottles of water to be shipped. Let y represent the number of medical kits. The plane's weight limitations can be described by the following inequality:

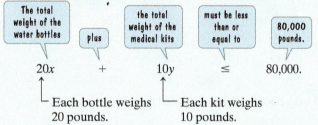

a. Graph the inequality. Because x and y must be nonnegative, limit the graph to quadrant I and its boundary only.

b. Select an ordered pair satisfying the inequality. What are its coordinates and what do they represent in this situation?

46. Bottled water and medical supplies are to be shipped to survivors of a hurricane by plane. Each plane can carry a total volume that does not exceed 6000 cubic feet. Each water bottle is 2 cubic feet and each medical kit has a volume of 1 cubic foot. Let x represent the number of bottles of water to be shipped. Let y represent the number of medical kits. The plane's volume limitations can be described by the following inequality:

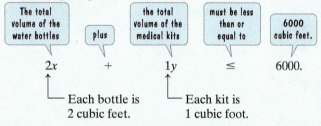

a. Graph the inequality. Because x and y must be nonnegative, limit the graph to quadrant I and its boundary only.

b. Select an ordered pair satisfying the inequality. What are its coordinates and what do they represent in this situation?

47. Many elevators have a capacity of 2000 pounds.

a. If a child averages 50 pounds and an adult 150 pounds, write an inequality that describes when x children and y adults will cause the elevator to be overloaded.

b. Graph the inequality. Because x and y must be nonnegative, limit the graph to quadrant I and its boundary only.

c. Select an ordered pair satisfying the inequality. What are its coordinates and what do they represent in this situation?

48. A patient is not allowed to have more than 330 milligrams of cholesterol per day from a diet of eggs and meat. Each egg provides 165 milligrams of cholesterol. Each ounce of meat provides 110 milligrams of cholesterol.

a. Write an inequality that describes the patient's dietary restrictions for x eggs and y ounces of meat.

b. Graph the inequality. Because x and y must be nonnegative, limit the graph to quadrant I and its boundary only.

c. Select an ordered pair satisfying the inequality. What are its coordinates and what do they represent in this situation?

Not all regions in rectangular coordinates are bounded by straight lines. Although these regions cannot be described using linear inequalities, they often have numerous applications. For example, the regions in the following two graphs indicate whether a person is obese, overweight, borderline overweight, normal weight, or underweight.

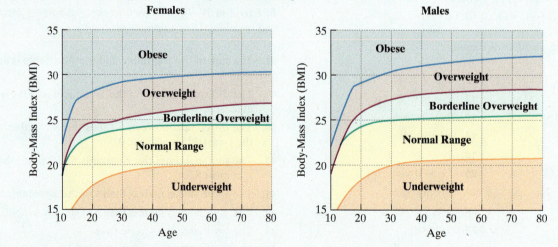

Source: Centers for Disease Control and Prevention

The horizontal axis of each graph on the previous page shows a person's age. The vertical axis shows that person's body-mass index (BMI), computed using the following formula:

$$\text{BMI} = \frac{703W}{H^2}.$$

The variable W represents weight, in pounds. The variable H represents height, in inches. Use this information to solve Exercises 49–50.

49. A man is 20 years old, 72 inches (6 feet) tall, and weighs 200 pounds.

   a. Compute the man's BMI. Round to the nearest tenth.

   b. Use the man's age and his BMI to locate this information as a point in the coordinate system for males shown on the previous page. Is this man obese, overweight, borderline overweight, normal weight, or underweight?

50. A woman is 25 years old, 66 inches (5 feet, 6 inches) tall, and weighs 105 pounds.

   a. Compute the woman's BMI. Round to the nearest tenth.

   b. Use the woman's age and her BMI to locate this information as a point in the coordinate system for females shown on the previous page. Is this woman obese, overweight, borderline overweight, normal weight, or underweight?

Explaining the Concepts

51. What is a linear inequality in two variables? Provide an example with your description.

52. How do you determine whether an ordered pair is a solution of an inequality in two variables, x and y?

53. What is a half-plane?

54. What does a solid line mean in the graph of an inequality?

55. What does a dashed line mean in the graph of an inequality?

56. Explain how to graph $2x - 3y < 6$.

57. Compare the graphs of $3x - 2y > 6$ and $3x - 2y \le 6$. Discuss similarities and differences between the graphs.

Critical Thinking Exercises

Make Sense?   In Exercises 58–61, determine whether each statement makes sense or does not make sense, and explain your reasoning.

58. By looking at a linear inequality in two variables, I can immediately determine whether the boundary line of its graph should be solid or dashed.

59. The inequality $2x - 3y < 6$ contains a "less than" symbol, so its graph lies below the boundary line.

60. When I write a linear inequality with y isolated on the left, $<$ indicates a region that lies below the boundary line.

61. I have less than $5.00 in nickels and dimes, so the linear inequality

$$0.05n + 0.10d < 5.00$$

models how many nickels, n, and how many dimes, d, that I might have.

In Exercises 62–65, determine whether each statement is true or false. If the statement is false, make the necessary change(s) to produce a true statement.

62. The ordered pair $(0, -3)$ satisfies $y > 2x - 3$.

63. The graph of $x < y + 1$ is the half-plane below the line $x = y + 1$.

64. In graphing $y \ge 4x$, a dashed line is used.

65. The graph of $x < 4$ is the half-plane to the left of the vertical line described by $x = 4$.

In Exercises 66–67, write an inequality that represents each graph.

66.   **67.**

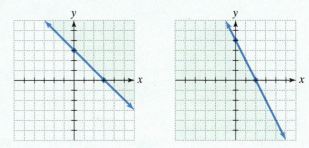

Technology Exercises

Graphing utilities can be used to shade regions in the rectangular coordinate system, thereby graphing an inequality in two variables. Read the section of the user's manual for your graphing utility that describes how to shade a region. Then use your graphing utility to graph the inequalities in Exercises 68–71.

68. $y \le 4x + 4$ **69.** $y \ge x - 2$

70. $y \ge \dfrac{1}{2}x + 4$ **71.** $y \le -\dfrac{1}{2}x + 4$

Review Exercises

72. Solve for h: $V = lwh$. (Section 2.4, Example 1)

73. Find the quotient: $\dfrac{2}{3} \div \left(-\dfrac{5}{4}\right)$.

   (Section 1.7, Example 5)

74. Evaluate $x^2 - 4$ for $x = -3$.

   (Section 1.8, Example 10)

Preview Exercises

Exercises 75–77 will help you prepare for the material covered in the first section of the next chapter.

75. Is $(4, -1)$ a solution of both $x + 2y = 2$ and $x - 2y = 6$?

76. Is $(-4, 3)$ a solution of both $x + 2y = 2$ and $x - 2y = 6$?

77. Determine the point of intersection of the graphs of $2x + 3y = 6$ and $2x + y = -2$ by graphing both equations in the same rectangular coordinate system.

Chapter 3 Summary

| **Definitions and Concepts** | **Examples** |

Section 3.1 Graphing Linear Equations in Two Variables

The rectangular coordinate system consists of a horizontal number line, the x-axis, and a vertical number line, the y-axis, intersecting at their zero points, the origin. Each point in the system corresponds to an ordered pair of real numbers (x, y). The first number in the pair is the x-coordinate; the second number is the y-coordinate.

Plot: $(2, 3), (-5, 4), (-4, -3)$, and $(5, -2)$.

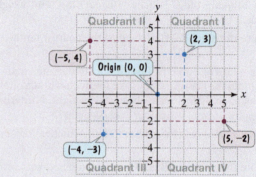

An ordered pair is a solution of an equation in two variables if replacing the variables by the coordinates of the ordered pair results in a true statement.

Is $(-1, 4)$ a solution of $2x + 5y = 18$?
$$2(-1) + 5 \cdot 4 \stackrel{?}{=} 18$$
$$-2 + 20 \stackrel{?}{=} 18$$
$$18 = 18, \text{ true}$$
Thus, $(-1, 4)$ is a solution.

One method for graphing an equation in two variables is point plotting. Find several ordered-pair solutions, plot them as points, and connect the points with a smooth curve or line.

Graph: $y = 2x + 1$.

x	$y = 2x + 1$	(x, y)
-2	$y = 2(-2) + 1 = -3$	$(-2, -3)$
-1	$y = 2(-1) + 1 = -1$	$(-1, -1)$
0	$y = 2 \cdot 0 + 1 = 1$	$(0, 1)$
1	$y = 2 \cdot 1 + 1 = 3$	$(1, 3)$
2	$y = 2 \cdot 2 + 1 = 5$	$(2, 5)$

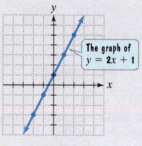

The graph of $y = 2x + 1$

Section 3.2 Graphing Linear Equations Using Intercepts

If a graph intersects the x-axis at $(a, 0)$, then a is an x-intercept.
If a graph intersects the y-axis at $(0, b)$, then b is a y-intercept.

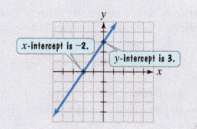

x-intercept is -2.

y-intercept is 3.

Definitions and Concepts	**Examples**

Section 3.2 Graphing Linear Equations Using Intercepts (continued)

An equation of the form $Ax + By = C$, where A, B, and C are integers, is called the standard form of the equation of a line. The graph of $Ax + By = C$ is a line that can be obtained using intercepts. To find the x-intercept, let $y = 0$ and solve for x. To find the y-intercept, let $x = 0$ and solve for y. Find a checkpoint, a third ordered-pair solution. Graph the equation by drawing a line through the three points.

Graph using intercepts: $4x + 3y = 12$.

x-intercept: $4x = 12$
            $x = 3$

y-intercept: $3y = 12$
            $y = 4$

Checkpoint: Let $x = 2$.
        $8 + 3y = 12$
        $3y = 4$
        $y = \frac{4}{3}$

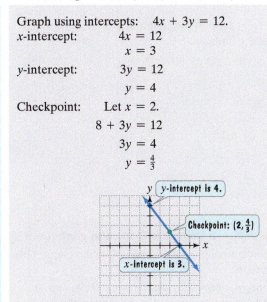

The graph of $Ax + By = 0$ is a line that passes through the origin. Find two other points by finding two other solutions of the equation. Graph the equation by drawing a line through the origin and these two points.

Graph: $x + 2y = 0$.

$x = 2$: $2 + 2y = 0$
        $2y = -2$
        $y = -1$

$y = 1$: $x + 2(1) = 0$
        $x = -2$

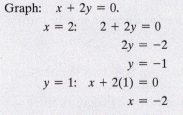

Horizontal and Vertical Lines

The graph of $y = b$ is a horizontal line. The y-intercept is b.
The graph of $x = a$ is a vertical line. The x-intercept is a.

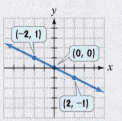

Section 3.3 Slope

The slope, m, of the line through the points (x_1, y_1) and (x_2, y_2) is

$$m = \frac{y_2 - y_1}{x_2 - x_1}, \quad x_2 - x_1 \neq 0$$

If the slope is positive, the line rises from left to right. If the slope is negative, the line falls from left to right.
The slope of a horizontal line is 0. The slope of a vertical line is undefined.
If two distinct nonvertical lines have the same slope, then the lines are parallel.
If the product of the slopes of two lines is -1, then the lines are perpendicular. Equivalently, if the slopes are negative reciprocals, then the lines are perpendicular.
The slope of a line represents its rate of change.

Find the slope of the line passing through the points shown.

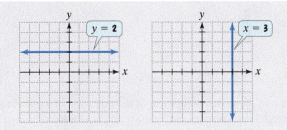

Let $(x_1, y_1) = (-1, 2)$ and $(x_2, y_2) = (2, -2)$.

$$m = \frac{y_2 - y_1}{x_2 - x_1} = \frac{-2 - 2}{2 - (-1)} = \frac{-4}{3} = -\frac{4}{3}$$

Definitions and Concepts	Examples

Section 3.4 The Slope-Intercept Form of the Equation of a Line

The slope-intercept form of the equation of a nonvertical line with slope m and y-intercept b is

$$y = mx + b.$$

Find the slope and the y-intercept of the line with the given equation.

- $y = -2x + 5$

  Slope is -2. y-intercept is 5.

- $2x + 3y = 9$ (Solve for y.)

  $3y = -2x + 9$ Subtract $2x$.

  $y = -\dfrac{2}{3}x + 3$ Divide by 3.

  Slope is $-\frac{2}{3}$. y-intercept is 3.

Graphing $y = mx + b$ Using the Slope and y-Intercept

1. Plot the point containing the y-intercept on the y-axis. This is the point $(0, b)$.
2. Use the slope, m, to obtain a second point. Write m as a fraction, and use rise over run, starting at the point on the y-axis, to plot this point.
3. Graph the equation by drawing a line through the two points.

Graph: $y = -\dfrac{3}{4}x + 1$.

Slope is $-\frac{3}{4}$. y-intercept is 1.

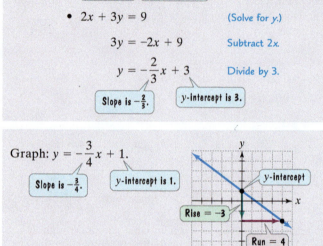

Rise $= -3$ Run $= 4$

Section 3.5 The Point-Slope Form of the Equation of a Line

The point-slope form of the equation of a nonvertical line with slope m that passes through the point (x_1, y_1) is

$$y - y_1 = m(x - x_1).$$

Slope $= -3$, passing through $(-1, 5)$

$m = -3$ $x_1 = -1$ $y_1 = 5$

The point-slope form of the line's equation is

$$y - 5 = -3[x - (-1)].$$

Simplify:

$$y - 5 = -3(x + 1).$$

To write the point-slope form of the equation of a line passing through two points, begin by using the points to compute the slope, m. Use either given point as (x_1, y_1) and write the point-slope form of the equation:

$$y - y_1 = m(x - x_1).$$

Solving this equation for y gives the slope-intercept form of the line's equation.

Write an equation in point-slope form and slope-intercept form of the line passing through $(-1, -3)$ and $(4, 2)$.

$$m = \frac{2 - (-3)}{4 - (-1)} = \frac{2 + 3}{4 + 1} = \frac{5}{5} = 1$$

Using $(4, 2)$ as (x_1, y_1), the point-slope form of the equation is

$$y - 2 = 1(x - 4).$$

Solve for y to obtain the slope-intercept form.

$$y = x - 2 \quad \text{Add 2 to both sides.}$$

Definitions and Concepts	**Examples**

Section 3.6 Linear Inequalities in Two Variables

A linear inequality in two variables can be written in one of the following forms:

$$Ax + By < C \quad Ax + By \leq C$$
$$Ax + By > C \quad Ax + By \geq C.$$

An ordered pair is a solution if replacing the variables by the coordinates of the ordered pair results in a true statement.

Is $(2, -6)$ a solution of $3x - 4y \leq 7$?

$$3 \cdot 2 - 4(-6) \overset{?}{\leq} 7$$
$$6 - (-24) \overset{?}{\leq} 7$$
$$6 + 24 \overset{?}{\leq} 7$$
$$30 \leq 7, \text{ false}$$

Thus, $(2, -6)$ is not a solution.

Graphing a Linear Inequality in Two Variables

1. Replace the inequality symbol with an equal sign and graph the boundary line. Use a solid line for $\leq$ or $\geq$ and a dashed line for $<$ or $>$.

2. Choose a test point not on the line and substitute its coordinates into the inequality.

3. If a true statement results, shade the half-plane containing the test point. If a false statement results, shade the half-plane not containing the test point.

Graph: $x - 2y \leq 4$.

1. Graph $x - 2y = 4$. Use a solid line because the inequality symbol is $\leq$.

2. Test $(0, 0)$.
$$0 - 2 \cdot 0 \overset{?}{\leq} 4$$
$$0 \leq 4, \text{ true}$$

3. The inequality is true. Shade the half-plane containing $(0, 0)$, shown in yellow at the right.

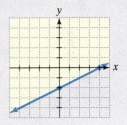

CHAPTER 3 REVIEW EXERCISES

3.1 *In Exercises 1–4, plot the given point in a rectangular coordinate system. Indicate in which quadrant each point lies.*

1. $(1, -5)$

2. $(4, -3)$

3. $\left(\dfrac{7}{2}, \dfrac{3}{2}\right)$

4. $(-5, 2)$

5. Give the ordered pairs that correspond to the points labeled in the figure.

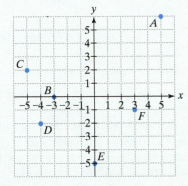

In Exercises 6–7, determine whether each ordered pair is a solution of the given equation.

6. $y = 3x + 6$ $(-3, 3), (0, 6), (1, 9)$

7. $3x - y = 12$ $(0, 4), (4, 0), (-1, 15)$

In Exercises 8–9,

a. *Find five solutions of each equation. Organize your work in a table of values.*

b. *Use the five solutions in the table to graph each equation.*

8. $y = 2x - 3$

9. $y = \frac{1}{2}x + 1$

3.2 *In Exercises 10–12, use the graph to identify the*

a. *x-intercept, or state that there is no x-intercept.*

b. *y-intercept, or state that there is no y-intercept.*

10.

11.

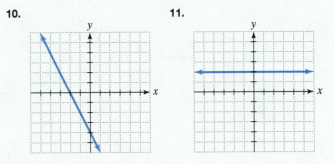

12.

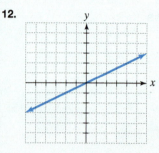

In Exercises 13–16, use intercepts to graph each equation.

13. $2x + y = 4$

14. $3x - 2y = 12$

15. $3x = 6 - 2y$

16. $3x - y = 0$

In Exercises 17–20, graph each equation.

17. $x = 3$

18. $y = -5$

19. $y + 3 = 5$

20. $2x = -8$

21. The graph shows the Fahrenheit temperature, y, x hours after noon.

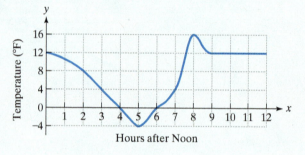

a. At what time did the minimum temperature occur? What is the minimum temperature?

b. At what time did the maximum temperature occur? What is the maximum temperature?

c. What are the x-intercepts? In terms of time and temperature, interpret the meaning of these intercepts.

d. What is the y-intercept? What does this mean in terms of time and temperature?

e. From 9 P.M. until midnight, the graph is shown as a horizontal line. What does this mean about the temperature over this period of time?

3.3 *In Exercises 22–25, calculate the slope of the line passing through the given points. If the slope is undefined, so state. Then indicate whether the line rises, falls, is horizontal, or is vertical.*

22. $(3, 2)$ and $(5, 1)$

23. $(-1, 2)$ and $(-3, -4)$

24. $(-3, 4)$ and $(6, 4)$

25. $(5, 3)$ and $(5, -3)$

In Exercises 26–29, find the slope of each line, or state that the slope is undefined.

26. **27.**

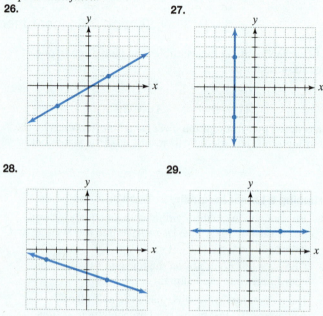

28. **29.**

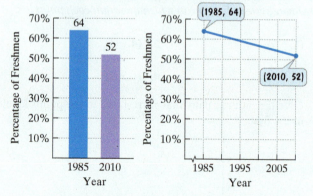

In Exercises 30–32, determine whether the lines through each pair of points are parallel, perpendicular, or neither.

30. $(-1, -3)$ and $(2, -8)$; $(8, -7)$ and $(9, 10)$

31. $(0, -4)$ and $(5, -1)$; $(-6, 8)$ and $(3, -7)$

32. $(5, 4)$ and $(9, 7)$; $(-6, 0)$ and $(-2, 3)$

33. In a 2010 survey of more than 200,000 freshmen at 279 colleges, only 52% rated their emotional health high or above average, a drop from 64% in 1985.

Percentage of U.S. College Freshmen Rating Their Emotional Health High or Above Average

Source: UCLA Higher Education Research Institute

a. Find the slope of the line passing through the two points shown by the voice ballons.

b. Use your answer from part (a) to complete this statement:
For each year from 1985 through 2010, the percentage of U.S. college freshmen rating their emotional health high or above average decreased by _____. The rate of change was _____ per _____.

3.4 *In Exercises 34–37, find the slope and the y-intercept of the line with the given equation.*

34. $y = 5x - 7$

35. $y = 6 - 4x$

36. $y = 3$

37. $2x + 3y = 6$

In Exercises 38–40, graph each linear equation using the slope and y-intercept.

38. $y = 2x - 4$

39. $y = \dfrac{1}{2}x - 1$

40. $y = -\dfrac{2}{3}x + 5$

In Exercises 41–42, write each equation in slope-intercept form. Then use the slope and y-intercept to graph the equation.

41. $y - 2x = 0$

42. $\dfrac{1}{3}x + y = 2$

43. Graph $y = -\dfrac{1}{2}x + 4$ and $y = -\dfrac{1}{2}x - 1$ in the same rectangular coordinate system. Are the lines parallel? If so, explain why.

44. The graphs show the percentage of Americans ages 18–25 who used marijuana regularly from 2008 through 2013.

Marijuana Use among 18- to 25-Year-Olds in the United States

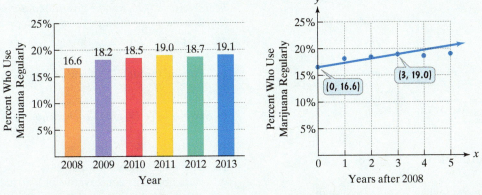

Source: National Survey on Drug Use and Health

a. Use the two points labeled by the voice balloons to find an equation in the form $y = mx + b$ that models the percentage of Americans ages 18–25 who used marijuana regularly, y, x years after 2008.

b. If trends shown by the graphs continue, use the model in part (a) to project the percentage of Americans ages 18–25 who will use marijuana regularly in 2018.

3.5 *Write the point-slope form of the equation of the line satisfying the conditions in Exercises 45–46. Then use the point-slope form of the equation to write the slope-intercept form.*

45. Slope = 6, passing through $(-4, 7)$

46. Passing through $(3, 4)$ and $(2, 1)$

47. The bar graph shows world population, in billions, for seven selected years from 1950 through 2010. Also shown is a scatter plot with a line passing through two of the data points.

World Population, 1950–2010

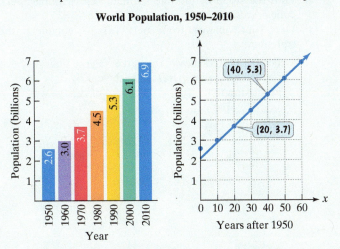

Source: U.S. Census Bureau, International Database

a. Use the two points whose coordinates are shown by the voice balloons to write the slope-intercept form of an equation that models world population, y, in billions, x years after 1950.

b. Use the model from part (a) to project world population in 2025.

3.6

48. Determine whether each ordered pair is a solution of $3x - 4y > 7$:

$$(0, 0), \ (3, -6), \ (-2, -5), \ (-3, 4).$$

In Exercises 49–56, graph each inequality.

49. $x - 2y > 6$

50. $4x - 6y \le 12$

51. $y > 3x + 2$

52. $y \le \dfrac{1}{3}x - 1$

53. $y < -\dfrac{1}{2}x$

54. $x < 4$

55. $y \ge -2$

56. $x + 2y \le 0$

1. Determine whether each ordered pair is a solution of $4x - 2y = 10$:
$$(0, -5), \quad (-2, 1), \quad (4, 3).$$

2. Find five solutions of $y = 3x + 1$. Organize your work in a table of values. Then use the five solutions in the table to graph the equation.

3. Use the graph to identify the
   a. x-intercept, or state that there is no x-intercept.
   b. y-intercept, or state that there is no y-intercept.

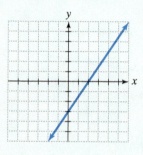

4. Use intercepts to graph $4x - 2y = -8$.

5. Graph $y = 4$ in a rectangular coordinate system.

In Exercises 6–7, calculate the slope of the line passing through the given points. If the slope is undefined, so state. Then indicate whether the line rises, falls, is horizontal, or is vertical.

6. $(-3, 4)$ and $(-5, -2)$

7. $(6, -1)$ and $(6, 3)$

8. Find the slope of the line in the figure shown or state that the slope is undefined.

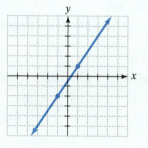

In Exercises 9–10 determine whether the lines through each pair of points are parallel, perpendicular, or neither.

9. $(-2, 10)$ and $(0, 2)$; $(-8, -7)$ and $(24, 1)$

10. $(2, 4)$ and $(6, 1)$; $(-3, 1)$ and $(1, -2)$

In Exercises 11–12, find the slope and the y-intercept of the line with the given equation.

11. $y = -x + 10$

12. $2x + y = 6$

In Exercises 13–14, graph each linear equation using the slope and y-intercept.

13. $y = \dfrac{2}{3}x - 1$

14. $y = -2x + 3$

In Exercises 15–16, use the given conditions to write an equation for each line in point-slope form and slope-intercept form.

15. Slope $= -2$, passing through $(-1, 4)$

16. Passing through $(2, 1)$ and $(-1, -8)$

In Exercises 17–19, graph each inequality.

17. $3x - 2y < 6$ 18. $y \geq 2x - 2$ 19. $x > -1$

20. The graph shows the percentage of college freshmen who were liberal or conservative in 1980 and in 2010.

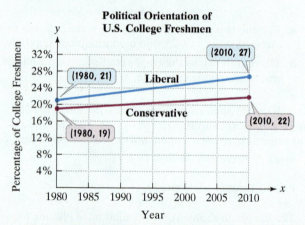

Political Orientation of U.S. College Freshmen

Source: The Higher Education Research Institute

   a. Find the slope of the line segment representing liberal college freshmen.
   b. Use your answer from part (a) to complete this statement:

   For the period shown, the percentage of liberal college freshmen increased each year by approximately _____. The rate of change was _____ per _____.

CUMULATIVE REVIEW EXERCISES   (CHAPTERS 1–3)

1. Perform the indicated operations:
$$\frac{10 - (-6)}{3^2 - (4 - 3)}.$$

2. Simplify: $6 - 2[3(x - 1) + 4]$.

3. List all the irrational numbers in this set:
$\left\{-3, 0, 1, \sqrt{4}, \sqrt{5}, \frac{11}{2}\right\}$.

In Exercises 4–5, solve each equation.

4. $6(2x - 1) - 6 = 11x + 7$

5. $x - \dfrac{3}{4} = \dfrac{1}{2}$

6. Solve for x: $y = mx + b$.

7. 120 is 15% of what number?

8. The formula $y = 4.5x - 46.7$ models the stopping distance, y, in feet, for a car traveling x miles per hour. If the stopping distance is 133.3 feet, how fast was the car traveling?

In Exercises 9–10, solve each inequality and graph the solution set on a number line.

9. $2 - 6x \geq 2(5 - x)$

10. $6(2 - x) > 12$

11. A plumber charged a customer $228, listing $18 for parts and the remainder for labor. If the cost of the labor is $35 per hour, how many hours did the plumber work?

12. The length of a rectangular football field is 14 meters more than twice the width. If the perimeter is 346 meters, find the field's dimensions.

13. After a 10% weight loss, a person weighed 180 pounds. What was the weight before the loss?

14. In a triangle, the measure of the second angle is 20° greater than the measure of the first angle. The measure of the third angle is twice that of the first. What is the measure of each angle?

15. Evaluate $x^2 - 10x$ for $x = -3$.

In Exercises 16–20, graph each equation or inequality in the rectangular coordinate system.

16. $2x - y = 4$

17. $x = -5$

18. $y = -4x + 3$

19. $3x - 2y < -6$

20. $y \geq -1$

Systems of Linear Equations and Inequalities

Television, movies, and magazines place great emphasis on physical beauty. Our culture emphasizes physical appearance to such an extent that it is a central factor in the perception and judgment of others. The modern emphasis on thinness as the ideal body shape has been suggested as a major cause of eating disorders among adolescent women.

What is the relationship between our changing cultural values of physical attractiveness and undernutrition? Given the importance of culture in setting standards of attractiveness, how can you establish a healthy weight range for your age and height? In this chapter, we will use systems of linear equations and inequalities to explore these skin-deep issues.

Here's where you'll find this application:

- The Blitzer Bonus on page 290 contains a graphical look at the shifting shape of that icon of American beauty, Miss America.
- You'll find a weight that fits you using the models (mathematical, not fashion) in Example 1 of Section 4.5 and Exercises 45–48 in Exercise Set 4.5.

4.1

Solving Systems of Linear Equations by Graphing

What am I supposed to learn?

After studying this section, you should be able to:

1 Decide whether an ordered pair is a solution of a linear system.

2 Solve systems of linear equations by graphing.

3 Use graphing to identify systems with no solution or infinitely many solutions.

4 Use graphs of linear systems to solve problems.

1 Decide whether an ordered pair is a solution of a linear system.

Visitors to the world's great bridges are frequently inspired by their beauty. This feeling is not always shared by daily commuters dealing with escalating toll costs and peak-hour congestion. For these frequent users, most bridge authorities provide the option of a fixed cost that reduces the toll. In this section, you will see how toll options in these situations can be modeled by two linear equations in two variables and their graphs.

Systems of Linear Equations and Their Solutions

We have seen that all equations in the form $Ax + By = C$ are straight lines when graphed. Two such equations are called a **system of linear equations** or a **linear system**. A **solution to a system of two linear equations in two variables** is an ordered pair that satisfies both equations in the system. For example, (3, 4) satisfies the system

$$\begin{cases} x + y = 7 & \text{(3 + 4 is, indeed, 7.)} \\ x - y = -1. & \text{(3 - 4 is, indeed, -1.)} \end{cases}$$

Thus, (3, 4) satisfies both equations and is a solution of the system. The solution can be described by saying that $x = 3$ and $y = 4$. The solution can also be described using the ordered pair (3, 4). The solution set of the system is {(3, 4)}—that is, the set consisting of the ordered pair (3, 4).

A system of linear equations can have exactly one solution, no solution, or infinitely many solutions. We begin with systems having exactly one solution.

EXAMPLE 1   Determining Whether Ordered Pairs Are Solutions of a System

Consider the system:

$$\begin{cases} x + 2y = 2 \\ x - 2y = 6. \end{cases}$$

Determine if each ordered pair is a solution of the system:

a. $(4, -1)$ **b.** $(-4, 3)$.

Solution

a. We begin by determining whether $(4, -1)$ is a solution. Because 4 is the x-coordinate and -1 is the y-coordinate of $(4, -1)$, we replace x with 4 and y with -1.

$$x + 2y = 2$$
$$4 + 2(-1) \stackrel{?}{=} 2$$
$$4 + (-2) \stackrel{?}{=} 2$$
$$2 = 2, \quad \text{true}$$

$$x - 2y = 6$$
$$4 - 2(-1) \stackrel{?}{=} 6$$
$$4 - (-2) \stackrel{?}{=} 6$$
$$4 + 2 \stackrel{?}{=} 6$$
$$6 = 6, \quad \text{true}$$

The pair $(4, -1)$ satisfies both equations: It makes each equation true. Thus, the ordered pair is a solution of the system.

b. To determine whether $(-4, 3)$ is a solution, we replace x with -4 and y with 3.

$$x + 2y = 2$$
$$-4 + 2 \cdot 3 \stackrel{?}{=} 2$$
$$-4 + 6 \stackrel{?}{=} 2$$
$$2 = 2, \quad \text{true}$$

$$x - 2y = 6$$
$$-4 - 2 \cdot 3 \stackrel{?}{=} 6$$
$$-4 - 6 \stackrel{?}{=} 6$$
$$-10 = 6, \quad \text{false}$$

The pair $(-4, 3)$ fails to satisfy *both* equations: It does not make both equations true. Thus, the ordered pair is not a solution of the system. ■

✓ **CHECK POINT 1** Consider the system:

$$\begin{cases} 2x - 3y = -4 \\ 2x + y = 4. \end{cases}$$

Determine if each ordered pair is a solution of the system:

a. $(1, 2)$ **b.** $(7, 6)$.

---

2 Solve systems of linear equations by graphing. ▶

Solving Linear Systems by Graphing

The solution of a system of two linear equations in two variables can be found by graphing both of the equations in the same rectangular coordinate system. For a system with one solution, **the coordinates of the point of intersection give the system's solution.** For example, the system in Example 1,

$$\begin{cases} x + 2y = 2 \\ x - 2y = 6 \end{cases}$$

is graphed in **Figure 4.1**. The solution of the system, $(4, -1)$, corresponds to the point of intersection of the lines.

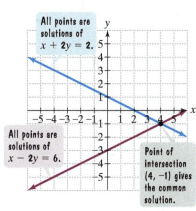

Figure 4.1 Visualizing a system's solution

Solving Systems of Two Linear Equations in Two Variables, *x* and *y*, by Graphing

1. Graph the first equation.
2. Graph the second equation on the same set of axes.
3. If the lines representing the two graphs intersect at a point, determine the coordinates of this point of intersection. The ordered pair is the solution of the system.
4. Check the solution in both equations.

EXAMPLE 2 Solving a Linear System by Graphing

Solve by graphing:

$$\begin{cases} 2x + 3y = 6 \\ 2x + y = -2. \end{cases}$$

Solution

Step 1. Graph the first equation. We use intercepts to graph $2x + 3y = 6$.

x-intercept (Set $y = 0$.)	**y-intercept** (Set $x = 0$.)
$2x + 3 \cdot 0 = 6$	$2 \cdot 0 + 3y = 6$
$2x = 6$	$3y = 6$
$x = 3$	$y = 2$

The x-intercept is 3, so the line passes through $(3, 0)$. The y-intercept is 2, so the line passes through $(0, 2)$. The graph of $2x + 3y = 6$ is shown as the red line in **Figure 4.2**.

Step 2. Graph the second equation on the same axes. We use intercepts to graph $2x + y = -2$.

x-intercept (Set $y = 0$.)	**y-intercept** (Set $x = 0$.)
$2x + 0 = -2$	$2 \cdot 0 + y = -2$
$2x = -2$	$y = -2$
$x = -1$	

The x-intercept is -1, so the line passes through $(-1, 0)$. The y-intercept is -2, so the line passes through $(0, -2)$. The graph of $2x + y = -2$ is shown as the blue line in **Figure 4.2**.

Step 3. Determine the coordinates of the intersection point. This ordered pair is the system's solution. Using **Figure 4.2**, it appears that the lines intersect at $(-3, 4)$. The "apparent" solution of the system is $(-3, 4)$.

Step 4. Check the solution in both equations.

Check $(-3, 4)$ in $2x + 3y = 6$:	Check $(-3, 4)$ in $2x + y = -2$:
$2(-3) + 3 \cdot 4 \stackrel{?}{=} 6$	$2(-3) + 4 \stackrel{?}{=} -2$
$-6 + 12 \stackrel{?}{=} 6$	$-6 + 4 \stackrel{?}{=} -2$
$6 = 6$, true	$-2 = -2$, true

Because both equations are satisfied, $(-3, 4)$ is the solution of the system and $\{(-3, 4)\}$ is the solution set. ■

☑ **CHECK POINT 2** Solve by graphing:

$$\begin{cases} 2x + y = 6 \\ 2x - y = -2. \end{cases}$$

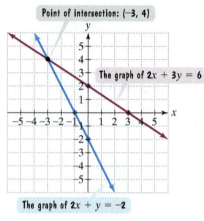

Point of intersection: (−3, 4)

The graph of 2x + 3y = 6

The graph of 2x + y = −2

Figure 4.2

Discover for Yourself

Must two lines intersect at exactly one point? Sketch two lines that have less than one intersection point. Now sketch two lines that have more than one intersection point. What does this say about each of these systems?

Great Question!

Can I use a rough sketch on scratch paper to solve a linear system by graphing?

No. When solving linear systems by graphing, neatly drawn graphs are essential for determining points of intersection.

- Use rectangular coordinate graph paper.
- Use a ruler or straightedge.
- Use a pencil with a sharp point.

EXAMPLE 3 Solving a Linear System by Graphing

Solve by graphing:

$$\begin{cases} y = -3x + 2 \\ y = 5x - 6. \end{cases}$$

Solution Each equation is in the form $y = mx + b$. Thus, we use the y-intercept, b, and the slope, m, to graph each line.

Step 1. Graph the first equation.

$$y = -3x + 2$$

The slope is -3. The y-intercept is 2.

The y-intercept is 2, so the line passes through $(0, 2)$. The slope is $-\frac{3}{1}$. Start at the y-intercept and move 3 units down (the rise) and 1 unit to the right (the run). The graph of $y = -3x + 2$ is shown as the red line in **Figure 4.3**.

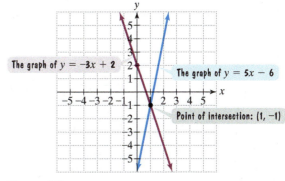

The graph of $y = -3x + 2$

The graph of $y = 5x - 6$

Point of intersection: $(1, -1)$

Figure 4.3

Step 2. Graph the second equation on the same axes.

$$y = 5x - 6$$

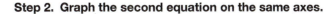

The slope is 5. The y-intercept is -6.

The y-intercept is -6, so the line passes through $(0, -6)$. The slope is $\frac{5}{1}$. Start at the y-intercept and move 5 units up (the rise) and 1 unit to the right (the run). The graph of $y = 5x - 6$ is shown as the blue line in **Figure 4.3**.

Step 3. Determine the coordinates of the intersection point. This ordered pair is the system's solution. Using **Figure 4.3**, it appears that the lines intersect at $(1, -1)$. The "apparent" solution of the system is $(1, -1)$.

Step 4. Check the solution in both equations.

Check $(1, -1)$ in
$y = -3x + 2$:
$-1 \stackrel{?}{=} -3 \cdot 1 + 2$
$-1 \stackrel{?}{=} -3 + 2$
$-1 = -1$,   true

Check $(1, -1)$ in
$y = 5x - 6$:
$-1 \stackrel{?}{=} 5 \cdot 1 - 6$
$-1 \stackrel{?}{=} 5 - 6$
$-1 = -1$,   true

Because both equations are satisfied, $(1, -1)$ is the solution and the solution set is $\{(1, -1)\}$. ■

☑ **CHECK POINT 3** Solve by graphing:

$$\begin{cases} y = -x + 6 \\ y = 3x - 6. \end{cases}$$

3 Use graphing to identify systems with no solution or infinitely many solutions. ▶

Linear Systems Having No Solution or Infinitely Many Solutions

We have seen that a system of linear equations in two variables represents a pair of lines. The lines either intersect at one point, are parallel, or are identical. Thus, there are three possibilities for the number of solutions to a system of two linear equations.

The Number of Solutions to a System of Two Linear Equations

The number of solutions to a system of two linear equations in two variables is given by one of the following. (See **Figure 4.4**.)

Number of Solutions	What This Means Graphically
Exactly one ordered-pair solution	The two lines intersect at one point.
No solution	The two lines are parallel.
Infinitely many solutions	The two lines are identical.

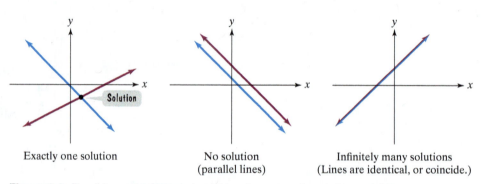

Exactly one solution   No solution (parallel lines)   Infinitely many solutions (Lines are identical, or coincide.)

Figure 4.4 Possible graphs for a system of two linear equations in two variables

A linear system with no solution is called an **inconsistent system**. If you attempt to solve such a system by graphing, you will obtain two parallel lines. The solution set is the empty set, $\varnothing$.

EXAMPLE 4   A System with No Solution

Solve by graphing:
$$\begin{cases} y = 2x - 1 \\ y = 2x + 3. \end{cases}$$

Solution   Compare the slopes and y-intercepts in the two equations.

> The lines have the same slope, **2**.

$$y = 2x - 1 \qquad y = 2x + 3$$

> The lines have different y-intercepts, **−1** and **3**.

Figure 4.5 shows the graphs of the two equations. Because both equations have the same slope, 2, but different y-intercepts, the lines are parallel. The system is inconsistent and has no solution. The solution set is the empty set, $\varnothing$. ∎

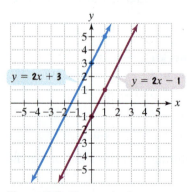

Figure 4.5 The graph of an inconsistent system

✓ **CHECK POINT 4**   Solve by graphing:
$$\begin{cases} y = 3x - 2 \\ y = 3x + 1. \end{cases}$$

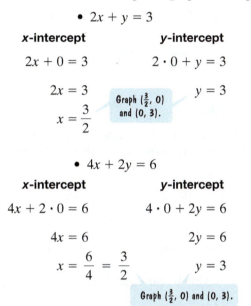

EXAMPLE 5   A System with Infinitely Many Solutions

Solve by graphing:
$$\begin{cases} 2x + y = 3 \\ 4x + 2y = 6. \end{cases}$$

Solution   We use intercepts to graph each equation.

- $2x + y = 3$

x-intercept

$2x + 0 = 3$

$2x = 3$

$x = \dfrac{3}{2}$

> Graph $\left(\frac{3}{2}, 0\right)$ and $(0, 3)$.

y-intercept

$2 \cdot 0 + y = 3$

$y = 3$

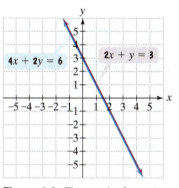

- $4x + 2y = 6$

x-intercept

$4x + 2 \cdot 0 = 6$

$4x = 6$

$x = \dfrac{6}{4} = \dfrac{3}{2}$

y-intercept

$4 \cdot 0 + 2y = 6$

$2y = 6$

$y = 3$

> Graph $\left(\frac{3}{2}, 0\right)$ and $(0, 3)$.

Figure 4.6   The graph of a system with infinitely many solutions

Both lines have the same x-intercept, $\frac{3}{2}$ or 1.5, and the same y-intercept, 3. Thus, the graphs of the two equations in the system are the same line, shown in **Figure 4.6**. The two equations have the same solutions. Any ordered pair that is a solution of one equation is a solution of the other, and, consequently, a solution of the system. The system has an infinite number of solutions, namely all points that are solutions of either line.

We express the solution set for the system in one of two equivalent ways:

$$\{(x, y) \mid 2x + y = 3\} \quad \text{or} \quad \{(x, y) \mid 4x + 2y = 6\}.$$

> The set of all ordered pairs (x, y) such that $2x + y = 3$

> The set of all ordered pairs (x, y) such that $4x + 2y = 6$ ■

Great Question!

The system in Example 5 has infinitely many solutions. Does that mean that any ordered pair of numbers is a solution?

No. Although the system in Example 5 has infinitely many solutions, this does not mean that any ordered pair of numbers you can form will be a solution. The ordered pair (x, y) must satisfy one of the system's equations, $2x + y = 3$ or $4x + 2y = 6$, and there are infinitely many such ordered pairs. Because the graphs are coinciding lines, the ordered pairs that are solutions of one of the equations are also solutions of the other equation.

Take a second look at the two equations, $2x + y = 3$ and $4x + 2y = 6$, in Example 5. If you multiply both sides of the first equation, $2x + y = 3$, by 2, you will obtain the second equation, $4x + 2y = 6$.

$2x + y = 3$	This is the first equation in the system.
$2(2x + y) = 2 \cdot 3$	Multiply both sides by 2.
$2 \cdot 2x + 2y = 2 \cdot 3$	Use the distributive property.
$4x + 2y = 6.$	Simplify.

> This is the second equation in the system.

Because $2x + y = 3$ and $4x + 2y = 6$ are different forms of the same equation, these equations are called dependent equations. In general, the equations in a linear system with infinitely many solutions are called **dependent equations**.

☑ **CHECK POINT 5** Solve by graphing:

$$\begin{cases} x + y = 3 \\ 2x + 2y = 6. \end{cases}$$

4 Use graphs of linear systems to solve problems. ▶

Applications

One advantage of solving linear systems by graphing is that it allows us to "see" the equations and visualize the solution. This idea is demonstrated in Example 6.

EXAMPLE 6 Modeling Options for a Toll

The toll to a bridge costs \$2.50. Commuters who use the bridge frequently have the option of purchasing a monthly discount pass for \$21.00. With the discount pass, the toll is reduced to \$1.00. The monthly cost, y, of using the bridge x times can be modeled by the following linear equations:

Without the discount pass:

$$y = 2.50x \qquad \text{The monthly cost, } y, \text{ is } \$2.50 \text{ times the number of times, } x, \text{ that the bridge is used.}$$

With the discount pass:

$$y = 21 + 1 \cdot x \qquad \text{The monthly cost, } y, \text{ is } \$21 \text{ for the pass plus } \$1 \text{ times the number of times, } x, \text{ that the bridge is used.}$$

$$y = 21 + x \qquad \text{Simplify: } 1 \cdot x = x.$$

Expressing 2.50 as $\frac{5}{2}$, the options for the toll can be modeled by the linear system

$$\begin{cases} y = \dfrac{5}{2}x & \text{Without discount pass} \\ y = x + 21. & \text{With discount pass} \end{cases}$$

a. Use graphing to solve the system.
b. Interpret the coordinates of the solution in practical terms.

Solution

a. Notice that x, the number of times the bridge is used in a month, and y, the monthly cost, are nonnegative. Thus, we will use only the first quadrant and its boundary to graph the system. We begin by identifying slopes and y-intercepts.

$$y = \frac{5}{2}x + 0 \qquad\qquad y = x + 21$$

The slope is $\frac{5}{2}$. The y-intercept is 0.   The slope is 1. The y-intercept is 21.

Using slopes and y-intercepts, we graph the two lines, as shown in **Figure 4.7**.

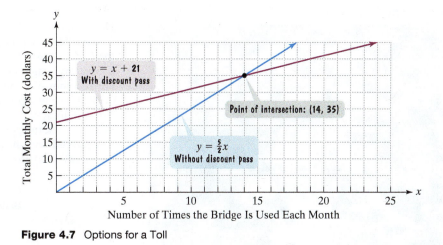

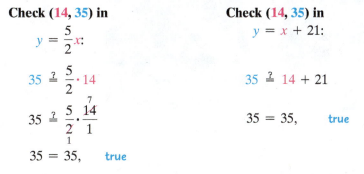

Figure 4.7 Options for a Toll

Using **Figure 4.7**, it appears that the lines intersect at $(14, 35)$. Let's check this solution in both equations.

Check (14, 35) in	**Check (14, 35) in**

$$\text{Check } (14, 35) \text{ in } \quad y = \frac{5}{2}x:$$
$$35 \stackrel{?}{=} \frac{5}{2} \cdot 14$$
$$35 \stackrel{?}{=} \frac{5}{2} \cdot \frac{\overset{7}{\cancel{14}}}{\underset{1}{1}}$$
$$35 = 35, \qquad \text{true}$$

$$\text{Check } (14, 35) \text{ in } \quad y = x + 21:$$
$$35 \stackrel{?}{=} 14 + 21$$
$$35 = 35, \qquad \text{true}$$

Because both equations are satisfied, $(14, 35)$ is the solution and the solution set is $\{(14, 35)\}$.

b. The graphs intersect at $(14, 35)$. This means that if the bridge is used 14 times in a month, the total monthly cost without the discount pass is the same as the total monthly cost with the discount pass, namely $35. ■

In **Figure 4.7**, look at the two graphs to the right of the intersection point $(14, 35)$. The red graph of $y = x + 21$ lies below the blue graph of $y = \frac{5}{2}x$. This means that if the bridge is used more than 14 times in a month $(x > 14)$, the (red) monthly cost, y, with the discount pass is cheaper than the (blue) monthly cost, y, without the discount pass.

✓ **CHECK POINT 6**   The toll to a bridge costs $2.00. If you use the bridge x times in a month, the monthly cost, y, is $y = 2x$. With a $10 discount pass, the toll is reduced to $1.00. The monthly cost, y, of using the bridge x times in a month with the discount pass is $y = x + 10$.

a. Solve the system by graphing:

$$\begin{cases} y = 2x \\ y = x + 10. \end{cases}$$

Suggestion: Let the x-axis extend from 0 to 20 and let the y-axis extend from 0 to 40.

b. Interpret the coordinates of the solution in practical terms.

Blitzer Bonus ▶

Missing America

Norma Smallwood, Miss America 1926

Here she is, Miss America, the icon of American beauty. Always thin, she is becoming more so. The scatter plot in the figure shows Miss America's body-mass index, a ratio comparing weight divided by the square of height. Two lines are also shown: a line that passes near the data points and a horizontal line representing the World Health Organization's cutoff point for undernutrition. The intersection point indicates that in approximately 1978, Miss America reached this cutoff. There she goes: If the trend continues, Miss America's body-mass index could reach zero in about 320 years.

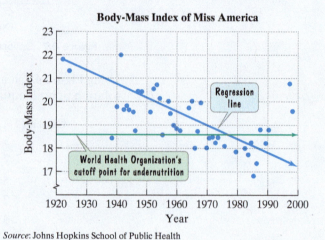

Source: Johns Hopkins School of Public Health

Achieving Success

Learn from your mistakes. Being human means making mistakes. By finding and understanding your errors, you will become a better math student.

Source of Error	Remedy
Not Understanding a Concept	Review the concept by finding a similar example in your textbook or class notes. Ask your professor questions to help clarify the concept.
Skipping Steps	Show clear step-by-step solutions. Detailed solution procedures help organize your thoughts and enhance understanding. Doing too many steps mentally often results in preventable mistakes.
Carelessness	Write neatly. Not being able to read your own math writing leads to errors. Avoid writing in pen so you won't have to put huge marks through incorrect work.

"You can achieve your goal if you persistently pursue it."

—Cha Sa-Soon, a 68-year-old South Korean woman who passed her country's written driver's-license exam on her 950th try (*Source: Newsweek*)

CONCEPT AND VOCABULARY CHECK

Fill in each blank so that the resulting statement is true.

1. A solution to a system of linear equations in two variables is an ordered pair that _____.

2. When solving a system of linear equations by graphing, the system's solution is determined by using _____.

3. A system of two linear equations that has no solution is called a/an _____ system. If you attempt to solve such a system by graphing, you will obtain two lines that are _____.

4. The equations in a system of two linear equations with infinitely many solutions are called _____ equations. If you attempt to solve such a system by graphing, you will obtain two lines that _____.

4.1 EXERCISE SET ▶ MyMathLab®

Practice Exercises

In Exercises 1–10, determine whether the given ordered pair is a solution of the system.

1. $(2, -3)$
$$\begin{cases} 2x + 3y = -5 \\ 7x - 3y = 23 \end{cases}$$

2. $(-2, -5)$
$$\begin{cases} 6x - 2y = -2 \\ 3x + y = -11 \end{cases}$$

3. $\left(\dfrac{2}{3}, \dfrac{1}{9}\right)$
$$\begin{cases} x + 3y = 1 \\ 4x + 3y = 3 \end{cases}$$

4. $\left(\dfrac{7}{25}, -\dfrac{1}{25}\right)$
$$\begin{cases} 4x + 3y = 1 \\ 3x - 4y = 1 \end{cases}$$

5. $(-5, 9)$
$$\begin{cases} 5x + 3y = 2 \\ x + 4y = 14 \end{cases}$$

6. $(10, 7)$
$$\begin{cases} 6x - 5y = 25 \\ 4x + 15y = 13 \end{cases}$$

7. $(1400, 450)$
$$\begin{cases} x - 2y = 500 \\ 0.03x + 0.02y = 51 \end{cases}$$

8. $(200, 700)$
$$\begin{cases} -4x + y = -100 \\ 0.05x - 0.06y = -32 \end{cases}$$

9. $(8, 5)$
$$\begin{cases} 5x - 4y = 20 \\ 3y = 2x + 1 \end{cases}$$

10. $(5, -2)$
$$\begin{cases} 4x - 3y = 26 \\ x = 15 - 5y \end{cases}$$

In Exercises 11–42, solve each system by graphing. If there is no solution or an infinite number of solutions, so state. Use set notation to express solution sets.

11. $\begin{cases} x + y = 6 \\ x - y = 2 \end{cases}$

12. $\begin{cases} x + y = 2 \\ x - y = 4 \end{cases}$

13. $\begin{cases} x + y = 1 \\ y - x = 3 \end{cases}$

14. $\begin{cases} x + y = 4 \\ y - x = 4 \end{cases}$

15. $\begin{cases} 2x - 3y = 6 \\ 4x + 3y = 12 \end{cases}$

16. $\begin{cases} x + 2y = 2 \\ x - y = 2 \end{cases}$

17. $\begin{cases} 4x + y = 4 \\ 3x - y = 3 \end{cases}$

18. $\begin{cases} 5x - y = 10 \\ 2x + y = 4 \end{cases}$

19. $\begin{cases} y = x + 5 \\ y = -x + 3 \end{cases}$

20. $\begin{cases} y = x + 1 \\ y = 3x - 1 \end{cases}$

21. $\begin{cases} y = 2x \\ y = -x + 6 \end{cases}$

22. $\begin{cases} y = 2x + 1 \\ y = -2x - 3 \end{cases}$

23. $\begin{cases} y = -2x + 3 \\ y = -x + 1 \end{cases}$

24. $\begin{cases} y = 3x - 4 \\ y = -2x + 1 \end{cases}$

25. $\begin{cases} y = 2x - 1 \\ y = 2x + 1 \end{cases}$

26. $\begin{cases} y = 3x - 1 \\ y = 3x + 2 \end{cases}$

27. $\begin{cases} x + y = 4 \\ x = -2 \end{cases}$

28. $\begin{cases} x + y = 6 \\ y = -3 \end{cases}$

29. $\begin{cases} x - 2y = 4 \\ 2x - 4y = 8 \end{cases}$

30. $\begin{cases} 2x + 3y = 6 \\ 4x + 6y = 12 \end{cases}$

31. $\begin{cases} y = 2x - 1 \\ x - 2y = -4 \end{cases}$

32. $\begin{cases} y = -2x - 4 \\ 4x - 2y = 8 \end{cases}$

33. $\begin{cases} x + y = 5 \\ 2x + 2y = 12 \end{cases}$

34. $\begin{cases} x - y = 2 \\ 3x - 3y = -6 \end{cases}$

35. $\begin{cases} x - y = 0 \\ y = x \end{cases}$

36. $\begin{cases} 2x - y = 0 \\ y = 2x \end{cases}$

37. $\begin{cases} x = 2 \\ y = 4 \end{cases}$

38. $\begin{cases} x = 3 \\ y = 5 \end{cases}$

39. $\begin{cases} x = 2 \\ x = -1 \end{cases}$

40. $\begin{cases} x = 3 \\ x = -2 \end{cases}$

41. $\begin{cases} y = 0 \\ y = 4 \end{cases}$

42. $\begin{cases} y = 0 \\ y = 5 \end{cases}$

Practice PLUS

In Exercises 43–50, find the slope and the y-intercept for the graph of each equation in the given system. Use this information (and not the equations' graphs) to determine if the system has no solution, one solution, or an infinite number of solutions.

43. $\begin{cases} y = \dfrac{1}{2}x - 3 \\ y = \dfrac{1}{2}x - 5 \end{cases}$

44. $\begin{cases} y = \dfrac{3}{4}x - 2 \\ y = \dfrac{3}{4}x + 1 \end{cases}$

45. $\begin{cases} y = -\dfrac{1}{2}x + 4 \\ 3x - y = -4 \end{cases}$

46. $\begin{cases} y = -\dfrac{1}{4}x + 3 \\ 4x - y = -3 \end{cases}$

47. $\begin{cases} 3x - y = 6 \\ x = \dfrac{y}{3} + 2 \end{cases}$

48. $\begin{cases} 2x - y = 4 \\ x = \dfrac{y}{2} + 2 \end{cases}$

49. $\begin{cases} 3x + y = 0 \\ y = -3x + 1 \end{cases}$

50. $\begin{cases} 2x + y = 0 \\ y = -2x + 1 \end{cases}$

Application Exercises

51. A rental company charges $40.00 a day plus $0.35 per mile to rent a moving truck. The total cost, y, for a day's rental if x miles are driven is described by $y = 0.35x + 40$. A second company charges $36.00 a day plus $0.45 per mile, so the daily cost, y, if x miles are driven is described by $y = 0.45x + 36$. The graphs of the two equations are shown in the same rectangular coordinate system.

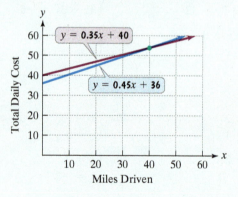

a. What is the x-coordinate of the intersection point of the graphs? Describe what this x-coordinate means in practical terms.

b. What is a reasonable estimate for the y-coordinate of the intersection point?

c. Substitute the x-coordinate of the intersection point into each of the equations and find the corresponding value for y. Describe what this value represents in practical terms. How close is this value to your estimate from part (b)?

52. A band plans to record a demo. Studio A rents for $100 plus $50 per hour. Studio B rents for $50 plus $75 per hour. The total cost, y, in dollars, of renting the studios for x hours can be modeled by the linear system

$\begin{cases} y = 50x + 100 \quad \boxed{\text{Studio A}} \\ y = 75x + 50. \quad \boxed{\text{Studio B}} \end{cases}$

a. Use graphing to solve the system. Extend the x-axis from 0 to 4 and let each tick mark represent 1 unit (one hour in a recording studio). Extend the y-axis from 0 to 400 and let each tick mark represent 100 units (a rental cost of $100).

b. Interpret the coordinates of the solution in practical terms.

53. You plan to start taking an aerobics class. Nonmembers pay $4 per class. Members pay a $10 monthly fee plus an additional $2 per class. The monthly cost, y, of taking x classes can be modeled by the linear system

$\begin{cases} y = 4x \quad \boxed{\text{Nonmembers}} \\ y = 2x + 10. \quad \boxed{\text{Members}} \end{cases}$

a. Use graphing to solve the system.

b. Interpret the coordinates of the solution in practical terms.

Explaining the Concepts

54. What is a system of linear equations? Provide an example with your description.

55. What is a solution of a system of linear equations?

56. Explain how to determine if an ordered pair is a solution of a system of linear equations.

57. Explain how to solve a system of linear equations by graphing.

58. What is an inconsistent system? What happens if you attempt to solve such a system by graphing?

59. Explain how a linear system can have infinitely many solutions.

60. What are dependent equations? Provide an example with your description.

61. The following system models the winning times, y, in seconds, in the Olympic 500-meter speed skating event x years after 1970:

$\begin{cases} y = -0.19x + 43.7 \quad \boxed{\text{Women}} \\ y = -0.16x + 39.9. \quad \boxed{\text{Men}} \end{cases}$

Use the slope of each model to explain why the system has a solution. What does this solution represent?

Critical Thinking Exercises

Make Sense? *In Exercises 62–65, determine whether each statement makes sense or does not make sense, and explain your reasoning.*

62. Each equation in a system of linear equations has infinitely many ordered-pair solutions.

63. Every linear system has infinitely many ordered-pair solutions.

64. In dependent systems, the two equations represent the same line.

65. When I use graphing to solve an inconsistent system, the lines should look parallel, and I can always use slope to confirm that they really are.

In Exercises 66–69, determine whether each statement is true or false. If the statement is false, make the necessary change(s) to produce a true statement.

66. If a linear system has graphs with equal slopes, the system must be inconsistent.

67. If a linear system has graphs with equal y-intercepts, the system must have infinitely many solutions.

68. If a linear system has two distinct points that are solutions, then the graphs of the system's equations have equal slopes and equal y-intercepts.

69. It is possible for a linear system with one solution to have graphs with equal slopes.

70. Write a system of linear equations whose solution is $(5, 1)$. How many different systems are possible? Explain.

71. Write a system of equations with one solution, a system of equations with no solution, and a system of equations with infinitely many solutions. Explain how you were able to think of these systems.

Technology Exercises

72. Verify your solutions to any five exercises from Exercises 11 through 36 by using a graphing utility to graph the two equations in the system in the same viewing rectangle. After entering the two equations, one as y_1 and the other as y_2, and graphing them, use the $\boxed{\text{INTERSECTION}}$ feature to find the system's solution. (It may first be necessary to solve the equations for y before entering them.)

Read Exercise 72. Then use a graphing utility to solve the systems in Exercises 73–80.

73. $\begin{cases} y = 2x + 2 \\ y = -2x + 6 \end{cases}$

74. $\begin{cases} y = -x + 5 \\ y = x - 7 \end{cases}$

75. $\begin{cases} x + 2y = 4 \\ x - y = 4 \end{cases}$

76. $\begin{cases} 2x - 3y = 10 \\ 4x + 3y = 20 \end{cases}$

77. $\begin{cases} 3x - y = 5 \\ -5x + 2y = -10 \end{cases}$

78. $\begin{cases} 2x - 3y = 7 \\ 3x + 5y = 1 \end{cases}$

79. $\begin{cases} y = \dfrac{1}{3}x + \dfrac{2}{3} \\ y = \dfrac{5}{7}x - 2 \end{cases}$

80. $\begin{cases} y = -\dfrac{1}{2}x + 2 \\ y = \dfrac{3}{4}x + 7 \end{cases}$

Review Exercises

In Exercises 81–83, perform the indicated operation.

81. $-3 + (-9)$ (Section 1.7, **Table 1.7**)

82. $-3 - (-9)$ (Section 1.7, **Table 1.7**)

83. $-3(-9)$ (Section 1.7, **Table 1.7**)

Preview Exercises

Exercises 84–86 will help you prepare for the material covered in the next section. In each exercise, solve the given equation.

84. $4x - 3(-x - 1) = 24$

85. $5(2y - 3) - 4y = 9$

86. $(5x - 1) + 1 = 5x + 5$

4.2

Solving Systems of Linear Equations by the Substitution Method

What am I supposed to learn?

After studying this section, you should be able to:

1. Solve linear systems by the substitution method.

2. Use the substitution method to identify systems with no solution or infinitely many solutions.

3. Solve problems using the substitution method.

1. Solve linear systems by the substitution method.

Great Question!

In the first step of the substitution method, how do I know which variable to isolate and in which equation?

You can choose both the variable and the equation. If possible, solve for a variable whose coefficient is 1 or −1 to avoid working with fractions.

Other than outrage, what is going on at the gas pumps? Is surging demand creating the increasing oil prices? Like all things in a free market economy, the price of a commodity is based on supply and demand. In this section, we use a second method for solving linear systems, the *substitution method*, to understand this economic phenomenon.

Eliminating a Variable Using the Substitution Method

Finding the solution of a linear system by graphing equations may not be easy to do. For example, a solution of $\left(-\frac{2}{3}, \frac{157}{29}\right)$ would be difficult to "see" as an intersection point on a graph.

Let's consider a method that does not depend on finding a system's solution visually: the substitution method. This method involves converting the system to one equation in one variable by an appropriate substitution.

Solving Linear Systems by Substitution

1. Solve either of the equations for one variable in terms of the other. (If one of the equations is already in this form, you can skip this step.)

2. Substitute the expression found in step 1 into the *other* equation. This will result in an equation in one variable.

3. Solve the equation containing one variable.

4. Back-substitute the value found in step 3 into the equation from step 1. Simplify and find the value of the remaining variable.

5. Check the proposed solution in both of the system's given equations.

EXAMPLE 1 Solving a System by Substitution

Solve by the substitution method:

$$\begin{cases} y = -x - 1 \\ 4x - 3y = 24. \end{cases}$$

Solution

Step 1. Solve either of the equations for one variable in terms of the other. This step has already been done for us. The first equation, $y = -x - 1$, is solved for y in terms of x.

Step 2. Substitute the expression from step 1 into the other equation. We substitute the expression $-x - 1$ for y into the second equation, $4x - 3y = 24$:

$$y = \boxed{-x - 1} \qquad 4x - 3\boxed{y} = 24. \qquad \text{Substitute } -x - 1 \text{ for } y.$$

This gives us an equation in one variable, namely

$$4x - 3(-x - 1) = 24.$$

The variable y has been eliminated.

Step 3. Solve the resulting equation containing one variable.

$$
\begin{array}{ll}
4x - 3(-x - 1) = 24 & \text{This is the equation containing one variable.} \\
4x + 3x + 3 = 24 & \text{Apply the distributive property.} \\
7x + 3 = 24 & \text{Combine like terms.} \\
7x = 21 & \text{Subtract 3 from both sides.} \\
x = 3 & \text{Divide both sides by 7.}
\end{array}
$$

Step 4. Back-substitute the obtained value into the equation from step 1. We now know that the x-coordinate of the solution is 3. To find the y-coordinate, we back-substitute the x-value in the equation from step 1,

$$y = -x - 1. \qquad \text{This is the equation from step 1.}$$

Substitute **3** for x.

$$
\begin{array}{l}
y = -3 - 1 \\
y = -4 \qquad \text{Simplify.}
\end{array}
$$

With $x = 3$ and $y = -4$, the proposed solution is $(3, -4)$.

Step 5. Check the proposed solution in both of the system's given equations. Replace x with 3 and y with -4.

$$
\begin{array}{ll}
y = -x - 1 & 4x - 3y = 24 \\
-4 \stackrel{?}{=} -3 - 1 & 4(3) - 3(-4) \stackrel{?}{=} 24 \\
-4 = -4, \quad \text{true} & 12 + 12 \stackrel{?}{=} 24 \\
& 24 = 24, \quad \text{true}
\end{array}
$$

The pair $(3, -4)$ satisfies both equations. The system's solution is $(3, -4)$ and the solution set is $\{(3, -4)\}$. ■

Using Technology

A graphing utility can be used to solve the system in Example 1. Graph each equation and use the intersection feature. The utility displays the solution $(3, -4)$ as $x = 3, y = -4$.

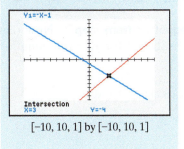

$[-10, 10, 1]$ by $[-10, 10, 1]$

☑ **CHECK POINT 1**   Solve by the substitution method:

$$
\begin{cases}
y = 5x - 13 \\
2x + 3y = 12.
\end{cases}
$$

EXAMPLE 2   Solving a System by Substitution

Solve by the substitution method:

$$
\begin{cases}
5x - 4y = 9 \\
x - 2y = -3.
\end{cases}
$$

Solution

Step 1. Solve either of the equations for one variable in terms of the other. We begin by isolating one of the variables in either of the equations. By solving for x in the second equation, which has a coefficient of 1, we can avoid fractions.

$$
\begin{array}{ll}
x - 2y = -3 & \text{This is the second equation in the given system.} \\
x = 2y - 3 & \text{Solve for } x \text{ by adding } 2y \text{ to both sides.}
\end{array}
$$

Step 2. Substitute the expression from step 1 into the other equation. We substitute $2y - 3$ for x in the first equation, $5x - 4y = 9$.

$$x = \boxed{2y - 3} \qquad 5\boxed{x} - 4y = 9$$

This gives us an equation in one variable, namely

$$5(2y - 3) - 4y = 9.$$

The variable x has been eliminated.

Step 3. Solve the resulting equation containing one variable.

$5(2y - 3) - 4y = 9$	This is the equation containing one variable.
$10y - 15 - 4y = 9$	Apply the distributive property.
$6y - 15 = 9$	Combine like terms.
$6y = 24$	Add 15 to both sides.
$y = 4$	Divide both sides by 6.

Step 4. Back-substitute the obtained value into the equation from step 1. Now that we have the y-coordinate of the solution, we back-substitute 4 for y in the equation $x = 2y - 3$.

$x = 2y - 3$	Use the equation obtained in step 1.
$x = 2(4) - 3$	Substitute 4 for y.
$x = 8 - 3$	Multiply.
$x = 5$	Subtract.

With $x = 5$ and $y = 4$, the proposed solution is $(5, 4)$.

Step 5. Check. Take a moment to show that $(5, 4)$ satisfies both given equations. The solution is $(5, 4)$ and the solution set is $\{(5, 4)\}$. ∎

> ✓ **CHECK POINT 2** Solve by the substitution method:
>
> $$\begin{cases} 3x + 2y = -1 \\ x - y = 3. \end{cases}$$

Great Question!

If my solution satisfies one of the equations in the system, do I have to check the solution in the other equation?

Yes. Get into the habit of checking ordered-pair solutions in both equations of the system.

2 Use the substitution method to identify systems with no solution or infinitely many solutions. ▶

The Substitution Method with Linear Systems Having No Solution or Infinitely Many Solutions

Recall that a linear system with no solution is called an **inconsistent system**. If you attempt to solve such a system by substitution, you will eliminate both variables. A false statement such as $0 = 5$ will be the result.

EXAMPLE 3 Using the Substitution Method on an Inconsistent System

Solve the system:

$$\begin{cases} y + 1 = 5(x + 1) \\ y = 5x - 1. \end{cases}$$

Using Technology

A graphing utility was used to graph the equations in Example 3. The lines are parallel and have no point of intersection. This verifies that the system is inconsistent.

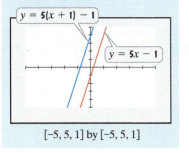

$[-5, 5, 1]$ by $[-5, 5, 1]$

Solution   The variable y is isolated in the second equation, $y = 5x - 1$. We use the substitution method and substitute the expression for y in the first equation, $y + 1 = 5(x + 1)$.

$$\boxed{y} + 1 = 5(x + 1) \qquad y = \boxed{5x - 1} \qquad \text{Substitute } 5x - 1 \text{ for } y.$$

$$(5x - 1) + 1 = 5(x + 1) \qquad \text{This substitution gives an equation in one variable.}$$

$$5x = 5x + 5 \qquad \text{Simplify on the left side. Use the distributive property on the right side.}$$

There are no values of x and y for which $0 = 5$. $0 = 5$ false   Subtract $5x$ from both sides.

The false statement $0 = 5$ indicates that the system is inconsistent and has no solution. The solution set is the empty set, $\varnothing$. ∎

✓ **CHECK POINT 3**   Solve the system:

$$\begin{cases} 3x + y = -5 \\ \quad\ y = -3x + 3. \end{cases}$$

Do you remember that the equations in a linear system with infinitely many solutions are called **dependent**? If you attempt to solve such a system by substitution, you will eliminate both variables. However, a true statement such as $6 = 6$ will be the result.

EXAMPLE 4   Using the Substitution Method on a System with Infinitely Many Solutions

Solve the system:

$$\begin{cases} \quad\ y = 3 - 2x \\ 4x + 2y = 6. \end{cases}$$

Solution   The variable y is isolated in the first equation. We use the substitution method and substitute the expression for y in the second equation.

$$y = \boxed{3 - 2x} \qquad 4x + 2\boxed{y} = 6 \qquad \text{Substitute } 3 - 2x \text{ for } y.$$

$$4x + 2(3 - 2x) = 6 \qquad \text{This substitution gives an equation in one variable.}$$

$$4x + 6 - 4x = 6 \qquad \text{Apply the distributive property:}$$

$$2(3 - 2x) = 2 \cdot 3 - 2 \cdot 2x = 6 - 4x.$$

$$6 = 6 \quad \text{true} \qquad \text{Simplify: } 4x - 4x = 0.$$

This true statement indicates that the system contains dependent equations and has infinitely many solutions. We express the solution set for the system in one of two equivalent ways:

$$\{(x, y) \mid y = 3 - 2x\} \quad \text{The set of all ordered pairs } (x, y) \text{ such that } y = 3 - 2x$$

$$\text{or } \{(x, y) \mid 4x + 2y = 6\}. \quad \text{The set of all ordered pairs } (x, y) \text{ such that } 4x + 2y = 6 \ \blacksquare$$

✓ **CHECK POINT 4**   Solve the system:

$$\begin{cases} \quad\ y = 3x - 4 \\ 9x - 3y = 12. \end{cases}$$

③ Solve problems using the substitution method. ⊙

Applications

An important application of systems of equations arises in connection with supply and demand. As the price of a product increases, the demand for that product decreases. However, at higher prices suppliers are willing to produce greater quantities of the product.

EXAMPLE 5   Supply and Demand Models

Table 4.1 shows the price of a slice of pizza. For each price, the table lists the number of slices that consumers are willing to buy and the number of slices that pizzerias are willing to supply.

Table 4.1   Supply and Demand for Pizza Slices

Price of a slice of pizza	Quantity demanded per day	Quantity supplied per day	Result
$0.50	300	100	There is a shortage from excess demand.
$1.00	250	150	
$2.00	150	250	There is a surplus from excess supply.
$3.00	50	350	

Source: O'Sullivan and Sheffrin, *Economics,* Prentice Hall, 2007.

The data in **Table 4.1** can be described by the following demand and supply models:

Demand Model	**Supply Model**
$p = -0.01x + 3.5$	$p = 0.01x - 0.5.$

Demand Model: Price per slice, Number of slices demanded per day

Supply Model: Price per slice, Number of slices supplied per day

The price at which supply and demand are equal is called the **equilibrium price**. The quantity supplied and demanded at that price is called the **equilibrium quantity**. Find the equilibrium quantity and the equilibrium price for pizza slices.

Solution   We can find the equilibrium quantity and the equilibrium price by solving the demand-supply linear system. We will use substitution, substituting $-0.01x + 3.5$ for p in the second equation.

$p = \boxed{-0.01x + 3.5}$ $\boxed{p} = 0.01x - 0.5$ Substitute $-0.01x + 3.5$ for p.

$-0.01x + 3.5 = 0.01x - 0.5$ The resulting equation contains only one variable.

$-0.02x + 3.5 = -0.5$ Subtract 0.01x from both sides.

$-0.02x = -4.0$ Subtract 3.5 from both sides: $-0.5 - 3.5 = -4.0$.

$x = 200$ Divide both sides by -0.02: $\dfrac{-4.0}{-0.02} = 200.$

Because $x = 200$, the equilibrium quantity is 200 slices per day. To find the equilibrium price, we back-substitute 200 for x into either the demand or the supply model. We'll use both models to make sure we get the same number in each case.

Demand Model	**Supply Model**
$p = -0.01x + 3.5$	$p = 0.01x - 0.5$
Substitute 200 for x.	Substitute 200 for x.
$p = -0.01(200) + 3.5$	$p = 0.01(200) - 0.5$
$= -2 + 3.5 = 1.5$	$= 2 - 0.5 = 1.5$

The equilibrium price is \$1.50 per slice. At that price, and only at that price, the quantity demanded and the quantity supplied are equal, at 200 slices per day. **Figure 4.8** shows the equilibrium point, (200, 1.50), as the intersection point for the graphs of the demand and supply models.

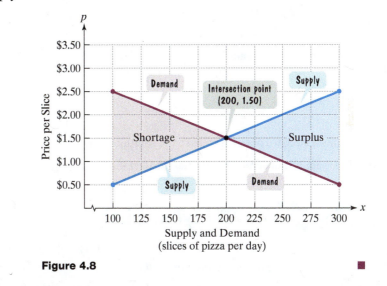

Figure 4.8

✓ **CHECK POINT 5**   The following models describe demand and supply for two-bedroom rental apartments.

Demand Model	**Supply Model**
$p = -30x + 1800$	$p = 30x$
Monthly rental price Number of apartments demanded, in thousands	Monthly rental price Number of apartments supplied, in thousands

a. Solve the system and find the equilibrium quantity and the equilibrium price.

b. Use your answer from part (a) to complete this statement:

When rents are _____ per month, consumers will demand _____ apartments and suppliers will offer _____ apartments for rent.

Achieving Success

In college, it is recommended that students study and do homework for at least two hours for each hour of class time. For many students, finding the necessary time to study is not always easy. In order to make education a top priority, efficient time management is essential. You can continue to improve your time management by identifying your biggest time-wasters. Do you really need to spend hours on the phone, overshare the details of your life on Facebook, or watch that sitcom rerun? My biggest time-wasters: _____

CONCEPT AND VOCABULARY CHECK

Fill in each blank so that the resulting statement is true.

1. When solving

$$\begin{cases} x = 5 - y \\ 2x - y = 7 \end{cases}$$

by the substitution method, we obtain $y = 1$, so the solution set is _____.

2. When solving

$$\begin{cases} x - 7y = -22 \\ 5x + 2y = 1 \end{cases}$$

by the substitution method, we obtain $x = -1$, so the solution set is _____.

3. When solving

$$\begin{cases} x + 2y = 4 \\ \quad y = -\frac{1}{2}x + 1 \end{cases}$$

by the substitution method, we obtain $2 = 4$, so the solution set is _____.

4. When solving

$$\begin{cases} 2x - 6y = 8 \\ \quad x = 3y + 4 \end{cases}$$

by the substitution method, we obtain $8 = 8$, so the solution set is _____.

5. The price at which supply and demand are equal is called the _____ price.

4.2 EXERCISE SET ▶ MyMathLab®

Practice Exercises

In Exercises 1–32, solve each system by the substitution method.
If there is no solution or an infinite number of solutions, so state.
Use set notation to express solution sets.

1. $\begin{cases} x + y = 4 \\ y = 3x \end{cases}$

2. $\begin{cases} x + y = 6 \\ y = 2x \end{cases}$

3. $\begin{cases} x + 3y = 8 \\ y = 2x - 9 \end{cases}$

4. $\begin{cases} 2x - 3y = -13 \\ y = 2x + 7 \end{cases}$

5. $\begin{cases} x + 3y = 5 \\ 4x + 5y = 13 \end{cases}$

6. $\begin{cases} x + 2y = 5 \\ 2x - y = -15 \end{cases}$

7. $\begin{cases} 2x - y = -5 \\ x + 5y = 14 \end{cases}$

8. $\begin{cases} 2x + 3y = 11 \\ x - 4y = 0 \end{cases}$

9. $\begin{cases} 2x - y = 3 \\ 5x - 2y = 10 \end{cases}$

10. $\begin{cases} -x + 3y = 10 \\ 2x + 8y = -6 \end{cases}$

11. $\begin{cases} -3x + y = -1 \\ x - 2y = 4 \end{cases}$

12. $\begin{cases} -4x + y = -11 \\ 2x - 3y = 5 \end{cases}$

13. $\begin{cases} x = 9 - 2y \\ x + 2y = 13 \end{cases}$

14. $\begin{cases} 6x + 2y = 7 \\ y = 2 - 3x \end{cases}$

15. $\begin{cases} y = 3x - 5 \\ 21x - 35 = 7y \end{cases}$

16. $\begin{cases} 9x - 3y = 12 \\ y = 3x - 4 \end{cases}$

17. $\begin{cases} 5x + 2y = 0 \\ x - 3y = 0 \end{cases}$

18. $\begin{cases} 4x + 3y = 0 \\ 2x - y = 0 \end{cases}$

19. $\begin{cases} 2x - y = 6 \\ 3x + 2y = 5 \end{cases}$

20. $\begin{cases} 2x - y = 4 \\ 3x - 5y = 2 \end{cases}$

21. $\begin{cases} 2(x - 1) - y = -3 \\ y = 2x + 3 \end{cases}$

22. $\begin{cases} x + y - 1 = 2(y - x) \\ y = 3x - 1 \end{cases}$

23. $\begin{cases} x = 2y + 9 \\ x = 7y + 10 \end{cases}$

24. $\begin{cases} x = 5y - 3 \\ x = 8y + 4 \end{cases}$

25. $\begin{cases} 4x - y = 100 \\ 0.05x - 0.06y = -32 \end{cases}$

26. $\begin{cases} x + 6y = 8000 \\ 0.3x - 0.6y = 0 \end{cases}$

27. $\begin{cases} y = \dfrac{1}{3}x + \dfrac{2}{3} \\ y = \dfrac{5}{7}x - 2 \end{cases}$

28. $\begin{cases} y = -\dfrac{1}{2}x + 2 \\ y = \dfrac{3}{4}x + 7 \end{cases}$

29. $\begin{cases} \dfrac{x}{6} - \dfrac{y}{2} = \dfrac{1}{3} \\ x + 2y = -3 \end{cases}$

30. $\begin{cases} \dfrac{x}{4} - \dfrac{y}{4} = -1 \\ x + 4y = -9 \end{cases}$

31. $\begin{cases} 2x - 3y = 8 - 2x \\ 3x + 4y = x + 3y + 14 \end{cases}$

32. $\begin{cases} 3x - 4y = x - y + 4 \\ 2x + 6y = 5y - 4 \end{cases}$

Practice PLUS

In Exercises 33–38, write a system of equations modeling the given conditions. Then solve the system by the substitution method and find the two numbers.

33. The sum of two numbers is 81. One number is 41 more than the other. Find the numbers.

34. The sum of two numbers is 62. One number is 12 more than the other. Find the numbers.

35. The difference between two numbers is 5. Four times the larger number is 6 times the smaller number. Find the numbers.

36. The difference between two numbers is 25. Two times the larger number is 12 times the smaller number. Find the numbers.

37. The difference between two numbers is 1. The sum of the larger number and twice the smaller number is 7. Find the numbers.

38. The difference between two numbers is 5. The sum of the larger number and twice the smaller number is 14. Find the numbers.

In Exercises 39–40, multiply each equation in the system by an appropriate number so that the coefficients are integers. Then solve the system by the substitution method.

39. $\begin{cases} 0.7x - 0.1y = 0.6 \\ 0.8x - 0.3y = -0.8 \end{cases}$

40. $\begin{cases} 1.25x - 0.01y = 4.5 \\ 0.5x - 0.02y = 1 \end{cases}$

Application Exercises

41. The table shows the price of a gallon of unleaded premium gasoline. For each price, the table lists the number of gallons per day that a gas station sells and the number of gallons per day that can be supplied.

Supply and Demand for Unleaded Premium Gasoline

Price per gallon	Gallons demanded per day	Gallons supplied per day
$3.20	1400	200
$3.60	1200	600
$4.40	800	1400
$4.80	600	1800

The data in the table are described by the following demand and supply models:

Demand Model   **Supply Model**

$p = -0.002x + 6$ $p = 0.001x + 3.$

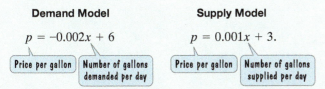

a. Solve the system and find the equilibrium quantity and the equilibrium price for a gallon of unleaded premium gasoline.

b. Use your answer from part (a) to complete this statement:

If unleaded premium gasoline is sold for _____ per gallon, there will be a demand for _____ gallons per day and _____ gallons will be supplied per day.

42. The table shows the price of a package of cookies. For each price, the table lists the number of packages that consumers are willing to buy and the number of packages that bakers are willing to supply.

Supply and Demand for Packages of Cookies

Price of a package of cookies	Quantity demanded (millions of packages) per week	Quantity supplied (millions of packages) per week
30¢	150	70
40¢	130	90
60¢	90	130
70¢	70	150

The data in the table can be described by the following demand and supply models:

Demand Model   **Supply Model**

$p = -0.5x + 105$ $p = 0.5x - 5.$

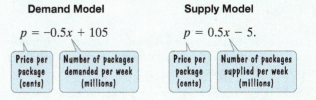

a. Solve the system and find the equilibrium quantity and the equilibrium price for a package of cookies.

b. Use your answer from part (a) to complete this statement:

If cookies are sold for _____ per package, there will be a demand for _____ million packages per week and bakers will supply _____ million packages per week.

Explaining the Concepts

43. Describe a problem that might arise when solving a system of equations using graphing. Assume that both equations in the system have been graphed correctly and the system has exactly one solution.

44. Explain how to solve a system of equations using the substitution method. Use $y = 3 - 3x$ and $3x + 4y = 6$ to illustrate your explanation.

45. When using the substitution method, how can you tell if a system of linear equations has no solution?

46. When using the substitution method, how can you tell if a system of linear equations has infinitely many solutions?

47. The law of supply and demand states that, in a free market economy, a commodity tends to be sold at its equilibrium price. At this price, the amount that the seller will supply is the same amount that the consumer will buy. Explain how systems of equations can be used to determine the equilibrium price.

Critical Thinking Exercises

Make Sense? *In Exercises 48–51, determine whether each statement makes sense or does not make sense, and explain your reasoning.*

48. The substitution method provides me with solutions that might be awkward to determine by graphing.

49. When using substitution to solve

$$\begin{cases} 5x - 4y = 9 \\ x - 2y = -3, \end{cases}$$

I find it easiest to solve for x in the first equation.

50. While solving a system using substitution, I eliminated both variables and obtained $6 = 6$, so the system has no solution.

51. I think of equilibrium as the point at which quantity demanded exceeds quantity supplied.

In Exercises 52–55, determine whether each statement is true or false. If the statement is false, make the necessary change(s) to produce a true statement.

52. Solving an inconsistent system by substitution results in a true statement.

53. The line passing through the intersection of the graphs of $x + y = 4$ and $x - y = 0$ with slope = 3 has an equation given by $y - 2 = 3(x - 2)$.

54. Unlike the graphing method, where solutions cannot be seen, the substitution method provides a way to visualize solutions as intersection points.

55. To solve the system

$$\begin{cases} 2x - y = 5 \\ 3x + 4y = 7 \end{cases}$$

by substitution, replace y in the second equation with $5 - 2x$.

56. If $x = 3 - y - z$, $2x + y - z = -6$, and $3x - y + z = 11$, find the values for $x, y,$ and z.

57. Find the value of m that makes

$$\begin{cases} y = mx + 3 \\ 5x - 2y = 7 \end{cases}$$

an inconsistent system.

Review Exercises

58. Graph: $4x + 6y = 12$. (Section 3.2, Example 4)

59. Solve: $4(x + 1) = 25 + 3(x - 3)$. (Section 2.3, Example 3)

60. List all the integers in this set: $\left\{ -73, -\dfrac{2}{3}, 0, \dfrac{3}{1}, \dfrac{3}{2}, \dfrac{\pi}{1} \right\}$. (Section 1.3, Example 5)

Preview Exercises

Exercises 61–63 will help you prepare for the material covered in the next section.

61. Use both equations in the system

$$\begin{cases} 3x + 2y = 48 \\ 9x - 8y = -24 \end{cases}$$

to find x for $y = 12$. What do you observe?

62. Solve: $-14y = -168$.

63. Multiply both sides of $x - 5y = 3$ by -4 and simplify.

SECTION

4.3

Solving Systems of Linear Equations by the Addition Method

What am I supposed to learn?

After studying this section, you should be able to:

1 Solve linear systems by the addition method. ▶

2 Use the addition method to identify systems with no solution or infinitely many solutions. ▶

3 Determine the most efficient method for solving a linear system. ▶

Figure 4.9
Source: Gerrig and Zimbardo, *Psychology and Life*, 18th Edition, Allyn and Bacon, 2008.

1 Solve linear systems by the addition method. ▶

Researchers identified college students who generally were procrastinators or nonprocrastinators. The students were asked to report throughout the semester how many symptoms of physical illness they had experienced. **Figure 4.9** shows that by late in the semester, all students experienced increases in symptoms. Early in the semester, procrastinators reported fewer symptoms, but late in the semester, as work came due, they reported more symptoms than their nonprocrastinating peers.

The data in **Figure 4.9** can be analyzed using a pair of linear models in two variables. The figure shows that by week 6, both groups reported the same number of symptoms of illness, an average of approximately 3.5 symptoms per group. In this section, you will learn a third method, called *addition*, that will reinforce this graphic observation, verifying $(6, 3.5)$ as the point of intersection.

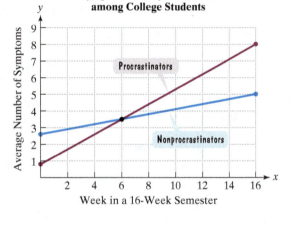

Symptoms of Physical Illness among College Students

Procrastinators

Nonprocrastinators

Average Number of Symptoms

Week in a 16-Week Semester

Eliminating a Variable Using the Addition Method

The substitution method is most useful if one of the given equations has an isolated variable. A third, and frequently the easiest, method for solving a linear system is the addition method. Like the substitution method, the addition method involves eliminating a variable and ultimately solving an equation containing only one variable. However, this time we eliminate a variable by adding the equations.

For example, consider the following system of linear equations:

$$\begin{cases} 3x - 4y = 11 \\ -3x + 2y = -7. \end{cases}$$

When we add these two equations, the x-terms are eliminated. This occurs because the coefficients of the x-terms, 3 and -3, are opposites (additive inverses) of each other.

$$\begin{cases} 3x - 4y = 11 \\ \underline{-3x + 2y = -7} \end{cases}$$

Add: $\qquad -2y = 4$ The sum is an equation in one variable.

$$y = -2 \qquad \text{Divide both sides by } -2 \text{ and solve for } y.$$

Now we can back-substitute -2 for y into one of the original equations to find x. It does not matter which equation you use; you will obtain the same value for x in either case. If we use either equation, we can show that $x = 1$. The solution $(1, -2)$ satisfies both equations in the system.

When we use the addition method, we want to obtain two equations whose sum is an equation containing only one variable. The key step is to **obtain, for one of the variables, coefficients that differ only in sign.** To do this, we may need to multiply one or both equations by some nonzero number so that the coefficients of one of the variables, x or y, become opposites. Then when the two equations are added, this variable is eliminated.

EXAMPLE 1 Solving a System by the Addition Method

Solve by the addition method:

$$\begin{cases} x + y = 4 \\ x - y = 6. \end{cases}$$

Solution The coefficients of y in the two equations, 1 and -1, differ only in sign and are opposites. Therefore, by adding the two left sides and the two right sides, we can eliminate the y-terms.

$$\begin{cases} x + y = 4 \\ \underline{x - y = 6} \end{cases}$$
Add: $2x + 0y = 10$
$ 2x = 10$ Simplify.

Now y is eliminated and we can solve $2x = 10$ for x.

$$2x = 10$$
$$x = 5 \quad \text{Divide both sides by 2 and solve for } x.$$

We back-substitute 5 for x into one of the original equations to find y. We will use both equations to show that we obtain the same value for y in either case.

Use the first equation:	Use the second equation:
$x + y = 4$	$x - y = 6$
$5 + y = 4$	$5 - y = 6$ Replace x with 5.
$y = -1.$	$-y = 1$ Solve for y.
	$y = -1.$

Thus, $x = 5$ and $y = -1$. The proposed solution, $(5, -1)$, can be shown to satisfy both equations in the system. Consequently, the solution is $(5, -1)$ and the solution set is $\{(5, -1)\}$. ∎

☑ **CHECK POINT 1** Solve by the addition method:

$$\begin{cases} x + y = 5 \\ x - y = 9. \end{cases}$$

EXAMPLE 2 Solving a System by the Addition Method

Solve by the addition method:

$$\begin{cases} 3x - y = 11 \\ 2x + 5y = 13. \end{cases}$$

Solution We must rewrite one or both equations in equivalent forms so that the coefficients of the same variable (either x or y) are opposites. Consider the terms in y in each equation, that is, $-1y$ and $5y$. To eliminate y, we can multiply each term of the first equation by 5 and then add the equations.

Great Question!

Isn't the addition method also called the elimination method?

Although the addition method is also known as the elimination method, variables are eliminated when using both the substitution and addition methods. The name *addition method* specifically tells us that the elimination of a variable is accomplished by adding two equations.

$$\begin{cases} 3x - y = 11 \\ 2x + 5y = 13 \end{cases} \xrightarrow[\text{No change}]{\text{Multiply by 5.}} \begin{cases} 15x - 5y = 55 \\ \underline{2x + 5y = 13} \end{cases}$$

$$\text{Add:}\quad 17x + 0y = 68$$

$$17x = 68 \qquad \text{Simplify.}$$

$$x = 4 \qquad \text{Divide both sides by 17 and solve for } x.$$

Thus, $x = 4$. To find y, we back-substitute 4 for x into either one of the given equations. We'll use the second equation.

$$2x + 5y = 13 \qquad \text{This is the second equation in the given system.}$$

$$2 \cdot 4 + 5y = 13 \qquad \text{Substitute 4 for } x.$$

$$8 + 5y = 13 \qquad \text{Multiply: } 2 \cdot 4 = 8.$$

$$5y = 5 \qquad \text{Subtract 8 from both sides.}$$

$$y = 1 \qquad \text{Divide both sides by 5.}$$

The solution is $(4, 1)$. Check to see that it satisfies both of the original equations in the system. The solution set is $\{(4, 1)\}$. ∎

✓ **CHECK POINT 2**   Solve by the addition method:

$$\begin{cases} 4x - y = 22 \\ 3x + 4y = 26. \end{cases}$$

Before considering additional examples, let's summarize the steps for solving linear systems by the addition method.

Solving Linear Systems by Addition

1. If necessary, rewrite both equations in the form $Ax + By = C$.
2. If necessary, multiply either equation or both equations by appropriate nonzero numbers so that the sum of the x-coefficients or the sum of the y-coefficients is 0.
3. Add the equations in step 2. The sum is an equation in one variable.
4. Solve the equation in one variable.
5. Back-substitute the value obtained in step 4 into either of the given equations and solve for the other variable.
6. Check the solution in both of the original equations.

EXAMPLE 3   Solving a System by the Addition Method

Solve by the addition method:

$$\begin{cases} 3x + 2y = 48 \\ 9x - 8y = -24. \end{cases}$$

Solution

Step 1. Rewrite both equations in the form $Ax + By = C$. Both equations are already in this form. Variable terms appear on the left and constants appear on the right.

Step 2. If necessary, multiply either equation or both equations by appropriate numbers so that the sum of the x-coefficients or the sum of the y-coefficients is 0. We can eliminate x or y. Let's eliminate x. Consider the terms in x in each equation, that is, $3x$ and $9x$. To eliminate x, we can multiply each term of the first equation by -3 and then add the equations.

$$\begin{cases} 3x + 2y = 48 \\ 9x - 8y = -24 \end{cases} \xrightarrow[\text{No change}]{\text{Multiply by } -3.} \begin{cases} -9x - 6y = -144 \\ 9x - 8y = \ \ -24 \end{cases}$$

Step 3. Add the equations.   Add: $-14y = -168$

Step 4. Solve the equation in one variable. We solve $-14y = -168$ by dividing both sides by -14.

$$\frac{-14y}{-14} = \frac{-168}{-14} \qquad \text{Divide both sides by } -14.$$
$$y = 12 \qquad \text{Simplify.}$$

Step 5. Back-substitute and find the value for the other variable. We can back-substitute 12 for y in either one of the given equations. We'll use the first one.

$$3x + 2y = 48 \qquad \text{This is the first equation in the given system.}$$
$$3x + 2(12) = 48 \qquad \text{Substitute 12 for } y.$$
$$3x + 24 = 48 \qquad \text{Multiply.}$$
$$3x = 24 \qquad \text{Subtract 24 from both sides.}$$
$$x = 8 \qquad \text{Divide both sides by 3.}$$

The solution is $(8, 12)$.

Step 6. Check. Take a minute or so to show that $(8, 12)$ satisfies both of the original equations in the system. The solution set is $\{(8, 12)\}$. ∎

✓ **CHECK POINT 3**   Solve by the addition method:

$$\begin{cases} 4x + 5y = 3 \\ 2x - 3y = 7. \end{cases}$$

Some linear systems have solutions that are not integers. If the value of one variable turns out to be a "messy" fraction, back-substitution might lead to cumbersome arithmetic. If this happens, you can return to the original system and use addition to find the value of the other variable.

EXAMPLE 4   Solving a System by the Addition Method

Solve by the addition method:

$$\begin{cases} 2x = 7y - 17 \\ 5y = 17 - 3x. \end{cases}$$

Solution

Step 1. Rewrite both equations in the form $Ax + By = C$. We first arrange the system so that variable terms appear on the left and constants appear on the right. We obtain

$$\begin{cases} 2x - 7y = -17 \qquad \text{Subtract 7y from both sides of the first equation.} \\ 3x + 5y = \ \ 17. \qquad \text{Add 3x to both sides of the second equation.} \end{cases}$$

Step 2. If necessary, multiply either equation or both equations by appropriate numbers so that the sum of the x-coefficients or the sum of the y-coefficients is 0. We can eliminate x or y. Let's eliminate x by multiplying the first equation by 3 and the second equation by -2.

$$\begin{cases} 2x - 7y = -17 \\ 3x + 5y = 17 \end{cases} \xrightarrow[\text{Multiply by } -2.]{\text{Multiply by } 3.} \begin{cases} 6x - 21y = -51 \\ -6x - 10y = -34 \end{cases}$$

Step 3. Add the equations.   Add: $-31y = -85$

Step 4. Solve the equation in one variable. We solve $-31y = -85$ by dividing both sides by -31.

$$\frac{-31y}{-31} = \frac{-85}{-31} \qquad \text{Divide both sides by } -31.$$

$$y = \frac{85}{31} \qquad \text{Simplify.}$$

Step 5. Back-substitute and find the value for the other variable. Back-substitution of $\frac{85}{31}$ for y into either of the given equations results in cumbersome arithmetic. Instead, let's use the addition method on the given system in the form $Ax + By = C$ to find the value for x. Thus, we eliminate y by multiplying the first equation by 5 and the second equation by 7.

$$\begin{cases} 2x - 7y = -17 \\ 3x + 5y = 17 \end{cases} \xrightarrow[\text{Multiply by } 7.]{\text{Multiply by } 5.} \begin{cases} 10x - 35y = -85 \\ 21x + 35y = 119 \end{cases}$$

Add: $31x = 34$

$$x = \frac{34}{31} \qquad \text{Divide both sides by 31.}$$

The solution is $\left(\dfrac{34}{31}, \dfrac{85}{31}\right)$.

Step 6. Check. For this system, a calculator is helpful in showing that $\left(\frac{34}{31}, \frac{85}{31}\right)$ satisfies both of the original equations in the system. The solution set is $\left\{\left(\frac{34}{31}, \frac{85}{31}\right)\right\}$. ■

☑ **CHECK POINT 4**   Solve by the addition method:

$$\begin{cases} 2x = 9 + 3y \\ 4y = 8 - 3x. \end{cases}$$

---

2 Use the addition method to identify systems with no solution or infinitely many solutions.

The Addition Method with Linear Systems Having No Solution or Infinitely Many Solutions

As with the substitution method, if the addition method results in a false statement, the linear system is inconsistent and has no solution.

EXAMPLE 5   Using the Addition Method on an Inconsistent System

Solve the system: $\begin{cases} 4x + 6y = 12 \\ 6x + 9y = 12. \end{cases}$

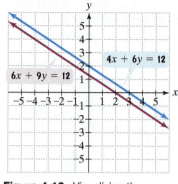

Figure 4.10 Visualizing the inconsistent system in Example 5

Solution We can eliminate x or y. Let's eliminate x by multiplying the first equation, $4x + 6y = 12$, by 3 and the second equation, $6x + 9y = 12$, by -2.

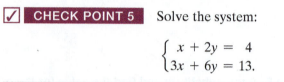

$$\begin{cases} 4x + 6y = 12 \xrightarrow{\text{Multiply by 3.}} \\ 6x + 9y = 12 \xrightarrow{\text{Multiply by } -2.} \end{cases} \begin{cases} 12x + 18y = 36 \\ -12x - 18y = -24 \end{cases}$$
$$\text{Add:} \qquad 0 = 12$$

> There are no values of x and y for which $0 = 12$.

The false statement $0 = 12$ indicates that the system is inconsistent and has no solution. The solution set is the empty set, $\varnothing$. The graphs of the system's equations are shown in **Figure 4.10**. The lines are parallel and have no point of intersection. ∎

✓ **CHECK POINT 5** Solve the system:

$$\begin{cases} x + 2y = 4 \\ 3x + 6y = 13. \end{cases}$$

If you use the addition method, how can you tell if a system has infinitely many solutions? As with the substitution method, you will eliminate both variables and obtain a true statement.

EXAMPLE 6 Using the Addition Method on a System with Infinitely Many Solutions

Solve by the addition method:

$$\begin{cases} 2x - y = 3 \\ -4x + 2y = -6. \end{cases}$$

Solution We can eliminate y by multiplying the first equation by 2.

$$\begin{cases} 2x - y = 3 \xrightarrow{\text{Multiply by 2.}} \\ -4x + 2y = -6 \xrightarrow{\text{No change}} \end{cases} \begin{cases} 4x - 2y = 6 \\ -4x + 2y = -6 \end{cases}$$
$$\text{Add:} \qquad 0 = 0$$

The true statement $0 = 0$ indicates that the system contains dependent equations and has infinitely many solutions. Any ordered pair that satisfies the first equation also satisfies the second equation. We express the solution set for the system in one of two equivalent ways:

$$\{(x, y) \mid 2x - y = 3\} \text{ or } \{(x, y) \mid -4x + 2y = -6\}. \quad ∎$$

✓ **CHECK POINT 6** Solve by the addition method:

$$\begin{cases} x - 5y = 7 \\ 3x - 15y = 21. \end{cases}$$

3 Determine the most efficient method for solving a linear system. ▶

Comparing the Three Solution Methods

Table 4.2 at the top of the next page compares the graphing, substitution, and addition methods for solving systems of linear equations. With increased practice, it will become easier for you to select the best method for solving a particular linear system.

Table 4.2	Comparing Solution Methods	
Method	**Advantages**	**Disadvantages**
Graphing	You can see the solutions.	If the solutions do not involve integers or are too large to be seen on the graph, it's impossible to tell exactly what the solutions are.
Substitution	Gives exact solutions. Easy to use if a variable is on one side by itself.	Solutions cannot be seen. Can introduce extensive work with fractions when no variable has a coefficient of 1 or -1.
Addition	Gives exact solutions. Easy to use even if no variable has a coefficient of 1 or -1.	Solutions cannot be seen.

Achieving Success

We have seen that **the best way to achieve success in math is through practice**. Keeping up with your homework, preparing for tests, asking questions of your professor, reading your textbook, and attending all classes will help you learn the material and boost your confidence. Use language in a proactive way that reflects a sense of responsibility for your own successes and failures.

Reactive Language	Proactive Language
I'll try.	I'll do it.
That's just the way I am.	I can do better than that.
There's not a thing I can do.	I have options for improvement.
I have to.	I choose to.
I can't.	I can find a way.

CONCEPT AND VOCABULARY CHECK

Fill in each blank so that the resulting statement is true.

1. When solving

$$\begin{cases} x + 5y = 18 \\ 3x + 2y = -11 \end{cases}$$

by the addition method, we can eliminate x by multiplying the first equation by _____ and then adding the equations.

2. When solving

$$\begin{cases} 2x + 10y = 9 \\ 8x + 5y = 7 \end{cases}$$

by the addition method, we can eliminate y by multiplying the second equation by _____ and then adding the equations.

3. When solving

$$\begin{cases} 2x + 5y = 7 \\ 3x + 4y = 9 \end{cases}$$

by the addition method, we can eliminate x by multiplying the first equation by 3 and the second equation by _____, and then adding the equations.

4. When solving

$$\begin{cases} 4x - 3y = 2 \\ -5x + 8y = 1 \end{cases}$$

by the addition method, we can eliminate y by multiplying the first equation by 8 and the second equation by _____, and then adding the equations.

4.3 EXERCISE SET ▶ MyMathLab®

Practice Exercises

In Exercises 1–44, solve each system by the addition method. If there is no solution or an infinite number of solutions, so state. Use set notation to express solution sets.

1. $\begin{cases} x + y = -3 \\ x - y = 11 \end{cases}$

2. $\begin{cases} x + y = 6 \\ x - y = -2 \end{cases}$

3. $\begin{cases} 2x + 3y = 6 \\ 2x - 3y = 6 \end{cases}$

4. $\begin{cases} 3x + 2y = 14 \\ 3x - 2y = 10 \end{cases}$

5. $\begin{cases} x + 2y = 7 \\ -x + 3y = 18 \end{cases}$

6. $\begin{cases} 2x + y = -2 \\ -2x - 3y = -6 \end{cases}$

7. $\begin{cases} 5x - y = 14 \\ -5x + 2y = -13 \end{cases}$

8. $\begin{cases} 7x - 4y = 13 \\ -7x + 6y = -11 \end{cases}$

9. $\begin{cases} 3x + y = 7 \\ 2x - 5y = -1 \end{cases}$

10. $\begin{cases} 3x - y = 11 \\ 2x + 5y = 13 \end{cases}$

11. $\begin{cases} x + 3y = 4 \\ 4x + 5y = 2 \end{cases}$

12. $\begin{cases} x + 2y = -1 \\ 4x - 5y = 22 \end{cases}$

13. $\begin{cases} -3x + 7y = 14 \\ 2x - y = -13 \end{cases}$

14. $\begin{cases} 2x - 5y = -1 \\ 3x + y = 7 \end{cases}$

15. $\begin{cases} 3x - 14y = 6 \\ 5x + 7y = 10 \end{cases}$

16. $\begin{cases} 5x - 4y = 19 \\ 3x + 2y = 7 \end{cases}$

17. $\begin{cases} 3x - 4y = 11 \\ 2x + 3y = -4 \end{cases}$

18. $\begin{cases} 2x + 3y = -16 \\ 5x - 10y = 30 \end{cases}$

19. $\begin{cases} 3x + 2y = -1 \\ -2x + 7y = 9 \end{cases}$

20. $\begin{cases} 5x + 3y = 27 \\ 7x - 2y = 13 \end{cases}$

21. $\begin{cases} 3x = 2y + 7 \\ 5x = 2y + 13 \end{cases}$

22. $\begin{cases} 9x = 25 + y \\ 2y = 4 - 9x \end{cases}$

23. $\begin{cases} 2x = 3y - 4 \\ -6x + 12y = 6 \end{cases}$

24. $\begin{cases} 5x = 4y - 8 \\ 3x + 7y = 14 \end{cases}$

25. $\begin{cases} 2x - y = 3 \\ 4x + 4y = -1 \end{cases}$

26. $\begin{cases} 3x - y = 22 \\ 4x + 5y = -21 \end{cases}$

27. $\begin{cases} 4x = 5 + 2y \\ 2x + 3y = 4 \end{cases}$

28. $\begin{cases} 3x = 4y + 1 \\ 4x + 3y = 1 \end{cases}$

29. $\begin{cases} 3x - y = 1 \\ 3x - y = 2 \end{cases}$

30. $\begin{cases} 4x - 9y = -2 \\ -4x + 9y = -2 \end{cases}$

31. $\begin{cases} x + 3y = 2 \\ 3x + 9y = 6 \end{cases}$

32. $\begin{cases} 4x - 2y = 2 \\ 2x - y = 1 \end{cases}$

33. $\begin{cases} 7x - 3y = 4 \\ -14x + 6y = -7 \end{cases}$

34. $\begin{cases} 2x + 4y = 5 \\ 3x + 6y = 6 \end{cases}$

35. $\begin{cases} 5x + y = 2 \\ 3x + y = 1 \end{cases}$

36. $\begin{cases} 2x - 5y = -1 \\ 2x - y = 1 \end{cases}$

37. $\begin{cases} x = 5 - 3y \\ 2x + 6y = 10 \end{cases}$

38. $\begin{cases} 4x = 36 + 8y \\ 3x - 6y = 27 \end{cases}$

39. $\begin{cases} 4(3x - y) = 0 \\ 3(x + 3) = 10y \end{cases}$

40. $\begin{cases} 2(2x + 3y) = 0 \\ 7x = 3(2y + 3) + 2 \end{cases}$

41. $\begin{cases} x + y = 11 \\ \dfrac{x}{5} + \dfrac{y}{7} = 1 \end{cases}$

42. $\begin{cases} x - y = -3 \\ \dfrac{x}{9} - \dfrac{y}{7} = -1 \end{cases}$

43. $\begin{cases} \dfrac{4}{5}x - y = -1 \\ \dfrac{2}{5}x + y = 1 \end{cases}$

44. $\begin{cases} \dfrac{x}{3} + y = 3 \\ \dfrac{x}{2} - \dfrac{y}{4} = 1 \end{cases}$

In Exercises 45–56, solve each system by the method of your choice. If there is no solution or an infinite number of solutions, so state. Use set notation to express solution sets. Explain why you selected one method over the other two.

45. $\begin{cases} 3x - 2y = 8 \\ x = -2y \end{cases}$

46. $\begin{cases} 2x - y = 10 \\ y = 3x \end{cases}$

47. $\begin{cases} 3x + 2y = -3 \\ 2x - 5y = 17 \end{cases}$

48. $\begin{cases} 2x - 7y = 17 \\ 4x - 5y = 25 \end{cases}$

49. $\begin{cases} 3x - 2y = 6 \\ y = 3 \end{cases}$

50. $\begin{cases} 2x + 3y = 7 \\ x = 2 \end{cases}$

51. $\begin{cases} y = 2x + 1 \\ y = 2x - 3 \end{cases}$

52. $\begin{cases} y = 2x + 4 \\ y = 2x - 1 \end{cases}$

53. $\begin{cases} 2(x + 2y) = 6 \\ 3(x + 2y - 3) = 0 \end{cases}$

54. $\begin{cases} 2(x + y) = 4x + 1 \\ 3(x - y) = x + y - 3 \end{cases}$

55. $\begin{cases} 3y = 2x \\ 2x + 9y = 24 \end{cases}$

56. $\begin{cases} 4y = -5x \\ 5x + 8y = 20 \end{cases}$

Practice PLUS

In Exercises 57–60, write a system of equations modeling the given conditions. Then solve the system by the addition method and find the two numbers.

57. Five times a first number increased by a second number is 14. The difference between four times the first number and the second number is 4. Find the numbers.

58. Three times a first number increased by twice a second number is 11. The difference between the first number and twice the second number is 9. Find the numbers.

59. If four times a first number is decreased by three times a second number, the result is 0. The sum of the numbers is −7. Find the numbers.

60. If three times a first number is decreased by six times a second number, the result is 15. The sum of the numbers is 2. Find the numbers.

In Exercises 61–68, solve each system or state that the system is inconsistent or dependent.

61. $\begin{cases} \dfrac{3x}{5} + \dfrac{4y}{5} = 1 \\ \dfrac{x}{4} - \dfrac{3y}{8} = -1 \end{cases}$

62. $\begin{cases} \dfrac{x}{3} - \dfrac{y}{2} = \dfrac{2}{3} \\ \dfrac{2x}{3} + y = \dfrac{4}{3} \end{cases}$

63. $\begin{cases} 5(x + 1) = 7(y + 1) - 7 \\ 6(x + 1) + 5 = 5(y + 1) \end{cases}$

64. $\begin{cases} 6x = 5(x + y + 3) - x \\ 3(x - y) + 4y = 5(y + 1) \end{cases}$

65. $\begin{cases} 0.4x + \quad y = 2.2 \\ 0.5x - 1.2y = 0.3 \end{cases}$

66. $\begin{cases} 1.25x - \quad 1.5y = \quad 2 \\ 3.5x - 1.75y = 10.5 \end{cases}$

67. $\begin{cases} \dfrac{x}{2} = \dfrac{y + 8}{3} \\ \dfrac{x + 2}{2} = \dfrac{y + 11}{3} \end{cases}$

68. $\begin{cases} \dfrac{x}{2} = \dfrac{y + 8}{4} \\ \dfrac{x + 3}{2} = \dfrac{y + 5}{4} \end{cases}$

Application Exercises

69. We opened this section with a study showing that late in the semester, procrastinating students reported more symptoms of physical illness than their nonprocrastinating peers. The data can be modeled by the following system of equations:

$$\begin{cases} -0.45x + y = 0.8 \\ -0.15x + y = 2.6. \end{cases}$$

The average number of symptoms for procrastinators, y, after x weeks

The average number of symptoms for nonprocrastinators, y, after x weeks

Use the addition method to determine by which week in the semester both groups report the same number of symptoms of physical illness. For that week, how many symptoms were reported by each group? How is this shown in **Figure 4.9** on page 303?

70. The bar graph shows the percentage of U.S. income received by the top 10 percent of Americans and the bottom 90 percent of Americans for seven selected years. Data displayed for 2020, 2025, and 2030 are projections.

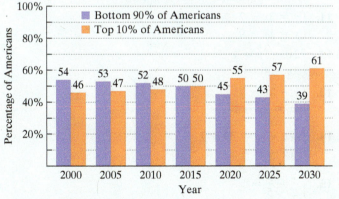

Share of Total U.S. Income

■ Bottom 90% of Americans
■ Top 10% of Americans

Source: motherjones.com/richest

a. In which year shown by the graph will the top 10 percent of Americans receive more than 60% of total U.S. income?

b. In which year shown by the graph did the top 10 percent of Americans and the bottom 90 percent of Americans receive the same percentage of total U.S. income? What percentage of total income did each group control?

c. The data displayed by the graph can be modeled by the following system of equations:

$$\begin{cases} x + 2y = 112 \\ -x + 2y = 88. \end{cases}$$

The percentage of U.S. income, y, received by the bottom 90% of Americans x years after 2000

The percentage of U.S. income, y, received by the top 10% of Americans x years after 2000

Use these models and the addition method to determine in which year the top 10 percent of Americans and the bottom 90 percent of Americans received the same percentage of total U.S. income. Use the models to find the percentage of total income that each group received.

d. How well do the models in part (c) describe the data displayed by the graphs that you used to answer part (b)?

Explaining the Concepts

71. Explain how to solve a system of equations using the addition method. Use $3x + 5y = -2$ and $2x + 3y = 0$ to illustrate your explanation.

72. When using the addition method, how can you tell if a system of linear equations has no solution?

73. When using the addition method, how can you tell if a system of linear equations has infinitely many solutions?

74. Take a second look at the data about the share of U.S. income controlled by the top 10 percent and the bottom 90 percent of Americans, shown in Exercise 70. Do you think that these trends will continue? Explain your answer.

75. The formula $3239x + 96y = 134{,}014$ models the number of daily evening newspapers, y, x years after 1980. The formula $-665x + 36y = 13{,}800$ models the number of daily morning newspapers, y, x years after 1980. What is the most efficient method for solving this system? Explain why. What does the solution mean in terms of the variables in the formulas? (It is not necessary to actually solve the system.)

Critical Thinking Exercises

Make Sense? *In Exercises 76–79, determine whether each statement makes sense or does not make sense, and explain your reasoning.*

76. When I use the addition method, I add the equations only if the x-coefficients or the y-coefficients are opposites.

77. Unlike substitution, the addition method lets me see solutions as intersection points of graphs.

78. When I use the addition method, I sometimes need to multiply more than one equation by a nonzero number before adding the equations.

79. I find it easiest to use the addition method when one of the equations has a variable on one side by itself.

In Exercises 80–83, determine whether each statement is true or false. If the statement is false, make the necessary change(s) to produce a true statement.

80. If x can be eliminated by the addition method, y cannot be eliminated by using the original equations of the system.

81. If $Ax + 2y = 2$ and $2x + By = 10$ have graphs that intersect at $(2, -2)$, then $A = -3$ and $B = 3$.

82. The equations $y = x - 1$ and $x = y + 1$ are dependent.

83. If the two equations in a linear system are $5x - 3y = 7$ and $4x + 9y = 11$, multiplying the first equation by 4, the second by 5, and then adding equations will eliminate x.

84. Solve by expressing x and y in terms of a and b:
$$\begin{cases} x - y = a \\ y = 2x + b. \end{cases}$$

85. The point of intersection of the graphs of the equations $Ax - 3y = 16$ and $3x + By = 7$ is $(5, -2)$. Find A and B.

Review Exercises

86. For which number is 5 times the number equal to the number increased by 40? (Section 2.5, Example 1)

87. In which quadrant is $\left(-\frac{3}{2}, 15\right)$ located? (Section 3.1, Example 1)

88. Solve: $29,700 + 150x = 5000 + 1100x$. (Section 2.2, Example 7)

Preview Exercises

Exercises 89–91 will help you prepare for the material covered in the next section.

89. The sum of two numbers, x and y, is 28. The difference between the numbers is 6.

  a. Write a system of linear equations that models these conditions.

  b. Solve the system and find the numbers.

90. If a slice of cheese contains x calories and a glass of wine contains y calories, write an algebraic expression for the number of calories in 3 slices of cheese and 2 glasses of wine.

91. A cellphone data plan has a monthly fee of $40 with a charge of $15 per gigabyte (GB).

  a. If you use 6 GB of data per month, what is the total cost of the plan?

  b. Write a formula that describes the total monthly cost of the plan, y, for x GB of data.

MID-CHAPTER CHECK POINT Section 4.1 – Section 4.3

What You Know: We learned how to solve systems of linear equations by graphing, by the substitution method, and by the addition method. We saw that some systems, called inconsistent systems, have no solution, whereas other systems, called dependent systems, have infinitely many solutions.

In Exercises 1–3, solve each system by graphing.

1. $\begin{cases} 3x + 2y = 6 \\ 2x - y = 4 \end{cases}$

2. $\begin{cases} y = 2x - 1 \\ y = 3x - 2 \end{cases}$

3. $\begin{cases} y = 2x - 1 \\ 6x - 3y = 12 \end{cases}$

In Exercises 4–15, solve each system by the method of your choice.

4. $\begin{cases} 5x - 3y = 1 \\ y = 3x - 7 \end{cases}$

5. $\begin{cases} 6x + 5y = 7 \\ 3x - 7y = 13 \end{cases}$

6. $\begin{cases} x = \dfrac{y}{3} - 1 \\ 6x + y = 21 \end{cases}$

7. $\begin{cases} 3x - 4y = 6 \\ 5x - 6y = 8 \end{cases}$

8. $\begin{cases} 3x - 2y = 32 \\ \dfrac{x}{5} + 3y = -1 \end{cases}$

9. $\begin{cases} x - y = 3 \\ 2x = 4 + 2y \end{cases}$

10. $\begin{cases} x = 2(y - 5) \\ 4x + 40 = y - 7 \end{cases}$

11. $\begin{cases} y = 3x - 2 \\ y = 2x - 9 \end{cases}$

12. $\begin{cases} 2x - 3y = 4 \\ 3x + 4y = 0 \end{cases}$

13. $\begin{cases} y - 2x = 7 \\ 4x = 2y - 14 \end{cases}$

14. $\begin{cases} 4(x + 3) = 3y + 7 \\ 2(y - 5) = x + 5 \end{cases}$

15. $\begin{cases} \dfrac{x}{2} - \dfrac{y}{5} = 1 \\ y - \dfrac{x}{3} = 8 \end{cases}$

Problem Solving Using Systems of Equations ▶

"Bad television is three things: a bullet train to a morally bankrupt youth, a slow spiral into an intellectual void, and of course, a complete blast to watch."

—*Dennis Miller*

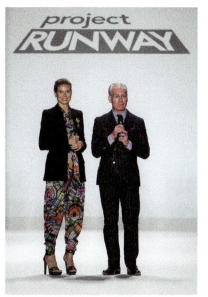

A blast, indeed. Americans spend more of their free time watching TV than any other activity. We open this section with a system of equations that models how we spend free time, including how many of those 1440 minutes in a day we spend glued to the tube.

A Strategy for Solving Word Problems Using Systems of Equations

When we solved problems in Chapter 2, we let *x* represent a quantity that was unknown. Problems in this section involve two unknown quantities. We will let *x* and *y* represent these quantities. We then translate from the verbal conditions of the problem to a *system* of linear equations.

EXAMPLE 1   Free Time

Figure 4.11 shows the average time per day on weekends that Americans spend on various leisure and sports activities.

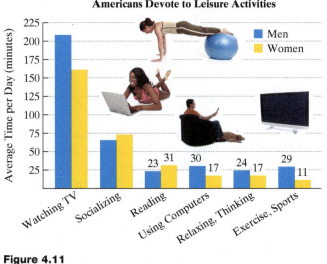

Average Time per Day on Weekends Americans Devote to Leisure Activities

Figure 4.11

Source: Bureau of Labor Statistics American Time Use Survey

It appears that TV is our preferred way of goofing off. Each weekend day, the sum of the average times spent watching TV for men and women is 369 minutes. (That's more than 6 hours.) The difference between the average times spent watching TV for men and women is 47 minutes. How many minutes per day on weekends do men and women devote to watching TV?

Solution

Step 1. Use variables to represent unknown quantities.

Let x = average time per day men watch TV.

Let y = average time per day women watch TV.

Step 2. Write a system of equations that models the problem's conditions.

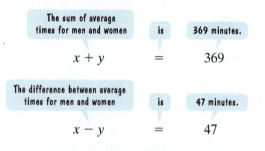

The sum of average times for men and women is 369 minutes.

$$x + y = 369$$

The difference between average times for men and women is 47 minutes.

$$x - y = 47$$

Step 3. Solve the system and answer the problem's question. The system

$$\begin{cases} x + y = 369 \\ x - y = 47 \end{cases}$$

can easily be solved by addition because the y-coefficients are opposites. Adding equations will eliminate y.

$$\begin{cases} x + y = 369 \\ \underline{x - y = 47} \end{cases}$$

Add: $2x = 416$

$x = 208$ Divide both sides by 2.

Because x represents average time per day men watch TV, we see that men devote 208 minutes per day on weekends (or 3 hours 28 minutes) to this leisure activity. Now we can find y, the time for women. We do so by back-substituting 208 for x in either of the system's equations.

$x + y = 369$ We'll use the first equation.

$208 + y = 369$ Back-substitute 208 for x.

$y = 161$ Subtract 208 from both sides.

Because y represents average time per day women watch TV, we see that women devote 161 minutes per day on weekends (or 2 hours 41 minutes) to this leisure activity.

Step 4. Check the proposed answers in the original wording of the problem. We check the times, 208 minutes for men and 161 minutes for women, in each of the problem's conditions.

The sum of the average times spent watching TV for men and women is 369 minutes:

$$208 \text{ minutes} + 161 \text{ minutes} = 369 \text{ minutes.}$$

The difference between the average times spent watching TV for men and women is 47 minutes:

$$208 \text{ minutes} - 161 \text{ minutes} = 47 \text{ minutes.}$$

This verifies that men spend 208 minutes per day on weekends watching TV and women spend 161 minutes per day. ■

✓ CHECK POINT 1 **Figure 4.11** on page 313 indicates that socializing is our second-favorite leisure activity. Each weekend day, the sum of the average times spent socializing for men and women is 138 minutes. The difference between the average times spent socializing for women and men is 8 minutes. How many minutes per day on weekends do men and women devote to socializing?

EXAMPLE 2   Cholesterol and Heart Disease

The verdict is in. After years of research, the nation's health experts agree that high cholesterol in the blood is a major contributor to heart disease. Thus, cholesterol intake should be limited to 300 milligrams or less each day. Fast foods provide a cholesterol carnival. All together, two McDonald's Quarter Pounders and three Burger King Whoppers with cheese contain 520 milligrams of cholesterol. Three Quarter Pounders and one Whopper with cheese exceed the suggested daily cholesterol intake by 53 milligrams. Determine the cholesterol content in each item.

Solution
Step 1. Use variables to represent the unknown quantities.

Let x = the cholesterol content, in milligrams, of a Quarter Pounder.

Let y = the cholesterol content, in milligrams, of a Whopper with cheese.

Step 2. Write a system of equations that models the problem's conditions.

The amount of cholesterol in 2 Quarter Pounders	plus	the amount of cholesterol in 3 Whoppers with cheese	is	520 mg.
$2x$	$+$	$3y$	$=$	520

The amount of cholesterol in 3 Quarter Pounders	plus	the amount of cholesterol in 1 Whopper with cheese	is	the suggested daily limit plus 53 mg.
$3x$	$+$	y	$=$	$300 + 53$

Step 3. Solve the system and answer the problem's question. The system

$$\begin{cases} 2x + 3y = 520 \\ 3x + y = 353 \end{cases}$$

can be solved by substitution or addition. We will use addition; if we multiply the second equation by -3, adding equations will eliminate y.

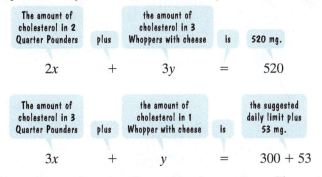

$$\begin{cases} 2x + 3y = 520 \quad \xrightarrow{\text{No change}} \\ 3x + y = 353 \quad \xrightarrow{\text{Multiply by } -3.} \end{cases} \begin{cases} 2x + 3y = 520 \\ -9x - 3y = -1059 \end{cases}$$

$$\text{Add: } -7x = -539$$

$$x = \frac{-539}{-7} = 77$$

Because x represents the cholesterol content of a Quarter Pounder, we see that a Quarter Pounder contains 77 milligrams of cholesterol. Now we can find y, the cholesterol content of a Whopper with cheese. We do so by back-substituting 77 for x in either of the system's equations.

$3x + y = 353$	We'll use the second equation.
$3(77) + y = 353$	Back-substitute 77 for x.
$231 + y = 353$	Multiply.
$y = 122$	Subtract 231 from both sides.

Because $x = 77$ and $y = 122$, a Quarter Pounder contains 77 milligrams of cholesterol and a Whopper with cheese contains 122 milligrams of cholesterol.

Step 4. Check the proposed answers in the original wording of the problem. Two Quarter Pounders and three Whoppers with cheese contain

$$2(77 \text{ mg}) + 3(122 \text{ mg}) = 520 \text{ mg},$$

which checks with the given conditions. Furthermore, three Quarter Pounders and one Whopper with cheese contain

$$3(77 \text{ mg}) + 1(122 \text{ mg}) = 353 \text{ mg},$$

which does exceed the daily limit of 300 milligrams by 53 milligrams. ∎

✓ **CHECK POINT 2** How do the Quarter Pounder and Whopper with cheese measure up in the calorie department? Actually, not too well. Two Quarter Pounders and three Whoppers with cheese provide 2607 calories. Even combining one of each provides enough calories to bring tears to Jenny Craig's eyes—9 calories in excess of what is allowed on a 1000-calorie-a-day diet. Find the caloric content of each item.

EXAMPLE 3 Fencing a Waterfront Lot

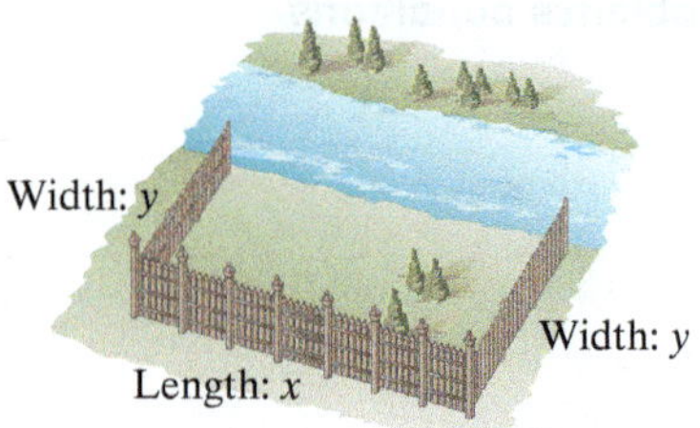

You just purchased a rectangular waterfront lot along a river's edge. The perimeter of the lot is 1000 feet. To create a sense of privacy, you decide to fence along three sides, excluding the side that fronts the river (see **Figure 4.12**). An expensive fencing along the lot's front length costs $25 per foot. An inexpensive fencing along the two side widths costs only $5 per foot. The total cost of the fencing along three sides comes to $9500.

a. What are the lot's dimensions?

b. You are considering using the expensive fencing on all three sides of the lot. However, you are limited by a $16,000 budget. Can this be done within your budget constraints?

Width: y
Width: y
Length: x

Figure 4.12

Solution

a. We begin by finding the lot's dimensions.

Step 1. Use variables to represent unknown quantities.

Let $x =$ the lot's length, in feet.

Let $y =$ the lot's width, in feet.

(If you prefer, you can use the variable l for length and w for width.)

Step 2. Write a system of equations that models the problem's conditions.

- The lot's perimeter is 1000 feet.

Twice the length	plus	twice the width	is	the perimeter.
$2x$	$+$	$2y$	$=$	1000

- The cost of fencing three sides of the lot is $9500.

Fencing along the front length	plus	fencing along the two side widths	costs	$9500.

Cost per foot	·	number of feet	+	cost per foot	·	number of feet	is	$9500.
25	·	x	$+$	5	·	$2y$	$=$	9500

Step 3. Solve the system and answer the problem's question. The system

$$\begin{cases} 2x + 2y = 1000 \\ 25x + 10y = 9500 \end{cases}$$

can be solved most easily by addition. If we multiply the first equation by −5, adding equations will eliminate y.

$$\begin{cases} 2x + 2y = 1000 \\ 25x + 10y = 9500 \end{cases} \xrightarrow[\text{No change}]{\text{Multiply by }-5} \begin{cases} -10x - 10y = -5000 \\ 25x + 10y = 9500 \end{cases}$$

$$\text{Add: } 15x = 4500$$

$$x = \frac{4500}{15} = 300$$

Because x represents length, we see that the lot is 300 feet long. Now we can find y, the lot's width. We do so by back-substituting 300 for x in either of the system's equations.

$2x + 2y = 1000$	We'll use the first equation.
$2 \cdot 300 + 2y = 1000$	Back-substitute 300 for x.
$600 + 2y = 1000$	Multiply: 2 · 300 = 600.
$2y = 400$	Subtract 600 from both sides.
$y = 200$	Divide both sides by 2.

Because $x = 300$ and $y = 200$, the lot is 300 feet long and 200 feet wide. Its dimensions are 300 feet by 200 feet.

Step 4. Check the proposed answers in the original wording of the problem. The perimeter should be 1000 feet:

$$2(300 \text{ ft}) + 2(200 \text{ ft}) = 600 \text{ ft} + 400 \text{ ft} = 1000 \text{ ft}.$$

The cost of fencing three sides of the lot should be $9500:

Fencing cost: front length   Fencing cost: two side widths

$$300 \cdot \$25 + 400 \cdot \$5 = \$7500 + \$2000 = \$9500.$$

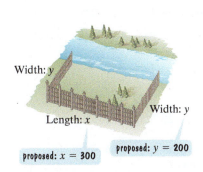

Width: y

Width: y

Length: x

proposed: $x = 300$

proposed: $y = 200$

Figure 4.12 (repeated)
Showing proposed values for x and y

b. Can you use the expensive fencing along all three sides of the lot and stay within a $16,000 budget? The fencing is to be placed along the length, 300 feet, and the two side widths, 2 · 200 feet, or 400 feet. Thus, you will fence 700 feet. The expensive fencing costs $25 per foot. The cost is

$$700 \text{ feet} \cdot \frac{\$25}{\text{feet}} = 700 \cdot \$25 = \$17,\!500.$$

Limited by a $16,000 budget, you cannot use the expensive fencing on three sides of the lot. ∎

✓ **CHECK POINT 3** A rectangular lot whose perimeter is 360 feet is fenced along three sides. An expensive fencing along the lot's length costs $20 per foot. An inexpensive fencing along the two side widths costs only $8 per foot. The total cost of the fencing along the three sides comes to $3280. What are the lot's dimensions?

EXAMPLE 4   Solar and Electric Heating Systems

The costs for two different kinds of heating systems for a two-bedroom home are given in the following table.

System	Cost to Install	Operating Cost/Year
Solar	$29,700	$150
Electric	$5000	$1100

After how many years will total costs for solar heating and electric heating be the same? What will be the cost at that time?

Solution

Step 1. Use variables to represent unknown quantities.

Let x = the number of years the heating system is used.

Let y = the total cost for the heating system.

Step 2. Write a system of equations that models the problem's conditions.

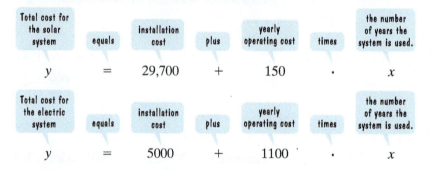

Total cost for the solar system	equals	installation cost	plus	yearly operating cost	times	the number of years the system is used.
y	$=$	$29{,}700$	$+$	150	$\cdot$	x

Total cost for the electric system	equals	installation cost	plus	yearly operating cost	times	the number of years the system is used.
y	$=$	5000	$+$	1100	$\cdot$	x

Step 3. Solve the system and answer the problem's question. We want to know after how many years the total costs for the two heating systems will be the same. We must solve the system of equations

$$\begin{cases} y = 29{,}700 + 150x \\ y = 5000 + 1100x. \end{cases}$$

Substitution works well because y is isolated in each equation.

$y = \boxed{29{,}700 + 150x} \qquad \boxed{y} = 5000 + 1100x$ Substitute 29,700 + 150x for y.

$29{,}700 + 150x = 5000 + 1100x$ This substitution gives an equation in one variable.

$29{,}700 = 5000 + 950x$ Subtract 150x from both sides.

$24{,}700 = 950x$ Subtract 5000 from both sides.

$26 = x$ Divide both sides by 950: $\dfrac{24{,}700}{950} = 26.$

Because x represents the number of years the heating system is used, we see that after 26 years, the total costs for the two systems will be the same. Now we can find y, the total cost. Back-substitute 26 for x in either of the system's equations. We will use the second equation, $y = 5000 + 1100x$.

$$y = 5000 + 1100 \cdot 26 = 5000 + 28{,}600 = 33{,}600$$

Because $x = 26$ and $y = 33{,}600$, after 26 years, the total costs for the two systems will be the same. The cost for each system at that time will be $33,600.

Step 4. Check the proposed answers in the original wording of the problem. Let's verify that after 26 years, the two systems will cost the same amount. The installation cost for the solar system is $29,700 and the yearly operating cost is $150. Thus, the total cost after 26 years is

$$\$29{,}700 + \$150(26) = \$29{,}700 + \$3900 = \$33{,}600.$$

The installation cost for the electric system is $5000 and the yearly operating cost is $1100. Thus, the total cost after 26 years is

$$\$5000 + \$1100(26) = \$5000 + \$28{,}600 = \$33{,}600.$$

This verifies that after 26 years the two systems will cost the same amount, $33,600. ■

The graphs in **Figure 4.13** give us a way of visualizing the solution to Example 4. The total cost of solar heating over 40 years is represented by the blue line. The total cost of electric heating over 40 years is represented by the red line. The lines intersect at (26, 33,600): After 26 years, the cost for each system is the same, $33,600.

Can you see that to the right of the intersection point, (26, 33,600), the blue graph representing solar costs lies below the red graph representing electric costs? Thus, after 26 years, or when $x > 26$, the cost for solar heating is less than the cost for electric heating.

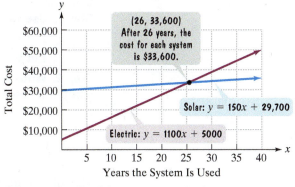

Figure 4.13 Visualizing total costs of solar and electric heating

✓ **CHECK POINT 4**   Costs for two different kinds of heating systems for a home are given in the following table.

System	Cost to Install	Operating Cost/Year
Electric	$5000	$1100
Gas	$12,000	$700

After how long will total costs for electric heating and gas heating be the same? What will be the cost at that time?

Next, we will solve problems involving investments and mixtures with systems of equations. We will continue using our four-step problem-solving strategy. We will also use tables to help organize the information in the problems.

2 Solve simple interest problems. ▶

Dual Investments with Simple Interest

Interest is the amount of money that we get paid for investing money or that we pay for borrowing money. Simple interest involves interest calculated only on the amount of money that we invest, called the **principal**.

Calculating Simple Interest for One Year

$$I = Pr$$

I is the interest earned for one year when the principal, P, is invested at an annual interest rate r. The rate, r, is expressed as a decimal when calculating simple interest.

For example, if you deposit $2000 at a rate of 6%, the interest at the end of the first year is

$$I = Pr = (2000)(0.06) = 120, \text{ or } \$120.$$

Dual investment problems involve different amounts of money in two or more investments, each paying a different rate.

EXAMPLE 5 Solving a Dual Investment Problem

Suppose that you invested $16,000, part at 6% simple interest and the remainder at 8%. If the total yearly interest from these investments was $1180, find the amount invested at each rate.

Solution

Step 1. Use variables to represent unknown quantities.

Let x = the amount invested at 6%.

Let y = the amount invested at 8%.

Step 2. Write a system of equations that models the problem's conditions. The total amount invested is $16,000.

The amount invested at 6%	plus	the amount invested at 8%	equals	$16,000.
x	$+$	y	$=$	$16,000$

The interest for the two investments combined is $1180. We can use a table to organize the information and obtain a second equation.

	Principal	×	Rate	=	Interest
6% Investment	x		0.06		0.06x
8% Investment	y		0.08		0.08y

The interest for the two investments combined is $1180.

Interest from 6% investment	plus	interest from 8% investment	is	$1180.
$0.06x$	$+$	$0.08y$	$=$	1180
rate times principal		rate times principal		

Step 3. Solve the system and answer the problem's question. The system

$$\begin{cases} x + y = 16{,}000 \\ 0.06x + 0.08y = 1180 \end{cases}$$

can be solved by substitution or addition. Substitution works well because both variables in the first equation have coefficients of 1. Addition also works well; if we multiply the first equation by -0.06 or -0.08, adding equations will eliminate a variable. We will use addition.

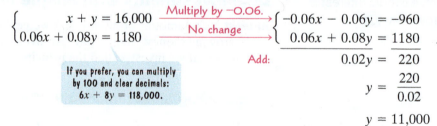

$$\begin{cases} x + y = 16{,}000 \quad \xrightarrow{\text{Multiply by } -0.06} \\ 0.06x + 0.08y = 1180 \quad \xrightarrow{\text{No change}} \end{cases}$$

$$\begin{cases} -0.06x - 0.06y = -960 \\ 0.06x + 0.08y = 1180 \end{cases}$$

Add:

$$0.02y = 220$$

$$y = \frac{220}{0.02}$$

$$y = 11{,}000$$

If you prefer, you can multiply by 100 and clear decimals:
$6x + 8y = 118{,}000.$

Because y represents the amount invested at 8%, $11,000 was invested at 8%. Now we can find x, the amount invested at 6%. We do so by back-substituting 11,000 for y in either of the system's equations.

$$\begin{array}{ll} x + y = 16{,}000 & \text{We'll use the first equation.} \\ x + 11{,}000 = 16{,}000 & \text{Back-substitute 11,000 for } y. \\ x = 5000 & \text{Subtract 11,000 from both sides.} \end{array}$$

Because x represents the amount invested at 6%, $5000 was invested at 6%.

With $x = 5000$ and $y = 11,000$, our proposed solution is that $5000 was invested at 6% and $11,000 was invested at 8%.

Step 4. Check the proposed answers in the original wording of the problem. The problem states that the total interest was $1180. The interest earned on $5000 at 6% is ($5000)(0.06), or $300. The interest earned on $11,000 at 8% is ($11,000)(0.08), or $880. The total interest is $300 + $880, or $1180, exactly the amount given in the problem. ∎

> ☑ **CHECK POINT 5**   Suppose that you invested $25,000, part at 9% simple interest and the remainder at 12%. If the total yearly interest from these investments was $2550, find the amount invested at each rate.

 3 Solve mixture problems.

Problems Involving Mixtures

Chemists and pharmacists often have to change the concentration of solutions and other mixtures. In these situations, the amount of a particular ingredient in the solution or mixture is expressed as a percentage of the total solution.

For example, if a 40-milliliter solution of acid in water contains 35% acid, the amount of acid in the solution is 35% of the total solution.

| Amount of acid in the solution | is | 35% | of | total number of milliliters in the solution. |

$$= (0.35) \cdot (40)$$
$$= 14$$

There are 14 milliliters of acid in the solution.

> **EXAMPLE 6**   Solving a Mixture Problem

A chemist needs to mix an 18% acid solution with a 45% acid solution to obtain 12 liters of a 36% acid solution. How many liters of each of the acid solutions must be used?

Solution

Step 1. Use variables to represent unknown quantities.

Let x = the number of liters of the 18% solution to be used in the mixture.

Let y = the number of liters of the 45% solution to be used in the mixture.

Step 2. Write a system of equations that models the problem's conditions. The situation is illustrated in **Figure 4.14**. The chemist needs 12 liters of a 36% acid solution. We form a table that shows the amount of acid in each of the three solutions.

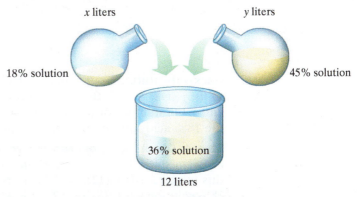

Figure 4.14 Obtaining 12 liters of a 36% acid mixture

	Number of Liters	×	Percent of Acid	=	Amount of Acid
18% Acid Solution	x		18% = 0.18		$0.18x$
45% Acid Solution	y		45% = 0.45		$0.45y$
36% Acid Mixture	12		36% = 0.36		0.36(12) = 4.32

The chemist needs to obtain a 12-liter mixture.

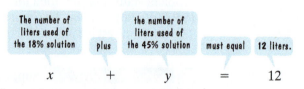

The number of liters used of the 18% solution	plus	the number of liters used of the 45% solution	must equal	12 liters.
x	$+$	y	$=$	12

The 12-liter mixture must be 36% acid. The amount of acid must be 36% of 12, or $(0.36)(12) = 4.32$ liters.

Amount of acid in the 18% solution	plus	amount of acid in the 45% solution	equals	amount of acid in the 36% mixture.
$0.18x$	$+$	$0.45y$	$=$	4.32

Step 3. Solve the system and answer the problem's question. The system

$$\begin{cases} x + y = 12 \\ 0.18x + 0.45y = 4.32 \end{cases}$$

can be solved by substitution or addition. Let's use substitution. The first equation can easily be solved for x or y. Solving for y, we obtain $y = 12 - x$.

$$y = \boxed{12 - x} \qquad 0.18x + 0.45\boxed{y} = 4.32$$

We substitute $12 - x$ for y in the second equation. This gives us an equation in one variable.

If you prefer, you can multiply by 100 and clear decimals:
$18x + 45(12 - x) = 432.$

$0.18x + 0.45(12 - x) = 4.32$ This equation contains one variable, x.

$0.18x + 5.4 - 0.45x = 4.32$ Apply the distributive property.

$-0.27x + 5.4 = 4.32$ Combine like terms.

$-0.27x = -1.08$ Subtract 5.4 from both sides.

$\dfrac{-0.27x}{-0.27} = \dfrac{-1.08}{-0.27}$ Divide both sides by -0.27.

$x = 4$ Simplify.

Back-substituting 4 for x in either of the system's equations ($x + y = 12$ is easier to use) gives $y = 8$. Because x represents the number of liters of the 18% solution and y the number of liters of the 45% solution, the chemist should mix 4 liters of the 18% acid solution with 8 liters of the 45% acid solution.

Step 4. Check the proposed answers in the original wording of the problem. The problem states that the 12-liter mixture should be 36% acid. The amount of acid in this mixture is 0.36(12), or 4.32 liters of acid. The amount of acid in 4 liters of the 18% solution is 0.18(4), or 0.72 liter. The amount of acid in 8 liters of the 45% solution is 0.45(8), or 3.6 liters. The amount of acid in the two solutions used in the mixture is 0.72 liter + 3.6 liters, or 4.32 liters, exactly as it should be. ■

✓ **CHECK POINT 6**   A chemist needs to mix a 10% acid solution with a 60% acid solution to obtain 50 milliliters of a 30% acid solution. How many milliliters of each of the acid solutions must be used?

Great Question!

Are there similarities between dual investment problems and mixture problems?

Problems involving dual investments and problems involving mixtures are both based on the same idea: The total amount times the rate gives the amount.

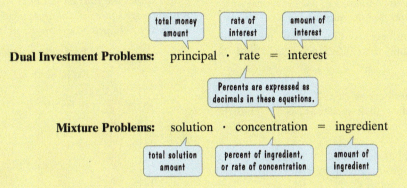

Our dual investment problem involved mixing two investments. Our mixture problem involved mixing two liquids. The equations in these problems are obtained from similar conditions:

Being aware of the similarities between dual investment and mixture problems should make you a better problem solver in a variety of situations that involve mixtures.

Achieving Success

Avoid asking for help on a problem that you have not thought about. This is basically asking your professor to do the work for you. First try solving the problem on your own!

CONCEPT AND VOCABULARY CHECK

Fill in each blank so that the resulting statement is true.

1. Suppose that x represents the cost of one apple and y represents the cost of one banana. The cost of five apples and six bananas is represented by _____.

2. Suppose that x represents the length of a rectangular lot, in feet, and y represents its width, in feet. The cost of fencing the entire lot at $10 per foot for the length and $15 per foot for the width is represented by _____.

3. A solar heating system costs $25,600 to install and has operating costs of $225 per year. The total cost for the solar system after x years is represented by _____.

4. The combined yearly interest for x dollars invested at 4% and y dollars invested at 5% is represented by _____.

5. The total amount of acid in x milliliters of a 7% acid solution and y milliliters of a 15% acid solution is represented by _____.

4.4 EXERCISE SET ▶ MyMathLab®

Practice Exercises

In Exercises 1–4, let x represent one number and let y represent the other number. Use the given conditions to write a system of equations. Solve the system and find the numbers.

1. The sum of two numbers is 17. If one number is subtracted from the other, their difference is −3. Find the numbers.

2. The sum of two numbers is 5. If one number is subtracted from the other, their difference is 13. Find the numbers.

3. Three times a first number decreased by a second number is −1. The first number increased by twice the second number is 23. Find the numbers.

4. The sum of three times a first number and twice a second number is 43. If the second number is subtracted from twice the first number, the result is −4. Find the numbers.

Application Exercises

The bar graph shows the average time per day that Americans devote to sprucing up. Exercises 5–6 are based on the graph.

Average Time per Day Americans Spend on Grooming

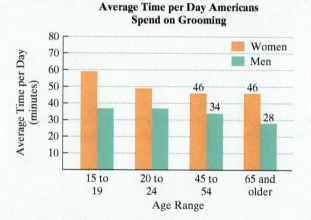

Source: Bureau of Labor Statistics American Time Use Survey

5. Each day, the sum of the average times spent on grooming for 20- to 24-year-old women and men is 86 minutes. The difference between grooming times for 20- to 24-year-old women and men is 12 minutes. How many minutes per day do 20- to 24-year-old women and men spend on grooming?

6. Each day, the sum of the average times spent on grooming for 15- to 19-year-old women and men is 96 minutes. The difference between grooming times for 15- to 19-year-old women and men is 22 minutes. How many minutes per day do 15- to 19-year-old women and men spend on grooming?

Looking for Mr. Goodbar? It's probably not a good idea if you want to look like Mr. Universe or Kate Winslet. The graph shows the four candy bars with the highest fat content, representing grams of fat and calories in each bar. Exercises 7–10 are based on the graph.

Candy Bars with the Highest Fat Content

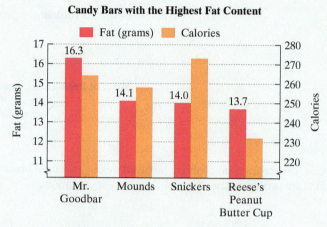

Source: Krantz and Sveum, *The World's Worsts*, HarperCollins, 2005.

7. One Mr. Goodbar and two Mounds bars contain 780 calories. Two Mr. Goodbars and one Mounds bar contain 786 calories. Find the caloric content of each candy bar.

(In Exercises 8–10 refer to the bar graph to visualize that your solution is reasonable.)

8. One Snickers bar and two Reese's Peanut Butter Cups contain 737 calories. Two Snickers bars and one Reese's Peanut Butter Cup contain 778 calories. Find the caloric content of each candy bar.

9. A collection of Halloween candy contains a total of five Mr. Goodbars and Mounds bars. Chew on this: The grams of fat in these candy bars exceed the daily maximum desirable fat intake of 70 grams by 7.1 grams. How many bars of each kind of candy are contained in the Halloween collection?

10. A collection of Halloween candy contains a total of 12 Snickers bars and Reese's Peanut Butter Cups. Chew on this: The grams of fat in these candy bars exceed twice the daily maximum desirable fat intake of 70 grams by 26.5 grams. How many bars of each kind of candy are contained in the Halloween collection?

11. In a discount clothing store, all sweaters are sold at one fixed price and all shirts are sold at another fixed price. If one sweater and three shirts cost $42, while three sweaters and two shirts cost $56, find the price of one sweater and the price of one shirt.

12. A restaurant purchased eight tablecloths and five napkins for $106. A week later, a tablecloth and six napkins were bought for $24. Find the cost of one tablecloth and the cost of one napkin, assuming the same prices for both purchases.

13. The perimeter of a badminton court is 128 feet. After a game of badminton, a player's coach estimates that the athlete has run a total of 444 feet, which is equivalent to six times the court's length plus nine times its width. What are the dimensions of a standard badminton court?

14. The perimeter of a tennis court is 228 feet. After a round of tennis, a player's coach estimates that the athlete has run a total of 690 feet, which is equivalent to seven times the court's length plus four times its width. What are the dimensions of a standard tennis court?

15. A rectangular lot whose perimeter is 320 feet is fenced along three sides. An expensive fencing along the lot's length costs $16 per foot. An inexpensive fencing along the two side widths costs only $5 per foot. The total cost of the fencing along the three sides comes to $2140. What are the lot's dimensions?

16. A rectangular lot whose perimeter is 1600 feet is fenced along three sides. An expensive fencing along the lot's length costs $20 per foot. An inexpensive fencing along the two side widths costs only $5 per foot. The total cost of the fencing along the three sides comes to $13,000. What are the lot's dimensions?

17. You are choosing between two cellphone plans. Data Plan A has a monthly fee of $40 with a charge of $15 per gigabyte (GB). Data Plan B has a monthly fee of $30 with a charge of $20 per GB.

    a. For how many GB of data will the costs for the two data plans be the same? What will be the cost for each plan?

    b. If you use 6 GB of data per month, which plan is the better deal and by how much?

18. You are choosing between two cellphone plans. Data Plan A has a monthly fee of $52 with a charge of $18 per gigabyte (GB). Data Plan B has a monthly fee of $32 with a charge of $22 per GB.

    a. For how many GB of data will the costs for the two data plans be the same? What will be the cost for each plan?

    b. If you use 6 GB of data per month, which plan is the better deal and by how much?

19. You are choosing between two plans at a discount warehouse. Plan A offers an annual membership fee of $100 and you pay 80% of the manufacturer's recommended list price. Plan B offers an annual membership fee of $40 and you pay 90% of the manufacturer's recommended list price. How many dollars of merchandise would you have to purchase in a year to pay the same amount under both plans? What will be the cost for each plan?

20. You are choosing between two plans at a discount warehouse. Plan A offers an annual membership fee of $300 and you pay 70% of the manufacturer's recommended list price. Plan B offers an annual membership fee of $40 and you pay 90% of the manufacturer's recommended list price. How many dollars of merchandise would you have to purchase in a year to pay the same amount under both plans? What will be the cost for each plan?

21. A community center sells a total of 301 tickets for a basketball game. An adult ticket costs $3. A student ticket costs $1. The sponsors collect $487 in ticket sales. Find the number of each type of ticket sold.

22. On a special day, tickets for a minor league baseball game cost $5 for adults and $1 for students. The attendance that day was 1281 and $3425 was collected. Find the number of each type of ticket sold.

23. At the 1987 annual meeting of the National Council of Teachers of Mathematics, the system below was found on a menu card at a restaurant in California. Find the cost of each item in column A and the cost of each item in column B.

24. The perimeter of the rectangle is 34 inches. The perimeter of the triangle is 30 inches.

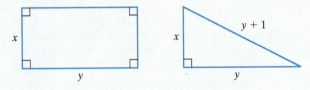

    a. Find the values of x and y.

    b. Find the area of the rectangle and the area of the triangle.

25. Nutritional information for macaroni and broccoli is given in the table. How many servings of each would it take to get exactly 14 grams of protein and 48 grams of carbohydrates?

	Macaroni	Broccoli
Protein (grams/serving)	3	2
Carbohydrates (grams/serving)	16	4

26. The calorie-nutrient information for an apple and an avocado is given in the table. How many of each should be eaten to get exactly 1000 calories and 100 grams of carbohydrates?

	One Apple	One Avocado
Calories	100	350
Carbohydrates (grams)	24	14

In Exercises 27–28, an isosceles triangle in which angles B and C have the same measure is shown. Find the measure of each angle whose degree measure is represented with variables.

27.

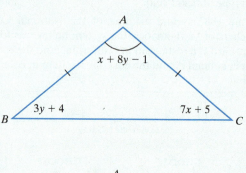

28.

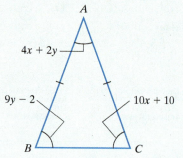

Exercises 29–36 involve dual investments.

29. You invest $20,000 in two accounts paying 7% and 8% annual interest. How much should be invested at each rate if the total interest earned for the year is to be $1520? Begin by filling in the missing entry in the following table. Then use the fact that the interest for the two investments combined must be $1520.

	Principal	×	Rate	=	Interest
7% Investment	x		0.07		0.07x
8% Investment	y		0.08		

30. You invested $20,000 in two accounts paying 7% and 9% annual interest. If the total interest earned for the year was $1550, how much was invested at each rate? Begin by filling in the missing entry in the following table. Then use the fact that the interest for the two investments combined was $1550.

	Principal	×	Rate	=	Interest
7% Investment	x		0.07		0.07x
9% Investment	y		0.09		

31. A bank loaned out $120,000, part of it at the rate of 8% annual mortgage interest and the rest at the rate of 18% annual credit card interest. The interest received on both loans totaled $10,000. How much was loaned at each rate? Organize your work in the following table.

	Principal	×	Rate	=	Interest
8% Loan	x		0.08		
18% Loan			0.18		

32. A bank loaned out $250,000, part of it at the rate of 8% annual mortgage interest and the rest at the rate of 18% annual credit card interest. The interest received on both loans totaled $23,000. How much was loaned at each rate? Organize your work in the following table.

	Principal	×	Rate	=	Interest
8% Loan	x		0.08		
18% Loan			0.18		

33. You invest $6000 in two accounts paying 6% and 9% annual interest. At the end of the year, the accounts earn the same interest. How much was invested at each rate?

34. You invest $7200 in two accounts paying 8% and 10% annual interest. At the end of the year, the accounts earn the same interest. How much was invested at each rate?

35. Your grandmother needs your help. She has $50,000 to invest. Part of this money is to be invested in noninsured bonds paying 15% annual interest. The rest of this money is to be invested in a government-insured certificate of deposit paying 7% annual interest. She told you that she requires a total of $6000 per year in extra income from these investments. How much money should be placed in each investment?

36. Things did not go quite as planned. You invested part of $8000 in an account that paid 12% annual interest. However, the rest of the money suffered a 5% loss. If the total annual income from both investments was $620, how much was invested at each rate?

Exercises 37–42 involve mixtures.

37. A lab technician needs to mix a 5% fungicide solution with a 10% fungicide solution to obtain a 50-liter mixture consisting of 8% fungicide. How many liters of each of the fungicide solutions must be used? Begin by filling in the missing entries in the table on the next page. Then use the fact that the amount of fungicide in the 5% solution plus the amount of fungicide in the 10% solution must equal the amount of fungicide in the 8% mixture.

	Number of Liters	×	Percent of Fungicide	=	Amount of Fungicide
5% Fungicide Solution	x		0.05		$0.05x$
10% Fungicide Solution	y		0.10		
8% Fungicide Mixture	50		0.08		

38. A candy company needs to mix a 20% fat-content chocolate with a 15% fat-content chocolate to obtain 50 kilograms of a 16% fat-content chocolate. How many kilograms of each kind of chocolate must be used? Begin by filling in the missing entries in the following table. Then use the fact that the amount of fat in the 20% fat-content chocolate plus the amount of fat in the 15% fat-content chocolate must equal the amount of fat in the 16% mixture.

	Number of Kilograms	×	Percent of Fat	=	Amount of Fat
20% Fat Chocolate	x		0.20		$0.20x$
15% Fat Chocolate	y		0.15		
16% Fat Mixture	50		0.16		

39. How many ounces of a 15% alcohol solution must be mixed with 4 ounces of a 20% alcohol solution to make a 17% alcohol solution?

40. How many ounces of a 50% alcohol solution must be mixed with 80 ounces of a 20% alcohol solution to make a 40% alcohol solution?

41. At the north campus of a performing arts school, 10% of the students are music majors. At the south campus, 90% of the students are music majors. The campuses are merged into one east campus. If 42% of the 1000 students at the east campus are music majors, how many students did each of the north and south campuses have before the merger?

42. At the north campus of a small liberal arts college, 10% of the students are women. At the south campus, 50% of the students are women. The campuses are merged into one east campus. If 40% of the 1200 students at the east campus are women, how many students did each of the north and south campuses have before the merger?

Explaining the Concepts

43. Describe the conditions in a problem that enable it to be solved using a system of linear equations.

44. Use **Figure 4.11** on page 313 to write a word problem similar to Example 1 or Check Point 1 that involves the leisure activity of reading. Then solve the problem.

45. Exercises 17–20 involve using systems of linear equations to compare costs of cellphone plans and plans at a discount warehouse. Describe another situation that involves choosing between two options that can be modeled and solved with a linear system.

46. Describe two similarities between interest and mixture problems.

47. Must the concentration of a mixture always be greater than the concentration of an ingredient in one of the solutions and less than the concentration of the ingredient in the other solution being mixed? Explain your answer.

Critical Thinking Exercises

Make Sense? *In Exercises 48–51, determine whether each statement makes sense or does not make sense, and explain your reasoning.*

48. The Sea Drift Hotel charges $650 for 7 nights and 10 meals, so I modeled these conditions using $7n + 10m = 650$, where n is the cost of one night and m is the cost of one meal.

49. A cellphone plan charges $40 per month plus $19 per gigabyte, so I modeled these conditions using $y = 19 + 40x$, where y is the monthly charge for x GB of data.

50. The perimeter of the large rectangle shown in the figure is 58 meters, so I modeled this condition using $2(x + 8) + 2(y + 10) = 58$.

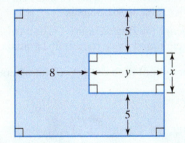

51. Angles A and B are supplementary, so I modeled this condition using

$$4x - 2y + 4 = 12x + 6y + 12 + 180.$$

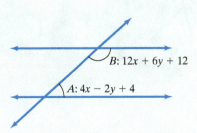

52. A set of identical twins can only be distinguished by the characteristic that one always tells the truth and the other always lies. One twin tells you of a lucky number pair: "When I multiply my first lucky number by 3 and my second lucky number by 6, the addition of the resulting numbers produces a sum of 12. When I add my first lucky number and twice my second lucky number, the sum is 5." Which twin is talking?

53. Tourist: "How many birds and lions do you have in your zoo?" Zookeeper: "There are 30 heads and 100 feet." Tourist: "I can't tell from that." Zookeeper: "Oh, yes, you can!" Can you? Find the number of each.

54. Find the measure of each angle whose degree measure is represented with a variable.

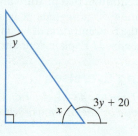

55. One apartment is directly above a second apartment. The resident living downstairs calls his neighbor living above him and states, "If one of you is willing to come downstairs, we'll have the same number of people in both apartments." The upstairs resident responds, "We're all too tired to move. Why don't one of you come up here? Then we'll have twice as many people up here as you've got down there." How many people are in each apartment?

56. In Lewis Carroll's *Through the Looking Glass*, the following dialogue takes place:

  Tweedledum (to Tweedledee): The sum of your weight and twice mine is 361 pounds.

  Tweedledee (to Tweedledum): Contrawise, the sum of your weight and twice mine is 362 pounds.

Find the weight of each of the two characters.

57. You have $70,000 to invest. Part of the money is to be placed in a certificate of deposit paying 8% per year. The rest is to be placed in corporate bonds paying 12% per year. If you wish to obtain an overall return of 9% per year, how much should you place in each investment?

Technology Exercise

58. Select any two problems that you solved from Exercises 5–26. Use a graphing utility to graph the system of equations that you wrote for that problem. Then use the [TRACE] or [INTERSECTION] feature to show the point on the graphs that corresponds to the problem's solution.

Review Exercises

59. Solve: $2(x + 3) = 24 - 2(x + 4)$.
(Section 2.3, Example 3)

60. Simplify: $5 + 6(x + 1)$.
(Section 1.7, Example 7)

61. Write the slope-intercept form of the equation of the line passing through $(-5, 6)$ and $(3, -10)$.
(Section 3.5, Example 2)

Preview Exercises

Exercises 62–64 will help you prepare for the material covered in the next section. In each exercise, graph the given inequality in a rectangular coordinate system.

62. $2x - y < 4$

63. $y \geq x + 1$

64. $x \geq 2$

SECTION

4.5

What am I supposed to learn?

After studying this section, you should be able to:

1 Use mathematical models involving systems of linear inequalities.

2 Graph the solution sets of systems of linear inequalities.

1 Use mathematical models involving systems of linear inequalities.

Systems of Linear Inequalities

We opened the chapter noting that the modern emphasis on thinness as the ideal body shape has been suggested as a major cause of eating disorders. In this section, as well as in the Exercise Set, we use systems of linear inequalities in two variables that will enable you to establish a healthy weight range for your height and age.

Systems of Linear Inequalities

In Section 3.6, we graphed the solutions of a linear inequality in two variables, such as $4.9x - y \geq 165$. Just as two linear equations make up a system of linear equations, two (or more) linear inequalities make up a **system of linear inequalities**. Here is an example of a system of linear inequalities:

$$\begin{cases} 4.9x - y \geq 165 \\ 3.7x - y \leq 125. \end{cases}$$

A **solution of a system of linear inequalities** in two variables is an ordered pair that satisfies each inequality in the system.

EXAMPLE 1 Does Your Weight Fit You?

The latest guidelines, which apply to both men and women, give healthy weight ranges, rather than specific weights, for your height. **Figure 4.15** shows the healthy weight region for various heights for people between the ages of 19 and 34, inclusive.

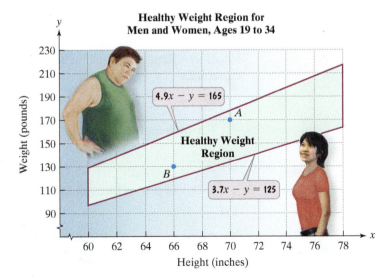

Figure 4.15

Source: U.S. Department of Health and Human Services

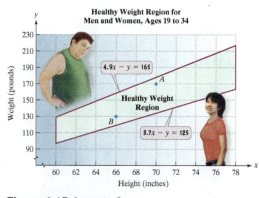

Figure 4.15 (repeated)

If x represents height, in inches, and y represents weight, in pounds, the healthy weight region in **Figure 4.15** can be modeled by the following system of linear inequalities:

$$\begin{cases} 4.9x - y \geq 165 \\ 3.7x - y \leq 125. \end{cases}$$

Show that point A in **Figure 4.15** is a solution of the system of inequalities that describes healthy weight.

Solution   Point A has coordinates $(70, 170)$. This means that if a person is 70 inches tall, or 5 feet 10 inches, and weighs 170 pounds, then that person's weight is within the healthy weight region. We can show that $(70, 170)$ satisfies the system of inequalities by substituting 70 for x and 170 for y in each inequality in the system.

$$4.9x - y \geq 165$$
$$4.9(70) - 170 \stackrel{?}{\geq} 165$$
$$343 - 170 \stackrel{?}{\geq} 165$$
$$173 \geq 165, \quad \text{true}$$

$$3.7x - y \leq 125$$
$$3.7(70) - 170 \stackrel{?}{\leq} 125$$
$$259 - 170 \stackrel{?}{\leq} 125$$
$$89 \leq 125, \quad \text{true}$$

The coordinates $(70, 170)$ make each inequality true. Thus, $(70, 170)$ satisfies the system for the healthy weight region and is a solution of the system. ■

✓ **CHECK POINT 1**   Show that point B in **Figure 4.15** is a solution of the system of inequalities that describes healthy weight.

---

② Graph the solution sets of systems of linear inequalities. ▶

Graphing Systems of Linear Inequalities

The **solution set of a system of linear inequalities in two variables** is the set of all ordered pairs that satisfy each inequality in the system. Thus, to graph a system of inequalities in two variables, begin by graphing each individual inequality in the same rectangular coordinate system. Then find the region, if there is one, that is common to every graph in the system. This region of intersection gives a picture of the system's solution set.

EXAMPLE 2   Graphing a System of Linear Inequalities

Graph the solution set of the system:

$$\begin{cases} 2x - y < 4 \\ x + y \geq -1. \end{cases}$$

Solution   Replacing each inequality symbol with an equal sign indicates that we need to graph $2x - y = 4$ and $x + y = -1$. We can use intercepts to graph these lines.

$2x - y = 4$

x-intercept: $2x - 0 = 4$
$2x = 4$
$x = 2$

> Set $y = 0$ in each equation.

$x + y = -1$

x-intercept: $x + 0 = -1$
$x = -1$

The line passes through $(2, 0)$.

The line passes through $(-1, 0)$.

y-intercept: $2 \cdot 0 - y = 4$
$-y = 4$
$y = -4$

> Set $x = 0$ in each equation.

y-intercept: $0 + y = -1$
$y = -1$

The line passes through $(0, -4)$.

The line passes through $(0, -1)$.

Now we are ready to graph the solution set of the system of linear inequalities.

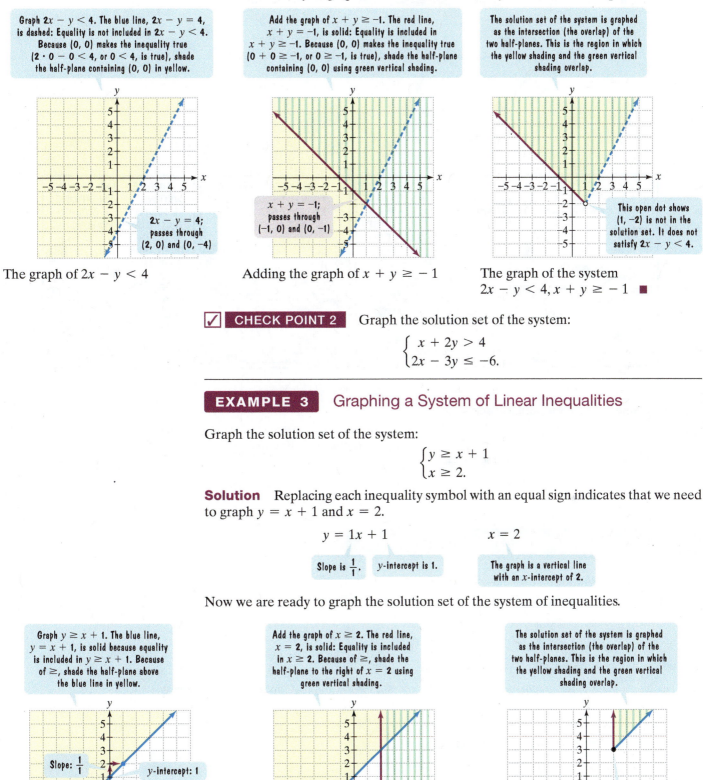

Graph $2x - y < 4$. The blue line, $2x - y = 4$, is dashed: Equality is not included in $2x - y < 4$. Because $(0, 0)$ makes the inequality true $(2 \cdot 0 - 0 < 4$, or $0 < 4$, is true), shade the half-plane containing $(0, 0)$ in yellow.

$2x - y = 4$; passes through $(2, 0)$ and $(0, -4)$

The graph of $2x - y < 4$

Add the graph of $x + y \geq -1$. The red line, $x + y = -1$, is solid: Equality is included in $x + y \geq -1$. Because $(0, 0)$ makes the inequality true $(0 + 0 \geq -1$, or $0 \geq -1$, is true), shade the half-plane containing $(0, 0)$ using green vertical shading.

$x + y = -1$; passes through $(-1, 0)$ and $(0, -1)$

Adding the graph of $x + y \geq -1$

The solution set of the system is graphed as the intersection (the overlap) of the two half-planes. This is the region in which the yellow shading and the green vertical shading overlap.

This open dot shows $(1, -2)$ is not in the solution set. It does not satisfy $2x - y < 4$.

The graph of the system
$2x - y < 4, x + y \geq -1$ ∎

✓ **CHECK POINT 2** Graph the solution set of the system:

$$\begin{cases} x + 2y > 4 \\ 2x - 3y \leq -6. \end{cases}$$

EXAMPLE 3 Graphing a System of Linear Inequalities

Graph the solution set of the system:

$$\begin{cases} y \geq x + 1 \\ x \geq 2. \end{cases}$$

Solution Replacing each inequality symbol with an equal sign indicates that we need to graph $y = x + 1$ and $x = 2$.

$$y = 1x + 1 \qquad\qquad x = 2$$

Slope is $\frac{1}{1}$. y-intercept is 1.

The graph is a vertical line with an x-intercept of 2.

Now we are ready to graph the solution set of the system of inequalities.

Graph $y \geq x + 1$. The blue line, $y = x + 1$, is solid because equality is included in $y \geq x + 1$. Because of $\geq$, shade the half-plane above the blue line in yellow.

Slope: $\frac{1}{1}$ y-intercept: 1

$y = x + 1$, with slope 1 and y-intercept 1

The graph of $y \geq x + 1$

Add the graph of $x \geq 2$. The red line, $x = 2$, is solid: Equality is included in $x \geq 2$. Because of $\geq$, shade the half-plane to the right of $x = 2$ using green vertical shading.

$x = 2$

Adding the graph of $x \geq 2$

The solution set of the system is graphed as the intersection (the overlap) of the two half-planes. This is the region in which the yellow shading and the green vertical shading overlap.

This closed dot shows $(2, 3)$ is in the solution set, satisfying $y \geq x + 1$ and $x \geq 2$.

The graph of the system
$y \geq x + 1, x \geq 2$ ∎

☑ **CHECK POINT 3**   Graph the solution set of the system:

$$\begin{cases} y \geq x + 2 \\ x \geq 1. \end{cases}$$

Achieving Success

FoxTrot

FOXTROT copyright © 2003, 2009 by Bill Amend/Distributed by Universal Uclick

A test-taking tip: Live for partial credit.

Always show your work. If worse comes to worst, write *something* down, anything, even if it's a formula that you think might solve a problem or a possible idea or procedure for solving the problem. As Bill Amend's *FoxTrot* cartoon indicates, partial credit has salvaged more than a few test scores.

CONCEPT AND VOCABULARY CHECK

Fill in each blank so that the resulting statement is true.

1. The solution set of the system

$$\begin{cases} x + y \geq 4 \\ x - y \leq 2 \end{cases}$$

   is the set of ordered pairs that satisfy _____ and _____.

2. The intersection of the graphs of $y \geq 2x + 1$ and $y \leq 4$ gives the graph of the solution set of the system

$$\begin{cases} \text{_____.} \end{cases}$$

3. True or false: The graph of the solution set of the system

$$\begin{cases} x - y \geq 1 \\ x - y < 3 \end{cases}$$

   includes the intersection point of $x - y = 1$ and $x - y = 3$. _____

4. True or false: The graph of the solution set of the system

$$\begin{cases} y \geq 2x + 1 \\ y \leq 4 \end{cases}$$

   includes the intersection point of $y = 2x + 1$ and $y = 4$. _____

4.5 EXERCISE SET ▶ MyMathLab®

Practice Exercises

In Exercises 1–36, graph the solution set of each system of linear inequalities.

1. $\begin{cases} x + y \le 4 \\ x - y \le 2 \end{cases}$ **2.** $\begin{cases} x + y \ge 4 \\ x - y \le 2 \end{cases}$

3. $\begin{cases} 2x - 4y \le 8 \\ x + y \ge -1 \end{cases}$ **4.** $\begin{cases} 4x + 3y \le 12 \\ x - 2y \le 4 \end{cases}$

5. $\begin{cases} x + 3y \le 6 \\ x - 2y \le 4 \end{cases}$ **6.** $\begin{cases} 2x + y \le 4 \\ 2x - y \le 6 \end{cases}$

7. $\begin{cases} x - 2y > 4 \\ 2x + y \ge 6 \end{cases}$ **8.** $\begin{cases} 3x + y < 6 \\ x + 2y \ge 2 \end{cases}$

9. $\begin{cases} x + y > 1 \\ x + y < 4 \end{cases}$ **10.** $\begin{cases} x - y > 1 \\ x - y < 3 \end{cases}$

11. $\begin{cases} y \ge 2x + 1 \\ y \le 4 \end{cases}$ **12.** $\begin{cases} y \ge \frac{1}{2}x + 2 \\ y \le 2 \end{cases}$

13. $\begin{cases} y > x - 1 \\ x > 5 \end{cases}$ **14.** $\begin{cases} y > x - 2 \\ x > 3 \end{cases}$

15. $\begin{cases} y \ge 2x - 3 \\ y \le 2x + 1 \end{cases}$ **16.** $\begin{cases} y \ge 3x - 2 \\ y \le 3x + 1 \end{cases}$

17. $\begin{cases} y > 2x - 3 \\ y \le -x + 6 \end{cases}$ **18.** $\begin{cases} y < -2x + 4 \\ y \le x - 4 \end{cases}$

19. $\begin{cases} x + 2y \le 4 \\ y \ge x - 3 \end{cases}$ **20.** $\begin{cases} x + y \le 4 \\ y \ge 2x - 4 \end{cases}$

21. $\begin{cases} x \le 3 \\ y \ge -2 \end{cases}$ **22.** $\begin{cases} x \le 4 \\ y \le -3 \end{cases}$

23. $\begin{cases} x \ge 3 \\ y < 2 \end{cases}$ **24.** $\begin{cases} x \ge -2 \\ y < -1 \end{cases}$

25. $\begin{cases} x \ge 0 \\ y \le 0 \end{cases}$ **26.** $\begin{cases} x \le 0 \\ y \ge 0 \end{cases}$

27. $\begin{cases} x \ge 0 \\ y > 0 \end{cases}$ **28.** $\begin{cases} x \le 0 \\ y < 0 \end{cases}$

29. $\begin{cases} x + y \le 5 \\ x \ge 0 \\ y \ge 0 \end{cases}$ **30.** $\begin{cases} 2x + y \le 4 \\ x \ge 0 \\ y \ge 0 \end{cases}$

31. $\begin{cases} 4x - 3y > 12 \\ x \ge 0 \\ y \le 0 \end{cases}$ **32.** $\begin{cases} 2x - 6y > 12 \\ x \le 0 \\ y \le 0 \end{cases}$

33. $\begin{cases} 0 \le x \le 3 \\ 0 \le y \le 3 \end{cases}$ **34.** $\begin{cases} 0 \le x \le 5 \\ 0 \le y \le 5 \end{cases}$

35. $\begin{cases} x - y \le 4 \\ x + 2y \le 4 \end{cases}$ **36.** $\begin{cases} x - y \le 3 \\ 2x + y \le 4 \end{cases}$

Practice PLUS

In Exercises 37–44, graph the solution set of each system of linear inequalities. If the system has no solutions, state this and explain why.

37. $\begin{cases} x + y \ge 1 \\ x - y \ge 1 \\ x \ge 4 \end{cases}$ **38.** $\begin{cases} x - y \le 3 \\ x + y \le 3 \\ x \ge -2 \end{cases}$

39. $\begin{cases} x + 2y < 6 \\ y > 2x - 2 \\ y \ge 2 \end{cases}$ **40.** $\begin{cases} 2x - 3y < 6 \\ 2x - 3y > -6 \\ -3 \le x \le 2 \end{cases}$

41. $\begin{cases} y \le -3x + 3 \\ y \ge -x - 1 \\ y < x + 7 \end{cases}$ **42.** $\begin{cases} y \ge -3x + 5 \\ y \ge -x + 3 \\ y \ge \frac{1}{2}x \\ x \ge 0 \\ y \ge 0 \end{cases}$

43. $\begin{cases} y \ge 2x + 2 \\ y < 2x - 3 \\ x \ge 2 \end{cases}$ **44.** $\begin{cases} y \ge -3x + 2 \\ y < -3x \\ x \ge 1 \end{cases}$

Application Exercises

The figure shows the healthy weight region for various heights for people ages 35 and older.

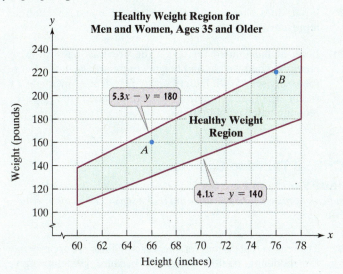

Healthy Weight Region for Men and Women, Ages 35 and Older

Source: U.S. Department of Health and Human Services

If x represents height, in inches, and y represents weight, in pounds, the healthy weight region can be modeled by the following system of linear inequalities:

$$\begin{cases} 5.3x - y \ge 180 \\ 4.1x - y \le 140. \end{cases}$$

Use this information to solve Exercises 45–48.

45. Show that point *A* is a solution of the system of inequalities that describes healthy weight for this age group.

(In Exercises 46–48, refer to the graph and the system of inequalities on the previous page.)

46. Show that point *B* is a solution of the system of inequalities that describes healthy weight for this age group.

47. Is a person in this age group who is 6 feet tall weighing 205 pounds within the healthy weight region?

48. Is a person in this age group who is 5 feet 8 inches tall weighing 135 pounds within the healthy weight region?

49. Suppose a patient is not allowed to have more than 330 milligrams of cholesterol per day from a diet of eggs and meat. Each egg provides 165 milligrams of cholesterol and each ounce of meat provides 110 milligrams of cholesterol. Thus, $165x + 110y \le 330$, where *x* is the number of eggs and *y* the number of ounces of meat. Furthermore, the patient must have at least 165 milligrams of cholesterol from the diet. Graph the system of inequalities in the first quadrant. Give the coordinates of any two points in the solution set. Describe what each set of coordinates means in terms of the variables in the problem.

Explaining the Concepts

50. What does the graph of a system of linear inequalities represent?

51. Look at the shaded region on the previous page showing recommended weight and height combinations in the figure for Exercises 45–48. Describe why a system of inequalities, rather than an equation, is better suited to give the recommended combinations.

Critical Thinking Exercises

Make Sense? *In Exercises 52–55, determine whether each statement makes sense or does not make sense, and explain your reasoning.*

52. The reason that systems of linear inequalities are appropriate for modeling healthy weight is because guidelines give healthy weight ranges, rather than specific weights, for various heights.

53. I graphed the solution set of $y \ge x + 2$ and $x \ge 1$ without using test points.

54. I graphed the solution set of $2x - y < 4$ and $x + y > -1$ without using test points.

55. I use two different colors to graph solution sets of systems of inequalities, selecting the region where the colors overlap.

In Exercises 56–57, write a system of inequalities whose solution set is shown by the region in which the yellow shading and the green vertical shading overlap.

56.

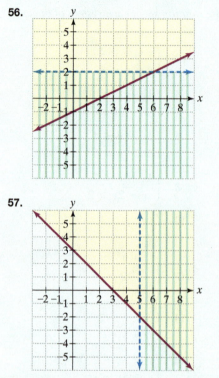

57.

58. Write a system of inequalities that has no solutions.

59. Write a system of inequalities that describes all the points in quadrant III of a rectangular coordinate system.

60. A person plans to invest no more than $15,000, placing the money in two investments. One investment is high risk, high yield; the other is low risk, low yield. At least $2000 is to be placed in the high-risk investment. Furthermore, the amount invested at low risk should be at least three times the amount invested at high risk. Write and graph a system of inequalities that describes all possibilities for placing the money in the high- and low-risk investments.

61. Promoters of a rock concert must sell at least 25,000 tickets priced at $35 and $50 per ticket. Furthermore, the promoters must take in at least $1,025,000 in ticket sales. Write and graph a system of inequalities that describes all possibilities for selling the $35 tickets and the $50 tickets.

Review Exercises

62. Find the slope of the line containing the points $(-6, 1)$ and $(2, -1)$. (Section 3.3, Example 1)

63. Add: $\dfrac{1}{5} + \left(-\dfrac{3}{4}\right)$. (Section 1.5, Example 4)

64. Solve: $7x = 10 + 6(11 - 2x)$.
(Section 2.3, Example 3)

Preview Exercises

Exercises 65–67 will help you prepare for the material covered in the first section of the next chapter.

65. Add: $5x^3 + 12x^3$.

66. Add: $-8x^2 + 6x^2$.

67. Subtract: $-9y^4 - (-2y^4)$.

Chapter 4 Summary

Definitions and Concepts	Examples

Section 4.1 Solving Systems of Linear Equations by Graphing

A system of linear equations in two variables, x and y, consists of two equations of the form $Ax + By = C$. A solution is an ordered pair of numbers that satisfies both equations.

Determine whether $(3, -1)$ is a solution of

$$\begin{cases} 2x + 5y = 1 \\ 4x + y = 11. \end{cases}$$

Replace x with 3 and y with -1 in both equations.

$$2x + 5y = 1 \qquad\qquad 4x + y = 11$$
$$2 \cdot 3 + 5(-1) \stackrel{?}{=} 1 \qquad\qquad 4 \cdot 3 + (-1) \stackrel{?}{=} 11$$
$$6 + (-5) \stackrel{?}{=} 1 \qquad\qquad 12 + (-1) \stackrel{?}{=} 11$$
$$1 = 1, \text{ true} \qquad\qquad 11 = 11, \text{ true}$$

Thus, $(3, -1)$ is a solution of the system.

To solve a linear system by graphing,

1. Graph the first equation.
2. Graph the second equation on the same set of axes.
3. If the graphs intersect, determine the coordinates of this point. The ordered pair is the solution of the system.
4. Check the solution in both equations.

If the graphs are parallel lines, the system has no solution and is called inconsistent. If the graphs are the same line, the system has infinitely many solutions. The equations are called dependent.

Solve by graphing:

$$\begin{cases} 2x + y = 4 \\ x + y = 2. \end{cases}$$

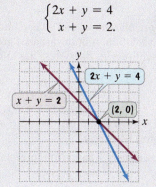

The solution is $(2, 0)$ and the solution set is $\{(2, 0)\}$.

Definitions and Concepts	**Examples**

Section 4.2 Solving Systems of Linear Equations by the Substitution Method

To solve a linear system by the substitution method,

1. Solve one equation for one variable in terms of the other.
2. Substitute the expression for that variable into the other equation. This will result in an equation in one variable.
3. Solve the equation in one variable.
4. Back-substitute the value of the variable found in step 3 in the equation from step 1. Simplify and find the value of the remaining variable.
5. Check the proposed solution in both of the system's given equations.

If both variables are eliminated and a false statement results, the system has no solution. If both variables are eliminated and a true statement results, the system has infinitely many solutions.

Solve by the substitution method:

$$\begin{cases} y = 2x + 3 \\ 7x - 5y = -18. \end{cases}$$

Substitute $2x + 3$ for y in the second equation.

$$7x - 5(2x + 3) = -18$$
$$7x - 10x - 15 = -18$$
$$-3x - 15 = -18$$
$$-3x = -3$$
$$x = 1$$

Find y. Substitute 1 for x in $y = 2x + 3$.

$$y = 2 \cdot 1 + 3 = 2 + 3 = 5$$

The solution, $(1, 5)$, checks. The solution set is $\{(1, 5)\}$.

Section 4.3 Solving Systems of Linear Equations by the Addition Method

To solve a linear system by the addition method,

1. Write the equations in $Ax + By = C$ form.
2. Multiply one or both equations by nonzero numbers so that coefficients of one of the variables are opposites.
3. Add the equations.
4. Solve the resulting equation for the remaining variable.
5. Back-substitute the value of the variable into either original equation and find the value of the other variable.
6. Check the proposed solution in both of the original equations.

If both variables are eliminated and a false statement results, the system has no solution. If both variables are eliminated and a true statement results, the system has infinitely many solutions.

Solve by the addition method:

$$\begin{cases} 3x + y = -11 \\ 6x - 2y = -2. \end{cases}$$

Eliminate y. Multiply both sides of the first equation by 2.

$$\begin{cases} 6x + 2y = -22 \\ 6x - 2y = \underline{-2} \end{cases}$$
$$\text{Add:} \quad 12x = -24$$
$$x = -2$$

Find y. Back-substitute -2 for x. We'll use the first equation.

$$3(-2) + y = -11$$
$$-6 + y = -11$$
$$y = -5$$

The solution, $(-2, -5)$, checks. The solution set is $\{(-2, -5)\}$.

Section 4.4 Problem Solving Using Systems of Equations

A Problem-Solving Strategy

1. Use variables, usually x and y, to represent unknown quantities.
2. Write a system of equations that models the problem's conditions.
3. Solve the system and answer the problem's question.
4. Check proposed answers in the problem's wording.

You invested $14,000 in two funds paying 7% and 9% interest. Total year-end interest was $1180. How much was invested at each rate?

Let x = amount invested at 7% and
y = amount invested at 9%.

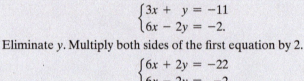

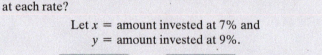

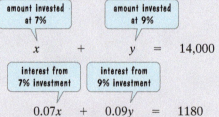

Solving by substitution or addition, $x = 4000$ and $y = 10,000$. Thus, $4000 was invested at 7% and $10,000 at 9%.

Definitions and Concepts	**Examples**

Section 4.5 Systems of Linear Inequalities

Two (or more) linear inequalities make up a system of linear inequalities. A solution is an ordered pair satisfying each of the inequalities in the system. The solution set is the set of all such ordered pairs. To graph the solution set of a system of inequalities, graph each inequality in the system in the same rectangular coordinate system. The overlapping region represents the solution set of the system.

Graph the solution set of the system:

$$\begin{cases} y \le -2x \\ x - y \ge 3. \end{cases}$$

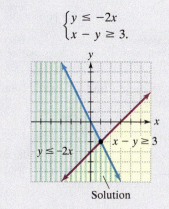

Solution

CHAPTER 4 REVIEW EXERCISES

4.1 *In Exercises 1–2, determine whether the given ordered pair is a solution of the system.*

1. $(1, -5)$

$$\begin{cases} 4x - y = 9 \\ 2x + 3y = -13 \end{cases}$$

2. $(-5, 2)$

$$\begin{cases} 2x + 3y = -4 \\ x - 4y = -10 \end{cases}$$

3. Does the graphing-utility screen show the solution for the following system? Explain.

$$\begin{cases} x + y = 2 \\ 2x + y = -5 \end{cases}$$

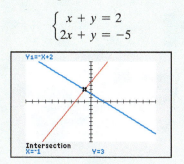

In Exercises 4–14, solve each system by graphing. If there is no solution or an infinite number of solutions, so state. Use set notation to express solution sets.

4. $\begin{cases} x + y = 2 \\ x - y = 6 \end{cases}$

5. $\begin{cases} 2x - 3y = 12 \\ -2x + y = -8 \end{cases}$

6. $\begin{cases} 3x + 2y = 6 \\ 3x - 2y = 6 \end{cases}$

7. $\begin{cases} y = \dfrac{1}{2}x \\ y = 2x - 3 \end{cases}$

8. $\begin{cases} x + 2y = 2 \\ y = x - 5 \end{cases}$

9. $\begin{cases} x + 2y = 8 \\ 3x + 6y = 12 \end{cases}$

10. $\begin{cases} 2x - 4y = 8 \\ x - 2y = 4 \end{cases}$

11. $\begin{cases} y = 3x - 1 \\ y = 3x + 2 \end{cases}$

12. $\begin{cases} x - y = 4 \\ x = -2 \end{cases}$

13. $\begin{cases} x = 2 \\ y = 5 \end{cases}$

14. $\begin{cases} x = 2 \\ x = 5 \end{cases}$

4.2 *In Exercises 15–23, solve each system by the substitution method. If there is no solution or an infinite number of solutions, so state. Use set notation to express solution sets.*

15. $\begin{cases} 2x - 3y = 7 \\ y = 3x - 7 \end{cases}$

16. $\begin{cases} 2x - y = 6 \\ x = 13 - 2y \end{cases}$

17. $\begin{cases} 2x - 5y = 1 \\ 3x + y = -7 \end{cases}$

18. $\begin{cases} 3x + 4y = -13 \\ 5y - x = -21 \end{cases}$

19. $\begin{cases} y = 39 - 3x \\ y = 2x - 61 \end{cases}$

20. $\begin{cases} 4x + y = 5 \\ 12x + 3y = 15 \end{cases}$

21. $\begin{cases} 4x - 2y = 10 \\ y = 2x + 3 \end{cases}$

22. $\begin{cases} x - 4 = 0 \\ 9x - 2y = 0 \end{cases}$

23. $\begin{cases} 8y = 4x \\ 7x + 2y = -8 \end{cases}$

24. The following models describe demand and supply for three-bedroom rental apartments.

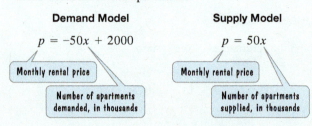

Demand Model

$$p = -50x + 2000$$

Monthly rental price

Number of apartments demanded, in thousands

Supply Model

$$p = 50x$$

Monthly rental price

Number of apartments supplied, in thousands

a. Solve the system and find the equilibrium quantity and the equilibrium price.

b. Use your answer from part (a) to complete this statement:

When rents are _____ per month, consumers will demand _____ apartments and suppliers will offer _____ apartments for rent.

4.3 *In Exercises 25–35, solve each system by the addition method. If there is no solution or an infinite number of solutions, so state. Use set notation to express solution sets.*

25. $\begin{cases} x + y = 6 \\ 2x + y = 8 \end{cases}$

26. $\begin{cases} 3x - 4y = 1 \\ 12x - y = -11 \end{cases}$

27. $\begin{cases} 3x - 7y = 13 \\ 6x + 5y = 7 \end{cases}$

28. $\begin{cases} 8x - 4y = 16 \\ 4x + 5y = 22 \end{cases}$

29. $\begin{cases} 5x - 2y = 8 \\ 3x - 5y = 1 \end{cases}$

30. $\begin{cases} 2x + 7y = 0 \\ 7x + 2y = 0 \end{cases}$

31. $\begin{cases} x + 3y = -4 \\ 3x + 2y = 3 \end{cases}$

32. $\begin{cases} 2x + y = 5 \\ 2x + y = 7 \end{cases}$

33. $\begin{cases} 3x - 4y = -1 \\ -6x + 8y = 2 \end{cases}$

34. $\begin{cases} 2x = 8y + 24 \\ 3x + 5y = 2 \end{cases}$

35. $\begin{cases} 5x - 7y = 2 \\ 3x = 4y \end{cases}$

In Exercises 36–41, solve each system by the method of your choice. If there is no solution or an infinite number of solutions, so state. Use set notation to express solution sets.

36. $\begin{cases} 3x + 4y = -8 \\ 2x + 3y = -5 \end{cases}$

37. $\begin{cases} 6x + 8y = 39 \\ y = 2x - 2 \end{cases}$

38. $\begin{cases} x + 2y = 7 \\ 2x + y = 8 \end{cases}$

39. $\begin{cases} y = 2x - 3 \\ y = -2x - 1 \end{cases}$

40. $\begin{cases} 3x - 6y = 7 \\ 3x = 6y \end{cases}$

41. $\begin{cases} y - 7 = 0 \\ 7x - 3y = 0 \end{cases}$

4.4

42. Talk about paintings by numbers: In 2007, this glittering Klimt set the record for the most ever paid for a painting, a title that had been held by Picasso's *Boy with a Pipe*. Combined, the two paintings sold for $239 million. The difference between the selling price for Klimt's work and the selling price for Picasso's work was $31 million. Find the amount paid for each painting.

Adele Bloch-Bauer I (1907), Gustav Klimt. Neue Galerie/Art Resource

Boy with a Pipe (1905), Pablo Picasso. © 2011 Picasso Estate/ARS

43. Health experts agree that cholesterol intake should be limited to 300 milligrams or less each day. Three ounces of shrimp and 2 ounces of scallops contain 156 milligrams of cholesterol. Five ounces of shrimp and 3 ounces of scallops contain 45 milligrams of cholesterol less than the suggested maximum daily intake. Determine the cholesterol content in an ounce of each item.

44. The perimeter of a table tennis top is 28 feet. The difference between 4 times the length and 3 times the width is 21 feet. Find the dimensions.

Width: *y*

Length: *x*

45. A rectangular garden has a perimeter of 24 yards. Fencing across the length costs $3 per yard and along the width $2 per yard. The total cost of the fencing is $62. Find the length and width of the garden.

46. A travel agent offers two package vacation plans. The first plan costs $360 and includes 3 days at a hotel and a rental car for 2 days. The second plan costs $500 and includes 4 days at a hotel and a rental car for 3 days. The daily charge for the room is the same under each plan, as is the daily charge for the car. Find the cost per day for the room and for the car.

47. You are choosing between two cellphone plans. Data Plan A offers a flat monthly rate of $20 per gigabyte (GB). Data Plan B has a monthly fee of $40 with a charge of $15 per GB. For how many GB of data will the costs for the two data plans be the same? What will be the cost for each plan?

48. Tickets for a touring production of *Sweeney Todd* cost $90 for orchestra seats and $60 for balcony seats. You purchase nine tickets, some in the orchestra and some in the balcony, for a total cost of $720. How many orchestra tickets and how many balcony tickets were purchased?

49. You invested $10,000 in two accounts paying 8% and 10% annual interest. At the end of the year, the total interest from these investments was $940. How much was invested at each rate?

50. A chemist needs to mix a 75% saltwater solution with a 50% saltwater solution to obtain 10 gallons of a 60% saltwater solution. How many gallons of each of the solutions must be used?

4.5 *In Exercises 51–57, graph the solution set of each system of linear inequalities.*

51. $\begin{cases} 3x - y \le 6 \\ x + y \ge 2 \end{cases}$

52. $\begin{cases} x + y < 4 \\ x - y < 4 \end{cases}$

53. $\begin{cases} y < 2x - 2 \\ x \ge 3 \end{cases}$

54. $\begin{cases} 4x + 6y \le 24 \\ y > 2 \end{cases}$

55. $\begin{cases} x \le 3 \\ y \ge -2 \end{cases}$

56. $\begin{cases} y \ge \dfrac{1}{2}x - 2 \\ y \le \dfrac{1}{2}x + 1 \end{cases}$

57. $\begin{cases} x \le 0 \\ y \ge 0 \end{cases}$

CHAPTER 4 TEST

Step-by-step test solutions are found on the Chapter Test Prep Videos available in MyMathLab® or on You Tube (search "BlitzerIntroAlg7e" and click on "Channels").

In Exercises 1–2, determine whether the given ordered pair is a solution of the system.

1. $(5, -5)$

$\begin{cases} 2x + y = 5 \\ x + 3y = -10 \end{cases}$

2. $(-3, 2)$

$\begin{cases} x + 5y = 7 \\ 3x - 4y = 1 \end{cases}$

In Exercises 3–4, solve each system by graphing. If there is no solution or an infinite number of solutions, so state. Use set notation to express solution sets.

3. $\begin{cases} x + y = 6 \\ 4x - y = 4 \end{cases}$

4. $\begin{cases} 2x + y = 8 \\ y = 3x - 2 \end{cases}$

In Exercises 5–7, solve each system by the substitution method. If there is no solution or an infinite number of solutions, so state. Use set notation to express solution sets.

5. $\begin{cases} x = y + 4 \\ 3x + 7y = -18 \end{cases}$

6. $\begin{cases} 2x - y = 7 \\ 3x + 2y = 0 \end{cases}$

7. $\begin{cases} 2x - 4y = 3 \\ x = 2y + 4 \end{cases}$

In Exercises 8–10, solve each system by the addition method. If there is no solution or an infinite number of solutions, so state. Use set notation to express solution sets.

8. $\begin{cases} 2x + y = 2 \\ 4x - y = -8 \end{cases}$

9. $\begin{cases} 2x + 3y = 1 \\ 3x + 2y = -6 \end{cases}$

10. $\begin{cases} 3x - 2y = 2 \\ -9x + 6y = -6 \end{cases}$

11. Our Dogs, Ourselves. Dogs rely on a dazzling sense of smell to interpret their world. The average number of scent receptors in dogs and humans combined is 225 million. The difference between the average number of scent receptors in dogs and in humans is 215 million. What is the average number of scent receptors in dogs and what is the average number in humans? (*Source: AARP Magazine*)

12. A rectangular garden has a perimeter of 34 yards. Fencing across the length costs $2 per yard and fencing along the width costs $1 per yard. The total cost of the fencing is $58. Find the length and width of the garden.

13. You are choosing between two cellphone plans. Data Plan A has a monthly fee of $46 with a charge of $17 per gigabyte (GB). Data Plan B has a monthly fee of $34 with a charge of $19 per GB. For how many GB of data will the costs for the two data plans be the same? What will be the cost for each plan?

14. You invested $6000 in two accounts paying 9% and 6% annual interest. At the end of the year, the total interest from these investments was $480. How much was invested at each rate?

15. A chemist needs to mix a 50% acid solution with an 80% acid solution to obtain 100 milliliters of a 68% acid solution. How many milliliters of each of the solutions must be used?

In Exercises 16–17, graph the solution set of each system of linear inequalities.

16. $\begin{cases} x - 3y > 6 \\ 2x + 4y \le 8 \end{cases}$

17. $\begin{cases} y \ge 2x - 4 \\ x < 2 \end{cases}$

CUMULATIVE REVIEW EXERCISES (CHAPTERS 1–4)

1. Perform the indicated operations:

$$-14 - [18 - (6 - 10)].$$

2. Simplify: $6(3x - 2) - (x - 1)$.

In Exercises 3–4, solve each equation.

3. $17(x + 3) = 13 + 4(x - 10)$

4. $\dfrac{x}{4} - 1 = \dfrac{x}{5}$

5. Solve for t: $A = P + Prt$.

6. Solve and graph the solution set on a number line: $2x - 5 < 5x - 11$.

In Exercises 7–9, graph each equation in the rectangular coordinate system.

7. $x - 3y = 6$

8. $y = 4$

9. $y = -\dfrac{3}{5}x + 2$

In Exercises 10–11, solve each linear system.

10. $\begin{cases} 3x - 4y = 8 \\ 4x + 5y = -10 \end{cases}$

11. $\begin{cases} 2x - 3y = 9 \\ y = 4x - 8 \end{cases}$

12. Find the slope of the line passing through $(5, -6)$ and $(6, -5)$.

13. Write the point-slope form and the slope-intercept form of the equation of the line passing through $(-1, 6)$ with slope $= -4$.

14. The area of a triangle is 80 square feet. Find the height if the base is 16 feet.

15. If 10 pens and 15 pads of paper cost $26, and 5 of the same pens and 10 of the same pads cost $16, find the cost of a pen and a pad.

16. List all the integers in this set:

$$\left\{ -93, -\dfrac{7}{3}, 0, \sqrt{3}, \dfrac{7}{1}, \sqrt{100} \right\}.$$

Gay Marriage. Gallup polls conducted in 2010 and 2014 asked a random sample of U.S. adults:

> *Should marriages between same-sex couples be recognized by the law as valid?*

The graphs show the percentage of American adults who answered "yes" or "no" in Gallup's 2010 and 2014 surveys.

Exercises 17–20 are based on these graphs.

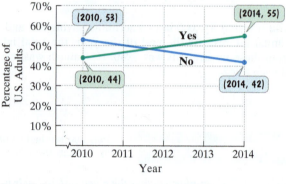

Should Marriages between Same-Sex Couples Be Recognized by the Law as Valid?

Source: Gallup

17. Which line has a positive slope? What does this mean in terms of the variables in this situation?

18. Which line has a negative slope? What does this mean in terms of the variables in this situation?

19. Find the slope of the line in Exercise 18. Then complete the following statement:

> In each year since 2010, the percentage of U.S. adults who disapproved of same-sex marriage _____ by approximately _____ percent per year.

20. Use the graph to estimate in which year, to the nearest whole year, the percentage of U.S. adults who approved of same-sex marriage was the same as the percentage of U.S. adults who disapproved of same-sex marriage. What percent, to the nearest whole percent, approved and what percent disapproved at that time?

Exponents and Polynomials

Surfing the web, you hear politicians discussing the problem of the national debt that exceeds $15 trillion. They state that the interest on the debt equals government spending on veterans, homeland security, education, and transportation combined. They make it seem like the national debt is a real problem, but later you realize that you don't really know what a number like 15 trillion means. If the national debt were evenly divided among all citizens of the country, how much would every man, woman, and child have to pay? Is economic doomsday about to arrive?

Here's where you'll find these application:

Literacy with numbers, called *numeracy*, is a prerequisite for functioning in a meaningful way personally, professionally, and as a citizen. In this chapter, you will learn to use exponents to provide a way of putting large and small numbers into perspective. The problem of the national debt appears as Example 10 in Section 5.7.

SECTION

5.1 Adding and Subtracting Polynomials

What am I supposed to learn?

After studying this section, you should be able to:

1. Understand the vocabulary used to describe polynomials.
2. Add polynomials.
3. Subtract polynomials.
4. Graph equations defined by polynomials of degree 2.

What? Me? Racist? More than 2 million people have tested their racial prejudice using an online version of the Implicit Association Test. Most groups' average scores fall between "slight" and "moderate" bias, but the differences among groups, by age and by political identification, are intriguing.

In this section's Exercise Set (Exercises 103 and 104), you will be working with models that measure bias:

$$S = 0.3x^3 - 2.8x^2 + 6.7x + 30$$
$$S = -0.03x^3 + 0.2x^2 + 2.3x + 24.$$

In each model, S represents the score on the Implicit Association Test. (Higher scores indicate stronger bias.) In the first model (see Exercise 103), x represents age group. In the second model (see Exercise 104), x represents political identification.

The algebraic expressions that appear on the right side of the models are examples of *polynomials*. A **polynomial** is a single term or the sum of two or more terms containing variables with whole-number exponents. These particular polynomials each contain four terms. Equations containing polynomials are used in such diverse areas as science, business, medicine, psychology, and sociology. In this section, we present basic ideas about polynomials. We then use our knowledge of combining like terms to find sums and differences of polynomials.

1. Understand the vocabulary used to describe polynomials.

Describing Polynomials

Consider the polynomial

$$7x^3 - 9x^2 + 13x - 6.$$

We can express this polynomial as

$$7x^3 + (-9x^2) + 13x + (-6).$$

The polynomial contains four terms. It is customary to write the terms in the order of descending powers of the variables. This is the **standard form** of a polynomial.

We begin this chapter by limiting our discussion to polynomials containing only one variable. Each term of such a polynomial in x is of the form ax^n. The **degree** of ax^n is n. For example, the degree of the term $7x^3$ is 3.

Great Question!

If the degree of a nonzero constant is 0, why doesn't the constant 0 also have degree 0?

We can express 0 in many ways, including $0x$, $0x^2$, and $0x^3$. It is impossible to assign a unique exponent to the variable. This is why 0 has no defined degree.

The Degree of ax^n

If $a \neq 0$ and n is a whole number, the degree of ax^n is n. The degree of a nonzero constant term is 0. The constant 0 has no defined degree.

Here is an example of a polynomial and the degree of each of its four terms:

$$6x^4 - 3x^3 - 2x - 5.$$

| degree 4 | degree 3 | degree 1 | degree of nonzero constant: 0 |

Notice that the exponent on x for the term $-2x$, meaning $-2x^1$, is understood to be 1. For this reason, the degree of $-2x$ is 1.

A polynomial is simplified when it contains no grouping symbols and no like terms. A simplified polynomial that has exactly one term is called a **monomial**. A **binomial** is a simplified polynomial that has two terms. A **trinomial** is a simplified polynomial with three terms. Simplified polynomials with four or more terms have no special names.

The **degree of a polynomial** is the greatest degree of all the terms of the polynomial. For example, $4x^2 + 3x$ is a binomial of degree 2 because the degree of the first term is 2, and the degree of the other term is less than 2. Also, $7x^5 - 2x^2 + 4$ is a trinomial of degree 5 because the degree of the first term is 5, and the degrees of the other terms are less than 5.

Up to now, we have used x to represent the variable in a polynomial. However, any letter can be used. For example,

- $7x^5 - 3x^3 + 8$ is a polynomial (in x) of degree 5. Because there are three terms, the polynomial is a trinomial.

- $6y^3 + 4y^2 - y + 3$ is a polynomial (in y) of degree 3. Because there are four terms, the polynomial has no special name.

- $z^7 + \sqrt{2}$ is a polynomial (in z) of degree 7. Because there are two terms, the polynomial is a binomial.

2 Add polynomials.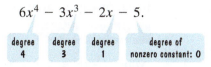

Adding Polynomials

Recall that *like terms* are terms containing exactly the same variables to the same powers. Polynomials are added by combining like terms. For example, we can add the monomials $-9x^3$ and $13x^3$ as follows:

$$-9x^3 + 13x^3 = (-9 + 13)x^3 = 4x^3.$$

| These like terms both contain x to the third power. | Add coefficients and keep the same variable factor, x^3. |

EXAMPLE 1 Adding Polynomials

Add: $(-9x^3 + 7x^2 - 5x + 3) + (13x^3 + 2x^2 - 8x - 6)$.

Solution The like terms are $-9x^3$ and $13x^3$, containing the same variable to the same power (x^3), as well as $7x^2$ and $2x^2$ (both containing x^2), $-5x$ and $-8x$ (both containing x), and the constant terms 3 and -6. We begin by grouping these pairs of like terms.

$$(-9x^3 + 7x^2 - 5x + 3) + (13x^3 + 2x^2 - 8x - 6)$$
$$= (-9x^3 + 13x^3) + (7x^2 + 2x^2) + (-5x - 8x) + (3 - 6) \quad \text{Group like terms.}$$
$$= 4x^3 + 9x^2 + (-13x) + (-3) \quad \text{Combine like terms.}$$
$$= 4x^3 + 9x^2 - 13x - 3 \quad \text{Express addition of opposites as subtraction.} \ \blacksquare$$

✓ **CHECK POINT 1** Add: $(-11x^3 + 7x^2 - 11x - 5) + (16x^3 - 3x^2 + 3x - 15)$.

Polynomials can also be added by arranging like terms in columns. Then combine like terms, column by column.

EXAMPLE 2 Adding Polynomials Vertically

Add: $(-9x^3 + 7x^2 - 5x + 3) + (13x^3 + 2x^2 - 8x - 6)$.

Solution

$$
\begin{array}{ll}
-9x^3 + 7x^2 - 5x + 3 & \text{Line up like terms vertically.} \\
\underline{13x^3 + 2x^2 - 8x - 6} & \\
4x^3 + 9x^2 - 13x - 3 & \text{Add the like terms in each column.}
\end{array}
$$

This is the same answer that we found in Example 1. ∎

☑ **CHECK POINT 2** Add the polynomials in Check Point 1 using a vertical format. Begin by arranging like terms in columns.

Great Question!

What's the advantage of using a vertical format to add polynomials instead of a horizontal format?

A vertical format often makes it easier to see the like terms.

---

3 Subtract polynomials.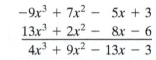

Subtracting Polynomials

We subtract real numbers by adding the opposite, or additive inverse, of the number being subtracted. For example,

$$8 - 3 = 8 + (-3) = 5.$$

Subtraction of polynomials also involves opposites. If the sum of two polynomials is 0, the polynomials are **opposites**, or **additive inverses**, of each other. Here is an example:

$$(4x^2 - 6x - 7) + (-4x^2 + 6x + 7) = 0.$$

The opposite of $4x^2 - 6x - 7$ is $-4x^2 + 6x + 7$, and vice versa.

Observe that the opposite of $4x^2 - 6x - 7$ can be obtained by changing the sign of each of its coefficients:

Polynomial		**Opposite**
$4x^2 - 6x - 7$	Change 4 to −4, change −6 to 6, and change −7 to 7.	$-4x^2 + 6x + 7.$

In general, **the opposite of a polynomial is that polynomial with the sign of every coefficient changed.** Just as we did with real numbers, we subtract one polynomial from another by adding the opposite of the polynomial being subtracted.

> ### Subtracting Polynomials
>
> To subtract two polynomials, add the first polynomial and the opposite of the polynomial being subtracted.

EXAMPLE 3 Subtracting Polynomials

Subtract: $(7x^2 + 3x - 4) - (4x^2 - 6x - 7)$.

Solution

$$(7x^2 + 3x - 4) - (4x^2 - 6x - 7)$$

Change the sign of each coefficient.

$$
\begin{aligned}
&= (7x^2 + 3x - 4) + (-4x^2 + 6x + 7) &&\text{Add the opposite of the polynomial being subtracted.} \\
&= (7x^2 - 4x^2) + (3x + 6x) + (-4 + 7) &&\text{Group like terms.} \\
&= 3x^2 + 9x + 3 &&\text{Combine like terms.} \quad ∎
\end{aligned}
$$

Great Question!

I'm confused by what it means to subtract one polynomial from a second polynomial. Which polynomial should I write first?

Be careful of the order in Example 4. For example, subtracting 2 from 5 means $5 - 2$. In general, subtracting B from A means $A - B$. The order of the resulting algebraic expression is not the same as the order in English. The polynomial following the word *from* is the one to write first.

✓ **CHECK POINT 3**   Subtract: $(9x^2 + 7x - 2) - (2x^2 - 4x - 6)$.

EXAMPLE 4   Subtracting Polynomials

Subtract $2x^3 - 6x^2 - 3x + 9$ from $7x^3 - 8x^2 + 9x - 6$.

Solution

$$(7x^3 - 8x^2 + 9x - 6) - (2x^3 - 6x^2 - 3x + 9)$$

Change the sign of each coefficient.

$$= (7x^3 - 8x^2 + 9x - 6) + (-2x^3 + 6x^2 + 3x - 9)$$ Add the opposite of the polynomial being subtracted.

$$= (7x^3 - 2x^3) + (-8x^2 + 6x^2) + (9x + 3x) + (-6 - 9)$$ Group like terms.

$$= 5x^3 + (-2x^2) + 12x + (-15)$$ Combine like terms.

$$= 5x^3 - 2x^2 + 12x - 15$$ Express addition of opposites as subtraction. ■

✓ **CHECK POINT 4**   Subtract $3x^3 - 8x^2 - 5x + 6$ from $10x^3 - 5x^2 + 7x - 2$.

Subtraction can also be performed in columns.

EXAMPLE 5   Subtracting Polynomials Vertically

Use the method of subtracting in columns to find

$$(12y^3 - 9y^2 - 11y - 3) - (4y^3 - 5y + 8).$$

Solution   Arrange like terms in columns.

$$
\begin{array}{l}
12y^3 - 9y^2 - 11y - 3 \\
-(4y^3 \qquad\ - 5y + 8)
\end{array}
$$

Add the opposite of the polynomial being subtracted.

Leave space for the missing term.

$$
\begin{array}{l}
\ \ 12y^3 - 9y^2 - 11y - 3 \\
+\ -4y^3 \qquad\ \ + 5y - 8 \\
\hline
\ \ \ 8y^3 - 9y^2\ - 6y - 11
\end{array}
$$

Change the sign of each coefficient of $4y^3 - 5y + 8$. Combine like terms. ■

✓ **CHECK POINT 5**   Use the method of subtracting in columns to find

$$(8y^3 - 10y^2 - 14y - 2) - (5y^3 - 3y + 6).$$

Graphing Equations Defined by Polynomials

Look at the picture of this gymnast. He has created a perfect balance in which the two halves of his body are mirror images of each other. Graphs of equations defined by polynomials of degree 2, such as $y = x^2 - 4$, have this mirrorlike quality. We can obtain their graphs, shaped like bowls or inverted bowls, using the point-plotting method for graphing an equation in two variables.

4 Graph equations defined by polynomials of degree 2.

EXAMPLE 6 Graphing an Equation Defined by a Polynomial of Degree 2

Graph the equation: $y = x^2 - 4$.

Solution The given equation involves two variables, x and y. However, because the variable x is squared, it is not a linear equation in two variables.

$$y = x^2 - 4$$

> This is not in the form $y = mx + b$ because x is squared.

Although the graph is not a line, it is still a picture of all the ordered-pair solutions of $y = x^2 - 4$. Thus, we can use the point-plotting method to obtain the graph.

Step 1. Find several ordered pairs that are solutions of the equation. To find some solutions of $y = x^2 - 4$, we select integers for x, starting with -3 and ending with 3.

> Start with x.   Compute y.   Form the ordered pair (x, y).

x	$y = x^2 - 4$	(x, y)
-3	$y = (-3)^2 - 4 = 9 - 4 = 5$	$(-3, 5)$
-2	$y = (-2)^2 - 4 = 4 - 4 = 0$	$(-2, 0)$
-1	$y = (-1)^2 - 4 = 1 - 4 = -3$	$(-1, -3)$
0	$y = 0^2 - 4 = 0 - 4 = -4$	$(0, -4)$
1	$y = 1^2 - 4 = 1 - 4 = -3$	$(1, -3)$
2	$y = 2^2 - 4 = 4 - 4 = 0$	$(2, 0)$
3	$y = 3^2 - 4 = 9 - 4 = 5$	$(3, 5)$

Step 2. Plot these ordered pairs as points in the rectangular coordinate system. The seven ordered pairs in the table of values are plotted in **Figure 5.1(a)**.

Step 3. Connect the points with a smooth curve. The seven points are joined with a smooth curve in **Figure 5.1(b)**. The graph of $y = x^2 - 4$ is a curve where the part of the graph to the right of the y-axis is a reflection of the part to the left of it, and vice versa. The arrows on both ends of the curve indicate that it extends indefinitely in both directions.

Great Question!

When I graphed lines, I used two, or possibly three, points. Why isn't this enough to use when graphing equations that are not linear?

If the graph of an equation is not a straight line, extra points are needed to get a better general idea of the graph's shape.

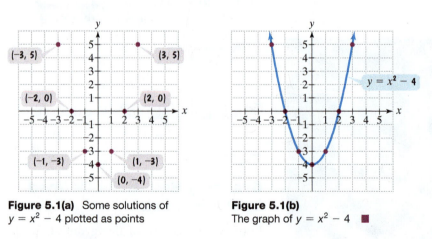

Figure 5.1(a) Some solutions of $y = x^2 - 4$ plotted as points

Figure 5.1(b) The graph of $y = x^2 - 4$ ∎

✓ **CHECK POINT 6** Graph the equation: $y = x^2 - 1$. Select integers for x, starting with -3 and ending with 3.

Blitzer Bonus ▶

Modeling *American Idol* with a Polynomial of Degree 2

The graph in **Figure 5.2** indicates that the ratings of *American Idol* from season 1 (2002) through season 12 (2013) have a mirrorlike quality. This suggests modeling the show's average number of viewers with a polynomial of degree 2.

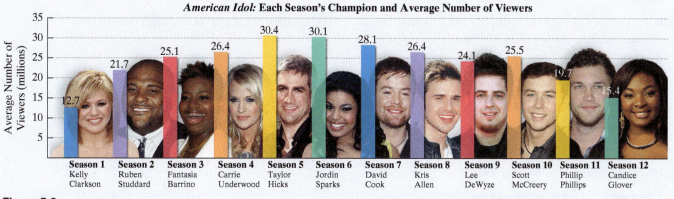

American Idol: Each Season's Champion and Average Number of Viewers

Figure 5.2

Source: The Nielsen Company

The equation

$$y = -0.48x^2 + 6.17x + 9.57$$

models *American Idol*'s average number of viewers, *y*, in millions, where *x* is the show's season number. The graph of the model in **Figure 5.3** is shaped like an inverted bowl. Can you see why projections based on this graph have the show's producers looking for a shake-up?

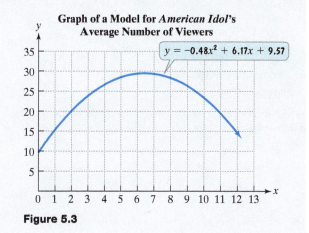

Graph of a Model for *American Idol*'s Average Number of Viewers

$$y = -0.48x^2 + 6.17x + 9.57$$

Figure 5.3

Achieving Success

Address your stress. Stress levels can help or hinder performance. The graph of the polynomial equation of degree 2 in **Figure 5.4** serves as a model that shows people under both low stress and high stress perform worse than their moderate-stress counterparts.

Figure 5.4

Source: Herbert Benson, *Your Maximum Mind,* Random House, 1987.

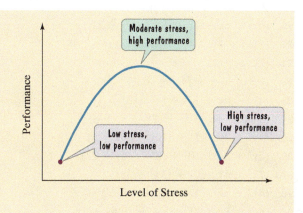

CONCEPT AND VOCABULARY CHECK

Fill in each blank so that the resulting statement is true.

1. A polynomial is a single term or the sum of two or more terms containing variables with exponents that are _____ numbers.

2. It is customary to write the terms of a polynomial in the order of descending powers of the variable. This is called the _____ form of a polynomial.

3. A simplified polynomial that has exactly one term is called a/an _____.

4. A simplified polynomial that has two terms is called a/an _____.

5. A simplified polynomial that has three terms is called a/an _____.

6. The degree of ax^n is _____, provided $a \neq 0$.

7. The degree of a polynomial is the _____ degree of all the terms of the polynomial.

8. Polynomials are added by combining _____ terms.

9. To subtract two polynomials, add the first polynomial and the _____ of the polynomial being subtracted.

5.1 EXERCISE SET ▶ MyMathLab®

Practice Exercises

In Exercises 1–16, identify each polynomial as a monomial, a binomial, or a trinomial. Give the degree of the polynomial.

1. $3x + 7$
2. $5x - 2$
3. $x^3 - 2x$
4. $x^5 - 7x$
5. $8x^2$
6. $10x^2$
7. 5
8. 9
9. $x^2 - 3x + 4$
10. $x^2 - 9x + 2$
11. $7y^2 - 9y^4 + 5$
12. $3y^2 - 14y^5 + 6$
13. $15x - 7x^3$
14. $9x - 5x^3$
15. $-9y^{23}$
16. $-11y^{26}$

In Exercises 17–38, add the polynomials.

17. $(9x + 8) + (-17x + 5)$
18. $(8x - 5) + (-13x + 9)$
19. $(4x^2 + 6x - 7) + (8x^2 + 9x - 2)$
20. $(11x^2 + 7x - 4) + (27x^2 + 10x - 20)$
21. $(7x^2 - 11x) + (3x^2 - x)$
22. $(-3x^2 + x) + (4x^2 + 8x)$
23. $(4x^2 - 6x + 12) + (x^2 + 3x + 1)$
24. $(-7x^2 + 8x + 3) + (2x^2 + x + 8)$
25. $(4y^3 + 7y - 5) + (10y^2 - 6y + 3)$
26. $(2y^3 + 3y + 10) + (3y^2 + 5y - 22)$
27. $(2x^2 - 6x + 7) + (3x^3 - 3x)$
28. $(4x^3 + 5x + 13) + (-4x^2 + 22)$
29. $(4y^2 + 8y + 11) + (-2y^3 + 5y + 2)$
30. $(7y^3 + 5y - 1) + (2y^2 - 6y + 3)$
31. $(-2y^6 + 3y^4 - y^2) + (-y^6 + 5y^4 + 2y^2)$
32. $(7r^4 + 5r^2 + 2r) + (-18r^4 - 5r^2 - r)$

33. $\left(9x^3 - x^2 - x - \dfrac{1}{3}\right) + \left(x^3 + x^2 + x + \dfrac{4}{3}\right)$

34. $\left(12x^3 - x^2 - x + \dfrac{4}{3}\right) + \left(x^3 + x^2 + x - \dfrac{1}{3}\right)$

35. $\left(\dfrac{1}{5}x^4 + \dfrac{1}{3}x^3 + \dfrac{3}{8}x^2 + 6\right) +$
    $\left(-\dfrac{3}{5}x^4 + \dfrac{2}{3}x^3 - \dfrac{1}{2}x^2 - 6\right)$

36. $\left(\dfrac{2}{5}x^4 + \dfrac{2}{3}x^3 + \dfrac{5}{8}x^2 + 7\right) +$
    $\left(-\dfrac{4}{5}x^4 + \dfrac{1}{3}x^3 - \dfrac{1}{4}x^2 - 7\right)$

37. $(0.03x^5 - 0.1x^3 + x + 0.03) +$
    $(-0.02x^5 + x^4 - 0.7x + 0.3)$

38. $(0.06x^5 - 0.2x^3 + x + 0.05) +$
    $(-0.04x^5 + 2x^4 - 0.8x + 0.5)$

In Exercises 39–54, use a vertical format to add the polynomials.

39. $5y^3 - 7y^2$
    $6y^3 + 4y^2$

40. $13x^4 - x^2$
    $7x^4 + 2x^2$

41. $3x^2 - 7x + 4$
    $-5x^2 + 6x - 3$

42. $7x^2 - 5x - 6$
    $-9x^2 + 4x + 6$

43. $\dfrac{1}{4}x^4 - \dfrac{2}{3}x^3 - 5$
    $-\dfrac{1}{2}x^4 + \dfrac{1}{5}x^3 + 4.7$

44. $\frac{1}{3}x^9 - \frac{1}{5}x^5 - 2.7$
$-\frac{3}{4}x^9 + \frac{2}{3}x^5 + 1$

45. $y^3 + 5y^2 - 7y - 3$
$-2y^3 + 3y^2 + 4y - 11$

46. $y^3 + y^2 - 7y + 9$
$-y^3 - 6y^2 - 8y + 11$

47. $4x^3 - 6x^2 + 5x - 7$
$-9x^3 \qquad - 4x + 3$

48. $-4y^3 + 6y^2 - 8y + 11$
$2y^3 \qquad + 9y - 3$

49. $7x^4 - 3x^3 + x^2$
$\qquad x^3 - x^2 + 4x - 2$

50. $7y^5 - 3y^3 + y^2$
$\qquad 2y^3 - y^2 - 4y - 3$

51. $7x^2 - 9x + 3$
$4x^2 + 11x - 2$
$-3x^2 + 5x - 6$

52. $7y^2 - 11y - 6$
$8y^2 + 3y + 4$
$-9y^2 - 5y + 2$

53. $1.2x^3 - 3x^2 + 9.1$
$7.8x^3 - 3.1x^2 + 8$
$1.2x^2 - 6$

54. $7.9x^3 - 6.8x^2 + 3.3$
$6.1x^3 - 2.2x^2 + 7$
$4.3x^2 - 5$

In Exercises 55–74, subtract the polynomials.

55. $(x - 8) - (3x + 2)$
56. $(x - 2) - (7x + 9)$
57. $(x^2 - 5x - 3) - (6x^2 + 4x + 9)$
58. $(3x^2 - 8x - 2) - (11x^2 + 5x + 4)$
59. $(x^2 - 5x) - (6x^2 - 4x)$
60. $(3x^2 - 2x) - (5x^2 - 6x)$
61. $(x^2 - 8x - 9) - (5x^2 - 4x - 3)$
62. $(x^2 - 5x + 3) - (x^2 - 6x - 8)$
63. $(y - 8) - (3y - 2)$
64. $(y - 2) - (7y - 9)$
65. $(6y^3 + 2y^2 - y - 11) - (y^2 - 8y + 9)$
66. $(5y^3 + y^2 - 3y - 8) - (y^2 - 8y + 11)$
67. $(7n^3 - n^7 - 8) - (6n^3 - n^7 - 10)$
68. $(2n^2 - n^7 - 6) - (2n^3 - n^7 - 8)$
69. $(y^6 - y^3) - (y^2 - y)$
70. $(y^5 - y^3) - (y^4 - y^2)$
71. $(7x^4 + 4x^2 + 5x) - (-19x^4 - 5x^2 - x)$
72. $(-3x^6 + 3x^4 - x^2) - (-x^6 + 2x^4 + 2x^2)$
73. $\left(\frac{3}{7}x^3 - \frac{1}{5}x - \frac{1}{3}\right) - \left(-\frac{2}{7}x^3 + \frac{1}{4}x - \frac{1}{3}\right)$
74. $\left(\frac{3}{8}x^2 - \frac{1}{3}x - \frac{1}{4}\right) - \left(-\frac{1}{8}x^2 + \frac{1}{2}x - \frac{1}{4}\right)$

In Exercises 75–88, use a vertical format to subtract the polynomials.

75. $7x + 1$
$-(3x - 5)$

76. $4x + 2$
$-(3x - 5)$

77. $7x^2 - 3$
$-(-3x^2 + 4)$

78. $9y^2 - 6$
$-(-5y^2 + 2)$

79. $7y^2 - 5y + 2$
$-(11y^2 + 2y - 3)$

80. $3x^5 - 5x^3 + 6$
$-(7x^5 + 4x^3 - 2)$

81. $7x^3 + 5x^2 - 3$
$-(-2x^3 - 6x^2 + 5)$

82. $3y^4 - 4y^2 + 7$
$-(-5y^4 - 6y^2 - 13)$

83. $5y^3 + 6y^2 - 3y + 10$
$-(6y^3 - 2y^2 - 4y - 4)$

84. $4y^3 + 5y^2 + 7y + 11$
$-(-5y^3 + 6y^2 - 9y - 3)$

85. $7x^4 - 3x^3 + 2x^2$
$-(\quad - x^3 - x^2 + x - 2)$

86. $5y^6 - 3y^3 + 2y^2$
$-(\quad - y^3 - y^2 - y - 1)$

87. $0.07x^3 - 0.01x^2 + 0.02x$
$-(0.02x^3 - 0.03x^2 - x)$

88. $0.04x^3 - 0.03x^2 + 0.05x$
$-(0.02x^3 - 0.06x^2 - x)$

Graph each equation in Exercises 89–94. Find seven solutions in your table of values for each equation by using integers for x, starting with −3 and ending with 3.

89. $y = x^2$ **90.** $y = x^2 - 2$
91. $y = x^2 + 1$ **92.** $y = x^2 + 2$
93. $y = 4 - x^2$ **94.** $y = 9 - x^2$

Practice PLUS

In Exercises 95–98, perform the indicated operations.

95. $[(4x^2 + 7x - 5) - (2x^2 - 10x + 3)] - (x^2 + 5x - 8)$

96. $[(10x^3 - 5x^2 + 4x + 3) - (-3x^3 - 4x^2 + x)] - (7x^3 - 5x + 4)$

97. $[(4y^2 - 3y + 8) - (5y^2 + 7y - 4)] - [(8y^2 + 5y - 7) + (-10y^2 + 4y + 3)]$

98. $[(7y^2 - 4y + 2) - (12y^2 + 3y - 5)] - [(5y^2 - 2y - 8) + (-7y^2 + 10y - 13)]$

99. Subtract $x^3 - 2x^2 + 2$ from the sum of $4x^3 + x^2$ and $-x^3 + 7x - 3$.

100. Subtract $-3x^3 - 7x + 5$ from the sum of $2x^2 + 4x - 7$ and $-5x^3 - 2x - 3$.

101. Subtract $-y^2 + 7y^3$ from the difference between $-5 + y^2 + 4y^3$ and $-8 - y + 7y^3$. Express the answer in standard form.

102. Subtract $-2y^2 + 8y^3$ from the difference between $-6 + y^2 + 5y^3$ and $-12 - y + 13y^3$. Express the answer in standard form.

Application Exercises

103. The bar graph shows the differences among age groups on the Implicit Association Test that measures levels of racial prejudice. Higher scores indicate stronger bias.

Measuring Racial Prejudice, by Age

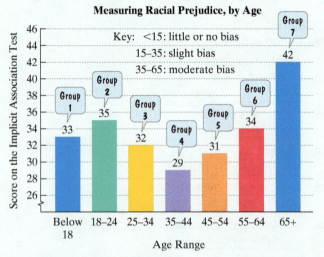

Source: The Race Implicit Association Test on the Project Implicit Demonstration Website

a. The data can be described by the following polynomial model of degree 3:

$$S = 0.2x^3 - 1.5x^2 + 3.4x + 25$$
$$+ (0.1x^3 - 1.3x^2 + 3.3x + 5).$$

In this polynomial model, S represents the score on the Implicit Association Test for age group x. Simplify the model.

b. Use the simplified form of the model from part (a) to find the score on the Implicit Association Test for the group in the 45–54 age range. How well does the model describe the score displayed by the bar graph?

c. Shown in a rectangular coordinate system is the graph of the polynomial model of degree 3 that describes scores on the Implicit Association Test, by age group. Identify your solution from part (b) as a point on the graph.

Graph of the Model for Measuring Racial Prejudice, by Age

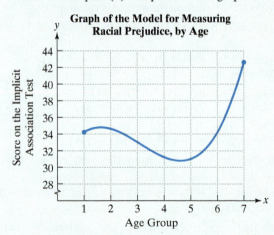

104. The bar graph at the top of the next column shows the differences among political identification groups on the Implicit Association Test that measures levels of racial prejudice. Higher scores indicate stronger bias.

Measuring Racial Prejudice, by Political Identification

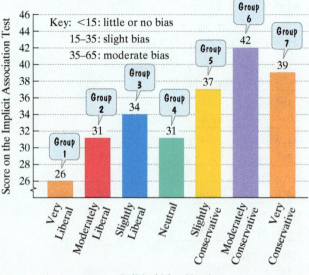

Source: The Race Implicit Association Test on the Project Implicit Demonstration Website

a. The data can be described by the following polynomial model of degree 3:

$$S = -0.02x^3 + 0.4x^2 + 1.2x + 22$$
$$+ (-0.01x^3 - 0.2x^2 + 1.1x + 2).$$

In this polynomial model, S represents the score on the Implicit Association Test for political identification group x. Simplify the model.

b. Use the simplified form of the model from part (a) to find the score on the Implicit Association Test for the slightly conservative political identification group. Does the model underestimate or overestimate the score displayed by the bar graph? By how much?

c. Shown in a rectangular coordinate system is the graph of the polynomial model of degree 3 that describes scores on the Implicit Association Test, by political identification. Identify your solution from part (b) as a point on the graph.

Graph of the Model Measuring Racial Prejudice, by Political Identification

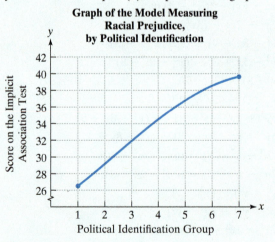

Explaining the Concepts

105. What is a polynomial?

106. What is a monomial? Give an example with your explanation.

107. What is a binomial? Give an example with your explanation.

108. What is a trinomial? Give an example with your explanation.

109. What is the degree of a polynomial? Provide an example with your explanation.

110. Explain how to add polynomials.

111. Explain how to subtract polynomials.

Critical Thinking Exercises

Make Sense?   *In Exercises 112–115, determine whether each statement makes sense or does not make sense, and explain your reasoning.*

112. I add like monomials by adding both their coefficients and the exponents that appear on their common variable factor.

113. By looking at the first term of a polynomial, I can determine its degree.

114. As long as I understand how to add and subtract polynomials, I can select the format, horizontal or vertical, that works better for me.

115. I used two points and a checkpoint to graph $y = x^2 - 4$.

In Exercises 116–119, determine whether each statement is true or false. If the statement is false, make the necessary change(s) to produce a true statement.

116. It is not possible to write a binomial with degree 0.

117. $\dfrac{1}{5x^2} + \dfrac{1}{3x}$ is a binomial.

118. $(2x^2 - 8x + 6) - (x^2 - 3x + 5) = x^2 - 5x + 1$ for any value of x.

119. In the polynomial $3x^2 - 5x + 13$, the coefficient of x is 5.

120. What polynomial must be subtracted from $5x^2 - 2x + 1$ so that the difference is $8x^2 - x + 3$?

121. The number of people who catch a cold t weeks after January 1 is $5t - 3t^2 + t^3$. The number of people who recover t weeks after January 1 is $t - t^2 + \frac{1}{3}t^3$. Write a polynomial in standard form for the number of people who are still ill with a cold t weeks after January 1.

122. Explain why it is not possible to add two polynomials of degree 3 and get a polynomial of degree 4.

Review Exercises

123. Simplify: $(-10)(-7) \div (1 - 8)$.
(Section 1.8, Example 8)

124. Subtract: $-4.6 - (-10.2)$.
(Section 1.6, Example 2)

125. Solve: $3(x - 2) = 9(x + 2)$.
(Section 2.3, Example 3)

Preview Exercises

Exercises 126–128 will help you prepare for the material covered in the next section.

126. Find the missing exponent, designated by the question mark, in the final step.
$$x^3 \cdot x^4 = (x \cdot x \cdot x) \cdot (x \cdot x \cdot x \cdot x) = x^?$$

127. Use the distributive property to multiply: $3x(x + 5)$.

128. Simplify: $x(x + 2) + 3(x + 2)$.

SECTION

5.2

Multiplying Polynomials

What am I supposed to learn?

After studying this section, you should be able to:

1. Use the product rule for exponents.

2. Use the power rule for exponents.

3. Use the products-to-powers rule.

4. Multiply monomials.

5. Multiply a monomial and a polynomial.

6. Multiply polynomials when neither is a monomial.

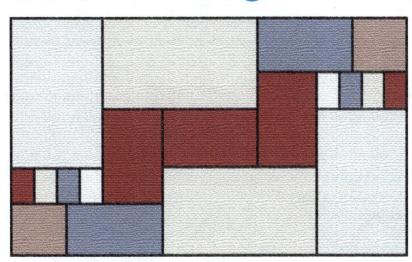

The ancient Greeks believed that the most visually pleasing rectangles have a ratio of length to width of approximately 1.618 to 1. With the exception of the squares on the lower left and the upper right, the interior of this geometric figure is filled entirely with these *golden rectangles*. Furthermore, the large rectangle is also a golden rectangle.

The total area of the large rectangle shown on the previous page can be found in many ways. This is because the area of any large rectangular region is related to the areas of the smaller rectangles that make up that region. In this section, we apply areas of rectangles as a way to picture the multiplication of polynomials. Before studying how polynomials are multiplied, we must develop some rules for working with exponents.

1 Use the product rule for exponents. ⊙

The Product Rule for Exponents

We have seen that exponents are used to indicate repeated multiplication. For example, 2^4, where 2 is the base and 4 is the exponent, indicates that 2 occurs as a factor four times:

$$2^4 = 2 \cdot 2 \cdot 2 \cdot 2.$$

Now consider the multiplication of two exponential expressions, such as $2^4 \cdot 2^3$. We are multiplying 4 factors of 2 and 3 factors of 2. We have a total of 7 factors of 2:

4 factors of 2   3 factors of 2

$$2^4 \cdot 2^3 = (2 \cdot 2 \cdot 2 \cdot 2) \cdot (2 \cdot 2 \cdot 2)$$

Total: 7 factors of 2

Thus, $2^4 \cdot 2^3 = 2^7.$ Caution: $2^4 \cdot 2^3$ is not equal to $2^{4 \cdot 3}$, or 2^{12}, as might be expected.

We can quickly find the exponent, 7, of the product by adding 4 and 3, the original exponents:

$$2^4 \cdot 2^3 = 2^{4+3} = 2^7.$$

This suggests the following rule:

> ## The Product Rule
>
> $$b^m \cdot b^n = b^{m+n}$$
>
> When multiplying exponential expressions with the same base, add the exponents. Use this sum as the exponent of the common base.

EXAMPLE 1   Using the Product Rule

Multiply each expression using the product rule:

a. $2^2 \cdot 2^3$ **b.** $x^7 \cdot x^9$ **c.** $y \cdot y^5$ **d.** $y^3 \cdot y^2 \cdot y^5.$

Great Question!

Can I use the product rule to multiply an expression such as $x^7 \cdot y^9$?

The product rule does not apply to exponential expressions with different bases:

$x^7 \cdot y^9$, or $x^7 y^9$, cannot be simplified.

Solution

a. $2^2 \cdot 2^3 = 2^{2+3} = 2^5$ or 32

b. $x^7 \cdot x^9 = x^{7+9} = x^{16}$

c. $y \cdot y^5 = y^1 \cdot y^5 = y^{1+5} = y^6$

d. $y^3 \cdot y^2 \cdot y^5 = y^{3+2+5} = y^{10}$ ∎

☑ **CHECK POINT 1**   Multiply each expression using the product rule:

a. $2^2 \cdot 2^4$ **b.** $x^6 \cdot x^4$ **c.** $y \cdot y^7$ **d.** $y^4 \cdot y^3 \cdot y^2.$

2 Use the power rule for exponents. ▶

The Power Rule for Exponents

The next property of exponents applies when an exponential expression is raised to a power. Here is an example:

$$(3^2)^4.$$

> The exponential expression 3^2 is raised to the fourth power.

There are 4 factors of 3^2. Thus,

$$(3^2)^4 = 3^2 \cdot 3^2 \cdot 3^2 \cdot 3^2 = 3^{2+2+2+2} = 3^8.$$

> Add exponents when multiplying with the same base.

We can obtain the answer, 3^8, by multiplying the exponents:

$$(3^2)^4 = 3^{2 \cdot 4} = 3^8.$$

By generalizing $(3^2)^4 = 3^{2 \cdot 4} = 3^8$, we obtain the following rule:

The Power Rule (Powers to Powers)

$$(b^m)^n = b^{mn}$$

When an exponential expression is raised to a power, multiply the exponents. Place the product of the exponents on the base and remove the parentheses.

EXAMPLE 2 Using the Power Rule

Simplify each expression using the power rule:

a. $(2^3)^5$ **b.** $(x^6)^4$ **c.** $[(-3)^7]^5.$

Solution

a. $(2^3)^5 = 2^{3 \cdot 5} = 2^{15}$

b. $(x^6)^4 = x^{6 \cdot 4} = x^{24}$

c. $[(-3)^7]^5 = (-3)^{7 \cdot 5} = (-3)^{35}$ ∎

✓ **CHECK POINT 2** Simplify each expression using the power rule:

a. $(3^4)^5$ **b.** $(x^9)^{10}$ **c.** $[(-5)^7]^3.$

Great Question!

Can you show me examples that illustrate the difference between the product rule and the power rule?

Do not confuse the product and power rules. Note the following differences:

- $x^4 \cdot x^7 = x^{4+7} = x^{11}$
- $(x^4)^7 = x^{4 \cdot 7} = x^{28}.$

3 Use the products-to-powers rule. ▶

The Products-to-Powers Rule for Exponents

The next property of exponents applies when we are raising a product to a power. Here is an example:

$$(2x)^4.$$

> The product $2x$ is raised to the fourth power.

There are four factors of $2x$. Thus,

$$(2x)^4 = 2x \cdot 2x \cdot 2x \cdot 2x = 2 \cdot 2 \cdot 2 \cdot 2 \cdot x \cdot x \cdot x \cdot x = 2^4 x^4.$$

We can obtain the answer, $2^4 x^4$, by raising each factor within the parentheses to the fourth power:

$$(2x)^4 = 2^4 x^4.$$

Generalizing from $(2x)^4 = 2^4x^4$ suggests the following rule:

Products to Powers

$$(ab)^n = a^n b^n$$

When a product is raised to a power, raise each factor to the power.

EXAMPLE 3   Using the Products-to-Powers Rule

Simplify each expression using the products-to-powers rule:

a. $(5x)^3$ **b.** $(-2y^4)^5$.

Solution

a. $(5x)^3 = 5^3x^3$ Raise each factor to the third power.

$\qquad\quad = 125x^3$ $5^3 = 5 \cdot 5 \cdot 5 = 125$

b. $(-2y^4)^5 = (-2)^5(y^4)^5$ Raise each factor to the fifth power.

$\qquad\qquad\quad = (-2)^5 y^{4 \cdot 5}$ To raise an exponential expression to a power, multiply exponents: $(b^m)^n = b^{mn}$.

$\qquad\qquad\quad = -32y^{20}$ $(-2)^5 = (-2)(-2)(-2)(-2)(-2) = -32$ ■

✓ **CHECK POINT 3**   Simplify each expression using the products-to-powers rule:

a. $(2x)^4$ **b.** $(-4y^2)^3$.

Great Question!

What are some common errors to avoid when simplifying exponential expressions?

Here's a partial list. The first column shows the correct simplification. The second column illustrates a common error.

Correct	Incorrect	Description of Error
$b^3 \cdot b^4 = b^{3+4} = b^7$	$b^3 \cdot b^4 = b^{12}$	Exponents should be added, not multiplied.
$3^2 \cdot 3^4 = 3^{2+4} = 3^6$	$3^2 \cdot 3^4 = 9^{2+4} = 9^6$	The common base should be retained, not multiplied.
$(x^5)^3 = x^{5 \cdot 3} = x^{15}$	$(x^5)^3 = x^{5+3} = x^8$	Exponents should be multiplied, not added, when raising a power to a power.
$(4x)^3 = 4^3x^3 = 64x^3$	$(4x)^3 = 4x^3$	Both factors should be cubed.

---

4   Multiply monomials. ▶

Multiplying Monomials

Now that we have developed three properties of exponents, we are ready to turn to polynomial multiplication. We begin with the product of two monomials, such as $-8x^6$ and $5x^3$. This product is obtained by multiplying the coefficients, -8 and 5, and then multiplying the variables using the product rule for exponents.

$$(-8x^6)(5x^3) = -8 \cdot 5 \cdot x^6 \cdot x^3 = -8 \cdot 5x^{6+3} = -40x^9$$

Multiply coefficients and add exponents.

Multiplying Monomials

To multiply monomials with the same variable base, multiply the coefficients and then multiply the variables. Use the product rule for exponents to multiply the variables: Keep the variable and add the exponents.

EXAMPLE 4 Multiplying Monomials

Multiply: **a.** $(2x)(4x^2)$ **b.** $(-10x^6)(6x^{10})$.

Solution

a. $(2x)(4x^2) = (2 \cdot 4)(x \cdot x^2)$ Multiply the coefficients and multiply the variables.
$\qquad\qquad\quad = 8x^{1+2}$ Add exponents: $b^m \cdot b^n = b^{m+n}$.
$\qquad\qquad\quad = 8x^3$ Simplify.

b. $(-10x^6)(6x^{10}) = (-10 \cdot 6)(x^6 \cdot x^{10})$ Multiply the coefficients and multiply the variables.
$\qquad\qquad\qquad\quad = -60x^{6+10}$ Add exponents: $b^m \cdot b^n = b^{m+n}$.
$\qquad\qquad\qquad\quad = -60x^{16}$ Simplify. ■

☑ **CHECK POINT 4** Multiply: **a.** $(7x^2)(10x)$ **b.** $(-5x^4)(4x^5)$.

Multiplying a Monomial and a Polynomial That Is Not a Monomial

We use the distributive property to multiply a monomial and a polynomial that is not a monomial. For example,

$$3x^2(2x^3 + 5x) = 3x^2 \cdot 2x^3 + 3x^2 \cdot 5x = 3 \cdot 2x^{2+3} + 3 \cdot 5x^{2+1} = 6x^5 + 15x^3.$$

Monomial   Binomial   Multiply coefficients and add exponents.

Multiplying a Monomial and a Polynomial That Is Not a Monomial

To multiply a monomial and a polynomial, use the distributive property to multiply each term of the polynomial by the monomial.

EXAMPLE 5 Multiplying a Monomial and a Polynomial

Multiply: **a.** $2x(x + 4)$ **b.** $3x^2(4x^3 - 5x + 2)$.

Solution

a. $2x(x + 4) = 2x \cdot x + 2x \cdot 4$ Use the distributive property.
$\qquad\qquad\quad = 2 \cdot 1x^{1+1} + 2 \cdot 4x$ To multiply the monomials, multiply coefficients and add exponents.
$\qquad\qquad\quad = 2x^2 + 8x$ Simplify.

b. $3x^2(4x^3 - 5x + 2)$
$\quad = 3x^2 \cdot 4x^3 - 3x^2 \cdot 5x + 3x^2 \cdot 2$ Use the distributive property.
$\quad = 3 \cdot 4x^{2+3} - 3 \cdot 5x^{2+1} + 3 \cdot 2x^2$ To multiply the monomials, multiply coefficients and add exponents.
$\quad = 12x^5 - 15x^3 + 6x^2$ Simplify. ■

Great Question!

Because monomials with the same base and different exponents can be multiplied, can they also be added?

No. Don't confuse adding and multiplying monomials.

Addition:

$5x^4 + 6x^4 = 11x^4$

Multiplication:

$(5x^4)(6x^4) = (5 \cdot 6)(x^4 \cdot x^4)$
$\qquad\qquad\quad = 30x^{4+4}$
$\qquad\qquad\quad = 30x^8$

Only like terms can be added or subtracted, but unlike terms may be multiplied.

Addition:

$5x^4 + 3x^2$ cannot be simplified.

Multiplication:

$(5x^4)(3x^2) = (5 \cdot 3)(x^4 \cdot x^2)$
$\qquad\qquad\quad = 15x^{4+2}$
$\qquad\qquad\quad = 15x^6$

5 Multiply a monomial and a polynomial. ▶

Rectangles often make it possible to visualize polynomial multiplication. For example, **Figure 5.5** shows a rectangle with length $2x$ and width $x + 4$. The area of the large rectangle is

$$2x(x + 4).$$

The sum of the areas of the two smaller rectangles is

$$2x^2 + 8x.$$

Conclusion:

$$2x(x + 4) = 2x^2 + 8x$$

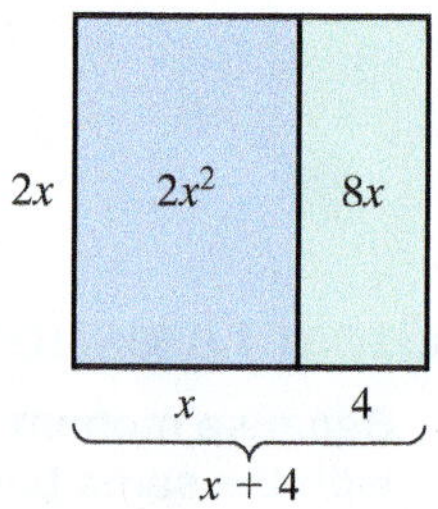

Figure 5.5

☑ **CHECK POINT 5** Multiply:

a. $3x(x + 5)$ **b.** $6x^2(5x^3 - 2x + 3)$.

6 Multiply polynomials when neither is a monomial. ▶

Multiplying Polynomials When Neither Is a Monomial

How do we multiply two polynomials if neither is a monomial? For example, consider

$$(2x + 3)(x^2 + 4x + 5).$$

Binomial   Trinomial

One way to perform this multiplication is to distribute $2x$ throughout the trinomial

$$2x(x^2 + 4x + 5)$$

and 3 throughout the trinomial

$$3(x^2 + 4x + 5).$$

Then combine the like terms that result. In general, the product of two polynomials is the polynomial obtained by multiplying each term of one polynomial by each term of the other polynomial and then combining like terms.

Using Technology

Graphic Connections
The graphs of
$$y_1 = (x + 3)(x + 2)$$
and $y_2 = x^2 + 5x + 6$
are the same.

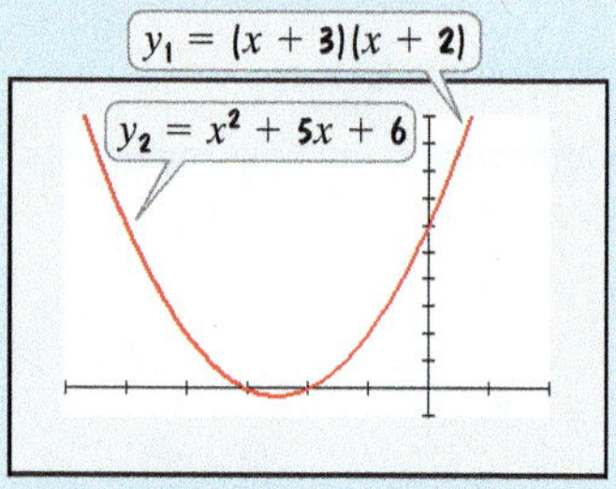

$y_1 = (x + 3)(x + 2)$

$y_2 = x^2 + 5x + 6$

$[-6, 2, 1]$ by $[-1, 10, 1]$

This verifies that
$(x + 3)(x + 2) = x^2 + 5x + 6$.

Multiplying Polynomials When Neither Is a Monomial

Multiply each term of one polynomial by each term of the other polynomial. Then combine like terms.

EXAMPLE 6   Multiplying Binomials

Multiply: **a.** $(x + 3)(x + 2)$ **b.** $(3x + 7)(2x - 4)$.

Solution   We begin by multiplying each term of the second binomial by each term of the first binomial.

a. $(x + 3)(x + 2)$

$= x(x + 2) + 3(x + 2)$ Multiply the second binomial by each term of the first binomial.

$= x \cdot x + x \cdot 2 + 3 \cdot x + 3 \cdot 2$ Use the distributive property.

$= x^2 + 2x + 3x + 6$ Multiply. Note that $x \cdot x = x^1 \cdot x^1 = x^{1+1} = x^2$.

$= x^2 + 5x + 6$ Combine like terms.

b. $(3x + 7)(2x - 4)$

$= 3x(2x - 4) + 7(2x - 4)$ Multiply the second binomial by each term of the first binomial.

$= 3x \cdot 2x - 3x \cdot 4 + 7 \cdot 2x - 7 \cdot 4$ Use the distributive property.

$= 6x^2 - 12x + 14x - 28$ Multiply.

$= 6x^2 + 2x - 28$ Combine like terms. ∎

☑ **CHECK POINT 6**   Multiply:

a. $(x + 4)(x + 5)$ **b.** $(5x + 3)(2x - 7)$.

You can visualize the polynomial multiplication in Example 6(a), $(x + 3)(x + 2) = x^2 + 5x + 6$, by analyzing the areas in **Figure 5.6**.

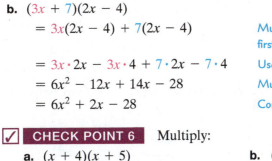

Area of the large rectangle $(x + 3)(x + 2)$

Sum of the areas of the four smaller rectangles inside the large rectangle $x^2 + 3x + 2x + 6$
$= x^2 + 5x + 6$

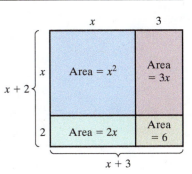

Figure 5.6

Conclusion:

$$(x + 3)(x + 2) = x^2 + 5x + 6$$

EXAMPLE 7   Multiplying a Binomial and a Trinomial

Multiply: $(2x + 3)(x^2 + 4x + 5)$.

Solution

$(2x + 3)(x^2 + 4x + 5)$

$= 2x(x^2 + 4x + 5) + 3(x^2 + 4x + 5)$ Multiply the trinomial by each term of the binomial.

$= 2x \cdot x^2 + 2x \cdot 4x + 2x \cdot 5 + 3x^2 + 3 \cdot 4x + 3 \cdot 5$ Use the distributive property.

$= 2x^3 + 8x^2 + 10x + 3x^2 + 12x + 15$ Multiply monomials: Multiply coefficients and add exponents.

$= 2x^3 + 11x^2 + 22x + 15$ Combine like terms: $8x^2 + 3x^2 = 11x^2$ and $10x + 12x = 22x$. ∎

☑ **CHECK POINT 7**   Multiply: $(5x + 2)(x^2 - 4x + 3)$.

Another method for solving Example 7 is to use a vertical format similar to that used for multiplying whole numbers.

$$
\begin{array}{r}
x^2 + 4x + 5 \\
2x + 3 \\
\hline
3x^2 + 12x + 15 \\
2x^3 + 8x^2 + 10x \\
\hline
2x^3 + 11x^2 + 22x + 15
\end{array}
$$

$3(x^2 + 4x + 5)$

$2x(x^2 + 4x + 5)$

Write like terms in the same column.

Combine like terms.

EXAMPLE 8   Multiplying Polynomials Using a Vertical Format

Multiply: $(2x^2 - 3x)(5x^3 - 4x^2 + 7x)$.

Solution To use the vertical format to find $(2x^2 - 3x)(5x^3 - 4x^2 + 7x)$, it is most convenient to write the polynomial with the greater number of terms in the top row.

$$5x^3 - 4x^2 + 7x$$
$$\underline{2x^2 - 3x}$$

We now multiply each term in the top polynomial by the last term in the bottom polynomial.

$$5x^3 \quad - 4x^2 \quad + 7x$$
$$\underline{2x^2 \quad - 3x}$$
$$-15x^4 + 12x^3 - 21x^2 \qquad -3x(5x^3 - 4x^2 + 7x)$$

Then we multiply each term in the top polynomial by $2x^2$, the first term in the bottom polynomial. Like terms are placed in columns because the final step involves combining them.

$$5x^3 \quad - 4x^2 \quad + 7x$$
$$\underline{2x^2 \quad - 3x}$$

Write like terms in the same column.

$$-15x^4 + 12x^3 - 21x^2 \qquad -3x(5x^3 - 4x^2 + 7x)$$
$$\underline{10x^5 \quad - 8x^4 + 14x^3} \qquad 2x^2(5x^3 - 4x^2 + 7x)$$
$$10x^5 - 23x^4 + 26x^3 - 21x^2 \qquad \text{Combine like terms, which are lined up in columns.}$$

✓ **CHECK POINT 8** Multiply using a vertical format: $(3x^2 - 2x)(2x^3 - 5x^2 + 4x)$.

Achieving Success

Think about finding a tutor.

If you're attending all lectures, taking good class notes, reading the textbook, and doing all the assigned homework, but still having difficulty in a math course, you might want to find a tutor. Many on-campus learning centers and math labs have trained people available to help you. Sometimes a TA who has previously taught the course is available. **Make sure the tutor is both good at math and familiar with the particular course you're taking.** Bring your textbook, class notes, the problems you've done, and information about course policy and tests to each meeting with your tutor. That way he or she can be sure the tutoring sessions address your exact needs.

CONCEPT AND VOCABULARY CHECK

Fill in each blank so that the resulting statement is true.

1. The product rule for exponents states that $b^m \cdot b^n = $ _____. When multiplying exponential expressions with the same base, _____ the exponents.

2. The power rule for exponents states that $(b^m)^n = $ _____. When an exponential expression is raised to a power, _____ the exponents.

3. The products-to-powers rule for exponents states that $(ab)^n = $ _____. When a product is raised to a power, raise each _____ to the power.

4. To multiply $2x^2(x^2 + 5x + 7)$, use the _____ property to multiply each term of the polynomial _____ by the monomial _____.

5. To multiply $(4x + 7)(x^2 + 8x + 3)$, begin by multiplying each term of $x^2 + 8x + 3$ by _____. Then multiply each term of $x^2 + 8x + 3$ by _____. Then combine _____ terms.

5.2 EXERCISE SET ▶ MyMathLab®

Practice Exercises

In Exercises 1–8, multiply each expression using the product rule.

1. $x^{15} \cdot x^3$ **2.** $x^{12} \cdot x^4$

3. $y \cdot y^{11}$ **4.** $y \cdot y^{19}$

5. $x^2 \cdot x^6 \cdot x^3$ **6.** $x^4 \cdot x^3 \cdot x^5$

7. $7^9 \cdot 7^{10}$ **8.** $8^7 \cdot 8^{10}$

In Exercises 9–14, simplify each expression using the power rule.

9. $(6^9)^{10}$ **10.** $(6^7)^{10}$

11. $(x^{15})^3$ **12.** $(x^{12})^4$

13. $[(-20)^3]^3$ **14.** $[(-50)^4]^4$

In Exercises 15–24, simplify each expression using the products-to-powers rule.

15. $(2x)^3$ **16.** $(4x)^3$

17. $(-5x)^2$ **18.** $(-6x)^2$

19. $(4x^3)^2$ **20.** $(6x^3)^2$

21. $(-2y^6)^4$ **22.** $(-2y^5)^4$

23. $(-2x^7)^5$ **24.** $(-2x^{11})^5$

In Exercises 25–34, multiply the monomials.

25. $(7x)(2x)$ **26.** $(8x)(3x)$

27. $(6x)(4x^2)$ **28.** $(10x)(3x^2)$

29. $(-5y^4)(3y^3)$ **30.** $(-6y^4)(2y^3)$

31. $\left(-\dfrac{1}{2}a^3\right)\left(-\dfrac{1}{4}a^2\right)$ **32.** $\left(-\dfrac{1}{3}a^4\right)\left(-\dfrac{1}{2}a^2\right)$

33. $(2x^2)(-3x)(8x^4)$ **34.** $(3x^3)(-2x)(5x^6)$

In Exercises 35–54, find each product of the monomial and the polynomial.

35. $4x(x + 3)$ **36.** $6x(x + 5)$

37. $x(x - 3)$ **38.** $x(x - 7)$

39. $2x(x - 6)$ **40.** $3x(x - 5)$

41. $-4y(3y + 5)$ **42.** $-5y(6y + 7)$

43. $4x^2(x + 2)$ **44.** $5x^2(x + 6)$

45. $2y^2(y^2 + 3y)$ **46.** $4y^2(y^2 + 2y)$

47. $2y^2(3y^2 - 4y + 7)$

48. $4y^2(5y^2 - 6y + 3)$

49. $(3x^3 + 4x^2)(2x)$

50. $(4x^3 + 5x^2)(2x)$

51. $(x^2 + 5x - 3)(-2x)$

52. $(x^3 - 2x + 2)(-4x)$

53. $-3x^2(-4x^2 + x - 5)$

54. $-6x^2(3x^2 - 2x - 7)$

In Exercises 55–78, find each product. In each case, neither factor is a monomial.

55. $(x + 3)(x + 5)$

56. $(x + 4)(x + 6)$

57. $(2x + 1)(x + 4)$

58. $(2x + 5)(x + 3)$

59. $(x + 3)(x - 5)$

60. $(x + 4)(x - 6)$

61. $(x - 11)(x + 9)$

62. $(x - 12)(x + 8)$

63. $(2x - 5)(x + 4)$

64. $(3x - 4)(x + 5)$

65. $\left(\dfrac{1}{4}x + 4\right)\left(\dfrac{3}{4}x - 1\right)$

66. $\left(\dfrac{1}{5}x + 5\right)\left(\dfrac{3}{5}x - 1\right)$

67. $(x + 1)(x^2 + 2x + 3)$

68. $(x + 2)(x^2 + x + 5)$

69. $(y - 3)(y^2 - 3y + 4)$

70. $(y - 2)(y^2 - 4y + 3)$

71. $(2a - 3)(a^2 - 3a + 5)$

72. $(2a - 1)(a^2 - 4a + 3)$

73. $(x + 1)(x^3 + 2x^2 + 3x + 4)$

74. $(x + 1)(x^3 + 4x^2 + 7x + 3)$

75. $\left(x - \dfrac{1}{2}\right)(4x^3 - 2x^2 + 5x - 6)$

76. $\left(x - \dfrac{1}{3}\right)(3x^3 - 6x^2 + 5x - 9)$

77. $(x^2 + 2x + 1)(x^2 - x + 2)$

78. $(x^2 + 3x + 1)(x^2 - 2x - 1)$

In Exercises 79–92, use a vertical format to find each product.

79. $\begin{array}{r} x^2 - 5x + 3 \\ \underline{x + 8} \end{array}$

80. $\begin{array}{r} x^2 - 7x + 9 \\ \underline{x + 4} \end{array}$

81. $\begin{array}{r} x^2 - 3x + 9 \\ \underline{2x - 3} \end{array}$

82. $\begin{array}{r} y^2 - 5y + 3 \\ \underline{4y - 5} \end{array}$

83. $\begin{array}{r} 2x^3 + x^2 + 2x + 3 \\ \underline{x + 4} \end{array}$

84. $\begin{array}{r} 3y^3 + 2y^2 + y + 4 \\ \underline{y + 3} \end{array}$

85. $\begin{array}{r} 4z^3 - 2z^2 + 5z - 4 \\ \underline{3z - 2} \end{array}$

86. $\begin{array}{r} 5z^3 - 3z^2 + 4z - 3 \\ \underline{2z - 4} \end{array}$

87. $\begin{array}{r} 7x^3 - 5x^2 + 6x \\ \underline{3x^2 - 4x} \end{array}$

88. $\begin{array}{r} 9y^3 - 7y^2 + 5y \\ \underline{3y^2 + 5y} \end{array}$

89. $\begin{array}{r} 2y^5 - 3y^3 + y^2 - 2y + 3 \\ \underline{2y - 1} \end{array}$

90. $\begin{array}{r} n^4 - n^3 + n^2 - n + 1 \\ \underline{2n + 3} \end{array}$

91. $x^2 + 7x - 3$
$\underline{x^2 - x - 1}$

92. $x^2 + 6x - 4$
$\underline{x^2 - x - 2}$

Practice PLUS

In Exercises 93–100, perform the indicated operations.

93. $(x + 4)(x - 5) - (x + 3)(x - 6)$

94. $(x + 5)(x - 6) - (x + 2)(x - 9)$

95. $4x^2(5x^3 + 3x - 2) - 5x^3(x^2 - 6)$

96. $3x^2(6x^3 + 2x - 3) - 4x^3(x^2 - 5)$

97. $(y + 1)(y^2 - y + 1) + (y - 1)(y^2 + y + 1)$

98. $(y + 1)(y^2 - y + 1) - (y - 1)(y^2 + y + 1)$

99. $(y + 6)^2 - (y - 2)^2$

100. $(y + 5)^2 - (y - 4)^2$

Application Exercises

101. Find a trinomial for the area of the rectangular rug shown below whose sides are $x + 5$ feet and $2x - 3$ feet.

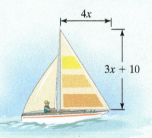

102. The base of a triangular sail is $4x$ feet and its height is $3x + 10$ feet. Write a binomial in terms of x for the area of the sail.

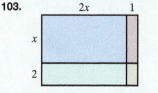

In Exercises 103–104,

a. *Express the area of the large rectangle as the product of two binomials.*

b. *Find the sum of the areas of the four smaller rectangles.*

c. *Use polynomial multiplication to show that your expressions for area in parts (a) and (b) are equal.*

103.

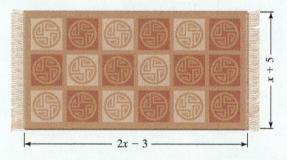

104.

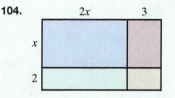

Explaining the Concepts

105. Explain the product rule for exponents. Use $2^3 \cdot 2^5$ in your explanation.

106. Explain the power rule for exponents. Use $(3^2)^4$ in your explanation.

107. Explain how to simplify an expression that involves a product raised to a power. Provide an example with your explanation.

108. Explain how to multiply monomials. Give an example.

109. Explain how to multiply a monomial and a polynomial that is not a monomial. Give an example.

110. Explain how to multiply polynomials when neither is a monomial. Give an example.

111. Explain the difference between performing these two operations:

$$2x^2 + 3x^2 \quad \text{and} \quad (2x^2)(3x^2).$$

112. Discuss situations in which a vertical format, rather than a horizontal format, is useful for multiplying polynomials.

Critical Thinking Exercises

Make Sense? *In Exercises 113–116, determine whether each statement makes sense or does not make sense, and explain your reasoning.*

113. I'm working with two monomials that I cannot add, although I can multiply them.

114. I'm working with two monomials that I can add, although I cannot multiply them.

115. Other than multiplying monomials, the distributive property is used to multiply other kinds of polynomials.

116. I used the product rule for exponents to multiply x^7 and y^9.

In Exercises 117–120, determine whether each statement is true or false. If the statement is false, make the necessary change(s) to produce a true statement.

117. $4x^3 \cdot 3x^4 = 12x^{12}$

118. $5x^2 \cdot 4x^6 = 9x^8$

119. $(y - 1)(y^2 + y + 1) = y^3 - 1$

120. Some polynomial multiplications can only be performed by using a vertical format.

121. Find a polynomial in descending powers of x representing the area of the shaded region.

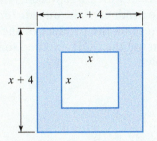

122. Find each of the products in parts (a)–(c).

   a. $(x - 1)(x + 1)$

   b. $(x - 1)(x^2 + x + 1)$

   c. $(x - 1)(x^3 + x^2 + x + 1)$

   d. Using the pattern found in parts (a)–(c), find $(x - 1)(x^4 + x^3 + x^2 + x + 1)$ without actually multiplying.

123. Find the missing factor.

$$(\underline{\qquad})\left(-\frac{1}{4}xy^3\right) = 2x^5y^3$$

Review Exercises

124. Solve: $4x - 7 > 9x - 2$. (Section 2.7, Example 7)

125. Graph $3x - 2y = 6$ using intercepts. (Section 3.2, Example 4)

126. Find the slope of the line passing through the points $(-2, 8)$ and $(1, 6)$. (Section 3.3, Example 1)

Preview Exercises

Exercises 127–129 will help you prepare for the material covered in the next section. In each exercise, find the indicated products. Then, if possible, state a fast method for finding these products. (You may already be familiar with some of these methods from a high school algebra course.)

127. **a.** $(x + 3)(x + 4)$

   b. $(x + 5)(x + 20)$

128. **a.** $(x + 3)(x - 3)$

   b. $(x + 5)(x - 5)$

129. **a.** $(x + 3)^2$

   b. $(x + 5)^2$

SECTION

5.3

Special Products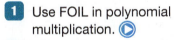

What am I supposed to learn?

After studying this section, you should be able to:

1 Use FOIL in polynomial multiplication.

2 Multiply the sum and difference of two terms.

3 Find the square of a binomial sum.

4 Find the square of a binomial difference.

Let's cut to the chase. Are there fast methods for finding products of polynomials? The answer is beepingly "yes." (Or should that be $(\text{BEEP})^2$ yes?) In this section, we'll cut to the chase by using the distributive property to develop patterns that will let you multiply certain binomials quite rapidly.

The Product of Two Binomials: FOIL

Frequently, we need to find the product of two binomials. One way to perform this multiplication is to distribute each term in the first binomial through the second binomial. For example, we can find the product of the binomials $3x + 2$ and $4x + 5$ as follows:

$$(3x + 2)(4x + 5) = 3x(4x + 5) + 2(4x + 5)$$

Distribute $3x$ over $4x + 5$.   Distribute 2 over $4x + 5$.

$$= 3x(4x) + 3x(5) + 2(4x) + 2(5)$$

$$= 12x^2 + 15x + 8x + 10.$$

> We'll combine these like terms later. For now, our interest is in how to obtain each of these four terms.

We can also find the product of $3x + 2$ and $4x + 5$ using a method called FOIL, which is based on our work shown on the previous page. Any two binomials can be quickly multiplied by using the FOIL method, in which **F** represents the product of the **first** terms in each binomial, **O** represents the product of the **outside** terms, **I** represents the product of the **inside** terms, and **L** represents the product of the **last**, or second, terms in each binomial. For example, we can use the FOIL method to find the product of the binomials $3x + 2$ and $4x + 5$ as follows:

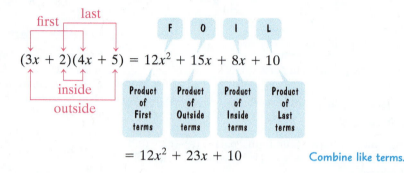

$$(3x + 2)(4x + 5) = 12x^2 + 15x + 8x + 10$$

$$= 12x^2 + 23x + 10 \qquad \text{Combine like terms.}$$

In general, here's how to use the FOIL method to find the product of $ax + b$ and $cx + d$:

1 Use FOIL in polynomial multiplication. ▶

Using the FOIL Method to Multiply Binomials

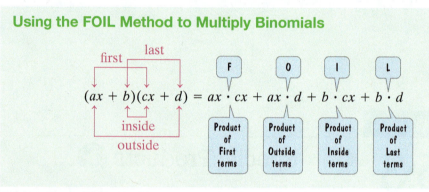

$$(ax + b)(cx + d) = ax \cdot cx + ax \cdot d + b \cdot cx + b \cdot d$$

EXAMPLE 1   Using the FOIL Method

Multiply: $(x + 3)(x + 4)$.

Solution

F: First terms $= x \cdot x = x^2$ $(x + 3)(x + 4)$

O: Outside terms $= x \cdot 4 = 4x$ $(x + 3)(x + 4)$

I: Inside terms $= 3 \cdot x = 3x$ $(x + 3)(x + 4)$

L: Last terms $= 3 \cdot 4 = 12$ $(x + 3)(x + 4)$

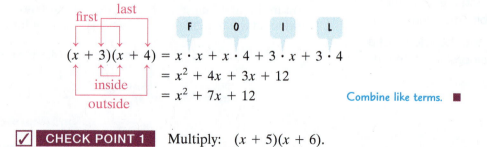

$$(x + 3)(x + 4) = x \cdot x + x \cdot 4 + 3 \cdot x + 3 \cdot 4$$

$$= x^2 + 4x + 3x + 12$$

$$= x^2 + 7x + 12 \qquad \text{Combine like terms.} \quad ■$$

✓ CHECK POINT 1   Multiply: $(x + 5)(x + 6)$.

EXAMPLE 2 Using the FOIL Method

Multiply: $(3x + 4)(5x - 3)$.

Solution

$$(3x + 4)(5x - 3) = 3x \cdot 5x + 3x(-3) + 4 \cdot 5x + 4(-3)$$
$$= 15x^2 - 9x + 20x - 12$$
$$= 15x^2 + 11x - 12 \qquad \text{Combine like terms.} \quad \blacksquare$$

✓ **CHECK POINT 2** Multiply: $(7x + 5)(4x - 3)$.

EXAMPLE 3 Using the FOIL Method

Multiply: $(2 - 5x)(3 - 4x)$.

Solution

$$(2 - 5x)(3 - 4x) = 2 \cdot 3 + 2(-4x) + (-5x)(3) + (-5x)(-4x)$$
$$= 6 - 8x - 15x + 20x^2$$
$$= 6 - 23x + 20x^2 \qquad \text{Combine like terms.}$$

The product can also be expressed in standard form as $20x^2 - 23x + 6$. $\blacksquare$

✓ **CHECK POINT 3** Multiply: $(4 - 2x)(5 - 3x)$.

---

② Multiply the sum and difference of two terms.

Multiplying the Sum and Difference of Two Terms

We can use the FOIL method to multiply $A + B$ and $A - B$ as follows:

$$(A + B)(A - B) = A^2 - AB + AB - B^2 = A^2 - B^2.$$

Notice that the outside and inside products have a sum of 0 and the terms cancel. The FOIL multiplication provides us with a quick rule for multiplying the sum and difference of two terms, referred to as a *special-product formula*.

> ### The Product of the Sum and Difference of Two Terms
>
> $$(A + B)(A - B) = A^2 - B^2$$
>
> The product of the sum and the difference of the same two terms **is** the square of the first term minus the square of the second term.

EXAMPLE 4 Finding the Product of the Sum and Difference of Two Terms

Multiply: **a.** $(4y + 3)(4y - 3)$ **b.** $(3x - 7)(3x + 7)$ **c.** $(5a^4 + 6)(5a^4 - 6)$.

Solution Use the special-product formula shown.

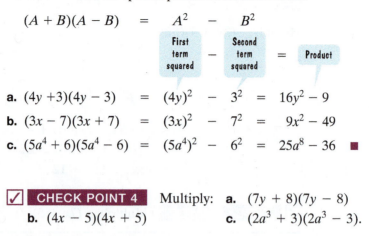

$$(A + B)(A - B) = A^2 - B^2$$

First term squared − Second term squared = Product

a. $(4y + 3)(4y - 3) = (4y)^2 - 3^2 = 16y^2 - 9$

b. $(3x - 7)(3x + 7) = (3x)^2 - 7^2 = 9x^2 - 49$

c. $(5a^4 + 6)(5a^4 - 6) = (5a^4)^2 - 6^2 = 25a^8 - 36$ ∎

☑ **CHECK POINT 4** Multiply: **a.** $(7y + 8)(7y - 8)$
b. $(4x - 5)(4x + 5)$ **c.** $(2a^3 + 3)(2a^3 - 3)$.

---

3 Find the square of a binomial sum.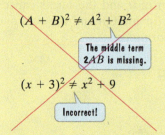

The Square of a Binomial

Let's now find $(A + B)^2$, the square of a binomial sum. To do so, we begin with the FOIL method and look for a general rule.

F   O   I   L

$$(A + B)^2 = (A + B)(A + B) = A \cdot A + A \cdot B + A \cdot B + B \cdot B$$
$$= A^2 + 2AB + B^2$$

This result implies the following rule, which is another example of a special-product formula:

The Square of a Binomial Sum

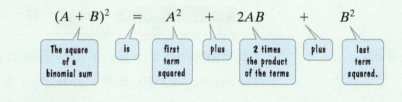

$$(A + B)^2 = A^2 + 2AB + B^2$$

The square of a binomial sum | is | first term squared | plus | 2 times the product of the terms | plus | last term squared.

EXAMPLE 5 Finding the Square of a Binomial Sum

Multiply:

a. $(x + 3)^2$ **b.** $(3x + 7)^2$.

Solution Use the special-product formula shown.

$(A + B)^2 = A^2 + 2AB + B^2$

	(First Term)^2	+	2 · Product of the Terms	+	(Last Term)^2	= Product
a. $(x + 3)^2 =$	x^2	+	$2 \cdot x \cdot 3$	+	3^2	$= x^2 + 6x + 9$
b. $(3x + 7)^2 =$	$(3x)^2$	+	$2(3x)(7)$	+	7^2	$= 9x^2 + 42x + 49$

∎

☑ **CHECK POINT 5** Multiply:

a. $(x + 10)^2$ **b.** $(5x + 4)^2$.

The formula for the square of a binomial sum can be interpreted geometrically by analyzing the areas in **Figure 5.7**.

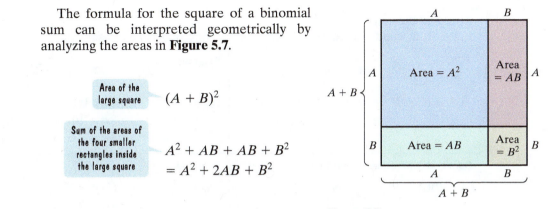

Area of the large square $(A + B)^2$

Sum of the areas of the four smaller rectangles inside the large square
$$A^2 + AB + AB + B^2$$
$$= A^2 + 2AB + B^2$$

Figure 5.7

Conclusion:

$$(A + B)^2 = A^2 + 2AB + B^2$$

4 Find the square of a binomial difference. ▶

A similar pattern occurs for $(A - B)^2$, the square of a binomial difference. Using the FOIL method on $(A - B)^2$, we obtain the following rule:

The Square of a Binomial Difference

$$(A - B)^2 \quad = \quad A^2 \quad - \quad 2AB \quad + \quad B^2$$

The square of a binomial difference | is | first term squared | minus | 2 times the product of the terms | plus | last term squared.

EXAMPLE 6   Finding the Square of a Binomial Difference

Multiply:

a. $(x - 4)^2$ **b.** $(5y - 6)^2$.

Solution   Use the special-product formula shown.

$$(A - B)^2 = A^2 - 2AB + B^2$$

	(First Term)^2	$-$	2 · Product of the Terms	$+$	(Last Term)^2	= Product
a. $(x - 4)^2 =$	x^2	$-$	$2 \cdot x \cdot 4$	$+$	4^2	$= x^2 - 8x + 16$
b. $(5y - 6)^2 =$	$(5y)^2$	$-$	$2(5y)(6)$	$+$	6^2	$= 25y^2 - 60y + 36$ ∎

☑ CHECK POINT 6   Multiply:

a. $(x - 9)^2$ **b.** $(7x - 3)^2$.

The table at the top of the next page summarizes the FOIL method and the three special products. The special products occur so frequently in algebra that it is convenient to memorize the form or pattern of these formulas.

FOIL and Special Products
Let *A*, *B*, *C*, and *D* be real numbers, variables, or algebraic expressions.

FOIL	**Example**
$(A + B)(C + D) = AC + AD + BC + BD$	$(2x + 3)(4x + 5) = (2x)(4x) + (2x)(5) + (3)(4x) + (3)(5)$ $= 8x^2 + 10x + 12x + 15$ $= 8x^2 + 22x + 15$
Sum and Difference of Two Terms $(A + B)(A - B) = A^2 - B^2$	**Example** $(2x + 3)(2x - 3) = (2x)^2 - 3^2$ $= 4x^2 - 9$
Square of a Binomial $(A + B)^2 = A^2 + 2AB + B^2$ $(A - B)^2 = A^2 - 2AB + B^2$	**Example** $(2x + 3)^2 = (2x)^2 + 2(2x)(3) + 3^2$ $= 4x^2 + 12x + 9$ $(2x - 3)^2 = (2x)^2 - 2(2x)(3) + 3^2$ $= 4x^2 - 12x + 9$

Achieving Success

Manage your time. Use a day planner such as the one shown below. (Go online and search "day planner" or "day scheduler" to find a schedule grid that you can print and use.)

Sample Day Planner

	Monday	Tuesday	Wednesday	Thursday	Friday	Saturday	Sunday
5:00 A.M.							
6:00 A.M.							
7:00 A.M.							
8:00 A.M.							
9:00 A.M.							
10:00 A.M.							
11:00 A.M.							
12:00 P.M.							
1:00 P.M.							
2:00 P.M.							
3:00 P.M.							
4:00 P.M.							
5:00 P.M.							
6:00 P.M.							
7:00 P.M.							
8:00 P.M.							
9:00 P.M.							
10:00 P.M.							
11:00 P.M.							
Midnight							

- On the Sunday before the week begins, fill in the time slots with fixed items such as school, extracurricular activities, work, etc.
- Because your education should be a top priority, decide what times you would like to study and do homework. Fill in these activities on the day planner.
- Plan other flexible activities such as exercise, socializing, etc. around the times already established.
- Be flexible. Things may come up that are unavoidable or some items may take longer than planned. However, be honest with yourself: Playing video games with friends is not "unavoidable."
- Stick to your schedule. At the end of the week, you will look back and be impressed at all the things you have accomplished.

CONCEPT AND VOCABULARY CHECK

Fill in each blank so that the resulting statement is true.

1. For $(x + 5)(2x + 3)$, the product of the first terms is _____, the product of the outside terms is _____, the product of the inside terms is _____, and the product of the last terms is _____.

2. $(A + B)(A - B) =$ _____. The product of the sum and the difference of the same two terms is the square of the first term _____ the square of the second term.

3. $(A + B)^2 =$ _____. The square of a binomial sum is the first term _____ plus 2 times the _____ plus the last term _____.

4. $(A - B)^2 =$ _____. The square of a binomial difference is the first term squared _____ 2 times the _____ plus the last term _____.

5. True or false: $(x + 5)(x - 5) = x^2 - 25.$ _____

6. True or false: $(x + 5)^2 = x^2 + 25.$ _____

5.3 EXERCISE SET ▶ MyMathLab®

Practice Exercises

In Exercises 1–24, use the FOIL method to find each product. Express the product in descending powers of the variable.

1. $(x + 4)(x + 6)$

2. $(x + 8)(x + 2)$

3. $(y - 7)(y + 3)$

4. $(y - 3)(y + 4)$

5. $(2x - 3)(x + 5)$

6. $(3x - 5)(x + 7)$

7. $(4y + 3)(y - 1)$

8. $(5y + 4)(y - 2)$

9. $(2x - 3)(5x + 3)$

10. $(2x - 5)(7x + 2)$

11. $(3y - 7)(4y - 5)$

12. $(4y - 5)(7y - 4)$

13. $(7 + 3x)(1 - 5x)$

14. $(2 + 5x)(1 - 4x)$

15. $(5 - 3y)(6 - 2y)$

16. $(7 - 2y)(10 - 3y)$

17. $(5x^2 - 4)(3x^2 - 7)$

18. $(7x^2 - 2)(3x^2 - 5)$

19. $(6x - 5)(2 - x)$

20. $(4x - 3)(2 - x)$

21. $(x + 5)(x^2 + 3)$

22. $(x + 4)(x^2 + 5)$

23. $(8x^3 + 3)(x^2 + 5)$

24. $(7x^3 + 5)(x^2 + 2)$

In Exercises 25–44, multiply using the rule for finding the product of the sum and difference of two terms.

25. $(x + 3)(x - 3)$

26. $(y + 5)(y - 5)$

27. $(3x + 2)(3x - 2)$

28. $(2x + 5)(2x - 5)$

29. $(3r - 4)(3r + 4)$

30. $(5z - 2)(5z + 2)$

31. $(3 + r)(3 - r)$

32. $(4 + s)(4 - s)$

33. $(5 - 7x)(5 + 7x)$

34. $(4 - 3y)(4 + 3y)$

35. $\left(2x + \dfrac{1}{2}\right)\left(2x - \dfrac{1}{2}\right)$

36. $\left(3y + \dfrac{1}{3}\right)\left(3y - \dfrac{1}{3}\right)$

37. $(y^2 + 1)(y^2 - 1)$

38. $(y^2 + 2)(y^2 - 2)$

39. $(r^3 + 2)(r^3 - 2)$

40. $(m^3 + 4)(m^3 - 4)$

41. $(1 - y^4)(1 + y^4)$

42. $(2 - s^5)(2 + s^5)$

43. $(x^{10} + 5)(x^{10} - 5)$

44. $(x^{12} + 3)(x^{12} - 3)$

In Exercises 45–62, multiply using the rules for the square of a binomial.

45. $(x + 2)^2$

46. $(x + 5)^2$

47. $(2x + 5)^2$

48. $(5x + 2)^2$

49. $(x - 3)^2$

50. $(x - 6)^2$

51. $(3y - 4)^2$

52. $(4y - 3)^2$

53. $(4x^2 - 1)^2$

54. $(5x^2 - 3)^2$

55. $(7 - 2x)^2$

56. $(9 - 5x)^2$

57. $\left(2x + \dfrac{1}{2}\right)^2$

58. $\left(3x + \dfrac{1}{3}\right)^2$

59. $\left(4y - \dfrac{1}{4}\right)^2$

60. $\left(2y - \dfrac{1}{2}\right)^2$

61. $(x^8 + 3)^2$

62. $(x^8 + 5)^2$

In Exercises 63–82, multiply using the method of your choice.

63. $(x - 1)(x^2 + x + 1)$

64. $(x + 1)(x^2 - x + 1)$

65. $(x - 1)^2$

66. $(x + 1)^2$

67. $(3y + 7)(3y - 7)$

68. $(4y + 9)(4y - 9)$

69. $3x^2(4x^2 + x + 9)$

70. $5x^2(7x^2 + x + 6)$

71. $(7y + 3)(10y - 4)$

72. $(8y + 3)(10y - 5)$

73. $(x^2 + 1)^2$

74. $(x^2 + 2)^2$

75. $(x^2 + 1)(x^2 + 2)$

76. $(x^2 + 2)(x^2 + 3)$

77. $(x^2 + 4)(x^2 - 4)$

78. $(x^2 + 5)(x^2 - 5)$

79. $(2 - 3x^5)^2$

80. $(2 - 3x^6)^2$

81. $\left(\dfrac{1}{4}x^2 + 12\right)\left(\dfrac{3}{4}x^2 - 8\right)$

82. $\left(\dfrac{1}{4}x^2 + 16\right)\left(\dfrac{3}{4}x^2 - 4\right)$

In Exercises 83–88, find the area of each shaded region. Write the answer as a polynomial in descending powers of x.

83.

$x + 1$

$x + 1$

84.

$x + 3$

$x + 3$

85.

$2x + 3$

$2x - 3$

86.

$4x + 3$

$4x - 3$

87.

$x + 9$

$x + 5$

$x + 3$

$x + 1$

88.

$x + 4$

$x + 2$

$x + 3$

$x + 1$

Practice PLUS

In Exercises 89–96, multiply by the method of your choice.

89. $[(2x + 3)(2x - 3)]^2$

90. $[(3x + 2)(3x - 2)]^2$

91. $(4x^2 + 1)[(2x + 1)(2x - 1)]$

92. $(9x^2 + 1)[(3x + 1)(3x - 1)]$

93. $(x + 2)^3$

94. $(x + 4)^3$

95. $[(x + 3) - y][(x + 3) + y]$

96. $[(x + 5) - y][(x + 5) + y]$

Application Exercises

The square garden shown in the figure measures x yards on each side. The garden is to be expanded so that one side is increased by 2 yards and an adjacent side is increased by 1 yard.

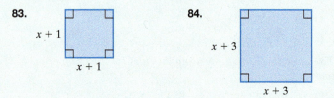

1 yard

$x + 1$

x yards

x yards

$x + 2$

2 yards

The graph shows the area of the expanded garden, y, in terms of the length of one of its original sides, x. Use this information to solve Exercises 97–100.

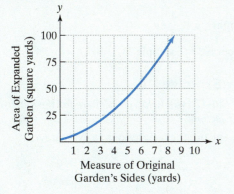

Measure of Original
Garden's Sides (yards)

97. Use the gardens pictured at the bottom of the previous page to write a product of two binomials that expresses the area of the larger garden.

98. Use the gardens pictured at the bottom of the previous page to write a polynomial in descending powers of x that expresses the area of the larger garden.

99. If the original garden measures 6 yards on a side, use your expression from Exercise 97 to find the area of the larger garden. Then identify your solution as a point on the graph shown above.

100. If the original garden measures 8 yards on a side, use your polynomial from Exercise 98 to find the area of the larger garden. Then identify your solution as a point on the graph shown above.

The square painting in the figure measures x inches on each side. The painting is uniformly surrounded by a frame that measures 1 inch wide. Use this information to solve Exercises 101–102.

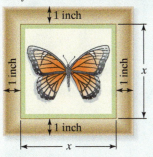

101. Write a polynomial in descending powers of x that expresses the area of the square that includes the painting and the frame.

102. Write an algebraic expression that describes the area of the frame. (*Hint:* The area of the frame is the area of the square that includes the painting and the frame minus the area of the painting.)

Explaining the Concepts

103. Explain how to multiply two binomials using the FOIL method. Give an example with your explanation.

104. Explain how to find the product of the sum and difference of two terms. Give an example with your explanation.

105. Explain how to square a binomial sum. Give an example with your explanation.

106. Explain how to square a binomial difference. Give an example with your explanation.

107. Explain why the graph for Exercises 97–100 is shown only in quadrant I.

Critical Thinking Exercises

Make Sense? *In Exercises 108–111, determine whether each statement makes sense or does not make sense, and explain your reasoning.*

108. Squaring a binomial sum is as simple as squaring each of the two terms and then writing their sum.

109. I can distribute the exponent 2 on each factor of $(5x)^2$, but I cannot do the same thing on each term of $(x + 5)^2$.

110. Instead of using the formula for the square of a binomial sum, I prefer to write the binomial sum twice and then apply the FOIL method.

111. Special-product formulas for $(A + B)(A - B)$, $(A + B)^2$, and $(A - B)^2$ have patterns that make their multiplications quicker than using the FOIL method.

In Exercises 112–115, determine whether each statement is true or false. If the statement is false, make the necessary change(s) to produce a true statement.

112. $(3 + 4)^2 = 3^2 + 4^2$

113. $(2y + 7)^2 = 4y^2 + 28y + 49$

114. $(3x^2 + 2)(3x^2 - 2) = 9x^2 - 4$

115. $(x - 5)^2 = x^2 - 5x + 25$

116. What two binomials must be multiplied using the FOIL method to give a product of $x^2 - 8x - 20$?

117. Express the volume of the box as a polynomial in standard form.

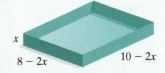

118. Express the area of the plane figure shown as a polynomial in standard form.

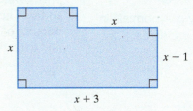

Technology Exercises

In Exercises 119–122, use a graphing utility to graph each side of the equation in the same viewing rectangle. (Call the left side y_1 and the right side y_2.) If the graphs coincide, verify that the multiplication has been performed correctly. If the graphs do not appear to coincide, this indicates that the multiplication is incorrect. In these exercises, correct the right side of the equation. Then graph the left side and the corrected right side to verify that the graphs coincide.

119. $(x + 1)^2 = x^2 + 1$; Use a $[-5, 5, 1]$ by $[0, 20, 1]$ viewing rectangle.

120. $(x + 2)^2 = x^2 + 2x + 4$; Use a $[-6, 5, 1]$ by $[0, 20, 1]$ viewing rectangle.

121. $(x + 1)(x - 1) = x^2 - 1$; Use a $[-6, 5, 1]$ by $[-2, 18, 1]$ viewing rectangle.

122. $(x - 2)(x + 2) + 4 = x^2$; Use a $[-6, 5, 1]$ by $[-2, 18, 1]$ viewing rectangle.

Review Exercises

In Exercises 123–124, solve each system by the method of your choice.

123. $\begin{cases} 2x + 3y = 1 \\ y = 3x - 7 \end{cases}$

(Section 4.2, Example 1)

124. $\begin{cases} 3x + 4y = 7 \\ 2x + 7y = 9 \end{cases}$

(Section 4.3, Example 3)

125. Graph: $y \le \dfrac{1}{3}x$.

(Section 3.6, Example 3)

Preview Exercises

Exercises 126–128 will help you prepare for the material covered in the next section.

126. Use the order of operations to evaluate

$$x^3y + 2xy^2 + 5x - 2$$

for $x = -2$ and $y = 3$.

127. Use the second step to combine the like terms.

$$5xy + 6xy = (5 + 6)xy = \,?$$

128. Multiply using FOIL: $(x + 2y)(3x + 5y)$.

SECTION

5.4 Polynomials in Several Variables

What am I supposed to learn?

After studying this section, you should be able to:

1. Evaluate polynomials in several variables.

2. Understand the vocabulary of polynomials in two variables.

3. Add and subtract polynomials in several variables.

4. Multiply polynomials in several variables.

The next time you visit a lumberyard and go rummaging through piles of wood, think *polynomials*, although polynomials a bit different from those we have encountered so far. The construction industry uses a polynomial in two variables to determine the number of board feet that can be manufactured from a tree with a diameter of x inches and a length of y feet. This polynomial is

$$\frac{1}{4}x^2y - 2xy + 4y.$$

We call a polynomial containing two or more variables a **polynomial in several variables**. These polynomials can be evaluated, added, subtracted, and multiplied just like polynomials that contain only one variable.

1 Evaluate polynomials in several variables.

Evaluating a Polynomial in Several Variables

Two steps can be used to evaluate a polynomial in several variables.

> ### Evaluating a Polynomial in Several Variables
>
> 1. Substitute the given value for each variable.
> 2. Perform the resulting computation using the order of operations.

> **EXAMPLE 1** Evaluating a Polynomial in Two Variables
>
> Evaluate $2x^3y + xy^2 + 7x - 3$ for $x = -2$ and $y = 3$.
>
> **Solution** We begin by substituting -2 for x and 3 for y in the polynomial.
>
> | | |
> |---|---|
> | $2x^3y + xy^2 + 7x - 3$ | This is the given polynomial. |
> | $= 2(-2)^3 \cdot 3 + (-2) \cdot 3^2 + 7(-2) - 3$ | Replace x with -2 and y with 3. |
> | $= 2(-8) \cdot 3 + (-2) \cdot 9 + 7(-2) - 3$ | Evaluate exponential expressions: $(-2)^3 = (-2)(-2)(-2) = -8$ and $3^2 = 3 \cdot 3 = 9$. |
> | $= -48 + (-18) + (-14) - 3$ | Perform the indicated multiplications. |
> | $= -83$ | Add from left to right. ■ |

> ✓ **CHECK POINT 1** Evaluate $3x^3y + xy^2 + 5y + 6$ for $x = -1$ and $y = 5$.

2 Understand the vocabulary of polynomials in two variables.

Describing Polynomials in Two Variables

In this section, we will limit our discussion of polynomials in several variables to two variables.

In general, a **polynomial in two variables**, x and y, contains the sum of one or more monomials in the form ax^ny^m. The constant, a, is the **coefficient**. The exponents, n and m, represent whole numbers. The **degree** of the monomial ax^ny^m is $n + m$. We'll use the polynomial from the construction industry to illustrate these ideas.

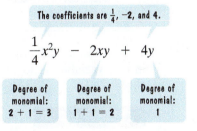

The coefficients are $\frac{1}{4}$, -2, and 4.

$$\frac{1}{4}x^2y \quad - \quad 2xy \quad + \quad 4y$$

| Degree of monomial: $2 + 1 = 3$ | Degree of monomial: $1 + 1 = 2$ | Degree of monomial: 1 |

The **degree of a polynomial in two variables** is the highest degree of all its terms. For the preceding polynomial, the degree is 3.

EXAMPLE 2 Using the Vocabulary of Polynomials

Determine the coefficient of each term, the degree of each term, and the degree of the polynomial:

$$7x^2y^3 - 17x^4y^2 + xy - 6y^2 + 9.$$

Solution

Term	Coefficient	Degree (Sum of Exponents on the Variables)
$7x^2y^3$	7	$2 + 3 = 5$
$-17x^4y^2$	-17	$4 + 2 = 6$
xy	1	$1 + 1 = 2$
$-6y^2$	-6	2
9	9	0

Think of xy as $1x^1y^1$.

The degree of the polynomial is the highest degree of all its terms, which is 6. ∎

✓ **CHECK POINT 2** Determine the coefficient of each term, the degree of each term, and the degree of the polynomial:

$$8x^4y^5 - 7x^3y^2 - x^2y - 5x + 11.$$

3 Add and subtract polynomials in several variables. ▶

Adding and Subtracting Polynomials in Several Variables

Polynomials in several variables are added by combining like terms. For example, we can add the monomials $-7xy^2$ and $13xy^2$ as follows:

$$-7xy^2 + 13xy^2 = (-7 + 13)xy^2 = 6xy^2.$$

These like terms both contain the variable factors x and y^2. *Add coefficients and keep the same variable factors, xy^2.*

EXAMPLE 3 Adding Polynomials in Two Variables

Add: $(6xy^2 - 5xy + 7) + (9xy^2 + 2xy - 6)$.

Solution
$$(6xy^2 - 5xy + 7) + (9xy^2 + 2xy - 6)$$
$$= (6xy^2 + 9xy^2) + (-5xy + 2xy) + (7 - 6)$$ Group like terms.
$$= 15xy^2 - 3xy + 1$$ Combine like terms by adding coefficients and keeping the same variable factors. ∎

✓ **CHECK POINT 3** Add: $(-8x^2y - 3xy + 6) + (10x^2y + 5xy - 10)$.

We subtract polynomials in two variables just as we did when subtracting polynomials in one variable. Add the first polynomial and the opposite of the polynomial being subtracted.

EXAMPLE 4 Subtracting Polynomials in Two Variables

Subtract:

$$(5x^3 - 9x^2y + 3xy^2 - 4) - (3x^3 - 6x^2y - 2xy^2 + 3).$$

Solution

$$(5x^3 - 9x^2y + 3xy^2 - 4) - (3x^3 - 6x^2y - 2xy^2 + 3)$$

Change the sign of each coefficient.

$$= (5x^3 - 9x^2y + 3xy^2 - 4) + (-3x^3 + 6x^2y + 2xy^2 - 3)$$

Add the opposite of the polynomial being subtracted.

$$= (5x^3 - 3x^3) + (-9x^2y + 6x^2y) + (3xy^2 + 2xy^2) + (-4 - 3)$$

Group like terms.

$$= 2x^3 - 3x^2y + 5xy^2 - 7$$

Combine like terms by adding coefficients and keeping the same variable factors. ∎

✓ **CHECK POINT 4**   Subtract:

$$(7x^3 - 10x^2y + 2xy^2 - 5) - (4x^3 - 12x^2y - 3xy^2 + 5).$$

---

4 Multiply polynomials in several variables. ▶

Multiplying Polynomials in Several Variables

The product of monomials forms the basis of polynomial multiplication. As with monomials in one variable, multiplication can be done mentally by multiplying coefficients and adding exponents on variables with the same base.

EXAMPLE 5   Multiplying Monomials

Multiply: $(7x^2y)(5x^3y^2)$.

Solution

$$(7x^2y)(5x^3y^2)$$
$$= (7 \cdot 5)(x^2 \cdot x^3)(y \cdot y^2)$$

This regrouping can be worked mentally.

$$= 35x^{2+3}y^{1+2}$$

Multiply coefficients and add exponents on variables with the same base.

$$= 35x^5y^3$$

Simplify. ∎

✓ **CHECK POINT 5**   Multiply: $(6xy^3)(10x^4y^2)$.

How do we multiply a monomial and a polynomial that is not a monomial? As we did with polynomials in one variable, multiply each term of the polynomial by the monomial.

EXAMPLE 6   Multiplying a Monomial and a Polynomial

Multiply: $3x^2y(4x^3y^2 - 6x^2y + 2)$.

Solution

$$3x^2y(4x^3y^2 - 6x^2y + 2)$$
$$= 3x^2y \cdot 4x^3y^2 - 3x^2y \cdot 6x^2y + 3x^2y \cdot 2$$

Use the distributive property.

$$= 12x^{2+3}y^{1+2} - 18x^{2+2}y^{1+1} + 6x^2y$$

Multiply coefficients and add exponents on variables with the same base.

$$= 12x^5y^3 - 18x^4y^2 + 6x^2y$$

Simplify. ∎

✓ **CHECK POINT 6**   Multiply: $6xy^2(10x^4y^5 - 2x^2y + 3)$.

FOIL and the special-products formulas can be used to multiply polynomials in several variables.

EXAMPLE 7   Multiplying Polynomials in Two Variables

Multiply: **a.** $(x + 4y)(3x - 5y)$ **b.** $(5x + 3y)^2$.

Solution   We will perform the multiplication in part (a) using the FOIL method. We will multiply in part (b) using the formula for the square of a binomial, $(A + B)^2$.

a. $(x + 4y)(3x - 5y)$

Multiply these binomials using the FOIL method.

F   0   I   L

$= (x)(3x) + (x)(-5y) + (4y)(3x) + (4y)(-5y)$

$= 3x^2 - 5xy + 12xy - 20y^2$

$= 3x^2 + 7xy - 20y^2$ Combine like terms.

$$(A + B)^2 = A^2 + 2 \cdot A \cdot B + B^2$$

b. $(5x + 3y)^2 = (5x)^2 + 2(5x)(3y) + (3y)^2$

$= 25x^2 + 30xy + 9y^2$ ■

✓ **CHECK POINT 7**   Multiply:

a. $(7x - 6y)(3x - y)$ **b.** $(2x + 4y)^2$.

EXAMPLE 8   Multiplying Polynomials in Two Variables

Multiply: **a.** $(4x^2y + 3y)(4x^2y - 3y)$ **b.** $(x + y)(x^2 - xy + y^2)$.

Solution   We perform the multiplication in part (a) using the formula for the product of the sum and difference of two terms. We perform the multiplication in part (b) by multiplying each term of the trinomial, $x^2 - xy + y^2$, by x and y, respectively, and then adding like terms.

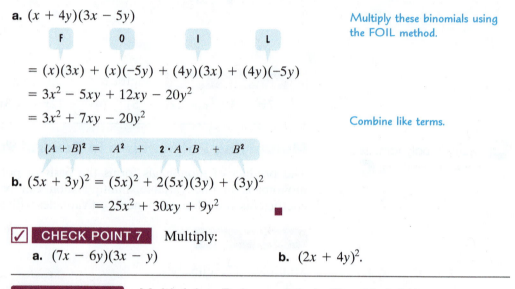

$$(A + B) \cdot (A - B) = A^2 - B^2$$

a. $(4x^2y + 3y)(4x^2y - 3y) = (4x^2y)^2 - (3y)^2$

$= 16x^4y^2 - 9y^2$

b. $(x + y)(x^2 - xy + y^2)$

$= x(x^2 - xy + y^2) + y(x^2 - xy + y^2)$ Multiply the trinomial by each term of the binomial.

$= x \cdot x^2 - x \cdot xy + x \cdot y^2 + y \cdot x^2 - y \cdot xy + y \cdot y^2$ Use the distributive property.

$= x^3 - x^2y + xy^2 + x^2y - xy^2 + y^3$ Add exponents on variables with the same base.

$= x^3 + y^3$ Combine like terms:
$-x^2y + x^2y = 0$ and
$xy^2 - xy^2 = 0$. ■

✓ **CHECK POINT 8**   Multiply:

a. $(6xy^2 + 5x)(6xy^2 - 5x)$ **b.** $(x - y)(x^2 + xy + y^2)$.

Achieving Success

Two Ways to Stay Sharp

- **Concentrate on one task at a time.** Do not multitask. Doing several things at once can cause confusion and can take longer to complete the tasks than tackling them sequentially.
- **Get enough sleep.** Fatigue impedes the ability to learn and do complex tasks.

CONCEPT AND VOCABULARY CHECK

Fill in each blank so that the resulting statement is true.

1. The coefficient of the monomial $-18x^4y^2$ is _____.

2. The degree of the monomial $-18x^4y^2$ is _____.

3. The coefficient of the monomial ax^ny^m is _____ and the degree is _____.

4. The degree of x^3y^2 is _____ and the degree of x^2y^7 is _____, so the degree of $x^3y^2 - 8x^2y^7$ is _____.

5. True or false: The monomials $7xy^2$ and $2x^2y$ can be added. _____

6. True or false: The monomials $7xy^2$ and $2x^2y$ can be multiplied. _____

5.4 EXERCISE SET ▶ MyMathLab®

Practice Exercises

In Exercises 1–6, evaluate each polynomial for $x = 2$ and $y = -3$.

1. $x^2 + 2xy + y^2$
2. $x^2 + 3xy + y^2$
3. $xy^3 - xy + 1$
4. $x^3y - xy + 2$
5. $2x^2y - 5y + 3$
6. $3x^2y - 4y + 5$

In Exercises 7–8, determine the coefficient of each term, the degree of each term, and the degree of the polynomial.

7. $x^3y^2 - 5x^2y^7 + 6y^2 - 3$
8. $12x^4y - 5x^3y^7 - x^2 + 4$

In Exercises 9–20, add or subtract as indicated.

9. $(5x^2y - 3xy) + (2x^2y - xy)$
10. $(-2x^2y + xy) + (4x^2y + 7xy)$
11. $(4x^2y + 8xy + 11) + (-2x^2y + 5xy + 2)$
12. $(7x^2y + 5xy + 13) + (-3x^2y + 6xy + 4)$
13. $(7x^4y^2 - 5x^2y^2 + 3xy) + (-18x^4y^2 - 6x^2y^2 - xy)$
14. $(6x^4y^2 - 10x^2y^2 + 7xy) + (-12x^4y^2 - 3x^2y^2 - xy)$
15. $(x^3 + 7xy - 5y^2) - (6x^3 - xy + 4y^2)$
16. $(x^4 - 7xy - 5y^3) - (6x^4 - 3xy + 4y^3)$
17. $(3x^4y^2 + 5x^3y - 3y) - (2x^4y^2 - 3x^3y - 4y + 6x)$
18. $(5x^4y^2 + 6x^3y - 7y) - (3x^4y^2 - 5x^3y - 6y + 8x)$
19. $(x^3 - y^3) - (-4x^3 - x^2y + xy^2 + 3y^3)$
20. $(x^3 - y^3) - (-6x^3 + x^2y - xy^2 + 2y^3)$

21. Add: $5x^2y^2 - 4xy^2 + 6y^2$
    $\phantom{\text{Add: }}-8x^2y^2 + 5xy^2 - y^2$

22. Add: $7a^2b^2 - 5ab^2 + 6b^2$
    $\phantom{\text{Add: }}-10a^2b^2 + 6ab^2 + 6b^2$

23. Subtract: $3a^2b^4 - 5ab^2 + 7ab$
    $\phantom{\text{Subtract: }}-(-5a^2b^4 - 8ab^2 - ab)$

24. Subtract: $13x^2y^4 - 17xy^2 + xy$
    $\phantom{\text{Subtract: }}-(-7x^2y^4 - 8xy^2 - xy)$

25. Subtract $11x - 5y$ from the sum of $7x + 13y$ and $-26x + 19y$.

26. Subtract $23x - 5y$ from the sum of $6x + 15y$ and $x - 19y$.

In Exercises 27–76, find each product.

27. $(5x^2y)(8xy)$
28. $(10x^2y)(5xy)$
29. $(-8x^3y^4)(3x^2y^5)$
30. $(7x^4y^5)(-10x^7y^{11})$
31. $9xy(5x + 2y)$
32. $7xy(8x + 3y)$
33. $5xy^2(10x^2 - 3y)$
34. $6x^2y(5x^2 - 9y)$
35. $4ab^2(7a^2b^3 + 2ab)$
36. $2ab^2(20a^2b^3 + 11ab)$
37. $-b(a^2 - ab + b^2)$
38. $-b(a^3 - ab + b^3)$
39. $(x + 5y)(7x + 3y)$
40. $(x + 9y)(6x + 7y)$
41. $(x - 3y)(2x + 7y)$

42. $(3x - y)(2x + 5y)$
43. $(3xy - 1)(5xy + 2)$
44. $(7xy + 1)(2xy - 3)$
45. $(2x + 3y)^2$
46. $(2x + 5y)^2$
47. $(xy - 3)^2$
48. $(xy - 5)^2$
49. $(x^2 + y^2)^2$
50. $(2x^2 + y^2)^2$
51. $(x^2 - 2y^2)^2$
52. $(x^2 - y^2)^2$
53. $(3x + y)(3x - y)$
54. $(x + 5y)(x - 5y)$
55. $(ab + 1)(ab - 1)$
56. $(ab + 2)(ab - 2)$
57. $(x + y^2)(x - y^2)$
58. $(x^2 + y)(x^2 - y)$
59. $(3a^2b + a)(3a^2b - a)$
60. $(5a^2b + a)(5a^2b - a)$
61. $(3xy^2 - 4y)(3xy^2 + 4y)$
62. $(7xy^2 - 10y)(7xy^2 + 10y)$
63. $(a + b)(a^2 - b^2)$
64. $(a - b)(a^2 + b^2)$
65. $(x + y)(x^2 + 3xy + y^2)$
66. $(x + y)(x^2 + 5xy + y^2)$
67. $(x - y)(x^2 - 3xy + y^2)$
68. $(x - y)(x^2 - 4xy + y^2)$
69. $(xy + ab)(xy - ab)$
70. $(xy + ab^2)(xy - ab^2)$
71. $(x^2 + 1)(x^4y + x^2 + 1)$
72. $(x^2 + 1)(xy^4 + y^2 + 1)$
73. $(x^2y^2 - 3)^2$
74. $(x^2y^2 - 5)^2$
75. $(x + y + 1)(x + y - 1)$
76. $(x + y + 1)(x - y + 1)$

In Exercises 77–80, write a polynomial in two variables that describes the total area of each region shaded in blue. Express each polynomial as the sum or difference of terms.

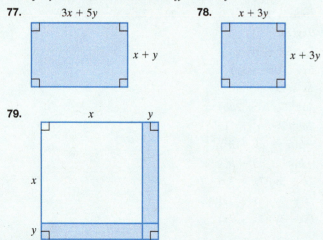

77. $3x + 5y$ / $x + y$
78. $x + 3y$ / $x + 3y$
79. x / y / x / y

80.

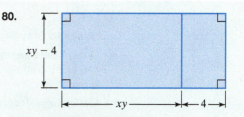

$xy - 4$ / xy / 4

Practice PLUS

In Exercises 81–86, find each product. As we said in the Section 5.3 opener, cut to the chase in each part of the polynomial multiplication: Use only the special-product formula for the sum and difference of two terms or the formulas for the square of a binomial.

81. $[(x^3y^3 + 1)(x^3y^3 - 1)]^2$
82. $[(1 - a^3b^3)(1 + a^3b^3)]^2$
83. $(xy - 3)^2(xy + 3)^2$ (Do not begin by squaring a binomial.)
84. $(ab - 4)^2(ab + 4)^2$ (Do not begin by squaring a binomial.)
85. $[x + y + z][x - (y + z)]$
86. $(a - b - c)(a + b + c)$

Application Exercises

87. The number of board feet, N, that can be manufactured from a tree with a diameter of x inches and a length of y feet is modeled by the formula

$$N = \frac{1}{4}x^2y - 2xy + 4y.$$

A building contractor estimates that 3000 board feet of lumber is needed for a job. The lumber company has just milled a fresh load of timber from 20 trees that averaged 10 inches in diameter and 16 feet in length. Is this enough to complete the job? If not, how many additional board feet of lumber is needed?

88. The storage shed shown in the figure has a volume given by the polynomial

$$2x^2y + \frac{1}{2}\pi x^2y.$$

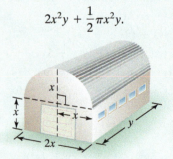

a. A small business is considering having a shed installed like the one shown in the figure. The shed's height, $2x$, is 26 feet and its length, y, is 27 feet. Using $x = 13$ and $y = 27$, find the volume of the storage shed.

b. The business requires at least 18,000 cubic feet of storage space. Should they construct the storage shed described in part (a)?

An object that is falling or vertically projected into the air has its height, in feet, above the ground given by

$$s = -16t^2 + v_0t + s_0,$$

where s is the height, in feet, v_0 is the original velocity of the object, in feet per second, t is the time the object is in motion, in seconds, and s_0 is the height, in feet, from which the object is dropped or projected. The figure shows that a ball is thrown straight up from a rooftop at an original velocity of 80 feet per second from a height of 96 feet. The ball misses the rooftop on its way down and eventually strikes the ground. Use the formula and this information to solve Exercises 89–91.

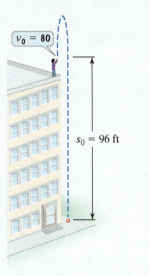

$v_0 = 80$

$s_0 = 96$ ft

89. How high above the ground will the ball be 2 seconds after being thrown?

90. How high above the ground will the ball be 4 seconds after being thrown?

91. How high above the ground will the ball be 6 seconds after being thrown? Describe what this means in practical terms.

The graph visually displays the information about the thrown ball described in Exercises 89–91. The horizontal axis represents the ball's time in motion, in seconds. The vertical axis represents the ball's height above the ground, in feet. Use the graph to solve Exercises 92–97.

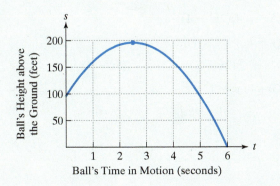

Ball's Time in Motion (seconds)

92. During which time period is the ball rising?

93. During which time period is the ball falling?

94. Identify your answer from Exercise 90 as a point on the graph.

95. Identify your answer from Exercise 89 as a point on the graph.

96. After how many seconds does the ball strike the ground?

97. After how many seconds does the ball reach its maximum height above the ground? What is a reasonable estimate of this maximum height?

Explaining the Concepts

98. What is a polynomial in two variables? Provide an example with your description.

99. Explain how to find the degree of a polynomial in two variables.

100. Suppose that you take up sky diving. Explain how to use the formula for Exercises 89–91 to determine your height above the ground at every instant of your fall.

Critical Thinking Exercises

Make Sense? *In Exercises 101–104, determine whether each statement makes sense or does not make sense, and explain your reasoning.*

101. I use the same procedures for operations with polynomials in two variables as I did when performing these operations with polynomials in one variable.

102. Adding polynomials in several variables is the same as adding like terms.

103. I used FOIL to find the product of $x + y$ and $x^2 - xy + y^2$.

104. I used FOIL to multiply $5xy$ and $3xy + 4$.

In Exercises 105–108, determine whether each statement is true or false. If the statement is false, make the necessary change(s) to produce a true statement.

105. The degree of $5x^{24} - 3x^{16}y^9 - 7xy^2 + 6$ is 24.

106. In the polynomial $4x^2y + x^3y^2 + 3x^2y^3 + 7y$, the term x^3y^2 has degree 5 and no numerical coefficient.

107. $(2x + 3 - 5y)(2x + 3 + 5y) = 4x^2 + 12x + 9 - 25y^2$

108. $(6x^2y - 7xy - 4) - (6x^2y + 7xy - 4) = 0$

In Exercises 109–110, find a polynomial in two variables that describes the area of the region of each figure shaded in blue. Write the polynomial as the sum or difference of terms.

109.

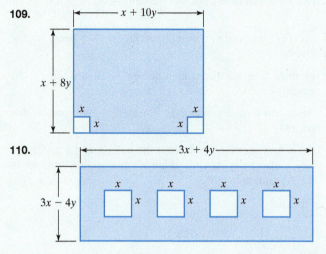

110.

111. Use the formulas for the volume of a rectangular solid and a cylinder to derive the polynomial in Exercise 88 that describes the volume of the storage building.

Review Exercises

112. Solve for W: $R = \dfrac{L + 3W}{2}$. (Section 2.4, Example 4)

113. Subtract: $-6.4 - (-10.2)$. (Section 1.6, Example 2)

114. Solve: $0.02(x - 5) = 0.03 - 0.03(x + 7)$. (Section 2.3, Example 5)

Preview Exercises

Exercises 115–117 will help you prepare for the material covered in the next section.

115. Find the missing exponent, designated by the question mark, in the final step.

$$\frac{x^7}{x^3} = \frac{x \cdot x \cdot x \cdot x \cdot x \cdot x \cdot x}{x \cdot x \cdot x} = x^?$$

116. Simplify: $\dfrac{(x^2)^3}{5^3}$.

117. Simplify: $\dfrac{(2a^3)^5}{(b^4)^5}$.

MID-CHAPTER CHECK POINT   Section 5.1–Section 5.4

✓ **What You Know:** We learned to add, subtract, and multiply polynomials. We used a number of fast methods for finding products of polynomials, including the FOIL method for multiplying binomials, a special-product formula for the product of the sum and difference of two terms $[(A + B)(A - B) = A^2 - B^2]$, and special-product formulas for squaring binomials $[(A + B)^2 = A^2 + 2AB + B^2; (A - B)^2 = A^2 - 2AB + B^2]$. Finally, we applied all of these operations to polynomials in several variables.

In Exercises 1–21, perform the indicated operations.

1. $(11x^2y^3)(-5x^2y^3)$

2. $11x^2y^3 - 5x^2y^3$

3. $(3x + 5)(4x - 7)$

4. $(3x + 5) - (4x - 7)$

5. $(2x - 5)(x^2 - 3x + 1)$

6. $(2x - 5) + (x^2 - 3x + 1)$

7. $(8x - 3)^2$

8. $(-10x^4)(-7x^5)$

9. $(x^2 + 2)(x^2 - 2)$

10. $(x^2 + 2)^2$

11. $(9a - 10b)(2a + b)$

12. $7x^2(10x^3 - 2x + 3)$

13. $(3a^2b^3 - ab + 4b^2) - (-2a^2b^3 - 3ab + 5b^2)$

14. $2(3y - 5)(3y + 5)$

15. $(-9x^3 + 5x^2 - 2x + 7) + (11x^3 - 6x^2 + 3x - 7)$

16. $10x^2 - 8xy - 3(y^2 - xy)$

17. $(-2x^5 + x^4 - 3x + 10) - (2x^5 - 6x^4 + 7x - 13)$

18. $(x + 3y)(x^2 - 3xy + 9y^2)$

19. $(5x^4 + 4)(2x^3 - 1)$

20. $(y - 6z)^2$

21. $(2x + 3)(2x - 3) - (5x + 4)(5x - 4)$

22. Graph: $y = 1 - x^2$.

SECTION

5.5

What am I supposed to learn?

After studying this section, you should be able to:

1️⃣ Use the quotient rule for exponents.

2️⃣ Use the zero-exponent rule.

3️⃣ Use the quotients-to-powers rule.

4️⃣ Divide monomials.

5️⃣ Check polynomial division.

6️⃣ Divide a polynomial by a monomial.

1️⃣ Use the quotient rule for exponents.

Dividing Polynomials

In the dramatic arts, ours is the era of the movies. As individuals and as a nation, we've grown up with them. Our images of love, war, family, country—even of things that terrify us—owe much to what we've seen on screen. In this section's Exercise Set, we'll model our love for movies with polynomials and polynomial division. Before discussing polynomial division, we must develop some additional rules for working with exponents.

The Quotient Rule for Exponents

Consider the quotient of two exponential expressions, such as the quotient of 2^7 and 2^3. We are dividing 7 factors of 2 by 3 factors of 2. We are left with 4 factors of 2:

$$\frac{2^7}{2^3} = \frac{\overbrace{2 \cdot 2 \cdot 2 \cdot 2 \cdot 2 \cdot 2 \cdot 2}^{\text{7 factors of 2}}}{\underbrace{2 \cdot 2 \cdot 2}_{\text{3 factors of 2}}} = \frac{\cancel{2} \cdot \cancel{2} \cdot \cancel{2} \cdot 2 \cdot 2 \cdot 2 \cdot 2}{\cancel{2} \cdot \cancel{2} \cdot \cancel{2}} = \overbrace{2 \cdot 2 \cdot 2 \cdot 2}^{\text{4 factors of 2}}$$

Divide out pairs of factors: $\frac{2}{2} = 1$.

Thus,

$$\frac{2^7}{2^3} = 2^4.$$

We can quickly find the exponent, 4, on the quotient by subtracting the original exponents:

$$\frac{2^7}{2^3} = 2^{7-3}.$$

This suggests the following rule:

The Quotient Rule

$$\frac{b^m}{b^n} = b^{m-n}, \quad b \neq 0$$

When dividing exponential expressions with the same nonzero base, subtract the exponent in the denominator from the exponent in the numerator. Use this difference as the exponent of the common base.

| **EXAMPLE 1** | Using the Quotient Rule |

Divide each expression using the quotient rule:

a. $\dfrac{2^8}{2^4}$ **b.** $\dfrac{x^{13}}{x^3}$ **c.** $\dfrac{y^{15}}{y}$.

Solution

a. $\dfrac{2^8}{2^4} = 2^{8-4} = 2^4$ or 16

b. $\dfrac{x^{13}}{x^3} = x^{13-3} = x^{10}$

c. $\dfrac{y^{15}}{y} = \dfrac{y^{15}}{y^1} = y^{15-1} = y^{14}$ ∎

| ✓ | **CHECK POINT 1** | Divide each expression using the quotient rule:

a. $\dfrac{5^{12}}{5^4}$ **b.** $\dfrac{x^9}{x^2}$ **c.** $\dfrac{y^{20}}{y}$.

---

2 Use the zero-exponent rule. ▶

Zero as an Exponent

A nonzero base can be raised to the 0 power. The quotient rule can be used to help determine what zero as an exponent should mean. Consider the quotient of b^4 and b^4, where b is not zero. We can determine this quotient in two ways.

$$\frac{b^4}{b^4} = 1 \qquad\qquad \frac{b^4}{b^4} = b^{4-4} = b^0$$

Any nonzero expression divided by itself is 1.   Use the quotient rule and subtract exponents.

This means that b^0 must equal 1.

> ### The Zero-Exponent Rule
>
> If b is any real number other than 0,
>
> $$b^0 = 1.$$

| **EXAMPLE 2** | Using the Zero-Exponent Rule |

Use the zero-exponent rule to simplify each expression:

a. 7^0 **b.** $(-5)^0$ **c.** -5^0 **d.** $10x^0$ **e.** $(10x)^0$.

Solution

a. $7^0 = 1$ Any nonzero number raised to the O power is 1.

b. $(-5)^0 = 1$ Any nonzero number raised to the O power is 1.

c. $-5^0 = -1$ $\qquad\qquad$ $-5^0 = -(5^0) = -1$

> Only 5 is raised to the 0 power.

d. $10x^0 = 10 \cdot 1 = 10$

> Only x is raised to the 0 power.

e. $(10x)^0 = 1$

> The entire expression, $10x$, is raised to the 0 power.

■

✓ **CHECK POINT 2**   Use the zero-exponent rule to simplify each expression:
$\qquad$ **a.** 14^0 $\qquad$ **b.** $(-10)^0$ $\qquad$ **c.** -10^0 $\qquad$ **d.** $20x^0$ $\qquad$ **e.** $(20x)^0$.

3 Use the quotients-to-powers rule. ⏵

The Quotients-to-Powers Rule for Exponents

We have seen that when a product is raised to a power, we raise every factor in the product to the power:

$$(ab)^n = a^n b^n.$$

There is a similar property for raising a quotient to a power.

> **Quotients to Powers**
>
> If a and b are real numbers and b is nonzero, then
>
> $$\left(\frac{a}{b}\right)^n = \frac{a^n}{b^n}.$$
>
> When a quotient is raised to a power, raise the numerator to the power and divide by the denominator raised to the power.

EXAMPLE 3   Using the Quotients-to-Powers Rule

Simplify each expression using the quotients-to-powers rule:

a. $\left(\dfrac{x}{4}\right)^2$ $\qquad$ **b.** $\left(\dfrac{x^2}{5}\right)^3$ $\qquad$ **c.** $\left(\dfrac{2a^3}{b^4}\right)^5$.

Solution

a. $\left(\dfrac{x}{4}\right)^2 = \dfrac{x^2}{4^2} = \dfrac{x^2}{16}$ $\qquad$ Square the numerator and the denominator.

b. $\left(\dfrac{x^2}{5}\right)^3 = \dfrac{(x^2)^3}{5^3} = \dfrac{x^{2\cdot3}}{5\cdot5\cdot5} = \dfrac{x^6}{125}$ $\qquad$ Cube the numerator and the denominator.

c. $\left(\dfrac{2a^3}{b^4}\right)^5 = \dfrac{(2a^3)^5}{(b^4)^5}$ $\qquad$ Raise the numerator and the denominator to the fifth power.

$\qquad = \dfrac{2^5(a^3)^5}{(b^4)^5}$ $\qquad$ Raise each factor in the numerator to the fifth power.

$\qquad = \dfrac{2^5 a^{3\cdot5}}{b^{4\cdot5}}$ $\qquad$ To raise exponential expressions to powers, multiply exponents: $(b^m)^n = b^{mn}$.

$\qquad = \dfrac{32a^{15}}{b^{20}}$ $\qquad$ Simplify. ■

✓ **CHECK POINT 3**   Simplify each expression using the quotients-to-powers rule:

a. $\left(\dfrac{x}{5}\right)^2$ **b.** $\left(\dfrac{x^4}{2}\right)^3$ **c.** $\left(\dfrac{2a^{10}}{b^3}\right)^4.$

Great Question!

What are some common errors to avoid when using the quotient rule or the zero-exponent rule?

Here's a partial list. The first column shows the correct simplification. The second column illustrates a common error.

Correct	Incorrect	Description of Error
$\dfrac{2^{20}}{2^4} = 2^{20-4} = 2^{16}$	~~$\dfrac{2^{20}}{2^4} = 2^5$~~	Exponents should be subtracted, not divided.
$-8^0 = -1$	~~$-8^0 = 1$~~	Only 8 is raised to the 0 power.
$\left(\dfrac{x}{5}\right)^2 = \dfrac{x^2}{5^2} = \dfrac{x^2}{25}$	~~$\left(\dfrac{x}{5}\right)^2 = \dfrac{x^2}{5}$~~	The numerator and denominator must both be squared.

④ Divide monomials. ▶

Dividing Monomials

Now that we have developed three additional properties of exponents, we are ready to turn to polynomial division. We begin with the quotient of two monomials, such as $16x^{14}$ and $8x^2$. This quotient is obtained by dividing the coefficients, 16 and 8, and then dividing the variables using the quotient rule for exponents.

$$\frac{16x^{14}}{8x^2} = \frac{16}{8}x^{14-2} = 2x^{12}$$

> Divide coefficients and subtract exponents.

Dividing Monomials

To divide monomials, divide the coefficients and then divide the variables. Use the quotient rule for exponents to divide the variables: Keep the variable and subtract the exponents.

Great Question!

I notice that the exponents on x are the same in Example 4(b). Do I have to subtract exponents since I know the quotient of x^3 and x^3 is 1?

No. Rather than subtracting exponents for division that results in a 0 exponent, you might prefer to divide out x^3.

$$\frac{2x^3}{8x^3} = \frac{2}{8} = \frac{1}{4}$$

EXAMPLE 4   Dividing Monomials

Divide: **a.** $\dfrac{-12x^8}{4x^2}$ **b.** $\dfrac{2x^3}{8x^3}$ **c.** $\dfrac{15x^5y^4}{3x^2y}.$

Solution

a. $\dfrac{-12x^8}{4x^2} = \dfrac{-12}{4}x^{8-2} = -3x^6$

b. $\dfrac{2x^3}{8x^3} = \dfrac{2}{8}x^{3-3} = \dfrac{1}{4}x^0 = \dfrac{1}{4}\cdot 1 = \dfrac{1}{4}$

c. $\dfrac{15x^5y^4}{3x^2y} = \dfrac{15}{3}x^{5-2}y^{4-1} = 5x^3y^3$ ■

✓ **CHECK POINT 4**   Divide:

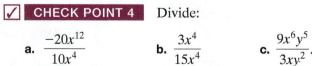

a. $\dfrac{-20x^{12}}{10x^4}$ **b.** $\dfrac{3x^4}{15x^4}$ **c.** $\dfrac{9x^6y^5}{3xy^2}.$

5 Check polynomial division.

Checking Division of Polynomial Problems

The answer to a division problem can be checked. For example, consider the following problem:

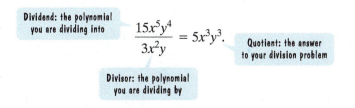

Dividend: the polynomial you are dividing into

$$\frac{15x^5y^4}{3x^2y} = 5x^3y^3.$$

Quotient: the answer to your division problem

Divisor: the polynomial you are dividing by

The quotient is correct if the product of the divisor and the quotient is the dividend. Is the quotient shown in the preceding equation correct?

$$(3x^2y)(5x^3y^3) = 3 \cdot 5x^{2+3}y^{1+3} = 15x^5y^4$$

Divisor   Quotient   This is the dividend.

Because the product of the divisor and the quotient is the dividend, the answer to the division problem is correct.

> ### Checking Division of Polynomials
>
> To check a quotient in a division problem, multiply the divisor and the quotient. If this product is the dividend, the quotient is correct.

6 Divide a polynomial by a monomial.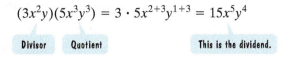

Great Question!

Can I cancel identical terms in the dividend and the divisor?

No. Try to avoid this common error:

Incorrect:

$$\frac{x^4 - \overset{1}{\cancel{x}}}{\underset{1}{\cancel{x}}} = \frac{x^4 - 1}{1} = x^4 - 1$$

Correct:

$$\frac{x^4 - x}{x} = \frac{x^4}{x} - \frac{x}{x}$$

$$= x^{4-1} - x^{1-1}$$

$$= x^3 - x^0$$

$$= x^3 - 1$$

Don't leave out the 1.

Dividing a Polynomial That Is Not a Monomial by a Monomial

To divide a polynomial by a monomial, we divide each term of the polynomial by the monomial. For example,

Polynomial dividend

$$\frac{10x^8 + 15x^6}{5x^3} = \frac{10x^8}{5x^3} + \frac{15x^6}{5x^3} = \frac{10}{5}x^{8-3} + \frac{15}{5}x^{6-3} = 2x^5 + 3x^3.$$

Monomial divisor

Divide the first term by $5x^3$.

Divide the second term by $5x^3$.

Is the quotient correct? Multiply the divisor and the quotient.

$$5x^3(2x^5 + 3x^3) = 5x^3 \cdot 2x^5 + 5x^3 \cdot 3x^3$$

$$= 5 \cdot 2x^{3+5} + 5 \cdot 3x^{3+3} = 10x^8 + 15x^6$$

Because this product gives the dividend, the quotient is correct.

> ### Dividing a Polynomial That Is Not a Monomial by a Monomial
>
> To divide a polynomial by a monomial, divide each term of the polynomial by the monomial.

EXAMPLE 5 Dividing a Polynomial by a Monomial

Find the quotient: $(-12x^8 + 4x^6 - 8x^3) \div 4x^2$.

Solution

$$\frac{-12x^8 + 4x^6 - 8x^3}{4x^2}$$ Rewrite the division in a vertical format.

$$= \frac{-12x^8}{4x^2} + \frac{4x^6}{4x^2} - \frac{8x^3}{4x^2}$$ Divide each term of the polynomial by the monomial.

$$= \frac{-12}{4}x^{8-2} + \frac{4}{4}x^{6-2} - \frac{8}{4}x^{3-2}$$ Divide coefficients and subtract exponents.

$$= -3x^6 + x^4 - 2x$$ Simplify.

To check the answer, multiply the divisor and the quotient.

$$4x^2(-3x^6 + x^4 - 2x) = 4x^2(-3x^6) + 4x^2 \cdot x^4 - 4x^2(2x)$$

Divisor   Quotient

$$= 4(-3)x^{2+6} + 4x^{2+4} - 4 \cdot 2x^{2+1}$$

$$= -12x^8 + 4x^6 - 8x^3$$ This is the dividend.

Because the product of the divisor and the quotient is the dividend, the answer—that is, the quotient—is correct. ∎

☑ **CHECK POINT 5** Find the quotient: $(-15x^9 + 6x^5 - 9x^3) \div 3x^2$.

EXAMPLE 6 Dividing a Polynomial by a Monomial

Divide: $\dfrac{16x^5 - 9x^4 + 8x^3}{2x^3}$.

Solution

$$\frac{16x^5 - 9x^4 + 8x^3}{2x^3}$$ This is the given polynomial division.

$$= \frac{16x^5}{2x^3} - \frac{9x^4}{2x^3} + \frac{8x^3}{2x^3}$$ Divide each term by $2x^3$.

$$= \frac{16}{2}x^{5-3} - \frac{9}{2}x^{4-3} + \frac{8}{2}x^{3-3}$$ Divide coefficients and subtract exponents. Did you immediately write the last term as 4?

$$= 8x^2 - \frac{9}{2}x + 4x^0$$ Simplify.

$$= 8x^2 - \frac{9}{2}x + 4$$ $x^0 = 1$, so $4x^0 = 4 \cdot 1 = 4$.

Check the answer by showing that the product of the divisor and the quotient is the dividend. ∎

☑ **CHECK POINT 6** Divide: $\dfrac{25x^9 - 7x^4 + 10x^3}{5x^3}$.

EXAMPLE 7 Dividing Polynomials in Two Variables

Divide: $(15x^5y^4 - 3x^3y^2 + 9x^2y) \div 3x^2y$.

Solution

$$\frac{15x^5y^4 - 3x^3y^2 + 9x^2y}{3x^2y}$$

Rewrite the division in a vertical format.

$$= \frac{15x^5y^4}{3x^2y} - \frac{3x^3y^2}{3x^2y} + \frac{9x^2y}{3x^2y}$$

Divide each term of the polynomial by the monomial.

$$= \frac{15}{3}x^{5-2}y^{4-1} - \frac{3}{3}x^{3-2}y^{2-1} + \frac{9}{3}x^{2-2}y^{1-1}$$

Divide coefficients and subtract exponents.

$$= 5x^3y^3 - xy + 3$$

Simplify.

Check the answer by showing that the product of the divisor and the quotient is the dividend. ∎

✓ **CHECK POINT 7** Divide: $(18x^7y^6 - 6x^2y^3 + 60xy^2) \div 6xy^2$.

CONCEPT AND VOCABULARY CHECK

Fill in each blank so that the resulting statement is true.

1. The quotient rule for exponents states that $\dfrac{b^m}{b^n} =$ _____, $b \neq 0$. When dividing exponential expressions with the same nonzero base, _____ the exponents.

2. If $b \neq 0$, $b^0 =$ _____.

3. The quotients-to-powers rule for exponents states that $\left(\dfrac{a}{b}\right)^n =$ _____, $b \neq 0$. When a quotient is raised to a power, raise both the _____ and the _____ to the power.

4. To divide monomials, _____ the coefficients and _____ the exponents.

5. Consider the following division problem:
$$\frac{20x^6y^4}{10x^2y} = 2x^4y^3.$$
The polynomial in the numerator, $20x^6y^4$, is called the _____. The polynomial in the denominator, $10x^2y$, is called the _____. The answer to the problem, $2x^4y^3$, is called the _____.

6. To check the answer to a division problem, multiply the _____ and the _____. If this product is the _____, the answer is correct.

7. To perform the division $\dfrac{20x^8 - 10x^4 + 6x^3}{2x^3}$, divide each term of _____ by _____.

5.5 EXERCISE SET ▶ MyMathLab®

Practice Exercises

In Exercises 1–10, divide each expression using the quotient rule. Express any numerical answers in exponential form.

1. $\dfrac{3^{20}}{3^5}$

2. $\dfrac{3^{30}}{3^{10}}$

3. $\dfrac{x^6}{x^2}$

4. $\dfrac{x^8}{x^4}$

5. $\dfrac{y^{13}}{y^5}$

6. $\dfrac{y^{19}}{y^6}$

7. $\dfrac{5^6 \cdot 2^8}{5^3 \cdot 2^4}$

8. $\dfrac{3^6 \cdot 2^8}{3^3 \cdot 2^4}$

9. $\dfrac{x^{100}y^{50}}{x^{25}y^{10}}$

10. $\dfrac{x^{200}y^{40}}{x^{25}y^{10}}$

In Exercises 11–24, use the zero-exponent rule to simplify each expression.

11. 2^0

12. 4^0

13. $(-2)^0$

14. $(-4)^0$

15. -2^0

16. -4^0

17. $100y^0$

18. $200y^0$

19. $(100y)^0$

20. $(200y)^0$

21. $-5^0 + (-5)^0$

22. $-6^0 + (-6)^0$

23. $-\pi^0 - (-\pi)^0$

24. $-\sqrt{3}^0 - (-\sqrt{3})^0$

In Exercises 25–36, simplify each expression using the quotients-to-powers rule. If possible, evaluate exponential expressions.

25. $\left(\dfrac{x}{3}\right)^2$

26. $\left(\dfrac{x}{5}\right)^2$

27. $\left(\dfrac{x^2}{4}\right)^3$

28. $\left(\dfrac{x^2}{3}\right)^3$

29. $\left(\dfrac{2x^3}{5}\right)^2$

30. $\left(\dfrac{3x^4}{7}\right)^2$

31. $\left(\dfrac{-4}{3a^3}\right)^3$

32. $\left(\dfrac{-5}{2a^3}\right)^3$

33. $\left(\dfrac{-2a^7}{b^4}\right)^5$

34. $\left(\dfrac{-2a^8}{b^3}\right)^5$

35. $\left(\dfrac{x^2y^3}{2z}\right)^4$

36. $\left(\dfrac{x^3y^2}{2z}\right)^4$

In Exercises 37–52, divide the monomials. Check each answer by showing that the product of the divisor and the quotient is the dividend.

37. $\dfrac{30x^{10}}{10x^5}$

38. $\dfrac{45x^{12}}{15x^4}$

39. $\dfrac{-8x^{22}}{4x^2}$

40. $\dfrac{-15x^{40}}{3x^4}$

41. $\dfrac{-9y^8}{18y^5}$

42. $\dfrac{-15y^{13}}{45y^9}$

43. $\dfrac{7y^{17}}{5y^5}$

44. $\dfrac{9y^{19}}{7y^{11}}$

45. $\dfrac{30x^7y^5}{5x^2y}$

46. $\dfrac{40x^9y^5}{2x^2y}$

47. $\dfrac{-18x^{14}y^2}{36x^2y^2}$

48. $\dfrac{-15x^{16}y^2}{45x^2y^2}$

49. $\dfrac{9x^{20}y^{20}}{7x^{20}y^{20}}$

50. $\dfrac{7x^{30}y^{30}}{15x^{30}y^{30}}$

51. $\dfrac{-5x^{10}y^{12}z^6}{50x^2y^3z^2}$

52. $\dfrac{-8x^{12}y^{10}z^4}{40x^2y^3z^2}$

In Exercises 53–78, divide the polynomial by the monomial. Check each answer by showing that the product of the divisor and the quotient is the dividend.

53. $\dfrac{10x^4 + 2x^3}{2}$

54. $\dfrac{20x^4 + 5x^3}{5}$

55. $\dfrac{14x^4 - 7x^3}{7x}$

56. $\dfrac{24x^4 - 8x^3}{8x}$

57. $\dfrac{y^7 - 9y^2 + y}{y}$

58. $\dfrac{y^8 - 11y^3 + y}{y}$

59. $\dfrac{24x^3 - 15x^2}{-3x}$

60. $\dfrac{10x^3 - 20x^2}{-5x}$

61. $\dfrac{18x^5 + 6x^4 + 9x^3}{3x^2}$

62. $\dfrac{18x^5 + 24x^4 + 12x^3}{6x^2}$

63. $\dfrac{12x^4 - 8x^3 + 40x^2}{4x}$

64. $\dfrac{49x^4 - 14x^3 + 70x^2}{-7x}$

65. $(4x^2 - 6x) \div x$

66. $(16y^2 - 8y) \div y$

67. $\dfrac{30z^3 + 10z^2}{-5z}$

68. $\dfrac{12y^4 - 42y^2}{-4y}$

69. $\dfrac{8x^3 + 6x^2 - 2x}{2x}$

70. $\dfrac{9x^3 + 12x^2 - 3x}{3x}$

71. $\dfrac{25x^7 - 15x^5 - 5x^4}{5x^3}$

72. $\dfrac{49x^7 - 28x^5 - 7x^4}{7x^3}$

73. $\dfrac{18x^7 - 9x^6 + 20x^5 - 10x^4}{-2x^4}$

74. $\dfrac{25x^8 - 50x^7 + 3x^6 - 40x^5}{-5x^5}$

75. $\dfrac{12x^2y^2 + 6x^2y - 15xy^2}{3xy}$

76. $\dfrac{18a^3b^2 - 9a^2b - 27ab^2}{9ab}$

77. $\dfrac{20x^7y^4 - 15x^3y^2 - 10x^2y}{-5x^2y}$

78. $\dfrac{8x^6y^3 - 12x^8y^2 - 4x^{14}y^6}{-4x^6y^2}$

Practice PLUS

In Exercises 79–82, simplify each expression.

79. $\dfrac{2x^3(4x + 2) - 3x^2(2x - 4)}{2x^2}$

80. $\dfrac{6x^3(3x - 1) + 5x^2(6x - 3)}{3x^2}$

81. $\left(\dfrac{18x^2y^4}{9xy^2}\right) - \left(\dfrac{15x^5y^6}{5x^4y^4}\right)$

82. $\left(\dfrac{9x^3 + 6x^2}{3x}\right) - \left(\dfrac{12x^2y^2 - 4xy^2}{2xy^2}\right)$

83. Divide the sum of $(y + 5)^2$ and $(y + 5)(y - 5)$ by $2y$.

84. Divide the sum of $(y + 4)^2$ and $(y + 4)(y - 4)$ by $2y$.

In Exercises 85–86, the variable n in each exponent represents a natural number. Divide the polynomial by the monomial. Then use polynomial multiplication to check the quotient.

85. $\dfrac{12x^{15n} - 24x^{12n} + 8x^{3n}}{4x^{3n}}$

86. $\dfrac{35x^{10n} - 15x^{8n} + 25x^{2n}}{5x^{2n}}$

Application Exercises

The bar graphs show U.S. film box-office receipts, in millions of dollars, and box-office admissions, in millions of tickets sold, for five selected years.

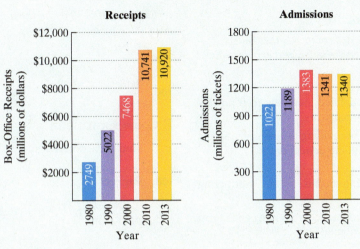

United States Film Box-Office Receipts and Admissions

Source: Motion Picture Association of America

The following polynomial models of degree 2 can be used to describe the data in the bar graphs:

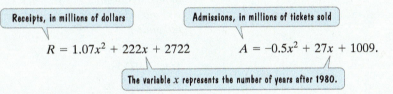

Receipts, in millions of dollars	Admissions, in millions of tickets sold
$R = 1.07x^2 + 222x + 2722$	$A = -0.5x^2 + 27x + 1009.$

The variable x represents the number of years after 1980.

Use this information and a calculator to solve Exercises 87–88.

87. a. Use the data displayed by the bar graphs to find the average admission charge for a film ticket in 1990. Round to two decimal places, or to the nearest cent.

b. Use the models to write an algebraic expression that describes the average admission charge for a film ticket x years after 1980.

c. Use the model from part (b) to find the average admission charge for a film ticket in 1990. Round to the nearest cent. Does the model underestimate or overestimate the actual average charge that you found in part (a)? By how much?

d. Can the polynomial division for the model in part (b) be performed using the methods that you learned in this section? Explain your answer.

88. a. Use the data displayed by the bar graphs to find the average admission charge for a film ticket in 2000. Round to two decimal places, or to the nearest cent.

b. Use the models to write an algebraic expression that describes the average admission charge for a film ticket x years after 1980.

c. Use the model from part (b) to find the average admission charge for a film ticket in 2000. Round to the nearest cent. Does the model underestimate or overestimate the actual average charge that you found in part (a)? By how much?

d. Can the polynomial division for the model in part (b) be performed using the methods that you learned in this section? Explain your answer.

Explaining the Concepts

89. Explain the quotient rule for exponents. Use $\dfrac{3^6}{3^2}$ in your explanation.

90. Explain how to find any nonzero number to the 0 power.

91. Explain the difference between $(-7)^0$ and -7^0.

92. Explain how to simplify an expression that involves a quotient raised to a power. Provide an example with your explanation.

93. Explain how to divide monomials. Give an example.

94. Explain how to divide a polynomial that is not a monomial by a monomial. Give an example.

95. Are the expressions

$$\frac{12x^2 + 6x}{3x} \quad \text{and} \quad 4x + 2$$

equal for every value of x? Explain.

Critical Thinking Exercises

Make Sense? *In Exercises 96–99, determine whether each statement makes sense or does not make sense, and explain your reasoning.*

96. Because division by 0 is undefined, numbers to 0 powers should not be written in denominators.

97. The quotient rule is applied by dividing the exponent in the numerator by the exponent in the denominator.

98. I divide monomials by dividing coefficients and subtracting exponents.

99. I divide a polynomial by a monomial by dividing each term of the monomial by the polynomial.

In Exercises 100–103, determine whether each statement is true or false. If the statement is false, make the necessary change(s) to produce a true statement.

100. $x^{10} \div x^2 = x^5$ for all nonzero real numbers x.

101. $\dfrac{12x^3 - 6x}{2x} = 6x^2 - 6x$

102. $\dfrac{x^2 + x}{x} = x$

103. If a polynomial in x of degree 6 is divided by a monomial in x of degree 2, the degree of the quotient is 4.

104. What polynomial, when divided by $3x^2$, yields the trinomial $6x^6 - 9x^4 + 12x^2$ as a quotient?

In Exercises 105–106, find the missing coefficients and exponents designated by question marks.

105. $\dfrac{?x^8 - ?x^6}{3x^?} = 3x^5 - 4x^3$

106. $\dfrac{3x^{14} - 6x^{12} - ?x^7}{?x^?} = -x^7 + 2x^5 + 3$

Review Exercises

107. Find the absolute value: $|-20.3|$. (Section 1.3, Example 8)

108. Express $\frac{7}{8}$ as a decimal. (Section 1.3, Example 4)

109. Graph: $y = \dfrac{1}{3}x + 2$. (Section 3.4, Example 3)

Preview Exercises

Exercises 110–112 will help you prepare for the material covered in the next section. In each exercise, perform the long division without using a calculator, and then state the quotient and the remainder.

110. $19\overline{)494}$

111. $24\overline{)2958}$

112. $98\overline{)25,187}$

SECTION

5.6

Dividing Polynomials by Binomials

What am I supposed to learn?

After studying this section, you should be able to:

1. Divide polynomials by binomials.

"He was an arithmetician rather than a mathematician. None of the humor, the music, or the mysticism of higher mathematics ever entered his head."

—*John Steinbeck*

"Arithmetic has a very great and elevating effect. He who can properly divide is to be considered a god."

—*Plato*

"You cannot ask us to take sides against arithmetic."

—*Winston Churchill*

So, what's the deal? Will performing the repetitive procedure of long division (don't reach for that calculator!) have an elevating effect? Or will it confine you to a computational box that allows neither humor nor music to enter? Forget the box: Mathematician Wilhelm Leibniz believed that music is nothing but unconscious arithmetic. But do think elevation, if not to the level of an ancient Greek god, then to new, algebraic highs. The bottom line: Understanding long division of whole numbers lays the foundation for performing the division of a polynomial by a binomial, such as

$$x + 3 \overline{)x^2 + 10x + 21}.$$

Divisor has two terms and is a binomial.

The polynomial dividend has three terms and is a trinomial.

In this section, you will learn how to perform such divisions.

The Steps in Dividing a Polynomial by a Binomial

Dividing a polynomial by a binomial may remind you of long division. Let's review long division of whole numbers by dividing 3983 by 26.

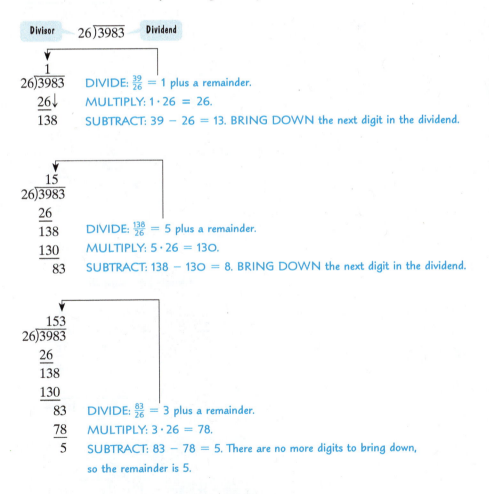

The quotient is 153 and the remainder is 5. This can be written as

Quotient $153 \dfrac{5}{26}$.   Remainder / Divisor

We see that $26\overline{)3983}$ is $153\frac{5}{26}$.

This answer can be checked. Multiply the divisor and the quotient. Then add the remainder. If the result is the dividend, the answer is correct. In this case, we have

$$26(153) \quad + \quad 5 \quad = \quad 3978 + 5 \quad = \quad 3983.$$

Divisor   Quotient   Remainder   This is the dividend.

Because we obtained the dividend, the answer to the division problem, $153\frac{5}{26}$, is correct.

1 Divide polynomials by binomials.

When a divisor is a binomial, the four steps used to divide whole numbers—**divide, multiply, subtract, bring down the next term**—form the repetitive procedure for dividing a polynomial by a binomial.

EXAMPLE 1   Dividing a Polynomial by a Binomial

Divide $x^2 + 10x + 21$ by $x + 3$.

Solution   The following steps illustrate how polynomial division is very similar to numerical division.

$$x + 3\overline{)x^2 + 10x + 21}$$

Arrange the terms of the dividend ($x^2 + 10x + 21$) and the divisor ($x + 3$) in descending powers of x.

$$\begin{array}{r} x \\ x + 3\overline{)x^2 + 10x + 21} \end{array}$$

DIVIDE x^2 (the first term in the dividend) by x (the first term in the divisor): $\frac{x^2}{x} = x$. Align like terms.

$$x(x + 3) = x^2 + 3x \qquad \begin{array}{r} x \\ x + 3\overline{)x^2 + 10x + 21} \\ x^2 + 3x \end{array}$$

MULTIPLY each term in the divisor ($x + 3$) by x, aligning terms of the product under like terms in the dividend.

$$\begin{array}{r} x \\ x + 3\overline{)x^2 + 10x + 21} \\ \ominus x^2 \ominus 3x \\ \hline 7x \end{array}$$

Change signs of the polynomial being subtracted.

SUBTRACT $x^2 + 3x$ from $x^2 + 10x$ by changing the sign of each term in the lower expression and adding.

$$\begin{array}{r} x \\ x + 3\overline{)x^2 + 10x + 21} \\ \underline{x^2 + 3x} \quad \downarrow \\ 7x + 21 \end{array}$$

BRING DOWN 21 from the original dividend and add algebraically to form a new dividend.

$$\begin{array}{r} x + 7 \\ x + 3\overline{)x^2 + 10x + 21} \\ \underline{x^2 + 3x} \\ 7x + 21 \end{array}$$

FIND the second term of the quotient. Divide the first term of $7x + 21$ by x, the first term of the divisor: $\frac{7x}{x} = 7$.

$$7(x + 3) = 7x + 21 \qquad \begin{array}{r} x + 7 \\ x + 3\overline{)x^2 + 10x + 21} \\ \underline{x^2 + 3x} \\ 7x + 21 \\ \ominus 7x \ominus 21 \\ \hline 0 \end{array}$$ Remainder

MULTIPLY the divisor ($x + 3$) by 7, aligning under like terms in the new dividend. Then subtract to obtain the remainder of 0.

The quotient is $x + 7$ and the remainder is 0. We will not list a remainder of 0 in the answer. Thus,

$$\frac{x^2 + 10x + 21}{x + 3} = x + 7. \quad \blacksquare$$

When dividing polynomials by binomials, the answer can be checked. Find the product of the divisor and the quotient and add the remainder. If the result is the dividend, the answer to the division problem is correct. For example, let's check our work in Example 1.

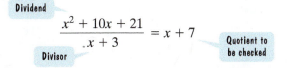

Multiply the divisor and the quotient and add the remainder, 0:

$$(x + 3)(x + 7) + 0 = x^2 + 7x + 3x + 21 + 0 = x^2 + 10x + 21.$$

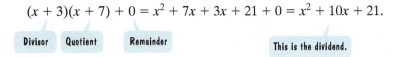

Because we obtained the dividend, the quotient is correct.

☑ **CHECK POINT 1**   Divide $x^2 + 14x + 45$ by $x + 9$.

Before considering additional examples, let's summarize the general procedure for dividing a polynomial by a binomial.

> ### Dividing a Polynomial by a Binomial
>
> 1. **Arrange the terms** of both the dividend and the divisor in descending powers of the variable.
> 2. **Divide** the first term in the dividend by the first term in the divisor. The result is the first term of the quotient.
> 3. **Multiply** every term in the divisor by the first term in the quotient. Write the resulting product beneath the dividend with like terms lined up.
> 4. **Subtract** the product from the dividend.
> 5. **Bring down** the next term in the original dividend and write it next to the remainder to form a new dividend.
> 6. Use this new expression as the dividend and repeat this process until the remainder can no longer be divided. This will occur when the degree of the remainder (the highest exponent on a variable in the remainder) is less than the degree of the divisor.

In our next division, we will obtain a nonzero remainder.

EXAMPLE 2   Dividing a Polynomial by a Binomial

Divide: $\dfrac{7x - 9 - 4x^2 + 4x^3}{2x - 1}$.

Solution We begin by writing the dividend, $7x - 9 - 4x^2 + 4x^3$, in descending powers of x.

> Think of 9 as $9x^0$. The powers descend from 3 to 0.

$$7x - 9 - 4x^2 + 4x^3 = 4x^3 - 4x^2 + 7x - 9$$

$$2x - 1 \overline{)4x^3 - 4x^2 + 7x - 9}$$
This is the problem with the dividend in descending powers of x.

$$\begin{array}{r} 2x^2 \\ 2x - 1 \overline{)4x^3 - 4x^2 + 7x - 9} \end{array}$$
DIVIDE: $\dfrac{4x^3}{2x} = 2x^2$.

$2x^2(2x - 1) = 4x^3 - 2x^2$

$$\begin{array}{r} 2x^2 \\ 2x - 1 \overline{)4x^3 - 4x^2 + 7x - 9} \\ 4x^3 - 2x^2 \end{array}$$
MULTIPLY: $2x^2(2x - 1) = 4x^3 - 2x^2$.

$$\begin{array}{r} 2x^2 \\ 2x - 1 \overline{)4x^3 - 4x^2 + 7x - 9} \\ \ominus 4x^3 \oplus 2x^2 \\ -2x^2 \end{array}$$
SUBTRACT: $4x^3 - 4x^2 - (4x^3 - 2x^2)$
$= 4x^3 - 4x^2 - 4x^3 + 2x^2$
$= -2x^2$.

Change signs of the polynomial being subtracted.

$$\begin{array}{r} 2x^2 \\ 2x - 1 \overline{)4x^3 - 4x^2 + 7x - 9} \\ 4x^3 - 2x^2 \downarrow \\ -2x^2 + 7x \end{array}$$
BRING DOWN $7x$. The new dividend is $-2x^2 + 7x$.

$$\begin{array}{r} 2x^2 - x \\ 2x - 1 \overline{)4x^3 - 4x^2 + 7x - 9} \\ 4x^3 - 2x^2 \\ -2x^2 + 7x \end{array}$$
DIVIDE: $\dfrac{-2x^2}{2x} = -x$.

$-x(2x - 1) = -2x^2 + x$

$$\begin{array}{r} 2x^2 - x \\ 2x - 1 \overline{)4x^3 - 4x^2 + 7x - 9} \\ 4x^3 - 2x^2 \\ -2x^2 + 7x \\ -2x^2 + x \end{array}$$
MULTIPLY: $-x(2x - 1) = -2x^2 + x$.

$$\begin{array}{r} 2x^2 - x \\ 2x - 1 \overline{)4x^3 - 4x^2 + 7x - 9} \\ 4x^3 - 2x^2 \\ -2x^2 + 7x \\ \oplus -2x^2 \ominus x \\ 6x \end{array}$$
SUBTRACT: $-2x^2 + 7x - (-2x^2 + x)$
$= -2x^2 + 7x + 2x^2 - x$
$= 6x$.

$$\begin{array}{r} 2x^2 - x \\ 2x - 1 \overline{)4x^3 - 4x^2 + 7x - 9} \\ 4x^3 - 2x^2 \\ -2x^2 + 7x \\ -2x^2 + x \\ 6x - 9 \end{array}$$
BRING DOWN -9. The new dividend is $6x - 9$.

$$
\begin{array}{r}
2x^2 - \ \ x + 3 \\
2x - 1\overline{\smash{)}4x^3 - 4x^2 + 7x - 9} \\
\underline{4x^3 - 2x^2} \\
-2x^2 + 7x \\
\underline{-2x^2 + \ \ x} \\
6x - 9
\end{array}
$$

DIVIDE: $\dfrac{6x}{2x} = 3.$

3(2x − 1) = 6x − 3

$$
\begin{array}{r}
2x^2 - \ \ x + 3 \\
2x - 1\overline{\smash{)}4x^3 - 4x^2 + 7x - 9} \\
\underline{4x^3 - 2x^2} \\
-2x^2 + 7x \\
\underline{-2x^2 + \ \ x} \\
6x - 9 \\
6x - 3
\end{array}
$$

MULTIPLY: $3(2x - 1) = 6x - 3.$

$$
\begin{array}{r}
2x^2 - \ \ x + 3 \\
2x - 1\overline{\smash{)}4x^3 - 4x^2 + 7x - 9} \\
\underline{4x^3 - 2x^2} \\
-2x^2 + 7x \\
\underline{-2x^2 + \ \ x} \\
6x - 9 \\
\underline{6x - 3} \\
-6
\end{array}
$$

SUBTRACT: $6x - 9 - (6x - 3)$
$ = 6x - 9 - 6x + 3$
$ = -6.$

Remainder

The quotient is $2x^2 - x + 3$ and the remainder is -6. When there is a nonzero remainder, as in this example, list the quotient, plus the remainder above the divisor. Thus,

$$
\frac{7x - 9 - 4x^2 + 4x^3}{2x - 1} = 2x^2 - x + 3 + \frac{-6}{2x - 1}
$$

Remainder above divisor

Quotient

or

$$
\frac{7x - 9 - 4x^2 + 4x^3}{2x - 1} = 2x^2 - x + 3 - \frac{6}{2x - 1}.
$$

Check this result by showing that the product of the divisor and the quotient,

$$
(2x - 1)(2x^2 - x + 3),
$$

plus the remainder, -6, is the dividend, $7x - 9 - 4x^2 + 4x^3$. ■

✓ **CHECK POINT 2**   Divide: $\dfrac{6x + 8x^2 - 12}{2x + 3}.$

If a power of the variable is missing in a dividend, add that power of the variable with a coefficient of 0 and then divide. In this way, like terms will be aligned as you carry out the division.

EXAMPLE 3 Dividing a Polynomial with Missing Terms

Divide: $\dfrac{8x^3 - 1}{2x - 1}$.

Solution We write the dividend, $8x^3 - 1$, as

$$8x^3 + 0x^2 + 0x - 1.$$

> Use a coefficient of 0 with missing terms.

By doing this, we will keep all like terms aligned.

$4x^2(2x - 1) = 8x^3 - 4x^2$

$$\begin{array}{r} 4x^2 \\ 2x - 1 \overline{\smash{\big)}\, 8x^3 + 0x^2 + 0x - 1} \\ 8x^3 - 4x^2 \\ \hline 4x^2 + 0x \end{array}$$

Divide $\left(\dfrac{4x^2}{2x} = 2x\right)$, multiply $[4x^2(2x - 1) = 8x^3 - 4x^2]$, subtract, and bring down the next term. The new dividend is $4x^2 + 0x$.

$2x(2x - 1) = 4x^2 - 2x$

$$\begin{array}{r} 4x^2 + 2x \\ 2x - 1 \overline{\smash{\big)}\, 8x^3 + 0x^2 + 0x - 1} \\ 8x^3 - 4x^2 \\ \hline 4x^2 + 0x \\ 4x^2 - 2x \\ \hline 2x - 1 \end{array}$$

Divide $\left(\dfrac{4x^2}{2x} = 2x\right)$, multiply $[2x(2x - 1) = 4x^2 - 2x]$, subtract, and bring down the next term. The new dividend is $2x - 1$.

$1(2x - 1) = 2x - 1$

$$\begin{array}{r} 4x^2 + 2x + 1 \\ 2x - 1 \overline{\smash{\big)}\, 8x^3 + 0x^2 + 0x - 1} \\ 8x^3 - 4x^2 \\ \hline 4x^2 + 0x \\ 4x^2 - 2x \\ \hline 2x - 1 \\ 2x - 1 \\ \hline 0 \end{array}$$

Divide $\left(\dfrac{2x}{2x} = 1\right)$, multiply $[1(2x - 1) = 2x - 1]$, and subtract. The remainder is 0.

Thus,

$$\frac{8x^3 - 1}{2x - 1} = 4x^2 + 2x + 1.$$

Check this result by showing that the product of the divisor and the quotient

$$(2x - 1)(4x^2 + 2x + 1)$$

plus the remainder, 0, is the dividend, $8x^3 - 1$. ∎

CHECK POINT 3 Divide: $\dfrac{x^3 - 1}{x - 1}$.

Using Technology

Graphic Connections

The graphs of $y_1 = \dfrac{8x^3 - 1}{2x - 1}$ and $y_2 = 4x^2 + 2x + 1$ are shown below.

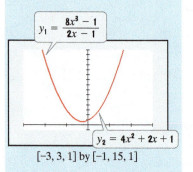

$y_1 = \dfrac{8x^3 - 1}{2x - 1}$

$y_2 = 4x^2 + 2x + 1$

[-3, 3, 1] by [-1, 15, 1]

The graphs coincide. Thus,

$$\frac{8x^3 - 1}{2x - 1} = 4x^2 + 2x + 1.$$

Achieving Success

Read your lecture notes before starting your homework.

Often homework problems, and later the test problems, are variations of the ones done by your professor in class.

CONCEPT AND VOCABULARY CHECK

Fill in each blank so that the resulting statement is true.

1. Consider the following long division problem:

$$2x + 3\overline{)4x^2 + 9 + 10x^3}.$$

We begin the division process by rewriting the dividend as _____.

2. Consider the following long division problem:

$$2x + 1\overline{)8x^2 + 10x - 1}.$$

We begin the division process by dividing _____ by _____. We obtain _____. We write this result above _____ in the dividend.

3. In the following long division problem, the first step has been completed:

$$3x - 2\overline{)\begin{array}{l}5x \\ 15x^2 - 22x + 19\end{array}}.$$

The next step is to multiply _____ and _____. We obtain _____. We write this result below _____.

4. In the following long division problem, the first steps have been completed:

$$3x - 5\overline{)\begin{array}{l}2x \\ 6x^2 + 8x - 4 \\ \underline{6x^2 - 10x}\end{array}}.$$

The next step is to subtract _____ from _____. We obtain _____. Then we bring down _____ and form the new dividend _____.

5. In the following long division problem, most of the steps have been completed:

$$x - 5\overline{)\begin{array}{l}x - 12 \\ x^2 - 17x + 74 \\ \underline{x^2 - 5x} \\ -12x + 74 \\ \underline{-12x + 60} \\ ?\end{array}}.$$

Completing the step designated by the question mark, we obtain _____. Thus, the quotient is _____ and the remainder is _____. The answer to this long division problem is _____.

5.6 EXERCISE SET   ▶ MyMathLab®

Practice Exercises

In Exercises 1–36, divide as indicated. Check each answer by showing that the product of the divisor and the quotient, plus the remainder, is the dividend.

1. $\dfrac{x^2 + 6x + 8}{x + 2}$

2. $\dfrac{x^2 + 7x + 10}{x + 5}$

3. $\dfrac{2x^2 + x - 10}{x - 2}$

4. $\dfrac{2x^2 + 13x + 15}{x + 5}$

5. $\dfrac{x^2 - 5x + 6}{x - 3}$

6. $\dfrac{x^2 - 2x - 24}{x + 4}$

7. $\dfrac{2y^2 + 5y + 2}{y + 2}$

8. $\dfrac{2y^2 - 13y + 21}{y - 3}$

9. $\dfrac{x^2 - 3x + 4}{x + 2}$

10. $\dfrac{x^2 - 7x + 5}{x + 3}$

11. $\dfrac{5y + 10 + y^2}{y + 2}$

12. $\dfrac{-8y + y^2 - 9}{y - 3}$

13. $\dfrac{x^3 - 6x^2 + 7x - 2}{x - 1}$

14. $\dfrac{x^3 + 3x^2 + 5x + 3}{x + 1}$

15. $\dfrac{12y^2 - 20y + 3}{2y - 3}$

16. $\dfrac{4y^2 - 8y - 5}{2y + 1}$

17. $\dfrac{4a^2 + 4a - 3}{2a - 1}$

18. $\dfrac{2b^2 - 9b - 5}{2b + 1}$

19. $\dfrac{3y - y^2 + 2y^3 + 2}{2y + 1}$

20. $\dfrac{9y + 18 - 11y^2 + 12y^3}{4y + 3}$

21. $\dfrac{6x^2 - 5x - 30}{2x - 5}$

22. $\dfrac{4y^2 + 8y + 3}{2y - 1}$

23. $\dfrac{x^3 + 4x - 3}{x - 2}$

24. $\dfrac{x^3 + 2x^2 - 3}{x - 2}$

25. $\dfrac{4y^3 + 8y^2 + 5y + 9}{2y + 3}$

26. $\dfrac{2y^3 - y^2 + 3y + 2}{2y + 1}$

27. $\dfrac{6y^3 - 5y^2 + 5}{3y + 2}$

28. $\dfrac{4y^3 + 3y + 5}{2y - 3}$

29. $\dfrac{27x^3 - 1}{3x - 1}$

30. $\dfrac{8x^3 + 27}{2x + 3}$

31. $\dfrac{81 - 12y^3 + 54y^2 + y^4 - 108y}{y - 3}$

32. $\dfrac{8y^3 + y^4 + 16 + 32y + 24y^2}{y + 2}$

33. $\dfrac{4y^2 + 6y}{2y - 1}$

34. $\dfrac{10x^2 - 3x}{x + 3}$

35. $\dfrac{y^4 - 2y^2 + 5}{y - 1}$

36. $\dfrac{y^4 - 6y^2 + 3}{y - 1}$

Practice PLUS

In Exercises 37–42, divide as indicated.

37. $\dfrac{4x^3 - 3x^2 + x + 1}{x^2 + 2}$

38. $\dfrac{3x^3 + 4x^2 + x + 7}{x^2 + 1}$

39. $\dfrac{x^3 - a^3}{x - a}$

40. $\dfrac{x^4 - a^4}{x - a}$

41. $\dfrac{6x^4 - 5x^3 - 8x^2 + 16x - 8}{3x^2 + 2x - 4}$

42. $\dfrac{2x^4 + 5x^3 - 11x^2 - 20x + 12}{x^2 + x - 6}$

43. Divide the difference between $4x^3 + x^2 - 2x + 7$ and $3x^3 - 2x^2 - 7x + 4$ by $x + 1$.

44. Divide the difference between $4x^3 + 2x^2 - x - 1$ and $2x^3 - x^2 + 2x - 5$ by $x + 2$.

Application Exercises

45. Write a simplified polynomial that represents the length of the rectangle.

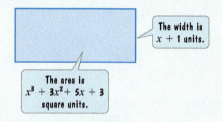

The width is $x + 1$ units.

The area is $x^3 + 3x^2 + 5x + 3$ square units.

46. Write a simplified polynomial that represents the measure of the base of the parallelogram.

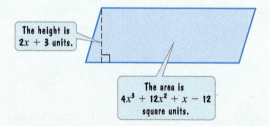

The height is $2x + 3$ units.

The area is $4x^3 + 12x^2 + x - 12$ square units.

You just signed a contract for a new job. The salary for the first year is $30,000 and there is to be a percent increase in your salary each year. The algebraic expression

$$\dfrac{30{,}000x^n - 30{,}000}{x - 1}$$

describes your total salary over n years, where x is the sum of 1 and the yearly percent increase, expressed as a decimal. Use this information to solve Exercises 47–48.

47. a. Use the given expression and write a quotient of polynomials that describes your total salary over three years.

   b. Simplify the expression in part (a) by performing the division.

c. Suppose you are to receive an increase of 5% per year. Thus, x is the sum of 1 and 0.05, or 1.05. Substitute 1.05 for x in the expression in part (a) as well as in the simplified form of the expression in part (b). Evaluate each expression. What is your total salary over the three-year period?

48. a. Use the expression given on the previous page and write a quotient of polynomials that describes your total salary over four years.

  b. Simplify the expression in part (a) by performing the division.

  c. Suppose you are to receive an increase of 8% per year. Thus, x is the sum of 1 and 0.08, or 1.08. Substitute 1.08 for x in the expression in part (a) as well as in the simplified form of the expression in part (b). Evaluate each expression. What is your total salary over the four-year period?

Explaining the Concepts

49. In your own words, explain how to divide a polynomial by a binomial. Use $\dfrac{x^2 + 4}{x + 2}$ in your explanation.

50. When dividing a polynomial by a binomial, explain when to stop dividing.

51. After dividing a polynomial by a binomial, explain how to check the answer.

52. When dividing a binomial into a polynomial with missing terms, explain the advantage of writing the missing terms with zero coefficients.

Critical Thinking Exercises

Make Sense? In Exercises 53–56, determine whether each statement makes sense or does not make sense, and explain your reasoning. Each statement applies to the division problem

$$\frac{x^3 + 1}{x + 1}.$$

53. The purpose of writing $x^3 + 1$ as $x^3 + 0x^2 + 0x + 1$ is to keep all like terms aligned.

54. Rewriting $x^3 + 1$ as $x^3 + 0x^2 + 0x + 1$ can change the value of the variable expression for certain values of x.

55. There's no need to apply the long-division process to this problem because I can work the problem in my head and see that the quotient must be $x^2 + 1$.

56. The degree of the quotient must be $3 - 1$.

In Exercises 57–60, determine whether each statement is true or false. If the statement is false, make the necessary change(s) to produce a true statement.

57. If $4x^2 + 25x - 3$ is divided by $4x + 1$, the remainder is 9.

58. If polynomial division results in a remainder of zero, then the product of the divisor and the quotient is the dividend.

59. A nonzero remainder indicates that the answer to a polynomial long-division problem is not a polynomial.

60. When a polynomial is divided by a binomial, the division process stops when the last term of the dividend is brought down.

61. When a certain polynomial is divided by $2x + 4$, the quotient is

$$x - 3 + \frac{17}{2x + 4}.$$

What is the polynomial?

62. Find the number k such that when $16x^2 - 2x + k$ is divided by $2x - 1$, the remainder is 0.

63. Describe the pattern that you observe in the following quotients and remainders.

$$\frac{x^3 - 1}{x + 1} = x^2 - x + 1 - \frac{2}{x + 1}$$

$$\frac{x^5 - 1}{x + 1} = x^4 - x^3 + x^2 - x + 1 - \frac{2}{x + 1}$$

Use this pattern to find $\dfrac{x^7 - 1}{x + 1}$. Verify your result by dividing.

Technology Exercises

In Exercises 64–68, use a graphing utility to determine whether the divisions have been performed correctly. Graph each side of the given equation in the same viewing rectangle. The graphs should coincide. If they do not, correct the expression on the right side by using polynomial division. Then use your graphing utility to show that the division has been performed correctly.

64. $\dfrac{x^2 - 4}{x - 2} = x + 2$

65. $\dfrac{x^2 - 25}{x - 5} = x - 5$

66. $\dfrac{2x^2 + 13x + 15}{x - 5} = 2x + 3$

67. $\dfrac{6x^2 + 16x + 8}{3x + 2} = 2x - 4$

68. $\dfrac{x^3 + 3x^2 + 5x + 3}{x + 1} = x^2 - 2x + 3$

Review Exercises

69. Graph the solution set of the system:

$$\begin{cases} 2x - y \ge \ \ 4 \\ \ \ x - y \le -1. \end{cases}$$

(Section 4.5, Example 2)

70. What is 6% of 20? (Section 2.4, Example 5)

71. Solve: $\dfrac{x}{3} + \dfrac{2}{5} = \dfrac{x}{5} - \dfrac{2}{5}$. (Section 2.3, Example 4)

Preview Exercises

Exercises 72–74 will help you prepare for the material covered in the next section.

72. a. Find the missing exponent, designated by the question mark, in each final step.

$$\frac{7^3}{7^5} = \frac{7 \cdot 7 \cdot 7}{7 \cdot 7 \cdot 7 \cdot 7 \cdot 7} = \frac{1}{7^?}$$

$$\frac{7^3}{7^5} = 7^{3-5} = 7^?$$

b. Based on your two results for $\dfrac{7^3}{7^5}$, what can you conclude?

73. Simplify: $\dfrac{(2x^3)^4}{x^{10}}$.

74. Simplify: $\left(\dfrac{x^5}{x^2}\right)^3$.

SECTION

5.7

Negative Exponents and Scientific Notation ▶

Bigger than the biggest thing ever and then some. Much bigger than that in fact, really amazingly immense, a totally stunning size, real 'wow, that's big', time ... Gigantic multiplied by colossal multiplied by staggeringly huge is the sort of concept we're trying to get across here.

—Douglas Adams, *The Restaurant at the End of the Universe*

What am I supposed to learn?

After studying this section, you should be able to:

1. Use the negative exponent rule. ▶

2. Simplify exponential expressions. ▶

3. Convert from scientific notation to decimal notation. ▶

4. Convert from decimal notation to scientific notation. ▶

5. Compute with scientific notation. ▶

6. Solve applied problems using scientific notation. ▶

Although Adams's description may not quite apply to this $15.2 trillion national debt, exponents can be used to explore the meaning of this "staggeringly huge" number. In this section, you will learn to use exponents to provide a way of putting large and small numbers in perspective.

Negative Integers as Exponents

A nonzero base can be raised to a negative power. The quotient rule can be used to help determine what a negative integer as an exponent should mean. Consider the quotient of b^3 and b^5, where b is not zero. We can determine this quotient in two ways.

$$\frac{b^3}{b^5} = \frac{1 \cdot \cancel{b} \cdot \cancel{b} \cdot \cancel{b}}{\cancel{b} \cdot \cancel{b} \cdot \cancel{b} \cdot b \cdot b} = \frac{1}{b^2}$$

After dividing out pairs of factors, we have two factors of b in the denominator.

$$\frac{b^3}{b^5} = b^{3-5} = b^{-2}$$

Use the quotient rule and subtract exponents.

Notice that $\dfrac{b^3}{b^5}$ equals both b^{-2} and $\dfrac{1}{b^2}$. This means that b^{-2} must equal $\dfrac{1}{b^2}$. This example is a special case of the **negative exponent rule**.

1 Use the negative exponent rule.

> ### The Negative Exponent Rule
>
> If b is any real number other than 0 and n is a natural number, then
>
> $$b^{-n} = \frac{1}{b^n}.$$

Great Question!

Does a negative exponent make the value of an expression negative?

No. For example,

$$7^{-2} = \frac{1}{7^2} = \frac{1}{49}$$

is positive. Avoid these common errors:

Incorrect!

$$7^{-2} = -7^2$$
$$7^{-2} = -\frac{1}{7^2}$$

| **EXAMPLE 1** | Using the Negative Exponent Rule |

Use the negative exponent rule to write each expression with a positive exponent. Then simplify the expression.

a. 7^{-2} **b.** 4^{-3} **c.** $(-2)^{-4}$ **d.** -2^{-4} **e.** 5^{-1}

Solution

a. $7^{-2} = \dfrac{1}{7^2} = \dfrac{1}{7 \cdot 7} = \dfrac{1}{49}$

b. $4^{-3} = \dfrac{1}{4^3} = \dfrac{1}{4 \cdot 4 \cdot 4} = \dfrac{1}{64}$

c. $(-2)^{-4} = \dfrac{1}{(-2)^4} = \dfrac{1}{(-2)(-2)(-2)(-2)} = \dfrac{1}{16}$

d. $-2^{-4} = -\dfrac{1}{2^4} = -\dfrac{1}{2 \cdot 2 \cdot 2 \cdot 2} = -\dfrac{1}{16}$

> The negative is not inside parentheses and is not taken to the −4 power.

e. $5^{-1} = \dfrac{1}{5^1} = \dfrac{1}{5}$ ∎

✓ **CHECK POINT 1** Use the negative exponent rule to write each expression with a positive exponent. Then simplify the expression.

a. 6^{-2} **b.** 5^{-3} **c.** $(-3)^{-4}$

d. -3^{-4} **e.** 8^{-1}

Negative exponents can also appear in denominators. For example,

$$\frac{1}{2^{-10}} = \frac{1}{\dfrac{1}{2^{10}}} = 1 \div \frac{1}{2^{10}} = 1 \cdot \frac{2^{10}}{1} = 2^{10}.$$

In general, if a negative exponent appears in a denominator, an expression can be written with a positive exponent using

$$\frac{1}{b^{-n}} = b^n.$$

For example,

$$\frac{1}{2^{-3}} = 2^3 = 8 \qquad \text{and} \qquad \frac{1}{(-6)^{-2}} = (-6)^2 = 36.$$

> Change only the sign of the exponent and not the sign of the base, −6.

Negative Exponents in Numerators and Denominators

If b is any real number other than 0 and n is a natural number, then

$$b^{-n} = \frac{1}{b^n} \quad \text{and} \quad \frac{1}{b^{-n}} = b^n.$$

When a negative number appears as an exponent, switch the position of the base (from numerator to denominator or from denominator to numerator) and make the exponent positive. The sign of the base does not change.

EXAMPLE 2 Using Negative Exponents

Write each expression with positive exponents only. Then simplify, if possible.

a. $\dfrac{4^{-3}}{5^{-2}}$ **b.** $\left(\dfrac{3}{4}\right)^{-2}$ **c.** $\dfrac{1}{4x^{-3}}$ **d.** $\dfrac{x^{-5}}{y^{-1}}$

Solution

a. $\dfrac{4^{-3}}{5^{-2}} = \dfrac{5^2}{4^3} = \dfrac{5 \cdot 5}{4 \cdot 4 \cdot 4} = \dfrac{25}{64}$

> Switch the position of the bases and make the exponents positive.

b. $\left(\dfrac{3}{4}\right)^{-2} = \dfrac{3^{-2}}{4^{-2}} = \dfrac{4^2}{3^2} = \dfrac{4 \cdot 4}{3 \cdot 3} = \dfrac{16}{9}$

> Switch the position of the bases and make the exponents positive.

c. $\dfrac{1}{4x^{-3}} = \dfrac{x^3}{4}$

> Switch the position of the base and make the exponent positive. Note that only x is raised to the -3 power.

d. $\dfrac{x^{-5}}{y^{-1}} = \dfrac{y^1}{x^5} = \dfrac{y}{x^5}$ ∎

☑ **CHECK POINT 2** Write each expression with positive exponents only. Then simplify, if possible.

a. $\dfrac{2^{-3}}{7^{-2}}$ **b.** $\left(\dfrac{4}{5}\right)^{-2}$ **c.** $\dfrac{1}{7y^{-2}}$ **d.** $\dfrac{x^{-1}}{y^{-8}}$

---

2 Simplify exponential expressions. ▶

Simplifying Exponential Expressions

Properties of exponents are used to simplify exponential expressions. An exponential expression is **simplified** when

- Each base occurs only once.
- No parentheses appear.
- No powers are raised to powers.
- No negative or zero exponents appear.

Simplifying Exponential Expressions

Example

1. If necessary, be sure that each base appears only once, using

$$b^m \cdot b^n = b^{m+n} \quad \text{or} \quad \frac{b^m}{b^n} = b^{m-n}.$$

$x^4 \cdot x^3 = x^{4+3} = x^7$

2. If necessary, remove parentheses using

$$(ab)^n = a^n b^n \quad \text{or} \quad \left(\frac{a}{b}\right)^n = \frac{a^n}{b^n}.$$

$(xy)^3 = x^3 y^3$

3. If necessary, simplify powers to powers using

$$(b^m)^n = b^{mn}.$$

$(x^4)^3 = x^{4 \cdot 3} = x^{12}$

4. If necessary, rewrite exponential expressions with zero powers as 1 ($b^0 = 1$). Furthermore, write the answer with positive exponents using

$$b^{-n} = \frac{1}{b^n} \quad \text{or} \quad \frac{1}{b^{-n}} = b^n.$$

$\dfrac{x^5}{x^8} = x^{5-8} = x^{-3} = \dfrac{1}{x^3}$

Great Question!

Is the procedure in Example 3 the only way to simplify $x^{-9} \cdot x^4$?

There is often more than one way to simplify an exponential expression. For example, you may prefer to simplify Example 3 as follows:

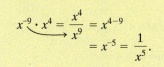

The following examples show how to simplify exponential expressions. In each example, assume that any variable in a denominator is not equal to zero.

EXAMPLE 3 Simplifying an Exponential Expression

Simplify: $x^{-9} \cdot x^4$.

Solution

$$x^{-9} \cdot x^4 = x^{-9+4} \qquad b^m \cdot b^n = b^{m+n}$$
$$= x^{-5} \qquad \text{The base, } x, \text{ now appears only once.}$$
$$= \frac{1}{x^5} \qquad b^{-n} = \frac{1}{b^n} \quad \blacksquare$$

✓ **CHECK POINT 3** Simplify: $x^{-12} \cdot x^2$.

EXAMPLE 4 Simplifying Exponential Expressions

Simplify:

a. $\dfrac{x^4}{x^{20}}$ **b.** $\dfrac{25x^6}{5x^8}$ **c.** $\dfrac{10y^7}{-2y^{10}}$.

Solution

a. $\dfrac{x^4}{x^{20}} = x^{4-20} = x^{-16} = \dfrac{1}{x^{16}}$

b. $\dfrac{25x^6}{5x^8} = \dfrac{25}{5} \cdot \dfrac{x^6}{x^8} = 5x^{6-8} = 5x^{-2} = \dfrac{5}{x^2}$

c. $\dfrac{10y^7}{-2y^{10}} = \dfrac{10}{-2} \cdot \dfrac{y^7}{y^{10}} = -5y^{7-10} = -5y^{-3} = -\dfrac{5}{y^3}$ $\blacksquare$

✓ **CHECK POINT 4** Simplify:

a. $\dfrac{x^2}{x^{10}}$ **b.** $\dfrac{75x^3}{5x^9}$ **c.** $\dfrac{50y^8}{-25y^{14}}$.

EXAMPLE 5   Simplifying an Exponential Expression

Simplify: $\dfrac{(5x^3)^2}{x^{10}}$.

Solution

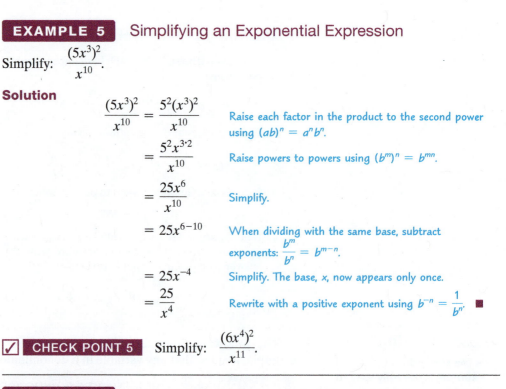

$$\frac{(5x^3)^2}{x^{10}} = \frac{5^2(x^3)^2}{x^{10}}$$ Raise each factor in the product to the second power using $(ab)^n = a^n b^n$.

$$= \frac{5^2 x^{3\cdot 2}}{x^{10}}$$ Raise powers to powers using $(b^m)^n = b^{mn}$.

$$= \frac{25 x^6}{x^{10}}$$ Simplify.

$$= 25 x^{6-10}$$ When dividing with the same base, subtract exponents: $\dfrac{b^m}{b^n} = b^{m-n}$.

$$= 25 x^{-4}$$ Simplify. The base, x, now appears only once.

$$= \frac{25}{x^4}$$ Rewrite with a positive exponent using $b^{-n} = \dfrac{1}{b^n}$.   ■

 **CHECK POINT 5**   Simplify: $\dfrac{(6x^4)^2}{x^{11}}$.

EXAMPLE 6   Simplifying an Exponential Expression

Simplify: $\left(\dfrac{x^5}{x^2}\right)^{-3}$.

Solution

Method 1. First perform the division within the parentheses.

$$\left(\frac{x^5}{x^2}\right)^{-3} = (x^{5-2})^{-3}$$ Within parentheses, divide by subtracting exponents: $\dfrac{b^m}{b^n} = b^{m-n}$.

$$= (x^3)^{-3}$$ Simplify. The base, x, now appears only once.

$$= x^{3(-3)}$$ Raise powers to powers: $(b^m)^n = b^{mn}$.

$$= x^{-9}$$ Simplify.

$$= \frac{1}{x^9}$$ Rewrite with a positive exponent using $b^{-n} = \dfrac{1}{b^n}$.

Method 2. Remove parentheses first by raising the numerator and the denominator to the -3 power.

$$\left(\frac{x^5}{x^2}\right)^{-3} = \frac{(x^5)^{-3}}{(x^2)^{-3}}$$ Use $\left(\dfrac{a}{b}\right)^n = \dfrac{a^n}{b^n}$ and raise the numerator and denominator to the -3 power.

$$= \frac{x^{5(-3)}}{x^{2(-3)}}$$ Raise powers to powers using $(b^m)^n = b^{mn}$.

$$= \frac{x^{-15}}{x^{-6}}$$ Simplify.

$$= x^{-15-(-6)}$$ When dividing with the same base, subtract the exponent in the denominator from the exponent in the numerator: $\dfrac{b^m}{b^n} = b^{m-n}$.

$$= x^{-9}$$ Subtract: $-15 - (-6) = -15 + 6 = -9$. The base, x, now appears only once.

$$= \frac{1}{x^9}$$ Rewrite with a positive exponent using $b^{-n} = \dfrac{1}{b^n}$.

Which method do you prefer?   ■

✓ CHECK POINT 6   Simplify: $\left(\dfrac{x^8}{x^4}\right)^{-5}$.

Scientific Notation

Earth is a 4.5-billion-year-old ball of rock orbiting the sun. Because a billion is 10^9 (see **Table 5.1**), the age of our world can be expressed as

$$4.5 \times 10^9.$$

The number 4.5×10^9 is written in a form called *scientific notation*.

Table 5.1	Names of Large Numbers
10^2	hundred
10^3	thousand
10^6	million
10^9	billion
10^{12}	trillion
10^{15}	quadrillion
10^{18}	quintillion
10^{21}	sextillion
10^{24}	septillion
10^{27}	octillion
10^{30}	nonillion
10^{100}	googol
10^{googol}	googolplex

Scientific Notation

A positive number is written in **scientific notation** when it is expressed in the form

$$a \times 10^n,$$

where a is a number greater than or equal to 1 and less than 10 ($1 \le a < 10$) and n is an integer.

It is customary to use the multiplication symbol, $\times$, rather than a dot, when writing a number in scientific notation.

Here are three examples of numbers in scientific notation:

- The universe is 1.375×10^{10} years old.
- In 2010, humankind generated 1.2×10^{21} bytes, or 1.2 zettabytes, of digital information. (Put into perspective, if all 6.8 billion (6.8×10^9) people on Earth joined Twitter and continuously tweeted for a century, they would crank out one zettabyte of data.) (*Source*: LiveScience.com)
- The length of the AIDS virus is 1.1×10^{-4} millimeter.

3 Convert from scientific notation to decimal notation. ▶

We can use n, the exponent on the 10 in $a \times 10^n$, to change a number in scientific notation to decimal notation. If n is **positive**, move the decimal point in a to the **right** n places. If n is **negative**, move the decimal point in a to the **left** $|n|$ places.

EXAMPLE 7   Converting from Scientific to Decimal Notation

Write each number in decimal notation:

a. 2.6×10^7 **b.** 1.1×10^{-4}.

Solution   In each case, we use the exponent on the 10 to move the decimal point. In part (a), the exponent is positive, so we move the decimal point to the right. In part (b), the exponent is negative, so we move the decimal point to the left.

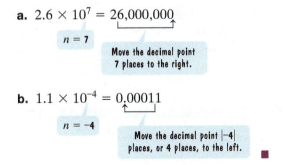

a. $2.6 \times 10^7 = 26,000,000$

$n = 7$

Move the decimal point 7 places to the right.

b. $1.1 \times 10^{-4} = 0.00011$

$n = -4$

Move the decimal point $|-4|$ places, or 4 places, to the left. ∎

> ✓ **CHECK POINT 7** Write each number in decimal notation:
> **a.** 7.4×10^9 **b.** 3.017×10^{-6}.

4 Convert from decimal notation to scientific notation.

To convert a positive number from decimal notation to scientific notation, we reverse the procedure of Example 7.

Converting from Decimal to Scientific Notation

Write the number in the form $a \times 10^n$.

- Determine a, the numerical factor. Move the decimal point in the given number to obtain a number greater than or equal to 1 and less than 10.
- Determine n, the exponent on 10^n. The absolute value of n is the number of places the decimal point was moved. The exponent n is positive if the given number is greater than 10 and negative if the given number is between 0 and 1.

> **EXAMPLE 8** Converting from Decimal Notation to Scientific Notation

Write each number in scientific notation:

a. 4,600,000 **b.** 0.000023.

Solution

a. $4{,}600{,}000 \ = \ 4.6 \ \times \ 10^6$

| This number is greater than 10, so n is positive in $a \times 10^n$. | Move the decimal point in 4,600,000 to get $1 \leq a < 10$. | The decimal point moved 6 places from 4,600,000 to 4.6. |

b. $0.000023 \ = \ 2.3 \ \times \ 10^{-5}$

| This number is less than 1, so n is negative in $a \times 10^n$. | Move the decimal point in 0.000023 to get $1 \leq a < 10$. | The decimal point moved 5 places from 0.000023 to 2.3. | ∎

> ✓ **CHECK POINT 8** Write each number in scientific notation:
> **a.** 7,410,000,000 **b.** 0.000000092.

5 Compute with scientific notation.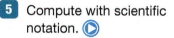

Computations with Scientific Notation

Properties of exponents are used to perform computations with numbers that are expressed in scientific notation.

Computations with Numbers in Scientific Notation

Multiplication

$$(a \times 10^n) \times (b \times 10^m) = (a \times b) \times 10^{n+m}$$

> Add the exponents on 10 and multiply the other parts of the numbers separately.

Division

$$\frac{a \times 10^n}{b \times 10^m} = \left(\frac{a}{b}\right) \times 10^{n-m}$$

> Subtract the exponents on 10 and divide the other parts of the numbers separately.

Exponentiation

$$(a \times 10^n)^m = a^n \times 10^{nm}$$

> Multiply exponents on 10 and raise the other part of the number to the power.

After the computation is completed, the answer may require an adjustment before it is expressed in scientific notation.

EXAMPLE 9   Computations with Scientific Notation

Perform the indicated computations, writing the answers in scientific notation:

a. $(4 \times 10^5)(2 \times 10^9)$ **b.** $\dfrac{1.2 \times 10^6}{4.8 \times 10^{-3}}$ **c.** $(5 \times 10^{-4})^3$.

Solution

a.
$$\begin{aligned}
(4 \times 10^5)(2 \times 10^9) &= (4 \times 2) \times (10^5 \times 10^9) && \text{Regroup.}\\
&= 8 \times 10^{5+9} && \text{Add the exponents on 10 and multiply}\\
&&& \text{the other parts.}\\
&= 8 \times 10^{14} && \text{Simplify.}
\end{aligned}$$

b.
$$\begin{aligned}
\frac{1.2 \times 10^6}{4.8 \times 10^{-3}} &= \left(\frac{1.2}{4.8}\right) \times \left(\frac{10^6}{10^{-3}}\right) && \text{Regroup.}\\
&= 0.25 \times 10^{6-(-3)} && \text{Subtract the exponents on 10 and divide the other}\\
&&& \text{parts.}\\
&= 0.25 \times 10^9 && \text{Simplify. Because 0.25 is not between 1 and 10, it}\\
&&& \text{must be written in scientific notation.}\\
&= 2.5 \times 10^{-1} \times 10^9 && 0.25 = 2.5 \times 10^{-1}\\
&= 2.5 \times 10^{-1+9} && \text{Add the exponents on 10.}\\
&= 2.5 \times 10^8 && \text{Simplify.}
\end{aligned}$$

c.
$$\begin{aligned}
(5 \times 10^{-4})^3 &= 5^3 \times (10^{-4})^3 && (ab)^n = a^n b^n. \text{ Cube each factor in parentheses.}\\
&= 125 \times 10^{-4 \cdot 3} && \text{Multiply the exponents and cube the other part of}\\
&&& \text{the number.}\\
&= 125 \times 10^{-12} && \text{Simplify. 125 must be written in scientific notation.}\\
&= 1.25 \times 10^2 \times 10^{-12} && 125 = 1.25 \times 10^2\\
&= 1.25 \times 10^{2+(-12)} && \text{Add the exponents on 10.}\\
&= 1.25 \times 10^{-10} && \text{Simplify.} \blacksquare
\end{aligned}$$

Using Technology

$(4 \times 10^5)(2 \times 10^9)$ On a Calculator:

Many Scientific Calculators

4 [EE] 5 [×] 2 [EE] 9 [=]

Display: 8. 14

Many Graphing Calculators

4 [EE] 5 [×] 2 [EE] 9 [ENTER]

Display: 8E14

✓ **CHECK POINT 9** Perform the indicated computations, writing the answers in scientific notation:

a. $(3 \times 10^8)(2 \times 10^2)$

b. $\dfrac{8.4 \times 10^7}{4 \times 10^{-4}}$

c. $(4 \times 10^{-2})^3$.

6 Solve applied problems using scientific notation. ▶

Applications: Putting Numbers in Perspective

Due to tax cuts and spending increases, the United States began accumulating large deficits in the 1980s. To finance the deficit, the government had borrowed $15.2 trillion as of December 2011. The graph in **Figure 5.8** shows the national debt increasing over time.

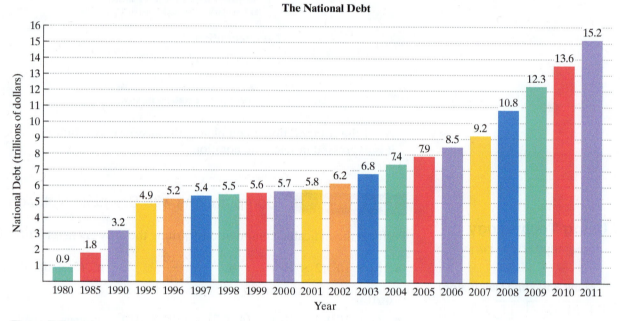

The National Debt

Figure 5.8

Source: Office of Management and Budget

Example 10 shows how we can use scientific notation to comprehend the meaning of a number such as 15.2 trillion.

EXAMPLE 10 The National Debt

As of December 2011, the national debt was $15.2 trillion, or 15.2×10^{12} dollars. At that time, the U.S. population was approximately 312,000,000 (312 million), or 3.12×10^8. If the national debt was evenly divided among every individual in the United States, how much would each citizen have to pay?

Solution The amount each citizen must pay is the total debt, 15.2×10^{12} dollars, divided by the number of citizens, 3.12×10^8.

$$\frac{15.2 \times 10^{12}}{3.12 \times 10^8} = \left(\frac{15.2}{3.12}\right) \times \left(\frac{10^{12}}{10^8}\right)$$

$$\approx 4.87 \times 10^{12-8}$$

$$= 4.87 \times 10^4$$

$$= 48{,}700$$

Every U.S. citizen would have to pay approximately $48,700 to the federal government to pay off the national debt. ■

Using Technology

Here is the keystroke sequence for solving Example 10 using a calculator:

15.2 EE 12 ÷ 3.12 EE 8.

The quotient is displayed by pressing = on a scientific calculator or ENTER on a graphing calculator. The answer can be displayed in scientific or decimal notation. Consult your manual.

If a number is written in scientific notation, $a \times 10^n$, the digits in a are called **significant digits**.

$$\text{National Debt: } 15.2 \times 10^{12} \qquad \text{U.S. Population: } 3.12 \times 10^8$$

Three significant digits   Three significant digits

Because these were the given numbers in Example 10, we rounded the answer, 4.87×10^4, to three significant digits. When multiplying or dividing in scientific notation where rounding is necessary and rounding instructions are not given, **round the scientific notation answer to the least number of significant digits found in any of the given numbers.**

✓ **CHECK POINT 10**   As of December 2011, the United States had spent $2.6 trillion for the wars in Iraq and Afghanistan. (Source: costsofwar.org) At that time, the U.S. population was approximately 312 million (3.12×10^8). If the cost of these wars was evenly divided among every individual in the United States, how much would each citizen have to pay?

Blitzer Bonus ⓒ

Seven Ways to Spend $1 Trillion

Confronting a national debt of $15.2 trillion starts with grasping just how colossal $1 trillion ($1 \times 10^{12}$) actually is. To help you wrap your head around this mind-boggling number, and to put the national debt in further perspective, consider what $1 trillion will buy:

- 40,816,326 new cars based on an average sticker price of $24,500 each
- 5,574,136 homes based on the national median price of $179,400 for existing single-family homes
- one year's salary for 14.7 million teachers based on the average teacher salary of $68,000 in California
- the annual salaries of all 535 members of Congress for the next 10,742 years based on current salaries of $174,000 per year
- the salary of basketball superstar LeBron James for 50,000 years based on an annual salary of $20 million
- annual base pay for 59.5 million U.S. privates (that's 100 times the total number of active-duty soldiers in the Army) based on basic pay of $16,794 per year
- salaries to hire all 2.8 million residents of the state of Kansas in full-time minimum-wage jobs for the next 23 years based on a federal minimum wage of $7.25 per hour

Source: Kiplinger.com

Image © photobank.kiev.ua, 2009

Achieving Success

Form a study group with other students in your class. Working in small groups often serves as an excellent way to learn and reinforce new material. Set up helpful procedures and guidelines for the group. "Talk" math by discussing and explaining the concepts and exercises to one another.

CONCEPT AND VOCABULARY CHECK

Fill in each blank so that the resulting statement is true.

1. The negative exponent rule states that $b^{-n} =$ _____, $b \neq 0$.

2. True or false: $4^{-2} = -4^2$ _____

3. True or false: $4^{-2} = \dfrac{1}{4^2}$ _____

4. Negative exponents in denominators can be evaluated using $\dfrac{1}{b^{-n}} =$ _____, $b \neq 0$.

5. True or false: $\dfrac{1}{5^{-2}} = 5^2$ _____

6. True or false: $\dfrac{1}{5^{-2}} = -5^2$ _____

7. A positive number is written in scientific notation when it is expressed in the form $a \times 10^n$, where a is _____ and n is a/an _____.

8. True or false: 4×10^3 is written in scientific notation. _____

9. True or false: 40×10^2 is written in scientific notation. _____

5.7 EXERCISE SET ▶ MyMathLab®

Practice Exercises

In Exercises 1–28, write each expression with positive exponents only. Then simplify, if possible.

1. 8^{-2}

2. 9^{-2}

3. 5^{-3}

4. 4^{-3}

5. $(-6)^{-2}$

6. $(-7)^{-2}$

7. -6^{-2}

8. -7^{-2}

9. 4^{-1}

10. 6^{-1}

11. $2^{-1} + 3^{-1}$

12. $3^{-1} - 6^{-1}$

13. $\dfrac{1}{3^{-2}}$

14. $\dfrac{1}{4^{-3}}$

15. $\dfrac{1}{(-3)^{-2}}$

16. $\dfrac{1}{(-2)^{-2}}$

17. $\dfrac{2^{-3}}{8^{-2}}$

18. $\dfrac{4^{-3}}{2^{-2}}$

19. $\left(\dfrac{1}{4}\right)^{-2}$

20. $\left(\dfrac{1}{5}\right)^{-2}$

21. $\left(\dfrac{3}{5}\right)^{-3}$

22. $\left(\dfrac{3}{4}\right)^{-3}$

23. $\dfrac{1}{6x^{-5}}$

24. $\dfrac{1}{8x^{-6}}$

25. $\dfrac{x^{-8}}{y^{-1}}$

26. $\dfrac{x^{-12}}{y^{-1}}$

27. $\dfrac{3}{(-5)^{-3}}$

28. $\dfrac{4}{(-3)^{-3}}$

In Exercises 29–78, simplify each exponential expression. Assume that variables represent nonzero real numbers.

29. $x^{-8} \cdot x^3$

30. $x^{-11} \cdot x^5$

31. $(4x^{-5})(2x^2)$

32. $(5x^{-7})(3x^3)$

33. $\dfrac{x^3}{x^9}$

34. $\dfrac{x^5}{x^{12}}$

35. $\dfrac{y}{y^{100}}$

36. $\dfrac{y}{y^{50}}$

37. $\dfrac{30z^5}{10z^{10}}$

38. $\dfrac{45z^4}{15z^{12}}$

39. $\dfrac{-8x^3}{2x^7}$

40. $\dfrac{-15x^4}{3x^9}$

41. $\dfrac{-9a^5}{27a^8}$

42. $\dfrac{-15a^8}{45a^{13}}$

43. $\dfrac{7w^5}{5w^{13}}$

44. $\dfrac{7w^8}{9w^{14}}$

45. $\dfrac{x^3}{(x^4)^2}$

46. $\dfrac{x^5}{(x^3)^2}$

47. $\dfrac{y^{-3}}{(y^4)^2}$

48. $\dfrac{y^{-5}}{(y^3)^2}$

49. $\dfrac{(4x^3)^2}{x^8}$

50. $\dfrac{(5x^3)^2}{x^7}$

51. $\dfrac{(6y^4)^3}{y^{-5}}$

52. $\dfrac{(4y^5)^3}{y^{-4}}$

53. $\left(\dfrac{x^4}{x^2}\right)^{-3}$

54. $\left(\dfrac{x^6}{x^2}\right)^{-3}$

55. $\left(\dfrac{4x^5}{2x^2}\right)^{-4}$

56. $\left(\dfrac{6x^7}{2x^2}\right)^{-4}$

57. $(3x^{-1})^{-2}$

58. $(4x^{-1})^{-2}$

59. $(-2y^{-1})^{-3}$

60. $(-3y^{-1})^{-3}$

61. $\dfrac{2x^5 \cdot 3x^7}{15x^6}$

62. $\dfrac{3x^3 \cdot 5x^{14}}{20x^{14}}$

63. $(x^3)^5 \cdot x^{-7}$

64. $(x^4)^3 \cdot x^{-5}$

65. $(2y^3)^4 y^{-6}$

66. $(3y^4)^3 y^{-7}$

67. $\dfrac{(y^3)^4}{(y^2)^7}$

68. $\dfrac{(y^2)^5}{(y^3)^4}$

69. $(y^{10})^{-5}$

70. $(y^{20})^{-5}$

71. $(a^4 b^5)^{-3}$

72. $(a^5 b^3)^{-4}$

73. $(a^{-2} b^6)^{-4}$

74. $(a^{-7} b^2)^{-5}$

75. $\left(\dfrac{x^2}{2}\right)^{-2}$

76. $\left(\dfrac{x^2}{2}\right)^{-3}$

77. $\left(\dfrac{x^2}{y^3}\right)^{-3}$

78. $\left(\dfrac{x^3}{y^2}\right)^{-4}$

In Exercises 79–90, write each number in decimal notation without the use of exponents.

79. 8.7×10^2

80. 2.75×10^3

81. 9.23×10^5

82. 7.24×10^4

83. 3.4×10^0

84. 9.115×10^0

85. 7.9×10^{-1}

86. 8.6×10^{-1}

87. 2.15×10^{-2}

88. 3.14×10^{-2}

89. 7.86×10^{-4}

90. 4.63×10^{-5}

In Exercises 91–106, write each number in scientific notation.

91. 32,400

92. 327,000

93. 220,000,000

94. 370,000,000,000

95. 713

96. 623

97. 6751

98. 9832

99. 0.0027

100. 0.00083

101. 0.0000202

102. 0.00000103

103. 0.005

104. 0.006

105. 3.14159

106. 2.71828

In Exercises 107–126, perform the indicated computations. Write the answers in scientific notation.

107. $(2 \times 10^3)(3 \times 10^2)$

108. $(3 \times 10^4)(3 \times 10^2)$

109. $(2 \times 10^5)(8 \times 10^3)$

110. $(4 \times 10^3)(5 \times 10^4)$

111. $\dfrac{12 \times 10^6}{4 \times 10^2}$

112. $\dfrac{20 \times 10^{20}}{10 \times 10^{10}}$

113. $\dfrac{15 \times 10^4}{5 \times 10^{-2}}$

114. $\dfrac{18 \times 10^2}{9 \times 10^{-3}}$

115. $\dfrac{15 \times 10^{-4}}{5 \times 10^2}$

116. $\dfrac{18 \times 10^{-2}}{9 \times 10^3}$

117. $\dfrac{180 \times 10^6}{2 \times 10^3}$

118. $\dfrac{180 \times 10^8}{2 \times 10^4}$

119. $\dfrac{3 \times 10^4}{12 \times 10^{-3}}$

120. $\dfrac{5 \times 10^2}{20 \times 10^{-3}}$

121. $(5 \times 10^2)^3$

122. $(4 \times 10^3)^2$

123. $(3 \times 10^{-2})^4$

124. $(2 \times 10^{-3})^5$

125. $(4 \times 10^6)^{-1}$

126. $(5 \times 10^4)^{-1}$

Practice PLUS

In Exercises 127–134, simplify each exponential expression. Assume that variables represent nonzero real numbers.

127. $\dfrac{(x^{-2}y)^{-3}}{(x^2 y^{-1})^3}$

128. $\dfrac{(xy^{-2})^{-2}}{(x^{-2}y)^{-3}}$

129. $(2x^{-3}yz^{-6})(2x)^{-5}$

130. $(3x^{-4}yz^{-7})(3x)^{-3}$

131. $\left(\dfrac{x^3 y^4 z^5}{x^{-3} y^{-4} z^{-5}}\right)^{-2}$

132. $\left(\dfrac{x^4 y^5 z^6}{x^{-4} y^{-5} z^{-6}}\right)^{-4}$

133. $\dfrac{(2^{-1}x^{-2}y^{-1})^{-2}(2x^{-4}y^3)^{-2}(16x^{-3}y^3)^0}{(2x^{-3}y^{-5})^2}$

134. $\dfrac{(2^{-1}x^{-3}y^{-1})^{-2}(2x^{-6}y^4)^{-2}(9x^3y^{-3})^0}{(2x^{-4}y^{-6})^2}$

In Exercises 135–138, perform the indicated computations. Express answers in scientific notation.

135. $(5 \times 10^3)(1.2 \times 10^{-4}) \div (2.4 \times 10^2)$

136. $(2 \times 10^2)(2.6 \times 10^{-3}) \div (4 \times 10^3)$

137. $\dfrac{(1.6 \times 10^4)(7.2 \times 10^{-3})}{(3.6 \times 10^8)(4 \times 10^{-3})}$

138. $\dfrac{(1.2 \times 10^6)(8.7 \times 10^{-2})}{(2.9 \times 10^6)(3 \times 10^{-3})}$

Application Exercises

In 2012, the United States government spent more than it had collected in taxes, resulting in a budget deficit of $1.09 trillion. In Exercises 139–142, you will use scientific notation to put a number like 1.09 trillion in perspective.

139. a. Express 1.09 trillion in scientific notation.

b. Express the 2012 U.S. population, 313 million, in scientific notation.

c. Use your scientific notation answers from parts (a) and (b) to answer this question: If the 2012 budget deficit was evenly divided among every individual in the United States, how much would each citizen have to pay? Express the answer in scientific and decimal notations.

140. a. Express 1.09 trillion in scientific notation.

b. A trip around the world at the Equator is approximately 25,000 miles. Express this number in scientific notation.

c. Use your scientific notation answers from parts (a) and (b) to answer this question: How many times can you circle the world at the Equator by traveling 1.09 trillion miles?

141. If there are approximately 3.2×10^7 seconds in a year, approximately how many years is 1.09 trillion seconds? (*Note*: 1.09 trillion seconds would take us back in time to a period when Neanderthals were using stones to make tools.)

142. The Washington Monument, overlooking the U.S. Capitol, stands about 555 feet tall. Stacked end to end, how many monuments would it take to reach 1.09 trillion feet? (*Note*: That's more than twice the distance from Earth to the Sun.)

Use the motion formula $d = rt$, distance equals rate times time, and the fact that light travels at the rate of 1.86×10^5 miles per second, to solve Exercises 143–144.

143. If the Moon is approximately 2.325×10^5 miles from Earth, how many seconds does it take moonlight to reach Earth?

144. If the sun is approximately 9.14×10^7 miles from Earth, how many seconds, to the nearest tenth of a second, does it take sunlight to reach Earth?

145. Refer to the Blitzer Bonus on page 407. Use scientific notation to verify any three of the bulleted items on ways to spend $1 trillion.

Explaining the Concepts

146. Explain the negative exponent rule and give an example.

147. How do you know if an exponential expression is simplified?

148. How do you know if a number is written in scientific notation?

149. Explain how to convert from scientific to decimal notation and give an example.

150. Explain how to convert from decimal to scientific notation and give an example.

151. Describe one advantage of expressing a number in scientific notation over decimal notation.

Critical Thinking Exercises

Make Sense?   In Exercises 152–155, determine whether each statement makes sense or does not make sense, and explain your reasoning.

152. There are many exponential expressions that are equal to $36x^{12}$, such as $(6x^6)^2$, $(6x^3)(6x^9)$, $36(x^3)^9$, and $6^2(x^2)^6$.

153. If 5^{-2} is raised to the third power, the result is a number between 0 and 1.

154. The population of Colorado is approximately 4.6×10^{12}.

155. I wrote a number where there is no advantage to using scientific notation instead of decimal notation.

In Exercises 156–163, determine whether each statement is true or false. If the statement is false, make the necessary change(s) to produce a true statement.

156. $4^{-2} < 4^{-3}$

157. $5^{-2} > 2^{-5}$

158. $(-2)^4 = 2^{-4}$

159. $5^2 \cdot 5^{-2} > 2^5 \cdot 2^{-5}$

160. $534.7 = 5.347 \times 10^3$

161. $\dfrac{8 \times 10^{30}}{4 \times 10^{-5}} = 2 \times 10^{25}$

162. $(7 \times 10^5) + (2 \times 10^{-3}) = 9 \times 10^2$

163. $(4 \times 10^3) + (3 \times 10^2) = 4.3 \times 10^3$

164. The mad Dr. Frankenstein has gathered enough bits and pieces (so to speak) for $2^{-1} + 2^{-2}$ of his creature-to-be. Write a fraction that represents the amount of his creature that must still be obtained.

Technology Exercises

165. Use a calculator in a fraction mode to check any five of your answers in Exercises 1–22.

166. Use a calculator to check any three of your answers in Exercises 79–90.

167. Use a calculator to check any three of your answers in Exercises 91–106.

168. Use a calculator with an $\boxed{\text{EE}}$ or $\boxed{\text{EXP}}$ key to check any four of your computations in Exercises 107–126. Display the result of the computation in scientific notation.

Review Exercises

169. Solve: $8 - 6x > 4x - 12$. (Section 2.7, Example 7)

170. Simplify: $24 \div 8 \cdot 3 + 28 \div (-7)$. (Section 1.8, Example 8)

171. List the whole numbers in this set:

$$\left\{ -4, -\frac{1}{5}, 0, \pi, \sqrt{16}, \sqrt{17} \right\}.$$

(Section 1.3, Example 5)

Preview Exercises

Exercises 172–174 will help you prepare for the material covered in the first section of the next chapter. In each exercise, find the product.

172. $4x^3(4x^2 - 3x + 1)$

173. $9xy(3xy^2 - y + 9)$

174. $(x + 3)(x^2 + 5)$

Chapter 5 Summary

Definitions and Concepts	**Examples**

Section 5.1 Adding and Subtracting Polynomials

A polynomial is a single term or the sum of two or more terms containing variables with whole number exponents. A monomial is a polynomial with exactly one term; a binomial has exactly two terms; a trinomial has exactly three terms. The degree of a polynomial is the highest power of all the terms. The standard form of a polynomial is written in descending powers of the variable.	Polynomials Monomial: $2x^5$ Degree is 5. Binomial: $6x^3 + 5x$ Degree is 3. Trinomial: $7x + 4x^2 - 5$ Degree is 2.
To add polynomials, add like terms.	$(6x^3 + 5x^2 - 7x) + (-9x^3 + x^2 + 6x)$ $= (6x^3 - 9x^3) + (5x^2 + x^2) + (-7x + 6x)$ $= -3x^3 + 6x^2 - x$
The opposite, or additive inverse, of a polynomial is that polynomial with the sign of every coefficient changed. To subtract two polynomials, add the first polynomial and the opposite of the polynomial being subtracted.	$(5y^3 - 9y^2 - 4) - (3y^3 - 12y^2 - 5)$ $= (5y^3 - 9y^2 - 4) + (-3y^3 + 12y^2 + 5)$ $= (5y^3 - 3y^3) + (-9y^2 + 12y^2) + (-4 + 5)$ $= 2y^3 + 3y^2 + 1$
The graphs of equations defined by polynomials of degree 2, shaped like bowls or inverted bowls, can be obtained using the point-plotting method.	Graph: $y = x^2 - 1$. 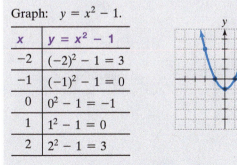

x	$y = x^2 - 1$
-2	$(-2)^2 - 1 = 3$
-1	$(-1)^2 - 1 = 0$
0	$0^2 - 1 = -1$
1	$1^2 - 1 = 0$
2	$2^2 - 1 = 3$

Definitions and Concepts	Examples

Section 5.2 Multiplying Polynomials

Properties of Exponents

Product Rule: $b^m \cdot b^n = b^{m+n}$
Power Rule: $(b^m)^n = b^{mn}$
Products to Powers: $(ab)^n = a^n b^n$

$$x^3 \cdot x^8 = x^{3+8} = x^{11}$$
$$(x^3)^8 = x^{3 \cdot 8} = x^{24}$$
$$(-5x^2)^3 = (-5)^3(x^2)^3 = -125x^6$$

To multiply monomials, multiply coefficients and add exponents.

$$(-6x^4)(3x^{10}) = -6 \cdot 3x^{4+10} = -18x^{14}$$

To multiply a monomial and a polynomial, multiply each term of the polynomial by the monomial.

$$2x^4(3x^2 - 6x + 5)$$
$$= 2x^4 \cdot 3x^2 - 2x^4 \cdot 6x + 2x^4 \cdot 5$$
$$= 6x^6 - 12x^5 + 10x^4$$

To multiply polynomials when neither is a monomial, multiply each term of one polynomial by each term of the other polynomial. Then combine like terms.

$$(2x + 3)(5x^2 - 4x + 2)$$
$$= 2x(5x^2 - 4x + 2) + 3(5x^2 - 4x + 2)$$
$$= 10x^3 - 8x^2 + 4x + 15x^2 - 12x + 6$$
$$= 10x^3 + 7x^2 - 8x + 6$$

Section 5.3 Special Products

The FOIL method may be used when multiplying two binomials: First terms multiplied. Outside terms multiplied. Inside terms multiplied. Last terms multiplied.

F	O	I	L

$$(3x + 7)(2x - 5) = 3x \cdot 2x + 3x(-5) + 7 \cdot 2x + 7(-5)$$
$$= 6x^2 - 15x + 14x - 35$$
$$= 6x^2 - x - 35$$

The Product of the Sum and Difference of Two Terms
$$(A + B)(A - B) = A^2 - B^2$$

$$(4x + 7)(4x - 7) = (4x)^2 - 7^2$$
$$= 16x^2 - 49$$

The Square of a Binomial Sum
$$(A + B)^2 = A^2 + 2AB + B^2$$

$$(x^2 + 6)^2 = (x^2)^2 + 2 \cdot x^2 \cdot 6 + 6^2$$
$$= x^4 + 12x^2 + 36$$

The Square of a Binomial Difference
$$(A - B)^2 = A^2 - 2AB + B^2$$

$$(9x - 3)^2 = (9x)^2 - 2 \cdot 9x \cdot 3 + 3^2$$
$$= 81x^2 - 54x + 9$$

Section 5.4 Polynomials in Several Variables

To evaluate a polynomial in several variables, substitute the given value for each variable and perform the resulting computation.

Evaluate $4x^2y + 3xy - 2x$ for $x = -1$ and $y = -3$.
$$4x^2y + 3xy - 2x$$
$$= 4(-1)^2(-3) + 3(-1)(-3) - 2(-1)$$
$$= 4(1)(-3) + 3(-1)(-3) - 2(-1)$$
$$= -12 + 9 + 2 = -1$$

Definitions and Concepts	**Examples**

Section 5.4 Polynomials in Several Variables (continued)

For a polynomial in two variables, the degree of a term is the sum of the exponents on its variables. The degree of the polynomial is the highest degree of all its terms.	$7x^2y + 12x^4y^3 - 17x^5 + 6$ degree: $2 + 1 = 3$ degree: $4 + 3 = 7$ degree: 5 degree: 0 Degree of polynomial $= 7$
Polynomials in several variables are added, subtracted, and multiplied using the same rules for polynomials in one variable.	$(5x^2y^3 - xy + 4y^2) - (8x^2y^3 - 6xy - 2y^2)$ $= (5x^2y^3 - xy + 4y^2) + (-8x^2y^3 + 6xy + 2y^2)$ $= (5x^2y^3 - 8x^2y^3) + (-xy + 6xy) + (4y^2 + 2y^2)$ $= -3x^2y^3 + 5xy + 6y^2$ **F** **O** **I** **L** $(3x - 2y)(x - y) = 3x \cdot x + 3x(-y) + (-2y)x + (-2y)(-y)$ $= 3x^2 - 3xy - 2xy + 2y^2$ $= 3x^2 - 5xy + 2y^2$

Section 5.5 Dividing Polynomials

Additional Properties of Exponents **Quotient Rule:** $\dfrac{b^m}{b^n} = b^{m-n}, \quad b \neq 0$ **Zero-Exponent Rule:** $b^0 = 1, \quad b \neq 0$ **Quotients to Powers:** $\left(\dfrac{a}{b}\right)^n = \dfrac{a^n}{b^n}, \quad b \neq 0$	$\dfrac{x^{12}}{x^4} = x^{12-4} = x^8$ $(-3)^0 = 1 \qquad -3^0 = -(3^0) = -1$ $\left(\dfrac{y^2}{4}\right)^3 = \dfrac{(y^2)^3}{4^3} = \dfrac{y^{2 \cdot 3}}{4 \cdot 4 \cdot 4} = \dfrac{y^6}{64}$
To divide monomials, divide coefficients and subtract exponents.	$\dfrac{-40x^{40}}{20x^{20}} = \dfrac{-40}{20}x^{40-20} = -2x^{20}$
To divide a polynomial by a monomial, divide each term of the polynomial by the monomial.	$\dfrac{8x^6 - 4x^3 + 10x}{2x}$ $= \dfrac{8x^6}{2x} - \dfrac{4x^3}{2x} + \dfrac{10x}{2x}$ $= 4x^{6-1} - 2x^{3-1} + 5x^{1-1} = 4x^5 - 2x^2 + 5$

Section 5.6 Dividing Polynomials by Binomials

To divide a polynomial by a binomial, begin by arranging the polynomial in descending powers of the variable. If a power of a variable is missing, add that power with a coefficient of 0. Repeat the four steps—divide, multiply, subtract, bring down the next term—until the degree of the remainder is less than the degree of the divisor.	Divide: $\dfrac{10x^2 + 13x + 8}{2x + 3}$. $5x - 1 + \dfrac{11}{2x + 3}$ $2x + 3\,\overline{)10x^2 + 13x + 8}$ $\underline{10x^2 + 15x}$ $-2x + 8$ $\underline{-2x - 3}$ 11

Definitions and Concepts	**Examples**

Section 5.7 Negative Exponents and Scientific Notation

Negative Exponents in Numerators and Denominators

If $b \neq 0$, $b^{-n} = \dfrac{1}{b^n}$ and $\dfrac{1}{b^{-n}} = b^n$.

$$6^{-2} = \frac{1}{6^2} = \frac{1}{36}$$

$$\frac{1}{(-2)^{-4}} = (-2)^4 = 16$$

$$\left(\frac{2}{3}\right)^{-3} = \frac{2^{-3}}{3^{-3}} = \frac{3^3}{2^3} = \frac{27}{8}$$

An exponential expression is simplified when
- Each base occurs only once.
- No parentheses appear.
- No powers are raised to powers.
- No negative or zero exponents appear.

Simplify: $\dfrac{(2x^4)^3}{x^{18}}$.

$$\frac{(2x^4)^3}{x^{18}} = \frac{2^3(x^4)^3}{x^{18}} = \frac{8x^{4\cdot3}}{x^{18}} = \frac{8x^{12}}{x^{18}} = 8x^{12-18} = 8x^{-6} = \frac{8}{x^6}$$

A positive number in scientific notation is expressed as $a \times 10^n$, where $1 \le a < 10$ and n is an integer.

Write 2.9×10^{-3} in decimal notation.

$$2.9 \times 10^{-3} = .0029 = 0.0029$$

Write 16,000 in scientific notation.

$$16{,}000 = 1.6 \times 10^4$$

Use properties of exponents with base 10

$$10^m \cdot 10^n = 10^{m+n}, \quad \frac{10^m}{10^n} = 10^{m-n}, \quad \text{and} \quad (10^m)^n = 10^{mn}$$

to perform computations with scientific notation.

$$(5 \times 10^3)(4 \times 10^{-8})$$
$$= 5 \cdot 4 \times 10^{3-8}$$
$$= 20 \times 10^{-5}$$
$$= 2 \times 10^1 \times 10^{-5} = 2 \times 10^{-4}$$

CHAPTER 5 REVIEW EXERCISES

5.1 In Exercises 1–3, identify each polynomial as a monomial, binomial, or trinomial. Give the degree of the polynomial.

1. $7x^4 + 9x$
2. $3x + 5x^2 - 2$
3. $16x$

In Exercises 4–8, add or subtract as indicated.

4. $(-6x^3 + 7x^2 - 9x + 3) + (14x^3 + 3x^2 - 11x - 7)$
5. $(9y^3 - 7y^2 + 5) + (4y^3 - y^2 + 7y - 10)$
6. $(5y^2 - y - 8) - (-6y^2 + 3y - 4)$
7. $(13x^4 - 8x^3 + 2x^2) - (5x^4 - 3x^3 + 2x^2 - 6)$
8. Subtract $x^4 + 7x^2 - 11x$ from $-13x^4 - 6x^2 + 5x$.

In Exercises 9–11, add or subtract as indicated.

9. Add. $\begin{array}{r} 7y^4 - 6y^3 + 4y^2 - 4y \\ y^3 - y^2 + 3y - 4 \\ \hline \end{array}$

10. Subtract. $\begin{array}{r} 7x^2 - 9x + 2 \\ -(4x^2 - 2x - 7) \\ \hline \end{array}$

11. Subtract. $\begin{array}{r} 5x^3 - 6x^2 - 9x + 14 \\ -(-5x^3 + 3x^2 - 11x + 3) \\ \hline \end{array}$

In Exercises 12–13, graph each equation.

12. $y = x^2 + 3$
13. $y = 1 - x^2$

5.2 *In Exercises 14–18, simplify each expression.*

14. $x^{20} \cdot x^3$

15. $y \cdot y^5 \cdot y^8$

16. $(x^{20})^5$

17. $(10y)^2$

18. $(-4x^{10})^3$

In Exercises 19–27, find each product.

19. $(5x)(10x^3)$

20. $(-12y^7)(3y^4)$

21. $(-2x^5)(-3x^4)(5x^3)$

22. $7x(3x^2 + 9)$

23. $5x^3(4x^2 - 11x)$

24. $3y^2(-7y^2 + 3y - 6)$

25. $2y^5(8y^3 - 10y^2 + 1)$

26. $(x + 3)(x^2 - 5x + 2)$

27. $(3y - 2)(4y^2 + 3y - 5)$

In Exercises 28–29, use a vertical format to find each product.

28. $y^2 - 4y + 7$
$$\underline{\ 3y - 5}$$

29. $4x^3 - 2x^2 - 6x - 1$
$$\underline{\ 2x + 3}$$

5.3 *In Exercises 30–42, find each product.*

30. $(x + 6)(x + 2)$

31. $(3y - 5)(2y + 1)$

32. $(4x^2 - 2)(x^2 - 3)$

33. $(5x + 4)(5x - 4)$

34. $(7 - 2y)(7 + 2y)$

35. $(y^2 + 1)(y^2 - 1)$

36. $(x + 3)^2$ 37. $(3y + 4)^2$

38. $(y - 1)^2$ 39. $(5y - 2)^2$

40. $(x^2 + 4)^2$

41. $(x^2 + 4)(x^2 - 4)$

42. $(x^2 + 4)(x^2 - 5)$

43. Write a polynomial in descending powers of x that represents the area of the shaded region.

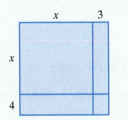

44. The parking garage shown in the figure measures 30 yards by 20 yards. The length and the width are each increased by a fixed amount, x yards. Write a trinomial that describes the area of the expanded garage.

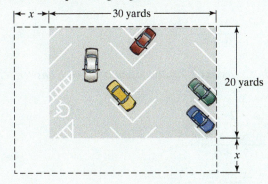

5.4

45. Evaluate $2x^3y - 4xy^2 + 5y + 6$ for $x = -1$ and $y = 2$.

46. Determine the coefficient of each term, the degree of each term, and the degree of the polynomial:

$$4x^2y + 9x^3y^2 - 17x^4 - 12.$$

In Exercises 47–56, perform the indicated operations.

47. $(7x^2 - 8xy + y^2) + (-8x^2 - 9xy + 4y^2)$

48. $(13x^3y^2 - 5x^2y - 9x^2) - (11x^3y^2 - 6x^2y - 3x^2 + 4)$

49. $(-7x^2y^3)(5x^4y^6)$

50. $5ab^2(3a^2b^3 - 4ab)$

51. $(x + 7y)(3x - 5y)$

52. $(4xy - 3)(9xy - 1)$

53. $(3x + 5y)^2$

54. $(xy - 7)^2$

55. $(7x + 4y)(7x - 4y)$

56. $(a - b)(a^2 + ab + b^2)$

5.5 *In Exercises 57–63, simplify each expression.*

57. $\dfrac{6^{40}}{6^{10}}$ 58. $\dfrac{x^{18}}{x^3}$

59. $(-10)^0$ 60. -10^0

61. $400x^0$ 62. $\left(\dfrac{x^4}{2}\right)^3$

63. $\left(\dfrac{-3}{2y^6}\right)^4$

In Exercises 64–68, divide and check each answer.

64. $\dfrac{-15y^8}{3y^2}$ 65. $\dfrac{40x^8y^6}{5xy^3}$

66. $\dfrac{18x^4 - 12x^2 + 36x}{6x}$

67. $\dfrac{30x^8 - 25x^7 - 40x^5}{-5x^3}$

68. $\dfrac{27x^3y^2 - 9x^2y - 18xy^2}{3xy}$

5.6 *In Exercises 69–72, divide and check each answer.*

69. $\dfrac{2x^2 + 3x - 14}{x - 2}$

70. $\dfrac{2x^3 - 5x^2 + 7x + 5}{2x + 1}$

71. $\dfrac{x^3 - 2x^2 - 33x - 7}{x - 7}$

72. $\dfrac{y^3 - 27}{y - 3}$

5.7 *In Exercises 73–77, write each expression with positive exponents only and then simplify.*

73. 7^{-2}

74. $(-4)^{-3}$

75. $2^{-1} + 4^{-1}$

76. $\dfrac{1}{5^{-2}}$

77. $\left(\dfrac{2}{5}\right)^{-3}$

In Exercises 78–86, simplify each exponential expression. Assume that variables in denominators do not equal zero.

78. $\dfrac{x^3}{x^9}$

79. $\dfrac{30y^6}{5y^8}$

80. $(5x^{-7})(6x^2)$

81. $\dfrac{x^4 \cdot x^{-2}}{x^{-6}}$

82. $\dfrac{(3y^3)^4}{y^{10}}$

83. $\dfrac{y^{-7}}{(y^4)^3}$

84. $(2x^{-1})^{-3}$

85. $\left(\dfrac{x^7}{x^4}\right)^{-2}$

86. $\dfrac{(y^3)^4}{(y^{-2})^4}$

In Exercises 87–89, write each number in decimal notation without the use of exponents.

87. 2.3×10^4

88. 1.76×10^{-3}

89. 9×10^{-1}

In Exercises 90–93, write each number in scientific notation.

90. 73,900,000

91. 0.00062

92. 0.38

93. 3.8

In Exercises 94–96, perform the indicated computation. Write the answers in scientific notation.

94. $(6 \times 10^{-3})(1.5 \times 10^6)$

95. $\dfrac{2 \times 10^2}{4 \times 10^{-3}}$

96. $(4 \times 10^{-2})^2$

97. The average salary of a professional baseball player is $3.4 million. (*Source:* Major League Baseball Player Association) Express this number in scientific notation.

98. The average salary of a nurse is $68,000. (*Source:* U.S. Department of Labor) Express this number in scientific notation.

99. Use your scientific notation answers from Exercises 97 and 98 to answer this question:

How many times greater is the average salary of a professional baseball player than the average salary of a nurse?

CHAPTER 5 TEST

Step-by-step test solutions are found on the Chapter Test Prep Videos available in MyMathLab® or on YouTube (search "BlitzerIntroAlg7e" and click on "Channels").

1. Identify $9x + 6x^2 - 4$ as a monomial, binomial, or trinomial. Give the degree of the polynomial.

In Exercises 2–3, add or subtract as indicated.

2. $(7x^3 + 3x^2 - 5x - 11) + (6x^3 - 2x^2 + 4x - 13)$

3. $(9x^3 - 6x^2 - 11x - 4) - (4x^3 - 8x^2 - 13x + 5)$

4. Graph the equation: $y = x^2 - 3$. Select integers for x, starting with -3 and ending with 3.

In Exercises 5–11, find each product.

5. $(-7x^3)(5x^8)$

6. $6x^2(8x^3 - 5x - 2)$

7. $(3x + 2)(x^2 - 4x - 3)$

8. $(3y + 7)(2y - 9)$

9. $(7x + 5)(7x - 5)$

10. $(x^2 + 3)^2$

11. $(5x - 3)^2$

12. Evaluate $4x^2y + 5xy - 6x$ for $x = -2$ and $y = 3$.

In Exercises 13–15, perform the indicated operations.

13. $(8x^2y^3 - xy + 2y^2) - (6x^2y^3 - 4xy - 10y^2)$

14. $(3a - 7b)(4a + 5b)$

15. $(2x + 3y)^2$

In Exercises 16–18, divide and check each answer.

16. $\dfrac{-25x^{16}}{5x^4}$

17. $\dfrac{15x^4 - 10x^3 + 25x^2}{5x}$

18. $\dfrac{2x^3 - 3x^2 + 4x + 4}{2x + 1}$

In Exercises 19–20, write each expression with positive exponents only and then simplify.

19. 10^{-2}

20. $\dfrac{1}{4^{-3}}$

In Exercises 21–26, simplify each expression.

21. $(-3x^2)^3$

22. $\dfrac{20x^3}{5x^8}$

23. $(-7x^{-8})(3x^2)$

24. $\dfrac{(2y^3)^4}{y^8}$

25. $(5x^{-4})^{-2}$

26. $\left(\dfrac{x^{10}}{x^5}\right)^{-3}$

27. Write 3.7×10^{-4} in decimal notation.

28. Write $7{,}600{,}000$ in scientific notation.

In Exercises 29–30, perform the indicated computation. Write the answers in scientific notation.

29. $(4.1 \times 10^2)(3 \times 10^{-5})$

30. $\dfrac{8.4 \times 10^6}{4 \times 10^{-2}}$

31. Write a polynomial in descending powers of x that represents the area of the figure.

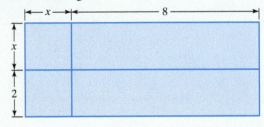

CUMULATIVE REVIEW EXERCISES (CHAPTERS 1–5)

In Exercises 1–2, perform the indicated operation or operations.

1. $(-7)(-5) \div (12 - 3)$

2. $(3 - 7)^2(9 - 11)^3$

3. What is the difference in elevation between a plane flying 14,300 feet above sea level and a submarine traveling 750 feet below sea level?

In Exercises 4–5, solve each equation.

4. $2(x + 3) + 2x = x + 4$

5. $\dfrac{x}{5} - \dfrac{1}{3} = \dfrac{x}{10} - \dfrac{1}{2}$

6. The length of a rectangular sign is 2 feet less than three times its width. If the perimeter of the sign is 28 feet, what are its dimensions?

7. Solve $7 - 8x \le -6x - 5$ and graph the solution set on a number line.

8. You invested $6000 in two accounts paying 12% and 14% annual interest. At the end of the year, the total interest from these investments was $772. How much was invested at each rate?

9. You need to mix a solution that is 70% antifreeze with one that is 30% antifreeze to obtain 20 liters of a mixture that is 60% antifreeze. How many liters of each of the solutions must be used?

10. Graph $y = -\frac{2}{5}x + 2$ using the slope and y-intercept.

11. Graph $x - 2y = 4$ using intercepts.

12. Find the slope of the line passing through the points $(-3, 2)$ and $(2, -4)$. Is the line rising, falling, horizontal, or vertical?

13. The slope of a line is -2 and the line passes through the point $(3, -1)$. Write the line's equation in point-slope form and slope-intercept form.

In Exercises 14–15, solve each system by the method of your choice.

14. $\begin{cases} 3x + 2y = 10 \\ 4x - 3y = -15 \end{cases}$

15. $\begin{cases} 2x + 3y = -6 \\ y = 3x - 13 \end{cases}$

16. The toll to a bridge costs $6.00. Commuters who frequently use the bridge have the option of purchasing a monthly discount pass for $30.00. With the discount pass, the toll is reduced to $4.00. For how many bridge crossings per month will the cost without the discount pass be the same as the cost with the discount pass? What will be the monthly cost for each option?

17. Graph the solution set for the system of linear inequalities:
$$\begin{cases} 2x + 5y \le 10 \\ x - y \ge 4. \end{cases}$$

18. Subtract: $(9x^5 - 3x^3 + 2x - 7) - (6x^5 + 3x^3 - 7x - 9)$.

19. Divide: $\dfrac{x^3 + 3x^2 + 5x + 3}{x + 1}$.

20. Simplify: $\dfrac{(3x^2)^4}{x^{10}}$.

Factoring Polynomials

Motion and change are the very essence of life. Moving air brushes against our faces; rain falls on our heads; birds fly past us; plants spring from the earth, grow, and then die; and rocks thrown upward reach a maximum height before falling to the ground. In this chapter, we analyze the where and when of moving objects by writing polynomials as products and using equations in which the highest exponent on the variable is 2.

Here's where you'll find this application:

▪ Motion is modeled in Example 7 of Section 6.6, as well as in the visual discussion that follows the example.

6.1

The Greatest Common Factor and Factoring by Grouping

What am I supposed to learn?

After studying this section, you should be able to:

1. Find the greatest common factor.

2. Factor out the greatest common factor of a polynomial.

3. Factor out the negative of the greatest common factor of a polynomial.

4. Factor by grouping.

A two-year-old boy is asked, "Do you have a brother?" He answers, "Yes." "What is your brother's name?" "Tom." Asked if Tom has a brother, the two-year-old replies, "No." The child can go in the direction from self to brother, but he cannot reverse this direction and move from brother back to self.

As our intellects develop, we learn to reverse the direction of our thinking. Reversibility of thought is found throughout algebra. For example, we can multiply polynomials and show that

$$5x(2x + 3) = 10x^2 + 15x.$$

We can also reverse this process and express the resulting polynomial as

$$10x^2 + 15x = 5x(2x + 3).$$

Factoring a polynomial containing the sum of monomials means finding an equivalent expression that is a product.

Factoring $10x^2 + 15x$

Sum of monomials	Equivalent expression that is a product

$$10x^2 + 15x = 5x(2x + 3)$$

The factors of $10x^2 + 15x$ are $5x$ and $2x + 3$.

In this chapter, we will be factoring over the set of integers, meaning that the coefficients in the factors are integers. Polynomials that cannot be factored using integer coefficients are called **prime polynomials** over the set of integers.

Factoring Out the Greatest Common Factor

We use the distributive property to multiply a monomial and a polynomial of two or more terms. When we factor, we reverse this process, expressing the polynomial as a product.

Multiplication	Factoring
$a(b + c) = ab + ac$	$ab + ac = a(b + c)$

Here is a specific example:

Multiplication	Factoring
$5x(2x + 3)$	$10x^2 + 15x$
$= 5x \cdot 2x + 5x \cdot 3$	$= 5x \cdot 2x + 5x \cdot 3$
$= 10x^2 + 15x$	$= 5x(2x + 3).$

In the process of finding an equivalent expression for $10x^2 + 15x$ that is a product, we used the fact that $5x$ is a factor of the monomials $10x^2$ and $15x$. The factoring on the right shows that $5x$ is a *common factor* for both terms of the binomial $10x^2 + 15x$.

1 Find the greatest common factor. ▶

In any factoring problem, the first step is to look for the *greatest common factor*. The **greatest common factor**, abbreviated GCF, is an expression of the highest degree that divides each term of the polynomial. Can you see that $5x$ is the greatest common factor of $10x^2 + 15x$? 5 is the greatest integer that divides both 10 and 15. Furthermore, x is the greatest expression that divides both x^2 and x.

The variable part of the greatest common factor always contains the smallest power of a variable that appears in all terms of the polynomial. For example, consider the polynomial

$$10x^2 + 15x.$$

> x^1, or x, is the variable raised to the smaller exponent.

We see that x is the variable part of the greatest common factor, $5x$.

EXAMPLE 1 Finding the Greatest Common Factor

Find the greatest common factor of each list of monomials:

a. $6x^3$ and $10x^2$ **b.** $15y^5$, $-9y^4$, and $27y^3$ **c.** x^5y^3, x^4y^4, and x^3y^2.

Solution Use numerical coefficients to determine the coefficient of the GCF. Use variable factors to determine the variable factor of the GCF.

> 2 is the greatest integer that divides 6 and 10.

a. $6x^3$ and $10x^2$

> x^2 is the variable raised to the smaller exponent.

We see that 2 is the coefficient of the GCF and x^2 is the variable factor of the GCF. Thus, the GCF of $6x^3$ and $10x^2$ is $2x^2$.

> 3 is the greatest integer that divides 15, -9, and 27.

b. $15y^5$, $-9y^4$,   and $27y^3$

> y^3 is the variable raised to the smallest exponent.

We see that 3 is the coefficient of the GCF and y^3 is the variable factor of the GCF. Thus, the GCF of $15y^5$, $-9y^4$, and $27y^3$ is $3y^3$.

> x^3 is the variable, x, raised to the smallest exponent.

c. x^5y^3, x^4y^4,   and x^3y^2

> y^2 is the variable, y, raised to the smallest exponent.

Because all terms have coefficients of 1, 1 is the greatest integer that divides these coefficients. Thus, 1 is the coefficient of the GCF. The voice balloons show that x^3 and y^2 are the variable factors of the GCF. Thus, the GCF of x^5y^3, x^4y^4, and x^3y^2 is x^3y^2. ■

✓ **CHECK POINT 1** Find the greatest common factor of each list of monomials:

a. $18x^3$ and $15x^2$ **b.** $-20x^2$, $12x^4$, and $40x^3$

c. x^4y, x^3y^2, and x^2y.

2 Factor out the greatest common factor of a polynomial. ▶

When we factor a monomial from a polynomial, we determine the greatest common factor of all terms in the polynomial. Sometimes there may not be a GCF other than 1. When a GCF other than 1 exists, we use the following procedure:

> **Factoring a Monomial from a Polynomial**
>
> 1. Determine the greatest common factor of all terms in the polynomial.
> 2. Express each term as the product of the GCF and its other factor.
> 3. Use the distributive property to factor out the GCF.

Great Question!

Is $5 \cdot x^2 + 5 \cdot 6$ a factorization of $5x^2 + 30$?

No. When we express $5x^2 + 30$ as $5 \cdot x^2 + 5 \cdot 6$, we have factored the *terms* of the binomial, but not the binomial itself. The factorization of the binomial is not complete until we write

$$5(x^2 + 6).$$

Now we have expressed the binomial *as a product*.

EXAMPLE 2 Factoring Out the Greatest Common Factor

Factor: $5x^2 + 30$.

Solution The GCF of $5x^2$ and 30 is 5.

$$5x^2 + 30$$
$$= 5 \cdot x^2 + 5 \cdot 6 \qquad \text{Express each term as the product of the GCF and its other factor.}$$
$$= 5(x^2 + 6) \qquad \text{Factor out the GCF.}$$

Because factoring reverses the process of multiplication, all factorizations can be checked by multiplying.

$$5(x^2 + 6) = 5 \cdot x^2 + 5 \cdot 6 = 5x^2 + 30$$

The factorization is correct because multiplication gives us the original polynomial. ■

☑ **CHECK POINT 2** Factor: $6x^2 + 18$.

EXAMPLE 3 Factoring Out the Greatest Common Factor

Factor: $18x^3 + 27x^2$.

Solution We begin by determining the greatest common factor.

9 is the greatest integer that divides 18 and 27.

$$18x^3 \qquad \text{and} \qquad 27x^2$$

x^2 is the variable raised to the smaller exponent.

The GCF of the two terms in the polynomial is $9x^2$.

$$18x^3 + 27x^2$$
$$= 9x^2 \cdot 2x + 9x^2 \cdot 3 \qquad \text{Express each term as the product of the GCF and its other factor.}$$
$$= 9x^2(2x + 3) \qquad \text{Factor out the GCF.}$$

We can check this factorization by multiplying $9x^2$ and $2x + 3$, obtaining the original polynomial as the answer. ■

Discover for Yourself

What happens if you factor out $3x^2$ rather than $9x^2$ from $18x^3 + 27x^2$? Although $3x^2$ is a common factor of the two terms, it is not the *greatest* common factor. Factor out $3x^2$ from $18x^3 + 27x^2$ and describe what happens with the second factor. Now factor again. Make the final result look like the factorization in Example 3. What is the advantage of factoring out the greatest common factor rather than just a common factor?

☑ **CHECK POINT 3** Factor: $25x^2 + 35x^3$.

EXAMPLE 4 Factoring Out the Greatest Common Factor

Factor: $16x^5 - 12x^4 + 4x^3$.

Solution First, determine the greatest common factor.

> 4 is the greatest integer that divides 16, −12, and 4.

$$16x^5, \quad -12x^4, \quad \text{and} \quad 4x^3$$

> x^3 is the variable raised to the smallest exponent.

The GCF of the three terms of the polynomial is $4x^3$.

$$16x^5 - 12x^4 + 4x^3$$
$$= 4x^3 \cdot 4x^2 - 4x^3 \cdot 3x + 4x^3 \cdot 1$$

Express each term as the product of the GCF and its other factor.

> You can obtain the factors shown in black by dividing each term of the given polynomial by $4x^3$, the GCF:
>
> $\dfrac{16x^5}{4x^3} = 4x^2 \quad \dfrac{12x^4}{4x^3} = 3x \quad \dfrac{4x^3}{4x^3} = 1.$

$$= 4x^3(4x^2 - 3x + 1)$$

Factor out the GCF. ∎

> Don't leave out the 1.

✓ **CHECK POINT 4** Factor: $15x^5 + 12x^4 - 27x^3$.

EXAMPLE 5 Factoring Out the Greatest Common Factor

Factor: $27x^2y^3 - 9xy^2 + 81xy$.

Solution First, determine the greatest common factor.

> 9 is the greatest integer that divides 27, −9, and 81.

$$27x^2y^3, \quad -9xy^2, \quad \text{and} \quad 81xy$$

> The variables raised to the smallest exponents are x and y.

The GCF of the three terms of the polynomial is $9xy$.

$$27x^2y^3 - 9xy^2 + 81xy$$
$$= 9xy \cdot 3xy^2 - 9xy \cdot y + 9xy \cdot 9$$

Express each term as the product of the GCF and its other factor.

> You can obtain the factors shown in black by dividing each term of the given polynomial by $9xy$, the GCF:
>
> $\dfrac{27x^2y^3}{9xy} = 3xy^2 \quad \dfrac{9xy^2}{9xy} = y \quad \dfrac{81xy}{9xy} = 9.$

$$= 9xy(3xy^2 - y + 9)$$

Factor out the GCF. ∎

✓ **CHECK POINT 5** Factor: $8x^3y^2 - 14x^2y + 2xy$.

3 Factor out the negative of the greatest common factor of a polynomial. ▶

Factoring with a Negative Coefficient in the First Term

Suppose we are interested in factoring

$$-5x^2 + 30.$$

> The first term has a negative coefficient.

When factoring polynomials, it is preferable to have a first term with a positive coefficient inside parentheses. We can do this with $-5x^2 + 30$ by factoring out -5, the negative of the GCF.

$$-5x^2 + 30$$
$$= -5 \cdot x^2 - 5(-6) \qquad \text{Express each term as the product of the negative of the GCF and its other factor.}$$
$$= -5(x^2 - 6) \qquad \text{Factor out the negative of the GCF.}$$

> **Factoring with a Negative Coefficient in the First Term**
>
> Express each term as the product of the negative of the GCF and its other factor. Then use the distributive property to factor out the negative of the GCF.

EXAMPLE 6 Factoring Out the Negative of the GCF

Factor: $-18a^4b^3 + 6a^2b^2 - 15a^3b$.

Solution First, determine the greatest common factor.

> 3 is the greatest integer that divides −18, 6, and −15.

$$-18a^4b^3, \qquad 6a^2b^2, \qquad \text{and} \qquad -15a^3b$$

> The variables raised to the smallest exponents are a^2 and b.

The GCF of the three terms of the polynomial is $3a^2b$. Because the polynomial has a negative coefficient in the first term, -18, we will factor out the negative of the GCF. Thus, we will factor out $-3a^2b$.

$$-18a^4b^3 + 6a^2b^2 - 15a^3b$$
$$= -3a^2b \cdot 6a^2b^2 - 3a^2b(-2b) - 3a^2b \cdot 5a \qquad \text{Express each term as the product of the negative of the GCF and its other factor.}$$
$$= -3a^2b(6a^2b^2 - 2b + 5a) \qquad \text{Factor out the negative of the GCF.} \quad \blacksquare$$

☑ **CHECK POINT 6** Factor: $-16a^4b^5 + 24a^3b^4 - 20ab^2$.

---

4 Factor by grouping. ▶

Factoring by Grouping

Up to now, we have factored a monomial from a polynomial. By contrast, in our next example, the greatest common factor of the polynomial is a binomial.

EXAMPLE 7 Factoring Out the Greatest Common Binomial Factor

Factor:

a. $x^2(x + 3) + 5(x + 3)$ **b.** $x(y + 1) - 2(y + 1)$.

Solution Let's identify the common binomial factor in each part of the problem.

$$x^2(x + 3) \quad \text{and} \quad 5(x + 3) \qquad\qquad x(y + 1) \quad \text{and} \quad -2(y + 1)$$

The GCF, a binomial, is $x + 3$.   The GCF, a binomial, is $y + 1$.

We factor out these common binomial factors as follows.

a. $x^2(x + 3) + 5(x + 3)$

$= (x + 3)x^2 + (x + 3)5$ Express each term as the product of the GCF and its other factor, in that order. Hereafter, we omit this step.

$= (x + 3)(x^2 + 5)$ Factor out the GCF, $x + 3$.

b. $x(y + 1) - 2(y + 1)$ The GCF is $y + 1$.

$= (y + 1)(x - 2)$ Factor out the GCF. ■

✓ **CHECK POINT 7**   Factor:

a. $x^2(x + 1) + 7(x + 1)$ **b.** $x(y + 4) - 7(y + 4)$.

Some polynomials have only a greatest common factor of 1. However, by a suitable grouping of the terms, it still may be possible to factor. This process, called **factoring by grouping**, is illustrated in Example 8.

EXAMPLE 8   Factoring by Grouping

Factor: $x^3 + 4x^2 + 3x + 12$.

Solution There is no factor other than 1 common to all four terms. However, we can group terms that have a common factor:

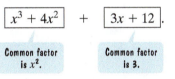

$$\boxed{x^3 + 4x^2} \quad + \quad \boxed{3x + 12}.$$

Common factor is x^2.   Common factor is 3.

Discover for Yourself

In Example 8, group the terms as follows:

$$(x^3 + 3x) + (4x^2 + 12).$$

Factor out the greatest common factor from each group and complete the factoring process. Describe what happens. What can you conclude?

We now factor the given polynomial as follows:

$x^3 + 4x^2 + 3x + 12$

$= (x^3 + 4x^2) + (3x + 12)$ Group terms with common factors.

$= x^2(x + 4) + 3(x + 4)$ Factor out the greatest common factor from the grouped terms. The remaining two terms have $x + 4$ as a common binomial factor.

$= (x + 4)(x^2 + 3).$ Factor out the GCF, $x + 4$.

Thus, $x^3 + 4x^2 + 3x + 12 = (x + 4)(x^2 + 3)$. Check the factorization by multiplying the right side of the equation using the FOIL method. Because the factorization is correct, you should obtain the original polynomial. ■

✓ **CHECK POINT 8**   Factor: $x^3 + 5x^2 + 2x + 10$.

Factoring by Grouping

1. Group terms that have a common monomial factor. There will usually be two groups. Sometimes the terms must be rearranged.

2. Factor out the common monomial factor from each group.

3. Factor out the remaining common binomial factor (if one exists).

EXAMPLE 9 Factoring by Grouping

Factor: $xy + 5x - 4y - 20$.

Solution There is no factor other than 1 common to all four terms. However, we can group terms that have a common factor:

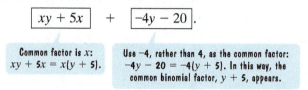

$$\boxed{xy + 5x} \quad + \quad \boxed{-4y - 20}.$$

Common factor is x:
$xy + 5x = x(y + 5)$.

Use -4, rather than 4, as the common factor: $-4y - 20 = -4(y + 5)$. In this way, the common binomial factor, $y + 5$, appears.

The voice balloons illustrate that it is sometimes necessary to factor out a negative number from a grouping to obtain a common binomial factor for the two groupings. We now factor the given polynomial as follows:

$$xy + 5x - 4y - 20$$
$$= x(y + 5) - 4(y + 5) \qquad \text{Factor } x \text{ and } -4, \text{ respectively, from each grouping.}$$
$$= (y + 5)(x - 4). \qquad \text{Factor out the GCF, } y + 5.$$

Thus, $xy + 5x - 4y - 20 = (y + 5)(x - 4)$. Using the commutative property of multiplication, the factorization can also be expressed as $(x - 4)(y + 5)$. Multiply these factors using the FOIL method to verify that, regardless of the order, these are the correct factors. ∎

☑ **CHECK POINT 9** Factor: $xy + 3x - 5y - 15$.

Achieving Success

When using your professor's office hours, show up prepared. If you are having difficulty with a concept or problem, bring your work so that your instructor can determine where you are having trouble. If you miss a lecture, read the appropriate section in the textbook, borrow class notes, and attempt the assigned homework before your office visit. Because this text has an accompanying video lesson for every section, you might find it helpful to view the video covering the material you missed. It is not realistic to expect your professor to rehash all or part of a class lecture during office hours.

CONCEPT AND VOCABULARY CHECK

Fill in each blank so that the resulting statement is true.

1. The process of writing a polynomial containing the sum of monomials as a product is called _____.

2. An expression of the highest degree that divides each term of a polynomial is called the _____. The variable part of this expression contains the _____ power of a variable that appears in all terms of the polynomial.

3. True or false: The factorization of $15x + 20$ is $5 \cdot 3x + 5 \cdot 4$. _____

4. True or false: The factorization of $x^2 + 3x + 5x + 15$ is $x(x + 3) + 5(x + 3)$. _____

6.1 EXERCISE SET ▶ MyMathLab®

Practice Exercises

In Exercises 1–12, find the greatest common factor of each list of monomials.

1. 4 and $8x$
2. 5 and $15x$
3. $12x^2$ and $8x$
4. $20x^2$ and $15x$
5. $-2x^4$ and $6x^3$
6. $-3x^4$ and $6x^3$
7. $9y^5$, $18y^2$, and $-3y$
8. $10y^5$, $20y^2$, and $-5y$
9. xy, xy^2, and xy^3
10. x^2y, $3x^3y$, and $6x^2$
11. $16x^5y^4$, $8x^6y^3$, and $20x^4y^5$
12. $18x^5y^4$, $6x^6y^3$, and $12x^4y^5$

In Exercises 13–48, factor each polynomial using the greatest common factor. If there is no common factor other than 1 and the polynomial cannot be factored, so state.

13. $8x + 8$
14. $9x + 9$
15. $4y - 4$
16. $5y - 5$
17. $5x + 30$
18. $10x + 30$
19. $30x - 12$
20. $32x - 24$
21. $x^2 + 5x$
22. $x^2 + 6x$
23. $18y^2 + 12$
24. $20y^2 + 15$
25. $14x^3 + 21x^2$
26. $6x^3 + 15x^2$
27. $13y^2 - 25y$
28. $11y^2 - 30y$
29. $9y^4 + 27y^6$
30. $10y^4 + 15y^6$
31. $8x^2 - 4x^4$
32. $12x^2 - 4x^4$
33. $12y^2 + 16y - 8$
34. $15y^2 - 3y + 9$
35. $9x^4 + 18x^3 + 6x^2$
36. $32x^4 + 2x^3 + 8x^2$
37. $100y^5 - 50y^3 + 100y^2$
38. $26y^5 - 13y^3 + 39y^2$
39. $10x - 20x^2 + 5x^3$
40. $6x - 4x^2 + 2x^3$
41. $11x^2 - 23$
42. $12x^2 - 25$
43. $6x^3y^2 + 9xy$
44. $4x^2y^3 + 6xy$
45. $30x^2y^3 - 10xy^2 + 20xy$
46. $27x^2y^3 - 18xy^2 + 45x^2y$
47. $32x^3y^2 - 24x^3y - 16x^2y$
48. $18x^3y^2 - 12x^3y - 24x^2y$

In Exercises 49–56, factor each polynomial using the negative of the greatest common factor.

49. $-12x^2 + 18$
50. $-15x^2 + 20$
51. $-8x^4 + 32x^3 + 16x^2$
52. $-18x^4 + 9x^3 + 6x^2$
53. $-4a^3b^2 + 6ab$
54. $-9a^2b^3 + 12ab$
55. $-12x^3y^2 - 18x^3y + 24x^2y$
56. $-24x^3y^2 - 32x^3y + 16x^2y$

In Exercises 57–68, factor each polynomial using the greatest common binomial factor.

57. $x(x + 5) + 3(x + 5)$
58. $x(x + 7) + 10(x + 7)$
59. $x(x + 2) - 4(x + 2)$
60. $x(x + 3) - 8(x + 3)$
61. $x(y + 6) - 7(y + 6)$
62. $x(y + 9) - 11(y + 9)$
63. $3x(x + y) - (x + y)$
64. $7x(x + y) - (x + y)$
65. $4x(3x + 1) + 3x + 1$
66. $5x(2x + 1) + 2x + 1$
67. $7x^2(5x + 4) + 5x + 4$
68. $9x^2(7x + 2) + 7x + 2$

In Exercises 69–86, factor by grouping.

69. $x^2 + 2x + 4x + 8$
70. $x^2 + 3x + 5x + 15$
71. $x^2 + 3x - 5x - 15$
72. $x^2 + 7x - 4x - 28$
73. $x^3 - 2x^2 + 5x - 10$
74. $x^3 - 3x^2 + 4x - 12$
75. $x^3 - x^2 + 2x - 2$
76. $x^3 + 6x^2 - 2x - 12$
77. $xy + 5x + 9y + 45$
78. $xy + 6x + 2y + 12$
79. $xy - x + 5y - 5$
80. $xy - x + 7y - 7$
81. $3x^2 - 6xy + 5xy - 10y^2$
82. $10x^2 - 12xy + 35xy - 42y^2$
83. $3x^3 - 2x^2 - 6x + 4$
84. $4x^3 - x^2 - 12x + 3$
85. $x^2 - ax - bx + ab$
86. $x^2 + ax + bx + ab$

Practice PLUS

In Exercises 87–94, factor each polynomial.

87. $24x^3y^3z^3 + 30x^2y^2z + 18x^2yz^2$
88. $16x^2y^2z^2 + 32x^2yz^2 + 24x^2yz$
89. $x^3 - 4 + 3x^3y - 12y$
90. $x^3 - 5 + 2x^3y - 10y$
91. $4x^5(x + 1) - 6x^3(x + 1) - 8x^2(x + 1)$

92. $8x^5(x + 2) - 10x^3(x + 2) - 2x^2(x + 2)$

93. $3x^5 - 3x^4 + x^3 - x^2 + 5x - 5$

94. $7x^5 - 7x^4 + x^3 - x^2 + 3x - 3$

The figures for Exercises 95–96 show one or more circles drawn inside a square. Write a polynomial that represents the shaded blue area in each figure. Then factor the polynomial.

95.   **96.**

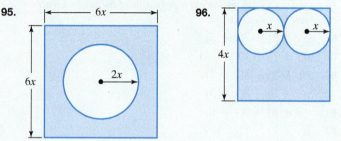

Application Exercises

97. An explosion causes debris to rise vertically with an initial velocity of 64 feet per second. The polynomial $64x - 16x^2$ describes the height of the debris above the ground, in feet, after x seconds.

   a. Find the height of the debris after 3 seconds.

   b. Factor the polynomial.

   c. Use the factored form of the polynomial in part (b) to find the height of the debris after 3 seconds. Do you get the same answer as you did in part (a)? If so, does this prove that your factorization is correct? Explain.

98. An explosion causes debris to rise vertically with an initial velocity of 72 feet per second. The polynomial $72x - 16x^2$ describes the height of the debris above the ground, in feet, after x seconds.

   a. Find the height of the debris after 4 seconds.

   b. Factor the polynomial.

   c. Use the factored form of the polynomial in part (b) to find the height of the debris after 4 seconds. Do you get the same answer as you did in part (a)? If so, does this prove that your factorization is correct? Explain.

In Exercises 99–100, write a polynomial for the length of each rectangle.

99.

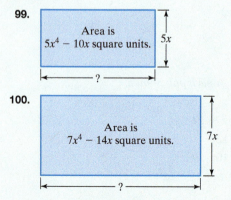

100.

Explaining the Concepts

101. What is factoring?

102. What is a prime polynomial?

103. Explain how to find the greatest common factor of a list of terms. Give an example with your explanation.

104. Use an example and explain how to factor out the greatest common factor of a polynomial.

105. Suppose that a polynomial contains four terms and can be factored by grouping. Explain how to obtain the factorization.

106. Write a sentence that uses the word "factor" as a noun. Then write a sentence that uses the word "factor" as a verb.

Critical Thinking Exercises

Make Sense?   In Exercises 107–110, determine whether each statement makes sense or does not make sense, and explain your reasoning.

107. After factoring $20x^3 + 8x^2$ and $20x^3 + 10x$, I noticed that I factored the monomial $20x^3$ in two different ways.

108. After I've factored a polynomial, my answer cannot always be checked by multiplication.

109. The word *greatest* in greatest common factor is helpful because it tells me to look for the greatest power of a variable appearing in all terms.

110. You grouped the polynomial's terms using different groupings than I did, yet we both obtained the same factorization.

In Exercises 111–114, determine whether each statement is true or false. If the statement is false, make the necessary change(s) to produce a true statement.

111. Since the GCF of $9x^3 + 6x^2 + 3x$ is $3x$, it is not necessary to write the 1 when $3x$ is factored from the last term.

112. $a(x - 7) + b(7 - x) = a(x - 7) + b(-1)(x - 7)$
$$= a(x - 7) - b(x - 7)$$
$$= (x - 7)(a - b)$$

113. $a^2 + b^2 = a^2 + ab - ab + b^2$
$$= a(a + b) - b(a + b)$$
$$= (a + b)(a - b)$$

114. $-4x^2 + 12x$ can be factored as $-4x(x - 3)$ or $4x(-x + 3)$.

115. Suppose you receive x dollars in January. Each month thereafter, you receive $100 more than you received the month before. Write a factored polynomial that describes the total dollar amount you receive from January through April.

In Exercises 116–117, write a polynomial that fits the given description. Do not use a polynomial that appears in this section or in the Exercise Set.

116. The polynomial has four terms and can be factored using a greatest common factor that has both a coefficient and a variable.

117. The polynomial has four terms and can be factored by grouping.

Technology Exercises

In Exercises 118–120, use a graphing utility to graph each side of the equation in the same viewing rectangle. Do the graphs coincide? If so, this means that the polynomial on the left side has been factored correctly. If not, factor the polynomial correctly and then use your graphing utility to verify the factorization.

118. $-3x - 6 = -3(x - 2)$

119. $x^2 - 2x + 5x - 10 = (x - 2)(x - 5)$

120. $x^2 + 2x + x + 2 = x(x + 2) + 1$

Review Exercises

121. Multiply: $(x + 7)(x + 10)$. (Section 5.3, Example 1)

122. Solve the system by graphing:
$$\begin{cases} 2x - y = -4 \\ x - 3y = 3. \end{cases}$$
(Section 4.1, Example 2)

123. Write the point-slope form of the equation of the line passing through $(-7, 2)$ and $(-4, 5)$. Then use the point-slope form of the equation to write the slope-intercept form of the equation. (Section 3.5, Example 2)

Preview Exercises

Exercises 124–126 will help you prepare for the material covered in the next section.

124. Find two factors of 8 whose sum is 6.

125. Find two factors of 6 whose sum is -5.

126. Find two factors of -35 whose sum is 2.

SECTION

6.2

Factoring Trinomials Whose Leading Coefficient Is 1 ▶

What am I supposed to learn?

After studying this section, you should be able to:

1 Factor trinomials of the form $x^2 + bx + c$. ▶

Not afraid of heights and cutting-edge excitement? How about sky diving? Behind your exhilarating experience is the world of algebra. After you jump from the airplane, your height above the ground at every instant of your fall can be described by a formula involving a variable that is squared. At a height of approximately 2000 feet, you'll need to open your parachute. How can you determine when you must do so?

The answer to this critical question involves using the factoring technique presented in this section. In Section 6.6, in which applications are discussed, this technique is applied to models involving the height of any free-falling object—in this case, you.

① Factor trinomials of the form $x^2 + bx + c$. ▶

A Strategy for Factoring $x^2 + bx + c$

In Section 5.3, we used the FOIL method to multiply two binomials. The product was often a trinomial. The following are some examples:

Factored Form	F	O	I	L	Trinomial Form
$(x + 3)(x + 4)$ =	x^2	$+ 4x$	$+ 3x$	$+ 12$	$= x^2 + 7x + 12$
$(x - 3)(x - 4)$ =	x^2	$- 4x$	$- 3x$	$+ 12$	$= x^2 - 7x + 12$
$(x + 3)(x - 5)$ =	x^2	$- 5x$	$+ 3x$	$- 15$	$= x^2 - 2x - 15.$

Observe that each trinomial is of the form $x^2 + bx + c$, where the coefficient of the squared term, called the **leading coefficient**, is 1. Our goal in this section is to start with the trinomial form and, assuming that it is factorable, return to the factored form.

The first FOIL multiplication shown above indicates that $(x + 3)(x + 4) = x^2 + 7x + 12$. Let's reverse the sides of this equation:

$$x^2 + 7x + 12 = (x + 3)(x + 4).$$

We can make several important observations about the factors on the right side.

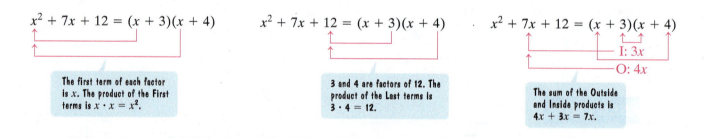

The first term of each factor is x. The product of the First terms is $x \cdot x = x^2$.

3 and 4 are factors of 12. The product of the Last terms is $3 \cdot 4 = 12$.

The sum of the Outside and Inside products is $4x + 3x = 7x$.

These observations provide us with a procedure for factoring $x^2 + bx + c$.

A Strategy for Factoring $x^2 + bx + c$

1. Enter x as the first term of each factor.

$$(x \qquad)(x \qquad) = x^2 + bx + c$$

2. List pairs of factors of the constant c.

3. Try various combinations of these factors as the second term in each set of parentheses. Select the combination in which the sum of the Outside and Inside products is equal to bx.

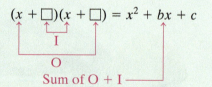

$$(x + \square)(x + \square) = x^2 + bx + c$$

I
O
Sum of O + I

4. Check your work by multiplying the factors using the FOIL method. You should obtain the original trinomial.

If none of the possible combinations yield an Outside product and an Inside product whose sum is equal to bx, the trinomial cannot be factored using integers and is called **prime** over the set of integers.

EXAMPLE 1 Factoring a Trinomial in $x^2 + bx + c$ Form

Factor: $x^2 + 6x + 8$.

Solution

Step 1. Enter x as the first term of each factor.

$$x^2 + 6x + 8 = (x \quad)(x \quad)$$

Step 2. List pairs of factors of the constant, 8.

Factors of 8	8, 1	4, 2	−8, −1	−4, −2

Step 3. Try various combinations of these factors. The correct factorization of $x^2 + 6x + 8$ is the one in which the sum of the Outside and Inside products is equal to $6x$. Here is a list of the possible factorizations:

Possible Factorizations of $x^2 + 6x + 8$	Sum of Outside and Inside Products (Should Equal 6x)	
$(x + 8)(x + 1)$	$x + 8x = 9x$	
$(x + 4)(x + 2)$	$2x + 4x = 6x$	This is the required middle term.
$(x - 8)(x - 1)$	$-x - 8x = -9x$	
$(x - 4)(x - 2)$	$-2x - 4x = -6x$	

Thus, $x^2 + 6x + 8 = (x + 4)(x + 2)$.

Step 4. Check this result by multiplying the right side using the FOIL method. You should obtain the original trinomial. Because of the commutative property, the factorization can also be expressed as

$$x^2 + 6x + 8 = (x + 2)(x + 4). \quad \blacksquare$$

Using Technology

Graphic and Numeric Connections

If a polynomial contains one variable, a graphing utility can be used to check its factorization. For example, the factorization in Example 1,

$$x^2 + 6x + 8 = (x + 4)(x + 2),$$

can be checked graphically or numerically.

Graphic Check
Use the GRAPH feature. Graph $y_1 = x^2 + 6x + 8$ and $y_2 = (x + 4)(x + 2)$ on the same screen. Because the graphs are identical, the factorization appears to be correct.

Numeric Check
Use the TABLE feature. Enter $y_1 = x^2 + 6x + 8$ and $y_2 = (x + 4)(x + 2)$ and press TABLE. Two columns of values are shown, one for y_1 and one for y_2. Because the corresponding values are equal regardless of how far up or down we scroll, the factorization is correct.

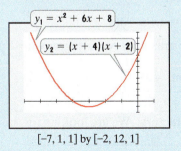

$y_1 = x^2 + 6x + 8$
$y_2 = (x + 4)(x + 2)$

$[-7, 1, 1]$ by $[-2, 12, 1]$

X	Y1	Y2
-3	-1	-1
-2	0	0
-1	3	3
0	8	8
1	15	15
2	24	24
3	35	35
4	48	48
5	63	63
6	80	80
7	99	99

X=-3

☑ **CHECK POINT 1** Factor: $x^2 + 5x + 6$.

EXAMPLE 2 Factoring a Trinomial in $x^2 + bx + c$ Form

Factor: $x^2 - 5x + 6$.

Solution

Step 1. Enter x as the first term of each factor.

$$x^2 - 5x + 6 = (x \quad)(x \quad)$$

Step 2. List pairs of factors of the constant, 6.

Factors of 6	6, 1	3, 2	−6, −1	−3, −2

Step 3. Try various combinations of these factors. The correct factorization of $x^2 - 5x + 6$ is the one in which the sum of the Outside and Inside products is equal to $-5x$. Here is a list of the possible factorizations:

Possible Factorizations of $x^2 - 5x + 6$	Sum of Outside and Inside Products (Should Equal $-5x$)
$(x + 6)(x + 1)$	$x + 6x = 7x$
$(x + 3)(x + 2)$	$2x + 3x = 5x$
$(x - 6)(x - 1)$	$-x - 6x = -7x$
$(x - 3)(x - 2)$	$-2x - 3x = -5x$

This is the required middle term.

Thus, $x^2 - 5x + 6 = (x - 3)(x - 2)$. Verify this result using the FOIL method. ∎

In factoring a trinomial of the form $x^2 + bx + c$, you can speed things up by listing the factors of c and then finding their sums. We are interested in a sum of b. For example, in factoring $x^2 - 5x + 6$, we are interested in the factors of 6 whose sum is -5.

Factors of 6	6, 1	3, 2	−6, −1	−3, −2
Sum of Factors	7	5	−7	−5

This is the desired sum.

Thus, $x^2 - 5x + 6 = (x - 3)(x - 2)$.

✅ **CHECK POINT 2** Factor: $x^2 - 6x + 8$.

Great Question!

Is there a way to eliminate some of the combinations of factors for $x^2 + bx + c$ when c is positive?

Yes. To factor $x^2 + bx + c$ when c is positive, find two numbers with the same sign as the middle term.

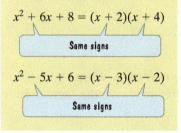

$x^2 + 6x + 8 = (x + 2)(x + 4)$

Same signs

$x^2 - 5x + 6 = (x - 3)(x - 2)$

Same signs

EXAMPLE 3 Factoring a Trinomial in $x^2 + bx + c$ Form

Factor: $x^2 + 2x - 35$.

Solution

Step 1. Enter x as the first term of each factor.

$$x^2 + 2x - 35 = (x \quad)(x \quad)$$

To find the second term of each factor, we must find two integers whose product is -35 and whose sum is 2.

Step 2. List pairs of factors of the constant, -35.

Factors of −35	−35, 1	−7, 5	35, −1	7, −5

Step 3. Try various combinations of these factors. We are looking for the pair of factors whose sum is 2.

Factors of −35	−35, 1	−7, 5	35, −1	7, −5
Sum of Factors	−34	−2	34	2

This is the desired sum.

Thus, $x^2 + 2x - 35 = (x + 7)(x - 5)$.

Step 4. Verify the factorization using the FOIL method.

$$(x + 7)(x - 5) = x^2 - 5x + 7x - 35 = x^2 + 2x - 35$$

Because the product of the factors is the original polynomial, the factorization is correct. ∎

✓ **CHECK POINT 3**   Factor: $x^2 + 3x - 10$.

___

EXAMPLE 4   Factoring a Trinomial Whose Leading Coefficient Is 1

Factor: $y^2 - 2y - 99$.

Solution

Step 1. Enter y as the first term of each factor.

$$y^2 - 2y - 99 = (y \quad)(y \quad)$$

To find the second term of each factor, we must find two integers whose product is -99 and whose sum is -2.

Step 2. List pairs of factors of the constant, -99.

Factors of -99	$-99, 1$	$-11, 9$	$-33, 3$	$99, -1$	$11, -9$	$33, -3$

Step 3. Try various combinations of these factors. In order to factor $y^2 - 2y - 99$, we are interested in the pair of factors of -99 whose sum is -2.

Factors of -99	$-99, 1$	$-11, 9$	$-33, 3$	$99, -1$	$11, -9$	$33, -3$
Sum of Factors	-98	-2	-30	98	2	30

This is the desired sum.

Thus, $y^2 - 2y - 99 = (y - 11)(y + 9)$. Verify this result using the FOIL method. ∎

✓ **CHECK POINT 4**   Factor: $y^2 - 6y - 27$.

___

EXAMPLE 5   Trying to Factor a Trinomial in $x^2 + bx + c$ Form

Factor: $x^2 + x - 5$.

Solution

Step 1. Enter x as the first term of each factor.

$$x^2 + x - 5 = (x \quad)(x \quad)$$

To find the second term of each factor, we must find two integers whose product is -5 and whose sum is 1.

Steps 2 and 3. List pairs of factors of the constant, -5, and try various combinations of these factors. We are interested in a pair of factors whose sum is 1.

Factors of -5	$-5, 1$	$5, -1$
Sum of Factors	-4	4

No pair gives the desired sum, 1.

Because neither pair has a sum of 1, $x^2 + x - 5$ cannot be factored using integers. This trinomial is prime. ∎

✓ **CHECK POINT 5**   Factor: $x^2 + x - 7$.

Great Question!

Is there a way to eliminate some of the combinations of factors for $x^2 + bx + c$ when c is negative?

Yes. To factor $x^2 + bx + c$ when c is negative, find two numbers with opposite signs whose sum is the coefficient of the middle term.

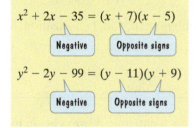

$$x^2 + 2x - 35 = (x + 7)(x - 5)$$

Negative   Opposite signs

$$y^2 - 2y - 99 = (y - 11)(y + 9)$$

Negative   Opposite signs

EXAMPLE 6   Factoring a Trinomial in Two Variables

Factor: $x^2 - 5xy + 6y^2$.

Solution

Step 1. Enter *x* as the first term of each factor. Because the last term of the trinomial contains y^2, the second term of each factor must contain y.

$$x^2 - 5xy + 6y^2 = (x \quad ?y)(x \quad ?y)$$

The question marks indicate that we are looking for the coefficients of y in each factor. To find these coefficients, we must find two integers whose product is 6 and whose sum is -5.

Steps 2 and 3. List pairs of factors of the coefficient of the last term, 6, and try various combinations of these factors. We are interested in the pair of factors whose sum is -5.

Factors of 6	6, 1	3, 2	−6, −1	−3, −2
Sum of Factors	7	5	−7	−5

This is the desired sum.

Thus, $x^2 - 5xy + 6y^2 = (x - 3y)(x - 2y)$.

Step 4. Verify the factorization using the FOIL method.

$$(x - 3y)(x - 2y) = x^2 - 2xy - 3xy + 6y^2 = x^2 - 5xy + 6y^2$$

Because the product of the factors is the original polynomial, the factorization is correct. ∎

✓ CHECK POINT 6   Factor: $x^2 - 4xy + 3y^2$.

___

Some polynomials can be factored using more than one technique. **Always begin by looking for common factors** and, if there are common factors, factor out the greatest common factor. A polynomial is **factored completely** when it is written as the product of prime polynomials.

EXAMPLE 7   Factoring Completely

Factor: $3x^3 - 15x^2 - 42x$.

Solution   The GCF of the three terms of the polynomial is $3x$. We begin by factoring out $3x$. Then we factor the remaining trinomial by the methods of this section.

$$3x^3 - 15x^2 - 42x$$
$$= 3x(x^2 - 5x - 14) \quad \text{Factor out the GCF.}$$
$$= 3x(x \quad)(x \quad) \quad \text{Begin factoring } x^2 - 5x - 14. \text{ Find two integers whose product is } -14 \text{ and whose sum is } -5.$$
$$= 3x(x - 7)(x + 2) \quad \text{The integers are } -7 \text{ and } 2.$$

Thus,

$$3x^3 - 15x^2 - 42x = 3x(x - 7)(x + 2).$$

Be sure to include the GCF in the factorization.

How can we check this factorization? We will multiply the binomials using the FOIL method. Then use the distributive property and multiply each term of this product by $3x$. If the factorization is correct, we should obtain the original polynomial.

$$3x(x - 7)(x + 2) = 3x(x^2 + 2x - 7x - 14) = 3x(x^2 - 5x - 14) = 3x^3 - 15x^2 - 42x$$

> Use the FOIL method on $(x - 7)(x + 2)$.

> This is the original polynomial.

The factorization is correct. ■

✓ **CHECK POINT 7**   Factor: $2x^3 + 6x^2 - 56x$.

EXAMPLE 8   Factoring Completely by Factoring Out the Negative of the GCF

Factor: $-16t^2 + 16t + 96$.

Solution   The GCF of the three terms of the polynomial is 16. Because the polynomial has a negative coefficient in the first term, we will factor out the negative of the GCF. Thus, we will factor out -16.

$$-16t^2 + 16t + 96$$
$$= -16(t^2 - t - 6) \qquad \text{Factor out the negative of the GCF.}$$
$$= -16(t \quad)(t \quad) \qquad \text{Begin factoring } t^2 - t - 6. \text{ Find two integers whose product is } -6 \text{ and whose sum is } -1.$$
$$= -16(t - 3)(t + 2) \qquad \text{The integers are } -3 \text{ and } 2.$$

Thus,

$$-16t^2 + 16t + 96 = -16(t - 3)(t + 2).$$

Verify this factorization using the FOIL method to multiply $t - 3$ and $t + 2$. Then use the distributive property and multiply each term of this product by -16. You should obtain the original polynomial. ■

✓ **CHECK POINT 8**   Factor: $-2y^2 - 10y + 28$.

CONCEPT AND VOCABULARY CHECK

Fill in each blank so that the resulting statement is true.

1. To factor $x^2 - 12x + 20$, we must find two integers whose product is _____ and whose sum is _____.

2. A polynomial is factored _____ when it is written as a product of prime polynomials.

3. $x^2 + 13x + 30 = (x + 3)(x$ ____$)$

4. $x^2 - 9x + 18 = (x - 3)(x$ ____$)$

5. $x^2 - x - 30 = (x - 6)(x$ ____$)$

6. $x^2 - 5x - 14 = (x + 2)(x$ ____$)$

7. $x^2 - 10xy + 16y^2 = (x - 8y)(x$ ____$)$

6.2 EXERCISE SET ▶ MyMathLab®

Practice Exercises

In Exercises 1–42, factor each trinomial, or state that the trinomial is prime. Check each factorization using FOIL multiplication.

1. $x^2 + 7x + 6$
2. $x^2 + 9x + 8$
3. $x^2 + 7x + 10$
4. $x^2 + 9x + 14$
5. $x^2 + 11x + 10$
6. $x^2 + 13x + 12$
7. $x^2 - 7x + 12$
8. $x^2 - 13x + 40$
9. $x^2 - 12x + 36$
10. $x^2 - 8x + 16$
11. $y^2 - 8y + 15$
12. $y^2 - 8y + 7$
13. $x^2 + 3x - 10$
14. $x^2 + 3x - 28$
15. $y^2 + 10y - 39$
16. $y^2 + 5y - 24$
17. $x^2 - 2x - 15$
18. $x^2 - 4x - 5$
19. $x^2 - 2x - 8$
20. $x^2 - 5x - 6$
21. $x^2 + 4x + 12$ 22. $x^2 + 4x + 5$
23. $y^2 - 16y + 48$
24. $y^2 - 10y + 21$
25. $x^2 - 3x + 6$ 26. $x^2 + 4x - 10$
27. $w^2 - 30w - 64$
28. $w^2 + 12w - 64$
29. $y^2 - 18y + 65$
30. $y^2 - 22y + 72$
31. $r^2 + 12r + 27$
32. $r^2 - 15r - 16$
33. $y^2 - 7y + 5$ 34. $y^2 - 15y + 5$
35. $x^2 + 7xy + 6y^2$
36. $x^2 + 6xy + 8y^2$
37. $x^2 - 8xy + 15y^2$
38. $x^2 - 9xy + 14y^2$
39. $x^2 - 3xy - 18y^2$
40. $x^2 - xy - 30y^2$
41. $a^2 - 18ab + 45b^2$
42. $a^2 - 18ab + 80b^2$

In Exercises 43–66, factor completely.

43. $3x^2 + 15x + 18$
44. $3x^2 + 21x + 36$
45. $4y^2 - 4y - 8$
46. $3y^2 + 3y - 18$
47. $10x^2 - 40x - 600$
48. $2x^2 + 10x - 48$
49. $3x^2 - 33x + 54$
50. $2x^2 - 14x + 24$
51. $2r^3 + 6r^2 + 4r$
52. $2r^3 + 8r^2 + 6r$
53. $4x^3 + 12x^2 - 72x$
54. $3x^3 - 15x^2 + 18x$
55. $2r^3 + 8r^2 - 64r$
56. $3r^3 - 9r^2 - 54r$
57. $y^4 + 2y^3 - 80y^2$
58. $y^4 - 12y^3 + 35y^2$
59. $x^4 - 3x^3 - 10x^2$
60. $x^4 - 22x^3 + 120x^2$
61. $2w^4 - 26w^3 - 96w^2$
62. $3w^4 + 54w^3 + 135w^2$
63. $15xy^2 + 45xy - 60x$
64. $20x^2y - 100xy + 120y$
65. $x^5 + 3x^4y - 4x^3y^2$
66. $x^3y - 2x^2y^2 - 3xy^3$

In Exercises 67–74, use the negative of the greatest common factor to factor completely.

67. $-16t^2 + 64t + 80$
68. $-16t^2 + 80t + 96$
69. $-5x^2 + 50x - 45$
70. $-3x^2 + 36x - 33$
71. $-x^2 - 3x + 40$
72. $-x^2 - 4x + 45$
73. $-2x^3 - 6x^2 + 8x$
74. $-3x^3 + 6x^2 + 24x$

Practice PLUS

In Exercises 75–82, factor completely.

75. $2x^2y^2 - 32x^2yz + 30x^2z^2$
76. $2x^2y^2 - 30x^2yz + 28x^2z^2$
77. $(a + b)x^2 + (a + b)x - 20(a + b)$
78. $(a + b)x^2 - 13(a + b)x + 36(a + b)$

(Hint for Exercises 79–82: Factors contain rational numbers.)

79. $x^2 + 0.5x + 0.06$
80. $x^2 - 0.5x - 0.06$
81. $x^2 - \dfrac{2}{5}x + \dfrac{1}{25}$
82. $x^2 + \dfrac{2}{3}x + \dfrac{1}{9}$

Application Exercises

83. You dive directly upward from a board that is 32 feet high. After t seconds, your height above the water is described by the polynomial

$$-16t^2 + 16t + 32.$$

a. Factor the polynomial completely.

b. Evaluate both the original polynomial and its factored form for $t = 2$. Do you get the same answer for each evaluation? Describe what this answer means.

84. You dive directly upward from a board that is 48 feet high. After t seconds, your height above the water is described by the polynomial

$$-16t^2 + 32t + 48.$$

a. Factor the polynomial completely.

b. Evaluate both the original polynomial and its factored form for $t = 3$. Do you get the same answer for each evaluation? Describe what this answer means.

Explaining the Concepts

85. Explain how to factor $x^2 + 8x + 15$.

86. Give two helpful suggestions for factoring $x^2 - 5x + 6$.

87. In factoring $x^2 + bx + c$, describe how the last terms in each factor are related to b and c.

88. Without actually factoring and without multiplying the given factors, explain why the following factorization is not correct:

$$x^2 + 46x + 513 = (x - 27)(x - 19).$$

Critical Thinking Exercises

Make Sense? *In Exercises 89–92, determine whether each statement makes sense or does not make sense, and explain your reasoning.*

89. If I have a correct factorization, switching the order of the factors can give me an incorrect factorization.

90. I began factoring $x^2 - 17x + 72$ by finding all number pairs with a sum of -17.

91. It's easy to factor $x^2 + x + 1$ because of the relatively small numbers for the constant term and the coefficient of x.

92. I factor $x^2 + bx + c$ by finding two numbers that have a product of c and a sum of b.

In Exercises 93–96, determine whether each statement is true or false. If the statement is false, make the necessary change(s) to produce a true statement.

93. One factor of $x^2 + x + 20$ is $x + 5$.

94. A trinomial can never have two identical factors.

95. One factor of $y^2 + 5y - 24$ is $y - 3$.

96. $x^2 + 4 = (x + 2)(x + 2)$

In Exercises 97–98, find all positive integers b so that the trinomial can be factored.

97. $x^2 + bx + 15$

98. $x^2 + 4x + b$

99. Write a trinomial of the form $x^2 + bx + c$ that is prime.

100. Factor: $x^{2n} + 20x^n + 99$.

101. Factor $x^3 + 3x^2 + 2x$. If x represents an integer, use the factorization to describe what the trinomial represents.

102. A box with no top is to be made from an 8-inch by 6-inch piece of metal by cutting identical squares from each corner and turning up the sides (see the figure). The volume of the box is modeled by the polynomial $4x^3 - 28x^2 + 48x$. Factor the polynomial completely. Then use the dimensions given on the box and show that its volume is equivalent to the factorization that you obtain.

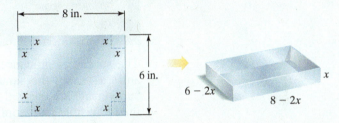

Technology Exercises

In Exercises 103–106, use the GRAPH *or* TABLE *feature of a graphing utility to determine if the polynomial on the left side of each equation has been correctly factored. If the graphs of y_1 and y_2 coincide, or if their corresponding table values are equal, this means that the polynomial on the left side has been correctly factored. If not, factor the trinomial correctly and then use your graphing utility to verify the factorization.*

103. $x^2 - 5x + 6 = (x - 2)(x - 3)$

104. $2x^2 + 2x - 12 = 2(x - 3)(x + 2)$

105. $x^2 - 2x + 1 = (x + 1)(x - 1)$

106. $2x^2 + 8x + 6 = (x + 3)(x + 1)$

Review Exercises

107. Solve: $4(x - 2) = 3x + 5$. (Section 2.3, Example 2)

108. Graph: $6x - 5y \leq 30$. (Section 3.6, Example 2)

109. Graph: $y = -\frac{1}{2}x + 2$. (Section 3.4, Example 3)

Preview Exercises

Exercises 110–112 will help you prepare for the material covered in the next section.

110. Multiply: $(2x + 3)(x - 2)$.

111. Multiply: $(3x + 4)(3x + 1)$.

112. Factor by grouping: $8x^2 - 2x - 20x + 5$.

SECTION

6.3　Factoring Trinomials Whose Leading Coefficient Is Not 1

What am I supposed to learn?

After studying this section, you should be able to:

1. Factor trinomials by trial and error. ▶

2. Factor trinomials by grouping. ▶

The False Mirror (*Le Faux Miroir*) (1928), René Magritte. © 2011 MoMA/Herscovici/ARS.

The special significance of the number 1 is reflected in our language. "One," "an," and "a" mean the same thing. The words "unit," "unity," "union," "unique," and "universal" are derived from the Latin word for "one." For the ancient Greeks, 1 was the indivisible unit from which all other numbers arose.

The Greeks' philosophy of 1 applies to our work in this section. Factoring trinomials whose leading coefficient is 1 is the basic technique from which other methods of factoring $ax^2 + bx + c$, where the leading coefficient a is not equal to 1, follow.

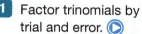

1. Factor trinomials by trial and error. ▶

Factoring by the Trial-and-Error Method

How do we factor a trinomial such as $3x^2 - 20x + 28$? Notice that the leading coefficient is 3. We must find two binomials whose product is $3x^2 - 20x + 28$. The product of the First terms must be $3x^2$:

$$(3x\quad)(x\quad).$$

From this point on, the factoring strategy is exactly the same as the one we use to factor trinomials whose leading coefficient is 1.

A Strategy for Factoring $ax^2 + bx + c$

Assume, for the moment, that there is no common factor other than 1.

1. Find two First terms whose product is ax^2:

$$(\square x + \quad)(\square x + \quad) = ax^2 + bx + c.$$

2. Find two Last terms whose product is c:

$$(\square x + \square)(\square x + \square) = ax^2 + bx + c.$$

3. By trial and error, perform steps 1 and 2 until the sum of the Outside product and the Inside product is bx:

$$(\square x + \square)(\square x + \square) = ax^2 + bx + c.$$
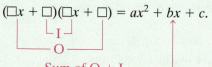

If no such combinations exist, the polynomial is prime.

Great Question!

Should I feel discouraged if it takes me a while to get the correct factorization?

The *error* part of the factoring strategy plays an important role in the process. If you do not get the correct factorization the first time, this is not a bad thing. This error is often helpful in leading you to the correct factorization.

| EXAMPLE 1 | Factoring a Trinomial Whose Leading Coefficient Is Not 1 |

Factor: $3x^2 - 20x + 28$.

Solution

Step 1. Find two First terms whose product is $3x^2$.

$$3x^2 - 20x + 28 = (3x\quad)(x\quad)$$

Step 2. Find two Last terms whose product is 28. The number 28 has pairs of factors that are either both positive or both negative. Because the middle term, $-20x$, is negative, both factors must be negative. The negative factorizations of 28 are $-1(-28)$, $-2(-14)$, and $-4(-7)$.

Step 3. Try various combinations of these factors. The correct factorization of $3x^2 - 20x + 28$ is the one in which the sum of the Outside and Inside products is equal to $-20x$. Here is a list of the possible factorizations:

Possible Factorizations of $3x^2 - 20x + 28$	Sum of Outside and Inside Products (Should Equal $-20x$)	
$(3x - 1)(x - 28)$	$-84x - x = -85x$	
$(3x - 28)(x - 1)$	$-3x - 28x = -31x$	
$(3x - 2)(x - 14)$	$-42x - 2x = -44x$	
$(3x - 14)(x - 2)$	$-6x - 14x = -20x$	This is the required middle term.
$(3x - 4)(x - 7)$	$-21x - 4x = -25x$	
$(3x - 7)(x - 4)$	$-12x - 7x = -19x$	

Thus,

$$3x^2 - 20x + 28 = (3x - 14)(x - 2) \quad \text{or} \quad (x - 2)(3x - 14).$$

Show that this factorization is correct by multiplying the factors using the FOIL method. You should obtain the original trinomial. ■

Great Question!

When factoring trinomials, must I list every possible factorization before getting the correct one?

With practice, you will find that it is not necessary to list every possible factorization of the trinomial. As you practice factoring, you will be able to narrow down the list of possible factors to just a few. When it comes to factoring, practice makes perfect.

 **CHECK POINT 1**   Factor: $5x^2 - 14x + 8$.

| EXAMPLE 2 | Factoring a Trinomial Whose Leading Coefficient Is Not 1 |

Factor: $8x^2 - 10x - 3$.

Solution

Step 1. Find two First terms whose product is $8x^2$.

$$8x^2 - 10x - 3 \overset{?}{=} (8x\quad)(x\quad)$$
$$8x^2 - 10x - 3 \overset{?}{=} (4x\quad)(2x\quad)$$

Step 2. Find two Last terms whose product is -3. The possible factorizations are $1(-3)$ and $-1(3)$.

Step 3. Try various combinations of these factors. The correct factorization of $8x^2 - 10x - 3$ is the one in which the sum of the Outside and Inside products is equal to $-10x$. Here is a list of the possible factorizations:

These four factorizations use $(8x\ \ )(x\ \ )$ with $1(-3)$ and $-1(3)$ as factorizations of -3.

These four factorizations use $(4x\ \ )(2x\ \ )$ with $1(-3)$ and $-1(3)$ as factorizations of -3.

Possible Factorizations of $8x^2 - 10x - 3$	Sum of Outside and Inside Products (Should Equal $-10x$)
$(8x + 1)(x - 3)$	$-24x + x = -23x$
$(8x - 3)(x + 1)$	$8x - 3x = 5x$
$(8x - 1)(x + 3)$	$24x - x = 23x$
$(8x + 3)(x - 1)$	$-8x + 3x = -5x$
$(4x + 1)(2x - 3)$	$-12x + 2x = -10x$
$(4x - 3)(2x + 1)$	$4x - 6x = -2x$
$(4x - 1)(2x + 3)$	$12x - 2x = 10x$
$(4x + 3)(2x - 1)$	$-4x + 6x = 2x$

This is the required middle term.

Thus,

$$8x^2 - 10x - 3 = (4x + 1)(2x - 3) \quad \text{or} \quad (2x - 3)(4x + 1).$$

Use FOIL multiplication to check either of these factorizations. ∎

✓ **CHECK POINT 2** Factor: $6x^2 + 19x - 7$.

EXAMPLE 3 Factoring a Trinomial in Two Variables

Factor: $2x^2 - 7xy + 3y^2$.

Solution

Step 1. Find two First terms whose product is $2x^2$.

$$2x^2 - 7xy + 3y^2 = (2x\ \ )(x\ \ )$$

Step 2. Find two Last terms whose product is $3y^2$. The possible factorizations are $(y)(3y)$ and $(-y)(-3y)$.

Step 3. Try various combinations of these factors. The correct factorization of $2x^2 - 7xy + 3y^2$ is the one in which the sum of the Outside and Inside products is equal to $-7xy$. Here is a list of possible factorizations:

Possible Factorizations of $2x^2 - 7xy + 3y^2$	Sum of Outside and Inside Products (Should Equal $-7xy$)
$(2x + 3y)(x + y)$	$2xy + 3xy = 5xy$
$(2x + y)(x + 3y)$	$6xy + xy = 7xy$
$(2x - 3y)(x - y)$	$-2xy - 3xy = -5xy$
$(2x - y)(x - 3y)$	$-6xy - xy = -7xy$

This is the required middle term.

Thus,

$$2x^2 - 7xy + 3y^2 = (2x - y)(x - 3y) \quad \text{or} \quad (x - 3y)(2x - y).$$

Use FOIL multiplication to check either of these factorizations. ∎

✓ **CHECK POINT 3** Factor: $3x^2 - 13xy + 4y^2$.

Great Question!

I zone out reading your long lists of possible factorizations. Are there any rules for shortening these lists?

Here are some suggestions for reducing the list of possible factorizations for $ax^2 + bx + c$:

1. If b is relatively small, avoid the larger factors of a.

2. If c is positive, the signs in both binomial factors must match the sign of b.

3. If the trinomial has no common factor, no binomial factor can have a common factor.

4. Reversing the signs in the binomial factors changes the sign of bx, the middle term.

② Factor trinomials by grouping.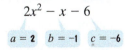

Factoring by the Grouping Method

A second method for factoring $ax^2 + bx + c, a \neq 0$, is called the **grouping method**. The method involves both trial and error, as well as grouping. The trial and error in factoring $ax^2 + bx + c$ depends on finding two numbers, p and q, for which $p + q = b$. Then we factor $ax^2 + px + qx + c$ using grouping.

Let's see how this works by looking at our factorization in Example 2:

$$8x^2 - 10x - 3 = (2x - 3)(4x + 1).$$

If we multiply using FOIL on the right, we obtain:

$$(2x - 3)(4x + 1) = 8x^2 + 2x - 12x - 3.$$

In this case, the desired numbers, p and q, are $p = 2$ and $q = -12$. Compare these numbers to ac and b in the given polynomial:

$$\boxed{a = 8} \quad \boxed{b = -10} \quad \boxed{c = -3}$$

$$8x^2 - 10x - 3.$$

$$\boxed{ac = 8(-3) = -24}$$

Can you see that p and q, 2 and -12, are factors of ac, or -24? Furthermore, p and q have a sum of b, namely -10. By expressing the middle term, $-10x$, in terms of p and q, we can factor by grouping as follows:

$$
\begin{aligned}
8x^2 - 10x - 3 & \\
= 8x^2 + (2x - 12x) - 3 & \qquad \text{Rewrite } -10x \text{ as } 2x - 12x. \\
= (8x^2 + 2x) + (-12x - 3) & \qquad \text{Group terms.} \\
= 2x(4x + 1) - 3(4x + 1) & \qquad \text{Factor from each group.} \\
= (4x + 1)(2x - 3) & \qquad \text{Factor out the common binomial factor.}
\end{aligned}
$$

As we obtained in Example 2,

$$8x^2 - 10x - 3 = (4x + 1)(2x - 3).$$

Generalizing from this example, here's how to factor a trinomial by grouping:

> ### Factoring $ax^2 + bx + c$ Using Grouping ($a \neq 1$)
>
> 1. Multiply the leading coefficient, a, and the constant, c.
> 2. Find factors of ac whose sum is b.
> 3. Rewrite the middle term, bx, as a sum or difference using the factors from step 2.
> 4. Factor by grouping.

EXAMPLE 4   Factoring by Grouping

Factor by grouping: $2x^2 - x - 6$.

Solution   The trinomial is of the form $ax^2 + bx + c$.

$$2x^2 - x - 6$$

$$\boxed{a = 2} \quad \boxed{b = -1} \quad \boxed{c = -6}$$

Step 1. Multiply the leading coefficient, a, and the constant, c. Using $a = 2$ and $c = -6$,

$$ac = 2(-6) = -12.$$

Step 2. Find factors of ac whose sum is b. We want factors of -12 whose sum is b, or -1. The factors of -12 whose sum is -1 are -4 and 3.

Discover for Yourself

In step 2, we discovered that the desired numbers were -4 and 3, and in step 3 we wrote $-x$ as $-4x + 3x$. What happens if we write $-x$ as $3x - 4x$? Use factoring by grouping on

$$2x^2 - x - 6$$
$$= 2x^2 + 3x - 4x - 6.$$

Describe what happens.

Step 3. Rewrite the middle term, $-x$, as a sum or difference using the factors from step 2, -4 and 3.

$$2x^2 - x - 6 = 2x^2 - 4x + 3x - 6$$

Step 4. Factor by grouping.

$$= (2x^2 - 4x) + (3x - 6) \quad \text{Group terms.}$$
$$= 2x(x - 2) + 3(x - 2) \quad \text{Factor from each group.}$$
$$= (x - 2)(2x + 3) \quad \text{Factor out the common binomial factor.}$$

Thus,

$$2x^2 - x - 6 = (x - 2)(2x + 3) \quad \text{or} \quad (2x + 3)(x - 2). \;\blacksquare$$

✓ **CHECK POINT 4**   Factor by grouping: $3x^2 - x - 10$.

EXAMPLE 5   Factoring by Grouping

Factor by grouping: $8x^2 - 22x + 5$.

Solution   The trinomial is of the form $ax^2 + bx + c$.

$$8x^2 - 22x + 5$$

$$a = 8 \quad b = -22 \quad c = 5$$

Step 1. Multiply the leading coefficient, *a*, and the constant, *c*. Using $a = 8$ and $c = 5$, $ac = 8 \cdot 5 = 40$.

Step 2. Find factors of *ac* whose sum is *b*. We want factors of 40 whose sum is b, or -22. The factors of 40 whose sum is -22 are -2 and -20.

Step 3. Rewrite the middle term, $-22x$, as a sum or difference using the factors from step 2, -2 and -20.

$$8x^2 - 22x + 5 = 8x^2 - 2x - 20x + 5$$

Step 4. Factor by grouping.

$$= (8x^2 - 2x) + (-20x + 5) \quad \text{Group terms.}$$
$$= 2x(4x - 1) - 5(4x - 1) \quad \text{Factor from each group.}$$
$$= (4x - 1)(2x - 5) \quad \text{Factor out the common binomial factor.}$$

Thus,

$$8x^2 - 22x + 5 = (4x - 1)(2x - 5) \quad \text{or} \quad (2x - 5)(4x - 1). \;\blacksquare$$

✓ **CHECK POINT 5**   Factor by grouping: $8x^2 - 10x + 3$.

Factoring Completely

Always begin the process of factoring a polynomial by looking for a greatest common factor other than 1. If there is one, **factor out the GCF first**. After doing this, you should attempt to factor the remaining polynomial, using one of the methods presented in this section, if appropriate.

EXAMPLE 6   Factoring Completely

Factor completely: $15y^4 + 26y^3 + 7y^2$.

Solution   We will first factor out a common monomial factor from the polynomial and then factor the resulting trinomial by the methods of this section. The GCF of the three terms is y^2.

$$15y^4 + 26y^3 + 7y^2 = y^2(15y^2 + 26y + 7) \quad \text{Factor out the GCF.}$$
$$= y^2(5y + 7)(3y + 1) \quad \text{Factor } 15y^2 + 26y + 7 \text{ using trial and error or grouping.}$$

Thus,

$$15y^4 + 26y^3 + 7y^2 = y^2(5y + 7)(3y + 1) \quad \text{or} \quad y^2(3y + 1)(5y + 7).$$

> Be sure to include the GCF, y^2, in the factorization.

☑ **CHECK POINT 6**   Factor completely: $5y^4 + 13y^3 + 6y^2$.

Great Question!

As I approach the next Exercise Set (it looks long!), can you give me any other hints for shortening those oppressive lists of possible factorizations?

Here is an observation that sometimes helps narrow down the list of possible factorizations. If a polynomial does not have a common factor other than 1 or if you have factored out the GCF, there will be no common factor within any of its binomial factors. Here is an example:

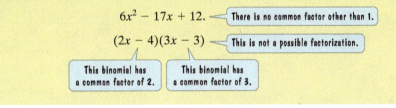

$6x^2 - 17x + 12.$ ── There is no common factor other than 1.

$(2x - 4)(3x - 3)$ ── This is not a possible factorization.

This binomial has a common factor of 2.   This binomial has a common factor of 3.

CONCEPT AND VOCABULARY CHECK

Fill in each blank so that the resulting statement is true.

1. We begin the process of factoring a polynomial by first factoring out the _____, assuming that there is a common factor other than 1.

2. $8x^2 - 10x - 3 = (4x + 1)(2x \underline{\quad})$

3. $12x^2 - x - 20 = (4x + 5)(3x \underline{\quad})$

4. $2x^2 - 5x + 3 = (x - 1)(\underline{\quad\quad})$

5. $6x^2 + 17x + 12 = (2x + 3)(\underline{\quad})$

6. $5x^2 - 8xy - 4y^2 = (5x + 2y)(\underline{\quad})$

6.3 EXERCISE SET   ▶ MyMathLab®

Practice Exercises

In Exercises 1–58, use the method of your choice to factor each trinomial, or state that the trinomial is prime. Check each factorization using FOIL multiplication.

1. $2x^2 + 5x + 3$
2. $3x^2 + 5x + 2$
3. $3x^2 + 13x + 4$
4. $2x^2 + 7x + 3$
5. $2x^2 + 11x + 12$
6. $2x^2 + 19x + 35$
7. $5y^2 - 16y + 3$
8. $5y^2 - 17y + 6$
9. $3y^2 + y - 4$
10. $3y^2 - y - 4$
11. $3x^2 + 13x - 10$
12. $3x^2 + 14x - 5$
13. $3x^2 - 22x + 7$
14. $3x^2 - 10x + 7$
15. $5y^2 - 16y + 3$
16. $5y^2 - 8y + 3$
17. $3x^2 - 17x + 10$
18. $3x^2 - 25x - 28$
19. $6w^2 - 11w + 4$
20. $6w^2 - 17w + 12$

21. $8x^2 + 33x + 4$

22. $7x^2 + 43x + 6$

23. $5x^2 + 33x - 14$

24. $3x^2 + 22x - 16$

25. $14y^2 + 15y - 9$

26. $6y^2 + 7y - 24$

27. $6x^2 - 7x + 3$

28. $9x^2 + 3x + 2$

29. $25z^2 - 30z + 9$

30. $9z^2 + 12z + 4$

31. $15y^2 - y - 2$

32. $15y^2 + 13y - 2$

33. $5x^2 + 2x + 9$

34. $3x^2 - 5x + 1$

35. $10y^2 + 43y - 9$

36. $16y^2 - 46y + 15$

37. $8x^2 - 2x - 1$

38. $8x^2 - 22x + 5$

39. $9y^2 - 9y + 2$

40. $9y^2 + 5y - 4$

41. $20x^2 + 27x - 8$

42. $15x^2 - 19x + 6$

43. $2x^2 + 3xy + y^2$

44. $3x^2 + 4xy + y^2$

45. $3x^2 + 5xy + 2y^2$

46. $3x^2 + 11xy + 6y^2$

47. $2x^2 - 9xy + 9y^2$

48. $3x^2 + 5xy - 2y^2$

49. $6x^2 - 5xy - 6y^2$

50. $6x^2 - 7xy - 5y^2$

51. $15x^2 + 11xy - 14y^2$

52. $15x^2 - 31xy + 10y^2$

53. $2a^2 + 7ab + 5b^2$

54. $2a^2 + 5ab + 2b^2$

55. $15a^2 - ab - 6b^2$

56. $3a^2 - ab - 14b^2$

57. $12x^2 - 25xy + 12y^2$

58. $12x^2 + 7xy - 12y^2$

In Exercises 59–88, factor completely.

59. $4x^2 + 26x + 30$

60. $4x^2 - 18x - 10$

61. $9x^2 - 6x - 24$

62. $12x^2 - 33x + 21$

63. $4y^2 + 2y - 30$

64. $36y^2 + 6y - 12$

65. $9y^2 + 33y - 60$

66. $16y^2 - 16y - 12$

67. $3x^3 + 4x^2 + x$

68. $3x^3 + 14x^2 + 8x$

69. $2x^3 - 3x^2 - 5x$

70. $6x^3 + 4x^2 - 10x$

71. $9y^3 - 39y^2 + 12y$

72. $10y^3 + 12y^2 + 2y$

73. $60z^3 + 40z^2 + 5z$

74. $80z^3 + 80z^2 - 60z$

75. $15x^4 - 39x^3 + 18x^2$

76. $24x^4 + 10x^3 - 4x^2$

77. $10x^5 - 17x^4 + 3x^3$

78. $15x^5 - 2x^4 - x^3$

79. $6x^2 - 3xy - 18y^2$

80. $4x^2 + 14xy + 10y^2$

81. $12x^2 + 10xy - 8y^2$

82. $24x^2 + 3xy - 27y^2$

83. $8x^2y + 34xy - 84y$

84. $6x^2y - 2xy - 60y$

85. $12a^2b - 46ab^2 + 14b^3$

86. $12a^2b - 34ab^2 + 14b^3$

87. $-32x^2y^4 + 20xy^4 + 12y^4$

88. $-10x^2y^4 + 14xy^4 + 12y^4$

Practice PLUS

In Exercises 89–90, factor completely.

89. $30(y + 1)x^2 + 10(y + 1)x - 20(y + 1)$

90. $6(y + 1)x^2 + 33(y + 1)x + 15(y + 1)$

91. a. Factor $2x^2 - 5x - 3$.

    b. Use the factorization in part (a) to factor
$$2(y + 1)^2 - 5(y + 1) - 3.$$
Then simplify each factor.

92. a. Factor $3x^2 + 5x - 2$.

    b. Use the factorization in part (a) to factor
$$3(y + 1)^2 + 5(y + 1) - 2.$$
Then simplify each factor.

93. Divide $3x^3 - 11x^2 + 12x - 4$ by $x - 2$. Use the quotient to factor $3x^3 - 11x^2 + 12x - 4$ completely.

94. Divide $2x^3 + x^2 - 13x + 6$ by $x - 2$. Use the quotient to factor $2x^3 + x^2 - 13x + 6$ completely.

Application Exercises

It is possible to construct geometric models for factorizations so that you can see the factoring. This idea is developed in Exercises 95–96.

95. Consider the following figure.

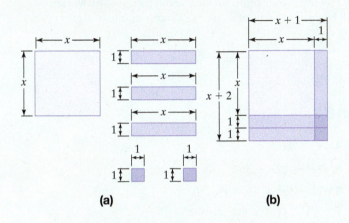

(a)　　　　　　　(b)

a. Write a trinomial that expresses the sum of the areas of the six rectangular pieces shown in figure (a).

b. Express the area of the large rectangle in figure (b) as the product of two binomials.

c. Are the pieces in figures (a) and (b) the same? Set the expressions that you wrote in parts (a) and (b) equal to each other. What factorization is illustrated?

96. Copy the figure and cut out the six pieces. Use the pieces to create a geometric model for the factorization
$$2x^2 + 3x + 1 = (2x + 1)(x + 1)$$
by forming a large rectangle using all the pieces.

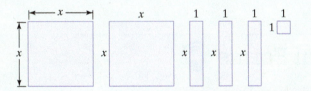

Explaining the Concepts

97. Explain how to factor $2x^2 - x - 1$.

98. Why is it a good idea to factor out the GCF first and then use other methods of factoring? Use $3x^2 - 18x + 15$ as an example. Discuss what happens if one first uses trial and error to factor as two binomials rather than first factoring out the GCF.

99. In factoring $3x^2 - 10x - 8$, a student lists $(3x - 2)(x + 4)$ as a possible factorization. Use FOIL multiplication to determine if this factorization is correct. If it is not correct, describe how the correct factorization can quickly be obtained using these factors.

100. Explain why $2x - 10$ cannot be one of the factors in the correct factorization of $6x^2 - 19x + 10$.

Critical Thinking Exercises

Make Sense? *In Exercises 101–104, determine whether each statement makes sense or does not make sense, and explain your reasoning.*

101. I'm often able to use an incorrect factorization to lead me to the correct factorization.

102. First factoring out the GCF makes it easier for me to determine how to factor the remaining factor, assuming it is not prime.

103. I'm working with a polynomial that has a GCF other than 1, but then it doesn't factor further, so the polynomial that I'm working with is prime.

104. My graphing calculator showed the same graphs for $y_1 = 4x^2 - 20x + 24$ and $y_2 = 4(x^2 - 5x + 6)$, so I can conclude that the complete factorization of $4x^2 - 20x + 24$ is $4(x^2 - 5x + 6)$.

In Exercises 105–108, determine whether each statement is true or false. If the statement is false, make the necessary change(s) to produce a true statement.

105. Once the GCF is factored from $18y^2 - 6y + 6$, the remaining trinomial factor is prime.

106. One factor of $12x^2 - 13x + 3$ is $4x + 3$.

107. One factor of $4y^2 - 11y - 3$ is $y + 3$.

108. The trinomial $3x^2 + 2x + 1$ has relatively small coefficients and therefore can be factored.

In Exercises 109–110, find all integers b so that the trinomial can be factored.

109. $3x^2 + bx + 2$

110. $2x^2 + bx + 3$

111. Factor: $3x^{10} - 4x^5 - 15$.

112. Factor: $2x^{2n} - 7x^n - 4$.

Review Exercises

113. Solve the system:
$$\begin{cases} 4x - y = 105 \\ x + 7y = -10. \end{cases} \text{(Section 4.3, Example 3)}$$

114. Write 0.00086 in scientific notation. (Section 5.7, Example 8)

115. Solve: $8x - \dfrac{x}{6} = \dfrac{1}{6} - 8$. (Section 2.3, Example 4)

Preview Exercises

Exercises 116–118 will help you prepare for the material covered in the next section. In each exercise, perform the indicated operation.

116. $(9x + 10)(9x - 10)$

117. $(4x + 5y)^2$

118. $(x + 2)(x^2 - 2x + 4)$

Achieving Success

In the next chapter, you will see how mathematical models involving quotients of polynomials describe environmental issues. Factoring is an essential skill for working with such models. **Success in mathematics cannot be achieved without a complete understanding of factoring.** Be sure to work all the exercises in the Mid-Chapter Check Point so that you can apply each of the factoring techniques discussed in the first half of the chapter. The more deeply you force your brain to think about factoring by working many exercises, the better will be your chances of achieving success in future algebra courses.

MID-CHAPTER CHECK POINT   Section 6.1–Section 6.3

✓ **What You Know:** We learned to factor out a polynomial's greatest common factor and to use grouping to factor polynomials with four terms. We factored polynomials with three terms, beginning with trinomials with leading coefficient 1 and moving on to $ax^2 + bx + c$, with $a \neq 1$. We saw that the factoring process should begin by looking for a GCF other than 1 and, if there is one, factoring it out first.

In Exercises 1–13, factor completely, or state that the polynomial is prime.

1. $x^5 + x^4$

2. $x^2 + 7x - 18$

3. $x^2y^3 - x^2y^2 + x^2y$

4. $x^2 - 2x + 4$

5. $7x^2 - 22x + 3$

6. $x^3 + 5x^2 + 3x + 15$

7. $2x^3 - 11x^2 + 5x$

8. $xy - 7x - 4y + 28$

9. $x^2 - 17xy + 30y^2$

10. $25x^2 - 25x - 14$

11. $16x^2 - 70x + 24$

12. $3x^2 + 10xy + 7y^2$

13. $-6x^3 + 8x^2 + 30x$

SECTION

6.4

Factoring Special Forms ▶

What am I supposed to learn?

After studying this section, you should be able to:

1. Factor the difference of two squares. ▶

2. Factor perfect square trinomials. ▶

3. Factor the sum or difference of two cubes. ▶

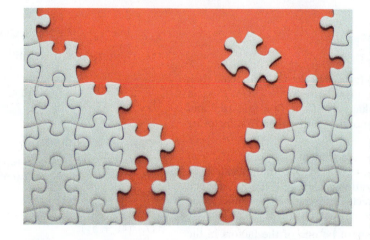

Do you enjoy solving puzzles? The process is a natural way to develop problem-solving skills that are important to every area of our lives. Engaging in problem solving for sheer pleasure releases chemicals in the brain that enhance our feeling of well-being. Perhaps this is why puzzles date back 12,000 years.

In this section, we develop factoring techniques by reversing the formulas for special products discussed in Chapter 5. These factorizations can be visualized by fitting pieces of a puzzle together to form rectangles.

1 Factor the difference of two squares. ▶

Factoring the Difference of Two Squares

A method for factoring the difference of two squares is obtained by reversing the special product for the sum and difference of two terms.

> ### The Difference of Two Squares
>
> If A and B are real numbers, variables, or algebraic expressions, then
>
> $$A^2 - B^2 = (A + B)(A - B).$$
>
> In words: The difference of the squares of two terms factors as the product of a sum and a difference of those terms.

EXAMPLE 1 Factoring the Difference of Two Squares

Factor:

a. $x^2 - 4$ **b.** $81x^2 - 49$.

Solution We must express each term as the square of some monomial. Then we use the formula for factoring $A^2 - B^2$.

a. $x^2 - 4 = x^2 - 2^2 = (x + 2)(x - 2)$

$$A^2 - B^2 = (A + B)(A - B)$$

b. $81x^2 - 49 = (9x)^2 - 7^2 = (9x + 7)(9x - 7)$ ∎

In order to apply the factoring formula for $A^2 - B^2$, each term must be the square of an integer or a polynomial.

- A number that is the square of an integer is called a **perfect square**. For example, 100 is a perfect square because $100 = 10^2$.
- Any monomial involving a perfect-square coefficient and variables to even powers is a perfect square. For example, $16x^{10}$ is a perfect square because $16x^{10} = (4x^5)^2$.

Great Question!

> **You mentioned that because 100 = 10^2, 100 is a perfect square. What are some other perfect squares that I should recognize?**
>
> It's helpful to identify perfect squares. Here are 16 perfect squares, printed in boldface.
>
> | | | | |
> |---|---|---|---|
> | **1** $= 1^2$ | **25** $= 5^2$ | **81** $= 9^2$ | **169** $= 13^2$ |
> | **4** $= 2^2$ | **36** $= 6^2$ | **100** $= 10^2$ | **196** $= 14^2$ |
> | **9** $= 3^2$ | **49** $= 7^2$ | **121** $= 11^2$ | **225** $= 15^2$ |
> | **16** $= 4^2$ | **64** $= 8^2$ | **144** $= 12^2$ | **256** $= 16^2$ |

☑ **CHECK POINT 1** Factor:

a. $x^2 - 81$ **b.** $36x^2 - 25$.

Be careful when determining whether or not to apply the factoring formula for the difference of two squares.

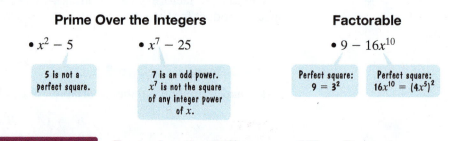

Prime Over the Integers

• $x^2 - 5$

• $x^7 - 25$

Factorable

• $9 - 16x^{10}$

5 is not a perfect square.

7 is an odd power. x^7 is not the square of any integer power of x.

Perfect square: $9 = 3^2$

Perfect square: $16x^{10} = (4x^5)^2$

EXAMPLE 2 Factoring the Difference of Two Squares

Factor:

a. $9 - 16x^{10}$ **b.** $25x^2 - 4y^2$.

Solution Begin by expressing each term as the square of some monomial. Then use the formula for factoring $A^2 - B^2$.

a. $9 - 16x^{10} = 3^2 - (4x^5)^2 = (3 + 4x^5)(3 - 4x^5)$

$$A^2 - B^2 = (A + B)(A - B)$$

b. $25x^2 - 4y^2 = (5x)^2 - (2y)^2 = (5x + 2y)(5x - 2y)$ ∎

☑ **CHECK POINT 2** Factor:

a. $25 - 4x^{10}$ **b.** $100x^2 - 9y^2$.

When factoring, always check first for common factors. If there are common factors other than 1, factor out the GCF and then factor the resulting polynomial.

EXAMPLE 3 Factoring Out the GCF and Then Factoring the Difference of Two Squares

Factor:

a. $12x^3 - 3x$ **b.** $80 - 125x^2$.

Solution

a. $12x^3 - 3x = 3x(4x^2 - 1) = 3x[(2x)^2 - 1^2] = 3x(2x + 1)(2x - 1)$

Factor out the GCF. $A^2 - B^2 = (A + B)(A - B)$

b. $80 - 125x^2 = 5(16 - 25x^2) = 5[4^2 - (5x)^2] = 5(4 + 5x)(4 - 5x)$ ∎

Great Question!

If I rewrite $80 - 125x^2$ as $-125x^2 + 80$, will I still get the same factorization?

Yes. Because the coefficient of $-125x^2$ is negative, factor out -5, the negative of the GCF.

$$-125x^2 + 80 = -5(25x^2 - 16) = -5[(5x)^2 - 4^2] = -5(5x + 4)(5x - 4)$$

Factor out the negative of the GCF.

To show that this is the same factorization as the one in the solution to part (b), factor -1 from $(5x - 4)$.

✓ CHECK POINT 3   Factor:

a. $18x^3 - 2x$ b. $72 - 18x^2$.

We have seen that a polynomial is factored completely when it is written as the product of prime polynomials. To be sure that you have factored completely, check to see whether any factors with more than one term in the factored polynomial can be factored further. If so, continue factoring.

EXAMPLE 4   A Repeated Factorization

Factor completely: $x^4 - 81$.

Solution

$$\begin{aligned} x^4 - 81 &= (x^2)^2 - 9^2 && \text{Express as the difference of two squares.} \\ &= (x^2 + 9)(x^2 - 9) && \text{The factors are the sum and the difference of} \\ &&& \text{the expressions being squared.} \\ &= (x^2 + 9)(x^2 - 3^2) && \text{The factor } x^2 - 9 \text{ is the difference of two} \\ &&& \text{squares and can be factored.} \\ &= (x^2 + 9)(x + 3)(x - 3) && \text{The factors of } x^2 - 9 \text{ are the sum and the} \\ &&& \text{difference of the expressions being squared.} \quad\blacksquare \end{aligned}$$

Great Question!

Why isn't factoring $x^4 - 81$ as $(x^2 + 9)(x^2 - 9)$ a complete factorization?

The second factor, $x^2 - 9$, is itself a difference of two squares and can be factored.

Are you tempted to factor $x^2 + 9$ further, the sum of two squares, in Example 4? Resist the temptation! **The sum of two squares, $A^2 + B^2$, with no common factor other than 1 is a prime polynomial over the integers.**

✓ CHECK POINT 4   Factor completely: $81x^4 - 16$.

In Examples 1–4, we used the formula for factoring the difference of two squares. Although we obtained the formula by reversing the special product for the sum and difference of two terms, it can also be obtained geometrically.

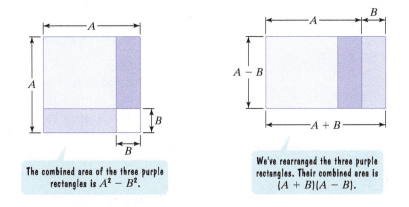

The combined area of the three purple rectangles is $A^2 - B^2$.

We've rearranged the three purple rectangles. Their combined area is $(A + B)(A - B)$.

Because the three purple rectangles make up the same combined area in both figures,

$$A^2 - B^2 = (A + B)(A - B).$$

2 Factor perfect square trinomials.

Factoring Perfect Square Trinomials

Our next factoring technique is obtained by reversing the special products for squaring binomials. The trinomials that are factored using this technique are called **perfect square trinomials**.

Factoring Perfect Square Trinomials

Let A and B be real numbers, variables, or algebraic expressions.

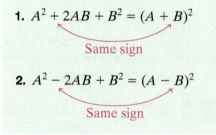

1. $A^2 + 2AB + B^2 = (A + B)^2$

Same sign

2. $A^2 - 2AB + B^2 = (A - B)^2$

Same sign

The two formulas in the box show that perfect square trinomials, $A^2 + 2AB + B^2$ and $A^2 - 2AB + B^2$, come in two forms: one in which the coefficient of the middle term is positive and one in which the coefficient of the middle term is negative. Here's how to recognize a perfect square trinomial:

1. The first and last terms are squares of monomials or integers.

2. The middle term is twice the product of the expressions being squared in the first and last terms.

EXAMPLE 5 Factoring Perfect Square Trinomials

Factor:

a. $x^2 + 6x + 9$ **b.** $x^2 - 16x + 64$ **c.** $25x^2 - 60x + 36$.

Solution

a. $x^2 + 6x + 9 = x^2 + 2 \cdot x \cdot 3 + 3^2 = (x + 3)^2$

             $A^2 + 2AB + B^2 = (A + B)^2$

The middle term has a positive sign.

b. $x^2 - 16x + 64 = x^2 - 2 \cdot x \cdot 8 + 8^2 = (x - 8)^2$

             $A^2 - 2AB + B^2 = (A - B)^2$

The middle term has a negative sign.

c. We suspect that $25x^2 - 60x + 36$ is a perfect square trinomial because $25x^2 = (5x)^2$ and $36 = 6^2$. The middle term can be expressed as twice the product of $5x$ and 6.

$$25x^2 - 60x + 36 = (5x)^2 - 2 \cdot 5x \cdot 6 + 6^2 = (5x - 6)^2$$

             $A^2 - 2AB + B^2 = (A - B)^2$ ∎

✓ CHECK POINT 5 Factor:

a. $x^2 + 14x + 49$ **b.** $x^2 - 6x + 9$

c. $16x^2 - 56x + 49$.

EXAMPLE 6 Factoring a Perfect Square Trinomial in Two Variables

Factor: $16x^2 + 40xy + 25y^2$.

Solution   Observe that $16x^2 = (4x)^2$, $25y^2 = (5y)^2$, and $40xy$ is twice the product of $4x$ and $5y$. Thus, we have a perfect square trinomial.

$$16x^2 + 40xy + 25y^2 = (4x)^2 + 2 \cdot 4x \cdot 5y + (5y)^2 = (4x + 5y)^2$$

$$\underbrace{A^2} \quad + \quad \underbrace{2AB} \quad + \quad \underbrace{B^2} \quad = \quad \underbrace{(A + B)^2}$$

✓ **CHECK POINT 6**   Factor: $4x^2 + 12xy + 9y^2$.

In Examples 5 and 6, we factored perfect square trinomials using $A^2 + 2AB + B^2 = (A + B)^2$. Although we obtained the formula by reversing the special product for $(A + B)^2$, it can also be obtained geometrically.

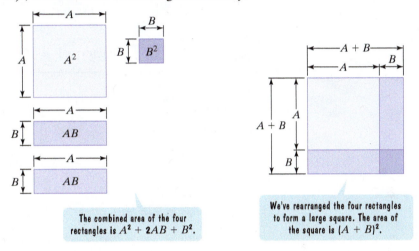

The combined area of the four rectangles is $A^2 + 2AB + B^2$.

We've rearranged the four rectangles to form a large square. The area of the square is $(A + B)^2$.

Because the four rectangles make up the same combined area in both figures,

$$A^2 + 2AB + B^2 = (A + B)^2.$$

3 Factor the sum or difference of two cubes. ▶

Factoring the Sum or Difference of Two Cubes

We can use the following formulas to factor the sum or the difference of two cubes:

Factoring the Sum or Difference of Two Cubes

1. Factoring the Sum of Two Cubes

$$A^3 + B^3 = (A + B)(A^2 - AB + B^2)$$

    Same signs   Opposite signs

2. Factoring the Difference of Two Cubes

$$A^3 - B^3 = (A - B)(A^2 + AB + B^2)$$

    Same signs   Opposite signs

EXAMPLE 7   Factoring the Sum of Two Cubes

Factor: $x^3 + 8$.

Solution   We must express each term as the cube of some monomial. Then we use the formula for factoring $A^3 + B^3$.

$$x^3 + 8 = x^3 + 2^3 = (x + 2)(x^2 - x \cdot 2 + 2^2) = (x + 2)(x^2 - 2x + 4)$$

$$\underbrace{A^3} \;+\; \underbrace{B^3} \;=\; \underbrace{(A + B)} \; \underbrace{(A^2 - AB + B^2)}$$

Great Question!

What are some cubes that I should be able to identify?

When factoring the sum or difference of cubes, it is helpful to recognize the following cubes:

$$1 = 1^3$$
$$8 = 2^3$$
$$27 = 3^3$$
$$64 = 4^3$$
$$125 = 5^3$$
$$216 = 6^3$$
$$1000 = 10^3.$$

✓ **CHECK POINT 7** Factor: $x^3 + 27$.

EXAMPLE 8 Factoring the Difference of Two Cubes

Factor: $27 - y^3$.

Solution Express each term as the cube of some monomial. Then use the formula for factoring $A^3 - B^3$.

$$27 - y^3 = 3^3 - y^3 = (3 - y)(3^2 + 3y + y^2) = (3 - y)(9 + 3y + y^2)$$

$$A^3 - B^3 = (A - B)(A^2 + AB + B^2)$$

■

✓ **CHECK POINT 8** Factor: $1 - y^3$.

EXAMPLE 9 Factoring the Sum of Two Cubes

Factor: $64x^3 + 125$.

Solution Express each term as the cube of some monomial. Then use the formula for factoring $A^3 + B^3$.

$$64x^3 + 125 = (4x)^3 + 5^3 = (4x + 5)[(4x)^2 - (4x)(5) + 5^2]$$

$$A^3 + B^3 = (A + B)(A^2 - AB + B^2)$$

$$= (4x + 5)(16x^2 - 20x + 25)$$

■

✓ **CHECK POINT 9** Factor: $125x^3 + 8$.

Great Question!

A Cube of SOAP

The formulas for factoring $A^3 + B^3$ and $A^3 - B^3$ are difficult to remember and easy to confuse. Can you help me out?

When factoring sums or differences of cubes, observe the sign patterns.

Same signs

$$A^3 + B^3 = (A + B)(A^2 - AB + B^2)$$

Opposite signs | Always positive

Same signs

$$A^3 - B^3 = (A - B)(A^2 + AB + B^2)$$

Opposite signs | Always positive

The word SOAP is a way to remember these patterns:

S   O   A   P.

Same signs | Opposite signs | Always Positive

Achieving Success

Create and maintain a civil and respectful environment on campus.

In "Civility on Community College Campuses: A Shared Responsibility" (*College Student Journal*, Spring 2014), author Alexander J. Popovics cites behaviors that are disrespectful in classrooms and should be avoided. These include:

1. Arriving late or leaving early from class.
2. Walking in and out of class while class is in session.
3. Carrying on personal conversations or in other ways interrupting classroom presentations.
4. Using language or gestures intended to offend others.
5. Sleeping in class.
6. Using beepers, cellphones, and other disruptive devices in class.

CONCEPT AND VOCABULARY CHECK

Fill in each blank so that the resulting statement is true.

1. The formula for factoring the difference of two squares is $A^2 - B^2 =$ _____.
2. A formula for factoring a perfect square trinomial is $A^2 + 2AB + B^2 =$ _____.
3. A formula for factoring a perfect square trinomial is $A^2 - 2AB + B^2 =$ _____.
4. The formula for factoring the sum of two cubes is $A^3 + B^3 =$ _____.
5. The formula for factoring the difference of two cubes is $A^3 - B^3 =$ _____.
6. $36x^2 - 49 = ($ _____ $+ 7)($ _____ $- 7)$
7. $x^2 - 12x + 36 = (x$ _____ $)^2$
8. $16x^2 - 24x + 9 = ($ _____ $- 3)^2$
9. $x^3 + 8 = (x$ _____ $)(x^2 - 2x + 4)$
10. $x^3 - 27 = (x$ _____ $)(x^2 + 3x$ _____ $)$
11. True or false: $x^2 - 8$ is the difference of two perfect squares. _____
12. True or false: $x^2 + 8x + 16$ is a perfect square trinomial. _____
13. True or false: $x^2 - 5x + 25$ is a perfect square trinomial. _____
14. True or false: $x^3 + 1000$ is the sum of two cubes. _____
15. True or false: $x^3 - 100$ is the difference of two cubes. _____

6.4 EXERCISE SET ▶ MyMathLab®

Practice Exercises

In Exercises 1–26, factor each difference of two squares.

1. $x^2 - 25$
2. $x^2 - 16$
3. $y^2 - 1$
4. $y^2 - 9$
5. $4x^2 - 9$
6. $9x^2 - 25$
7. $25 - x^2$
8. $16 - x^2$
9. $1 - 49x^2$
10. $1 - 64x^2$
11. $9 - 25y^2$
12. $16 - 49y^2$
13. $x^4 - 9$
14. $x^4 - 25$
15. $49y^4 - 16$
16. $49y^4 - 25$
17. $x^{10} - 9$

18. $x^{10} - 1$

19. $25x^2 - 16y^2$

20. $9x^2 - 25y^2$

21. $x^4 - y^{10}$

22. $x^{14} - y^4$

23. $x^4 - 16$

24. $x^4 - 1$

25. $16x^4 - 81$

26. $81x^4 - 1$

In Exercises 27–44, factor completely, or state that the polynomial is prime.

27. $2x^2 - 18$

28. $5x^2 - 45$

29. $2x^3 - 72x$

30. $2x^3 - 8x$

31. $x^2 + 36$

32. $x^2 + 4$

33. $3x^3 + 27x$

34. $3x^3 + 15x$

35. $18 - 2y^2$

36. $32 - 2y^2$

37. $3y^3 - 48y$

38. $3y^3 - 75y$

39. $18x^3 - 2x$

40. $20x^3 - 5x$

41. $-3x^2 + 75$

42. $-4x^2 + 4$

43. $-5y^3 + 20y$

44. $-54y^3 + 6y$

In Exercises 45–66, factor any perfect square trinomials, or state that the polynomial is prime.

45. $x^2 + 2x + 1$

46. $x^2 + 4x + 4$

47. $x^2 - 14x + 49$

48. $x^2 - 10x + 25$

49. $x^2 - 2x + 1$

50. $x^2 - 4x + 4$

51. $x^2 + 22x + 121$

52. $x^2 + 24x + 144$

53. $4x^2 + 4x + 1$

54. $9x^2 + 6x + 1$

55. $25y^2 - 10y + 1$

56. $64y^2 - 16y + 1$

57. $x^2 - 10x + 100$

58. $x^2 - 7x + 49$

59. $x^2 + 14xy + 49y^2$

60. $x^2 + 16xy + 64y^2$

61. $x^2 - 12xy + 36y^2$

62. $x^2 - 18xy + 81y^2$

63. $x^2 - 8xy + 64y^2$

64. $x^2 + 4xy + 16y^2$

65. $16x^2 - 40xy + 25y^2$

66. $9x^2 + 48xy + 64y^2$

In Exercises 67–78, factor completely.

67. $12x^2 - 12x + 3$

68. $18x^2 + 24x + 8$

69. $9x^3 + 6x^2 + x$

70. $25x^3 - 10x^2 + x$

71. $2y^2 - 4y + 2$

72. $2y^2 - 40y + 200$

73. $2y^3 + 28y^2 + 98y$

74. $50y^3 + 20y^2 + 2y$

75. $-6x^2 + 24x - 24$

76. $-5x^2 + 30x - 45$

77. $-16y^3 - 16y^2 - 4y$

78. $-45y^3 - 30y^2 - 5y$

In Exercises 79–96, factor using the formula for the sum or difference of two cubes.

79. $x^3 + 1$

80. $x^3 + 64$

81. $x^3 - 27$

82. $x^3 - 64$

83. $8y^3 - 1$

84. $27y^3 - 1$

85. $27x^3 + 8$

86. $125x^3 + 8$

87. $x^3y^3 - 64$

88. $x^3y^3 - 27$

89. $27y^4 + 8y$

90. $64y - y^4$

91. $54 - 16y^3$

92. $128 - 250y^3$

93. $64x^3 + 27y^3$

94. $8x^3 + 27y^3$

95. $125x^3 - 64y^3$

96. $125x^3 - y^3$

Practice PLUS

In Exercises 97–104, factor completely. (Hint for Exercises 97–102: Factors contain rational numbers.)

97. $25x^2 - \dfrac{4}{49}$

98. $16x^2 - \dfrac{9}{25}$

99. $y^4 - \dfrac{y}{1000}$

100. $y^4 - \dfrac{y}{8}$

101. $0.25x - x^3$

102. $0.64x - x^3$

103. $(x + 1)^2 - 25$

104. $(x + 2)^2 - 49$

105. Divide $x^3 - x^2 - 5x - 3$ by $x - 3$. Use the quotient to factor $x^3 - x^2 - 5x - 3$ completely.

106. Divide $x^3 + 4x^2 - 3x - 18$ by $x - 2$. Use the quotient to factor $x^3 + 4x^2 - 3x - 18$ completely.

Application Exercises

In Exercises 107–110, find the formula for the area of the shaded blue region and express it in factored form.

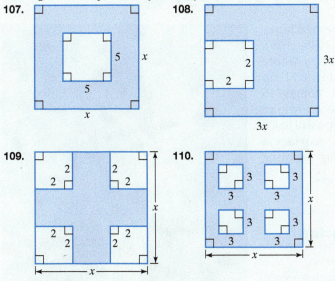

107.

108.

109.

110.

Explaining the Concepts

111. Explain how to factor the difference of two squares. Provide an example with your explanation.

112. What is a perfect square trinomial and how is it factored?

113. Explain why $x^2 - 1$ is factorable, but $x^2 + 1$ is not.

114. Explain how to factor $x^3 + 1$.

Critical Thinking Exercises

Make Sense? *In Exercises 115–118, determine whether each statement makes sense or does not make sense, and explain your reasoning.*

115. I factored $9x^2 - 36$ completely and obtained $(3x + 6)(3x - 6)$.

116. Although I can factor the difference of squares and perfect square trinomials using trial and error, recognizing these special forms shortens the process.

117. I factored $9 - 25x^2$ as $(3 + 5x)(3 - 5x)$ and then applied the commutative property to rewrite the factorization as $(5x + 3)(5x - 3)$.

118. I compared the factorization for the sum of cubes with the factorization for the difference of cubes and noticed that the only difference between them is the positive and negative signs.

In Exercises 119–122, determine whether each statement is true or false. If the statement is false, make the necessary change(s) to produce a true statement.

119. Because $x^2 - 25 = (x + 5)(x - 5)$, then $x^2 + 25 = (x - 5)(x + 5)$.

120. All perfect square trinomials are squares of binomials.

121. Any polynomial that is the sum of two squares is prime.

122. The polynomial $16x^2 + 20x + 25$ is a perfect square trinomial.

123. Where is the error in this "proof" that $2 = 0$?

$a = b$	Suppose that a and b are any equal real numbers.
$a^2 = b^2$	Square both sides of the equation.
$a^2 - b^2 = 0$	Subtract b^2 from both sides.
$2(a^2 - b^2) = 2 \cdot 0$	Multiply both sides by 2.
$2(a^2 - b^2) = 0$	On the right side, $2 \cdot 0 = 0$.
$2(a + b)(a - b) = 0$	Factor $a^2 - b^2$.
$2(a + b) = 0$	Divide both sides by $a - b$.
$2 = 0$	Divide both sides by $a + b$.

In Exercises 124–127, factor each polynomial.

124. $x^2 - y^2 + 3x + 3y$

125. $x^{2n} - 25y^{2n}$

126. $4x^{2n} + 12x^n + 9$

127. $(x + 3)^2 - 2(x + 3) + 1$

In Exercises 128–129, find all integers k so that the trinomial is a perfect square trinomial.

128. $9x^2 + kx + 1$

129. $64x^2 - 16x + k$

Technology Exercises

In Exercises 130–133, use the $\boxed{\text{GRAPH}}$ or $\boxed{\text{TABLE}}$ feature of a graphing utility to determine if the polynomial on the left side of each equation has been correctly factored. If the graphs of y_1 and y_2 coincide, or if their corresponding table values are equal, this means that the polynomial on the left side has been correctly factored. If not, factor the polynomial correctly and then use your graphing utility to verify the factorization.

130. $4x^2 - 9 = (4x + 3)(4x - 3)$

131. $x^2 - 6x + 9 = (x - 3)^2$

132. $4x^2 - 4x + 1 = (4x - 1)^2$

133. $x^3 - 1 = (x - 1)(x^2 - x + 1)$

Review Exercises

134. Simplify: $(2x^2y^3)^4(5xy^2)$. (Section 5.7, Example 5)

135. Subtract: $(10x^2 - 5x + 2) - (14x^2 - 5x - 1)$. (Section 5.1, Example 3)

136. Divide: $\dfrac{6x^2 + 11x - 10}{3x - 2}$. (Section 5.6, Example 1)

Preview Exercises

Exercises 137–139 will help you prepare for the material covered in the next section. In each exercise, factor completely.

137. $3x^3 - 75x$

138. $2x^2 - 20x + 50$

139. $x^3 - 2x^2 - x + 2$

6.5

A General Factoring Strategy

What am I supposed to learn?

After studying this section, you should be able to:

1 Recognize the appropriate method for factoring a polynomial.

2 Use a general strategy for factoring polynomials.

1 Recognize the appropriate method for factoring a polynomial.

Yogi Berra, catcher and renowned hitter for the New York Yankees (1946–1963), said it best: "If you don't know where you're going, you'll probably end up someplace else." When it comes to factoring, it's easy to know where you're going. Why? In this section, you will learn a step-by-step strategy that provides a plan and direction for solving factoring problems.

A Strategy for Factoring Polynomials

It is important to practice factoring a wide variety of polynomials so that you can quickly select the appropriate technique. The polynomial is factored completely when all its polynomial factors, except possibly the monomial factor, are prime. Because of the commutative property, the order of the factors does not matter.

Man with a Bowler Hat (1964), René Magritte. © 2011 Herscovici/ARS

Here is a general strategy for factoring polynomials:

A Strategy for Factoring a Polynomial

1. If there is a common factor other than 1, factor out the GCF.
2. Determine the number of terms in the polynomial and try factoring as follows:

   a. If there are two terms, can the binomial be factored by one of the following special forms?

   Difference of two squares: $A^2 - B^2 = (A + B)(A - B)$
   Sum of two cubes: $A^3 + B^3 = (A + B)(A^2 - AB + B^2)$
   Difference of two cubes: $A^3 - B^3 = (A - B)(A^2 + AB + B^2)$

   b. If there are three terms, is the trinomial a perfect square trinomial? If so, factor by one of the following special forms:

   $$A^2 + 2AB + B^2 = (A + B)^2$$
   $$A^2 - 2AB + B^2 = (A - B)^2.$$

   If the trinomial is not a perfect square trinomial, try factoring by trial and error or grouping.

   c. If there are four or more terms, try factoring by grouping.

3. Check to see if any factors with more than one term in the factored polynomial can be factored further. If so, factor completely.
4. Check by multiplying.

2 Use a general strategy for factoring polynomials. ▶

The following examples and those in the Exercise Set are similar to the previous factoring problems. One difference is that although these polynomials may be factored using the techniques we have studied in this chapter, most must be factored using at least two techniques. Also different is that these factorizations are not all of the same type. They are intentionally mixed to promote the development of a general factoring strategy.

EXAMPLE 1 Factoring a Polynomial

Factor: $4x^4 - 16x^2$.

Solution

Step 1. If there is a common factor, factor out the GCF. Because $4x^2$ is common to both terms, we factor it out.

$$4x^4 - 16x^2 = 4x^2(x^2 - 4) \qquad \text{Factor out the GCF.}$$

Step 2. Determine the number of terms and factor accordingly. The factor $x^2 - 4$ has two terms. It is the difference of two squares: $x^2 - 2^2$. We factor using the special form for the difference of two squares and rewrite the GCF.

$$4x^4 - 16x^2 = 4x^2(x + 2)(x - 2) \qquad \text{Use } A^2 - B^2 = (A + B)(A - B) \text{ on } x^2 - 4\text{: } A = x \text{ and } B = 2.$$

Step 3. Check to see if any factors with more than one term can be factored further. No factor with more than one term can be factored further, so we have factored completely.

Step 4. Check by multiplying.

$$4x^2(x + 2)(x - 2) = 4x^2(x^2 - 4) = 4x^4 - 16x^2$$

This is the original polynomial, so the factorization is correct. ∎

☑ **CHECK POINT 1** Factor: $5x^4 - 45x^2$.

EXAMPLE 2 Factoring a Polynomial

Factor: $3x^2 - 6x - 45$.

Solution

Step 1. If there is a common factor, factor out the GCF. Because 3 is common to all terms, we factor it out.

$$3x^2 - 6x - 45 = 3(x^2 - 2x - 15) \qquad \text{Factor out the GCF.}$$

Step 2. Determine the number of terms and factor accordingly. The factor $x^2 - 2x - 15$ has three terms, but it is not a perfect square trinomial. We factor it using trial and error.

$$3x^2 - 6x - 45 = 3(x^2 - 2x - 15) = 3(x - 5)(x + 3)$$

Step 3. Check to see if factors can be factored further. In this case, they cannot, so we have factored completely.

Step 4. Check by multiplying.

$$3(x - 5)(x + 3) = 3(x^2 - 2x - 15) = 3x^2 - 6x - 45$$

FOIL

This is the original polynomial, so the factorization is correct. ∎

✓ **CHECK POINT 2**   Factor: $4x^2 - 16x - 48$.

EXAMPLE 3   Factoring a Polynomial

Factor: $7x^5 - 7x$.

Solution

Step 1. If there is a common factor, factor out the GCF. Because $7x$ is common to both terms, we factor it out.

$$7x^5 - 7x = 7x(x^4 - 1) \qquad \text{Factor out the GCF.}$$

Step 2. Determine the number of terms and factor accordingly. The factor $x^4 - 1$ has two terms. This binomial can be expressed as $(x^2)^2 - 1^2$, so it can be factored as the difference of two squares.

$$7x^5 - 7x = 7x(x^4 - 1) = 7x(x^2 + 1)(x^2 - 1) \qquad \begin{array}{l}\text{Use } A^2 - B^2 = (A + B)(A - B)\\ \text{on } x^4 - 1: A = x^2 \text{ and } B = 1.\end{array}$$

Step 3. Check to see if factors can be factored further. We note that $(x^2 - 1)$ is also the difference of two squares, $x^2 - 1^2$, so we continue factoring.

$$7x^5 - 7x = 7x(x^2 + 1)(x + 1)(x - 1) \qquad \begin{array}{l}\text{Factor } x^2 - 1 \text{ as the difference}\\ \text{of two squares.}\end{array}$$

Step 4. Check by multiplying.

$$7x(x^2 + 1)(x + 1)(x - 1) = 7x(x^2 + 1)(x^2 - 1) = 7x(x^4 - 1) = 7x^5 - 7x$$

We obtain the original polynomial, so the factorization is correct. ■

✓ **CHECK POINT 3**   Factor: $4x^5 - 64x$.

EXAMPLE 4   Factoring a Polynomial

Factor: $x^3 - 5x^2 - 4x + 20$.

Solution

Step 1. If there is a common factor, factor out the GCF. Other than 1, there is no common factor.

Step 2. Determine the number of terms and factor accordingly. There are four terms. We try factoring by grouping.

$$
\begin{aligned}
&x^3 - 5x^2 - 4x + 20 \\
&= (x^3 - 5x^2) + (-4x + 20) && \text{Group terms with common factors.} \\
&= x^2(x - 5) - 4(x - 5) && \text{Factor from each group.} \\
&= (x - 5)(x^2 - 4) && \text{Factor out the common binomial factor, } x - 5.
\end{aligned}
$$

Step 3. Check to see if factors can be factored further. We note that $(x^2 - 4)$ is the difference of two squares, $x^2 - 2^2$, so we continue factoring.

$$x^3 - 5x^2 - 4x + 20 = (x - 5)(x + 2)(x - 2) \qquad \begin{array}{l}\text{Factor } x^2 - 4 \text{ as the difference}\\ \text{of two squares.}\end{array}$$

We have factored completely because no factor with more than one term can be factored further.

Step 4. Check by multiplying.

$$\text{F} \quad \text{O} \quad \text{I} \quad \text{L}$$

$$(x - 5)(x + 2)(x - 2) = (x - 5)(x^2 - 4) = x^3 - 4x - 5x^2 + 20$$

$$= x^3 - 5x^2 - 4x + 20$$

We obtain the original polynomial, so the factorization is correct.   ■

✓ **CHECK POINT 4**   Factor: $x^3 - 4x^2 - 9x + 36$.

EXAMPLE 5   Factoring a Polynomial

Factor: $2x^3 - 24x^2 + 72x$.

Solution

Step 1. If there is a common factor, factor out the GCF. Because $2x$ is common to all terms, we factor it out.

$$2x^3 - 24x^2 + 72x = 2x(x^2 - 12x + 36) \qquad \text{Factor out the GCF.}$$

Step 2. Determine the number of terms and factor accordingly. The factor $x^2 - 12x + 36$ has three terms. Is it a perfect square trinomial? Yes. The first term, x^2, is the square of a monomial. The last term, 36 or 6^2, is the square of an integer. The middle term involves twice the product of x and 6. We factor using $A^2 - 2AB + B^2 = (A - B)^2$.

$$2x^3 - 24x^2 + 72x = 2x(x^2 - 12x + 36)$$

$$= 2x(\underbrace{x^2 - 2 \cdot x \cdot 6 + 6^2}_{A^2 \;-\; 2AB \;+\; B^2}) \qquad \text{The second factor is a perfect square trinomial.}$$

$$= 2x(x - 6)^2 \qquad A^2 - 2AB + B^2 = (A - B)^2$$

Step 3. Check to see if factors can be factored further. In this problem, they cannot, so we have factored completely.

Step 4. Check by multiplying. Let's verify that $2x^3 - 24x^2 + 72x = 2x(x - 6)^2$.

$$2x(x - 6)^2 = 2x(x^2 - 12x + 36) = 2x^3 - 24x^2 + 72x$$

We obtain the original polynomial, so the factorization is correct.   ■

✓ **CHECK POINT 5**   Factor: $3x^3 - 30x^2 + 75x$.

EXAMPLE 6   Factoring a Polynomial

Factor: $3x^5 + 24x^2$.

Solution

Step 1. If there is a common factor, factor out the GCF. Because $3x^2$ is common to both terms, we factor it out.

$$3x^5 + 24x^2 = 3x^2(x^3 + 8) \qquad \text{Factor out the GCF.}$$

Step 2. Determine the number of terms and factor accordingly. Our factorization up to this point, $3x^5 + 24x^2 = 3x^2(x^3 + 8)$, is not complete. The factor $x^3 + 8$ has two terms. This binomial can be expressed as $x^3 + 2^3$, so it can be factored as the sum of two cubes.

$$3x^5 + 24x^2 = 3x^2(\underbrace{x^3 + 2^3}_{A^3\ +\ B^3})$$

Express $x^3 + 8$ as the sum of two cubes.

$$= 3x^2\underbrace{(x + 2)(x^2 - 2x + 4)}_{(A\ +\ B)\ (A^2\ -\ AB\ +\ B^2)}$$

Factor the sum of two cubes.

Step 3. Check to see if factors can be factored further. In this problem, they cannot, so we have factored completely.

Step 4. Check by multiplying.

$$3x^2(x + 2)(x^2 - 2x + 4) = 3x^2[x(x^2 - 2x + 4) + 2(x^2 - 2x + 4)]$$
$$= 3x^2(x^3 - 2x^2 + 4x + 2x^2 - 4x + 8)$$
$$= 3x^2(x^3 + 8) = 3x^5 + 24x^2$$

We obtain the original polynomial, so the factorization is correct. ∎

Discover for Yourself

In Examples 1–6, substitute 1 for the variable in both the given polynomial and in its factored form. Evaluate each expression. What do you observe? Do this for a second value of the variable. Is this a complete check or only a partial check of the factorization? Explain.

✅ **CHECK POINT 6**   Factor: $2x^5 + 54x^2$.

EXAMPLE 7   Factoring a Polynomial in Two Variables

Factor: $32x^4y - 2y^5$.

Solution

Step 1. If there is a common factor, factor out the GCF. Because $2y$ is common to both terms, we factor it out.

$$32x^4y - 2y^5 = 2y(16x^4 - y^4)$$ Factor out the GCF.

Step 2. Determine the number of terms and factor accordingly. The factor $16x^4 - y^4$ has two terms. It is the difference of two squares: $(4x^2)^2 - (y^2)^2$. We factor using the special form for the difference of two squares.

$$32x^4y - 2y^5 = 2y[\underbrace{(4x^2)^2 - (y^2)^2}_{A^2\ -\ B^2}]$$

Express $16x^4 - y^4$ as the difference of two squares.

$$= 2y\underbrace{(4x^2 + y^2)}_{(A\ +\ B)}\underbrace{(4x^2 - y^2)}_{(A\ -\ B)}$$

$A^2 - B^2 = (A + B)(A - B)$

Step 3. Check to see if factors can be factored further. We note that the last factor, $4x^2 - y^2$, is also the difference of two squares, $(2x)^2 - y^2$, so we continue factoring.

$$32x^4y - 2y^5 = 2y(4x^2 + y^2)(2x + y)(2x - y)$$

Step 4. Check by multiplying. Multiply the factors in the factorization and verify that you obtain the original polynomial. ∎

✅ **CHECK POINT 7**   Factor: $3x^4y - 48y^5$.

EXAMPLE 8   Factoring a Polynomial in Two Variables

Factor: $18x^3 + 48x^2y + 32xy^2$.

Solution

Step 1. If there is a common factor, factor out the GCF. Because $2x$ is common to all terms, we factor it out.

$$18x^3 + 48x^2y + 32xy^2 = 2x(9x^2 + 24xy + 16y^2)$$

Step 2. Determine the number of terms and factor accordingly. The factor $9x^2 + 24xy + 16y^2$ has three terms. Is it a perfect square trinomial? Yes. The first term, $9x^2$ or $(3x)^2$, and the last term, $16y^2$ or $(4y)^2$, are squares of monomials. The middle term, $24xy$, is twice the product of $3x$ and $4y$. We factor using $A^2 + 2AB + B^2 = (A + B)^2$.

$$18x^3 + 48x^2y + 32xy^2 = 2x(9x^2 + 24xy + 16y^2)$$
$$= 2x[(3x)^2 + 2 \cdot 3x \cdot 4y + (4y)^2]$$

The second factor is a perfect square trinomial.

$$\underbrace{A^2 + 2 \cdot A \cdot B + B^2}$$

$$= 2x(3x + 4y)^2 \qquad\qquad A^2 + 2AB + B^2 = (A + B)^2$$

Step 3. Check to see if factors can be factored further. In this problem, they cannot, so we have factored completely.

Step 4. Check by multiplication. Multiply the factors in the factorization and verify that you obtain the original polynomial. ■

☑ **CHECK POINT 8**   Factor: $12x^3 + 36x^2y + 27xy^2$.

CONCEPT AND VOCABULARY CHECK

Here is a list of the factoring techniques that we have discussed.

a. Factoring out the GCF

b. Factoring out the negative of the GCF

c. Factoring by grouping

d. Factoring trinomials by trial and error or grouping

e. Factoring the difference of two squares
$$A^2 - B^2 = (A + B)(A - B)$$

f. Factoring perfect square trinomials
$$A^2 + 2AB + B^2 = (A + B)^2$$
$$A^2 - 2AB + B^2 = (A - B)^2$$

g. Factoring the sum of two cubes
$$A^3 + B^3 = (A + B)(A^2 - AB + B^2)$$

h. Factoring the difference of two cubes
$$A^3 - B^3 = (A - B)(A^2 + AB + B^2)$$

Fill in each blank by writing the letter of the technique (a through h) for factoring the polynomial.

1. $-3x^2 + 21x$ _____

2. $16x^2 - 25$ _____

3. $27x^3 - 1$ _____

4. $x^2 + 7x + xy + 7y$ _____

5. $4x^2 + 8x + 3$ _____

6. $9x^2 + 24x + 16$ _____

7. $5x^2 + 10x$ _____

8. $x^3 + 1000$ _____

6.5 EXERCISE SET ▶ MyMathLab®

Practice Exercises

Before getting to multiple-step factorizations, let's be sure that you are comfortable with exercises requiring only one of the factoring techniques. In Exercises 1–16, factor each polynomial.

1. $-7x^2 + 35x$
2. $-6x^2 + 24x$
3. $25x^2 - 49$
4. $100x^2 - 81$
5. $27x^3 - 1$
6. $64x^3 - 1$
7. $5x + 5y + x^2 + xy$
8. $7x + 7y + x^2 + xy$
9. $14x^2 - 9x + 1$
10. $3x^2 + 2x - 5$
11. $x^2 - 2x + 1$
12. $x^2 - 4x + 4$
13. $27x^3y^3 + 8$
14. $216x^3y^3 + 125$
15. $6x^2 + x - 15$
16. $4x^2 - x - 5$

Now let's move on to factorizations that may require two or more techniques. In Exercises 17–80, factor completely, or state that the polynomial is prime. Check factorizations using multiplication or a graphing utility.

17. $5x^3 - 20x$
18. $4x^3 - 100x$
19. $7x^3 + 7x$
20. $6x^3 + 24x$
21. $5x^2 - 5x - 30$
22. $5x^2 - 15x - 50$
23. $2x^4 - 162$
24. $7x^4 - 7$
25. $x^3 + 2x^2 - 9x - 18$
26. $x^3 + 3x^2 - 25x - 75$
27. $3x^3 - 24x^2 + 48x$
28. $5x^3 - 20x^2 + 20x$
29. $2x^5 + 2x^2$
30. $2x^5 + 128x^2$
31. $6x^2 + 8x$
32. $21x^2 - 35x$
33. $-2y^2 + 2y + 112$
34. $-6x^2 + 6x + 12$
35. $7y^4 + 14y^3 + 7y^2$
36. $2y^4 + 28y^3 + 98y^2$
37. $y^2 + 8y - 16$
38. $y^2 - 18y - 81$
39. $16y^2 - 4y - 2$
40. $32y^2 + 4y - 6$
41. $r^2 - 25r$
42. $3r^2 - 27r$
43. $4w^2 + 8w - 5$
44. $35w^2 - 2w - 1$
45. $x^3 - 4x$
46. $9x^3 - 9x$
47. $x^2 + 64$
48. $y^2 + 36$
49. $9y^2 + 13y + 4$
50. $20y^2 + 12y + 1$
51. $y^3 + 2y^2 - 4y - 8$
52. $y^3 + 2y^2 - y - 2$
53. $16y^2 + 24y + 9$
54. $25y^2 + 20y + 4$
55. $-4y^3 + 28y^2 - 40y$
56. $-7y^3 + 21y^2 - 14y$
57. $y^5 - 81y$
58. $y^5 - 16y$
59. $20a^4 - 45a^2$
60. $48a^4 - 3a^2$
61. $9x^4 + 18x^3 + 6x^2$
62. $10x^4 + 20x^3 + 15x^2$
63. $12y^2 - 11y + 2$
64. $21x^2 - 25x - 4$
65. $9y^2 - 64$
66. $100y^2 - 49$
67. $9y^2 + 64$
68. $100y^2 + 49$
69. $2y^3 + 3y^2 - 50y - 75$
70. $12y^3 + 16y^2 - 3y - 4$
71. $2r^3 + 30r^2 - 68r$
72. $3r^3 - 27r^2 - 210r$
73. $8x^5 - 2x^3$
74. $y^9 - y^5$
75. $3x^2 + 243$
76. $27x^2 + 75$
77. $x^4 + 8x$
78. $x^4 + 27x$
79. $2y^5 - 2y^2$
80. $2y^5 - 128y^2$

Exercises 81–112 contain polynomials in several variables. Factor each polynomial completely and check using multiplication.

81. $6x^2 + 8xy$
82. $21x^2 - 35xy$
83. $xy - 7x + 3y - 21$
84. $xy - 5x + 2y - 10$
85. $x^2 - 3xy - 4y^2$
86. $x^2 - 4xy - 12y^2$
87. $72a^3b^2 + 12a^2 - 24a^4b^2$
88. $24a^4b + 60a^3b^2 + 150a^2b^3$
89. $3a^2 + 27ab + 54b^2$
90. $3a^2 + 15ab + 18b^2$
91. $48x^4y - 3x^2y$
92. $16a^3b^2 - 4ab^2$
93. $6a^2b + ab - 2b$

94. $16a^2 - 32ab + 12b^2$
95. $7x^5y - 7xy^5$
96. $3x^4y^2 - 3x^2y^2$
97. $10x^3y - 14x^2y^2 + 4xy^3$
98. $18x^3y + 57x^2y^2 + 30xy^3$
99. $2bx^2 + 44bx + 242b$
100. $3xz^2 - 72xz + 432x$
101. $15a^2 + 11ab - 14b^2$
102. $25a^2 + 25ab + 6b^2$
103. $-36x^3y + 62x^2y^2 - 12xy^3$
104. $-10a^4b^2 + 15a^3b^3 + 25a^2b^4$
105. $a^2y - b^2y - a^2x + b^2x$
106. $bx^2 - 4b + ax^2 - 4a$
107. $9ax^3 + 15ax^2 - 14ax$
108. $4ay^3 - 12ay^2 + 9ay$
109. $2x^4 + 6x^3y + 2x^2y^2$
110. $3x^4 - 9x^3y + 3x^2y^2$
111. $81x^4y - y^5$
112. $16x^4y - y^5$

Practice PLUS

In Exercises 113–122, factor completely.

113. $10x^2(x + 1) - 7x(x + 1) - 6(x + 1)$

114. $12x^2(x - 1) - 4x(x - 1) - 5(x - 1)$

115. $6x^4 + 35x^2 - 6$
116. $7x^4 + 34x^2 - 5$
117. $(x - 7)^2 - 4a^2$
118. $(x - 6)^2 - 9a^2$
119. $x^2 + 8x + 16 - 25a^2$
120. $x^2 + 14x + 49 - 16a^2$
121. $y^7 + y$
122. $(y + 1)^3 + 1$

Application Exercises

123. A rock is dropped from the top of a 256-foot cliff. The height, in feet, of the rock above the water after t seconds is modeled by the polynomial $256 - 16t^2$. Factor this expression completely.

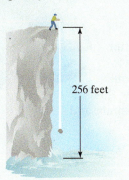

256 feet

124. The building shown in the figure has a height represented by x feet. The building's base is a square and the building's volume is $x^3 - 60x^2 + 900x$ cubic feet. Express the building's dimensions in terms of x.

x

125. Express the area of the blue shaded ring shown in the figure in terms of π. Then factor this expression completely.

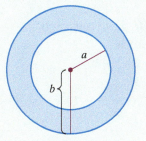

a

b

Explaining the Concepts

126. Describe a strategy that can be used to factor polynomials.

127. Describe some of the difficulties in factoring polynomials. What suggestions can you offer to overcome these difficulties?

128. You are about to take a great picture of fog rolling into San Francisco from the middle of the Golden Gate Bridge, 400 feet above the water. Whoops! You accidently lean too far over the safety rail and drop your camera. The height, in feet, of the camera after t seconds is modeled by the polynomial $400 - 16t^2$. The factored form of the polynomial is $16(5 + t)(5 - t)$. Describe something about your falling camera that is easier to see from the factored form, $16(5 + t)(5 - t)$, than from the form $400 - 16t^2$.

Critical Thinking Exercises

Make Sense? *In Exercises 129–132, determine whether each statement makes sense or does not make sense, and explain your reasoning.*

129. It takes a great deal of practice to get good at factoring a wide variety of polynomials.

130. Multiplying polynomials is relatively mechanical, but factoring often requires a great deal of thought.

131. The factorable trinomial $4x^2 + 8x + 3$ and the prime trinomial $4x^2 + 8x + 1$ are in the form $ax^2 + bx + c$, but $b^2 - 4ac$ is a perfect square only in the case of the factorable trinomial.

132. When a factorization requires two factoring techniques, I'm less likely to make errors if I show one technique at a time rather than combining the two factorizations into one step.

In Exercises 133–136, determine whether each statement is true or false. If the statement is false, make the necessary change(s) to produce a true statement.

133. $x^2 - 9 = (x - 3)^2$ for any real number x.

134. The polynomial $4x^2 + 100$ is the sum of two squares and therefore cannot be factored.

135. If the general factoring strategy is used to factor a polynomial, at least two factorizations are necessary before the given polynomial is factored completely.

136. Once a common monomial factor is removed from $3xy^3 + 9xy^2 + 21xy$, the remaining trinomial factor cannot be factored further.

In Exercises 137–141, factor completely.

137. $3x^5 - 21x^3 - 54x$

138. $5y^5 - 5y^4 - 20y^3 + 20y^2$

139. $4x^4 - 9x^2 + 5$

140. $(x + 5)^2 - 20(x + 5) + 100$

141. $3x^{2n} - 27y^{2n}$

Technology Exercises

In Exercises 142–146, use the GRAPH *or* TABLE *feature of a graphing utility to determine if the polynomial on the left side*

of each equation has been correctly factored. If not, factor the polynomial correctly and then use your graphing utility to verify the factorization.

142. $4x^2 - 12x + 9 = (4x - 3)^2$; $[-5, 5, 1]$ by $[0, 20, 1]$

143. $3x^3 - 12x^2 - 15x = 3x(x + 5)(x - 1)$; $[-5, 7, 1]$ by $[-80, 80, 10]$

144. $6x^2 + 10x - 4 = 2(3x - 1)(x + 2)$; $[-5, 5, 1]$ by $[-20, 20, 2]$

145. $x^4 - 16 = (x^2 + 4)(x + 2)(x - 2)$; $[-5, 5, 1]$ by $[-20, 20, 2]$

146. $2x^3 + 10x^2 - 2x - 10 = 2(x + 5)(x^2 + 1)$; $[-8, 4, 1]$ by $[-100, 100, 10]$

Review Exercises

147. Factor: $9x^2 - 16$. (Section 6.4, Example 1)

148. Graph using intercepts: $5x - 2y = 10$. (Section 3.2, Example 4)

149. The second angle of a triangle measures three times that of the first angle's measure. The third angle measures $80°$ more than the first. Find the measure of each angle. (Section 2.6, Example 6)

Preview Exercises

Exercises 150–152 will help you prepare for the material covered in the next section.

150. Evaluate $(3x - 1)(x + 2)$ for $x = \dfrac{1}{3}$.

151. Evaluate $2x^2 + 7x - 4$ for $x = \dfrac{1}{2}$.

152. Factor: $(x - 2)(x + 3) - 6$.

SECTION

6.6 Solving Quadratic Equations by Factoring ▶

What am I supposed to learn?

After studying this section, you should be able to:

1. Use the zero-product principle. ▶

2. Solve quadratic equations by factoring. ▶

3. Solve problems using quadratic equations. ▶

The alligator, at one time an endangered species, was the subject of a protection program at Florida's Everglades National Park. Park rangers used the formula

$$P = -10x^2 + 475x + 3500$$

to estimate the alligator population, P, after x years of the protection program. Their goal was to bring the population up to 7250. To find out how long the program had to be continued for this to happen, we substitute 7250 for P in the formula and solve for x:

$$7250 = -10x^2 + 475x + 3500.$$

Do you see how this equation differs from a linear equation? The highest exponent on x is 2. Solving such an equation involves finding the numbers that will make the equation a true statement. In this section, we use factoring to solve equations in the form $ax^2 + bx + c = 0$. We also look at applications of these equations.

The Standard Form of a Quadratic Equation

We begin by defining a quadratic equation.

Definition of a Quadratic Equation

A **quadratic equation** in x is an equation that can be written in the **standard form**

$$ax^2 + bx + c = 0,$$

where a, b, and c are real numbers, with $a \neq 0$. A quadratic equation in x is also called a **second-degree polynomial equation** in x.

Here is an example of a quadratic equation in standard form:

$$x^2 - 7x + 10 = 0.$$

$a = 1 \quad b = -7 \quad c = 10$

1 Use the zero-product principle. ⊙

Solving Quadratic Equations by Factoring

We can factor the left side of the quadratic equation $x^2 - 7x + 10 = 0$. We obtain $(x - 5)(x - 2) = 0$. If a quadratic equation has zero on one side and a factored expression on the other side, it can be solved using the **zero-product principle**.

The Zero-Product Principle

If the product of two algebraic expressions is zero, then at least one of the factors is equal to zero.

If $AB = 0$, then $A = 0$ or $B = 0$.

For example, consider the equation $(x - 5)(x - 2) = 0$. According to the zero-product principle, this product can be zero only if at least one of the factors is zero. We set each individual factor equal to zero and solve each resulting equation for x.

$$(x - 5)(x - 2) = 0$$
$$x - 5 = 0 \quad \text{or} \quad x - 2 = 0$$
$$x = 5 \qquad\qquad x = 2$$

We can check each of the proposed solutions, 5 and 2, in the original quadratic equation, $x^2 - 7x + 10 = 0$. Substitute each one separately for x in the equation.

Check 5:	Check 2:
$x^2 - 7x + 10 = 0$	$x^2 - 7x + 10 = 0$
$5^2 - 7 \cdot 5 + 10 \stackrel{?}{=} 0$	$2^2 - 7 \cdot 2 + 10 \stackrel{?}{=} 0$
$25 - 35 + 10 \stackrel{?}{=} 0$	$4 - 14 + 10 \stackrel{?}{=} 0$
$0 = 0, \quad$ true	$0 = 0, \quad$ true

The resulting true statements indicate that the solutions are 5 and 2. Note that with a quadratic equation, we can have two solutions, compared to the linear equation that usually had one.

EXAMPLE 1 Using the Zero-Product Principle

Solve the equation: $(3x - 1)(x + 2) = 0$.

Solution   The product $(3x - 1)(x + 2)$ is equal to zero. By the zero-product principle, the only way that this product can be zero is if at least one of the factors is zero. Thus,

$$3x - 1 = 0 \quad \text{or} \quad x + 2 = 0.$$
$$3x = 1 \qquad\qquad x = -2 \quad \text{Solve each equation for } x.$$
$$x = \frac{1}{3}$$

Because each linear equation has a solution, the original equation, $(3x - 1)(x + 2) = 0$, has two solutions, $\frac{1}{3}$ and -2. Check these solutions by substituting each one separately into the given equation. The equation's solution set is $\left\{-2, \frac{1}{3}\right\}$. ∎

✓ **CHECK POINT 1**   Solve the equation: $(2x + 1)(x - 4) = 0$.

---

2 Solve quadratic equations by factoring. ▶

    In Example 1 and Check Point 1, the given equations were in factored form. Here is a procedure for solving a quadratic equation when we must first do the factoring.

> ### Solving a Quadratic Equation by Factoring
> 1. If necessary, rewrite the equation in the standard form $ax^2 + bx + c = 0$, moving all terms to one side, thereby obtaining zero on the other side.
> 2. Factor.
> 3. Apply the zero-product principle, setting each factor equal to zero.
> 4. Solve the equations formed in step 3.
> 5. Check the solutions in the original equation.

EXAMPLE 2 Solving a Quadratic Equation by Factoring

Solve: $2x^2 + 7x - 4 = 0$.

Solution

Step 1. Move all terms to one side and obtain zero on the other side. All terms are already on the left and zero is on the other side, so we can skip this step.

Step 2. Factor.

$$2x^2 + 7x - 4 = 0$$
$$(2x - 1)(x + 4) = 0$$

Steps 3 and 4. Set each factor equal to zero and solve each resulting equation.

$$2x - 1 = 0 \quad \text{or} \quad x + 4 = 0$$
$$2x = 1 \qquad\qquad x = -4$$
$$x = \frac{1}{2}$$

Step 5. Check the solutions in the original equation.

Check $\frac{1}{2}$:

$$2x^2 + 7x - 4 = 0$$

$$2\left(\frac{1}{2}\right)^2 + 7\left(\frac{1}{2}\right) - 4 \overset{?}{=} 0$$

$$2\left(\frac{1}{4}\right) + 7\left(\frac{1}{2}\right) - 4 \overset{?}{=} 0$$

$$\frac{1}{2} + \frac{7}{2} - 4 \overset{?}{=} 0$$

$$4 - 4 \overset{?}{=} 0$$

$$0 = 0, \quad \text{true}$$

Check -4:

$$2x^2 + 7x - 4 = 0$$

$$2(-4)^2 + 7(-4) - 4 \overset{?}{=} 0$$

$$2(16) + 7(-4) - 4 \overset{?}{=} 0$$

$$32 + (-28) - 4 \overset{?}{=} 0$$

$$4 - 4 \overset{?}{=} 0$$

$$0 = 0, \quad \text{true}$$

The solutions are -4 and $\frac{1}{2}$, and the solution set is $\left\{-4, \frac{1}{2}\right\}$. ■

✓ **CHECK POINT 2** Solve: $x^2 - 6x + 5 = 0$.

Great Question!

After factoring a polynomial, should I set each factor equal to zero?

No. Do not confuse factoring a polynomial with solving a quadratic equation by factoring.

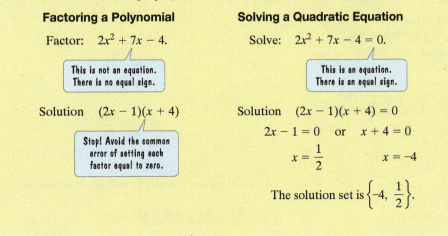

Factoring a Polynomial

Factor: $2x^2 + 7x - 4$.

> This is not an equation. There is no equal sign.

Solution $(2x - 1)(x + 4)$

> Stop! Avoid the common error of setting each factor equal to zero.

Solving a Quadratic Equation

Solve: $2x^2 + 7x - 4 = 0$.

> This is an equation. There is an equal sign.

Solution $(2x - 1)(x + 4) = 0$

$2x - 1 = 0$ or $x + 4 = 0$

$x = \frac{1}{2}$ $x = -4$

The solution set is $\left\{-4, \frac{1}{2}\right\}$.

Using Technology

Graphic Connections

You can use a graphing utility to check the real number solutions of a quadratic equation. **The solutions of $ax^2 + bx + c = 0$ correspond to the x-intercepts for the graph of $y = ax^2 + bx + c$.** For example, to check the solutions of $2x^2 + 7x - 4 = 0$, graph $y = 2x^2 + 7x - 4$. The U-shaped, bowl-like graph is shown below. The x-intercepts are -4 and $\frac{1}{2}$, verifying -4 and $\frac{1}{2}$ as the solutions.

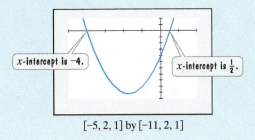

x-intercept is -4.

x-intercept is $\frac{1}{2}$.

$[-5, 2, 1]$ by $[-11, 2, 1]$

EXAMPLE 3 Solving a Quadratic Equation by Factoring

Solve: $3x^2 = 2x$.

Solution

Step 1. Move all terms to one side and obtain zero on the other side. Subtract $2x$ from both sides and write the equation in standard form.

$$3x^2 - 2x = 2x - 2x$$

$$3x^2 - 2x = 0$$

Step 2. Factor. We factor out x from the two terms on the left side.

$$3x^2 - 2x = 0$$

$$x(3x - 2) = 0$$

Steps 3 and 4. Set each factor equal to zero and solve the resulting equations.

$$x = 0 \quad \text{or} \quad 3x - 2 = 0$$

$$3x = 2$$

$$x = \frac{2}{3}$$

Step 5. Check the solutions in the original equation.

Great Question!

Can I simplify $3x^2 = 2x$ by dividing both sides by x?

No. If you divide both sides of $3x^2 = 2x$ by x, you will obtain $3x = 2$ and, consequently, $x = \frac{2}{3}$. The other solution, 0, is lost. We can divide both sides of an equation by any *nonzero* real number. If x is zero, we lose the second solution.

Check 0:

$$3x^2 = 2x$$

$$3 \cdot 0^2 \overset{?}{=} 2 \cdot 0$$

$$0 = 0, \quad \text{true}$$

Check $\frac{2}{3}$:

$$3x^2 = 2x$$

$$3\left(\frac{2}{3}\right)^2 \overset{?}{=} 2\left(\frac{2}{3}\right)$$

$$3\left(\frac{4}{9}\right) \overset{?}{=} 2\left(\frac{2}{3}\right)$$

$$\frac{4}{3} = \frac{4}{3}, \quad \text{true}$$

The solutions are 0 and $\frac{2}{3}$, and the solution set is $\left\{0, \frac{2}{3}\right\}$. ∎

✓ **CHECK POINT 3** Solve: $4x^2 = 2x$.

EXAMPLE 4 Solving a Quadratic Equation by Factoring

Solve: $x^2 = 6x - 9$.

Solution

Step 1. Move all terms to one side and obtain zero on the other side. To obtain zero on the right, we subtract $6x$ and add 9 on both sides.

$$x^2 - 6x + 9 = 6x - 6x - 9 + 9$$

$$x^2 - 6x + 9 = 0$$

Step 2. Factor. The trinomial on the left side is a perfect square trinomial: $x^2 - 6x + 9 = x^2 - 2 \cdot x \cdot 3 + 3^2$. We factor using $A^2 - 2AB + B^2 = (A - B)^2$: $A = x$ and $B = 3$.

$$x^2 - 6x + 9 = 0$$

$$(x - 3)^2 = 0$$

Using Technology

Graphic Connections
The graph of
$y = x^2 - 6x + 9$ is shown
below. Notice that there
is only one x-intercept,
namely 3, verifying that the
solution of
$$x^2 - 6x + 9 = 0$$
is 3.

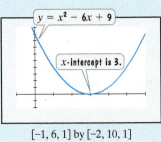

$[-1, 6, 1]$ by $[-2, 10, 1]$

Steps 3 and 4. Set each factor equal to zero and solve the resulting equations. Because both factors are the same, it is only necessary to set one of them equal to zero.

$$x - 3 = 0$$
$$x = 3$$

Step 5. Check the solution in the original equation.

Check 3:

$$x^2 = 6x - 9$$
$$3^2 \stackrel{?}{=} 6 \cdot 3 - 9$$
$$9 \stackrel{?}{=} 18 - 9$$
$$9 = 9, \quad \text{true}$$

The solution is 3, and the solution set is {3}. ∎

✓ **CHECK POINT 4**   Solve: $x^2 = 10x - 25$.

EXAMPLE 5   Solving a Quadratic Equation by Factoring

Solve: $9x^2 = 16$.

Solution
Step 1. Move all terms to one side and obtain zero on the other side. Subtract 16 from both sides and write the equation in standard form.

$$9x^2 - 16 = 16 - 16$$
$$9x^2 - 16 = 0$$

Step 2. Factor. The binomial on the left side is the difference of two squares: $9x^2 - 16 = (3x)^2 - 4^2$. We factor using $A^2 - B^2 = (A + B)(A - B)$: $A = 3x$ and $B = 4$.

$$9x^2 - 16 = 0$$
$$(3x + 4)(3x - 4) = 0$$

Steps 3 and 4. Set each factor equal to zero and solve the resulting equations. We use the zero-product principle to solve $(3x + 4)(3x - 4) = 0$.

$$3x + 4 = 0 \quad \text{or} \quad 3x - 4 = 0$$
$$3x = -4 \qquad\qquad 3x = 4$$
$$x = -\frac{4}{3} \qquad\qquad x = \frac{4}{3}$$

Step 5. Check the solutions in the original equation. Do this now and verify that the solutions of $9x^2 = 16$ are $-\frac{4}{3}$ and $\frac{4}{3}$. The equation's solution set is $\left\{ -\frac{4}{3}, \frac{4}{3} \right\}$. ∎

✓ **CHECK POINT 5**   Solve: $16x^2 = 25$.

EXAMPLE 6   Solving a Quadratic Equation by Factoring

Solve: $(x - 2)(x + 3) = 6$.

Solution

Step 1. Move all terms to one side and obtain zero on the other side. We write $(x - 2)(x + 3) = 6$ in standard form by multiplying out the product on the left side and then subtracting 6 from both sides.

$$(x - 2)(x + 3) = 6 \qquad \text{This is the given equation.}$$
$$x^2 + 3x - 2x - 6 = 6 \qquad \text{Use the FOIL method.}$$
$$x^2 + x - 6 = 6 \qquad \text{Simplify.}$$
$$x^2 + x - 6 - 6 = 6 - 6 \qquad \text{Subtract 6 from both sides.}$$
$$x^2 + x - 12 = 0 \qquad \text{Simplify.}$$

Step 2. Factor.

$$x^2 + x - 12 = 0$$
$$(x + 4)(x - 3) = 0$$

Steps 3 and 4. Set each factor equal to zero and solve the resulting equations.

$$x + 4 = 0 \quad \text{or} \quad x - 3 = 0$$
$$x = -4 \qquad\qquad x = 3$$

Step 5. Check the solutions in the original equation. Do this now and verify that the solutions are -4 and 3. The equation's solution set is $\{-4, 3\}$. ■

✓ **CHECK POINT 6**   Solve: $(x - 5)(x - 2) = 28$.

3 Solve problems using quadratic equations. ▶

Applications of Quadratic Equations

Solving quadratic equations by factoring can be used to answer questions about variables contained in mathematical models.

EXAMPLE 7   Modeling Motion

You throw a ball straight up from a rooftop 160 feet high with an initial speed of 48 feet per second. The formula

$$h = -16t^2 + 48t + 160$$

describes the ball's height above the ground, h, in feet, t seconds after you throw it. The ball misses the rooftop on its way down and eventually strikes the ground. The situation is illustrated in **Figure 6.1**. How long will it take for the ball to hit the ground?

Figure 6.1

Solution   The ball hits the ground when h, its height above the ground, is 0 feet. Thus, we substitute 0 for h in the given formula and solve for t.

$$h = -16t^2 + 48t + 160 \qquad \text{This is the formula that models the ball's height.}$$
$$0 = -16t^2 + 48t + 160 \qquad \text{Substitute 0 for } h.$$
$$0 = -16(t^2 - 3t - 10) \qquad \text{Factor out the negative of the GCF.}$$
$$0 = -16(t - 5)(t + 2) \qquad \text{Factor the trinomial.}$$

> Do not set the constant, -16, equal to zero: $-16 \neq 0$.

$$t - 5 = 0 \text{ or } t + 2 = 0 \qquad \text{Set each variable factor equal to 0.}$$
$$t = 5 \qquad\quad t = -2 \qquad \text{Solve for } t.$$

Because we begin describing the ball's height at $t = 0$, we discard the solution $t = -2$. The ball hits the ground after 5 seconds. ■

Great Question!

You began solving 0 = −16t² + 48t + 160 by factoring out −16, the negative of the GCF. What happens if I begin by multiplying both sides of the equation by −1?

You will still get the same solutions, 5 and −2. Here's how it works:

$$-1 \cdot 0 = -1(-16t^2 + 48t + 160)$$ Multiply both sides of
    0 = −16t² + 48t + 160 by −1.

$$0 = 16t^2 - 48t - 160$$ Multiply each term on the right side by −1.

$$0 = 16(t^2 - 3t - 10)$$ Factor out the GCF, 16.

$$0 = 16(t - 5)(t + 2)$$ Factor the trinomial.

$$t - 5 = 0 \text{ or } t + 2 = 0$$ Set each variable factor equal to 0.

$$t = 5 \qquad t = -2.$$ Solve for t.

Figure 6.2 shows the graph of the formula $h = -16t^2 + 48t + 160$. The horizontal axis is labeled t, for the ball's time in motion. The vertical axis is labeled h, for the ball's height above the ground. Because time and height are both positive, the model is graphed in quadrant I and its boundaries only.

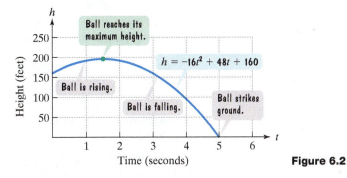

Figure 6.2

The graph visually shows what we discovered algebraically: The ball hits the ground after 5 seconds. The graph also reveals that the ball reaches its maximum height, nearly 200 feet, after 1.5 seconds. Then the ball begins to fall.

✓ **CHECK POINT 7**   Use the formula $h = -16t^2 + 48t + 160$ to determine when the ball's height is 192 feet. Identify your solutions as points on the graph in **Figure 6.2**.

In our next example, we use our five-step strategy for solving word problems.

EXAMPLE 8   Solving a Problem About a Rectangle's Area

An architect is allowed no more than 15 square meters to add a small bedroom to a house. Because of the room's design in relationship to the existing structure, the width of its rectangular floor must be 7 meters less than two times the length. Find the precise length and width of the rectangular floor of maximum area that the architect is permitted.

Solution

Step 1. Let x represet one of the quantities. We know something about the width: It must be 7 meters less than two times the length. We will let

$$x = \text{the length of the floor.}$$

Step 2. Represent other unknown quantities in terms of x. Because the width must be 7 meters less than two times the length, let

$$2x - 7 = \text{the width of the floor.}$$

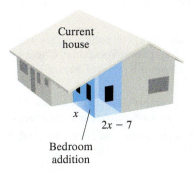

Figure 6.3

The problem is illustrated in **Figure 6.3**.

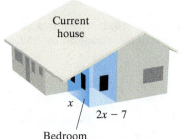

Current house

x

$2x - 7$

Bedroom addition

Figure 6.3 (repeated)

Step 3. Write an equation that models the conditions. Because the architect is allowed no more than 15 square meters, an area of 15 square meters is the maximum area permitted. The area of a rectangle is the product of its length and its width.

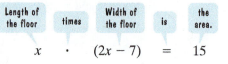

| Length of the floor | times | Width of the floor | is | the area. |

$$x \cdot (2x - 7) = 15$$

Step 4. Solve the equation and answer the question.

$$x(2x - 7) = 15 \quad \text{This is the equation that models the problem's conditions.}$$

$$2x^2 - 7x = 15 \quad \text{Use the distributive property.}$$

$$2x^2 - 7x - 15 = 0 \quad \text{Subtract 15 from both sides.}$$

$$(2x + 3)(x - 5) = 0 \quad \text{Factor.}$$

$$2x + 3 = 0 \quad \text{or} \quad x - 5 = 0 \quad \text{Set each factor equal to zero.}$$

$$2x = -3 \quad \quad x = 5 \quad \text{Solve the resulting equations.}$$

$$x = -\frac{3}{2}$$

A rectangle cannot have a negative length. Thus,

$$\text{Length} = x = 5$$

$$\text{Width} = 2x - 7 = 2 \cdot 5 - 7 = 10 - 7 = 3.$$

The architect is permitted a room of maximum area whose length is 5 meters and whose width is 3 meters.

Step 5. Check the proposed solution in the original wording of the problem. The area of the floor using the dimensions that we found is

$$A = lw = (5 \text{ meters})(3 \text{ meters}) = 15 \text{ square meters.}$$

Because the problem's wording tells us that the maximum area permitted is 15 square meters, our dimensions are correct. ∎

☑ **CHECK POINT 8** The length of a rectangular sign is 3 feet longer than the width. If the sign's area is 54 square feet, find its length and width.

Achieving Success

Be sure to use the Chapter Test Prep on YouTube for each chapter test. The Chapter Test Prep videos provide step-by-step solutions to every exercise in the test and let you review any exercises you miss.

Are you using any of the other textbook supplements for help and additional study? These include:

- The Student Solutions Manual. This contains fully worked solutions to the odd-numbered section exercises plus all Check Points, Concept and Vocabulary Checks, Review/Preview Exercises, Mid-Chapter Check Points, Chapter Reviews, Chapter Tests, and Cumulative Reviews.

- Lecture Videos. These interactive lectures highlight key examples from every section of the textbook. The interface allows you to navigate to sections, objectives, and examples.

- MyMathLab is a text-specific online course. Math XL is an online homework, tutorial, and assessment system. Ask your instructor whether these are available to you.

CONCEPT AND VOCABULARY CHECK

Fill in each blank so that the resulting statement is true.

1. An equation that can be written in the standard form $ax^2 + bx + c = 0$, $a \neq 0$, is called a/an _____.

2. The zero-product principle states that if $AB = 0$, then _____.

3. The solutions of $ax^2 + bx + c = 0$ correspond to the _____ for the graph of $y = ax^2 + bx + c$.

4. The equation $3x^2 = 5x$ can be written in standard form by _____ on both sides.

5. The equation $9x^2 = 30x - 25$ can be written in standard form by _____ and _____ on both sides.

6.6 EXERCISE SET ▶ MyMathLab®

Practice Exercises

In Exercises 1–8, solve each equation using the zero-product principle.

1. $x(x + 7) = 0$
2. $x(x - 3) = 0$
3. $(x - 6)(x + 4) = 0$
4. $(x - 3)(x + 8) = 0$
5. $(x - 9)(5x + 4) = 0$
6. $(x + 7)(3x - 2) = 0$
7. $10(x - 4)(2x + 9) = 0$
8. $8(x - 5)(3x + 11) = 0$

In Exercises 9–56, use factoring to solve each quadratic equation. Check by substitution or by using a graphing utility and identifying x-intercepts.

9. $x^2 + 8x + 15 = 0$
10. $x^2 + 5x + 6 = 0$
11. $x^2 - 2x - 15 = 0$
12. $x^2 + x - 42 = 0$
13. $x^2 - 4x = 21$
14. $x^2 + 7x = 18$
15. $x^2 + 9x = -8$
16. $x^2 - 11x = -10$
17. $x^2 + 4x = 0$
18. $x^2 - 6x = 0$
19. $x^2 - 5x = 0$
20. $x^2 + 3x = 0$
21. $x^2 = 4x$
22. $x^2 = 8x$
23. $2x^2 = 5x$
24. $3x^2 = 5x$
25. $3x^2 = -5x$
26. $2x^2 = -3x$
27. $x^2 + 4x + 4 = 0$
28. $x^2 + 6x + 9 = 0$
29. $x^2 = 12x - 36$
30. $x^2 = 14x - 49$
31. $4x^2 = 12x - 9$
32. $9x^2 = 30x - 25$
33. $2x^2 = 7x + 4$
34. $3x^2 = x + 4$
35. $5x^2 = 18 - x$
36. $3x^2 = 15 + 4x$
37. $x^2 - 49 = 0$
38. $x^2 - 25 = 0$
39. $4x^2 - 25 = 0$
40. $9x^2 - 100 = 0$
41. $81x^2 = 25$
42. $25x^2 = 49$
43. $x(x - 4) = 21$
44. $x(x - 3) = 18$
45. $4x(x + 1) = 15$
46. $x(3x + 8) = -5$
47. $(x - 1)(x + 4) = 14$
48. $(x - 3)(x + 8) = -30$
49. $(x + 1)(2x + 5) = -1$
50. $(x + 3)(3x + 5) = 7$
51. $y(y + 8) = 16(y - 1)$
52. $y(y + 9) = 4(2y + 5)$
53. $4y^2 + 20y + 25 = 0$
54. $4y^2 + 44y + 121 = 0$
55. $64w^2 = 48w - 9$
56. $25w^2 = 80w - 64$

Practice PLUS

In Exercises 57–66, solve each equation and check your solutions.

57. $(x - 4)(x^2 + 5x + 6) = 0$
58. $(x - 5)(x^2 - 3x + 2) = 0$
59. $x^3 - 36x = 0$
60. $x^3 - 4x = 0$
61. $y^3 + 3y^2 + 2y = 0$
62. $y^3 + 2y^2 - 3y = 0$
63. $2(x - 4)^2 + x^2 = x(x + 50) - 46x$
64. $(x - 4)(x - 5) + (2x + 3)(x - 1) = x(2x - 25) - 13$
65. $(x - 2)^2 - 5(x - 2) + 6 = 0$
66. $(x - 3)^2 + 2(x - 3) - 8 = 0$

Application Exercises

A ball is thrown straight up from a rooftop 300 feet high. The formula

$$h = -16t^2 + 20t + 300$$

describes the ball's height above the ground, h, in feet, t seconds after it was thrown. The ball misses the rooftop on its way down and eventually strikes the ground. The graph of the formula is shown, with tick marks omitted along the horizontal axis. Use the formula to solve Exercises 67–69.

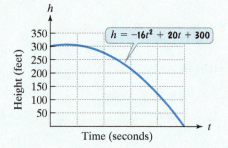

67. How long will it take for the ball to hit the ground? Use this information to provide tick marks with appropriate numbers along the horizontal axis in the figure shown.

68. When will the ball's height be 304 feet? Identify the solution as a point on the graph.

69. When will the ball's height be 276 feet? Identify the solution as a point on the graph.

An explosion causes debris to rise vertically with an initial speed of 72 feet per second. The formula

$$h = -16t^2 + 72t$$

describes the height of the debris above the ground, h, in feet, t seconds after the explosion. Use this information to solve Exercises 70–71.

70. How long will it take for the debris to hit the ground?

71. When will the debris be 32 feet above the ground?

The formula

$$S = 2x^2 - 12x + 82$$

models spending by international travelers to the United States, S, in billions of dollars, x years after 2000. Use this formula to solve Exercises 72–73.

72. In which years did international travelers spend $72 billion?

73. In which years did international travelers spend $66 billion?

The graph of the formula modeling spending by international travelers is shown below. Use the graph to solve Exercises 74–75.

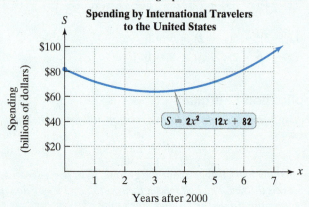

Spending by International Travelers to the United States

Source: Travel Industry Association of America

74. Identify your solutions from Exercise 72 as points on the graph.

75. Identify your solutions from Exercise 73 as points on the graph.

The alligator, at one time an endangered species, is the subject of a protection program. The formula

$$P = -10x^2 + 475x + 3500$$

models the alligator population, P, after x years of the protection program, where $0 \le x \le 12$. Use the formula to solve Exercises 76–77.

76. After how long is the population up to 5990?

77. After how long is the population up to 7250?

The graph of the alligator population is shown over time. Use the graph to solve Exercises 78–79.

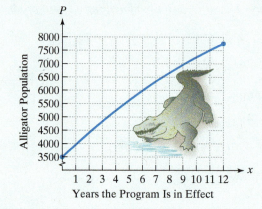

Years the Program Is in Effect

78. Identify your solution from Exercise 76 as a point on the graph.

79. Identify your solution from Exercise 77 as a point on the graph.

The formula

$$N = \frac{t^2 - t}{2}$$

describes the number of football games, N, that must be played in a league with t teams if each team is to play every other team once. Use this information to solve Exercises 80–81.

80. If a league has 36 games scheduled, how many teams belong to the league, assuming that each team plays every other team once?

81. If a league has 45 games scheduled, how many teams belong to the league, assuming that each team plays every other team once?

82. The length of a rectangular garden is 5 feet greater than the width. The area of the rectangle is 300 square feet. Find the length and the width.

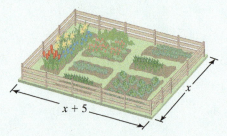

83. A rectangular parking lot has a length that is 3 yards greater than the width. The area of the parking lot is 180 square yards. Find the length and the width.

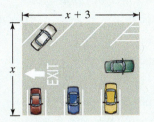

84. Each end of a glass prism is a triangle with a height that is 1 inch shorter than twice the base. If the area of the triangle is 60 square inches, how long are the base and height?

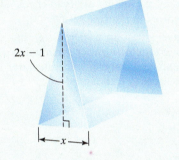

85. Great white sharks have triangular teeth with a height that is 1 centimeter longer than the base. If the area of one tooth is 15 square centimeters, find its base and height.

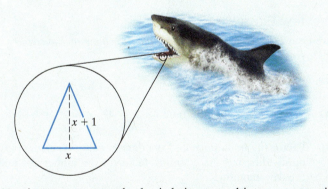

86. A vacant rectangular lot is being turned into a community vegetable garden measuring 15 meters by 12 meters. A path of uniform width is to surround the garden. If the area of the lot is 378 square meters, find the width of the path surrounding the garden.

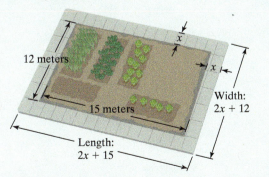

87. As part of a landscaping project, you put in a flower bed measuring 10 feet by 12 feet. You plan to surround the bed with a uniform border of low-growing plants.

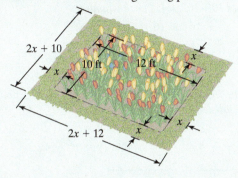

a. Write a polynomial that describes the area of the uniform border that surrounds your flower bed. (*Hint:* The area of the border is the area of the large rectangle shown in the figure minus the area of the flower bed.)

b. The low-growing plants surrounding the flower bed require 1 square foot each when mature. If you have 168 of these plants, how wide a strip around the flower bed should you prepare for the border?

Explaining the Concepts

88. What is a quadratic equation?

89. Explain how to solve $x^2 + 6x + 8 = 0$ using factoring and the zero-product principle.

90. If $(x + 2)(x - 4) = 0$ indicates that $x + 2 = 0$ or $x - 4 = 0$, explain why $(x + 2)(x - 4) = 6$ does not mean $x + 2 = 6$ or $x - 4 = 6$. Could we solve the equation using $x + 2 = 3$ and $x - 4 = 2$ because $3 \cdot 2 = 6$?

Critical Thinking Exercises

Make Sense? In Exercises 91–94, determine whether each statement makes sense or does not make sense, and explain your reasoning.

91. When solving $4(x - 3)(x + 2) = 0$ and $4x(x + 2) = 0$, I can ignore the monomial factors.

92. I set the quadratic equation $2x^2 - 5x = 12$ equal to zero and obtained $2x^2 - 5x = 0$.

93. Because some trinomials are prime, some quadratic equations cannot be solved by factoring.

94. I'm looking at a graph with one x-intercept, so it must be the graph of a linear equation.

In Exercises 95–98, determine whether each statement is true or false. If the statement is false, make the necessary change(s) to produce a true statement.

95. If $(x + 3)(x - 4) = 2$, then $x + 3 = 0$ or $x - 4 = 0$.

96. The solutions of the equation $4(x - 5)(x + 3) = 0$ are 4, 5, and −3.

97. Equations solved by factoring always have two different solutions.

98. Both 0 and $-\pi$ are solutions of the equation $x(x + \pi) = 0$.

99. Write a quadratic equation in standard form whose solutions are -3 and 5.

In Exercises 100–102, solve each equation.

100. $x^3 - x^2 - 16x + 16 = 0$

101. $3^{x^2 - 9x + 20} = 1$

102. $(x^2 - 5x + 5)^3 = 1$

In Exercises 103–106, match each equation with its graph. The graphs are labeled (a) through (d).

103. $y = x^2 - x - 2$

104. $y = x^2 + x - 2$

105. $y = x^2 - 4$

106. $y = x^2 - 4x$

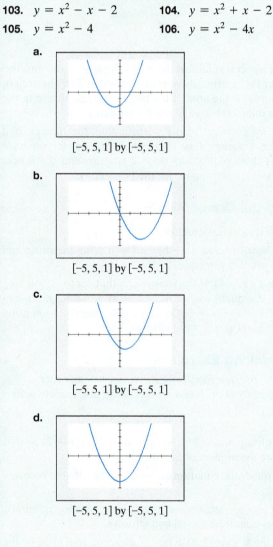

a.

$[-5, 5, 1]$ by $[-5, 5, 1]$

b.

$[-5, 5, 1]$ by $[-5, 5, 1]$

c.

$[-5, 5, 1]$ by $[-5, 5, 1]$

d.

$[-5, 5, 1]$ by $[-5, 5, 1]$

Technology Exercises

In Exercises 107–110, use the x-intercepts for the graph in a $[-10, 10, 1]$ by $[-13, 10, 1]$ viewing rectangle to solve the quadratic equation. Check by substitution.

107. Use the graph of $y = x^2 + 3x - 4$ to solve
$$x^2 + 3x - 4 = 0.$$

108. Use the graph of $y = x^2 + x - 6$ to solve
$$x^2 + x - 6 = 0.$$

109. Use the graph of $y = (x - 2)(x + 3) - 6$ to solve
$$(x - 2)(x + 3) - 6 = 0.$$

110. Use the graph of $y = x^2 - 2x + 1$ to solve
$$x^2 - 2x + 1 = 0.$$

111. Use the technique of identifying x-intercepts on a graph generated by a graphing utility to check any five equations that you solved in Exercises 9–56.

112. If you have access to a calculator that solves quadratic equations, consult the owner's manual to determine how to use this feature. Then use your calculator to solve any five of the equations in Exercises 9–56.

Review Exercises

113. Graph: $y > -\frac{2}{3}x + 1$. (Section 3.6, Example 3)

114. Simplify: $\left(\frac{8x^4}{4x^7}\right)^2$. (Section 5.7, Example 6)

115. Solve: $5x + 28 = 6 - 6x$. (Section 2.2, Example 7)

Preview Exercises

Exercises 116–118 will help you prepare for the material covered in the first section of the next chapter.

116. Evaluate $\frac{250x}{100 - x}$ for $x = 60$.

117. Why is $\frac{6x + 12}{7x - 28}$ undefined for $x = 4$?

118. Factor the numerator and the denominator. Then simplify by dividing out the common factor in the numerator and the denominator.
$$\frac{x^2 + 6x + 5}{x^2 - 25}$$

Chapter 6 Summary

Definitions and Concepts	Examples

Section 6.1 The Greatest Common Factor and Factoring by Grouping

Factoring a polynomial containing the sum of monomials means finding an equivalent expression that is a product. The greatest common factor, GCF, is an expression that divides every term of the polynomial. The variable part of the GCF contains the smallest power of a variable that appears in all terms of the polynomial.

Find the GCF of $16x^2y$, $20x^3y^2$, and $8x^2y^3$.

The GCF of 16, 20, and 8 is 4.

The GCF of x^2, x^3, and x^2 is x^2.

The GCF of y, y^2, and y^3 is y.

$$\text{GCF} = 4 \cdot x^2 \cdot y = 4x^2y$$

To factor a monomial from a polynomial, express each term as the product of the GCF and its other factor. Then use the distributive property to factor out the GCF.

$$16x^2y + 20x^3y^2 + 8x^2y^3$$
$$= 4x^2y \cdot 4 + 4x^2y \cdot 5xy + 4x^2y \cdot 2y^2$$
$$= 4x^2y(4 + 5xy + 2y^2)$$

To factor a monomial from a polynomial with a negative coefficient in the first term, express each term as the product of the negative of the GCF and its other factor. Then use the distributive property to factor out the negative of the GCF.

$$-20x^4y^3 + 10x^2y^2 - 15x^3y$$
$$= -5x^2y \cdot 4x^2y^2 - 5x^2y(-2y) - 5x^2y \cdot 3x$$

The negative of the GCF is $-5x^2y$.

$$= -5x^2y(4x^2y^2 - 2y + 3x)$$

To factor by grouping, factor out the GCF from each group. Then factor out the remaining common factor.

$$xy + 5x - 3y - 15$$
$$= x(y + 5) - 3(y + 5)$$
$$= (y + 5)(x - 3)$$

Section 6.2 Factoring Trinomials Whose Leading Coefficient Is 1

To factor a trinomial of the form $x^2 + bx + c$, find two numbers whose product is c and whose sum is b. The factorization is

$$(x + \text{one number})(x + \text{other number}).$$

Factor: $x^2 + 9x + 20$.

Find two numbers whose product is 20 and whose sum is 9. The numbers are 4 and 5.

$$x^2 + 9x + 20 = (x + 4)(x + 5)$$

Section 6.3 Factoring Trinomials Whose Leading Coefficient Is Not 1

To factor $ax^2 + bx + c$ by trial and error, try various combinations of factors of ax^2 and c until a middle term of bx is obtained for the sum of outside and inside products.

Factor: $3x^2 + 7x - 6$.

Factors of $3x^2$: $3x, x$

Factors of -6: 1 and -6, -1 and 6, 2 and -3, -2 and 3

A possible combination of these factors is

$$(3x - 2)(x + 3).$$

Sum of outside and inside products should equal $7x$.

$$9x - 2x = 7x$$

Thus, $3x^2 + 7x - 6 = (3x - 2)(x + 3)$.

To factor $ax^2 + bx + c$ by grouping, find factors of ac whose sum is b. Write bx using these factors. Then factor by grouping.

Factor: $3x^2 + 7x - 6$.

Find factors of $3(-6)$, or -18, whose sum is 7. They are 9 and -2.

$$3x^2 + 7x - 6$$
$$= 3x^2 + 9x - 2x - 6$$
$$= 3x(x + 3) - 2(x + 3) = (x + 3)(3x - 2)$$

Definitions and Concepts	Examples

Section 6.4 Factoring Special Forms

The Difference of Two Squares

$$A^2 - B^2 = (A + B)(A - B)$$

$$9x^2 - 25y^2$$
$$= (3x)^2 - (5y)^2 = (3x + 5y)(3x - 5y)$$

Perfect Square Trinomials

$$A^2 + 2AB + B^2 = (A + B)^2$$
$$A^2 - 2AB + B^2 = (A - B)^2$$

$$x^2 + 16x + 64 = x^2 + 2 \cdot x \cdot 8 + 8^2 = (x + 8)^2$$
$$25x^2 - 30x + 9 = (5x)^2 - 2 \cdot 5x \cdot 3 + 3^2 = (5x - 3)^2$$

Sum or Difference of Cubes

$$A^3 + B^3 = (A + B)(A^2 - AB + B^2)$$
$$A^3 - B^3 = (A - B)(A^2 + AB + B^2)$$

$$8x^3 - 125 = (2x)^3 - 5^3$$
$$= (2x - 5)[(2x)^2 + 2x \cdot 5 + 5^2]$$
$$= (2x - 5)(4x^2 + 10x + 25)$$

Section 6.5 A General Factoring Strategy

A Factoring Strategy

1. Factor out the GCF.
2. **a.** If two terms, try
   $$A^2 - B^2 = (A + B)(A - B)$$
   $$A^3 + B^3 = (A + B)(A^2 - AB + B^2)$$
   $$A^3 - B^3 = (A - B)(A^2 + AB + B^2).$$
   b. If three terms, try
   $$A^2 + 2AB + B^2 = (A + B)^2$$
   $$A^2 - 2AB + B^2 = (A - B)^2.$$

   If not a perfect square trinomial, try trial and error or grouping.
   c. If four terms, try factoring by grouping.
3. See if any factors can be factored further.
4. Check by multiplying.

Factor: $2x^4 + 10x^3 - 8x^2 - 40x$.
The GCF is $2x$.

$$2x^4 + 10x^3 - 8x^2 - 40x$$
$$= 2x(x^3 + 5x^2 - 4x - 20)$$

> Four terms: Try grouping.

$$= 2x[x^2(x + 5) - 4(x + 5)]$$
$$= 2x(x + 5)(x^2 - 4)$$

> This can be factored further.

$$= 2x(x + 5)(x + 2)(x - 2)$$

Section 6.6 Solving Quadratic Equations by Factoring

The Zero-Product Principle

If $AB = 0$, then $A = 0$ or $B = 0$.

Solve: $(x - 6)(x + 10) = 0$.
$$x - 6 = 0 \quad \text{or} \quad x + 10 = 0$$
$$x = 6 \qquad\qquad x = -10$$
The solutions are -10 and 6, and the solution set is $\{-10, 6\}$.

A quadratic equation in x is an equation that can be written in the standard form

$$ax^2 + bx + c = 0, \quad a \neq 0.$$

To solve by factoring, write the equation in standard form, factor, set each factor equal to zero, and solve each resulting equation. Check proposed solutions in the original equation.

Solve: $4x^2 + 9x = 9$.
$$4x^2 + 9x - 9 = 0$$
$$(4x - 3)(x + 3) = 0$$
$$4x - 3 = 0 \quad \text{or} \quad x + 3 = 0$$
$$x = \frac{3}{4} \qquad\qquad x = -3$$
The solutions are -3 and $\frac{3}{4}$, and the solution set is $\left\{-3, \frac{3}{4}\right\}$.

CHAPTER 6 REVIEW EXERCISES

6.1 *In Exercises 1–5, factor each polynomial using the greatest common factor. If there is no common factor other than 1 and the polynomial cannot be factored, so state.*

1. $30x - 45$
2. $-12x^3 - 16x^2 + 400x$
3. $30x^4y + 15x^3y + 5x^2y$
4. $7(x + 3) - 2(x + 3)$
5. $7x^2(x + y) - (x + y)$

In Exercises 6–9, factor by grouping.

6. $x^3 + 3x^2 + 2x + 6$
7. $xy + y + 4x + 4$
8. $x^3 + 5x + x^2 + 5$
9. $xy + 4x - 2y - 8$

6.2 *In Exercises 10–17, factor completely, or state that the trinomial is prime.*

10. $x^2 - 3x + 2$
11. $x^2 - x - 20$
12. $x^2 + 19x + 48$
13. $x^2 - 6xy + 8y^2$
14. $x^2 + 5x - 9$
15. $x^2 + 16xy - 17y^2$
16. $3x^2 + 6x - 24$
17. $3x^3 - 36x^2 + 33x$

6.3 *In Exercises 18–26, factor completely, or state that the trinomial is prime.*

18. $3x^2 + 17x + 10$
19. $5y^2 - 17y + 6$
20. $4x^2 + 4x - 15$
21. $5y^2 + 11y + 4$
22. $-8x^2 - 8x + 6$
23. $2x^3 + 7x^2 - 72x$
24. $12y^3 + 28y^2 + 8y$
25. $2x^2 - 7xy + 3y^2$
26. $5x^2 - 6xy - 8y^2$

6.4 *In Exercises 27–30, factor each difference of two squares completely.*

27. $4x^2 - 1$
28. $81 - 100y^2$
29. $25a^2 - 49b^2$
30. $z^4 - 16$

In Exercises 31–34, factor completely, or state that the polynomial is prime.

31. $2x^2 - 18$
32. $x^2 + 1$
33. $9x^3 - x$
34. $18xy^2 - 8x$

In Exercises 35–41, factor any perfect square trinomials, or state that the polynomial is prime.

35. $x^2 + 22x + 121$
36. $x^2 - 16x + 64$
37. $9y^2 + 48y + 64$
38. $16x^2 - 40x + 25$
39. $25x^2 + 15x + 9$
40. $36x^2 + 60xy + 25y^2$
41. $25x^2 - 40xy + 16y^2$

In Exercises 42–45, factor using the formula for the sum or difference of two cubes.

42. $x^3 - 27$
43. $64x^3 + 1$
44. $54x^3 - 16y^3$
45. $27x^3y + 8y$

In Exercises 46–47, find the formula for the area of the blue shaded region and express it in factored form.

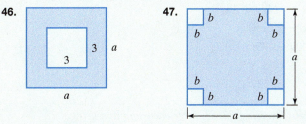

48. The figure shows a geometric interpretation of a factorization. Use the sum of the areas of the four pieces on the left and the area of the square on the right to write the factorization that is illustrated.

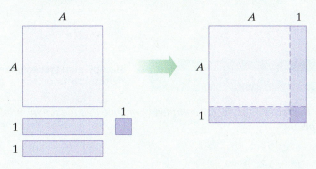

6.5 *In Exercises 49–81, factor completely, or state that the polynomial is prime.*

49. $x^3 - 8x^2 + 7x$
50. $10y^2 + 9y + 2$
51. $128 - 2y^2$
52. $9x^2 + 6x + 1$
53. $-20x^7 + 36x^3$
54. $x^3 - 3x^2 - 9x + 27$
55. $y^2 + 16$
56. $2x^3 + 19x^2 + 35x$
57. $3x^3 - 30x^2 + 75x$

58. $3x^5 - 24x^2$

59. $4y^4 - 36y^2$

60. $5x^2 + 20x - 105$

61. $9x^2 + 8x - 3$

62. $-10x^5 + 44x^4 - 16x^3$

63. $100y^2 - 49$

64. $9x^5 - 18x^4$

65. $x^4 - 1$

66. $2y^3 - 16$

67. $x^3 + 64$

68. $6x^2 + 11x - 10$

69. $3x^4 - 12x^2$

70. $x^2 - x - 90$

71. $25x^2 + 25xy + 6y^2$

72. $x^4 + 125x$

73. $32y^3 + 32y^2 + 6y$

74. $-2y^2 + 16y - 32$

75. $x^2 - 2xy - 35y^2$

76. $x^2 + 7x + xy + 7y$

77. $9x^2 + 24xy + 16y^2$

78. $2x^4y - 2x^2y$

79. $100y^2 - 49z^2$

80. $x^2 + xy + y^2$

81. $3x^4y^2 - 12x^2y^4$

6.6 *In Exercises 82–83, solve each equation using the zero-product principle.*

82. $x(x - 12) = 0$

83. $3(x - 7)(4x + 9) = 0$

In Exercises 84–92, use factoring to solve each quadratic equation.

84. $x^2 + 5x - 14 = 0$

85. $5x^2 + 20x = 0$

86. $2x^2 + 15x = 8$

87. $x(x - 4) = 32$

88. $(x + 3)(x - 2) = 50$

89. $x^2 = 14x - 49$

90. $9x^2 = 100$

91. $3x^2 + 21x + 30 = 0$

92. $3x^2 = 22x - 7$

93. You dive from a board that is 32 feet above the water. The formula
$$h = -16t^2 + 16t + 32$$
describes your height above the water, h, in feet, t seconds after you dive. How long will it take you to hit the water?

94. The length of a rectangular sign is 3 feet longer than the width. If the sign has space for 40 square feet of advertising, find its length and its width.

95. The square lot shown here is being turned into a garden with a 3-meter path at one end. If the area of the garden is 88 square meters, find the dimensions of the square lot.

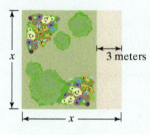

In Exercises 1–21, factor completely, or state that the polynomial is prime.

1. $x^2 - 9x + 18$

2. $x^2 - 14x + 49$

3. $15y^4 - 35y^3 + 10y^2$

4. $x^3 + 2x^2 + 3x + 6$

5. $x^2 - 9x$

6. $x^3 + 6x^2 - 7x$

7. $14x^2 + 64x - 30$

8. $25x^2 - 9$ 9. $x^3 + 8$

10. $x^2 - 4x - 21$

11. $x^2 + 4$

12. $6y^3 + 9y^2 + 3y$

13. $4y^2 - 36$

14. $16x^2 + 48x + 36$

15. $2x^4 - 32$

16. $36x^2 - 84x + 49$

17. $7x^2 - 50x + 7$

18. $x^3 + 2x^2 - 5x - 10$

19. $-12y^3 + 12y^2 + 45y$

20. $y^3 - 125$

21. $5x^2 - 5xy - 30y^2$

In Exercises 22–27, solve each quadratic equation.

22. $x^2 + 2x - 24 = 0$

23. $3x^2 - 5x = 2$

24. $x(x - 6) = 16$

25. $6x^2 = 21x$

26. $16x^2 = 81$

27. $(5x + 4)(x - 1) = 2$

28. Find a formula for the area of the shaded blue region and express it in factored form.

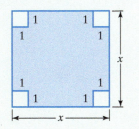

29. A model rocket is launched from a height of 96 feet. The formula

$$h = -16t^2 + 80t + 96$$

describes the rocket's height, h, in feet, t seconds after it was launched. How long will it take the rocket to reach the ground?

30. The length of a rectangular garden is 6 feet longer than its width. If the area of the garden is 55 square feet, find its length and its width.

CUMULATIVE REVIEW EXERCISES   (CHAPTERS 1–6)

1. Simplify: $6[5 + 2(3 - 8) - 3]$.

2. Solve: $4(x - 2) = 2(x - 4) + 3x$.

3. Solve: $\dfrac{x}{2} - 1 = \dfrac{x}{3} + 1$.

4. Solve and graph the solution set on a number line.

$$5 - 5x > 2(5 - x) + 1$$

5. Find the measures of the angles of a triangle whose two base angles have equal measure and whose third angle is $10°$ less than three times the measure of a base angle.

6. A dinner for six people cost $159, including 6% tax. What was the dinner cost before tax?

7. Graph using the slope and y-intercept: $y = -\frac{3}{5}x + 3$.

8. Write the point-slope form of the equation of the line passing through $(2, -4)$ and $(3, 1)$. Then use the point-slope form of the equation to write the slope-intercept form of the equation.

9. Graph: $5x - 6y > 30$.

10. Solve the system:

$$\begin{cases} 5x + 2y = 13 \\ y = 2x - 7. \end{cases}$$

11. Solve the system:

$$\begin{cases} 2x + 3y = 5 \\ 3x - 2y = -4. \end{cases}$$

12. Subtract: $\dfrac{4}{5} - \dfrac{9}{8}$.

In Exercises 13–15, perform the indicated operations.

13. $\dfrac{6x^5 - 3x^4 + 9x^2 + 27x}{3x}$

14. $(3x - 5y)(2x + 9y)$

15. $\dfrac{6x^3 + 5x^2 - 34x + 13}{3x - 5}$

16. Write 0.0071 in scientific notation.

In Exercises 17–19, factor completely.

17. $3x^2 + 11x + 6$

18. $y^5 - 16y$

19. $4x^2 + 12x + 9$

20. The length of a rectangle is 2 feet greater than its width. If the rectangle's area is 24 square feet, find its dimensions.

Rational Expressions

We are plagued by questions about the environment. Will we run out of oil? How hot will it get? Will there be neighborhoods where the air is pristine? Can we make garbage disappear? Will there be any wilderness left? Which wild animals will become extinct? How much will it cost to clean up toxic wastes from our rivers so that they can safely provide food, recreation, and enjoyment of wildlife for the millions who live along and visit their banks?

When making decisions on public policies dealing with the environment, algebraic fractions play an important role in modeling the costs. By learning to work with these fractional expressions, you will gain new insights into phenomena as diverse as the dosage of drugs prescribed for children, the digital transformation of the music industry, the cost of environmental cleanup, and even the shape of our heads.

Here's where you'll find these applications:

- Children and drug dosage: Exercise Set 7.4, Exercises 93–100
- The digital transformation of the music industry: Exercise Set 7.5, Exercises 51–52
- Costs of environmental cleanup: Section 7.1 opener; Section 7.1, page 490; Exercise Set 7.1, Exercise 86; Section 7.6, Example 6; Exercise Set 7.6, Exercises 57–58
- Shape of our heads: Section 7.3 opener; Exercise Set 7.3, Exercise 73.

7.1

Rational Expressions and Their Simplification

What am I supposed to learn?

After studying this section, you should be able to:

1 Find numbers for which a rational expression is undefined. ▶

2 Simplify rational expressions. ▶

3 Solve applied problems involving rational expressions. ▶

Discover for Yourself

What happens if you try substituting 100 for x in

$$\frac{250x}{100 - x}?$$

What does this tell you about the cost of cleaning up all of the river's pollutants?

How do we describe the costs of reducing environmental pollution? We often use algebraic expressions involving quotients of polynomials. For example, the algebraic expression

$$\frac{250x}{100 - x}$$

describes the cost, in millions of dollars, to remove x percent of the pollutants that are discharged into a river. Removing a modest percentage of pollutants, say 40%, is far less costly than removing a substantially greater percentage, such as 95%. We see this by evaluating the algebraic expression for $x = 40$ and $x = 95$.

$$\text{Evaluating } \frac{250x}{100 - x} \text{ for}$$

$$x = 40: \qquad\qquad x = 95:$$

$$\text{Cost is } \frac{250(40)}{100 - 40} \approx 167. \qquad \text{Cost is } \frac{250(95)}{100 - 95} = 4750.$$

The cost increases from approximately \$167 million to a possibly prohibitive \$4750 million, or \$4.75 billion. Costs spiral upward as the percentage of removed pollutants increases.

Many algebraic expressions that describe costs of environmental projects are examples of *rational expressions*. In this section, we introduce rational expressions and their simplification.

1 Find numbers for which a rational expression is undefined. ▶

Excluding Numbers from Rational Expressions

A **rational expression** is the quotient of two polynomials. Some examples are

$$\frac{x - 2}{4}, \quad \frac{4}{x - 2}, \quad \frac{x}{x^2 - 1}, \quad \text{and} \quad \frac{x^2 + 1}{x^2 + 2x - 3}.$$

Rational expressions indicate division and division by zero is undefined. This means that **we must exclude any value or values of the variable that make a denominator zero.** For example, consider the rational expression

$$\frac{4}{x - 2}.$$

When x is replaced with 2, the denominator is 0 and the expression is undefined.

$$\text{If } x = 2: \quad \frac{4}{x - 2} = \frac{4}{2 - 2} = \frac{4}{0}. \quad \boxed{\text{Division by zero is undefined.}}$$

Notice that if x is replaced by a number other than 2, such as 1, the expression is defined because the denominator is nonzero.

$$\text{If } x = 1: \quad \frac{4}{x - 2} = \frac{4}{1 - 2} = \frac{4}{-1} = -4.$$

Thus, only 2 must be excluded as a replacement for x in the rational expression $\dfrac{4}{x - 2}$.

Using Technology

We can use the $\boxed{\text{TABLE}}$ feature of a graphing utility to verify our work with $\dfrac{4}{x - 2}$. Enter

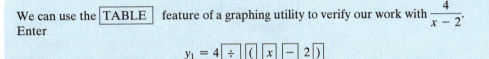

and press $\boxed{\text{TABLE}}$.

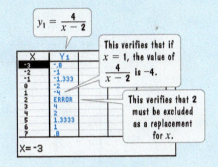

Excluding Values from Rational Expressions

If a variable in a rational expression is replaced by a number that causes the denominator to be 0, that number must be excluded as a replacement for the variable. The rational expression is undefined at any value that produces a denominator of 0.

How do we determine the value or values of the variable for which a rational expression is undefined? Set the denominator equal to 0 and then solve the resulting equation for the variable.

EXAMPLE 1   Determining Numbers for Which Rational Expressions Are Undefined

Find all the numbers for which the rational expression is undefined:

a. $\dfrac{6x + 12}{7x - 28}$ **b.** $\dfrac{2x + 6}{x^2 + 3x - 10}$.

Solution   In each case, we set the denominator equal to 0 and solve.

$$\frac{6x + 12}{7x - 28} \qquad \begin{array}{c} \text{Exclude values of } x \\ \text{that make these} \\ \text{denominators 0.} \end{array} \qquad \frac{2x + 6}{x^2 + 3x - 10}$$

a. $7x - 28 = 0$ Set the denominator of $\dfrac{6x + 12}{7x - 28}$ equal to O.

        $7x = 28$ Add 28 to both sides.

         $x = 4$ Divide both sides by 7.

Thus, $\dfrac{6x + 12}{7x - 28}$ is undefined for $x = 4$.

b. $x^2 + 3x - 10 = 0$ Set the denominator of $\dfrac{2x + 6}{x^2 + 3x - 10}$ equal to O.

$(x + 5)(x - 2) = 0$ Factor.

$x + 5 = 0$ or $x - 2 = 0$ Set each factor equal to O.

$x = -5$ $x = 2$ Solve the resulting equations.

Thus, $\dfrac{2x + 6}{x^2 + 3x - 10}$ is undefined for $x = -5$ and $x = 2$. ■

Using Technology

Graphic Connections

When using a graphing utility to graph an equation containing a rational expression, remember to enclose both the numerator and denominator in parentheses. Compare these two graphs of $y = \dfrac{6x + 12}{7x - 28}$.

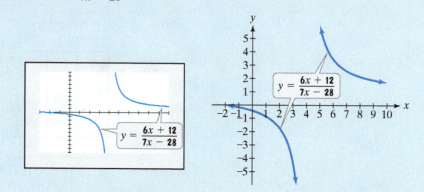

The graph on the left was obtained using a $[-3, 10, 1]$ by $[-10, 10, 1]$ viewing rectangle. Examine the behavior of the graph near $x = 4$, the number for which the rational expression is undefined. The values of the rational expression are decreasing as the values of x get closer to 4 on the left and increasing as the values of x get closer to 4 on the right. However, there is no point on the graph corresponding to $x = 4$. Some graphing utilities incorrectly connect the two pieces of the graph shown above, resulting in a nearly vertical line appearing on the screen. If your graphing utility does this, we recommend that you use $\boxed{\text{DOT}}$ mode. Consult your owner's manual.

☑ **CHECK POINT 1**   Find all the numbers for which the rational expression is undefined:

a. $\dfrac{7x - 28}{8x - 40}$ **b.** $\dfrac{8x - 40}{x^2 + 3x - 28}$.

Is every rational expression undefined for at least one number? No. Consider

$$\frac{x - 2}{x^2 + 1}.$$

Because the denominator, $x^2 + 1$, is not zero for any real-number replacement for x, the rational expression is defined for all real numbers. Thus, it is not necessary to exclude any values for x.

2 Simplify rational expressions.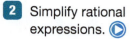

Simplifying Rational Expressions

A rational expression is **simplified** if its numerator and denominator have no common factors other than 1 or −1. The following principle is used to simplify a rational expression:

Fundamental Principle of Rational Expressions

If P, Q, and R are polynomials, and Q and R are not 0,

$$\frac{PR}{QR} = \frac{P}{Q}.$$

As you read the Fundamental Principle, can you see why $\dfrac{PR}{QR}$ is not simplified? The numerator and denominator have a common factor, the polynomial R. By dividing the numerator and the denominator by the common factor, R, we obtain the simplified form $\dfrac{P}{Q}$. This is often shown as follows:

$$\frac{P\overset{1}{R}}{Q\underset{1}{R}} = \frac{P}{Q}.$$

Observe that
$$\frac{PR}{QR} = \frac{P}{Q} \cdot \frac{R}{R} = \frac{P}{Q} \cdot 1 = \frac{P}{Q}.$$

The following procedure can be used to simplify rational expressions:

Simplifying Rational Expressions

1. Factor the numerator and the denominator completely.
2. Divide both the numerator and the denominator by any common factors.

EXAMPLE 2   Simplifying a Rational Expression

Simplify: $\dfrac{5x + 35}{20x}$.

Solution

$$\frac{5x + 35}{20x} = \frac{5(x + 7)}{5 \cdot 4x}$$ Factor the numerator and denominator. Because the denominator is 20x, $x \neq 0$.

$$= \frac{\overset{1}{5}(x + 7)}{\underset{1}{5} \cdot 4x}$$ Divide out the common factor of 5.

$$= \frac{x + 7}{4x} \quad \blacksquare$$

✓ CHECK POINT 2   Simplify: $\dfrac{7x + 28}{21x}$.

EXAMPLE 3   Simplifying a Rational Expression

Simplify: $\dfrac{x^3 + x^2}{x + 1}$.

Solution

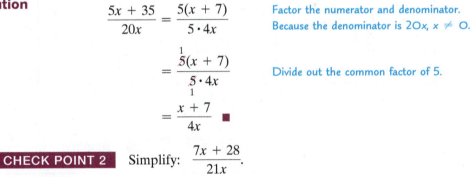

$$\frac{x^3 + x^2}{x + 1} = \frac{x^2(x + 1)}{x + 1}$$ Factor the numerator. Because the denominator is $x + 1$, $x \neq -1$.

$$= \frac{x^2(\overset{1}{\cancel{x + 1}})}{\underset{1}{\cancel{x + 1}}}$$ Divide out the common factor of $x + 1$.

$$= x^2 \quad \blacksquare$$

Simplifying a rational expression can change the numbers that make it undefined. For example, we just showed that

$$\frac{x^3 + x^2}{x + 1} = x^2.$$

This is undefined for $x = -1$.

This simplified form is defined for all real numbers.

Thus, to equate the two expressions, we must restrict the values for x in the simplified expression to exclude -1. We can write

$$\frac{x^3 + x^2}{x + 1} = x^2, \quad x \neq -1.$$

Hereafter, we will assume that the simplified rational expression is equal to the original rational expression for all real numbers except those for which either denominator is 0.

Using Technology

Graphic and Numeric Connections
A graphing utility can be used to verify that

$$\frac{x^3 + x^2}{x + 1} = x^2, \quad x \neq -1.$$

Enter $y_1 = \dfrac{x^3 + x^2}{x + 1}$ and $y_2 = x^2$.

Graphic Check

The screen shows the graph of y_1. Notice the gap in the graph at $x = -1$, where the rational function is undefined.

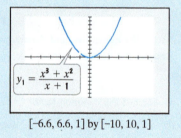

$[-6.6, 6.6, 1]$ by $[-10, 10, 1]$

This screen shows y_1 and y_2 graphed simultaneously. The red graph of y_2 completely covers the blue graph of y_1, but notice that the red graph does not have a gap at $x = -1$.

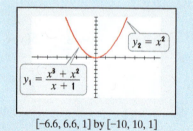

$[-6.6, 6.6, 1]$ by $[-10, 10, 1]$

Numeric Check

No matter how far up or down we scroll, if $x \neq -1$, $y_1 = y_2$. If $x = -1$, y_1 is undefined, although the value of y_2 is 1.

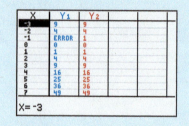

☑ **CHECK POINT 3** Simplify: $\dfrac{x^3 - x^2}{7x - 7}$.

EXAMPLE 4 Simplifying a Rational Expression

Simplify: $\dfrac{x^2 + 6x + 5}{x^2 - 25}$.

Solution

$$\frac{x^2 + 6x + 5}{x^2 - 25} = \frac{(x + 5)(x + 1)}{(x + 5)(x - 5)}$$

Factor the numerator and denominator. Because the denominator is $(x + 5)(x - 5)$, $x \neq -5$ and $x \neq 5$.

$$= \frac{\overset{1}{\cancel{(x + 5)}}(x + 1)}{\underset{1}{\cancel{(x + 5)}}(x - 5)}$$

Divide out the common factor of $x + 5$.

$$= \frac{x + 1}{x - 5} \quad \blacksquare$$

Great Question!

Can I simplify $\dfrac{x+1}{x-5}$ **by dividing the numerator and the denominator by** x**?**

No. When simplifying rational expressions, only *factors* that are common to the *entire numerator* and the *entire denominator* can be divided out. **It is incorrect to divide out common terms from the numerator and denominator.**

Incorrect!

$$\frac{\overset{1}{\cancel{x}}+1}{\underset{1}{\cancel{x}}-5} = \frac{1}{-5} = -\frac{1}{5} \qquad \frac{x^2-\overset{1}{\cancel{4}}}{\underset{1}{4}} = x^2 - 1 \qquad \frac{\overset{x}{\cancel{x^2}}-\overset{3}{\cancel{9}}}{\underset{1}{\cancel{x}}-\underset{1}{\cancel{3}}} = x-3$$

The first two expressions, $\dfrac{x+1}{x-5}$ and $\dfrac{x^2-4}{4}$, have no common factors in their numerators and denominators. Thus, these rational expressions are in simplified form. The rational expression $\dfrac{x^2-9}{x-3}$ can be simplified as follows:

Correct

$$\frac{x^2-9}{x-3} = \frac{(x+3)\overset{1}{\cancel{(x-3)}}}{\underset{1}{\cancel{x-3}}} = x+3.$$

> Divide out the common factor, $x - 3$.

✓ **CHECK POINT 4** Simplify: $\dfrac{x^2-1}{x^2+2x+1}$.

Factors That Are Opposites

How do we simplify rational expressions that contain factors in the numerator and denominator that are opposites, or additive inverses? Here is an example of such an expression:

$$\frac{x-3}{3-x}.$$

> The numerator and denominator are opposites. They differ only in their signs.

Factor out -1 from either the numerator or the denominator. Then divide out the common factor.

$$\frac{x-3}{3-x} = \frac{-1(-x+3)}{3-x}$$
    Factor -1 from the numerator. Notice how the sign of each term in the polynomial $x - 3$ changes.

$$= \frac{-1(3-x)}{3-x}$$
    In the numerator, use the commutative property to rewrite $-x + 3$ as $3 - x$.

$$= \frac{-1\overset{1}{\cancel{(3-x)}}}{\underset{1}{\cancel{3-x}}}$$
    Divide out the common factor of $3 - x$.

$$= -1$$

Our result, -1, suggests the following useful property:

Simplifying Rational Expressions with Opposite Factors in the Numerator and Denominator

The quotient of two polynomials that have opposite signs and are additive inverses is -1.

EXAMPLE 5 Simplifying a Rational Expression

Simplify: $\dfrac{4x^2 - 25}{15 - 6x}$.

Solution

$$\dfrac{4x^2 - 25}{15 - 6x} = \dfrac{(2x + 5)(2x - 5)}{3(5 - 2x)}$$ Factor the numerator and denominator.

$$= \dfrac{(2x + 5)\overset{-1}{\cancel{(2x - 5)}}}{3\,\cancel{(5 - 2x)}}$$ The quotient of polynomials with opposite signs is −1.

$$= \dfrac{-(2x + 5)}{3} \quad \text{or} \quad -\dfrac{2x + 5}{3} \quad \text{or} \quad \dfrac{-2x - 5}{3}$$

Each of these forms is an acceptable answer. ■

✓ **CHECK POINT 5** Simplify: $\dfrac{9x^2 - 49}{28 - 12x}$.

3 Solve applied problems involving rational expressions.

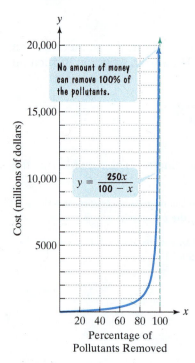

Figure 7.1

Applications

The equation

$$y = \dfrac{250x}{100 - x}$$

models the cost, in millions of dollars, to remove x percent of the pollutants that are discharged into a river. This equation contains the rational expression that we looked at in the opening to this section. Do you remember how costs were spiraling upward as the percentage of removed pollutants increased?

Is it possible to clean up the river completely? To do this, we must remove 100% of the pollutants. The problem is that the rational expression is undefined for $x = 100$.

$$y = \dfrac{250x}{100 - x}$$ If $x = 100$, the value of the denominator is 0.

Notice how the graph of $y = \dfrac{250x}{100 - x}$, shown in **Figure 7.1**, approaches but never touches the dashed vertical line $x = 100$, where 100 is the value of x for which $\dfrac{250x}{100 - x}$ is undefined. The graph continues to rise more and more steeply, visually showing the escalating costs. By never touching the dashed vertical line, the graph illustrates that no amount of money will be enough to remove all pollutants from the river.

Achieving Success

Analyze the errors you make on quizzes and tests.

For each error, write out the correct solution along with a description of the concept needed to solve the problem correctly. Do your mistakes indicate gaps in understanding concepts or do you at times believe that you are just not a good test taker? Are you repeatedly making the same kinds of mistakes on tests? Keeping track of errors should increase your understanding of the material, resulting in improved test scores.

CONCEPT AND VOCABULARY CHECK

Fill in each blank so that the resulting statement is true.

1. A rational expression is the quotient of two _____.

2. A rational expression is undefined for values that make the denominator _____.

3. We simplify a rational expression by _____ the numerator and the denominator completely. Then we divide the numerator and the denominator by any _____.

4. The rational expression $\dfrac{x-7}{x-7}$ simplifies to _____.

5. The rational expression $\dfrac{x-7}{7-x}$ simplifies to _____.

6. True or false: $\dfrac{x+5}{x+10}$ can be simplified. _____

7. True or false: $\dfrac{5x}{x+10}$ can be simplified. _____

8. True or false: $\dfrac{-x-10}{x+10}$ can be simplified. _____

9. True or false: The rational expression $\dfrac{x-2}{7x}$ is undefined for $x=2$. _____

7.1 EXERCISE SET   ▶ MyMathLab®

Practice Exercises

In Exercises 1–20, find all numbers for which each rational expression is undefined. If the rational expression is defined for all real numbers, so state.

1. $\dfrac{5}{2x}$

2. $\dfrac{11}{3x}$

3. $\dfrac{x}{x-8}$

4. $\dfrac{x}{x-6}$

5. $\dfrac{13}{5x-20}$

6. $\dfrac{17}{6x-30}$

7. $\dfrac{x+3}{(x+9)(x-2)}$

8. $\dfrac{x+5}{(x+7)(x-9)}$

9. $\dfrac{4x}{(3x-17)(x+3)}$

10. $\dfrac{8x}{(4x-19)(x+2)}$

11. $\dfrac{x+5}{x^2+x-12}$

12. $\dfrac{7x-14}{x^2-9x+20}$

13. $\dfrac{x+5}{5}$

14. $\dfrac{x+7}{7}$

15. $\dfrac{y+3}{4y^2+y-3}$

16. $\dfrac{y+8}{6y^2-y-2}$

17. $\dfrac{y+5}{y^2-25}$

18. $\dfrac{y+7}{y^2-49}$

19. $\dfrac{5}{x^2+1}$

20. $\dfrac{8}{x^2+4}$

In Exercises 21–76, simplify each rational expression. If the rational expression cannot be simplified, so state.

21. $\dfrac{14x^2}{7x}$

22. $\dfrac{9x^2}{6x}$

23. $\dfrac{5x-15}{25}$

24. $\dfrac{7x+21}{49}$

25. $\dfrac{2x-8}{4x}$

26. $\dfrac{3x-9}{6x}$

27. $\dfrac{3}{3x-9}$

28. $\dfrac{12}{6x-18}$

29. $\dfrac{-15}{3x-9}$

30. $\dfrac{-21}{7x-14}$

31. $\dfrac{3x+9}{x+3}$

32. $\dfrac{5x-10}{x-2}$

33. $\dfrac{x+5}{x^2-25}$

34. $\dfrac{x+4}{x^2-16}$

35. $\dfrac{2y-10}{3y-15}$

36. $\dfrac{6y+18}{11y+33}$

37. $\dfrac{x+1}{x^2-2x-3}$

38. $\dfrac{x+2}{x^2-x-6}$

39. $\dfrac{4x-8}{x^2-4x+4}$

40. $\dfrac{x^2-12x+36}{4x-24}$

41. $\dfrac{y^2-3y+2}{y^2+7y-18}$

42. $\dfrac{y^2+5y+4}{y^2-4y-5}$

43. $\dfrac{2y^2-7y+3}{2y^2-5y+2}$

44. $\dfrac{3y^2+4y-4}{6y^2-y-2}$

45. $\dfrac{2x+3}{2x+5}$

46. $\dfrac{3x+7}{3x+10}$

47. $\dfrac{x^2+12x+36}{x^2-36}$

48. $\dfrac{x^2-14x+49}{x^2-49}$

49. $\dfrac{x^3-2x^2+x-2}{x-2}$

50. $\dfrac{x^3+4x^2-3x-12}{x+4}$

51. $\dfrac{x^3-8}{x-2}$

52. $\dfrac{x^3-125}{x^2-25}$

53. $\dfrac{(x-4)^2}{x^2-16}$

54. $\dfrac{(x+5)^2}{x^2-25}$

55. $\dfrac{x}{x+1}$

56. $\dfrac{x}{x+7}$

57. $\dfrac{x+4}{x^2+16}$

58. $\dfrac{x+5}{x^2+25}$

59. $\dfrac{x-5}{5-x}$

60. $\dfrac{x-7}{7-x}$

61. $\dfrac{2x-3}{3-2x}$

62. $\dfrac{5x-4}{4-5x}$

63. $\dfrac{x-5}{x+5}$

64. $\dfrac{x-7}{x+7}$

65. $\dfrac{4x-6}{3-2x}$

66. $\dfrac{9x-15}{5-3x}$

67. $\dfrac{4-6x}{3x^2-2x}$

68. $\dfrac{9-15x}{5x^2-3x}$

69. $\dfrac{x^2-1}{1-x}$

70. $\dfrac{x^2-4}{2-x}$

71. $\dfrac{y^2-y-12}{4-y}$

72. $\dfrac{y^2-7y+12}{3-y}$

73. $\dfrac{x^2y-x^2}{x^3-x^3y}$

74. $\dfrac{xy-2x}{3y-6}$

75. $\dfrac{x^2+2xy-3y^2}{2x^2+5xy-3y^2}$

76. $\dfrac{x^2+3xy-10y^2}{3x^2-7xy+2y^2}$

Practice PLUS

In Exercises 77–84, simplify each rational expression.

77. $\dfrac{x^2-9x+18}{x^3-27}$

78. $\dfrac{x^3-8}{x^2+2x-8}$

79. $\dfrac{9-y^2}{y^2-3(2y-3)}$

80. $\dfrac{16-y^2}{y(y-8)+16}$

81. $\dfrac{xy+2y+3x+6}{x^2+5x+6}$

82. $\dfrac{xy+4y-7x-28}{x^2+11x+28}$

83. $\dfrac{8x^2+4x+2}{1-8x^3}$

84. $\dfrac{x^3-3x^2+9x}{x^3+27}$

Application Exercises

85. The rational expression

    $$\frac{130x}{100-x}$$

    describes the cost, in millions of dollars, to inoculate x percent of the population against a particular strain of flu.

    a. Evaluate the expression for $x = 40$, $x = 80$, and $x = 90$. Describe the meaning of each evaluation in terms of percentage inoculated and cost.

    b. For what value of x is the expression undefined?

    c. What happens to the cost as x approaches 100%? How can you interpret this observation?

86. The rational expression

    $$\frac{60{,}000x}{100-x}$$

    describes the cost, in dollars, to remove x percent of the air pollutants in the smokestack emissions of a utility company that burns coal to generate electricity.

    a. Evaluate the expression for $x = 20$, $x = 50$, and $x = 80$. Describe the meaning of each evaluation in terms of percentage of pollutants removed and cost.

    b. For what value of x is the expression undefined?

    c. What happens to the cost as x approaches 100%? How can you interpret this observation?

Doctors use the rational expression

$$\frac{DA}{A+12}$$

to determine the dosage of a drug prescribed for children. In this expression, A = the child's age and D = the adult dosage. Use the expression to solve Exercises 87–88.

87. If the normal adult dosage of medication is 1000 milligrams, what dosage should an 8-year-old child receive?

88. If the normal adult dosage of medication is 1000 milligrams, what dosage should a 4-year-old child receive?

89. A company that manufactures bicycles has costs given by the equation

$$C = \frac{100x + 100,000}{x},$$

in which x is the number of bicycles manufactured and C is the cost to manufacture each bicycle.

   a. Find the cost per bicycle when manufacturing 500 bicycles.

   b. Find the cost per bicycle when manufacturing 4000 bicycles.

   c. Does the cost per bicycle increase or decrease as more bicycles are manufactured? Explain why this happens.

90. A company that manufactures small canoes has costs given by the equation

$$C = \frac{20x + 20,000}{x},$$

in which x is the number of canoes manufactured and C is the cost to manufacture each canoe.

   a. Find the cost per canoe when manufacturing 100 canoes.

   b. Find the cost per canoe when manufacturing 10,000 canoes.

   c. Does the cost per canoe increase or decrease as more canoes are manufactured? Explain why this happens.

A drug is injected into a patient and the concentration of the drug in the bloodstream is monitored. The drug's concentration, y, in milligrams per liter, after x hours is modeled by

$$y = \frac{5x}{x^2 + 1}.$$

The graph of this equation, obtained with a graphing utility, is shown in the figure in a $[0, 10, 1]$ by $[0, 3, 1]$ viewing rectangle. Use this information to solve Exercises 91–92.

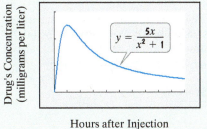

Hours after Injection
$[0, 10, 1]$ by $[0, 3, 1]$

91. Use the equation to find the drug's concentration after 3 hours. Then identify the point on the equation's graph that conveys this information.

92. Use the graph of the equation to find after how many hours the drug reaches its maximum concentration. Then use the equation to find the drug's concentration at this time.

Explaining the Concepts

93. What is a rational expression? Give an example with your explanation.

94. Explain how to find the number or numbers, if any, for which a rational expression is undefined.

95. Explain how to simplify a rational expression.

96. Explain how to simplify a rational expression with opposite factors in the numerator and denominator.

97. Use the graph shown for Exercises 91–92 to write a description of the drug's concentration over time. In your description, try to convey as much information as possible that is displayed visually by the graph.

Critical Thinking Exercises

Make Sense? *In Exercises 98–101, determine whether each statement makes sense or does not make sense, and explain your reasoning.*

98. Simplifying rational expressions is similar to reducing fractions.

99. I cannot simplify rational expressions without knowing how to factor polynomials.

100. The rational expressions

$$\frac{7}{14x} \quad \text{and} \quad \frac{7}{14 + x}$$

can both be simplified by dividing each numerator and each denominator by 7.

101. I evaluated $\dfrac{3x - 3}{4x(x - 1)}$ for $x = 1$ and obtained 0.

In Exercises 102–105, determine whether each statement is true or false. If the statement is false, make the necessary change(s) to produce a true statement.

102. $\dfrac{3x + 1}{3} = x + 1$

103. $\dfrac{x^2 + 3}{3} = x^2 + 1$

104. The expression $\dfrac{-3y - 6}{y + 2}$ simplifies to the consecutive integer that follows -4.

105. $\dfrac{3x + 7}{3x + 10} = \dfrac{8}{11}$

106. Write a rational expression that cannot be simplified.

107. Write a rational expression that is undefined for $x = -4$.

108. Write a rational expression with $x^2 - x - 6$ in the numerator that can be simplified to $x - 3$.

Technology Exercises

In Exercises 109–111, use the GRAPH *or* TABLE *feature of a graphing utility to determine if the rational expression has been correctly simplified. If the simplification is wrong, correct it and then verify your answer using the graphing utility.*

109. $\dfrac{3x + 15}{x + 5} = 3, \ x \neq -5$

110. $\dfrac{2x^2 - x - 1}{x - 1} = 2x^2 - 1, x \neq 1$

111. $\dfrac{x^2 - x}{x} = x^2 - 1, \ x \neq 0$

112. Use a graphing utility to verify the graph in **Figure 7.1** on page 490. TRACE along the graph as x approaches 100. What do you observe?

Review Exercises

113. Multiply: $\dfrac{5}{6} \cdot \dfrac{9}{25}$. (Section 1.2, Example 5)

114. Divide: $\dfrac{2}{3} \div 4$. (Section 1.2, Example 6)

115. Solve by the addition method:

$$\begin{cases} 2x - 5y = -2 \\ 3x + 4y = 20. \end{cases} \text{(Section 4.3, Example 3)}$$

Preview Exercises

Exercises 116–118 will help you prepare for the material covered in the next section. In each exercise, perform the indicated operation.

116. $\dfrac{2}{5} \cdot \dfrac{3}{7}$

117. $\dfrac{3}{4} \div \dfrac{1}{2}$

118. $\dfrac{5}{4} \div \dfrac{15}{8}$

SECTION

7.2

Multiplying and Dividing Rational Expressions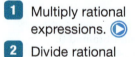

What am I supposed to learn?

After studying this section, you should be able to:

1. Multiply rational expressions.

2. Divide rational expressions.

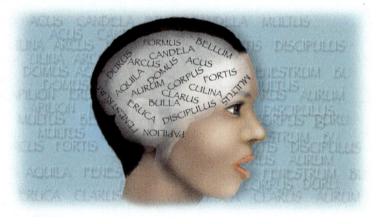

Your psychology class is learning various techniques to double what we remember over time. At the beginning of the course, students memorize 40 words in Latin, a language with which they are not familiar. The rational expression

$$\frac{5t + 30}{t}$$

models the class average for the number of words remembered after t days, where $t \geq 1$. If the techniques are successful, what will be the new memory model?

The new model can be found by multiplying the given rational expression by 2. In this section, you will see that we multiply rational expressions in the same way that we multiply rational numbers. Thus, we multiply numerators and multiply denominators. The rational expression for doubling what the class remembers over time is

$$\frac{2}{1} \cdot \frac{5t + 30}{t} = \frac{2(5t + 30)}{1 \cdot t} = \frac{2 \cdot 5t + 2 \cdot 30}{t} = \frac{10t + 60}{t}.$$

1 Multiply rational expressions.

Multiplying Rational Expressions

The product of two rational expressions is the product of their numerators divided by the product of their denominators.

> ### Multiplying Rational Expressions
>
> If P, Q, R, and S are polynomials, where $Q \neq 0$ and $S \neq 0$, then
>
> $$\frac{P}{Q} \cdot \frac{R}{S} = \frac{PR}{QS}.$$

EXAMPLE 1   Multiplying Rational Expressions

Multiply: $\dfrac{7}{x+3} \cdot \dfrac{x-2}{5}$.

Solution

$$\frac{7}{x+3} \cdot \frac{x-2}{5} = \frac{7(x-2)}{(x+3)5}$$ Multiply numerators. Multiply denominators. ($x \neq -3$)

$$= \frac{7x-14}{5x+15}$$ ■

☑ **CHECK POINT 1**   Multiply: $\dfrac{9}{x+4} \cdot \dfrac{x-5}{2}$.

Here is a step-by-step procedure for multiplying rational expressions. Before multiplying, divide out any factors common to both a numerator and a denominator.

> ### Multiplying Rational Expressions
>
> 1. Factor all numerators and denominators completely.
> 2. Divide numerators and denominators by common factors.
> 3. Multiply the remaining factors in the numerators and multiply the remaining factors in the denominators.

EXAMPLE 2   Multiplying Rational Expressions

Multiply: $\dfrac{x-3}{x+5} \cdot \dfrac{10x+50}{7x-21}$.

Solution

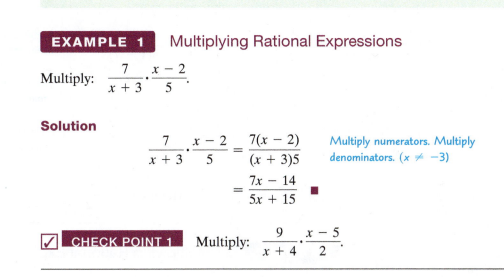

$$\frac{x-3}{x+5} \cdot \frac{10x+50}{7x-21}$$

$$= \frac{x-3}{x+5} \cdot \frac{10(x+5)}{7(x-3)}$$ Factor as many numerators and denominators as possible.

$$= \frac{x-\overset{1}{\cancel{3}}}{\cancel{x+5}} \cdot \frac{10(\overset{1}{\cancel{x+5}})}{7(\cancel{x-3})}$$ Divide numerators and denominators by common factors.

$$= \frac{10}{7}$$ Multiply the remaining factors in the numerators and in the denominators. ■

✓ **CHECK POINT 2** Multiply: $\dfrac{x+4}{x-7} \cdot \dfrac{3x-21}{8x+32}$.

EXAMPLE 3 Multiplying Rational Expressions

Multiply: $\dfrac{x-7}{x-1} \cdot \dfrac{x^2-1}{3x-21}$.

Solution

$$\frac{x-7}{x-1} \cdot \frac{x^2-1}{3x-21}$$

$$= \frac{x-7}{x-1} \cdot \frac{(x+1)(x-1)}{3(x-7)} \qquad \text{Factor as many numerators and denominators as possible.}$$

$$= \frac{\overset{1}{\cancel{x-7}}}{\underset{1}{\cancel{x-1}}} \cdot \frac{(x+1)\overset{1}{\cancel{(x-1)}}}{3\cancel{(x-7)}} \qquad \text{Divide numerators and denominators by common factors.}$$

$$= \frac{x+1}{3} \qquad \text{Multiply the remaining factors in the numerators and in the denominators.} \ ■$$

✓ **CHECK POINT 3** Multiply: $\dfrac{x-5}{x-2} \cdot \dfrac{x^2-4}{9x-45}$.

EXAMPLE 4 Multiplying Rational Expressions

Multiply: $\dfrac{4x+8}{6x-3x^2} \cdot \dfrac{3x^2-4x-4}{9x^2-4}$.

Solution

$$\frac{4x+8}{6x-3x^2} \cdot \frac{3x^2-4x-4}{9x^2-4}$$

$$= \frac{4(x+2)}{3x(2-x)} \cdot \frac{(3x+2)(x-2)}{(3x+2)(3x-2)} \qquad \text{Factor as many numerators and denominators as possible.}$$

$$= \frac{4(x+2)}{3x\underset{1}{\cancel{(2-x)}}} \cdot \frac{\overset{1}{\cancel{(3x+2)}}\overset{-1}{\cancel{(x-2)}}}{\underset{1}{\cancel{(3x+2)}}(3x-2)} \qquad \text{Divide numerators and denominators by common factors. Because } 2-x \text{ and } x-2 \text{ have opposite signs, their quotient is } -1.$$

$$= \frac{-4(x+2)}{3x(3x-2)} \quad \text{or} \quad -\frac{4(x+2)}{3x(3x-2)} \qquad \text{Multiply the remaining factors in the numerators and in the denominators.} \ ■$$

It is not necessary to carry out these multiplications.

✓ **CHECK POINT 4** Multiply: $\dfrac{5x+5}{7x-7x^2} \cdot \dfrac{2x^2+x-3}{4x^2-9}$.

---

2 Divide rational expressions. ▶

Dividing Rational Expressions

The quotient of two rational expressions is the product of the first expression and the multiplicative inverse, or reciprocal, of the second. The reciprocal is found by interchanging the numerator and the denominator.

Dividing Rational Expressions

If P, Q, R, and S are polynomials, where $Q \neq 0$, $R \neq 0$, and $S \neq 0$, then

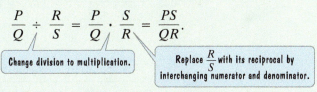

$$\frac{P}{Q} \div \frac{R}{S} = \frac{P}{Q} \cdot \frac{S}{R} = \frac{PS}{QR}.$$

Change division to multiplication.

Replace $\frac{R}{S}$ with its reciprocal by interchanging numerator and denominator.

Thus, **we find the quotient of two rational expressions by inverting the divisor and multiplying**. For example,

$$\frac{x}{7} \div \frac{6}{y} = \frac{x}{7} \cdot \frac{y}{6} = \frac{xy}{42}.$$

Change the division to multiplication.

Replace $\frac{6}{y}$ with its reciprocal by interchanging numerator and denominator.

Great Question!

If I'm multiplying or dividing rational expressions and an expression appears without a denominator, what should I do?

When performing operations with rational expressions, if a rational expression is written without a denominator, it is helpful to write the expression with a denominator of 1. In Example 5, we wrote $x + 5$ as

$$\frac{x + 5}{1}.$$

EXAMPLE 5 Dividing Rational Expressions

Divide: $(x + 5) \div \dfrac{x - 2}{x + 9}$.

Solution

$$(x + 5) \div \frac{x - 2}{x + 9} = \frac{x + 5}{1} \cdot \frac{x + 9}{x - 2} \qquad \text{Invert the divisor and multiply.}$$

$$= \frac{(x + 5)(x + 9)}{x - 2} \qquad \begin{array}{l}\text{Multiply the factors in the numerators and}\\ \text{in the denominators. We need not carry out}\\ \text{the multiplication in the numerator.} \ \blacksquare\end{array}$$

☑ **CHECK POINT 5** Divide: $(x + 3) \div \dfrac{x - 4}{x + 7}$.

EXAMPLE 6 Dividing Rational Expressions

Divide: $\dfrac{x^2 - 2x - 8}{x^2 - 9} \div \dfrac{x - 4}{x + 3}$.

Solution

$$\frac{x^2 - 2x - 8}{x^2 - 9} \div \frac{x - 4}{x + 3}$$

$$= \frac{x^2 - 2x - 8}{x^2 - 9} \cdot \frac{x + 3}{x - 4} \qquad \text{Invert the divisor and multiply.}$$

$$= \frac{(x - 4)(x + 2)}{(x + 3)(x - 3)} \cdot \frac{x + 3}{x - 4} \qquad \begin{array}{l}\text{Factor as many numerators and}\\ \text{denominators as possible.}\end{array}$$

$$= \frac{\overset{1}{\cancel{(x - 4)}}(x + 2)}{\underset{1}{\cancel{(x + 3)}}(x - 3)} \cdot \frac{\overset{1}{\cancel{(x + 3)}}}{\underset{1}{\cancel{(x - 4)}}} \qquad \begin{array}{l}\text{Divide numerators and denominators}\\ \text{by common factors.}\end{array}$$

$$= \frac{x + 2}{x - 3} \qquad \begin{array}{l}\text{Multiply the remaining factors in the}\\ \text{numerators and in the denominators.} \ \blacksquare\end{array}$$

☑ **CHECK POINT 6** Divide: $\dfrac{x^2 + 5x + 6}{x^2 - 25} \div \dfrac{x + 2}{x + 5}$.

EXAMPLE 7   Dividing Rational Expressions

Divide: $\dfrac{y^2 + 7y + 12}{y^2 + 9} \div (7y^2 + 21y)$.

Solution

$$\dfrac{y^2 + 7y + 12}{y^2 + 9} \div \dfrac{7y^2 + 21y}{1}$$ It is helpful to write the divisor with a denominator of 1.

$$= \dfrac{y^2 + 7y + 12}{y^2 + 9} \cdot \dfrac{1}{7y^2 + 21y}$$ Invert the divisor and multiply.

$$= \dfrac{(y + 4)(y + 3)}{y^2 + 9} \cdot \dfrac{1}{7y(y + 3)}$$ Factor as many numerators and denominators as possible.

$$= \dfrac{(y + 4)\overset{1}{(y + 3)}}{y^2 + 9} \cdot \dfrac{1}{7y\underset{1}{(y + 3)}}$$ Divide numerators and denominators by common factors.

$$= \dfrac{y + 4}{7y(y^2 + 9)}$$ Multiply the remaining factors in the numerators and in the denominators. ∎

✓ **CHECK POINT 7**   Divide: $\dfrac{y^2 + 3y + 2}{y^2 + 1} \div (5y^2 + 10y)$.

CONCEPT AND VOCABULARY CHECK

Fill in each blank so that the resulting statement is true.

1. The product of two rational expressions is the product of their _____ divided by the product of their _____:
   $\dfrac{P}{Q} \cdot \dfrac{R}{S} =$ _____ , $Q \neq 0, S \neq 0$.

2. The quotient of two rational expressions is the product of the first expression and the _____ of the second:
   $\dfrac{P}{Q} \div \dfrac{R}{S} = \dfrac{P}{Q} \cdot$ _____ = _____ , $Q \neq 0, R \neq 0, S \neq 0$.

3. $\dfrac{x}{5} \cdot \dfrac{x}{3} =$ _____

4. $\dfrac{x}{5} \div \dfrac{x}{3} =$ _____ , $x \neq 0$

7.2 EXERCISE SET   ▶ MyMathLab®

Practice Exercises

In Exercises 1–32, multiply as indicated.

1. $\dfrac{4}{x + 3} \cdot \dfrac{x - 5}{9}$

2. $\dfrac{8}{x - 2} \cdot \dfrac{x + 5}{3}$

3. $\dfrac{x}{3} \cdot \dfrac{12}{x + 5}$

4. $\dfrac{x}{5} \cdot \dfrac{30}{x - 4}$

5. $\dfrac{3}{x} \cdot \dfrac{4x}{15}$

6. $\dfrac{7}{x} \cdot \dfrac{5x}{35}$

7. $\dfrac{x - 3}{x + 5} \cdot \dfrac{4x + 20}{9x - 27}$

8. $\dfrac{x - 2}{x + 9} \cdot \dfrac{5x + 45}{2x - 4}$

9. $\dfrac{x^2 + 9x + 14}{x + 7} \cdot \dfrac{1}{x + 2}$

10. $\dfrac{x^2 + 9x + 18}{x + 6} \cdot \dfrac{1}{x + 3}$

11. $\dfrac{x^2 - 25}{x^2 - 3x - 10} \cdot \dfrac{x + 2}{x}$

12. $\dfrac{x^2 - 49}{x^2 - 4x - 21} \cdot \dfrac{x + 3}{x}$

13. $\dfrac{4y + 30}{y^2 - 3y} \cdot \dfrac{y - 3}{2y + 15}$

14. $\dfrac{9y + 21}{y^2 - 2y} \cdot \dfrac{y - 2}{3y + 7}$

15. $\dfrac{y^2 - 7y - 30}{y^2 - 6y - 40} \cdot \dfrac{2y^2 + 5y + 2}{2y^2 + 7y + 3}$

16. $\dfrac{3y^2 + 17y + 10}{3y^2 - 22y - 16} \cdot \dfrac{y^2 - 4y - 32}{y^2 - 8y - 48}$

17. $(y^2 - 9) \cdot \dfrac{4}{y - 3}$

18. $(y^2 - 16) \cdot \dfrac{3}{y - 4}$

19. $\dfrac{x^2 - 5x + 6}{x^2 - 2x - 3} \cdot \dfrac{x^2 - 1}{x^2 - 4}$

20. $\dfrac{x^2 + 5x + 6}{x^2 + x - 6} \cdot \dfrac{x^2 - 9}{x^2 - x - 6}$

21. $\dfrac{x^3 - 8}{x^2 - 4} \cdot \dfrac{x + 2}{3x}$

22. $\dfrac{x^2 + 6x + 9}{x^3 + 27} \cdot \dfrac{1}{x + 3}$

23. $\dfrac{(x - 2)^3}{(x - 1)^3} \cdot \dfrac{x^2 - 2x + 1}{x^2 - 4x + 4}$

24. $\dfrac{(x + 4)^3}{(x + 2)^3} \cdot \dfrac{x^2 + 4x + 4}{x^2 + 8x + 16}$

25. $\dfrac{6x + 2}{x^2 - 1} \cdot \dfrac{1 - x}{3x^2 + x}$

26. $\dfrac{8x + 2}{x^2 - 9} \cdot \dfrac{3 - x}{4x^2 + x}$

27. $\dfrac{25 - y^2}{y^2 - 2y - 35} \cdot \dfrac{y^2 - 8y - 20}{y^2 - 3y - 10}$

28. $\dfrac{2y}{3y - y^2} \cdot \dfrac{2y^2 - 9y + 9}{8y - 12}$

29. $\dfrac{x^2 - y^2}{x} \cdot \dfrac{x^2 + xy}{x + y}$

30. $\dfrac{4x - 4y}{x} \cdot \dfrac{x^2 + xy}{x^2 - y^2}$

31. $\dfrac{x^2 + 2xy + y^2}{x^2 - 2xy + y^2} \cdot \dfrac{4x - 4y}{3x + 3y}$

32. $\dfrac{x^2 - y^2}{x + y} \cdot \dfrac{x + 2y}{2x^2 - xy - y^2}$

In Exercises 33–64, divide as indicated.

33. $\dfrac{x}{7} \div \dfrac{5}{3}$

34. $\dfrac{x}{3} \div \dfrac{3}{8}$

35. $\dfrac{3}{x} \div \dfrac{12}{x}$

36. $\dfrac{x}{5} \div \dfrac{20}{x}$

37. $\dfrac{15}{x} \div \dfrac{3}{2x}$

38. $\dfrac{9}{x} \div \dfrac{3}{4x}$

39. $\dfrac{x + 1}{3} \div \dfrac{3x + 3}{7}$

40. $\dfrac{x + 5}{7} \div \dfrac{4x + 20}{9}$

41. $\dfrac{7}{x - 5} \div \dfrac{28}{3x - 15}$

42. $\dfrac{4}{x - 6} \div \dfrac{40}{7x - 42}$

43. $\dfrac{x^2 - 4}{x} \div \dfrac{x + 2}{x - 2}$

44. $\dfrac{x^2 - 4}{x - 2} \div \dfrac{x + 2}{4x - 8}$

45. $(y^2 - 16) \div \dfrac{y^2 + 3y - 4}{y^2 + 4}$

46. $(y^2 + 4y - 5) \div \dfrac{y^2 - 25}{y + 7}$

47. $\dfrac{y^2 - y}{15} \div \dfrac{y - 1}{5}$

48. $\dfrac{y^2 - 2y}{15} \div \dfrac{y - 2}{5}$

49. $\dfrac{4x^2 + 10}{x - 3} \div \dfrac{6x^2 + 15}{x^2 - 9}$

50. $\dfrac{x^2 + x}{x^2 - 4} \div \dfrac{x^2 - 1}{x^2 + 5x + 6}$

51. $\dfrac{x^2 - 25}{2x - 2} \div \dfrac{x^2 + 10x + 25}{x^2 + 4x - 5}$

52. $\dfrac{x^2 - 4}{x^2 + 3x - 10} \div \dfrac{x^2 + 5x + 6}{x^2 + 8x + 15}$

53. $\dfrac{y^3 + y}{y^2 - y} \div \dfrac{y^3 - y^2}{y^2 - 2y + 1}$

54. $\dfrac{3y^2 - 12}{y^2 + 4y + 4} \div \dfrac{y^3 - 2y^2}{y^2 + 2y}$

55. $\dfrac{y^2 + 5y + 4}{y^2 + 12y + 32} \div \dfrac{y^2 - 12y + 35}{y^2 + 3y - 40}$

56. $\dfrac{y^2 + 4y - 21}{y^2 + 3y - 28} \div \dfrac{y^2 + 14y + 48}{y^2 + 4y - 32}$

57. $\dfrac{2y^2 - 128}{y^2 + 16y + 64} \div \dfrac{y^2 - 6y - 16}{3y^2 + 30y + 48}$

58. $\dfrac{3y + 12}{y^2 + 3y} \div \dfrac{y^2 + y - 12}{9y - y^3}$

59. $\dfrac{2x + 2y}{3} \div \dfrac{x^2 - y^2}{x - y}$

60. $\dfrac{5x + 5y}{7} \div \dfrac{x^2 - y^2}{x - y}$

61. $\dfrac{x^2 - y^2}{8x^2 - 16xy + 8y^2} \div \dfrac{4x - 4y}{x + y}$

62. $\dfrac{4x^2 - y^2}{x^2 + 4xy + 4y^2} \div \dfrac{4x - 2y}{3x + 6y}$

63. $\dfrac{xy - y^2}{x^2 + 2x + 1} \div \dfrac{2x^2 + xy - 3y^2}{2x^2 + 5xy + 3y^2}$

64. $\dfrac{x^2 - 4y^2}{x^2 + 3xy + 2y^2} \div \dfrac{x^2 - 4xy + 4y^2}{x + y}$

Practice PLUS

In Exercises 65–72, perform the indicated operation or operations.

65. $\left(\dfrac{y - 2}{y^2 - 9y + 18} \cdot \dfrac{y^2 - 4y - 12}{y + 2} \right) \div \dfrac{y^2 - 4}{y^2 + 5y + 6}$

66. $\left(\dfrac{6y^2 + 31y + 18}{3y^2 - 20y + 12} \cdot \dfrac{2y^2 - 15y + 18}{6y^2 + 35y + 36} \right) \div \dfrac{2y^2 - 13y + 15}{9y^2 + 15y + 4}$

67. $\dfrac{3x^2 + 3x - 60}{2x - 8} \div \left(\dfrac{30x^2}{x^2 - 7x + 10} \cdot \dfrac{x^3 + 3x^2 - 10x}{25x^3} \right)$

68. $\dfrac{5x^2 - x}{3x + 2} \div \left(\dfrac{6x^2 + x - 2}{10x^2 + 3x - 1} \cdot \dfrac{2x^2 - x - 1}{2x^2 - x} \right)$

69. $\dfrac{x^2 + xz + xy + yz}{x - y} \div \dfrac{x + z}{x + y}$

70. $\dfrac{x^2 - xz + xy - yz}{x - y} \div \dfrac{x - z}{y - x}$

71. $\dfrac{3xy + ay + 3xb + ab}{9x^2 - a^2} \div \dfrac{y^3 + b^3}{6x - 2a}$

72. $\dfrac{5xy - ay - 5xb + ab}{25x^2 - a^2} \div \dfrac{y^3 - b^3}{15x + 3a}$

Application Exercises

In Exercises 73–76, write a simplified rational expression for the area of each figure. Assume that all measures are in inches.

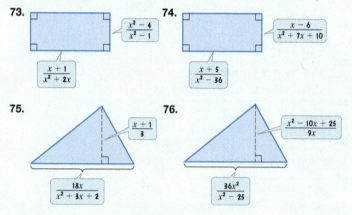

73. $\dfrac{x^2 - 4}{x^2 - 1}$ $\dfrac{x + 1}{x^2 + 2x}$

74. $\dfrac{x - 6}{x^2 + 7x + 10}$ $\dfrac{x + 5}{x^2 - 36}$

75. $\dfrac{x + 1}{3}$ $\dfrac{18x}{x^2 + 3x + 2}$

76. $\dfrac{x^2 - 10x + 25}{9x}$ $\dfrac{36x^2}{x^2 - 25}$

Explaining the Concepts

77. Explain how to multiply rational expressions.

78. Explain how to divide rational expressions.

79. In dividing polynomials

$$\frac{P}{Q} \div \frac{R}{S},$$

why is it necessary to state that polynomial R is not equal to 0?

Critical Thinking Exercises

Make Sense? *In Exercises 80–83, determine whether each statement makes sense or does not make sense, and explain your reasoning.*

80. When dividing rational expressions, I multiply by the reciprocal of the divisor, just as I did when dividing rational numbers.

81. When opposite factors appear in the numerator and the denominator of a multiplication problem, I can put their quotient, -1, in either the numerator or the denominator.

82. When performing the division

$$\frac{7x}{x + 3} \div \frac{(x + 3)^2}{x - 5},$$

I began by dividing the numerator and the denominator by the common factor, $x + 3$.

83. The quotient

$$\frac{x + 2}{x - 5} \div \frac{x - 4}{x + 3}$$

is undefined for $x = 5$, $x = -3$, and $x = 4$.

In Exercises 84–87, determine whether each statement is true or false. If the statement is false, make the necessary change(s) to produce a true statement.

84. $5 \div x = \dfrac{1}{5} \cdot x$ for any nonzero number x.

85. $\dfrac{4}{x} \div \dfrac{x - 2}{x} = \dfrac{4}{x - 2}$ if $x \neq 0$ and $x \neq 2$.

86. $\dfrac{x - 5}{6} \cdot \dfrac{3}{5 - x} = \dfrac{1}{2}$ for any value of x except 5.

87. The quotient of two rational expressions can be found by inverting each expression and multiplying.

88. Find the missing polynomials: $\dfrac{\rule{1cm}{2mm}}{\rule{1cm}{2mm}} \cdot \dfrac{3x - 12}{2x} = \dfrac{3}{2}$.

89. Find the missing polynomials: $-\dfrac{1}{2x - 3} \div \dfrac{\rule{1cm}{2mm}}{\rule{1cm}{2mm}} = \dfrac{1}{3}$.

90. Divide:

$$\frac{9x^2 - y^2 + 15x - 5y}{3x^2 + xy + 5x} \div \frac{3x + y}{9x^3 + 6x^2y + xy^2}.$$

Technology Exercises

In Exercises 91–94, use the [GRAPH] or [TABLE] feature of a graphing utility to determine if the multiplication or division has been performed correctly. If the answer is wrong, correct it and then verify your correction using the graphing utility.

91. $\dfrac{x^2 + x}{3x} \cdot \dfrac{6x}{x + 1} = 2x$

92. $\dfrac{x^3 - 25x}{x^2 - 3x - 10} \cdot \dfrac{x + 2}{x} = x + 5$

93. $\dfrac{x^2 - 9}{x + 4} \div \dfrac{x - 3}{x + 4} = x - 3$

94. $(x - 5) \div \dfrac{2x^2 - 11x + 5}{4x^2 - 1} = 2x - 1$

Review Exercises

95. Solve: $2x + 3 < 3(x - 5)$. (Section 2.7, Example 8)

96. Factor completely: $3x^2 - 15x - 42$. (Section 6.5, Example 2)

97. Solve: $x(2x + 9) = 5$. (Section 6.6, Example 6)

Preview Exercises

Exercises 98–100 will help you prepare for the material covered in the next section.

98. Subtract: $\dfrac{7}{9} - \dfrac{1}{9}$.

99. Add: $\dfrac{2x}{3} + \dfrac{x}{3}$.

100. Simplify: $\dfrac{x^2 - 6x + 9}{x^2 - 9}$.

SECTION

7.3

Adding and Subtracting Rational Expressions with the Same Denominator

What am I supposed to learn?

After studying this section, you should be able to:

1. Add rational expressions with the same denominator. ▶

2. Subtract rational expressions with the same denominator. ▶

3. Add and subtract rational expressions with opposite denominators. ▶

1. Add rational expressions with the same denominator. ▶

Are you long, medium, or round? Your skull, that is? The varying shapes of the human skull create physical diversity in the human species. By learning to add and subtract rational expressions with the same denominator, you will obtain an expression that models this diversity. (See Exercise 73.)

Addition When Denominators Are the Same

To add rational numbers having the same denominators, such as $\frac{2}{9}$ and $\frac{5}{9}$, we add the numerators and place the sum over the common denominator:

$$\frac{2}{9} + \frac{5}{9} = \frac{2+5}{9} = \frac{7}{9}.$$

We add rational expressions with the same denominator in an identical manner.

Adding Rational Expressions with Common Denominators

If $\dfrac{P}{R}$ and $\dfrac{Q}{R}$ are rational expressions, then

$$\frac{P}{R} + \frac{Q}{R} = \frac{P+Q}{R}.$$

To add rational expressions with the same denominator, add numerators and place the sum over the common denominator. If possible, factor and simplify the result.

EXAMPLE 1   Adding Rational Expressions When Denominators Are the Same

Add: $\dfrac{2x-1}{3} + \dfrac{x+4}{3}$.

Solution

$$\frac{2x-1}{3} + \frac{x+4}{3} = \frac{2x-1+x+4}{3}$$

Add numerators. Place this sum over the common denominator.

$$= \frac{3x+3}{3}$$

Combine like terms.

$$= \frac{\overset{1}{3}(x+1)}{\underset{1}{3}}$$

Factor and simplify.

$$= x+1 \quad \blacksquare$$

 **CHECK POINT 1** Add: $\dfrac{3x-2}{5}+\dfrac{2x+12}{5}$.

EXAMPLE 2 Adding Rational Expressions When Denominators Are the Same

Add: $\dfrac{x^2}{x^2-9}+\dfrac{9-6x}{x^2-9}$.

Solution

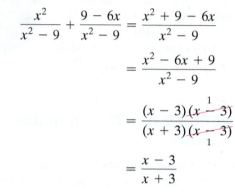

$$\dfrac{x^2}{x^2-9}+\dfrac{9-6x}{x^2-9}=\dfrac{x^2+9-6x}{x^2-9}$$ Add numerators. Place this sum over the common denominator.

$$=\dfrac{x^2-6x+9}{x^2-9}$$ Write the numerator in descending powers of x.

$$=\dfrac{(x-3)\overset{1}{\cancel{(x-3)}}}{(x+3)\underset{1}{\cancel{(x-3)}}}$$ Factor and simplify. What values of x are not permitted?

$$=\dfrac{x-3}{x+3}$$ ∎

 **CHECK POINT 2** Add: $\dfrac{x^2}{x^2-25}+\dfrac{25-10x}{x^2-25}$.

---

2 Subtract rational expressions with the same denominator. ◉

Subtraction When Denominators Are the Same

The following box shows how to subtract rational expressions with the same denominator:

> ### Subtracting Rational Expressions with Common Denominators
>
> If $\dfrac{P}{R}$ and $\dfrac{Q}{R}$ are rational expressions, then
>
> $$\dfrac{P}{R}-\dfrac{Q}{R}=\dfrac{P-Q}{R}.$$
>
> To subtract rational expressions with the same denominator, subtract numerators and place the difference over the common denominator. If possible, factor and simplify the result.

EXAMPLE 3 Subtracting Rational Expressions When Denominators Are the Same

Subtract:

a. $\dfrac{2x+3}{x+1}-\dfrac{x}{x+1}$ **b.** $\dfrac{5x+1}{x^2-9}-\dfrac{4x-2}{x^2-9}$.

Solution

a. $\dfrac{2x + 3}{x + 1} - \dfrac{x}{x + 1} = \dfrac{2x + 3 - x}{x + 1}$

Subtract numerators. Place this difference over the common denominator.

$= \dfrac{x + 3}{x + 1}$

Combine like terms.

b. $\dfrac{5x + 1}{x^2 - 9} - \dfrac{4x - 2}{x^2 - 9} = \dfrac{5x + 1 - (4x - 2)}{x^2 - 9}$

Subtract numerators and include parentheses to indicate that both terms are subtracted. Place this difference over the common denominator.

$= \dfrac{5x + 1 - 4x + 2}{x^2 - 9}$

Remove parentheses and then change the sign of each term.

$= \dfrac{x + 3}{x^2 - 9}$

Combine like terms.

$= \dfrac{\overset{1}{\cancel{(x + 3)}}}{\underset{1}{\cancel{(x + 3)}}(x - 3)}$

Factor and simplify ($x \neq -3$, and $x \neq 3$).

$= \dfrac{1}{x - 3}$ ∎

✓ **CHECK POINT 3** Subtract:

a. $\dfrac{4x + 5}{x + 7} - \dfrac{x}{x + 7}$

b. $\dfrac{3x^2 + 4x}{x - 1} - \dfrac{11x - 4}{x - 1}.$

Great Question!

When subtracting a numerator containing more than one term, do I really need to insert parentheses like you did in Example 3(b)?

Yes. When a numerator is being subtracted, we need to be sure that we **subtract every term in that expression**. Parentheses indicate that every term inside is being subtracted.

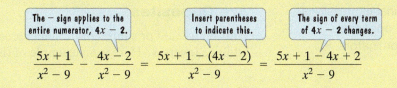

The − sign applies to the entire numerator, 4x − 2. Insert parentheses to indicate this. The sign of every term of 4x − 2 changes.

$$\dfrac{5x + 1}{x^2 - 9} - \dfrac{4x - 2}{x^2 - 9} = \dfrac{5x + 1 - (4x - 2)}{x^2 - 9} = \dfrac{5x + 1 - 4x + 2}{x^2 - 9}$$

The entire numerator of the second rational expression must be subtracted. Avoid the common error of subtracting only the first term.

Incorrect!

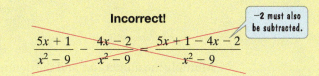

−2 must also be subtracted.

$$\dfrac{5x + 1}{x^2 - 9} - \dfrac{4x - 2}{x^2 - 9} \quad \dfrac{5x + 1 - 4x - 2}{x^2 - 9}$$

EXAMPLE 4 Subtracting Rational Expressions When Denominators Are the Same

Subtract: $\dfrac{20y^2 + 5y + 1}{6y^2 + y - 2} - \dfrac{8y^2 - 12y - 5}{6y^2 + y - 2}.$

Solution

$\dfrac{20y^2 + 5y + 1}{6y^2 + y - 2} - \dfrac{8y^2 - 12y - 5}{6y^2 + y - 2}$

> Don't forget the parentheses.

$= \dfrac{20y^2 + 5y + 1 - (8y^2 - 12y - 5)}{6y^2 + y - 2}$

Subtract numerators. Place this difference over the common denominator.

$= \dfrac{20y^2 + 5y + 1 - 8y^2 + 12y + 5}{6y^2 + y - 2}$

Remove parentheses and then change the sign of each term.

$= \dfrac{(20y^2 - 8y^2) + (5y + 12y) + (1 + 5)}{6y^2 + y - 2}$

Group like terms. This step is usually performed mentally.

$= \dfrac{12y^2 + 17y + 6}{6y^2 + y - 2}$

Combine like terms.

$= \dfrac{\overset{1}{\cancel{(3y + 2)}}(4y + 3)}{\underset{1}{\cancel{(3y + 2)}}(2y - 1)}$

Factor and simplify.

$= \dfrac{4y + 3}{2y - 1}$ ∎

✓ **CHECK POINT 4** Subtract: $\dfrac{y^2 + 3y - 6}{y^2 - 5y + 4} - \dfrac{4y - 4 - 2y^2}{y^2 - 5y + 4}.$

---

③ Add and subtract rational expressions with opposite denominators. ▶

Addition and Subtraction When Denominators Are Opposites

How do we add or subtract rational expressions when denominators are opposites, or additive inverses? Here is an example of this type of addition problem:

$$\dfrac{x^2}{x - 5} + \dfrac{4x + 5}{5 - x}.$$

> These denominators are opposites. They differ only in their signs.

Multiply the numerator and the denominator of either of the rational expressions by −1. Then they will both have the same denominator.

EXAMPLE 5 Adding Rational Expressions When Denominators Are Opposites

Add: $\dfrac{x^2}{x-5} + \dfrac{4x+5}{5-x}$.

Solution

$$\dfrac{x^2}{x-5} + \dfrac{4x+5}{5-x}$$

$$= \dfrac{x^2}{x-5} + \dfrac{(-1)}{(-1)} \cdot \dfrac{4x+5}{5-x}$$ Multiply the numerator and denominator of the second rational expression by -1.

$$= \dfrac{x^2}{x-5} + \dfrac{-4x-5}{-5+x}$$ Perform the multiplications by -1 by changing every term's sign.

$$= \dfrac{x^2}{x-5} + \dfrac{-4x-5}{x-5}$$ Rewrite $-5+x$ as $x-5$. Both rational expressions have the same denominator.

$$= \dfrac{x^2 + (-4x-5)}{x-5}$$ Add numerators. Place this sum over the common denominator.

$$= \dfrac{x^2 - 4x - 5}{x-5}$$ Remove parentheses.

$$= \dfrac{\overset{1}{\cancel{(x-5)}}(x+1)}{\underset{1}{\cancel{x-5}}}$$ Factor and simplify.

$$= x + 1 \quad \blacksquare$$

✓ **CHECK POINT 5** Add: $\dfrac{x^2}{x-7} + \dfrac{4x+21}{7-x}$.

Adding and Subtracting Rational Expressions with Opposite Denominators

When one denominator is the opposite, or additive inverse, of the other, first multiply either rational expression by $\dfrac{-1}{-1}$ to obtain a common denominator.

EXAMPLE 6 Subtracting Rational Expressions When Denominators Are Opposites

Subtract: $\dfrac{5x-x^2}{x^2-4x-3} - \dfrac{3x-x^2}{3+4x-x^2}$.

Solution We note that $x^2 - 4x - 3$ and $3 + 4x - x^2$ are opposites. We multiply the second rational expression by $\dfrac{-1}{-1}$.

$$\dfrac{(-1)}{(-1)} \cdot \dfrac{3x-x^2}{3+4x-x^2} = \dfrac{-3x+x^2}{-3-4x+x^2}$$ Multiply the numerator and denominator by -1 by changing every term's sign.

$$= \dfrac{x^2-3x}{x^2-4x-3}$$ Write the numerator and the denominator in descending powers of x.

We now return to the original subtraction problem.

$$\frac{5x - x^2}{x^2 - 4x - 3} - \frac{3x - x^2}{3 + 4x - x^2}$$ This is the given problem.

$$= \frac{5x - x^2}{x^2 - 4x - 3} - \frac{x^2 - 3x}{x^2 - 4x - 3}$$ Replace the second rational expression by the form obtained through multiplication by $\frac{-1}{-1}$.

$$= \frac{5x - x^2 - (x^2 - 3x)}{x^2 - 4x - 3}$$ Subtract numerators. Place this difference over the common denominator. Don't forget parentheses!

$$= \frac{5x - x^2 - x^2 + 3x}{x^2 - 4x - 3}$$ Remove parentheses and then change the sign of each term.

$$= \frac{-2x^2 + 8x}{x^2 - 4x - 3}$$ Combine like terms in the numerator. Although the numerator can be factored, further simplification is not possible. ∎

✓ **CHECK POINT 6** Subtract: $\dfrac{7x - x^2}{x^2 - 2x - 9} - \dfrac{5x - 3x^2}{9 + 2x - x^2}$.

CONCEPT AND VOCABULARY CHECK

Fill in each blank so that the resulting statement is true.

1. $\dfrac{P}{R} + \dfrac{Q}{R} =$ _____: To add rational expressions with the same denominator, add _____ and place the sum over the _____.

2. $\dfrac{P}{R} - \dfrac{Q}{R} =$ _____: To subtract rational expressions with the same denominator, subtract _____ and place the difference over the _____.

3. When adding and subtracting rational expressions with opposite denominators, multiply either expression by _____ to obtain a common denominator.

4. $\dfrac{x}{3} + \dfrac{5}{3} =$ _____

5. $\dfrac{x}{3} - \dfrac{5}{3} =$ _____

6. $\dfrac{x}{3} - \dfrac{5 - y}{3} =$ _____

7.3 EXERCISE SET   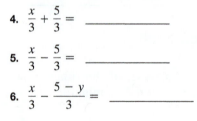 MyMathLab®

Practice Exercises

In Exercises 1–38, add or subtract as indicated. Simplify the result, if possible.

1. $\dfrac{7x}{13} + \dfrac{2x}{13}$

2. $\dfrac{3x}{17} + \dfrac{8x}{17}$

3. $\dfrac{8x}{15} + \dfrac{x}{15}$

4. $\dfrac{9x}{24} + \dfrac{x}{24}$

5. $\dfrac{x - 3}{12} + \dfrac{5x + 21}{12}$

6. $\dfrac{x + 4}{9} + \dfrac{2x - 25}{9}$

7. $\dfrac{4}{x} + \dfrac{2}{x}$

8. $\dfrac{5}{x} + \dfrac{13}{x}$

9. $\dfrac{8}{9x} + \dfrac{13}{9x}$

10. $\dfrac{4}{9x} + \dfrac{11}{9x}$

11. $\dfrac{5}{x + 3} + \dfrac{4}{x + 3}$

12. $\dfrac{8}{x + 6} + \dfrac{10}{x + 6}$

13. $\dfrac{x}{x - 3} + \dfrac{4x + 5}{x - 3}$

14. $\dfrac{x}{x - 4} + \dfrac{9x + 7}{x - 4}$

15. $\dfrac{4x + 1}{6x + 5} + \dfrac{8x + 9}{6x + 5}$

16. $\dfrac{3x + 2}{3x + 4} + \dfrac{3x + 6}{3x + 4}$

17. $\dfrac{y^2 + 7y}{y^2 - 5y} + \dfrac{y^2 - 4y}{y^2 - 5y}$

18. $\dfrac{y^2 - 2y}{y^2 + 3y} + \dfrac{y^2 + y}{y^2 + 3y}$

19. $\dfrac{4y - 1}{5y^2} + \dfrac{3y + 1}{5y^2}$

20. $\dfrac{y + 2}{6y^3} + \dfrac{3y - 2}{6y^3}$

21. $\dfrac{x^2 - 2}{x^2 + x - 2} + \dfrac{2x - x^2}{x^2 + x - 2}$

22. $\dfrac{x^2 + 9x}{4x^2 - 11x - 3} + \dfrac{3x - 5x^2}{4x^2 - 11x - 3}$

23. $\dfrac{x^2 - 4x}{x^2 - x - 6} + \dfrac{4x - 4}{x^2 - x - 6}$

24. $\dfrac{x}{2x + 7} - \dfrac{2}{2x + 7}$

25. $\dfrac{3x}{5x - 4} - \dfrac{4}{5x - 4}$

26. $\dfrac{x}{x - 1} - \dfrac{1}{x - 1}$

27. $\dfrac{4x}{4x - 3} - \dfrac{3}{4x - 3}$

28. $\dfrac{2y + 1}{3y - 7} - \dfrac{y + 8}{3y - 7}$

29. $\dfrac{14y}{7y + 2} - \dfrac{7y - 2}{7y + 2}$

30. $\dfrac{2x + 3}{3x - 6} - \dfrac{3 - x}{3x - 6}$

31. $\dfrac{3x + 1}{4x - 2} - \dfrac{x + 1}{4x - 2}$

32. $\dfrac{x^3 - 3}{2x^4} - \dfrac{7x^3 - 3}{2x^4}$

33. $\dfrac{3y^2 - 1}{3y^3} - \dfrac{6y^2 - 1}{3y^3}$

34. $\dfrac{y^2 + 3y}{y^2 + y - 12} - \dfrac{y^2 - 12}{y^2 + y - 12}$

35. $\dfrac{4y^2 + 5}{9y^2 - 64} - \dfrac{y^2 - y + 29}{9y^2 - 64}$

36. $\dfrac{2y^2 + 6y + 8}{y^2 - 16} - \dfrac{y^2 - 3y - 12}{y^2 - 16}$

37. $\dfrac{6y^2 + y}{2y^2 - 9y + 9} - \dfrac{2y + 9}{2y^2 - 9y + 9} - \dfrac{4y - 3}{2y^2 - 9y + 9}$

38. $\dfrac{3y^2 - 2}{3y^2 + 10y - 8} - \dfrac{y + 10}{3y^2 + 10y - 8} - \dfrac{y^2 - 6y}{3y^2 + 10y - 8}$

In Exercises 39–64, denominators are opposites, or additive inverses. Add or subtract as indicated. Simplify the result, if possible.

39. $\dfrac{4}{x - 3} + \dfrac{2}{3 - x}$

40. $\dfrac{6}{x - 5} + \dfrac{2}{5 - x}$

41. $\dfrac{6x + 7}{x - 6} + \dfrac{3x}{6 - x}$

42. $\dfrac{6x + 5}{x - 2} + \dfrac{4x}{2 - x}$

43. $\dfrac{5x - 2}{3x - 4} + \dfrac{2x - 3}{4 - 3x}$

44. $\dfrac{9x - 1}{7x - 3} + \dfrac{6x - 2}{3 - 7x}$

45. $\dfrac{x^2}{x - 2} + \dfrac{4}{2 - x}$

46. $\dfrac{x^2}{x - 3} + \dfrac{9}{3 - x}$

47. $\dfrac{y - 3}{y^2 - 25} + \dfrac{y - 3}{25 - y^2}$

48. $\dfrac{y - 7}{y^2 - 16} + \dfrac{7 - y}{16 - y^2}$

49. $\dfrac{6}{x - 1} - \dfrac{5}{1 - x}$

50. $\dfrac{10}{x - 2} - \dfrac{6}{2 - x}$

51. $\dfrac{10}{x + 3} - \dfrac{2}{-x - 3}$

52. $\dfrac{11}{x + 7} - \dfrac{5}{-x - 7}$

53. $\dfrac{y}{y - 1} - \dfrac{1}{1 - y}$

54. $\dfrac{y}{y - 4} - \dfrac{4}{4 - y}$

55. $\dfrac{3 - x}{x - 7} - \dfrac{2x - 5}{7 - x}$

56. $\dfrac{4 - x}{x - 9} - \dfrac{3x - 8}{9 - x}$

57. $\dfrac{x - 2}{x^2 - 25} - \dfrac{x - 2}{25 - x^2}$

58. $\dfrac{x - 8}{x^2 - 16} - \dfrac{x - 8}{16 - x^2}$

59. $\dfrac{x}{x - y} + \dfrac{y}{y - x}$

60. $\dfrac{2x - y}{x - y} + \dfrac{x - 2y}{y - x}$

61. $\dfrac{2x}{x^2 - y^2} + \dfrac{2y}{y^2 - x^2}$

62. $\dfrac{2y}{x^2 - y^2} + \dfrac{2x}{y^2 - x^2}$

63. $\dfrac{x^2 - 2}{x^2 + 6x - 7} + \dfrac{19 - 4x}{7 - 6x - x^2}$

64. $\dfrac{2x + 3}{x^2 - x - 30} + \dfrac{x - 2}{30 + x - x^2}$

Practice PLUS

In Exercises 65–72, perform the indicated operation or operations. Simplify the result, if possible.

65. $\dfrac{6b^2 - 10b}{16b^2 - 48b + 27} + \dfrac{7b^2 - 20b}{16b^2 - 48b + 27} - \dfrac{6b - 3b^2}{16b^2 - 48b + 27}$

66. $\dfrac{22b + 15}{12b^2 + 52b - 9} + \dfrac{30b - 20}{12b^2 + 52b - 9} - \dfrac{4 - 2b}{12b^2 + 52b - 9}$

67. $\dfrac{2y}{y - 5} - \left(\dfrac{2}{y - 5} + \dfrac{y - 2}{y - 5} \right)$

68. $\dfrac{3x}{(x + 1)^2} - \left[\dfrac{5x + 1}{(x + 1)^2} - \dfrac{3x + 2}{(x + 1)^2} \right]$

69. $\dfrac{b}{ac + ad - bc - bd} - \dfrac{a}{ac + ad - bc - bd}$

70. $\dfrac{y}{ax + bx - ay - by} - \dfrac{x}{ax + bx - ay - by}$

71. $\dfrac{(y - 3)(y + 2)}{(y + 1)(y - 4)} - \dfrac{(y + 2)(y + 3)}{(y + 1)(4 - y)} - \dfrac{(y + 5)(y - 1)}{(y + 1)(4 - y)}$

72. $\dfrac{(y + 1)(2y - 1)}{(y - 2)(y - 3)} + \dfrac{(y + 2)(y - 1)}{(y - 2)(y - 3)} - \dfrac{(y + 5)(2y + 1)}{(3 - y)(2 - y)}$

Application Exercises

73. Anthropologists and forensic scientists classify skulls using

$$\frac{L + 60W}{L} - \frac{L - 40W}{L},$$

where L is the skull's length and W is its width.

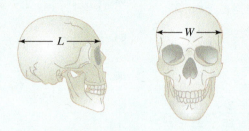

a. Express the classification as a single rational expression.

b. If the value of the rational expression in part (a) is less than 75, a skull is classified as long. A medium skull has a value between 75 and 80, and a round skull has a value over 80. Use your rational expression from part (a) to classify a skull that is 5 inches wide and 6 inches long.

74. The temperature, in degrees Fahrenheit, of a dessert placed in a freezer for t hours is modeled by

$$\frac{t + 30}{t^2 + 4t + 1} - \frac{t - 50}{t^2 + 4t + 1}.$$

a. Express the temperature as a single rational expression.

b. Use your rational expression from part (a) to find the temperature of the dessert, to the nearest hundredth of a degree, after 1 hour and after 2 hours.

In Exercises 75–76, find the perimeter of each rectangle.

75.

$\frac{5}{x + 3}$ meters

$\frac{5x + 10}{x + 3}$ meters

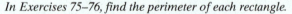

76.

$\frac{7}{x + 4}$ inches

$\frac{4x + 9}{x + 4}$ inches

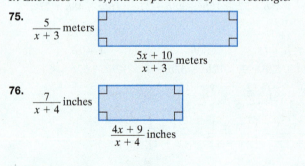

Explaining the Concepts

77. Explain how to add rational expressions when denominators are the same. Give an example with your explanation.

78. Explain how to subtract rational expressions when denominators are the same. Give an example with your explanation.

79. Describe two similarities between the following problems:

$$\frac{3}{8} + \frac{1}{8} \quad \text{and} \quad \frac{x}{x^2 - 1} + \frac{1}{x^2 - 1}.$$

80. Explain how to add rational expressions when denominators are opposites. Use an example to support your explanation.

Critical Thinking Exercises

Make Sense? *In Exercises 81–84, determine whether each statement makes sense or does not make sense, and explain your reasoning.*

81. After adding $\frac{3x + 1}{4}$ and $\frac{x + 2}{4}$, I simplified the sum by dividing the numerator and the denominator by 4.

82. I use similar procedures to find each of the following sums:

$$\frac{3}{8} + \frac{1}{8} \quad \text{and} \quad \frac{x}{x^2 - 1} + \frac{1}{x^2 - 1}.$$

83. I subtracted $\frac{3x - 5}{x - 1}$ from $\frac{x - 3}{x - 1}$ and obtained a constant.

84. I added $\frac{5}{x - 7}$ and $\frac{3}{7 - x}$ by first multiplying the second rational expression by -1.

In Exercises 85–88, determine whether each statement is true or false. If the statement is false, make the necessary change(s) to produce a true statement.

85. The sum of two rational expressions with the same denominator can be found by adding numerators, adding denominators, and then simplifying.

86. $\frac{4}{b} - \frac{2}{-b} = -\frac{2}{b}$

87. The difference between two rational expressions with the same denominator can always be simplified.

88. $\frac{2x + 1}{x - 7} + \frac{3x + 1}{x - 7} - \frac{5x + 2}{x - 7} = 0$

In Exercises 89–90, perform the indicated operations. Simplify the result, if possible.

89. $\left(\frac{3x - 1}{x^2 + 5x - 6} - \frac{2x - 7}{x^2 + 5x - 6} \right) \div \frac{x + 2}{x^2 - 1}$

90. $\left(\frac{3x^2 - 4x + 4}{3x^2 + 7x + 2} - \frac{10x + 9}{3x^2 + 7x + 2} \right) \div \frac{x - 5}{x^2 - 4}$

In Exercises 91–95, find the missing expression.

91. $\frac{2x}{x + 3} + \frac{\rule{0.5cm}{0.3cm}}{x + 3} = \frac{4x + 1}{x + 3}$

92. $\frac{3x}{x + 2} - \frac{\rule{0.5cm}{0.3cm}}{x + 2} = \frac{6 - 17x}{x + 2}$

93. $\frac{6}{x - 2} + \frac{\rule{0.5cm}{0.3cm}}{2 - x} = \frac{13}{x - 2}$

94. $\frac{a^2}{a - 4} - \frac{\rule{0.5cm}{0.3cm}}{a - 4} = a + 3$

95. $\frac{3x}{x - 5} + \frac{\rule{0.5cm}{0.3cm}}{5 - x} = \frac{7x + 1}{x - 5}$

Technology Exercises

In Exercises 96–98, use the GRAPH *or* TABLE *feature of a graphing utility to determine if the subtraction has been performed correctly. If the answer is wrong, correct it and then verify your correction using the graphing utility.*

96. $\dfrac{3x + 6}{2} - \dfrac{x}{2} = x + 3$

97. $\dfrac{x^2 + 4x + 3}{x + 2} - \dfrac{5x + 9}{x + 2} = x - 2, x \neq -2$

98. $\dfrac{x^2 - 13}{x + 4} - \dfrac{3}{x + 4} = x + 4, x \neq -4$

Review Exercises

99. Subtract: $\dfrac{13}{15} - \dfrac{8}{45}$. (Section 1.2, Example 9)

100. Factor completely: $81x^4 - 1$. (Section 6.4, Example 4)

101. Divide: $\dfrac{3x^3 + 2x^2 - 26x - 15}{x + 3}$. (Section 5.6, Example 2)

Preview Exercises

Exercises 102–104 will help you prepare for the material covered in the next section.
In Exercises 102–103, perform the indicated operation. Where possible, reduce the answer to its lowest terms.

102. $\dfrac{1}{2} + \dfrac{2}{3}$

103. $\dfrac{1}{8} - \dfrac{5}{6}$

104. Simplify: $\dfrac{(y + 2)y - 2 \cdot 4}{4y(y + 4)}$.

SECTION

7.4

Adding and Subtracting Rational Expressions with Different Denominators

What am I supposed to learn?

After studying this section, you should be able to:

1. Find the least common denominator. ▶

2. Add and subtract rational expressions with different denominators. ▶

1 Find the least common denominator. ▶

When my aunt asked how I liked my five-year-old nephew, I replied "medium rare." Unfortunately, my little joke did not get me out of baby sitting for the Dennis the Menace of our family. Now the little squirt doesn't want to go to bed because his head hurts. Does my aunt have any Tylenol®? What is the proper dosage for a child his age?

In this section's Exercise Set (Exercises 93–100), you will use two formulas that model drug dosage for children. Before working with these models, we continue drawing on your experience from arithmetic to add and subtract rational expressions that have different denominators.

Finding the Least Common Denominator

We can gain insight into adding rational expressions with different denominators by looking closely at what we do when adding fractions with different denominators. For example, suppose that we want to add $\frac{1}{2}$ and $\frac{2}{3}$. We must first write the fractions with the same denominator. We look for the smallest number that contains both 2 and 3 as factors. This number, 6, is then used as the *least common denominator*, or LCD.

The **least common denominator** of several rational expressions is a polynomial consisting of the product of all prime factors in the denominators, with each factor raised to the greatest power of its occurrence in any denominator.

Finding the Least Common Denominator

1. Factor each denominator completely.
2. List the factors of the first denominator.
3. Add to the list in step 2 any factors of the second denominator that do not appear in the list.
4. Form the product of the factors from the list in step 3. This product is the least common denominator.

EXAMPLE 1 Finding the Least Common Denominator

Find the LCD of $\dfrac{7}{6x^2}$ and $\dfrac{2}{9x}$.

Solution

Step 1. Factor each denominator completely.

$$6x^2 = 3 \cdot 2 \cdot x \cdot x$$
$$9x = 3 \cdot 3 \cdot x$$

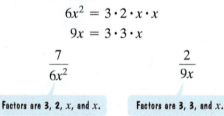

$$\dfrac{7}{6x^2} \qquad\qquad \dfrac{2}{9x}$$

Factors are **3, 2,** x, and x.   Factors are **3, 3,** and x.

Step 2. List the factors of the first denominator.

$$3, 2, x, x$$

Step 3. Add any unlisted factors from the second denominator. Two factors from $9x$ are already in our list. These factors include one factor of 3 and x. We add the other factor of 3 to our list. We have

$$3, 3, 2, x, x.$$

Step 4. The least common denominator is the product of all factors in the final list. Thus,

$$3 \cdot 3 \cdot 2 \cdot x \cdot x,$$

or $18x^2$ is the least common denominator. ∎

✓ **CHECK POINT 1**   Find the LCD of $\dfrac{3}{10x^2}$ and $\dfrac{7}{15x}$.

EXAMPLE 2 Finding the Least Common Denominator

Find the LCD of $\dfrac{3}{x + 1}$ and $\dfrac{5}{x - 1}$.

Solution

Step 1. Factor each denominator completely.

$$x + 1 = 1(x + 1)$$
$$x - 1 = 1(x - 1)$$

$$\dfrac{3}{x + 1} \qquad\qquad \dfrac{5}{x - 1}$$

Factors are 1 and $x + 1$.   Factors are 1 and $x - 1$.

Step 2. List the factors of the first denominator.

$$1, x + 1$$

Step 3. Add any unlisted factors from the second denominator. One factor of $x - 1$ is already in our list. That factor is 1. However, the other factor, $x - 1$, is not listed in step 2. We add $x - 1$ to the list. We have

$$1, x + 1, x - 1.$$

Step 4. The least common denominator is the product of all factors in the final list. Thus

$$1(x + 1)(x - 1)$$

or $(x + 1)(x - 1)$ is the least common denominator of $\dfrac{3}{x + 1}$ and $\dfrac{5}{x - 1}$. ∎

☑ **CHECK POINT 2**   Find the LCD of $\dfrac{2}{x + 3}$ and $\dfrac{4}{x - 3}$.

EXAMPLE 3   Finding the Least Common Denominator

Find the LCD of

$$\frac{7}{5x^2 + 15x} \quad \text{and} \quad \frac{9}{x^2 + 6x + 9}.$$

Solution

Step 1. Factor each denominator completely.

$$5x^2 + 15x = 5x(x + 3)$$

$$x^2 + 6x + 9 = (x + 3)^2 \quad \text{or} \quad (x + 3)(x + 3)$$

$$\underbrace{\frac{7}{5x^2 + 15x}}_{\text{Factors are 5, } x, \text{ and } x + 3.} \qquad \underbrace{\frac{9}{x^2 + 6x + 9}}_{\text{Factors are } x + 3 \text{ and } x + 3.}$$

Step 2. List the factors of the first denominator.

$$5, x, x + 3$$

Step 3. Add any unlisted factors from the second denominator. One factor of $x^2 + 6x + 9$ is already in our list. That factor is $x + 3$. However, the other factor of $x + 3$ is not listed in step 2. We add $x + 3$ to the list. We have

$$5, x, x + 3, x + 3.$$

Step 4. The least common denominator is the product of all factors in the final list. Thus,

$$5x(x + 3)(x + 3) \quad \text{or} \quad 5x(x + 3)^2$$

is the least common denominator. ∎

☑ **CHECK POINT 3**   Find the LCD of $\dfrac{9}{7x^2 + 28x}$ and $\dfrac{11}{x^2 + 8x + 16}$.

---

2 Add and subtract rational expressions with different denominators. ▶

Adding and Subtracting Rational Expressions with Different Denominators

Finding the least common denominator for two (or more) rational expressions is the first step needed to add or subtract the expressions. For example, to add $\frac{1}{2}$ and $\frac{2}{3}$, we first determine that the LCD is 6. Then we write each fraction in terms of the LCD.

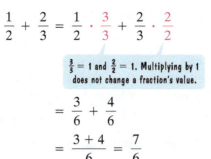

$$\frac{1}{2} + \frac{2}{3} = \frac{1}{2} \cdot \frac{3}{3} + \frac{2}{3} \cdot \frac{2}{2}$$

Multiply the numerator and denominator of each fraction by whatever extra factors are required to form 6, the LCD.

$\frac{3}{3} = 1$ and $\frac{2}{2} = 1$. Multiplying by 1 does not change a fraction's value.

$$= \frac{3}{6} + \frac{4}{6}$$

$$= \frac{3 + 4}{6} = \frac{7}{6}$$

Add numerators. Place this sum over the LCD.

We follow the same steps in adding or subtracting rational expressions with different denominators.

> ### Adding and Subtracting Rational Expressions That Have Different Denominators
>
> 1. Find the LCD of the rational expressions.
> 2. Rewrite each rational expression as an equivalent expression whose denominator is the LCD. To do so, multiply the numerator and the denominator of each rational expression by any factor(s) needed to convert the denominator into the LCD.
> 3. Add or subtract numerators, placing the resulting expression over the LCD.
> 4. If possible, simplify the resulting rational expression.

EXAMPLE 4 Adding Rational Expressions with Different Denominators

Add: $\dfrac{7}{6x^2} + \dfrac{2}{9x}$.

Solution

Step 1. Find the least common denominator. In Example 1, we found that the LCD for these rational expressions is $18x^2$.

Step 2. Write equivalent expressions with the LCD as denominators. We must rewrite each rational expression with a denominator of $18x^2$.

$$\frac{7}{6x^2} \cdot \frac{3}{3} = \frac{21}{18x^2} \qquad \frac{2}{9x} \cdot \frac{2x}{2x} = \frac{4x}{18x^2}$$

Multiply the numerator and denominator by 3 to get $18x^2$, the LCD.

Multiply the numerator and denominator by $2x$ to get $18x^2$, the LCD.

Great Question!

Can't I just simplify things and add rational expressions by adding numerators and adding denominators?

No. It is incorrect to add rational expressions by adding numerators and adding denominators. Avoid this common error.

Incorrect!

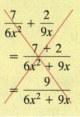

Because $\dfrac{3}{3} = 1$ and $\dfrac{2x}{2x} = 1$, we are not changing the value of either rational expression, only its appearance.

Now we are ready to perform the indicated addition.

$$\frac{7}{6x^2} + \frac{2}{9x}$$

This is the given problem. The LCD is $18x^2$.

$$= \frac{7}{6x^2} \cdot \frac{3}{3} + \frac{2}{9x} \cdot \frac{2x}{2x}$$

Write equivalent expressions with the LCD.

$$= \frac{21}{18x^2} + \frac{4x}{18x^2}$$

Steps 3 and 4. Add numerators, putting this sum over the LCD. Simplify, if possible.

$$= \frac{21 + 4x}{18x^2} \quad \text{or} \quad \frac{4x + 21}{18x^2}$$

The numerator is prime and further simplification is not possible. ∎

✓ **CHECK POINT 4** Add: $\dfrac{3}{10x^2} + \dfrac{7}{15x}$.

EXAMPLE 5 Adding Rational Expressions with Different Denominators

Add: $\dfrac{3}{x+1} + \dfrac{5}{x-1}$.

Solution

Step 1. Find the least common denominator. The factors of the denominators are 1, $x + 1$, and $x - 1$. In Example 2, we found that the LCD is $(x + 1)(x - 1)$.

Step 2. Write equivalent expressions with the LCD as denominators.

$$\dfrac{3}{x+1} + \dfrac{5}{x-1}$$

$$= \dfrac{3(x-1)}{(x+1)(x-1)} + \dfrac{5(x+1)}{(x+1)(x-1)}$$

Multiply each numerator and denominator by the extra factor required to form $(x+1)(x-1)$, the LCD.

Steps 3 and 4. Add numerators, putting this sum over the LCD. Simplify, if possible.

$$= \dfrac{3(x-1) + 5(x+1)}{(x+1)(x-1)}$$

$$= \dfrac{3x - 3 + 5x + 5}{(x+1)(x-1)}$$

Use the distributive property to multiply and remove grouping symbols.

$$= \dfrac{8x + 2}{(x+1)(x-1)}$$

Combine like terms: $3x + 5x = 8x$ and $-3 + 5 = 2$. ∎

We can factor 2 from the numerator of the answer in Example 5 to obtain

$$\dfrac{2(4x + 1)}{(x+1)(x-1)}.$$

Because the numerator and denominator do not have any common factors, further simplification is not possible. In this section, unless there is a common factor in the numerator and denominator, we will leave an answer's numerator in unfactored form and the denominator in factored form.

✓ **CHECK POINT 5** Add: $\dfrac{2}{x+3} + \dfrac{4}{x-3}$.

EXAMPLE 6 Subtracting Rational Expressions with Different Denominators

Subtract: $\dfrac{x}{x+3} - 1$.

Solution

Step 1. Find the least common denominator. We know that 1 means $\frac{1}{1}$. The factors of the first denominator are 1 and $x + 3$:

$$1, x + 3.$$

The factor of the second denominator, 1, is already in this list. Thus, the LCD is $1(x + 3)$, or $x + 3$.

Step 2. Write equivalent expressions with the LCD as denominators.

$$\frac{x}{x + 3} - 1$$

$$= \frac{x}{x + 3} - \frac{1}{1} \qquad \text{Write 1 as } \tfrac{1}{1}.$$

$$= \frac{x}{x + 3} - \frac{1(x + 3)}{1(x + 3)} \qquad \text{Multiply the numerator and denominator of } \tfrac{1}{1} \text{ by the extra factor required to form } x + 3, \text{ the LCD.}$$

Steps 3 and 4. Subtract numerators, putting this difference over the LCD. Simplify, if possible.

$$= \frac{x - (x + 3)}{x + 3}$$

$$= \frac{x - x - 3}{x + 3} \qquad \text{Remove parentheses and then change the sign of each term.}$$

$$= \frac{-3}{x + 3} \quad \text{or} \quad -\frac{3}{x + 3} \qquad \text{Simplify.} \ \blacksquare$$

✓ **CHECK POINT 6** Subtract: $\dfrac{x}{x + 5} - 1.$

EXAMPLE 7 Subtracting Rational Expressions with Different Denominators

Subtract: $\dfrac{y + 2}{4y + 16} - \dfrac{2}{y^2 + 4y}.$

Solution

Step 1. Find the least common denominator. Start by factoring the denominators.

$$4y + 16 = 4(y + 4)$$
$$y^2 + 4y = y(y + 4)$$

The factors of the first denominator are 4 and $y + 4$. The only factor from the second denominator that is unlisted is y. Thus, the least common denominator is $4y(y + 4)$.

Step 2. Write equivalent expressions with the LCD as denominators.

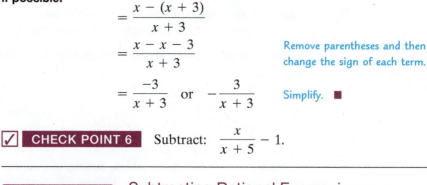

$$\frac{y + 2}{4y + 16} - \frac{2}{y^2 + 4y}$$

$$= \frac{y + 2}{4(y + 4)} - \frac{2}{y(y + 4)} \qquad \text{Factor denominators. The LCD is } 4y(y + 4).$$

$$= \frac{(y + 2)y}{4y(y + 4)} - \frac{2 \cdot 4}{4y(y + 4)} \qquad \text{Multiply each numerator and denominator by the extra factor required to form } 4y(y + 4), \text{ the LCD.}$$

Steps 3 and 4. Subtract numerators, putting this difference over the LCD. Simplify, if possible.

$$= \frac{(y + 2)y - 2 \cdot 4}{4y(y + 4)}$$

$$= \frac{y^2 + 2y - 8}{4y(y + 4)} \qquad \text{Use the distributive property.}$$

$$= \frac{\overset{1}{\cancel{(y + 4)}}(y - 2)}{4y\,\underset{1}{\cancel{(y + 4)}}} \qquad \text{Factor and simplify.}$$

$$= \frac{y - 2}{4y} \ \blacksquare$$

☑ **CHECK POINT 7** Subtract: $\dfrac{5}{y^2 - 5y} - \dfrac{y}{5y - 25}$.

In some situations, after factoring denominators, a factor in one denominator is the opposite of a factor in the other denominator. When this happens, we can use the following procedure:

> ### Adding and Subtracting Rational Expressions When Denominators Contain Opposite Factors
>
> When one denominator contains the opposite factor of the other, first multiply either rational expression by $\frac{-1}{-1}$. Then apply the procedure for adding or subtracting rational expressions that have different denominators to the rewritten problem.

EXAMPLE 8 ### Adding Rational Expressions with Opposite Factors in the Denominators

Add: $\dfrac{x^2 - 2}{2x^2 - x - 3} + \dfrac{x - 2}{3 - 2x}$.

Solution

Step 1. Find the least common denominator. Start by factoring the denominators.

$$2x^2 - x - 3 = (2x - 3)(x + 1)$$

$$3 - 2x = 1(3 - 2x)$$

Do you see that $2x - 3$ and $3 - 2x$ are opposite factors? Thus, we multiply either rational expression by $\frac{-1}{-1}$. We will use the second rational expression, resulting in $2x - 3$ in the denominator.

$$\frac{x^2 - 2}{2x^2 - x - 3} + \frac{x - 2}{3 - 2x}$$

$$= \frac{x^2 - 2}{(2x - 3)(x + 1)} + \frac{(-1)}{(-1)} \cdot \frac{x - 2}{3 - 2x}$$ Factor the first denominator. Multiply the second rational expression by $\frac{-1}{-1}$.

$$= \frac{x^2 - 2}{(2x - 3)(x + 1)} + \frac{-x + 2}{-3 + 2x}$$ Perform the multiplications by -1 by changing every term's sign.

$$= \frac{x^2 - 2}{(2x - 3)(x + 1)} + \frac{2 - x}{2x - 3}$$ Rewrite $-3 + 2x$ as $2x - 3$.

The LCD of our rewritten addition problem is $(2x - 3)(x + 1)$.

Step 2. Write equivalent expressions with the LCD as denominators.

$$= \frac{x^2 - 2}{(2x - 3)(x + 1)} + \frac{(2 - x)(x + 1)}{(2x - 3)(x + 1)}$$ Multiply the numerator and denominator of the second rational expression by the extra factor required to form $(2x - 3)(x + 1)$, the LCD.

Here is the original addition problem and our equivalent expressions with the LCD:

$$\frac{x^2 - 2}{2x^2 - x - 3} + \frac{x - 2}{3 - 2x}$$

$$= \frac{x^2 - 2}{(2x - 3)(x + 1)} + \frac{(2 - x)(x + 1)}{(2x - 3)(x + 1)}.$$

Discover for Yourself

In Example 8, the denominators can be factored as follows:

$2x^2 - x - 3 = (2x - 3)(x + 1)$
$3 - 2x = -1(2x - 3).$

Using these factorizations, what is the LCD? Solve Example 8 by obtaining this LCD in each rational expression. Then combine the expressions. How does your solution compare with the one shown on the right?

Steps 3 and 4. Add numerators, putting this sum over the LCD. Simplify, if possible.

$$= \frac{x^2 - 2 + (2 - x)(x + 1)}{(2x - 3)(x + 1)}$$

$$= \frac{x^2 - 2 + 2x + 2 - x^2 - x}{(2x - 3)(x + 1)}$$
Use the FOIL method to multiply $(2 - x)(x + 1)$.

$$= \frac{(x^2 - x^2) + (2x - x) + (-2 + 2)}{(2x - 3)(x + 1)}$$
Group like terms.

$$= \frac{x}{(2x - 3)(x + 1)}$$
Combine like terms. ■

✓ **CHECK POINT 8**   Add: $\dfrac{4x}{x^2 - 25} + \dfrac{3}{5 - x}.$

Achieving Success

Take some time to consider how you are doing in this course. Check your performance by answering the following questions:

- Are you attending all lectures?
- For each hour of class time, are you spending at least two hours outside of class completing all homework assignments, checking answers, correcting errors, and using all resources to get the help that you need?
- Are you reviewing for quizzes and tests?
- Are you reading the textbook? In all college courses, you are responsible for the information in the text, whether or not it is covered in class.
- Are you keeping an organized notebook? Does each page have the appropriate section number from the text on top? Do the pages contain examples your instructor works during lecture and other relevant class notes? Have you included your worked-out homework exercises? Do you keep a special section for graded exams?
- Are you analyzing your mistakes and learning from your errors?
- Are there ways you can improve how you are doing in the course?

CONCEPT AND VOCABULARY CHECK

Fill in each blank so that the resulting statement is true.

1. The first step in finding the least common denominator of $\dfrac{7}{x^2 - 1}$ and $\dfrac{x}{x^2 - 2x + 1}$ is to _____.

2. Consider the following addition problem:

$$\frac{7}{x(x + 3)} + \frac{10}{(x + 3)(x + 5)}.$$

The factors of the first denominator are _____. The factors of the second denominator are _____. The LCD is _____.

3. An equivalent expression for $\dfrac{7}{8x}$ with a denominator of $24x$ can be obtained by multiplying the numerator and denominator by _____.

4. An equivalent expression for $\dfrac{x}{x + 5}$ with a denominator of $x^2 - 25$ can be obtained by multiplying the numerator and denominator by _____.

5. An equivalent expression for $\dfrac{7}{9 - 5x}$ with a denominator of $5x - 9$ can be obtained by multiplying the numerator and denominator by _____.

7.4 EXERCISE SET   ▶ MyMathLab®

Practice Exercises

In Exercises 1–16, find the least common denominator of the rational expressions.

1. $\dfrac{7}{15x^2}$ and $\dfrac{13}{24x}$

2. $\dfrac{11}{25x^2}$ and $\dfrac{17}{35x}$

3. $\dfrac{8}{15x^2}$ and $\dfrac{5}{6x^5}$

4. $\dfrac{7}{15x^2}$ and $\dfrac{11}{24x^5}$

5. $\dfrac{4}{x-3}$ and $\dfrac{7}{x+1}$

6. $\dfrac{2}{x-5}$ and $\dfrac{3}{x+7}$

7. $\dfrac{5}{7(y+2)}$ and $\dfrac{10}{y}$

8. $\dfrac{8}{11(y+5)}$ and $\dfrac{12}{y}$

9. $\dfrac{17}{x+4}$ and $\dfrac{18}{x^2-16}$

10. $\dfrac{3}{x-6}$ and $\dfrac{4}{x^2-36}$

11. $\dfrac{8}{y^2-9}$ and $\dfrac{14}{y(y+3)}$

12. $\dfrac{14}{y^2-49}$ and $\dfrac{12}{y(y-7)}$

13. $\dfrac{7}{y^2-1}$ and $\dfrac{y}{y^2-2y+1}$

14. $\dfrac{9}{y^2-25}$ and $\dfrac{y}{y^2-10y+25}$

15. $\dfrac{3}{x^2-x-20}$ and $\dfrac{x}{2x^2+7x-4}$

16. $\dfrac{7}{x^2-5x-6}$ and $\dfrac{x}{x^2-4x-5}$

In Exercises 17–82, add or subtract as indicated. Simplify the result, if possible.

17. $\dfrac{3}{x}+\dfrac{5}{x^2}$

18. $\dfrac{4}{x}+\dfrac{8}{x^2}$

19. $\dfrac{2}{9x}+\dfrac{11}{6x}$

20. $\dfrac{5}{6x}+\dfrac{7}{8x}$

21. $\dfrac{4}{x}+\dfrac{7}{2x^2}$

22. $\dfrac{10}{x}+\dfrac{3}{5x^2}$

23. $6+\dfrac{1}{x}$

24. $3+\dfrac{1}{x}$

25. $\dfrac{2}{x}+9$

26. $\dfrac{7}{x}+4$

27. $\dfrac{x-1}{6}+\dfrac{x+2}{3}$

28. $\dfrac{x+3}{2}+\dfrac{x+5}{4}$

29. $\dfrac{4}{x}+\dfrac{3}{x-5}$

30. $\dfrac{3}{x}+\dfrac{4}{x-6}$

31. $\dfrac{2}{x-1}+\dfrac{3}{x+2}$

32. $\dfrac{3}{x-2}+\dfrac{4}{x+3}$

33. $\dfrac{2}{y+5}+\dfrac{3}{4y}$

34. $\dfrac{3}{y+1}+\dfrac{2}{3y}$

35. $\dfrac{x}{x+7}-1$

36. $\dfrac{x}{x+6}-1$

37. $\dfrac{7}{x+5}-\dfrac{4}{x-5}$

38. $\dfrac{8}{x+6}-\dfrac{2}{x-6}$

39. $\dfrac{2x}{x^2-16}+\dfrac{x}{x-4}$

40. $\dfrac{4x}{x^2-25}+\dfrac{x}{x+5}$

41. $\dfrac{5y}{y^2-9}-\dfrac{4}{y+3}$

42. $\dfrac{8y}{y^2-16}-\dfrac{5}{y+4}$

43. $\dfrac{7}{x-1}-\dfrac{3}{(x-1)^2}$

44. $\dfrac{5}{x+3}-\dfrac{2}{(x+3)^2}$

45. $\dfrac{3y}{4y-20}+\dfrac{9y}{6y-30}$

46. $\dfrac{4y}{5y-10}+\dfrac{3y}{10y-20}$

47. $\dfrac{y+4}{y}-\dfrac{y}{y+4}$

48. $\dfrac{y}{y-5}-\dfrac{y-5}{y}$

49. $\dfrac{2x+9}{x^2-7x+12}-\dfrac{2}{x-3}$

50. $\dfrac{3x+7}{x^2-5x+6}-\dfrac{3}{x-3}$

51. $\dfrac{3}{x^2-1}+\dfrac{4}{(x+1)^2}$

52. $\dfrac{6}{x^2-4}+\dfrac{2}{(x+2)^2}$

53. $\dfrac{3x}{x^2+3x-10}-\dfrac{2x}{x^2+x-6}$

54. $\dfrac{x}{x^2-2x-24}-\dfrac{x}{x^2-7x+6}$

55. $\dfrac{y}{y^2+2y+1}+\dfrac{4}{y^2+5y+4}$

56. $\dfrac{y}{y^2 + 5y + 6} + \dfrac{4}{y^2 - y - 6}$

57. $\dfrac{x - 5}{x + 3} + \dfrac{x + 3}{x - 5}$

58. $\dfrac{x - 7}{x + 4} + \dfrac{x + 4}{x - 7}$

59. $\dfrac{5}{2y^2 - 2y} - \dfrac{3}{2y - 2}$

60. $\dfrac{7}{5y^2 - 5y} - \dfrac{2}{5y - 5}$

61. $\dfrac{4x + 3}{x^2 - 9} - \dfrac{x + 1}{x - 3}$

62. $\dfrac{2x - 1}{x + 6} - \dfrac{6 - 5x}{x^2 - 36}$

63. $\dfrac{y^2 - 39}{y^2 + 3y - 10} - \dfrac{y - 7}{y - 2}$

64. $\dfrac{y^2 - 6}{y^2 + 9y + 18} - \dfrac{y - 4}{y + 6}$

65. $4 + \dfrac{1}{x - 3}$

66. $7 + \dfrac{1}{x - 5}$

67. $3 - \dfrac{3y}{y + 1}$

68. $7 - \dfrac{4y}{y + 5}$

69. $\dfrac{9x + 3}{x^2 - x - 6} + \dfrac{x}{3 - x}$

70. $\dfrac{x^2 + 9x}{x^2 - 2x - 3} + \dfrac{5}{3 - x}$

71. $\dfrac{x + 3}{x^2 + x - 2} - \dfrac{2}{x^2 - 1}$

72. $\dfrac{x}{x^2 - 10x + 25} - \dfrac{x - 4}{2x - 10}$

73. $\dfrac{y + 3}{5y^2} - \dfrac{y - 5}{15y}$

74. $\dfrac{y - 7}{3y^2} - \dfrac{y - 2}{12y}$

75. $\dfrac{x + 3}{3x + 6} + \dfrac{x}{4 - x^2}$

76. $\dfrac{x + 7}{4x + 12} + \dfrac{x}{9 - x^2}$

77. $\dfrac{y}{y^2 - 1} + \dfrac{2y}{y - y^2}$

78. $\dfrac{y}{y^2 - 1} + \dfrac{5y}{y - y^2}$

79. $\dfrac{x - 1}{x} + \dfrac{y + 1}{y}$

80. $\dfrac{x + 2}{y} + \dfrac{y - 2}{x}$

81. $\dfrac{3x}{x^2 - y^2} - \dfrac{2}{y - x}$

82. $\dfrac{7x}{x^2 - y^2} - \dfrac{3}{y - x}$

Practice PLUS

In Exercises 83–92, perform the indicated operation or operations. Simplify the result, if possible.

83. $\dfrac{x + 6}{x^2 - 4} - \dfrac{x + 3}{x + 2} + \dfrac{x - 3}{x - 2}$

84. $\dfrac{x + 8}{x^2 - 9} - \dfrac{x + 2}{x + 3} + \dfrac{x - 2}{x - 3}$

85. $\dfrac{5}{x^2 - 25} + \dfrac{4}{x^2 - 11x + 30} - \dfrac{3}{x^2 - x - 30}$

86. $\dfrac{3}{x^2 - 49} + \dfrac{2}{x^2 - 15x + 56} - \dfrac{5}{x^2 - x - 56}$

87. $\dfrac{x + 6}{x^3 - 27} - \dfrac{x}{x^3 + 3x^2 + 9x}$

88. $\dfrac{x + 8}{x^3 - 8} - \dfrac{x}{x^3 + 2x^2 + 4x}$

89. $\dfrac{9y + 3}{y^2 - y - 6} + \dfrac{y}{3 - y} + \dfrac{y - 1}{y + 2}$

90. $\dfrac{7y - 2}{y^2 - y - 12} + \dfrac{2y}{4 - y} + \dfrac{y + 1}{y + 3}$

91. $\dfrac{3}{x^2 + 4xy + 3y^2} - \dfrac{5}{x^2 - 2xy - 3y^2} + \dfrac{2}{x^2 - 9y^2}$

92. $\dfrac{5}{x^2 + 3xy + 2y^2} - \dfrac{7}{x^2 - xy - 2y^2} + \dfrac{4}{x^2 - 4y^2}$

Application Exercises

Two formulas that approximate the dosage of a drug prescribed for children are

$$\text{Young's rule:} \quad C = \frac{DA}{A + 12}$$

$$\text{and Cowling's rule:} \quad C = \frac{D(A + 1)}{24}.$$

In each formula, A = the child's age, in years, D = an adult dosage, and C = the proper child's dosage. The formulas apply for ages 2 through 13, inclusive. Use the formulas to solve Exercises 93–96.

93. Use Young's rule to find the difference in a child's dosage for an 8-year-old child and a 3-year-old child. Express the answer as a single rational expression in terms of D. Then describe what your answer means in terms of the variables in the model.

94. Use Young's rule to find the difference in a child's dosage for a 10-year-old child and a 3-year-old child. Express the answer as a single rational expression in terms of D. Then describe what your answer means in terms of the variables in the model.

95. For a 12-year-old child, what is the difference in the dosage given by Cowling's rule and Young's rule? Express the answer as a single rational expression in terms of D. Then describe what your answer means in terms of the variables in the models.

96. Use Cowling's rule to find the difference in a child's dosage for a 12-year-old child and a 10-year-old child. Express the answer as a single rational expression in terms of D. Then describe what your answer means in terms of the variables in the model.

The graphs illustrate Young's rule and Cowling's rule when the dosage of a drug prescribed for an adult is 1000 milligrams. Use the graphs to solve Exercises 97–100.

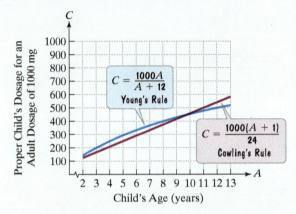

97. Does either formula consistently give a smaller dosage than the other? If so, which one?

98. Is there an age at which the dosage given by one formula becomes greater than the dosage given by the other? If so, what is a reasonable estimate of that age?

99. For what age under 11 is the difference in dosage given by the two formulas the greatest?

100. For what age over 11 is the difference in dosage given by the two formulas the greatest?

In Exercises 101–102, express the perimeter of each rectangle as a single rational expression.

101.

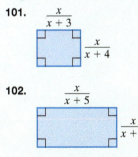

102.

$\dfrac{x}{x + 5}$

$\dfrac{x}{x + 6}$

Explaining the Concepts

103. Explain how to find the least common denominator for denominators of $x^2 - 100$ and $x^2 - 20x + 100$.

104. Explain how to add rational expressions that have different denominators. Use $\dfrac{3}{x + 5} + \dfrac{7}{x + 2}$ in your explanation.

Explain the error in Exercises 105–106. Then rewrite the right side of the equation to correct the error that now exists.

105. $\dfrac{1}{x} + \dfrac{2}{5} = \dfrac{3}{x + 5}$

106. $\dfrac{1}{x} + 7 = \dfrac{1}{x + 7}$

107. The formulas in Exercises 93–96 relate the dosage of a drug prescribed for children to the child's age. Describe another factor that might be used when determining a child's dosage. Is this factor more or less important than age? Explain why.

Critical Thinking Exercises

Make Sense? *In Exercises 108–111, determine whether each statement makes sense or does not make sense, and explain your reasoning.*

108. It takes me more steps to find $\dfrac{2}{x + 5} + \dfrac{7}{x - 3}$ than it does to find $\dfrac{2x^3}{x + 5} + \dfrac{7x^3}{x + 5}$.

109. The reason I can rewrite rational expressions with a common denominator is that 1 is the multiplicative identity.

110. The fastest way for me to find $\dfrac{4}{x - 5} + \dfrac{9}{5 - x}$ is by using $(x - 5)(5 - x)$ as the LCD.

111. After adding rational expressions with different denominators, I factored the numerator and found no common factors in the numerator and denominator, so my final answer is incorrect if I leave the numerator in factored form.

In Exercises 112–115, determine whether each statement is true or false. If the statement is false, make the necessary change(s) to produce a true statement.

112. $x - \dfrac{1}{5} = \dfrac{4}{5}x$

113. The LCD of $\dfrac{1}{x}$ and $\dfrac{2x}{x - 1}$ is $x^2 - 1$.

114. $\dfrac{1}{x} + \dfrac{x}{1} = \dfrac{1}{\frac{x}{1}} + \dfrac{\frac{x}{1}}{1} = 1 + 1 = 2$

115. $\dfrac{2}{x} + 1 = \dfrac{2 + x}{x}, x \neq 0$

In Exercises 116–117, perform the indicated operations. Simplify the result, if possible.

116. $\dfrac{y^2 + 5y + 4}{y^2 + 2y - 3} \cdot \dfrac{y^2 + y - 6}{y^2 + 2y - 3} - \dfrac{2}{y - 1}$

117. $\left(\dfrac{1}{x + h} - \dfrac{1}{x} \right) \div h$

In Exercises 118–119, find the missing rational expression.

118. $\dfrac{2}{x - 1} + \dfrac{\blacksquare}{\blacksquare} = \dfrac{2x^2 + 3x - 1}{x^2(x - 1)}$

119. $\dfrac{4}{x - 2} - \dfrac{\blacksquare}{\blacksquare} = \dfrac{2x + 8}{(x - 2)(x + 1)}$

Review Exercises

120. Multiply: $(3x + 5)(2x - 7)$. (Section 5.3, Example 2)

121. Graph: $3x - y < 3$. (Section 3.6, Example 2)

122. Write the slope-intercept form of the equation of the line passing through $(-3, -4)$ and $(1, 0)$. (Section 3.5, Example 2)

Preview Exercises

Exercises 123–125 will help you prepare for the material covered in the next section.

123. a. Add: $\dfrac{1}{3} + \dfrac{2}{5}$.

   b. Subtract: $\dfrac{2}{5} - \dfrac{1}{3}$.

   c. Use your answers from parts (a) and (b) to find

$$\left(\frac{1}{3} + \frac{2}{5}\right) \div \left(\frac{2}{5} - \frac{1}{3}\right).$$

124. a. Add: $\dfrac{1}{x} + \dfrac{1}{y}$.

   b. Use your answer from part (a) to find

$$\frac{1}{xy} \div \left(\frac{1}{x} + \frac{1}{y}\right).$$

125. Multiply and simplify: $xy\left(\dfrac{1}{x} + \dfrac{1}{y}\right)$.

MID-CHAPTER CHECK POINT   Section 7.1–Section 7.4

✓ **What You Know:** We learned that it is necessary to exclude any value or values of a variable that make the denominator of a rational expression zero. We learned to simplify rational expressions by dividing the numerator and the denominator by common factors. We performed a variety of operations with rational expressions, including multiplication, division, addition, and subtraction.

1. Find all numbers for which $\dfrac{x^2 - 4}{x^2 - 2x - 8}$ is undefined.

In Exercises 2–4, simplify each rational expression.

2. $\dfrac{3x^2 - 7x + 2}{6x^2 + x - 1}$

3. $\dfrac{9 - 3y}{y^2 - 5y + 6}$

4. $\dfrac{16w^3 - 24w^2}{8w^4 - 12w^3}$

In Exercises 5–20, perform the indicated operations. Simplify the result, if possible.

5. $\dfrac{7x - 3}{x^2 + 3x - 4} - \dfrac{3x + 1}{x^2 + 3x - 4}$

6. $\dfrac{x + 2}{2x - 4} \cdot \dfrac{8}{x^2 - 4}$

7. $1 + \dfrac{7}{x - 2}$

8. $\dfrac{2x^2 + x - 1}{2x^2 - 7x + 3} \div \dfrac{x^2 - 3x - 4}{x^2 - x - 6}$

9. $\dfrac{1}{x^2 + 2x - 3} + \dfrac{1}{x^2 + 5x + 6}$

10. $\dfrac{17}{x - 5} + \dfrac{x + 8}{5 - x}$

11. $\dfrac{4y^2 - 1}{9y - 3y^2} \cdot \dfrac{y^2 - 7y + 12}{2y^2 - 7y - 4}$

12. $\dfrac{y}{y + 1} - \dfrac{2y}{y + 2}$

13. $\dfrac{w^2 + 6w + 5}{7w^2 - 63} \div \dfrac{w^2 + 10w + 25}{7w + 21}$

14. $\dfrac{2z}{z^2 - 9} - \dfrac{5}{z^2 + 4z + 3}$

15. $\dfrac{z + 2}{3z - 1} + \dfrac{5}{(3z - 1)^2}$

16. $\dfrac{8}{x^2 + 4x - 21} + \dfrac{3}{x + 7}$

17. $\dfrac{x^4 - 27x}{x^2 - 9} \cdot \dfrac{x + 3}{x^2 + 3x + 9}$

18. $\dfrac{x - 1}{x^2 - x - 2} - \dfrac{x + 2}{x^2 + 4x + 3}$

19. $\dfrac{x^2 - 2xy + y^2}{x + y} \div \dfrac{x^2 - xy}{5x + 5y}$

20. $\dfrac{5}{x + 5} + \dfrac{x}{x - 4} - \dfrac{11x - 8}{x^2 + x - 20}$

7.5

Complex Rational Expressions

What am I supposed to learn?

After studying this section, you should be able to:

1 Simplify complex rational expressions by dividing.

2 Simplify complex rational expressions by multiplying by the LCD.

Do you drive to and from campus each day? If the one-way distance of your round-trip commute is d, then your average rate, or speed, is given by the expression

$$\frac{2d}{\dfrac{d}{r_1} + \dfrac{d}{r_2}},$$

in which r_1 and r_2 are your average rates on the outgoing and return trips, respectively. Do you notice anything unusual about this expression? It has two separate rational expressions in its denominator.

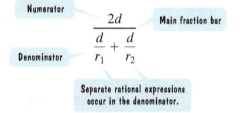

Complex rational expressions, also called **complex fractions**, have numerators or denominators containing one or more rational expressions. Here is another example of such an expression:

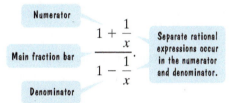

In this section, we study two methods for *simplifying* complex rational expressions. A complex rational expression is **simplified** when it is in the form $\frac{P}{Q}$, where P and Q are polynomials that have no common factors.

1 Simplify complex rational expressions by dividing.

Simplifying by Rewriting Complex Rational Expressions as a Quotient of Two Rational Expressions

One method for simplifying a complex rational expression is to combine its numerator into a single expression and combine its denominator into a single expression. Then perform the division by inverting the denominator and multiplying.

Simplifying a Complex Rational Expression by Dividing

1. If necessary, add or subtract to get a single rational expression in the numerator.
2. If necessary, add or subtract to get a single rational expression in the denominator.
3. Perform the division indicated by the main fraction bar: Invert the denominator of the complex rational expression and multiply.
4. If possible, simplify.

The following examples illustrate the use of this first method.

EXAMPLE 1 Simplifying a Complex Rational Expression

Simplify:

$$\dfrac{\dfrac{1}{3}+\dfrac{2}{5}}{\dfrac{2}{5}-\dfrac{1}{3}}.$$

Solution Let's first identify the parts of this complex rational expression.

Numerator

Main fraction bar $\quad$ Denominator

$$\dfrac{\dfrac{1}{3}+\dfrac{2}{5}}{\dfrac{2}{5}-\dfrac{1}{3}}$$

Step 1. Add to get a single rational expression in the numerator.

$$\frac{1}{3}+\frac{2}{5}=\frac{1\cdot 5}{3\cdot 5}+\frac{2\cdot 3}{5\cdot 3}=\frac{5}{15}+\frac{6}{15}=\frac{11}{15}$$

The LCD is 3 · 5, or 15.

Step 2. Subtract to get a single rational expression in the denominator.

$$\frac{2}{5}-\frac{1}{3}=\frac{2\cdot 3}{5\cdot 3}-\frac{1\cdot 5}{3\cdot 5}=\frac{6}{15}-\frac{5}{15}=\frac{1}{15}$$

The LCD is 15.

Steps 3 and 4. Perform the division indicated by the main fraction bar: Invert and multiply. If possible, simplify.

$$\dfrac{\dfrac{1}{3}+\dfrac{2}{5}}{\dfrac{2}{5}-\dfrac{1}{3}}=\dfrac{\dfrac{11}{15}}{\dfrac{1}{15}}=\frac{11}{15}\cdot\frac{15}{1}=\frac{11}{\cancel{15}}\cdot\frac{\overset{1}{\cancel{15}}}{1}=11$$

Invert and multiply. ∎

☑ **CHECK POINT 1** Simplify: $\dfrac{\dfrac{1}{4}+\dfrac{2}{3}}{\dfrac{2}{3}-\dfrac{1}{4}}.$

EXAMPLE 2   Simplifying a Complex Rational Expression

Simplify:

$$\frac{1 + \dfrac{1}{x}}{1 - \dfrac{1}{x}}.$$

Solution

Step 1. Add to get a single rational expression in the numerator.

$$1 + \frac{1}{x} = \frac{1}{1} + \frac{1}{x} = \frac{1 \cdot x}{1 \cdot x} + \frac{1}{x} = \frac{x}{x} + \frac{1}{x} = \frac{x + 1}{x}$$

The LCD is 1 · x, or x.

Step 2. Subtract to get a single rational expression in the denominator.

$$1 - \frac{1}{x} = \frac{1}{1} - \frac{1}{x} = \frac{1 \cdot x}{1 \cdot x} - \frac{1}{x} = \frac{x}{x} - \frac{1}{x} = \frac{x - 1}{x}$$

The LCD is 1 · x, or x.

Steps 3 and 4. Perform the division indicated by the main fraction bar: Invert and multiply. If possible, simplify.

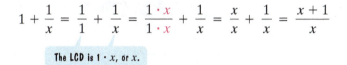

$$\frac{1 + \dfrac{1}{x}}{1 - \dfrac{1}{x}} = \frac{\dfrac{x + 1}{x}}{\dfrac{x - 1}{x}} = \frac{x + 1}{x} \cdot \frac{x}{x - 1} = \frac{x + 1}{\overset{1}{\cancel{x}}} \cdot \frac{\overset{1}{\cancel{x}}}{x - 1} = \frac{x + 1}{x - 1} \ \blacksquare$$

Invert and multiply.

✓ **CHECK POINT 2**   Simplify: $\dfrac{2 - \dfrac{1}{x}}{2 + \dfrac{1}{x}}.$

EXAMPLE 3   Simplifying a Complex Rational Expression

Simplify:

$$\frac{\dfrac{1}{xy}}{\dfrac{1}{x} + \dfrac{1}{y}}.$$

Solution

Step 1. Get a single rational expression in the numerator. The numerator, $\dfrac{1}{xy}$, already contains a single rational expression, so we can skip this step.

Step 2. Add to get a single rational expression in the denominator.

$$\frac{1}{x} + \frac{1}{y} = \frac{1 \cdot y}{x \cdot y} + \frac{1 \cdot x}{y \cdot x} = \frac{y}{xy} + \frac{x}{xy} = \frac{y + x}{xy}$$

The LCD is xy.

Steps 3 and 4. Perform the division indicated by the main fraction bar: Invert and multiply. If possible, simplify.

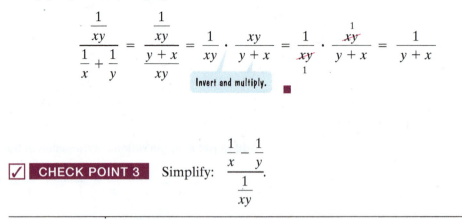

$$\frac{\dfrac{1}{xy}}{\dfrac{1}{x}+\dfrac{1}{y}}=\frac{\dfrac{1}{xy}}{\dfrac{y+x}{xy}}=\frac{1}{xy}\cdot\frac{xy}{y+x}=\frac{1}{\cancel{xy}}\cdot\frac{\overset{1}{\cancel{xy}}}{y+x}=\frac{1}{y+x}$$

Invert and multiply. ∎

✓ **CHECK POINT 3**   Simplify: $\dfrac{\dfrac{1}{x}-\dfrac{1}{y}}{\dfrac{1}{xy}}$.

2 Simplify complex rational expressions by multiplying by the LCD. ▶

Simplifying Complex Rational Expressions by Multiplying by the LCD

A second method for simplifying a complex rational expression is to find the least common denominator of all the rational expressions in its numerator and denominator. Then multiply each term in its numerator and denominator by this least common denominator. Because we are multiplying by a form of 1, we will obtain an equivalent expression that does not contain fractions in its numerator or denominator.

> ### Simplifying a Complex Rational Expression by Multiplying by the LCD
>
> 1. Find the LCD of all rational expressions within the complex rational expression.
> 2. Multiply both the numerator and the denominator of the complex rational expression by this LCD.
> 3. Use the distributive property and multiply each term in the numerator and denominator by this LCD. Simplify each term. No fractional expressions should remain.
> 4. If possible, factor and simplify.

We now rework Examples 1, 2, and 3 using the method of multiplying by the LCD. Compare the two simplification methods to see if there is one method that you prefer.

EXAMPLE 4   Simplifying a Complex Rational Expression by the LCD Method

Simplify:

$$\frac{\dfrac{1}{3}+\dfrac{2}{5}}{\dfrac{2}{5}-\dfrac{1}{3}}.$$

Solution The denominators in the complex rational expression are 3, 5, 5, and 3. The LCD is $3 \cdot 5$, or 15. Multiply both the numerator and denominator of the complex rational expression by 15.

$$\frac{\frac{1}{3} + \frac{2}{5}}{\frac{2}{5} - \frac{1}{3}} = \frac{15}{15} \cdot \frac{\left(\frac{1}{3} + \frac{2}{5}\right)}{\left(\frac{2}{5} - \frac{1}{3}\right)} = \frac{15 \cdot \frac{1}{3} + 15 \cdot \frac{2}{5}}{15 \cdot \frac{2}{5} - 15 \cdot \frac{1}{3}} = \frac{5 + 6}{6 - 5} = \frac{11}{1} = 11$$

$\frac{15}{15} = 1$, so we are not changing the complex fraction's value. ∎

✓ **CHECK POINT 4** Simplify by the LCD method: $\dfrac{\frac{1}{4} + \frac{2}{3}}{\frac{2}{3} - \frac{1}{4}}$.

EXAMPLE 5 Simplifying a Complex Rational Expression by the LCD Method

Simplify:

$$\frac{1 + \frac{1}{x}}{1 - \frac{1}{x}}.$$

Solution The denominators in the complex rational expression are $1, x, 1$, and x.

$$\frac{1 + \frac{1}{x}}{1 - \frac{1}{x}} = \frac{\frac{1}{1} + \frac{1}{x}}{\frac{1}{1} - \frac{1}{x}}$$

Denominators

The LCD is $1 \cdot x$, or x. Multiply both the numerator and denominator of the complex rational expression by x.

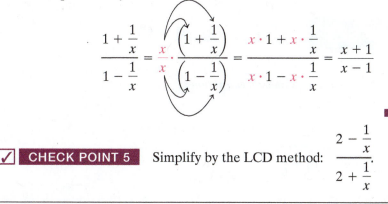

$$\frac{1 + \frac{1}{x}}{1 - \frac{1}{x}} = \frac{x}{x} \cdot \frac{\left(1 + \frac{1}{x}\right)}{\left(1 - \frac{1}{x}\right)} = \frac{x \cdot 1 + x \cdot \frac{1}{x}}{x \cdot 1 - x \cdot \frac{1}{x}} = \frac{x + 1}{x - 1}$$

∎

✓ **CHECK POINT 5** Simplify by the LCD method: $\dfrac{2 - \frac{1}{x}}{2 + \frac{1}{x}}$.

EXAMPLE 6 Simplifying a Complex Rational Expression by the LCD Method

Simplify:

$$\frac{\frac{1}{xy}}{\frac{1}{x} + \frac{1}{y}}.$$

Solution The denominators in the complex rational expression are xy, x, and y. The LCD is xy. Multiply both the numerator and denominator of the complex rational expression by xy.

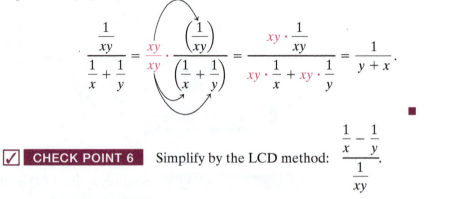

$$\frac{\dfrac{1}{xy}}{\dfrac{1}{x}+\dfrac{1}{y}} = \frac{xy}{xy}\cdot\frac{\left(\dfrac{1}{xy}\right)}{\left(\dfrac{1}{x}+\dfrac{1}{y}\right)} = \frac{xy\cdot\dfrac{1}{xy}}{xy\cdot\dfrac{1}{x}+xy\cdot\dfrac{1}{y}} = \frac{1}{y+x}.$$

■

☑ **CHECK POINT 6**   Simplify by the LCD method: $\dfrac{\dfrac{1}{x}-\dfrac{1}{y}}{\dfrac{1}{xy}}$.

CONCEPT AND VOCABULARY CHECK

Fill in each blank so that the resulting statement is true.

1. A rational expression whose numerator or denominator or both contains rational expressions is called a/an _____ rational expression or a/an _____ fraction.

2. $\dfrac{\dfrac{3}{4}+\dfrac{1}{2}}{\dfrac{3}{8}-\dfrac{1}{6}} = \dfrac{24}{24}\cdot\dfrac{\left(\dfrac{3}{4}+\dfrac{1}{2}\right)}{\left(\dfrac{3}{8}-\dfrac{1}{6}\right)} = \dfrac{24\cdot\dfrac{3}{4}+24\cdot\dfrac{1}{2}}{24\cdot\dfrac{3}{8}-24\cdot\dfrac{1}{6}} = \dfrac{\underline{}+\underline{}}{\underline{}-\underline{}} = \dfrac{\underline{}}{\underline{}} = \underline{}$

3. $\dfrac{\dfrac{5}{x}+\dfrac{x}{3}}{\dfrac{4}{x}-\dfrac{x}{6}} = \dfrac{6x}{6x}\cdot\dfrac{\left(\dfrac{5}{x}+\dfrac{x}{3}\right)}{\left(\dfrac{4}{x}-\dfrac{x}{6}\right)} = \dfrac{6x\cdot\dfrac{5}{x}+6x\cdot\dfrac{x}{3}}{6x\cdot\dfrac{4}{x}-6x\cdot\dfrac{x}{6}} = \dfrac{\underline{}+\underline{}}{\underline{}-\underline{}}$

4. $\dfrac{\dfrac{2}{x}+\dfrac{3}{x^2}}{\dfrac{5}{x}+1} = \dfrac{x^2}{x^2}\cdot\dfrac{\left(\dfrac{2}{x}+\dfrac{3}{x^2}\right)}{\left(\dfrac{5}{x}+1\right)} = \dfrac{x^2\cdot\dfrac{2}{x}+x^2\cdot\dfrac{3}{x^2}}{x^2\cdot\dfrac{5}{x}+x^2\cdot1} = \dfrac{\underline{}+\underline{}}{\underline{}+\underline{}}$

7.5 EXERCISE SET   ▶ MyMathLab®

Practice Exercises

In Exercises 1–40, simplify each complex rational expression by the method of your choice.

1. $\dfrac{\dfrac{1}{2}+\dfrac{1}{4}}{\dfrac{1}{2}+\dfrac{1}{3}}$

2. $\dfrac{\dfrac{1}{3}+\dfrac{1}{4}}{\dfrac{1}{3}+\dfrac{1}{6}}$

3. $\dfrac{5+\dfrac{2}{5}}{7-\dfrac{1}{10}}$

4. $\dfrac{1+\dfrac{3}{5}}{2-\dfrac{1}{4}}$

5. $\dfrac{\dfrac{2}{5}-\dfrac{1}{3}}{\dfrac{2}{3}-\dfrac{3}{4}}$

6. $\dfrac{\dfrac{1}{2}-\dfrac{1}{4}}{\dfrac{3}{8}+\dfrac{1}{16}}$

7. $\dfrac{\dfrac{3}{4}-x}{\dfrac{3}{4}+x}$

8. $\dfrac{\dfrac{2}{3}-x}{\dfrac{2}{3}+x}$

9. $\dfrac{7-\dfrac{2}{x}}{5+\dfrac{1}{x}}$

10. $\dfrac{8+\dfrac{3}{x}}{1-\dfrac{7}{x}}$

11. $\dfrac{2+\dfrac{3}{y}}{1-\dfrac{7}{y}}$

12. $\dfrac{4-\dfrac{7}{y}}{3-\dfrac{2}{y}}$

13. $\dfrac{\dfrac{1}{y}-\dfrac{3}{2}}{\dfrac{1}{y}+\dfrac{3}{4}}$

14. $\dfrac{\dfrac{1}{y}-\dfrac{3}{4}}{\dfrac{1}{y}+\dfrac{2}{3}}$

15. $\dfrac{\dfrac{x}{5} - \dfrac{5}{x}}{\dfrac{1}{5} + \dfrac{1}{x}}$

16. $\dfrac{\dfrac{3}{x} + \dfrac{x}{3}}{\dfrac{x}{3} - \dfrac{3}{x}}$

17. $\dfrac{1 + \dfrac{1}{x}}{1 - \dfrac{1}{x^2}}$

18. $\dfrac{1 + \dfrac{2}{x}}{1 - \dfrac{4}{x^2}}$

19. $\dfrac{\dfrac{1}{7} - \dfrac{1}{y}}{\dfrac{7 - y}{7}}$

20. $\dfrac{\dfrac{1}{9} - \dfrac{1}{y}}{\dfrac{9 - y}{9}}$

21. $\dfrac{x + \dfrac{2}{y}}{\dfrac{x}{y}}$

22. $\dfrac{x - \dfrac{2}{y}}{\dfrac{x}{y}}$

23. $\dfrac{\dfrac{1}{x} + \dfrac{1}{y}}{xy}$

24. $\dfrac{\dfrac{1}{x} + \dfrac{1}{y}}{x + y}$

25. $\dfrac{\dfrac{x}{y} + \dfrac{1}{x}}{\dfrac{y}{x} + \dfrac{1}{x}}$

26. $\dfrac{\dfrac{1}{x} + \dfrac{1}{y}}{\dfrac{1}{x} - \dfrac{1}{y}}$

27. $\dfrac{\dfrac{1}{y} + \dfrac{2}{y^2}}{\dfrac{2}{y} + 1}$

28. $\dfrac{\dfrac{1}{y} + \dfrac{3}{y^2}}{\dfrac{3}{y} + 1}$

29. $\dfrac{\dfrac{12}{x^2} - \dfrac{3}{x}}{\dfrac{15}{x} - \dfrac{9}{x^2}}$

30. $\dfrac{\dfrac{8}{x^2} - \dfrac{2}{x}}{\dfrac{10}{x} - \dfrac{6}{x^2}}$

31. $\dfrac{2 + \dfrac{6}{y}}{1 - \dfrac{9}{y^2}}$

32. $\dfrac{3 + \dfrac{12}{y}}{1 - \dfrac{16}{y^2}}$

33. $\dfrac{\dfrac{1}{x + 2}}{1 + \dfrac{1}{x + 2}}$

34. $\dfrac{\dfrac{1}{x - 2}}{1 - \dfrac{1}{x - 2}}$

35. $\dfrac{x - 5 + \dfrac{3}{x}}{x - 7 + \dfrac{2}{x}}$

36. $\dfrac{x + 9 - \dfrac{7}{x}}{x - 6 + \dfrac{4}{x}}$

37. $\dfrac{\dfrac{3}{xy^2} + \dfrac{2}{x^2 y}}{\dfrac{1}{x^2 y} + \dfrac{2}{xy^3}}$

38. $\dfrac{\dfrac{2}{x^3 y} + \dfrac{5}{xy^4}}{\dfrac{5}{x^3 y} - \dfrac{3}{xy}}$

39. $\dfrac{\dfrac{3}{x + 1} - \dfrac{3}{x - 1}}{\dfrac{5}{x^2 - 1}}$

40. $\dfrac{\dfrac{3}{x + 2} - \dfrac{3}{x - 2}}{\dfrac{5}{x^2 - 4}}$

Practice PLUS

In Exercises 41–48, simplify each complex rational expression.

41. $\dfrac{\dfrac{6}{x^2 + 2x - 15} - \dfrac{1}{x - 3}}{\dfrac{1}{x + 5} + 1}$

42. $\dfrac{\dfrac{1}{x - 2} - \dfrac{6}{x^2 + 3x - 10}}{1 + \dfrac{1}{x - 2}}$

43. $\dfrac{y^{-1} - (y + 5)^{-1}}{5}$

44. $\dfrac{y^{-1} - (y + 2)^{-1}}{2}$

45. $\dfrac{\dfrac{1}{1 - \dfrac{1}{x}} - 1}{}$

46. $\dfrac{\dfrac{1}{1 - \dfrac{1}{x + 1}} - 1}{}$

47. $\dfrac{1}{1 + \dfrac{1}{1 + \dfrac{1}{x}}}$

48. $\dfrac{1}{1 + \dfrac{1}{1 + \dfrac{1}{2}}}$

Application Exercises

49. The average rate on a round-trip commute having a one-way distance d is given by the complex rational expression

$$\dfrac{2d}{\dfrac{d}{r_1} + \dfrac{d}{r_2}},$$

in which r_1 and r_2 are the average rates on the outgoing and return trips, respectively. Simplify the expression. Then find your average rate if you drive to campus averaging 40 miles per hour and return home on the same route averaging 30 miles per hour.

50. If two electrical resistors with resistances R_1 and R_2 are connected in parallel (see the figure), then the total resistance in the circuit is given by the complex rational expression

$$\dfrac{1}{\dfrac{1}{R_1} + \dfrac{1}{R_2}}.$$

Simplify the expression. Then find the total resistance if $R_1 = 10$ ohms and $R_2 = 20$ ohms.

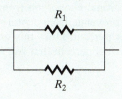

With MP3s and streaming sites, the Internet has transformed the music industry. The bar graph shows the revenue, in billions of dollars, for the U.S. music industry from digital music downloads and total revenue for the music industry for seven selected years.

U.S. Music Industry Revenue

Source: Recording Industry Association of America

The fraction of revenue from digital sales can be modeled by the complex rational expression

$$\frac{1 + \dfrac{10}{x} + \dfrac{135}{x^2}}{9 - \dfrac{85}{x} + \dfrac{525}{x^2}},$$

where x is the number of years after 2007. Use this information to solve Exercises 51–52.

51. a. Simplify the complex rational expression for the fraction of revenue from digital sales. _____

    b. Use the data displayed by the bar graph to find the percentage of revenue from digital sales in 2013. Round to the nearest percent.

    c. Use the simplified rational expression from part (a) to find the percentage of revenue from digital sales in 2013. Round to the nearest percent. Does this underestimate or overestimate the actual percent that you found in part (b)? By how much?

52. a. Simplify the complex rational expression for the fraction of revenue from digital sales. _____

    b. Use the data displayed by the bar graph to find the percentage of revenue from digital sales in 2010. Round to the nearest percent.

    c. Use the simplified rational expression from part (a) to find the percentage of revenue from digital sales in 2010. Round to the nearest percent. Does this underestimate or overestimate the actual percent that you found in part (b)? By how much?

Explaining the Concepts

53. What is a complex rational expression? Give an example with your explanation.

54. Describe two ways to simplify $\dfrac{\dfrac{3}{x} + \dfrac{2}{x^2}}{\dfrac{1}{x^2} + \dfrac{2}{x}}$.

55. Which method do you prefer for simplifying complex rational expressions? Why?

Critical Thinking Exercises

Make Sense? In Exercises 56–59, determine whether each statement makes sense or does not make sense, and explain your reasoning.

56. I simplified $\dfrac{\dfrac{1}{2} + \dfrac{x}{3}}{4}$ by multiplying the numerator by 6.

57. I added 1 to $\dfrac{1}{1 + \dfrac{1}{2}}$ and obtained $\dfrac{5}{3}$.

58. I used the LCD method to simplify
$$\frac{3 - \dfrac{6}{x}}{1 + \dfrac{7}{y}}$$

and obtained $\dfrac{3 - 6y}{1 + 7x}$.

59. Because x^{-1} means $\dfrac{1}{x}$ and y^{-1} means $\dfrac{1}{y}$, I simplified $\dfrac{x^{-1} + y^{-1}}{x^{-1} - y^{-1}}$ and obtained $\dfrac{x - y}{x + y}$.

In Exercises 60–63, determine whether each statement is true or false. If the statement is false, make the necessary change(s) to produce a true statement.

60. The fraction $\dfrac{31{,}729{,}546}{72{,}578{,}112}$ is a complex rational expression.

61. $\dfrac{y - \dfrac{1}{2}}{y + \dfrac{3}{4}} = \dfrac{4y - 2}{4y + 3}$ for any value of y except $-\dfrac{3}{4}$.

62. $\dfrac{\dfrac{1}{4} - \dfrac{1}{3}}{\dfrac{1}{3} + \dfrac{1}{6}} = \dfrac{1}{12} \div \dfrac{3}{6} = \dfrac{1}{6}$

63. Some complex rational expressions cannot be simplified by both methods discussed in this section.

64. In one short sentence, five words or less, explain what

$$\frac{\dfrac{1}{x} + \dfrac{1}{x^2} + \dfrac{1}{x^3}}{\dfrac{1}{x^4} + \dfrac{1}{x^5} + \dfrac{1}{x^6}}$$

does to each number x.

In Exercises 65–66, simplify completely.

65. $\dfrac{2y}{2 + \dfrac{2}{y}} + \dfrac{y}{1 + \dfrac{1}{y}}$

66. $\dfrac{1 + \dfrac{1}{y} - \dfrac{6}{y^2}}{1 - \dfrac{5}{y} + \dfrac{6}{y^2}} - \dfrac{1 - \dfrac{1}{y}}{1 - \dfrac{2}{y} - \dfrac{3}{y^2}}$

Technology Exercises

In Exercises 67–69, use the $\boxed{\text{GRAPH}}$ *or* $\boxed{\text{TABLE}}$ *feature of a graphing utility to determine if the simplification is correct. If the answer is wrong, correct it and then verify your corrected simplification using the graphing utility.*

67. $\dfrac{x - \dfrac{1}{2x + 1}}{1 - \dfrac{x}{2x + 1}} = 2x - 1$

68. $\dfrac{\dfrac{1}{x} + 1}{\dfrac{1}{x}} = 2$

69. $\dfrac{\dfrac{1}{x} + \dfrac{1}{3}}{\dfrac{1}{3x}} = x + \dfrac{1}{3}$

Review Exercises

70. Factor completely: $2x^3 - 20x^2 + 50x$. (Section 6.5, Example 2)

71. Solve: $2 - 3(x - 2) = 5(x + 5) - 1$. (Section 2.3, Example 3)

72. Multiply: $(x + y)(x^2 - xy + y^2)$. (Section 5.2, Example 7)

Preview Exercises

Exercises 73–75 will help you prepare for the material covered in the next section.

73. Solve: $\dfrac{x}{3} + \dfrac{x}{2} = \dfrac{5}{6}$.

74. Solve: $\dfrac{2x}{3} = \dfrac{14}{3} - \dfrac{x}{2}$.

75. Solve: $2x^2 + 2 = 5x$.

SECTION

7.6

Solving Rational Equations

What am I supposed to learn?

After studying this section, you should be able to:

1 Solve rational equations. ▶

2 Solve problems involving formulas with rational expressions. ▶

The time has come to clean up the river. Suppose that the government has committed \$375 million for this project. We know that

$$y = \frac{250x}{100 - x}$$

models the cost, y, in millions of dollars, to remove x percent of the river's pollutants. What percentage of pollutants can be removed for \$375 million?

In order to determine the percentage, we use the given model. The government has committed \$375 million, so substitute 375 for y:

$$375 = \frac{250x}{100 - x} \quad \text{or} \quad \frac{250x}{100 - x} = 375.$$

> The equation contains a rational expression.

Now we need to solve the equation and find the value for x. This variable represents the percentage of pollutants that can be removed for \$375 million.

A **rational equation**, also called a **fractional equation**, is an equation containing one or more rational expressions. The preceding equation, $\frac{250x}{100 - x} = 375$, is an example of a rational equation. Do you see that there is a variable in the denominator? This is a characteristic of many rational equations. In this section, you will learn a procedure for solving such equations.

① Solve rational equations. ▶

Solving Rational Equations

We have seen that the LCD is used to add and subtract rational expressions. By contrast, when solving rational equations, **the LCD is used as a multiplier that clears an equation of fractions**.

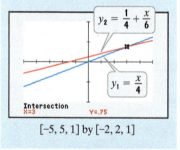
EXAMPLE 1 Solving a Rational Equation

Solve: $\dfrac{x}{4} = \dfrac{1}{4} + \dfrac{x}{6}$.

Solution The denominators are 4, 4, and 6. The least common denominator is 12. To clear the equation of fractions, we multiply both sides by 12.

$$\frac{x}{4} = \frac{1}{4} + \frac{x}{6} \qquad \text{This is the given equation.}$$

$$12\left(\frac{x}{4}\right) = 12\left(\frac{1}{4} + \frac{x}{6}\right) \qquad \text{Multiply both sides by 12, the LCD of all the fractions in the equation.}$$

$$12 \cdot \frac{x}{4} = 12 \cdot \frac{1}{4} + 12 \cdot \frac{x}{6} \qquad \text{Use the distributive property on the right side.}$$

$$3x = 3 + 2x \qquad \text{Simplify: } = \frac{\overset{3}{\cancel{12}}}{1} \cdot \frac{x}{\cancel{4}} = 3x;\ \frac{\overset{3}{\cancel{12}}}{1} \cdot \frac{1}{\cancel{4}} = 3;\ \frac{\overset{2}{\cancel{12}}}{1} \cdot \frac{x}{\cancel{6}} = 2x.$$

$$x = 3 \qquad \text{Subtract 2x from both sides.}$$

Substitute 3 for x in the original equation. You should obtain the true statement $\frac{3}{4} = \frac{3}{4}$. This verifies that the solution is 3 and the solution set is {3}. ■

☑ **CHECK POINT 1** Solve: $\dfrac{x}{6} = \dfrac{1}{6} + \dfrac{x}{8}$.

In Example 1, we solved a rational equation with constants in the denominators. Now, let's consider an equation such as

$$\frac{1}{x} = \frac{1}{5} + \frac{3}{2x}.$$

Can you see how this equation differs from the rational equation that we solved earlier? The variable, x, appears in two of the denominators. The procedure for solving this equation still involves multiplying each side by the least common denominator. However, we must avoid any values of the variable that make a denominator zero. For example, examine the denominators in the equation:

$$\frac{1}{x} = \frac{1}{5} + \frac{3}{2x}.$$

This denominator would equal zero if $x = 0$.   This denominator would equal zero if $x = 0$.

We see that x cannot equal zero. With this in mind, let's solve the equation.

EXAMPLE 2 Solving a Rational Equation

Solve: $\dfrac{1}{x} = \dfrac{1}{5} + \dfrac{3}{2x}$.

Solution The denominators are x, 5, and $2x$. The least common denominator is $10x$. We begin by multiplying both sides of the equation by $10x$. We will also write the restriction that x cannot equal zero to the right of the equation.

$$\frac{1}{x} = \frac{1}{5} + \frac{3}{2x}, \quad x \neq 0 \qquad \text{This is the given equation.}$$

$$10x \cdot \frac{1}{x} = 10x\left(\frac{1}{5} + \frac{3}{2x}\right) \qquad \text{Multiply both sides by 10x.}$$

$$10x \cdot \frac{1}{x} = 10x \cdot \frac{1}{5} + 10x \cdot \frac{3}{2x} \qquad \begin{array}{l}\text{Use the distributive property. Be sure to} \\ \text{multiply all terms by 10x.}\end{array}$$

$$10x \cdot \frac{1}{\cancel{x}} = \overset{2}{\cancel{10x}} \cdot \frac{1}{\underset{1}{\cancel{5}}} + \overset{5}{\cancel{10x}} \cdot \frac{3}{\underset{1}{\cancel{2x}}} \qquad \begin{array}{l}\text{Divide out common factors in the} \\ \text{multiplications.}\end{array}$$

$$10 = 2x + 15 \qquad \text{Simplify.}$$

Observe that the resulting equation,

$$10 = 2x + 15$$

is now cleared of fractions. With the variable term, $2x$, already on the right, we will collect constant terms on the left by subtracting 15 from both sides.

$$-5 = 2x \qquad \text{Subtract 15 from both sides.}$$

$$-\frac{5}{2} = x \qquad \text{Divide both sides by 2.}$$

We check our solution by substituting $-\frac{5}{2}$ into the original equation or by using a calculator. With a calculator, evaluate each side of the equation for $x = -\frac{5}{2}$, or for $x = -2.5$. Note that the original restriction that $x \neq 0$ is met. The solution is $-\frac{5}{2}$ and the solution set is $\left\{-\frac{5}{2}\right\}$. ∎

✓ **CHECK POINT 2** Solve: $\dfrac{5}{2x} = \dfrac{17}{18} - \dfrac{1}{3x}$.

The following steps may be used to solve a rational equation:

Solving Rational Equations

1. List restrictions on the variable. Avoid any values of the variable that make a denominator zero.
2. Clear the equation of fractions by multiplying both sides by the LCD of all rational expressions in the equation.
3. Solve the resulting equation.
4. Reject any proposed solution that is in the list of restrictions on the variable. Check other proposed solutions in the original equation.

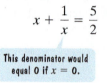

EXAMPLE 3   Solving a Rational Equation

Solve: $x + \dfrac{1}{x} = \dfrac{5}{2}$.

Solution

Step 1. List restrictions on the variable.

$$x + \dfrac{1}{x} = \dfrac{5}{2}$$

> This denominator would
> equal 0 if $x = 0$.

The restriction is $x \neq 0$.

Step 2. Multiply both sides by the LCD. The denominators are x and 2. Thus, the LCD is $2x$. We multiply both sides by $2x$.

$x + \dfrac{1}{x} = \dfrac{5}{2}, \quad x \neq 0$ This is the given equation.

$2x\left(x + \dfrac{1}{x}\right) = 2x\left(\dfrac{5}{2}\right)$ Multiply both sides by the LCD.

$2x \cdot x + 2x \cdot \dfrac{1}{x} = 2x \cdot \dfrac{5}{2}$ Use the distributive property on the left side.

$2x^2 + 2 = 5x$ Simplify.

Step 3. Solve the resulting equation. Can you see that we have a quadratic equation? Write the equation in standard form and solve for x.

$2x^2 - 5x + 2 = 0$ Subtract $5x$ from both sides.

$(2x - 1)(x - 2) = 0$ Factor.

$2x - 1 = 0 \quad \text{or} \quad x - 2 = 0$ Set each factor equal to O.

$2x = 1 \qquad\qquad x = 2$ Solve the resulting equations.

$x = \dfrac{1}{2}$

Step 4. Check proposed solutions in the original equation. The proposed solutions, $\frac{1}{2}$ and 2, are not part of the restriction that $x \neq 0$. Neither makes a denominator in the original equation equal to zero.

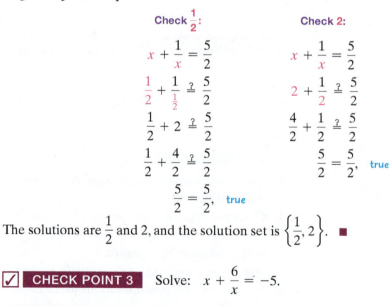

The solutions are $\dfrac{1}{2}$ and 2, and the solution set is $\left\{\dfrac{1}{2}, 2\right\}$. ∎

✓ **CHECK POINT 3**   Solve: $x + \dfrac{6}{x} = -5$.

EXAMPLE 4   Solving a Rational Equation

Solve: $\dfrac{3x}{x^2 - 9} + \dfrac{1}{x - 3} = \dfrac{3}{x + 3}$.

Solution

Step 1. List restrictions on the variable. By factoring denominators, it makes it easier to see values that make denominators zero.

$$\underbrace{\frac{3x}{(x + 3)(x - 3)}} + \underbrace{\frac{1}{x - 3}} = \underbrace{\frac{3}{x + 3}}$$

This denominator is zero if $x = -3$ or $x = 3$.   This denominator is zero if $x = 3$.   This denominator is zero if $x = -3$.

The restrictions are $x \neq -3$ and $x \neq 3$.

Step 2. Multiply both sides by the LCD. The LCD is $(x + 3)(x - 3)$.

$$\frac{3x}{(x + 3)(x - 3)} + \frac{1}{x - 3} = \frac{3}{x + 3}, \quad x \neq -3, x \neq 3$$ This is the given equation with a denominator factored.

$$(x + 3)(x - 3)\left[\frac{3x}{(x + 3)(x - 3)} + \frac{1}{x - 3}\right] = (x + 3)(x - 3) \cdot \frac{3}{x + 3}$$ Multiply both sides by the LCD.

$$(x + 3)(x - 3) \cdot \frac{3x}{(x + 3)(x - 3)} + (x + 3)(x - 3) \cdot \frac{1}{x - 3}$$

$$= (x + 3)(x - 3) \cdot \frac{3}{x + 3}$$ Use the distributive property on the left side.

$$3x + (x + 3) = 3(x - 3)$$ Simplify.

Step 3. Solve the resulting equation.

$$3x + (x + 3) = 3(x - 3)$$ This is the equation cleared of fractions.

$$4x + 3 = 3x - 9$$ Combine like terms on the left side.
Use the distributive property on the right side.

$$x + 3 = -9$$ Subtract $3x$ from both sides.

$$x = -12$$ Subtract 3 from both sides.

Step 4. Check proposed solutions in the original equation. The proposed solution, -12, is not part of the restriction that $x \neq -3$ and $x \neq 3$. Substitute -12 for x in the given equation and show that -12 is the solution. The equation's solution set is $\{-12\}$. ■

✓ **CHECK POINT 4**   Solve: $\dfrac{11}{x^2 - 25} + \dfrac{4}{x + 5} = \dfrac{3}{x - 5}$.

EXAMPLE 5   Solving a Rational Equation

Solve: $\dfrac{8x}{x + 1} = 4 - \dfrac{8}{x + 1}$.

Solution

Step 1. List restrictions on the variable.

This denominator is zero if $x = -1$. $\dfrac{8x}{x + 1} = 4 - \dfrac{8}{x + 1}$ This denominator is zero if $x = -1$.

The restriction is $x \neq -1$.

Step 2. Multiply both sides by the LCD. The LCD is $x + 1$.

$$\frac{8x}{x + 1} = 4 - \frac{8}{x + 1}, \quad x \neq -1$$ This is the given equation.

$$(x + 1) \cdot \frac{8x}{x + 1} = (x + 1)\left[4 - \frac{8}{x + 1}\right]$$ Multiply both sides by the LCD.

$$(x + 1) \cdot \frac{8x}{x + 1} = (x + 1) \cdot 4 - (x + 1) \cdot \frac{8}{x + 1}$$ Use the distributive property on the right side.

$$8x = 4(x + 1) - 8$$ Simplify.

Step 3. Solve the resulting equation.

$$8x = 4(x + 1) - 8$$ This is the equation cleared of fractions.

$$8x = 4x + 4 - 8$$ Use the distributive property on the right side.

$$8x = 4x - 4$$ Simplify.

$$4x = -4$$ Subtract $4x$ from both sides.

$$x = -1$$ Divide both sides by 4.

Step 4. Check proposed solutions. The proposed solution, -1, is *not* a solution because of the restriction that $x \neq -1$. Notice that -1 makes both of the denominators zero in the original equation. There is *no solution to this equation*. The solution set is $\varnothing$, the empty set. ■

> ## Great Question!
>
> **Give me the bottom line: When do I get rid of proposed solutions in rational equations?**
>
> Reject any proposed solution that causes any denominator in a rational equation to equal 0.

✓ **CHECK POINT 5** Solve: $\dfrac{x}{x - 3} = \dfrac{3}{x - 3} + 9.$

> ## Great Question!
>
> **What's the difference between adding or subtracting rational expressions and solving equations containing the addition or subtraction of rational expressions?**
>
> We simplify rational expressions. We solve rational equations. Notice the differences between the procedures.
>
> **Adding Rational Expressions**
>
> Simplify: $\dfrac{5}{3x} + \dfrac{3}{x}.$
>
>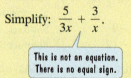
> This is not an equation. There is no equal sign.
>
> **Solving a Rational Equation**
>
> Solve: $\dfrac{5}{3x} + \dfrac{3}{x} = 1.$
>
> This is an equation. There is an equal sign.
>
> **Solution**
>
> The LCD is $3x$. Rewrite each expression with this LCD and retain the LCD.
>
> $$\frac{5}{3x} + \frac{3}{x}$$
> $$= \frac{5}{3x} + \frac{3}{x} \cdot \frac{3}{3}$$
> $$= \frac{5}{3x} + \frac{9}{3x}$$
> $$= \frac{5 + 9}{3x}$$
> $$= \frac{14}{3x}$$
>
> **Solution**
>
> The LCD is $3x$. Multiply both sides by this LCD and clear the fractions.
>
> $$3x\left(\frac{5}{3x} + \frac{3}{x}\right) = 3x \cdot 1$$
> $$3x \cdot \frac{5}{3x} + 3x \cdot \frac{3}{x} = 3x$$
> $$5 + 9 = 3x$$
> $$14 = 3x$$
> $$\frac{14}{3} = x$$
>
>
>
> The solution set is $\left\{\dfrac{14}{3}\right\}$.
>
> You only eliminate the denominators when solving a rational equation with an equal sign. You should never begin by eliminating the denominators when simplifying a rational expression involving addition or subtraction with no equal sign.

2 Solve problems involving formulas with rational expressions. ▶

Applications of Rational Equations

Rational equations can be solved to answer questions about variables contained in mathematical models.

EXAMPLE 6   A Government-Funded Cleanup

The formula

$$y = \frac{250x}{100 - x}$$

models the cost, y, in millions of dollars, to remove x percent of a river's pollutants. If the government commits $375 million for this project, what percentage of pollutants can be removed?

Solution   Substitute 375 for y and solve the resulting rational equation for x.

$$375 = \frac{250x}{100 - x} \qquad \text{The LCD is } 100 - x.$$

$$(100 - x)375 = (100 - x) \cdot \frac{250x}{100 - x} \qquad \text{Multiply both sides by the LCD.}$$

$$375(100 - x) = 250x \qquad \text{Simplify.}$$

$$37{,}500 - 375x = 250x \qquad \text{Use the distributive property on the left side.}$$

$$37{,}500 = 625x \qquad \text{Add } 375x \text{ to both sides.}$$

$$\frac{37{,}500}{625} = \frac{625x}{625} \qquad \text{Divide both sides by 625.}$$

$$60 = x \qquad \text{Simplify.}$$

If the government spends $375 million, 60% of the river's pollutants can be removed. ∎

☑ **CHECK POINT 6**   Use the model in Example 6 to answer this question: If government funding is increased to $750 million, what percentage of pollutants can be removed?

Achieving Success

Watching lectures on YouTube should not be a substitute for going to class.

For many students, it's much easier to get to class at designated times than to watch lectures at home, where no time ever seems to be the right time. If working from the web or MyMathLab, avoid the temptation of listening to multiple lectures in a single sitting. This cramming of information is never a good way to learn. It's risky business to think you can rely on the web as a substitute for the valuable insights and interactions that occur during classroom lecture.

CONCEPT AND VOCABULARY CHECK

Fill in each blank so that the resulting statement is true.

1. We clear a rational equation of fractions by multiplying both sides by the _____ of all rational expressions in the equation.

2. We reject any proposed solution of a rational equation that causes a denominator to equal _____.

3. The first step in solving

$$\frac{8}{x} + \frac{1}{4x} = \frac{17}{8}, x \neq 0,$$

is to multiply both sides by _____.

4. The first step in solving

$$\frac{x + 1}{3(x + 3)} + \frac{x}{2(x + 3)} = \frac{5}{4(x + 3)}, x \neq -3,$$

is to multiply both sides by _____.

5.

$$\frac{1}{3x} - \frac{1}{3} = x$$

$$3x\left(\frac{1}{3x} - \frac{1}{3}\right) = 3x \cdot x$$

The resulting equation cleared of fractions is _____.

6.

$$\frac{6}{(x + 1)(x - 1)} = \frac{5}{x - 1} + \frac{3}{x + 1}$$

$$(x + 1)(x - 1) \cdot \frac{6}{(x + 1)(x - 1)} = (x + 1)(x - 1)\left[\frac{5}{x - 1} + \frac{3}{x + 1}\right]$$

The resulting equation cleared of fractions is _____.

7. The restriction on the variable in the rational equation

$$\frac{2}{x + 2} + 2 = \frac{7}{x + 2}$$

is _____.

8. The restrictions on the variable in the rational equation

$$\frac{4}{x - 3} - \frac{3}{x - 2} = \frac{4x + 9}{(x - 3)(x - 2)}$$

are _____ and _____.

9. True or false: A rational equation can have no solution. _____

10. True or false: There are no restrictions on the variable in the rational equation $\frac{5}{6} + \frac{3}{x} = \frac{1}{4}$. _____

7.6 EXERCISE SET ▶ MyMathLab®

Practice Exercises

In Exercises 1–46, solve each rational equation.

1. $\frac{x}{3} = \frac{x}{2} - 2$

2. $\frac{x}{5} = \frac{x}{6} + 1$

3. $\frac{4x}{3} = \frac{x}{18} - \frac{x}{6}$

4. $\frac{5x}{4} = \frac{x}{12} - \frac{x}{2}$

5. $2 - \frac{8}{x} = 6$

6. $1 - \frac{9}{x} = 4$

7. $\frac{2}{x} + \frac{1}{3} = \frac{4}{x}$

8. $\frac{5}{x} + \frac{1}{3} = \frac{6}{x}$

9. $\frac{2}{x} + 3 = \frac{5}{2x} + \frac{13}{4}$

10. $\frac{7}{2x} = \frac{5}{3x} + \frac{22}{3}$

11. $\frac{2}{3x} + \frac{1}{4} = \frac{11}{6x} - \frac{1}{3}$

12. $\frac{5}{2x} - \frac{8}{9} = \frac{1}{18} - \frac{1}{3x}$

13. $\frac{6}{x + 3} = \frac{4}{x - 3}$

14. $\frac{7}{x + 1} = \frac{5}{x - 3}$

15. $\dfrac{x-2}{2x} + 1 = \dfrac{x+1}{x}$

16. $\dfrac{7x-4}{5x} = \dfrac{9}{5} - \dfrac{4}{x}$

17. $x + \dfrac{6}{x} = -7$

18. $x + \dfrac{7}{x} = -8$

19. $\dfrac{x}{5} - \dfrac{5}{x} = 0$

20. $\dfrac{x}{4} - \dfrac{4}{x} = 0$

21. $x + \dfrac{3}{x} = \dfrac{12}{x}$

22. $x + \dfrac{3}{x} = \dfrac{19}{x}$

23. $\dfrac{4}{y} - \dfrac{y}{2} = \dfrac{7}{2}$

24. $\dfrac{4}{3y} - \dfrac{1}{3} = y$

25. $\dfrac{x-4}{x} = \dfrac{15}{x+4}$

26. $\dfrac{x-1}{2x+3} = \dfrac{6}{x-2}$

27. $\dfrac{2}{x^2-1} = \dfrac{4}{x+1}$

28. $\dfrac{3}{x+1} = \dfrac{1}{x^2-1}$

29. $\dfrac{1}{x-1} + 5 = \dfrac{11}{x-1}$

30. $\dfrac{3}{x+4} - 7 = \dfrac{-4}{x+4}$

31. $\dfrac{8y}{y+1} = 4 - \dfrac{8}{y+1}$

32. $\dfrac{2}{y-2} = \dfrac{y}{y-2} - 2$

33. $\dfrac{3}{x-1} + \dfrac{8}{x} = 3$

34. $\dfrac{2}{x-2} + \dfrac{4}{x} = 2$

35. $\dfrac{3y}{y-4} - 5 = \dfrac{12}{y-4}$

36. $\dfrac{10}{y+2} = 3 - \dfrac{5y}{y+2}$

37. $\dfrac{1}{x} + \dfrac{1}{x-3} = \dfrac{x-2}{x-3}$

38. $\dfrac{1}{x-1} + \dfrac{2}{x} = \dfrac{x}{x-1}$

39. $\dfrac{x+1}{3x+9} + \dfrac{x}{2x+6} = \dfrac{2}{4x+12}$

40. $\dfrac{3}{2y-2} + \dfrac{1}{2} = \dfrac{2}{y-1}$

41. $\dfrac{4y}{y^2-25} + \dfrac{2}{y-5} = \dfrac{1}{y+5}$

42. $\dfrac{1}{x+4} + \dfrac{1}{x-4} = \dfrac{22}{x^2-16}$

43. $\dfrac{1}{x-4} - \dfrac{5}{x+2} = \dfrac{6}{x^2-2x-8}$

44. $\dfrac{6}{x+3} - \dfrac{5}{x-2} = \dfrac{-20}{x^2+x-6}$

45. $\dfrac{2}{x+3} - \dfrac{2x+3}{x-1} = \dfrac{6x-5}{x^2+2x-3}$

46. $\dfrac{x-3}{x-2} + \dfrac{x+1}{x+3} = \dfrac{2x^2-15}{x^2+x-6}$

Practice PLUS

In Exercises 47–54, solve or simplify, whichever is appropriate.

47. $\dfrac{x^2-10}{x^2-x-20} = 1 + \dfrac{7}{x-5}$

48. $\dfrac{x^2+4x-2}{x^2-2x-8} = 1 + \dfrac{4}{x-4}$

49. $\dfrac{x^2-10}{x^2-x-20} - 1 - \dfrac{7}{x-5}$

50. $\dfrac{x^2+4x-2}{x^2-2x-8} - 1 - \dfrac{4}{x-4}$

51. $5y^{-2} + 1 = 6y^{-1}$

52. $3y^{-2} + 1 = 4y^{-1}$

53. $\dfrac{3}{y+1} - \dfrac{1}{1-y} = \dfrac{10}{y^2-1}$

54. $\dfrac{4}{y-2} - \dfrac{1}{2-y} = \dfrac{25}{y+6}$

Application Exercises

A company that manufactures wheelchairs has fixed costs of $500,000. The average cost per wheelchair, C, for the company to manufacture x wheelchairs per month is modeled by the formula

$$C = \dfrac{400x + 500{,}000}{x}.$$

Use this mathematical model to solve Exercises 55–56.

55. How many wheelchairs per month can be produced at an average cost of $450 per wheelchair?

56. How many wheelchairs per month can be produced at an average cost of $405 per wheelchair?

In Palo Alto, California, a government agency ordered computer-related companies to contribute to a pool of money to clean up underground water supplies. (The companies had stored toxic chemicals in leaking underground containers.) The formula

$$C = \dfrac{2x}{100 - x}$$

models the cost, C, in millions of dollars, for removing x percent of the contaminants. Use this mathematical model to solve Exercises 57–58.

57. What percentage of the contaminants can be removed for $2 million?

58. What percentage of the contaminants can be removed for $8 million?

We have seen that Young's rule

$$C = \dfrac{DA}{A + 12}$$

can be used to approximate the dosage of a drug prescribed for children. In this formula, A = the child's age, in years, D = an adult dosage, and C = the proper child's dosage. Use this formula to solve Exercises 59–60.

59. When the adult dosage is 1000 milligrams, a child is given 300 milligrams. What is that child's age? Round to the nearest year.

60. When the adult dosage is 1000 milligrams, a child is given 500 milligrams. What is that child's age?

A grocery store sells 4000 cases of canned soup per year. By averaging costs to purchase soup and to pay storage costs, the owner has determined that if x cases are ordered at a time, the yearly inventory cost, C, can be modeled by

$$C = \frac{10,000}{x} + 3x.$$

The graph of this model is shown below. Use this information to solve Exercises 61–62.

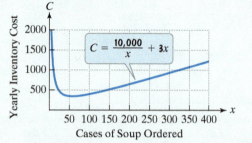

61. How many cases should be ordered at a time for yearly inventory costs to be $350? Identify your solutions as points on the graph.

62. How many cases should be ordered at a time for yearly inventory costs to be $790? Identify your solutions as points on the graph.

Explaining the Concepts

63. What is a rational equation?

64. Explain how to solve a rational equation.

65. Explain how to find restrictions on the variable in a rational equation.

66. Why should restrictions on the variable in a rational equation be listed before you begin solving the equation?

67. Describe similarities and differences between the procedures needed to solve the following problems:

$$\text{Add: } \frac{2}{x} + \frac{3}{4}. \quad \text{Solve: } \frac{2}{x} + \frac{3}{4} = 1.$$

Critical Thinking Exercises

Make Sense? *In Exercises 68–71, determine whether each statement makes sense or does not make sense, and explain your reasoning.*

68. I added two rational expressions and found the solution set.

69. I can solve the equation $\frac{6}{x+3} = \frac{4}{x-3}$ by multiplying both sides by the LCD.

70. I'm solving a rational equation that became a quadratic equation, so my rational equation must have two solutions.

71. I must have made an error if a rational equation produces no solution.

In Exercises 72–75, determine whether each statement is true or false. If the statement is false, make the necessary change(s) to produce a true statement.

72. $\frac{1}{x} + \frac{1}{6} = 6x\left(\frac{1}{x} + \frac{1}{6}\right) = 6 + x$

73. If a is any real number, the equation $\frac{a}{x} + 1 = \frac{a}{x}$ has no solution.

74. All real numbers satisfy the equation $\frac{3}{x} - \frac{1}{x} = \frac{2}{x}$.

75. To solve $\frac{5}{3x} + \frac{3}{x} = 1$, we must first add the rational expressions on the left side.

76. Solve for f: $\frac{1}{p} + \frac{1}{q} = \frac{1}{f}$.

77. Solve for f_2: $f = \frac{f_1 f_2}{f_1 + f_2}$.

In Exercises 78–79, solve each rational equation.

78. $\frac{x+1}{2x^2 - 11x + 5} = \frac{x-7}{2x^2 + 9x - 5} - \frac{2x-6}{x^2 - 25}$

79. $\left(\frac{x+1}{x+7}\right)^2 \div \left(\frac{x+1}{x+7}\right)^4 = 0.$

80. Find b so that the solution of

$$\frac{7x+4}{b} + 13 = x$$

is -6.

Technology Exercises

In Exercises 81–83, use a graphing utility to solve each rational equation. Graph each side of the equation in the given viewing rectangle. The first coordinate of each point of intersection is a solution. Check by direct substitution.

81. $\frac{x}{2} + \frac{x}{4} = 6$

$[-5, 10, 1]$ by $[-5, 10, 1]$

82. $\frac{50}{x} = 2x$

$[-10, 10, 1]$ by $[-20, 20, 2]$

83. $x + \frac{6}{x} = -5$

$[-10, 10, 1]$ by $[-10, 10, 1]$

Review Exercises

84. Factor completely: $x^4 + 2x^3 - 3x - 6$. (Section 6.1, Example 8)

85. Simplify: $(3x^2)(-4x^{-10})$. (Section 5.7, Example 3)

86. Simplify: $-5[4(x - 2) - 3]$. (Section 1.8, Example 11)

Preview Exercises

Exercises 87–89 will help you prepare for the material covered in the next section.

87. Solve: $\dfrac{15}{8 + x} = \dfrac{9}{8 - x}$.

88. If you can complete a job in 5 hours, what fractional part of the job can you complete in 1 hour? in 3 hours? in x hours?

89. Write as an equation, where x represents the number: The quotient of 63 and a number is equal to the quotient of 7 and 5.

SECTION

7.7

Applications Using Rational Equations and Proportions ▶

What am I supposed to learn?

After studying this section, you should be able to:

1 Solve problems involving motion. ▶

2 Solve problems involving work. ▶

3 Solve problems involving proportions. ▶

4 Solve problems involving similar triangles. ▶

1 Solve problems involving motion. ▶

The possibility of seeing a blue whale, the largest mammal ever to grace the earth, increases the excitement of gazing out over the ocean's swell of waves. Blue whales were hunted to near extinction in the last half of the nineteenth and the first half of the twentieth centuries. Using a method for estimating wildlife populations that we discuss in this section, by the mid-1960s it was determined that the world population of blue whales was less than 1000. This led the International Whaling Commission to ban the killing of blue whales to prevent their extinction. A dramatic increase in blue whale sightings indicates an ongoing increase in their population and the success of the killing ban.

Problems Involving Motion

Suppose that you ride your bike at an average speed of 12 miles per hour. What distance do you cover in 2 hours? Your distance is the product of your speed and the time that you travel:

$$\frac{12 \text{ miles}}{\text{hour}} \times 2 \text{ hours} = 24 \text{ miles}.$$

Your distance is 24 miles. Notice how the hour units cancel. The distance is expressed in miles.

In general, the distance covered by any moving body is the product of its average speed, or rate, and its time in motion:

A Formula for Motion

$$d = rt$$

Distance equals rate times time.

Rational expressions appear in motion problems when the conditions of the problem involve the time traveled. We can obtain an expression for t, the time traveled, by dividing both sides of $d = rt$ by r.

$$d = rt \qquad \text{Distance equals rate times time.}$$

$$\frac{d}{r} = \frac{rt}{r} \qquad \text{Divide both sides by } r.$$

$$\frac{d}{r} = t \qquad \text{Simplify.}$$

Time in Motion

$$t = \frac{d}{r}$$

$$\text{Time traveled} = \frac{\text{Distance traveled}}{\text{Rate of travel}}$$

Downstream (with the current)

←——15 miles——→

Upstream (against the current)

←—— 9 miles ——→

Traveling 15 miles downstream takes the same time as traveling 9 miles upstream.

EXAMPLE 1 A Motion Problem Involving Time

In still water, your small boat averages 8 miles per hour. It takes you the same amount of time to travel 15 miles downstream, with the current, as 9 miles upstream, against the current. What is the rate of the water's current?

Solution

Step 1. Let x represent one of the quantities. Let

$$x = \text{the rate of the current.}$$

Step 2. Represent other unknown quantities in terms of x. We still need expressions for the rate of your boat with the current and the rate against the current. Traveling with the current, the boat's rate in still water, 8 miles per hour, is increased by the current's rate, x miles per hour. Thus,

$$8 + x = \text{the boat's rate with the current.}$$

Traveling against the current, the boat's rate in still water, 8 miles per hour, is decreased by the current's rate, x miles per hour. Thus,

$$8 - x = \text{the boat's rate against the current.}$$

Step 3. Write an equation that models the conditions. By reading the problem again, we discover that the crucial idea is that the time spent going 15 miles with the current equals the time spent going 9 miles against the current. This information is summarized in the following table.

	Distance	Rate	Time $= \dfrac{\text{Distance}}{\text{Rate}}$
With the current	15	$8 + x$	$\dfrac{15}{8 + x}$
Against the current	9	$8 - x$	$\dfrac{9}{8 - x}$

These two times are equal.

We are now ready to write an equation that models the problem's conditions.

The time spent going 15 miles with the current $\qquad \dfrac{15}{8 + x} = \dfrac{9}{8 - x} \qquad$ the time spent going 9 miles against the current.

equals

Step 4. Solve the equation and answer the question.

$$\frac{15}{8 + x} = \frac{9}{8 - x}$$

This is the equation that models the problem's conditions.

$$(8 + x)(8 - x) \cdot \frac{15}{8 + x} = (8 + x)(8 - x) \cdot \frac{9}{8 - x}$$

Multiply both sides by the LCD, $(8 + x)(8 - x)$.

$$15(8 - x) = 9(8 + x)$$

Simplify.

$$120 - 15x = 72 + 9x$$

Use the distributive property.

$$120 = 72 + 24x$$

Add 15x to both sides.

$$48 = 24x$$

Subtract 72 from both sides.

$$2 = x$$

Divide both sides by 24.

The rate of the water's current is 2 miles per hour.

Step 5. Check the proposed solution in the original wording of the problem. Does it take you the same amount of time to travel 15 miles downstream as 9 miles upstream if the current is 2 miles per hour? Keep in mind that your rate in still water is 8 miles per hour.

Time required to travel 15 miles with the current $= \dfrac{\text{Distance}}{\text{Rate}} = \dfrac{15}{8 + 2} = \dfrac{15}{10} = 1\dfrac{1}{2}$ hours

Time required to travel 9 miles against the current $= \dfrac{\text{Distance}}{\text{Rate}} = \dfrac{9}{8 - 2} = \dfrac{9}{6} = 1\dfrac{1}{2}$ hours

These times are the same, which checks with the original conditions of the problem. ∎

✓ **CHECK POINT 1**   Forget the small boat! This time we have you canoeing on the Colorado River. In still water, your average canoeing rate is 3 miles per hour. It takes you the same amount of time to travel 10 miles downstream, with the current, as 2 miles upstream, against the current. What is the rate of the water's current?

2 Solve problems involving work. ▶

Problems Involving Work

You are thinking of designing your own Web site. You estimate that it will take 30 hours to do the job. In 1 hour, $\dfrac{1}{30}$ of the job is completed. In 2 hours, $\dfrac{2}{30}$, or $\dfrac{1}{15}$, of the job is completed. In 3 hours, the fractional part of the job done is $\dfrac{3}{30}$, or $\dfrac{1}{10}$. In x hours, the fractional part of the job that you can complete is $\dfrac{x}{30}$.

Your friend, who has experience developing Web sites, took 20 hours working on his own to design an impressive site. You wonder about the possibility of working together. How long would it take both of you to design your Web site?

Problems involving work usually have two people working together to complete a job. The amount of time it takes each person to do the job working alone is frequently known. The question deals with how long it will take both people working together to complete the job.

In work problems, **the number 1 represents one whole job completed**. For example, the completion of your Web site is represented by 1. Equations in work problems are based on the following condition:

| Fractional part of the job done by the first person | + | fractional part of the job done by the second person | = | 1 (one whole job completed). |

EXAMPLE 2 Solving a Problem Involving Work

You can design a Web site in 30 hours. Your friend can design the same site in 20 hours. How long will it take to design the Web site if you both work together?

Solution

Step 1. Let *x* represent one of the quantities. Let $x =$ the time, in hours, for you and your friend to design the Web site together.

Step 2. Represent other unknown quantities in terms of *x*. Because there are no other unknown quantities, we can skip this step.

Step 3. Write an equation that models the conditions. We construct a table to help find the fractional part of the task completed by you and your friend in x hours.

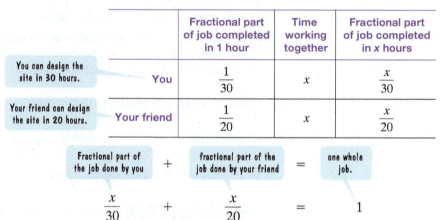

		Fractional part of job completed in 1 hour	Time working together	Fractional part of job completed in *x* hours
You can design the site in 30 hours.	You	$\dfrac{1}{30}$	x	$\dfrac{x}{30}$
Your friend can design the site in 20 hours.	Your friend	$\dfrac{1}{20}$	x	$\dfrac{x}{20}$

Fractional part of the job done by you $+$ fractional part of the job done by your friend $=$ one whole job.

$$\frac{x}{30} + \frac{x}{20} = 1$$

Step 4. Solve the equation and answer the question.

$$\frac{x}{30} + \frac{x}{20} = 1 \qquad \text{This is the equation that models the problem's conditions.}$$

$$60\left(\frac{x}{30} + \frac{x}{20}\right) = 60 \cdot 1 \qquad \text{Multiply both sides by 60, the LCD.}$$

$$60 \cdot \frac{x}{30} + 60 \cdot \frac{x}{20} = 60 \qquad \text{Use the distributive property on the left side.}$$

$$2x + 3x = 60 \qquad \text{Simplify: } \frac{\overset{2}{\cancel{60}}}{1} \cdot \frac{x}{\underset{1}{\cancel{30}}} = 2x \text{ and } \frac{\overset{3}{\cancel{60}}}{1} \cdot \frac{x}{\underset{1}{\cancel{20}}} = 3x.$$

$$5x = 60 \qquad \text{Combine like terms.}$$

$$x = 12 \qquad \text{Divide both sides by 5.}$$

If you both work together, you can design your Web site in 12 hours.

Step 5. Check the proposed solution in the original wording of the problem. Will you both complete the job in 12 hours? Because you can design the site in 30 hours, in 12 hours, you can complete $\frac{12}{30}$, or $\frac{2}{5}$, of the job. Because your friend can design the site in 20 hours, in 12 hours, he can complete $\frac{12}{20}$, or $\frac{3}{5}$, of the job. Notice that $\frac{2}{5} + \frac{3}{5} = 1$, which represents the completion of the entire job, or one whole job. ∎

Great Question!

Is there an equation that models all the work problems in this section?

Yes. Let

$a =$ the time it takes person A to do a job working alone

$b =$ the time it takes person B to do the same job working alone.

If x represents the time it takes for A and B to complete the entire job working together, then the situation can be modeled by the rational equation

$$\frac{x}{a} + \frac{x}{b} = 1.$$

✓ **CHECK POINT 2** One person can paint the outside of a house in 8 hours. A second person can do it in 4 hours. How long will it take them to do the job if they work together?

3 Solve problems involving proportions.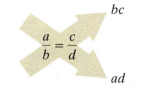

Problems Involving Proportions

A **ratio** compares quantities by division. For example, this year's entering class at a medical school contains 60 women and 30 men. The ratio of women to men is $\frac{60}{30}$. We can express this ratio as a fraction reduced to lowest terms:

$$\frac{60}{30} = \frac{30 \cdot 2}{30 \cdot 1} = \frac{2}{1}.$$

This ratio can be expressed as 2:1, or 2 to 1.

A **proportion** is a statement that says that two ratios are equal. If the ratios are $\frac{a}{b}$ and $\frac{c}{d}$, then the proportion is

$$\frac{a}{b} = \frac{c}{d}.$$

We can clear this equation of fractions by multiplying both sides by bd, the least common denominator:

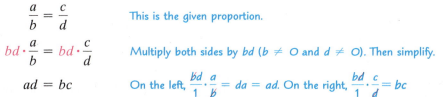

We see that the following principle is true for any proportion:

$\dfrac{a}{b} \underset{ad}{\overset{bc}{\rightleftarrows}} \dfrac{c}{d}$

The cross-products principle:
$ad = bc$

The Cross-Products Principle for Proportions

If $\dfrac{a}{b} = \dfrac{c}{d}$, then $ad = bc$. ($b \neq 0$ and $d \neq 0$)

The cross products ad and bc are equal.

For example, since $\frac{2}{3} = \frac{6}{9}$, we see that $2 \cdot 9 = 3 \cdot 6$, or $18 = 18$. We can also use $\frac{2}{3} = \frac{6}{9}$ and conclude that $3 \cdot 6 = 2 \cdot 9$. When using the cross-products principle, it does not matter on which side of the equation each product is placed.

Here is a procedure for solving problems involving proportions:

Solving Applied Problems Using Proportions

1. Read the problem and represent the unknown quantity by x (or any letter).
2. Set up a proportion by listing the given ratio on one side and the ratio with the unknown quantity on the other side. Each respective quantity should occupy the same corresponding position on each side of the proportion.
3. Drop units and apply the cross-products principle.
4. Solve for x and answer the question.

EXAMPLE 3 Applying Proportions: Calculating Taxes

The property tax on a house with an assessed value of $480,000 is $5760. Determine the property tax on a house with an assessed value of $600,000, assuming the same tax rate.

Great Question!

Are there other proportions that I can use in step 2 to model the problem's conditions?

Yes. Here are three other correct proportions you can use:

- $\dfrac{\$480{,}000 \text{ value}}{\$5760 \text{ tax}} = \dfrac{\$600{,}000 \text{ value}}{\$x \text{ tax}}$

- $\dfrac{\$480{,}000 \text{ value}}{\$600{,}000 \text{ value}} = \dfrac{\$5760 \text{ tax}}{\$x \text{ tax}}$

- $\dfrac{\$600{,}000 \text{ value}}{\$480{,}000 \text{ value}} = \dfrac{\$x \text{ tax}}{\$5760 \text{ tax}}$.

Each proportion gives the same cross product obtained in step 3.

Solution

Step 1. Represent the unknown by x. Let x = the tax on the $600,000 house.

Step 2. Set up a proportion. We will set up a proportion comparing taxes to assessed value.

$$\underbrace{\frac{\text{Tax on \$480,000 house}}{\text{Assessed value (\$480,000)}}}_{} \ \text{equals} \ \underbrace{\frac{\text{Tax on \$600,000 house}}{\text{Assessed value (\$600,000)}}}_{}$$

$$\text{Given ratio}\begin{cases} \dfrac{\$5760}{\$480{,}000} \end{cases} = \dfrac{\$x \ \leftarrow \text{Unknown}}{\$600{,}000 \ \leftarrow \text{Given quantity}}$$

Step 3. Drop the units and apply the cross-products principle. We drop the dollar signs and begin to solve for x.

$$\frac{5760}{480{,}000} = \frac{x}{600{,}000} \qquad \text{This is the proportion that models the problem's conditions.}$$

$$480{,}000x = (5760)(600{,}000) \qquad \text{Apply the cross-products principle.}$$

$$480{,}000x = 3{,}456{,}000{,}000 \qquad \text{Multiply.}$$

Step 4. Solve for x and answer the question.

$$\frac{480{,}000x}{480{,}000} = \frac{3{,}456{,}000{,}000}{480{,}000} \qquad \text{Divide both sides by 480,000.}$$

$$x = 7200 \qquad \text{Simplify.}$$

The property tax on the $600,000 house is $7200. ∎

☑ **CHECK POINT 3** The property tax on a house with an assessed value of $250,000 is $3500. Determine the property tax on a house with an assessed value of $420,000, assuming the same tax rate.

Sampling in Nature

The method that was used to estimate the blue whale population described in the section opener is called the **capture-recapture method**. Because it is impossible to count each individual animal within a population, wildlife biologists randomly catch and tag a given number of animals. Sometime later they recapture a second sample of animals and count the number of recaptured tagged animals. The total size of the wildlife population is then estimated using the following proportion:

$$\underset{\substack{\text{Initially} \\ \text{unknown} \\ (x)}}{\underbrace{\frac{\text{Original number of tagged animals}}{\text{Total number of animals in the population}}}} = \left.\frac{\text{Number of recaptured tagged animals}}{\text{Number of animals in second sample}}\right\} \begin{matrix}\text{Known} \\ \text{ratio}\end{matrix}$$

Although this is called the capture-recapture method, it is not necessary to recapture animals to observe whether or not they are tagged. This could be done from a distance, with binoculars for instance.

EXAMPLE 4 Applying Proportions: Estimating Wildlife Population

Wildlife biologists catch, tag, and then release 135 deer back into a wildlife refuge. Two weeks later they observe a sample of 140 deer, 30 of which are tagged. Assuming the ratio of tagged deer in the sample holds for all deer in the refuge, approximately how many deer are in the refuge?

Solution

Step 1. Represent the unknown by x. Let x = the total number of deer in the refuge.

Step 2. Set up a proportion. Recall that the biologists tagged 135 deer. They later observed 30 tagged deer in a sample of 140 animals.

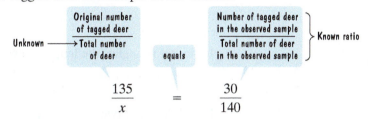

$$\frac{135}{x} = \frac{30}{140}$$

Steps 3 and 4. Apply the cross-products principle, solve, and answer the question.

$$\frac{135}{x} = \frac{30}{140} \qquad \text{This is the proportion that models the problem's conditions.}$$

$$(135)(140) = 30x \qquad \text{Apply the cross-products principle.}$$

$$18{,}900 = 30x \qquad \text{Multiply.}$$

$$\frac{18{,}900}{30} = \frac{30x}{30} \qquad \text{Divide both sides by 30.}$$

$$630 = x \qquad \text{Simplify.}$$

There are approximately 630 deer in the refuge. ∎

✓ **CHECK POINT 4** Wildlife biologists catch, tag, and then release 120 deer back into a wildlife refuge. Two weeks later they observe a sample of 150 deer, 25 of which are tagged. Assuming the ratio of tagged deer in the sample holds for all deer in the refuge, approximately how many deer are in the refuge?

4 Solve problems involving similar triangles. ▶

Pedestrian Crossing

Similar Triangles

Shown in the margin is an international road sign. This sign is shaped just like the actual sign, although its size is smaller. Figures that have the same shape, but not the same size, are used in **scale drawings**. A scale drawing always pictures the exact shape of the object that the drawing represents. Architects, engineers, landscape gardeners, and interior decorators use scale drawings in planning their work.

Figures that have the same shape, but not necessarily the same size, are called **similar figures**. In **Figure 7.2**, triangles ABC and DEF are similar. Angles A and D measure the same number of degrees and are called **corresponding angles**. Angles C and F are corresponding angles, as are angles B and E. Angles with the same number of tick marks in **Figure 7.2** are the corresponding angles.

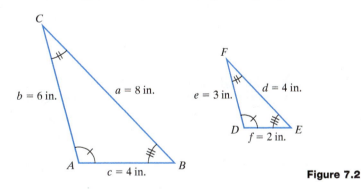

Figure 7.2

The sides opposite the corresponding angles are called **corresponding sides**. Although the measures of corresponding angles are equal, corresponding sides may or may not be the same length. For the triangles in **Figure 7.2**, each side in the smaller triangle is half the length of the corresponding side in the larger triangle.

The triangles in **Figure 7.2** illustrate what it means to be **similar triangles. Corresponding angles have the same measure and the ratios of the lengths of the corresponding sides are equal.** For the triangles in **Figure 7.2**, each of these ratios is equal to 2:

$$\frac{a}{d} = \frac{8}{4} = 2 \qquad \frac{b}{e} = \frac{6}{3} = 2 \qquad \frac{c}{f} = \frac{4}{2} = 2.$$

In similar triangles, the lengths of the corresponding sides are proportional. Thus,

$$\frac{a}{d} = \frac{b}{e} = \frac{c}{f}.$$

If we know that two triangles are similar, we can set up a proportion to solve for the length of an unknown side.

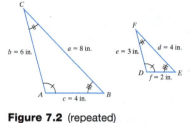

Figure 7.2 (repeated)

EXAMPLE 5 Using Similar Triangles

The triangles in **Figure 7.3** are similar. Find the missing length, x.

Figure 7.3

Solution Because the triangles in **Figure 7.3** are similar, their corresponding sides are proportional.

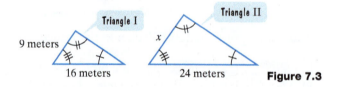

We solve this rational equation by multiplying both sides by the LCD, $24x$. (You can also apply the cross-products principle for solving proportions.)

$$24x \cdot \frac{9}{x} = 24x \cdot \frac{16}{24} \qquad \text{Multiply both sides by the LCD, } 24x.$$
$$24 \cdot 9 = 16x \qquad \text{Simplify.}$$
$$216 = 16x \qquad \text{Multiply: } 24 \cdot 9 = 216.$$
$$13.5 = x \qquad \text{Divide both sides by 16.}$$

The missing length, x, is 13.5 meters. ∎

✓ **CHECK POINT 5** The similar triangles in the figure are shown with corresponding sides in the same relative position. Find the missing length, x.

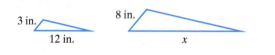

How can we quickly determine if two triangles are similar? **If the measures of two angles of one triangle are equal to those of two angles of a second triangle, then the two triangles are similar.** If the triangles are similar, then their corresponding sides are proportional.

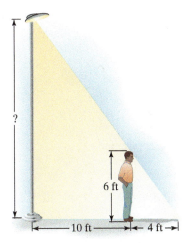

Figure 7.4

> **EXAMPLE 6**   Problem Solving Using Similar Triangles

A man who is 6 feet tall is standing 10 feet from the base of a lamppost (see **Figure 7.4**). The man's shadow has a length of 4 feet. How tall is the lamppost?

Solution   The drawing in **Figure 7.5** makes the similarity of the triangles easier to see. The large triangle with the lamppost on the left and the small triangle with the man on the left both contain 90° angles. They also share an angle. Thus, two angles of the large triangle are equal in measure to two angles of the small triangle. This means that the triangles are similar and their corresponding sides are proportional. We begin by letting x represent the height of the lamppost, in feet. Because corresponding sides of similar triangles are proportional,

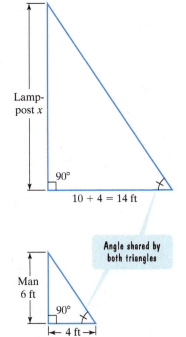

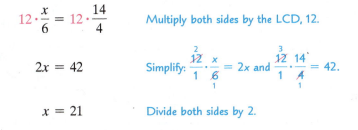

We solve for x by multiplying both sides by the LCD, 12.

Figure 7.5

$$12 \cdot \frac{x}{6} = 12 \cdot \frac{14}{4} \qquad \text{Multiply both sides by the LCD, 12.}$$

$$2x = 42 \qquad \text{Simplify: } \frac{\overset{2}{\cancel{12}}}{1} \cdot \frac{x}{\cancel{6}} = 2x \text{ and } \frac{\overset{3}{\cancel{12}}}{1} \cdot \frac{14}{\cancel{4}} = 42.$$

$$x = 21 \qquad \text{Divide both sides by 2.}$$

The lamppost is 21 feet tall. ∎

> ✓ **CHECK POINT 6**   Find the height of the lookout tower shown in **Figure 7.6** using the figure that lines up the top of the tower with the top of a stick that is 2 yards long.

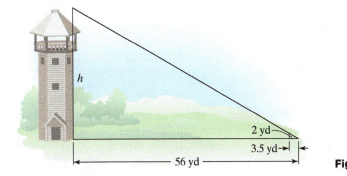

Figure 7.6

CONCEPT AND VOCABULARY CHECK

Fill in each blank so that the resulting statement is true.

1. The formula $t = \dfrac{d}{r}$ states that time traveled is _____ divided by _____.

2. In work problems, the number _____ represents the whole job completed.

3. If you can complete a job in 5 hours, the fractional part of the job that you can complete in x hours is represented by _____.

4. The cross-products principle for proportions states that if $\dfrac{a}{b} = \dfrac{c}{d}$, then _____, $b \neq 0$ and $d \neq 0$.

5. Triangles that have the same shape, but not necessarily the same size, are called _____ triangles.

7.7 EXERCISE SET ▶ MyMathLab®

Practice and Application Exercises

Use rational equations to solve Exercises 1–10. Each exercise is a problem involving motion.

1. How bad is the heavy traffic? You can walk 10 miles in the same time that it takes to travel 15 miles by car. If the car's rate is 3 miles per hour faster than your walking rate, find the average rate of each.

	Distance	Rate	Time = $\dfrac{\text{Distance}}{\text{Rate}}$
Walking	10	x	$\dfrac{10}{x}$
Car in Heavy Traffic	15	$x + 3$	$\dfrac{15}{x + 3}$

2. You can travel 40 miles on your motorcycle in the same time that it takes to travel 15 miles on your bicycle. If your motorcycle's rate is 20 miles per hour faster than your bicycle's, find the average rate for each.

	Distance	Rate	Time = $\dfrac{\text{Distance}}{\text{Rate}}$
Motorcycle	40	$x + 20$	$\dfrac{40}{x + 20}$
Bicycle	15	x	$\dfrac{15}{x}$

3. A jogger runs 4 miles per hour faster downhill than uphill. If the jogger can run 5 miles downhill in the same time that it takes to run 3 miles uphill, find the jogging rate in each direction.

4. A truck can travel 120 miles in the same time that it takes a car to travel 180 miles. If the truck's rate is 20 miles per hour slower than the car's, find the average rate for each.

5. In still water, a boat averages 15 miles per hour. It takes the same amount of time to travel 20 miles downstream, with the current, as 10 miles upstream, against the current. What is the rate of the water's current?

6. In still water, a boat averages 18 miles per hour. It takes the same amount of time to travel 33 miles downstream, with the current, as 21 miles upstream, against the current. What is the rate of the water's current?

7. As part of an exercise regimen, you walk 2 miles on an indoor track. Then you jog at twice your walking speed for another 2 miles. If the total time spent walking and jogging is 1 hour, find the walking and jogging rates.

8. The joys of the Pacific Coast! You drive 90 miles along the Pacific Coast Highway and then take a 5-mile run along a hiking trail in Point Reyes National Seashore. Your driving rate is nine times that of your running rate. If the total time for driving and running is 3 hours, find the average rate driving and the average rate running.

9. The water's current is 2 miles per hour. A boat can travel 6 miles downstream, with the current, in the same amount of time it travels 4 miles upstream, against the current. What is the boat's average rate in still water?

10. The water's current is 2 miles per hour. A canoe can travel 6 miles downstream, with the current, in the same amount of time it travels 2 miles upstream, against the current. What is the canoe's average rate in still water?

Use a rational equation to solve Exercises 11–16. Each exercise is a problem involving work.

11. You must leave for campus in 10 minutes or you will be late for class. Unfortunately, you are snowed in. You can shovel the driveway in 20 minutes and your brother claims he can do it in 15 minutes. If you shovel together, how long will it take to clear the driveway? Will this give you enough time before you have to leave?

12. You promised your parents that you would wash the family car. You have not started the job and they are due home in 16 minutes. You can wash the car in 40 minutes and your sister claims she can do it in 30 minutes. If you work together, how long will it take to do the job? Will this give you enough time before your parents return?

13. The MTV crew will arrive in one week and begin filming the city for *The Real World Kalamazoo*. The mayor is desperate to clean the city streets before filming begins. Two teams are available, one that requires 400 hours and one that requires 300 hours. If the teams work together, how long will it take to clean all of Kalamazoo's streets? Is this enough time before the cameras begin rolling?

14. A hurricane strikes and a rural area is without food or water. Three crews arrive. One can dispense needed supplies in 10 hours, a second in 15 hours, and a third in 20 hours. How long will it take all three crews working together to dispense food and water?

15. A pool can be filled by one pipe in 4 hours and by a second pipe in 6 hours. How long will it take using both pipes to fill the pool?

16. A pool can be filled by one pipe in 3 hours and by a second pipe in 6 hours. How long will it take using both pipes to fill the pool?

Use a proportion to solve each problem in Exercises 17–24.

17. The tax on a property with an assessed value of $65,000 is $720. Find the tax on a property with an assessed value of $162,500.

18. The maintenance bill for a shopping center containing 180,000 square feet is $45,000. What is the bill for a store in the center that is 4800 square feet?

19. St. Paul Island in Alaska has 12 fur seal rookeries (breeding places). In 1961, to estimate the fur seal pup population in the Gorbath rookery, 4963 fur seal pups were tagged in early August. In late August, a sample of 900 pups was observed and 218 of these were found to have been previously tagged. Estimate the total number of fur seal pups in this rookery.

20. To estimate the number of bass in a lake, wildlife biologists tagged 50 bass and released them in the lake. Later they netted 108 bass and found that 27 of them were tagged. Approximately how many bass are in the lake?

21. According to the authors of *Number Freaking*, in a global village of 200 people, 28 suffer from malnutrition. How many people of the world's 6.9 billion people (2010 population) suffer from malnutrition? Round to the nearest hundredth of a billion.

22. According to the authors of *Number Freaking*, in a global village of 200 people, 9 get drunk every day. How many of the world's 6.9 billion people (2010 population) get drunk every day? Round to the nearest hundredth of a billion.

23. Height is proportional to foot length. A person whose foot length is 10 inches is 67 inches tall. In 1951, photos of large footprints were published. Some believed that these footprints were made by the "Abominable Snowman." Each footprint was 23 inches long. If indeed they belonged to the Abominable Snowman, how tall is the critter?

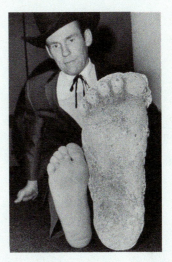

Roger Patterson comparing his foot with a plaster cast of a footprint of the purported "Bigfoot" that Mr. Patterson said he sighted in a California forest in 1967.

24. A person's hair length is proportional to the number of years it has been growing. After 2 years, a person's hair grows 8 inches. The longest moustache on record was grown by Kalyan Sain of India. Sain grew his moustache for 17 years. How long was each side of the moustache?

In Exercises 25–30, use similar triangles and the fact that corresponding sides are proportional to find the length of the side marked with an x.

25.
26.

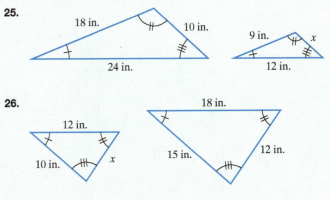

27.

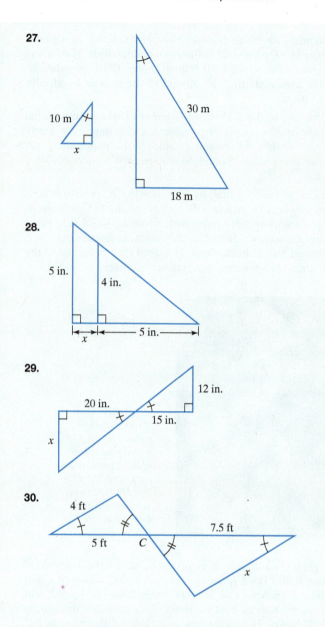

28.

29.

30.

Use similar triangles to solve Exercises 31–32.

31. A tree casts a shadow 12 feet long. At the same time, a vertical rod 8 feet high casts a shadow 6 feet long. How tall is the tree?

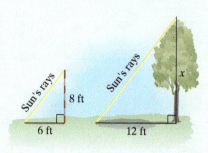

32. A person who is 5 feet tall is standing 80 feet from the base of a tree. The tree casts an 86-foot shadow. The person's shadow is 6 feet in length. What is the tree's height?

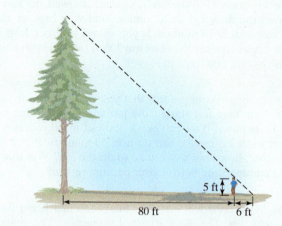

Explaining the Concepts

33. What is the relationship among time traveled, distance traveled, and rate of travel?

34. If you know how many hours it takes for you to do a job, explain how to find the fractional part of the job you can complete in *x* hours.

35. If you can do a job in 6 hours and your friend can do the same job in 3 hours, explain how to find how long it takes to complete the job working together. It is not necessary to solve the problem.

36. When two people work together to complete a job, describe one factor that can result in more or less time than the time given by the rational equations we have been using.

37. What is a proportion? Give an example with your description.

38. What are similar triangles?

39. If the ratio of the corresponding sides of two similar triangles is 1 to 1 $\left(\frac{1}{1}\right)$, what must be true about the triangles?

40. What are corresponding angles in similar triangles?

41. Describe how to identify the corresponding sides in similar triangles.

Critical Thinking Exercises

Make Sense?   In Exercises 42–45, determine whether each statement makes sense or does not make sense, and explain your reasoning.

42. I can solve $\frac{x}{9} = \frac{4}{6}$ by using the cross-products principle or by multiplying both sides by 18, the least common denominator.

43. It took me the same amount of time to travel 10 miles with the current as it did to travel 15 miles against the current.

44. I used $\dfrac{a}{d} = \dfrac{b}{e}$ to show that corresponding sides of similar triangles are proportional, but I could also use $\dfrac{a}{b} = \dfrac{d}{e}$ or $\dfrac{d}{a} = \dfrac{e}{b}$.

45. I can clean my house in 3 hours and my sloppy friend can completely mess it up in 6 hours, so if we both "work" together, the time, x, it takes to clean the house can be modeled by $\dfrac{x}{3} - \dfrac{x}{6} = 1$.

46. Two skiers begin skiing along a trail at the same time. The faster skier averages 9 miles per hour and the slower skier averages 6 miles per hour. The faster skier completes the trail $\frac{1}{4}$ hour before the slower skier. How long is the trail?

47. A snowstorm causes a bus driver to decrease the usual average rate along a 60-mile route by 15 miles per hour. As a result, the bus takes two hours longer than usual to complete the route. At what average rate does the bus usually cover the 60-mile route?

48. One pipe can fill a swimming pool in 2 hours, a second can fill the pool in 3 hours, and a third pipe can fill the pool in 4 hours. How many minutes, to the nearest minute, would it take to fill the pool with all three pipes operating?

49. Ben can prepare a company report in 3 hours. Shane can prepare a report in 4.2 hours. How long will it take them, working together, to prepare *four* company reports?

50. An experienced carpenter can panel a room 3 times faster than an apprentice can. Working together, they can panel the room in 6 hours. How long would it take each person working alone to do the job?

51. It normally takes 2 hours to fill a swimming pool. The pool has developed a slow leak. If the pool were full, it would take 10 hours for all the water to leak out. If the pool is empty, how long will it take to fill it?

52. Two investments have interest rates that differ by 1%. An investment for 1 year at the lower rate earns $175. The same principal invested for a year at the higher rate earns $200. What are the two interest rates?

Review Exercises

53. Factor: $25x^2 - 81$. (Section 6.4, Example 1)

54. Solve: $x^2 - 12x + 36 = 0$. (Section 6.6, Example 4)

55. Graph: $y = -\dfrac{2}{3}x + 4$. (Section 3.4, Example 3)

Preview Exercises

Exercises 56–58 will help you prepare for the material covered in the next section.

56. a. If $y = kx$, find the value of k using $x = 2$ and $y = 64$.

b. Substitute the value for k into $y = kx$ and write the resulting equation.

c. Use the equation from part (b) to find y when $x = 5$.

57. If $B = kW$, find the value of k, in decimal form, using $B = 5$ and $W = 160$.

58. a. If $y = \dfrac{k}{x}$, find the value of k using $x = 8$ and $y = 12$.

b. Substitute the value for k into $y = \dfrac{k}{x}$ and write the resulting equation.

c. Use the equation from part (b) to find y when $x = 3$.

SECTION

7.8

Modeling Using Variation ▶

What am I supposed to learn?

After studying this section, you should be able to:

1 Solve direct variation problems. ▶

2 Solve inverse variation problems. ▶

The vampire legend is death as seducer: He/she sucks our blood to take us to a perverse immortality. The vampire resembles us, but appears hidden among mortals. Can we offer an algebraic application that is particularly seductive to these legendary creatures of the night? Among the examples you'll be reading in this section, there is one that promises to have Dracula panting with anticipatory delight.

Horror films offer the pleasure of vicarious terror, of being safely scared.

1 Solve direct variation problems.

Direct Variation

Certain formulas occur so frequently in applied situations that they are given special names. Variation formulas show how one quantity changes in relation to another quantity. Quantities can vary directly or inversely. We begin by discussing direct variation.

Because light travels faster than sound, during a thunderstorm we see lightning before we hear thunder. The formula

$$d = 1080t$$

describes your distance, d, in feet, from the storm's center if it takes you t seconds to hear thunder after seeing lightning. Thus,

If $t = 1$, $d = 1080 \cdot 1 = 1080$:   If it takes 1 second to hear thunder, the storm's center is 1080 feet away.

If $t = 2$, $d = 1080 \cdot 2 = 2160$:   If it takes 2 seconds to hear thunder, the storm's center is 2160 feet away.

If $t = 3$, $d = 1080 \cdot 3 = 3240$:   If it takes 3 seconds to hear thunder, the storm's center is 3240 feet away.

The formula $d = 1080t$ illustrates that distance to the storm's center is a constant multiple of how long it takes to hear the thunder. If the time is doubled, the storm's distance is doubled; if the time is tripled, the storm's distance is tripled; and so on. Because of this, the distance is said to **vary directly** as the time. The **equation of variation** is

$$d = 1080t.$$

Generalizing, we obtain the following statement:

> ### Direct Variation
>
> If a situation is described by an equation in the form
>
> $$y = kx,$$
>
> where k is a constant, we say that **y varies directly as x** or **y is directly proportional to x**. The number k is called the **constant of variation**.

Can you see that the direct variation equation, $y = kx$, is a special case of the slope-intercept form of a line's equation, $y = mx + b$? When $m = k$ and $b = 0$, $y = mx + b$ becomes $y = kx$. Thus, the slope of a direct variation equation is k, the constant of variation. Because b, the y-intercept, is 0, the graph of a direct variation equation is a line through the origin. This is illustrated in **Figure 7.7**, which shows the graph of $d = 1080t$: Distance to a storm's center varies directly as the time it takes to hear thunder.

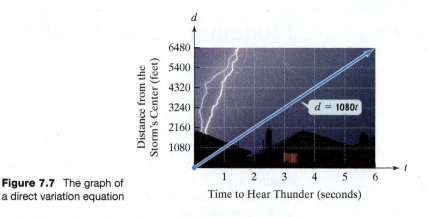

Figure 7.7 The graph of a direct variation equation

Problems involving direct variation can be solved using the procedure at the top of the next page. This procedure applies to direct variation problems, as well as to inverse variation problems, the other kind of variation problem that we will discuss.

Solving Variation Problems

1. Write an equation that describes the given English statement.
2. Substitute the given pair of values into the equation in step 1 and find the value of k, the constant of variation.
3. Substitute the value of k into the equation in step 1.
4. Use the equation from step 3 to answer the problem's question.

EXAMPLE 1   Solving a Direct Variation Problem

The volume of blood, B, in a person's body varies directly as body weight, W. A person who weighs 160 pounds has approximately 5 quarts of blood. Estimate the number of quarts of blood in a person who weighs 200 pounds.

Solution

Step 1. Write an equation. We know that *y varies directly as x* is expressed as

$$y = kx.$$

By changing letters, we can write an equation that describes the following English statement: The volume of blood, B, varies directly as body weight, W.

$$B = kW$$

Step 2. Use the given values to find k. A person who weighs 160 pounds has approximately 5 quarts of blood. Substitute 160 for W and 5 for B in the direct variation equation. Then solve for k.

$B = kW$	The volume of blood varies directly as body weight.
$5 = k \cdot 160$	Substitute 160 for W and 5 for B.
$\dfrac{5}{160} = \dfrac{k \cdot 160}{160}$	Divide both sides by 160.
$0.03125 = k$	Express $\frac{5}{160}$, or $\frac{1}{32}$, in decimal form.

Step 3. Substitute the value of k into the equation.

$B = kW$	Use the equation from step 1.
$B = 0.03125W$	Replace k, the constant of variation, with 0.03125.

Step 4. Answer the problem's question. We are interested in estimating the number of quarts of blood in a person who weighs 200 pounds. Substitute 200 for W in $B = 0.03125W$ and solve for B.

$B = 0.03125W$	This is the equation from step 3.
$B = 0.03125(200)$	Substitute 200 for W.
$= 6.25$	Multiply.

A person who weighs 200 pounds has approximately 6.25 quarts of blood. ∎

✓ CHECK POINT 1   The number of gallons of water, W, used when taking a shower varies directly as the time, t, in minutes, in the shower. A shower lasting 5 minutes uses 30 gallons of water. How much water is used in a shower lasting 8 minutes?

2 Solve inverse variation problems.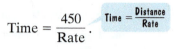

Inverse Variation

The distance from Atlanta, Georgia, to Orlando, Florida, is 450 miles. The time that it takes to drive from Atlanta to Orlando depends on the rate at which one drives and is given by

$$\text{Time} = \frac{450}{\text{Rate}}.$$

Time $=\frac{\text{Distance}}{\text{Rate}}$

For example, if you average 45 miles per hour, the time for the drive is

$$\text{Time} = \frac{450}{45} = 10$$

or 10 hours. If you average 75 miles per hour, the time for the drive is

$$\text{Time} = \frac{450}{75} = 6$$

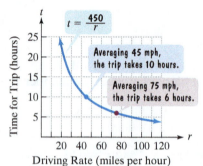

Figure 7.8

or 6 hours. As your rate (or speed) increases, the time for the trip decreases, and vice versa. This is illustrated by the graph in **Figure 7.8**. Observe the graph's shape. It shows that as the rate increases, the time for the trip decreases quite rapidly.

We can express the time for the Atlanta–Orlando trip using t for time and r for rate:

$$t = \frac{450}{r}.$$

This equation is an example of an **inverse variation** equation. Time, t, **varies inversely** as rate, r. When two quantities vary inversely, one quantity increases as the other decreases, and vice versa.

Generalizing, we obtain the following statement:

Inverse Variation

If a situation is described by an equation in the form

$$y = \frac{k}{x},$$

where k is a constant, we say that **y varies inversely as x** or **y is inversely proportional to x**. The number k is called the **constant of variation**.

We use the same procedure to solve inverse variation problems as we did to solve direct variation problems. Example 2 illustrates this procedure.

EXAMPLE 2 Solving an Inverse Variation Problem

The rate at which heat is lost, L, through a window pane varies inversely as the thickness, T, of the pane. A normal $\frac{1}{8}$-inch-thick pane loses 400 calories per hour. How many calories per hour are lost through a $\frac{3}{8}$-inch-thick window pane?

Solution

Step 1. Write an equation. We know that y *varies inversely as x* is expressed as

$$y = \frac{k}{x}.$$

By changing letters, we can write an equation that describes the following English statement: The rate at which heat is lost, L, varies inversely as the window pane's thickness, T.

$$L = \frac{k}{T}$$

Step 2. Use the given values to find k. A normal $\frac{1}{8}$-inch-thick pane loses 400 calories per hour. Substitute $\frac{1}{8}$ for T and 400 for L in the inverse variation equation. Then solve for k.

$$L = \frac{k}{T} \qquad \text{Heat loss varies inversely as thickness.}$$

$$400 = \frac{k}{\frac{1}{8}} \qquad \text{A } \frac{1}{8}\text{-inch-thick pane loses 400 calories per hour:} \\ \text{If } T = \frac{1}{8}, L = 400.$$

$$400 = 8k \qquad \frac{k}{\frac{1}{8}} = \frac{k}{1} \div \frac{1}{8} = \frac{k}{1} \cdot \frac{8}{1} = 8k$$

$$\frac{400}{8} = \frac{8k}{8} \qquad \text{Divide both sides by 8.}$$

$$50 = k \qquad \text{Simplify.}$$

Step 3. Substitute the value of k into the equation.

$$L = \frac{k}{T} \qquad \text{Use the equation from step 1.}$$

$$L = \frac{50}{T} \qquad \text{Replace } k, \text{ the constant of variation, with 50.}$$

Step 4. Answer the problem's question. We need to find the number of calories per hour lost through a $\frac{3}{8}$-inch-thick window pane. Substitute $\frac{3}{8}$ for T and solve for L.

$$L = \frac{50}{T} \qquad \text{This is the equation from step 3.}$$

$$L = \frac{50}{\frac{3}{8}} = 50 \div \frac{3}{8} = \frac{50}{1} \cdot \frac{8}{3} = \frac{400}{3} = 133\tfrac{1}{3}$$

A $\frac{3}{8}$-inch-thick window pane loses $133\frac{1}{3}$ calories per hour.   ∎

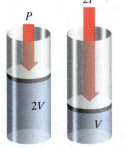

2*P*

P

2*V*

V

Doubling the pressure halves the volume.

✓ **CHECK POINT 2**   When you use a spray can and press the valve at the top, you decrease the pressure of the gas in the can. This decrease of pressure causes the volume of the gas in the can to increase. Because the gas needs more room than is provided in the can, it expands in spray form through the small hole near the valve. In general, if the temperature is constant, the pressure, P, of a gas in a container varies inversely as the volume, V, of the container. The pressure of a gas sample in a container whose volume is 8 cubic inches is 12 pounds per square inch. If the sample expands to a volume of 20 cubic inches, what is the new pressure of the gas?

CONCEPT AND VOCABULARY CHECK

Fill in each blank so that the resulting statement is true.

1. y varies directly as x can be modeled by the equation _____, where k is called the _____.

2. y varies inversely as x can be modeled by the equation _____, where k is called the _____.

3. In the equation $L = \dfrac{50}{T}$, L varies _____ as T.

4. In the equation $B = 0.03125W$, B varies _____ as W.

7.8 EXERCISE SET ▶ MyMathLab®

Practice Exercises

In Exercises 1–4, write an equation that expresses each relationship. Use k as the constant of variation.

1. g varies directly as h.   **2.** v varies directly as r.

3. w varies inversely as v.   **4.** a varies inversely as b.

In Exercises 5–8, determine the constant of variation for each stated condition.

5. y varies directly as x, and $y = 80$ when $x = 4$.

6. y varies directly as x, and $y = 108$ when $x = 12$.

7. W varies inversely as r, and $W = 600$ when $r = 10$.

8. T varies inversely as n, and $T = 4$ when $n = 24$.

Use the four-step procedure for solving variation problems given on page 553 to solve Exercises 9–12.

9. y varies directly as x. $y = 35$ when $x = 5$. Find y when $x = 12$.

10. y varies directly as x. $y = 55$ when $x = 5$. Find y when $x = 13$.

11. y varies inversely as x. $y = 10$ when $x = 5$. Find y when $x = 2$.

12. y varies inversely as x. $y = 5$ when $x = 3$. Find y when $x = 9$.

Application Exercises

13. A person's fingernail growth, G, in inches, varies directly as the number of weeks it has been growing, W.

   a. Write an equation that expresses this relationship.

   b. Fingernails grow at a rate of about 0.02 inch per week. Substitute 0.02 for k, the constant of variation, in the equation in part (a) and write the equation for fingernail growth.

   c. Substitute 52 for W to determine your fingernail growth at the end of one year if for some bizarre reason you decided not to cut them and they did not break.

14. A person's salary, S, varies directly as the number of hours worked, h.

   a. Write an equation that expresses this relationship.

   b. For a 40-hour work week, Gloria earned $1400. Substitute 1400 for S and 40 for h in the equation from part (a) and find k, the constant of variation.

   c. Substitute the value of k into your equation in part (a) and write the equation that describes Gloria's salary in terms of the number of hours she works.

   d. Use the equation from part (c) to find Gloria's salary for 25 hours of work.

Use the four-step procedure for solving variation problems given on page 553 to solve Exercises 15–24.

15. The cost, C, of an airplane ticket varies directly as the number of miles, M, in the trip. A 3000-mile trip costs $400. What is the cost of a 450-mile trip?

16. An object's weight on the moon, M, varies directly as its weight on Earth, E. A person who weighs 55 kilograms on Earth weighs 8.8 kilograms on the moon. What is the moon weight of a person who weighs 90 kilograms on Earth?

17. The Mach number is a measurement of speed named after the man who suggested it, Ernst Mach (1838–1916). The speed of an aircraft varies directly as its Mach number. Shown here are two aircraft. Use the figures for the Concorde to determine the Blackbird's speed.

Concorde
Mach 2.03
Speed = 1502.2 miles per hour

SR-71 Blackbird
Mach 3.3
Speed = ?

18. A golf ball's bounce height, B, in inches, varies directly as its drop height, d, in inches. A golf ball bounces 36 inches when dropped from a height of 40 inches. What is the ball's bounce height if the drop height is increased to 50 inches?

19. The time that it takes to get to campus varies inversely as your driving rate. Averaging 20 miles per hour in terrible traffic, it takes you 1.5 hours to get to campus. How long would the trip take averaging 60 miles per hour?

20. The weight that can be supported by a 2-inch by 4-inch piece of pine (called a 2-by-4) varies inversely as its length. A 10-foot 2-by-4 can support 500 pounds. What weight can be supported by a 5-foot 2-by-4?

21. The volume of a gas in a container at a constant temperature varies inversely as the pressure. If the volume is 32 cubic centimeters at a pressure of 8 pounds per square centimeter, find the pressure when the volume is 40 cubic centimeters.

22. The current in a circuit varies inversely as the resistance. The current is 20 amperes when the resistance is 5 ohms. Find the current for a resistance of 16 ohms.

23. The number of pens sold varies inversely as the price per pen. If 4000 pens are sold at a price of $1.50 each, find the number of pens sold at a price of $1.20 each.

24. The time required to accomplish a task varies inversely as the number of people working on the task. It takes 6 hours for 20 people to put a new roof on a porch. How long would it take 30 people to do the job?

Explaining the Concepts

25. What does it mean if two quantities vary directly?

26. In your own words, explain how to solve a variation problem.

27. What does it mean if two quantities vary inversely?

28. Explain the meaning of this statement: A company's monthly sales vary directly as its advertising budget.

29. Explain the meaning of this statement: A company's monthly sales vary inversely as the price of its product.

Critical Thinking Exercises

Make Sense? In Exercises 30–33, determine whether each statement makes sense or does not make sense, and explain your reasoning.

30. The time it takes me to drive to campus varies directly as my rate of travel.

31. It seems reasonable that a student's grade in a course varies directly as the number of hours spent studying.

32. It seems reasonable that a student's grade in a course varies directly as the number of hours spent watching TV.

33. It seems reasonable that a student's exam score varies inversely as the number of missed assignments.

34. The intensity of radiation from a machine used to treat tumors varies inversely as the square of the distance from the machine. If the intensity is 140.5 milliroentgens per hour at 2 meters, what is the intensity at a distance of 3 meters?

Review Exercises

35. Solve:
$$8(2 - x) = -5x.$$
(Section 2.3, Example 2)

36. Divide:
$$\frac{27x^3 - 8}{3x + 2}.$$
(Section 5.6, Example 3)

37. Factor:
$$6x^3 - 6x^2 - 120x.$$
(Section 6.5, Example 2)

Preview Exercises

Exercises 38–40 will help you prepare for the material covered in the first section of the next chapter.

38. Evaluate $\sqrt{x - 1}$ for $x = 17$.

39. Evaluate $4\sqrt{x} + 30$ for $x = 25$.

40. Simplify: $(-2)^5 - (-1)^3$.

Chapter 7 Summary

Definitions and Concepts	Examples

Section 7.1 Rational Expressions and Their Simplification

A rational expression is the quotient of two polynomials. To find values for which a rational expression is undefined, set the denominator equal to 0 and solve.	Find all numbers for which $$\frac{7x}{x^2 - 3x - 4}$$ is undefined. $$x^2 - 3x - 4 = 0$$ $$(x - 4)(x + 1) = 0$$ $$x - 4 = 0 \quad \text{or} \quad x + 1 = 0$$ $$x = 4 \qquad\qquad x = -1$$ The expression is undefined for 4 and -1.
To simplify a rational expression: 1. Factor the numerator and the denominator completely. 2. Divide the numerator and the denominator by any common factors. If factors in the numerator and denominator are opposites, their quotient is -1.	Simplify: $\dfrac{3x + 18}{x^2 - 36}$. $$\frac{3x + 18}{x^2 - 36} = \frac{3\overset{1}{\cancel{(x + 6)}}}{\underset{1}{\cancel{(x + 6)}}(x - 6)} = \frac{3}{x - 6}$$

Definitions and Concepts	**Examples**

Section 7.2 Multiplying and Dividing Rational Expressions

Multiplying Rational Expressions

1. Factor completely.

2. Divide numerators and denominators by common factors.

3. Multiply the remaining factors in the numerators and multiply the remaining factors in the denominators.

$$\frac{x^2 + 3x - 10}{x^2 - 2x} \cdot \frac{x^2}{x^2 - 25}$$

$$= \frac{\overset{1}{\cancel{(x+5)}}\overset{1}{\cancel{(x-2)}}}{x\underset{1}{\cancel{(x-2)}}} \cdot \frac{x \cdot x}{\underset{1}{\cancel{(x+5)}}(x-5)}$$

$$= \frac{x}{x-5}$$

Dividing Rational Expressions

Invert the divisor and multiply.

$$\frac{3y + 3}{(y+2)^2} \div \frac{y^2 - 1}{y+2}$$

$$= \frac{3y + 3}{(y+2)^2} \cdot \frac{y+2}{y^2 - 1}$$

$$= \frac{3\overset{1}{\cancel{(y+1)}}}{(y+2)\underset{1}{\cancel{(y+2)}}} \cdot \frac{\overset{1}{\cancel{(y+2)}}}{\underset{1}{\cancel{(y+1)}}(y-1)}$$

$$= \frac{3}{(y+2)(y-1)}$$

Section 7.3 Adding and Subtracting Rational Expressions with the Same Denominator

To add or subtract rational expressions with the same denominator, add or subtract the numerators and place the result over the common denominator. If possible, factor and simplify the resulting expression.

$$\frac{y^2 - 3y + 4}{y^2 + 8y + 15} - \frac{y^2 - 5y - 2}{y^2 + 8y + 15}$$

$$= \frac{y^2 - 3y + 4 - (y^2 - 5y - 2)}{y^2 + 8y + 15}$$

$$= \frac{y^2 - 3y + 4 - y^2 + 5y + 2}{y^2 + 8y + 15}$$

$$= \frac{2y + 6}{(y+5)(y+3)}$$

$$= \frac{2\overset{1}{\cancel{(y+3)}}}{(y+5)\underset{1}{\cancel{(y+3)}}} = \frac{2}{y+5}$$

To add or subtract rational expressions with opposite denominators, multiply either rational expression by $\frac{-1}{-1}$ to obtain a common denominator.

$$\frac{7}{x-6} + \frac{x+4}{6-x}$$

$$= \frac{7}{x-6} + \frac{(-1)}{(-1)} \cdot \frac{x+4}{6-x}$$

$$= \frac{7}{x+6} + \frac{-x-4}{x-6}$$

$$= \frac{7 - x - 4}{x-6} = \frac{3-x}{x-6}$$

Definitions and Concepts	**Examples**

Section 7.4 Adding and Subtracting Rational Expressions with Different Denominators

Finding the Least Common Denominator (LCD)

1. Factor denominators completely.
2. List factors of the first denominator.
3. Add to the list any factors of the second denominator that are not already in the list.
4. The LCD is the product of factors in step 3.

Find the LCD of

$$\frac{x+1}{2x-2} \quad \text{and} \quad \frac{2x}{x^2+2x-3}.$$

$$2x - 2 = 2(x-1)$$

$$x^2 + 2x - 3 = (x-1)(x+3)$$

Factors of first denominator: $2, x - 1$

Factors of second denominator not in the list: $x + 3$

LCD: $2(x-1)(x+3)$

Adding and Subtracting Rational Expressions with Different Denominators

1. Find the LCD.
2. Rewrite each rational expression as an equivalent expression with the LCD.
3. Add or subtract numerators, placing the resulting expression over the LCD.
4. If possible, simplify.

$$\frac{x+1}{2x-2} - \frac{2x}{x^2+2x-3}$$

$$= \frac{x+1}{2(x-1)} - \frac{2x}{(x-1)(x+3)}$$

LCD is $2(x-1)(x+3)$.

$$= \frac{(x+1)(x+3)}{2(x-1)(x+3)} - \frac{2x \cdot 2}{2(x-1)(x+3)}$$

$$= \frac{x^2 + 4x + 3 - 4x}{2(x-1)(x+3)}$$

$$= \frac{x^2 + 3}{2(x-1)(x+3)}$$

Section 7.5 Complex Rational Expressions

Complex rational expressions have numerators or denominators containing one or more rational expressions. Complex rational expressions can be simplified by obtaining single expressions in the numerator and denominator and then dividing. They can also be simplified by multiplying the numerator and denominator by the LCD of all rational expressions within the complex rational expression.

Simplify by dividing: $\dfrac{\dfrac{1}{x} + 5}{\dfrac{1}{x} - \dfrac{1}{3}}$.

$$= \frac{\dfrac{1}{x} + \dfrac{5x}{x}}{\dfrac{3}{3x} - \dfrac{x}{3x}} = \frac{\dfrac{1+5x}{x}}{\dfrac{3-x}{3x}} = \frac{1+5x}{\overset{1}{\cancel{x}}} \cdot \frac{\overset{1}{\cancel{3x}}}{3-x}$$

$$= \frac{3(1+5x)}{3-x} \quad \text{or} \quad \frac{3+15x}{3-x}$$

Simplify by the LCD method: $\dfrac{\dfrac{1}{x} + 5}{\dfrac{1}{x} - \dfrac{1}{3}}$.

LCD is $3x$.

$$\frac{3x}{3x} \cdot \frac{\left(\dfrac{1}{x} + 5\right)}{\left(\dfrac{1}{x} - \dfrac{1}{3}\right)} = \frac{3x \cdot \dfrac{1}{x} + 3x \cdot 5}{3x \cdot \dfrac{1}{x} - 3x \cdot \dfrac{1}{3}}$$

$$= \frac{3+15x}{3-x}$$

Definitions and Concepts	**Examples**

Section 7.6 Solving Rational Equations

A rational equation is an equation containing one or more rational expressions.

Solving Rational Equations

1. List restrictions on the variable.
2. Clear fractions by multiplying both sides by the LCD.
3. Solve the resulting equation.
4. Reject any proposed solution in the list of restrictions. Check other proposed solutions in the original equation.

Solve: $\dfrac{7x}{x^2 - 4} + \dfrac{5}{x - 2} = \dfrac{2x}{x^2 - 4}$.

$$\frac{7x}{(x + 2)(x - 2)} + \frac{5}{x - 2} = \frac{2x}{(x + 2)(x - 2)}$$

> Denominators would equal 0 if $x = -2$ or $x = 2$.
> Restrictions: $x \neq -2$ and $x \neq 2$.

LCD is $(x + 2)(x - 2)$.

$$(x + 2)(x - 2)\left[\frac{7x}{(x + 2)(x - 2)} + \frac{5}{x - 2}\right]$$

$$= (x + 2)(x - 2) \cdot \frac{2x}{(x + 2)(x - 2)}$$

$$7x + 5(x + 2) = 2x$$
$$7x + 5x + 10 = 2x$$
$$12x + 10 = 2x$$
$$10 = -10x$$
$$-1 = x$$

The proposed solution, -1, is not part of the restriction $x \neq -2$ and $x \neq 2$. It checks. The solution is -1 and the solution set is $\{-1\}$.

Section 7.7 Applications Using Rational Equations and Proportions

Motion problems involving time are solved using

$$t = \frac{d}{r}.$$

$$\text{Time traveled} = \frac{\text{Distance traveled}}{\text{Rate of travel}}$$

It takes a cyclist who averages 16 miles per hour in still air the same time to travel 48 miles with the wind as 16 miles against the wind. What is the wind's rate?

$$x = \text{wind's rate}$$
$$16 + x = \text{cyclist's rate with wind}$$
$$16 - x = \text{cyclist's rate against wind}$$

	Distance	Rate	Time = $\dfrac{\text{Distance}}{\text{Rate}}$
With wind	48	$16 + x$	$\dfrac{48}{16 + x}$
Against wind	16	$16 - x$	$\dfrac{16}{16 - x}$

Two times are equal.

$$\frac{48}{16 + x} = \frac{16}{16 - x}$$

$$(16 + x)(16 - x) \cdot \frac{48}{16 + x} = \frac{16}{16 - x} \cdot (16 + x)(16 - x)$$

$$48(16 - x) = 16(16 + x)$$

Solving this equation, $x = 8$.
The wind's rate is 8 miles per hour.

Definitions and Concepts	**Examples**

Section 7.7 Applications Using Rational Equations and Proportions (continued)

Work problems are solved using the following condition:

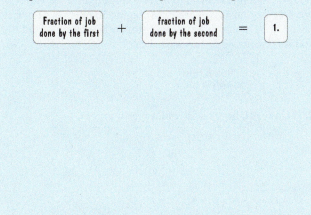

One pipe fills a pool in 20 hours and a second pipe in 15 hours. How long will it take to fill the pool using both pipes?

$$x = \text{time using both pipes}$$

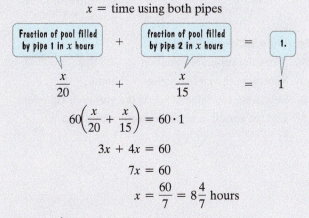

$$60\left(\frac{x}{20} + \frac{x}{15}\right) = 60 \cdot 1$$

$$3x + 4x = 60$$

$$7x = 60$$

$$x = \frac{60}{7} = 8\frac{4}{7} \text{ hours}$$

It will take $8\frac{4}{7}$ hours for both pipes to fill the pool.

A proportion is a statement in the form $\dfrac{a}{b} = \dfrac{c}{d}$.

The cross-products principle states that if $\dfrac{a}{b} = \dfrac{c}{d}$, then $ad = bc$ ($b \neq 0$ and $d \neq 0$).

Solving Applied Problems Using Proportions

1. Read the problem and represent the unknown quantity by x (or any letter).
2. Set up a proportion by listing the given ratio on one side and the ratio with the unknown quantity on the other side.
3. Drop units and apply the cross-products principle.
4. Solve for x and anwer the question.

30 elk are tagged and released. Sometime later, a sample of 80 elk is observed and 10 are tagged. How many elk are there?

$$x = \text{number of elk}$$

$$\text{Tagged} \rightarrow \frac{30}{x} = \frac{10}{80} \leftarrow \text{Total}$$

$$10x = 30 \cdot 80$$

$$10x = 2400$$

$$x = 240$$

There are 240 elk.

Similar triangles have the same shape, but not necessarily the same size. Corresponding angles have the same measure, and corresponding sides are proportional. If the measures of two angles of one triangle are equal to those of two angles of a second triangle, then the two triangles are similar.

Find x for these similar triangles.

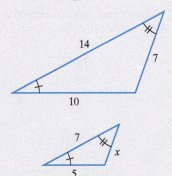

Corresponding sides are proportional:

$$\frac{7}{x} = \frac{10}{5}. \quad \left(\text{or } \frac{7}{x} = \frac{14}{7}\right)$$

$$5x \cdot \frac{7}{x} = \frac{10}{5} \cdot 5x$$

$$35 = 10x$$

$$x = \frac{35}{10} = 3.5$$

Definitions and Concepts	**Examples**

Section 7.8 Modeling Using Variation

y varies directly as x: $y = kx$.

y varies inversely as x: $y = \dfrac{k}{x}$. — constant of variation

Solving a Variation Problem

1. Write the variation equation.
2. Substitute the pair of values and find k.
3. Substitute the value of k into the variation equation.
4. Use the variation equation with k to answer the problem's question.

The time that it takes you to drive a certain distance varies inversely as your driving rate. Averaging 30 miles per hour, it takes you 10 hours to drive the distance. How long would the trip take averaging 50 miles per hour?

1. $t = \dfrac{k}{r}$ — Time, t, varies inversely as rate, r.

2. It takes 10 hours at 30 miles per hour.
$$10 = \frac{k}{30}$$
$$k = 10(30) = 300$$

3. $t = \dfrac{300}{r}$

4. How long at 50 miles per hour? Substitute 50 for r.
$$t = \frac{300}{50} = 6$$

It takes 6 hours at 50 miles per hour.

CHAPTER 7 REVIEW EXERCISES

7.1 *In Exercises 1–4, find all numbers for which each rational expression is undefined. If the rational expression is defined for all real numbers, so state.*

1. $\dfrac{5x}{6x - 24}$

2. $\dfrac{x + 3}{(x - 2)(x + 5)}$

3. $\dfrac{x^2 + 3}{x^2 - 3x + 2}$

4. $\dfrac{7}{x^2 + 81}$

In Exercises 5–12, simplify each rational expression. If the rational expression cannot be simplified, so state.

5. $\dfrac{16x^2}{12x}$

6. $\dfrac{x^2 - 4}{x - 2}$

7. $\dfrac{x^3 + 2x^2}{x + 2}$

8. $\dfrac{x^2 + 3x - 18}{x^2 - 36}$

9. $\dfrac{x^2 - 4x - 5}{x^2 + 8x + 7}$

10. $\dfrac{y^2 + 2y}{y^2 + 4y + 4}$

11. $\dfrac{x^2}{x^2 + 4}$

12. $\dfrac{2x^2 - 18y^2}{3y - x}$

7.2 *In Exercises 13–17, multiply as indicated.*

13. $\dfrac{x^2 - 4}{12x} \cdot \dfrac{3x}{x + 2}$

14. $\dfrac{5x + 5}{6} \cdot \dfrac{3x}{x^2 + x}$

15. $\dfrac{x^2 + 6x + 9}{x^2 - 4} \cdot \dfrac{x - 2}{x + 3}$

16. $\dfrac{y^2 - 2y + 1}{y^2 - 1} \cdot \dfrac{2y^2 + y - 1}{5y - 5}$

17. $\dfrac{2y^2 + y - 3}{4y^2 - 9} \cdot \dfrac{3y + 3}{5y - 5y^2}$

In Exercises 18–22, divide as indicated.

18. $\dfrac{x^2 + x - 2}{10} \div \dfrac{2x + 4}{5}$

19. $\dfrac{6x + 2}{x^2 - 1} \div \dfrac{3x^2 + x}{x - 1}$

20. $\dfrac{1}{y^2 + 8y + 15} \div \dfrac{7}{y + 5}$

21. $\dfrac{y^2 + y - 42}{y - 3} \div \dfrac{y + 7}{(y - 3)^2}$

22. $\dfrac{8x + 8y}{x^2} \div \dfrac{x^2 - y^2}{x^2}$

7.3 *In Exercises 23–28, add or subtract as indicated. Simplify the result, if possible.*

23. $\dfrac{4x}{x + 5} + \dfrac{20}{x + 5}$

24. $\dfrac{8x - 5}{3x - 1} + \dfrac{4x + 1}{3x - 1}$

25. $\dfrac{3x^2 + 2x}{x - 1} - \dfrac{10x - 5}{x - 1}$

26. $\dfrac{6y^2 - 4y}{2y - 3} - \dfrac{12 - 3y}{2y - 3}$

27. $\dfrac{x}{x - 2} + \dfrac{x - 4}{2 - x}$

28. $\dfrac{x + 5}{x - 3} - \dfrac{x}{3 - x}$

7.4   *In Exercises 29–31, find the least common denominator of the rational expressions.*

29. $\dfrac{7}{9x^3}$ and $\dfrac{5}{12x}$

30. $\dfrac{3}{x^2(x-1)}$ and $\dfrac{11}{x(x-1)^2}$

31. $\dfrac{x}{x^2+4x+3}$ and $\dfrac{17}{x^2+10x+21}$

In Exercises 32–42, add or subtract as indicated. Simplify the result, if possible.

32. $\dfrac{7}{3x}+\dfrac{5}{2x^2}$

33. $\dfrac{5}{x+1}+\dfrac{2}{x}$

34. $\dfrac{7}{x+3}+\dfrac{4}{(x+3)^2}$

35. $\dfrac{6y}{y^2-4}-\dfrac{3}{y+2}$

36. $\dfrac{y-1}{y^2-2y+1}-\dfrac{y+1}{y-1}$

37. $\dfrac{x+y}{y}-\dfrac{x-y}{x}$

38. $\dfrac{2x}{x^2+2x+1}+\dfrac{x}{x^2-1}$

39. $\dfrac{5x}{x+1}-\dfrac{2x}{1-x^2}$

40. $\dfrac{4}{x^2-x-6}-\dfrac{4}{x^2-4}$

41. $\dfrac{7}{x+3}+2$

42. $\dfrac{2y-5}{6y+9}-\dfrac{4}{2y^2+3y}$

7.5   *In Exercises 43–47, simplify each complex rational expression.*

43. $\dfrac{\dfrac{1}{2}+\dfrac{3}{8}}{\dfrac{3}{4}-\dfrac{1}{2}}$

44. $\dfrac{\dfrac{1}{x}}{1-\dfrac{1}{x}}$

45. $\dfrac{\dfrac{1}{x}+\dfrac{1}{y}}{\dfrac{1}{xy}}$

46. $\dfrac{\dfrac{1}{x}-\dfrac{1}{2}}{\dfrac{1}{3}-\dfrac{x}{6}}$

47. $\dfrac{3+\dfrac{12}{x}}{1-\dfrac{16}{x^2}}$

7.6   *In Exercises 48–55, solve each rational equation.*

48. $\dfrac{3}{x}-\dfrac{1}{6}=\dfrac{1}{x}$

49. $\dfrac{3}{4x}=\dfrac{1}{x}+\dfrac{1}{4}$

50. $x+5=\dfrac{6}{x}$

51. $4-\dfrac{x}{x+5}=\dfrac{5}{x+5}$

52. $\dfrac{2}{x-3}=\dfrac{4}{x+3}+\dfrac{8}{x^2-9}$

53. $\dfrac{2}{x}=\dfrac{2}{3}+\dfrac{x}{6}$

54. $\dfrac{13}{y-1}-3=\dfrac{1}{y-1}$

55. $\dfrac{1}{x+3}-\dfrac{1}{x-1}=\dfrac{x+1}{x^2+2x-3}$

56. Park rangers introduce 50 elk into a wildlife preserve. The formula
$$P=\dfrac{250(3t+5)}{t+25}$$
models the elk population, P, after t years. How many years will it take for the population to increase to 125 elk?

57. The formula
$$S=\dfrac{C}{1-r}$$
describes the selling price, S, of a product in terms of its cost to the retailer, C, and its markup, r, usually expressed as a percent. A small television cost a retailer \$140 and was sold for \$200. Find the markup. Express the answer as a percent.

7.7

58. In still water, a paddle boat averages 20 miles per hour. It takes the boat the same amount of time to travel 72 miles downstream, with the current, as 48 miles upstream, against the current. What is the rate of the water's current?

59. A car travels 60 miles in the same time that a car traveling 10 miles per hour faster travels 90 miles. What is the rate of each car?

60. A painter can paint a fence around a house in 6 hours. Working alone, the painter's apprentice can paint the same fence in 12 hours. How many hours would it take them to do the job if they worked together?

61. If a school board determines that there should be 3 teachers for every 50 students, how many teachers are needed for an enrollment of 5400 students?

62. To determine the number of trout in a lake, a conservationist catches 112 trout, tags them, and returns them to the lake. Later, 82 trout are caught, and 32 of them are found to be tagged. How many trout are in the lake?

63. The triangles shown in the figure are similar. Find the length of the side marked with an x.

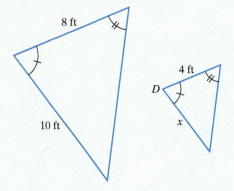

64. Find the height of the lamppost in the figure.

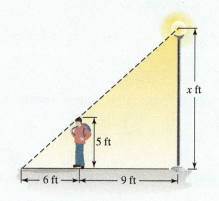

5 ft

x ft

← 6 ft → ← 9 ft →

7.8

65. An electric bill varies directly as the amount of electricity used. The bill for 1400 kilowatts of electricity is $98. What is the bill for 2200 kilowatts of electricity?

66. The time it takes to drive a certain distance varies inversely as the rate of travel. If it takes 4 hours at 50 miles per hour to drive the distance, how long will it take at 40 miles per hour?

CHAPTER 7 TEST

Step-by-step test solutions are found on the Chapter Test Prep Videos available in MyMathLab® or on YouTube (search "BlitzerIntroAlg7e" and click on "Channels").

1. Find all numbers for which

$$\frac{x + 7}{x^2 + 5x - 36}$$

is undefined.

In Exercises 2–3, simplify each rational expression.

2. $\dfrac{x^2 + 2x - 3}{x^2 - 3x + 2}$

3. $\dfrac{4x^2 - 20x}{x^2 - 4x - 5}$

In Exercises 4–16, perform the indicated operations. Simplify the result, if possible.

4. $\dfrac{x^2 - 16}{10} \cdot \dfrac{5}{x + 4}$

5. $\dfrac{x^2 - 7x + 12}{x^2 - 4x} \cdot \dfrac{x^2}{x^2 - 9}$

6. $\dfrac{2x + 8}{x - 3} \div \dfrac{x^2 + 5x + 4}{x^2 - 9}$

7. $\dfrac{5y + 5}{(y - 3)^2} \div \dfrac{y^2 - 1}{y - 3}$

8. $\dfrac{2y^2 + 5}{y + 3} + \dfrac{6y - 5}{y + 3}$

9. $\dfrac{y^2 - 2y + 3}{y^2 + 7y + 12} - \dfrac{y^2 - 4y - 5}{y^2 + 7y + 12}$

10. $\dfrac{x}{x + 3} + \dfrac{5}{x - 3}$

11. $\dfrac{2}{x^2 - 4x + 3} + \dfrac{6}{x^2 + x - 2}$

12. $\dfrac{4}{x - 3} + \dfrac{x + 5}{3 - x}$

13. $1 + \dfrac{3}{x - 1}$

14. $\dfrac{2x + 3}{x^2 - 7x + 12} - \dfrac{2}{x - 3}$

15. $\dfrac{8y}{y^2 - 16} - \dfrac{4}{y - 4}$

16. $\dfrac{(x - y)^2}{x + y} \div \dfrac{x^2 - xy}{3x + 3y}$

In Exercises 17–18, simplify each complex rational expression.

17. $\dfrac{5 + \dfrac{5}{x}}{2 + \dfrac{1}{x}}$

18. $\dfrac{\dfrac{1}{x} - \dfrac{1}{y}}{\dfrac{1}{x}}$

In Exercises 19–21, solve each rational equation.

19. $\dfrac{5}{x} + \dfrac{2}{3} = 2 - \dfrac{2}{x} - \dfrac{1}{6}$

20. $\dfrac{3}{y + 5} - 1 = \dfrac{4 - y}{2y + 10}$

21. $\dfrac{2}{x - 1} = \dfrac{3}{x^2 - 1} + 1$

22. In still water, a boat averages 30 miles per hour. It takes the boat the same amount of time to travel 16 miles downstream, with the current, as 14 miles upstream, against the current. What is the rate of the water's current?

23. One pipe can fill a hot tub in 20 minutes and a second pipe can fill it in 30 minutes. If the hot tub is empty, how long will it take both pipes to fill it?

24. Park rangers catch, tag, and release 200 tule elk back into a wildlife refuge. Two weeks later they observe a sample of 150 elk, of which 5 are tagged. Assuming that the ratio of tagged elk in the sample holds for all elk in the refuge, how many elk are there in the park?

25. The triangles in the figure are similar. Find the length of the side marked with an *x*.

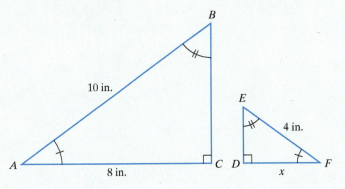

26. The amount of current flowing in an electrical circuit varies inversely as the resistance in the circuit. When the resistance in a particular circuit is 5 ohms, the current is 42 amperes. What is the current when the resistance is 4 ohms?

CUMULATIVE REVIEW EXERCISES   (CHAPTERS 1–7)

In Exercises 1–6, solve each equation, inequality, or system of equations.

1. $2(x - 3) + 5x = 8(x - 1)$

2. $-3(2x - 4) > 2(6x - 12)$

3. $x^2 + 3x = 18$

4. $\dfrac{2x}{x^2 - 4} + \dfrac{1}{x - 2} = \dfrac{2}{x + 2}$

5. $\begin{cases} y = 2x - 3 \\ x + 2y = 9 \end{cases}$

6. $\begin{cases} 3x + 2y = -2 \\ -4x + 5y = 18 \end{cases}$

In Exercises 7–9, graph each equation or inequality in a rectangular coordinate system.

7. $3x - 2y = 6$

8. $y > -2x + 3$

9. $y = -3$

In Exercises 10–12, simplify each expression.

10. $-21 - 16 - 3(2 - 8)$

11. $\left(\dfrac{4x^5}{2x^2}\right)^3$

12. $\dfrac{\dfrac{1}{x} - 2}{4 - \dfrac{1}{x}}$

In Exercises 13–15, factor completely.

13. $4x^2 - 13x + 3$

14. $4x^2 - 20x + 25$

15. $3x^2 - 75$

In Exercises 16–18, perform the indicated operations.

16. $(4x^2 - 3x + 2) - (5x^2 - 7x - 6)$

17. $\dfrac{-8x^6 + 12x^4 - 4x^2}{4x^2}$

18. $\dfrac{x + 6}{x - 2} + \dfrac{2x + 1}{x + 3}$

19. You invested $4000, part at 5% and the remainder at 9% annual interest. At the end of the year, the total interest from these investments was $311. How much was invested at each rate?

20. A 68-inch board is to be cut into two pieces. If one piece must be three times as long as the other, find the length of each piece.

Roots and Radicals

The popular comic strip *FoxTrot* follows the off-the-wall lives of the Fox family. Youngest son Jason is forever obsessed by his love of math. In the math-themed strip shown below, Jason shares his opinion in a coded message about the mathematical abilities of his sister Paige.

Decoding Jason's message requires solving problems A through Z. Problems A ($\sqrt{121}$), E ($\sqrt[3]{1000}$), I ($\sqrt{13} \times \sqrt{13}$), N ($\sqrt{400}$), P ($4^{\sqrt{4}}$), T ($\sqrt{144}$), and Y ($\sqrt{49} \times \sqrt{9}$) involve roots and radicals, our focus in this chapter. Of course, we'll move beyond numerical problems and show you how models with roots and radicals describe an array of unique applications, ranging from gender imbalance on campus to marriage inequality to sexual abstinence among teens and young adults.

Here's where you'll find these applications:

- We'll ask you to decode Jason's message, with a few hints along the way, in Exercise 121 of Exercise Set 8.2.
- We revisit the issue of gender imbalance on college campuses in Example 4 of Section 8.1.
- A model related to marriage inequality is developed in Exercises 59–60 of Exercise Set 8.5.
- Sexual abstinence among young people is the application for the model in Example 4 of Section 8.6.

8.1

Finding Roots ▶

What is the maximum speed at which a racing cyclist can turn a corner without tipping over? The answer, in miles per hour, is given by the algebraic expression $4\sqrt{x}$, where x is the radius of the corner, in feet. Algebraic expressions containing roots describe phenomena as diverse as the percentage of bachelor's degrees awarded to women, a wild animal's territorial area, and Albert Einstein's bizarre concept of how an astronaut moving close to the speed of light would barely age relative to friends watching from Earth. No description of your world can be complete without roots and radicals. In this section, we develop the basics of radical expressions through a notation that takes us from a number raised to a power back to the number itself.

What am I supposed to learn?

After studying this section, you should be able to:

1 Find square roots. ▶

2 Evaluate models containing square roots. ▶

3 Use a calculator to find decimal approximations for irrational square roots. ▶

4 Use a calculator to evaluate models containing square roots. ▶

5 Find higher roots. ▶

1 Find square roots. ▶

Square Roots

From our earlier work with exponents, we are aware that the square of both 5 and -5 is 25:

$$5^2 = 25 \quad \text{and} \quad (-5)^2 = 25.$$

The reverse operation of squaring a number is finding the *square root* of the number. For example,

- One square root of 25 is 5 because $5^2 = 25$.
- Another square root of 25 is -5 because $(-5)^2 = 25$.

In general, **if $b^2 = a$, then b is a square root of a.**

The symbol $\sqrt{}$ is used to denote the nonnegative or *principal square root* of a number. For example,

- $\sqrt{25} = 5$ because $5^2 = 25$ and 5 is positive.
- $\sqrt{100} = 10$ because $10^2 = 100$ and 10 is positive.

The symbol $\sqrt{}$ that we use to denote the principal square root is called a **radical sign**. The number under the radical sign is called the **radicand**. Together we refer to the radical sign and its radicand as a **radical**.

Radical sign $\quad \sqrt{a} \quad$ Radicand

Radical

Definition of the Principal Square Root

If a is a nonnegative real number, the nonnegative number b such that $b^2 = a$, denoted by $b = \sqrt{a}$, is the **principal square root** of a.

The symbol $-\sqrt{}$ is used to denote the negative square root of a number. For example,

- $-\sqrt{25} = -5$ because $(-5)^2 = 25$ and -5 is negative.
- $-\sqrt{100} = -10$ because $(-10)^2 = 100$ and -10 is negative.

EXAMPLE 1   Evaluating Expressions Containing Radicals

Evaluate:

a. $\sqrt{64}$ **b.** $-\sqrt{49}$ **c.** $\sqrt{\dfrac{1}{4}}$ **d.** $\sqrt{9+16}$ **e.** $\sqrt{9} + \sqrt{16}$.

Solution

a. $\sqrt{64} = 8$ The principal square root of 64 is 8. Check: $8^2 = 64$.

b. $-\sqrt{49} = -7$ The negative square root of 49 is -7.
Check: $(-7)^2 = 49$.

c. $\sqrt{\dfrac{1}{4}} = \dfrac{1}{2}$ The principal square root of $\frac{1}{4}$ is $\frac{1}{2}$. Check: $\left(\frac{1}{2}\right)^2 = \frac{1}{4}$.

d. $\sqrt{9+16} = \sqrt{25}$ First simplify the expression under the radical sign.
$\phantom{\sqrt{9+16}} = 5$ Then take the principal square root of 25, which is 5.

e. $\sqrt{9} + \sqrt{16} = 3 + 4$ $\sqrt{9} = 3$ because $3^2 = 9$. $\sqrt{16} = 4$ because $4^2 = 16$.
$\phantom{\sqrt{9} + \sqrt{16}} = 7$ ∎

> **Great Question!**
>
> **Is $\sqrt{a+b}$ equal to $\sqrt{a} + \sqrt{b}$?**
>
> No. In Example 1, parts (d) and (e), observe that $\sqrt{9+16}$ is not equal to $\sqrt{9} + \sqrt{16}$. In general,
> $$\sqrt{a+b} \neq \sqrt{a} + \sqrt{b}$$
> and
> $$\sqrt{a-b} \neq \sqrt{a} - \sqrt{b}.$$

✓ **CHECK POINT 1**   Evaluate:

a. $\sqrt{81}$ **b.** $-\sqrt{9}$ **c.** $\sqrt{\dfrac{1}{25}}$

d. $\sqrt{36+64}$ **e.** $\sqrt{36} + \sqrt{64}$.

2 Evaluate models containing square roots. ▶

EXAMPLE 2   An Application: A Mathematical Model Containing a Radical

Racing cyclists use the formula

$$v = 4\sqrt{r}$$

to determine the maximum velocity, v, in miles per hour, to turn a corner with radius r, in feet, without tipping over. What is the maximum velocity that a cyclist should travel around a corner of radius 9 feet without tipping over?

Solution

$v = 4\sqrt{r}$ This is the given formula.

$v = 4\sqrt{9}$ The radius of the corner is 9 feet, so substitute 9 for r.

$v = 4 \cdot 3$ $\sqrt{9} = 3$ because $3^2 = 9$.

$ = 12$ Multiply: $4 \cdot 3 = 12$.

The maximum velocity that a cyclist should travel around a corner of radius 9 feet without tipping over is 12 miles per hour. ∎

> **Great Question!**
>
> **Should I know the square roots of certain numbers by heart?**
>
> Some square roots occur so frequently that you may want to memorize them.
>
> $\sqrt{1} = 1$ $\sqrt{4} = 2$
> $\sqrt{9} = 3$ $\sqrt{16} = 4$
> $\sqrt{25} = 5$ $\sqrt{36} = 6$
> $\sqrt{49} = 7$ $\sqrt{64} = 8$
> $\sqrt{81} = 9$ $\sqrt{100} = 10$
> $\sqrt{121} = 11$ $\sqrt{144} = 12$

☑ **CHECK POINT 2**   Use the formula $v = 4\sqrt{r}$, described in Example 2, to determine the maximum velocity that a cyclist should travel around a corner of radius 16 feet without tipping over.

The Graph of $y = \sqrt{x}$

We graph $y = \sqrt{x}$ by first selecting integers for x. It is easiest to choose **perfect squares**, numbers that are squares of integers. **Table 8.1** shows five ordered pairs that are solutions of $y = \sqrt{x}$. We plot these ordered pairs as points in the rectangular coordinate system and connect the points with a smooth curve. The graph of $y = \sqrt{x}$ is shown in **Figure 8.1**.

Table 8.1		
x	**$y = \sqrt{x}$**	**(x, y)**
0	$y = \sqrt{0} = 0$	$(0, 0)$
1	$y = \sqrt{1} = 1$	$(1, 1)$
4	$y = \sqrt{4} = 2$	$(4, 2)$
9	$y = \sqrt{9} = 3$	$(9, 3)$
16	$y = \sqrt{16} = 4$	$(16, 4)$

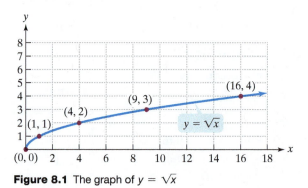

Figure 8.1 The graph of $y = \sqrt{x}$

Is it possible to choose values of x in **Table 8.1** that are not perfect squares? Yes. For example, we can let $x = 3$. Thus, $y = \sqrt{x} = \sqrt{3}$. Because 3 is not a perfect square, $\sqrt{3}$ is an irrational number, one that cannot be expressed as a quotient of integers. We can use a calculator to find a decimal approximation for $\sqrt{3}$.

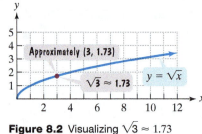

Figure 8.2 Visualizing $\sqrt{3} \approx 1.73$

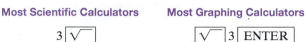

Most Scientific Calculators	Most Graphing Calculators
3 √☐	√☐ 3 ENTER

Rounding the displayed number to two decimal places, $\sqrt{3} \approx 1.73$. This information is shown visually as a point, approximately $(3, 1.73)$, on the graph of $y = \sqrt{x}$ in **Figure 8.2**.

Did you notice that in **Figure 8.1** we graphed $y = \sqrt{x}$ in quadrant I, where $x \geq 0$? What happens if x is negative? Is the square root of a negative number a real number? For example, consider $\sqrt{-25}$. Is there a real number whose square is -25? No. Thus, $\sqrt{-25}$ is not a real number. In general, **a square root of a negative number is not a real number**.

3 Use a calculator to find decimal approximations for irrational square roots. ⊙

EXAMPLE 3   Finding a Decimal Approximation for an Irrational Square Root

Use a calculator to find a decimal approximation for $\sqrt{11}$. Round to two decimal places. How is this shown on the graph of $y = \sqrt{x}$?

Solution   To approximate $\sqrt{11}$, use one of the following keystrokes:

Scientific Calculator	Graphing Calculator
11 √☐	√☐ 11 ENTER

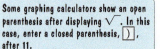

Some graphing calculators show an open parenthesis after displaying √. In this case, enter a closed parenthesis, ⟨⟩, after 11.

Figure 8.3

Figure 8.3 shows the graphing calculator screen with a decimal approximation for $\sqrt{11}$. Rounding the displayed number to two decimal places,

$$\sqrt{11} \approx 3.32.$$

This information is shown visually as a point, approximately $(11, 3.32)$ on the graph of $y = \sqrt{x}$ in **Figure 8.4**.

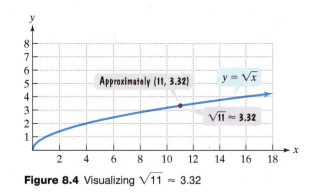

Figure 8.4 Visualizing $\sqrt{11} \approx 3.32$

✓ **CHECK POINT 3** Use a calculator to find a decimal approximation for $\sqrt{7}$. Round to two decimal places. How is this shown on the graph of $y = \sqrt{x}$ in **Figure 8.4**?

4 Use a calculator to evaluate models containing square roots. ▶

Modeling Data with Square Roots

We have seen that the graph of $y = \sqrt{x}$ is increasing from left to right. However, the rate of increase is slowing down as the graph moves to the right. For this reason, equations with square roots are often used to model growing phenomena with growth that is leveling off.

For example, let's revisit our discussion from Section 1.7 with a new set of data related to gender imbalance on U.S. college campuses. The bar graph in **Figure 8.5(a)** displays the percentage of bachelor's degrees awarded to women for five selected years from 1975 through 2010. The data are displayed as a set of five points in the scatter plot in **Figure 8.5(b)**.

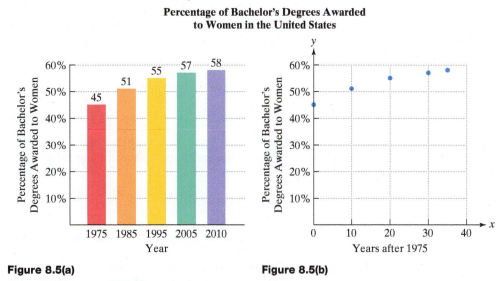

Percentage of Bachelor's Degrees Awarded to Women in the United States

Figure 8.5(a)

Figure 8.5(b)

Source: U.S. Department of Education

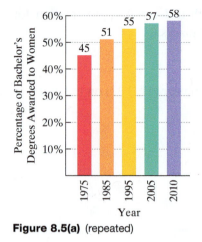

Figure 8.5(a) (repeated)

5 Find higher roots.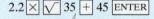

Can you see that the percentage of degrees awarded to women is slowing down for the period shown? Based on the shape of the graph of $y = \sqrt{x}$, square roots can be used to model the data.

EXAMPLE 4 A Matter of Degree: Modeling Gender Imbalance on U.S. College Campuses

The mathematical model

$$P = 2.2\sqrt{t} + 45$$

describes the percentage of bachelor's degrees, P, awarded to women in U.S. colleges t years after 1975. Use the formula to find the percentage, to the nearest percent, of degrees awarded to women in 2010.

Solution Because 2010 is 35 years after 1975, we substitute 35 for t in the given formula. Then we use a calculator to find P, the percentage of degrees awarded to women.

$P = 2.2\sqrt{t} + 45$	Use the given formula.
$P = 2.2\sqrt{35} + 45$	Substitute 35 for t.
≈ 58	Use a calculator. See the Using Technology box in the margin.

Rounding to the nearest percent, approximately 58% of bachelor's degrees were awarded to women in 2010. **Figure 8.5(a)** also shows that 58% of bachelor's degrees were awarded to women in 2010, so the model provides an excellent description of the 2010 data. ∎

✓ **CHECK POINT 4** Use the mathematical model in Example 4 to find the percentage, to the nearest percent, of degrees awarded to women in 2005. How does this approximation compare with the actual percentage displayed by **Figure 8.5(a)**?

Higher Roots

Finding the square root of a number reverses the process of squaring a number. Similarly, the cube root of a number reverses the process of cubing a number. For example, $2^3 = 8$, and so the cube root of 8 is 2. The notation that we use is $\sqrt[3]{8} = 2$.

We define the **principal nth root** of a real number a, symbolized by $\sqrt[n]{a}$, as follows:

Definition of the Principal nth Root of a Real Number

$$\sqrt[n]{a} = b \text{ means that } b^n = a,$$

where a and b are of the same sign. The natural number n is called the **index**. The index for square roots, 2, is usually omitted.

For example,

$$\sqrt[3]{64} = 4 \text{ because } 4^3 = 64 \text{ and } \sqrt[5]{-32} = -2 \text{ because } (-2)^5 = -32.$$

The same vocabulary that we learned for square roots applies to nth roots. The symbol $\sqrt[n]{a}$ is called a **radical** and a is called the **radicand**.

Table 8.2 shows how various roots reverse raising numbers to powers.

Table 8.2	Reversing nth Powers with nth Roots		
	Powers	**Roots**	**Vocabulary**
Cube Roots	$4^3 = 64$	$\sqrt[3]{64} = 4$	3 is the index of the radical.
	$(-2)^3 = -8$	$\sqrt[3]{-8} = -2$	
	$5^3 = 125$	$\sqrt[3]{125} = 5$	
Fourth Roots	$1^4 = 1$	$\sqrt[4]{1} = 1$	Index = 4
	$3^4 = 81$	$\sqrt[4]{81} = 3$	
Fifth Roots	$2^5 = 32$	$\sqrt[5]{32} = 2$	Index = 5
nth Roots	$b^n = a$	$\sqrt[n]{a} = b$	Index = n

One of the entries in the table involves a negative radicand: $\sqrt[3]{-8} = -2$. By contrast to the square root of a negative number, the cube root of a negative number is a real number. In general, **if the index is even,** such as $\sqrt{}$, $\sqrt[4]{}$, $\sqrt[6]{}$, and so on, **the radicand must be nonnegative for the root to be a real number.** For example,

$$\sqrt[4]{81} = 3, \text{ but } \sqrt[4]{-81} \text{ is not a real number.}$$
$$\sqrt[6]{64} = 2, \text{ but } \sqrt[6]{-64} \text{ is not a real number.}$$

Furthermore, if the index is even, the principal nth root is nonnegative.

Great Question!

Should I know the higher roots of certain numbers by heart?

Some higher roots occur so frequently that you may want to memorize them.

Cube Roots	Fourth Roots
$\sqrt[3]{1} = 1$	$\sqrt[4]{1} = 1$
$\sqrt[3]{8} = 2$	$\sqrt[4]{16} = 2$
$\sqrt[3]{27} = 3$	$\sqrt[4]{81} = 3$
$\sqrt[3]{64} = 4$	$\sqrt[4]{256} = 4$
$\sqrt[3]{125} = 5$	$\sqrt[4]{625} = 5$
$\sqrt[3]{216} = 6$	

Fifth Roots
$\sqrt[5]{1} = 1$
$\sqrt[5]{32} = 2$
$\sqrt[5]{243} = 3$

EXAMPLE 5   Finding Cube Roots

Find the cube roots:

a. $\sqrt[3]{27}$ **b.** $\sqrt[3]{-1}$ **c.** $\sqrt[3]{\dfrac{1}{8}}$.

Solution

a. $\sqrt[3]{27} = 3$ because $3^3 = 27$.

b. $\sqrt[3]{-1} = -1$ because $(-1)^3 = -1$.

c. $\sqrt[3]{\dfrac{1}{8}} = \dfrac{1}{2}$ because $\left(\dfrac{1}{2}\right)^3 = \dfrac{1}{8}$. ■

☑ **CHECK POINT 5**   Find the cube roots:

a. $\sqrt[3]{1}$ **b.** $\sqrt[3]{-27}$ **c.** $\sqrt[3]{\dfrac{1}{125}}$.

EXAMPLE 6   Finding Higher Roots

Find the indicated root, or state that the expression is not a real number:

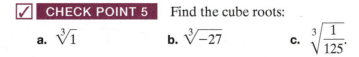

a. $\sqrt[4]{16}$ **b.** $\sqrt[4]{-16}$ **c.** $-\sqrt[4]{16}$ **d.** $\sqrt[5]{-32}$.

Solution

a. $\sqrt[4]{16} = 2$ because $2^4 = 16$.

b. $\sqrt[4]{-16}$ is not a real number because the index, 4, is even and the radicand, -16, is negative. No real number equals $\sqrt[4]{-16}$ because any real number raised to the fourth power gives a nonnegative result.

c. $-\sqrt[4]{16} = -2$

Copy the negative sign and use the fact that $\sqrt[4]{16} = 2$.

d. $\sqrt[5]{-32} = -2$ because $(-2)^5 = -32$. ■

✓ **CHECK POINT 6** Find the indicated root, or state that the expression is not a real number:

a. $\sqrt[4]{81}$

b. $\sqrt[4]{-81}$

c. $-\sqrt[4]{81}$

d. $\sqrt[5]{-\dfrac{1}{32}}$.

CONCEPT AND VOCABULARY CHECK

Fill in each blank so that the resulting statement is true.

1. The symbol $\sqrt{}$ is used to denote the nonnegative, or _____, square root of a number.

2. $\sqrt{81} = 9$ because _____ $= 81$.

3. In the expression $\sqrt[3]{125}$, the number 3 is called the _____ and the number 125 is called the _____.

4. $\sqrt[3]{64} = 4$ because _____ $= 64$.

5. $\sqrt[4]{16} = 2$ because _____ $= 16$.

6. True or false: Equations with square roots are often used to model growing phenomena with growth that is leveling off. _____

7. True or false: $-\sqrt{25}$ is a real number. _____

8. True or false: $\sqrt{-25}$ is a real number. _____

9. True or false: $\sqrt[3]{-1}$ is a real number. _____

10. True or false: $\sqrt[4]{-1}$ is a real number. _____

8.1 EXERCISE SET ▶ MyMathLab®

Practice Exercises

In Exercises 1–26, evaluate each expression, or state that the expression is not a real number.

1. $\sqrt{36}$

2. $\sqrt{16}$

3. $-\sqrt{36}$

4. $-\sqrt{16}$

5. $\sqrt{-36}$

6. $\sqrt{-16}$

7. $\sqrt{\dfrac{1}{9}}$

8. $\sqrt{\dfrac{1}{49}}$

9. $\sqrt{\dfrac{1}{100}}$

10. $\sqrt{\dfrac{1}{81}}$

11. $-\sqrt{\dfrac{1}{36}}$

12. $-\sqrt{\dfrac{1}{121}}$

13. $\sqrt{-\dfrac{1}{36}}$

14. $\sqrt{-\dfrac{1}{121}}$

15. $\sqrt{0.04}$

16. $\sqrt{0.64}$

17. $\sqrt{33 - 8}$

18. $\sqrt{51 + 13}$

19. $\sqrt{2 \cdot 32}$

20. $\sqrt{\dfrac{75}{3}}$

21. $\sqrt{144 + 25}$

22. $\sqrt{25 - 16}$

23. $\sqrt{144} + \sqrt{25}$

24. $\sqrt{25} - \sqrt{16}$

25. $\sqrt{25 - 144}$

26. $\sqrt{16 - 25}$

In Exercises 27–28, graph each equation. Begin by filling in the table and finding five solutions of the equation. Then plot these ordered pairs as points in the rectangular coordinate system and connect the points with a smooth curve.

27. $y = \sqrt{x - 1}$

Because the radicand cannot be negative, the graph begins at the point with this x-coordinate.

x	$y = \sqrt{x-1}$	(x, y)
1		
2		
5		
10		
17		

28. $y = \sqrt{x + 2}$

Because the radicand cannot be negative, the graph begins at the point with this x-coordinate.

x	$y = \sqrt{x+2}$	(x, y)
-2		
-1		
2		
7		
14		

29. Describe one similarity and one difference between your graph in Exercise 27 and the graph of $y = \sqrt{x}$, shown in **Figure 8.1** on page 570.

30. Describe one similarity and one difference between your graph in Exercise 28 and the graph of $y = \sqrt{x}$, shown in **Figure 8.1** on page 570.

In Exercises 31–36, use a calculator to approximate each square root. Round to two decimal places.

31. $\sqrt{7}$

32. $\sqrt{11}$

33. $\sqrt{23}$

34. $\sqrt{97}$

35. $-\sqrt{65}$

36. $-\sqrt{83}$

In Exercises 37–46, use a calculator to approximate each expression. Round to two decimal places. If the expression is not a real number and an approximation is not possible, so state.

37. $12 + \sqrt{11}$

38. $14 + \sqrt{13}$

39. $\dfrac{12 + \sqrt{11}}{2}$

40. $\dfrac{14 + \sqrt{13}}{2}$

41. $\dfrac{-5 + \sqrt{321}}{6}$

42. $\dfrac{-7 + \sqrt{839}}{5}$

43. $\sqrt{13 - 5}$

44. $\sqrt{21 - 4}$

45. $\sqrt{5 - 13}$

46. $\sqrt{4 - 21}$

In Exercises 47–56, find each cube root.

47. $\sqrt[3]{64}$

48. $\sqrt[3]{27}$

49. $\sqrt[3]{-27}$

50. $\sqrt[3]{-64}$

51. $-\sqrt[3]{8}$

52. $-\sqrt[3]{27}$

53. $\sqrt[3]{\dfrac{1}{125}}$

54. $\sqrt[3]{\dfrac{1}{1000}}$

55. $\sqrt[3]{-1000}$

56. $\sqrt[3]{-125}$

In Exercises 57–74, find the indicated root, or state that the expression is not a real number.

57. $\sqrt[4]{1}$

58. $\sqrt[5]{1}$

59. $\sqrt[4]{16}$

60. $\sqrt[4]{81}$

61. $-\sqrt[4]{16}$

62. $-\sqrt[4]{81}$

63. $\sqrt[4]{-16}$

64. $\sqrt[4]{-81}$

65. $\sqrt[5]{-1}$

66. $\sqrt[7]{-1}$

67. $\sqrt[6]{-1}$

68. $\sqrt[8]{-1}$

69. $-\sqrt[4]{256}$

70. $-\sqrt[4]{10,000}$

71. $\sqrt[6]{64}$

72. $\sqrt[5]{32}$

73. $-\sqrt[5]{32}$

74. $-\sqrt[6]{64}$

Practice PLUS

In Exercises 75–84, for which values of x is each radical expression a real number? Express your answer as an inequality or write "all real numbers."

75. $\sqrt{2x}$

76. $\sqrt{3x}$

77. $\sqrt{x - 2}$

78. $\sqrt{x - 3}$

79. $\sqrt{2 - x}$

80. $\sqrt{3 - x}$

81. $\sqrt{x^2 + 2}$

82. $\sqrt{x^2 + 3}$

83. $\sqrt{12 - 2x}$

84. $\sqrt{8 - 2x}$

Application Exercises

85. Racing cyclists use the formula $v = 4\sqrt{r}$ to determine the maximum velocity, v, in miles per hour, to turn a corner with radius r, in feet, without tipping over. What is the maximum velocity that a cyclist should travel around a corner of radius 4 feet without tipping over?

86. The formula $v = \sqrt{2.5r}$ models the maximum safe speed, v, in miles per hour, at which a car can travel on a curved road with radius of curvature r, in feet. A highway crew measures the radius of curvature at an exit ramp on a highway as 360 feet. What is the maximum safe speed?

Police use the formula $v = \sqrt{20L}$ to estimate the speed of a car, v, in miles per hour, based on the length, L, in feet, of its skid marks upon sudden braking on a dry asphalt road. Use the formula to solve Exercises 87–88.

87. A motorist is involved in an accident. A police officer measures the car's skid marks to be 245 feet long. Estimate the speed at which the motorist was traveling before braking. If the posted speed limit is 50 miles per hour and the motorist tells the officer he was not speeding, should the officer believe him? Explain.

88. A motorist is involved in an accident. A police officer measures the car's skid marks to be 45 feet long. Use the formula described at the bottom of the previous page to estimate the speed at which the motorist was traveling before braking. If the posted speed limit is 35 miles per hour and the motorist tells the officer she was not speeding, should the officer believe her? Explain.

Autism is a neurological disorder that impedes language and derails social and emotional development. New findings suggest that the condition is not a sudden calamity that strikes children at the age of 2 or 3, but a developmental problem linked to abnormally rapid brain growth during infancy. The graphs show that the heads of severely autistic children start out smaller than average and then go through a period of explosive growth. Exercises 89–90 involve mathematical models for the data shown by the graphs.

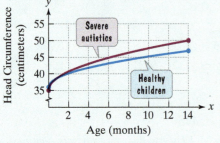

Developmental Differences between Healthy Children and Severe Autistics

Source: The Journal of the American Medical Association

89. The data for one of the two groups shown by the graphs can be modeled by

$$y = 2.9\sqrt{x} + 36,$$

where y is the head circumference, in centimeters, at age x months, $0 \le x \le 14$.

a. According to the model, what is the head circumference at birth?

b. According to the model, what is the head circumference at 9 months?

c. According to the model, what is the head circumference at 14 months? Use a calculator and round to the nearest tenth of a centimeter.

d. Use the values that you obtained in parts (a) through (c) and the graphs shown above to determine if the given model describes healthy children or severe autistics.

90. The data for one of the two groups shown by the graphs can be modeled by

$$y = 4\sqrt{x} + 35,$$

where y is the head circumference, in centimeters, at age x months, $0 \le x \le 14$.

a. According to the model, what is the head circumference at birth?

b. According to the model, what is the head circumference at 9 months?

c. According to the model, what is the head circumference at 14 months? Use a calculator and round to the nearest centimeter.

d. Use the values that you obtained in parts (a) through (c) and the graphs shown in the previous column to determine if the given model describes healthy children or severe autistics.

91. The bar graph shows the percentage of Americans ages 25 and older who had completed at least four years of high school in each of seven selected years. The data can be modeled by an equation in the form

$$y = a\sqrt{x} + b,$$

where y is the percentage of Americans ages 25 and over who had completed at least four years of high school x years after 1980.

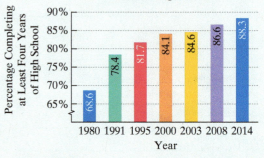

Percentage of Americans Ages 25 and Older Completing at Least Four Years of High School

Source: U.S. Census Bureau

In this exercise, you will use the data from 1980 and 2014 to find values for a and b. This will give you an equation of a square root model for educational attainment.

a. In 1980, 68.6% of Americans ages 25 and over had completed at least four years of high school. Substitute 0 for x (1980 is 0 years after 1980) and 68.6 for y in $y = a\sqrt{x} + b$. Find the value for b and rewrite the model using this value.

b. In 2014, 88.3% of Americans ages 25 and over had completed at least four years of high school. Substitute 34 for x (2014 is 34 years after 1980) and 88.3 for y in your model from part (a). Find the value of a, rounded to one decimal place, and rewrite the model using this value.

c. According to your model from part (b), what percentage of Americans ages 25 and over had completed at least four years of high school in 2003? Use a calculator and round to the nearest tenth of a percent. Does this underestimate or overestimate the actual percentage given by the bar graph? By how much?

Explaining the Concepts

92. What are the square roots of 36? Explain why each of these numbers is a square root.

93. What does the symbol $\sqrt{}$ denote? Which of your answers in Exercise 92 is given by this symbol? Write the symbol needed to obtain the other answer.

94. Explain why $\sqrt{-1}$ is not a real number.

95. Explain why $\sqrt[3]{8}$ is 2. Then describe what is meant by $\sqrt[n]{a} = b$.

96. Explain the meaning of the words *radical, radicand,* and *index.* Give an example with your explanation.

97. Suppose that a motorist first lightly applies the brakes and then forcefully applies the brakes after a few seconds. Does the model given in Exercises 87–88 underestimate or overestimate the speed at which the motorist was traveling before braking? Explain your answer.

98. The formula $d = \sqrt{1.5h}$ models the distance, d, in miles, that can be seen to the horizon from a height h, in feet. Write a word problem using the formula. Then solve the problem.

Critical Thinking Exercises

Make Sense? *In Exercises 99–102, determine whether each statement makes sense or does not make sense, and explain your reasoning.*

99. Because of the negative sign, $\sqrt{-9}$ and $-\sqrt{9}$ are not real numbers.

100. A negative number has an nth root only if n is odd.

101. Using my calculator, I determined that $6^7 = 279,936$, so 6 must be a seventh root of 279,936.

102. Using my calculator to approximate $\dfrac{12 + \sqrt{11}}{2}$, I pressed $\boxed{=}$ after entering $12 + \sqrt{11}$ before I divided by 2, so in this case my calculator is not following the order of operations.

In Exercises 103–106, determine whether each statement is true or false. If the statement is false, make the necessary change(s) to produce a true statement.

103. $\sqrt{9} + \sqrt{16} = \sqrt{25}$

104. $\dfrac{\sqrt{64}}{2} = \sqrt{32}$

105. $\sqrt[3]{-27}$ is not a real number.

106. $\sqrt{\dfrac{1}{4}} + \sqrt{\dfrac{1}{9}} = \sqrt{\dfrac{25}{36}}$

In Exercises 107–109, simplify each expression.

107. $\sqrt{\sqrt[3]{64}}$

108. $\sqrt[3]{-\sqrt{1}}$

109. $\sqrt{\sqrt{16}} - \sqrt[3]{\sqrt{64}}$

110. Between which two consecutive integers is $-\sqrt{47}$?

Technology Exercises

111. Use a graphing utility to graph $y_1 = \sqrt{x}$, $y_2 = \sqrt{x + 4}$, and $y_3 = \sqrt{x - 3}$ in the same $[-5, 10, 1]$ by $[0, 6, 1]$ viewing rectangle. Describe one similarity and one difference that you observe among the graphs. Use the word "shift" in your response.

112. Use a graphing utility to graph $y_1 = \sqrt{x}$, $y_2 = \sqrt{x} + 4$, and $y_3 = \sqrt{x} - 3$ in the same $[-1, 10, 1]$ by $[-10, 10, 1]$ viewing rectangle. Describe one similarity and one difference that you observe among the graphs.

Review Exercises

113. Graph: $4x - 5y = 20$. (Section 3.2, Example 4)

114. Solve and graph the solution set on a number line: $2(x - 3) > 4x + 10$. (Section 2.7, Example 8)

115. Divide: $\dfrac{1}{x^2 - 17x + 30} \div \dfrac{1}{x^2 + 7x - 18}$.

(Section 7.2, Example 6)

Preview Exercises

Exercises 116–118 will help you prepare for the material covered in the next section.

116. a. Find $\sqrt{25} \cdot \sqrt{4}$.

    b. Find $\sqrt{25 \cdot 4}$.

    c. Based on your answers to parts (a) and (b), what can you conclude?

117. a. Use a calculator to approximate $\sqrt{500}$ to three decimal places.

    b. Use a calculator to approximate $10\sqrt{5}$ to three decimal places.

    c. Based on your answers to parts (a) and (b), what can you conclude?

118. a. Find $\dfrac{\sqrt{64}}{\sqrt{4}}$.

    b. Find $\sqrt{\dfrac{64}{4}}$.

    c. Based on your answers to parts (a) and (b), what can you conclude?

8.2

Multiplying and Dividing Radicals

What am I supposed to learn?

After studying this section, you should be able to:

1. Multiply square roots.
2. Simplify square roots.
3. Use the quotient rule for square roots.
4. Use the product and quotient rules for other roots.

This photograph shows mathematical models used by Albert Einstein at a lecture on relativity. Notice the radicals that appear in many of the formulas. Among these models, there is one describing how an astronaut in a moving spaceship ages more slowly than friends who remain on Earth. In this section, in addition to learning how to multiply and divide radicals, you will see how radicals model time dilation for a futuristic high-speed trip to a nearby star.

The Product Rule for Square Roots

A rule for multiplying square roots can be generalized by comparing $\sqrt{25} \cdot \sqrt{4}$ and $\sqrt{25 \cdot 4}$. Notice that

$$\sqrt{25} \cdot \sqrt{4} = 5 \cdot 2 = 10 \quad \text{and} \quad \sqrt{25 \cdot 4} = \sqrt{100} = 10.$$

Because we obtain 10 in both situations, the original radical expressions must be equal. That is,

$$\sqrt{25} \cdot \sqrt{4} = \sqrt{25 \cdot 4}.$$

1. Multiply square roots.

This result is a special case of the **product rule for square roots** that can be generalized as follows:

> **The Product Rule for Square Roots**
>
> If a and b represent nonnegative real numbers, then
>
> $$\sqrt{a} \cdot \sqrt{b} = \sqrt{ab} \quad \text{and} \quad \sqrt{ab} = \sqrt{a} \cdot \sqrt{b}.$$
>
> The product of two square roots is the square root of the product of the radicands.
>
> The square root of a product is the product of the square roots.

EXAMPLE 1   Using the Product Rule to Multiply Square Roots

Use the product rule for square roots to find each product:

a. $\sqrt{2} \cdot \sqrt{5}$

b. $\sqrt{7x} \cdot \sqrt{11y}$

c. $\sqrt{7} \cdot \sqrt{7}$

d. $\sqrt{\dfrac{2}{5}} \cdot \sqrt{\dfrac{3}{7}}.$

Solution Because $\sqrt{a} \cdot \sqrt{b} = \sqrt{ab}$, we find each product by multiplying the radicands.

a. $\sqrt{2} \cdot \sqrt{5} = \sqrt{2 \cdot 5} = \sqrt{10}$

b. $\sqrt{7x} \cdot \sqrt{11y} = \sqrt{7x \cdot 11y} = \sqrt{77xy}$ Assume that $x \geq 0$ and $y \geq 0$.

c. $\sqrt{7} \cdot \sqrt{7} = \sqrt{7 \cdot 7} = \sqrt{49} = 7$

d. $\sqrt{\dfrac{2}{5}} \cdot \sqrt{\dfrac{3}{7}} = \sqrt{\dfrac{2}{5} \cdot \dfrac{3}{7}} = \sqrt{\dfrac{6}{35}}$ ∎

✓ **CHECK POINT 1**   Use the product rule for square roots to find each product:

a. $\sqrt{3} \cdot \sqrt{10}$ **b.** $\sqrt{2x} \cdot \sqrt{13y}$

c. $\sqrt{5} \cdot \sqrt{5}$ **d.** $\sqrt{\dfrac{3}{2}} \cdot \sqrt{\dfrac{5}{11}}$.

2 Simplify square roots. ⏵

Using the Product Rule to Simplify Square Roots

We have seen that a number that is the square of an integer is called a **perfect square**. For example, 100 is a perfect square because $100 = 10^2$. Thus, $\sqrt{100} = 10$.

A square root is **simplified** when its radicand has no factors other than 1 that are perfect squares. For example, $\sqrt{500}$ is not simplified because it can be expressed as $\sqrt{100 \cdot 5}$ and 100 is a perfect square. We can use the product rule in the form

$$\sqrt{ab} = \sqrt{a}\sqrt{b}$$

to simplify $\sqrt{500}$. We factor 500 so that one of its factors is the largest perfect square possible.

$$\begin{aligned} \sqrt{500} &= \sqrt{100 \cdot 5} & \text{Factor 500. 100 is the largest perfect square factor.} \\ &= \sqrt{100}\sqrt{5} & \text{Use the product rule: } \sqrt{ab} = \sqrt{a}\sqrt{b}. \\ &= 10\sqrt{5} & \text{Write } \sqrt{100} \text{ as 10. We read } 10\sqrt{5} \text{ as} \\ & & \text{"ten times the square root of 5."} \end{aligned}$$

Great Question!

When simplifying square roots, what happens if I use a perfect square factor that isn't the greatest perfect square factor possible?

You'll need to simplify even further. For example, consider the following factorization:

$$\sqrt{500} = \sqrt{25 \cdot 20} = \sqrt{25}\sqrt{20} = 5\sqrt{20}.$$

25 is a perfect square factor of 500, but not the greatest perfect square factor.

Because 20 contains a perfect square factor, 4, the simplification is not complete.

$$5\sqrt{20} = 5\sqrt{4 \cdot 5} = 5\sqrt{4}\sqrt{5} = 5 \cdot 2\sqrt{5} = 10\sqrt{5}$$

Although the result checks with our simplification using $\sqrt{500} = \sqrt{100 \cdot 5}$, more work is required when the greatest perfect square factor is not used.

EXAMPLE 2   Using the Product Rule to Simplify Square Roots

Simplify:

a. $\sqrt{75}$ **b.** $\sqrt{80}$ **c.** $\sqrt{38}$.

Solution In each case, try to factor the radicand as a product of the greatest perfect square factor and another factor.

a. $\sqrt{75} = \sqrt{25 \cdot 3}$ 25 is the greatest perfect square that is a factor of 75.

    $= \sqrt{25}\sqrt{3}$ $\sqrt{ab} = \sqrt{a}\sqrt{b}$

    $= 5\sqrt{3}$ Write $\sqrt{25}$ as 5.

b. $\sqrt{80} = \sqrt{16 \cdot 5}$ 16 is the greatest perfect square that is a factor of 80.

    $= \sqrt{16}\sqrt{5}$ $\sqrt{ab} = \sqrt{a}\sqrt{b}$

    $= 4\sqrt{5}$ Write $\sqrt{16}$ as 4.

c. Finally, we consider $\sqrt{38}$. Although the radicand, 38, can be factored as $2 \cdot 19$, neither of these factors is a perfect square. Because 38 has no perfect square factors other than 1, $\sqrt{38}$ cannot be simplified. ∎

Using Technology

Numeric Connections

You can use a calculator to provide numerical support that your simplified answer is correct. For example, to support $\sqrt{75} = 5\sqrt{3}$, find decimal approximations for $\sqrt{75}$ and $5\sqrt{3}$. These approximations should be equal.

	$\sqrt{75}$:	$5\sqrt{3}$:
Scientific Calculator	75 $\sqrt{}$	5 × 3 $\sqrt{}$ =
Graphing Calculator	$\sqrt{}$ 75 ENTER	5 $\sqrt{}$ 3 ENTER

Correct to two decimal places, $\sqrt{75} \approx 8.66$ and $5\sqrt{3} \approx 8.66$. Use this technique to support the numerical results for the answers in this section. Caution: **A simplified square root does not mean a decimal approximation.**

☑ **CHECK POINT 2**   Simplify, if possible:

a. $\sqrt{12}$ b. $\sqrt{60}$ c. $\sqrt{55}$.

Simplifying Square Roots Containing Variables

Square roots can also contain variables. Here are three examples:

$$\sqrt{x^2} \qquad \sqrt{x^4} \qquad \sqrt{x^6}.$$

We know that the square root of a negative number is not a real number. Thus, we want to avoid values for variables in the radicand that would make the radicand negative. To avoid negative radicands, we assume throughout this chapter that **if a variable appears in the radicand of a radical expression, it represents nonnegative numbers only**.

Do you see that we can simplify each of the three square root expressions just shown?

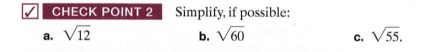

$$\sqrt{x^2} = x \quad \text{Because } (x)^2 = x^2 \qquad \sqrt{x^4} = x^2 \quad \text{Because } (x^2)^2 = x^4 \qquad \sqrt{x^6} = x^3 \quad \text{Because } (x^3)^2 = x^6$$

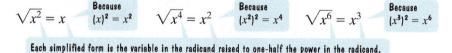

Each simplified form is the variable in the radicand raised to one-half the power in the radicand.

To simplify square roots when the radicand contains x to an even power, we can use the following rule:

Simplifying Square Roots with Variables to Even Powers

$$\sqrt{x^{2n}} = x^n$$

The square root of a variable raised to an even power equals the variable raised to one-half that power.

For example,

$$\sqrt{x^{10}} = x^5 \qquad \text{and} \qquad \sqrt{y^{34}} = y^{17}.$$

$$\tfrac{1}{2} \cdot 10 = 5 \qquad\qquad\qquad \tfrac{1}{2} \cdot 34 = 17$$

EXAMPLE 3 Simplifying Square Roots with Variables to Even Powers

Simplify: $\sqrt{72x^{14}}$.

Solution We write the radicand of $\sqrt{72x^{14}}$ as the product of the greatest perfect square factor and another factor. Variables to even powers are part of the perfect square factor.

$$\sqrt{72x^{14}} = \sqrt{36x^{14} \cdot 2} \qquad 36x^{14} \text{ is the greatest perfect square factor of } 72x^{14}.$$
$$= \sqrt{36x^{14}}\sqrt{2} \qquad \text{Use the product rule: } \sqrt{ab} = \sqrt{a}\,\sqrt{b}.$$
$$= 6x^7\sqrt{2} \text{ or } 6\sqrt{2}x^7 \qquad \sqrt{36} = 6 \text{ and, using } \sqrt{x^{2n}} = x^n,\ \sqrt{x^{14}} = x^7. \ \blacksquare$$

✓ **CHECK POINT 3** Simplify: $\sqrt{40x^{16}}$.

How do we simplify the square root of a radicand containing a variable raised to an odd power, such as $\sqrt{x^5}$? Express the variable as the product of two factors, one of which has an even power. Then use the product rule to simplify. For example,

$$\sqrt{x^5} = \sqrt{x^4 \cdot x} = \sqrt{x^4}\sqrt{x} = x^2\sqrt{x}.$$

$$\sqrt{x^{2n}} = x^n$$

EXAMPLE 4 Simplifying Square Roots with Variables to Odd Powers

Simplify: $\sqrt{50x^7}$.

Solution

$$\sqrt{50x^7} = \sqrt{25x^6 \cdot 2x} \qquad 25x^6 \text{ is the greatest perfect square factor of } 50x^7.$$
$$= \sqrt{25x^6}\sqrt{2x} \qquad \text{Use the product rule: } \sqrt{ab} = \sqrt{a}\,\sqrt{b}.$$
$$= 5x^3\sqrt{2x} \qquad \sqrt{25} = 5 \text{ and, using } \sqrt{x^{2n}} = x^n,\ \sqrt{x^6} = x^3. \ \blacksquare$$

✓ **CHECK POINT 4** Simplify: $\sqrt{27x^9}$.

Now we look at an example where we use the product rule to multiply radicals and then simplify.

EXAMPLE 5 Multiplying and Simplifying Square Roots

Multiply and then simplify:

$$\sqrt{6x^8} \cdot \sqrt{2x^3}.$$

Solution

$$\sqrt{6x^8} \cdot \sqrt{2x^3} = \sqrt{6x^8 \cdot 2x^3} \qquad \text{Use the product rule: } \sqrt{a}\sqrt{b} = \sqrt{ab}.$$
$$= \sqrt{12x^{11}} \qquad \text{Multiply in the radicand: Multiply coefficients and add exponents.}$$
$$= \sqrt{4x^{10} \cdot 3x} \qquad 4x^{10} \text{ is the greatest perfect square factor of } 12x^{11}.$$
$$= \sqrt{4x^{10}}\sqrt{3x} \qquad \text{Use the product rule: } \sqrt{ab} = \sqrt{a}\sqrt{b}.$$
$$= 2x^5\sqrt{3x} \qquad \sqrt{4} = 2 \text{ and, using } \sqrt{x^{2n}} = x^n, \sqrt{x^{10}} = x^5. \quad \blacksquare$$

☑ **CHECK POINT 5** Multiply and then simplify: $\sqrt{15x^6} \cdot \sqrt{3x^7}$.

3 Use the quotient rule for square roots. ▶

The Quotient Rule for Square Roots

A rule for dividing square roots can be generalized by comparing

$$\sqrt{\frac{64}{4}} \quad \text{and} \quad \frac{\sqrt{64}}{\sqrt{4}}.$$

Note that

$$\sqrt{\frac{64}{4}} = \sqrt{16} = 4 \quad \text{and} \quad \frac{\sqrt{64}}{\sqrt{4}} = \frac{8}{2} = 4.$$

Because we obtain 4 in both situations, the original radical expressions must be equal:

$$\sqrt{\frac{64}{4}} = \frac{\sqrt{64}}{\sqrt{4}}.$$

This result is a special case of the **quotient rule for square roots** that is generalized as follows:

> ### The Quotient Rule for Square Roots
>
> If a and b represent nonnegative real numbers and $b \neq 0$, then
>
> $$\frac{\sqrt{a}}{\sqrt{b}} = \sqrt{\frac{a}{b}} \quad \text{and} \quad \sqrt{\frac{a}{b}} = \frac{\sqrt{a}}{\sqrt{b}}.$$
>
>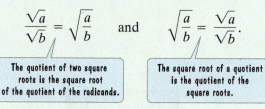
>
> The quotient of two square roots is the square root of the quotient of the radicands.   The square root of a quotient is the quotient of the square roots.

EXAMPLE 6 Using the Quotient Rule to Simplify Square Roots

Simplify:

a. $\sqrt{\dfrac{100}{9}}$ **b.** $\dfrac{\sqrt{48x^3}}{\sqrt{6x}}$.

Solution

a. $\sqrt{\dfrac{100}{9}} = \dfrac{\sqrt{100}}{\sqrt{9}} = \dfrac{10}{3}$

b. $\dfrac{\sqrt{48x^3}}{\sqrt{6x}} = \sqrt{\dfrac{48x^3}{6x}} = \sqrt{8x^2} = \sqrt{4x^2}\sqrt{2} = 2x\sqrt{2}$ $\quad\blacksquare$

☑ **CHECK POINT 6**   Simplify: **a.** $\sqrt{\dfrac{49}{25}}$ **b.** $\dfrac{\sqrt{48x^5}}{\sqrt{3x}}$.

4 Use the product and quotient rules for other roots. ▶

The Product and Quotient Rules for Other Roots

The product and quotient rules apply to cube roots, fourth roots, and all higher roots.

The Product and Quotient Rules for nth Roots

For all real numbers, where the indicated roots represent real numbers,

$$\sqrt[n]{a} \cdot \sqrt[n]{b} = \sqrt[n]{ab} \quad \text{and} \quad \sqrt[n]{ab} = \sqrt[n]{a} \cdot \sqrt[n]{b}$$

The product of two nth roots is the nth root of the product of the radicands.

The nth root of a product is the product of the nth roots.

$$\frac{\sqrt[n]{a}}{\sqrt[n]{b}} = \sqrt[n]{\frac{a}{b}}, \, b \neq 0 \quad \text{and} \quad \sqrt[n]{\frac{a}{b}} = \frac{\sqrt[n]{a}}{\sqrt[n]{b}}, \, b \neq 0.$$

The quotient of two nth roots is the nth root of the quotient of the radicands.

The nth root of a quotient is the quotient of the nth roots.

EXAMPLE 7   Simplifying, Multiplying, and Dividing Higher Roots

Simplify:

a. $\sqrt[3]{24}$ **b.** $\sqrt[4]{8} \cdot \sqrt[4]{4}$ **c.** $\sqrt[4]{\dfrac{81}{16}}$.

Solution

a. $\sqrt[3]{24} = \sqrt[3]{8 \cdot 3}$ Find the greatest perfect cube that is a factor of 24. $2^3 = 8$, so 8 is a perfect cube and is the greatest perfect cube factor of 24.

$\phantom{\sqrt[3]{24}} = \sqrt[3]{8} \cdot \sqrt[3]{3}$ $\sqrt[n]{ab} = \sqrt[n]{a} \cdot \sqrt[n]{b}$

$\phantom{\sqrt[3]{24}} = 2\sqrt[3]{3}$ $\sqrt[3]{8} = 2$

b. $\sqrt[4]{8} \cdot \sqrt[4]{4} = \sqrt[4]{8 \cdot 4}$ $\sqrt[n]{a} \cdot \sqrt[n]{b} = \sqrt[n]{ab}$

$\phantom{\sqrt[4]{8} \cdot \sqrt[4]{4}} = \sqrt[4]{32}$

$\phantom{\sqrt[4]{8} \cdot \sqrt[4]{4}} = \sqrt[4]{16 \cdot 2}$ Find the greatest perfect fourth power that is a factor of 32. $2^4 = 16$, so 16 is a perfect fourth power and is the largest perfect fourth power that is a factor of 32.

$\phantom{\sqrt[4]{8} \cdot \sqrt[4]{4}} = \sqrt[4]{16} \cdot \sqrt[4]{2}$ $\sqrt[n]{ab} = \sqrt[n]{a} \cdot \sqrt[n]{b}$

$\phantom{\sqrt[4]{8} \cdot \sqrt[4]{4}} = 2\sqrt[4]{2}$ $\sqrt[4]{16} = 2$

c. $\sqrt[4]{\dfrac{81}{16}} = \dfrac{\sqrt[4]{81}}{\sqrt[4]{16}}$ $\sqrt[n]{\dfrac{a}{b}} = \dfrac{\sqrt[n]{a}}{\sqrt[n]{b}}$

$\phantom{\sqrt[4]{\dfrac{81}{16}}} = \dfrac{3}{2}$ $\sqrt[4]{81} = 3$ because $3^4 = 81$, and $\sqrt[4]{16} = 2$ because $2^4 = 16$. ∎

☑ **CHECK POINT 7**   Simplify:

a. $\sqrt[3]{40}$ **b.** $\sqrt[5]{8} \cdot \sqrt[5]{8}$ **c.** $\sqrt[3]{\dfrac{125}{27}}$.

Blitzer Bonus ⊙

A Radical Idea: Time Is Relative

Salvador Dali "The Persistence of Memory" 1931, oil on canvas, $9\frac{1}{2} \times 13$ in. (24.1 × 33 cm). The Museum of Modern Art/Licensed by Scala-Art Resource, NY. © 1999 Demart Pro Arte, Geneva/Artists Rights Society (ARS), New York.

What does travel in space have to do with radicals? Imagine that in the future we will be able to travel at velocities approaching the speed of light (approximately 186,000 miles per second). According to Einstein's theory of special relativity, time would pass more quickly on Earth than it would in the moving spaceship.

The special-relativity equation

$$R_a = R_f \sqrt{1 - \left(\frac{v}{c}\right)^2}$$

gives the aging rate of an astronaut, R_a, relative to the aging rate of a friend, R_f, on Earth. In this formula, v is the astronaut's speed and c is the speed of light. As the astronaut's speed approaches the speed of light, we can substitute c for v.

$$R_a = R_f \sqrt{1 - \left(\frac{v}{c}\right)^2}$$

Einstein's equation gives the aging rate of an astronaut, R_a, relative to the aging rate of a friend, R_f, on Earth.

$$R_a = R_f \sqrt{1 - \left(\frac{c}{c}\right)^2}$$

The velocity, v, is approaching the speed of light, c, so let $v = c$.

$$= R_f \sqrt{1 - 1} \qquad \left(\frac{c}{c}\right)^2 = 1^2 = 1 \cdot 1 = 1$$

$$= R_f \sqrt{0} \qquad \text{Simplify the radicand: } 1 - 1 = 0.$$

$$= R_f \cdot 0 \qquad \sqrt{0} = 0$$

$$= 0 \qquad \text{Multiply: } R_f \cdot 0 = 0.$$

Close to the speed of light, the astronaut's aging rate, R_a, relative to a friend, R_f, on Earth is nearly 0. What does this mean? As we age here on Earth, the space traveler would barely get older. The space traveler would return to an unknown futuristic world in which friends and loved ones would be long gone.

Achieving Success

Warm up your brain before starting the assigned homework. Researchers say the mind can be strengthened, just like your muscles, with regular training and rigorous practice. Think of the book's Exercise Sets as brain calisthenics. If you're feeling a bit sluggish before any of your mental workouts, try this warmup:

In the list below say the color the word is printed in, not the word itself. Once you can do this in 15 seconds without an error, the warmup is over and it's time to move on to the assigned exercises.

Blue Yellow Red Green Yellow Green Blue Red Yellow Red

CONCEPT AND VOCABULARY CHECK

Fill in each blank so that the resulting statement is true.

1. If $\sqrt{a}$ and $\sqrt{b}$ are real numbers, then $\sqrt{a \cdot b} =$ _____.

2. $\sqrt{49 \cdot 6} = \sqrt{\underline{}} \cdot \sqrt{\underline{}} = \underline{} \cdot \sqrt{\underline{}} =$ _____

3. Square roots with variables to even powers can be simplified using $\sqrt{x^{2n}} =$ _____. The square root of a variable raised to an even power equals the variable raised to _____ that power.

4. $\sqrt{x^{12}} = x^{\frac{1}{2}\cdot\underline{}} = x^{\underline{}}$

5. If $\sqrt{a}$ and $\sqrt{b}$ are real numbers, then $\sqrt{\dfrac{a}{b}} = $ _____, $b \neq 0$.

6. $\sqrt{\dfrac{25}{4}} = \dfrac{\sqrt{}}{\sqrt{}} = $ _____

7. $\sqrt[4]{16 \cdot 3} = \sqrt[4]{} \cdot \sqrt[4]{} = \underline{} \cdot \sqrt[4]{} = $ _____

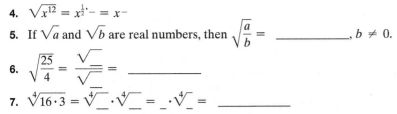

8.2 EXERCISE SET   ▶ MyMathLab®

Remember that throughout this chapter, variable expressions in radicands represent nonnegative real numbers.

Practice Exercises

In Exercises 1–14, use the product rule for square roots to find each product.

1. $\sqrt{2} \cdot \sqrt{7}$
2. $\sqrt{2} \cdot \sqrt{17}$
3. $\sqrt{3x} \cdot \sqrt{5y}$
4. $\sqrt{7x} \cdot \sqrt{11y}$
5. $\sqrt{5} \cdot \sqrt{5}$
6. $\sqrt{10} \cdot \sqrt{10}$
7. $\sqrt{\dfrac{2}{3}} \cdot \sqrt{\dfrac{5}{7}}$
8. $\sqrt{\dfrac{3}{5}} \cdot \sqrt{\dfrac{5}{7}}$
9. $\sqrt{0.1x} \cdot \sqrt{5y}$
10. $\sqrt{0.2x} \cdot \sqrt{3y}$
11. $\sqrt{\dfrac{1}{5}a} \cdot \sqrt{\dfrac{1}{5}b}$
12. $\sqrt{\dfrac{1}{7}a} \cdot \sqrt{\dfrac{1}{7}b}$
13. $\sqrt{\dfrac{2x}{9}} \cdot \sqrt{\dfrac{9}{2}}$
14. $\sqrt{\dfrac{5x}{11}} \cdot \sqrt{\dfrac{11}{5}}$

In Exercises 15–54, simplify each expression. If the expression cannot be simplified, so state.

15. $\sqrt{50}$
16. $\sqrt{27}$
17. $\sqrt{45}$
18. $\sqrt{28}$
19. $\sqrt{200}$
20. $\sqrt{300}$
21. $\sqrt{75x}$
22. $\sqrt{40x}$
23. $\sqrt{9x}$
24. $\sqrt{25x}$
25. $\sqrt{35}$
26. $\sqrt{22}$
27. $\sqrt{y^2}$
28. $\sqrt{z^2}$
29. $\sqrt{64x^2}$
30. $\sqrt{36x^2}$
31. $\sqrt{11x^2}$
32. $\sqrt{17x^2}$
33. $\sqrt{8x^2}$
34. $\sqrt{12x^2}$
35. $\sqrt{x^{20}}$
36. $\sqrt{x^{30}}$
37. $\sqrt{25y^{10}}$
38. $\sqrt{36y^{10}}$
39. $\sqrt{20x^6}$
40. $\sqrt{24x^8}$
41. $\sqrt{72y^{100}}$
42. $\sqrt{32y^{200}}$
43. $\sqrt{x^3}$
44. $\sqrt{y^3}$
45. $\sqrt{x^7}$
46. $\sqrt{x^9}$
47. $\sqrt{y^{17}}$
48. $\sqrt{y^{19}}$
49. $\sqrt{25x^5}$
50. $\sqrt{49x^5}$
51. $\sqrt{8x^{17}}$
52. $\sqrt{12x^7}$
53. $\sqrt{90y^{19}}$
54. $\sqrt{600y^{23}}$

In Exercises 55–68, multiply and, if possible, simplify.

55. $\sqrt{3} \cdot \sqrt{15}$
56. $\sqrt{3} \cdot \sqrt{6}$
57. $\sqrt{5x} \cdot \sqrt{10y}$
58. $\sqrt{8x} \cdot \sqrt{10y}$
59. $\sqrt{12x} \cdot \sqrt{3x}$
60. $\sqrt{20x} \cdot \sqrt{5x}$
61. $\sqrt{15x^2} \cdot \sqrt{3x}$
62. $\sqrt{2x^2} \cdot \sqrt{10x}$
63. $\sqrt{15x^4} \cdot \sqrt{5x^9}$
64. $\sqrt{50x^9} \cdot \sqrt{4x^4}$
65. $\sqrt{7x} \cdot \sqrt{3y}$
66. $\sqrt{5x} \cdot \sqrt{11y}$
67. $\sqrt{50xy} \cdot \sqrt{4xy^2}$
68. $\sqrt{5xy} \cdot \sqrt{10xy^2}$

In Exercises 69–92, simplify using the quotient rule for square roots.

69. $\sqrt{\dfrac{49}{16}}$
70. $\sqrt{\dfrac{100}{9}}$
71. $\sqrt{\dfrac{3}{4}}$
72. $\sqrt{\dfrac{7}{25}}$
73. $\sqrt{\dfrac{x^2}{36}}$
74. $\sqrt{\dfrac{x^2}{49}}$
75. $\sqrt{\dfrac{7}{x^4}}$
76. $\sqrt{\dfrac{13}{x^6}}$
77. $\sqrt{\dfrac{72}{y^{20}}}$
78. $\sqrt{\dfrac{300}{y^{30}}}$
79. $\dfrac{\sqrt{54}}{\sqrt{6}}$
80. $\dfrac{\sqrt{75}}{\sqrt{3}}$
81. $\dfrac{\sqrt{24}}{\sqrt{3}}$
82. $\dfrac{\sqrt{60}}{\sqrt{5}}$
83. $\dfrac{\sqrt{75}}{\sqrt{15}}$
84. $\dfrac{\sqrt{21}}{\sqrt{3}}$
85. $\dfrac{\sqrt{48x}}{\sqrt{3x}}$
86. $\dfrac{\sqrt{27x}}{\sqrt{3x}}$
87. $\dfrac{\sqrt{32x^3}}{\sqrt{8x}}$
88. $\dfrac{\sqrt{75x^3}}{\sqrt{3x}}$
89. $\dfrac{\sqrt{150x^4}}{\sqrt{3x}}$
90. $\dfrac{\sqrt{400x^4}}{\sqrt{2x}}$
91. $\dfrac{\sqrt{400x^{10}}}{\sqrt{10x^3}}$
92. $\dfrac{\sqrt{800x^{12}}}{\sqrt{10x^3}}$

In Exercises 93–108, simplify each radical expression.

93. $\sqrt[3]{16}$
94. $\sqrt[3]{32}$
95. $\sqrt[3]{54}$
96. $\sqrt[3]{250}$
97. $\sqrt[4]{32}$
98. $\sqrt[4]{48}$
99. $\sqrt[3]{4} \cdot \sqrt[3]{2}$
100. $\sqrt[3]{3} \cdot \sqrt[3]{9}$
101. $\sqrt[3]{9} \cdot \sqrt[3]{6}$
102. $\sqrt[3]{12} \cdot \sqrt[3]{4}$
103. $\sqrt[4]{4} \cdot \sqrt[4]{8}$
104. $\sqrt[5]{16} \cdot \sqrt[5]{4}$

105. $\sqrt[3]{\dfrac{27}{8}}$ **106.** $\sqrt[3]{\dfrac{125}{8}}$

107. $\sqrt[3]{\dfrac{3}{8}}$ **108.** $\sqrt[3]{\dfrac{5}{27}}$

Practice PLUS

In Exercises 109–114, simplify each expression.

109. $\sqrt{90(x+4)^3}$

110. $\sqrt{150(x+8)^3}$

111. $\sqrt{x^2-6x+9}$ **112.** $\sqrt{x^2-10x+25}$

113. $\sqrt{2^{43}x^{104}y^{13}}$ **114.** $\sqrt{3^{41}x^{102}y^{17}}$

In Exercises 115–116, use the fact that $\sqrt[3]{x^3}=x$ (why?) to simplify each expression.

115. $\sqrt[3]{24x^5}$ **116.** $\sqrt[3]{16x^4}$

Application Exercises

In Exercises 117–118, express the area of each rectangle as a square root in simplified form.

117.

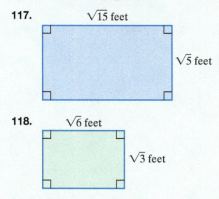

$\sqrt{15}$ feet

$\sqrt{5}$ feet

118.

$\sqrt{6}$ feet

$\sqrt{3}$ feet

The mathematical model

$$A = \dfrac{\sqrt{h} \cdot \sqrt{w}}{56}$$

describes adult body surface area, A, in square meters, where h is the person's height, in inches, and w is the adult's weight, in pounds. Use this model to solve Exercises 119–120.

119. Consider an adult who is 68 inches tall and weighs 200 pounds.

     a. Determine this person's body surface area, A, in simplified radical form. Begin by simplifying each radical factor in the numerator of A using the given values for h and w.

     b. Use a calculator to approximate the surface area in part (a) correct to the nearest hundredth of a square meter.

120. Consider an adult who is 75 inches tall and weighs 220 pounds.

     a. Determine this person's body surface area, A, in simplified radical form. Begin by simplifying each radical factor in the numerator of A using the given values for h and w.

     b. Use a calculator to approximate the surface area in part (a) correct to the nearest hundredth of a square meter.

121. Refer to the *FoxTrot* cartoon on page 567. Solve problems A through Z in the left panel. Then decode Jason Fox's message involving his opinion about the mathematical abilities of his sister Paige shown on the first line.

Hints: Here is the solution for problem C and partial solutions for problems Q and U.

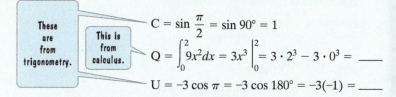

These are from trigonometry. | This is from calculus.

$C = \sin\dfrac{\pi}{2} = \sin 90° = 1$

$Q = \displaystyle\int_0^2 9x^2\,dx = 3x^3 \Big|_0^2 = 3\cdot 2^3 - 3\cdot 0^3 = \underline{\quad}$

$U = -3\cos\pi = -3\cos 180° = -3(-1) = \underline{\quad}$

Note: The comic strip *FoxTrot* is now printed in more than one thousand newspapers. What made cartoonist Bill Amend, a college physics major, put math in the comic? "I always try to use math in the strip to make the joke accessible to anyone," he said. "But if you understand math, hopefully you'll like it that much more!" We highly recommend the math humor in Amend's *FoxTrot* collection *Math, Science, and Unix Underpants* (Andrews McMeel Publishing, 2009).

Explaining the Concepts

122. Use words to state the product rule for square roots, $\sqrt{a} \cdot \sqrt{b} = \sqrt{ab}$. Give an example with your description.

123. Explain why $\sqrt{50}$ is not simplified. What do we mean when we say that a square root is simplified?

124. Explain how to simplify square roots with variables to even powers.

125. Explain how to simplify square roots with variables to odd powers.

126. Explain how to simplify $\sqrt{10} \cdot \sqrt{5}$.

127. Use words to state the quotient rule for square roots, $\dfrac{\sqrt{a}}{\sqrt{b}} = \sqrt{\dfrac{a}{b}}$. Give an example with your description.

128. Read the Blitzer Bonus on page 584. The future is now: You have the opportunity to explore the cosmos in a starship traveling near the speed of light. The experience will enable you to understand the mysteries of the universe in deeply personal ways, transporting you to unimagined levels of knowing and being. The downside: You return from your two-year journey to a futuristic world in which friends and loved ones are long gone. Do you explore space or stay here on Earth? What are the reasons for your choice?

Critical Thinking Exercises

Make Sense? *In Exercises 129–132, determine whether each statement makes sense or does not make sense, and explain your reasoning.*

129. I know that I've simplified a radical expression when it contains a single radical.

130. When multiplying $\sqrt{68}$ and $\sqrt{200}$, I find it easier to simplify each radical before multiplying rather than multiplying first and then simplifying the result.

131. Because the product rule for radicals applies when $\sqrt[n]{a}$ and $\sqrt[n]{b}$ are real numbers, I can use the rule to find $\sqrt[3]{3} \cdot \sqrt[3]{-2}$, but not to find $\sqrt{3} \cdot \sqrt{-2}$.

132. I multiply nth roots by taking the nth root of the product of the radicands.

In Exercises 133–136, determine whether each statement is true or false. If the statement is false, make the necessary change(s) to produce a true statement.

133. $\sqrt{20} = 4\sqrt{5}$

134. If $y \geq 0$, $\sqrt{y^9} = y^3$.

135. $\sqrt{2x}\sqrt{6y} = 2\sqrt{3xy}$ if x and y are nonnegative real numbers.

136. $\sqrt{2} \cdot \sqrt{8} = 16$

In Exercises 137–138, fill in the missing coefficients and exponents to make each statement true.

137. $\sqrt{\blacksquare \, x^{\blacksquare}} = 5x^7$

138. $\sqrt{15x^{\blacksquare}} \cdot \sqrt{\blacksquare \, x^5} = 3x^3\sqrt{10x}$

Technology Exercises

139. Use a calculator to provide numerical support for your simplifications in Exercises 15–20. In each case, find a decimal approximation for the given expression. Then find a decimal approximation for your simplified expression. The approximations should be the same.

In Exercises 140–142, determine if each simplification is correct by graphing each side of the equation with your graphing utility. Use the given viewing rectangle. The graphs should be the same. If they are not, correct the right side of the equation and then use your graphing utility to verify the simplification.

140. $\sqrt{x^4} = x^2$ [0, 5, 1] by [0, 20, 1]

141. $\sqrt{8x^2} = 4x\sqrt{2}$ [0, 5, 1] by [0, 20, 1]

142. $\sqrt{x^3} = x\sqrt{x}$ [0, 5, 1] by [0, 10, 1]

Review Exercises

143. Solve the system:
$$\begin{cases} 4x + 3y = 18 \\ 5x - 9y = 48. \end{cases}$$
(Section 4.3, Example 3)

144. Subtract: $\dfrac{6x}{x^2 - 4} - \dfrac{3}{x + 2}$.
(Section 7.4, Example 7)

145. Factor completely: $2x^3 - 16x^2 + 30x$.
(Section 6.5, Example 2)

Preview Exercises

Exercises 146–148 will help you prepare for the material covered in the next section.

146. **a.** Simplify: $7x + 5x$.
    b. Simplify: $7\sqrt{2} + 5\sqrt{2}$.

147. Simplify: $4\sqrt{50x}$.

148. **a.** Multiply: $3(x + 5)$.
    b. Multiply: $\sqrt{3}(\sqrt{7} + \sqrt{5})$.

SECTION

8.3

Operations with Radicals ▶

What am I supposed to learn?

After studying this section, you should be able to:

1 Add and subtract radicals. ▶

2 Multiply radical expressions with more than one term. ▶

3 Multiply conjugates. ▶

If you are a full-time college student with a job, you're part of the data for a radical model. The formula

$$J = 1.4\sqrt{x} + 55 - (20 - 1.2\sqrt{x})$$

describes the percentage of full-time college students with jobs, J, x years after 1975. Using operations with radicals discussed in this section, you will see that this model can be simplified to a more manageable form. We will return to the model and its graph in the Exercise Set.

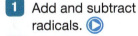

1 Add and subtract radicals. ⊙

Adding and Subtracting Like Radicals

We know that like terms have exactly the same variable factors and can be combined. For example,

$$7x + 6x = (7 + 6)x = 13x.$$

Two or more square roots that have the same radicands are called **like radicals**. Like radicals are combined using the distributive property in exactly the same way that we combine like terms. For example,

$$7\sqrt{11} + 6\sqrt{11} = (7 + 6)\sqrt{11} = 13\sqrt{11}.$$

7 square roots of 11 plus 6 square roots of 11 result in 13 square roots of 11.

EXAMPLE 1 Adding and Subtracting Like Radicals

Add or subtract as indicated:

a. $7\sqrt{2} + 5\sqrt{2}$ **b.** $\sqrt{5x} - 7\sqrt{5x}.$

Solution

a. $7\sqrt{2} + 5\sqrt{2} = (7 + 5)\sqrt{2}$ Apply the distributive property.

$= 12\sqrt{2}$ Simplify.

b. $\sqrt{5x} - 7\sqrt{5x} = 1\sqrt{5x} - 7\sqrt{5x}$ Write $\sqrt{5x}$ as $1\sqrt{5x}$.

$= (1 - 7)\sqrt{5x}$ Apply the distributive property.

$= -6\sqrt{5x}$ Simplify. ■

☑ **CHECK POINT 1** Add or subtract as indicated:

a. $8\sqrt{13} + 9\sqrt{13}$ **b.** $\sqrt{17x} - 20\sqrt{17x}.$

In some cases, radicals can be combined once they have been simplified. For example, to add $\sqrt{2}$ and $\sqrt{8}$, we can write $\sqrt{8}$ as $\sqrt{4 \cdot 2}$ because 4 is a perfect square factor of 8.

$$\sqrt{2} + \sqrt{8} = \sqrt{2} + \sqrt{4 \cdot 2} = 1\sqrt{2} + 2\sqrt{2} = (1 + 2)\sqrt{2} = 3\sqrt{2}$$

Always begin by simplifying radical terms. This makes it possible to identify and combine any like radicals.

EXAMPLE 2 Combining Radicals That First Require Simplification

Add or subtract as indicated:

a. $7\sqrt{3} + \sqrt{12}$ **b.** $4\sqrt{50x} - 6\sqrt{32x}.$

Solution

a. $7\sqrt{3} + \sqrt{12}$

$= 7\sqrt{3} + \sqrt{4 \cdot 3}$ Split 12 into two factors such that one is a perfect square.

$= 7\sqrt{3} + 2\sqrt{3}$ $\sqrt{4 \cdot 3} = \sqrt{4}\sqrt{3} = 2\sqrt{3}$

$= (7 + 2)\sqrt{3}$ Apply the distributive property. You will find that this step is usually done mentally.

$= 9\sqrt{3}$ Simplify.

b. $4\sqrt{50x} - 6\sqrt{32x}$

$= 4\sqrt{25 \cdot 2x} - 6\sqrt{16 \cdot 2x}$ *25 is the greatest perfect square factor of 50x and 16 is the greatest perfect square factor of 32x.*

$= 4 \cdot 5\sqrt{2x} - 6 \cdot 4\sqrt{2x}$ $\sqrt{25 \cdot 2x} = \sqrt{25}\sqrt{2x} = 5\sqrt{2x}$ *and* $\sqrt{16 \cdot 2x} = \sqrt{16}\sqrt{2x} = 4\sqrt{2x}.$

$= 20\sqrt{2x} - 24\sqrt{2x}$ *Multiply:* $4 \cdot 5 = 20$ *and* $6 \cdot 4 = 24.$

$= (20 - 24)\sqrt{2x}$ *Apply the distributive property.*

$= -4\sqrt{2x}$ *Simplify.* ∎

✓ **CHECK POINT 2**   Add or subtract as indicated:

a. $5\sqrt{27} + \sqrt{12}$ **b.** $6\sqrt{18x} - 4\sqrt{8x}.$

---

2 Multiply radical expressions with more than one term.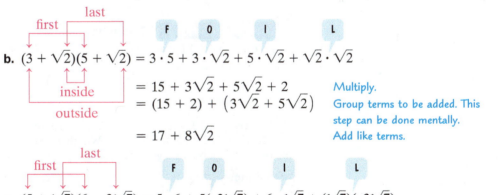

Multiplying Radical Expressions with More Than One Term

Radical expressions with more than one term are multiplied in much the same way that polynomials with more than one term are multiplied. Example 3 uses the distributive property and the FOIL method to perform multiplications.

EXAMPLE 3   Multiplying Radicals

Multiply:

a. $\sqrt{3}(\sqrt{7} + \sqrt{5})$ **b.** $(3 + \sqrt{2})(5 + \sqrt{2})$ **c.** $(5 + \sqrt{7})(6 - 3\sqrt{7}).$

Solution

a. $\sqrt{3}(\sqrt{7} + \sqrt{5}) = \sqrt{3} \cdot \sqrt{7} + \sqrt{3} \cdot \sqrt{5}$ *Use the distributive property.*

$\phantom{\sqrt{3}(\sqrt{7} + \sqrt{5})} = \sqrt{21} + \sqrt{15}$ *Use the product rule:* $\sqrt{a}\sqrt{b} = \sqrt{ab}.$

b. $(3 + \sqrt{2})(5 + \sqrt{2}) = 3 \cdot 5 + 3 \cdot \sqrt{2} + 5 \cdot \sqrt{2} + \sqrt{2} \cdot \sqrt{2}$

$\phantom{(3 + \sqrt{2})(5 + \sqrt{2})} = 15 + 3\sqrt{2} + 5\sqrt{2} + 2$ *Multiply.*

$\phantom{(3 + \sqrt{2})(5 + \sqrt{2})} = (15 + 2) + (3\sqrt{2} + 5\sqrt{2})$ *Group terms to be added. This step can be done mentally.*

$\phantom{(3 + \sqrt{2})(5 + \sqrt{2})} = 17 + 8\sqrt{2}$ *Add like terms.*

c. $(5 + \sqrt{7})(6 - 3\sqrt{7}) = 5 \cdot 6 + 5(-3\sqrt{7}) + 6 \cdot \sqrt{7} + (\sqrt{7})(-3\sqrt{7})$

$\phantom{(5 + \sqrt{7})(6 - 3\sqrt{7})} = 30 - 15\sqrt{7} + 6\sqrt{7} - 21$ *Multiply. For L, or Last terms:* $(\sqrt{7})(-3\sqrt{7}) = -3\sqrt{49} = -3 \cdot 7 = -21.$

$\phantom{(5 + \sqrt{7})(6 - 3\sqrt{7})} = (30 - 21) + (-15\sqrt{7} + 6\sqrt{7})$ *Group like terms.*

$\phantom{(5 + \sqrt{7})(6 - 3\sqrt{7})} = 9 - 9\sqrt{7}$ *Combine like terms.* ∎

✓ **CHECK POINT 3**   Multiply:

a. $\sqrt{2}(\sqrt{5} + \sqrt{11})$

b. $(4 + \sqrt{3})(2 + \sqrt{3})$

c. $(3 + \sqrt{5})(8 - 4\sqrt{5}).$

3 Multiply conjugates. ▶

Radical expressions that involve the sum and difference of the same two terms are called **conjugates**. For example,

$$5 + \sqrt{2} \quad \text{and} \quad 5 - \sqrt{2}$$

are conjugates of each other. Although you can multiply these conjugates using the FOIL method, there is an even faster way. Use the special-product formula for the sum and difference of the same two terms:

$$(A + B)(A - B) = A^2 - B^2.$$

Thus,

$$(A + B)(A - B) = A^2 - B^2$$

$$(5 + \sqrt{2})(5 - \sqrt{2}) = 5^2 - (\sqrt{2})^2 = 25 - 2 = 23.$$

EXAMPLE 4 Multiplying Conjugates

Multiply:

a. $\left(2 + \sqrt{7}\right)\left(2 - \sqrt{7}\right)$ **b.** $\left(\sqrt{5} - \sqrt{3}\right)\left(\sqrt{5} + \sqrt{3}\right).$

Solution Use the special-product formula

$$(A + B)(A - B) = A^2 - B^2.$$

$$\boxed{\text{First term squared}} - \boxed{\text{second term squared}} = \boxed{\text{product}}$$

a. $(2 + \sqrt{7})(2 - \sqrt{7}) = 2^2 - (\sqrt{7})^2 = 4 - 7 = -3$

b. $\left(\sqrt{5} - \sqrt{3}\right)\left(\sqrt{5} + \sqrt{3}\right) = \left(\sqrt{5}\right)^2 - \left(\sqrt{3}\right)^2 = 5 - 3 = 2$ ∎

In the next section, we will use conjugates to simplify quotients.

✓ CHECK POINT 4 Multiply:

a. $\left(3 + \sqrt{11}\right)\left(3 - \sqrt{11}\right)$ **b.** $\left(\sqrt{7} - \sqrt{2}\right)\left(\sqrt{7} + \sqrt{2}\right).$

Achieving Success

In Exercise Set 8.3, you will be working with a model that describes the percentage of full-time college students with jobs. Whether or not you're part of the nearly 50% of students with jobs, it's a good idea to **analyze the effects of working while in college**.

Advantages

- General and career-specific experience
- Developing skills that can be useful after graduation
- Contacts who might be helpful in the future
- Time commitment that requires effective time management
- Extra money to spend and invest
- Confidence builder

Disadvantages

- Time commitment that reduces available study time
- Reduced opportunity for social and extracurricular activities
- Mentally shifting gears from work to classroom

CONCEPT AND VOCABULARY CHECK

Fill in each blank so that the resulting statement is true.

1. Two or more square roots that have the same radicands are called _____.

2. $8\sqrt{3} + 10\sqrt{3} = (_ + _)\sqrt{3} = $ _____

3. $7\sqrt{5} + \sqrt{5} = 7\sqrt{5} + _\sqrt{5} = (_ + _)\sqrt{5} = $ _____

4. $\sqrt{50} + \sqrt{32} = \sqrt{25 \cdot 2} + \sqrt{16 \cdot 2} = _\sqrt{2} + _\sqrt{2} = $ _____

5. Consider the following multiplication problem:

$$(7 + \sqrt{2})(3 + \sqrt{2}).$$

Using the FOIL method, the product of the first terms is _____, the product of the outside terms is _____, the product of the inside terms is _____, and the product of the last terms is _____.

6. The conjugate of $8 + \sqrt{6}$ is _____.

7. $(6 + \sqrt{5})(6 - \sqrt{5}) = _ - (_)^2 = _ - _ = $ _____

EXERCISE SET 8.3   ▶ MyMathLab®

Practice Exercises

In Exercises 1–22, add or subtract as indicated. If terms are not like radicals and cannot be combined, so state.

1. $8\sqrt{3} + 5\sqrt{3}$
2. $9\sqrt{5} + 6\sqrt{5}$
3. $17\sqrt{6} - 2\sqrt{6}$
4. $19\sqrt{7} - 2\sqrt{7}$
5. $3\sqrt{13} - 8\sqrt{13}$
6. $5\sqrt{17} - 11\sqrt{17}$
7. $12\sqrt{x} + 3\sqrt{x}$
8. $9\sqrt{x} + 2\sqrt{x}$
9. $70\sqrt{y} - 76\sqrt{y}$
10. $8\sqrt{y} - 28\sqrt{y}$
11. $7\sqrt{10x} + 2\sqrt{10x}$
12. $4\sqrt{6x} + 3\sqrt{6x}$
13. $7\sqrt{5y} - \sqrt{5y}$
14. $8\sqrt{3y} - \sqrt{3y}$
15. $\sqrt{5} + \sqrt{5}$
16. $\sqrt{3} + \sqrt{3}$
17. $4\sqrt{2} + 3\sqrt{2} + 5\sqrt{2}$
18. $6\sqrt{3} + 2\sqrt{3} + 5\sqrt{3}$
19. $4\sqrt{7} - 5\sqrt{7} + 8\sqrt{7}$
20. $5\sqrt{7} - 6\sqrt{7} + 10\sqrt{7}$
21. $4\sqrt{11} - 6\sqrt{11} + 2\sqrt{11}$
22. $7\sqrt{17} - 10\sqrt{17} + 3\sqrt{17}$

In Exercises 23–42, add or subtract as indicated. You will need to simplify terms before they can be combined. If terms cannot be simplified so that they can be combined, so state.

23. $\sqrt{5} + \sqrt{20}$
24. $\sqrt{3} + \sqrt{27}$
25. $\sqrt{8} - \sqrt{2}$
26. $\sqrt{50} - \sqrt{2}$
27. $\sqrt{50} + \sqrt{18}$
28. $\sqrt{28} + \sqrt{63}$

29. $7\sqrt{12} + \sqrt{75}$
30. $5\sqrt{12} + \sqrt{75}$
31. $3\sqrt{27} - 2\sqrt{18}$
32. $5\sqrt{27} - 3\sqrt{18}$
33. $2\sqrt{45x} - 2\sqrt{20x}$
34. $2\sqrt{50x} - 2\sqrt{18x}$
35. $\sqrt{8} + \sqrt{16} + \sqrt{18} + \sqrt{25}$
36. $\sqrt{6} + \sqrt{9} + \sqrt{24} + \sqrt{25}$
37. $\sqrt{2} + \sqrt{11}$
38. $\sqrt{3} + \sqrt{13}$
39. $2\sqrt{80} + 3\sqrt{75}$
40. $2\sqrt{75} + 3\sqrt{125}$
41. $3\sqrt{54} - 2\sqrt{20} + 4\sqrt{45} - \sqrt{24}$
42. $4\sqrt{8} - \sqrt{128} + 2\sqrt{48} + 3\sqrt{18}$

In Exercises 43–78, multiply as indicated. If possible, simplify any square roots that appear in the product.

43. $\sqrt{2}(\sqrt{3} + \sqrt{5})$
44. $\sqrt{5}(\sqrt{3} + \sqrt{6})$
45. $\sqrt{7}(\sqrt{6} - \sqrt{10})$
46. $\sqrt{7}(\sqrt{5} - \sqrt{11})$
47. $\sqrt{3}(5 + \sqrt{3})$
48. $\sqrt{6}(7 + \sqrt{6})$
49. $\sqrt{3}(\sqrt{6} - \sqrt{3})$
50. $\sqrt{6}(\sqrt{6} - \sqrt{2})$

51. $(5 + \sqrt{2})(6 + \sqrt{2})$

52. $(7 + \sqrt{2})(8 + \sqrt{2})$

53. $(4 + \sqrt{5})(10 - 3\sqrt{5})$

54. $(6 + \sqrt{5})(9 - 4\sqrt{5})$

55. $(6 - 3\sqrt{7})(2 - 5\sqrt{7})$

56. $(7 - 2\sqrt{7})(5 - 3\sqrt{7})$

57. $(\sqrt{10} - 3)(\sqrt{10} - 5)$

58. $(\sqrt{10} - 4)(\sqrt{10} - 6)$

59. $(\sqrt{3} + \sqrt{6})(\sqrt{3} + 2\sqrt{6})$

60. $(\sqrt{6} + \sqrt{3})(\sqrt{6} + 5\sqrt{3})$

61. $(\sqrt{2} + 1)(\sqrt{3} - 6)$

62. $(\sqrt{5} + 3)(\sqrt{2} - 8)$

63. $(3 + \sqrt{5})(3 - \sqrt{5})$

64. $(4 + \sqrt{7})(4 - \sqrt{7})$

65. $(1 - \sqrt{6})(1 + \sqrt{6})$

66. $(1 - \sqrt{5})(1 + \sqrt{5})$

67. $(\sqrt{11} + 5)(\sqrt{11} - 5)$

68. $(\sqrt{11} + 6)(\sqrt{11} - 6)$

69. $(\sqrt{7} - \sqrt{5})(\sqrt{7} + \sqrt{5})$

70. $(\sqrt{10} - \sqrt{7})(\sqrt{10} + \sqrt{7})$

71. $(2\sqrt{3} + 7)(2\sqrt{3} - 7)$

72. $(5\sqrt{3} + 6)(5\sqrt{3} - 6)$

73. $(2\sqrt{3} + \sqrt{5})(2\sqrt{3} - \sqrt{5})$

74. $(4\sqrt{5} + \sqrt{2})(4\sqrt{5} - \sqrt{2})$

75. $(\sqrt{2} + \sqrt{3})^2$

76. $(\sqrt{3} + \sqrt{5})^2$

77. $(\sqrt{x} - \sqrt{10})^2$

78. $(\sqrt{x} - \sqrt{11})^2$

Practice PLUS

In Exercises 79–86, add or subtract as indicated. You will need to simplify terms before they can be combined. Assume all variables represent nonnegative real numbers.

79. $5\sqrt{27x^3} - 3x\sqrt{12x}$

80. $7\sqrt{32x^3} - 3x\sqrt{50x}$

81. $6y^2\sqrt{x^5y} + 2x^2\sqrt{xy^5}$

82. $9y^2\sqrt{x^5y} + x^2\sqrt{xy^5}$

83. $3\sqrt[3]{54} - 4\sqrt[3]{16}$

84. $7\sqrt[3]{24} - 5\sqrt[3]{81}$

85. $x\sqrt[3]{32x} + 9\sqrt[3]{4x^4}$ Hint: $\sqrt[3]{x^3} = x$

86. $x\sqrt[3]{48x} + 11\sqrt[3]{6x^4}$ Hint: $\sqrt[3]{x^3} = x$

Application Exercises

In Exercises 87–92, write expressions for the perimeter and area of each figure. Then simplify these expressions. Assume that all measures are given in inches.

87.

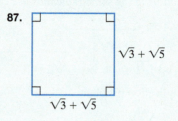

$\sqrt{3} + \sqrt{5}$

$\sqrt{3} + \sqrt{5}$

88.

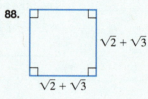

$\sqrt{2} + \sqrt{3}$

$\sqrt{2} + \sqrt{3}$

89.

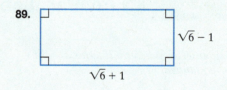

$\sqrt{6} - 1$

$\sqrt{6} + 1$

90.

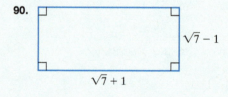

$\sqrt{7} - 1$

$\sqrt{7} + 1$

91.

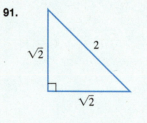

$\sqrt{2}$ 2

$\sqrt{2}$

92.

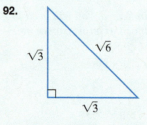

$\sqrt{3}$ $\sqrt{6}$

$\sqrt{3}$

There is a formula for adding $\sqrt{a}$ and $\sqrt{b}$. The formula is $\sqrt{a} + \sqrt{b} = \sqrt{(a + b) + 2\sqrt{ab}}$. Use this formula to add the radicals in Exercises 93–94. Then work the problem again by the methods discussed in this section. Which method do you prefer? Why?

93. $\sqrt{2} + \sqrt{8}$

94. $\sqrt{5} + \sqrt{20}$

The bar graph shows the percentage of full-time college students in the United States who had jobs for five selected years. The data can be described by the mathematical model

$$J = 1.4\sqrt{x} + 55 - (20 - 1.2\sqrt{x}),$$

where J is the percentage of full-time college students with jobs x years after 1975. Use this information to solve Exercises 95–96.

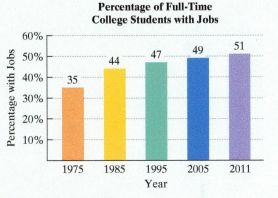

Source: Department of Education

95. a. Simplify the mathematical model for J.

b. Use the simplified form of the model to find the percentage of full-time college students who had jobs in 2011. Round to the nearest percent. How does this compare with the actual percentage displayed by the bar graph?

96. a. Simplify the mathematical model for J.

b. Use the simplified form of the model to find the percentage of full-time college students who had jobs in 2005. Round to the nearest percent. How does this compare with the actual percentage displayed by the bar graph?

The graph of the model for the percentage of full-time college students with jobs is shown in a rectangular coordinate system. Use the graph to solve Exercises 97–98.

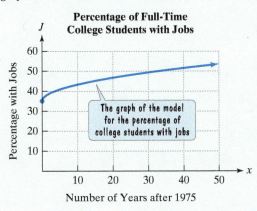

97. a. Identify your solution from Exercise 95(b) as a point on the graph.

b. Use the graph to estimate the percentage of full-time college students who had jobs in 2015.

98. a. Identify your solution from Exercise 96(b) as a point on the graph.

b. Use the graph to estimate the percentage of full-time college students who will have jobs in 2025.

Explaining the Concepts

99. What are like radicals? Give an example with your explanation.

100. Explain how to add like radicals. Give an example with your explanation.

101. If only like radicals can be added, why is it possible to add $\sqrt{2}$ and $\sqrt{8}$?

102. Explain how to perform this multiplication:
$$\sqrt{2}(\sqrt{7} + \sqrt{10}).$$

103. Explain how to perform this multiplication:
$$(2 + \sqrt{3})(4 + \sqrt{3}).$$

104. What are conjugates? Give an example with your explanation.

105. Describe how to multiply conjugates.

Critical Thinking Exercises

Make Sense? *In Exercises 106–109, determine whether each statement makes sense or does not make sense, and explain your reasoning.*

106. The unlike radicals $5\sqrt{2}$ and $4\sqrt{3}$ remind me of the unlike terms $5x$ and $4y$ that cannot be combined by addition or subtraction.

107. I simplified the terms of $2\sqrt{20} + 4\sqrt{75}$, and then I was able to identify and add the like radicals.

108. I use the same ideas to find $(3 + \sqrt{2})(5 + \sqrt{2})$ that I did to find the binomial product $(3 + x)(5 + x)$.

109. In some cases when I multiply a radical expression and its conjugate, the simplified product contains a radical.

In Exercises 110–113, determine whether each statement is true or false. If the statement is false, make the necessary change(s) to produce a true statement.

110. $\sqrt{2} + \sqrt{8} = \sqrt{10}$

111. $(\sqrt{5} + \sqrt{3})^2 = 5 + 3$

112. $4\sqrt{3} + 5\sqrt{3} = 9\sqrt{6}$

113. $(\sqrt{7} + \sqrt{3})(\sqrt{7} - \sqrt{3}) = 40$

114. Simplify: $\sqrt{5} \cdot \sqrt{15} + 6\sqrt{3}$.

115. Multiply: $(\sqrt[3]{4} + 1)(\sqrt[3]{2} - 3)$.

116. Fill in the boxes to make the statement true:
$(5 + \sqrt{\blacksquare})(5 - \sqrt{\blacksquare}) = 22$.

117. Multiply: $(4\sqrt{3x} + \sqrt{2y})(4\sqrt{3x} - \sqrt{2y})$.

Technology Exercises

In Exercises 118–121, determine if each operation is performed correctly by graphing each side of the equation with your graphing utility. Use the given viewing rectangle. The graphs should be the same. If they are not, correct the right side of the equation and then use your graphing utility to verify the correction.

118. $\sqrt{4x} + \sqrt{9x} = 5\sqrt{x}$

[0, 5, 1] by [0, 10, 1]

119. $\sqrt{16x} - \sqrt{9x} = \sqrt{7x}$

[0, 5, 1] by [0, 5, 1]

120. $(\sqrt{x} - 1)(\sqrt{x} - 1) = x + 1$

[0, 5, 1] by [-1, 2, 1]

121. $(\sqrt{x} + 2)(\sqrt{x} - 2) = x^2 - 4$

[-10, 10, 1] by [-10, 10, 1]

Review Exercises

122. Multiply: $(5x + 3)(5x - 3)$.
(Section 5.3, Example 4)

123. Factor completely: $64x^3 - x$.
(Section 6.5, Example 3)

124. Graph: $y = -\dfrac{1}{4}x + 3$.
(Section 3.4, Example 3)

Preview Exercises

Exercises 125–127 will help you prepare for the material covered in the next section.

125. Multiply and simplify: $\dfrac{\sqrt{3}}{\sqrt{5}} \cdot \dfrac{\sqrt{5}}{\sqrt{5}}$.

126. Multiply: $(5 + \sqrt{3})(5 - \sqrt{3})$.

127. Multiply: $(\sqrt{6} - \sqrt{2})(\sqrt{6} + \sqrt{2})$.

MID-CHAPTER CHECK POINT   Section 8.1–Section 8.3

✓ **What You Know:** We learned to find roots of numbers. We saw that if an index is even, the radicand must be nonnegative for a root to be a real number. We learned to simplify radicals. We performed various operations with radicals including multiplication, division, addition, and subtraction.

In Exercises 1–20, simplify the given expression or perform the indicated operation and, if possible, simplify. Assume that all variables represent positive real numbers.

1. $\sqrt{50} \cdot \sqrt{6}$

2. $\sqrt{6} + 9\sqrt{6}$

3. $\sqrt{96x^3}$

4. $\sqrt[5]{\dfrac{4}{32}}$

5. $\sqrt{27} + 3\sqrt{12}$

6. $(\sqrt{10} + \sqrt{3})(\sqrt{10} - \sqrt{3})$

7. $\dfrac{\sqrt{5}}{2}(4\sqrt{3} - 6\sqrt{20})$

8. $-\sqrt{32x^{21}}$

9. $\sqrt{6x^3} \cdot \sqrt{2x^4}$

10. $\dfrac{\sqrt[3]{32}}{\sqrt[3]{2}}$

11. $-3\sqrt{90} - 5\sqrt{40}$

12. $(2 - \sqrt{3})(5 + 2\sqrt{3})$

13. $\dfrac{\sqrt{56x^5}}{\sqrt{7x^3}}$

14. $-\sqrt[4]{32}$

15. $(\sqrt{2} + \sqrt{7})^2$

16. $\sqrt[3]{\dfrac{1}{2}} \cdot \sqrt[3]{32}$

17. $\sqrt{5} + \sqrt{20} + \sqrt{45}$

18. $\dfrac{1}{3}\sqrt{\dfrac{90}{16}}$

19. $3\sqrt{2}(\sqrt{2} + \sqrt{5})$

20. $(5 - \sqrt{2})(5 + \sqrt{2})$

8.4

What am I supposed to learn?

After studying this section, you should be able to:

1 Rationalize denominators containing one term.

2 Rationalize denominators containing two terms.

1 Rationalize denominators containing one term.

```
1/√3
              .5773502692
√3/3
              .5773502692
```

Figure 8.6 The calculator screen shows approximate values for $\frac{1}{\sqrt{3}}$ and $\frac{\sqrt{3}}{3}$.

Rationalizing the Denominator

Some cartoonists use their comic art to humanize complicated mathematical topics. In this *Peanuts* cartoon, Woodstock moves from $\frac{7\sqrt{2}}{\sqrt{6}}$, an expression with a radical in its denominator, to $\frac{7}{3}\sqrt{3}$, a number without a radical in the denominator.

The technique, called *rationalizing the denominator*, is discussed in this section. We return to *Peanuts* in the Exercise Set (Exercise 83), as well as presenting a general discussion of comic art and algebra in the Blitzer Bonus on pages 598–599.

PEANUTS © United Feature Syndicate, Inc.

Rationalizing Denominators Containing One Term

A calculator makes it fairly easy to find an approximation for divisions involving radicals, such as $\frac{1}{\sqrt{3}}$. However, it is sometimes easier to work with radical expressions if the denominators do not contain any radicals.

The calculator screen in **Figure 8.6** shows the approximate values for $\frac{1}{\sqrt{3}}$ and $\frac{\sqrt{3}}{3}$. The two approximations are the same. This is not a coincidence:

$$\frac{1}{\sqrt{3}} = \frac{1}{\sqrt{3}} \cdot \boxed{\frac{\sqrt{3}}{\sqrt{3}}} = \frac{\sqrt{3}}{\sqrt{9}} = \frac{\sqrt{3}}{3}.$$

Any nonzero number divided by itself is 1. Multiplication by 1 does not change the value of $\frac{1}{\sqrt{3}}$.

This process involves rewriting a radical expression as an equivalent expression in which the denominator no longer contains any radicals. The process is called **rationalizing the denominator**. If the denominator contains the square root of a natural number that is not a perfect square, **multiply the numerator and the denominator by the smallest number that produces the square root of a perfect square in the denominator**.

EXAMPLE 1 Rationalizing Denominators

Rationalize the denominator:

a. $\dfrac{15}{\sqrt{6}}$ **b.** $\sqrt{\dfrac{3}{5}}$.

Great Question!

What exactly does rationalizing a denominator do to an irrational number in the denominator?

Rationalizing a numerical denominator makes that denominator a rational number.

Solution

a. If we multiply the numerator and denominator of $\dfrac{15}{\sqrt{6}}$ by $\sqrt{6}$, the denominator becomes $\sqrt{6} \cdot \sqrt{6} = \sqrt{36} = 6$. Therefore, we multiply by 1, choosing $\dfrac{\sqrt{6}}{\sqrt{6}}$ for 1.

$$\dfrac{15}{\sqrt{6}} = \dfrac{15}{\sqrt{6}} \cdot \dfrac{\sqrt{6}}{\sqrt{6}}$$
Multiply the numerator and denominator by $\sqrt{6}$ to remove the square root in the denominator.

$$= \dfrac{15\sqrt{6}}{6}$$
$\sqrt{6} \cdot \sqrt{6} = \sqrt{36} = 6$

$$= \dfrac{5\sqrt{6}}{2}$$
Simplify, dividing the numerator and denominator by 3.

b. $\sqrt{\dfrac{3}{5}} = \dfrac{\sqrt{3}}{\sqrt{5}}$
The square root of a quotient is the quotient of the square roots.

$$= \dfrac{\sqrt{3}}{\sqrt{5}} \cdot \dfrac{\sqrt{5}}{\sqrt{5}}$$
Because $\sqrt{5}$ is the smallest factor that will produce a perfect square in the denominator, multiply by 1, choosing $\dfrac{\sqrt{5}}{\sqrt{5}}$ for 1.

$$= \dfrac{\sqrt{15}}{5}$$
$\sqrt{5} \cdot \sqrt{5} = \sqrt{25} = 5$ ■

☑ **CHECK POINT 1** Rationalize the denominator:

a. $\dfrac{25}{\sqrt{10}}$ **b.** $\sqrt{\dfrac{2}{7}}.$

It is a good idea to simplify a radical expression before attempting to rationalize the denominator.

EXAMPLE 2 Simplifying and Then Rationalizing Denominators

Rationalize the denominator:

a. $\dfrac{12}{\sqrt{8}}$ **b.** $\sqrt{\dfrac{7x}{75}}.$

Solution

a. We begin by simplifying $\sqrt{8}$.

$$\dfrac{12}{\sqrt{8}} = \dfrac{12}{\sqrt{4 \cdot 2}}$$
4 is the greatest perfect square factor of 8.

$$= \dfrac{12}{2\sqrt{2}}$$
$\sqrt{4 \cdot 2} = \sqrt{4}\sqrt{2} = 2\sqrt{2}$

$$= \dfrac{6}{\sqrt{2}}$$
Simplify, dividing the numerator and denominator by 2.

Discover for Yourself

Rationalize the denominator in Example 2(a) without first simplifying. Multiply the numerator and the denominator by $\sqrt{8}$. Do this again by multiplying by $\dfrac{\sqrt{2}}{\sqrt{2}}$.

Which method do you prefer?

$$= \dfrac{6}{\sqrt{2}} \cdot \dfrac{\sqrt{2}}{\sqrt{2}}$$
Rationalize the denominator.

$$= \dfrac{6\sqrt{2}}{2}$$
$\sqrt{2} \cdot \sqrt{2} = \sqrt{4} = 2$

$$= 3\sqrt{2}$$
Simplify.

Great Question!

Can I simplify

$$\frac{\sqrt{21x}}{15}$$

by dividing the numerator and the denominator by 3?

No. Although 3 is a factor of 15, it is *not* a factor of $\sqrt{21x}$. (The factor is $\sqrt{3}$.) Thus, the answer in Example 2(b) *cannot* be written as

$$\frac{\sqrt{7x}}{5}.$$

b.

$$\sqrt{\frac{7x}{75}} = \frac{\sqrt{7x}}{\sqrt{75}}$$

The square root of a quotient is the quotient of the square roots.

$$= \frac{\sqrt{7x}}{\sqrt{25 \cdot 3}}$$

Simplify the denominator. 25 is the greatest perfect square factor of 75.

$$= \frac{\sqrt{7x}}{5\sqrt{3}}$$

$$\sqrt{25 \cdot 3} = \sqrt{25}\sqrt{3} = 5\sqrt{3}$$

$$= \frac{\sqrt{7x}}{5\sqrt{3}} \cdot \frac{\sqrt{3}}{\sqrt{3}}$$

Rationalize the denominator, choosing $\dfrac{\sqrt{3}}{\sqrt{3}}$ for 1.

$$= \frac{\sqrt{21x}}{5 \cdot 3}$$

$$\sqrt{3} \cdot \sqrt{3} = \sqrt{9} = 3$$

$$= \frac{\sqrt{21x}}{15} \quad \blacksquare$$

✓ **CHECK POINT 2**   Rationalize the denominator:

a. $\dfrac{15}{\sqrt{18}}$ **b.** $\sqrt{\dfrac{7x}{20}}.$

---

2 Rationalize denominators containing two terms.

Rationalizing Denominators Containing Two Terms

How can we rationalize a denominator if the denominator contains two terms with one or more square roots? **Multiply the numerator and the denominator by the conjugate of the denominator.** Here are two examples of such expressions:

$$\frac{7}{5 + \sqrt{3}} \qquad \frac{2}{\sqrt{6} - \sqrt{2}}.$$

The conjugate of the denominator is $5 - \sqrt{3}$.

The conjugate of the denominator is $\sqrt{6} + \sqrt{2}$.

The product of the denominator and its conjugate is found using the formula

$$(A + B)(A - B) = A^2 - B^2.$$

The simplified product will not contain a radical.

EXAMPLE 3   Rationalizing a Denominator Containing Two Terms

Rationalize the denominator: $\dfrac{7}{5 + \sqrt{3}}.$

Solution   The conjugate of the denominator is $5 - \sqrt{3}$. If we multiply the numerator and the denominator by $5 - \sqrt{3}$, the denominator will not contain a radical. Therefore, we multiply by 1, choosing $\dfrac{5 - \sqrt{3}}{5 - \sqrt{3}}$ for 1.

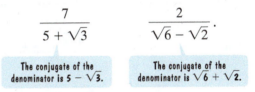

$$\frac{7}{5 + \sqrt{3}} = \frac{7}{5 + \sqrt{3}} \cdot \frac{5 - \sqrt{3}}{5 - \sqrt{3}}$$

Multiply by 1.

$$= \frac{7(5 - \sqrt{3})}{5^2 - (\sqrt{3})^2}$$

$$(A + B)(A - B) = A^2 - B^2$$

$$= \frac{7(5 - \sqrt{3})}{25 - 3}$$

$$(\sqrt{3})^2 = \sqrt{3} \cdot \sqrt{3} = \sqrt{9} = 3$$

$$= \frac{7(5 - \sqrt{3})}{22} \quad \text{or} \quad \frac{35 - 7\sqrt{3}}{22} \quad \blacksquare$$

☑ **CHECK POINT 3** Rationalize the denominator: $\dfrac{8}{4 + \sqrt{5}}$.

EXAMPLE 4 Rationalizing a Denominator Containing Two Terms

Rationalize the denominator: $\dfrac{2}{\sqrt{6} - \sqrt{2}}$.

Solution The conjugate of the denominator is $\sqrt{6} + \sqrt{2}$. If we multiply the numerator and the denominator by $\sqrt{6} + \sqrt{2}$, the denominator will not contain a radical. Therefore, we multiply by 1, choosing $\dfrac{\sqrt{6} + \sqrt{2}}{\sqrt{6} + \sqrt{2}}$ for 1.

$$\frac{2}{\sqrt{6} - \sqrt{2}} = \frac{2}{\sqrt{6} - \sqrt{2}} \cdot \frac{\sqrt{6} + \sqrt{2}}{\sqrt{6} + \sqrt{2}} \qquad \text{Multiply by 1.}$$

$$= \frac{2(\sqrt{6} + \sqrt{2})}{(\sqrt{6})^2 - (\sqrt{2})^2} \qquad (A - B)(A + B) = A^2 - B^2$$

$$= \frac{2(\sqrt{6} + \sqrt{2})}{6 - 2} \qquad \begin{array}{l} (\sqrt{6})^2 = \sqrt{6} \cdot \sqrt{6} = \sqrt{36} = 6 \\ \text{and } (\sqrt{2})^2 = \sqrt{2} \cdot \sqrt{2} = \sqrt{4} = 2 \end{array}$$

$$= \frac{2(\sqrt{6} + \sqrt{2})}{4}$$

$$= \frac{\overset{1}{2}(\sqrt{6} + \sqrt{2})}{\underset{2}{4}} \qquad \begin{array}{l} \text{Divide the numerator and denominator by} \\ \text{the common factor, 2.} \end{array}$$

$$= \frac{\sqrt{6} + \sqrt{2}}{2} \qquad \blacksquare$$

☑ **CHECK POINT 4** Rationalize the denominator: $\dfrac{8}{\sqrt{7} - \sqrt{3}}$.

Blitzer Bonus ⊙

Comic Art and Algebra

For an unusual presentation of introductory algebra using graphics and humor, read *The Cartoon Guide to Algebra* (HarperCollins, 2015) by cartoonist and Harvard-educated mathematician Larry Gonick. (The material on roots and radicals discussed in this chapter can be found on pages 181–192 of the comical guide.) Do you find Gonick's attempt to teach algebra, using cartoon characters talking, commenting, and joking, helpful or distracting?

If you enjoy the use of comic art to humanize the world of algebra, check out *Super Graphic* (Chronicle Books, San Francisco) by Tim Leong. The author uses algebraic and statistical graphs to present a guide to the comic book universe. You haven't fully studied comics until you've seen Bruce Wayne's (*Batman*) net worth graphed over time as a linear equation or Archie Andrews' (*Archie*) acquaintances represented as points in rectangular coordinates (shown, in part, in **Figure 8.7** on the next page). *Super Graphic's* compendium illustrating comic-book-world factoids will give infographic lovers much to enjoy.

Comic Art and Algebra (continued)

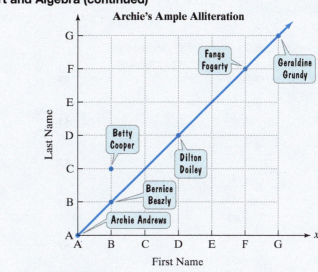

Figure 8.7

Source: Super Graphic

The comic-book character Archie Andrews is known for being in the middle of a love triangle between Betty, the girl next door, and Veronica, the one-percenter. Archie's acquaintances are vast, with names that are alliterations (two or more words starting with the same letter).

CONCEPT AND VOCABULARY CHECK

Fill in each blank so that the resulting statement is true.

1. The process of rewriting a radical expression as an equivalent expression in which the denominator no longer contains any radicals is called _____.

2. The number $\sqrt{\dfrac{2}{7}}$ can be rewritten without a radical in the denominator by multiplying the numerator and denominator by _____.

3. The number $\dfrac{5}{\sqrt{12}}$ can be rewritten without a radical in the denominator by first simplifying $\sqrt{12}$ to _____ and then multiplying the numerator and denominator by _____.

4. The number $\dfrac{4}{7 + \sqrt{5}}$ can be rewritten without a radical in the denominator by multiplying the numerator and denominator by _____.

5. The number $\dfrac{15}{\sqrt{11} - \sqrt{3}}$ can be rewritten without a radical in the denominator by multiplying the numerator and denominator by _____.

EXERCISE SET 8.4 ▶ MyMathLab®

Practice Exercises

In Exercises 1–24, rationalize each denominator. If possible, simplify the rationalized expression by dividing the numerator and denominator by the greatest common factor.

1. $\dfrac{1}{\sqrt{10}}$

2. $\dfrac{1}{\sqrt{2}}$

3. $\dfrac{5}{\sqrt{5}}$

4. $\dfrac{7}{\sqrt{7}}$

5. $\dfrac{2}{\sqrt{6}}$

6. $\dfrac{4}{\sqrt{6}}$

7. $\dfrac{28}{\sqrt{7}}$

8. $\dfrac{40}{\sqrt{5}}$

9. $\sqrt{\dfrac{3}{5}}$

10. $\sqrt{\dfrac{3}{7}}$

11. $\sqrt{\dfrac{7}{3}}$

12. $\sqrt{\dfrac{5}{2}}$

13. $\sqrt{\dfrac{x^2}{3}}$

14. $\sqrt{\dfrac{x^2}{7}}$

15. $\sqrt{\dfrac{11}{x}}$

16. $\sqrt{\dfrac{6}{x}}$

17. $\sqrt{\dfrac{x}{y}}$

18. $\sqrt{\dfrac{a}{b}}$

19. $\sqrt{\dfrac{x^4}{2}}$

20. $\sqrt{\dfrac{x^4}{3}}$

21. $\dfrac{\sqrt{7}}{\sqrt{5}}$

22. $\dfrac{\sqrt{11}}{\sqrt{5}}$

23. $\dfrac{\sqrt{3x}}{\sqrt{14}}$

24. $\dfrac{\sqrt{2x}}{\sqrt{17}}$

In Exercises 25–52, begin by simplifying the expression. Then rationalize the denominator using the simplified expression.

25. $\dfrac{1}{\sqrt{20}}$

26. $\dfrac{1}{\sqrt{18}}$

27. $\dfrac{12}{\sqrt{32}}$

28. $\dfrac{15}{\sqrt{50}}$

29. $\dfrac{15}{\sqrt{12}}$

30. $\dfrac{13}{\sqrt{40}}$

31. $\sqrt{\dfrac{5}{18}}$

32. $\sqrt{\dfrac{7}{12}}$

33. $\sqrt{\dfrac{x}{32}}$

34. $\sqrt{\dfrac{x}{40}}$

35. $\sqrt{\dfrac{1}{45}}$

36. $\sqrt{\dfrac{1}{54}}$

37. $\dfrac{\sqrt{7}}{\sqrt{12}}$

38. $\dfrac{\sqrt{5}}{\sqrt{18}}$

39. $\dfrac{8x}{\sqrt{8}}$

40. $\dfrac{27x}{\sqrt{27}}$

41. $\dfrac{\sqrt{7y}}{\sqrt{8}}$

42. $\dfrac{\sqrt{3y}}{\sqrt{125}}$

43. $\sqrt{\dfrac{7x}{12}}$

44. $\sqrt{\dfrac{11x}{18}}$

45. $\sqrt{\dfrac{45}{x}}$

46. $\sqrt{\dfrac{27}{x}}$

47. $\dfrac{5}{\sqrt{x^3}}$

48. $\dfrac{11}{\sqrt{x^3}}$

49. $\sqrt{\dfrac{27}{y^3}}$

50. $\sqrt{\dfrac{45}{y^3}}$

51. $\dfrac{\sqrt{50x^2}}{\sqrt{12y^3}}$

52. $\dfrac{\sqrt{27x^2}}{\sqrt{12y^3}}$

In Exercises 53–74, rationalize each denominator. Simplify, if possible.

53. $\dfrac{1}{4+\sqrt{3}}$

54. $\dfrac{1}{5+\sqrt{2}}$

55. $\dfrac{9}{2-\sqrt{7}}$

56. $\dfrac{12}{2-\sqrt{7}}$

57. $\dfrac{16}{\sqrt{11}+3}$

58. $\dfrac{15}{\sqrt{7}+2}$

59. $\dfrac{18}{3-\sqrt{3}}$

60. $\dfrac{40}{5-\sqrt{5}}$

61. $\dfrac{\sqrt{2}}{\sqrt{2}+1}$

62. $\dfrac{\sqrt{3}}{\sqrt{3}+1}$

63. $\dfrac{\sqrt{10}}{\sqrt{10}-\sqrt{7}}$

64. $\dfrac{\sqrt{5}}{\sqrt{5}-\sqrt{3}}$

65. $\dfrac{6}{\sqrt{6}+\sqrt{3}}$

66. $\dfrac{8}{\sqrt{7}+\sqrt{3}}$

67. $\dfrac{2}{\sqrt{5}-\sqrt{3}}$

68. $\dfrac{5}{\sqrt{7}-\sqrt{2}}$

69. $\dfrac{2}{4+\sqrt{x}}$

70. $\dfrac{6}{4-\sqrt{x}}$

71. $\dfrac{2\sqrt{3}}{\sqrt{15}+2}$

72. $\dfrac{3\sqrt{2}}{\sqrt{10}+2}$

73. $\dfrac{\sqrt{5}+\sqrt{2}}{\sqrt{5}-\sqrt{2}}$

74. $\dfrac{\sqrt{5}+\sqrt{3}}{\sqrt{5}-\sqrt{3}}$

Practice PLUS

In Exercises 75–82, rationalize each denominator. Simplify, if possible.

75. $\dfrac{\sqrt{36x^2y^5}}{\sqrt{2x^3y}}$

76. $\dfrac{\sqrt{100x^5y^2}}{\sqrt{2xy^3}}$

77. $\dfrac{2}{\sqrt{x+2} - \sqrt{x}}$

78. $\dfrac{3}{\sqrt{x+3} - \sqrt{x}}$

79. $\dfrac{\sqrt{2}}{\sqrt{3}} + \dfrac{\sqrt{3}}{\sqrt{2}}$

80. $\dfrac{\sqrt{2}}{\sqrt{7}} + \dfrac{\sqrt{7}}{\sqrt{2}}$

81. $\dfrac{2x + 4 - 2h}{\sqrt{x+2} - h}$

82. $\dfrac{4x + 12 - 4h}{\sqrt{x+3} - h}$

Application Exercises

83. Refer to the *Peanuts* cartoon on page 595. In the last frame, Woodstock appears to be working steps mentally. Fill in the missing steps that show how to go from $\dfrac{7\sqrt{2\cdot 2\cdot 3}}{6}$ to $\dfrac{7}{3}\sqrt{3}$.

84. Do you expect to pay more taxes than were withheld? Would you be surprised to know that the percentage of taxpayers who receive a refund and the percentage of taxpayers who pay more taxes vary according to age? The formula

$$P = \frac{x(13 + \sqrt{x})}{5\sqrt{x}}$$

models the percentage, P, of taxpayers who are x years old who must pay more taxes.

a. What percentage of 25-year-olds must pay more taxes?

b. Rewrite the formula by rationalizing the denominator.

c. Use the rationalized form of the formula from part (b) to find the percentage of 25-year-olds who must pay more taxes. Do you get the same answer as you did in part (a)? If so, does this prove that you correctly rationalized the denominator? Explain.

85. The early Greeks believed that the most pleasing of all rectangles were *golden rectangles*, whose ratio of width to height is

$$\frac{w}{h} = \frac{2}{\sqrt{5} - 1}.$$

Rationalize the denominator for this ratio and then use a calculator to approximate the answer correct to the nearest hundredth.

Explaining the Concepts

86. Describe what it means to rationalize a denominator. Use both $\dfrac{1}{\sqrt{5}}$ and $\dfrac{1}{5 + \sqrt{5}}$ in your explanation.

87. When a radical expression has its denominator rationalized, we change the denominator so that it no longer contains a radical. Doesn't this change the value of the radical expression? Explain.

88. Square the real number $\dfrac{2}{\sqrt{3}}$. Observe that the radical is eliminated from the denominator. Explain whether this process is equivalent to rationalizing the denominator.

89. Use the model in Exercise 84 and a calculator to find the percentage of taxpayers ages 30, 40, and 50 who expect to pay more taxes. Describe the trend that you observe. What explanation can you offer for this trend?

Critical Thinking Exercises

Make Sense? *In Exercises 90–93, determine whether each statement makes sense or does not make sense, and explain your reasoning.*

90. I rationalized a numerical denominator and the simplified denominator still contained an irrational number.

91. Without using a calculator and knowing that $\sqrt{2} \approx 1.4142$, rationalizing the denominator of $\dfrac{1}{\sqrt{2}}$ makes division to obtain a decimal approximation for $\dfrac{1}{\sqrt{2}}$ easier to perform.

92. Because 10 and 8 share a common factor of 2, I simplified $\dfrac{\sqrt{10}}{8}$ to $\dfrac{\sqrt{5}}{4}$.

93. I use the fact that 1 is the multiplicative identity to both rationalize denominators and rewrite rational expressions with a common denominator.

In Exercises 94–97, determine whether each statement is true or false. If the statement is false, make the necessary change(s) to produce a true statement.

94. $\dfrac{4 + 8\sqrt{3}}{4} = 1 + 8\sqrt{3}$

95. $\dfrac{3\sqrt{x}}{x\sqrt{6}} = \dfrac{\sqrt{6x}}{2x}$ for $x > 0$

96. Conjugates are used to rationalize the denominator of $\dfrac{2 - \sqrt{5}}{\sqrt{3}}$.

97. Radical expressions with rationalized denominators require less space to write than before they are rationalized.

98. Rationalize the denominator: $\dfrac{1}{\sqrt[3]{2}}$.

99. Simplify: $\sqrt{2} + \sqrt{\dfrac{1}{2}}$.

100. Simplify: $\sqrt{13 + \sqrt{2} + \dfrac{7}{3 + \sqrt{2}}}$.

101. Fill in the box to make the statement true:
$$\dfrac{4}{2 + \sqrt{\blacksquare}} = 8 - 4\sqrt{3}.$$

Review Exercises

102. Solve: $6x^2 - 11x + 5 = 0$. (Section 6.6, Example 2)

103. Simplify: $(2x^2)^{-3}$. (Section 5.7, Example 6)

104. Multiply: $\dfrac{x^2 - 6x + 9}{12} \cdot \dfrac{3}{x^2 - 9}$.
(Section 7.2, Example 3)

Preview Exercises

Exercises 105–107 will help you prepare for the material covered in the next section.

105. Solve: $7 = -2.5x + 17$.

106. Solve: $2x - 1 = x^2 - 4x + 4$.

107. Which solution of the equation in Exercise 106 is a solution of $\sqrt{2x - 1} + 2 = x$?

Radical Equations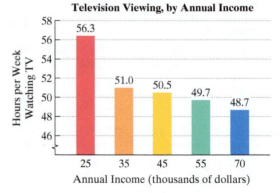

What am I supposed to learn?

After studying this section, you should be able to:

1 Solve radical equations. ▶

2 Solve problems involving square-root models. ▶

In Chapter 4, we looked at data that showed Americans spend more of their free time watching TV than any other activity. **Figure 8.8** shows the average number of hours per week spent watching TV, by annual income.

Television Viewing, by Annual Income

Bar graph. Vertical axis: Hours per Week Watching TV, from 46 to 58. Horizontal axis: Annual Income (thousands of dollars).

Annual Income	Hours per Week
25	56.3
35	51.0
45	50.5
55	49.7
70	48.7

Figure 8.8
Source: Nielsen Media Research

The bar graph indicates that television viewing decreases with increasing income. However, the rate of decrease is slowing down. For this reason, a model in the form $y = a\sqrt{x} + b$, where a is *negative*, is appropriate for describing the data. The formula

$$H = -2.3\sqrt{I} + 67.6$$

models weekly television viewing time, H, in hours, by annual income, I, in thousands of dollars.

Suppose that we are interested in the annual income that corresponds to 44.6 hours per week spent watching TV. Substitute 44.6 for H in the formula and solve for I:

$$44.6 = -2.3\sqrt{I} + 67.6.$$

The resulting equation contains a variable in the radicand and is called a *radical equation*. A **radical equation** is an equation in which the variable occurs in a square root, cube root, or any higher root.

$$44.6 = -2.3\sqrt{I} + 67.6$$

> The variable occurs in a square root.

In this section, you will learn how to solve radical equations with square roots. Solving such equations will enable you to solve new kinds of problems using square-root models, such as the model for television viewing, by annual income. We will return to this model at the end of the section.

① Solve radical equations. ▶

Radical Equations with Square Roots

Consider the following radical equation:

$$\sqrt{x} = 5.$$

We solve the equation by squaring both sides:

> Squaring both sides eliminates the square root.

$$(\sqrt{x})^2 = 5^2$$
$$x = 25.$$

The proposed solution, 25, can be checked in the original equation, $\sqrt{x} = 5$. Because $\sqrt{25} = 5$, the solution is 25.

In general, we solve radical equations with square roots by squaring both sides of the equation. Unfortunately, all the solutions of the squared equation may not be solutions of the original equation. Consider, for example, the equation

$$x = 4.$$

If we square both sides, we obtain

$$x^2 = 16.$$

This new equation has two solutions, -4 and 4. By contrast, only 4 is a solution of the original equation, $x = 4$. For this reason, we must **always check proposed solutions in the original equation when we square both sides while solving**.

Here is a general method for solving radical equations with square roots:

Solving Radical Equations Containing Square Roots

1. If necessary, arrange terms so that one radical is isolated on one side of the equation.
2. Square both sides of the equation to eliminate the square root.
3. Solve the resulting equation.
4. Check all proposed solutions in the original equation.

EXAMPLE 1 Solving a Radical Equation

Solve: $\sqrt{3x + 4} = 8$.

Solution

Step 1. Isolate the radical. The radical, $\sqrt{3x + 4}$, is already isolated on the left side of the equation, so we can skip this step.

Step 2. Square both sides.

$$\sqrt{3x + 4} = 8 \qquad \text{This is the given equation.}$$

$$(\sqrt{3x + 4})^2 = 8^2 \qquad \text{Square both sides to eliminate the radical.}$$

$$3x + 4 = 64 \qquad \text{Simplify.}$$

Step 3. Solve the resulting equation.

$$3x + 4 = 64 \qquad \text{The resulting equation is a linear equation.}$$

$$3x = 60 \qquad \text{Subtract 4 from both sides.}$$

$$x = 20 \qquad \text{Divide both sides by 3.}$$

Step 4. Check the proposed solution in the original equation.

Check 20:

$$\sqrt{3x + 4} = 8$$

$$\sqrt{3 \cdot 20 + 4} \stackrel{?}{=} 8$$

$$\sqrt{60 + 4} \stackrel{?}{=} 8$$

$$\sqrt{64} \stackrel{?}{=} 8$$

$$8 = 8, \quad \text{true}$$

The solution is 20 and the solution set is $\{20\}$. ∎

Using Technology

Graphic Connections
We can use a graphing utility to verify the solution to the equation in Example 1:

$$\sqrt{3x + 4} = 8.$$

Graph each side of the equation:

$$y_1 = \sqrt{3x + 4}$$
$$y_2 = 8.$$

$\boxed{\text{TRACE}}$ along the curves or use the utility's intersection feature. The solution, as shown below, is the first coordinate of the intersection point. Thus, the solution is 20.

$[0, 22, 1]$ by $[0, 10, 1]$

✓ CHECK POINT 1 Solve: $\sqrt{2x + 3} = 5$.

EXAMPLE 2 Solving a Radical Equation

Solve: $\sqrt{3x + 9} - 2\sqrt{x} = 0$.

Solution

Step 1. Isolate each radical. We can get each radical by itself on one side of the equation by adding $2\sqrt{x}$ to both sides.

$$\sqrt{3x + 9} - 2\sqrt{x} = 0 \qquad \text{This is the given equation.}$$

$$\sqrt{3x + 9} - 2\sqrt{x} + 2\sqrt{x} = 0 + 2\sqrt{x} \qquad \text{Add } 2\sqrt{x} \text{ to both sides.}$$

$$\sqrt{3x + 9} = 2\sqrt{x} \qquad \text{Simplify.}$$

Step 2. Square both sides.

$$(\sqrt{3x + 9})^2 = (2\sqrt{x})^2$$

$$3x + 9 = 4x \qquad \text{Simplify.}$$

Step 3. Solve the resulting equation.

$$3x + 9 = 4x \qquad \text{The resulting equation is a linear equation.}$$

$$9 = x \qquad \text{Subtract } 3x \text{ from both sides.}$$

Step 4. Check the proposed solution in the original equation.

Check 9:

$$\sqrt{3x + 9} - 2\sqrt{x} = 0$$
$$\sqrt{3 \cdot 9 + 9} - 2\sqrt{9} \stackrel{?}{=} 0$$
$$\sqrt{36} - 2\sqrt{9} \stackrel{?}{=} 0$$
$$6 - 2 \cdot 3 \stackrel{?}{=} 0$$
$$6 - 6 \stackrel{?}{=} 0$$
$$0 = 0, \quad \text{true}$$

The solution is 9 and the solution set is {9}. ∎

✅ **CHECK POINT 2**   Solve: $\sqrt{x + 32} - 3\sqrt{x} = 0$.

EXAMPLE 3   Solving a Radical Equation

Solve: $\sqrt{x} + 5 = 0$.

Solution

Step 1. Isolate the radical. We isolate the radical, $\sqrt{x}$, on the left side by subtracting 5 from both sides.

$$\sqrt{x} + 5 = 0 \qquad \text{This is the given equation.}$$
$$\sqrt{x} = -5 \qquad \text{Subtract 5 from both sides.}$$

> A principal square root cannot be negative. This equation has no solution. Let's continue the solution procedure to see what happens.

Using Technology

Graphic Connections
The graphs of

$$y_1 = \sqrt{x}$$

and

$$y_2 = -5$$

do not intersect, as shown below. Thus, $\sqrt{x} = -5$ has no real number solution.

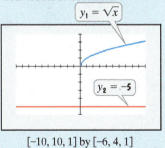

$[-10, 10, 1]$ by $[-6, 4, 1]$

Step 2. Square both sides.

$$(\sqrt{x})^2 = (-5)^2$$
$$x = 25 \qquad \text{Simplify.}$$

Step 3. Solve the resulting equation. We immediately see that 25 is the proposed solution.

Step 4. Check the proposed solution in the original equation.

Check 25:

$$\sqrt{x} + 5 = 0$$
$$\sqrt{25} + 5 \stackrel{?}{=} 0$$
$$5 + 5 \stackrel{?}{=} 0$$
$$10 = 0, \quad \text{false}$$

This false statement indicates that 25 is not a solution. Thus, the equation has no solution. The solution set is ∅, the empty set. ∎

Example 3 illustrates that extra solutions may be introduced when you raise both sides of a radical equation to an even power. Such solutions, which are not solutions of the given equation, are called **extraneous solutions**. Thus, 25 is an extraneous solution of the equation $\sqrt{x} + 5 = 0$.

✅ **CHECK POINT 3**   Solve: $\sqrt{x} + 1 = 0$.

EXAMPLE 4   Solving a Radical Equation

Solve: $\sqrt{2x - 1} + 2 = x$.

Solution

Step 1. Isolate the radical. We isolate the radical, $\sqrt{2x-1}$, by subtracting 2 from both sides.

$\sqrt{2x-1} + 2 = x$	This is the given equation.
$\sqrt{2x-1} = x - 2$	Subtract 2 from both sides.

Step 2. Square both sides.

$(\sqrt{2x-1})^2 = (x-2)^2$	Simplify. Use the formula
$2x - 1 = x^2 - 4x + 4$	$(A-B)^2 = A^2 - 2AB + B^2$ on the right side.

Step 3. Solve the resulting equation. Because of the x^2-term, the resulting equation is a quadratic equation. We can obtain 0 on the left side by subtracting $2x$ and adding 1 on both sides.

$2x - 1 = x^2 - 4x + 4$	The resulting equation is quadratic.
$0 = x^2 - 6x + 5$	Write in standard form, subtracting $2x$ and adding 1 on both sides.
$0 = (x-1)(x-5)$	Factor.
$x - 1 = 0$ or $x - 5 = 0$	Set each factor equal to O.
$x = 1$ $x = 5$	Solve the resulting equations.

Step 4. Check the proposed solutions in the original equation.

Check 1:	**Check 5:**
$\sqrt{2x-1} + 2 = x$	$\sqrt{2x-1} + 2 = x$
$\sqrt{2 \cdot 1 - 1} + 2 \overset{?}{=} 1$	$\sqrt{2 \cdot 5 - 1} + 2 \overset{?}{=} 5$
$\sqrt{1} + 2 \overset{?}{=} 1$	$\sqrt{9} + 2 \overset{?}{=} 5$
$1 + 2 \overset{?}{=} 1$	$3 + 2 \overset{?}{=} 5$
$3 = 1$, false	$5 = 5$, true

Thus, 1 is an extraneous solution. The only solution is 5 and the solution set is {5}. ∎

Using Technology

Numeric Connections

A graphing utility's $\boxed{\text{TABLE}}$ feature provides a numeric check that 1 is not a solution and 5 is a solution of $\sqrt{2x-1} + 2 = x$.

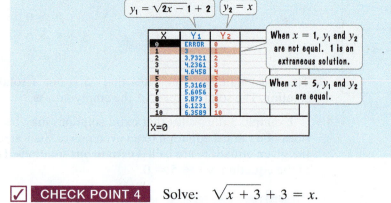

Great Question!

If there are two proposed solutions, will one of them always be extraneous?

No. If a radical equation leads to a quadratic equation with two solutions, there will not always be one solution that checks and one that does not. Both solutions of the quadratic equation can satisfy the original radical equation. It is also possible that neither solution of the quadratic equation satisfies the given radical equation.

✓ **CHECK POINT 4** Solve: $\sqrt{x+3} + 3 = x$.

2 Solve problems involving square-root models. ▶

Applications of Radical Equations

Radical equations can be solved to answer questions about variables contained in square-root models.

EXAMPLE 5 Using a Square-Root Model

The formula

$$H = -2.3\sqrt{I} + 67.6$$

models weekly television viewing time, H, in hours, by annual income, I, in thousands of dollars. What annual income corresponds to 44.6 hours per week watching TV?

Solution Because we are interested in finding the annual income that corresponds to 44.6 hours of weekly viewing time, we substitute 44.6 for H in the given formula. Then we solve for I, the annual income, in thousands of dollars.

$H = -2.3\sqrt{I} + 67.6$	This is the given formula.
$44.6 = -2.3\sqrt{I} + 67.6$	Substitute 44.6 for H.
$-23 = -2.3\sqrt{I}$	Subtract 67.6 from both sides.
$\dfrac{-23}{-2.3} = \dfrac{-2.3\sqrt{I}}{-2.3}$	Divide both sides by -2.3.
$10 = \sqrt{I}$	Simplify: $\dfrac{-23}{-2.3}$
$10^2 = (\sqrt{I})^2$	Square both sides.
$100 = I$	Simplify.

The model indicates that an annual income of 100 thousand dollars, or $100,000, corresponds to 44.6 hours of weekly viewing time. ∎

✓ CHECK POINT 5 Use the formula in Example 5 to find the annual income that corresponds to 33.1 hours per week watching TV.

CONCEPT AND VOCABULARY CHECK

Fill in each blank so that the resulting statement is true.

1. An equation in which the variable occurs in a square root, cube root, or any higher root is called a/an _____ equation.

2. Solutions of a squared equation that are not solutions of the original equation are called _____ solutions.

3. Consider the equation

$$\sqrt{x + 3} = x - 3.$$

Squaring the left side and simplifying results in _____. Squaring the right side and simplifying results in _____.

4. True or false: 36 is a solution of $\sqrt{x} + 6 = 0$. _____

5. True or false: -3 is a solution of $\sqrt{x + 7} = x + 5$. _____

6. True or false: 8 is a solution of $\sqrt{x + 1} = \sqrt{x - 4} - 5$. _____

EXERCISE SET 8.5   ▶ MyMathLab®

Practice Exercises

In Exercises 1–44, solve each radical equation.

1. $\sqrt{x} = 5$
2. $\sqrt{x} = 6$
3. $\sqrt{x} - 4 = 0$
4. $\sqrt{x} - 9 = 0$
5. $\sqrt{x + 2} = 3$
6. $\sqrt{x - 6} = 5$
7. $\sqrt{x - 3} - 11 = 0$
8. $\sqrt{x + 5} - 8 = 0$
9. $\sqrt{3x - 5} = 4$
10. $\sqrt{5x - 6} = 8$
11. $\sqrt{x + 5} + 2 = 5$
12. $\sqrt{x + 6} + 1 = 3$
13. $\sqrt{x + 3} = \sqrt{4x - 3}$
14. $\sqrt{x + 8} = \sqrt{5x - 4}$
15. $\sqrt{6x - 2} = \sqrt{4x + 4}$
16. $\sqrt{5x - 4} = \sqrt{3x + 6}$
17. $11 = 6 + \sqrt{x + 1}$
18. $7 = 5 + \sqrt{x + 1}$

19. $\sqrt{x} + 10 = 0$

20. $\sqrt{x} + 8 = 0$

21. $\sqrt{x-1} = -3$

22. $\sqrt{x-2} = -5$

23. $3\sqrt{x} = \sqrt{8x+16}$

24. $3\sqrt{x-2} = \sqrt{7x+4}$

25. $\sqrt{2x-3} + 5 = 0$

26. $\sqrt{2x-8} + 4 = 0$

27. $\sqrt{3x+4} - 2 = 3$

28. $\sqrt{5x-4} - 2 = 4$

29. $3\sqrt{x-1} = \sqrt{3x+3}$

30. $2\sqrt{3x+4} = \sqrt{5x+9}$

31. $\sqrt{x+7} = x+5$

32. $\sqrt{x+10} = x-2$

33. $\sqrt{2x+13} = x+7$

34. $\sqrt{6x+1} = x-1$

35. $\sqrt{9x^2 + 2x - 4} = 3x$

36. $\sqrt{9x^2 - 2x + 8} = 3x$

37. $x = \sqrt{2x-2} + 1$

38. $x = \sqrt{3x+7} - 3$

39. $x = \sqrt{8-7x} + 2$

40. $x = \sqrt{1-8x} + 2$

41. $\sqrt{3x+10} = x+4$

42. $\sqrt{x-3} = x-9$

43. $3\sqrt{x} + 5 = 2$

44. $3\sqrt{x} + 8 = 5$

Practice PLUS

45. Two added to the square root of the product of 4 and a number is equal to 10. Find the number.

46. Five added to the square root of the product of 6 and a number is equal to 8. Find the number.

47. A number is 4 more than the principal square root of twice the number. Find the number.

48. A number is 6 more than the principal square root of 3 times the number. Find the number.

49. Solve for h: $v = \sqrt{2gh}$.

50. Solve for l: $t = \dfrac{\pi}{2}\sqrt{\dfrac{l}{2}}$.

Solve the equations in Exercises 51–54. You will need to square both sides of each equation twice.

51. $\sqrt{x} + 2 = \sqrt{x+8}$

52. $\sqrt{x} + 6 = \sqrt{x+72}$

53. $\sqrt{x-8} = \sqrt{x} - 2$

54. $\sqrt{x-4} = \sqrt{x} - 2$

Application Exercises

When firefighters are working to put out a fire, the rate at which they spray water on the fire depends on the nozzle pressure. The formula

$$f = 120\sqrt{p}$$

models the water's flow rate, f, in gallons per minute, in terms of the nozzle pressure, p, in pounds per square inch. Use this formula to solve Exercises 55–56.

55. What nozzle pressure is needed to achieve a water flow rate of 840 gallons per minute?

56. What nozzle pressure is needed to achieve a water flow rate of 720 gallons per minute?

Use the graph of the firefighting formula to solve Exercises 57–58.

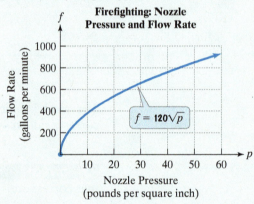

57. How is your answer to Exercise 55 shown on the graph?

58. How is your answer to Exercise 56 shown on the graph?

Long before the Supreme Court decision legalizing marriage between people of the same sex, a substantial percentage of Americans were in favor of laws prohibiting interracial marriage. Since 1972, the General Social Survey has asked

Do you think that there should be laws against marriages between negroes/blacks/African Americans and whites?

In 1972, 39.3% of the adult U.S. population was in favor of such laws. The bar graph shows the percentage of Americans in favor of legislation prohibiting interracial marriage for five selected years from 1993 through 2002, the last year the data were collected.

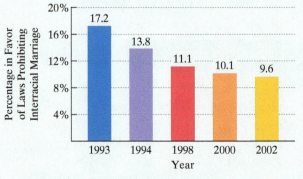

Source: Ben Schott, Schott's Almanac 2007, Donnelley and Sons

The formula

$$p = -2.5\sqrt{t} + 17$$

models the percentage of Americans, p, in favor of legislation prohibiting interracial marriage t years after 1993. Use this mathematical model to solve Exercises 59–60.

59. a. According to the formula, what percentage of Americans were in favor of laws prohibiting interracial marriage in 2002?

   b. Does your answer in part (a) underestimate or overestimate the actual percentage given by the bar graph at the bottom of the previous page? By how much?

   c. If trends indicated by the model continue, in what year did the percentage favoring such legislation decrease to 7%?

60. a. According to the formula, what percentage of Americans were in favor of laws prohibiting interracial marriage in 2000? Round to the nearest tenth of a percent.

   b. Does your answer in part (a) underestimate or overestimate the percentage given by the bar graph at the bottom of the previous page? By how much?

   c. If trends indicated by the model continue, in what year will only 4.5% of Americans favor such legislation?

Explaining the Concepts

61. What is a radical equation?

62. In solving $\sqrt{2x - 1} + 2 = x$, why is it a good idea to isolate the radical term? What if we don't do this and simply square each side? Describe what happens.

63. What is an extraneous solution of a radical equation?

64. Explain why $\sqrt{x} = -1$ has no solution.

Critical Thinking Exercises

Make Sense?   *In Exercises 65–68, determine whether each statement makes sense or does not make sense, and explain your reasoning.*

65. When checking a radical equation's proposed solution, I can substitute into the original equation or any equation that is part of the solution process.

66. After squaring both sides of a radical equation, the only solution that I obtained was extraneous, so $\varnothing$ must be the solution set of the original equation.

67. Whenever a radical equation leads to a quadratic equation with two solutions, one solution satisfies the original radical equation and the other does not.

68. I squared both sides of $\sqrt{x + 7} = x + 5$ and obtained $x + 7 = x^2 + 25$.

In Exercises 69–72, determine whether each statement is true or false. If the statement is false, make the necessary change(s) to produce a true statement.

69. The equation $y^2 = 25$ has the same solutions as the equation $y = 5$.

70. The equation $\sqrt{x^2 + 2x} = -1$ has no real number solution.

71. The first step in solving $\sqrt{x} + 3 = 4$ is to square each side.

72. When an extraneous solution is substituted into an equation with radicals, a denominator of zero results.

73. The square root of the sum of two consecutive integers is one less than the smaller integer. Find the integers.

74. If $w = 2$, find x, y, and z if $y = \sqrt{x - 2} + 2$, $z = \sqrt{y - 2} + 2$, and $w = \sqrt{z - 2} + 2$.

Technology Exercises

In Exercises 75–79, use a graphing utility to solve each radical equation. Graph each side of the equation in the given viewing rectangle. The equation's solution is given by the x-coordinate of the point(s) of intersection. Check by substitution.

75. $\sqrt{2x + 2} = \sqrt{3x - 5}$

   $[-1, 10, 1]$ by $[-1, 5, 1]$

76. $\sqrt{x} + 3 = 5$

   $[-1, 6, 1]$ by $[-1, 6, 1]$

77. $\sqrt{x^2 + 3} = x + 1$

   $[-1, 6, 1]$ by $[-1, 6, 1]$

78. $4\sqrt{x} = x + 3$

   $[-1, 10, 1]$ by $[-1, 13, 1]$

79. $\sqrt{x} + 4 = 2$

   $[-2, 18, 1]$ by $[0, 10, 1]$

80. Use a graphing utility's TABLE feature to provide a numeric check for the solutions you obtained in any three radical equations from Exercises 29–38.

Review Exercises

81. A total of $9000 was invested for 1 year, part at 6% and the remainder at 4% simple interest. At the end of the year, the investments earned $500 in interest. How much was invested at each rate? (Section 4.4, Example 5)

82. Producers of *Breaking Bad the Musical* are a bit worried about their basic concept and decide to sell tickets for previews at cut-rate prices. If four orchestra and two mezzanine seats sell for $22, while two orchestra and three mezzanine seats sell for $16, what is the price of an orchestra seat? (Section 4.4, Example 2)

83. Solve by graphing:

$$\begin{cases} 2x + y = -4 \\ x + y = -3. \end{cases}$$

   (Section 4.1, Example 2)

Preview Exercises

Exercises 84–86 will help you prepare for the material covered in the next section. In each exercise, evaluate the given expression.

84. $-\left(\sqrt[5]{32}\right)^4$

85. $\left(\sqrt[3]{27}\right)^2$

86. $\dfrac{1}{\left(\sqrt[4]{81}\right)^3}$

SECTION

8.6

Rational Exponents

What am I supposed to learn?

After studying this section, you should be able to:

1 Evaluate expressions with rational exponents.

2 Solve problems using models with rational exponents.

1 Evaluate expressions with rational exponents.

Animals in the wild have regions to which they confine their movement, called their territorial area. Territorial area, in square miles, is related to an animal's body weight. If an animal weighs W pounds, its territorial area is

$$W^{\frac{141}{100}}$$

square miles.

W to the *what* power?! How can we interpret the information given by this algebraic expression? In this section, we turn our attention to rational exponents such as $\frac{141}{100}$ and their relationship to roots of real numbers.

Defining Rational Exponents

We define rational exponents so that their properties are the same as the properties for integer exponents. For example, we know that exponents are multiplied when an exponential expression is raised to a power. For this to be true,

$$\left(7^{\frac{1}{2}}\right)^2 = 7^{\frac{1}{2} \cdot 2} = 7^1 = 7.$$

We also know that

$$\left(\sqrt{7}\right)^2 = \sqrt{7} \cdot \sqrt{7} = \sqrt{49} = 7.$$

Can you see that the square of both $7^{\frac{1}{2}}$ and $\sqrt{7}$ is 7? It is reasonable to conclude that

$$7^{\frac{1}{2}} \text{ means } \sqrt{7}.$$

We can generalize this idea with the following definition:

The Definition of $a^{\frac{1}{n}}$

If $\sqrt[n]{a}$ represents a real number and $n \geq 2$ is an integer, then

$$a^{\frac{1}{n}} = \sqrt[n]{a}.$$

The denominator of the rational exponent is the radical's index.

EXAMPLE 1 Using the Definition of $a^{\frac{1}{n}}$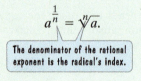

Simplify:

a. $64^{\frac{1}{2}}$ **b.** $125^{\frac{1}{3}}$ **c.** $-16^{\frac{1}{4}}$ **d.** $(-27)^{\frac{1}{3}}$.

Solution

a. $64^{\frac{1}{2}} = \sqrt{64} = 8$

b. $125^{\frac{1}{3}} = \sqrt[3]{125} = 5$

The denominator is the index.

c. $-16^{\frac{1}{4}} = -(\sqrt[4]{16}) = -2$

The base is 16 and the negative sign is not affected by the exponent.

d. $(-27)^{\frac{1}{3}} = \sqrt[3]{-27} = -3$

Parentheses show that the base is −27 and that the negative sign is affected by the exponent. ∎

✓ **CHECK POINT 1**   Simplify:

a. $25^{\frac{1}{2}}$ **b.** $8^{\frac{1}{3}}$ **c.** $-81^{\frac{1}{4}}$ **d.** $(-8)^{\frac{1}{3}}$.

In Example 1 and Check Point 1, each rational exponent had a numerator of 1. If the numerator is some other integer, we still want to multiply exponents when raising a power to a power. For this reason,

$$a^{\frac{2}{3}} = \left(a^{\frac{1}{3}}\right)^2 \quad \text{and} \quad a^{\frac{2}{3}} = \left(a^2\right)^{\frac{1}{3}}.$$

This means $(\sqrt[3]{a})^2$.   This means $\sqrt[3]{a^2}$.

Thus,

$$a^{\frac{2}{3}} = \left(\sqrt[3]{a}\right)^2 = \sqrt[3]{a^2}.$$

Do you see that the denominator, 3, of the rational exponent is the same as the index of the radical? The numerator, 2, of the rational exponent serves as an exponent in each of the two radical forms. We generalize these ideas with the following definition:

The Definition of $a^{\frac{m}{n}}$

If $\sqrt[n]{a}$ represents a real number, and $\dfrac{m}{n}$ is a positive rational number, $n \geq 2$, then

$$a^{\frac{m}{n}} = \left(\sqrt[n]{a}\right)^m.$$

First take the nth root of a.

Also,

$$a^{\frac{m}{n}} = \sqrt[n]{a^m}.$$

First raise a to the m power.

The first form of the definition, shown again below, involves taking the root first. This form is often preferable because smaller numbers are involved. Notice that the rational exponent consists of two parts, indicated by the following voice balloons:

The numerator is the exponent.

$$a^{\frac{m}{n}} = \left(\sqrt[n]{a}\right)^m.$$

The denominator is the radical's index.

EXAMPLE 2 Using the Definition of $a^{\frac{m}{n}}$

Simplify:

a. $27^{\frac{2}{3}}$ b. $9^{\frac{3}{2}}$ c. $-32^{\frac{4}{5}}$.

Solution

a. $27^{\frac{2}{3}} = \left(\sqrt[3]{27}\right)^2 = 3^2 = 9$

b. $9^{\frac{3}{2}} = \left(\sqrt{9}\right)^3 = 3^3 = 27$

c. $-32^{\frac{4}{5}} = -\left(\sqrt[5]{32}\right)^4 = -2^4 = -16$

The base is 32 and the negative sign is not affected by the exponent. ∎

✓ **CHECK POINT 2** Simplify:

a. $27^{\frac{4}{3}}$ b. $4^{\frac{3}{2}}$ c. $-16^{\frac{3}{4}}$.

Can a rational exponent be negative? Yes. The way that negative rational exponents are defined is similar to the way that negative integer exponents are defined:

> **The Definition of $a^{-\frac{m}{n}}$**
>
> If $a^{-\frac{m}{n}}$ is a nonzero real number, then
>
> $$a^{-\frac{m}{n}} = \frac{1}{a^{\frac{m}{n}}}.$$

EXAMPLE 3 Using the Definition of $a^{-\frac{m}{n}}$

Simplify:

a. $100^{-\frac{1}{2}}$ b. $27^{-\frac{1}{3}}$ c. $81^{-\frac{3}{4}}$.

Solution

a. $100^{-\frac{1}{2}} = \frac{1}{100^{\frac{1}{2}}} = \frac{1}{\sqrt{100}} = \frac{1}{10}$

b. $27^{-\frac{1}{3}} = \frac{1}{27^{\frac{1}{3}}} = \frac{1}{\sqrt[3]{27}} = \frac{1}{3}$

c. $81^{-\frac{3}{4}} = \frac{1}{81^{\frac{3}{4}}} = \frac{1}{\left(\sqrt[4]{81}\right)^3} = \frac{1}{3^3} = \frac{1}{27}$ ∎

✓ **CHECK POINT 3** Simplify:

a. $25^{-\frac{1}{2}}$ b. $64^{-\frac{1}{3}}$ c. $32^{-\frac{4}{5}}$.

---

2 Solve problems using models with rational exponents.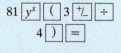

Applications

Now that you know the meaning of rational exponents, you can work with mathematical models that contain these exponents.

EXAMPLE 4 Sexual Abstinence among Teens and Young Adults

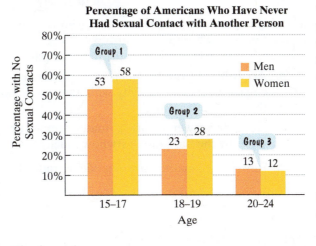

Percentage of Americans Who Have Never Had Sexual Contact with Another Person

Figure 8.9

Source: National Center for Health Statistics (2008 data)

The bar graph in **Figure 8.9** shows the percentage of American teens and young adults who have never had sexual contact with another person.

The formula

$$M = 54x^{-\frac{13}{10}}$$

models the percentage of men, M, who have never engaged in sexual activity with another person, where x is the group number designating the age ranges shown on the horizontal axis in **Figure 8.9**.

a. According to the formula, what percentage of men ages 18–19 have never engaged in sexual activity with another person? Use a calculator and round to the nearest percent.

b. Does the answer in part (a) underestimate or overestimate the actual percentage given by the bar graph? By how much?

c. Rewrite the mathematical model in radical notation.

Solution

a. We are interested in men ages 18–19. The voice balloon in **Figure 8.9** shows that this age range is designated by group 2. We substitute 2 for x in the given formula. Then we use a calculator to find M, the percentage of sexually abstinent men.

$$M = 54x^{-\frac{13}{10}}$$ This is the given formula.

$$M = 54 \cdot 2^{-\frac{13}{10}}$$ Substitute 2 for x.

$$M \approx 22$$ Use a calculator. See the Using Technology box in the margin.

The mathematical model indicates that approximately 22% of American men ages 18–19 have never had sexual contact with another person.

b. The bar graph in **Figure 8.9** shows that 23% of men ages 18–19 are sexually abstinent. Because we obtained 22% in part (a), the model underestimates the percentage by 1%.

c. Now we are ready to express the mathematical model in radical notation.

$$M = 54x^{-\frac{13}{10}}$$ This is the given formula.

$$M = \frac{54}{x^{\frac{13}{10}}}$$ $a^{-\frac{m}{n}} = \frac{1}{a^{\frac{m}{n}}}$

$$M = \frac{54}{\left(\sqrt[10]{x}\right)^{13}} \text{ or } \frac{54}{\sqrt[10]{x^{13}}}$$ $a^{\frac{m}{n}} = \left(\sqrt[n]{a}\right)^m$ or $\sqrt[n]{a^m}$ ■

Using Technology

Here are the calculator keystroke sequences for $54 \cdot 2^{-\frac{13}{10}}$:

Many Scientific Calculators

54 × 2 y^x (13 +/− ÷ 10) =

Many Graphing Calculators

54 × 2 ^ ((−) 13 ÷ 10) ENTER

✓ **CHECK POINT 4** The formula

$$W = 62x^{-\frac{7}{5}}$$

models the percentage of women, W, who have never engaged in sexual activity with another person, where x is the group number designating the age ranges shown on the horizontal axis in **Figure 8.9** on the previous page.

a. According to the formula, what percentage of women ages 20–24 have never engaged in sexual activity with another person? Use a calculator and round to the nearest percent.

b. Does the answer in part (a) underestimate or overestimate the actual percentage given by the bar graph? By how much?

c. Rewrite the mathematical model in radical notation.

Achieving Success

Do you get to choose your seat during an exam? If so, select a desk that minimizes distractions and puts you in the right frame of mind. Many students can focus better if they do not sit near friends. If possible, sit near a window, next to a wall, or in the front row. In these locations, you can glance up from time to time without looking at another student's work.

CONCEPT AND VOCABULARY CHECK

Fill in each blank so that the resulting statement is true.

1. $36^{\frac{1}{2}} = \sqrt{36} = $ _____

2. $a^{\frac{1}{2}} = $ _____, $a \geq 0$

3. $8^{\frac{1}{3}} = \sqrt[3]{_} = $ _____

4. $a^{\frac{1}{3}} = $ _____

5. $16^{\frac{3}{4}} = (\sqrt[4]{_})^3 = (_____)^3 = $ _____

6. $a^{\frac{4}{5}} = $ _____

7. $16^{-\frac{3}{2}} = \dfrac{1}{16^{\frac{3}{2}}} = \dfrac{1}{(\sqrt{16})^-} = \dfrac{1}{(\_)^-} = \dfrac{1}{\_}$

8.6 EXERCISE SET ▶ MyMathLab®

Practice Exercises

In Exercises 1–48, simplify by first writing the expression in radical form. If applicable, use a calculator to verify your answer.

1. $49^{\frac{1}{2}}$
2. $100^{\frac{1}{2}}$
3. $121^{\frac{1}{2}}$
4. $1^{\frac{1}{2}}$
5. $27^{\frac{1}{3}}$
6. $64^{\frac{1}{3}}$
7. $-125^{\frac{1}{3}}$
8. $-27^{\frac{1}{3}}$
9. $16^{\frac{1}{4}}$
10. $81^{\frac{1}{4}}$
11. $-32^{\frac{1}{5}}$
12. $-243^{\frac{1}{5}}$
13. $\left(\dfrac{1}{9}\right)^{\frac{1}{2}}$
14. $\left(\dfrac{1}{25}\right)^{\frac{1}{2}}$
15. $\left(\dfrac{27}{64}\right)^{\frac{1}{3}}$
16. $\left(\dfrac{64}{125}\right)^{\frac{1}{3}}$
17. $81^{\frac{3}{2}}$
18. $25^{\frac{3}{2}}$
19. $125^{\frac{2}{3}}$
20. $1000^{\frac{2}{3}}$
21. $9^{\frac{3}{2}}$
22. $16^{\frac{3}{2}}$
23. $(-32)^{\frac{3}{5}}$
24. $(-27)^{\frac{2}{3}}$
25. $9^{-\frac{1}{2}}$
26. $49^{-\frac{1}{2}}$
27. $125^{-\frac{1}{3}}$
28. $27^{-\frac{1}{3}}$
29. $32^{-\frac{1}{5}}$
30. $243^{-\frac{1}{5}}$

31. $\left(\dfrac{1}{4}\right)^{-\frac{1}{2}}$ **32.** $\left(\dfrac{1}{9}\right)^{-\frac{1}{2}}$

33. $16^{-\frac{3}{4}}$ **34.** $625^{-\frac{3}{4}}$

35. $81^{-\frac{5}{4}}$ **36.** $32^{-\frac{4}{5}}$

37. $8^{-\frac{2}{3}}$ **38.** $625^{-\frac{5}{4}}$

39. $\left(\dfrac{4}{25}\right)^{-\frac{1}{2}}$ **40.** $\left(\dfrac{8}{27}\right)^{-\frac{1}{3}}$

41. $\left(\dfrac{8}{125}\right)^{-\frac{1}{3}}$ **42.** $\left(\dfrac{9}{100}\right)^{-\frac{1}{2}}$

43. $(-8)^{-\frac{2}{3}}$ **44.** $(-64)^{-\frac{2}{3}}$

45. $27^{\frac{2}{3}} + 16^{\frac{3}{4}}$ **46.** $4^{\frac{5}{2}} - 8^{\frac{2}{3}}$

47. $25^{\frac{3}{2}} \cdot 81^{\frac{1}{4}}$ **48.** $16^{-\frac{3}{4}} \cdot 16^{\frac{3}{2}}$

Practice PLUS

In Exercises 49–56, simplify each expression. Write answers in exponential form with positive exponents only. Assume that all variables represent positive real numbers.

49. $x^{\frac{1}{3}} \cdot x^{\frac{1}{4}}$ **50.** $x^{\frac{1}{4}} \cdot x^{\frac{1}{5}}$

51. $\dfrac{x^{\frac{1}{6}}}{x^{\frac{5}{6}}}$ **52.** $\dfrac{x^{\frac{1}{4}}}{x^{\frac{3}{4}}}$

53. $\left(x^{\frac{1}{4}}y^3\right)^{\frac{2}{3}}$ **54.** $\left(x^{\frac{1}{6}}y^{15}\right)^{\frac{3}{5}}$

55. $\left(\dfrac{x^{\frac{2}{5}}}{x^{\frac{6}{5}} \cdot x^{\frac{3}{5}}}\right)^5$ **56.** $\left(\dfrac{x^{\frac{4}{7}}}{x^{\frac{3}{7}} \cdot x^{\frac{2}{7}}}\right)^{49}$

Application Exercises

57. It is difficult to measure the height of a tall tree, particularly when it is growing in a dense forest. However, it is relatively easy to measure its base diameter. The formula

$$h = 0.84d^{\frac{2}{3}}$$

models a tree's height, h, in meters, in terms of its base diameter, d, in centimeters. (*Source:* Thomas McMahon, *Scientific American*, July, 1975)

  a. The largest known sequoia, the General Sherman in California, has a base diameter of 985 centimeters (about the size of a small house). Use a calculator to approximate the height of the General Sherman to the nearest tenth of a meter.

  b. Rewrite the formula in radical notation.

58. The formula

$$A = 1.83w^{\frac{2}{3}}$$

models the area of an egg shell, A, in square centimeters, in terms of its weight, w, in grams.

  a. An ostrich egg weighs approximately 1600 grams. Use a calculator to approximate the shell's area to the nearest tenth of a square centimeter.

  b. Rewrite the formula in radical notation.

Artificial gravity can be created in a space station to prevent bone and muscle loss for astronauts on lengthy space voyages. A rotating bedlike apparatus is used to create the artificial gravity. The required rate of rotation is given by the following formula:

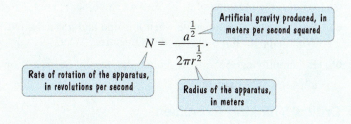

$$N = \frac{a^{\frac{1}{2}}}{2\pi r^{\frac{1}{2}}}.$$

Rate of rotation of the apparatus, in revolutions per second

Artificial gravity produced, in meters per second squared

Radius of the apparatus, in meters

Use this formula to solve Exercises 59–60.

59. **a.** How fast, in revolutions per second, would an apparatus with a radius of 1.7 meters have to rotate to simulate gravity on Earth, 9.8 meters per second squared? Express the answer correct to three decimal places.

  b. The required rotation that you obtained in part (a) is expressed in revolutions per second. Express the rotation needed for the apparatus in revolutions per minute.

  c. Rewrite the formula in radical notation without a radical in the denominator.

60. **a.** How fast, in revolutions per second, would an apparatus with a radius of 2.4 meters have to rotate to simulate gravity on Earth, 9.8 meters per second squared? Express the answer correct to three decimal places.

  b. The required rotation that you obtained in part (a) is expressed in revolutions per second. Express the rotation needed for the apparatus in revolutions per minute.

  c. Rewrite the formula in radical notation without a radical in the denominator.

Explaining the Concepts

61. What is the meaning of $a^{\frac{1}{n}}$? Give an example to support your explanation.

62. What is the meaning of $a^{\frac{m}{n}}$? Give an example.

63. What is the meaning of $a^{-\frac{m}{n}}$? Give an example.

64. Explain why $a^{\frac{1}{n}}$ is negative when n is odd and a is negative. What happens if n is even and a is negative? Why?

65. In simplifying $36^{\frac{3}{2}}$, is it better to use $a^{\frac{m}{n}} = \sqrt[n]{a^m}$ or $a^{\frac{m}{n}} = (\sqrt[n]{a})^m$? Explain.

Critical Thinking Exercises

Make Sense? In Exercises 66–69, determine whether each statement makes sense or does not make sense, and explain your reasoning.

66. When I used my calculator to approximate $5^{\frac{2}{3}}$, I found it easier to first rewrite the expression in radical form, using the radical form for the keystroke sequence.

67. By adding the exponents, I simplified $5^{\frac{1}{2}} \cdot 5^{\frac{1}{2}}$ and obtained 25.

68. When I use the definition for $a^{\frac{m}{n}}$, I prefer to first raise a to the m power because smaller numbers are involved.

69. There's no question that $(-64)^{\frac{1}{3}} = -64^{\frac{1}{3}}$, so I can conclude that $(-64)^{\frac{1}{2}} = -64^{\frac{1}{2}}$.

In Exercises 70–77, determine whether each statement is true or false. If the statement is false, make the necessary change(s) to produce a true statement.

70. $2^{\frac{1}{2}} \cdot 2^{\frac{1}{2}} = 4^{\frac{1}{2}}$

71. $8^{-\frac{1}{2}} = \frac{1}{4}$

72. $25^{-\frac{1}{2}} = -5$

73. $-3^{-2} = \frac{1}{9}$

74. $2^{\frac{1}{2}} \cdot 2^{\frac{3}{2}} = \left(\frac{1}{4}\right)^{-1}$

75. $81^{\frac{1}{4}} \cdot 125^{\frac{1}{3}}$ is an integer.

76. $16^{-\frac{1}{4}} = -2$

77. $(-8)^{\frac{1}{3}} = -8^{\frac{1}{3}}$

Without using a calculator, simplify the expressions in Exercises 78–79 completely.

78. $25^{\frac{1}{4}} \cdot 25^{-\frac{3}{4}}$

79. $\dfrac{3^{-1} \cdot 3^{\frac{1}{2}}}{3^{-\frac{3}{2}}}$

Technology Exercises

80. The territorial area of an animal in the wild is the area of the region to which the animal confines its movements. The formula
$$T = W^{1.41} = W^{\frac{141}{100}} = \sqrt[100]{W^{141}}$$
models the territorial area, T, in square miles, in terms of an animal's body weight, W, in pounds.

   a. Use a calculator to fill in the table of values, rounding T to the nearest whole square mile.

W	0	25	50	150	200	250	300
T = W^{1.41}							

b. Use the table of values to graph $T = W^{1.41}$. What does the shape of the graph indicate about the relationship between body weight and territorial area?

c. Verify your hand-drawn graph by using a graphing utility to graph the model.

81. If A is the surface area of a cube and V is its volume, then $A = 6V^{\frac{2}{3}}$.

   a. Graph the equation $\left(y = 6x^{\frac{2}{3}}\right)$ relating a cube's surface area and volume using a graphing utility and a $[0, 30, 3]$ by $[0, 60, 3]$ viewing rectangle.

   b. TRACE or ZOOM IN along the curve and verify the numbers shown in the figure below. In particular, show that a cube whose volume is 27 cubic units has a surface area of 54 square units.

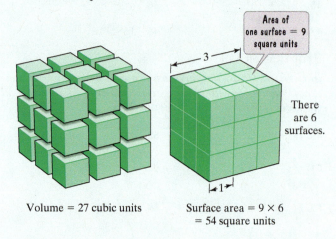

Volume = 27 cubic units

Area of one surface = 9 square units

There are 6 surfaces.

Surface area = 9 × 6 = 54 square units

c. TRACE or ZOOM IN along the curve and find the surface area of a cube whose volume is 8 cubic units.

Review Exercises

82. Solve the system:
$$\begin{cases} 7x - 3y = -14 \\ y = 3x + 6. \end{cases}$$
(Section 4.2, Example 1)

83. Graph the solution set of the system:
$$\begin{cases} -3x + 4y \le 12 \\ x \ge 2. \end{cases}$$
(Section 4.5, Example 3)

84. Simplify: $\dfrac{(2x)^5}{x^3}$. (Section 5.7, Example 5)

Preview Exercises

Exercises 85–87 will help you prepare for the material covered in the first section of the next chapter.

85. Use substitution to determine if $-\sqrt{5}$ is a solution of the quadratic equation $4x^2 = 20$.

86. Use substitution to determine if $1 + \sqrt{3}$ is a solution of $(x - 1)^2 = 5$.

87. Simplify: $\sqrt{[6 - (-4)]^2 + [2 - (-3)]^2}$.

Chapter 8 Summary

Definitions and Concepts	Examples

Section 8.1 Finding Roots

If $b^2 = a$, then b is a square root of a. The principal square root of a, designated $\sqrt{a}$, is the nonnegative number satisfying $b^2 = a$. The negative square root of a is written $-\sqrt{a}$. A square root of a negative number is not a real number.

- $\sqrt{100} = 10$ because $10^2 = 100$.
- $-\sqrt{100} = -10$
- $\sqrt{-100}$ is not a real number.

The principal nth root of a real number a, symbolized by $\sqrt[n]{a}$, is defined as follows:

$$\sqrt[n]{a} = b \text{ means that } b^n = a.$$

The natural number n is called the index, the symbol $\sqrt[n]{}$ is called a radical sign, and the expression under the radical sign is called the radicand. If the index is even, the radicand must be nonnegative for the root to be a real number and the principal nth root is nonnegative.

- $\sqrt[3]{8} = 2$ because $2^3 = 8$.
- $\sqrt[3]{-125} = -5$ because $(-5)^3 = -125$.
- $\sqrt[4]{81} = 3$ because $3^4 = 81$.
- $\sqrt[4]{-81}$ is not a real number.

Section 8.2 Multiplying and Dividing Radicals

The Product Rule for Roots

$$\sqrt[n]{a} \cdot \sqrt[n]{b} = \sqrt[n]{ab} \text{ and } \sqrt[n]{ab} = \sqrt[n]{a} \cdot \sqrt[n]{b}$$

(The roots represent real numbers.)

- $\sqrt{11} \cdot \sqrt{7} = \sqrt{11 \cdot 7} = \sqrt{77}$
- $\sqrt[3]{4} \cdot \sqrt[3]{2} = \sqrt[3]{4 \cdot 2} = \sqrt[3]{8} = 2$

A square root is simplified when its radicand has no factors other than 1 that are perfect squares. To simplify a square root, factor the radicand so that one of its factors is a perfect square. Simplify square roots with a variable to an even power using $\sqrt{x^{2n}} = x^n$. Simplify square roots with a variable to an odd power by expressing the variable as the product of two factors, one of which has an even power.

- $\sqrt{63} = \sqrt{9 \cdot 7} = \sqrt{9} \cdot \sqrt{7} = 3\sqrt{7}$
- $\sqrt{x^{26}} = x^{13}$
- $\sqrt{x^{27}} = \sqrt{x^{26} \cdot x}$

$$= \sqrt{x^{26}} \cdot \sqrt{x} = x^{13}\sqrt{x}$$

- $\sqrt{45x^3} = \sqrt{9x^2 \cdot 5x}$

$$= \sqrt{9x^2} \cdot \sqrt{5x} = 3x\sqrt{5x}$$

- $\sqrt{8} \cdot \sqrt{3} = \sqrt{24} = \sqrt{4 \cdot 6}$

$$= \sqrt{4} \cdot \sqrt{6} = 2\sqrt{6}$$

The Quotient Rule for Roots

$$\frac{\sqrt[n]{a}}{\sqrt[n]{b}} = \sqrt[n]{\frac{a}{b}} \text{ and } \sqrt[n]{\frac{a}{b}} = \frac{\sqrt[n]{a}}{\sqrt[n]{b}}$$

(The roots represent real numbers and no denominators are 0.)

- $\sqrt{\dfrac{36}{x^2}} = \dfrac{\sqrt{36}}{\sqrt{x^2}} = \dfrac{6}{x}$

- $\sqrt[3]{\dfrac{2}{125}} = \dfrac{\sqrt[3]{2}}{\sqrt[3]{125}} = \dfrac{\sqrt[3]{2}}{5}$

- $\dfrac{\sqrt{50x^4}}{\sqrt{2x}} = \sqrt{\dfrac{50x^4}{2x}} = \sqrt{25x^3}$

$$= \sqrt{25x^2 \cdot x} = 5x\sqrt{x}$$

Definitions and Concepts	**Examples**

Section 8.3 Operations with Radicals

Square roots with the same radicand are like radicals. Like radicals can be added or subtracted using the distributive property. In some cases, radicals can be combined once they have been simplified.

- $7\sqrt{2} + 9\sqrt{2} = (7 + 9)\sqrt{2}$
  $$= 16\sqrt{2}$$
- $6\sqrt{8} - \sqrt{2} = 6\sqrt{4 \cdot 2} - \sqrt{2}$
  $$= 6\sqrt{4}\sqrt{2} - \sqrt{2}$$
  $$= 6 \cdot 2\sqrt{2} - \sqrt{2}$$
  $$= 12\sqrt{2} - \sqrt{2}$$
  $$= (12 - 1)\sqrt{2}$$
  $$= 11\sqrt{2}$$

Radical expressions with more than one term are multiplied in much the same way that polynomials with more than one term are multiplied.

- $\sqrt{5}(\sqrt{2} + \sqrt{5}) = \sqrt{10} + \sqrt{25}$
  $$= \sqrt{10} + 5$$
- $(2 + \sqrt{3})(4 + 5\sqrt{3})$

  $\boxed{F} \quad \boxed{O} \quad \boxed{I} \quad \boxed{L}$

  $= 2 \cdot 4 + 2(5\sqrt{3}) + 4\sqrt{3} + \sqrt{3}(5\sqrt{3})$
  $= 8 + 10\sqrt{3} + 4\sqrt{3} + 15$
  $= 23 + 14\sqrt{3}$

Radical expressions that involve the sum and difference of the same two terms are called conjugates. Use the special product

$$(A + B)(A - B) = A^2 - B^2$$

to multiply conjugates.

$(\sqrt{11} + 5)(\sqrt{11} - 5)$

$= (\sqrt{11})^2 - 5^2 = 11 - 25 = -14$

Section 8.4 Rationalizing the Denominator

The process of rewriting a radical expression as an equivalent expression in which the denominator no longer contains any radicals is called rationalizing the denominator. If the denominator contains the square root of a natural number that is not a perfect square, multiply the numerator and the denominator by the smallest number that produces the square root of a perfect square in the denominator.

- Rationalize the denominator:

  $$\frac{4}{\sqrt{7}}.$$

  $$\frac{4}{\sqrt{7}} = \frac{4}{\sqrt{7}} \cdot \frac{\sqrt{7}}{\sqrt{7}} = \frac{4\sqrt{7}}{7}$$

If the denominator contains two terms, rationalize the denominator by multiplying the numerator and the denominator by the conjugate of the denominator.

$$\frac{9}{7 - \sqrt{5}} = \frac{9}{7 - \sqrt{5}} \cdot \frac{7 + \sqrt{5}}{7 + \sqrt{5}}$$

$$= \frac{9(7 + \sqrt{5})}{7^2 - (\sqrt{5})^2}$$

$$= \frac{9(7 + \sqrt{5})}{49 - 5} = \frac{9(7 + \sqrt{5})}{44}$$

Definitions and Concepts	Examples

Section 8.5 Radical Equations

A radical equation is an equation in which the variable occurs in a radicand.

Solving Radical Equations Containing Square Roots

1. Isolate the radical.
2. Square both sides.
3. Solve the resulting equation.
4. Check proposed solutions in the original equation. Solutions of the squared equation, but not the original equation, are called extraneous solutions.

Solve: $\sqrt{2x + 1} - x = -7$.

$\sqrt{2x + 1} = x - 7$ Isolate the radical.

$(\sqrt{2x + 1})^2 = (x - 7)^2$ Square both sides.

$2x + 1 = x^2 - 14x + 49$ Use $(A - B)^2 = A^2 - 2AB + B^2$.

$0 = x^2 - 16x + 48$ Write in standard form.

$0 = (x - 12)(x - 4)$ Factor.

$x - 12 = 0$ or $x - 4 = 0$ Set each factor equal to 0.

$x = 12$ $x = 4$ Solve the resulting equations.

Check both proposed solutions. 12 checks, but 4 is extraneous. The only solution is 12 and the solution set is {12}.

Section 8.6 Rational Exponents

- $a^{\frac{1}{n}} = \sqrt[n]{a}$

- $a^{\frac{m}{n}} = (\sqrt[n]{a})^m$ or $\sqrt[n]{a^m}$

- $a^{-\frac{m}{n}} = \dfrac{1}{a^{\frac{m}{n}}}$

- $16^{\frac{1}{2}} = \sqrt{16} = 4$

- $27^{\frac{1}{3}} = \sqrt[3]{27} = 3$

- $8^{\frac{5}{3}} = (\sqrt[3]{8})^5 = 2^5 = 32$

- $81^{-\frac{3}{4}} = \dfrac{1}{81^{\frac{3}{4}}} = \dfrac{1}{(\sqrt[4]{81})^3} = \dfrac{1}{3^3} = \dfrac{1}{27}$

CHAPTER 8 REVIEW EXERCISES

8.1 *In Exercises 1–6, find the indicated root, or state that the expression is not a real number.*

1. $\sqrt{121}$

2. $-\sqrt{121}$

3. $\sqrt{-121}$

4. $\sqrt[3]{\dfrac{8}{125}}$

5. $\sqrt[5]{-32}$

6. $-\sqrt[4]{81}$

In Exercises 7–8, use a calculator to approximate each square root. Round to three decimal places.

7. $\sqrt{75}$

8. $\sqrt{398} - 5$

9. The formula $M = 0.7\sqrt{t} + 12.5$ models the average number of nonprogram minutes in an hour of prime-time cable TV t years after 1996. (Nonprogram minutes consist of commercials and plugs for other shows.) According to the model, how many nonprogram minutes disrupted an hour of cable TV action in 2005?

10. Use the model in Exercise 9 to describe what is happening to the number of minutes disrupting an hour of cable TV over time.

11. The formula

$$d = \sqrt{\dfrac{3h}{2}}$$

models the distance, d, in miles, that you can see to the horizon at a height of h feet. The height of the Willis Tower in Chicago is 1450 feet. How far to the horizon can visitors see from the top of the building? Use a calculator and round to the nearest mile.

8.2 *In Exercises 12–19, simplify each expression.*

12. $\sqrt{54}$

13. $6\sqrt{20}$

14. $\sqrt{63x^2}$

15. $\sqrt{48x^3}$

16. $\sqrt{x^8}$

17. $\sqrt{75x^9}$

18. $\sqrt{45x^{23}}$

19. $\sqrt[3]{24}$

In Exercises 20–25, multiply and, if possible, simplify.

20. $\sqrt{7} \cdot \sqrt{11}$

21. $\sqrt{3} \cdot \sqrt{12}$

22. $\sqrt{5x} \cdot \sqrt{10x}$

23. $\sqrt{3x^2} \cdot \sqrt{4x^3}$

24. $\sqrt[3]{6} \cdot \sqrt[3]{9}$

25. $\sqrt{\frac{5}{2}} \cdot \sqrt{\frac{3}{8}}$

In Exercises 26–33, simplify using the quotient rule.

26. $\sqrt{\frac{121}{4}}$

27. $\sqrt{\frac{7x}{25}}$

28. $\sqrt{\frac{18}{x^2}}$

29. $\frac{\sqrt{200}}{\sqrt{2}}$

30. $\frac{\sqrt{96}}{\sqrt{3}}$

31. $\frac{\sqrt{72x^8}}{\sqrt{x^3}}$

32. $\sqrt[3]{\frac{5}{64}}$

33. $\sqrt[3]{\frac{40}{27}}$

8.3   *In Exercises 34–39, add or subtract as indicated.*

34. $7\sqrt{5} + 13\sqrt{5}$

35. $\sqrt{8} + \sqrt{50}$

36. $\sqrt{75} - \sqrt{48}$

37. $2\sqrt{80} + 3\sqrt{45}$

38. $4\sqrt{72} - 2\sqrt{48}$

39. $2\sqrt{18} + 3\sqrt{27} - \sqrt{12}$

In Exercises 40–47, multiply as indicated and, if possible, simplify.

40. $\sqrt{10}(\sqrt{5} + \sqrt{6})$

41. $\sqrt{3}(\sqrt{6} - \sqrt{12})$

42. $(9 + \sqrt{2})(10 + \sqrt{2})$

43. $(1 + 3\sqrt{7})(4 - \sqrt{7})$

44. $(\sqrt{3} + 2)(\sqrt{6} - 4)$

45. $(2 + \sqrt{7})(2 - \sqrt{7})$

46. $(\sqrt{11} - \sqrt{5})(\sqrt{11} + \sqrt{5})$

47. $(1 + \sqrt{2})^2$

8.4   *In Exercises 48–56, rationalize each denominator and, if possible, simplify.*

48. $\frac{30}{\sqrt{5}}$

49. $\frac{13}{\sqrt{50}}$

50. $\sqrt{\frac{2}{3}}$

51. $\sqrt{\frac{3}{8}}$

52. $\sqrt{\frac{17}{x}}$

53. $\frac{11}{\sqrt{5} + 2}$

54. $\frac{21}{4 - \sqrt{3}}$

55. $\frac{12}{\sqrt{5} + \sqrt{3}}$

56. $\frac{7\sqrt{2}}{\sqrt{2} - 4}$

8.5   *In Exercises 57–63, solve each radical equation.*

57. $\sqrt{x + 3} = 4$

58. $\sqrt{2x + 3} = 5$

59. $3\sqrt{x} = \sqrt{6x + 15}$

60. $\sqrt{5x + 1} = x + 1$

61. $\sqrt{x + 1} + 5 = x$

62. $\sqrt{x - 2} + 5 = 1$

63. $x = \sqrt{x^2 + 4x + 4}$

64. Paleontologists use the formula

$$W = 4\sqrt{2x}$$

to estimate the walking speed of a dinosaur, W, in feet per second, where x is the length, in feet, of the dinosaur's leg. What was the leg length of a dinosaur whose walking speed was 16 feet per second?

65. By the end of 2010, women made up more than half of the labor force in the United States, for the first time in the country's history. The graphs show the percentage of jobs in the U.S. labor force held by men and by women from 1972 through 2009.

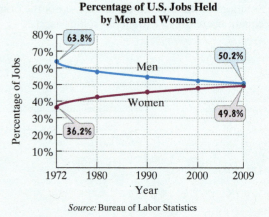

Percentage of U.S. Jobs Held by Men and Women

Source: Bureau of Labor Statistics

The formula

$$p = 2.2\sqrt{t} + 36.2$$

models the percentage of jobs in the U.S. labor force, p, held by women t years after 1972.

a. Use the appropriate graph to estimate the percentage of jobs in the U.S. labor force held by women in 2000. Give your estimation to the nearest percent.

b. Use the mathematical model to determine the percentage of jobs in the U.S. labor force held by women in 2000. Round to the nearest tenth of a percent.

c. According to the formula, when will 52% of jobs in the U.S. labor force be held by women? Round to the nearest year.

8.6   *In Exercises 66–71, simplify by first writing the expression in radical form.*

66. $16^{\frac{1}{2}}$

67. $125^{\frac{1}{3}}$

68. $64^{\frac{2}{3}}$

69. $25^{-\frac{1}{2}}$

70. $27^{-\frac{1}{3}}$

71. $(-8)^{-\frac{4}{3}}$

72. The formula $S = 28.6A^{\frac{1}{3}}$ models the number of plant species, S, on the various islands of the Galápagos chain in terms of the area, A, in square miles, of a particular island. Approximately how many species of plants are there on a Galápagos island whose area is 8 square miles?

CHAPTER 8 TEST

Step-by-step test solutions are found on the Chapter Test Prep Videos available in MyMathLab® or on YouTube (search "BlitzerIntroAlg7e" and click on "Channels").

In Exercises 1–2, find the indicated root, or state that the expression is not a real number.

1. $-\sqrt{64}$

2. $\sqrt[3]{64}$

In Exercises 3–8, simplify each expression.

3. $\sqrt{48}$

4. $\sqrt{72x^3}$

5. $\sqrt{x^{29}}$

6. $\sqrt{\dfrac{25}{x^2}}$

7. $\sqrt{\dfrac{75}{27}}$

8. $\sqrt[3]{\dfrac{5}{8}}$

In Exercises 9–18, perform the indicated operation and, if possible, simplify.

9. $\dfrac{\sqrt{80x^4}}{\sqrt{2x^2}}$

10. $\sqrt{10} \cdot \sqrt{5}$

11. $\sqrt{6x} \cdot \sqrt{6y}$

12. $\sqrt{10x^2} \cdot \sqrt{2x^3}$

13. $\sqrt{24} + 3\sqrt{54}$

14. $7\sqrt{8} - 2\sqrt{32}$

15. $\sqrt{3}(\sqrt{10} + \sqrt{3})$

16. $(7 - \sqrt{5})(10 + 3\sqrt{5})$

17. $(\sqrt{6} + 2)(\sqrt{6} - 2)$

18. $(3 + \sqrt{7})^2$

In Exercises 19–20, rationalize each denominator and, if possible, simplify.

19. $\dfrac{4}{\sqrt{5}}$

20. $\dfrac{5}{4 + \sqrt{3}}$

In Exercises 21–22, solve each radical equation.

21. $\sqrt{3x} + 5 = 11$

22. $\sqrt{2x - 1} = x - 2$

23. The less income people have, the more likely they are to report that their health is fair or poor. The formula

$$p = -4.4\sqrt{x} + 38$$

models the percentage of Americans reporting fair or poor health, p, in terms of annual income, x, in thousands of dollars. According to the formula, what is the income, in thousands of dollars, for the group in which 16% report fair or poor health?

In Exercises 24–25, simplify by first writing the expression in radical form.

24. $8^{\frac{2}{3}}$

25. $9^{-\frac{1}{2}}$

CUMULATIVE REVIEW EXERCISES   (Chapters 1–8)

In Exercises 1–6, solve each equation or system of equations.

1. $2x + 3x - 5 + 7 = 10x + 3 - 6x - 4$

2. $2x^2 + 5x = 12$

3. $\begin{cases} 8x - 5y = -4 \\ 2x + 15y = -66 \end{cases}$

4. $\dfrac{15}{x} - 4 = \dfrac{6}{x} + 3$

5. $-3x - 7 = 8$

6. $\sqrt{2x - 1} - x = -2$

In Exercises 7–11, simplify each expression.

7. $\dfrac{8x^3}{-4x^7}$

8. $6\sqrt{75} - 4\sqrt{12}$

9. $\dfrac{\dfrac{1}{x} - \dfrac{1}{2}}{\dfrac{1}{3} - \dfrac{x}{6}}$

10. $\dfrac{4 - x^2}{3x^2 - 5x - 2}$

11. $-5 - (-8) - (4 - 6)$

In Exercises 12–13, factor completely.

12. $x^2 - 18x + 77$

13. $x^3 - 25x$

In Exercises 14–18, perform the indicated operations. If possible, simplify the answer.

14. $\dfrac{6x^3 - 19x^2 + 16x - 4}{x - 2}$

15. $(2x - 3)(4x^2 + 6x + 9)$

16. $\dfrac{3x}{x^2 + x - 2} - \dfrac{2}{x + 2}$

17. $\dfrac{5x^2 - 6x + 1}{x^2 - 1} \div \dfrac{16x^2 - 9}{4x^2 + 7x + 3}$

18. $\sqrt{12} - 4\sqrt{75}$

In Exercises 19–21, graph each equation or inequality in a rectangular coordinate system.

19. $2x - y = 4$

20. $y = -\dfrac{2}{3}x$

21. $x \geq -1$

22. Find the slope of the line through $(-1, 5)$ and $(2, -3)$.

23. Write the point-slope form of the equation of the line with slope 5, passing through $(-2, -3)$. Then use the point-slope equation to write the slope-intercept form of the line's equation.

24. Seven subtracted from five times a number is 208. Find the number.

25. Park rangers catch, tag, and then release 318 deer back into a state park. Two weeks later, they select a sample of 168 deer, 56 of which are tagged. Assuming the ratio of tagged deer in the sample holds for all deer in the park, approximately how many deer are in the park?

Quadratic Equations and Introduction to Functions

America is a nation of immigrants. Since 1820, more than 40 million people have immigrated to the United States from all over the world. They chose to come for various reasons, such as to live in freedom, to practice religion without persecution, to escape poverty or oppression, and to make better lives for themselves and their children. As a result, in 2012, 12.9% of the United States population was foreign-born. How can we use mathematical models to project when this percentage might grow to one in every five Americans, or even more?

Here's where you'll find this application:

In this chapter, you will learn techniques for solving quadratic equations,

$$ax^2 + bx + c = 0,$$

that will give you new ways of exploring mathematical models. The United States foreign-born model appears as Example 3 in Section 9.3.

SECTION

9.1

Solving Quadratic Equations by the Square Root Property ▶

What am I supposed to learn?

After studying this section, you should be able to:

1. Solve quadratic equations using the square root property. ▶

2. Solve problems using the Pythagorean Theorem. ▶

3. Find the distance between two points. ▶

Shown here is Renaissance artist Raphael Sanzio's (1483–1520) image of Pythagoras from *The School of Athens* mural. Detail of left side.

Raphael (1483–1520) "School of Athens" detail of left side. Stanza della Segnatura, Vatican Palace, Vatican State. Scala/Art Resource, NY

Pythagoras

For the followers of the Greek mathematician Pythagoras in the sixth century B.C., numbers took on a life-and-death importance. The "Pythagorean Brotherhood" was a secret group whose members were convinced that properties of whole numbers were the key to understanding the universe. Members of the Brotherhood (which admitted women) thought that all numbers that were not whole numbers could be represented as the ratio of whole numbers. A crisis occurred for the Pythagoreans when they discovered the existence of a number that was not rational. Because the Pythagoreans viewed numbers with reverence and awe, the punishment for speaking about this irrational number was death. However, a member of the Brotherhood revealed the secret of the irrational number's existence. When he later died in a shipwreck, his death was viewed as punishment from the gods.

In this section, you will work with the triangle that led the Pythagoreans to the discovery of irrational numbers. You will find the lengths of one of the triangle's sides using a property other than factoring that can be used to solve quadratic equations.

1 Solve quadratic equations using the square root property. ▶

The Square Root Property

Let's begin with a relatively simple quadratic equation:

$$x^2 = 9.$$

The value of x must be a number whose square is 9. There are two numbers whose square is 9:

$$x = \sqrt{9} = 3 \quad \text{or} \quad x = -\sqrt{9} = -3.$$

Thus, the solutions of $x^2 = 9$ are 3 and -3. This is an example of the **square root property**.

The Square Root Property

If u is an algebraic expression and d is a positive real number, then $u^2 = d$ has exactly two solutions:

$$\text{If } u^2 = d, \quad \text{then } u = \sqrt{d} \text{ or } u = -\sqrt{d}.$$

Equivalently,

$$\text{If } u^2 = d, \quad \text{then } u = \pm\sqrt{d}.$$

Notice that $u = \pm\sqrt{d}$ is a shorthand notation to indicate that $u = \sqrt{d}$ or $u = -\sqrt{d}$. Although we usually read $u = \pm\sqrt{d}$ as "u equals plus or minus the square root of d," we actually mean that u is the positive square root of d or the negative square root of d.

EXAMPLE 1 Solving Quadratic Equations by the Square Root Property

Solve by the square root property:

a. $x^2 = 49$ **b.** $4x^2 = 20$ **c.** $2x^2 - 5 = 0$.

Solution

a.

$x^2 = 49$	This is the original equation.
$x = \sqrt{49}$ or $x = -\sqrt{49}$	Apply the square root property. You can also write $x = \pm\sqrt{49}$.
$x = 7$ or $x = -7$	In abbreviated notation, $x = \pm 7$.

Substitute both values into the original equation and confirm that the solutions are 7 and -7. The solution set is $\{\pm 7\}$.

b. To apply the square root property, we need a squared expression by itself on one side of the equation.

$$4x^2 = 20$$

> We want x^2 by itself.

We can get x^2 by itself if we divide both sides by 4.

$4x^2 = 20$	This is the original equation.
$\dfrac{4x^2}{4} = \dfrac{20}{4}$	Divide both sides by 4.
$x^2 = 5$	Simplify.
$x = \sqrt{5}$ or $x = -\sqrt{5}$	Apply the square root property.

Now let's check these proposed solutions in the original equation. Because the equation has an x^2-term and no x-term, we can check both values, $\pm\sqrt{5}$, at once.

Check $\sqrt{5}$ and $-\sqrt{5}$:

$4x^2 = 20$	This is the original equation.
$4(\pm\sqrt{5})^2 \stackrel{?}{=} 20$	Substitute the proposed solutions.
$4 \cdot 5 \stackrel{?}{=} 20$	$(\pm\sqrt{5})^2 = 5$
$20 = 20,$	true

The solutions are $-\sqrt{5}$ and $\sqrt{5}$. The solution set is $\{-\sqrt{5}, \sqrt{5}\}$ or $\{\pm\sqrt{5}\}$.

c. To solve $2x^2 - 5 = 0$ by the square root property, we must isolate the squared expression by itself on one side of the equation.

$$2x^2 - 5 = 0$$

> We want x^2 by itself.

$$2x^2 - 5 = 0 \qquad \text{This is the original equation.}$$

$$2x^2 = 5 \qquad \text{Add 5 to both sides.}$$

$$x^2 = \frac{5}{2} \qquad \text{Divide both sides by 2.}$$

$$x = \sqrt{\frac{5}{2}} \quad \text{or} \quad x = -\sqrt{\frac{5}{2}} \qquad \text{Apply the square root property.}$$

In this section, we will express irrational solutions in simplified radical form, rationalizing denominators when possible. Because the proposed solutions are opposites, we can rationalize both denominators at once:

$$\pm\sqrt{\frac{5}{2}} = \pm\frac{\sqrt{5}}{\sqrt{2}} = \pm\frac{\sqrt{5}}{\sqrt{2}} \cdot \frac{\sqrt{2}}{\sqrt{2}} = \pm\frac{\sqrt{10}}{2}.$$

Substitute these values into the original equation and verify that the solutions are $-\dfrac{\sqrt{10}}{2}$ and $\dfrac{\sqrt{10}}{2}$. The solution set is $\left\{-\dfrac{\sqrt{10}}{2}, \dfrac{\sqrt{10}}{2}\right\}$ or $\left\{\pm\dfrac{\sqrt{10}}{2}\right\}$. ■

✓ **CHECK POINT 1** Solve by the square root property:

a. $x^2 = 36$ **b.** $5x^2 = 15$

c. $2x^2 - 7 = 0$.

Can we solve an equation such as $(x - 5)^2 = 16$ using the square root property? Yes. The equation is in the form $u^2 = d$, where u^2, the squared expression, is by itself on the left side.

$$(x - 5)^2 = 16$$

This is u^2 in $u^2 = d$ with $u = x - 5$.

This is d in $u^2 = d$ with $d = 16$.

EXAMPLE 2 Solving a Quadratic Equation by the Square Root Property

Solve by the square root property: $(x - 5)^2 = 16$.

Solution

$$(x - 5)^2 = 16 \qquad \text{This is the original equation.}$$

$$x - 5 = \sqrt{16} \quad \text{or} \quad x - 5 = -\sqrt{16} \qquad \text{Apply the square root property.}$$

$$x - 5 = 4 \quad \text{or} \quad x - 5 = -4 \qquad \text{Simplify.}$$

$$x = 9 \qquad\qquad x = 1 \qquad \text{Add 5 to both sides in each equation.}$$

Substitute both values into the original equation and confirm that the solutions are 9 and 1. The solutions are visually confirmed in the Using Technology box on the next page. The solution set is {1, 9}. ■

Using Technology

Graphic Connections

We can use graphs to confirm that the solution set of $(x - 5)^2 = 16$ is $\{1, 9\}$.

The graph of $y = (x - 5)^2 - 16$ has x-intercepts at 1 and 9. The solutions of $(x - 5)^2 - 16 = 0$, or $(x - 5)^2 = 16$, are 1 and 9.

Another option is to graph each side of the equation and find the x-coordinates of the intersection points. Graphing

$$y_1 = (x - 5)^2$$

and

$$y_2 = 16,$$

the x-coordinates of the intersection points are 1 and 9.

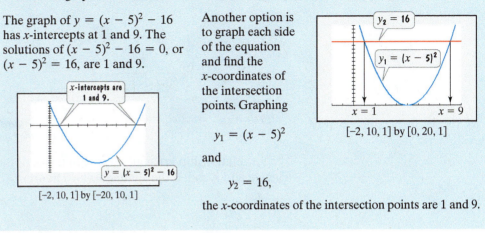

x-intercepts are 1 and 9.

$y = (x - 5)^2 - 16$

$[-2, 10, 1]$ by $[-20, 10, 1]$

$y_2 = 16$

$y_1 = (x - 5)^2$

$x = 1$ $x = 9$

$[-2, 10, 1]$ by $[0, 20, 1]$

✓ **CHECK POINT 2** Solve by the square root property: $(x - 3)^2 = 25$.

EXAMPLE 3 Solving a Quadratic Equation by the Square Root Property

Solve by the square root property: $(x - 1)^2 = 5$.

Solution

$$(x - 1)^2 = 5 \qquad \text{This is the original equation.}$$

$$x - 1 = \sqrt{5} \quad \text{or} \quad x - 1 = -\sqrt{5} \qquad \text{Apply the square root property.}$$

$$x = 1 + \sqrt{5} \qquad x = 1 - \sqrt{5} \qquad \text{Add 1 to both sides in each equation.}$$

Check $1 + \sqrt{5}$:

$$(x - 1)^2 = 5$$
$$(1 + \sqrt{5} - 1)^2 \overset{?}{=} 5$$
$$(\sqrt{5})^2 \overset{?}{=} 5$$
$$5 = 5, \quad \text{true}$$

Check $1 - \sqrt{5}$:

$$(x - 1)^2 = 5$$
$$(1 - \sqrt{5} - 1)^2 \overset{?}{=} 5$$
$$(-\sqrt{5})^2 \overset{?}{=} 5$$
$$5 = 5, \quad \text{true}$$

The solutions are $1 + \sqrt{5}$ and $1 - \sqrt{5}$, expressed in abbreviated notation as $1 \pm \sqrt{5}$. The solution set is $\left\{1 \pm \sqrt{5}\right\}$. ∎

✓ **CHECK POINT 3** Solve by the square root property: $(x - 2)^2 = 7$.

---

2 Solve problems using the Pythagorean Theorem. ▶

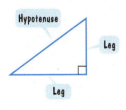

Hypotenuse

Leg

Leg

The Pythagorean Theorem and the Square Root Property

The ancient Greek philosopher and mathematician Pythagoras (approximately 582–500 B.C.) founded a school whose motto was "All is number." Pythagoras is best remembered for his work with the **right triangle**, a triangle with one angle measuring 90°. The side opposite the 90° angle is called the **hypotenuse**. The other sides are called **legs**.

Pythagoras found that if he constructed squares on each of the legs, as well as a larger square on the hypotenuse, the sum of the areas of the smaller squares is equal to the area of the larger square. This is illustrated in **Figure 9.1**.

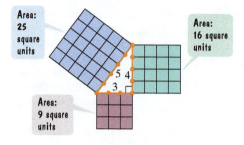

Area:
25
square
units

Area:
16 square
units

Area:
9 square
units

Figure 9.1 The area of the large square equals the sum of the areas of the smaller squares.

This relationship is usually stated in terms of the lengths of the three sides of a right triangle and is called the **Pythagorean Theorem**.

The Pythagorean Theorem

The sum of the squares of the lengths of the legs of a right triangle equals the square of the length of the hypotenuse.

If the legs have lengths a and b, and the hypotenuse has length c, then

$$a^2 + b^2 = c^2.$$

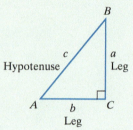

EXAMPLE 4   Screen Math

Did you know that the size of a television screen refers to the length of its diagonal? If the length of the HDTV screen in **Figure 9.2** is 28 inches and its width is 15.7 inches, what is the size of the screen to the nearest inch?

28 in.

15.7 in.

Figure 9.2

Solution   **Figure 9.2** shows that the length, width, and diagonal of the screen form a right triangle. The diagonal is the hypotenuse of the triangle. We use the Pythagorean Theorem with $a = 28$, $b = 15.7$, and solve for the screen size, c.

$$a^2 + b^2 = c^2$$ This is the Pythagorean Theorem.

$$28^2 + 15.7^2 = c^2$$ Let $a = 28$ and $b = 15.7$.

$$784 + 246.49 = c^2$$ $28^2 = 28 \cdot 28 = 784$ and $15.7^2 = 15.7 \cdot 15.7 = 246.49$.

$$c^2 = 1030.49$$ Add: $784 + 246.49 = 1030.49$. We also reversed the two sides.

$$c = \sqrt{1030.49} \quad \text{or} \quad c = -\sqrt{1030.49}$$ Apply the square root property.

$$c \approx 32 \quad \text{or} \quad c \approx -32$$ Use a calculator and round to the nearest inch.

1030.49 $\boxed{\sqrt{}}$ $\boxed{=}$ or $\boxed{\sqrt{}}$ 1030.49 $\boxed{\text{ENTER}}$

Because c represents the size of the screen, this dimension must be positive. We reject -32. Thus, the screen size of the HDTV is 32 inches. ■

Great Question!

Why did you include a decimal like 15.7 in the screen math example? Squaring 15.7 is awkward. Why not just use 16 inches for the length of the screen?

We wanted to use the actual dimensions for a 32-inch HDTV screen. In the Check Point that follows, we use the exact dimensions of an "old" TV screen (prior to HDTV). A calculator is helpful in squaring each of these dimensions.

✓ **CHECK POINT 4**   **Figure 9.3** shows the dimensions of an old TV screen. What is the size of the screen?

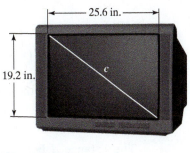

Figure 9.3

Blitzer Bonus ⏵

Screen Math

That new 32-inch HDTV you want: How much larger than your old 32-incher is it? Actually, based on Example 4 and Check Point 4, it's smaller! **Figure 9.4** compares the screen area of the old 32-inch TV in Check Point 4 with the 32-inch HDTV in Example 4.

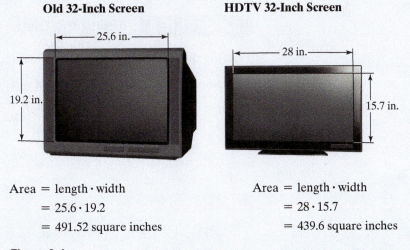

Old 32-Inch Screen

Area = length · width
    = 25.6 · 19.2
    = 491.52 square inches

HDTV 32-Inch Screen

Area = length · width
    = 28 · 15.7
    = 439.6 square inches

Figure 9.4

To make sure your HDTV has the same screen area as your old TV, it needs to have a diagonal measure, or screen size, that is 6% larger. Equivalently, take the screen size of the old TV and multiply by 1.06. If you have a 32-inch regular TV, this means the HDTV needs a 34-inch screen ($32 \times 1.06 = 33.92 \approx 34$) if you don't want your new TV picture to be smaller than the old one.

3 Find the distance between two points.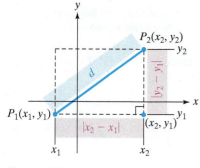

Figure 9.5

The Distance Formula

Using the Pythagorean Theorem, we can find the distance between the two points $P_1(x_1, y_1)$ and $P_2(x_2, y_2)$ in the rectangular coordinate system. The two points are illustrated in **Figure 9.5**.

The distance that we need to find is represented by d and shown in blue. Notice that the distance between two points on the dashed horizontal line is the absolute value of the difference between the x-coordinates of the two points. This distance, $|x_2 - x_1|$, is shown in pink. Similarly, the distance between two points on the dashed vertical line is the absolute value of the difference between the y-coordinates of the two points. This distance, $|y_2 - y_1|$, is also shown in pink.

Because the dashed lines are horizontal and vertical, a right triangle is formed. Thus, we can use the Pythagorean Theorem to find the distance d. Squaring the lengths of the triangle's sides results in positive numbers, so absolute value notation is not necessary.

$$d^2 = (x_2 - x_1)^2 + (y_2 - y_1)^2$$ Apply the Pythagorean Theorem to the right triangle in Figure 9.5.

$$d = \pm\sqrt{(x_2 - x_1)^2 + (y_2 - y_1)^2}$$ Apply the square root property.

$$d = \sqrt{(x_2 - x_1)^2 + (y_2 - y_1)^2}$$ Because distance is nonnegative, write only the principal square root.

This result is called the **distance formula**.

> ### The Distance Formula
>
> The distance, d, between the points (x_1, y_1) and (x_2, y_2) in the rectangular coordinate system is
>
> $$d = \sqrt{(x_2 - x_1)^2 + (y_2 - y_1)^2}.$$
>
> To compute the distance between two points, find the square of the difference between the x-coordinates plus the square of the difference between the y-coordinates. The principal square root of this sum is the distance.

When using the distance formula, it does not matter which point you call (x_1, y_1) and which you call (x_2, y_2).

EXAMPLE 5   Using the Distance Formula

Find the distance between $(-4, -3)$ and $(6, 2)$.

Solution   We will let $(x_1, y_1) = (-4, -3)$ and $(x_2, y_2) = (6, 2)$.

$$d = \sqrt{(x_2 - x_1)^2 + (y_2 - y_1)^2}$$ Use the distance formula.

$$= \sqrt{[6 - (-4)]^2 + [2 - (-3)]^2}$$ Substitute the given values.

$$= \sqrt{10^2 + 5^2}$$ Perform operations within grouping symbols: $6 - (-4) = 6 + 4 = 10$ and $2 - (-3) = 2 + 3 = 5$.

Caution! This is not equal to $\sqrt{100} + \sqrt{25}$.

$$= \sqrt{100 + 25}$$ Square 10 and 5.

$$= \sqrt{125}$$ Add.

$$= 5\sqrt{5} \approx 11.18$$ $\sqrt{125} = \sqrt{25 \cdot 5} = \sqrt{25}\sqrt{5} = 5\sqrt{5}$

The distance between the given points is $5\sqrt{5}$ units, or approximately 11.18 units. The situation is illustrated in **Figure 9.6**. ∎

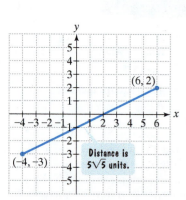

Figure 9.6 Finding the distance between two points

 ✓ **CHECK POINT 5**   Find the distance between $(-4, 9)$ and $(1, -3)$.

CONCEPT AND VOCABULARY CHECK

Fill in each blank so that the resulting statement is true.

1. The square root property states that if $u^2 = d$, then $u = $ _____, $d > 0$.

2. A triangle with one angle measuring 90° is called a/an _____ triangle. The side opposite the 90° angle is called the _____. The other sides are called _____.

3. The Pythagorean Theorem states that in any _____ triangle, the sum of the squares of the lengths of the _____ equals _____.

4. The distance, d, between the points (x_1, y_1) and (x_2, y_2) in the rectangular coordinate system is given by the formula $d = $ _____.

9.1 EXERCISE SET   ▶ MyMathLab®

Practice Exercises

In Exercises 1–30, solve each quadratic equation by the square root property. If possible, simplify radicals or rationalize denominators.

1. $x^2 = 16$
2. $x^2 = 100$
3. $y^2 = 81$
4. $y^2 = 144$
5. $x^2 = 7$
6. $x^2 = 13$
7. $x^2 = 50$
8. $x^2 = 27$
9. $5x^2 = 20$
10. $3x^2 = 75$
11. $4y^2 = 49$
12. $16y^2 = 25$
13. $2x^2 + 1 = 51$
14. $3x^2 - 1 = 47$
15. $3x^2 - 2 = 0$
16. $3x^2 - 5 = 0$
17. $5z^2 - 7 = 0$
18. $5z^2 - 2 = 0$
19. $(x - 3)^2 = 16$
20. $(x - 2)^2 = 25$
21. $(x + 5)^2 = 121$
22. $(x + 6)^2 = 144$
23. $(3x + 2)^2 = 9$
24. $(2x + 1)^2 = 49$
25. $(x - 5)^2 = 3$
26. $(x - 3)^2 = 15$
27. $(y + 8)^2 = 11$
28. $(y + 7)^2 = 5$
29. $(z - 4)^2 = 18$
30. $(z - 6)^2 = 12$

In Exercises 31–40, solve each quadratic equation by first factoring the perfect square trinomial on the left side. Then apply the square root property. Simplify radicals, if possible.

31. $x^2 + 4x + 4 = 16$
32. $x^2 + 4x + 4 = 25$
33. $x^2 - 6x + 9 = 36$
34. $x^2 - 6x + 9 = 49$
35. $x^2 - 10x + 25 = 2$
36. $x^2 - 10x + 25 = 3$
37. $x^2 + 2x + 1 = 5$
38. $x^2 + 2x + 1 = 7$
39. $y^2 - 14y + 49 = 12$
40. $y^2 - 14y + 49 = 18$

In Exercises 41–48, use the Pythagorean Theorem to find the missing length in each right triangle. Express the answer in radical form and simplify, if possible.

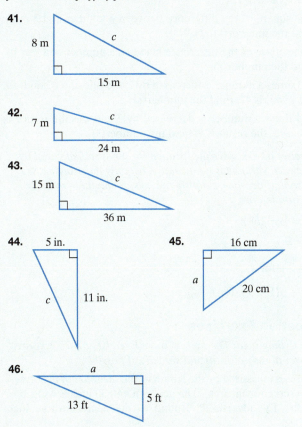

41.
8 m c
15 m

42.
7 m c
24 m

43.
15 m c
36 m

44.
5 in.
c 11 in.

45.
16 cm
a 20 cm

46.
a
13 ft   5 ft

47.

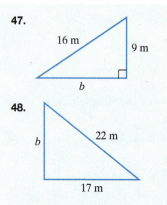

16 m · 9 m · b

48.

b · 22 m · 17 m

In Exercises 49–58, find the distance between each pair of points. Express answers in simplified radical form and, if necessary, round to two decimal places.

49. (3, 5) and (4, 1)

50. (1, 5) and (6, 2)

51. (−4, 2) and (4, 17)

52. (2, −2) and (5, 2)

53. (6, −1) and (9, 5)

54. (−4, −1) and (2, −3)

55. (−7, −5) and (−2, −1)

56. (−8, −4) and (−3, −8)

57. $(-2\sqrt{7}, 10)$ and $(4\sqrt{7}, 8)$

58. $(-\sqrt{3}, 4\sqrt{6})$ and $(2\sqrt{3}, \sqrt{6})$

Practice PLUS

59. The square of the difference between a number and 3 is 25. Find the number(s).

60. The square of the difference between a number and 7 is 16. Find the number(s).

61. If 3 times a number is increased by 2 and this sum is squared, the result is 49. Find the number(s).

62. If 4 times a number is decreased by 3 and this difference is squared, the result is 9. Find the number(s).

In Exercises 63–66, solve the formula for the specified variable. Because each variable is nonnegative, list only the principal square root. If possible, simplify radicals or eliminate radicals from denominators.

63. $A = \pi r^2$ for r

64. $ax^2 - b = 0$ for x

65. $I = \dfrac{k}{d^2}$ for d

66. $A = p(1 + r)^2$ for r

Application Exercises

Use the Pythagorean Theorem to solve Exercises 67–72. Express the answer in radical form and simplify, if possible.

67. A TV is measured by the length of the diagonal of the screen. If a screen measures 20 inches high and 30 inches wide, how is the TV advertised? Round to the nearest whole inch.

68. A TV is measured by the length of the diagonal of the screen. If a screen measures 21 inches high and 24 inches wide, how is the TV advertised? Round to the nearest whole inch.

69. Find the length of the ladder.

? · 8 ft · 10 ft

70. How high is the airplane above the ground?

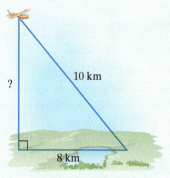

? · 10 km · 8 km

71. A baseball diamond is actually a square with 90-foot sides. What is the distance from home plate to second base?

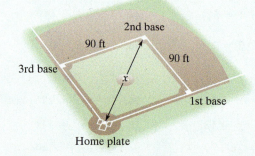

2nd base · 90 ft · 90 ft · 3rd base · x · 1st base · Home plate

72. The base of a 20-foot ladder is 15 feet from the house. How far up the house does the ladder reach?

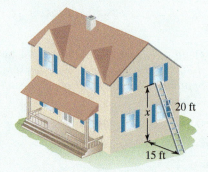

20 ft · x · 15 ft

Use the formula for the area of a circle, $A = \pi r^2$, to solve Exercises 73–74.

73. If the area of a circle is 36π square inches, find its radius.

74. If the area of a circle is 49π square inches, find its radius.

The weight of a human fetus is modeled by the formula $W = 3t^2$, where W is the weight, in grams, and t is the time, in weeks, with $0 \le t \le 39$. Use this formula to solve Exercises 75–76.

75. After how many weeks does the fetus weigh 108 grams?

76. After how many weeks does the fetus weigh 192 grams?

The distance, d, in feet, that an object falls in t seconds is modeled by the formula $d = 16t^2$. Use this formula to solve Exercises 77–78.

77. If you drop a rock from a cliff 400 feet above the water, how long will it take for the rock to hit the water?

78. If you drop a rock from a cliff 576 feet above the water, how long will it take for the rock to hit the water?

79. A square flower bed is to be enlarged by adding 2 meters on each side. If the larger square has an area of 144 square meters, what is the length of the original square?

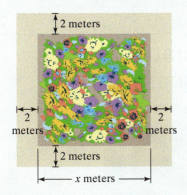

80. A square flower bed is to be enlarged by adding 3 feet on each side. If the larger square has an area of 169 square feet, what is the length of the original square?

81. A machine produces open boxes using square sheets of metal. The figure illustrates that the machine cuts equal-sized squares measuring 2 inches on a side from the corners and then shapes the metal into an open box by turning up the sides. If each box must have a volume of 200 cubic inches, find the size of the length and width of the open box.

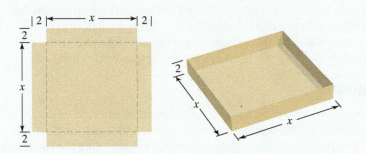

82. A machine produces open boxes using square sheets of metal. The machine cuts equal-sized squares measuring 3 inches on a side from the corners and then shapes the metal into an open box by turning up the sides. If each box must have a volume of 75 cubic inches, find the size of the length and width of the open box.

Explaining the Concepts

83. What is the square root property?

84. Explain how to solve $(x - 1)^2 = 16$ using the square root property.

85. In your own words, state the Pythagorean Theorem.

86. In the 1939 movie *The Wizard of Oz*, upon being presented with a Th.D. (Doctor of Thinkology), the Scarecrow proudly exclaims, "The sum of the square roots of any two sides of an isosceles triangle is equal to the square root of the remaining side." Did the Scarecrow get the Pythagorean Theorem right? In particular, describe four errors in the Scarecrow's statement.

Critical Thinking Exercises

Make Sense? *In Exercises 87–90, determine whether each statement makes sense or does not make sense, and explain your reasoning.*

87. I have a graphing calculator, so I used the x-coordinates of the intersection points of the graphs of $y_1 = (x - 2)^2$ and $y_2 = 25$ to verify the solution set for $(x - 2)^2 = 25$.

88. When I'm given the picture of a right triangle, the hypotenuse is always the side on top.

89. I've noticed that in mathematics there is a connection between topics, such as using the Pythagorean Theorem to derive the distance formula.

90. When I use the square root property to determine the length of a right triangle's side, I don't even bother to list the negative square root.

In Exercises 91–94, determine whether each statement is true or false. If the statement is false, make the necessary change(s) to produce a true statement.

91. The equation $(x + 5)^2 = 8$ is equivalent to $x + 5 = 2\sqrt{2}$.

92. The equation $x^2 = 0$ has no solution.

93. The equation $x^2 = -1$ has no solutions that are real numbers.

94. The solutions of $3x^2 - 5 = 0$ are $\dfrac{\sqrt{5}}{3}$ and $-\dfrac{\sqrt{5}}{3}$.

95. Find the value(s) of x if the distance between $(-3, -2)$ and $(x, -5)$ is 5 units.

Technology Exercises

96. Use a graphing utility to solve $4 - (x + 1)^2 = 0$. Graph $y = 4 - (x + 1)^2$ in a $[-5, 5, 1]$ by $[-5, 5, 1]$ viewing rectangle. The equation's solutions are the graph's x-intercepts. Check by substitution in the given equation.

97. Use a graphing utility to solve $(x - 1)^2 - 9 = 0$. Graph $y = (x - 1)^2 - 9$ in a $[-5, 5, 1]$ by $[-9, 3, 1]$ viewing rectangle. The equation's solutions are the graph's x-intercepts. Check by substitution in the given equation.

Review Exercises

98. Factor completely: $12x^2 + 14x - 6$.
(Section 6.5, Example 2)

99. Divide: $\dfrac{x^2 - x - 6}{3x - 3} \div \dfrac{x^2 - 4}{x - 1}$.
(Section 7.2, Example 6)

100. Solve: $4(x - 5) = 22 + 2(6x + 3)$.
(Section 2.3, Example 3)

Preview Exercises

Exercises 101–103 will help you prepare for the material covered in the next section.

101. Factor: $x^2 + 8x + 16$.

102. Factor: $x^2 - 14x + 49$.

103. Factor: $x^2 + 5x + \dfrac{25}{4}$.

9.2

Solving Quadratic Equations by Completing the Square

What am I supposed to learn?

After studying this section, you should be able to:

1 Complete the square of a binomial.

2 Solve quadratic equations by completing the square. ▶

Rind (1955), M. C. Escher © 2011 M. C. Escher Company, Holland.

There is a lack of completion in both the Escher image and the unfinished square on the left. Completion for the geometric figure can be obtained by adding a small square to its upper-right-hand corner. Understanding this process algebraically will give you a new method, appropriately called *completing the square*, for solving quadratic equations.

1 Complete the square a binomial.

Completing the Square

How do we solve a quadratic equation, $ax^2 + bx + c = 0$, if the trinomial $ax^2 + bx + c$ cannot be factored? We can convert the equation into an equivalent equation that can be solved using the square root property. This is accomplished by **completing the square**.

Completing the Square

If $x^2 + bx$ is a binomial, then by adding $\left(\dfrac{b}{2}\right)^2$, which is the square of half the coefficient of x, a perfect square trinomial will result. That is,

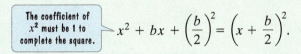

$$x^2 + bx + \left(\frac{b}{2}\right)^2 = \left(x + \frac{b}{2}\right)^2.$$

The coefficient of x^2 must be 1 to complete the square.

EXAMPLE 1 Completing the Square

Complete the square for each binomial. Then factor the resulting perfect square trinomial:

a. $x^2 + 8x$ **b.** $x^2 - 14x$ **c.** $x^2 + 5x.$

Solution To complete the square, we must add a term to each binomial. The term that should be added is the square of half the coefficient of x.

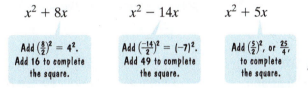

$$x^2 + 8x \qquad x^2 - 14x \qquad x^2 + 5x$$

Add $\left(\frac{8}{2}\right)^2 = 4^2$. Add 16 to complete the square.

Add $\left(\frac{-14}{2}\right)^2 = (-7)^2$. Add 49 to complete the square.

Add $\left(\frac{5}{2}\right)^2$, or $\frac{25}{4}$, to complete the square.

a. The coefficient of the x-term of $x^2 + 8x$ is 8. Half of 8 is 4, and $4^2 = 16$. Add 16.

$$x^2 + 8x + 16 = (x + 4)^2$$

b. The coefficient of the x-term of $x^2 - 14x$ is -14. Half of -14 is -7, and $(-7)^2 = 49$. Add 49.

$$x^2 - 14x + 49 = (x - 7)^2$$

c. The coefficient of the x-term of $x^2 + 5x$ is 5. Half of 5 is $\frac{5}{2}$, and $\left(\frac{5}{2}\right)^2 = \frac{25}{4}$. Add $\frac{25}{4}$.

$$x^2 + 5x + \frac{25}{4} = \left(x + \frac{5}{2}\right)^2 \quad \blacksquare$$

Geometric figures make it possible to visualize completing the square.

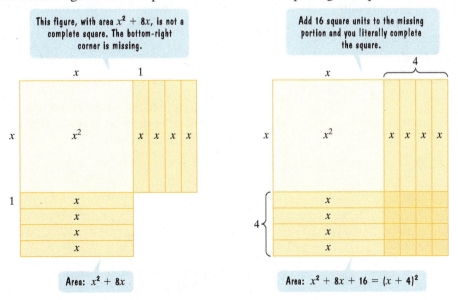

This figure, with area $x^2 + 8x$, is not a complete square. The bottom-right corner is missing.

Add 16 square units to the missing portion and you literally complete the square.

Area: $x^2 + 8x$

Area: $x^2 + 8x + 16 = (x + 4)^2$

✅ **CHECK POINT 1**   Complete the square for each binomial. Then factor the resulting perfect square trinomial:

a. $x^2 + 10x$

b. $x^2 - 6x$

c. $x^2 + 3x$.

---

2 Solve quadratic equations by completing the square.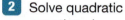

Solving Quadratic Equations by Completing the Square

We can solve *any* quadratic equation by completing the square. If the coefficient of the x^2-term is one, we add the square of half the coefficient of x to both sides of the equation. **When you add a constant term to one side of the equation to complete the square, be certain to add the same constant to the other side of the equation.** These ideas are illustrated in Example 2.

EXAMPLE 2   Solving Quadratic Equations by Completing the Square

Solve by completing the square:

a. $x^2 + 8x = -15$ **b.** $x^2 - 6x + 2 = 0$.

Great Question!

When I solve a quadratic equation by completing the square, does this result in a new equation? Why are the solutions of this new equation the same as those of the given equation?

When you complete the square for the binomial expression $x^2 + bx$, you obtain a different polynomial. When you solve a quadratic equation by completing the square, you obtain an equation with the same solution set because the constant needed to complete the square is added to *both sides*.

Solution

a. Our first equation is $x^2 + 8x = -15$. To complete the square on the binomial $x^2 + 8x$, we take half of 8, which is 4, and square 4, giving 16. We add 16 to both sides of the equation. This makes the left side a perfect square trinomial.

$$x^2 + 8x = -15$$ This is the given equation.

$$x^2 + 8x + 16 = -15 + 16$$ Add 16 to both sides to complete the square.

$$(x + 4)^2 = 1$$ Factor and simplify.

$$x + 4 = \sqrt{1} \quad \text{or} \quad x + 4 = -\sqrt{1}$$ Apply the square root property.

$$x + 4 = 1 \qquad\qquad x + 4 = -1$$ Simplify.

$$x = -3 \qquad\qquad x = -5$$ Subtract 4 from both sides in each equation.

The solutions are -5 and -3. The solution set is $\{-5, -3\}$.

b. To solve $x^2 - 6x + 2 = 0$ by completing the square, we first subtract 2 from both sides. This is done to isolate the binomial $x^2 - 6x$ so that we can complete the square.

$$x^2 - 6x + 2 = 0$$ This is the original equation.

$$x^2 - 6x = -2$$ Subtract 2 from both sides.

Next, we complete the square. Find half the coefficient of the x-term and square it. The coefficient of the x-term is -6. Half of -6 is -3, and $(-3)^2 = 9$. Thus, we add 9 to both sides of the equation.

$$x^2 - 6x + 9 = -2 + 9$$ Add 9 to both sides to complete the square.

$$(x - 3)^2 = 7$$ Factor and simplify.

$$x - 3 = \sqrt{7} \quad \text{or} \quad x - 3 = -\sqrt{7}$$ Apply the square root property.

$$x = 3 + \sqrt{7} \qquad x = 3 - \sqrt{7}$$ Add 3 to both sides in each equation.

The solutions are $3 + \sqrt{7}$ and $3 - \sqrt{7}$, expressed in abbreviated notation as $3 \pm \sqrt{7}$. The solution set is $\{3 \pm \sqrt{7}\}$. ∎

If you solve a quadratic equation by completing the square and the solutions are rational numbers, the equation can also be solved by factoring. By contrast, quadratic equations with irrational solutions cannot be solved by factoring. However, all quadratic equations can be solved by completing the square.

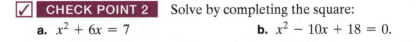

☑ **CHECK POINT 2**   Solve by completing the square:

a. $x^2 + 6x = 7$ **b.** $x^2 - 10x + 18 = 0$.

We have seen that the leading coefficient must be 1 in order to complete the square. If the coefficient of the x^2-term in a quadratic equation is not 1, you must divide each side of the equation by this coefficient before completing the square. For example, to solve $2x^2 + 5x - 4 = 0$ by completing the square, first divide every term by 2:

$$\frac{2x^2}{2} + \frac{5x}{2} - \frac{4}{2} = \frac{0}{2}$$

$$x^2 + \frac{5}{2}x - 2 = 0.$$

Now that the coefficient of the x^2-term is 1, we can solve by completing the square.

EXAMPLE 3   Solving a Quadratic Equation by Completing the Square

Solve by completing the square: $2x^2 + 5x - 4 = 0$.

Solution

$$2x^2 + 5x - 4 = 0$$ This is the original equation.

$$x^2 + \frac{5}{2}x - 2 = 0$$ Divide both sides by 2.

$$x^2 + \frac{5}{2}x = 2$$ Add 2 to both sides and isolate the binomial.

$$x^2 + \frac{5}{2}x + \frac{25}{16} = 2 + \frac{25}{16}$$ Complete the square: Half of $\frac{5}{2}$ is $\frac{5}{4}$ and $\left(\frac{5}{4}\right)^2 = \frac{25}{16}$.

$$\left(x + \frac{5}{4}\right)^2 = \frac{57}{16}$$ Factor and simplify. On the right: $2 + \frac{25}{16} = \frac{32}{16} + \frac{25}{16} = \frac{57}{16}$.

$$x + \frac{5}{4} = \sqrt{\frac{57}{16}} \quad \text{or} \quad x + \frac{5}{4} = -\sqrt{\frac{57}{16}}$$ Apply the square root property.

$$x + \frac{5}{4} = \frac{\sqrt{57}}{4} \qquad x + \frac{5}{4} = -\frac{\sqrt{57}}{4}$$ $\sqrt{\frac{57}{16}} = \frac{\sqrt{57}}{\sqrt{16}} = \frac{\sqrt{57}}{4}$

$$x = -\frac{5}{4} + \frac{\sqrt{57}}{4} \qquad x = -\frac{5}{4} - \frac{\sqrt{57}}{4}$$ Solve the equations, subtracting $\frac{5}{4}$ from both sides.

$$x = \frac{-5 + \sqrt{57}}{4} \qquad x = \frac{-5 - \sqrt{57}}{4}$$ Express solutions with a common denominator.

The solutions are $\dfrac{-5 \pm \sqrt{57}}{4}$ and the solution set is $\left\{\dfrac{-5 \pm \sqrt{57}}{4}\right\}$. ∎

Using Technology

Graphic Connections
Obtain a decimal approximation for each solution of $2x^2 + 5x - 4 = 0$, the equation in Example 3:

$$\frac{-5 + \sqrt{57}}{4} \approx 0.6$$

$$\frac{-5 - \sqrt{57}}{4} \approx -3.1.$$

The x-intercepts of $y = 2x^2 + 5x - 4$ verify these solutions.

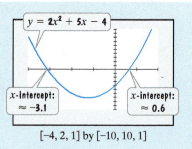

$[-4, 2, 1]$ by $[-10, 10, 1]$

☑ **CHECK POINT 3** Solve by completing the square: $2x^2 - 10x - 1 = 0$.

CONCEPT AND VOCABULARY CHECK

Fill in each blank so that the resulting statement is true.

1. To complete the square on $x^2 + 18x$, add _____,

2. To complete the square on $x^2 - 6x$, add _____.

3. To complete the square on $x^2 + 7x$, add _____.

4. To complete the square on $x^2 - \frac{5}{2}x$, add _____.

5. To complete the square on $x^2 + bx$, add _____.

Fill in each blank with the number needed to make each binominal a perfect square trinominal.

6. $x^2 + 20x +$ _____

7. $x^2 - 2x +$ _____

8. $x^2 + x +$ _____

9.2 EXERCISE SET   ▶ MyMathLab®

Practice Exercises

In Exercises 1–12, complete the square for each binomial. Then factor the resulting perfect square trinomial.

1. $x^2 + 10x$

2. $x^2 + 12x$

3. $x^2 - 2x$

4. $x^2 - 4x$

5. $x^2 + 5x$

6. $x^2 + 3x$

7. $x^2 - 7x$

8. $x^2 - x$

9. $x^2 + \frac{1}{2}x$

10. $x^2 + \frac{1}{3}x$

11. $x^2 - \frac{4}{3}x$

12. $x^2 - \frac{4}{5}x$

In Exercises 13–34, solve each quadratic equation by completing the square.

13. $x^2 + 4x = 5$

14. $x^2 + 6x = -8$

15. $x^2 - 10x = -24$

16. $x^2 - 2x = 8$

17. $x^2 - 2x = 5$

18. $x^2 - 4x = -2$

19. $x^2 + 4x + 1 = 0$

20. $x^2 + 6x - 5 = 0$

21. $x^2 - 3x = 28$

22. $x^2 - 5x = -6$

23. $x^2 + 3x - 1 = 0$

24. $x^2 - 3x - 5 = 0$

25. $x^2 = 7x - 3$

26. $x^2 = 5x - 3$

27. $2x^2 - 2x - 6 = 0$

28. $2x^2 - 4x - 2 = 0$

29. $2x^2 - 3x + 1 = 0$

30. $2x^2 - x - 1 = 0$

31. $2x^2 + 10x + 11 = 0$

32. $2x^2 + 8x + 5 = 0$

33. $4x^2 - 2x - 3 = 0$

34. $3x^2 - 2x - 4 = 0$

Practice PLUS

In Exercises 35–40, solve each quadratic equation by completing the square.

35. $\dfrac{x^2}{6} - \dfrac{x}{3} - 1 = 0$

36. $\dfrac{x^2}{6} + x - \dfrac{3}{2} = 0$

37. $(x + 2)(x - 3) = 1$

38. $(x - 5)(x - 3) = 1$

39. $x^2 + 4bx = 5b^2$

40. $x^2 + 6bx = 7b^2$

Explaining the Concepts

41. Explain how to complete the square for a binomial. Use $x^2 + 6x$ to illustrate your explanation.

42. Explain how to solve $x^2 + 6x + 8 = 0$ by completing the square.

Critical Thinking Exercises

Make Sense?   *In Exercises 43–46, determine whether each statement makes sense or does not make sense, and explain your reasoning.*

43. When I complete the square, I convert a quadratic equation into an equivalent equation that can be solved by the square root property.

44. When the coefficient of the *x*-term in a quadratic equation is negative and I'm solving by completing the square, I add a negative constant to each side of the equation.

45. When I complete the square for the binomial $x^2 + bx$, I obtain a different polynomial, but when I solve a quadratic equation by completing the square, I obtain an equation with the same solution set.

46. I solved $4x^2 + 10x = 0$ by completing the square and added 25 to both sides of the equation.

In Exercises 47–50, determine whether each statement is true or false. If the statement is false, make the necessary change(s) to produce a true statement.

47. Completing the square is a method for finding the area and perimeter of a square.

48. The trinomial $x^2 - 3x + 9$ is a perfect square trinomial.

49. Although not every quadratic equation can be solved by completing the square, they can all be solved using factoring.

50. In completing the square for $x^2 - 7x = 5$, we should add $\frac{49}{4}$ to both sides.

51. Write a perfect square trinomial whose *x*-term is $-20x$.

52. Solve by completing the square: $x^2 + x + c = 0$.

53. Solve by completing the square: $x^2 + bx + c = 0$.

Technology Exercises

54. Use the technique shown in the Using Technology box on page 638 to verify the solutions of any two quadratic equations in Exercises 17–20.

Review Exercises

In Exercises 55–56, perform the indicated operations. If possible, simplify the answer.

55. $\dfrac{2x + 3}{x^2 - 7x + 12} - \dfrac{2}{x - 3}$ (Section 7.4, Example 7)

56. $\dfrac{x - \dfrac{1}{3}}{3 - \dfrac{1}{x}}$ (Section 7.5, Example 2 or Example 5)

57. Solve: $\sqrt{2x + 3} = 2x - 3$. (Section 8.5, Example 4)

Preview Exercises

Exercises 58–60 will help you prepare for the material covered in the next section. In each exercise, evaluate

$$\frac{-b \pm \sqrt{b^2 - 4ac}}{2a}$$

for the given values of a, b, and c. Where necessary, express answers in simplified radical form.

58. $a = 2, b = 9, c = -5$

59. $a = 9, b = -12, c = 4$

60. $a = 1, b = -2, c = -6$

9.3

The Quadratic Formula

What am I supposed to learn?

After studying this section, you should be able to:

1. Solve quadratic equations using the quadratic formula.

2. Determine the most efficient method to use when solving a quadratic equation.

3. Solve problems using quadratic equations.

In the chapter opener, we observed that a substantial percentage of the United States population is foreign-born. In this section, we will explore the foreign-born model using a formula that will enable you to solve quadratic equations more quickly than using the method of completing the square. We begin by deriving this formula.

Solving Quadratic Equations Using the Quadratic Formula

We can apply the method of completing the square to derive a formula that can be used to solve all quadratic equations. The derivation given here also shows a particular quadratic equation, $3x^2 - 2x - 4 = 0$, to specifically illustrate each of the steps.

Deriving the Quadratic Formula

Standard Form of a Quadratic Equation	Comment	A Specific Example
$ax^2 + bx + c = 0, a > 0$	This is the given equation.	$3x^2 - 2x - 4 = 0$
$x^2 + \dfrac{b}{a}x + \dfrac{c}{a} = 0$	Divide both sides by a so that the coefficient of x^2 is 1.	$x^2 - \dfrac{2}{3}x - \dfrac{4}{3} = 0$
$x^2 + \dfrac{b}{a}x = -\dfrac{c}{a}$	Isolate the binomial by adding $-\dfrac{c}{a}$ on both sides of the equation.	$x^2 - \dfrac{2}{3}x = \dfrac{4}{3}$
$x^2 + \dfrac{b}{a}x + \left(\dfrac{b}{2a}\right)^2 = -\dfrac{c}{a} + \left(\dfrac{b}{2a}\right)^2$ (half)^2 $x^2 + \dfrac{b}{a}x + \dfrac{b^2}{4a^2} = -\dfrac{c}{a} + \dfrac{b^2}{4a^2}$	Complete the square. Add the square of half the coefficient of x to both sides.	$x^2 - \dfrac{2}{3}x + \left(-\dfrac{1}{3}\right)^2 = \dfrac{4}{3} + \left(-\dfrac{1}{3}\right)^2$ (half)^2 $x^2 - \dfrac{2}{3}x + \dfrac{1}{9} = \dfrac{4}{3} + \dfrac{1}{9}$
$\left(x + \dfrac{b}{2a}\right)^2 = -\dfrac{c}{a}\cdot\dfrac{4a}{4a} + \dfrac{b^2}{4a^2}$	Factor on the left side and obtain a common denominator on the right side.	$\left(x - \dfrac{1}{3}\right)^2 = \dfrac{4}{3}\cdot\dfrac{3}{3} + \dfrac{1}{9}$
$\left(x + \dfrac{b}{2a}\right)^2 = \dfrac{-4ac + b^2}{4a^2}$ $\left(x + \dfrac{b}{2a}\right)^2 = \dfrac{b^2 - 4ac}{4a^2}$	Add fractions on the right side.	$\left(x - \dfrac{1}{3}\right)^2 = \dfrac{12 + 1}{9}$ $\left(x - \dfrac{1}{3}\right)^2 = \dfrac{13}{9}$
$x + \dfrac{b}{2a} = \pm\sqrt{\dfrac{b^2 - 4ac}{4a^2}}$	Apply the square root property.	$x - \dfrac{1}{3} = \pm\sqrt{\dfrac{13}{9}}$
$x + \dfrac{b}{2a} = \pm\dfrac{\sqrt{b^2 - 4ac}}{2a}$	Take the square root of the quotient, simplifying the denominator.	$x - \dfrac{1}{3} = \pm\dfrac{\sqrt{13}}{3}$
$x = \dfrac{-b}{2a} \pm \dfrac{\sqrt{b^2 - 4ac}}{2a}$	Solve for x by subtracting $\dfrac{b}{2a}$ from both sides.	$x = \dfrac{1}{3} \pm \dfrac{\sqrt{13}}{3}$
$x = \dfrac{-b \pm \sqrt{b^2 - 4ac}}{2a}$	Combine fractions on the right side.	$x = \dfrac{1 \pm \sqrt{13}}{3}$

The formula shown at the bottom of the left column on the previous page is called the *quadratic formula*. A similar proof shows that the same formula can be used to solve quadratic equations if a, the coefficient of the x^2-term, is negative.

1 Solve quadratic equations using the quadratic formula. ▶

The Quadratic Formula

The solutions of a quadratic equation in standard form $ax^2 + bx + c = 0$, with $a \neq 0$, are given by the **quadratic formula**

$$x = \frac{-b \pm \sqrt{b^2 - 4ac}}{2a}.$$

> *x* equals negative *b* plus or minus the square root of $b^2 - 4ac$, all divided by 2*a*.

To use the quadratic formula, begin by writing the quadratic equation in standard form, if necessary. Then determine the numerical values for a (the coefficient of the x^2-term), b (the coefficient of the x-term), and c (the constant term). Substitute the values of $a, b,$ and c into the quadratic formula and evaluate the expression. The $\pm$ sign indicates that if $b^2 - 4ac$ is not zero, there are two solutions of the equation.

EXAMPLE 1   Solving a Quadratic Equation Using the Quadratic Formula

Solve using the quadratic formula: $2x^2 + 9x - 5 = 0$.

Solution   The given equation is in standard form. Begin by identifying the values for $a, b,$ and c.

$$2x^2 + 9x - 5 = 0$$

$$a = 2 \qquad b = 9 \qquad c = -5$$

Substituting these values into the quadratic formula and simplifying gives the equation's solutions.

$$x = \frac{-b \pm \sqrt{b^2 - 4ac}}{2a} \qquad \text{Use the quadratic formula.}$$

$$x = \frac{-9 \pm \sqrt{9^2 - 4(2)(-5)}}{2(2)} \qquad \begin{array}{l}\text{Substitute the values for } a, b, \text{ and } c. \\ a = 2, b = 9, \text{ and } c = -5.\end{array}$$

$$= \frac{-9 \pm \sqrt{81 + 40}}{4} \qquad 9^2 - 4(2)(-5) = 81 - (-40) = 81 + 40$$

$$= \frac{-9 \pm \sqrt{121}}{4} \qquad \text{Add under the radical sign.}$$

$$= \frac{-9 \pm 11}{4} \qquad \sqrt{121} = 11$$

Using Technology

Graphic Connections
The graph of
$y = 2x^2 + 9x - 5$ has
x-intercepts at -5 and $\frac{1}{2}$.
This verifies that -5 and $\frac{1}{2}$
are the solutions of

$$2x^2 + 9x - 5 = 0.$$

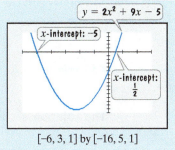

$y = 2x^2 + 9x - 5$

x-intercept: -5

x-intercept: $\frac{1}{2}$

$[-6, 3, 1]$ by $[-16, 5, 1]$

Now we will evaluate this expression in two different ways to obtain the two solutions. At the left, we will *add* 11 to -9. At the right, we will *subtract* 11 from -9.

$$x = \frac{-9 + 11}{4} \quad \text{or} \quad x = \frac{-9 - 11}{4}$$

$$= \frac{2}{4} = \frac{1}{2} \qquad\qquad = \frac{-20}{4} = -5$$

The solutions are -5 and $\frac{1}{2}$, and the solution set is $\left\{-5, \frac{1}{2}\right\}$. ∎

In Example 1, the solutions of $2x^2 + 9x - 5 = 0$, -5 and $\frac{1}{2}$, are rational numbers. This means that the equation can also be solved using factoring. The reason that the solutions are rational numbers is that $b^2 - 4ac$, the radicand in the quadratic formula, is 121, which is a perfect square.

✓ **CHECK POINT 1**   Solve using the quadratic formula: $8x^2 + 2x - 1 = 0$.

EXAMPLE 2   Solving a Quadratic Equation Using the Quadratic Formula

Solve using the quadratic formula: $2x^2 = 6x - 1$.

Solution   The quadratic equation must be in standard form to identify the values for a, b, and c. We need to move all terms to one side and obtain zero on the other side. To obtain zero on the right, we subtract $6x$ and add 1 on both sides. Then we can identify the values for a, b, and c.

$$2x^2 = 6x - 1 \qquad \text{This is the given equation.}$$

$$2x^2 - 6x + 1 = 6x - 6x - 1 + 1 \qquad \text{This step is usually performed mentally.}$$

$$2x^2 - 6x + 1 = 0 \qquad \text{Identify } a, \text{ the } x^2\text{-coefficient, } b, \text{ the } x\text{-coefficient, and } c, \text{ the constant.}$$

$$\boxed{a = 2} \quad \boxed{b = -6} \quad \boxed{c = 1}$$

Substituting these values into the quadratic formula and simplifying gives the equation's solutions.

$$x = \frac{-b \pm \sqrt{b^2 - 4ac}}{2a} \qquad \text{Use the quadratic formula.}$$

$$x = \frac{-(-6) \pm \sqrt{(-6)^2 - 4(2)(1)}}{2 \cdot 2} \qquad \text{Substitute the values for } a, b, \text{ and } c\text{:} \\ a = 2, b = -6, \text{ and } c = 1.$$

$$= \frac{6 \pm \sqrt{36 - 8}}{4} \qquad -(-6) = 6 \text{ and } (-6)^2 = (-6)(-6) = 36.$$

$$= \frac{6 \pm \sqrt{28}}{4} \qquad \text{Complete the subtraction under the radical.}$$

$$= \frac{6 \pm 2\sqrt{7}}{4} \qquad \sqrt{28} = \sqrt{4 \cdot 7} = \sqrt{4}\sqrt{7} = 2\sqrt{7}$$

$$= \frac{2(3 \pm \sqrt{7})}{4} \qquad \text{Factor out 2 from the numerator.}$$

$$= \frac{3 \pm \sqrt{7}}{2} \qquad \text{Divide the numerator and denominator by 2.}$$

The solutions are $\dfrac{3 + \sqrt{7}}{2}$ and $\dfrac{3 - \sqrt{7}}{2}$, abbreviated $\dfrac{3 \pm \sqrt{7}}{2}$. The solution set is $\left\{ \dfrac{3 \pm \sqrt{7}}{2} \right\}$. ∎

In Example 2, the solutions of $2x^2 = 6x - 1$ are irrational numbers. This means that the equation cannot be solved using factoring. The reason that the solutions are irrational numbers is that $b^2 - 4ac$, the radicand in the quadratic formula, is 28, which is not a perfect square.

Great Question!

The simplification of the irrational solutions in Example 2 was kind of tricky. Any suggestions to guide the process?

Many students use the quadratic formula correctly until the last step, where they make an error in simplifying the solutions. Be sure to factor the numerator before dividing the numerator and denominator by the greatest common factor:

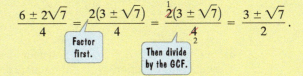

You cannot divide just one term in the numerator and the denominator by their greatest common factor.

Incorrect!

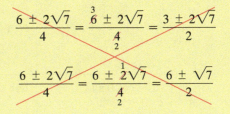

Great Question!

Can all irrational solutions of quadratic equations be simplified?

No. The following solutions cannot be simplified.

$$\frac{5 \pm 2\sqrt{7}}{2} \qquad\qquad \frac{-4 \pm 3\sqrt{7}}{2}$$

Other than 1, terms in each numerator have no common factor.

> ✓ **CHECK POINT 2**   Solve using the quadratic formula: $x^2 = 6x - 4$.

---

2 Determine the most efficient method to use when solving a quadratic equation. ▶

Determining Which Method to Use

All quadratic equations can be solved by the quadratic formula. However, if an equation is in the form $u^2 = d$, such as $x^2 = 5$ or $(2x + 3)^2 = 8$, it is faster to use the square root property, taking the square root of both sides. If the equation is not in the form $u^2 = d$, write the quadratic equation in standard form $(ax^2 + bx + c = 0)$. Try to solve the equation by factoring. If $ax^2 + bx + c$ cannot be factored, then solve the quadratic equation by the quadratic formula.

Because we used the method of completing the square to derive the quadratic formula, we no longer need it for solving quadratic equations. However, you will use completing the square in more advanced algebra courses to help graph certain kinds of equations.

Table 9.1 summarizes our observations about which technique to use when solving a quadratic equation.

Table 9.1 Determining the Most Efficient Technique to Use When Solving a Quadratic Equation		
Description and Form of the Quadratic Equation	**Most Efficient Solution Method**	**Example**
$ax^2 + bx + c = 0$ and $ax^2 + bx + c$ can be factored easily.	Factor and use the zero-product principle.	$3x^2 + 5x - 2 = 0$ $(3x - 1)(x + 2) = 0$ $3x - 1 = 0$ or $x + 2 = 0$ $x = \dfrac{1}{3}$ $\qquad x = -2$
$ax^2 + c = 0$ The quadratic equation has no x-term. $(b = 0)$	Solve for x^2 and apply the square root property.	$4x^2 - 7 = 0$ $4x^2 = 7$ $x^2 = \dfrac{7}{4}$ $x = \dfrac{\pm\sqrt{7}}{2}$
$u^2 = d$; u is a first-degree polynomial.	Use the square root property.	$(x + 4)^2 = 5$ $x + 4 = \pm\sqrt{5}$ $x = -4 \pm\sqrt{5}$
$ax^2 + bx + c = 0$ and $ax^2 + bx + c$ cannot be factored or the factoring is too difficult.	Use the quadratic formula: $x = \dfrac{-b \pm \sqrt{b^2 - 4ac}}{2a}$.	$x^2 - 2x - 6 = 0$ $a = 1 \quad b = -2 \quad c = -6$ $x = \dfrac{-(-2) \pm \sqrt{(-2)^2 - 4(1)(-6)}}{2(1)}$ $= \dfrac{2 \pm \sqrt{4 + 24}}{2}$ $= \dfrac{2 \pm \sqrt{28}}{2} = \dfrac{2 \pm \sqrt{4}\sqrt{7}}{2}$ $= \dfrac{2 \pm 2\sqrt{7}}{2} = \dfrac{2(1 \pm \sqrt{7})}{2}$ $= 1 \pm \sqrt{7}$

3 Solve problems using quadratic equations. ▶

Applications

Quadratic equations can be solved by any efficient method to answer questions about variables contained in mathematical models.

EXAMPLE 3	Making Predictions About the U.S. Foreign-Born Population

A substantial percentage of the United States population is foreign-born. The graph in **Figure 9.7** shows the percentage of foreign-born Americans for selected years from 1920 through 2012.

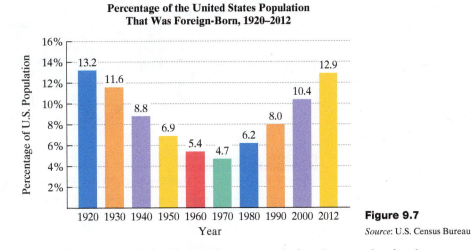

Percentage of the United States Population That Was Foreign-Born, 1920–2012

Figure 9.7

Source: U.S. Census Bureau

The percentage, p, of the United States population that was foreign-born x years after 1920 can be modeled by the formula

$$p = 0.004x^2 - 0.35x + 13.9.$$

According to this model, in which year will 19% of the United States population be foreign-born?

Solution   Because we are interested in when the foreign-born percentage will reach 19%, we substitute 19 for p in the given formula. Then we solve for x, the number of years after 1920.

$p = 0.004x^2 - 0.35x + 13.9$	This is the given formula.
$19 = 0.004x^2 - 0.35x + 13.9$	Substitute 19 for p.
$0 = 0.004x^2 - 0.35x - 5.1$	Subtract 19 from both sides and write the quadratic equation in standard form.

$$a = 0.004 \quad b = -0.35 \quad c = -5.1$$

Because the trinomial on the right side of the equation, $0.004x^2 - 0.35x - 5.1$, is not easily factored, if it can be factored at all, we solve using the quadratic formula.

$x = \dfrac{-b \pm \sqrt{b^2 - 4ac}}{2a}$	Use the quadratic formula.
$x = \dfrac{-(-0.35) \pm \sqrt{(-0.35)^2 - 4(0.004)(-5.1)}}{2(0.004)}$	Substitute the values for a, b, and c: $a = 0.004$, $b = -0.35$, and $c = -5.1$.
$= \dfrac{0.35 \pm \sqrt{0.2041}}{0.008}$	Perform the indicated computations. Use a calculator to simplify the radicand.

$$x = \frac{0.35 + \sqrt{0.2041}}{0.008} \quad \text{or} \quad x = \frac{0.35 - \sqrt{0.2041}}{0.008}$$

$$x \approx 100 \qquad\qquad\qquad x \approx -13$$

Use a calculator and round to the nearest whole number.

Reject this solution. The model applies to years *after*, not *before*, 1920.

According to the model, approximately 100 years after 1920, or in 2020, 19% of the United States population will be foreign-born. ∎

Using Technology

Numeric Connections

A graphing utility's ⌈TABLE⌋ feature can be used to verify the solution to Example 3.

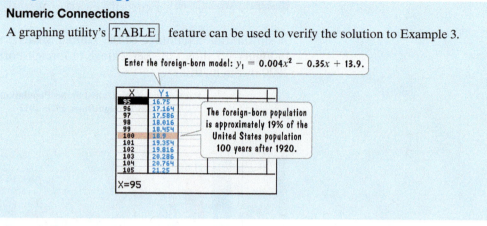

Enter the foreign-born model: $y_1 = 0.004x^2 - 0.35x + 13.9$.

X	Y₁
95	16.75
96	17.164
97	17.586
98	18.016
99	18.454
100	18.9
101	19.354
102	19.816
103	20.286
104	20.764
105	21.25

X=95

The foreign-born population is approximately 19% of the United States population 100 years after 1920.

✓ **CHECK POINT 3**   Use the model in Example 3 to determine in which year 25% of the United States population will be foreign-born.

Blitzer Bonus ◉

Art, Nature, and Quadratic Equations

A **golden rectangle** can be a rectangle of any size, but its long side must be Φ times as long as its short side, where Φ ≈ 1.6. Artists often use golden rectangles in their work because they are considered to be more visually pleasing than other rectangles.

If a golden rectangle is divided into a square and a rectangle, as in **Figure 9.8(a)**, the smaller rectangle is a golden rectangle. If the smaller golden rectangle is divided again, the same is true of the yet smaller rectangle, and so on. The process of repeatedly dividing each golden rectangle in this manner is illustrated in **Figure 9.8(b)**. We've also created a spiral by connecting the opposite corners of all the squares with a smooth curve. This spiral matches the spiral shape of the chambered nautilus shell shown in **Figure 9.8(c)**. The shell spirals out at an ever-increasing rate that is governed by this geometry.

In *Bathers at Asnières*, by the French impressionist Georges Seurat (1859–1891), the artist positions parts of the painting as though they were inside golden rectangles.
Bathers at Asnières (1884), George Seurat Art Resource

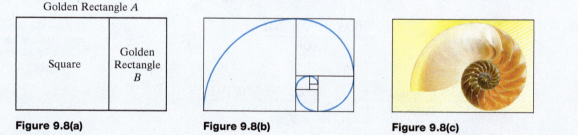

Golden Rectangle *A*

| Square | Golden Rectangle *B* |

Figure 9.8(a)   **Figure 9.8(b)**   **Figure 9.8(c)**

In the Exercise Set that follows, you will use the golden rectangles in **Figure 9.8(a)** to obtain an exact value of Φ, the ratio of the long side to the short side in a golden rectangle of any size. Your model will involve a quadratic equation that can be solved by the quadratic formula. (See Exercise 63.)

Achieving Success

Here are some strategies to help you prepare for your final exam:

- Review your previous exams. Be sure you understand any errors that you made. Seek help with any concepts that are still unclear.
- Ask your professor if there are additional materials to help students review for the final. This includes review sheets and final exams from previous semesters.
- Attend any review sessions conducted by your professor or by the math department.
- Use the strategy first introduced on page 269: Imagine that your professor will permit two 3 by 5 index cards of notes on the final. Organize and create such a two-card summary for the most vital information in the course, including all important formulas. Refer to the chapter summaries in the textbook to prepare your personalized summary.
- For further review, work the relevant exercises in the Cumulative Review at the end of this chapter.
- Write your own final exam with detailed solutions for each item. You can use test questions from previous exams in mixed order, worked examples from the textbook's chapter summaries, exercises in the Cumulative Reviews, and problems from course handouts. Use your test as a practice final exam.

CONCEPT AND VOCABULARY CHECK

Fill in each blank so that the resulting statement is true.

1. The solutions of a quadratic equation in standard form $ax^2 + bx + c = 0$, $a \neq 0$, are given by the quadratic formula

   $x = $ _____.

2. In order to solve $3x^2 + 10x - 8 = 0$ by the quadratic formula, we use $a = $ _____, $b = $ _____, and $c = $ _____.

3. In order to solve $x^2 + 8x - 15 = 14$ by the quadratic formula, we use $a = $ _____, $b = $ _____, and $c = $ _____.

4. In order to solve $6x^2 = 3x + 4$ by the quadratic formula, we use $a = $ _____, $b = $ _____, and $c = $ _____.

5. $x = \dfrac{-(-5) \pm \sqrt{(-5)^2 - 4(3)(-2)}}{2(3)}$ simplifies to $x = $ _____ and $x = $ _____.

6. $x = \dfrac{-8 \pm \sqrt{8^2 - 4(1)(-29)}}{2(1)}$ simplifies to $x = $ _____.

7. The most efficient technique for solving $(x + 3)^2 = 7$ is by using _____.

8. The most efficient technique for solving $x^2 - 2x - 3 = 0$ is by using _____.

9. The most efficient technique for solving $x^2 - 5x - 10 = 0$ is by using _____.

9.3 EXERCISE SET ▶ MyMathLab®

Practice Exercises

In Exercises 1–22, solve each equation using the quadratic formula. Simplify irrational solutions, if possible.

1. $x^2 + 5x + 6 = 0$
2. $x^2 + 7x + 10 = 0$
3. $x^2 + 5x + 3 = 0$
4. $x^2 + 5x + 2 = 0$
5. $x^2 + 4x - 6 = 0$
6. $x^2 + 2x - 4 = 0$
7. $x^2 + 4x - 7 = 0$
8. $x^2 + 4x + 1 = 0$
9. $x^2 - 3x - 18 = 0$
10. $x^2 - 3x - 10 = 0$

11. $6x^2 - 5x - 6 = 0$ **12.** $9x^2 - 12x - 5 = 0$

13. $x^2 - 2x - 10 = 0$ **14.** $x^2 + 6x - 10 = 0$

15. $x^2 - x = 14$

16. $x^2 - 5x = 10$

17. $6x^2 + 6x + 1 = 0$

18. $3x^2 - 5x + 1 = 0$

19. $9x^2 - 12x + 4 = 0$

20. $4x^2 + 12x + 9 = 0$

21. $4x^2 = 2x + 7$

22. $3x^2 = 6x - 1$

In Exercises 23–44, solve each equation by the method of your choice. Simplify irrational solutions, if possible.

23. $2x^2 - x = 1$ **24.** $3x^2 - 4x = 4$

25. $5x^2 + 2 = 11x$ **26.** $5x^2 = 6 - 13x$

27. $3x^2 = 60$ **28.** $2x^2 = 250$

29. $x^2 - 2x = 1$ **30.** $2x^2 + 3x = 1$

31. $(2x + 3)(x + 4) = 1$ **32.** $(2x - 5)(x + 1) = 2$

33. $(3x - 4)^2 = 16$ **34.** $(2x + 7)^2 = 25$

35. $3x^2 - 12x + 12 = 0$ **36.** $9 - 6x + x^2 = 0$

37. $4x^2 - 16 = 0$ **38.** $3x^2 - 27 = 0$

39. $x^2 + 9x = 0$ **40.** $x^2 - 6x = 0$

41. $\frac{3}{4}x^2 - \frac{5}{2}x - 2 = 0$ **42.** $\frac{1}{3}x^2 - \frac{1}{2}x - \frac{3}{2} = 0$

43. $(3x - 2)^2 = 10$ **44.** $(4x - 1)^2 = 15$

Practice PLUS

In Exercises 45–52, solve each equation by the method of your choice. Simplify irrational solutions, if possible.

45. $\dfrac{x^2}{x + 7} - \dfrac{3}{x + 7} = 0$ **46.** $\dfrac{x^2}{x + 9} - \dfrac{11}{x + 9} = 0$

47. $(x + 2)^2 + x(x + 1) = 4$

48. $(x - 1)(3x + 2) = -7(x - 1)$

49. $2x^2 - 9x - 3 = 9 - 9x$

50. $3x^2 - 6x - 3 = 12 - 6x$

51. $\dfrac{1}{x} + \dfrac{1}{x + 3} = \dfrac{1}{4}$ **52.** $\dfrac{1}{x} + \dfrac{2}{x + 3} = \dfrac{1}{4}$

Application Exercises

53. A football is kicked straight up from a height of 4 feet with an initial speed of 60 feet per second. The formula

$$h = -16t^2 + 60t + 4$$

describes the ball's height above the ground, *h*, in feet, *t* seconds after it is kicked. How long will it take for the football to hit the ground? Use a calculator and round to the nearest tenth of a second.

54. Standing on a platform 50 feet high, a person accidentally fires a gun straight into the air. The formula

$$h = -16t^2 + 100t + 50$$

describes the bullet's height above the ground, *h*, in feet, *t* seconds after the gun is fired. How long will it take for the bullet to hit the ground? Use a calculator and round to the nearest tenth of a second.

The height of the arch supporting the bridge shown in the figure is modeled by

$$h = -0.05x^2 + 27,$$

where x is the distance, in feet, from the center of the arch. Use this formula to solve Exercises 55–56.

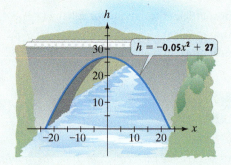

55. How far to the right of the center is the height 22 feet?

56. How far to the right of the center is the height 7 feet?

The bar graph shows undocumented immigrants living in the United States as a percentage of the foreign-born population.

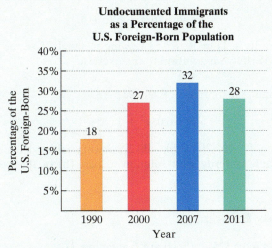

Undocumented Immigrants as a Percentage of the U.S. Foreign-Born Population

Source: Pew Hispanic Center

The data can be modeled by the formula

$$p = -0.05x^2 + 1.5x + 18,$$

where p is the percentage of the U.S. foreign-born who were undocumented immigrants x years after 1990. Use this information to solve Exercises 57–58.

57. a. According to the model, what percentage of the U.S. foreign-born were undocumented in 2000? Does this underestimate or overestimate the percent displayed by the bar graph? By how much?

  b. If trends shown by the model continue, in which year after 1990 will 18% of the U.S. foreign-born be undocumented?

58. a. According to the model, what percentage of the U.S. foreign-born were undocumented in 2011? Round to the nearest percent. Does your rounded answer underestimate or overestimate the percent displayed by the bar graph? By how much?

  b. If trends shown by the model continue, in which year will 13% of the U.S. foreign-born be undocumented?

59. The length of a rectangle is 3 meters longer than the width. If the area is 36 square meters, find the rectangle's dimensions. Round to the nearest tenth of a meter.

60. The length of a rectangle is 2 meters longer than the width. If the area is 10 square meters, find the rectangle's dimensions. Round to the nearest tenth of a meter.

61. The hypotenuse of a right triangle is 4 feet long. One leg is 1 foot longer than the other. Find the lengths of the legs. Round to the nearest tenth of a foot.

62. The hypotenuse of a right triangle is 6 feet long. One leg is 1 foot shorter than the other. Find the lengths of the legs. Round to the nearest tenth of a foot.

63. If you have not yet done so, read the Blitzer Bonus on page 646. In this exercise, you will use the golden rectangles shown to obtain an exact value for Φ, the ratio of the long side to the short side in a golden rectangle of any size.

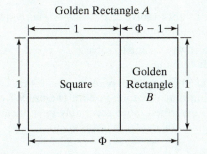

Golden Rectangle *A*

  a. The golden ratio in rectangle *A*, or the ratio of the long side to the short side, can be modeled by $\dfrac{\Phi}{1}$. Write a fractional expression that models the golden ratio in rectangle *B*.

  b. Set the expression for the golden ratio in rectangle *A* equal to the expression for the golden ratio in rectangle *B*. Solve the resulting proportion using the quadratic formula. Express Φ as an exact value in simplified radical form.

  c. Use your solution from part (b) to complete this statement: The ratio of the long side to the short side in a golden rectangle of any size is _____ to 1.

64. In Exercise Set 7.4, we considered two formulas that approximate the dosage of a drug prescribed for children:

$$\text{Young's rule:}\quad C = \frac{DA}{A + 12}$$

$$\text{Cowling's rule:}\quad C = \frac{D(A + 1)}{24}.$$

In each formula, $A =$ the child's age, in years, $D =$ an adult dosage, and $C =$ the proper child's dosage. The formulas apply for ages 2 through 13, inclusive. At which age, to the nearest tenth of a year, do the two formulas give the same dosage?

Explaining the Concepts

65. What is the quadratic formula and why is it useful?

66. Without going into specific details for each step, describe how the quadratic formula is derived.

67. Explain how to solve $x^2 + 6x + 8 = 0$ using the quadratic formula.

68. If you are given a quadratic equation, how do you determine which method to use to solve it?

Critical Thinking Exercises

Make Sense? *In Exercises 69–72, determine whether each statement makes sense or does not make sense, and explain your reasoning.*

69. Because I want to solve $25x^2 - 49 = 0$ fairly quickly, I'll use the quadratic formula.

70. The fastest way for me to solve $x^2 - x - 2 = 0$ is to use the quadratic formula.

71. The data showing the percentage of the United States population that was foreign-born (**Figure 9.7** on page 645) could be described more accurately by a linear model than by the quadratic model that was given.

72. I simplified $\dfrac{3 + 2\sqrt{3}}{2}$ to $3 + \sqrt{3}$ because 2 is a factor of $2\sqrt{3}$.

In Exercises 73–76, determine whether each statement is true or false. If the statement is false, make the necessary change(s) to produce a true statement.

73. The quadratic formula can be expressed as
$$x = -b \pm \frac{\sqrt{b^2 - 4ac}}{2a}.$$

74. The solutions $\dfrac{4 \pm \sqrt{3}}{2}$ can be simplified to $2 \pm \sqrt{3}$.

75. For the quadratic equation $-2x^2 + 3x = 0$, we have $a = -2, b = 3,$ and $c = 0$.

76. Any quadratic equation that can be solved by completing the square can be solved by the quadratic formula.

77. The radicand of the quadratic formula, $b^2 - 4ac$, can be used to determine whether $ax^2 + bx + c = 0$ has solutions that are rational, irrational, or not real numbers. Explain how this works. Is it possible to determine the kinds of answers that one will obtain to a quadratic equation without actually solving the equation? Explain.

78. Solve: $x^2 + 2\sqrt{3}x - 9 = 0$.

79. A rectangular vegetable garden is 5 feet wide and 9 feet long. The garden is to be surrounded by a tile border of uniform width. If there are 40 square feet of tile for the border, how wide, to the nearest tenth of a foot, should it be?

Technology Exercises

80. Graph the formula in Exercise 53,
$$y = -16x^2 + 60x + 4,$$
in a [0, 4, 1] by [0, 65, 5] viewing rectangle. Use the graph to verify your solution to the exercise.

81. Graph the formula in Exercise 57,
$$y = -0.05x^2 + 1.5x + 18,$$
in a [0, 40, 2] by [0, 40, 2] viewing rectangle. Use the graph to verify your solution to part (b) of the exercise.

Review Exercises

82. Evaluate: $125^{-\frac{2}{3}}$. (Section 8.6, Example 3)

83. Rationalize the denominator: $\dfrac{12}{3 + \sqrt{5}}$.
(Section 8.4, Example 3)

84. Multiply: $(x - y)(x^2 + xy + y^2)$.
(Section 5.2, Example 7)

Preview Exercises

Exercises 85–87 will help you prepare for the material covered in the next section.

85. Can squaring a real number result in a negative number? Based on your answer, are $\sqrt{-1}$ and $\sqrt{-4}$ real numbers?

In Exercises 86–87, list the numbers from each set that are:
a. *rational numbers;* **b.** *irrational numbers;* **c.** *real numbers;*
d. *not real numbers. (Hint: Your answer to each question in Exercise 85 should be "no.")*

86. $\{-\sqrt{5}, -\sqrt{4}, \sqrt{-1}, \sqrt{-4}, \sqrt{1}, \sqrt{5}\}$

87. $\{-\sqrt{9}, -\sqrt{7}, \sqrt{-9}, \sqrt{-7}, \sqrt{0}, \sqrt{7}, \sqrt{9}\}$

MID-CHAPTER CHECK POINT   Section 9.1–Section 9.3

What You Know: We saw that not all quadratic equations can be solved by factoring. We learned three new methods for solving these equations—the square root property, completing the square, and the quadratic formula. We also learned to determine the most efficient technique to use when solving a quadratic equation.

In Exercises 1–12, solve each equation by the method of your choice. Simplify irrational solutions, if possible.

1. $(3x - 2)^2 = 100$

2. $15x^2 = 5x$

3. $x^2 - 2x - 10 = 0$

4. $x^2 - 8x + 16 = 7$

5. $3x^2 - x - 2 = 0$

6. $6x^2 = 10x - 3$

7. $x^2 + (x + 1)^2 = 25$

8. $(x + 5)^2 = 40$

9. $2(x^2 - 8) = 11 - x^2$

10. $2x^2 + 5x + 1 = 0$

11. $(x - 8)(2x - 3) = 34$

12. $x + \dfrac{16}{x} = 8$

13. Solve by completing the square: $x^2 + 14x - 32 = 0$.

14. Find the missing length in the right triangle. Express the answer in simplified radical form.

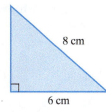

8 cm

6 cm

15. Find the distance between $(-3, 2)$ and $(9, -3)$.

16. The figure shows a right triangle whose hypotenuse measures 20 inches and whose leg measurements are in the ratio 3:4. Find the length of each leg.

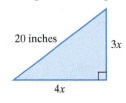

20 inches $3x$

$4x$

SECTION

9.4   Imaginary Numbers as Solutions of Quadratic Equations

What am I supposed to learn?

After studying this section, you should be able to:

1 Express square roots of negative numbers in terms of i.

2 Solve quadratic equations with imaginary solutions.

1 Express square roots of negative numbers in terms of i.

Who is this kid warning us about our eyeballs turning black if we attempt to find the square root of -9? Don't believe what you hear on the street. Although square roots of negative numbers are not real numbers, they do play a significant role in algebra. In this section, we move beyond the real numbers and discuss square roots with negative radicands.

The Imaginary Unit i

Throughout this chapter, we have avoided quadratic equations that have no real numbers as solutions. A fairly simple example is the equation

$$x^2 = -1.$$

Because the square of a real number is never negative, there is no real number x such that $x^2 = -1$. To

THE KID WHO LEARNED ABOUT MATH ON THE STREET

If you divide 6,973 by 0, you die

Once, this guy tried to find the square root of -9, and his eyeballs turned black.

This girl my brother knows found out exactly what π equals, but she went nuts.

R. Chast

provide a setting in which such equations have solutions, mathematicians invented an expanded system of numbers, the *complex numbers*. The *imaginary number i*, defined to be a solution to the equation $x^2 = -1$, is the basis of this new set.

The Imaginary Unit *i*

The **imaginary unit** *i* is defined as

$$i = \sqrt{-1}, \quad \text{where} \quad i^2 = -1.$$

Using the imaginary unit *i*, we can express the square root of any negative number as a real multiple of *i*. For example,

$$\sqrt{-25} = \sqrt{25(-1)} = \sqrt{25}\sqrt{-1} = 5i.$$

We can check this result by squaring $5i$ and obtaining -25.

$$(5i)^2 = 5^2 i^2 = 25(-1) = -25$$

The Square Root of a Negative Number

If *b* is a positive real number, then

$$\sqrt{-b} = \sqrt{b(-1)} = \sqrt{b}\sqrt{-1} = \sqrt{b}\,i \quad \text{or} \quad i\sqrt{b}.$$

EXAMPLE 1   Expressing Square Roots of Negative Numbers as Multiples of *i*

Write as a multiple of *i*:

a. $\sqrt{-9}$ **b.** $\sqrt{-7}$ **c.** $\sqrt{-8}$.

Solution

a. $\sqrt{-9} = \sqrt{9(-1)} = \sqrt{9}\sqrt{-1} = 3i$

b. $\sqrt{-7} = \sqrt{7(-1)} = \sqrt{7}\sqrt{-1} = \sqrt{7}i$

c. $\sqrt{-8} = \sqrt{8(-1)} = \sqrt{8}\sqrt{-1} = \sqrt{4 \cdot 2}\sqrt{-1} = 2\sqrt{2}i$

> Be sure not to write *i* under the radical.

In order to avoid writing *i* under a radical, let's agree to write *i* *before* any radical. Consequently, we express the multiple of *i* in part (b) as $i\sqrt{7}$ and the multiple of *i* in part (c) as $2i\sqrt{2}$. ∎

Great Question!

Now that we've introduced square roots of negative numbers, can I still use the rule $\sqrt{ab} = \sqrt{a}\sqrt{b}$?

We allow the use of the product rule

$$\sqrt{ab} = \sqrt{a}\sqrt{b}$$

when *a* is positive and *b* is −1. However, you cannot use $\sqrt{ab} = \sqrt{a}\sqrt{b}$ when both *a* and *b* are negative.

✓ **CHECK POINT 1**   Write as a multiple of *i*:

a. $\sqrt{-16}$ **b.** $\sqrt{-5}$ **c.** $\sqrt{-50}$.

A new system of numbers, called *complex numbers*, is based on adding multiples of *i*, such as $5i$, to the real numbers.

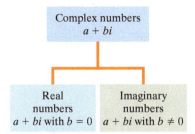

Figure 9.9 The complex number system

Complex Numbers and Imaginary Numbers

The set of all numbers in the form

$$a + bi$$

with real numbers a and b, and i, the imaginary unit, is called the set of **complex numbers**. The real number a is called the **real part** and the real number b is called the **imaginary part** of the complex number $a + bi$. If $b \neq 0$, then the complex number is called an **imaginary number** (see **Figure 9.9**).

Here are some examples of complex numbers. Each number can be written in the form $a + bi$.

$$-4 + 6i \qquad\qquad 2i = 0 + 2i \qquad\qquad 3 = 3 + 0i$$

| a, the real part, is -4. | b, the imaginary part, is 6. | a, the real part, is 0. | b, the imaginary part, is 2. | a, the real part, is 3. | b, the imaginary part, is 0. |

Can you see that b, the imaginary part, is not zero in the first two complex numbers? Because $b \neq 0$, these complex numbers are imaginary numbers. By contrast, the imaginary part of the complex number on the right, $3 + 0i$, is zero. This complex number is not an imaginary number. The number 3, or $3 + 0i$, is a real number.

2 Solve quadratic equations with imaginary solutions.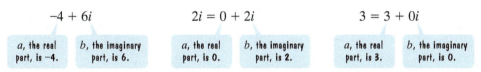

Solving Quadratic Equations with Imaginary Solutions

The equation $x^2 = -25$ has no real solutions, but it does have imaginary solutions.

$$x^2 = -25 \qquad \text{No real number squared results in a negative number.}$$
$$x = \pm\sqrt{-25} \qquad \text{Apply the square root property.}$$
$$x = \pm 5i \qquad \sqrt{-25} = \sqrt{25(-1)} = \sqrt{25}\sqrt{-1} = 5i$$

The solutions are $5i$ and $-5i$. The next examples involve quadratic equations that have no real solutions, but do have imaginary solutions.

> **EXAMPLE 2**   Solving a Quadratic Equation Using the Square Root Property

Solve: $(x + 4)^2 = -36$.

Solution

$$(x + 4)^2 = -36 \qquad \text{This is the given equation.}$$
$$x + 4 = \sqrt{-36} \quad \text{or} \quad x + 4 = -\sqrt{-36} \qquad \text{Apply the square root property.}$$
$$\text{Equivalently, } x + 4 = \pm\sqrt{-36}.$$
$$x + 4 = 6i \qquad\qquad x + 4 = -6i \qquad \sqrt{-36} = \sqrt{36(-1)} = \sqrt{36}\sqrt{-1} = 6i$$
$$x = -4 + 6i \qquad\qquad x = -4 - 6i \qquad \text{Solve the equations by subtracting 4 from both sides.}$$

Check $-4 + 6i$:

$$(x + 4)^2 = -36$$
$$(-4 + 6i + 4)^2 \stackrel{?}{=} -36$$
$$(6i)^2 \stackrel{?}{=} -36$$
$$36i^2 \stackrel{?}{=} -36$$
$$36(-1) \stackrel{?}{=} -36$$
$$-36 = -36, \quad \text{true}$$

Check $-4 - 6i$:

$$(x + 4)^2 = -36$$
$$(-4 - 6i + 4)^2 \stackrel{?}{=} -36$$
$$(-6i)^2 \stackrel{?}{=} -36$$
$$36i^2 \stackrel{?}{=} -36$$
$$36(-1) \stackrel{?}{=} -36$$
$$-36 = -36, \quad \text{true}$$

The imaginary solutions are $-4 + 6i$ and $-4 - 6i$. The solution set is $\{-4 + 6i, -4 - 6i\}$ or $\{-4 \pm 6i\}$. ■

✓ **CHECK POINT 2**   Solve: $(x + 2)^2 = -25$.

EXAMPLE 3   Solving a Quadratic Equation Using the Quadratic Formula

Solve: $x^2 - 2x + 2 = 0$.

Solution   Because the trinomial on the left side is prime, we solve using the quadratic formula.

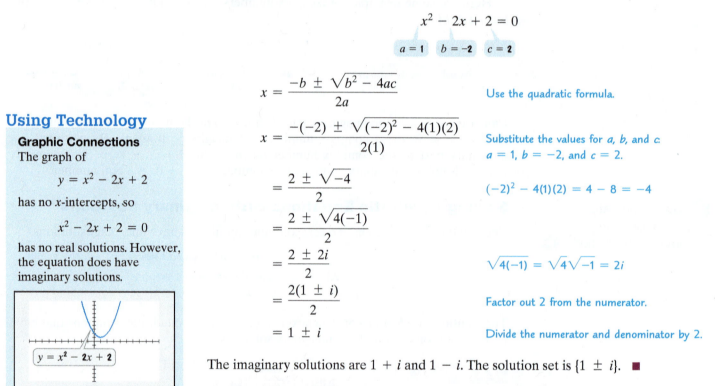

$$x^2 - 2x + 2 = 0$$

$$a = 1 \quad b = -2 \quad c = 2$$

$$x = \frac{-b \pm \sqrt{b^2 - 4ac}}{2a}$$ Use the quadratic formula.

$$x = \frac{-(-2) \pm \sqrt{(-2)^2 - 4(1)(2)}}{2(1)}$$ Substitute the values for a, b, and c: $a = 1$, $b = -2$, and $c = 2$.

$$= \frac{2 \pm \sqrt{-4}}{2}$$ $(-2)^2 - 4(1)(2) = 4 - 8 = -4$

$$= \frac{2 \pm \sqrt{4(-1)}}{2}$$

$$= \frac{2 \pm 2i}{2}$$ $\sqrt{4(-1)} = \sqrt{4}\sqrt{-1} = 2i$

$$= \frac{2(1 \pm i)}{2}$$ Factor out 2 from the numerator.

$$= 1 \pm i$$ Divide the numerator and denominator by 2.

The imaginary solutions are $1 + i$ and $1 - i$. The solution set is $\{1 \pm i\}$. ∎

Using Technology

Graphic Connections
The graph of

$$y = x^2 - 2x + 2$$

has no x-intercepts, so

$$x^2 - 2x + 2 = 0$$

has no real solutions. However, the equation does have imaginary solutions.

$y = x^2 - 2x + 2$

$[-10, 10, 1]$ by $[-10, 10, 1]$

✓ **CHECK POINT 3**   Solve: $x^2 + 6x + 13 = 0$.

CONCEPT AND VOCABULARY CHECK

Fill in each blank so that the resulting statement is true.

1. The imaginary unit i is defined as $i = $ _____, where $i^2 = $ _____.

2. $\sqrt{-25} = \sqrt{25(-1)} = \sqrt{25}\sqrt{-1} = $ _____

3. The set of all numbers in the form $a + bi$ is called the set of _____ numbers. If $b \neq 0$, then the number is also called a/an _____ number. If $b = 0$, then the number is also called a/an _____ number.

4. $x = \dfrac{4 \pm \sqrt{-4}}{2}$ simplifies to $x = $ _____.

9.4 EXERCISE SET   ▶ MyMathLab®

Practice Exercises

In Exercises 1–16, express each number in terms of i.

1. $\sqrt{-36}$ **2.** $\sqrt{-49}$

3. $\sqrt{-13}$ **4.** $\sqrt{-19}$

5. $\sqrt{-50}$ **6.** $\sqrt{-12}$

7. $\sqrt{-20}$ **8.** $\sqrt{-300}$

9. $-\sqrt{-28}$ **10.** $-\sqrt{-150}$

11. $7 + \sqrt{-16}$ **12.** $9 + \sqrt{-4}$

13. $10 + \sqrt{-3}$ **14.** $5 + \sqrt{-5}$

15. $6 - \sqrt{-98}$ **16.** $6 - \sqrt{-18}$

In Exercises 17–24, solve each quadratic equation using the square root property. Express imaginary solutions in a + bi form.

17. $(x - 3)^2 = -9$

18. $(x - 5)^2 = -36$

19. $(x + 7)^2 = -64$

20. $(x + 12)^2 = -100$

21. $(x - 2)^2 = -7$

22. $(x - 1)^2 = -13$

23. $(y + 3)^2 = -18$

24. $(y + 4)^2 = -48$

In Exercises 25–36, solve each quadratic equation using the quadratic formula.

25. $x^2 + 4x + 5 = 0$

26. $x^2 + 2x + 2 = 0$

27. $x^2 - 6x + 13 = 0$

28. $x^2 - 6x + 10 = 0$

29. $x^2 - 12x + 40 = 0$

30. $x^2 - 4x + 29 = 0$

31. $x^2 = 10x - 27$

32. $x^2 = 4x - 7$

33. $5x^2 = 2x - 3$

34. $6x^2 = -2x - 1$

35. $2y^2 = 4y - 5$

36. $5y^2 = 6y - 7$

Practice PLUS

In Exercises 37–42, solve each equation by the method of your choice.

37. $12x^2 + 35 = 8x^2 + 15$ **38.** $8x^2 - 9 = 5x^2 - 30$

39. $\dfrac{x + 3}{5} = \dfrac{x - 2}{x}$ **40.** $\dfrac{x + 4}{4} = \dfrac{x - 5}{x - 2}$

41. $\dfrac{1}{x + 1} - \dfrac{1}{2} = \dfrac{1}{x}$ **42.** $\dfrac{4}{x} = \dfrac{8}{x^2} + 1$

Application Exercises

43. The personnel manager of a roller skate company knows that the company's weekly revenue, R, in thousands of dollars, can be modeled by the formula

$$R = -2x^2 + 36x,$$

where x is the price of a pair of skates, in dollars. A job applicant promises the personnel manager an advertising campaign guaranteed to generate \$200,000 in weekly revenue. Substitute 200 for R in the given formula and solve the equation. Are the solutions real numbers? Explain why the applicant will or will not be hired in the advertising department.

44. A football is kicked straight up from a height of 4 feet with an initial speed of 60 feet per second. The formula

$$h = -16t^2 + 60t + 4$$

describes the ball's height above the ground, h, in feet, t seconds after it is kicked. Will the ball reach a height of 80 feet? Substitute 80 for h in the given formula and solve the equation. Are the solutions real numbers? Explain why the ball will or will not reach 80 feet.

Explaining the Concepts

45. What is the imaginary unit i?

46. Explain how to write $\sqrt{-64}$ as a multiple of i.

47. What is a complex number?

48. What is an imaginary number?

49. Why is every real number also a complex number?

50. Explain each of the three jokes in the cartoon on page 651.

Critical Thinking Exercises

Make Sense?   *In Exercises 51–54, determine whether each statement makes sense or does not make sense, and explain your reasoning.*

51. The word *imaginary* in imaginary numbers tells me that these numbers are undefined.

52. Writing i before any radical helps me to avoid placing i in the radicand.

53. By writing the imaginary number $6i$, I can immediately see that 6 is the constant and i is the variable.

54. When I use the quadratic formula to solve a quadratic equation and $b^2 - 4ac$ is negative, I can be certain that the equation has two imaginary solutions.

In Exercises 55–58, determine whether each statement is true or false. If the statement is false, make the necessary change(s) to produce a true statement.

55. $-\sqrt{-9} = -(-3) = 3$

56. The complex number $a + 0i$ is the real number a.

57. $2 + \sqrt{-4} = 2 - 2i$

58. $\dfrac{2 \pm 4i}{2} = 1 \pm 4i$

59. Show that $1 + i$ is a solution of $x^2 - 2x + 2 = 0$ by substituting $1 + i$ for x. You should obtain

$$(1 + i)^2 - 2(1 + i) + 2.$$

Square $1 + i$ as you would a binomial. Distribute -2 as indicated. Then simplify the resulting expression by combining like terms and replacing i^2 with -1. You should obtain 0. Use this procedure to show that $1 - i$ is the equation's other solution.

60. Prove that there is no real number such that when twice the number is subtracted from its square, the difference is -5.

Technology Exercises

61. Reread Exercise 43. Use your graphing utility to illustrate the problem's solution by graphing $y = -2x^2 + 36x$ and $y = 200$ in a $[0, 20, 1]$ by $[0, 210, 30]$ viewing rectangle. Explain how the two graphs show that weekly revenue will not reach $200,000 and, therefore, the applicant will not be hired.

62. Reread Exercise 44. Use your graphing utility to illustrate the problem's solution by graphing $y = -16x^2 + 60x + 4$ and $y = 80$ in a $[0, 4, 1]$ by $[0, 100, 10]$ viewing rectangle. Explain how the graphs show that the ball will not reach a height of 80 feet.

Review Exercises

In Exercises 63–65, graph each equation in a rectangular coordinate system.

63. $y = \dfrac{1}{3}x - 2$ (Section 3.4, Example 3)

64. $2x - 3y = 6$ (Section 3.2, Example 4)

65. $x = -2$ (Section 3.2, Example 8)

Preview Exercises

Exercises 66–68 will help you prepare for the material covered in the next section.

66. Use point plotting to graph $y = x^2 + 4x + 3$. Select integers from -5 to 1, inclusive, for x.

67. Replace y with 0 and find the x-intercepts for the graph of $y = x^2 - 2x - 3$.

68. Replace x with 0 and find the y-intercept for the graph of $y = x^2 - 2x - 3$.

SECTION

9.5

Graphs of Quadratic Equations

Many sports involve objects that are thrown, kicked, or hit, and then proceed with no additional force of their own. Such objects are called **projectiles**. Paths of projectiles, as well as their heights over time, can be modeled by quadratic equations. In this section, you will learn to use graphs of quadratic equations to gain a visual understanding of the algebra that describes football, baseball, basketball, the shot put, and other projectile sports.

What am I supposed to learn?

After studying this section, you should be able to:

1. Understand the characteristics of graphs of quadratic equations.

2. Find a parabola's intercepts.

3. Find a parabola's vertex.

4. Graph quadratic equations.

5. Solve problems using a parabola's vertex.

1 Understand the characteristics of graphs of quadratic equations.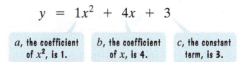

Graphs of Quadratic Equations

The graph of any quadratic equation

$$y = ax^2 + bx + c, \quad a \neq 0$$

is called a **parabola**. Parabolas are shaped like bowls or inverted bowls, as shown in **Figure 9.10**. If the coefficient of x^2 (the value of a in $ax^2 + bx + c$) is positive, the parabola opens upward. If the coefficient of x^2 is negative, the parabola opens downward. The **vertex** (or turning point) of the parabola is the lowest point, or minimum point, on the graph when it opens upward and the highest point, or maximum point, on the graph when it opens downward.

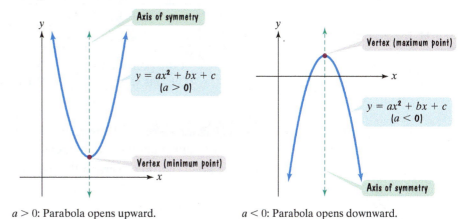

$a > 0$: Parabola opens upward. $a < 0$: Parabola opens downward.

Figure 9.10 Characteristics of graphs of quadratic equations

Look at the unusual image of the word *mirror* shown on the right. The artist, Scott Kim, has created the image so that the two halves of the whole are mirror images of each other. A parabola shares this kind of symmetry. A "mirror line" through the vertex, called the **axis of symmetry**, divides the figure in half. If a parabola is folded along its axis of symmetry, the two halves match exactly.

Copyright © 2011 Scott Kim, Scottkim.com

EXAMPLE 1   Using Point Plotting to Graph a Parabola

Consider the equation $y = x^2 + 4x + 3$.

a. Is the graph a parabola that opens upward or downward?

b. Use point plotting to graph the parabola. Select integers from -5 to 1, inclusive, for x.

Solution

a. To determine whether a parabola opens upward or downward, we begin by identifying a, the coefficient of x^2. The following voice balloons show the values for a, b, and c in $y = x^2 + 4x + 3$. Notice that we wrote x^2 as $1x^2$.

$$y = 1x^2 + 4x + 3$$

a, the coefficient of x^2, is 1. b, the coefficient of x, is 4. c, the constant term, is 3.

When a is greater than 0, a parabola opens upward. When a is less than 0, a parabola opens downward. Because $a = 1$, which is greater than 0, the parabola opens upward.

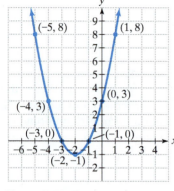

Figure 9.11 The graph of $y = x^2 + 4x + 3$

b. To use point plotting to graph the parabola, we first make a table of x- and y-coordinates.

x	$y = x^2 + 4x + 3$	(x, y)
-5	$y = (-5)^2 + 4(-5) + 3 = 8$	$(-5, 8)$
-4	$y = (-4)^2 + 4(-4) + 3 = 3$	$(-4, 3)$
-3	$y = (-3)^2 + 4(-3) + 3 = 0$	$(-3, 0)$
-2	$y = (-2)^2 + 4(-2) + 3 = -1$	$(-2, -1)$
-1	$y = (-1)^2 + 4(-1) + 3 = 0$	$(-1, 0)$
0	$y = 0^2 + 4(0) + 3 = 3$	$(0, 3)$
1	$y = 1^2 + 4(1) + 3 = 8$	$(1, 8)$

Then we plot the points and connect them with a smooth curve. The graph of $y = x^2 + 4x + 3$ is shown in **Figure 9.11**. ■

✓ **CHECK POINT 1**   Consider the equation $y = x^2 - 6x + 8$.

a. Is the graph a parabola that opens upward or downward?

b. Use point plotting to graph the parabola. Select integers from 0 to 6, inclusive, for x.

Several points are important when graphing a quadratic equation. These points, labeled in **Figure 9.12**, are the x-intercepts (although not every parabola has two x-intercepts), the y-intercept, and the vertex. Let's see how we can locate these points.

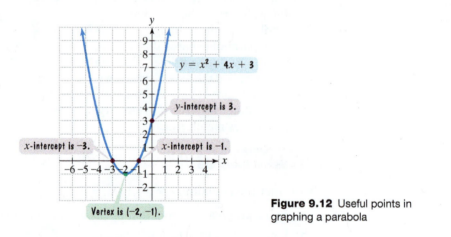

Figure 9.12 Useful points in graphing a parabola

2 Find a parabola's intercepts. ▶

Finding a Parabola's x-Intercepts

At each point where a parabola crosses the x-axis, the value of y equals 0. Thus, the x-intercepts can be found by replacing y with 0 in $y = ax^2 + bx + c$. Use factoring or the quadratic formula to solve the resulting quadratic equation for x.

EXAMPLE 2   Finding a Parabola's x-Intercepts

Find the x-intercepts for the parabola whose equation is $y = x^2 + 4x + 3$.

Solution   Replace y with 0 in $y = x^2 + 4x + 3$. We obtain $0 = x^2 + 4x + 3$, or $x^2 + 4x + 3 = 0$. We can solve this equation by factoring.

$$x^2 + 4x + 3 = 0$$
$$(x + 3)(x + 1) = 0$$
$$x + 3 = 0 \quad \text{or} \quad x + 1 = 0$$
$$x = -3 \qquad x = -1$$

Thus, the x-intercepts are -3 and -1. The parabola passes through $(-3, 0)$ and $(-1, 0)$, as shown in **Figure 9.12**. ∎

☑ **CHECK POINT 2**   Find the x-intercepts for the parabola whose equation is $y = x^2 - 6x + 8$.

Finding a Parabola's y-Intercept

At the point where a parabola crosses the y-axis, the value of x equals 0. Thus, the y-intercept can be found by replacing x with 0 in $y = ax^2 + bx + c$. Simple arithmetic will produce a value for y, which is the y-intercept.

EXAMPLE 3   Finding a Parabola's y-Intercept

Find the y-intercept for the parabola whose equation is $y = x^2 + 4x + 3$.

Solution   Replace x with 0 in $y = x^2 + 4x + 3$.

$$y = 0^2 + 4 \cdot 0 + 3 = 0 + 0 + 3 = 3$$

The y-intercept is 3. The parabola passes through $(0, 3)$, as shown in **Figure 9.12**. ∎

☑ **CHECK POINT 3**   Find the y-intercept for the parabola whose equation is $y = x^2 - 6x + 8$.

3 Find a parabola's vertex. ⊙

Finding a Parabola's Vertex

Keep in mind that a parabola's vertex is its turning point. The x-coordinate of the vertex for the parabola in **Figure 9.12**, -2, is midway between the x-intercepts, -3 and -1. If a parabola has two x-intercepts, they are found by solving $ax^2 + bx + c = 0$. The solutions of this equation,

$$x = \frac{-b - \sqrt{b^2 - 4ac}}{2a} \quad \text{and} \quad x = \frac{-b + \sqrt{b^2 - 4ac}}{2a},$$

are the x-intercepts. The value of x midway between these intercepts is $x = \dfrac{-b}{2a}$. This equation can be used to find the x-coordinate of the vertex even when no x-intercepts exist.

> **The Vertex of a Parabola**
>
> For a parabola whose equation is $y = ax^2 + bx + c$,
>
> **1.** The x-coordinate of the vertex is $\dfrac{-b}{2a}$.
>
> **2.** The y-coordinate of the vertex is found by substituting the x-coordinate into the parabola's equation and evaluating.

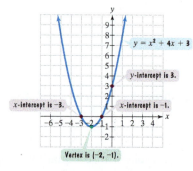

Figure 9.12 (repeated)

EXAMPLE 4 Finding a Parabola's Vertex

Find the vertex for the parabola whose equation is $y = x^2 + 4x + 3$.

Solution In the equation $y = x^2 + 4x + 3$, $a = 1$ and $b = 4$.

$$x\text{-coordinate of vertex} = \frac{-b}{2a} = \frac{-4}{2 \cdot 1} = \frac{-4}{2} = -2$$

To find the y-coordinate of the vertex, we substitute -2 for x in $y = x^2 + 4x + 3$ and then evaluate.

$$y\text{-coordinate of vertex} = (-2)^2 + 4(-2) + 3 = 4 + (-8) + 3 = -1$$

The vertex is $(-2, -1)$, as shown in **Figure 9.12**. ■

☑ **CHECK POINT 4** Find the vertex for the parabola whose equation is $y = x^2 - 6x + 8$.

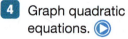

4 Graph quadratic equations.

A Strategy for Graphing Quadratic Equations

Here is a procedure to sketch the graph of the quadratic equation $y = ax^2 + bx + c$:

Graphing Quadratic Equations

The graph of $y = ax^2 + bx + c$, called a parabola, can be graphed using the following steps:

1. Determine whether the parabola opens upward or downward. If $a > 0$, it opens upward. If $a < 0$, it opens downward.

2. Determine the vertex of the parabola. The x-coordinate is $\frac{-b}{2a}$.

   The y-coordinate is found by substituting the x-coordinate into the parabola's equation and evaluating.

3. Find any x-intercepts by replacing y with 0. Solve the resulting quadratic equation for x. The real solutions are the x-intercepts.

4. Find the y-intercept by replacing x with 0. Because $y = a \cdot 0^2 + b \cdot 0 + c$ simplifies to $y = c$, the y-intercept is c, the constant term, and the parabola passes through $(0, c)$.

5. Plot the intercepts and the vertex.

6. Connect these points with a smooth curve.

EXAMPLE 5 Graphing a Parabola

Graph the quadratic equation: $y = x^2 - 2x - 3$.

Solution We can graph this equation by following the steps in the box.

Step 1. Determine how the parabola opens. Note that a, the coefficient of x^2, is 1. Thus, $a > 0$; this positive value tells us that the parabola opens upward.

Step 2. Find the vertex. We know that the x-coordinate of the vertex is $\frac{-b}{2a}$. Let's identify the numbers $a, b,$ and c in the given equation, which is in the form $y = ax^2 + bx + c$.

$$y = x^2 - 2x - 3$$

$$a = 1 \qquad b = -2 \qquad c = -3$$

Now we substitute the values of a and b, $a = 1$ and $b = -2$, into the expression for the x-coordinate:

$$x\text{-coordinate of vertex} = \frac{-b}{2a} = \frac{-(-2)}{2(1)} = \frac{2}{2} = 1.$$

The x-coordinate of the vertex is 1. We can substitute 1 for x in the equation $y = x^2 - 2x - 3$ to find the y-coordinate:

$$y\text{-coordinate of vertex} = 1^2 - 2 \cdot 1 - 3 = 1 - 2 - 3 = -4.$$

The vertex is $(1, -4)$.

Step 3. Find the x-intercepts. Replace y with 0 in $y = x^2 - 2x - 3$. We obtain $0 = x^2 - 2x - 3$ or $x^2 - 2x - 3 = 0$. We can solve this equation by factoring.

$$x^2 - 2x - 3 = 0$$
$$(x - 3)(x + 1) = 0$$
$$x - 3 = 0 \quad \text{or} \quad x + 1 = 0$$
$$x = 3 \qquad\qquad x = -1$$

The x-intercepts are 3 and -1. The parabola passes through $(3, 0)$ and $(-1, 0)$.

Step 4. Find the y-intercept. Replace x with 0 in $y = x^2 - 2x - 3$:

$$y = 0^2 - 2 \cdot 0 - 3 = 0 - 0 - 3 = -3.$$

The y-intercept is -3. The parabola passes through $(0, -3)$.

Steps 5 and 6. Plot the intercepts and the vertex. Connect these points with a smooth curve. The intercepts and the vertex are shown as the four labeled points in **Figure 9.13**. Also shown is the graph of the quadratic equation, obtained by connecting the points with a smooth curve. ∎

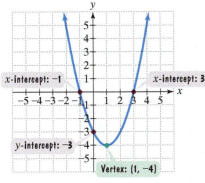

Figure 9.13 The graph of $y = x^2 - 2x - 3$

✓ **CHECK POINT 5**   Graph the quadratic equation: $y = x^2 + 6x + 5$.

EXAMPLE 6   Graphing a Parabola

Graph the quadratic equation: $y = -x^2 + 4x - 1$.

Solution

Step 1. Determine how the parabola opens. Note that a, the coefficient of x^2, is -1. Thus, $a < 0$; this negative value tells us that the parabola opens downward.

Step 2. Find the vertex. The x-coordinate of the vertex is $\dfrac{-b}{2a}$.

$$y = -x^2 + 4x - 1$$

$$a = -1 \quad b = 4 \quad c = -1$$

$$x\text{-coordinate of vertex} = \frac{-b}{2a} = \frac{-4}{2(-1)} = \frac{-4}{-2} = 2.$$

Substitute 2 for x in $y = -x^2 + 4x - 1$ to find the y-coordinate:

$$y\text{-coordinate of vertex} = -2^2 + 4 \cdot 2 - 1 = -4 + 8 - 1 = 3.$$

The vertex is $(2, 3)$.

Step 3. Find the *x*-intercepts. Replace *y* with 0 in $y = -x^2 + 4x - 1$. We obtain $0 = -x^2 + 4x - 1$ or $-x^2 + 4x - 1 = 0$. This equation cannot be solved by factoring. We will use the quadratic formula to solve it.

$$a = -1, \qquad b = 4, \qquad c = -1$$

$$x = \frac{-b \pm \sqrt{b^2 - 4ac}}{2a} = \frac{-4 \pm \sqrt{4^2 - 4(-1)(-1)}}{2(-1)} = \frac{-4 \pm \sqrt{16 - 4}}{-2}$$

> To locate the *x*-intercepts, we need decimal approximations. Thus, there is no need to simplify the radical form of the solutions.

$$x = \frac{-4 + \sqrt{12}}{-2} \approx 0.3 \quad \text{or} \quad x = \frac{-4 - \sqrt{12}}{-2} \approx 3.7$$

The *x*-intercepts are approximately 0.3 and 3.7. The parabola passes through (0.3, 0) and (3.7, 0).

Step 4. Find the *y*-intercept. Replace *x* with 0 in $y = -x^2 + 4x - 1$:

$$y = -0^2 + 4 \cdot 0 - 1 = -1.$$

The *y*-intercept is -1. The parabola passes through (0, −1).

Steps 5 and 6. Plot the intercepts and the vertex. Connect these points with a smooth curve. The intercepts and the vertex are shown as the four labeled points in **Figure 9.14**. Also shown is the graph of the quadratic equation, obtained by connecting the points with a smooth curve.

Vertex: (2, 3)

x-intercept: 0.3

x-intercept: 3.7

y-intercept: −1

Figure 9.14 The graph of $y = -x^2 + 4x - 1$ ■

✓ **CHECK POINT 6** Graph the quadratic equation: $y = -x^2 - 2x + 5$.

5 Solve problems using a parabola's vertex. ▶

Applications

Many applied problems involve finding the maximum or minimum value of equations in the form $y = ax^2 + bx + c$. The vertex of the graph is the point of interest. If $a < 0$, the parabola opens downward and the vertex is its highest point. If $a > 0$, the parabola opens upward and the vertex is its lowest point.

EXAMPLE 7 The Parabolic Path of a Punted Football

Figure 9.15 shows that when a football is kicked, the nearest defensive player is 6 feet from the point of impact with the kicker's foot. The height of the punted football, *y*, in feet, can be modeled by

$$y = -0.01x^2 + 1.18x + 2,$$

where *x* is the ball's horizontal distance, in feet, from the point of impact with the kicker's foot.

a. What is the maximum height of the punt and how far from the point of impact does this occur?

b. How far must the nearest defensive player, who is 6 feet from the kicker's point of impact, reach to block the punt?

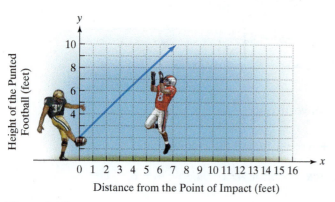

Height of the Punted Football (feet)

Distance from the Point of Impact (feet)

Figure 9.15

c. If the ball is not blocked by the defensive player, how far down the field will it go before hitting the ground?

d. Graph the equation that models the football's parabolic path.

Solution

a. We begin by identifying the numbers a, b, and c in the given model.

$$y = -0.01x^2 + 1.18x + 2$$

$$a = -0.01 \qquad b = 1.18 \qquad c = 2$$

Because the coefficient of x^2, -0.01, is negative, the parabola opens downward and the vertex is the highest point on the graph. The y-coordinate of the vertex gives the maximum height of the punt and the x-coordinate reveals how far from the point of impact this occurs.

$$x\text{-coordinate of vertex} = \frac{-b}{2a} = \frac{-1.18}{2(-0.01)} = \frac{-1.18}{-0.02} = 59$$

Substitute 59 for x in $y = -0.01x^2 + 1.18x + 2$ to find the y-coordinate.

$$y\text{-coordinate of vertex} = -0.01(59)^2 + 1.18(59) + 2$$
$$= -34.81 + 69.62 + 2 = 36.81$$

The vertex is $(59, 36.81)$. The maximum height of the punt is 36.81 feet and this occurs 59 feet from the kicker's point of impact.

b. **Figure 9.15** shows that the defensive player is 6 feet from the kicker's point of impact. To block the punt, he must touch the football along its parabolic path. This means that we must find the height of the ball 6 feet from the kicker. Replace x with 6 in the given model, $y = -0.01x^2 + 1.18x + 2$.

$$y = -0.01(6)^2 + 1.18(6) + 2 = -0.36 + 7.08 + 2 = 8.72$$

The defensive player must reach 8.72 feet above the ground to block the punt.

c. Assuming that the ball is not blocked by the defensive player, we are interested in how far down the field it will go before hitting the ground. We are looking for the ball's horizontal distance, x, when its height above the ground, y, is 0 feet. To find this x-intercept, replace y with 0 in $y = -0.01x^2 + 1.18x + 2$. We obtain $0 = -0.01x^2 + 1.18x + 2$, or $-0.01x^2 + 1.18x + 2 = 0$. The equation cannot be solved by factoring. We will use the quadratic formula to solve it.

> Use a calculator to evaluate the radicand.

$$x = \frac{-b \pm \sqrt{b^2 - 4ac}}{2a} = \frac{-1.18 \pm \sqrt{(1.18)^2 - 4(-0.01)(2)}}{2(-0.01)} = \frac{-1.18 \pm \sqrt{1.4724}}{-0.02}$$

$$x = \frac{-1.18 + \sqrt{1.4724}}{-0.02} \qquad \text{or} \qquad x = \frac{-1.18 - \sqrt{1.4724}}{-0.02}$$

$$x \approx -1.7 \qquad\qquad\qquad x \approx 119.7$$

> Use a calculator and round to the nearest tenth.

> Reject this value. We are interested in the football's height corresponding to horizontal distances from its point of impact onward, or $x \geq 0$.

If the football is not blocked by the defensive player, it will go approximately 119.7 feet down the field before hitting the ground.

d. In terms of graphing the model for the football's parabolic path, $y = -0.01x^2 + 1.18x + 2$, we have already determined the vertex and the approximate x-intercept.

vertex: $(59, 36.81)$  The ball's maximum height, 36.81 feet, occurs at a horizontal distance of 59 feet.

x-intercept: 119.7 | The ball's maximum horizontal distance is approximately 119.7 feet.

Figure 9.15 indicates that the y-intercept is 2, meaning that the ball is kicked from a height of 2 feet. Let's verify this value by replacing x with 0 in $y = -0.01x^2 + 1.18x + 2$.

$$y = -0.01 \cdot 0^2 + 1.18 \cdot 0 + 2 = 0 + 0 + 2 = 2$$

Using the vertex, $(59, 36.81)$, the x-intercept, 119.7, and the y-intercept, 2, the graph of the equation that models the football's parabolic path is shown in **Figure 9.16**. The graph is shown only for $x \geq 0$, indicating horizontal distances that begin at the football's impact with the kicker's foot and end with the ball hitting the ground.

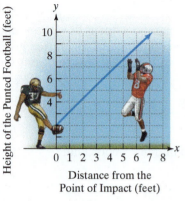

Figure 9.15 (repeated)

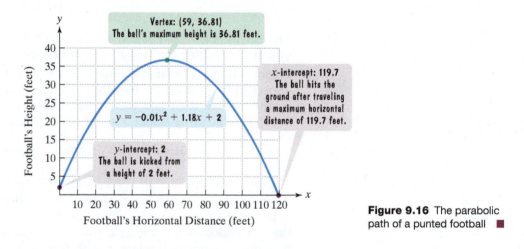

Figure 9.16 The parabolic path of a punted football ■

☑ **CHECK POINT 7**   An archer's arrow follows a parabolic path. The height of the arrow, y, in feet, can be modeled by

$$y = -0.005x^2 + 2x + 5,$$

where x is the arrow's horizontal distance, in feet.

a. What is the maximum height of the arrow and how far from its release does this occur?

b. Find the horizontal distance the arrow travels before it hits the ground. Round to the nearest foot.

c. Graph the equation that models the arrow's parabolic path.

CONCEPT AND VOCABULARY CHECK

Fill in each blank so that the resulting statement is true.

1. The graph of any quadratic equation $y = ax^2 + bx + c, a \neq 0$, is called a/an _____.

2. The graph of $y = ax^2 + bx + c$ opens upward if _____ and opens downward if _____.

3. The x-intercepts for the graph of $y = ax^2 + bx + c$ can be found by determining the real solutions of the equation _____.

4. The y-intercept for the graph of $y = ax^2 + bx + c$ can be determined by replacing x with _____.

5. The x-coordinate of the vertex of the graph of $y = ax^2 + bx + c$ is _____. The y-coordinate of the vertex is found by substituting _____ into the equation and evaluating.

9.5 EXERCISE SET ▶ MyMathLab®

Practice Exercises

In Exercises 1–4, determine if the parabola whose equation is given opens upward or downward.

1. $y = x^2 - 4x + 3$
2. $y = x^2 - 6x + 5$
3. $y = -2x^2 + x + 6$
4. $y = -2x^2 - 4x + 6$

In Exercises 5–10, find the x-intercepts for the parabola whose equation is given. If the x-intercepts are irrational numbers, round your answers to the nearest tenth.

5. $y = x^2 - 4x + 3$
6. $y = x^2 - 6x + 5$
7. $y = -x^2 + 8x - 12$
8. $y = -x^2 - 2x + 3$
9. $y = x^2 + 2x - 4$
10. $y = x^2 + 8x + 14$

In Exercises 11–18, find the y-intercept for the parabola whose equation is given.

11. $y = x^2 - 4x + 3$
12. $y = x^2 - 6x + 5$
13. $y = -x^2 + 8x - 12$
14. $y = -x^2 - 2x + 3$
15. $y = x^2 + 2x - 4$
16. $y = x^2 + 8x + 14$
17. $y = x^2 + 6x$
18. $y = x^2 + 8x$

In Exercises 19–24, find the vertex for the parabola whose equation is given.

19. $y = x^2 - 4x + 3$
20. $y = x^2 - 6x + 5$
21. $y = 2x^2 + 4x - 6$
22. $y = -2x^2 - 4x - 2$
23. $y = x^2 + 6x$
24. $y = x^2 + 8x$

In Exercises 25–36, graph the parabola whose equation is given.

25. $y = x^2 + 8x + 7$
26. $y = x^2 + 10x + 9$
27. $y = x^2 - 2x - 8$
28. $y = x^2 + 4x - 5$
29. $y = -x^2 + 4x - 3$
30. $y = -x^2 + 2x + 3$
31. $y = x^2 - 1$
32. $y = x^2 - 4$
33. $y = x^2 + 2x + 1$
34. $y = x^2 - 2x + 1$
35. $y = -2x^2 + 4x + 5$
36. $y = -3x^2 + 6x - 2$

Practice PLUS

In Exercises 37–44, find the vertex for the parabola whose equation is given by first writing the equation in the form $y = ax^2 + bx + c$.

37. $y = (x - 3)^2 + 2$
38. $y = (x - 4)^2 + 3$
39. $y = (x + 5)^2 - 4$
40. $y = (x + 6)^2 - 5$
41. $y = 2(x - 1)^2 - 3$
42. $y = 2(x - 1)^2 - 4$
43. $y = -3(x + 2)^2 + 5$
44. $y = -3(x + 4)^2 + 6$
45. Generalize your work in Exercises 37–44 and complete the following statement: For a parabola whose equation is $y = a(x - h)^2 + k$, the vertex is the point _____.

Application Exercises

An athlete whose event is the shot put releases the shot with the same initial velocity, but at different angles. The figure shows the parabolic paths for shots released at angles of 35° and 65°. Exercises 46–47 are based on the equations that model the parabolic paths.

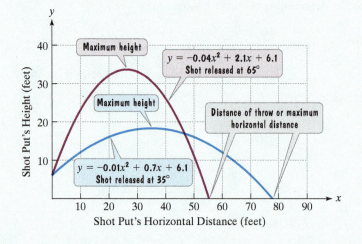

46. When the shot is released at an angle of 65°, its height, y, in feet, can be modeled by

$$y = -0.04x^2 + 2.1x + 6.1,$$

where x is the shot's horizontal distance, in feet, from its point of release. Use this model to solve parts (a) through (c) and verify your answers using the red graph.

a. What is the maximum height, to the nearest tenth of a foot, of the shot and how far from its point of release does this occur?

b. What is the shot's maximum horizontal distance, to the nearest tenth of a foot, or the distance of the throw?

c. From what height was the shot released?

47. When the shot is released at an angle of 35°, its height, y, in feet, can be modeled by

$$y = -0.01x^2 + 0.7x + 6.1,$$

where x is the shot's horizontal distance, in feet, from its point of release. Use this model to solve parts (a) through (c) and verify your answers using the blue graph.

a. What is the maximum height of the shot and how far from its point of release does this occur?

b. What is the shot's maximum horizontal distance, to the nearest tenth of a foot, or the distance of the throw?

c. From what height was the shot released?

48. A ball is thrown upward and outward from a height of 6 feet. The height of the ball, y, in feet, can be modeled by

$$y = -0.8x^2 + 2.4x + 6,$$

where x is the ball's horizontal distance, in feet, from where it was thrown.

a. What is the maximum height of the ball and how far from where it was thrown does this occur?

b. How far does the ball travel horizontally before hitting the ground? Round to the nearest tenth of a foot.

c. Graph the equation that models the ball's parabolic path.

49. A ball is thrown upward and outward from a height of 6 feet. The height of the ball, y, in feet, can be modeled by

$$y = -0.8x^2 + 3.2x + 6,$$

where x is the ball's horizontal distance, in feet, from where it was thrown.

a. What is the maximum height of the ball and how far from where it was thrown does this occur?

b. How far does the ball travel horizontally before hitting the ground? Round to the nearest tenth of a foot.

c. Graph the equation that models the ball's parabolic path.

50. You have 120 feet of fencing to enclose a rectangular plot that borders on a river. If you do not fence the side along the river, find the length and width of the plot that will maximize the area. What is the largest area that can be enclosed?

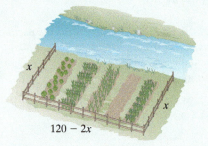

$120 - 2x$

51. You have 100 yards of fencing to enclose a rectangular area, as shown in the figure. Find the dimensions of the rectangle that maximize the enclosed area. What is the maximum area?

$50 - x$ x

Explaining the Concepts

52. What is a parabola? Describe its shape.

53. Explain how to decide whether a parabola opens upward or downward.

54. If a parabola has two x-intercepts, explain how to find them.

55. Explain how to find a parabola's y-intercept.

56. Describe how to find a parabola's vertex.

57. A parabola that opens upward has its vertex at $(1, 2)$. Describe as much as you can about the parabola based on this information. Include in your discussion the number of x-intercepts, if any, for the parabola.

Critical Thinking Exercises

Make Sense? In Exercises 58–61, determine whether each statement makes sense or does not make sense, and explain your reasoning.

58. I must have made an error when graphing this parabola because it is symmetric with respect to the y-axis.

59. Parabolas that open up appear to form smiles ($a > 0$), while parabolas that open down frown ($a < 0$).

60. I threw a baseball vertically upward and its path was a parabola.

61. **Figure 9.15** on page 662 shows that a linear model provides a better description of the football's path than a quadratic model.

In Exercises 62–65, determine whether each statement is true or false. If the statement is false, make the necessary change(s) to produce a true statement.

62. The x-coordinate of the vertex of the parabola whose equation is $y = ax^2 + bx + c$ is $\dfrac{b}{2a}$.

63. If a parabola has only one x-intercept, then the x-intercept is also the vertex.

64. There is no relationship between the graph of $y = ax^2 + bx + c$ and the number of real solutions of the equation $ax^2 + bx + c = 0$.

65. If $y = 4x^2 - 40x + 4$, then the vertex is the highest point on the graph.

66. Find two numbers whose sum is 200 and whose product is a maximum.

67. Graph $y = 2x^2 - 8$ and $y = -2x^2 + 8$ in the same rectangular coordinate system. What are the coordinates of the points of intersection?

68. A parabola has x-intercepts at 3 and 7, a y-intercept at -21, and $(5, 4)$ for its vertex. Write the parabola's equation.

Technology Exercises

69. Use a graphing utility to verify any five of your hand-drawn graphs in Exercises 25–36.

70. a. Use a graphing utility to graph $y = 2x^2 - 82x + 720$ in a standard viewing rectangle. What do you observe?

b. Find the coordinates of the vertex for the given quadratic equation.

c. The answer to part (b) is $(20.5, -120.5)$. Because the leading coefficient, 2, of $y = 2x^2 - 82x + 720$ is positive, the vertex is a minimum point on the graph. Use this fact to help find a viewing rectangle that will give a relatively complete picture of the parabola. With an axis of symmetry at $x = 20.5$, the setting for x should extend past this, so try Xmin = 0 and Xmax = 30. The setting for y should include (and probably go below) the y-coordinate of the graph's minimum point, so try Ymin = -130. Experiment with Ymax until your utility shows the parabola's major features.

d. In general, explain how knowing the coordinates of a parabola's vertex can help determine a reasonable viewing rectangle on a graphing utility for obtaining a complete picture of the parabola.

In Exercises 71–74, find the vertex for each parabola. Then determine a reasonable viewing rectangle on your graphing utility and use it to graph the parabola.

71. $y = -0.25x^2 + 40x$

72. $y = -4x^2 + 20x + 160$

73. $y = 5x^2 + 40x + 600$

74. $y = 0.01x^2 + 0.6x + 100$

Review Exercises

In Exercises 75–77, solve each equation or system of equations.

75. $7(x - 2) = 10 - 2(x + 3)$ (Section 2.3, Example 3)

76. $\dfrac{7}{x + 2} + \dfrac{2}{x + 3} = \dfrac{1}{x^2 + 5x + 6}$ (Section 7.6, Example 4)

77. $\begin{cases} 5x - 3y = -13 \text{ (Section 4.2, Example 1)} \\ x = 2 - 4y \end{cases}$

Preview Exercises

Exercises 78–80 will help you prepare for the material covered in the next section.

78. Here are two sets of ordered pairs:

$$\text{set 1: } \{(1, 5), (2, 5)\}$$
$$\text{set 2: } \{(5, 1), (5, 2)\}.$$

In which set is each x-coordinate paired with one and only one y-coordinate?

79. Here are two graphs:

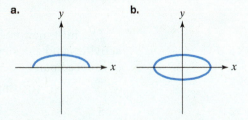

In which graph is each x-coordinate paired with one and only one y-coordinate?

80. Evaluate $x^2 + 3x + 5$ for $x = -3$.

SECTION

9.6

Introduction to Functions ▶

What am I supposed to learn?

After studying this section, you should be able to:

1. Find the domain and range of a relation. ▶
2. Determine whether a relation is a function. ▶
3. Evaluate a function. ▶
4. Use the vertical line test to identify functions. ▶
5. Find function values for functions that model data. ▶

1. Find the domain and range of a relation. ▶

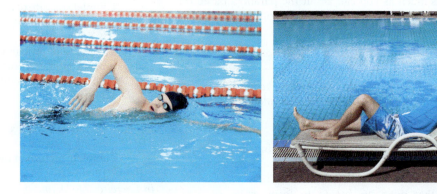

The number of calories that you burn per hour depends on the activity in which you are engaged. The average cost of cellphone use depends on the year in which this average is computed. In both of these situations, the relationship between variables can be illustrated with the notion of a *function*. Understanding this concept will give you a new perspective on many ordinary situations. Much of your work in subsequent algebra courses will be devoted to the important topic of functions and how they model your world.

Relations

Studies show that exercise can promote good long-term health no matter how much you weigh. A brisk half-hour walk each day is enough to get the benefits. Combined with a healthy diet, it also helps to stave off obesity. How many calories does your workout burn? The graph in **Figure 9.17** shows the calories burned per hour in six activities.

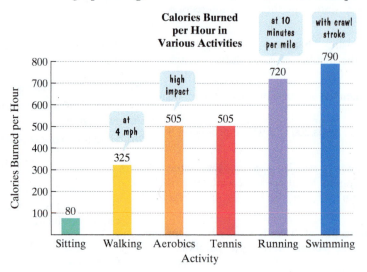

Figure 9.17
Counting calories

Source: FITRESOURCE.COM

The information shown in the bar graph indicates a correspondence between the activities and calories burned per hour. We can write this correspondence using a set of ordered pairs:

{(sitting, 80), (walking, 325), (aerobics, 505), (tennis, 505), (running, 720), (swimming, 790)}.

These braces indicate that we are representing a set.

The mathematical term for a set of ordered pairs is a **relation**.

Definition of a Relation

A **relation** is any set of ordered pairs. The set of all first components of the ordered pairs is called the **domain** of the relation and the set of all second components is called the **range** of the relation.

EXAMPLE 1   Finding the Domain and Range of a Relation

Find the domain and range of the relation:

{(sitting, 80), (walking, 325), (aerobics, 505), (tennis, 505),
(running, 720), (swimming, 790)}.

Solution   The domain is the set of all first components. Thus, the domain is

{sitting, walking, aerobics, tennis, running, swimming}.

> Parentheses and square brackets are not used to represent sets.

The range is the set of all second components. Thus, the range is

{80, 325, 505, 720, 790}.

> Although both aerobics and tennis burn 505 calories per hour, it is not necessary to list 505 twice.

☑ **CHECK POINT 1**   The following set shows calories burned per hour in activities not included in **Figure 9.17**. Find the domain and range of the relation:

{(golf, 250), (lawn mowing, 325), (water skiing, 430),
(hiking, 430), (bicycling, 720)}.

> with cart

> at 15 mph

---

2 Determine whether a relation is a function.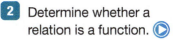

Functions

Shown in the margin are the calories burned per hour for the activities in the bar graph in **Figure 9.17**. We've used this information to define two relations. **Figure 9.18(a)** shows a correspondence between activities and calories burned. **Figure 9.18(b)** shows a correspondence between calories burned and activities.

Activity	Calories Burned per hour
Sitting	80
Walking	325
Aerobics	505
Tennis	505
Running	720
Swimming	790

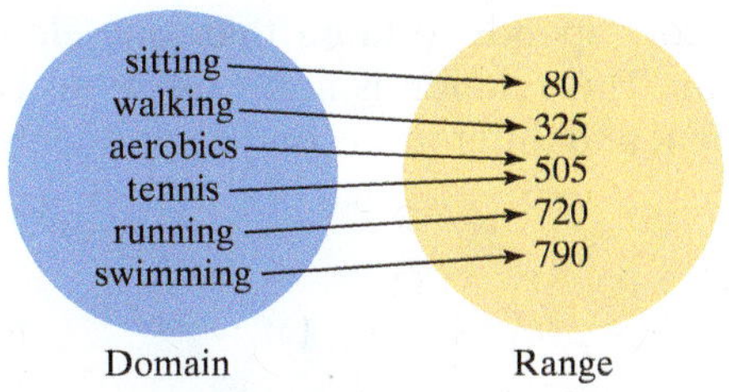

Figure 9.18(a) Activities correspond to calories burned.

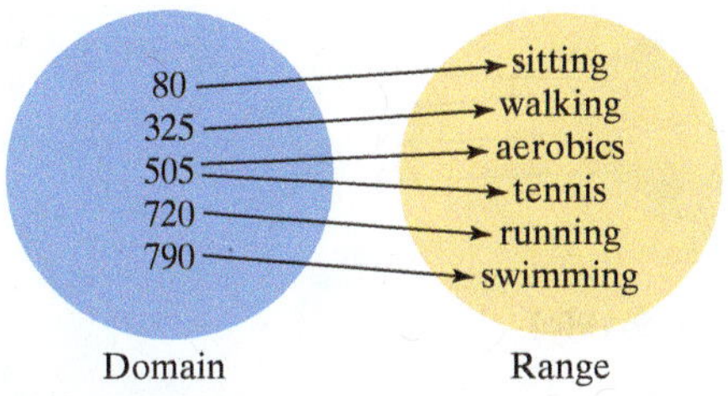

Figure 9.18(b) Calories burned correspond to activities.

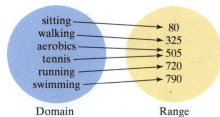

Figure 9.18(a) (repeated)

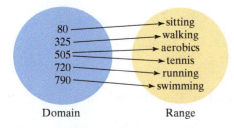

Figure 9.18(b) (repeated)

A relation in which each member of the domain corresponds to exactly one member of the range is a **function**. Can you see that the relation in **Figure 9.18(a)** is a function? Each activity in the domain corresponds to exactly one number representing calories burned per hour in the range. If we know the activity, we know the calories burned per hour. Notice that more than one element in the domain can correspond to the same element in the range: Aerobics and tennis both burn 505 calories per hour.

Is the relation in **Figure 9.18(b)** a function? Does each member of the domain correspond to precisely one member of the range? This relation is not a function because there is a member of the domain that corresponds to two members of the range:

$$(505, \text{aerobics}) \quad (505, \text{tennis}).$$

The member of the domain, 505, corresponds to both aerobics and tennis. If we know the calories burned per hour, 505, we cannot be sure of the activity. Because **a function is a relation in which no two ordered pairs have the same first component and different second components**, the ordered pairs (505, aerobics) and (505, tennis) are not ordered pairs of a function.

Same first component

$$(505, \text{aerobics}) \quad (505, \text{tennis})$$

Different second components

Definition of a Function

A **function** is a relation in which each member of the domain corresponds to exactly one member of the range. No two ordered pairs of a function can have the same first component and different second components.

EXAMPLE 2   Determining Whether a Relation Is a Function

Determine whether each relation is a function:

a. $\{(1, 6), (2, 6), (3, 8), (4, 9)\}$ **b.** $\{(6, 1), (6, 2), (8, 3), (9, 4)\}$.

Solution   We begin by making a figure for each relation that shows the domain and the range (**Figure 9.19**).

a. **Figure 9.19(a)** shows that every element in the domain corresponds to exactly one element in the range. The element 1 in the domain corresponds to the element 6 in the range. Furthermore, 2 corresponds to 6, 3 corresponds to 8, and 4 corresponds to 9. No two ordered pairs in the given relation have the same first component and different second components. Thus, the relation is a function.

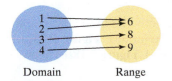

Figure 9.19(a)

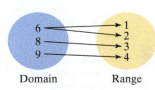

Figure 9.19(b)

b. **Figure 9.19(b)** shows that 6 corresponds to both 1 and 2. If any element in the domain corresponds to more than one element in the range, the relation is not a function. This relation is not a function: Two ordered pairs have the same first component and different second components.

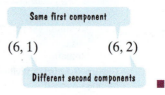

Same first component

$$(6, 1) \qquad (6, 2)$$

Different second components

Look at **Figure 9.19(a)** on the previous page again. The fact that 1 and 2 in the domain correspond to the same number, 6, in the range does not violate the definition of a function. **A function can have two different first components with the same second component.** By contrast, a relation is not a function when two different ordered pairs have the same first component and different second components. Thus, the relation in **Figure 9.19(b)** is not a function.

✅ **CHECK POINT 2**   Determine whether each relation is a function:

   a. $\{(1, 2), (3, 4), (5, 6), (5, 8)\}$

   b. $\{(1, 2), (3, 4), (6, 5), (8, 5)\}$.

---

3 Evaluate a function. ▶

Functions as Equations and Function Notation

Functions are usually given in terms of equations rather than as sets of ordered pairs. For example, here is an equation that shows a woman's cholesterol level as a function of her age:

$$y = 1.17x + 147.15.$$

The variable x represents a woman's age, in years. The variable y represents her cholesterol level. For each age, x, the equation gives one and only one cholesterol level, y. The variable y is a function of the variable x.

When an equation represents a function, the function is often named by a letter such as f, g, h, F, G, or H. Any letter can be used to name a function. Suppose that f names a function. Think of the domain as the set of the function's inputs and the range as the set of the function's outputs. As shown in **Figure 9.20**, the input is represented by x and the output by $f(x)$. The special notation $f(x)$, read "f of x" or "f at x," represents the **value of the function at the number x.**

Let's make this clearer by considering a specific example. We know the equation

$$y = 1.17x + 147.15$$

defines y as a function of x. We'll name the function f. Now, we can apply our new function notation.

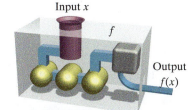

Input x
f
Output $f(x)$

Figure 9.20 A function as a machine with inputs and outputs

Input	Output	Equation
x	$f(x)$	$f(x) = 1.17x + 147.15$

We read this equation as "f of x equals $1.17x + 147.15$."

Suppose we are interested in finding $f(30)$, the function's output when the input is 30. To find the value of the function at 30, we substitute 30 for x. We are **evaluating the function** at 30.

$f(x) = 1.17x + 147.15$ This is the given function.

$f(30) = 1.17(30) + 147.15$ The input is 30.

$\quad\quad = 35.1 + 147.15$ Multiply.

$\quad\quad = 182.25$ Add.

The statement $f(30) = 182.25$, read "f of 30 equals 182.25," tells us that the value of the function at 30 is 182.25. When the function's input is 30, its output is 182.25. The model gives us exactly one cholesterol level for a 30-year-old woman, namely 182.25. **Figure 9.21** illustrates the input and output in terms of a function machine.

Great Question!

Doesn't $f(x)$ indicate that I need to multiply f and x?

The notation $f(x)$ does *not* mean "f times x." The notation describes the value of the function at x.

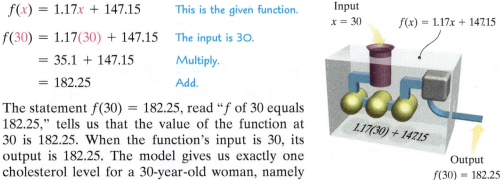

Input
$x = 30$ $f(x) = 1.17x + 147.15$

$1.17(30) + 147.15$

Output
$f(30) = 182.25$

Figure 9.21 A function machine at work

EXAMPLE 3 Evaluating a Function

If $f(x) = 3x + 1$, find each of the following:

a. $f(5)$ **b.** $f(-4)$ **c.** $f(0)$.

Solution We substitute 5, -4, and 0, respectively, for x in the function's equation, $f(x) = 3x + 1$.

a. $f(5) = 3 \cdot 5 + 1 = 15 + 1 = 16$ *f of 5 equals 16.*

b. $f(-4) = 3(-4) + 1 = -12 + 1 = -11$ *f of -4 equals -11.*

c. $f(0) = 3 \cdot 0 + 1 = 0 + 1 = 1$ *f of O equals 1.* ∎

The function $f(x) = 3x + 1$ is an example of a *linear function*. **Linear functions** have equations of the form $f(x) = mx + b$.

☑ **CHECK POINT 3**   If $f(x) = 4x + 3$, find each of the following:
a. $f(5)$ **b.** $f(-2)$ **c.** $f(0)$.

EXAMPLE 4 Evaluating a Function

If $g(x) = x^2 + 3x + 5$, find each of the following:

a. $g(2)$ **b.** $g(-3)$ **c.** $g(0)$.

Solution We substitute 2, -3, and 0, respectively, for x in the function's equation, $g(x) = x^2 + 3x + 5$.

a. $g(2) = 2^2 + 3 \cdot 2 + 5 = 4 + 6 + 5 = 15$ *g of 2 is 15.*

b. $g(-3) = (-3)^2 + 3(-3) + 5 = 9 + (-9) + 5 = 5$ *g of -3 is 5.*

c. $g(0) = 0^2 + 3 \cdot 0 + 5 = 0 + 0 + 5 = 5$ *g of O is 5.* ∎

The function in Example 4 is a *quadratic function*. **Quadratic functions** have equations of the form $f(x) = ax^2 + bx + c, a \neq 0$.

☑ **CHECK POINT 4**   If $g(x) = x^2 + 4x + 3$, find each of the following:
a. $g(5)$ **b.** $g(-4)$ **c.** $g(0)$.

4 Use the vertical line test to identify functions. ▶

Graphs of Functions and the Vertical Line Test

The **graph of a function** is the graph of its ordered pairs. For example, the graph of $f(x) = 3x + 1$ is the set of points (x, y) in the rectangular coordinate system satisfying the equation $y = 3x + 1$. Thus, the graph of f is a line with slope 3 and y-intercept 1. Similarly, the graph of $f(x) = x^2 + 3x + 5$ is the set of points (x, y) in the rectangular coordinate system satisfying the equation $y = x^2 + 3x + 5$. Thus, the graph of g is a parabola.

Not every graph in the rectangular coordinate system is the graph of a function. The definition of a function specifies that no value of x can be paired with two or more

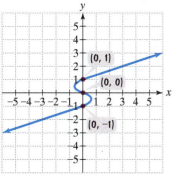

Figure 9.22 *y* is not a function of *x* because 0 is paired with three values of *y*, namely, 1, 0, and −1.

different values of *y*. Consequently, if a graph contains two or more different points with the same first coordinate, the graph cannot represent a function. This is illustrated in **Figure 9.22**. Observe that points sharing a common first coordinate are vertically above or below each other.

This observation is the basis of a useful test for determining whether a graph defines *y* as a function of *x*. The test is called the **vertical line test**.

The Vertical Line Test for Functions

If any vertical line intersects a graph in more than one point, the graph does not define *y* as a function of *x*.

EXAMPLE 5   Using the Vertical Line Test

Use the vertical line test to identify graphs in which *y* is a function of *x*.

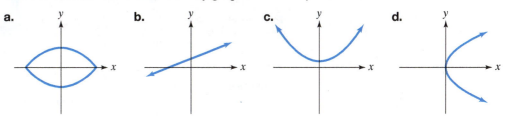

Solution   *y* is a function of *x* for the graphs in (b) and (c).

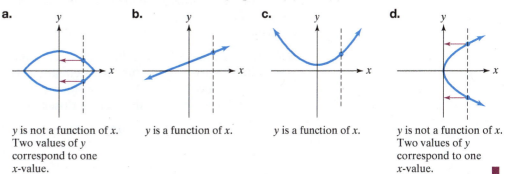

a. *y* is not a function of *x*. Two values of *y* correspond to one *x*-value.

b. *y* is a function of *x*.

c. *y* is a function of *x*.

d. *y* is not a function of *x*. Two values of *y* correspond to one *x*-value.

☑ **CHECK POINT 5**   Use the vertical line test to identify graphs in which *y* is a function of *x*.

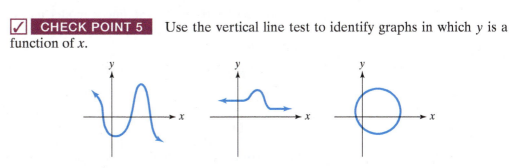

5 Find function values for functions that model data. ▶

Applications

Like formulas, functions can be obtained from verbal conditions or from actual data. Throughout your next algebra course, you'll have lots of practice doing this. For now, let's make sure that we can find and interpret function values for functions that were obtained from modeling data.

EXAMPLE 6 The Importance of Being Well-Off Financially for College Freshmen

Researchers have surveyed college freshmen every year since 1969. **Figure 9.23** shows the freshman class of 2013 was more interested in making money than the freshmen of 1980 had been.

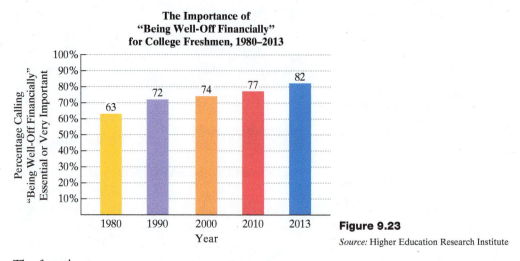

Figure 9.23

Source: Higher Education Research Institute

The functions

$$f(x) = 0.5x + 65$$

$$\text{and} \quad g(x) = -0.005x^2 + 0.67x + 64$$

model the percentage of college freshmen x years after 1980 who considered being well-off financially essential or very important. Which model serves as a better description for the freshman class of 2010?

Solution Because 2010 is 30 years after 1980, we evaluate each function at 30 by substituting 30 for x in the function's equation. The graph in **Figure 9.23** indicates that in 2010, 77% of college freshmen considered being well-off financially essential or very important. The evaluation that comes closer to 77 will serve as the better description.

$$f(x) = 0.5x + 65 \qquad \text{This is the given linear function.}$$

$$f(30) = 0.5(30) + 65 \qquad \text{Replace with 30.}$$

$$= 80 \qquad \text{Use a calculator.}$$

We see that $f(30) = 80$. The linear model indicates that 30 years after 1980, or in 2010, 80% of college freshmen considered being well-off financially essential or very important.

$$g(x) = -0.005x^2 + 0.67x + 64 \qquad \text{This is the given quadratic function.}$$

$$g(30) = -0.005(30)^2 + 0.67(30) + 64 \qquad \text{Replace each occurrence of with 30.}$$

$$= 79.6 \qquad \text{Use a calculator.}$$

We see that $g(30) = 79.6$. The quadratic model indicates that 30 years after 1980, or in 2010, 79.6% of college freshmen considered being well-off financially essential or very important.

- $f(30) = 80$

  Linear function

- $g(30) = 79.6$

  Quadratic function

- In 2010, 77% considered being well-off financially essential or very important.

  Actual data

We see that the quadratic function g serves as a slightly better description of the data. ■

Using Technology

Graphing utilities can be used to evaluate functions. The screens below show the evaluation of

$$f(x) = 0.5x + 65 \quad \text{and} \quad g(x) = -0.005x^2 + 0.67x + 64$$

at 30 on a TI-84 Plus C graphing calculator. The function f is named Y_1 and the function g is named Y_2.

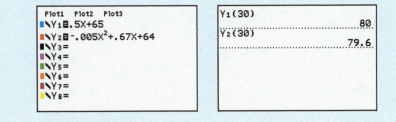

✓ **CHECK POINT 6**   Use the functions in Example 6 to solve this exercise.

a. According to the linear function, what percentage of college freshmen considered being well-off financially essential or very important in 2000?

b. According to the quadratic function, what percentage of college freshmen considered being well-off financially essential or very important in 2000?

c. Which model serves as a better description for the freshman class of 2000 in **Figure 9.23**?

Achieving Success

A recent government study cited in *Math: A Rich Heritage* (Globe Fearon Educational Publisher) found this simple fact: **The more college mathematics courses you take, the greater your earning potential will be.** Even jobs that do not require a college degree require mathematical thinking that involves attending to precision, making sense of complex problems, and persevering in solving them. No other discipline comes close to math in offering a more extensive set of tools for application and intellectual development. Take as much math as possible as you continue your journey into higher education.

CONCEPT AND VOCABULARY CHECK

Fill in each blank so that the resulting statement is true.

1. Any set of ordered pairs is called a/an _____. The set of all first components of the ordered pairs is called the _____. The set of all second components of the ordered pairs is called the _____.

2. A set of ordered pairs in which each first component corresponds to exactly one second component is called a/an _____.

3. The notation $f(x)$ describes the value of _____ at _____.

4. A function of the form $f(x) = mx + b$ is called a/an _____ function.

5. A function of the form $f(x) = ax^2 + bx + c, a \neq 0$, is called a/an _____ function.

6. If any vertical line intersects a graph _____, the graph does not define y as a/an _____ of x.

9.6 EXERCISE SET ▶ MyMathLab®

Practice Exercises

In Exercises 1–8, determine whether each relation is a function. Give the domain and range for each relation.

1. $\{(1, 2), (3, 4), (5, 5)\}$

2. $\{(4, 5), (6, 7), (8, 8)\}$

3. $\{(3, 4), (3, 5), (4, 4), (4, 5)\}$

4. $\{(5, 6), (5, 7), (6, 6), (6, 7)\}$

5. $\{(-3, -3), (-2, -2), (-1, -1), (0, 0)\}$

6. $\{(-7, -7), (-5, -5), (-3, -3), (0, 0)\}$

7. $\{(1, 4), (1, 5), (1, 6)\}$

8. $\{(4, 1), (5, 1), (6, 1)\}$

In Exercises 9–24, evaluate each function at the given values.

9. $f(x) = x + 5$
   a. $f(7)$ b. $f(-6)$ c. $f(0)$

10. $f(x) = x + 6$
    a. $f(4)$ b. $f(-8)$ c. $f(0)$

11. $f(x) = 7x$
    a. $f(10)$ b. $f(-4)$ c. $f(0)$

12. $f(x) = 9x$
    a. $f(10)$ b. $f(-5)$ c. $f(0)$

13. $f(x) = 8x - 3$
    a. $f(12)$ b. $f\left(-\frac{1}{2}\right)$ c. $f(0)$

14. $f(x) = 6x - 5$
    a. $f(12)$ b. $f\left(-\frac{1}{2}\right)$ c. $f(0)$

15. $g(x) = x^2 + 3x$
    a. $g(2)$ b. $g(-2)$ c. $g(0)$

16. $g(x) = x^2 + 7x$
    a. $g(2)$ b. $g(-2)$ c. $g(0)$

17. $h(x) = x^2 - 2x + 3$
    a. $h(4)$ b. $h(-4)$ c. $h(0)$

18. $h(x) = x^2 - 4x + 5$
    a. $h(4)$ b. $h(-4)$ c. $h(0)$

19. $f(x) = 5$
    a. $f(9)$ b. $f(-9)$ c. $f(0)$

20. $f(x) = 7$
    a. $f(10)$ b. $f(-10)$ c. $f(0)$

21. $f(r) = \sqrt{r + 6} + 3$
    a. $f(-6)$ b. $f(10)$

22. $f(r) = \sqrt{25 - r} - 6$
    a. $f(16)$ b. $f(-24)$

23. $f(x) = \dfrac{x}{|x|}$
    a. $f(6)$ b. $f(-6)$

24. $f(x) = \dfrac{x}{|x|}$
    a. $f(5)$ b. $f(-5)$

In Exercises 25–32, use the vertical line test to identify graphs in which y is a function of x.

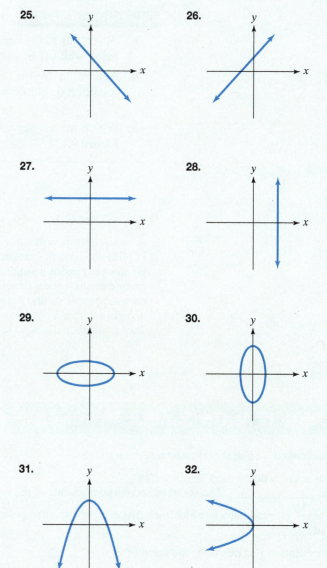

25.

26.

27.

28.

29.

30.

31.

32.

Practice PLUS

In Exercises 33–36, express each function as a set of ordered pairs.

33. $f(x) = 2x + 3$; domain: $\{-1, 0, 1\}$

34. $f(x) = 3x + 5$; domain: $\{-1, 0, 1\}$

35. $g(x) = x - x^2$; domain: the set of integers from -2 to 2, inclusive

36. $g(x) = x - |x|$; domain: the set of integers from -2 to 2, inclusive

In Exercises 37–40, find

$$\frac{f(x) - f(h)}{x - h}$$

and simplify.

37. $f(x) = 6x + 7$

38. $f(x) = 8x + 9$

39. $f(x) = x^2 - 1$

40. $f(x) = x^3 - 1$

Application Exercises

The bar graph shows minimum legal ages for sex and marriage in five selected countries. Use this information to solve Exercises 41–42. (We did not include data for the United States because the legal age of sexual consent varies according to state law. Furthermore, women are allowed to marry younger than men: 16 for women and 18 for men.)

Minimum Legal Ages for Sex and Marriage, for Heterosexual Relationships

Source: Mitchell Beazley, *Snapshot: The Visual Almanac for Our World Today,* Octopus Publishing, 2009

41. **a.** Write a set of five ordered pairs in which countries correspond to the minimum legal age of sexual consent. Each ordered pair should be in the form
    (country, minimum legal age of sexual consent).

    b. Is the relation in part (a) a function? Explain your answer.

    c. Write a set of five ordered pairs in which the minimum legal age of sexual consent corresponds to a country. Each ordered pair should be in the form
    (minimum legal age of sexual consent, country).

    d. Is the relation in part (c) a function?

42. **a.** Write a set of five ordered pairs in which countries correspond to the minimum legal age of marriage with parental consent. Each ordered pair should be in the form
    (country, minimum legal age of marriage with parental consent).

    b. Is the relation in part (a) a function? Explain your answer.

    c. Write a set of five ordered pairs in which the minimum legal age of marriage with parental consent corresponds to a country. Each ordered pair should be in the form
    (minimum legal age of marriage with parental consent, country).

    d. Is the relation in part (c) a function?

The function $f(x) = 0.76x + 171.4$ models the cholesterol level of an American man as a function of his age, x, in years. Use the function to solve Exercises 43–44.

43. Find and interpret $f(20)$.

44. Find and interpret $f(50)$.

The bar graph shows the percentage of college freshmen who spent six or more hours per week studying and doing homework during their high school senior year. The graph of a quadratic function that models the data is shown on the right. Use this information to solve Exercises 45–46.

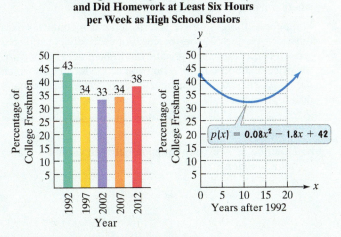

Percentage of College Freshmen Who Studied and Did Homework at Least Six Hours per Week as High School Seniors

$$p(x) = 0.08x^2 - 1.8x + 42$$

Source: UCLA Higher Education Research Institute

(In Exercises 45–46, be sure to refer to the graphs at the bottom of the previous page.)

45. The function

$$p(x) = 0.08x^2 - 1.8x + 42$$

models the percentage of college freshmen, $p(x)$, who studied and did homework at least six hours per week as high school seniors x years after 1992.

a. Find $p(20)$ and interpret the result.

b. How well does $p(20)$ describe the appropriate data displayed by the bar graph?

c. How is $p(20)$ shown on the graph of p?

46. The function

$$p(x) = 0.08x^2 - 1.8x + 42$$

models the percentage of college freshmen, $p(x)$, who studied and did homework at least six hours per week as high school seniors x years after 1992.

a. Find $p(15)$ and interpret the result.

b. Does $p(15)$ underestimate or overestimate the appropriate data displayed by the bar graph? By how much?

c. How is $p(15)$ shown on the graph of p?

Explaining the Concepts

47. If a relation is represented by a set of ordered pairs, explain how to determine whether the relation is a function.

48. Your friend heard that functions are studied in intermediate and college algebra courses. He asks you what a function is. Provide him with a clear, relatively concise response.

49. Does $f(x)$ mean f times x when referring to a function f? If not, what does $f(x)$ mean? Provide an example with your explanation.

50. Explain how the vertical line test is used to determine whether a graph is a function.

51. For people filing a single return, federal income tax is a function of adjusted gross income. For each value of adjusted gross income, there is a specific tax to be paid. By contrast, the price of a house is not a function of the lot size on which the house sits. Houses on same-sized lots can sell for many different prices.

a. Describe an everyday situation between variables that is a function.

b. Describe an everyday situation between variables that is not a function.

Critical Thinking Exercises

Make Sense? *In Exercises 52–55, determine whether each statement makes sense or does not make sense, and explain your reasoning.*

52. My body temperature is a function of the time of day.

53. My height is a function of my age.

54. Using $f(x) = 3x + 2$, I found $f(50)$ by applying the distributive property to $(3x + 2)50$.

55. A function models how a woman's cholesterol level depends on her age, so the domain is the set of various cholesterol levels.

In Exercises 56–59, determine whether each statement is true or false. If the statement is false, make the necessary change(s) to produce a true statement.

56. All relations are functions.

57. No two ordered pairs of a function can have the same second component and different first components.

58. The graph of every line is a function.

59. A horizontal line can intersect the graph of a function at more than one point.

60. Write a linear function, $f(x) = mx + b$, satisfying the following conditions:

$$f(0) = 7 \quad \text{and} \quad f(1) = 10.$$

61. If $f(x) = ax^2 + bx + c$ and $r = \dfrac{-b + \sqrt{b^2 - 4ac}}{2a}$, find $f(r)$ without doing any algebra and explain how you arrived at your result.

62. A car was purchased for \$22,500. The value of the car decreases by \$3200 per year for the first seven years. Write a function V that describes the value of the car after x years, where $0 \le x \le 7$. Then find and interpret $V(3)$.

Review Exercises

63. Write 0.00397 in scientific notation. (Section 5.7, Example 8)

64. Divide: $\dfrac{x^3 + 7x^2 - 2x + 3}{x - 2}$. (Section 5.6, Example 2)

65. Solve:

$$\begin{cases} 3x + 2y = 6 \\ 8x - 3y = 1 \end{cases}$$

(Section 4.3, Example 4)

Chapter 9 Summary

Definitions and Concepts	Examples

Section 9.1 Solving Quadratic Equations by the Square Root Property

The Square Root Property

If u is an algebraic expression and d is a positive real number, then

$$\text{If } u^2 = d, \text{ then } u = \sqrt{d} \text{ or } u = -\sqrt{d}.$$

Equivalently,

$$\text{If } u^2 = d, \text{ then } u = \pm\sqrt{d}.$$

Solve:

$$(x - 1)^2 = 7.$$
$$x - 1 = \pm\sqrt{7}$$
$$x = 1 \pm \sqrt{7}$$

The solution set is $\{1 \pm \sqrt{7}\}$.

The Pythagorean Theorem

The sum of the squares of the lengths of the legs of a right triangle equals the square of the length of the hypotenuse.

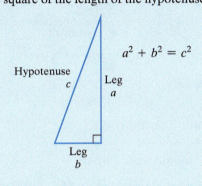

$$a^2 + b^2 = c^2$$

Find a.

$$a^2 + b^2 = c^2$$
$$a^2 + 2^2 = 6^2$$
$$a^2 + 4 = 36$$
$$a^2 = 32$$

a must be positive, so do not use $a = -\sqrt{32}.$

$$a = \sqrt{32} = \sqrt{16 \cdot 2} = 4\sqrt{2}$$

The Distance Formula

The distance, d, between the points (x_1, y_1) and (x_2, y_2) is given by

$$d = \sqrt{(x_2 - x_1)^2 + (y_2 - y_1)^2}.$$

Find the distance between $(-3, -5)$ and $(6, -2)$.

$$d = \sqrt{[6 - (-3)]^2 + [-2 - (-5)]^2}$$
$$= \sqrt{9^2 + 3^2} = \sqrt{81 + 9} = \sqrt{90} = \sqrt{9 \cdot 10} = 3\sqrt{10} \approx 9.49$$

Section 9.2 Solving Quadratic Equations by Completing the Square

Completing the Square

If $x^2 + bx$ is a binomial, then by adding $\left(\dfrac{b}{2}\right)^2$, the square of half the coefficient of x, you will obtain a perfect square trinomial. That is,

$$x^2 + bx + \left(\frac{b}{2}\right)^2 = \left(x + \frac{b}{2}\right)^2.$$

Complete the square: $x^2 + 10x$.

Add $\left(\frac{10}{2}\right)^2 = 5^2$, or **25**.

$$x^2 + 10x + 25 = (x + 5)^2$$

Definitions and Concepts	**Examples**

Section 9.2 Solving Quadratic Equations by Completing the Square (continued)

Solving Quadratic Equations by Completing the Square

1. If the coefficient of x^2 is not 1, divide both sides by this coefficient.

2. Isolate variable terms on one side.

3. Complete the square by adding the square of half the coefficient of x to both sides.

4. Factor the perfect square trinomial.

5. Solve by applying the square root property.

Solve by completing the square:

$$2x^2 + 12x - 4 = 0.$$

$$\frac{2x^2}{2} + \frac{12x}{2} - \frac{4}{2} = \frac{0}{2} \qquad \text{Divide by 2.}$$

$$x^2 + 6x - 2 = 0 \qquad \text{Simplify.}$$

$$x^2 + 6x = 2 \qquad \text{Add 2.}$$

The coefficient of x is 6. Half of 6 is 3, and $3^2 = 9$. Add 9 to both sides.

$$x^2 + 6x + 9 = 2 + 9$$

$$(x + 3)^2 = 11$$

$$x + 3 = \pm\sqrt{11}$$

$$x = -3 \pm \sqrt{11}$$

The solution set is $\{-3 \pm \sqrt{11}\}$.

Section 9.3 The Quadratic Formula

The solutions of a quadratic equation in standard form

$$ax^2 + bx + c = 0, \quad a \neq 0,$$

are given by the quadratic formula

$$x = \frac{-b \pm \sqrt{b^2 - 4ac}}{2a}.$$

Solve by the quadratic formula:

$$2x^2 + 4x = 5.$$

First write in standard form by subtracting 5 from both sides.

$$2x^2 + 4x - 5 = 0$$

$$\boxed{a = 2} \quad \boxed{b = 4} \quad \boxed{c = -5}$$

$$x = \frac{-4 \pm \sqrt{4^2 - 4 \cdot 2(-5)}}{2 \cdot 2}$$

$$= \frac{-4 \pm \sqrt{16 - (-40)}}{4}$$

$$= \frac{-4 \pm \sqrt{56}}{4} = \frac{-4 \pm \sqrt{4 \cdot 14}}{4}$$

$$= \frac{-4 \pm 2\sqrt{14}}{4}$$

$$= \frac{2(-2 \pm \sqrt{14})}{2 \cdot 2} = \frac{-2 \pm \sqrt{14}}{2}$$

Section 9.4 Imaginary Numbers as Solutions of Quadratic Equations

The imaginary unit i is defined as

$$i = \sqrt{-1}, \quad \text{where} \quad i^2 = -1.$$

The set of numbers in the form $a + bi$ is called the set of complex numbers. If $b = 0$, the complex number is a real number. If $b \neq 0$, the complex number is called an imaginary number.

- $\sqrt{-36} = \sqrt{36(-1)} = \sqrt{36}\sqrt{-1} = 6i$
- $\sqrt{-50} = \sqrt{50(-1)} = \sqrt{25 \cdot 2}\sqrt{-1} = 5i\sqrt{2}$

Definitions and Concepts	**Examples**

Section 9.4 Imaginary Numbers as Solutions of Quadratic Equations (continued)

Some quadratic equations have complex solutions that are imaginary numbers.

Solve: $x^2 - 2x + 2 = 0$.

$a = 1$ $b = -2$ $c = 2$

$$x = \frac{-b \pm \sqrt{b^2 - 4ac}}{2a} = \frac{-(-2) \pm \sqrt{(-2)^2 - 4 \cdot 1 \cdot 2}}{2 \cdot 1}$$

$$= \frac{2 \pm \sqrt{4 - 8}}{2} = \frac{2 \pm \sqrt{-4}}{2} = \frac{2 \pm \sqrt{4(-1)}}{2}$$

$$= \frac{2 \pm 2i}{2} = \frac{2(1 \pm i)}{2} = 1 \pm i$$

The solution set is $\{1 \pm i\}$.

Section 9.5 Graphs of Quadratic Equations

The graph of $y = ax^2 + bx + c$, called a parabola, can be graphed using the following steps:

1. If $a > 0$, the parabola opens upward. If $a < 0$, it opens downward.
2. Find the vertex, the lowest point if the parabola opens upward and the highest point if it opens downward. The x-coordinate of the vertex is $\frac{-b}{2a}$. Substitute this value into the parabola's equation to find the y-coordinate.
3. Find any x-intercepts by letting $y = 0$ and solving the resulting equation.
4. Find the y-intercept by letting $x = 0$.
5. Plot the intercepts and the vertex.
6. Connect these points with a smooth curve.

Graph: $y = x^2 - 2x - 8$.

$a = 1$ $b = -2$ $c = -8$

- $a > 0$, so the parabola opens upward.
- Vertex: x-coordinate $= \dfrac{-b}{2a} = \dfrac{-(-2)}{2 \cdot 1} = 1$

  y-coordinate $= 1^2 - 2 \cdot 1 - 8 = -9$

  Vertex is $(1, -9)$.
- x-intercepts: Let $y = 0$.

  $$x^2 - 2x - 8 = 0$$
  $$(x - 4)(x + 2) = 0$$
  $$x - 4 = 0 \quad \text{or} \quad x + 2 = 0$$
  $$x = 4 \qquad\qquad x = -2$$

  The parabola passes through $(4, 0)$ and $(-2, 0)$.
- y-intercept: Let $x = 0$.

  $$y = 0^2 - 2 \cdot 0 - 8 = 0 - 0 - 8 = -8$$

  The parabola passes through $(0, -8)$.

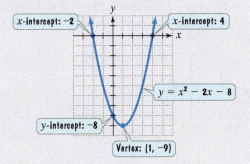

Definitions and Concepts	Examples

Section 9.6 Introduction to Functions

A relation is any set of ordered pairs. The set of first components is the domain and the set of second components is the range. A function is a relation in which each member of the domain corresponds to exactly one member of the range. No two ordered pairs of a function can have the same first component and different second components.

The domain of the relation $\{(1, 2), (3, 4), (3, 7)\}$ is $\{1, 3\}$. The range is $\{2, 4, 7\}$. The relation is not a function: 3, in the domain, corresponds to both 4 and 7 in the range.

If a function is defined as an equation, the notation $f(x)$, read "f of x" or "f at x," describes the value of the function at the number x.

If $f(x) = x^2 - 5x + 4$, find $f(3)$.
$$f(3) = 3^2 - 5 \cdot 3 + 4 = 9 - 15 + 4 = -2$$
Thus, f of 3 is -2.

The Vertical Line Test for Functions

If any vertical line intersects a graph in more than one point, the graph does not define y as a function of x.

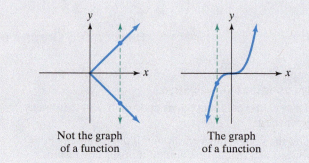

Not the graph
of a function

The graph
of a function

CHAPTER 9 REVIEW EXERCISES

9.1   *In Exercises 1–8, solve each quadratic equation by the square root property. If possible, simplify radicals or rationalize denominators.*

1. $x^2 = 64$

2. $x^2 = 17$

3. $2x^2 = 150$

4. $(x - 3)^2 = 9$

5. $(y + 4)^2 = 5$

6. $3y^2 - 5 = 0$

7. $(2x - 7)^2 = 25$

8. $(x + 5)^2 = 12$

In Exercises 9–11, use the Pythagorean Theorem to find the missing length in each right triangle. Express the answer in radical form and simplify, if possible.

9.

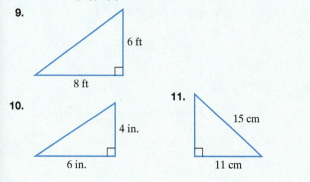

6 ft

8 ft

10.

4 in.

6 in.

11.

15 cm

11 cm

Use the Pythagorean Theorem to solve Exercises 12–13.

12. How far away from the building shown in the figure is the bottom of the ladder?

20 ft   25 ft

?

13. A vertical pole is to be supported by three wires. Each wire is 13 yards long and is anchored 5 yards from the base of the pole. How far up the pole will the wires be attached?

14. The weight of a human fetus is modeled by the formula $W = 3t^2$, where W is the weight, in grams, and t is the time, in weeks, $0 \leq t \leq 39$. After how many weeks does the fetus weigh 1200 grams?

15. The distance, d, in feet, that an object falls in t seconds is modeled by the formula $d = 16t^2$. If you dive from a height of 100 feet, how long will it take to hit the water?

In Exercises 16–17, find the distance between each pair of points. Express answers in simplified radical form and, if necessary, round to two decimal places.

16. $(-3, -2)$ and $(1, -5)$

17. $(3, 8)$ and $(5, 4)$

9.2 *In Exercises 18–21, complete the square for each binomial. Then factor the resulting perfect square trinomial.*

18. $x^2 + 16x$

19. $x^2 - 6x$

20. $x^2 + 3x$

21. $x^2 - 5x$

In Exercises 22–24, solve each quadratic equation by completing the square.

22. $x^2 - 12x + 27 = 0$

23. $x^2 - 6x + 4 = 0$

24. $3x^2 - 12x + 11 = 0$

9.3 *In Exercises 25–27, solve each equation using the quadratic formula. Simplify irrational solutions, if possible.*

25. $2x^2 + 5x - 3 = 0$

26. $x^2 = 2x + 4$

27. $3x^2 + 5 = 9x$

In Exercises 28–32, solve each equation by the method of your choice. Simplify irrational solutions, if possible.

28. $2x^2 - 11x + 5 = 0$

29. $(3x + 5)(x - 3) = 5$

30. $3x^2 - 7x + 1 = 0$

31. $x^2 - 9 = 0$

32. $(2x - 3)^2 = 5$

33. Most people in the United States are not prepared if a disaster strikes. The bar graph shows the percentage of Americans with a disaster-supply kit in their homes for four selected years.

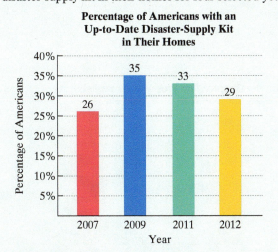

Percentage of Americans with an Up-to-Date Disaster-Supply Kit in Their Homes

Source: Federal Emergency Management Agency

The formula

$$p = -1.3x^2 + 6.9x + 26$$

models the percentage of Americans with a disaster-supply kit in their homes, p, x years after 2007.

a. According to the model, what percentage of Americans had a disaster-supply kit in their homes in 2012? Does this underestimate or overestimate the percent displayed by the graph? By how much?

b. According to the graph, in which year did 35% of Americans have a disaster-supply kit in their homes?

c. According to the mathematical model, in which year(s) did 35% of Americans have a disaster-supply kit in their homes?

9.4 *In Exercises 34–37, express each number in terms of i.*

34. $\sqrt{-81}$

35. $\sqrt{-23}$

36. $\sqrt{-48}$

37. $3 + \sqrt{-49}$

Exercises 38–42 involve quadratic equations with imaginary solutions. Solve each equation.

38. $x^2 = -100$

39. $5x^2 = -125$

40. $(2x + 1)^2 = -8$

41. $x^2 - 4x + 13 = 0$

42. $3x^2 - x + 2 = 0$

9.5 *In Exercises 43–46,*

a. *Determine if the parabola whose equation is given opens upward or downward.*

b. *Find the parabola's x-intercepts. If they are irrational, round to the nearest tenth.*

c. *Find the parabola's y-intercept.*

d. *Find the parabola's vertex.*

e. *Graph the parabola.*

43. $y = x^2 - 6x - 7$

44. $y = -x^2 - 2x + 3$

45. $y = -3x^2 + 6x + 1$

46. $y = x^2 - 4x$

47. Fireworks are launched into the air. The formula

$$y = -16x^2 + 200x + 4$$

models the fireworks' height, y, in feet, x seconds after they are launched. When should the fireworks explode so that they go off at the greatest height? What is that height?

48. A quarterback tosses a football to a receiver 40 yards downfield. The formula

$$y = -0.025x^2 + x + 6$$

models the football's height, y, in feet, when it is x yards from the quarterback.

a. How many yards from the quarterback does the football reach its greatest height? What is that height?

b. If a defender is 38 yards from the quarterback, how far must he reach to deflect or catch the ball?

c. If the football is neither deflected by the defender nor caught by the receiver, how many yards will it go, to the nearest tenth of a yard, before hitting the ground?

d. Graph the equation that models the football's parabolic path.

9.6 *In Exercises 49–51, determine whether each relation is a function. Give the domain and range for each relation.*

49. $\{(2, 7), (3, 7), (5, 7)\}$

50. $\{(1, 10), (2, 500), (3, \pi)\}$

51. $\{(12, 13), (14, 15), (12, 19)\}$

In Exercises 52–53, evaluate each function at the given values.

52. $f(x) = 3x - 4$

    a. $f(-5)$ **b.** $f(6)$ **c.** $f(0)$

53. $g(x) = x^2 - 5x + 2$

    a. $g(-4)$ **b.** $g(3)$ **c.** $g(0)$

In Exercises 54–57, use the vertical line test to identify graphs in which y is a function of x.

54.   **55.**

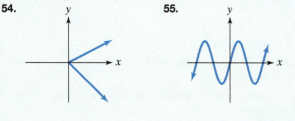

56.   **57.**

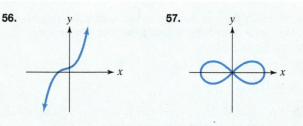

58. Whether on the slopes or at the shore, people are exposed to harmful amounts of the sun's skin-damaging ultraviolet (UV) rays. The quadratic function

$$D(x) = 0.8x^2 - 17x + 109$$

models the average time in which skin damage begins for burn-prone people, $D(x)$, in minutes, where x is the UV index, or measure of the sun's UV intensity. The graph of D is shown for a UV index from 1 (low) to 11 (high).

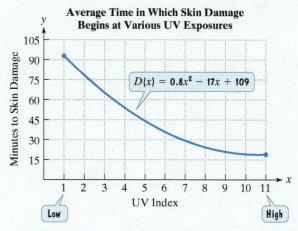

Average Time in Which Skin Damage Begins at Various UV Exposures

$D(x) = 0.8x^2 - 17x + 109$

Source: National Oceanic and Atmospheric Administration

a. Find and interpret $D(1)$. How is this shown on the graph of D?

b. Find and interpret $D(10)$. How is this shown on the graph of D?

CHAPTER 9 TEST   Step-by-step test solutions are found on the Chapter Test Prep Videos available in MyMathLab® or on You Tube (search "BlitzerIntroAlg" and click on "Channels").

In Exercises 1–2, solve by the square root property.

1. $3x^2 = 48$

2. $(x - 3)^2 = 5$

3. To find the distance across a lake, a surveyor inserts poles at P and Q, measuring the respective distances to point R, as shown in the figure. Use the surveyor's measurements given in the figure to find the distance PQ across the lake in simplified radical form.

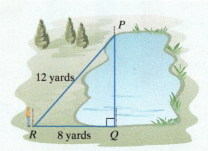

4. Find the distance between $(3, -2)$ and $(-4, 1)$. Express the answer in radical form and then round to two decimal places.

5. Solve by completing the square: $x^2 + 4x - 3 = 0$.

In Exercises 6–10, solve each equation by the method of your choice. Simplify irrational solutions, if possible.

6. $3x^2 + 5x + 1 = 0$

7. $(3x - 5)(x + 2) = -6$

8. $(2x + 1)^2 = 36$

9. $2x^2 = 6x - 1$

10. $2x^2 + 9x = 5$

In Exercises 11–12, express each number in terms of i.

11. $\sqrt{-121}$ **12.** $\sqrt{-75}$

In Exercises 13–15, solve each quadratic equation. Express imaginary solutions in a + bi form.

13. $x^2 + 36 = 0$ **14.** $(x - 5)^2 = -25$

15. $x^2 - 2x + 5 = 0$

In Exercises 16–17, graph each parabola whose equation is given. Label the x-intercepts, the y-intercept, and the vertex.

16. $y = x^2 + 2x - 8$

17. $y = -2x^2 + 16x - 24$

A batter hits a baseball into the air. The formula

$$y = -16x^2 + 64x + 5$$

models the baseball's height above the ground, y, in feet, x seconds after it is hit. Use the formula to solve Exercises 18–19.

18. When does the baseball reach its maximum height? What is that height?

19. After how many seconds does the baseball hit the ground? Round to the nearest tenth of a second.

In Exercises 20–21, determine whether each relation is a function. Give the domain and range for each relation.

20. $\{(1, 2), (3, 4), (5, 6), (6, 6)\}$

21. $\{(2, 1), (4, 3), (6, 5), (6, 6)\}$

22. If $f(x) = 7x - 3$, find $f(10)$.

23. If $g(x) = x^2 - 3x + 7$, find $g(-2)$.

In Exercises 24–25, identify the graph or graphs in which y is a function of x.

24.

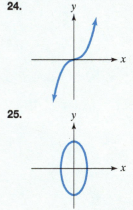

25.

26. In a round-robin chess tournament, each player is paired with every other player once. The function

$$f(x) = \frac{x^2 - x}{2}$$

models the number of chess games, $f(x)$, that must be played in a round-robin tournament with x chess players. Find and interpret $f(9)$.

CUMULATIVE REVIEW EXERCISES (CHAPTERS 1–9)

In Exercises 1–10, solve each equation, inequality, or system of equations.

1. $2 - 4(x + 2) = 5 - 3(2x + 1)$

2. $\frac{x}{2} - 3 = \frac{x}{5}$

3. $3x + 9 \geq 5(x - 1)$

4. $\begin{cases} 2x + 3y = 6 \\ x + 2y = 5 \end{cases}$

5. $\begin{cases} 3x - 2y = 1 \\ y = 10 - 2x \end{cases}$

6. $\frac{3}{x + 5} - 1 = \frac{4 - x}{2x + 10}$

7. $x + \frac{6}{x} = -5$

8. $x - 5 = \sqrt{x + 7}$

9. $(x - 2)^2 = 20$

10. $3x^2 - 6x + 2 = 0$

11. Solve for t: $A = \frac{5r + 2}{t}$.

In Exercises 12–24, perform the indicated operations. If possible, simplify the answer.

12. $\dfrac{12x^3}{3x^{12}}$

13. $4 \cdot 6 \div 2 \cdot 3 + (-5)$

14. $(6x^2 - 8x + 3) - (-4x^2 + x - 1)$

15. $(7x + 4)(3x - 5)$

16. $(5x - 2)^2$

17. $(x + y)(x^2 - xy + y^2)$

18. $\dfrac{x^2 + 6x + 8}{x^2} \div (3x^2 + 6x)$

19. $\dfrac{x}{x^2 + 2x - 3} - \dfrac{x}{x^2 - 5x + 4}$

20. $\dfrac{x - \dfrac{1}{5}}{5 - \dfrac{1}{x}}$

21. $3\sqrt{20} + 2\sqrt{45}$

22. $\sqrt{3x} \cdot \sqrt{6x}$

23. $\dfrac{2}{\sqrt{3}}$

24. $\dfrac{8}{3 - \sqrt{5}}$

In Exercises 25–30, factor completely.

25. $4x^2 - 49$

26. $x^3 + 3x^2 - x - 3$

27. $2x^2 + 8x - 42$

28. $x^5 - 16x$

29. $x^3 - 10x^2 + 25x$

30. $x^3 - 8$

31. Evaluate: $8^{-\frac{2}{3}}$.

In Exercises 32–37, graph each equation, inequality, or system of inequalities in a rectangular coordinate system.

32. $y = \dfrac{1}{3}x - 1$

33. $3x + 2y = -6$

34. $y = -2$

35. $3x - 4y \leq 12$

36. $y = x^2 - 2x - 3$

37. $\begin{cases} 2x + y < 4 \\ x > 2 \end{cases}$

38. Find the slope of the line passing through the points $(-1, 3)$ and $(2, -3)$.

39. Write the point-slope form of the equation of the line passing through the points $(1, 2)$ and $(3, 6)$. Then use the point-slope equation to write the slope-intercept form of the line's equation.

In Exercises 40–50, use an equation or a system of equations to solve each problem.

40. Seven subtracted from five times a number is 208. Find the number.

41. After a 20% reduction, a digital camera sold for $256. What was the price before the reduction?

42. A rectangular field is three times as long as it is wide. If the perimeter of the field is 400 yards, what are the field's dimensions?

43. You invested $20,000 in two accounts paying 7% and 9% annual interest. If the total interest earned for the year is $1550, how much was invested at each rate?

44. A chemist needs to mix a 40% acid solution with a 70% acid solution to obtain 12 liters of a 50% acid solution. How many liters of each solution should be used?

45. You can paint a room in 6 hours. Your friend can paint the same room in 12 hours. How long will it take to paint the room if you both work together?

46. A university with 176 people on the faculty wants to maintain a student-to-faculty ratio of 23:2. How many students should they enroll to maintain that ratio?

47. A sailboat has a triangular sail with an area of 120 square feet and a base that is 15 feet long. Find the height of the sail.

48. In a triangle, the measure of the first angle is 10° more than the measure of the second angle. The measure of the third angle is 20° more than four times that of the second angle. What is the measure of each angle?

49. A salesperson works in the TV and stereo department of an electronics store. One day she sold 3 TVs and 4 stereos for $2530. The next day, she sold 4 of the same TVs and 3 of the same stereos for $2510. Find the price of a TV and a stereo.

50. The length of a rectangle is 6 meters more than the width. The area is 55 square meters. Find the rectangle's dimensions.

Mean, Median, and Mode

One way to analyze a list of data items is to determine a single value that represents what is "average" or "typical" of the data. Such values are known as **measures of central tendency** because they are located toward the center of the data. Three such measures are discussed in this appendix: the *mean* (or *average*), the *median*, and the *mode*.

The Mean

By far the most commonly used measure of central tendency is the *mean*. The **mean** is obtained by adding all the data items and then dividing the sum by the number of items. The Greek letter sigma, Σ, called **a symbol of summation**, is used to indicate the sum of data items. The notation Σx, read "the sum of x," means to add all the data items in a given data set. We can use this symbol to give a formula for calculating the mean.

The Mean

The **mean** is the sum of the data items divided by the number of items.

$$\text{Mean} = \frac{\Sigma x}{n},$$

where Σx represents the sum of all the data items and n represents the number of items.

EXAMPLE 1   Calculating the Mean

Table A.1 shows the ten highest-earning TV actors and the ten highest-earning TV actresses for the 2010-2011 television season. Find the mean earnings, in millions of dollars, for the ten highest-earning actors.

Table A.1   Highest-Earning TV Actors and Actresses, 2010–2011

Actor	Earnings (millions of dollars)	Actress	Earnings (millions of dollars)
Charlie Sheen	$40	Eva Longoria	$13
Ray Romano	$20	Tina Fey	$13
Steve Carell	$15	Marcia Cross	$10
Mark Harmon	$13	Mariska Hargitay	$10
Jon Cryer	$11	Marg Helgenberger	$10
Laurence Fishburne	$11	Teri Hatcher	$9
Patrick Dempsey	$10	Felicity Huffman	$9
Simon Baker	$9	Courteney Cox	$7
Hugh Laurie	$9	Ellen Pompeo	$7
Chris Meloni	$9	Julianna Margulies	$7

Source: Forbes

Solution We find the mean, $\bar{x}$, by adding the earnings for the actors and dividing this sum by 10, the number of data items.

$$\bar{x} = \frac{\sum x}{n} = \frac{40 + 20 + 15 + 13 + 11 + 11 + 10 + 9 + 9 + 9}{10} = \frac{147}{10} = 14.7$$

The mean earnings of the ten highest-earning actors is $14.7 million. ■

One and only one mean can be calculated for any group of numerical data. The mean may or may not be one of the actual data items. In Example 1, the mean was 14.7, although no data item is 14.7.

☑ **CHECK POINT 1** Use **Table A.1** on the previous page to find the mean earnings, $\bar{x}$, in millions of dollars, for the ten highest-earning actresses.

2 Determine the median of a set of data items. ▶

The Median

The *median* age in the United States is 37.2. The oldest state by median age is Maine (42.7) and the youngest state is Utah (29.2). To find these values, researchers begin with appropriate random samples. The data items—that is, the ages—are arranged from youngest to oldest. The median age is the data item in the middle of each set of ranked, or ordered, data.

The Median

To find the **median** of a group of data items,

1. Arrange the data items in order, from smallest to largest.
2. If the number of data items is odd, the median is the data item in the middle of the list.
3. If the number of data items is even, the median is the mean of the two middle data items.

EXAMPLE 2 Finding the Median

Find the median for each of the following groups of data:

a. 84, 90, 98, 95, 88
b. 68, 74, 7, 13, 15, 25, 28, 59, 34, 47.

Solution

a. Arrange the data items in order, from smallest to largest. The number of data items in the list, five, is odd. Thus, the median is the middle number.

84, 88, 90, 95, 98

Middle data item

The median is 90. Notice that two data items lie above 90 and two data items lie below 90.

b. Arrange the data items in order, from smallest to largest. The number of data items in the list, ten, is even. Thus, the median is the mean of the two middle data items.

$$7, 13, 15, 25, 28, 34, 47, 59, 68, 74$$

Middle data items
are 28 and 34.

$$\text{Median} = \frac{28 + 34}{2} = \frac{62}{2} = 31$$

The median is 31. Five data items lie above 31 and five data items lie below 31.

7   13   15   25   28   31   34   47   59   68   74

Five data items lie below 31.   Five data items lie above 31.

Median is 31.

Great Question!

What exactly does the median do with the data?

The median splits the data items down the middle, like the median strip in a road.

✓ **CHECK POINT 2** Find the median for each of the following groups of data:

a. 28, 42, 40, 25, 35

b. 72, 61, 85, 93, 79, 87.

Statisticians generally use the median, rather than the mean, when reporting income. Why? Our next example will help to answer this question.

EXAMPLE 3 Comparing the Median and the Mean

Five employees in the assembly section of a television manufacturing company earn salaries of $19,700, $20,400, $21,500, $22,600, and $23,000 annually. The section manager has an annual salary of $95,000.

a. Find the median annual salary for the six people.

b. Find the mean annual salary for the six people.

Solution

a. To compute the median, first arrange the salaries in order:

$$\$19,700, \quad \$20,400, \quad \$21,500, \quad \$22,600, \quad \$23,000, \quad \$95,000.$$

Because the list contains an even number of data items, six, the median is the mean of the two middle items.

$$\text{Median} = \frac{\$21,500 + \$22,600}{2} = \frac{\$44,100}{2} = \$22,050$$

The median annual salary is $22,050.

b. We find the mean annual salary by adding the six annual salaries and dividing by 6.

$$\text{Mean} = \frac{\$19,700 + \$20,400 + \$21,500 + \$22,600 + \$23,000 + \$95,000}{6}$$

$$= \frac{\$202,200}{6} = \$33,700$$

The mean annual salary is $33,700. ∎

In Example 3, the median annual salary is $22,050 and the mean annual salary is $33,700. Why such a big difference between these two measures of central tendency? The relatively high annual salary of the section manager, $95,000, pulls the mean salary to a value considerably higher than the median salary. When one or more data items are much greater than the other items, these extreme values can greatly influence the mean. In cases like this, the median is often more representative of the data.

✓ **CHECK POINT 3**   **Table A.2** shows the net worth, in millions of 2010 dollars, for ten U.S. presidents from Kennedy through Obama.

Table A.2   Net Worth for Ten U.S. Presidents

President	Net Worth (millions of dollars)	President	Net Worth (millions of dollars)
Kennedy	$1000 (i.e., $1 billion)	Reagan	$13
Johnson	$98	Bush	$23
Nixon	$15	Clinton	$38
Ford	$7	Bush	$20
Carter	$7	Obama	$5

Source: Time

a. Find the mean net worth, in millions of dollars, for the ten presidents.

b. Find the median net worth, in millions of dollars, for the ten presidents.

c. Descibe why one of the measures of central tendency is greater than the other.

3   Determine the mode of a set of data items. ▶

The Mode

A value that occurs most often in a set of data items is called the *mode*.

> **The Mode**
>
> The **mode** is the data value that occurs most often in a data set. If more than one data value occurs most often, then each of these data values is a mode. If there is no data value that occurs most often, then the data set has no mode.

EXAMPLE 4   Finding the Mode

Find the mode for each of the following groups of data:

a. 7, 2, 4, 7, 8, 10
b. 2, 1, 4, 5, 3
c. 3, 3, 4, 5, 6, 6.

Solution

a. 7, 2, 4, 7, 8, 10

7 occurs most often.

The mode is 7.

b. 2, 1, 4, 5, 3

Each data item occurs the same number of times.

There is no mode.

c. 3, 3, 4, 5, 6, 6

Both 3 and 6 occur most often.

The modes are 3 and 6. ■

✓ **CHECK POINT 4**   Find the mode for each of the following groups of data:

a. 3, 8, 5, 8, 9, 10

b. 3, 8, 5, 8, 9, 3

c. 3, 8, 5, 6, 9, 10.

EXAMPLE 5   Finding Three Measures of Central Tendency

Suppose your six exam grades in a course are

$$52, \quad 69, \quad 75, \quad 86, \quad 86, \quad \text{and} \quad 92.$$

Compute your final course grade ($90-100 = A, 80-89 = B, 70-79 = C, 60-69 = D$, below $60 = F$) using the

a. mean.   **b.** median.   **c.** mode.

Solution

a. The mean is the sum of the data items divided by the number of items, 6.

$$\text{Mean} = \frac{52 + 69 + 75 + 86 + 86 + 92}{6} = \frac{460}{6} \approx 76.67$$

Using the mean, your final course grade is C.

b. The six data items, 52, 69, 75, 86, 86, and 92, are arranged in order. Because the number of data items is even, the median is the mean of the two middle items.

$$\text{Median} = \frac{75 + 86}{2} = \frac{161}{2} = 80.5$$

Using the median, your final course grade is B.

c. The mode is the data value that occurs most frequently. Because 86 occurs most often, the mode is 86. Using the mode, your final course grade is B. ∎

✓ **CHECK POINT 5**   *Consumer Reports* magazine gave the following data for the number of calories in a meat hot dog for each of eight brands:

$$107, \quad 136, \quad 138, \quad 138, \quad 172, \quad 173, \quad 190, \quad 191.$$

Find the mean, median, and mode for the number of calories in a meat hot dog for the eight brands. If necessary, round answers to the nearest tenth of a calorie.

APPENDIX EXERCISE SET   ▶ MyMathLab®

Practice Exercises

In Exercises 1–8, find the mean for each group of data items.

1. 7, 4, 3, 2, 8, 5, 1, 3

2. 11, 6, 4, 0, 2, 1, 12, 0, 0

3. 91, 95, 99, 97, 93, 95

4. 100, 100, 90, 30, 70, 100

5. 100, 40, 70, 40, 60

6. 1, 3, 5, 10, 8, 5, 6, 8

7. 1.6, 3.8, 5.0, 2.7, 4.2, 4.2, 3.2, 4.7, 3.6, 2.5, 2.5

8. 1.4, 2.1, 1.6, 3.0, 1.4, 2.2, 1.4, 9.0, 9.0, 1.8

In Exercises 9–16, find the median for each group of data items.

9. 7, 4, 3, 2, 8, 5, 1, 3

10. 11, 6, 4, 0, 2, 1, 12, 0, 0

11. 91, 95, 99, 97, 93, 95

12. 100, 100, 90, 30, 70, 100

13. 100, 40, 70, 40, 60

14. 1, 3, 5, 10, 8, 5, 6, 8

15. 1.6, 3.8, 5.0, 2.7, 4.2, 4.2, 3.2, 4.7, 3.6, 2.5, 2.5

16. 1.4, 2.1, 1.6, 3.0, 1.4, 2.2, 1.4, 9.0, 9.0, 1.8

In Exercises 17–24, find the mode for each group of data items. If there is no mode, so state.

17. 7, 4, 3, 2, 8, 5, 1, 3

18. 11, 6, 4, 0, 2, 1, 12, 0, 0

19. 91, 95, 99, 97, 93, 95

20. 100, 100, 90, 30, 70, 100

21. 100, 40, 70, 40, 60

22. 1, 3, 5, 10, 8, 5, 6, 8

23. 1.6, 3.8, 5.0, 2.7, 4.2, 4.2, 3.2, 4.7, 3.6, 2.5, 2.5

24. 1.4, 2.1, 1.6, 3.0, 1.4, 2.2, 1.4, 9.0, 9.0, 1.8

Application Exercises

Exercises 25–26 present data related to age. For each data set described in boldface, find the mean, median, and mode (or state that there is no mode). If necessary, round answers to the nearest tenth.

25. Ages of the Six Youngest U.S. Presidents at Inauguration

T. Roosevelt (42), Kennedy (43), Clinton (46), Grant (46), Cleveland (47), Obama (47)

26. Ages of the First Six U.S. Presidents at Inauguration

Washington (57), J. Adams (61), Jefferson (57), Madison (57), Monroe (58), J. Q. Adams (57)

27. The annual salaries of four salespeople and the owner of a bookstore are

$17,500, $19,000, $22,000, $27,500, $98,500.

Find the mean and the median. Is the mean or the median more representative of the five annual salaries? Briefly explain your answer.

28. In one common system for finding a grade-point average, or GPA,

A = 4, B = 3, C = 2, D = 1, F = 0.

The GPA is calculated by multiplying the number of credit hours for a course and the number assigned to each grade, and then adding these products. Then divide this sum by the total number of credit hours. Because each course grade is weighted according to the number of credits of the course, GPA is called a *weighted mean*. Calculate the GPA, rounded to the nearest tenth, for this transcript:

Sociology: 3 cr. A; Biology: 3.5 cr. C; Music: 1 cr. B;
Math: 4 cr. B; English: 3 cr. C.

Answers to Selected Exercises

CHAPTER 1

Section 1.1 Check Point Exercises

1. a. 26 **b.** 32 **2. a.** 37 **b.** 2 **3. a.** $6x$ **b.** $4 + x$ **c.** $3x + 5$ **d.** $12 - 2x$ **e.** $\dfrac{15}{x}$ **4. a.** not a solution
b. solution **5. a.** $\dfrac{x}{6} = 5$ **b.** $7 - 2x = 1$ **6. a.** 65% **b.** 60% **c.** overestimates by 5%

Concept and Vocabulary Check

1. variable **2.** expression **3.** substituting; evaluating **4.** equation; solution **5.** formula; modeling; models

Exercise Set 1.1

1. 12 **3.** 8 **5.** 20 **7.** 7 **9.** 17 **11.** 18 **13.** 5 **15.** 19 **17.** 24 **19.** 13 **21.** 10 **23.** 5 **25.** $x + 4$ **27.** $x - 4$
29. $x + 4$ **31.** $x - 9$ **33.** $9 - x$ **35.** $3x - 5$ **37.** $12x - 1$ **39.** $\dfrac{10}{x} + \dfrac{x}{10}$ **41.** $\dfrac{x}{30} + 6$ **43.** solution **45.** solution
47. not a solution **49.** solution **51.** not a solution **53.** solution **55.** solution **57.** not a solution **59.** $4x = 28$ **61.** $\dfrac{14}{x} = \dfrac{1}{2}$
63. $20 - x = 5$ **65.** $2x + 6 = 16$ **67.** $3x - 5 = 7$ **69.** $4x + 5 = 33$ **71.** $4(x + 5) = 33$ **73.** $5x = 24 - x$ **75.** 8 **77.** 59
79. a. 14 **b.** yes **81. a.** 6 **b.** no **83. a.** \$4.43; overestimates by \$0.20 **b.** underestimates by \$0.26 **85.** overestimates by 0.8%
87. a. 44 **b.** 164 **99.** makes sense **101.** makes sense **103.** true **105.** true **107.** Choices of variables may vary. Example:
$h = $ hours worked, $s = $ salary; $s = 20h$ **109.** $\dfrac{6}{35}$ **110.** $\dfrac{10}{21}$ **111.** $\dfrac{4}{17}$

Section 1.2 Check Point Exercises

1. $\dfrac{21}{8}$ **2.** $1\dfrac{2}{3}$ **3.** $2 \cdot 2 \cdot 3 \cdot 3$ **4. a.** $\dfrac{2}{3}$ **b.** $\dfrac{7}{4}$ **c.** $\dfrac{13}{15}$ **d.** $\dfrac{1}{5}$ **5. a.** $\dfrac{8}{33}$ **b.** $\dfrac{18}{5}$ or $3\dfrac{3}{5}$ **c.** $\dfrac{2}{7}$ **d.** $\dfrac{51}{10}$ or $5\dfrac{1}{10}$ **6. a.** $\dfrac{10}{3}$ or $3\dfrac{1}{3}$
b. $\dfrac{2}{9}$ **c.** $\dfrac{3}{2}$ or $1\dfrac{1}{2}$ **7. a.** $\dfrac{5}{11}$ **b.** $\dfrac{2}{3}$ **c.** $\dfrac{9}{4}$ or $2\dfrac{1}{4}$ **8.** $\dfrac{14}{21}$ **9. a.** $\dfrac{11}{10}$ or $1\dfrac{1}{10}$ **b.** $\dfrac{7}{12}$ **c.** $\dfrac{5}{4}$ or $1\dfrac{1}{4}$ **10.** $\dfrac{53}{60}$
11. a. solution **b.** solution **12. a.** $\dfrac{2}{3}(x - 6)$ **b.** $\dfrac{3}{4}x - 2 = \dfrac{1}{5}x$ **13.** 25°C

Concept and Vocabulary Check

1. numerator; denominator **2.** mixed; improper **3.** 5; 3; 2; 5 **4.** natural **5.** prime **6.** factors; product **7.** $\dfrac{a}{b}$ **8.** $\dfrac{a \cdot c}{b \cdot d}$
9. reciprocals **10.** $\dfrac{d}{c}$ **11.** $\dfrac{a + c}{b}$ **12.** least common denominator

Exercise Set 1.2

1. $\dfrac{19}{8}$ **3.** $\dfrac{38}{5}$ **5.** $\dfrac{135}{16}$ **7.** $4\dfrac{3}{5}$ **9.** $8\dfrac{4}{9}$ **11.** $35\dfrac{11}{20}$ **13.** $2 \cdot 11$ **15.** $2 \cdot 2 \cdot 5$ **17.** prime **19.** $2 \cdot 2 \cdot 3 \cdot 3$ **21.** $2 \cdot 2 \cdot 5 \cdot 7$
23. prime **25.** $3 \cdot 3 \cdot 3 \cdot 3$ **27.** $2 \cdot 2 \cdot 2 \cdot 2 \cdot 3 \cdot 5$ **29.** $\dfrac{5}{8}$ **31.** $\dfrac{5}{6}$ **33.** $\dfrac{7}{10}$ **35.** $\dfrac{2}{5}$ **37.** $\dfrac{22}{25}$ **39.** $\dfrac{60}{43}$ **41.** $\dfrac{2}{15}$ **43.** $\dfrac{21}{88}$
45. $\dfrac{36}{7}$ **47.** $\dfrac{1}{12}$ **49.** $\dfrac{15}{14}$ **51.** 6 **53.** $\dfrac{15}{16}$ **55.** $\dfrac{9}{5}$ **57.** $\dfrac{5}{9}$ **59.** 3 **61.** $\dfrac{7}{10}$ **63.** $\dfrac{1}{2}$ **65.** 6 **67.** $\dfrac{6}{11}$ **69.** $\dfrac{2}{3}$ **71.** $\dfrac{5}{4}$
73. $\dfrac{1}{6}$ **75.** 2 **77.** $\dfrac{7}{10}$ **79.** $\dfrac{9}{10}$ **81.** $\dfrac{19}{24}$ **83.** $\dfrac{7}{18}$ **85.** $\dfrac{7}{12}$ **87.** $\dfrac{41}{80}$ **89.** $1\dfrac{5}{12}$ or $\dfrac{17}{12}$ **91.** solution **93.** solution
95. not a solution **97.** solution **99.** not a solution **101.** solution **103.** $\dfrac{1}{5}x$ **105.** $x - \dfrac{1}{4}x$ **107.** $x - \dfrac{1}{4} = \dfrac{1}{2}x$ **109.** $\dfrac{1}{7}x + \dfrac{1}{8}x = 12$
111. $\dfrac{2}{3}(x + 6)$ **113.** $\dfrac{2}{3}x + 6 = x - 3$ **115.** $\dfrac{3a}{20}$ **117.** $\dfrac{20}{x}$ **119.** $\dfrac{4}{15}$ **121.** not a solution **123.** 20°C **125. a.** 140 beats per minute
b. 160 beats per minute **127. a.** $H = \dfrac{9}{10}(220 - a)$ **b.** 162 beats per minute **129. a.** $\dfrac{4}{5}$ **b.** $\dfrac{4}{5} = \dfrac{4}{5}$; They are equal. **c.** $\dfrac{49}{50}$

141. makes sense   **143.** makes sense   **145.** false   **147.** true
149.

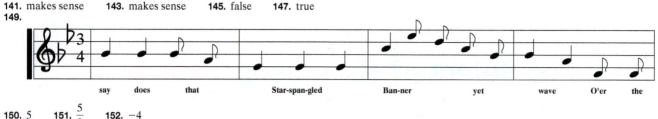

say   does   that   Star-span-gled   Ban-ner   yet   wave   O'er   the

150. 5   **151.** $\dfrac{5}{2}$ **152.** -4

Section 1.3 Check Point Exercises

1. a. -500 **b.** -282 **2.** (a)(b)(c) [number line: dots at -4, 0, 3]   **3.** (b)(a) [number line: dots at -2, 4]

4. a. 0.375 **b.** $0.\overline{45}$ **5. a.** $\sqrt{9}$ **b.** $0, \sqrt{9}$ **c.** $-9, 0, \sqrt{9}$ **d.** $-9, -1.3, 0, 0.\overline{3}, \sqrt{9}$ **e.** $\dfrac{\pi}{2}, \sqrt{10}$ **f.** $-9, -1.3, 0, 0.\overline{3}, \dfrac{\pi}{2}, \sqrt{9}, \sqrt{10}$

6. a. $>$ **b.** $<$ **c.** $<$ **d.** $<$ **7. a.** true   **b.** true   **c.** false   **8. a.** 4   **b.** 6   **c.** $\sqrt{2}$

Concept and Vocabulary Check

1. natural   **2.** whole   **3.** integers   **4.** rational   **5.** irrational   **6.** rational; irrational   **7.** left   **8.** absolute value; $|a|$

Exercise Set 1.3

1. -20 **3.** 8   **5.** -3000 **7.** -4 billion   **9–19.** [number line: -5, $-\frac{16}{5}$, -1.8, 2, $3\frac{1}{2}$, $\frac{11}{3}$]   **21.** 0.75 **23.** 0.35 **25.** 0.875 **27.** $0.\overline{81}$

29. -0.5 **31.** $-0.8\overline{3}$ **33. a.** $\sqrt{100}$ **b.** $0, \sqrt{100}$ **c.** $-9, 0, \sqrt{100}$ **d.** $-9, -\dfrac{4}{5}, 0, 0.25, 9.2, \sqrt{100}$ **e.** $\sqrt{3}$

f. $-9, -\dfrac{4}{5}, 0, 0.25, \sqrt{3}, 9.2, \sqrt{100}$ **35. a.** $\sqrt{64}$ **b.** $0, \sqrt{64}$ **c.** $-11, 0, \sqrt{64}$ **d.** $-11, -\dfrac{5}{6}, 0, 0.75, \sqrt{64}$ **e.** $\sqrt{5}, \pi$

f. $-11, -\dfrac{5}{6}, 0, 0.75, \sqrt{5}, \pi, \sqrt{64}$ **37.** 0   **39.** Answers will vary; $\dfrac{1}{2}$ is an example.   **41.** Answers will vary; 6 is an example.

43. Answers will vary; π is an example.   **45.** $<$ **47.** $>$ **49.** $>$ **51.** $<$ **53.** $>$ **55.** $<$ **57.** $<$ **59.** $>$ **61.** $>$ **63.** true

65. true   **67.** true   **69.** false   **71.** 6   **73.** 7   **75.** $\dfrac{5}{6}$ **77.** $\sqrt{11}$ **79.** $>$ **81.** $=$ **83.** $<$ **85.** $=$ **87.** rational numbers

89. integers   **91.** all real numbers   **93.** whole numbers
95. a. [number line: -25, -17, -16, -2, 12]   **b.** Rhode Island, Georgia, Louisiana, Florida, Hawaii

109. does not make sense   **111.** makes sense   **113.** false   **115.** false   **117.** false   **119.** $-\dfrac{1}{2}d$ **121.** -3.464; -4 and -3

123. -0.236; -1 and 0 **124.** $27; 27$; Both expressions have the same value.   **125.** $32; 32$; Both expressions have the same value.
126. $28; 28$; Both expressions have the same value.

Section 1.4 Check Point Exercises

1. a. 3 terms   **b.** 6   **c.** 11   **d.** $6x$ and $2x$ **2. a.** $14 + x$ **b.** $y7$ **3. a.** $17 + 5x$ **b.** $x5 + 17$ **4. a.** $20 + x$ or $x + 20$ **b.** $30x$
5. $12 + x$ or $x + 12$ **6.** $5x + 15$ **7.** $24y + 42$ **8. a.** $10x$ **b.** $5a$ **9. a.** $18x + 10$ **b.** $14x + 8y$ **10.** $25x + 21$ **11.** $38x + 23y$

Concept and Vocabulary Check

1. like   **2.** $b + a$ **3.** ab **4.** $a + (b + c)$ **5.** $(ab)c$ **6.** $ab + ac$ **7.** simplified

Exercise Set 1.4

1. a. 2   **b.** 3   **c.** 5   **d.** no   **3. a.** 3   **b.** 1   **c.** 2   **d.** yes; x and $5x$ **5. a.** 3   **b.** 4   **c.** 1   **d.** no   **7.** $4 + y$ **9.** $3x + 5$
11. $5y + 4x$ **13.** $5(3 + x)$ **15.** $x9$ **17.** $x + 6y$ **19.** $x7 + 23$ **21.** $(x + 3)5$ **23.** $(7 + 5) + x = 12 + x$ **25.** $(7 \cdot 4)x = 28x$

27. $3x + 15$ **29.** $16x + 24$ **31.** $4 + 2r$ **33.** $5x + 5y$ **35.** $3x - 6$ **37.** $8x - 10$ **39.** $\dfrac{5}{2}x - 6$ **41.** $8x + 28$ **43.** $6x + 18 + 12y$

45. $15x - 10 + 20y$ **47.** $17x$ **49.** $8a$ **51.** $14 + x$ **53.** $11y + 3$ **55.** $9x + 1$ **57.** $14a + 14$ **59.** $15x + 6$ **61.** $15x + 2$
63. $41a + 24b$ **65.** Distributive property; Commutative property of addition; Associative property of addition; Commutative property of addition
67. $7x + 2x; 9x$ **69.** $12x - 3x; 9x$ **71.** $6(4x); 24x$ **73.** $6(4 + x); 24 + 6x$ **75.** $8 + 5(x - 1); 5x + 3$ **77. a.** $U = 10n + 137$
b. 257 million; exact value   **89.** does not make sense   **91.** makes sense   **93.** false   **95.** false   **97.** 60   **98.** -60 **99.** -5

Mid-Chapter Check Point Exercises

1. 62   **2.** 2   **3.** 43   **4.** $\dfrac{1}{4}x - 2$ **5.** $\dfrac{x}{6} + 5 = 19$ **6.** not a solution   **7.** not a solution   **8. a.** \$48; underestimates by \$2   **b.** \$72

9. $\dfrac{1}{6}$ **10.** $\dfrac{1}{2}$ **11.** $\dfrac{25}{66}$ **12.** $\dfrac{2}{3}$ **13.** $\dfrac{1}{5}$ **14.** $\dfrac{25}{27}$ **15.** $\dfrac{37}{18}$ or $2\dfrac{1}{8}$ **16.** 10°C   **17.** <   **18.** $0.\overline{09}$ **19.** 19.3

20. $-11, -\dfrac{3}{7}, 0, 0.45, \sqrt{25}$ **21.** $(x + 3)5$ **22.** $5(3 + x)$ **23.** $5x + 15$ **24.** $65x + 21$ **25.** $10x + 34y$

Section 1.5 Check Point Exercises

1. $4 + (-7) = -3$    **2. a.** $-1 + (-3) = -4$

b. $-5 + 3 = -2$ **3. a.** -35 **b.** -1.5 **c.** $-\dfrac{5}{6}$ **4. a.** -13 **b.** 1.2 **c.** $-\dfrac{1}{2}$ **5. a.** $-17x$

b. $-7y + 6z$ **c.** $-20x + 15$ **6.** down 3 ft

Concept and Vocabulary Check

1. additive inverses   **2.** zero   **3.** negative number   **4.** positive number   **5.** 0   **6.** negative number   **7.** positive number   **8.** 0

Exercise Set 1.5

1. 4   **3.** -7 **5.** -4

7. 0   **9.** -7 **11.** 0   **13.** -60 **15.** -18 **17.** -1.3 **19.** -1 **21.** -5 **23.** 4   **25.** -3

27. -1.5 **29.** -5.7 **31.** $\dfrac{3}{10}$ **33.** $\dfrac{1}{8}$ **35.** $-\dfrac{43}{35}$ **37.** -8 **39.** 62   **41.** 8   **43.** -21

45. 22.1   **47.** $-8x$ **49.** $13y$ **51.** $-23a$ **53.** $-6y + 4z$ **55.** $-8b + 4$ **57.** $-2x + 14y$ **59.** $-3y + 24$ **61.** 12   **63.** -30

65. >   **67.** $-6x + (-13x); -19x$ **69.** $\dfrac{-20}{x} + \dfrac{3}{x}; \dfrac{-17}{x}$ **71.** 44°F   **73.** 600 ft below sea level   **75.** 3°F   **77.** the 25-yard line

79. a. $-1300; -\$1300$ billion   **b.** $-\$680$ billion   **c.** $-\$1980$ billion   **89.** makes sense   **91.** makes sense   **93.** true   **95.** false

97. negative   **99.** positive   **102. a.** $\sqrt{4}$ **b.** $0, \sqrt{4}$ **c.** $-6, 0, \sqrt{4}$ **d.** $-6, 0, 0.\overline{7}, \sqrt{4}$ **e.** $-\pi, \sqrt{3}$ **f.** $-6, -\pi, 0, 0.\overline{7}, \sqrt{3}, \sqrt{4}$

103. true   **104.** solution   **105.** -3 **106.** -21 **107.** 5

Section 1.6 Check Point Exercises

1. a. -8 **b.** 9   **c.** -5 **2. a.** 9.2   **b.** $-\dfrac{14}{15}$ **c.** 7π **3.** 15   **4.** $-6, 4a, -7ab$ **5. a.** $4 - 7x$ **b.** $-9x + 4y$

6. a. 8 years   **b.** 14 years

Concept and Vocabulary Check

1. (-14) **2.** 14   **3.** 14   **4.** $-8; (-14)$ **5.** $3; (-12); (-23)$ **6.** $(-4y); 6$ **7.** three; addition

Exercise Set 1.6

1. a. -12 **b.** $5 + (-12)$ **3. a.** 7   **b.** $5 + 7$ **5.** 6   **7.** -6 **9.** 23   **11.** 11   **13.** -11 **15.** -38 **17.** 0   **19.** 0   **21.** 26

23. -13 **25.** 13   **27.** $-\dfrac{2}{7}$ **29.** $\dfrac{4}{5}$ **31.** -1 **33.** $-\dfrac{3}{5}$ **35.** $\dfrac{3}{4}$ **37.** $\dfrac{1}{4}$ **39.** 7.6   **41.** -2 **43.** 2.6   **45.** 0   **47.** 3π **49.** 13π

51. 19   **53.** -3 **55.** -15 **57.** 0   **59.** -52 **61.** -187 **63.** $\dfrac{7}{6}$ **65.** -4.49 **67.** $-\dfrac{3}{8}$ **69.** $-3x, -8y$ **71.** $12x, -5xy, -4$

73. $-6x$ **75.** $4 - 10y$ **77.** $5 - 7a$ **79.** $-4 - 9b$ **81.** $24 + 11x$ **83.** $3y - 8x$ **85.** 9   **87.** $\dfrac{7}{8}$ **89.** -2 **91.** $6x - (-5x); 11x$

93. $\dfrac{-5}{x} - \left(\dfrac{-2}{x}\right); \dfrac{-3}{x}$ **95.** 19,757 ft   **97.** shrink by 5 years   **99.** shrink by 11 years   **101.** no change   **103.** 25 years   **105.** 1 year

107. 2892 meters   **109.** -2100 meters   **117.** makes sense   **119.** makes sense   **121.** false   **123.** true   **125.** negative   **127.** positive

131. not a solution   **132.** $15x + 40y$ **133.** Answers will vary; -1 is an example.   **134.** -12 **135.** -9 **136.** 12

Section 1.7 Check Point Exercises

1. a. -40 **b.** $-\dfrac{4}{21}$ **c.** 36   **d.** 5.5   **e.** 0   **2. a.** 24   **b.** -30 **3. a.** $\dfrac{1}{7}$ **b.** 8   **c.** $-\dfrac{1}{6}$ **d.** $-\dfrac{13}{7}$ **4. a.** -4 **b.** 8

5. a. 8   **b.** $-\dfrac{8}{15}$ **c.** -7.3 **d.** 0   **6. a.** $-20x$ **b.** $10x$ **c.** $-b$ **d.** $-21x + 28$ **e.** $-7y + 6$ **7.** $-y - 26$

8. not a solution   **9.** 49.4%; overestimates by 0.4%

Concept and Vocabulary Check

1. positive   **2.** negative   **3.** negative   **4.** positive   **5.** 0   **6.** negative   **7.** negative   **8.** positive   **9.** 0
10. undefined   **11.** positive

Exercise Set 1.7

1. -45 **3.** 24   **5.** -21 **7.** 19   **9.** 0   **11.** -12 **13.** 9   **15.** $\dfrac{12}{35}$ **17.** $-\dfrac{14}{27}$ **19.** -3.6 **21.** 0.12   **23.** 30   **25.** -72

27. 24   **29.** -27 **31.** 90   **33.** 0   **35.** $\dfrac{1}{4}$ **37.** 5   **39.** $-\dfrac{1}{10}$ **41.** $-\dfrac{5}{2}$ **43. a.** $-32 \cdot \left(\dfrac{1}{4}\right)$ **b.** -8 **45. a.** $-60 \cdot \left(-\dfrac{1}{5}\right)$
b. 12   **47.** -3 **49.** -7 **51.** 30   **53.** 0   **55.** undefined   **57.** -5 **59.** -12 **61.** 6   **63.** 0   **65.** undefined   **67.** -4.3
69. $\dfrac{5}{6}$ **71.** $-\dfrac{16}{9}$ **73.** -1 **75.** -15 **77.** $-10x$ **79.** $3y$ **81.** $9x$ **83.** $-4x$ **85.** $-b$ **87.** $3y$ **89.** $-8x + 12$
91. $6x - 12$ **93.** $-2y + 5$ **95.** $y - 14$ **97.** solution   **99.** solution   **101.** not a solution   **103.** solution   **105.** not a solution
107. solution   **109.** $4(-10) + 8 = -32$ **111.** $(-9)(-3) - (-2) = 29$ **113.** $\dfrac{-18}{-15 + 12} = 6$ **115.** $-6 - \left(\dfrac{12}{-4}\right) = -3$ **117.** $-30°C$
119. a. overestimates by 0.5%   **b.** 15%   **121. a.** 34%, although answers may vary by ± 1.   **b.** 34%; It's the same.
c. $R = \dfrac{-0.7n + 41}{0.6n + 51}$ **d.** $\dfrac{34}{57}$ **131.** does not make sense   **133.** makes sense   **135.** false   **137.** false   **139.** $5x$ **141.** $\dfrac{x}{12}$
145. $1.144x + 2.5$ **147.** -9 **148.** -3 **149.** 2   **150.** 36   **151.** -125 **152.** 16

Section 1.8 Check Point Exercises

1. a. 36   **b.** -64 **c.** 1   **d.** -1 **2. a.** $21x^2$ **b.** $8x^3$ **c.** cannot simplify   **3.** 15   **4.** 32   **5. a.** 36   **b.** 12   **6.** $\dfrac{3}{4}$ **7.** -40
8. -31 **9.** $\dfrac{5}{7}$ **10.** -5 **11.** $7x^2 + 15$ **12.** 2662 calories; underestimates by 38 calories   **13. a.** $330   **b.** $60   **c.** $33

Concept and Vocabulary Check

1. base; exponent   **2.** b to the nth power   **3.** multiply   **4.** add   **5.** divide   **6.** subtract   **7.** multiply

Exercise Set 1.8

1. 81   **3.** 64   **5.** 16   **7.** -64 **9.** 625   **11.** -625 **13.** -100 **15.** $19x^2$ **17.** $15x^3$ **19.** $9x^4$ **21.** $-x^2$ **23.** x^3
25. cannot be simplified   **27.** 0   **29.** 25   **31.** 27   **33.** 12   **35.** 5   **37.** 45   **39.** -24 **41.** 300   **43.** 0   **45.** -32 **47.** 64
49. 30   **51.** $\dfrac{4}{3}$ **53.** 2   **55.** 2   **57.** 3   **59.** 88   **61.** -60 **63.** -36 **65.** 14   **67.** $-\dfrac{3}{4}$ **69.** $-\dfrac{9}{40}$ **71.** $-\dfrac{37}{36}$ or $-1\dfrac{1}{36}$
73. 24   **75.** 28   **77.** 9   **79.** -7 **81.** $15x - 27$ **83.** $15 - 3y$ **85.** $16y - 25$ **87.** $-2x^2 - 9$ **89.** $-10 - (-2)^3 = -2$
91. $[2(7 - 10)]^2 = 36$ **93.** $x - (5x + 8) = -4x - 8$ **95.** $5(x^3 - 4) = 5x^3 - 20$ **97.** 1924 calories; underestimates by 76 calories
99. a. $1323; underestimates by $7   **b.** $985; underestimates by $1   **c.** $338; It's less than the difference shown by the graph.
101. a. $2,000,000 (or $200 tens of thousands)   **b.** $8,000,000 (or $800 tens of thousands)   **c.** Cost inceases.   **107.** does not make sense
109. makes sense   **111.** false   **113.** false   **115.** $(2 \cdot 3 + 3) \cdot 5 = 45$ **117.** 6   **118.** -24 **119.** Answers will vary; -3 is an example.
120. solution   **121.** solution   **122.** not a solution

Review Exercises

1. 40   **2.** 50   **3.** 3   **4.** 84   **5.** $7x - 6$ **6.** $\dfrac{x}{5} - 2 = 18$ **7.** $9 - 2x = 14$ **8.** $3(x + 7)$ **9.** not a solution   **10.** solution
11. solution   **12.** 11, although answers may vary by ± 1 **13.** 11 Latin words; very well; It is the same.   **14. a.** $29,070; underestimates by $221
b. $33,228   **15.** $\dfrac{23}{7}$ **16.** $\dfrac{64}{11}$ **17.** $1\dfrac{8}{9}$ **18.** $5\dfrac{2}{5}$ **19.** $2 \cdot 2 \cdot 3 \cdot 5$ **20.** $3 \cdot 3 \cdot 7$ **21.** prime   **22.** $\dfrac{5}{11}$ **23.** $\dfrac{8}{15}$ **24.** $\dfrac{21}{50}$ **25.** $\dfrac{8}{3}$
26. $\dfrac{1}{4}$ **27.** $\dfrac{2}{3}$ **28.** $\dfrac{29}{18}$ **29.** $\dfrac{37}{60}$ **30.** solution   **31.** not a solution   **32.** $2 - \dfrac{1}{2}x = \dfrac{1}{4}x$ **33.** $\dfrac{3}{5}(x + 6)$ **34.** 152 beats per minute
35.
36.
37. 0.625   **38.** $0.\overline{27}$ **39. a.** $\sqrt{81}$ **b.** $0, \sqrt{81}$
c. $-17, 0, \sqrt{81}$ **d.** $-17, -\dfrac{9}{13}, 0, 0.75, \sqrt{81}$ **e.** $\sqrt{2}, \pi$
f. $-17, -\dfrac{9}{13}, 0, 0.75, \sqrt{2}, \pi, \sqrt{81}$

40. Answers will vary; -2 is an example.   **41.** Answers will vary; $\dfrac{1}{2}$ is an example.   **42.** Answers will vary; π is an example.   **43.** <   **44.** >
45. >   **46.** <   **47.** false   **48.** true   **49.** 58   **50.** 2.75   **51.** $13y + 7$ **52.** $(x + 7)9$ **53.** $(6 + 4) + y = 10 + y$
54. $(7 \cdot 10)x = 70x$ **55.** $24x - 12 + 30y$ **56.** $7a + 2$ **57.** $28x + 19$
58. 2
   59. -3 **60.** $-\dfrac{11}{20}$ **61.** -7 **62.** $-4x + 5y$ or $5y - 4x$ **63.** $-10y + 40$ or $40 - 10y$
64. 800 ft below sea level   **65.** 23 ft   **66.** $9 + (-13)$ **67.** 4

68. $-\dfrac{6}{5}$ **69.** -1.5 **70.** -7 **71.** -3 **72.** $-5 - 8a$ **73.** $27{,}150$ ft   **74.** 84 **75.** $-\dfrac{3}{11}$ **76.** -120 **77.** -9 **78.** undefined   **79.** 2 **80.** $3x$ **81.** $-x - 1$ **82.** not a solution   **83.** solution   **84.** 2000 and 2014 **85.** 36 **86.** -36 **87.** -32 **88.** $6x^3$ **89.** cannot be simplified   **90.** -16 **91.** -16 **92.** 10 **93.** -2 **94.** 17 **95.** -88 **96.** 14 **97.** $-\dfrac{20}{3}$ **98.** $-\dfrac{2}{5}$ **99.** 6 **100.** 10 **101.** $28a - 20$ **102.** $6y - 12$ **103. a.** overestimates by $\$17$ **b.** $\$41{,}459$

Chapter Test

1. 4 **2.** -11 **3.** -51 **4.** $\dfrac{1}{5}$ **5.** $-\dfrac{35}{6}$ or $-5\dfrac{5}{6}$ **6.** -5 **7.** 1 **8.** -4 **9.** -32 **10.** 1 **11.** $4x + 4$ **12.** $13x - 19y$ **13.** $10 - 6x$ **14.** $-7, -\dfrac{4}{5}, 0, 0.25, \sqrt{4}, \dfrac{22}{7}$ **15.** $>$ **16.** 12.8 **17.** -15 **18.** 150 **19.** $2(3 + x)$ **20.** $(-6 \cdot 4)x = -24x$ **21.** $35x - 7 + 14y$ **22.** $17{,}030$ ft   **23.** not a solution   **24.** solution   **25.** $\dfrac{1}{4}x - 5 = 32$ **26.** $5(x + 4) - 7$ **27.** underestimates by 2% **28.** 144 beats per minute, although answers may vary by ± 1 **29.** 144 beats per minute; very well; It is the same.

CHAPTER 2

Section 2.1 Check Point Exercises

1. 17 or $\{17\}$ **2.** 2.29 or $\{2.29\}$ **3.** $\dfrac{1}{4}$ or $\left\{\dfrac{1}{4}\right\}$ **4.** 13 or $\{13\}$ **5.** 12 or $\{12\}$ **6.** 11 or $\{11\}$ **7.** 2100 words

Concept and Vocabulary Check

1. solving   **2.** linear   **3.** equivalent   **4.** $b + c$ **5.** subtract; solution   **6.** adding 7 **7.** subtracting $6x$

Exercise Set 2.1

1. linear   **3.** not linear   **5.** not linear   **7.** linear   **9.** not linear   **11.** 23 or $\{23\}$ **13.** -20 or $\{-20\}$ **15.** -16 or $\{-16\}$ **17.** -12 or $\{-12\}$ **19.** 4 or $\{4\}$ **21.** -11 or $\{-11\}$ **23.** 2 or $\{2\}$ **25.** $-\dfrac{17}{12}$ or $\left\{-\dfrac{17}{12}\right\}$ **27.** $\dfrac{21}{4}$ or $\left\{\dfrac{21}{4}\right\}$ **29.** $-\dfrac{11}{20}$ or $\left\{-\dfrac{11}{20}\right\}$ **31.** 4.3 or $\{4.3\}$ **33.** $-\dfrac{21}{4}$ or $\left\{-\dfrac{21}{4}\right\}$ **35.** 18 or $\{18\}$ **37.** $\dfrac{9}{10}$ or $\left\{\dfrac{9}{10}\right\}$ **39.** -310 or $\{-310\}$ **41.** 4.3 or $\{4.3\}$ **43.** 0 or $\{0\}$ **45.** 11 or $\{11\}$ **47.** 5 or $\{5\}$ **49.** -13 or $\{-13\}$ **51.** 6 or $\{6\}$ **53.** -12 or $\{-12\}$ **55.** $x = \triangle + \square$ **57.** $\triangle - \square = x$ **59.** $x - 12 = -2; 10$ **61.** $\dfrac{2}{5}x - 8 = \dfrac{7}{5}x; -8$ **63.** $\$1700$ **65. a.** 49%; overestimates by 1% **b.** 57% **67. a.** 55 **b.** 55; very well   **73.** does not make sense   **75.** makes sense   **77.** false   **79.** true   **81.** Answers will vary; example: $x - 100 = -101$.   **83.** 2.7529 or $\{2.7529\}$ **84.** $\dfrac{9}{x} - 4x$ **85.** -12 **86.** $6 - 9x$ **87.** x **88.** y **89.** yes

Section 2.2 Check Point Exercises

1. 36 or $\{36\}$ **2. a.** 21 or $\{21\}$ **b.** -4 or $\{-4\}$ **c.** -3.1 or $\{-3.1\}$ **3. a.** 24 or $\{24\}$ **b.** -16 or $\{-16\}$ **4. a.** -5 or $\{-5\}$ **b.** 3 or $\{3\}$ **5.** 6 or $\{6\}$ **6.** -10 or $\{-10\}$ **7.** 6 or $\{6\}$ **8. a.** underestimates by $\$11$ **b.** 35 years after 1980, or in 2015

Concept and Vocabulary Check

1. bc **2.** divide   **3.** multiplying; 7 **4.** dividing/multiplying; $-8/-\dfrac{1}{8}$ **5.** multiplying; $\dfrac{5}{3}$ **6.** multiplying/dividing; -1
7. subtracting 2; dividing; 5

Exercise Set 2.2

1. 30 or $\{30\}$ **3.** -33 or $\{-33\}$ **5.** 7 or $\{7\}$ **7.** -9 or $\{-9\}$ **9.** $-\dfrac{7}{2}$ or $\left\{-\dfrac{7}{2}\right\}$ **11.** 6 or $\{6\}$ **13.** $-\dfrac{3}{4}$ or $\left\{-\dfrac{3}{4}\right\}$ **15.** 0 or $\{0\}$ **17.** 18 or $\{18\}$ **19.** -8 or $\{-8\}$ **21.** -17 or $\{-17\}$ **23.** 47 or $\{47\}$ **25.** 45 or $\{45\}$ **27.** -5 or $\{-5\}$ **29.** 5 or $\{5\}$ **31.** 6 or $\{6\}$ **33.** -1 or $\{-1\}$ **35.** -2 or $\{-2\}$ **37.** $\dfrac{9}{4}$ or $\left\{\dfrac{9}{4}\right\}$ **39.** -6 or $\{-6\}$ **41.** -3 or $\{-3\}$ **43.** -3 or $\{-3\}$ **45.** 4 or $\{4\}$ **47.** $-\dfrac{3}{2}$ or $\left\{-\dfrac{3}{2}\right\}$ **49.** 2 or $\{2\}$ **51.** -4 or $\{-4\}$ **53.** -6 or $\{-6\}$ **55.** $x = \square \cdot \triangle$ **57.** $-\triangle = x$ **59.** $6x = 10; \dfrac{5}{3}$ **61.** $\dfrac{x}{-9} = 5; -45$ **63.** $4x - 8 = 56; 16$ **65.** $-3x + 15 = -6; 7$ **67.** 10 sec   **69.** 1502.2 mph   **71. a.** underestimates by $\$5$
b. 45 years after 1980, or in 2025 **77.** does not make sense   **79.** does not make sense   **81.** false   **83.** true   **85.** Answers will vary;
example: $\dfrac{5}{4}x = -20$.   **87.** 6.5 or $\{6.5\}$ **88.** 100 **88.** -100 **90.** 3 **91.** $-3x + 7$ **92.** yes   **93.** $2x - 78$

Section 2.3 Check Point Exercises

1. 6 or {6}   **2.** 2 or {2}   **3.** 5 or {5}   **4.** −2 or {−2}   **5.** −15 or {−15}   **6.** no solution or ∅   **7.** all real numbers or {x|x is a real number}   **8.** 3.7; shown as the point whose corresponding value on the vertical axis is 10 and whose value on the horizontal axis is 3.7

Concept and Vocabulary Check

1. simplify each side; combine like terms   **2.** 30   **3.** 100   **4.** inconsistent   **5.** identity   **6.** inconsistent   **7.** identity

Exercise Set 2.3

1. 3 or {3}   **3.** −1 or {−1}   **5.** 4 or {4}   **7.** 4 or {4}   **9.** $\frac{7}{2}$ or $\left\{\frac{7}{2}\right\}$ **11.** −3 or {−3}   **13.** 6 or {6}   **15.** 8 or {8}   **17.** 4 or {4}
19. 1 or {1}   **21.** −4 or {−4}   **23.** 5 or {5}   **25.** 6 or {6}   **27.** 1 or {1}   **29.** −57 or {−57}   **31.** −10 or {−10}   **33.** 18 or {18}
35. $\frac{7}{4}$ or $\left\{\frac{7}{4}\right\}$ **37.** 1 or {1}   **39.** 24 or {24}   **41.** −6 or {−6}   **43.** 20 or {20}   **45.** −7 or {−7}   **47.** 9 or {9}   **49.** 30 or {30}
51. 25 or {25}   **53.** 20 or {20}   **55.** −1 or {−1}   **57.** 7700 or {7700}   **59.** no solution or ∅   **61.** all real numbers or {x|x is a real number}
63. $\frac{2}{3}$ or $\left\{\frac{2}{3}\right\}$ **65.** all real numbers or {x|x is a real number}   **67.** no solution or ∅   **69.** 0 or {0}   **71.** no solution or ∅   **73.** 0 or {0}
75. $\frac{4}{3}$ or $\left\{\frac{4}{3}\right\}$ **77.** all real numbers or {x|x is a real number}   **79.** $x = \Box\$ - \Box\triangle$ **81.** 240   **83.** $\frac{x}{5} + \frac{x}{3} = 16; 30$
85. $\frac{3x}{4} - 3 = \frac{x}{2}; 12$ **87.** 85 mph   **89.** 142 lb; 13 lb   **91.** 409.2 ft   **99.** makes sense   **101.** does not make sense   **103.** false   **105.** false
107. 3 or {3}   **109.** <   **110.** <   **111.** −10   **112. a.** $T - D = pm$ **b.** $\frac{T-D}{p} = m$ **113.** 16 or {16}   **114.** 0.05 or {0.05}

Section 2.4 Check Point Exercises

1. $l = \frac{A}{w}$ **2.** $l = \frac{P-2w}{2}$ **3.** $m = \frac{T-D}{p}$ **4.** $x = 15 + 12y$ **5.** 4.5   **6.** 15   **7.** 36%   **8.** 35%   **9. a.** $1152   **b.** 4% decrease

Concept and Vocabulary Check

1. isolated on one side   **2.** $A = lw$ **3.** $P = 2l + 2w$ **4.** $A = PB$ **5.** subtract b; divide by m

Exercise Set 2.4

1. $r = \frac{d}{t}$ **3.** $P = \frac{I}{rt}$ **5.** $r = \frac{C}{2\pi}$ **7.** $m = \frac{E}{c^2}$ **9.** $m = \frac{y-b}{x}$ **11.** $D = T - pm$ **13.** $b = \frac{2A}{h}$ **15.** $n = 5M$ **17.** $c = 4F - 160$
19. $a = 2A - b$ **21.** $r = \frac{S-P}{Pt}$ **23.** $b = \frac{2A}{h} - a$ **25.** $x = \frac{C-By}{A}$ **27.** 6   **29.** 7.2   **31.** 5   **33.** 170   **35.** 20%   **37.** 12%
39. 60%   **41.** 75%   **43.** $x = \frac{y}{a+b}$ **45.** $x = \frac{y-5}{a-b}$ **47.** $x = \frac{y}{c+d}$ **49.** $x = \frac{y+C}{A-B}$ **51. a.** $z = 3A - x - y$ **b.** 96%
53. a. $t = \frac{d}{r}$ **b.** 2.5 hr   **55.** 522   **57. a.** 117 million households   **b.** $184,978   **59.** 36%   **61.** 12.5%   **63.** $9   **65. a.** $1008
b. $17,808   **67. a.** $103.20   **b.** $756.80   **69.** 15%   **71.** no; 2% loss   **75.** makes sense   **77.** does not make sense   **79.** false
81. false   **83.** $C = \frac{100M}{Q}$ **84.** 12 or {12}   **85.** 20 or {20}   **86.** 0.7x   **87.** $\frac{13}{x} - 7x$ **88.** $8(x + 14)$ **89.** $9(x - 5)$

Mid-Chapter Check Point Exercises

1. 16 or {16}   **2.** −3 or {−3}   **3.** $C = \frac{825H}{E}$ **4.** 8.4   **5.** 30 or {30}   **6.** $-\frac{8}{9}$ or $\left\{-\frac{8}{9}\right\}$ **7.** $r = \frac{S}{2\pi h}$ **8.** 40
9. 20 or {20}   **10.** 5.25 or {5.25}   **11.** −1 or {−1}   **12.** $x = \frac{By+C}{A}$ **13.** no solution or ∅   **14.** 40 or {40}   **15.** 12.5%
16. $-\frac{6}{5}$ or $\left\{-\frac{6}{5}\right\}$ **17.** 25%   **18.** all real numbers or {x|x is a real number}   **19. a.** underestimates by 2%   **b.** 24

Section 2.5 Check Point Exercises

1. 12   **2.** English: $41 thousand; computer science; $59 thousand   **3.** pages 72 and 73   **4.** 32; 4 mi   **5.** 40 ft wide and 120 ft long   **6.** $940

Concept and Vocabulary Check

1. $4x - 6$ **2.** $x + 215$ **3.** $x + 1$ **4.** $125 + 0.15x$ **5.** $2x + 2 \cdot 4x$ or $2 \cdot 4x + 2x$ or $10x$ **6.** $x - 0.35x$ or $0.65x$

Exercise Set 2.5

1. $x + 60 = 410$; 350   **3.** $x - 23 = 214$; 237   **5.** $7x = 126$; 18   **7.** $\frac{x}{19} = 5$; 95   **9.** $4 + 2x = 56$; 26   **11.** $5x - 7 = 178$; 37

13. $x + 5 = 2x$; 5   **15.** $2(x + 4) = 36$; 14   **17.** $9x = 3x + 30$; 5   **19.** $\frac{3x}{5} + 4 = 34$; 50   **21.** TV: 9 years; sleeping: 28 years

23. Bachelor's: \$81 thousand; Master's: \$92 thousand   **25.** pages 314 and 315   **27.** Ruth: 59; Maris: 61   **29.** 800 mi

31. 7 years after 2014; 2021   **33.** 50 yd wide and 200 yd long   **35.** 160 ft wide and 360 ft long   **37.** 12 ft long and 4 ft high   **39.** \$400
41. \$46,500   **43.** \$22,500   **45.** 11 hr   **51.** does not make sense   **53.** makes sense   **55.** false   **57.** true   **59.** 5 ft 7 in.

61. The uncle is 60 years old, and the woman is 20 years old.   **63.** -20 or $\{-20\}$ **64.** 0 or $\{0\}$ **65.** $w = \frac{3V}{lh}$ **66.** 5   **67.** 91   **68.** 64 or $\{64\}$

Section 2.6 Check Point Exercises

1. 12 ft   **2.** 400π ft$^2 \approx$ 1256 ft^2 or 1257 ft^2; 40π ft $\approx$ 126 ft   **3.** large pizza   **4.** 2 times   **5.** no; About 32 more cubic inches are needed.
6. first: 120°; second: 40°; third: 20°   **7.** 60°

Concept and Vocabulary Check

1. $A = \frac{1}{2}bh$ **2.** $A = \pi r^2$ **3.** $C = 2\pi r$ **4.** radius; diameter   **5.** $V = lwh$ **6.** $V = \pi r^2 h$ **7.** 180°   **8.** complementary
9. supplementary   **10.** $90 - x$; $180 - x$

Exercise Set 2.6

1. 18 m; 18 m^2 **3.** 56 in.^{2} **5.** 91 m^2 **7.** 50 ft   **9.** 8 ft   **11.** 50 cm   **13.** 16π cm$^2 \approx$ 50 cm^2; 8π cm $\approx$ 25 cm
15. 36π yd$^2 \approx$ 113 yd^2; 12π yd $\approx$ 38 yd   **17.** 7 in.; 14 in.   **19.** 36 in.^{3} **21.** 150π cm$^3 \approx$ 471 cm^3 **23.** 972π cm$^3 \approx$ 3052 or 3054 cm^3
25. 48π m$^3 \approx$ 151 m^3 **27.** $h = \frac{V}{\pi r^2}$ **29.** 9 times   **31.** $x = 50$; $x + 30 = 80$; 50°, 50°, 80°   **33.** $4x = 76$; $3x + 4 = 61$; $2x + 5 = 43$; 76°, 61°, 43°
35. 40°, 80°, 60°   **37.** 32°   **39.** 2°   **41.** 48°   **43.** 90°   **45.** 75°   **47.** 135°   **49.** 50°   **51.** 72 m^2 **53.** 70.5 cm^2 **55.** 448 in.^{3}
57. \$698.18   **59.** large pizza   **61.** \$2261 or \$2262   **63.** approx 19.7 ft   **65.** 21,000 yd^3 **67.** the can with diameter of 6 in. and height of 5 in.
69. Yes, the water tank is a little over one cubic foot too small.   **79.** does not make sense   **81.** does not make sense   **83.** true   **85.** false
87. 2.25 times   **89.** Volume increases 8 times.   **91.** 35°   **92.** $s = \frac{P - b}{2}$ **93.** 8 or $\{8\}$ **94.** 0   **95.** yes   **96.** yes   **97.** 2 or $\{2\}$

Section 2.7 Check Point Exercises

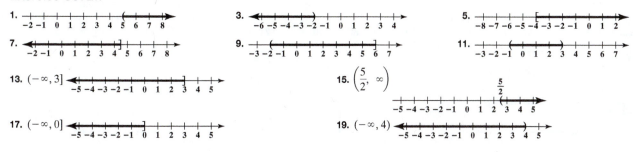

11. at least 83%   **12.** at most 43

Concept and Vocabulary Check

1. $(-\infty, 5)$ **2.** $[2, \infty)$ **3.** $< b + c$ **4.** $< bc$ **5.** $> bc$ **6.** subtracting 4; dividing; -3; direction; $>$; $<$
7. $\varnothing$ or the empty set   **8.** $(-\infty, \infty)$

Exercise Set 2.7

21. $(7, \infty)$ or $\{x \mid x > 7\}$

23. $(-\infty, 6]$ or $\{x \mid x \leq 6\}$

25. $(-\infty, 2)$ or $\{y \mid y < 2\}$

27. $(-\infty, 3]$ or $\{x \mid x \leq 3\}$

29. $(-\infty, 16)$ or $\{x \mid x < 16\}$

31. $(4, \infty)$ or $\{x \mid x > 4\}$

33. $\left(\dfrac{7}{6}, \infty\right)$ or $\left\{x \mid x > \dfrac{7}{6}\right\}$

35. $\left(-\infty, -\dfrac{3}{8}\right]$ or $\left\{y \mid y \leq -\dfrac{3}{8}\right\}$

37. $(0, \infty)$ or $\{y \mid y > 0\}$

39. $(-\infty, 8)$ or $\{x \mid x < 8\}$

41. $(-6, \infty)$ or $\{x \mid x > -6\}$

43. $(-\infty, 5)$ or $\{x \mid x < 5\}$

45. $[-7, \infty)$ or $\{x \mid x \geq -7\}$

47. $(-5, \infty)$ or $\{x \mid x > -5\}$

49. $(-\infty, -5]$ or $\{x \mid x \leq -5\}$

51. $(-\infty, 3)$ or $\{x \mid x < 3\}$

53. $\left[-\dfrac{1}{8}, \infty\right)$ or $\left\{y \mid y \geq -\dfrac{1}{8}\right\}$

55. $(-4, \infty)$ or $\{x \mid x > -4\}$

57. $(5, \infty)$ or $\{x \mid x > 5\}$

59. $(-\infty, 5)$ or $\{x \mid x < 5\}$

61. $[-2, \infty)$ or $\{x \mid x \geq -2\}$

63. $(-3, \infty)$ or $\{x \mid x > -3\}$

65. $[4, \infty)$ or $\{x \mid x \geq 4\}$

67. $\left(\dfrac{11}{3}, \infty\right)$ or $\left\{x \mid x > \dfrac{11}{3}\right\}$

69. $(2, \infty)$ or $\{y \mid y > 2\}$

71. $(-\infty, 2)$ or $\{y \mid y < 2\}$

73. $(-\infty, 3)$ or $\{x \mid x < 3\}$

75. $\left(\dfrac{5}{3}, \infty\right)$ or $\left\{x \mid x > \dfrac{5}{3}\right\}$

77. $[9, \infty)$ or $\{x \mid x \geq 9\}$

79. $(-\infty, -6)$ or $\{x \mid x < -6\}$

81. no solution or $\varnothing$ **83.** $(-\infty, \infty)$ or $\{x \mid x$ is a real number$\}$ **85.** no solution or $\varnothing$ **87.** $(-\infty, \infty)$ or $\{x \mid x$ is a real number$\}$
89. $(-\infty, 0]$ or $\{x \mid x \leq 0\}$ **91.** $x > \dfrac{b - a}{3}$ **93.** $\dfrac{y - b}{m} \geq x$ **95.** x is between -2 and 2; $|x| < 2$
97. x is greater than 2 or less than -2; $|x| > 2$ **99.** weird, cemetery, accommodation   **101.** supersede, inoculate   **103.** harass
105. a. 4%; It's the same.   **b.** 37 years; from 2017 onward   **107. a.** at least 96   **b.** if you get less than 66 on the final
109. up to 1280 mi   **111.** up to 29 bags of cement   **117.** makes sense   **119.** makes sense   **121.** false   **123.** false   **125.** more than 720 mi
127. $(-\infty, 0.4)$ or $\{x \mid x < 0.4\}$ **129.** 20   **130.** length: 11 in.; width: 6 in.   **131.** 4 or $\{4\}$ **132.** yes   **133.** no   **134.** -3

Review Exercises

1. 32 or $\{32\}$ **2.** -22 or $\{-22\}$ **3.** 12 or $\{12\}$ **4.** -22 or $\{-22\}$ **5.** 5 or $\{5\}$ **6.** 80 or $\{80\}$ **7.** -56 or $\{-56\}$ **8.** 11 or $\{11\}$
9. 4 or $\{4\}$ **10.** -15 or $\{-15\}$ **11.** -12 or $\{-12\}$ **12.** -25 or $\{-25\}$ **13.** 10 or $\{10\}$ **14.** 6 or $\{6\}$ **15.** -5 or $\{-5\}$ **16.** -10 or $\{-10\}$
17. 2 or $\{2\}$ **18.** 1 or $\{1\}$ **19. a.** underestimates by 0.1%   **b.** 2017   **20.** -1 or $\{-1\}$ **21.** 12 or $\{12\}$ **22.** -13 or $\{-13\}$ **23.** -3 or $\{-3\}$
24. -10 or $\{-10\}$ **25.** 2 or $\{2\}$ **26.** 2 or $\{2\}$ **27.** 9 or $\{9\}$ **28.** 4 or $\{4\}$ **29.** no solution or $\varnothing$ **30.** all real numbers or
$\{x \mid x$ is a real number$\}$ **31.** 30 years old   **32.** $r = \dfrac{I}{P}$ **33.** $h = \dfrac{3V}{B}$ **34.** $w = \dfrac{P - 2l}{2}$ **35.** $B = 2A - C$ **36.** $m = \dfrac{T - D}{p}$
37. 9.6   **38.** 200   **39.** 48%   **40.** 100%   **41.** 40%   **42.** 12.5%   **43.** no; 1%   **44. a.** $h = 7r$ **b.** 5 ft 3 in.   **45.** 350 gallons
46. 10   **47.** Gates: $81 billion; Buffett: $67 billion   **48.** pages 46 and 47   **49.** females: 49%; males: 51%   **50.** 15 years; 2016   **51.** 18 checks
52. length: 150 yd; width: 50 yd   **53.** $240   **54.** 32.5 ft^2 **55.** 50 cm^2 **56.** 135 yd^2 **57.** 7608 m^2 **58.** 20π m $\approx$ 63 m; 100π m^2 $\approx$ 314 m^2
59. 6 ft   **60.** 156 ft^2 **61.** $1890   **62.** medium pizza   **63.** 60 cm^3 **64.** 128π yd^3 $\approx$ 402 yd^3 **65.** 288π m^3 $\approx$ 905 m^3 **66.** 4800 m^3
67. 16 fish   **68.** $x = 30, 3x = 90, 2x = 60$; 30°, 60°, 90°   **69.** 85°, 35°, 60°   **70.** 33°   **71.** 105°   **72.** 57.5°   **73.** 45° and 135°

74.

75.

76. $\left[\dfrac{3}{2}, \infty\right)$

77. $(-\infty, 0)$

78. $(-\infty, 4)$ or $\{x \mid x < 4\}$

79. $(-8, \infty)$ or $\{x \mid x > -8\}$

80. $[-3, \infty)$ or $\{x \mid x \geq -3\}$ 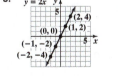 (number line from −5 to 5, closed bracket at −3)

81. $(6, \infty)$ or $\{x \mid x > 6\}$ (number line from −2 to 8, open paren at 6)

82. $[4, \infty)$ or $\{x \mid x \geq 4\}$ (number line from −2 to 8, closed bracket at 4)

83. $(-\infty, 2]$ or $\{x \mid x \leq 2\}$ (number line from −5 to 5, closed bracket at 2)

84. $(-\infty, \infty)$ or $\{x \mid x \text{ is a real number}\}$ **85.** no solution or $\varnothing$ **86.** at least 64   **87.** at most 30

Chapter Test

1. $\frac{9}{2}$ or $\left\{\frac{9}{2}\right\}$ **2.** -5 or $\{-5\}$ **3.** $-\frac{4}{3}$ or $\left\{-\frac{4}{3}\right\}$ **4.** 2 or $\{2\}$ **5.** -20 or $\{-20\}$ **6.** $-\frac{5}{3}$ or $\left\{-\frac{5}{3}\right\}$ **7.** -5 or $\{-5\}$ **8.** 60 yr; 2020

9. $h = \dfrac{V}{\pi r^2}$ **10.** $w = \dfrac{P - 2l}{2}$ **11.** 8.4   **12.** 150   **13.** 5%   **14.** 63   **15.** Americans: 13 days; Italians: 42 days   **16.** 600 messages

17. length: 150 yd; width: 75 yd   **18.** \$35   **19.** 517 m^2 **20.** 525 in.^{2} **21.** 66 ft^2 **22.** 18 in.^{3} **23.** 175π cm$^3 \approx 550$ cm^3

24. \$650   **25.** 14 ft   **26.** 126°, 42°, 12°   **27.** 53°

28. $(-2, \infty)$ (number line from −5 to 5, open paren at −2)   **29.** $(-\infty, 3]$ (number line from −5 to 5, closed bracket at 3)

30. $(-\infty, -6)$ or $\{x \mid x < -6\}$ (number line from −8 to 2, open paren at −6)   **31.** $(-\infty, -3]$ or $\{x \mid x \leq -3\}$ (number line from −5 to 5, closed bracket at −3)

32. $(7, \infty)$ or $\{x \mid x > 7\}$ (number line from −2 to 8, open paren at 7)   **33.** at least 92   **34.** widths greater than 8 in.

Cumulative Review Exercises

1. -4 **2.** -2 **3.** -128 **4.** $-103 - 20x$ **5.** $-4, -\dfrac{1}{3}, 0, \sqrt{4}, 1063$ **6.** $\dfrac{5}{x} - (x + 2)$ **7.** $<$ **8.** $24x - 6 - 30y$

9. 80%; It's the same.   **10.** 2020   **11.** 1 or $\{1\}$ **12.** -15 or $\{-15\}$ **13.** $A = \dfrac{3V}{h}$ **14.** 160   **15.** length: 130 yd; width: 70 yd

16. 75,000 gal   **17.** $\left(-\infty, \dfrac{1}{2}\right]$ (number line from −5 to 5, closed bracket at $\frac{1}{2}$)   **18.** $(-\infty, -3)$ or $\{x \mid x < -3\}$ (number line from −5 to 5, open paren at −3)

19. $[-6, \infty)$ or $\{x \mid x \geq -6\}$ (number line from −8 to 2, closed bracket at −6)   **20.** more than \$47,500

CHAPTER 3

Section 3.1 Check Point Exercises

1. 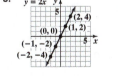 (graph with points $A(-2, 4)$, $C(-3, 0)$, $D(0, -3)$, $B(4, -2)$)   **2.** $E(-4, -2), F(-2, 0), G(6, 0)$ **3. a.** solution   **b.** not a solution   **4.** $(-2, -4), (-1, -1), (0, 2), (1, 5),$ and $(2, 8)$

5. 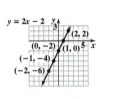 ($y = 2x$; points $(2, 4), (1, 2), (0, 0), (-1, -2), (-2, -4)$)   **6.** ($y = 2x - 2$; points $(2, 2), (1, 0), (0, -2), (-1, -4), (-2, -6)$)   **7.** 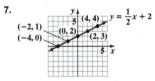 ($y = \frac{1}{2}x + 2$; points $(4, 4), (2, 3), (0, 2), (-2, 1), (-4, 0)$)

8. a. $(0, 1), (5, 8), (10, 15),$ and $(15, 22)$ **b.** 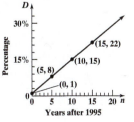 (graph of Percentage vs Years after 1995 with points $(0, 1), (5, 8), (10, 15), (15, 22)$)   **c.** approximately 29%, although answers may vary by $\pm 1\%$ **d.** 29%

Concept and Vocabulary Check

1. x-axis   **2.** y-axis   **3.** origin   **4.** quadrants; four   **5.** x-coordinate; y-coordinate   **6.** solution; satisfies   **7.** a/one   **8.** $mx + b$

Exercise Set 3.1

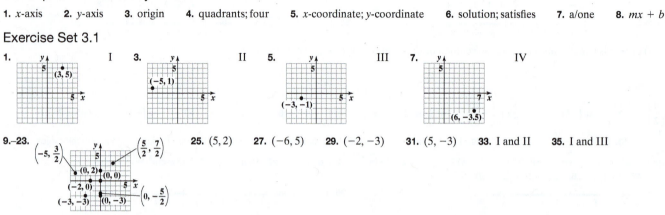

1. I   **3.** II   **5.** III   **7.** IV

9.–23.

25. $(5, 2)$ **27.** $(-6, 5)$ **29.** $(-2, -3)$ **31.** $(5, -3)$ **33.** I and II   **35.** I and III

37. $(2, 3)$ and $(3, 2)$ are not solutions; $(-4, -12)$ is a solution.   **39.** $(-5, -20)$ is not a solution; $(0, 0)$ and $(9, -36)$ are solutions.
41. $(2, -2)$ is not a solution; $(0, 6)$ and $(-3, 0)$ are solutions.   **43.** $(0, 5)$ is not a solution; $(-5, 6)$ and $(10, -3)$ are solutions.

45. $\left(1, \dfrac{1}{3}\right)$ is not a solution; $(0, 0)$ and $\left(2, -\dfrac{2}{3}\right)$ are solutions.   **47.** $(3, 4)$ and $(0, -4)$ are not solutions; $(4, 7)$ is a solution.

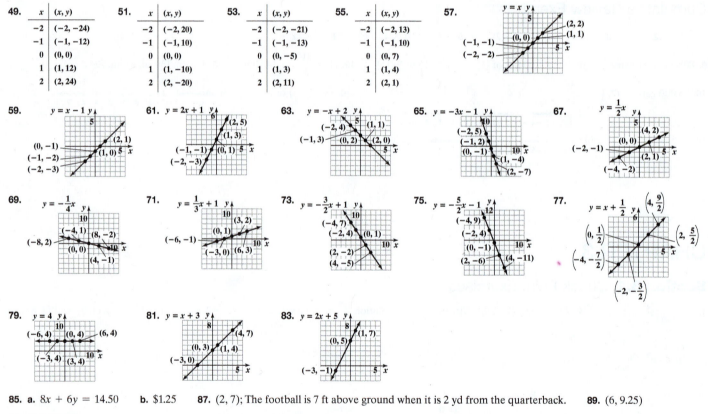

85. a. $8x + 6y = 14.50$ **b.** $1.25   **87.** $(2, 7)$; The football is 7 ft above ground when it is 2 yd from the quarterback.   **89.** $(6, 9.25)$

91. 12 ft; 15 yd

93. a. $(0, 31)$, $(5, 43)$, $(10, 55)$, $(15, 67)$, and $(20, 79)$

b.   **c.** $74\% \pm 1\%$ **d.** 74.2%

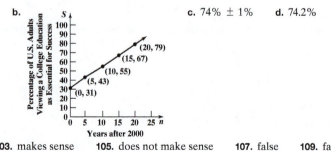

103. makes sense   **105.** does not make sense   **107.** false   **109.** false

111. a.

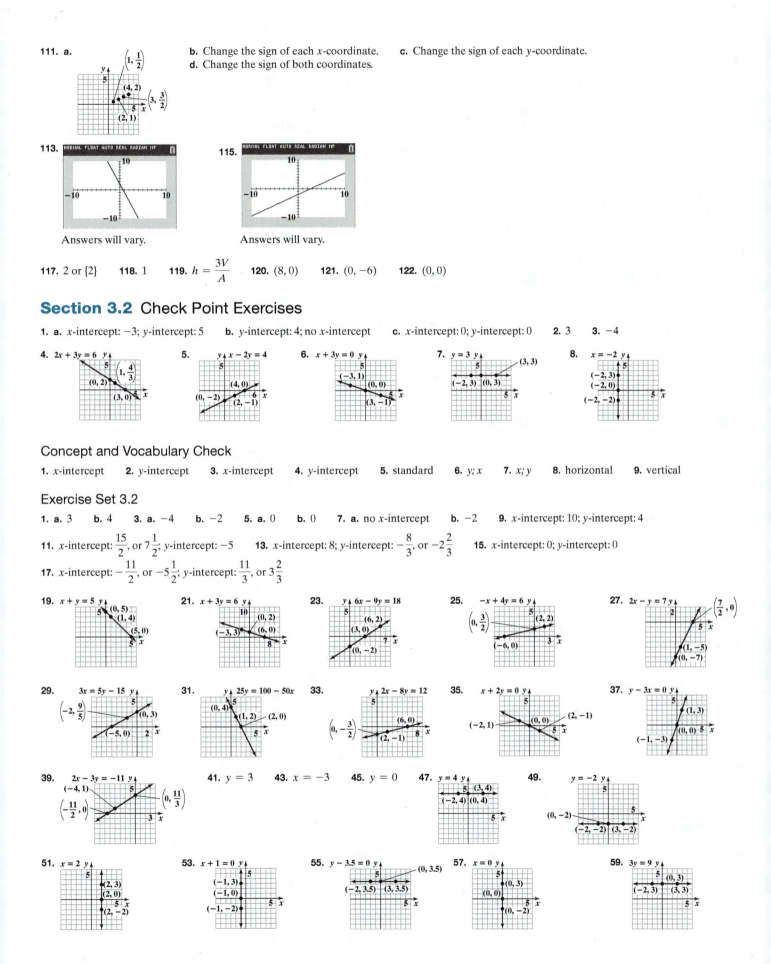

b. Change the sign of each x-coordinate.   **c.** Change the sign of each y-coordinate.
d. Change the sign of both coordinates.

113.   **115.**

Answers will vary.   Answers will vary.

117. 2 or {2}   **118.** 1   **119.** $h = \dfrac{3V}{A}$ **120.** $(8, 0)$ **121.** $(0, -6)$ **122.** $(0, 0)$

Section 3.2 Check Point Exercises

1. a. x-intercept: -3; y-intercept: 5   **b.** y-intercept: 4; no x-intercept   **c.** x-intercept: 0; y-intercept: 0   **2.** 3   **3.** -4

4. $2x + 3y = 6$ **5.** $x - 2y = 4$ **6.** $x + 3y = 0$ **7.** $y = 3$ **8.** $x = -2$

Concept and Vocabulary Check

1. x-intercept   **2.** y-intercept   **3.** x-intercept   **4.** y-intercept   **5.** standard   **6.** y; x **7.** x; y **8.** horizontal   **9.** vertical

Exercise Set 3.2

1. a. 3   **b.** 4   **3. a.** -4 **b.** -2 **5. a.** 0   **b.** 0   **7. a.** no x-intercept   **b.** -2 **9.** x-intercept: 10; y-intercept: 4

11. x-intercept: $\dfrac{15}{2}$, or $7\dfrac{1}{2}$; y-intercept: -5 **13.** x-intercept: 8; y-intercept: $-\dfrac{8}{3}$, or $-2\dfrac{2}{3}$ **15.** x-intercept: 0; y-intercept: 0

17. x-intercept: $-\dfrac{11}{2}$, or $-5\dfrac{1}{2}$; y-intercept: $\dfrac{11}{3}$, or $3\dfrac{2}{3}$

19. $x + y = 5$ **21.** $x + 3y = 6$ **23.** $6x - 9y = 18$ **25.** $-x + 4y = 6$ **27.** $2x - y = 7$

29. $3x = 5y - 15$ **31.** $25y = 100 - 50x$ **33.** $2x - 8y = 12$ **35.** $x + 2y = 0$ **37.** $y - 3x = 0$

39. $2x - 3y = -11$ **41.** $y = 3$ **43.** $x = -3$ **45.** $y = 0$ **47.** $y = 4$ **49.** $y = -2$

51. $x = 2$ **53.** $x + 1 = 0$ **55.** $y - 3.5 = 0$ **57.** $x = 0$ **59.** $3y = 9$

61. $12 - 3x = 0$ 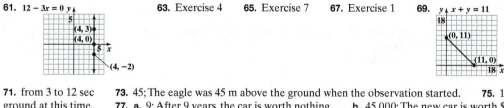   **63.** Exercise 4   **65.** Exercise 7   **67.** Exercise 1   **69.**

71. from 3 to 12 sec   **73.** 45; The eagle was 45 m above the ground when the observation started.   **75.** 12, 13, 14, 15, 16; The eagle is on the ground at this time.   **77. a.** 9; After 9 years, the car is worth nothing.   **b.** 45,000; The new car is worth $45,000.

c. 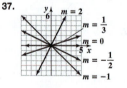   **d.** $20,000; Estimates will vary.   **87.** makes sense   **89.** makes sense   **91.** 2; 5

95.   **97.**

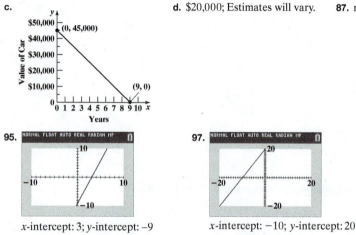

x-intercept: 3; y-intercept: –9 x-intercept: -10; y-intercept: 20

98. 13.4   **99.** $4x + 5$ **100.** $[-5, \infty)$ or $\{x \mid x \geq -5\}$ **101.** 2   **102.** -1 **103.** 0

Section 3.3 Check Point Exercises

1. a. 6   **b.** $-\frac{7}{5}$ **2. a.** 0   **b.** undefined   **3.** Both slopes equal 2, so the lines are parallel.   **4.** Product of slopes is $-\frac{1}{2}(2) = -1$, so the lines are perpendicular.   **5.** $m \approx 0.29$; The number of men living alone increased by 0.29 million per year. The rate of change is 0.29 million men per year.

Concept and Vocabulary Check

1. $\frac{y_2 - y_1}{x_2 - x_1}$ **2.** $y; x$ **3.** positive   **4.** negative   **5.** 0   **6.** undefined   **7.** parallel   **8.** perpendicular

Exercise Set 3.3

1. $\frac{3}{4}$; rises   **3.** $\frac{1}{4}$; rises   **5.** 0; horizontal   **7.** -5; falls   **9.** undefined; vertical   **11.** $\frac{1}{2}$ **13.** $-\frac{1}{3}$ **15.** $-\frac{1}{2}$ **17.** $-\frac{4}{3}$
19. 0   **21.** undefined   **23.** parallel   **25.** not parallel   **27.** perpendicular   **29.** not perpendicular   **31.** parallel
33. neither   **35.** perpendicular

37.   **39.** Slopes of corresponding opposite sides are equal: $-\frac{2}{5}$ and $\frac{4}{3}$.   **41.** 2
43. 4   **45. a.** -0.5 **b.** 0.5; -0.5; year of aging   **47.** 14.9%; green circle
49. 8.3%   **57.** does not make sense   **59.** makes sense   **61.** false   **63.** false

65. m_1, m_3, m_2, m_4 **67.** $m = 2$ **69.** $m = -\frac{1}{2}$ **71.** The line's slope is the coefficient of x.

72. 12 in. and 24 in.   **73.** -42 **74.** $(-\infty, 4]$ or $\{x \mid x \leq 4\}$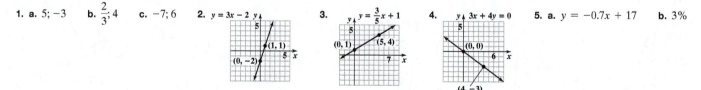

75. $(1, 1)$ **76.** $(3, -1)$ **77.** $y = -\frac{2}{5}x$

Section 3.4 Check Point Exercises

1. a. 5; -3 **b.** $\frac{2}{3}$; 4   **c.** -7; 6   **2.** $y = 3x - 2$ **3.**   **4.**   **5. a.** $y = -0.7x + 17$ **b.** 3%

Concept and Vocabulary Check

1. $y = mx + b$; slope; y-intercept   **2.** $(0, 3); 2; 5$ **3.** y

Exercise Set 3.4

1. $3; 2$ **3.** $3; -5$ **5.** $-\frac{1}{2}; 5$ **7.** $7; 0$ **9.** $0; 10$ **11.** $-1; 4$ **13.** $y = 5x + 7; 5; 7$ **15.** $y = -x + 6; -1; 6$ **17.** $y = -6x; -6; 0$

19. $y = 2x; 2; 0$ **21.** $y = -\frac{2}{7}x; -\frac{2}{7}; 0$ **23.** $y = -\frac{3}{2}x + \frac{3}{2}; -\frac{3}{2}, \frac{3}{2}$ **25.** $y = \frac{3}{4}x - 3; \frac{3}{4}, -3$

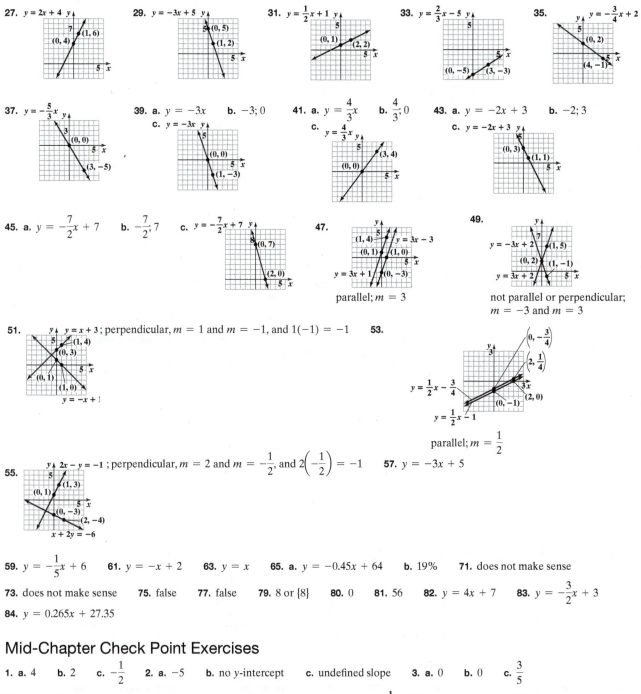

39. a. $y = -3x$ **b.** $-3; 0$ **41. a.** $y = \frac{4}{3}x$ **b.** $\frac{4}{3}; 0$ **43. a.** $y = -2x + 3$ **b.** $-2; 3$

45. a. $y = -\frac{7}{2}x + 7$ **b.** $-\frac{7}{2}; 7$ **c.** (graph)   **47.** parallel; $m = 3$ **49.** not parallel or perpendicular; $m = -3$ and $m = 3$

51. $y = x + 3$; perpendicular, $m = 1$ and $m = -1$, and $1(-1) = -1$ **53.** parallel; $m = \frac{1}{2}$

55. $2x - y = -1$; perpendicular, $m = 2$ and $m = -\frac{1}{2}$, and $2\left(-\frac{1}{2}\right) = -1$ **57.** $y = -3x + 5$

59. $y = -\frac{1}{5}x + 6$ **61.** $y = -x + 2$ **63.** $y = x$ **65. a.** $y = -0.45x + 64$ **b.** 19%   **71.** does not make sense

73. does not make sense   **75.** false   **77.** false   **79.** 8 or {8}   **80.** 0   **81.** 56   **82.** $y = 4x + 7$ **83.** $y = -\frac{3}{2}x + 3$

84. $y = 0.265x + 27.35$

Mid-Chapter Check Point Exercises

1. a. 4   **b.** 2   **c.** $-\frac{1}{2}$ **2. a.** -5 **b.** no y-intercept   **c.** undefined slope   **3. a.** 0   **b.** 0   **c.** $\frac{3}{5}$

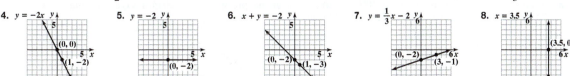

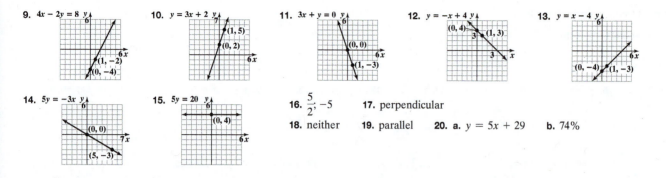

9. $4x - 2y = 8$ **10.** $y = 3x + 2$ **11.** $3x + y = 0$ **12.** $y = -x + 4$ **13.** $y = x - 4$

14. $5y = -3x$ **15.** $5y = 20$ **16.** $\dfrac{5}{2}; -5$ **17.** perpendicular

18. neither **19.** parallel **20. a.** $y = 5x + 29$ **b.** 74%

Section 3.5 Check Point Exercises

1. $y + 5 = 6(x - 2); y = 6x - 17$ **2. a.** $y + 1 = -5(x + 2)$ or $y + 6 = -5(x + 1)$ **b.** $y = -5x - 11$ **3.** $y = 0.28x + 27.2; 41.2$

Concept and Vocabulary Check

1. $y - y_1 = m(x - x_1)$ **2.** standard **3.** slope-intercept **4.** point-slope **5.** horizontal **6.** vertical

Exercise Set 3.5

1. $y - 5 = 3(x - 2); y = 3x - 1$ **3.** $y - 6 = 5(x + 2); y = 5x + 16$ **5.** $y + 2 = -8(x + 3); y = -8x - 26$

7. $y - 0 = -12(x + 8); y = -12x - 96$ **9.** $y + 2 = -1\left(x + \dfrac{1}{2}\right); y = -x - \dfrac{5}{2}$ **11.** $y - 0 = \dfrac{1}{2}(x - 0); y = \dfrac{1}{2}x$

13. $y + 2 = -\dfrac{2}{3}(x - 6); y = -\dfrac{2}{3}x + 2$ **15.** $y - 2 = 2(x - 1)$ or $y - 10 = 2(x - 5); y = 2x$

17. $y - 0 = 1(x + 3)$ or $y - 3 = 1(x - 0); y = x + 3$ **19.** $y + 1 = 1(x + 3)$ or $y - 4 = 1(x - 2); y = x + 2$

21. $y + 1 = \dfrac{5}{7}(x + 4)$ or $y - 4 = \dfrac{5}{7}(x - 3); y = \dfrac{5}{7}x + \dfrac{13}{7}$ **23.** $y + 1 = 0(x + 3)$ or $y + 1 = 0(x - 4); y = -1$

25. $y - 4 = 1(x - 2)$ or $y - 0 = 1(x + 2); y = x + 2$ **27.** $y - 0 = 8\left(x + \dfrac{1}{2}\right)$ or $y - 4 = 8(x - 0); y = 8x + 4$ **29.** $y = 4x + 14$

31. $y = -3x - 8$ **33.** $y = -2x + 1$ **35.** $y = 3x - 2$ **37.** $y = -5x - 20$ **39. a.** $y = 1100x + 16,700$ **b.** \$38,700

43. does not make sense **45.** makes sense **47.** false **49.** false **51.** $E = 2.4M - 20$ **53. b.**

c. $a = 0.1358315098; b = 22.94070022; r = 0.9897376983$ **d.**

54. at most 12 sheets of paper **55.** $1, \sqrt{4}$ **56.** **57.** yes **58.** no **59.** no

Section 3.6 Check Point Exercises

1. a. solution **b.** not a solution **2.** **3.** $y \geq \dfrac{1}{2}x$ **4. a.** **b.**

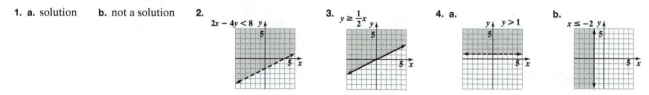

5. a. $B(60, 20)$; A region that has an average annual temperature of 60°F and an average annual precipitation of 20 inches is a grassland.
b. $5(60) - 7(20) \geq 70, 160 \geq 70$ true; $3(60) - 35(20) \leq -140, -520 \leq -140$ true

Concept and Vocabulary Check

1. solution; *x*; *y*; 9 > 2   **2.** graph   **3.** half-plane   **4.** false   **5.** true   **6.** false

Exercise Set 3.6

1. no; yes; yes   **3.** yes; yes; no   **5.** yes; yes; no   **7.** yes; no; yes

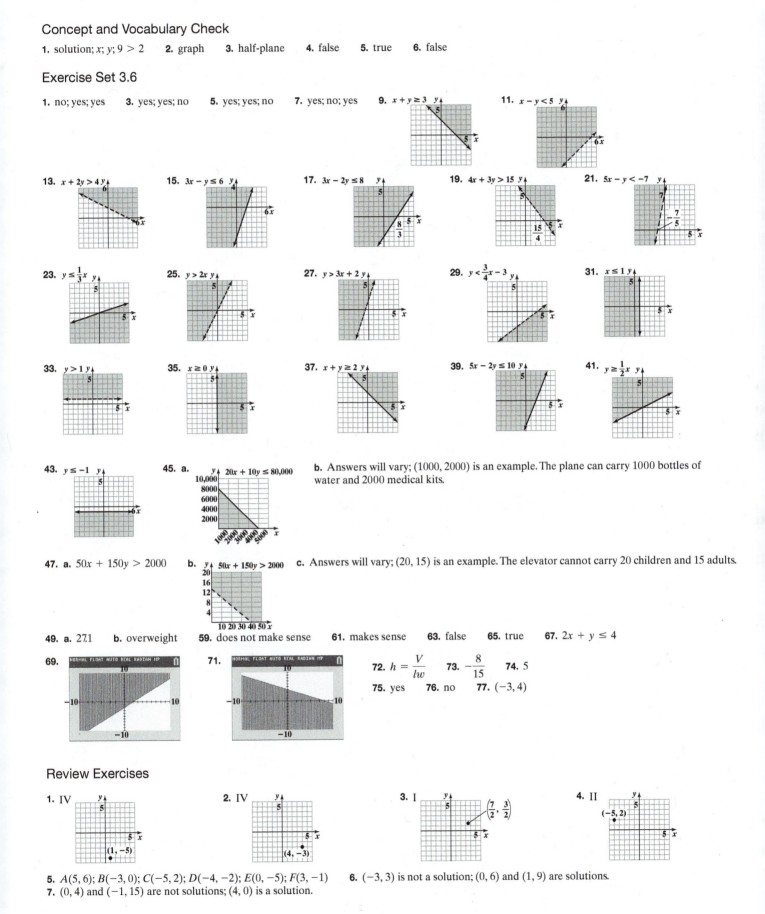

9. $x + y \geq 3$ **11.** $x - y < 5$

13. $x + 2y > 4$ **15.** $3x - y \leq 6$ **17.** $3x - 2y \leq 8$ **19.** $4x + 3y > 15$ **21.** $5x - y < -7$

23. $y \leq \frac{1}{3}x$ **25.** $y > 2x$ **27.** $y > 3x + 2$ **29.** $y < \frac{3}{4}x - 3$ **31.** $x \leq 1$

33. $y > 1$ **35.** $x \geq 0$ **37.** $x + y \geq 2$ **39.** $5x - 2y \leq 10$ **41.** $y \geq \frac{1}{2}x$

43. $y \leq -1$ **45. a.** $20x + 10y \leq 80{,}000$ **b.** Answers will vary; (1000, 2000) is an example. The plane can carry 1000 bottles of water and 2000 medical kits.

47. a. $50x + 150y > 2000$ **b.** $50x + 150y > 2000$ **c.** Answers will vary; (20, 15) is an example. The elevator cannot carry 20 children and 15 adults.

49. a. 27.1   **b.** overweight   **59.** does not make sense   **61.** makes sense   **63.** false   **65.** true   **67.** $2x + y \leq 4$

69.   **71.**

72. $h = \dfrac{V}{lw}$ **73.** $-\dfrac{8}{15}$ **74.** 5   **75.** yes   **76.** no   **77.** $(-3, 4)$

Review Exercises

1. IV   **2.** IV   **3.** I   **4.** II

$(1, -5)$ $(4, -3)$ $\left(\frac{7}{2}, \frac{3}{2}\right)$ $(-5, 2)$

5. $A(5, 6)$; $B(-3, 0)$; $C(-5, 2)$; $D(-4, -2)$; $E(0, -5)$; $F(3, -1)$ **6.** $(-3, 3)$ is not a solution; $(0, 6)$ and $(1, 9)$ are solutions.

7. $(0, 4)$ and $(-1, 15)$ are not solutions; $(4, 0)$ is a solution.

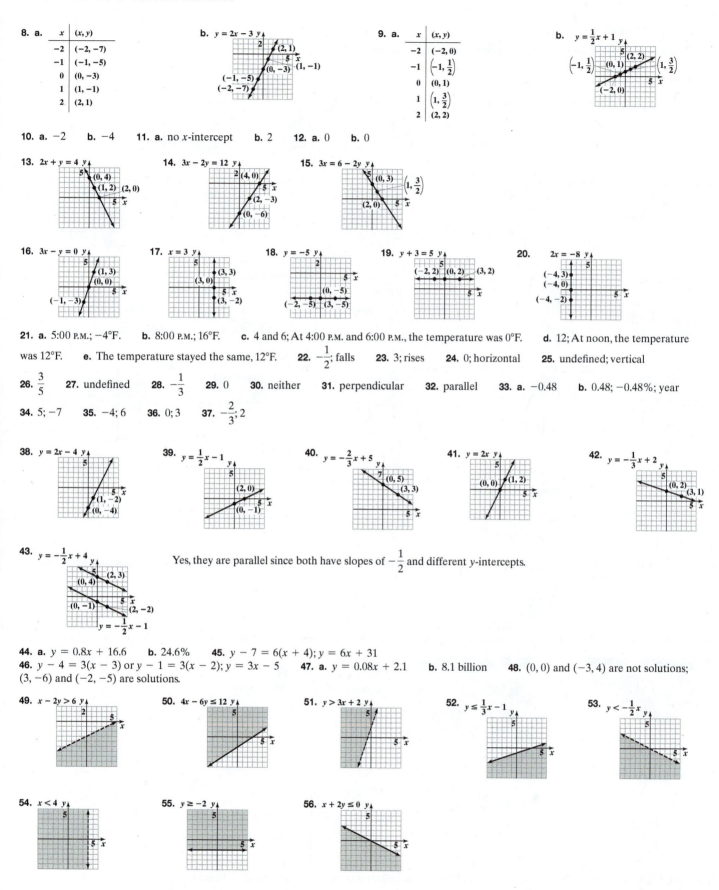

8. a.

x	(x, y)
-2	$(-2, -7)$
-1	$(-1, -5)$
0	$(0, -3)$
1	$(1, -1)$
2	$(2, 1)$

b. $y = 2x - 3$

9. a.

x	(x, y)
-2	$(-2, 0)$
-1	$\left(-1, \frac{1}{2}\right)$
0	$(0, 1)$
1	$\left(1, \frac{3}{2}\right)$
2	$(2, 2)$

b. $y = \frac{1}{2}x + 1$

10. a. -2 **b.** -4 **11. a.** no x-intercept   **b.** 2 **12. a.** 0 **b.** 0

13. $2x + y = 4$ **14.** $3x - 2y = 12$ **15.** $3x = 6 - 2y$

16. $3x - y = 0$ **17.** $x = 3$ **18.** $y = -5$ **19.** $y + 3 = 5$ **20.** $2x = -8$

21. a. 5:00 P.M.; $-4°$F.   **b.** 8:00 P.M.; $16°$F.   **c.** 4 and 6; At 4:00 P.M. and 6:00 P.M., the temperature was $0°$F.   **d.** 12; At noon, the temperature was $12°$F.   **e.** The temperature stayed the same, $12°$F.   **22.** $-\frac{1}{2}$; falls   **23.** 3; rises   **24.** 0; horizontal   **25.** undefined; vertical

26. $\frac{3}{5}$ **27.** undefined   **28.** $-\frac{1}{3}$ **29.** 0   **30.** neither   **31.** perpendicular   **32.** parallel   **33. a.** -0.48 **b.** 0.48; -0.48%; year

34. $5; -7$ **35.** $-4; 6$ **36.** $0; 3$ **37.** $-\frac{2}{3}; 2$

38. $y = 2x - 4$ **39.** $y = \frac{1}{2}x - 1$ **40.** $y = -\frac{2}{3}x + 5$ **41.** $y = 2x$ **42.** $y = -\frac{1}{3}x + 2$

43. $y = -\frac{1}{2}x + 4$ Yes, they are parallel since both have slopes of $-\frac{1}{2}$ and different y-intercepts.

44. a. $y = 0.8x + 16.6$ **b.** 24.6%   **45.** $y - 7 = 6(x + 4)$; $y = 6x + 31$
46. $y - 4 = 3(x - 3)$ or $y - 1 = 3(x - 2)$; $y = 3x - 5$ **47. a.** $y = 0.08x + 2.1$ **b.** 8.1 billion   **48.** $(0, 0)$ and $(-3, 4)$ are not solutions; $(3, -6)$ and $(-2, -5)$ are solutions.

49. $x - 2y > 6$ **50.** $4x - 6y \leq 12$ **51.** $y > 3x + 2$ **52.** $y \leq \frac{1}{3}x - 1$ **53.** $y < -\frac{1}{2}x$

54. $x < 4$ **55.** $y \geq -2$ **56.** $x + 2y \leq 0$

Chapter Test

1. $(-2, 1)$ is not a solution; $(0, -5)$ and $(4, 3)$ are solutions

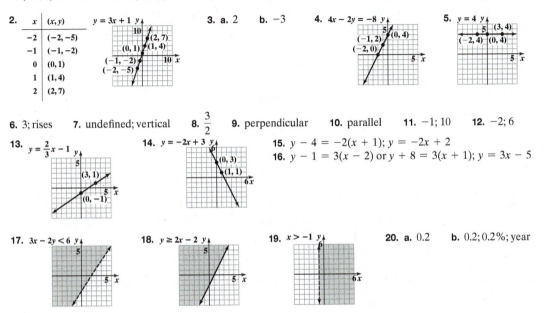

2.

x	(x, y)
-2	$(-2, -5)$
-1	$(-1, -2)$
0	$(0, 1)$
1	$(1, 4)$
2	$(2, 7)$

3. a. 2   **b.** -3 **4.** $4x - 2y = -8$ **5.** $y = 4$

6. 3; rises   **7.** undefined; vertical   **8.** $\dfrac{3}{2}$ **9.** perpendicular   **10.** parallel   **11.** $-1; 10$ **12.** $-2; 6$

13. $y = \dfrac{2}{3}x - 1$ **14.** $y = -2x + 3$ **15.** $y - 4 = -2(x + 1); y = -2x + 2$

16. $y - 1 = 3(x - 2)$ or $y + 8 = 3(x + 1); y = 3x - 5$

17. $3x - 2y < 6$ **18.** $y \geq 2x - 2$ **19.** $x > -1$ **20. a.** 0.2   **b.** 0.2; 0.2%; year

Cumulative Review Exercises

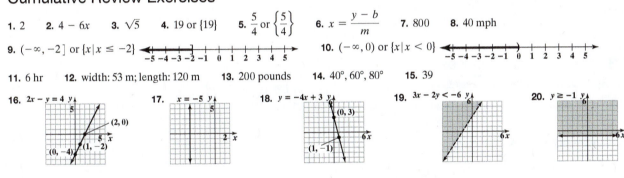

1. 2   **2.** $4 - 6x$ **3.** $\sqrt{5}$ **4.** 19 or $\{19\}$ **5.** $\dfrac{5}{4}$ or $\left\{\dfrac{5}{4}\right\}$ **6.** $x = \dfrac{y - b}{m}$ **7.** 800   **8.** 40 mph

9. $(-\infty, -2]$ or $\{x \mid x \leq -2\}$ **10.** $(-\infty, 0)$ or $\{x \mid x < 0\}$

11. 6 hr   **12.** width: 53 m; length: 120 m   **13.** 200 pounds   **14.** $40°, 60°, 80°$ **15.** 39

16. $2x - y = 4$ **17.** $x = -5$ **18.** $y = -4x + 3$ **19.** $3x - 2y < -6$ **20.** $y \geq -1$

CHAPTER 4

Section 4.1 Check Point Exercises

1. a. solution   **b.** not a solution   **2.** $(1, 4)$ or $\{(1, 4)\}$ **3.** $(3, 3)$ or $\{(3, 3)\}$ **4.** no solution or $\varnothing$
5. infinitely many solutions; $\{(x, y) \mid x + y = 3\}$ or $\{(x, y) \mid 2x + 2y = 6\}$ **6. a.** $(10, 20)$ or $\{(10, 20)\}$ **b.** If the bridge is used 10 times in a month, the total monthly cost without the discount pass is the same as the monthly cost with the discount pass, namely $20.

Concept and Vocabulary Check

1. satisfies both equations in the system   **2.** the intersection point   **3.** inconsistent; parallel   **4.** dependent; identical or coincide

Exercise Set 4.1

1. solution   **3.** solution   **5.** not a solution   **7.** solution   **9.** not a solution   **11.** $(4, 2)$ or $\{(4, 2)\}$ **13.** $(-1, 2)$ or $\{(-1, 2)\}$
15. $(3, 0)$ or $\{(3, 0)\}$ **17.** $(1, 0)$ or $\{(1, 0)\}$ **19.** $(-1, 4)$ or $\{(-1, 4)\}$ **21.** $(2, 4)$ or $\{(2, 4)\}$ **23.** $(2, -1)$ or $\{(2, -1)\}$ **25.** no solution or $\varnothing$
27. $(-2, 6)$ or $\{(-2, 6)\}$ **29.** infinitely many solutions; $\{(x, y) \mid x - 2y = 4\}$ or $\{(x, y) \mid 2x - 4y = 8\}$ **31.** $(2, 3)$ or $\{(2, 3)\}$ **33.** no solution or $\varnothing$
35. infinitely many solutions; $\{(x, y) \mid x - y = 0\}$ or $\{(x, y) \mid y = x\}$ **37.** $(2, 4)$ or $\{(2, 4)\}$ **39.** no solution or $\varnothing$ **41.** no solution or $\varnothing$
43. $m = \dfrac{1}{2}; b = -3$ and -5; no solution   **45.** $m = -\dfrac{1}{2}$ and 3; $b = 4$; one solution   **47.** $m = 3; b = -6$; infinite number of solutions

49. $m = -3; b = 0$ and 1; no solution   **51. a.** 40; Both companies charge the same for 40 miles driven.   **b.** Answers will vary: 55.
c. 54; Both companies charge $54 for 40 miles driven.   **53. a.** (5, 20) or {(5, 20)}   **b.** Nonmembers and members pay the same amount per month
for taking 5 classes, namely $20.   **63.** does not make sense   **65.** makes sense   **67.** false   **69.** false   **73.** (1, 4) or {(1, 4)}
75. (4, 0) or {(4, 0)}   **77.** (0, −5) or {(0, −5)}   **79.** (7, 3) or {(7, 3)}   **81.** −12   **82.** 6   **83.** 27   **84.** 3 or {3}   **85.** 4 or {4}
86. no solution or ∅

Section 4.2 Check Point Exercises

1. (3, 2) or {(3, 2)}   **2.** (1, −2) or {(1, −2)}   **3.** no solution or ∅   **4.** infinitely many solutions; $\{(x, y)|y = 3x - 4\}$ or $\{(x, y)|9x - 3y = 12\}$
5. a. (30, 900) or {(30, 900)}; equilibrium quantity: 30,000; equilibrium price: $900   **b.** $900; 30,000; 30,000

Concept and Vocabulary Check

1. {(4, 1)}   **2.** {(−1, 3)}   **3.** ∅   **4.** $\{(x, y)|2x - 6y = 8\}$ or $\{(x, y)|x = 3y + 4\}$ **5.** equilibrium

Exercise Set 4.2

1. (1, 3) or {(1, 3)}   **3.** (5, 1) or {(5, 1)}   **5.** (2, 1) or {(2, 1)}   **7.** (−1, 3) or {(−1, 3)}   **9.** (4, 5) or {(4, 5)}
11. $\left(-\frac{2}{5}, -\frac{11}{5}\right)$ or $\left\{\left(-\frac{2}{5}, -\frac{11}{5}\right)\right\}$ **13.** no solution or ∅   **15.** infinitely many solutions; $\{(x, y)|y = 3x - 5\}$ or $\{(x, y)|21x - 35 = 7y\}$
17. (0, 0) or {(0, 0)}   **19.** $\left(\frac{17}{7}, -\frac{8}{7}\right)$ or $\left\{\left(\frac{17}{7}, -\frac{8}{7}\right)\right\}$ **21.** no solution or ∅   **23.** $\left(\frac{43}{5}, -\frac{1}{5}\right)$ or $\left\{\left(\frac{43}{5}, -\frac{1}{5}\right)\right\}$ **25.** (200, 700) or {(200, 700)}
27. (7, 3) or {(7, 3)}   **29.** (−1, −1) or {(−1, −1)}   **31.** (5, 4) or {(5, 4)}   **33.** $x + y = 81, y = x + 41$; 20 and 61
35. $x - y = 5, 4x = 6y$; 10 and 15   **37.** $x - y = 1, x + 2y = 7$; 2 and 3   **39.** (2, 8) or {(2, 8)}   **41. a.** 1000 gallons; $4   **b.** $4; 1000; 1000
49. does not make sense   **51.** does not make sense   **53.** true   **55.** false   **57.** $m = \frac{5}{2}$

58. $4x + 6y = 12$
59. 12 or {12}   **60.** $-73, 0, \frac{3}{1}$
61. 8; The same value of x is obtained in both cases, so (8, 12) is the solution.
62. 12 or {12}   **63.** $-4x + 20y = -12$

Section 4.3 Check Point Exercises

1. (7, −2) or {(7, −2)}   **2.** (6, 2) or {(6, 2)}   **3.** (2, −1) or {(2, −1)}   **4.** $\left(\frac{60}{17}, -\frac{11}{17}\right)$ or $\left\{\left(\frac{60}{17}, -\frac{11}{17}\right)\right\}$
5. no solution or ∅   **6.** infinitely many solutions; $\{(x, y)|x - 5y = 7\}$ or $\{(x, y)|3x - 15y = 21\}$

Concept and Vocabulary Check

1. −3   **2.** −2   **3.** −2   **4.** 3

Exercise Set 4.3

1. (4, −7) or {(4, −7)}   **3.** (3, 0) or {(3, 0)}   **5.** (−3, 5) or {(−3, 5)}   **7.** (3, 1) or {(3, 1)}   **9.** (2, 1) or {(2, 1)}   **11.** (−2, 2) or {(−2, 2)}
13. (−7, −1) or {(−7, −1)}   **15.** (2, 0) or {(2, 0)}   **17.** (1, −2) or {(1, −2)}   **19.** (−1, 1) or {(−1, 1)}   **21.** (3, 1) or {(3, 1)}
23. (−5, −2) or {(−5, −2)}   **25.** $\left(\frac{11}{12}, -\frac{7}{6}\right)$ or $\left\{\left(\frac{11}{12}, -\frac{7}{6}\right)\right\}$ **27.** $\left(\frac{23}{16}, \frac{3}{8}\right)$ or $\left\{\left(\frac{23}{16}, \frac{3}{8}\right)\right\}$ **29.** no solution or ∅

31. infinitely many solutions; $\{(x, y)|x + 3y = 2\}$ or $\{(x, y)|3x + 9y = 6\}$ **33.** no solution or ∅   **35.** $\left(\frac{1}{2}, -\frac{1}{2}\right)$ or $\left\{\left(\frac{1}{2}, -\frac{1}{2}\right)\right\}$

37. infinitely many solutions; $\{(x, y)|x = 5 - 3y\}$ or $\{(x, y)|2x + 6y = 10\}$ **39.** $\left(\frac{1}{3}, 1\right)$ or $\left\{\left(\frac{1}{3}, 1\right)\right\}$ **41.** (−10, 21) or {(−10, 21)}
43. (0, 1) or {(0, 1)}   **45.** (2, −1) or {(2, −1)}   **47.** (1, −3) or {(1, −3)}   **49.** (4, 3) or {(4, 3)}   **51.** no solution or ∅
53. infinitely many solutions; $\{(x, y)|2(x + 2y) = 6\}$ or $\{(x, y)|3(x + 2y - 3) = 0\}$ **55.** (3, 2) or {(3, 2)}   **57.** $5x + y = 14, 4x - y = 4$; 2 and 4
59. $4x - 3y = 0, x + y = -7$; −3 and −4   **61.** (−1, 2) or {(−1, 2)}   **63.** (−1, 0) or {(−1, 0)}   **65.** (3, 1) or {(3, 1)}   **67.** dependent
69. week 6; 3.5 symptoms; by the intersection point (6, 3.5)   **77.** does not make sense   **79.** does not make sense   **81.** false   **83.** false
85. $A = 2, B = 4$ **86.** 10   **87.** II   **88.** 26 or {26}   **89. a.** $x + y = 28; x - y = 6$ **b.** 17 and 11   **90.** $3x + 2y$ **91. a.** $130
b. $y = 15x + 40$

Mid-Chapter Check Point Exercises

1. (2, 0) or {(2, 0)}   **2.** (1, 1) or {(1, 1)}   **3.** no solution or ∅   **4.** (5, 8) or {(5, 8)}   **5.** (2, −1) or {(2, −1)}   **6.** (2, 9) or {(2, 9)}
7. (−2, −3) or {(−2, −3)}   **8.** (10, −1) or {(10, −1)}   **9.** no solution or ∅   **10.** (−12, −1) or {(−12, −1)}   **11.** (−7, −23) or {(−7, −23)}
12. $\left(\frac{16}{17}, -\frac{12}{17}\right)$ or $\left\{\left(\frac{16}{17}, -\frac{12}{17}\right)\right\}$ **13.** infinitely many solutions; $\{(x, y)|y - 2x = 7\}$ or $\{(x, y)|4x = 2y - 14\}$ **14.** (7, 11) or {(7, 11)}
15. (6, 10) or {(6, 10)}

Section 4.4 Check Point Exercises

1. men: 65 minutes per day; women: 73 minutes per day   **2.** Quarter Pounder: 420 calories; Whopper with cheese: 589 calories
3. length: 100 ft; width: 80 ft   **4.** 17.5 yr; $24,250   **5.** $15,000 at 9%; $10,000 at 12%   **6.** 30 ml of 10%; 20 ml of 60% acid solution

Concept and Vocabulary Check

1. $5x + 6y$ **2.** $10 \cdot 2x + 15 \cdot 2y$ or $20x + 30y$ **3.** $25{,}600 + 225x$ **4.** $0.04x + 0.05y$ **5.** $0.07x + 0.15y$

Exercise Set 4.4

1. $x + y = 17, x - y = -3$; 7 and 10   **3.** $3x - y = -1, x + 2y = 23$; 3 and 10   **5.** women: 49 minutes; men: 37 minutes
7. Mr. Goodbar: 264 calories; Mounds bar: 258 calories   **9.** 3 Mr. Goodbars and 2 Mounds bars   **11.** sweater: $12; shirt: $10
13. 44 ft by 20 ft   **15.** 90 ft by 70 ft   **17. a.** 2 GB; $70   **b.** Plan A by $20   **19.** $600 of merchandise; $580
21. adult: 93 tickets; student: 208 tickets   **23.** each item in column A: $3.99; each item in column B: $1.50
25. 2 servings of macaroni and 4 servings of broccoli   **27.** A: 100°; B: 40°; C: 40°   **29.** $0.08y$; $8000 at 7% and $12,000 at 8%
31. $0.08x$; y; $0.18y$; $116,000 at 8% and $4000 at 18%   **33.** $3600 at 6% and $2400 at 9%   **35.** $31,250 at 15% and $18,750 at 7%
37. $0.10y$; $0.08(50)$; 20 liters of 5% and 30 liters of 10%   **39.** 6 oz   **41.** The north had 600 students, and the south had 400 students.
49. does not make sense   **51.** does not make sense   **53.** 10 birds and 20 lions   **55.** There are 5 people downstairs and 7 people upstairs.
57. $52,500 at 8% and $17,500 at 12%   **59.** $\dfrac{5}{2}$ or $\left\{\dfrac{5}{2}\right\}$ **60.** $6x + 11$ **61.** $y = -2x - 4$

62. $2x - y < 4$ **63.** $y \geq x + 1$ **64.** $x \geq 2$

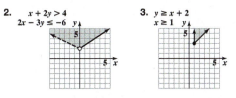

Section 4.5 Check Point Exercises

1. Point $B = (66, 130)$; $4.9(66) - 130 \geq 165$, or $193.4 \geq 165$, is true; $3.7(66) - 130 \leq 125$, or $114.2 \leq 125$, is true.

2. $x + 2y > 4$
$2x - 3y \leq -6$
3. $y \geq x + 2$
$x \geq 1$

Concept and Vocabulary Check

1. $x + y \geq 4$; $x - y \leq 2$ **2.** $\begin{cases} y \geq 2x + 1 \\ y \leq 4 \end{cases}$ **3.** false   **4.** true

Exercise Set 4.5

1. $x + y \leq 4$
$x - y \leq 2$
3. $2x - 4y \leq 8$
$x + y \geq -1$
5. $x + 3y \leq 6$
$x - 2y \leq 4$
7. $x - 2y > 4$
$2x + y \geq 6$
9. $x + y > 1$
$x + y < 4$

11. $y \geq 2x + 1$
$y \leq 4$
13. $y > x - 1$
$x > 5$
15. $y \geq 2x - 3$
$y \leq 2x + 1$
17. $y > 2x - 3$
$y \leq -x + 6$
19. $x + 2y \leq 4$
$y \geq x - 3$

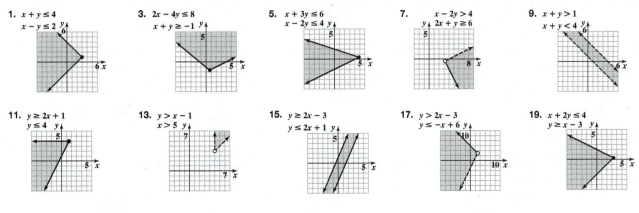

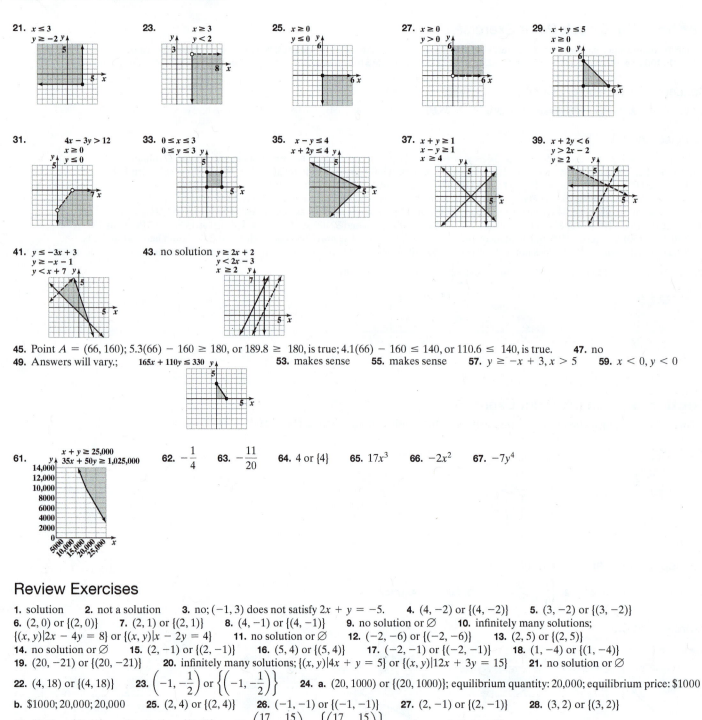

21. $x \le 3$
$y \ge -2$

23. $x \ge 3$
$y < 2$

25. $x \ge 0$
$y \le 0$

27. $x \ge 0$
$y > 0$

29. $x + y \le 5$
$x \ge 0$
$y \ge 0$

31. $4x - 3y > 12$
$x \ge 0$
$y \le 0$

33. $0 \le x \le 3$
$0 \le y \le 3$

35. $x - y \le 4$
$x + 2y \le 4$

37. $x + y \ge 1$
$x - y \ge 1$
$x \ge 4$

39. $x + 2y < 6$
$y > 2x - 2$
$y \ge 2$

41. $y \le -3x + 3$
$y \ge -x - 1$
$y < x + 7$

43. no solution $y \ge 2x + 2$
$y < 2x - 3$
$x \ge 2$

45. Point $A = (66, 160)$; $5.3(66) - 160 \ge 180$, or $189.8 \ge 180$, is true; $4.1(66) - 160 \le 140$, or $110.6 \le 140$, is true.   **47.** no

49. Answers will vary.; $165x + 110y \le 330$ **53.** makes sense   **55.** makes sense   **57.** $y \ge -x + 3, x > 5$ **59.** $x < 0, y < 0$

61. $x + y \ge 25{,}000$
$35x + 50y \ge 1{,}025{,}000$

62. $-\dfrac{1}{4}$ **63.** $-\dfrac{11}{20}$ **64.** 4 or {4}   **65.** $17x^3$ **66.** $-2x^2$ **67.** $-7y^4$

Review Exercises

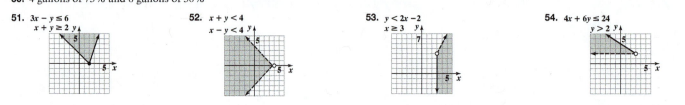

1. solution   **2.** not a solution   **3.** no; $(-1, 3)$ does not satisfy $2x + y = -5$.   **4.** $(4, -2)$ or $\{(4, -2)\}$ **5.** $(3, -2)$ or $\{(3, -2)\}$
6. $(2, 0)$ or $\{(2, 0)\}$ **7.** $(2, 1)$ or $\{(2, 1)\}$ **8.** $(4, -1)$ or $\{(4, -1)\}$ **9.** no solution or $\varnothing$ **10.** infinitely many solutions;
$\{(x, y)|2x - 4y = 8\}$ or $\{(x, y)|x - 2y = 4\}$ **11.** no solution or $\varnothing$ **12.** $(-2, -6)$ or $\{(-2, -6)\}$ **13.** $(2, 5)$ or $\{(2, 5)\}$
14. no solution or $\varnothing$ **15.** $(2, -1)$ or $\{(2, -1)\}$ **16.** $(5, 4)$ or $\{(5, 4)\}$ **17.** $(-2, -1)$ or $\{(-2, -1)\}$ **18.** $(1, -4)$ or $\{(1, -4)\}$
19. $(20, -21)$ or $\{(20, -21)\}$ **20.** infinitely many solutions; $\{(x, y)|4x + y = 5\}$ or $\{(x, y)|12x + 3y = 15\}$ **21.** no solution or $\varnothing$
22. $(4, 18)$ or $\{(4, 18)\}$ **23.** $\left(-1, -\dfrac{1}{2}\right)$ or $\left\{\left(-1, -\dfrac{1}{2}\right)\right\}$ **24. a.** $(20, 1000)$ or $\{(20, 1000)\}$; equilibrium quantity: 20,000; equilibrium price: $1000
b. $1000; 20,000; 20,000   **25.** $(2, 4)$ or $\{(2, 4)\}$ **26.** $(-1, -1)$ or $\{(-1, -1)\}$ **27.** $(2, -1)$ or $\{(2, -1)\}$ **28.** $(3, 2)$ or $\{(3, 2)\}$
29. $(2, 1)$ or $\{(2, 1)\}$ **30.** $(0, 0)$ or $\{(0, 0)\}$ **31.** $\left(\dfrac{17}{7}, -\dfrac{15}{7}\right)$ or $\left\{\left(\dfrac{17}{7}, -\dfrac{15}{7}\right)\right\}$ **32.** no solution or $\varnothing$
33. infinitely many solutions; $\{(x, y)|3x - 4y = -1\}$ or $\{(x, y)|-6x + 8y = 2\}$ **34.** $(4, -2)$ or $\{(4, -2)\}$ **35.** $(-8, -6)$ or $\{(-8, -6)\}$
36. $(-4, 1)$ or $\{(-4, 1)\}$ **37.** $\left(\dfrac{5}{2}, 3\right)$ or $\left\{\left(\dfrac{5}{2}, 3\right)\right\}$ **38.** $(3, 2)$ or $\{(3, 2)\}$ **39.** $\left(\dfrac{1}{2}, -2\right)$ or $\left\{\left(\dfrac{1}{2}, -2\right)\right\}$ **40.** no solution or $\varnothing$
41. $(3, 7)$ or $\{(3, 7)\}$ **42.** Klimt: $135 million; Picasso: $104 million   **43.** shrimp: 42 mg; scallops: 15 mg   **44.** 9 ft by 5 ft   **45.** 7 yd by 5 yd
46. room: $80; car: $60   **47.** 8 GB; $160   **48.** orchestra: 6 tickets; balcony: 3 tickets   **49.** $3000 at 8% and $7000 at 10%
50. 4 gallons of 75% and 6 gallons of 50%

51. $3x - y \le 6$
$x + y \ge 2$

52. $x + y < 4$
$x - y < 4$

53. $y < 2x - 2$
$x \ge 3$

54. $4x + 6y \le 24$
$y > 2$

55. $x \leq 3$
$y \geq -2$
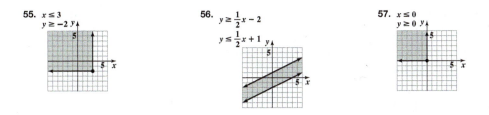

56. $y \geq \frac{1}{2}x - 2$
$y \leq \frac{1}{2}x + 1$

57. $x \leq 0$
$y \geq 0$

Chapter Test

1. solution **2.** not a solution **3.** (2, 4) or {(2, 4)} **4.** (2, 4) or {(2, 4)} **5.** (1, −3) or {(1, −3)} **6.** (2, −3) or {(2, −3)}
7. no solution or ∅ **8.** (−1, 4) or {(−1, 4)} **9.** (−4, 3) or {(−4, 3)} **10.** infinitely many solutions; {(x, y)|3x − 2y = 2} or
{(x, y)|−9x + 6y = −6} **11.** dogs: 220 million; humans: 5 million **12.** 12 yd by 5 yd **13.** 6 GB; $148
14. $4000 at 9% and $2000 at 6% **15.** 40 ml of 50% and 60 ml of 80%

16. $x - 3y > 6$
$2x + 4y \leq 8$

17. $y \geq 2x - 4$
$x < 2$

Cumulative Review Exercises

1. −36 **2.** 17x − 11 **3.** −6 or {−6} **4.** 20 or {20} **5.** $t = \dfrac{A - P}{Pr}$ **6.** (2, ∞) or {x|x > 2};

7. $x - 3y = 6$
(6, 0)
(0, −2)

8. $y = 4$

9. $y = -\frac{3}{5}x + 2$
(0, 2)

10. (0, −2) or {(0, −2)} **11.** $\left(\frac{3}{2}, -2\right)$ or $\left\{\left(\frac{3}{2}, -2\right)\right\}$ **12.** 1 **13.** y − 6 = −4(x + 1); y = −4x + 2 **14.** 10 ft

15. pen: $0.80; pad: $1.20 **16.** $-93, 0, \frac{7}{1}, \sqrt{100}$ **17.** The "yes" line has a positive slope; the percentage who answered "yes"

increased between 2010 and 2014. **18.** The "no" line has a negative slope; the percentage who answered "no" decreased between 2010 and 2014.
19. decreased; 2.75 **20.** 2012; 49%

CHAPTER 5

Section 5.1 Check Point Exercises

1. $5x^3 + 4x^2 - 8x - 20$ **2.** $5x^3 + 4x^2 - 8x - 20$ **3.** $7x^2 + 11x + 4$ **4.** $7x^3 + 3x^2 + 12x - 8$ **5.** $3y^3 - 10y^2 - 11y - 8$
6. $y = x^2 - 1$
(−3, 8) (3, 8)
(−2, 3) (2, 3)
(−1, 0) (1, 0)
(0, −1)

Concept and Vocabulary Check

1. whole **2.** standard **3.** monomial **4.** binomial **5.** trinomial **6.** n **7.** greatest **8.** like **9.** opposite

Exercise Set 5.1

1. binomial, 1 **3.** binomial, 3 **5.** monomial, 2 **7.** monomial, 0 **9.** trinomial, 2 **11.** trinomial, 4 **13.** binomial, 3
15. monomial, 23 **17.** $-8x + 13$ **19.** $12x^2 + 15x - 9$ **21.** $10x^2 - 12x$ **23.** $5x^2 - 3x + 13$ **25.** $4y^3 + 10y^2 + y - 2$
27. $3x^3 + 2x^2 - 9x + 7$ **29.** $-2y^3 + 4y^2 + 13y + 13$ **31.** $-3y^6 + 8y^4 + y^2$ **33.** $10x^3 + 1$ **35.** $-\frac{2}{5}x^4 + x^3 - \frac{1}{8}x^2$
37. $0.01x^5 + x^4 - 0.1x^3 + 0.3x + 0.33$ **39.** $11y^3 - 3y^2$ **41.** $-2x^2 - x + 1$ **43.** $-\frac{1}{4}x^4 + \frac{7}{15}x^3 - 0.3$ **45.** $-y^3 + 8y^2 - 3y - 14$

47. $-5x^3 - 6x^2 + x - 4$ **49.** $7x^4 - 2x^3 + 4x - 2$ **51.** $8x^2 + 7x - 5$ **53.** $9x^3 - 4.9x^2 + 11.1$ **55.** $-2x - 10$ **57.** $-5x^2 - 9x - 12$

59. $-5x^2 - x$ **61.** $-4x^2 - 4x - 6$ **63.** $-2y - 6$ **65.** $6y^3 + y^2 + 7y - 20$ **67.** $n^3 + 2$ **69.** $y^6 - y^3 - y^2 + y$

71. $26x^4 + 9x^2 + 6x$ **73.** $\frac{5}{7}x^3 - \frac{9}{20}x$ **75.** $4x + 6$ **77.** $10x^2 - 7$ **79.** $-4y^2 - 7y + 5$ **81.** $9x^3 + 11x^2 - 8$

83. $-y^3 + 8y^2 + y + 14$ **85.** $7x^4 - 2x^3 + 3x^2 - x + 2$ **87.** $0.05x^3 + 0.02x^2 + 1.02x$

89.   **91.**   **93.**

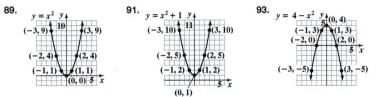

95. $x^2 + 12x$ **97.** $y^2 - 19y + 16$ **99.** $2x^3 + 3x^2 + 7x - 5$ **101.** $-10y^3 + 2y^2 + y + 3$

103. a. $S = 0.3x^3 - 2.8x^2 + 6.7x + 30$ **b.** 31; It's the same.   **c.** $(5, 31)$ **113.** does not make sense   **115.** does not make sense

117. false   **119.** false   **121.** $\frac{2}{3}t^3 - 2t^2 + 4t$ **123.** -10 **124.** 5.6   **125.** -4 or $\{-4\}$ **126.** 7

127. $3x^2 + 15x$ **128.** $x^2 + 5x + 6$

Section 5.2 Check Point Exercises

1. a. 2^6 or 64   **b.** x^{10} **c.** y^8 **d.** y^9 **2. a.** 3^{20} **b.** x^{90} **c.** $(-5)^{21}$ **3. a.** $16x^4$ **b.** $-64y^6$ **4. a.** $70x^3$ **b.** $-20x^9$

5. a. $3x^2 + 15x$ **b.** $30x^5 - 12x^3 + 18x^2$ **6. a.** $x^2 + 9x + 20$ **b.** $10x^2 - 29x - 21$ **7.** $5x^3 - 18x^2 + 7x + 6$

8. $6x^5 - 19x^4 + 22x^3 - 8x^2$

Concept and Vocabulary Check

1. b^{m+n}; add   **2.** b^{mn}; multiply   **3.** $a^n b^n$; factor   **4.** distributive; $x^2 + 5x + 7; 2x^2$ **5.** $4x; 7$; like

Exercise Set 5.2

1. x^{18} **3.** y^{12} **5.** x^{11} **7.** 7^{19} **9.** 6^{90} **11.** x^{45} **13.** $(-20)^9$ **15.** $8x^3$ **17.** $25x^2$ **19.** $16x^6$ **21.** $16y^{24}$ **23.** $-32x^{35}$

25. $14x^2$ **27.** $24x^3$ **29.** $-15y^7$ **31.** $\frac{1}{8}a^5$ **33.** $-48x^7$ **35.** $4x^2 + 12x$ **37.** $x^2 - 3x$ **39.** $2x^2 - 12x$ **41.** $-12y^2 - 20y$

43. $4x^3 + 8x^2$ **45.** $2y^4 + 6y^3$ **47.** $6y^4 - 8y^3 + 14y^2$ **49.** $6x^4 + 8x^3$ **51.** $-2x^3 - 10x^2 + 6x$ **53.** $12x^4 - 3x^3 + 15x^2$

55. $x^2 + 8x + 15$ **57.** $2x^2 + 9x + 4$ **59.** $x^2 - 2x - 15$ **61.** $x^2 - 2x - 99$ **63.** $2x^2 + 3x - 20$ **65.** $\frac{3}{16}x^2 + \frac{11}{4}x - 4$

67. $x^3 + 3x^2 + 5x + 3$ **69.** $y^3 - 6y^2 + 13y - 12$ **71.** $2a^3 - 9a^2 + 19a - 15$ **73.** $x^4 + 3x^3 + 5x^2 + 7x + 4$

75. $4x^4 - 4x^3 + 6x^2 - \frac{17}{2}x + 3$ **77.** $x^4 + x^3 + x^2 + 3x + 2$ **79.** $x^3 + 3x^2 - 37x + 24$ **81.** $2x^3 - 9x^2 + 27x - 27$

83. $2x^4 + 9x^3 + 6x^2 + 11x + 12$ **85.** $12z^4 - 14z^3 + 19z^2 - 22z + 8$ **87.** $21x^5 - 43x^4 + 38x^3 - 24x^2$

89. $4y^6 - 2y^5 - 6y^4 + 5y^3 - 5y^2 + 8y - 3$ **91.** $x^4 + 6x^3 - 11x^2 - 4x + 3$ **93.** $2x - 2$ **95.** $15x^5 + 42x^3 - 8x^2$

97. $2y^3$ **99.** $16y + 32$ **101.** $2x^2 + 7x - 15$ ft^2 **103. a.** $(2x + 1)(x + 2)$ **b.** $2x^2 + 5x + 2$ **c.** $(2x + 1)(x + 2) = 2x^2 + 5x + 2$

113. makes sense   **115.** makes sense   **117.** false   **119.** true   **121.** $8x + 16$ **123.** $-8x^4$ **124.** $(-\infty, -1)$ or $\{x \mid x < -1\}$

125. $3x - 2y = 6$ **126.** $-\frac{2}{3}$ **127. a.** $x^2 + 7x + 12$ **b.** $x^2 + 25x + 100$ **128. a.** $x^2 - 9$ **b.** $x^2 - 25$

129. a. $x^2 + 6x + 9$ **b.** $x^2 + 10x + 25$

Section 5.3 Check Point Exercises

1. $x^2 + 11x + 30$ **2.** $28x^2 - x - 15$ **3.** $6x^2 - 22x + 20$ **4. a.** $49y^2 - 64$ **b.** $16x^2 - 25$ **c.** $4a^6 - 9$ **5. a.** $x^2 + 20x + 100$

b. $25x^2 + 40x + 16$ **6. a.** $x^2 - 18x + 81$ **b.** $49x^2 - 42x + 9$

Concept and Vocabulary Check

1. $2x^2; 3x; 10x; 15$ **2.** $A^2 - B^2$; minus   **3.** $A^2 + 2AB + B^2$; squared; product of the terms; squared

4. $A^2 - 2AB + B^2$; minus; product of the terms; squared   **5.** true   **6.** false

Exercise Set 5.3

1. $x^2 + 10x + 24$ **3.** $y^2 - 4y - 21$ **5.** $2x^2 + 7x - 15$ **7.** $4y^2 - y - 3$ **9.** $10x^2 - 9x - 9$ **11.** $12y^2 - 43y + 35$

13. $-15x^2 - 32x + 7$ **15.** $6y^2 - 28y + 30$ **17.** $15x^4 - 47x^2 + 28$ **19.** $-6x^2 + 17x - 10$ **21.** $x^3 + 5x^2 + 3x + 15$

23. $8x^5 + 40x^3 + 3x^2 + 15$ **25.** $x^2 - 9$ **27.** $9x^2 - 4$ **29.** $9r^2 - 16$ **31.** $9 - r^2$ **33.** $25 - 49x^2$ **35.** $4x^2 - \frac{1}{4}$ **37.** $y^4 - 1$

39. $r^6 - 4$ **41.** $1 - y^8$ **43.** $x^{20} - 25$ **45.** $x^2 + 4x + 4$ **47.** $4x^2 + 20x + 25$ **49.** $x^2 - 6x + 9$ **51.** $9y^2 - 24y + 16$

53. $16x^4 - 8x^2 + 1$ **55.** $49 - 28x + 4x^2$ **57.** $4x^2 + 2x + \frac{1}{4}$ **59.** $16y^2 - 2y + \frac{1}{16}$ **61.** $x^{16} + 6x^8 + 9$ **63.** $x^3 - 1$

65. $x^2 - 2x + 1$ **67.** $9y^2 - 49$ **69.** $12x^4 + 3x^3 + 27x^2$ **71.** $70y^2 + 2y - 12$ **73.** $x^4 + 2x^2 + 1$ **75.** $x^4 + 3x^2 + 2$
77. $x^4 - 16$ **79.** $4 - 12x^5 + 9x^{10}$ **81.** $\dfrac{3}{16}x^4 + 7x^2 - 96$ **83.** $x^2 + 2x + 1$ **85.** $4x^2 - 9$ **87.** $6x + 22$
89. $16x^4 - 72x^2 + 81$ **91.** $16x^4 - 1$ **93.** $x^3 + 6x^2 + 12x + 8$ **95.** $x^2 + 6x + 9 - y^2$ **97.** $(x + 1)(x + 2)$ yd^2 **99.** 56 yd^2; (6, 56)
101. $(x^2 + 4x + 4)$ in.^{2} **109.** makes sense   **111.** makes sense, although answers may vary   **113.** true   **115.** false
117. $4x^3 - 36x^2 + 80x$ **119.** Change $x^2 + 1$ to $x^2 + 2x + 1$.   **121.** Graphs coincide.   **123.** $(2, -1)$ or $\{(2, -1)\}$
124. $(1, 1)$ or $\{(1, 1)\}$

125. $y \le \dfrac{1}{3}x$
   126. -72 **127.** $11xy$ **128.** $3x^2 + 11xy + 10y^2$

Section 5.4 Check Point Exercises

1. -9 **2.** polynomial degree: 9;

Term	Coefficient	Degree
$8x^4y^5$	8	9
$-7x^3y^2$	-7	5
$-x^2y$	-1	3
$-5x$	-5	1
11	11	0

3. $2x^2y + 2xy - 4$ **4.** $3x^3 + 2x^2y + 5xy^2 - 10$ **5.** $60x^5y^5$ **6.** $60x^5y^7 - 12x^3y^3 + 18xy^2$
7. a. $21x^2 - 25xy + 6y^2$ **b.** $4x^2 + 16xy + 16y^2$ **8. a.** $36x^2y^4 - 25x^2$ **b.** $x^3 - y^3$

Concept and Vocabulary Check

1. -18 **2.** 6   **3.** $a; n + m$ **4.** $5; 9; 9$ **5.** false   **6.** true

Exercise Set 5.4

1. 1   **3.** -47 **5.** -6 **7.** polynomial degree: 9;

Term	Coefficient	Degree
x^3y^2	1	5
$-5x^2y^7$	-5	9
$6y^2$	6	2
-3	-3	0

9. $7x^2y - 4xy$ **11.** $2x^2y + 13xy + 13$ **13.** $-11x^4y^2 - 11x^2y^2 + 2xy$ **15.** $-5x^3 + 8xy - 9y^2$ **17.** $x^4y^2 + 8x^3y + y - 6x$
19. $5x^3 + x^2y - xy^2 - 4y^3$ **21.** $-3x^2y^2 + xy^2 + 5y^2$ **23.** $8a^2b^4 + 3ab^2 + 8ab$ **25.** $-30x + 37y$ **27.** $40x^3y^2$ **29.** $-24x^5y^9$
31. $45x^2y + 18xy^2$ **33.** $50x^3y^2 - 15xy^3$ **35.** $28a^3b^5 + 8a^2b^3$ **37.** $-a^2b + ab^2 - b^3$ **39.** $7x^2 + 38xy + 15y^2$ **41.** $2x^2 + xy - 21y^2$
43. $15x^2y^2 + xy - 2$ **45.** $4x^2 + 12xy + 9y^2$ **47.** $x^2y^2 - 6xy + 9$ **49.** $x^4 + 2x^2y^2 + y^4$ **51.** $x^4 - 4x^2y^2 + 4y^4$ **53.** $9x^2 - y^2$
55. $a^2b^2 - 1$ **57.** $x^2 - y^4$ **59.** $9a^4b^2 - a^2$ **61.** $9x^2y^4 - 16y^2$ **63.** $a^3 - ab^2 + a^2b - b^3$ **65.** $x^3 + 4x^2y + 4xy^2 + y^3$
67. $x^3 - 4x^2y + 4xy^2 - y^3$ **69.** $x^2y^2 - a^2b^2$ **71.** $x^6y + x^4y + x^4 + 2x^2 + 1$ **73.** $x^4y^4 - 6x^2y^2 + 9$ **75.** $x^2 + 2xy + y^2 - 1$
77. $3x^2 + 8xy + 5y^2$ **79.** $2xy + y^2$ **81.** $x^{12}y^{12} - 2x^6y^6 + 1$ **83.** $x^4y^4 - 18x^2y^2 + 81$ **85.** $x^2 - y^2 - 2yz - z^2$
87. no; need 120 more board feet   **89.** 192 ft   **91.** 0 ft; The ball hits the ground.   **93.** 2.5 to 6 sec   **95.** (2, 192)
97. 2.5 sec; 196 ft   **101.** makes sense   **103.** does not make sense   **105.** false   **107.** true   **109.** $-x^2 + 18xy + 80y^2$
112. $W = \dfrac{2R - L}{3}$ **113.** 3.8   **114.** -1.6 or $\{-1.6\}$ **115.** 4   **116.** $\dfrac{x^6}{125}$ **117.** $\dfrac{32a^{15}}{b^{20}}$

Mid-Chapter Check Point Exercises

1. $-55x^4y^6$ **2.** $6x^2y^3$ **3.** $12x^2 - x - 35$ **4.** $-x + 12$ **5.** $2x^3 - 11x^2 + 17x - 5$ **6.** $x^2 - x - 4$ **7.** $64x^2 - 48x + 9$ **8.** $70x^9$
9. $x^4 - 4$ **10.** $x^4 + 4x^2 + 4$ **11.** $18a^2 - 11ab - 10b^2$ **12.** $70x^5 - 14x^3 + 21x^2$ **13.** $5a^2b^3 + 2ab - b^2$ **14.** $18y^2 - 50$

15. $2x^3 - x^2 + x$ **16.** $10x^2 - 5xy - 3y^2$ **17.** $-4x^5 + 7x^4 - 10x + 23$ **18.** $x^3 + 27y^3$ **19.** $10x^7 - 5x^4 + 8x^3 - 4$

20. $y^2 - 12yz + 36z^2$ **21.** $-21x^2 + 7$

22.

$y = 1 - x^2$

Section 5.5 Check Point Exercises

1. a. 5^8 **b.** x^7 **c.** y^{19} **2. a.** 1 **b.** 1 **c.** -1 **d.** 20 **e.** 1 **3. a.** $\dfrac{x^2}{25}$ **b.** $\dfrac{x^{12}}{8}$ **c.** $\dfrac{16a^{40}}{b^{12}}$ **4. a.** $-2x^8$ **b.** $\dfrac{1}{5}$

c. $3x^5y^3$ **5.** $-5x^7 + 2x^3 - 3x$ **6.** $5x^6 - \dfrac{7}{5}x + 2$ **7.** $3x^6y^4 - xy + 10$

Concept and Vocabulary Check

1. b^{m-n}; subtract   **2.** 1 **3.** $\dfrac{a^n}{b^n}$; numerator; denominator   **4.** divide; subtract   **5.** dividend; divisor; quotient

6. divisor; quotient; dividend   **7.** $20x^8 - 10x^4 + 6x^3; 2x^3$

Exercise Set 5.5

1. 3^{15} **3.** x^4 **5.** y^8 **7.** $5^3 \cdot 2^4$ **9.** $x^{75}y^{40}$ **11.** 1 **13.** 1 **15.** -1 **17.** 100 **19.** 1 **21.** 0 **23.** -2 **25.** $\dfrac{x^2}{9}$ **27.** $\dfrac{x^6}{64}$

29. $\dfrac{4x^6}{25}$ **31.** $-\dfrac{64}{27a^9}$ **33.** $-\dfrac{32a^{35}}{b^{20}}$ **35.** $\dfrac{x^8y^{12}}{16z^4}$ **37.** $3x^5$ **39.** $-2x^{20}$ **41.** $-\dfrac{1}{2}y^3$ **43.** $\dfrac{7}{5}y^{12}$ **45.** $6x^5y^4$ **47.** $-\dfrac{1}{2}x^{12}$ **49.** $\dfrac{9}{7}$

51. $-\dfrac{1}{10}x^8y^9z^4$ **53.** $5x^4 + x^3$ **55.** $2x^3 - x^2$ **57.** $y^6 - 9y + 1$ **59.** $-8x^2 + 5x$ **61.** $6x^3 + 2x^2 + 3x$ **63.** $3x^3 - 2x^2 + 10x$

65. $4x - 6$ **67.** $-6z^2 - 2z$ **69.** $4x^2 + 3x - 1$ **71.** $5x^4 - 3x^2 - x$ **73.** $-9x^3 + \dfrac{9}{2}x^2 - 10x + 5$ **75.** $4xy + 2x - 5y$

77. $-4x^5y^3 + 3xy + 2$ **79.** $4x^2 - x + 6$ **81.** $-xy^2$ **83.** $y + 5$ **85.** $3x^{12n} - 6x^{9n} + 2$

87. a. $\$4.22$ **b.** $\dfrac{1.07x^2 + 222x + 2722}{-0.5x^2 + 27x + 1009}$ **c.** $\$4.11$; underestimates by $\$0.11$ **d.** no; The divisor is not a monomial.

97. does not make sense   **99.** does not make sense   **101.** false   **103.** true   **105.** $\dfrac{9x^8 - 12x^6}{3x^3}$ **107.** 20.3 **108.** 0.875

109. $y = \dfrac{1}{3}x + 2$

110. quotient: 26; remainder: 0 **111.** quotient: 123; remainder: 6 **112.** quotient: 257; remainder: 1

Section 5.6 Check Point Exercises

1. $x + 5$ **2.** $4x - 3 - \dfrac{3}{2x + 3}$ **3.** $x^2 + x + 1$

Concept and Vocabulary Check

1. $10x^3 + 4x^2 + 0x + 9$ **2.** $8x^2; 2x; 4x; 10x$ **3.** $5x; 3x - 2; 15x^2 - 10x; 15x^2 - 22x$ **4.** $6x^2 - 10x; 6x^2 + 8x; 18x; -4; 18x - 4$

5. $14; x - 12; 14; x - 12 + \dfrac{14}{x - 5}$

Exercise Set 5.6

1. $x + 4$ **3.** $2x + 5$ **5.** $x - 2$ **7.** $2y + 1$ **9.** $x - 5 + \dfrac{14}{x + 2}$ **11.** $y + 3 + \dfrac{4}{y + 2}$ **13.** $x^2 - 5x + 2$ **15.** $6y - 1$

17. $2a + 3$ **19.** $y^2 - y + 2$ **21.** $3x + 5 - \dfrac{5}{2x - 5}$ **23.** $x^2 + 2x + 8 + \dfrac{13}{x - 2}$ **25.** $2y^2 + y + 1 + \dfrac{6}{2y + 3}$

27. $2y^2 - 3y + 2 + \dfrac{1}{3y + 2}$ **29.** $9x^2 + 3x + 1$ **31.** $y^3 - 9y^2 + 27y - 27$ **33.** $2y + 4 + \dfrac{4}{2y - 1}$ **35.** $y^3 + y^2 - y - 1 + \dfrac{4}{y - 1}$

37. $4x - 3 + \dfrac{-7x + 7}{x^2 + 2}$ **39.** $x^2 + ax + a^2$ **41.** $2x^2 - 3x + 2$ **43.** $x^2 + 2x + 3$ **45.** $x^2 + 2x + 3$ units   **47. a.** $\dfrac{30{,}000x^3 - 30{,}000}{x - 1}$

b. $30{,}000x^2 + 30{,}000x + 30{,}000$ **c.** $\$94{,}575$ **53.** makes sense   **55.** does not make sense   **57.** false   **59.** true   **61.** $2x^2 - 2x + 5$

63. Answers will vary; $x^6 - x^5 + x^4 - x^3 + x^2 - x + 1 - \dfrac{2}{x + 1}$ **65.** $x - 5$ should be $x + 5$.   **67.** $2x - 4$ should be $2x + 4$.

69. $2x - y \geq 4$
$x + y \leq -1$

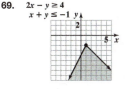

70. 1.2   **71.** -6 or $\{-6\}$ **72. a.** $2; -2$ **b.** $\dfrac{1}{7^2} = 7^{-2}$ **73.** $16x^2$ **74.** x^9

Section 5.7 Check Point Exercises

1. a. $\dfrac{1}{6^2} = \dfrac{1}{36}$ **b.** $\dfrac{1}{5^3} = \dfrac{1}{125}$ **c.** $\dfrac{1}{(-3)^4} = \dfrac{1}{81}$ **d.** $-\dfrac{1}{3^4} = -\dfrac{1}{81}$ **e.** $\dfrac{1}{8^1} = \dfrac{1}{8}$ **2. a.** $\dfrac{7^2}{2^3} = \dfrac{49}{8}$ **b.** $\dfrac{5^2}{4^2} = \dfrac{25}{16}$ **c.** $\dfrac{y^2}{7}$ **d.** $\dfrac{y^8}{x^1} = \dfrac{y^8}{x}$

3. $\dfrac{1}{x^{10}}$ **4. a.** $\dfrac{1}{x^8}$ **b.** $\dfrac{15}{x^6}$ **c.** $-\dfrac{2}{y^6}$ **5.** $\dfrac{36}{x^3}$ **6.** $\dfrac{1}{x^{20}}$ **7. a.** 7,400,000,000   **b.** 0.000003017   **8. a.** 7.41×10^9 **b.** 9.2×10^{-8}

9. a. 6×10^{10} **b.** 2.1×10^{11} **c.** 6.4×10^{-5} **10.** \$8300

Concept and Vocabulary Check

1. $\dfrac{1}{b^n}$ **2.** false   **3.** true   **4.** b^n **5.** true   **6.** false   **7.** a number greater than or equal to 1 and less than 10; integer

8. true   **9.** false

Exercise Set 5.7

1. $\dfrac{1}{8^2} = \dfrac{1}{64}$ **3.** $\dfrac{1}{5^3} = \dfrac{1}{125}$ **5.** $\dfrac{1}{(-6)^2} = \dfrac{1}{36}$ **7.** $-\dfrac{1}{6^2} = -\dfrac{1}{36}$ **9.** $\dfrac{1}{4^1} = \dfrac{1}{4}$ **11.** $\dfrac{1}{2^1} + \dfrac{1}{3^1} = \dfrac{1}{2} + \dfrac{1}{3} = \dfrac{5}{6}$ **13.** $3^2 = 9$

15. $(-3)^2 = 9$ **17.** $\dfrac{8^2}{2^3} = 8$ **19.** $\dfrac{4^2}{1^2} = 16$ **21.** $\dfrac{5^3}{3^3} = \dfrac{125}{27}$ **23.** $\dfrac{x^5}{6}$ **25.** $\dfrac{y^1}{x^8} = \dfrac{y}{x^8}$ **27.** $3 \cdot (-5)^3 = -375$ **29.** $\dfrac{1}{x^5}$ **31.** $\dfrac{8}{x^3}$

33. $\dfrac{1}{x^6}$ **35.** $\dfrac{1}{y^{99}}$ **37.** $\dfrac{3}{z^5}$ **39.** $-\dfrac{4}{x^4}$ **41.** $-\dfrac{1}{3a^3}$ **43.** $\dfrac{7}{5w^8}$ **45.** $\dfrac{1}{x^5}$ **47.** $\dfrac{1}{y^{11}}$ **49.** $\dfrac{16}{x^2}$ **51.** $216y^{17}$ **53.** $\dfrac{1}{x^6}$ **55.** $\dfrac{1}{16x^{12}}$

57. $\dfrac{x^2}{9}$ **59.** $-\dfrac{y^3}{8}$ **61.** $\dfrac{2x^6}{5}$ **63.** x^8 **65.** $16y^6$ **67.** $\dfrac{1}{y^2}$ **69.** $\dfrac{1}{y^{50}}$ **71.** $\dfrac{1}{a^{12}b^{15}}$ **73.** $\dfrac{a^8}{b^{24}}$ **75.** $\dfrac{4}{x^4}$ **77.** $\dfrac{y^9}{x^6}$ **79.** 870

81. 923,000   **83.** 3.4   **85.** 0.79   **87.** 0.0215   **89.** 0.000786   **91.** 3.24×10^4 **93.** 2.2×10^8 **95.** 7.13×10^2

97. 6.751×10^3 **99.** 2.7×10^{-3} **101.** 2.02×10^{-5} **103.** 5×10^{-3} **105.** 3.14159×10^0 **107.** 6×10^5 **109.** 1.6×10^9

111. 3×10^4 **113.** 3×10^6 **115.** 3×10^{-6} **117.** 9×10^4 **119.** 2.5×10^6 **121.** 1.25×10^8 **123.** 8.1×10^{-7}

125. 2.5×10^{-7} **127.** 1   **129.** $\dfrac{y}{16x^8z^6}$ **131.** $\dfrac{1}{x^{12}y^{16}z^{20}}$ **133.** $\dfrac{x^{18}y^6}{4}$ **135.** 2.5×10^{-3} **137.** 8×10^{-5} **139. a.** 1.09×10^{12}

b. 3.13×10^8 **c.** \$3.48 \times 10^3$; \$3480   **141.** 34,000 years   **143.** 1.25 sec   **153.** makes sense   **155.** makes sense   **157.** true

159. false   **161.** false   **163.** true   **169.** $(-\infty, 2)$ or $\{x \mid x < 2\}$ **170.** 5   **171.** $0, \sqrt{16}$ **172.** $16x^5 - 12x^4 + 4x^3$

173. $27x^2y^3 - 9xy^2 + 81xy$ **174.** $x^3 + 3x^2 + 5x + 15$

Review Exercises

1. binomial, 4   **2.** trinomial, 2   **3.** monomial, 1   **4.** $8x^3 + 10x^2 - 20x - 4$ **5.** $13y^3 - 8y^2 + 7y - 5$ **6.** $11y^2 - 4y - 4$
7. $8x^4 - 5x^3 + 6$ **8.** $-14x^4 - 13x^2 + 16x$ **9.** $7y^4 - 5y^3 + 3y^2 - y - 4$ **10.** $3x^2 - 7x + 9$ **11.** $10x^3 - 9x^2 + 2x + 11$

12.   **13.**

$y = x^2 + 3$
$(-2, 7)$ $(2, 7)$
$(-1, 4)$ $(1, 4)$
$(0, 3)$

$(0, 1)$
$(-1, 0)$ $(1, 0)$
$(-2, -3)$ $(2, -3)$
$y = 1 - x^2$

14. x^{23} **15.** y^{14} **16.** x^{100} **17.** $100y^2$ **18.** $-64x^{30}$ **19.** $50x^4$ **20.** $-36y^{11}$ **21.** $30x^{12}$ **22.** $21x^3 + 63x$ **23.** $20x^5 - 55x^4$
24. $-21y^4 + 9y^3 - 18y^2$ **25.** $16y^8 - 20y^7 + 2y^5$ **26.** $x^3 - 2x^2 - 13x + 6$ **27.** $12y^3 + y^2 - 21y + 10$ **28.** $3y^3 - 17y^2 + 41y - 35$
29. $8x^4 + 8x^3 - 18x^2 - 20x - 3$ **30.** $x^2 + 8x + 12$ **31.** $6y^2 - 7y - 5$ **32.** $4x^4 - 14x^2 + 6$ **33.** $25x^2 - 16$ **34.** $49 - 4y^2$
35. $y^4 - 1$ **36.** $x^2 + 6x + 9$ **37.** $9y^2 + 24y + 16$ **38.** $y^2 - 2y + 1$ **39.** $25y^2 - 20y + 4$ **40.** $x^4 + 8x^2 + 16$ **41.** $x^4 - 16$
42. $x^4 - x^2 - 20$ **43.** $x^2 + 7x + 12$ **44.** $x^2 + 50x + 600 \text{ yd}^2$ **45.** 28
46. polynomial degree: 5;

Term	Coefficient	Degree
$4x^2y$	4	3
$9x^3y^2$	9	5
$-17x^4$	-17	4
-12	-12	0

47. $-x^2 - 17xy + 5y^2$ **48.** $2x^3y^2 + x^2y - 6x^2 - 4$ **49.** $-35x^6y^9$ **50.** $15a^3b^5 - 20a^2b^3$ **51.** $3x^2 + 16xy - 35y^2$

52. $36x^2y^2 - 31xy + 3$ **53.** $9x^2 + 30xy + 25y^2$ **54.** $x^2y^2 - 14xy + 49$ **55.** $49x^2 - 16y^2$ **56.** $a^3 - b^3$ **57.** 6^{30} **58.** x^{15}

59. 1 **60.** -1 **61.** 400 **62.** $\dfrac{x^{12}}{8}$ **63.** $\dfrac{81}{16y^{24}}$ **64.** $-5y^6$ **65.** $8x^7y^3$ **66.** $3x^3 - 2x + 6$ **67.** $-6x^5 + 5x^4 + 8x^2$

68. $9x^2y - 3x - 6y$ **69.** $2x + 7$ **70.** $x^2 - 3x + 5$ **71.** $x^2 + 5x + 2 + \dfrac{7}{x - 7}$ **72.** $y^2 + 3y + 9$ **73.** $\dfrac{1}{7^2} = \dfrac{1}{49}$

74. $\dfrac{1}{(-4)^3} = -\dfrac{1}{64}$ **75.** $\dfrac{1}{2^1} + \dfrac{1}{4^1} = \dfrac{1}{2} + \dfrac{1}{4} = \dfrac{3}{4}$ **76.** $5^2 = 25$ **77.** $\dfrac{5^3}{2^3} = \dfrac{125}{8}$ **78.** $\dfrac{1}{x^6}$ **79.** $\dfrac{6}{y^2}$ **80.** $\dfrac{30}{x^5}$ **81.** x^8 **82.** $81y^2$

83. $\dfrac{1}{y^{19}}$ **84.** $\dfrac{x^3}{8}$ **85.** $\dfrac{1}{x^6}$ **86.** y^{20} **87.** 23,000 **88.** 0.00176 **89.** 0.9 **90.** 7.39×10^7 **91.** 6.2×10^{-4} **92.** 3.8×10^{-1}

93. 3.8×10^0 **94.** 9×10^3 **95.** 5×10^4 **96.** 1.6×10^{-3} **97.** 3.4×10^6 **98.** 6.8×10^4 **99.** 50 times

Chapter Test

1. trinomial, 2 **2.** $13x^3 + x^2 - x - 24$ **3.** $5x^3 + 2x^2 + 2x - 9$ **4.**

5. $-35x^{11}$ **6.** $48x^5 - 30x^3 - 12x^2$ **7.** $3x^3 - 10x^2 - 17x - 6$ **8.** $6y^2 - 13y - 63$ **9.** $49x^2 - 25$ **10.** $x^4 + 6x^2 + 9$

11. $25x^2 - 30x + 9$ **12.** 30 **13.** $2x^2y^3 + 3xy + 12y^2$ **14.** $12a^2 - 13ab - 35b^2$ **15.** $4x^2 + 12xy + 9y^2$ **16.** $-5x^{12}$

17. $3x^3 - 2x^2 + 5x$ **18.** $x^2 - 2x + 3 + \dfrac{1}{2x + 1}$ **19.** $\dfrac{1}{10^2} = \dfrac{1}{100}$ **20.** $4^3 = 64$ **21.** $-27x^6$ **22.** $\dfrac{4}{x^5}$ **23.** $-\dfrac{21}{x^6}$ **24.** $16y^4$

25. $\dfrac{x^8}{25}$ **26.** $\dfrac{1}{x^{15}}$ **27.** 0.00037 **28.** 7.6×10^6 **29.** 1.23×10^{-2} **30.** 2.1×10^8 **31.** $x^2 + 10x + 16$

Cumulative Review Exercises

1. $\dfrac{35}{9}$ **2.** -128 **3.** 15,050 ft **4.** $-\dfrac{2}{3}$ or $\left\{-\dfrac{2}{3}\right\}$ **5.** $-\dfrac{5}{3}$ or $\left\{-\dfrac{5}{3}\right\}$ **6.** 4 ft by 10 ft

7. $[6, \infty)$ or $\{x \mid x \geq 6\}$;

 8. $3400 at 12\% and $2600 at 14\%

9. 15 liters of 70% and 5 liters of 30% **10.**

 11.

12. $-\dfrac{6}{5}$; falling **13.** $y + 1 = -2(x - 3)$; $y = -2x + 5$ **14.** $(0, 5)$ or $\{(0, 5)\}$ **15.** $(3, -4)$ or $\{(3, -4)\}$ **16.** 15 crossings; $90

17.

 18. $3x^5 - 6x^3 + 9x + 2$ **19.** $x^2 + 2x + 3$ **20.** $\dfrac{81}{x^2}$

CHAPTER 6

Section 6.1 Check Point Exercises

1. a. $3x^2$ **b.** $4x^2$ **c.** x^2y **2.** $6(x^2 + 3)$ **3.** $5x^2(5 + 7x)$ **4.** $3x^3(5x^2 + 4x - 9)$ **5.** $2xy(4x^2y - 7x + 1)$
6. $-4ab^2(4a^3b^3 - 6a^2b^2 + 5)$ **7. a.** $(x + 1)(x^2 + 7)$ **b.** $(y + 4)(x - 7)$ **8.** $(x + 5)(x^2 + 2)$ **9.** $(y + 3)(x - 5)$

Concept and Vocabulary Check

1. factoring **2.** greatest common factor; smallest/least **3.** false **4.** false

Exercise Set 6.1

1. 4 **3.** $4x$ **5.** $2x^3$ **7.** $3y$ **9.** xy **11.** $4x^4y^3$ **13.** $8(x + 1)$ **15.** $4(y - 1)$ **17.** $5(x + 6)$ **19.** $6(5x - 2)$ **21.** $x(x + 5)$
23. $6(3y^2 + 2)$ **25.** $7x^2(2x + 3)$ **27.** $y(13y - 25)$ **29.** $9y^4(1 + 3y^2)$ **31.** $4x^2(2 - x^2)$ **33.** $4(3y^2 + 4y - 2)$ **35.** $3x^2(3x^2 + 6x + 2)$

37. $50y^2(2y^3 - y + 2)$ **39.** $5x(2 - 4x + x^2)$ **41.** cannot be factored **43.** $3xy(2x^2y + 3)$ **45.** $10xy(3xy^2 - y + 2)$
47. $8x^2y(4xy - 3x - 2)$ **49.** $-6(2x^2 - 3)$ **51.** $-8x^2(x^2 - 4x - 2)$ **53.** $-2ab(2a^2b - 3)$ **55.** $-6x^2y(2xy + 3x - 4)$ **57.** $(x + 5)(x + 3)$
59. $(x + 2)(x - 4)$ **61.** $(y + 6)(x - 7)$ **63.** $(x + y)(3x - 1)$ **65.** $(3x + 1)(4x + 1)$ **67.** $(5x + 4)(7x^2 + 1)$ **69.** $(x + 2)(x + 4)$
71. $(x - 5)(x + 3)$ **73.** $(x^2 + 5)(x - 2)$ **75.** $(x^2 + 2)(x - 1)$ **77.** $(y + 5)(x + 9)$ **79.** $(y - 1)(x + 5)$ **81.** $(x - 2y)(3x + 5y)$
83. $(3x - 2)(x^2 - 2)$ **85.** $(x - a)(x - b)$ **87.** $6x^2yz(4xy^2z^2 + 5y + 3z)$ **89.** $(x^3 - 4)(1 + 3y)$
91. $2x^2(x + 1)(2x^3 - 3x - 4)$ **93.** $(x - 1)(3x^4 + x^2 + 5)$ **95.** $36x^2 - 4\pi x^2; 4x^2(9 - \pi)$ **97. a.** 48 ft **b.** $16x(4 - x)$
c. 48 ft; yes; no; Answers will vary. **99.** $x^3 - 2$ **107.** makes sense **109.** does not make sense **111.** false
113. false **115.** $4(x + 150)$ **119.** $(x - 2)(x - 5)$ should be $(x - 2)(x + 5)$ **121.** $x^2 + 17x + 70$ **122.** $(-3, -2)$ or $\{(-3, -2)\}$
123. $y - 2 = 1(x + 7)$ or $y - 5 = 1(x + 4); y = x + 9$ **124.** 2 and 4 **125.** -3 and -2 **126.** -5 and 7

Section 6.2 Check Point Exercises

1. $(x + 2)(x + 3)$ **2.** $(x - 2)(x - 4)$ **3.** $(x + 5)(x - 2)$ **4.** $(y - 9)(y + 3)$ **5.** cannot factor over the integers; prime
6. $(x - 3y)(x - y)$ **7.** $2x(x - 4)(x + 7)$ **8.** $-2(y - 2)(y + 7)$

Concept and Vocabulary Check

1. $20; -12$ **2.** completely **3.** $+10$ **4.** -6 **5.** $+5$ **6.** -7 **7.** $-2y$

Exercise Set 6.2

1. $(x + 6)(x + 1)$ **3.** $(x + 2)(x + 5)$ **5.** $(x + 1)(x + 10)$ **7.** $(x - 4)(x - 3)$ **9.** $(x - 6)(x - 6)$ **11.** $(y - 3)(y - 5)$
13. $(x + 5)(x - 2)$ **15.** $(y + 13)(y - 3)$ **17.** $(x - 5)(x + 3)$ **19.** $(x - 4)(x + 2)$ **21.** prime **23.** $(y - 4)(y - 12)$ **25.** prime
27. $(w - 32)(w + 2)$ **29.** $(y - 5)(y - 13)$ **31.** $(r + 3)(r + 9)$ **33.** prime **35.** $(x + 6y)(x + y)$ **37.** $(x - 3y)(x - 5y)$
39. $(x - 6y)(x + 3y)$ **41.** $(a - 15b)(a - 3b)$ **43.** $3(x + 2)(x + 3)$ **45.** $4(y - 2)(y + 1)$ **47.** $10(x - 10)(x + 6)$
49. $3(x - 2)(x - 9)$ **51.** $2r(r + 2)(r + 1)$ **53.** $4x(x + 6)(x - 3)$ **55.** $2r(r + 8)(r - 4)$ **57.** $y^2(y + 10)(y - 8)$
59. $x^2(x - 5)(x + 2)$ **61.** $2w^2(w - 16)(w + 3)$ **63.** $15x(y - 1)(y + 4)$ **65.** $x^3(x - y)(x + 4y)$ **67.** $-16(t - 5)(t + 1)$
69. $-5(x - 9)(x - 1)$ **71.** $-(x + 8)(x - 5)$ **73.** $-2x(x + 4)(x - 1)$ **75.** $2x^2(y - 15z)(y - z)$ **77.** $(a + b)(x + 5)(x - 4)$
79. $(x + 0.3)(x + 0.2)$ **81.** $\left(x - \dfrac{1}{5}\right)\left(x - \dfrac{1}{5}\right)$ **83. a.** $-16(t - 2)(t + 1)$ **b.** 0; yes; After 2 seconds, you hit the water.

89. does not make sense **91.** does not make sense **93.** false **95.** true **97.** 8, 16 **101.** $x(x + 1)(x + 2)$; the product
of three consecutive integers **103.** correctly factored **105.** $(x + 1)(x - 1)$ should be $(x - 1)(x - 1)$. **107.** 13 or $\{13\}$
108.  **109.** **110.** $2x^2 - x - 6$ **111.** $9x^2 + 15x + 4$ **112.** $(4x - 1)(2x - 5)$

Section 6.3 Check Point Exercises

1. $(5x - 4)(x - 2)$ **2.** $(3x - 1)(2x + 7)$ **3.** $(3x - y)(x - 4y)$ **4.** $(3x + 5)(x - 2)$ **5.** $(2x - 1)(4x - 3)$ **6.** $y^2(5y + 3)(y + 2)$

Concept and Vocabulary Check

1. greatest common factor **2.** -3 **3.** -4 **4.** $2x - 3$ **5.** $3x + 4$ **6.** $x - 2y$

Exercise Set 6.3

1. $(2x + 3)(x + 1)$ **3.** $(3x + 1)(x + 4)$ **5.** $(2x + 3)(x + 4)$ **7.** $(5y - 1)(y - 3)$ **9.** $(3y + 4)(y - 1)$ **11.** $(3x - 2)(x + 5)$
13. $(3x - 1)(x - 7)$ **15.** $(5y - 1)(y - 3)$ **17.** $(3x - 2)(x - 5)$ **19.** $(3w - 4)(2w - 1)$ **21.** $(8x + 1)(x + 4)$
23. $(5x - 2)(x + 7)$ **25.** $(7y - 3)(2y + 3)$ **27.** prime **29.** $(5z - 3)(5z - 3)$ **31.** $(3y + 1)(5y - 2)$ **33.** prime
35. $(5y - 1)(2y + 9)$ **37.** $(4x + 1)(2x - 1)$ **39.** $(3y - 1)(3y - 2)$ **41.** $(5x + 8)(4x - 1)$ **43.** $(2x + y)(x + y)$
45. $(3x + 2y)(x + y)$ **47.** $(2x - 3y)(x - 3y)$ **49.** $(2x - 3y)(3x + 2y)$ **51.** $(3x - 2y)(5x + 7y)$ **53.** $(2a + 5b)(a + b)$
55. $(3a - 2b)(5a + 3b)$ **57.** $(3x - 4y)(4x - 3y)$ **59.** $2(2x + 3)(x + 5)$ **61.** $3(3x + 4)(x - 2)$ **63.** $2(2y - 5)(y + 3)$
65. $3(3y - 4)(y + 5)$ **67.** $x(3x + 1)(x + 1)$ **69.** $x(2x - 5)(x + 1)$ **71.** $3y(3y - 1)(y - 4)$ **73.** $5z(6z + 1)(2z + 1)$
75. $3x^2(5x - 3)(x - 2)$ **77.** $x^3(2x - 3)(5x - 1)$ **79.** $3(2x + 3y)(x - 2y)$ **81.** $2(2x - y)(3x + 4y)$ **83.** $2y(4x - 7)(x + 6)$
85. $2b(2a - 7b)(3a - b)$ **87.** $-4y^4(8x + 3)(x - 1)$ **89.** $10(y + 1)(x + 1)(3x - 2)$ **91. a.** $(2x + 1)(x - 3)$ **b.** $(2y + 3)(y - 2)$
93. $(x - 2)(3x - 2)(x - 1)$ **95. a.** $x^2 + 3x + 2$ **b.** $(x + 2)(x + 1)$ **c.** $x^2 + 3x + 2 = (x + 2)(x + 1)$ **101.** makes sense
103. does not make sense **105.** true **107.** false **109.** $5, 7, -5, -7$ **111.** $(3x^5 + 5)(x^5 - 3)$ **113.** $(25, -5)$ or $\{(25, -5)\}$
114. 8.6×10^{-4} **115.** -1 or $\{-1\}$ **116.** $81x^2 - 100$ **117.** $16x^2 + 40xy + 25y^2$ **118.** $x^3 + 8$

Mid-Chapter Check Point Exercises

1. $x^4(x + 1)$ **2.** $(x + 9)(x - 2)$ **3.** $x^2y(y^2 - y + 1)$ **4.** prime **5.** $(7x - 1)(x - 3)$ **6.** $(x^2 + 3)(x + 5)$ **7.** $x(2x - 1)(x - 5)$
8. $(x - 4)(y - 7)$ **9.** $(x - 15y)(x - 2y)$ **10.** $(5x + 2)(5x - 7)$ **11.** $2(8x - 3)(x - 4)$ **12.** $(3x + 7y)(x + y)$ **13.** $-2x(3x + 5)(x - 3)$

Section 6.4 Check Point Exercises

1. a. $(x + 9)(x - 9)$ **b.** $(6x + 5)(6x - 5)$ **2. a.** $(5 + 2x^5)(5 - 2x^5)$ **b.** $(10x + 3y)(10x - 3y)$ **3. a.** $2x(3x + 1)(3x - 1)$
b. $18(2 + x)(2 - x)$ or $-18(x + 2)(x - 2)$ **4.** $(9x^2 + 4)(3x + 2)(3x - 2)$ **5. a.** $(x + 7)^2$ **b.** $(x - 3)^2$ **c.** $(4x - 7)^2$ **6.** $(2x + 3y)^2$
7. $(x + 3)(x^2 - 3x + 9)$ **8.** $(1 - y)(1 + y + y^2)$ **9.** $(5x + 2)(25x^2 - 10x + 4)$

Concept and Vocabulary Check

1. $(A + B)(A - B)$ **2.** $(A + B)^2$ **3.** $(A - B)^2$ **4.** $(A + B)(A^2 - AB + B^2)$ **5.** $(A - B)(A^2 + AB + B^2)$
6. $6x$; $6x$ **7.** -6 **8.** $4x$ **9.** $+2$ **10.** -3; $+9$ **11.** false **12.** true **13.** false **14.** true **15.** false

Exercise Set 6.4

1. $(x + 5)(x - 5)$ **3.** $(y + 1)(y - 1)$ **5.** $(2x + 3)(2x - 3)$ **7.** $(5 + x)(5 - x)$ **9.** $(1 + 7x)(1 - 7x)$ **11.** $(3 + 5y)(3 - 5y)$
13. $(x^2 + 3)(x^2 - 3)$ **15.** $(7y^2 + 4)(7y^2 - 4)$ **17.** $(x^5 + 3)(x^5 - 3)$ **19.** $(5x + 4y)(5x - 4y)$ **21.** $(x^2 + y^5)(x^2 - y^5)$
23. $(x^2 + 4)(x + 2)(x - 2)$ **25.** $(4x^2 + 9)(2x + 3)(2x - 3)$ **27.** $2(x + 3)(x - 3)$ **29.** $2x(x + 6)(x - 6)$ **31.** prime **33.** $3x(x^2 + 9)$
35. $2(3 + y)(3 - y)$ **37.** $3y(y + 4)(y - 4)$ **39.** $2x(3x + 1)(3x - 1)$ **41.** $-3(x + 5)(x - 5)$ **43.** $-5y(y + 2)(y - 2)$
45. $(x + 1)^2$ **47.** $(x - 7)^2$ **49.** $(x - 1)^2$ **51.** $(x + 11)^2$ **53.** $(2x + 1)^2$ **55.** $(5y - 1)^2$ **57.** prime **59.** $(x + 7y)^2$ **61.** $(x - 6y)^2$
63. prime **65.** $(4x - 5y)^2$ **67.** $3(2x - 1)^2$ **69.** $x(3x + 1)^2$ **71.** $2(y - 1)^2$ **73.** $2y(y + 7)^2$ **75.** $-6(x - 2)^2$ **77.** $-4y(2y + 1)^2$
79. $(x + 1)(x^2 - x + 1)$ **81.** $(x - 3)(x^2 + 3x + 9)$ **83.** $(2y - 1)(4y^2 + 2y + 1)$ **85.** $(3x + 2)(9x^2 - 6x + 4)$
87. $(xy - 4)(x^2y^2 + 4xy + 16)$ **89.** $y(3y + 2)(9y^2 - 6y + 4)$ **91.** $2(3 - 2y)(9 + 6y + 4y^2)$ **93.** $(4x + 3y)(16x^2 - 12xy + 9y^2)$

95. $(5x - 4y)(25x^2 + 20xy + 16y^2)$ **97.** $\left(5x + \dfrac{2}{7}\right)\left(5x - \dfrac{2}{7}\right)$ **99.** $y\left(y - \dfrac{1}{10}\right)\left(y^2 + \dfrac{y}{10} + \dfrac{1}{100}\right)$ **101.** $x(0.5 + x)(0.5 - x)$

103. $(x + 6)(x - 4)$ **105.** $(x - 3)(x + 1)^2$ **107.** $x^2 - 25 = (x + 5)(x - 5)$ **109.** $x^2 - 16 = (x + 4)(x - 4)$
115. does not make sense **117.** does not make sense **119.** false **121.** false **123.** $a - b = 0$ and division by 0 is not permitted.
125. $(x^n + 5y^n)(x^n - 5y^n)$ **127.** $[(x + 3) - 1]^2$ or $(x + 2)^2$ **129.** 1 **131.** correctly factored
133. $(x - 1)(x^2 - x + 1)$ should be $(x - 1)(x^2 + x + 1)$. **134.** $80x^9y^{14}$ **135.** $-4x^2 + 3$
136. $2x + 5$ **137.** $3x(x + 5)(x - 5)$ **138.** $2(x - 5)^2$ **139.** $(x - 2)(x + 1)(x - 1)$

Section 6.5 Check Point Exercises

1. $5x^2(x + 3)(x - 3)$ **2.** $4(x - 6)(x + 2)$ **3.** $4x(x^2 + 4)(x + 2)(x - 2)$ **4.** $(x - 4)(x + 3)(x - 3)$ **5.** $3x(x - 5)^2$
6. $2x^2(x + 3)(x^2 - 3x + 9)$ **7.** $3y(x^2 + 4y^2)(x + 2y)(x - 2y)$ **8.** $3x(2x + 3y)^2$

Concept and Vocabulary Check

1. b **2.** e **3.** h **4.** c **5.** d **6.** f **7.** a **8.** g

Exercise Set 6.5

1. $-7x(x - 5)$ **3.** $(5x + 7)(5x - 7)$ **5.** $(3x - 1)(9x^2 + 3x + 1)$ **7.** $(5 + x)(x + y)$ **9.** $(7x - 1)(2x - 1)$ **11.** $(x - 1)^2$
13. $(3xy + 2)(9x^2y^2 - 6xy + 4)$ **15.** $(3x + 5)(2x - 3)$ **17.** $5x(x + 2)(x - 2)$ **19.** $7x(x^2 + 1)$ **21.** $5(x - 3)(x + 2)$
23. $2(x^2 + 9)(x + 3)(x - 3)$ **25.** $(x + 2)(x + 3)(x - 3)$ **27.** $3x(x - 4)^2$ **29.** $2x^2(x + 1)(x^2 - x + 1)$ **31.** $2x(3x + 4)$
33. $-2(y - 8)(y + 7)$ **35.** $7y^2(y + 1)^2$ **37.** prime **39.** $2(4y + 1)(2y - 1)$ **41.** $r(r - 25)$ **43.** $(2w + 5)(2w - 1)$
45. $x(x + 2)(x - 2)$ **47.** prime **49.** $(9y + 4)(y + 1)$ **51.** $(y + 2)(y + 2)(y - 2)$ **53.** $(4y + 3)^2$ **55.** $-4y(y - 5)(y - 2)$
57. $y(y^2 + 9)(y + 3)(y - 3)$ **59.** $5a^2(2a + 3)(2a - 3)$ **61.** $3x^2(3x^2 + 6x + 2)$ **63.** $(4y - 1)(3y - 2)$ **65.** $(3y + 8)(3y - 8)$ **67.** prime
69. $(2y + 3)(y + 5)(y - 5)$ **71.** $2r(r + 17)(r - 2)$ **73.** $2x^3(2x + 1)(2x - 1)$ **75.** $3(x^2 + 81)$ **77.** $x(x + 2)(x^2 - 2x + 4)$
79. $2y^2(y - 1)(y^2 + y + 1)$ **81.** $2x(3x + 4y)$ **83.** $(y - 7)(x + 3)$ **85.** $(x - 4y)(x + y)$ **87.** $12a^2(6ab^2 + 1 - 2a^2b^2)$
89. $3(a + 6b)(a + 3b)$ **91.** $3x^2y(4x + 1)(4x - 1)$ **93.** $b(3a + 2)(2a - 1)$ **95.** $7xy(x^2 + y^2)(x + y)(x - y)$ **97.** $2xy(5x - 2y)(x - y)$
99. $2b(x + 11)^2$ **101.** $(5a + 7b)(3a - 2b)$ **103.** $-2xy(9x - 2y)(2x - 3y)$ **105.** $(y - x)(a + b)(a - b)$ **107.** $ax(3x + 7)(3x - 2)$
109. $2x^2(x^2 + 3xy + y^2)$ **111.** $y(9x^2 + y^2)(3x + y)(3x - y)$ **113.** $(x + 1)(5x - 6)(2x + 1)$ **115.** $(x^2 + 6)(6x^2 - 1)$
117. $(x - 7 + 2a)(x - 7 - 2a)$ **119.** $(x + 4 + 5a)(x + 4 - 5a)$ **121.** $y(y^2 + 1)(y^4 - y^2 + 1)$ **123.** $16(4 + t)(4 - t)$
125. $\pi b^2 - \pi a^2$; $\pi(b + a)(b - a)$ **129.** makes sense **131.** makes sense **133.** false **135.** false **137.** $3x(x^2 + 2)(x + 3)(x - 3)$
139. $(4x^2 - 5)(x + 1)(x - 1)$ **141.** $3(x^n + 3y^n)(x^n - 3y^n)$ **143.** $3x(x + 5)(x - 1)$ should be $3x(x - 5)(x + 1)$.
145. correctly factored **147.** $(3x + 4)(3x - 4)$ **148.** $5x - 2y = 10$ **149.** $20°, 60°, 100°$ **150.** 0 **151.** 0 **152.** $(x + 4)(x - 3)$

Section 6.6 Check Point Exercises

1. $-\dfrac{1}{2}$ and 4, or $\left\{-\dfrac{1}{2}, 4\right\}$ **2.** 1 and 5, or $\{1, 5\}$ **3.** 0 and $\dfrac{1}{2}$, or $\left\{0, \dfrac{1}{2}\right\}$ **4.** 5 or $\{5\}$ **5.** $-\dfrac{5}{4}$ and $\dfrac{5}{4}$, or $\left\{-\dfrac{5}{4}, \dfrac{5}{4}\right\}$
6. -2 and 9, or $\{-2, 9\}$ **7.** 1 sec and 2 sec; (1, 192) and (2, 192) **8.** length: 9ft; width: 6ft

Concept and Vocabulary Check

1. quadratic equation **2.** $A = 0$ or $B = 0$ **3.** x-intercepts **4.** subtracting $5x$ **5.** subtracting $30x$; adding 25

Exercise Set 6.6

1. -7 and 0, or $\{-7, 0\}$ **3.** -4 and 6, or $\{-4, 6\}$ **5.** $-\dfrac{4}{5}$ and 9, or $\left\{-\dfrac{4}{5}, 9\right\}$ **7.** $-\dfrac{9}{2}$ and 4, or $\left\{-\dfrac{9}{2}, 4\right\}$ **9.** -5 and -3, or $\{-5, -3\}$

11. -3 and 5, or $\{-3, 5\}$ **13.** -3 and 7, or $\{-3, 7\}$ **15.** -8 and -1, or $\{-8, -1\}$ **17.** -4 and 0, or $\{-4, 0\}$ **19.** 0 and 5, or $\{0, 5\}$

21. 0 and 4, or $\{0, 4\}$ **23.** 0 and $\dfrac{5}{2}$, or $\left\{0, \dfrac{5}{2}\right\}$ **25.** $-\dfrac{5}{3}$ and 0, or $\left\{-\dfrac{5}{3}, 0\right\}$ **27.** -2 or $\{-2\}$ **29.** 6 or $\{6\}$ **31.** $\dfrac{3}{2}$ or $\left\{\dfrac{3}{2}\right\}$

33. $-\dfrac{1}{2}$ and 4, or $\left\{-\dfrac{1}{2}, 4\right\}$ **35.** -2 and $\dfrac{9}{5}$, or $\left\{-2, \dfrac{9}{5}\right\}$ **37.** -7 and 7, or $\{-7, 7\}$ **39.** $-\dfrac{5}{2}$ and $\dfrac{5}{2}$, or $\left\{-\dfrac{5}{2}, \dfrac{5}{2}\right\}$ **41.** $-\dfrac{5}{9}$ and $\dfrac{5}{9}$, or $\left\{-\dfrac{5}{9}, \dfrac{5}{9}\right\}$

43. -3 and 7, or $\{-3, 7\}$ **45.** $-\dfrac{5}{2}$ and $\dfrac{3}{2}$, or $\left\{-\dfrac{5}{2}, \dfrac{3}{2}\right\}$ **47.** -6 and 3, or $\{-6, 3\}$ **49.** -2 and $-\dfrac{3}{2}$, or $\left\{-2, -\dfrac{3}{2}\right\}$ **51.** 4 or $\{4\}$

53. $-\dfrac{5}{2}$ or $\left\{-\dfrac{5}{2}\right\}$ **55.** $\dfrac{3}{8}$ or $\left\{\dfrac{3}{8}\right\}$ **57.** $-3, -2$, and 4, or $\{-3, -2, 4\}$ **59.** $-6, 0$, and 6, or $\{-6, 0, 6\}$ **61.** $-2, -1$, and 0, or $\{-2, -1, 0\}$

63. 2 and 8, or $\{2, 8\}$ **65.** 4 and 5, or $\{4, 5\}$ **67.** 5 sec; Each tick represents one second.   **69.** 2 sec; $(2, 276)$ **71.** $\dfrac{1}{2}$ sec and 4 sec

73. 2002 and 2004   **75.** $(2, 66)$ and $(4, 66)$ **77.** 10 yr   **79.** $(10, 7250)$ **81.** 10 teams   **83.** length: 15 yd; width: 12 yd
85. base: 5 cm; height: 6 cm   **87. a.** $4x^2 + 44x$ **b.** 3 ft   **91.** does not make sense   **93.** makes sense   **95.** false   **97.** false
99. $x^2 - 2x - 15 = 0$ **101.** 4 and 5, or $\{4, 5\}$ **103.** c   **105.** d   **107.** -4 and 1, or $\{-4, 1\}$ **109.** -4 and 3, or $\{-4, 3\}$

113. $y > -\dfrac{2}{3}x + 1$

114. $\dfrac{4}{x^6}$ **115.** -2 or $\{-2\}$ **116.** 375   **117.** When x is replaced with 4, the denominator is 0.

118. $\dfrac{(x+5)(x+1)}{(x+5)(x-5)}, \dfrac{x+1}{x-5}$

Review Exercises

1. $15(2x - 3)$ **2.** $-4x(3x^2 + 4x - 100)$ **3.** $5x^2y(6x^2 + 3x + 1)$ **4.** $5(x + 3)$ **5.** $(7x^2 - 1)(x + y)$ **6.** $(x^2 + 2)(x + 3)$
7. $(x + 1)(y + 4)$ **8.** $(x^2 + 5)(x + 1)$ **9.** $(x - 2)(y + 4)$ **10.** $(x - 2)(x - 1)$ **11.** $(x - 5)(x + 4)$ **12.** $(x + 3)(x + 16)$
13. $(x - 4y)(x - 2y)$ **14.** prime   **15.** $(x + 17y)(x - y)$ **16.** $3(x + 4)(x - 2)$ **17.** $3x(x - 11)(x - 1)$ **18.** $(x + 5)(3x + 2)$
19. $(y - 3)(5y - 2)$ **20.** $(2x + 5)(2x - 3)$ **21.** prime   **22.** $-2(2x + 3)(2x - 1)$ **23.** $x(2x - 9)(x + 8)$ **24.** $4y(3y + 1)(y + 2)$
25. $(2x - y)(x - 3y)$ **26.** $(5x + 4y)(x - 2y)$ **27.** $(2x + 1)(2x - 1)$ **28.** $(9 + 10y)(9 - 10y)$ **29.** $(5a + 7b)(5a - 7b)$
30. $(z^2 + 4)(z + 2)(z - 2)$ **31.** $2(x + 3)(x - 3)$ **32.** prime   **33.** $x(3x + 1)(3x - 1)$ **34.** $2x(3y + 2)(3y - 2)$ **35.** $(x + 11)^2$
36. $(x - 8)^2$ **37.** $(3y + 8)^2$ **38.** $(4x - 5)^2$ **39.** prime   **40.** $(6x + 5y)^2$ **41.** $(5x - 4y)^2$ **42.** $(x - 3)(x^2 + 3x + 9)$
43. $(4x + 1)(16x^2 - 4x + 1)$ **44.** $2(3x - 2y)(9x^2 + 6xy + 4y^2)$ **45.** $y(3x + 2)(9x^2 - 6x + 4)$ **46.** $(a + 3)(a - 3)$
47. $(a + 2b)(a - 2b)$ **48.** $A^2 + 2A + 1 = (A + 1)^2$ **49.** $x(x - 7)(x - 1)$ **50.** $(5y + 2)(2y + 1)$ **51.** $2(8 + y)(8 - y)$
52. $(3x + 1)^2$ **53.** $-4x^3(5x^4 - 9)$ **54.** $(x - 3)^2(x + 3)$ **55.** prime   **56.** $x(2x + 5)(x + 7)$ **57.** $3x(x - 5)^2$
58. $3x^2(x - 2)(x^2 + 2x + 4)$ **59.** $4y^2(y + 3)(y - 3)$ **60.** $5(x + 7)(x - 3)$ **61.** prime   **62.** $-2x^3(5x - 2)(x - 4)$
63. $(10y + 7)(10y - 7)$ **64.** $9x^4(x - 2)$ **65.** $(x^2 + 1)(x + 1)(x - 1)$ **66.** $2(y - 2)(y^2 + 2y + 4)$ **67.** $(x + 4)(x^2 - 4x + 16)$
68. $(3x - 2)(2x + 5)$ **69.** $3x^2(x + 2)(x - 2)$ **70.** $(x - 10)(x + 9)$ **71.** $(5x + 2y)(5x + 3y)$ **72.** $x(x + 5)(x^2 - 5x + 25)$
73. $2y(4y + 3)(4y + 1)$ **74.** $-2(y - 4)^2$ **75.** $(x + 5y)(x - 7y)$ **76.** $(x + y)(x + 7)$ **77.** $(3x + 4y)^2$ **78.** $2x^2y(x + 1)(x - 1)$

79. $(10y + 7z)(10y - 7z)$ **80.** prime   **81.** $3x^2y^2(x + 2y)(x - 2y)$ **82.** 0 and 12, or $\{0, 12\}$ **83.** $-\dfrac{9}{4}$ and 7, or $\left\{-\dfrac{9}{4}, 7\right\}$

84. -7 and 2, or $\{-7, 2\}$ **85.** -4 and 0, or $\{-4, 0\}$ **86.** -8 and $\dfrac{1}{2}$, or $\left\{-8, \dfrac{1}{2}\right\}$ **87.** -4 and 8, or $\{-4, 8\}$ **88.** -8 and 7, or $\{-8, 7\}$

89. 7 or $\{7\}$ **90.** $-\dfrac{10}{3}$ and $\dfrac{10}{3}$, or $\left\{-\dfrac{10}{3}, \dfrac{10}{3}\right\}$ **91.** -5 and -2, or $\{-5, -2\}$ **92.** $\dfrac{1}{3}$ and 7, or $\left\{\dfrac{1}{3}, 7\right\}$

93. 2 sec   **94.** width: 5 ft; length: 8 ft   **95.** 11 m by 11 m

Chapter Test

1. $(x - 3)(x - 6)$ **2.** $(x - 7)^2$ **3.** $5y^2(3y - 1)(y - 2)$ **4.** $(x^2 + 3)(x + 2)$ **5.** $x(x - 9)$ **6.** $x(x + 7)(x - 1)$
7. $2(7x - 3)(x + 5)$ **8.** $(5x + 3)(5x - 3)$ **9.** $(x + 2)(x^2 - 2x + 4)$ **10.** $(x + 3)(x - 7)$ **11.** prime   **12.** $3y(2y + 1)(y + 1)$
13. $4(y + 3)(y - 3)$ **14.** $4(2x + 3)^2$ **15.** $2(x^2 + 4)(x + 2)(x - 2)$ **16.** $(6x - 7)^2$ **17.** $(7x - 1)(x - 7)$ **18.** $(x^2 - 5)(x + 2)$
19. $-3y(2y + 3)(2y - 5)$ **20.** $(y - 5)(y^2 + 5y + 25)$ **21.** $5(x - 3y)(x + 2y)$ **22.** -6 and 4, or $\{-6, 4\}$
23. $-\dfrac{1}{3}$ and 2, or $\left\{-\dfrac{1}{3}, 2\right\}$ **24.** -2 and 8, or $\{-2, 8\}$ **25.** 0 and $\dfrac{7}{2}$, or $\left\{0, \dfrac{7}{2}\right\}$ **26.** $-\dfrac{9}{4}$ and $\dfrac{9}{4}$, or $\left\{-\dfrac{9}{4}, \dfrac{9}{4}\right\}$
27. -1 and $\dfrac{6}{5}$, or $\left\{-1, \dfrac{6}{5}\right\}$ **28.** $x^2 - 4 = (x + 2)(x - 2)$ **29.** 6 sec   **30.** width: 5 ft; length: 11 ft

Cumulative Review Exercises

1. -48 **2.** 0 or $\{0\}$ **3.** 12 or $\{12\}$ **4.** $(-\infty, -2)$ or $\{x\,|\,x < -2\}$;   **5.** $38°, 38°, 104°$ **6.** $150

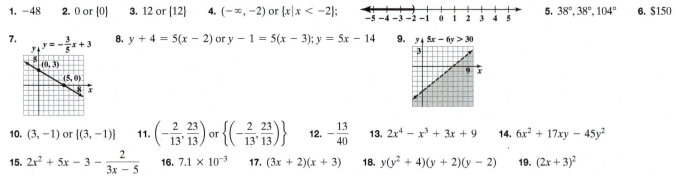

7.   **8.** $y + 4 = 5(x - 2)$ or $y - 1 = 5(x - 3)$; $y = 5x - 14$ **9.**

10. $(3, -1)$ or $\{(3, -1)\}$ **11.** $\left(-\dfrac{2}{13}, \dfrac{23}{13}\right)$ or $\left\{\left(-\dfrac{2}{13}, \dfrac{23}{13}\right)\right\}$ **12.** $-\dfrac{13}{40}$ **13.** $2x^4 - x^3 + 3x + 9$ **14.** $6x^2 + 17xy - 45y^2$

15. $2x^2 + 5x - 3 - \dfrac{2}{3x - 5}$ **16.** 7.1×10^{-3} **17.** $(3x + 2)(x + 3)$ **18.** $y(y^2 + 4)(y + 2)(y - 2)$ **19.** $(2x + 3)^2$

20. width: 4 ft; length: 6 ft

CHAPTER 7

Section 7.1 Check Point Exercises

1. a. $x = 5$ **b.** $x = -7$ and $x = 4$ **2.** $\dfrac{x + 4}{3x}$ **3.** $\dfrac{x^2}{7}$ **4.** $\dfrac{x - 1}{x + 1}$ **5.** $-\dfrac{3x + 7}{4}$

Concept and Vocabulary Check

1. polynomials   **2.** 0   **3.** factoring; common factors   **4.** 1   **5.** -1 **6.** false   **7.** false   **8.** true   **9.** false

Exercise Set 7.1

1. $x = 0$ **3.** $x = 8$ **5.** $x = 4$ **7.** $x = -9$ and $x = 2$ **9.** $x = \dfrac{17}{3}$ and $x = -3$ **11.** $x = -4$ and $x = 3$

13. defined for all real numbers   **15.** $y = -1$ and $y = \dfrac{3}{4}$ **17.** $y = -5$ and $y = 5$ **19.** defined for all real numbers   **21.** $2x$

23. $\dfrac{x - 3}{5}$ **25.** $\dfrac{x - 4}{2x}$ **27.** $\dfrac{1}{x - 3}$ **29.** $-\dfrac{5}{x - 3}$ **31.** 3   **33.** $\dfrac{1}{x - 5}$ **35.** $\dfrac{2}{3}$ **37.** $\dfrac{1}{x - 3}$ **39.** $\dfrac{4}{x - 2}$ **41.** $\dfrac{y - 1}{y + 9}$

43. $\dfrac{y - 3}{y - 2}$ **45.** cannot be simplified   **47.** $\dfrac{x + 6}{x - 6}$ **49.** $x^2 + 1$ **51.** $x^2 + 2x + 4$ **53.** $\dfrac{x - 4}{x + 4}$ **55.** cannot be simplified

57. cannot be simplified   **59.** -1 **61.** -1 **63.** cannot be simplified   **65.** -2 **67.** $-\dfrac{2}{x}$ **69.** $-x - 1$ **71.** $-y - 3$ **73.** $-\dfrac{1}{x}$

75. $\dfrac{x - y}{2x - y}$ **77.** $\dfrac{x - 6}{x^2 + 3x + 9}$ **79.** $\dfrac{3 + y}{3 - y}$ **81.** $\dfrac{y + 3}{x + 3}$ **83.** $\dfrac{2}{1 - 2x}$ **85. a.** It costs $86.67 million to inoculate 40% of the population. It costs $520 million to inoculate 80% of the population. It costs $1170 million to inoculate 90% of the population.   **b.** $x = 100$

c. The cost keeps rising; No amount of money will be enough to inoculate 100% of the population.   **87.** 400 mg

89. a. $300   **b.** $125   **c.** decrease   **91.** 1.5 mg per liter; $(3, 1.5)$ **99.** makes sense   **101.** does not make sense   **103.** false   **105.** false

109. correctly simplified   **111.** $x^2 - 1$ should be $x - 1$.   **113.** $\dfrac{3}{10}$ **114.** $\dfrac{1}{6}$ **115.** $(4, 2)$ or $\{(4, 2)\}$ **116.** $\dfrac{6}{35}$ **117.** $\dfrac{3}{2}$ **118.** $\dfrac{2}{3}$

Section 7.2 Check Point Exercises

1. $\dfrac{9x - 45}{2x + 8}$ **2.** $\dfrac{3}{8}$ **3.** $\dfrac{x + 2}{9}$ **4.** $-\dfrac{5(x + 1)}{7x(2x - 3)}$ **5.** $\dfrac{(x + 3)(x + 7)}{x - 4}$ **6.** $\dfrac{x + 3}{x - 5}$ **7.** $\dfrac{y + 1}{5y(y^2 + 1)}$

Concept and Vocabulary Check

1. numerators; denominators; $\dfrac{PR}{QS}$ **2.** multiplicative inverse/reciprocal; $\dfrac{S}{R}$; $\dfrac{PS}{QR}$ **3.** $\dfrac{x^2}{15}$ **4.** $\dfrac{3}{5}$

Exercise Set 7.2

1. $\dfrac{4x - 20}{9x + 27}$ **3.** $\dfrac{4x}{x + 5}$ **5.** $\dfrac{4}{5}$ **7.** $\dfrac{4}{9}$ **9.** 1   **11.** $\dfrac{x + 5}{x}$ **13.** $\dfrac{2}{y}$ **15.** $\dfrac{y + 2}{y + 4}$ **17.** $4(y + 3)$ **19.** $\dfrac{x - 1}{x + 2}$ **21.** $\dfrac{x^2 + 2x + 4}{3x}$

23. $\dfrac{x - 2}{x - 1}$ **25.** $-\dfrac{2}{x(x + 1)}$ **27.** $-\dfrac{y - 10}{y - 7}$ **29.** $(x - y)(x + y)$ **31.** $\dfrac{4(x + y)}{3(x - y)}$ **33.** $\dfrac{3x}{35}$ **35.** $\dfrac{1}{4}$ **37.** 10   **39.** $\dfrac{7}{9}$ **41.** $\dfrac{3}{4}$

43. $\dfrac{(x - 2)^2}{x}$ **45.** $\dfrac{(y - 4)(y^2 + 4)}{y - 1}$ **47.** $\dfrac{y}{3}$ **49.** $\dfrac{2(x + 3)}{3}$ **51.** $\dfrac{x - 5}{2}$ **53.** $\dfrac{y^2 + 1}{y^2}$ **55.** $\dfrac{y + 1}{y - 7}$ **57.** 6   **59.** $\dfrac{2}{3}$

61. $\dfrac{(x+y)^2}{32(x-y)^2}$ **63.** $\dfrac{y(x+y)}{(x+1)^2}$ **65.** $\dfrac{y+3}{y-3}$ **67.** $\dfrac{5(x-5)}{4}$ **69.** $\dfrac{(x+y)^2}{x-y}$ **71.** $\dfrac{2}{y^2-by+b^2}$ **73.** $\dfrac{x-2}{x(x-1)}$ sq in.   **75.** $\dfrac{3x}{x+2}$ sq in.

81. makes sense   **83.** makes sense   **85.** true   **87.** false   **89.** numerator: -3; denominator: $2x-3$ **91.** correct answer

93. $x-3$ should be $x+3$.   **95.** $(18, \infty)$ or $\{x|x>18\}$ **96.** $3(x-7)(x+2)$ **97.** -5 and $\dfrac{1}{2}$, or $\left\{-5, \dfrac{1}{2}\right\}$ **98.** $\dfrac{2}{3}$ **99.** x **100.** $\dfrac{x-3}{x+3}$

Section 7.3 Check Point Exercises

1. $x+2$ **2.** $\dfrac{x-5}{x+5}$ **3. a.** $\dfrac{3x+5}{x+7}$ **b.** $3x-4$ **4.** $\dfrac{3y+2}{y-4}$ **5.** $x+3$ **6.** $\dfrac{-4x^2+12x}{x^2-2x-9}$

Concept and Vocabulary Check

1. $\dfrac{P+Q}{R}$; numerators; common denominator   **2.** $\dfrac{P-Q}{R}$; numerators; common denominator

3. $\dfrac{-1}{-1}$ **4.** $\dfrac{x+5}{3}$ **5.** $\dfrac{x-5}{3}$ **6.** $\dfrac{x-5+y}{3}$

Exercise Set 7.3

1. $\dfrac{9x}{13}$ **3.** $\dfrac{3x}{5}$ **5.** $\dfrac{x+3}{2}$ **7.** $\dfrac{6}{x}$ **9.** $\dfrac{7}{3x}$ **11.** $\dfrac{9}{x+3}$ **13.** $\dfrac{5x+5}{x-3}$ **15.** 2 **17.** $\dfrac{2y+3}{y-5}$ **19.** $\dfrac{7}{5y}$ **21.** $\dfrac{2}{x+2}$ **23.** $\dfrac{x-2}{x-3}$

25. $\dfrac{3x-4}{5x-4}$ **27.** 1 **29.** 1 **31.** $\dfrac{x}{2x-1}$ **33.** $-\dfrac{1}{y}$ **35.** $\dfrac{y+3}{3y+8}$ **37.** $\dfrac{3y+2}{y-3}$ **39.** $\dfrac{2}{x-3}$ **41.** $\dfrac{3x+7}{x-6}$ **43.** $\dfrac{3x+1}{3x-4}$

45. $x+2$ **47.** 0 **49.** $\dfrac{11}{x-1}$ **51.** $\dfrac{12}{x+3}$ **53.** $\dfrac{y+1}{y-1}$ **55.** $\dfrac{x-2}{x-7}$ **57.** $\dfrac{2x-4}{x^2-25}$ **59.** 1 **61.** $\dfrac{2}{x+y}$ **63.** $\dfrac{x-3}{x-1}$

65. $\dfrac{4b}{4b-3}$ **67.** $\dfrac{y}{y-5}$ **69.** $-\dfrac{1}{c+d}$ **71.** $\dfrac{3y^2+8y-5}{(y+1)(y-4)}$ **73. a.** $\dfrac{100W}{L}$ **b.** round   **75.** 10 m   **81.** does not make sense

83. makes sense   **85.** false   **87.** false   **89.** $\dfrac{x+1}{x+2}$ **91.** $2x+1$ **93.** -7 **95.** $-4x-1$ **97.** $x-2$ should be $x-3$.

99. $\dfrac{31}{45}$ **100.** $(9x^2+1)(3x+1)(3x-1)$ **101.** $3x^2-7x-5$ **102.** $\dfrac{7}{6}$ **103.** $-\dfrac{17}{24}$ **104.** $\dfrac{y-2}{4y}$

Section 7.4 Check Point Exercises

1. $30x^2$ **2.** $(x+3)(x-3)$ **3.** $7x(x+4)(x+4)$ or $7x(x+4)^2$ **4.** $\dfrac{9+14x}{30x^2}$ **5.** $\dfrac{6x+6}{(x+3)(x-3)}$ **6.** $-\dfrac{5}{x+5}$ **7.** $-\dfrac{5+y}{5y}$

8. $\dfrac{x-15}{(x+5)(x-5)}$

Concept and Vocabulary Check

1. factor denominators   **2.** x and $x+3$; $x+3$ and $x+5$; $x(x+3)(x+5)$ **3.** 3 **4.** $x-5$ **5.** -1

Exercise Set 7.4

1. $120x^2$ **3.** $30x^5$ **5.** $(x-3)(x+1)$ **7.** $7y(y+2)$ **9.** $(x+4)(x-4)$ **11.** $y(y+3)(y-3)$ **13.** $(y+1)(y-1)(y-1)$

15. $(x-5)(x+4)(2x-1)$ **17.** $\dfrac{3x+5}{x^2}$ **19.** $\dfrac{37}{18x}$ **21.** $\dfrac{8x+7}{2x^2}$ **23.** $\dfrac{6x+1}{x}$ **25.** $\dfrac{2+9x}{x}$ **27.** $\dfrac{x+1}{2}$ **29.** $\dfrac{7x-20}{x(x-5)}$

31. $\dfrac{5x+1}{(x-1)(x+2)}$ **33.** $\dfrac{11y+15}{4y(y+5)}$ **35.** $-\dfrac{7}{x+7}$ **37.** $\dfrac{3x-55}{(x+5)(x-5)}$ **39.** $\dfrac{x^2+6x}{(x-4)(x+4)}$ **41.** $\dfrac{y+12}{(y+3)(y-3)}$

43. $\dfrac{7x-10}{(x-1)(x-1)}$ **45.** $\dfrac{9y}{4(y-5)}$ **47.** $\dfrac{8y+16}{y(y+4)}$ **49.** $\dfrac{17}{(x-3)(x-4)}$ **51.** $\dfrac{7x-1}{(x+1)(x+1)(x-1)}$ **53.** $\dfrac{x^2-x}{(x+3)(x-2)(x+5)}$

55. $\dfrac{y^2+8y+4}{(y+4)(y+1)(y+1)}$ **57.** $\dfrac{2x^2-4x+34}{(x+3)(x-5)}$ **59.** $\dfrac{5-3y}{2y(y-1)}$ **61.** $-\dfrac{x^2}{(x+3)(x-3)}$ **63.** $\dfrac{2}{y+5}$ **65.** $\dfrac{4x-11}{x-3}$ **67.** $\dfrac{3}{y+1}$

69. $\dfrac{-x^2+7x+3}{(x-3)(x+2)}$ **71.** $\dfrac{x^2+2x-1}{(x+1)(x-1)(x+2)}$ **73.** $\dfrac{-y^2+8y+9}{15y^2}$ **75.** $\dfrac{x^2-2x-6}{3(x+2)(x-2)}$ **77.** $\dfrac{-y-2}{(y+1)(y-1)}$

79. $\dfrac{x+2xy-y}{xy}$ **81.** $\dfrac{5x+2y}{(x+y)(x-y)}$ **83.** $\dfrac{-x+6}{(x+2)(x-2)}$ **85.** $\dfrac{6x+5}{(x-5)(x+5)(x-6)}$ **87.** $\dfrac{9}{(x-3)(x^2+3x+9)}$

89. $\dfrac{3}{y-3}$ **91.** $-\dfrac{22y}{(x+3y)(x+y)(x-3y)}$ **93.** $\dfrac{D}{5}$ **95.** $\dfrac{D}{24}$ **97.** no   **99.** 5 yr   **101.** $\dfrac{4x^2+14x}{(x+3)(x+4)}$

109. makes sense   **111.** does not make sense   **113.** false   **115.** true   **117.** $-\dfrac{1}{x(x+h)}$ **119.** $\dfrac{2}{x+1}$ **120.** $6x^2-11x-35$

121. $3x - y < 3$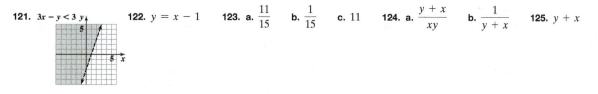

122. $y = x - 1$

123. a. $\dfrac{11}{15}$ **b.** $\dfrac{1}{15}$ **c.** 11

124. a. $\dfrac{y + x}{xy}$ **b.** $\dfrac{1}{y + x}$

125. $y + x$

Mid-Chapter Check Point Exercises

1. $x = -2$ and $x = 4$ **2.** $\dfrac{x - 2}{2x + 1}$ **3.** $\dfrac{-3}{y - 2}$ or $\dfrac{3}{2 - y}$ **4.** $\dfrac{2}{w}$ **5.** $\dfrac{4}{x + 4}$ **6.** $\dfrac{4}{(x - 2)^2}$ **7.** $\dfrac{x + 5}{x - 2}$ **8.** $\dfrac{x + 2}{x - 4}$

9. $\dfrac{2x + 1}{(x + 2)(x + 3)(x - 1)}$ **10.** $\dfrac{9 - x}{x - 5}$ **11.** $-\dfrac{2y - 1}{3y}$ or $\dfrac{1 - 2y}{3y}$ **12.** $\dfrac{-y^2}{(y + 1)(y + 2)}$ **13.** $\dfrac{w + 1}{(w - 3)(w + 5)}$

14. $\dfrac{2z^2 - 3z + 15}{(z + 3)(z - 3)(z + 1)}$ **15.** $\dfrac{3z^2 + 5z + 3}{(3z - 1)^2}$ **16.** $\dfrac{3x - 1}{(x + 7)(x - 3)}$ **17.** x **18.** $\dfrac{2x + 1}{(x + 1)(x - 2)(x + 3)}$ **19.** $\dfrac{5x - 5y}{x}$ **20.** $\dfrac{x + 3}{x + 5}$

Section 7.5 Check Point Exercises

1. $\dfrac{11}{5}$ **2.** $\dfrac{2x - 1}{2x + 1}$ **3.** $y - x$ **4.** $\dfrac{11}{5}$ **5.** $\dfrac{2x - 1}{2x + 1}$ **6.** $y - x$

Concept and Vocabulary Check

1. complex; complex **2.** $\dfrac{18 + 12}{9 - 4} = \dfrac{30}{5} = 6$ **3.** $\dfrac{30 + 2x^2}{24 - x^2}$ **4.** $\dfrac{2x + 3}{5x + x^2}$

Exercise Set 7.5

1. $\dfrac{9}{10}$ **3.** $\dfrac{18}{23}$ **5.** $-\dfrac{4}{5}$ **7.** $\dfrac{3 - 4x}{3 + 4x}$ **9.** $\dfrac{7x - 2}{5x + 1}$ **11.** $\dfrac{2y + 3}{y - 7}$ **13.** $\dfrac{4 - 6y}{4 + 3y}$ **15.** $x - 5$ **17.** $\dfrac{x}{x - 1}$ **19.** $-\dfrac{1}{y}$

21. $\dfrac{xy + 2}{x}$ **23.** $\dfrac{y + x}{x^2 y^2}$ **25.** $\dfrac{x^2 + y}{y(y + 1)}$ **27.** $\dfrac{1}{y}$ **29.** $\dfrac{4 - x}{5x - 3}$ **31.** $\dfrac{2y}{y - 3}$ **33.** $\dfrac{1}{x + 3}$ **35.** $\dfrac{x^2 - 5x + 3}{x^2 - 7x + 2}$ **37.** $\dfrac{2y^2 + 3xy}{y^2 + 2x}$

39. $-\dfrac{6}{5}$ **41.** $\dfrac{1 - x}{(x - 3)(x + 6)}$ **43.** $\dfrac{1}{y(y + 5)}$ **45.** $\dfrac{1}{x - 1}$ **47.** $\dfrac{x + 1}{2x + 1}$ **49.** $\dfrac{2r_1 r_2}{r_1 + r_2}; 34\dfrac{2}{7}$ mph **51. a.** $\dfrac{x^2 + 10x + 135}{9x^2 - 85x + 525}$

b. 67% **c.** 68%; overestimates by 1% **57.** makes sense **59.** does not make sense **61.** true **63.** false **65.** $\dfrac{2y^2}{y + 1}$

67. correct answer **69.** $x + \dfrac{1}{3}$ should be $3 + x$. **70.** $2x(x - 5)^2$ **71.** -2 or $\{-2\}$ **72.** $x^3 + y^3$ **73.** 1 or $\{1\}$ **74.** 4 or $\{4\}$

75. $\dfrac{1}{2}$ and 2, or $\left\{\dfrac{1}{2}, 2\right\}$

Section 7.6 Check Point Exercises

1. 4 or $\{4\}$ **2.** 3 or $\{3\}$ **3.** -3 and -2, or $\{-3, -2\}$ **4.** 24 or $\{24\}$ **5.** no solution or $\varnothing$ **6.** 75%

Concept and Vocabulary Check

1. LCD **2.** 0 **3.** $8x$ **4.** $12(x + 3)$ **5.** $1 - x = 3x^2$ **6.** $6 = 5(x + 1) + 3(x - 1)$ **7.** $x \neq -2$
8. $x \neq 3$ and $x \neq 2$ **9.** true **10.** false

Exercise Set 7.6

1. 12 or $\{12\}$ **3.** 0 or $\{0\}$ **5.** -2 or $\{-2\}$ **7.** 6 or $\{6\}$ **9.** -2 or $\{-2\}$ **11.** 2 or $\{2\}$ **13.** 15 or $\{15\}$ **15.** 4 or $\{4\}$
17. -6 and -1, or $\{-6, -1\}$ **19.** -5 and 5, or $\{-5, 5\}$ **21.** -3 and 3, or $\{-3, 3\}$ **23.** -8 and 1, or $\{-8, 1\}$ **25.** -1 and 16, or $\{-1, 16\}$

27. $\dfrac{3}{2}$ or $\left\{\dfrac{3}{2}\right\}$ **29.** 3 or $\{3\}$ **31.** no solution or $\varnothing$ **33.** $\dfrac{2}{3}$ and 4, or $\left\{\dfrac{2}{3}, 4\right\}$ **35.** no solution or $\varnothing$ **37.** 1 or $\{1\}$ **39.** $\dfrac{1}{5}$ or $\left\{\dfrac{1}{5}\right\}$

41. -3 or $\{-3\}$ **43.** no solution or $\varnothing$ **45.** -6 and $-\dfrac{1}{2}$, or $\left\{-6, -\dfrac{1}{2}\right\}$ **47.** -3 or $\{-3\}$ **49.** $\dfrac{-6x - 18}{(x - 5)(x + 4)}$

51. 1 and 5, or $\{1, 5\}$ **53.** 3 or $\{3\}$ **55.** 10,000 wheelchairs **57.** 50% **59.** 5 yr old

61. either 50 or approx 67 cases; $(50, 350)$ or $\left(66\dfrac{2}{3}, 350\right)$ **69.** makes sense **71.** does not make sense **73.** true **75.** false

77. $f_2 = \dfrac{-ff_1}{f - f_1}$ or $f_2 = \dfrac{ff_1}{f_1 - f}$ **79.** no solution or $\varnothing$ **81.** 8 or $\{8\}$ **83.** -3 and -2, or $\{-3, -2\}$ **84.** $(x^3 - 3)(x + 2)$

85. $-\dfrac{12}{x^8}$ **86.** $-20x + 55$ **87.** 2 or $\{2\}$ **88.** $\dfrac{1}{5}, \dfrac{3}{5}, \dfrac{x}{5}$ **89.** $\dfrac{63}{x} = \dfrac{7}{5}$

Section 7.7 Check Point Exercises

1. 2 mph   **2.** $2\frac{2}{3}$ hr or 2 hr 40 min   **3.** $5880   **4.** 720 deer   **5.** 32 in.   **6.** 32 yd

Concept and Vocabulary Check

1. distance traveled; rate of travel   **2.** 1   **3.** $\frac{x}{5}$ **4.** $ad = bc$ **5.** similar

Exercise Set 7.7

1. walking rate: 6 mph; car rate: 9 mph   **3.** downhill rate: 10 mph; uphill rate: 6 mph   **5.** 5 mph
7. walking rate: 3 mph; jogging rate: 6 mph   **9.** 10 mph   **11.** It will take about 8.6 min, which is enough time.
13. It will take about 171.4 hr, which is not enough time.   **15.** 2.4 hr, or 2 hr 24 min   **17.** $1800   **19.** 20,489 fur seal pups   **21.** 0.97 billion
23. 154.1 in.   **25.** 5 in.   **27.** 6 m   **29.** 16 in.   **31.** 16 ft   **43.** does not make sense   **45.** makes sense   **47.** 30 mph
49. 7 hr   **51.** 2.5 hr   **53.** $(5x + 9)(5x - 9)$ **54.** 6 or {6}
55. 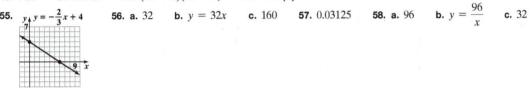 $y = -\frac{2}{3}x + 4$ **56. a.** 32   **b.** $y = 32x$ **c.** 160   **57.** 0.03125   **58. a.** 96   **b.** $y = \frac{96}{x}$ **c.** 32

Section 7.8 Check Point Exercises

1. 48 gal   **2.** 4.8 pounds per square inch

Concept and Vocabulary Check

1. $y = kx$; constant of variation   **2.** $y = \frac{k}{x}$; constant of variation   **3.** inversely   **4.** directly

Exercise Set 7.8

1. $g = kh$ **3.** $w = \frac{k}{v}$ **5.** 20   **7.** 6000   **9.** 84   **11.** 25   **13. a.** $G = kW$ **b.** $G = 0.02W$ **c.** 1.04 in.   **15.** $60

17. 2442 mph   **19.** 0.5 hr or 30 min   **21.** 6.4 lb per cm^2 **23.** 5000 pens   **31.** makes sense   **33.** makes sense

35. $\frac{16}{3}$ or $\left\{\frac{16}{3}\right\}$ **36.** $9x^2 - 6x + 4 - \frac{16}{3x + 2}$ **37.** $6x(x - 5)(x + 4)$ **38.** 4   **39.** 50   **40.** -31

Review Exercises

1. $x = 4$ **2.** $x = 2$ and $x = -5$ **3.** $x = 1$ and $x = 2$ **4.** defined for all real numbers   **5.** $\frac{4x}{3}$ **6.** $x + 2$ **7.** x^2 **8.** $\frac{x - 3}{x - 6}$
9. $\frac{x - 5}{x + 7}$ **10.** $\frac{y}{y + 2}$ **11.** cannot be simplified   **12.** $-2(x + 3y)$ **13.** $\frac{x - 2}{4}$ **14.** $\frac{5}{2}$ **15.** $\frac{x + 3}{x + 2}$ **16.** $\frac{2y - 1}{5}$ **17.** $\frac{-3(y + 1)}{5y(2y - 3)}$
18. $\frac{x - 1}{4}$ **19.** $\frac{2}{x(x + 1)}$ **20.** $\frac{1}{7(y + 3)}$ **21.** $(y - 3)(y - 6)$ **22.** $\frac{8}{x - y}$ **23.** 4   **24.** 4   **25.** $3x - 5$ **26.** $3y + 4$
27. $\frac{4}{x - 2}$ **28.** $\frac{2x + 5}{x - 3}$ **29.** $36x^3$ **30.** $x^2(x - 1)^2$ **31.** $(x + 3)(x + 1)(x + 7)$ **32.** $\frac{14x + 15}{6x^2}$ **33.** $\frac{7x + 2}{x(x + 1)}$ **34.** $\frac{7x + 25}{(x + 3)^2}$
35. $\frac{3}{y - 2}$ **36.** $-\frac{y}{y - 1}$ **37.** $\frac{x^2 + y^2}{xy}$ **38.** $\frac{3x^2 - x}{(x + 1)^2(x - 1)}$ **39.** $\frac{5x^2 - 3x}{(x + 1)(x - 1)}$ **40.** $\frac{4}{(x + 2)(x - 2)(x - 3)}$ **41.** $\frac{2x + 13}{x + 3}$
42. $\frac{y - 4}{3y}$ **43.** $\frac{7}{2}$ **44.** $\frac{1}{x - 1}$ **45.** $x + y$ **46.** $\frac{3}{x}$ **47.** $\frac{3x}{x - 4}$ **48.** 12 or {12}   **49.** -1 or {-1}   **50.** -6 and 1, or {-6, 1}
51. no solution or ∅   **52.** 5 or {5}   **53.** -6 and 2, or {-6, 2}   **54.** 5 or {5}   **55.** -5 or {-5}   **56.** 3 yr   **57.** 30%   **58.** 4 mph
59. Slower car's rate is 20 mph; faster car's rate is 30 mph.   **60.** 4 hr   **61.** 324 teachers   **62.** 287 trout   **63.** 5 ft   **64.** $12\frac{1}{2}$ ft
65. $154   **66.** 5 hr

Chapter Test

1. $x = -9$ and $x = 4$ **2.** $\frac{x + 3}{x - 2}$ **3.** $\frac{4x}{x + 1}$ **4.** $\frac{x - 4}{2}$ **5.** $\frac{x}{x + 3}$ **6.** $\frac{2x + 6}{x + 1}$ **7.** $\frac{5}{(y - 1)(y - 3)}$ **8.** $2y$
9. $\frac{2}{y + 3}$ **10.** $\frac{x^2 + 2x + 15}{(x + 3)(x - 3)}$ **11.** $\frac{8x - 14}{(x + 2)(x - 1)(x - 3)}$ **12.** $\frac{-x - 1}{x - 3}$ **13.** $\frac{x + 2}{x - 1}$ **14.** $\frac{11}{(x - 3)(x - 4)}$ **15.** $\frac{4}{y + 4}$
16. $\frac{3x - 3y}{x}$ **17.** $\frac{5x + 5}{2x + 1}$ **18.** $\frac{y - x}{y}$ **19.** 6 or {6}   **20.** -8 or {-8}   **21.** 0 and 2, or {0, 2}   **22.** 2 mph

23. 12 min   **24.** 6000 tule elk   **25.** 3.2 in.   **26.** 52.5 amp

Cumulative Review Exercises

1. 2 or {2}   **2.** $(-\infty, 2)$ or $\{x|x < 2\}$ **3.** -6 and 3, or $\{-6, 3\}$ **4.** -6 or $\{-6\}$ **5.** $(3, 3)$ or $\{(3, 3)\}$ **6.** $(-2, 2)$ or $\{(-2, 2)\}$

7. $3x - 2y = 6$ **8.** $y > -2x + 3$ **9.** $y = -3$

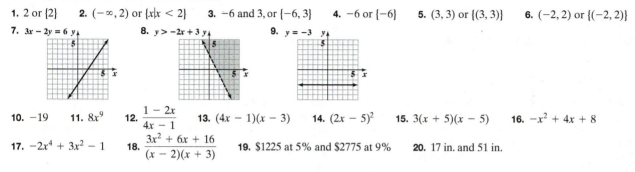

10. -19 **11.** $8x^9$ **12.** $\dfrac{1 - 2x}{4x - 1}$ **13.** $(4x - 1)(x - 3)$ **14.** $(2x - 5)^2$ **15.** $3(x + 5)(x - 5)$ **16.** $-x^2 + 4x + 8$

17. $-2x^4 + 3x^2 - 1$ **18.** $\dfrac{3x^2 + 6x + 16}{(x - 2)(x + 3)}$ **19.** \$1225 at 5% and \$2775 at 9%   **20.** 17 in. and 51 in.

CHAPTER 8

Section 8.1 Check Point Exercises

1. a. 9   **b.** -3 **c.** $\dfrac{1}{5}$ **d.** 10   **e.** 14   **2.** 16 miles per hour   **3.** 2.65; approximately by the point $(7, 2.65)$ **4.** 57%; It's the same.

5. a. 1   **b.** -3 **c.** $\dfrac{1}{5}$ **6. a.** 3   **b.** not a real number   **c.** -3 **d.** $-\dfrac{1}{2}$

Concept and Vocabulary Check

1. principal   **2.** 9^2 **3.** index; radicand   **4.** 4^3 **5.** 2^4 **6.** true   **7.** true   **8.** false   **9.** true   **10.** false

Exercise Set 8.1

1. 6   **3.** -6 **5.** not a real number   **7.** $\dfrac{1}{3}$ **9.** $\dfrac{1}{10}$ **11.** $-\dfrac{1}{6}$ **13.** not a real number   **15.** 0.2   **17.** 5   **19.** 8   **21.** 13

23. 17   **25.** not a real number   **27.** $y = \sqrt{x - 1}$ **29.** Answers will vary; example: Exercise 27 graph is shaped like $y = \sqrt{x}$, but shifted 1 unit to the right.   **31.** 2.65   **33.** 4.80   **35.** -8.06 **37.** 15.32   **39.** 7.66   **41.** 2.15   **43.** 2.83   **45.** not a real number   **47.** 4   **49.** -3 **51.** -2 **53.** $\dfrac{1}{5}$ **55.** -10 **57.** 1   **59.** 2   **61.** -2 **63.** not a real number

65. -1 **67.** not a real number   **69.** -4 **71.** 2   **73.** -2 **75.** $x \geq 0$ **77.** $x \geq 2$ **79.** $x \leq 2$ **81.** all real numbers   **83.** $x \leq 6$
85. 8 miles per hour   **87.** 70 mph; He was speeding.   **89. a.** 36 cm   **b.** 44.7 cm   **c.** 46.9 cm   **d.** describes healthy children
91. a. $y = a\sqrt{x} + 68.6$ **b.** $y = 3.4\sqrt{x} + 68.6$ **c.** 84.9%; overestimates by 0.3%   **99.** does not make sense
101. makes sense   **103.** false   **105.** false   **107.** 2   **109.** 0

111.   **113.** $4x - 5y = 20$ **114.** $(-\infty, -8)$ or $\{x|x < -8\}$

115. $\dfrac{x + 9}{x - 15}$ **116. a.** 10   **b.** 10   **c.** $\sqrt{25} \cdot \sqrt{4} = \sqrt{25 \cdot 4}$ **117. a.** 22.361   **b.** 22.361   **c.** $\sqrt{500} = 10\sqrt{5}$ **118. a.** 4
b. 4   **c.** $\dfrac{\sqrt{64}}{\sqrt{4}} = \sqrt{\dfrac{64}{4}}$

Section 8.2 Check Point Exercises

1. a. $\sqrt{30}$ **b.** $\sqrt{26xy}$ **c.** 5 **d.** $\sqrt{\dfrac{15}{22}}$ **2. a.** $2\sqrt{3}$ **b.** $2\sqrt{15}$ **c.** cannot be simplified **3.** $2x^8\sqrt{10}$ **4.** $3x^4\sqrt{3x}$

5. $3x^6\sqrt{5x}$ **6. a.** $\dfrac{7}{5}$ **b.** $4x^2$ **7. a.** $2\sqrt[3]{5}$ **b.** $2\sqrt[5]{2}$ **c.** $\dfrac{5}{3}$

Concept and Vocabulary Check

1. $\sqrt{a}\cdot\sqrt{b}$ **2.** $\sqrt{49}\cdot\sqrt{6}=7\cdot\sqrt{6}=7\sqrt{6}$ **3.** x^n; one-half **4.** 12; 6 **5.** $\dfrac{\sqrt{a}}{\sqrt{b}}$ **6.** $\dfrac{\sqrt{25}}{\sqrt{4}}=\dfrac{5}{2}$ **7.** $\sqrt[4]{16}\cdot\sqrt[4]{3}=2\cdot\sqrt[4]{3}=2\sqrt[4]{3}$

Exercise Set 8.2

1. $\sqrt{14}$ **3.** $\sqrt{15xy}$ **5.** 5 **7.** $\sqrt{\dfrac{10}{21}}$ **9.** $\sqrt{0.5xy}$ **11.** $\dfrac{1}{5}\sqrt{ab}$ **13.** $\sqrt{x}$ **15.** $5\sqrt{2}$ **17.** $3\sqrt{5}$ **19.** $10\sqrt{2}$ **21.** $5\sqrt{3x}$

23. $3\sqrt{x}$ **25.** cannot be simplified **27.** y **29.** $8x$ **31.** $x\sqrt{11}$ **33.** $2x\sqrt{2}$ **35.** x^{10} **37.** $5y^5$ **39.** $2x^3\sqrt{5}$ **41.** $6y^{50}\sqrt{2}$

43. $x\sqrt{x}$ **45.** $x^3\sqrt{x}$ **47.** $y^8\sqrt{y}$ **49.** $5x^2\sqrt{x}$ **51.** $2x^8\sqrt{2x}$ **53.** $3y^9\sqrt{10y}$ **55.** $3\sqrt{5}$ **57.** $5\sqrt{2xy}$ **59.** $6x$ **61.** $3x\sqrt{5x}$

63. $5x^6\sqrt{3x}$ **65.** $\sqrt{21xy}$ **67.** $10xy\sqrt{2y}$ **69.** $\dfrac{7}{4}$ **71.** $\dfrac{\sqrt{3}}{2}$ **73.** $\dfrac{x}{6}$ **75.** $\dfrac{\sqrt{7}}{x^2}$ **77.** $\dfrac{6\sqrt{2}}{y^{10}}$ **79.** 3 **81.** $2\sqrt{2}$ **83.** $\sqrt{5}$

85. 4 **87.** $2x$ **89.** $5x\sqrt{2x}$ **91.** $2x^3\sqrt{10x}$ **93.** $2\sqrt[3]{2}$ **95.** $3\sqrt[3]{2}$ **97.** $2\sqrt[4]{2}$ **99.** 2 **101.** $3\sqrt[3]{2}$ **103.** $2\sqrt[4]{2}$ **105.** $\dfrac{3}{2}$

107. $\dfrac{\sqrt[3]{3}}{2}$ **109.** $3(x+4)\sqrt{10(x+4)}$ **111.** $x-3$ **113.** $2^{21}x^{52}y^6\sqrt{2y}$ **115.** $2x\sqrt[3]{3x^2}$ **117.** $5\sqrt{3}$ sq ft

119. a. $\dfrac{5\sqrt{34}}{14}$ m^2 **b.** 2.08 m^2 **121.** Paige Fox is bad at math. **129.** does not make sense **131.** makes sense **133.** false **135.** true

137. $\sqrt{25x^{14}}$ **141.** $4x\sqrt{2}$ should be $2x\sqrt{2}$. **143.** $(6,-2)$ or $\{(6,-2)\}$ **144.** $\dfrac{3}{x-2}$ **145.** $2x(x-3)(x-5)$

146. a. $12x$ **b.** $12\sqrt{2}$ **147.** $20\sqrt{2x}$ **148. a.** $3x+15$ **b.** $\sqrt{21}+\sqrt{15}$

Section 8.3 Check Point Exercises

1. a. $17\sqrt{13}$ **b.** $-19\sqrt{17x}$ **2. a.** $17\sqrt{3}$ **b.** $10\sqrt{2x}$ **3. a.** $\sqrt{10}+\sqrt{22}$ **b.** $11+6\sqrt{3}$ **c.** $4-4\sqrt{5}$ **4. a.** -2 **b.** 5

Concept and Vocabulary Check

1. like radicals **2.** $(8+10)\sqrt{3}=18\sqrt{3}$ **3.** $7\sqrt{5}+1\sqrt{5}=(7+1)\sqrt{5}=8\sqrt{5}$ **4.** $5\sqrt{2}+4\sqrt{2}=9\sqrt{2}$ **5.** $21;7\sqrt{2};3\sqrt{2};2$
6. $8-\sqrt{6}$ **7.** $6^2-(\sqrt{5})^2=36-5=31$

Exercise Set 8.3

1. $13\sqrt{3}$ **3.** $15\sqrt{6}$ **5.** $-5\sqrt{13}$ **7.** $15\sqrt{x}$ **9.** $-6\sqrt{y}$ **11.** $9\sqrt{10x}$ **13.** $6\sqrt{5y}$ **15.** $2\sqrt{5}$ **17.** $12\sqrt{2}$ **19.** $7\sqrt{7}$
21. 0 **23.** $3\sqrt{5}$ **25.** $\sqrt{2}$ **27.** $8\sqrt{2}$ **29.** $19\sqrt{3}$ **31.** $9\sqrt{3}-6\sqrt{2}$ **33.** $2\sqrt{5x}$ **35.** $9+5\sqrt{2}$ **37.** cannot be combined
39. $8\sqrt{5}+15\sqrt{3}$ **41.** $7\sqrt{6}+8\sqrt{5}$ **43.** $\sqrt{6}+\sqrt{10}$ **45.** $\sqrt{42}-\sqrt{70}$ **47.** $5\sqrt{3}+3$ **49.** $3\sqrt{2}-3$ **51.** $32+11\sqrt{2}$
53. $25-2\sqrt{5}$ **55.** $117-36\sqrt{7}$ **57.** $25-8\sqrt{10}$ **59.** $15+9\sqrt{2}$ **61.** $\sqrt{6}-6\sqrt{2}+\sqrt{3}-6$ **63.** 4 **65.** -5 **67.** -14
69. 2 **71.** -37 **73.** 7 **75.** $5+2\sqrt{6}$ **77.** $x-2\sqrt{10x}+10$ **79.** $9x\sqrt{3x}$ **81.** $8x^2y^2\sqrt{xy}$ **83.** $\sqrt[3]{2}$ **85.** $11x\sqrt[4]{4x}$
87. perimeter $=4(\sqrt{3}+\sqrt{5})=(4\sqrt{3}+4\sqrt{5})$ in.; area $=(\sqrt{3}+\sqrt{5})^2=(8+2\sqrt{15})$ sq in.
89. perimeter $=2(\sqrt{6}-1)+2(\sqrt{6}+1)=4\sqrt{6}$ in.; area $=(\sqrt{6}+1)(\sqrt{6}-1)=5$ sq in.
91. perimeter $=\sqrt{2}+\sqrt{2}+2=(2+2\sqrt{2})$ in.; area $=\dfrac{1}{2}(\sqrt{2})(\sqrt{2})=1$ sq in. **93.** $3\sqrt{2}$; Answers will vary.
95. a. $J=2.6\sqrt{x}+35$ **b.** 51%; It's the same. **97. a.** $(36,51)$ **b.** 51% and 52% are reasonable estimates.
107. does not make sense **109.** does not make sense **111.** false **113.** false **115.** $-1-3\sqrt[3]{4}+\sqrt[3]{2}$ **117.** $48x-2y$
119. $\sqrt{7x}$ should be $\sqrt{x}$. **121.** x^2-4 should be $x-4$. **122.** $25x^2-9$ **123.** $x(8x+1)(8x-1)$

124. $y=-\dfrac{1}{4}x+3$ **125.** $\dfrac{\sqrt{15}}{5}$ **126.** 22 **127.** 4

Mid-Chapter Check Point Exercises

1. $10\sqrt{3}$ **2.** $10\sqrt{6}$ **3.** $4x\sqrt{6x}$ **4.** $\dfrac{\sqrt[5]{4}}{2}$ **5.** $9\sqrt{3}$ **6.** 7 **7.** $2\sqrt{15}-30$ **8.** $-4x^{10}\sqrt{2x}$ **9.** $2x^3\sqrt{3x}$ **10.** $2\sqrt[3]{2}$

11. $-19\sqrt{10}$ **12.** $4-\sqrt{3}$ **13.** $2x\sqrt{2}$ **14.** $-2\sqrt[4]{2}$ **15.** $9+2\sqrt{14}$ **16.** $2\sqrt[3]{2}$ **17.** $6\sqrt{5}$ **18.** $\dfrac{\sqrt{10}}{4}$ **19.** $6+3\sqrt{10}$ **20.** 23

Section 8.4 Check Point Exercises

1. a. $\frac{5\sqrt{10}}{2}$ b. $\frac{\sqrt{14}}{7}$ 2. a. $\frac{5\sqrt{2}}{2}$ b. $\frac{\sqrt{35x}}{10}$ 3. $\frac{32-8\sqrt{5}}{11}$ 4. $2\sqrt{7}+2\sqrt{3}$

Concept and Vocabulary Check

1. rationalizing the denominator 2. $\sqrt{7}$ 3. $2\sqrt{3}; \sqrt{3}$ 4. $7-\sqrt{5}$ 5. $\sqrt{11}+\sqrt{3}$

Exercise Set 8.4

1. $\frac{\sqrt{10}}{10}$ 3. $\sqrt{5}$ 5. $\frac{\sqrt{6}}{3}$ 7. $4\sqrt{7}$ 9. $\frac{\sqrt{15}}{5}$ 11. $\frac{\sqrt{21}}{3}$ 13. $\frac{x\sqrt{3}}{3}$ 15. $\frac{\sqrt{11x}}{x}$ 17. $\frac{\sqrt{xy}}{y}$ 19. $\frac{x^2\sqrt{2}}{2}$ 21. $\frac{\sqrt{35}}{5}$

23. $\frac{\sqrt{42x}}{14}$ 25. $\frac{\sqrt{5}}{10}$ 27. $\frac{3\sqrt{2}}{2}$ 29. $\frac{5\sqrt{3}}{2}$ 31. $\frac{\sqrt{10}}{6}$ 33. $\frac{\sqrt{2x}}{8}$ 35. $\frac{\sqrt{5}}{15}$ 37. $\frac{\sqrt{21}}{6}$ 39. $2x\sqrt{2}$ 41. $\frac{\sqrt{14y}}{4}$ 43. $\frac{\sqrt{21x}}{6}$

45. $\frac{3\sqrt{5x}}{x}$ 47. $\frac{5\sqrt{x}}{x^2}$ 49. $\frac{3\sqrt{3y}}{y^2}$ 51. $\frac{5x\sqrt{6y}}{6y^2}$ 53. $\frac{4-\sqrt{3}}{13}$ 55. $-6-3\sqrt{7}$ 57. $8\sqrt{11}-24$ 59. $9+3\sqrt{3}$

61. $2-\sqrt{2}$ 63. $\frac{10+\sqrt{70}}{3}$ 65. $2\sqrt{6}-2\sqrt{3}$ 67. $\sqrt{5}+\sqrt{3}$ 69. $\frac{8-2\sqrt{x}}{16-x}$ 71. $\frac{6\sqrt{5}-4\sqrt{3}}{11}$ 73. $\frac{7+2\sqrt{10}}{3}$

75. $\frac{3y^2\sqrt{2x}}{x}$ 77. $\sqrt{x+2}+\sqrt{x}$ 79. $\frac{5\sqrt{6}}{6}$ 81. $2\sqrt{x+2}-h$ 83. $\frac{7\sqrt{2\cdot2\cdot3}}{6}=\frac{7\cdot2\sqrt{3}}{6}=\frac{14\sqrt{3}}{6}=\frac{7\sqrt{3}}{3}=\frac{7}{3}\sqrt{3}$

85. $\frac{\sqrt{5}+1}{2}\approx1.62$ 89. about 20%, 24%, and 28% 91. makes sense 93. makes sense 95. true 97. false 99. $\frac{3\sqrt{2}}{2}$

101. $\frac{4}{2+\sqrt{3}}$ 102. $\frac{5}{6}$ and 1, or $\left\{\frac{5}{6},1\right\}$ 103. $\frac{1}{8x^6}$ 104. $\frac{x-3}{4(x+3)}$ 105. 4 or {4} 106. 1 and 5, or {1, 5} 107. 5

Section 8.5 Check Point Exercises

1. 11 or {11} 2. 4 or {4} 3. no solution or ∅ 4. 6 or {6} 5. $225,000

Concept and Vocabulary Check

1. radical 2. extraneous 3. $x+3; x^2-6x+9$ 4. false 5. true 6. false

Exercise Set 8.5

1. 25 or {25} 3. 16 or {16} 5. 7 or {7} 7. 124 or {124} 9. 7 or {7} 11. 4 or {4} 13. 2 or {2} 15. 3 or {3} 17. 24 or {24}
19. no solution or ∅ 21. no solution or ∅ 23. 16 or {16} 25. no solution or ∅ 27. 7 or {7} 29. 2 or {2} 31. −3 or {−3}
33. −6 or {−6} 35. 2 or {2} 37. 1 and 3, or {1, 3} 39. no solution or ∅ 41. 12 or {12} 43. no solution or ∅ 45. 16
47. 8 49. $h=\frac{v^2}{2g}$ 51. 1 or {1} 53. 9 or {9} 55. 49 pounds per square inch 57. by the point (49, 840) 59. a. 9.5%
b. underestimates by 0.1% c. 2009 65. does not make sense 67. does not make sense 69. false 71. false 73. 4 and 5
75. 7 or {7} 77. 1 or {1} 79. no solution or ∅ 81. $7000 at 6% and $2000 at 4% 82. $4.25 83. (−1, −2) or {(−1, −2)}
84. −16 85. 9 86. $\frac{1}{27}$

Section 8.6 Check Point Exercises

1. a. 5 b. 2 c. −3 d. −2 2. a. 81 b. 8 c. −8 3. a. $\frac{1}{5}$ b. $\frac{1}{4}$ c. $\frac{1}{16}$ 4. a. 13%

b. overestimates by 1% c. $W=\frac{62}{(\sqrt[5]{x})^7}$ or $W=\frac{62}{\sqrt[5]{x^7}}$

Concept and Vocabulary Check

1. 6 2. $\sqrt{a}$ 3. $\sqrt[3]{8}=2$ 4. $\sqrt[3]{a}$ 5. $(\sqrt[4]{16})^3=(2)^3=8$ 6. $(\sqrt[5]{a})^4$ 7. $\frac{1}{16^{\frac{3}{2}}}=\frac{1}{(\sqrt{16})^3}=\frac{1}{(4)^3}=\frac{1}{64}$

Exercise Set 8.6

1. 7 3. 11 5. 3 7. −5 9. 2 11. −2 13. $\frac{1}{3}$ 15. $\frac{3}{4}$ 17. 729 19. 25 21. 27 23. −8 25. $\frac{1}{3}$ 27. $\frac{1}{5}$
29. $\frac{1}{2}$ 31. 2 33. $\frac{1}{8}$ 35. $\frac{1}{243}$ 37. $\frac{1}{4}$ 39. $\frac{5}{2}$ 41. $\frac{5}{2}$ 43. $\frac{1}{4}$ 45. 17 47. 375 49. $x^{\frac{7}{12}}$ 51. $\frac{1}{x^{\frac{2}{3}}}$ 53. $x^{\frac{1}{6}}y^2$ 55. $\frac{1}{x^7}$

57. a. 83.2 m b. $h=0.84(\sqrt[3]{d})^2$ or $h=0.84\sqrt[3]{d^2}$ 59. a. approximately 0.382 revolution per second

b. approximately 22.92 revolutions per minute c. $N=\frac{\sqrt{ar}}{2\pi r}$ 67. does not make sense 69. does not make sense 71. false

73. false 75. true 77. true 79. 3

81. a–b. 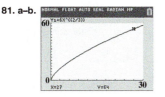   **c.** 24 square units   **82.** $(-2, 0)$ or $\{(-2, 0)\}$ **83.** 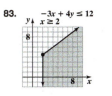   **84.** $32x^2$

85. solution   **86.** not a solution   **87.** $5\sqrt{5}$

Review Exercises

1. 11   **2.** -11 **3.** not a real number   **4.** $\dfrac{2}{5}$ **5.** -2 **6.** -3 **7.** 8.660   **8.** 19.824   **9.** 14.6 min

10. Answers will vary. Example: The number of disruptive minutes is increasing, but the rate of increase is slowing down.   **11.** 47 mi

12. $3\sqrt{6}$ **13.** $12\sqrt{5}$ **14.** $3x\sqrt{7}$ **15.** $4x\sqrt{3x}$ **16.** x^4 **17.** $5x^4\sqrt{3x}$ **18.** $3x^{11}\sqrt{5x}$ **19.** $2\sqrt[3]{3}$ **20.** $\sqrt{77}$ **21.** 6

22. $5x\sqrt{2}$ **23.** $2x^2\sqrt{3x}$ **24.** $3\sqrt[3]{2}$ **25.** $\dfrac{\sqrt{15}}{4}$ **26.** $\dfrac{11}{2}$ **27.** $\dfrac{\sqrt{7x}}{5}$ **28.** $\dfrac{3\sqrt{2}}{x}$ **29.** 10   **30.** $4\sqrt{2}$ **31.** $6x^2\sqrt{2x}$ **32.** $\dfrac{\sqrt[3]{5}}{4}$

33. $\dfrac{2\sqrt[3]{5}}{3}$ **34.** $20\sqrt{5}$ **35.** $7\sqrt{2}$ **36.** $\sqrt{3}$ **37.** $17\sqrt{5}$ **38.** $24\sqrt{2} - 8\sqrt{3}$ **39.** $6\sqrt{2} + 7\sqrt{3}$ **40.** $5\sqrt{2} + 2\sqrt{15}$ **41.** $3\sqrt{2} - 6$

42. $92 + 19\sqrt{2}$ **43.** $-17 + 11\sqrt{7}$ **44.** $3\sqrt{2} - 4\sqrt{3} + 2\sqrt{6} - 8$ **45.** -3 **46.** 6   **47.** $3 + 2\sqrt{2}$ **48.** $6\sqrt{5}$ **49.** $\dfrac{13\sqrt{2}}{10}$

50. $\dfrac{\sqrt{6}}{3}$ **51.** $\dfrac{\sqrt{6}}{4}$ **52.** $\dfrac{\sqrt{17x}}{x}$ **53.** $11\sqrt{5} - 22$ **54.** $\dfrac{84 + 21\sqrt{3}}{13}$ **55.** $6\sqrt{5} - 6\sqrt{3}$ **56.** $-1 - 2\sqrt{2}$ **57.** 13 or $\{13\}$

58. 11 or $\{11\}$ **59.** 5 or $\{5\}$ **60.** 0 and 3, or $\{0, 3\}$ **61.** 8 or $\{8\}$ **62.** no solution or $\varnothing$ **63.** no solution or $\varnothing$ **64.** 8 ft

65. a. $48\% \pm 1\%$ **b.** 47.8%   **c.** 2024   **66.** $\sqrt{16} = 4$ **67.** $\sqrt[3]{125} = 5$ **68.** $(\sqrt[3]{64})^2 = 16$ **69.** $\dfrac{1}{\sqrt{25}} = \dfrac{1}{5}$

70. $\dfrac{1}{\sqrt[3]{27}} = \dfrac{1}{3}$ **71.** $\dfrac{1}{(\sqrt[3]{-8})^4} = \dfrac{1}{16}$ **72.** approx 57 species

Chapter Test

1. -8 **2.** 4   **3.** $4\sqrt{3}$ **4.** $6x\sqrt{2x}$ **5.** $x^{14}\sqrt{x}$ **6.** $\dfrac{5}{x}$ **7.** $\dfrac{5}{3}$ **8.** $\dfrac{\sqrt[3]{5}}{2}$ **9.** $2x\sqrt{10}$ **10.** $5\sqrt{2}$ **11.** $6\sqrt{xy}$

12. $2x^2\sqrt{5x}$ **13.** $11\sqrt{6}$ **14.** $6\sqrt{2}$ **15.** $\sqrt{30} + 3$ **16.** $55 + 11\sqrt{5}$ **17.** 2   **18.** $16 + 6\sqrt{7}$ **19.** $\dfrac{4\sqrt{5}}{5}$ **20.** $\dfrac{20 - 5\sqrt{3}}{13}$

21. 12 or $\{12\}$ **22.** 5 or $\{5\}$ **23.** \$25 thousand   **24.** $(\sqrt[3]{8})^2 = 4$ **25.** $\dfrac{1}{\sqrt{9}} = \dfrac{1}{3}$

Cumulative Review Exercises

1. -3 or $\{-3\}$ **2.** -4 and $\dfrac{3}{2}$, or $\left\{-4, \dfrac{3}{2}\right\}$ **3.** $(-3, -4)$ or $\{(-3, -4)\}$ **4.** $\dfrac{9}{7}$ or $\left\{\dfrac{9}{7}\right\}$ **5.** -5 or $\{-5\}$ **6.** 5 or $\{5\}$

7. $-\dfrac{2}{x^4}$ **8.** $22\sqrt{3}$ **9.** $\dfrac{3}{x}$ **10.** $-\dfrac{2 + x}{3x + 1}$ **11.** 5   **12.** $(x - 7)(x - 11)$ **13.** $x(x + 5)(x - 5)$ **14.** $6x^2 - 7x + 2$ **15.** $8x^3 - 27$

16. $\dfrac{1}{x - 1}$ **17.** $\dfrac{5x - 1}{4x - 3}$ **18.** $-18\sqrt{3}$

19. $2x - y = 4$ **20.** $y = -\dfrac{2}{3}x$ **21.** $x \geq -1$

22. $-\dfrac{8}{3}$ **23.** $y + 3 = 5(x + 2); y = 5x + 7$ **24.** 43   **25.** 954 deer

CHAPTER 9

Section 9.1 Check Point Exercises

1. a. -6 and 6, or $\{\pm 6\}$ **b.** $-\sqrt{3}$ and $\sqrt{3}$, or $\{\pm\sqrt{3}\}$ **c.** $\pm\dfrac{\sqrt{14}}{2}$ or $\left\{\pm\dfrac{\sqrt{14}}{2}\right\}$ **2.** -2 and 8, or $\{-2, 8\}$ **3.** $2 \pm \sqrt{7}$ or $\{2 \pm \sqrt{7}\}$
4. 32 inches   **5.** 13 units

Concept and Vocabulary Check

1. $\pm\sqrt{d}$ **2.** right; hypotenuse; legs   **3.** right; legs; the square of the length of the hypotenuse   **4.** $\sqrt{(x_2 - x_1)^2 + (y_2 - y_1)^2}$

Exercise Set 9.1

1. ± 4 or $\{\pm 4\}$ **3.** ± 9 or $\{\pm 9\}$ **5.** $\pm\sqrt{7}$ or $\{\pm\sqrt{7}\}$ **7.** $\pm 5\sqrt{2}$ or $\{\pm 5\sqrt{2}\}$ **9.** ± 2 or $\{\pm 2\}$ **11.** $\pm\dfrac{7}{2}$ or $\left\{\pm\dfrac{7}{2}\right\}$ **13.** ± 5 or $\{\pm 5\}$

15. $\pm\dfrac{\sqrt{6}}{3}$ or $\left\{\pm\dfrac{\sqrt{6}}{3}\right\}$ **17.** $\pm\dfrac{\sqrt{35}}{5}$ or $\left\{\pm\dfrac{\sqrt{35}}{5}\right\}$ **19.** -1 and 7, or $\{-1, 7\}$ **21.** -16 and 6, or $\{-16; 6\}$ **23.** $-\dfrac{5}{3}$ and $\dfrac{1}{3}$, or $\left\{-\dfrac{5}{3}, \dfrac{1}{3}\right\}$

25. $5 \pm \sqrt{3}$ or $\{5 \pm \sqrt{3}\}$ **27.** $-8 \pm \sqrt{11}$ or $\{-8 \pm \sqrt{11}\}$ **29.** $4 \pm 3\sqrt{2}$ or $\{4 \pm 3\sqrt{2}\}$ **31.** -6 and 2, or $\{-6, 2\}$ **33.** -3 and 9, or $\{-3, 9\}$

35. $5 \pm \sqrt{2}$ or $\{5 \pm \sqrt{2}\}$ **37.** $-1 \pm \sqrt{5}$ or $\{-1 \pm \sqrt{5}\}$ **39.** $7 \pm 2\sqrt{3}$ or $\{7 \pm 2\sqrt{3}\}$ **41.** 17 m **43.** 39 m **45.** 12 cm **47.** $5\sqrt{7}$ m

49. $\sqrt{17}$ or 4.12 units **51.** 17 units **53.** $3\sqrt{5}$ or 6.71 units **55.** $\sqrt{41}$ or 6.40 units **57.** 16 units **59.** -2 and 8 **61.** -3 and $\dfrac{5}{3}$

63. $r = \dfrac{\sqrt{A\pi}}{\pi}$ **65.** $d = \dfrac{\sqrt{kI}}{I}$ **67.** 36 inches **69.** $2\sqrt{41}$ ft **71.** $90\sqrt{2}$ ft **73.** 6 in. **75.** 6 weeks **77.** 5 sec **79.** 8 m **81.** 10 in.

87. makes sense **89.** makes sense **91.** false **93.** true **95.** -7 and 1 **97.** -2 and 4, or $\{-2, 4\}$ **98.** $2(2x + 3)(3x - 1)$

99. $\dfrac{x - 3}{3(x - 2)}$ **100.** -6 or $\{-6\}$ **101.** $(x + 4)^2$ **102.** $(x - 7)^2$ **103.** $\left(x + \dfrac{5}{2}\right)^2$

Section 9.2 Check Point Exercises

1. a. $x^2 + 10x + 25 = (x + 5)^2$ **b.** $x^2 - 6x + 9 = (x - 3)^2$ **c.** $x^2 + 3x + \dfrac{9}{4} = \left(x + \dfrac{3}{2}\right)^2$

2. a. -7 and 1, or $\{-7, 1\}$ **b.** $5 \pm\sqrt{7}$ or $\{5 \pm\sqrt{7}\}$ **3.** $\dfrac{5 \pm 3\sqrt{3}}{2}$ or $\left\{\dfrac{5 \pm 3\sqrt{3}}{2}\right\}$

Concept and Vocabulary Check

1. 81 **2.** 9 **3.** $\dfrac{49}{4}$ **4.** $\dfrac{25}{16}$ **5.** $\left(\dfrac{b}{2}\right)^2$ or $\dfrac{b^2}{4}$ **6.** 100 **7.** 1 **8.** $\dfrac{1}{4}$

Exercise Set 9.2

1. $x^2 + 10x + 25 = (x + 5)^2$ **3.** $x^2 - 2x + 1 = (x - 1)^2$ **5.** $x^2 + 5x + \dfrac{25}{4} = \left(x + \dfrac{5}{2}\right)^2$ **7.** $x^2 - 7x + \dfrac{49}{4} = \left(x - \dfrac{7}{2}\right)^2$

9. $x^2 + \dfrac{1}{2}x + \dfrac{1}{16} = \left(x + \dfrac{1}{4}\right)^2$ **11.** $x^2 - \dfrac{4}{3}x + \dfrac{4}{9} = \left(x - \dfrac{2}{3}\right)^2$ **13.** -5 and 1, or $\{-5, 1\}$ **15.** 4 and 6, or $\{4, 6\}$

17. $1 \pm\sqrt{6}$ or $\{1 \pm\sqrt{6}\}$ **19.** $-2 \pm\sqrt{3}$ or $\{-2 \pm\sqrt{3}\}$ **21.** -4 and 7, or $\{-4, 7\}$ **23.** $\dfrac{-3 \pm\sqrt{13}}{2}$ or $\left\{\dfrac{-3 \pm\sqrt{13}}{2}\right\}$

25. $\dfrac{7 \pm\sqrt{37}}{2}$ or $\left\{\dfrac{7 \pm\sqrt{37}}{2}\right\}$ **27.** $\dfrac{1 \pm\sqrt{13}}{2}$ or $\left\{\dfrac{1 \pm\sqrt{13}}{2}\right\}$ **29.** $\dfrac{1}{2}$ and 1, or $\left\{\dfrac{1}{2}, 1\right\}$ **31.** $\dfrac{-5 \pm\sqrt{3}}{2}$ or $\left\{\dfrac{-5 \pm\sqrt{3}}{2}\right\}$

33. $\dfrac{1 \pm\sqrt{13}}{4}$ or $\left\{\dfrac{1 \pm\sqrt{13}}{4}\right\}$ **35.** $1 \pm\sqrt{7}$ or $\{1 \pm\sqrt{7}\}$ **37.** $\dfrac{1 \pm\sqrt{29}}{2}$ or $\left\{\dfrac{1 \pm\sqrt{29}}{2}\right\}$ **39.** $-5b$ and b, or $\{-5b, b\}$

43. makes sense **45.** makes sense **47.** false **49.** false **51.** $x^2 - 20x + 100$

53. $-\dfrac{b}{2} \pm\sqrt{\dfrac{b^2}{4} - c}$, $\dfrac{-b \pm\sqrt{b^2 - 4ac}}{2}$, $\left\{-\dfrac{b}{2} \pm\sqrt{\dfrac{b^2}{4} - c}\right\}$, or $\left\{\dfrac{-b \pm\sqrt{b^2 - 4ac}}{2}\right\}$ **55.** $\dfrac{11}{(x - 3)(x - 4)}$ **56.** $\dfrac{x}{3}$ **57.** 3 or $\{3\}$

58. $\dfrac{1}{2}$; -5 **59.** $\dfrac{2}{3}$ **60.** $1 + \sqrt{7}$; $1 - \sqrt{7}$

Section 9.3 Check Point Exercises

1. $-\dfrac{1}{2}$ and $\dfrac{1}{4}$, or $\left\{-\dfrac{1}{2}, \dfrac{1}{4}\right\}$ **2.** $3 \pm\sqrt{5}$ or $\{3 \pm\sqrt{5}\}$ **3.** 2032

Concept and Vocabulary Check

1. $\dfrac{-b \pm\sqrt{b^2 - 4ac}}{2a}$ **2.** 3; 10; -8 **3.** 1; 8; -29 **4.** 6; -3; -4 **5.** $-\dfrac{1}{3}$; 2 **6.** $-4 \pm 3\sqrt{5}$

7. the square root property **8.** factoring and the zero-product principle **9.** the quadratic formula

Exercise Set 9.3

1. -3 and -2, or $\{-3, -2\}$ **3.** $\dfrac{-5 \pm\sqrt{13}}{2}$ or $\left\{\dfrac{-5 \pm\sqrt{13}}{2}\right\}$ **5.** $-2 \pm\sqrt{10}$ or $\{-2 \pm\sqrt{10}\}$ **7.** $-2 \pm\sqrt{11}$ or $\{-2 \pm\sqrt{11}\}$

9. -3 and 6, or $\{-3, 6\}$ **11.** $-\dfrac{2}{3}$ and $\dfrac{3}{2}$, or $\left\{-\dfrac{2}{3}, \dfrac{3}{2}\right\}$ **13.** $1 \pm\sqrt{11}$ or $\{1 \pm\sqrt{11}\}$ **15.** $\dfrac{1 \pm\sqrt{57}}{2}$ or $\left\{\dfrac{1 \pm\sqrt{57}}{2}\right\}$

17. $\dfrac{-3 \pm\sqrt{3}}{6}$ or $\left\{\dfrac{-3 \pm\sqrt{3}}{6}\right\}$ **19.** $\dfrac{2}{3}$ or $\left\{\dfrac{2}{3}\right\}$ **21.** $\dfrac{1 \pm\sqrt{29}}{4}$ or $\left\{\dfrac{1 \pm\sqrt{29}}{4}\right\}$ **23.** $-\dfrac{1}{2}$ and 1, or $\left\{-\dfrac{1}{2}, 1\right\}$ **25.** $\dfrac{1}{5}$ and 2, or $\left\{\dfrac{1}{5}, 2\right\}$

27. $\pm2\sqrt{5}$ or $\{\pm2\sqrt{5}\}$ **29.** $1\pm\sqrt{2}$ or $\{1\pm\sqrt{2}\}$ **31.** $\dfrac{-11\pm\sqrt{33}}{4}$ or $\left\{\dfrac{-11\pm\sqrt{33}}{4}\right\}$ **33.** 0 and $\dfrac{8}{3}$, or $\left\{0,\dfrac{8}{3}\right\}$ **35.** 2 or $\{2\}$

37. -2 and 2, or $\{-2,2\}$ **39.** -9 and 0, or $\{-9,0\}$ **41.** $-\dfrac{2}{3}$ and 4, or $\left\{-\dfrac{2}{3},4\right\}$ **43.** $\dfrac{2\pm\sqrt{10}}{3}$ or $\left\{\dfrac{2\pm\sqrt{10}}{3}\right\}$ **45.** $\pm\sqrt{3}$ or $\{\pm\sqrt{3}\}$

47. $-\dfrac{5}{2}$ and 0, or $\left\{-\dfrac{5}{2},0\right\}$ **49.** $\pm\sqrt{6}$ or $\{\pm\sqrt{6}\}$ **51.** $\dfrac{5\pm\sqrt{73}}{2}$ or $\left\{\dfrac{5\pm\sqrt{73}}{2}\right\}$ **53.** about 3.8 sec   **55.** 10 ft

57. a. 28%; overestimates by 1%   **b.** 30 years after 1990, or 2020   **59.** width: 4.7 m; length: 7.7 m   **61.** 2.3 and 3.3 ft

63. a. $\dfrac{1}{\Phi-1}$ **b.** $\Phi=\dfrac{1+\sqrt{5}}{2}$ **c.** $\dfrac{1+\sqrt{5}}{2}$ **69.** does not make sense   **71.** does not make sense   **73.** false   **75.** true   **79.** about 1.2 feet

81.

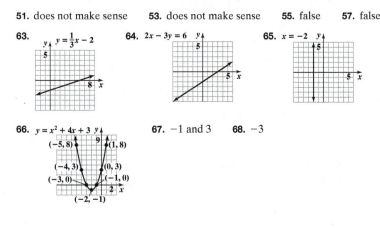

82. $\dfrac{1}{25}$ **83.** $9-3\sqrt{5}$ **84.** x^3-y^3 **85.** no; no   **86. a.** $-\sqrt{4},\sqrt{1}$ **b.** $-\sqrt{5},\sqrt{5}$ **c.** $-\sqrt{5},-\sqrt{4},\sqrt{1},\sqrt{5}$ **d.** $\sqrt{-4},\sqrt{-1}$
87. a. $-\sqrt{9},\sqrt{0},\sqrt{9}$ **b.** $-\sqrt{7},\sqrt{7}$ **c.** $-\sqrt{9},-\sqrt{7},\sqrt{0},\sqrt{9},\sqrt{7}$ **d.** $\sqrt{-9},\sqrt{-7}$

Mid-Chapter Check Point Exercises

1. $-\dfrac{8}{3}$ and 4, or $\left\{-\dfrac{8}{3},4\right\}$ **2.** 0 and $\dfrac{1}{3}$, or $\left\{0,\dfrac{1}{3}\right\}$ **3.** $1\pm\sqrt{11}$ or $\{1\pm\sqrt{11}\}$ **4.** $4\pm\sqrt{7}$ or $\{4\pm\sqrt{7}\}$ **5.** $-\dfrac{2}{3}$ and 1, or $\left\{-\dfrac{2}{3},1\right\}$

6. $\dfrac{5\pm\sqrt{7}}{6}$ or $\left\{\dfrac{5\pm\sqrt{7}}{6}\right\}$ **7.** -4 and 3, or $\{-4,3\}$ **8.** $-5\pm2\sqrt{10}$ or $\{-5\pm2\sqrt{10}\}$ **9.** -3 and 3, or $\{-3,3\}$

10. $\dfrac{-5\pm\sqrt{17}}{4}$ or $\left\{\dfrac{-5\pm\sqrt{17}}{4}\right\}$ **11.** $-\dfrac{1}{2}$ and 10, or $\left\{-\dfrac{1}{2},10\right\}$ **12.** 4 or $\{4\}$ **13.** -16 and 2, or $\{-16,2\}$ **14.** $2\sqrt{7}$ cm
15. 13 units   **16.** 12 in. and 16 in.

Section 9.4 Check Point Exercises

1. a. $4i$ **b.** $i\sqrt{5}$ **c.** $5i\sqrt{2}$ **2.** $-2+5i$ and $-2-5i$, or $\{-2\pm5i\}$ **3.** $-3\pm2i$ or $\{-3\pm2i\}$

Concept and Vocabulary Check

1. $\sqrt{-1}$; -1 **2.** $5i$ **3.** complex; imaginary; real   **4.** $2\pm i$

Exercise Set 9.4

1. $6i$ **3.** $i\sqrt{13}$ **5.** $5i\sqrt{2}$ **7.** $2i\sqrt{5}$ **9.** $-2i\sqrt{7}$ **11.** $7+4i$ **13.** $10+i\sqrt{3}$ **15.** $6-7i\sqrt{2}$ **17.** $3\pm3i$ or $\{3\pm3i\}$
19. $-7\pm8i$ or $\{-7\pm8i\}$ **21.** $2\pm i\sqrt{7}$ or $\{2\pm i\sqrt{7}\}$ **23.** $-3\pm3i\sqrt{2}$ or $\{-3\pm3i\sqrt{2}\}$ **25.** $-2\pm i$ or $\{-2\pm i\}$

27. $3\pm2i$ or $\{3\pm2i\}$ **29.** $6\pm2i$ or $\{6\pm2i\}$ **31.** $5\pm i\sqrt{2}$ or $\{5\pm i\sqrt{2}\}$ **33.** $\dfrac{1\pm i\sqrt{14}}{5}$ or $\left\{\dfrac{1\pm i\sqrt{14}}{5}\right\}$

35. $\dfrac{2\pm i\sqrt{6}}{2}$ or $\left\{\dfrac{2\pm i\sqrt{6}}{2}\right\}$ **37.** $\pm i\sqrt{5}$ or $\{\pm i\sqrt{5}\}$ **39.** $1\pm3i$ or $\{1\pm3i\}$ **41.** $\dfrac{-1\pm i\sqrt{7}}{2}$ or $\left\{\dfrac{-1\pm i\sqrt{7}}{2}\right\}$

43. $9\pm i\sqrt{19}$ are not real numbers. Applicant will not be hired. Answers will vary.

51. does not make sense   **53.** does not make sense   **55.** false   **57.** false

63. $y=\dfrac{1}{3}x-2$

64. $2x-3y=6$

65. $x=-2$

66. $y=x^2+4x+3$
$(-5,8)$ $(1,8)$ $(-4,3)$ $(0,3)$ $(-3,0)$ $(-1,0)$ $(-2,-1)$

67. -1 and 3 **68.** -3

Section 9.5 Check Point Exercises

1. a. upward **b.** $y = x^2 - 6x + 8$ **2.** 2 and 4 **3.** 8 **4.** $(3, -1)$

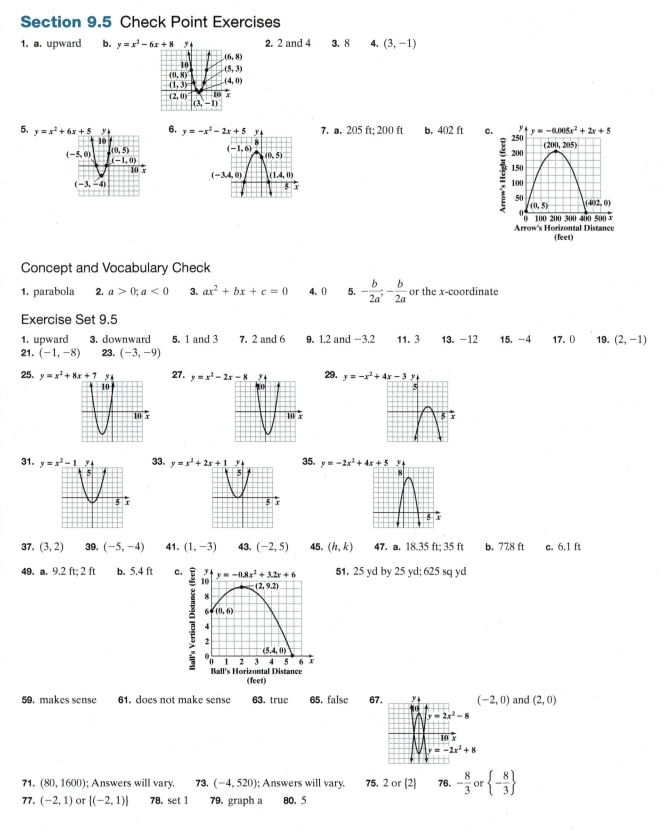

Concept and Vocabulary Check

1. parabola **2.** $a > 0; a < 0$ **3.** $ax^2 + bx + c = 0$ **4.** 0 **5.** $-\dfrac{b}{2a}; -\dfrac{b}{2a}$ or the x-coordinate

Exercise Set 9.5

1. upward **3.** downward **5.** 1 and 3 **7.** 2 and 6 **9.** 1.2 and -3.2 **11.** 3 **13.** -12 **15.** -4 **17.** 0 **19.** $(2, -1)$
21. $(-1, -8)$ **23.** $(-3, -9)$

37. $(3, 2)$ **39.** $(-5, -4)$ **41.** $(1, -3)$ **43.** $(-2, 5)$ **45.** (h, k) **47. a.** 18.35 ft; 35 ft **b.** 77.8 ft **c.** 6.1 ft

49. a. 9.2 ft; 2 ft **b.** 5.4 ft **c.** **51.** 25 yd by 25 yd; 625 sq yd

59. makes sense **61.** does not make sense **63.** true **65.** false **67.** $(-2, 0)$ and $(2, 0)$

71. $(80, 1600)$; Answers will vary. **73.** $(-4, 520)$; Answers will vary. **75.** 2 or {2} **76.** $-\dfrac{8}{3}$ or $\left\{-\dfrac{8}{3}\right\}$
77. $(-2, 1)$ or $\{(-2, 1)\}$ **78.** set 1 **79.** graph a **80.** 5

Section 9.6 Check Point Exercises

1. domain: {golf, lawn mowing, water skiing, hiking, bicycling }; range: {250, 325, 430, 720} **2. a.** not a function
b. function **3. a.** 23 **b.** -5 **c.** 3 **4. a.** 48 **b.** 3 **c.** 3 **5. a.** function **b.** function **c.** not a function
6. a. 75% **b.** 75.4% **c.** function f, the linear function

Concept and Vocabulary Check

1. relation; domain; range **2.** function **3.** $f; x$ **4.** linear **5.** quadratic **6.** more than once; function

Exercise Set 9.6

1. function; domain: $\{1, 3, 5\}$; range: $\{2, 4, 5\}$ **3.** not a function; domain: $\{3, 4\}$; range: $\{4, 5\}$ **5.** function; domain: $\{-3, -2, -1, 0\}$;
range: $\{-3, -2, -1, 0\}$ **7.** not a function; domain: $\{1\}$; range: $\{4, 5, 6\}$ **9. a.** 12 **b.** -1 **c.** 5 **11. a.** 70 **b.** -28 **c.** 0
13. a. 93 **b.** -7 **c.** -3 **15. a.** 10 **b.** -2 **c.** 0 **17. a.** 11 **b.** 27 **c.** 3 **19. a.** 5 **b.** 5 **c.** 5 **21. a.** 3 **b.** 7
23. a. 1 **b.** -1 **25.** function **27.** function **29.** not a function **31.** function **33.** $\{(-1, 1), (0, 3), (1, 5)\}$
35. $\{(-2, -6), (-1, -2), (0, 0), (1, 0), (2, -2)\}$ **37.** 6 **39.** $x + h$ **41. a.** $\{$(Philippines, 12), (Spain, 13), (Italy, 14), (Germany, 14), (Russia, 16)$\}$
b. Yes; Each country corresponds to exactly one age. **c.** $\{$(12, Philippines), (13, Spain), (14, Italy), (14, Germany), (16, Russia)$\}$
d. No; 14 in the domain corresponds to two members of the range, Italy and Germany. **43.** 186.6; At age 20, an American man's cholesterol level
is 186.6, or approx 187. **45. a.** $p(20) = 38$; in 2012, or 20 years after 1992, 38% of college freshmen had spent six or more hours per week studying
and doing homework during their high school senior year. **b.** It's the same. **c.** by the point (20, 38) **53.** makes sense
55. does not make sense **57.** false **59.** true **61.** $f(r) = 0$; r is a solution of the quadratic equation $ax^2 + bx + c = 0$. **63.** 3.97×10^{-3}
64. $x^2 + 9x + 16 + \dfrac{35}{x - 2}$ **65.** $\left(\dfrac{4}{5}, \dfrac{9}{5}\right)$ or $\left\{\left(\dfrac{4}{5}, \dfrac{9}{5}\right)\right\}$

Review Exercises

1. -8 and 8, or $\{-8, 8\}$ **2.** $\pm\sqrt{17}$ or $\{\pm\sqrt{17}\}$ **3.** $\pm 5\sqrt{3}$ or $\{\pm 5\sqrt{3}\}$ **4.** 0 and 6, or $\{0, 6\}$ **5.** $-4 \pm \sqrt{5}$ or $\{-4 \pm \sqrt{5}\}$
6. $\pm\dfrac{\sqrt{15}}{3}$ or $\left\{\pm\dfrac{\sqrt{15}}{3}\right\}$ **7.** 1 and 6, or $\{1, 6\}$ **8.** $-5 \pm 2\sqrt{3}$ or $\{-5 \pm 2\sqrt{3}\}$ **9.** 10 ft **10.** $2\sqrt{13}$ in. **11.** $2\sqrt{26}$ cm
12. 15 ft **13.** 12 yd **14.** 20 weeks **15.** 2.5 sec **16.** 5 units **17.** $2\sqrt{5}$ or 4.47 units **18.** $x^2 + 16x + 64 = (x + 8)^2$
19. $x^2 - 6x + 9 = (x - 3)^2$ **20.** $x^2 + 3x + \dfrac{9}{4} = \left(x + \dfrac{3}{2}\right)^2$ **21.** $x^2 - 5x + \dfrac{25}{4} = \left(x - \dfrac{5}{2}\right)^2$ **22.** 3 and 9, or $\{3, 9\}$
23. $3 \pm \sqrt{5}$ or $\{3 \pm \sqrt{5}\}$ **24.** $\dfrac{6 \pm \sqrt{3}}{3}$ or $\left\{\dfrac{6 \pm \sqrt{3}}{3}\right\}$ **25.** -3 and $\dfrac{1}{2}$, or $\left\{-3, \dfrac{1}{2}\right\}$ **26.** $1 \pm \sqrt{5}$ or $\{1 \pm \sqrt{5}\}$
27. $\dfrac{9 \pm \sqrt{21}}{6}$ or $\left\{\dfrac{9 \pm \sqrt{21}}{6}\right\}$ **28.** $\dfrac{1}{2}$ and 5, or $\left\{\dfrac{1}{2}, 5\right\}$ **29.** -2 and $\dfrac{10}{3}$, or $\left\{-2, \dfrac{10}{3}\right\}$ **30.** $\dfrac{7 \pm \sqrt{37}}{6}$ or $\left\{\dfrac{7 \pm \sqrt{37}}{6}\right\}$ **31.** -3 and 3, or $\{-3, 3\}$
32. $\dfrac{3 \pm \sqrt{5}}{2}$ or $\left\{\dfrac{3 \pm \sqrt{5}}{2}\right\}$ **33. a.** 28%; underestimates by 1% **b.** 2009 **c.** 2 or 3 years after 2007, or in 2009 or 2010 **34.** $9i$
35. $i\sqrt{23}$ **36.** $4i\sqrt{3}$ **37.** $3 + 7i$ **38.** $\pm 10i$ or $\{\pm 10i\}$ **39.** $\pm 5i$ or $\{\pm 5i\}$ **40.** $\dfrac{-1 \pm 2i\sqrt{2}}{2}$ or $\left\{\dfrac{-1 \pm 2i\sqrt{2}}{2}\right\}$
41. $2 \pm 3i$ or $\{2 \pm 3i\}$ **42.** $\dfrac{1 \pm i\sqrt{23}}{6}$ or $\left\{\dfrac{1 \pm i\sqrt{23}}{6}\right\}$

43. a. upward **b.** -1 and 7 **c.** -7 **d.** $(3, -16)$ **e.**
44. a. downward **b.** -3 and 1 **c.** 3 **d.** $(-1, 4)$ **e.**
45. a. downward **b.** -0.2 and 2.2 **c.** 1 **d.** $(1, 4)$ **e.**
46. a. upward **b.** 0 and 4 **c.** 0 **d.** $(2, -4)$ **e.**

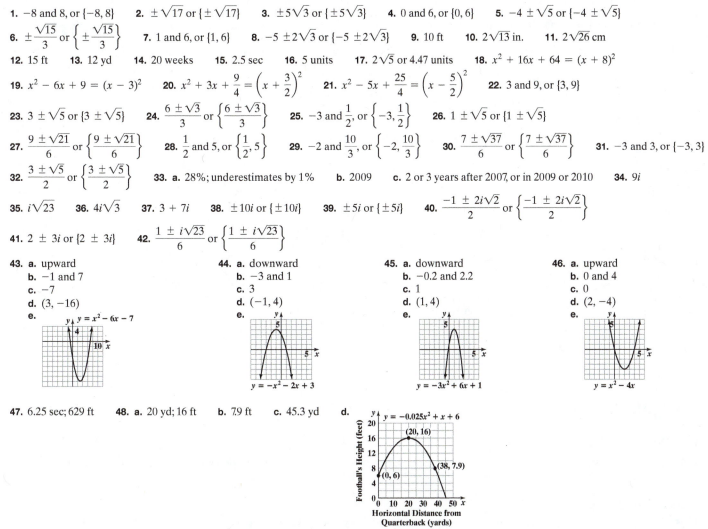

47. 6.25 sec; 629 ft **48. a.** 20 yd; 16 ft **b.** 7.9 ft **c.** 45.3 yd **d.**

49. function; domain: $\{2, 3, 5\}$; range: $\{7\}$ **50.** function; domain: $\{1, 2, 3\}$; range: $\{10, 500, \pi\}$ **51.** not a function; domain: $\{12, 14\}$;
range: $\{13, 15, 19\}$ **52. a.** -19 **b.** 14 **c.** -4 **53. a.** 38 **b.** -4 **c.** 2 **54.** not a function **55.** function **56.** function
57. not a function **58. a.** $D(1) = 92.8$; Skin damage begins for burn-prone people after 92.8 minutes when the sun's UV index is 1.; by the
point (1, 92.8) **b.** $D(10) = 19$; Skin damage begins for burn-prone people after 19 minutes when the sun's UV index is 10.; by the point (10, 19)

Chapter Test

1. −4 and 4, or {−4, 4}   **2.** $3 \pm \sqrt{5}$ or $\{3 \pm \sqrt{5}\}$ **3.** $4\sqrt{5}$ yd   **4.** $\sqrt{58}$ or 7.62 units   **5.** $-2 \pm \sqrt{7}$ or $\{-2 \pm \sqrt{7}\}$

6. $\dfrac{-5 \pm \sqrt{13}}{6}$ or $\left\{\dfrac{-5 \pm \sqrt{13}}{6}\right\}$ **7.** $-\dfrac{4}{3}$ and 1, or $\left\{-\dfrac{4}{3}, 1\right\}$ **8.** $-\dfrac{7}{2}$ and $\dfrac{5}{2}$, or $\left\{-\dfrac{7}{2}, \dfrac{5}{2}\right\}$ **9.** $\dfrac{3 \pm \sqrt{7}}{2}$ or $\left\{\dfrac{3 \pm \sqrt{7}}{2}\right\}$

10. -5 and $\dfrac{1}{2}$, or $\left\{-5, \dfrac{1}{2}\right\}$ **11.** $11i$ **12.** $5i\sqrt{3}$ **13.** $\pm 6i$ or $\{\pm 6i\}$ **14.** $5 \pm 5i$ or $\{5 \pm 5i\}$ **15.** $1 \pm 2i$ or $\{1 \pm 2i\}$

16.

$y = x^2 + 2x - 8$; (−4, 0), (2, 0), (0, −8), (−1, −9)

17.

$y = -2x^2 + 16x - 24$; (4, 8), (2, 0), (6, 0)

18. 2 sec; 69 ft   **19.** 4.1 sec   **20.** function; domain: {1, 3, 5, 6}; range: {2, 4, 6}   **21.** not a function; domain: {2, 4, 6}; range: {1, 3, 5, 6}

22. 67   **23.** 17   **24.** function   **25.** not a function   **26.** 36; In a tournament with 9 chess players, 36 games must be played.

Cumulative Review Exercises

1. 4 or {4}   **2.** 10 or {10}   **3.** $(-\infty, 7]$ or $\{x \mid x \le 7\}$ **4.** $(-3, 4)$ or $\{(-3, 4)\}$ **5.** $(3, 4)$ or $\{(3, 4)\}$ **6.** −8 or {−8}

7. −3 and −2, or {−3, −2}   **8.** 9 or {9}   **9.** $2 \pm 2\sqrt{5}$ or $\{2 \pm 2\sqrt{5}\}$ **10.** $\dfrac{3 \pm \sqrt{3}}{3}$ or $\left\{\dfrac{3 \pm \sqrt{3}}{3}\right\}$ **11.** $t = \dfrac{5r + 2}{A}$ **12.** $\dfrac{4}{x^9}$

13. 31   **14.** $10x^2 - 9x + 4$ **15.** $21x^2 - 23x - 20$ **16.** $25x^2 - 20x + 4$ **17.** $x^3 + y^3$ **18.** $\dfrac{x + 4}{3x^3}$ **19.** $\dfrac{-7x}{(x + 3)(x - 1)(x - 4)}$

20. $\dfrac{x}{5}$ **21.** $12\sqrt{5}$ **22.** $3x\sqrt{2}$ **23.** $\dfrac{2\sqrt{3}}{3}$ **24.** $6 + 2\sqrt{5}$ **25.** $(2x + 7)(2x - 7)$ **26.** $(x + 3)(x + 1)(x - 1)$ **27.** $2(x - 3)(x + 7)$

28. $x(x^2 + 4)(x + 2)(x - 2)$ **29.** $x(x - 5)^2$ **30.** $(x - 2)(x^2 + 2x + 4)$ **31.** $\dfrac{1}{4}$

32. $y = \dfrac{1}{3}x - 1$

33. $3x + 2y = -6$

34. $y = -2$

35. $3x - 4y \le 12$

36. $y = x^2 - 2x - 3$

37. $2x + y < 4$
$x > 2$

38. −2   **39.** $y - 2 = 2(x - 1)$ or $y - 6 = 2(x - 3)$; $y = 2x$ **40.** 43   **41.** \$320   **42.** length: 150 yd; width: 50 yd
43. \$12,500 at 7% and \$7500 at 9%   **44.** 8 l of 40% and 4 l of 70%   **45.** 4 hr   **46.** 2024 students   **47.** 16 ft
48. 35°, 25°, 120°   **49.** TV: \$350; stereo: \$370   **50.** length: 11 m; width: 5 m

APPENDIX

Check Point Exercises

1. \$9.5 million   **2. a.** 35   **b.** 82   **3. a.** \$122.6 million   **b.** \$17.5 million   **c.** Kennedy's net worth was much greater than the other presidents'.
4. a. 8   **b.** 3 and 8   **c.** no mode   **5.** mean: 155.6; median: 155; mode: 138

Appendix Exercise Set

1. 4.125   **3.** 95   **5.** 62   **7.** ≈ 3.45 **9.** 3.5   **11.** 95   **13.** 60   **15.** 3.6   **17.** 3   **19.** 95   **21.** 40   **23.** 2.5 and 4.2
25. mean: 45.2; median: 46; mode: 46, 47   **27.** mean: \$36,900; median: \$22,000; The median is more representative.; Answers will vary.

Applications Index

Subject Index

Credits

Definitions, Rules, and Formulas

The Real Numbers

Natural Numbers: $\{1, 2, 3, ...\}$

Whole Numbers: $\{0, 1, 2, 3, ...\}$

Integers: $\{..., -3, -2, -1, 0, 1, 2, 3, ...\}$

Rational Numbers: $\left\{\frac{a}{b} \mid a \text{ and } b \text{ are integers}, b \neq 0\right\}$

Irrational Numbers: $\{x \mid x \text{ is real and not rational}\}$

Set-Builder Notation and Graphs

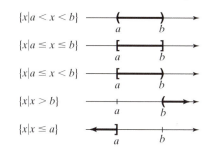

$\{x \mid a < x < b\}$

$\{x \mid a \leq x \leq b\}$

$\{x \mid a \leq x < b\}$

$\{x \mid x > b\}$

$\{x \mid x \leq a\}$

Basic Rules of Algebra

Commutative: $a + b = b + a; ab = ba$

Associative: $(a + b) + c = a + (b + c);$
$(ab)c = a(bc)$

Distributive: $a(b + c) = ab + ac;$
$a(b - c) = ab - ac$

Identity: $a + 0 = a; a \cdot 1 = a$

Inverse: $a + (-a) = 0; a \cdot \frac{1}{a} = 1 (a \neq 0)$

Multiplication Properties: $(-1)a = -a;$
$(-1)(-a) = a; a \cdot 0 = 0; (-a)(b) = (a)(-b) = -ab;$
$(-a)(-b) = ab$

Slope Formula

$$\text{slope}(m) = \frac{\text{Change in } y}{\text{Change in } x} = \frac{y_2 - y_1}{x_2 - x_1}, \quad x_2 - x_1 \neq 0$$

1. If $m > 0$, the line rises from left to right. If $m < 0$, the line falls from left to right.

2. The slope of a horizontal line is 0. The slope of a vertical line is undefined.

3. Parallel lines have equal slopes.

Order of Operations

1. Perform operations above and below any fraction bar, following steps (2) through (5).

2. Perform operations inside grouping symbols, innermost grouping symbols first, following steps (3) through (5).

3. Simplify exponential expressions.

4. Do multiplication and division as they occur, working from left to right.

5. Do addition and subtraction as they occur, working from left to right.

Equations of Lines

1. *Standard form:* $Ax + By = C$

2. *Slope-intercept form:* $y = mx + b$
   m is the line's slope and b is its y-intercept.

3. *Point-slope form:* $y - y_1 = m(x - x_1)$
   m is the line's slope and (x_1, y_1) is a fixed point on the line.

4. *Horizontal line parallel to the x-axis:* $y = b$

5. *Vertical line parallel to the y-axis:* $x = a$

Definitions, Rules, and Formulas (continued)

Properties of Exponents

1. $b^m \cdot b^n = b^{m+n}$
2. $(b^m)^n = b^{mn}$
3. $(ab)^m = a^m b^m$
4. $\dfrac{b^m}{b^n} = b^{m-n}$
5. $\left(\dfrac{a}{b}\right)^m = \dfrac{a^m}{b^m}$
6. $b^0 = 1$, where $b \neq 0$
7. $b^{-n} = \dfrac{1}{b^n}$ and $\dfrac{1}{b^{-n}} = b^n$, where $b \neq 0$

Special Factorizations

1. *Difference of two squares:*
$$A^2 - B^2 = (A + B)(A - B)$$
2. *Perfect square trinomials:*
$$A^2 + 2AB + B^2 = (A + B)^2$$
$$A^2 - 2AB + B^2 = (A - B)^2$$
3. *Sum of two cubes:*
$$A^3 + B^3 = (A + B)(A^2 - AB + B^2)$$
4. *Difference of two cubes:*
$$A^3 - B^3 = (A - B)(A^2 + AB + B^2)$$

Variation

English Statement	Equation
y varies directly as x.	$y = kx$
y varies inversely as x.	$y = \dfrac{k}{x}$

Properties of Radicals

All roots represent real numbers.

1. The product rule:
$$\sqrt[n]{a} \cdot \sqrt[n]{b} = \sqrt[n]{ab}$$
2. The quotient rule:
$$\dfrac{\sqrt[n]{a}}{\sqrt[n]{b}} = \sqrt[n]{\dfrac{a}{b}}, b \neq 0$$

Triangles

1. The sum of the measures of the interior angles of a triangle is $180°$.
2. Similar triangles have corresponding angles with the same measure and corresponding sides that are proportional. Two triangles are similar if two angles of one are equal in measure to two corresponding angles of the other.

Rational Exponents

1. $a^{\frac{1}{n}} = \sqrt[n]{a}$
2. $a^{\frac{m}{n}} = \left(\sqrt[n]{a}\right)^m = \sqrt[n]{a^m}$
3. $a^{-\frac{m}{n}} = \dfrac{1}{a^{\frac{m}{n}}} = \dfrac{1}{\left(\sqrt[n]{a}\right)^m} = \dfrac{1}{\sqrt[n]{a^m}}$

The Quadratic Formula

The solutions of $ax^2 + bx + c = 0$ with $a \neq 0$ are
$$x = \dfrac{-b \pm \sqrt{b^2 - 4ac}}{2a}.$$

Imaginary and Complex Numbers

1. $i = \sqrt{-1}$ and $i^2 = -1$
2. The set of numbers in the form $a + bi$ is the set of complex numbers. If $b = 0$, the complex number $a + bi$ is a real number. If $b \neq 0$, the complex number $a + bi$ is an imaginary number.

The Graph of $y = ax^2 + bx + c$

1. The graph of $y = ax^2 + bx + c$ is called a parabola, shaped like a bowl. If $a > 0$, the parabola opens upward, and if $a < 0$, the parabola opens downward. The turning point of the parabola is the vertex.
2. Graph $y = ax^2 + bx + c$ by finding any x-intercepts (replace y with 0), the y-intercept (replace x with 0), the vertex, and additional points near the vertex and intercepts. The x-coordinate of the vertex is $\dfrac{-b}{2a}$.

   The y-coordinate is found by substituting $\dfrac{-b}{2a}$ for x in $y = ax^2 + bx + c$ and solving for y.
3. The vertex of $y = ax^2 + bx + c$ is a minimum point when $a > 0$ and a maximum point when $a < 0$.

3. The Pythagorean Theorem

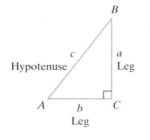

In any right triangle with hypotenuse of length c and legs of length a and b, $c^2 = a^2 + b^2$.